Heating with Renewable Energy

Practical, Hydronic-based Combisystems for Residential and Light Commercial Buildings

John Siegenthaler, P.E.

Associate Professor Emeritus

Mohawk Valley Community College

Utica, New York

CENGAGE Learning

Australia • Brazil • Canada • Mexico • Singapore • United Kingdom • United States

Heating with Renewable Energy
John Siegenthaler, P.E.

Senior Director, Development, Global Product Management, Skills: Dawn Gerrain

Product Team Manager: James Devoe

Senior Director Development: Marah Bellegarde

Senior Product Development Manager: Larry Main

Content Developer: Jenn Alverson

Product Assistant: Jason Koumourdas

Senior Production Director: Wendy A. Troeger

Senior Content Project Manager: Glenn Castle

Art Director: Bethany Casey

Technology Project Manager: Joe Pliss

Cover image(s): ©sjgh/Shutterstock.com
©NIKSPHOTO dot COM/Shutterstock.com
©gualtiero boffi/Shutterstock.com
Courtesy of Aimee Wetmore, MREA
Courtesy of ReVision Energy
Courtesy of C. Mark Sakry, CGD/Northern GroundSource Inc.
Courtesy of Mark Odell

Library of Congress Control Number: 2015946690

ISBN-13: 978-1-285-07560-0

Cengage Learning
200 First Stamford Place, 4th Floor
Stamford, CT 06902
USA

Cengage Learning is a leading provider of customized learning solutions with office locations around the globe, including Singapore, the United Kingdom, Australia, Mexico, Brazil, and Japan. Locate your local office at: **www.cengage.com/global**

To learn more about Cengage Learning, visit **www.cengage.com**

Purchase any of our products at your local college store or at our preferred online store **www.cengagebrain.com**

Notice to the Reader

Neither the publisher nor the author warrant or guarantee any of the products described herein or perform any independent analysis in connection with any of the product information contained herein. Neither the publisher nor the author assume, and expressly disclaims, any obligation to obtain and include information other than that provided to it by the manufacturer. The reader is expressly warned to consider and adopt all safety precautions that might be indicated by the activities described herein and to avoid all potential hazards. By following the instructions contained herein, the reader willingly assumes all risks in connection with such instructions. Neither the publisher nor the author make representations or warranties of any kind, including but not limited to, the warranties of fitness for particular purpose or merchantability, nor are any such representations implied with respect to the material set forth herein, and neither the publisher nor the author take responsibility with respect to such material. Neither the publisher nor the author shall be liable for any special, consequential, or exemplary damages resulting, in whole or part, from the readers' use of, or reliance upon, this material.

Printed in the United States of America
Print Number: 02 Print Year: 2021

Table of Contents

Preface

This text was written to help designers create efficient, reliable, and economically sustainable heating systems that incorporate renewable energy heat sources. The heat sources discussed include solar thermal collectors, air-to-water heat pumps, geothermal heat pumps, and boilers fueled by cordwood, chips, and pellets.

The framework for these systems is modern hydronics technology (e.g., water based conveyance of thermal energy). This technology provides a versatile and highly energy efficient means of controlling and distributing heat gathered by renewable energy heat sources. It allows the best renewable energy technology for *producing* heat to be combined with the best technology available for *delivering* heat.

Modern hydronics technology can be thought of as the "glue" that holds together many systems supplied by renewable energy heat sources. If you removed the renewable heat source from one of these systems, what remains are pipes, circulators, valves, tanks, heat emitters, and controllers. Much of this hardware is the same as that used in hydronic systems supplied by *conventional* heat sources. Still, the ways in which this hardware is applied, and the logic by which it is controlled, must respect the unique characteristics of each renewable heat source.

The concepts and systems presented in this text demonstrate the "commonality" of hydronic design elements across a wide spectrum of heat sources. For example, a home run distribution system serving low-temperature heat emitters, and powered by a pressure regulated circulator, could be supplied by any of the renewable energy heat sources presented in this text. So can a subassembly for "on-demand" domestic water heating. An outdoor reset controller can be used as a means of signaling when a thermal storage tank, supplied by any of the renewable energy heat sources discussed, can no longer supply the heating load. When these common design elements are understood, they greatly expand the possibilities for creative design, while also ensuring proven operating characteristics.

IT'S NOT JUST ABOUT THE HEAT SOURCE

The motivation for writing this text came from watching many aspiring designers focus almost exclusively on the promised benefits of a particular renewable energy heat source, and then attempt to apply that heat source, without a solid understanding of the remaining portions of the system. Unfortunately, some of these designers thought that the balance of system was not as important as the heat source. Others likely assumed that the renewable heat source was somehow universally compatible with *any* hydronic distribution system. This is analogous to thinking that installing a high-performance Ferrari engine in a Cub Cadet® *lawn tractor* will produce a competitive Formula 1 racing vehicle. Such a notion seems absurd, even to those not interested in auto racing. Still, mismatches between the "engine" and its associated "drive train" take place every day as under-informed individuals try to match renewable heat sources to incompatible distribution systems. This text was written to help avoid such situations.

This book does not focus on innumerable energy usage statistics, vague concepts, and oversimplified / incomplete diagrams. Instead, it focuses on methodology and details that allow designers to create *complete systems*. It covers the path of energy flow from its source to its delivery. It discusses and illustrates the "nuts and bolts" that designers need to specify and installers need to assemble, to create systems that meet expectations and provide years of low-maintenance operation.

In the highly competitive residential / light commercial HVAC market, heating systems using renewable energy heat sources can only succeed if they provide comfort, operate reliably, and are economically sustainable. The latter term implies that such systems must appeal to the "mass market" of North American building owners, rather than just renewable energy enthusiasts. These systems must be replicable. Furthermore, they must do this *without long-term subsidies from government or utilities.*

Experience has shown that simply labeling systems using renewable energy heat sources as "green" technology will not motivate most North American consumers to look beyond the higher installation cost of these systems. Doing so requires economic justification combined with assurances of uncompromised comfort, reliability, a high degree of automatic operation, and serviceability when required.

FUNDAMENTALS FIRST, SYSTEMS SECOND

This text begins with a discussion of energy fundamentals and the "building blocks" needed in any type of hydronic system. It emphasizes energy conservation as the prerequisite for successful use of renewable energy heat sources. You will be introduced to hardware and design principles that are *not* strictly associated with renewable energy. For example, there is a discussion of boilers that operate on *conventional fuels*. This information is not presented to simply fill pages, but because proper selection and application of such boilers is essential in creating systems that deliver uncompromised comfort, regardless of the status of their renewable energy heat source.

Once these underlying principles and building blocks are presented, the text moves on to discuss the construction, operation, and proper application of several renewable energy heat sources. These include solar thermal collectors, air-to-water heat pumps, geothermal heat pumps, and wood-fueled boilers. Each of these heat sources has unique operating characteristics. Good application of the hydronic building blocks will complement rather than compromise these characteristics. Finally, the text shows how to assemble the building blocks into many complete and fully documented systems. Systems that can be built today, using readily available nonproprietary components, and systems that are based on proven hydronics technologies.

ORGANIZATION

Chapter 1 covers the fundamentals of thermal energy use in residential and light commercial buildings. It gives a brief history of how renewable heat sources have been used in North America over the last several decades. It covers the basics of human thermal comfort, emphasizing that such comfort is a prerequisite to widespread acceptance of renewable energy heating systems. It also addresses domestic water heating, cooling, and heat recovery ventilation as elements that expand a "heating only" approach into a complete system that delivers high indoor environmental quality.

Chapter 2 discusses heating load calculations. Experience indicates that such calculations can be a stumbling block for students, as well as for those in the trade, who want to jump right into design and layout without first determining what the system needs to provide. Without proper load information, any type of heating system can fail to deliver the required comfort. A complete method for determining design heating loads is presented. Software-based load calculations are also discussed. The degree-day method is presented as a means of estimating annual heating energy used. Simple methods for comparing the cost of various fuel options on a unit cost basis are also given.

Chapter 3 is an overview of the essential hydronic design principles used in all the systems discussed in later chapters. It also provides the mathematical tools needed to quantify fluid flow in piping circuits.

Chapter 4 is a "show & tell" chapter that covers a wide range of hardware used in hydronic systems. It discusses the proper application and sizing of conventional boilers, piping, valves, circulators, expansion tanks, and heat exchangers. Such components are the building blocks, and proper selection is the mortar that bonds them together as the foundation of a complete system.

Chapter 5 surveys a range of low-temperature hydronic heat emitters. All of these heat emitters are compatible with the renewable heat sources discussed in the text. A given building might be supplied by just one type of these heat emitters, or a combination of them depending on factors such as aesthetics, available wall space, floor coverings, and recovery time from setbacks. My recommendation is to select all heat emitters in systems using renewable energy heat sources to provide the building's *design heating load* using a supply water temperature no higher than 120 °F. Even lower supply water temperatures are preferred when possible.

Chapter 6 describes control principles and devices that orchestrate the thermal and hydraulic processes taking place in the system. Emphasis is given to using nonproprietary control devices in unique ways that complement the operating characteristic of each renewable energy heat source.

Chapter 7 narrows the focus to the fundamentals of solar energy. It discusses past approaches to solar heating, and then presents methods for determining the sun's position and energy availability.

Chapter 8 provides detailed discussions on the common types of solar thermal collectors that are used in residential and light commercial combisystem applications. It covers construction of flat plate and evacuated tube collectors, standards for determining their thermal performance, and the details of how they are properly installed.

Chapter 9 covers thermal storage tank options. Storage is an integral part of most systems using renewable energy sources. Well-designed thermal storage can stabilize system operation both thermally and hydraulically. Proper sizing of thermal storage is essential to good system performance and affordability.

Chapter 10 is the first of several chapters that brings previously discussed principles and hardware together into complete systems. In this chapter, those systems are built around antifreeze-protected solar thermal collectors. Ancillary issues such a protecting system fluids during collector stagnation, and properly sizing the heat exchanger between the collector array and thermal storage tank are discussed.

Chapter 11 continues with the topic of solar thermal combisystems, but changes the method used to protect solar collectors from freezing. This chapter explains the simple elegance and multiple benefits of drainback-protected collector arrays. I have lived with a drainback-protected solar combisystem system, in a harsh winter climate, for over three decades, and can testify that such systems work well when properly designed and installed. This chapter lays out the necessary principles and details.

Chapter 12 focuses on the f-chart method of estimating the thermal performance of solar combisystems. It presents the f-chart method in a form suitable for manual calculations or implementation by spreadsheet. It also shows examples of how commercially available F-CHART software can be used to expedite calculations and quickly assess "what if" questions related to a specific system configuration.

Chapter 13 switches the renewable energy heat source from solar collectors to air-to-water heat pumps. Although currently not as well known as geothermal heat pumps, at least in North America, air-to-water heat pumps are extensively used in Europe and Asia. The market for these products is growing in North America, and those wanting to heat buildings with renewable energy should be familiar with their operation and the benefits they offer. This chapter covers the fundamentals of heat pump construction and operation. It also discusses how operating conditions can significantly affect the heating capacity of a heat pump. This is also the first chapter that discusses systems that provide both heating and cooling. The latter load is handled by chilled water produced by the heat pump. Several complete combisystems are presented that are built around air-to-water heat pumps.

Chapter 14 moves from air-source heat pumps to geothermal water-to-water heat pumps. The principal difference between these heat pumps is how low-temperature heat is gathered from soil or ground water, rather than from outside air. Both open-loop and closed-loop water sources are discussed. Methods are given to calculate the necessary size of horizontal earth loops. Heat pump performance factors are given and factored into earth loop sizing. Several complete combisystems are presented that are built around water-to-water geothermal heat pumps.

Chapter 15 describes how to combine modern hydronics technology with wood-fired and pellet-fired heat sources. This is a large sector of the rapidly expanding "biomass" heating market, and one appropriate for residential and light commercial buildings. This chapter covers a range of technology from outdoor wood-fired hydronic heaters, to high-efficiency wood gasification and pellet-fired boilers. As in other chapters, it shows how to combine readily available hydronics hardware in specific ways that address the operating characteristics of wood-fired and pellet-fired boilers. Several complete system designs are presented at the end of the chapter.

Chapter 16 rounds out the designer's tool box by presenting several methods for evaluating the financial merit of a proposed system. It discusses ways to frame a situation that often occurs with systems using renewable heat sources: higher installation cost and lower operating cost. The methods presented were selected to be understandable by designers, as well as clients who need to factor them into buying decisions.

This text also includes an extensive glossary, and appendices of data and other reference information frequently used in the design process.

HYDRONICS DESIGN STUDIO

This last page of this text references a website where the demo version of the Hydronics Design Studio software can be freely downloaded. This Windows®-based software is built around some of the analytical methods presented in the text. It allows rapid simulation of flow in user-defined piping systems. It also includes modules for room heat loss estimating, expansion tank sizing, and simulation of buffer tanks. Several examples of how the software can be used to evaluate long numerical calculations are given within the chapters. A professional version of the Hydronics Design Studio with increased capabilities is also available. Visit the website www.hydronicpros.com for additional information.

ACKNOWLEDGMENTS

My appreciation is extended to the many individuals, manufacturers, and organizations who have supplied images and information for this text. Their cooperation and willingness to accommodate what have often been tight schedules have been truly outstanding.

Further appreciation is extended to the following individuals who reviewed the manuscript: Dr. Benjamin Ballard, Dr. Phil Hofmeyer, Max Rohr, and Bob Rohr. Their generosity and collective comments have unquestionably improved this text.

Finally, I also thank the Cengage Learning production team, and especially John Fisher and Richard Hall, for their diligence and patience as this text has moved from proposal to hardcover.

DEDICATION

This text is dedicated to Bob "Hot Rod" Rohr, a long time friend and professional associate. Bob and I are of the same vintage, and have shared similar paths in the hydronics and renewable energy industry over the last three decades. We have frequently huddled, often by e-mail, to discuss new ideas and potential solutions to design and installation issues. I have valued his insight, ingenuity, and artistic craftsmanship over many years. My hope is that this text inspires others to become as passionate about hydronics and renewable energy as "Hot Rod" has been, and continues to be.

John Siegenthaler, P.E.

INSTRUCTOR COMPANION WEBSITE

This is an educational resource that creates a truly electronic classroom. It is a website containing tools and instructional resources that enrich your classroom and make your preparation time shorter. The elements of the Instructor Companion Website link directly to the text and tie together to provide a unified instructional system. With the Instructor Companion Website you can spend your time teaching, not preparing to teach. The website contains an Instructor Guide with answers to the text's review questions and exercises, chapter presentations in PowerPoint, and an Image Library.

CENGAGE LEARNING TESTING POWERED BY COGNERO is a flexible, online system that allows you to:

- author, edit, and manage test bank content from multiple Cengage Learning solutions
- create multiple test versions in an instant
- deliver tests from your LMS, your classroom, or wherever you want

Start right away!

Cengage Learning Testing Powered by Cognero works on any operating system or browser.

- No special installs or downloads needed
- Create tests from school, home, the coffee shop—anywhere with Internet access

What will you find?

- *Simplicity at every step.* A desktop-inspired interface features drop-down menus and familiar, intuitive tools that take you through content creation and management with ease.
- *Full-featured test generator.* Create ideal assessments with your choice of 15 question types (including true/false, multiple choice, opinion scale/Likert, and essay). Multi-language support, an equation editor, and unlimited metadata help ensure your tests are complete and compliant.
- *Cross-compatible capability.* Import and export content into other systems.

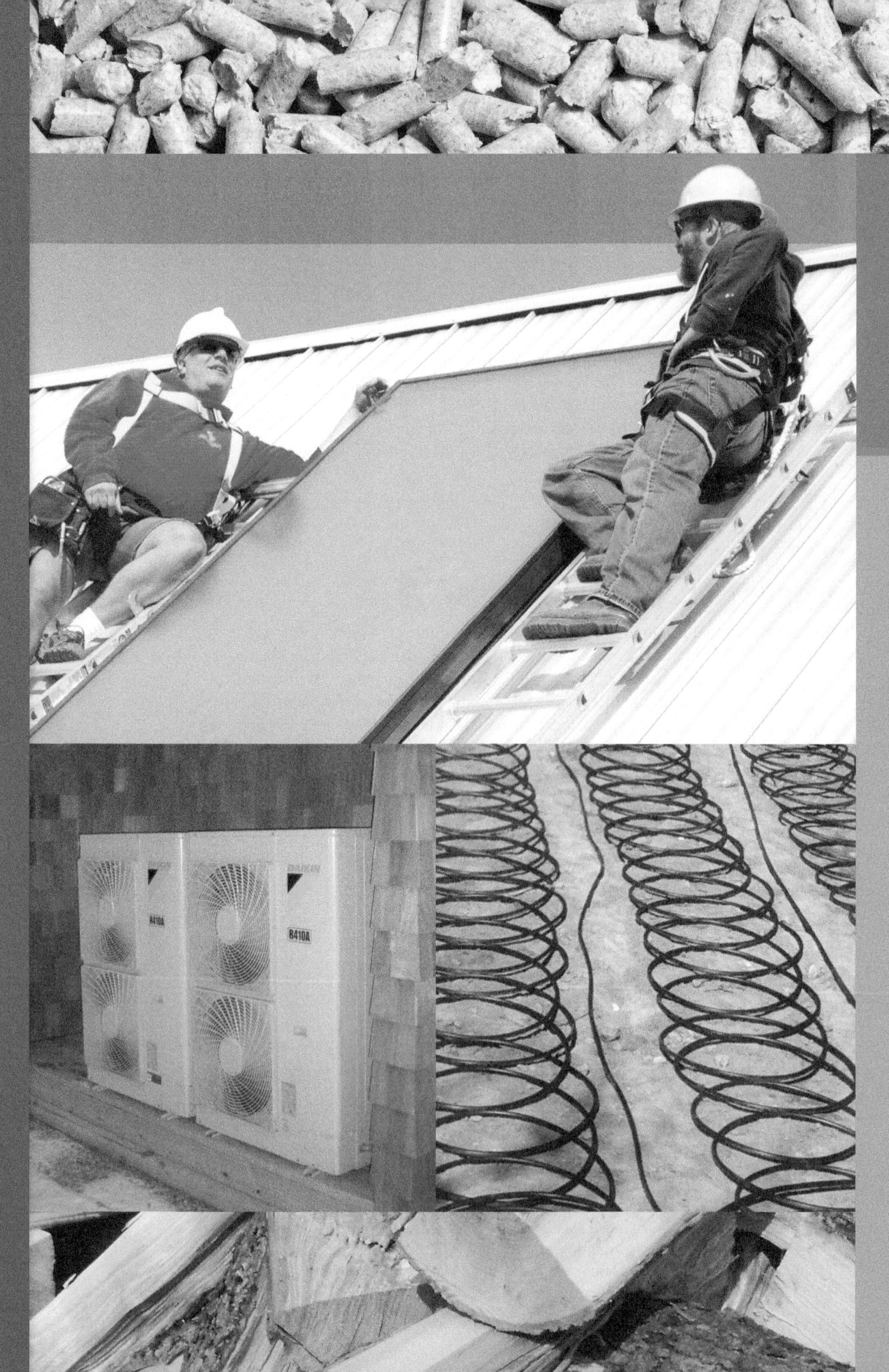
R410A
R410A

CHAPTER 1

Residential Energy Basics

OBJECTIVES

After studying this chapter, you should be able to:

- Describe factors that influence human comfort.
- Define heat and describe how it is measured.
- Describe three methods by which heat travels.
- Explain thermal equilibrium within a hydronic heating system.
- Explain why systems with a renewable energy heat source often need an auxiliary heat source.
- Understand the importance of energy conservation versus energy supply.
- Describe concepts and construction details that reduce the energy used for space heating.
- Explain several methods for reducing the energy required for domestic water heating.
- Calculate the unit price of thermal energy delivered from several different fuels.
- Understand the need for cooling and ventilation in addition to space heating.
- Discuss the operation and benefits of ductless cooling systems.
- Describe balanced heat recovery ventilation systems.

1.1 Introduction

In an ideal climate, prevailing weather would allow buildings to remain comfortable every day of the year. There would be no need to "artificially" add or remove heat from a building's interior. Operable windows would provide fresh air. The interior temperature would remain somewhere between 68 ºF and 72 ºF during the day, and perhaps drop slightly during sleeping hours. The indoor relative humidity would remain in the range of 30 to 40 percent throughout the year.

There are locations within North America where such conditions do exist, at least for a few days in a typical year. However, as nearly all readers can attest, very few locations in North America can claim such conditions *throughout* the year. Hence, to maintain human comfort, nearly all residential and light commercial buildings in North America require a means of generating heat in cooler weather, and often a means of rejecting excess heat during warmer weather.

Beyond space heating and cooling, most buildings designed for human occupancy must also supply domestic hot water for washing, cleaning, and bathing. In most residential buildings, the energy required to heat domestic water is second only to that used for space heating (in cool climates), or space cooling (in warmer climates). Thus, domestic water heating is a load that deserves serious and accurate consideration.

The vast majority of North American buildings provide the energy needed for space heating, cooling, and domestic water heating using conventional energy sources such as natural gas, propane, fuel oil, and electricity. Occupants of such buildings expect indoor conditions to remain comfortable, regardless of outside conditions. Still, most of these occupants pay little attention to the systems that are expected to maintain these conditions. Most simply accept the heating, cooling, and domestic water-heating systems that have been specified by a designer, or installed by a builder. This is especially true when the perceived cost of heating and cooling is low.

Relatively low oil prices were the norm through much of the American industrial expansion years spanning from the late 1800s into the early 1970s. Then, in 1973, American consumers were jolted with rapid increases in energy costs. This situation was informally known as the **Arab oil embargo**. Figure 1-1 shows how the price of oil spiked at this time, in comparison to previous decades of relative price stability.

America's first experience with petroleum price shocks generated strong interest in developing **alternate energy sources** that could either replace or supplement conventional energy. At the time, "alternate energy source" meant an alternative to fossil fuels such as oil, natural gas, and propane as well as electricity generated using these fuels or nuclear energy. Government expenditures for research, and subsidies for installation of alternate energy equipment, helped this fledgling market expand. Products such as solar thermal collectors, wind turbines, and wood-burning heaters quickly appeared on the market, and many early adapters eagerly installed them, often without guidance on proper system design or installation.

Performance results were mixed. While some systems proved successful in reducing energy costs, others quickly failed due to poor design, inadequate materials, or lack of installation standards. As time passed, more systems developed problems, and many became "orphans" due to the lack of trained technicians to service them. Inevitably, systems that performed marginally, or failed all together, influenced public opinion on the viability of alternate energy sources.

Skepticism of such systems grew into the mid 1980s. During this time, the price of crude oil also dropped by over 50 percent. These factors quickly dampened public interest in renewable energy. Many of the companies that supplied renewable energy products in the late 1970s abandoned further market development, or simply went out of business, as Americans returned to a status quo with conventional energy sources. From the mid 1980s through the 1990s and the early part of the new century, residential use of renewable energy sources was often perceived as either a "hobby" or an environmental statement. Strong economic growth and increased disposable income altered

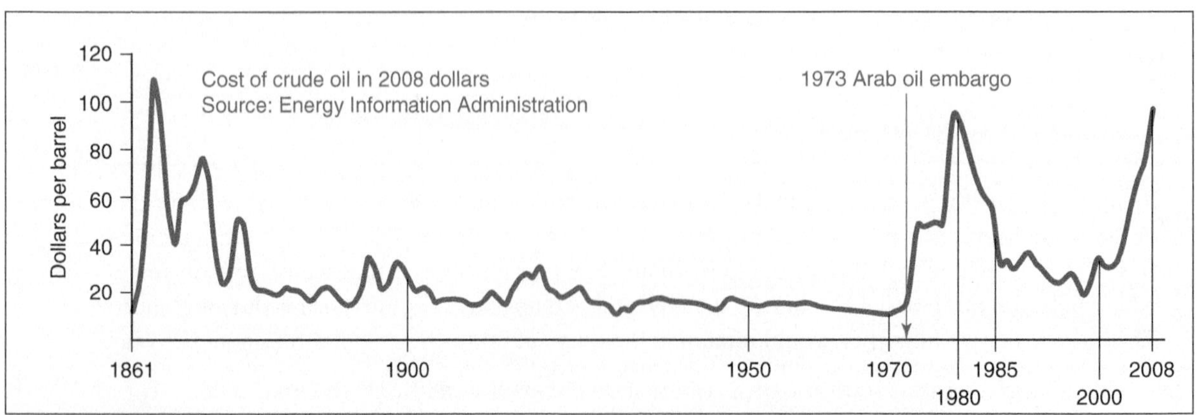

Figure 1-1 | The price of crude oil in America from 1861 to 2008.

the priorities of those investing in new homes and other buildings. Larger floor plans and interior amenities usually claimed construction budgets long before decisions on heating or cooling were made. Thus, many homes and light commercial buildings of this era simply got the heating system that the builder's HVAC subcontractor preferred.

Then, as it often has in the case of energy use, history began repeating itself. Worldwide demand for conventional energy increased dramatically beginning in 2007. The economics of supply and demand created new price spikes for conventional energy. This, along with rising concern over global climate change, and a return of government incentives, rekindled interest in a broader use of what is now commonly called **renewable energy**: energy from sources that are naturally replenished on a human timescale. Examples include sunlight, solar energy temporarily stored in the atmosphere or soil, chemical energy stored in plants and trees (e.g., **biomass**), energy extracted by water moving from higher to lower elevations (e.g., **hydropower**), energy extracted from wind, and energy extracted from tides or waves.

As this text is being written, this interest continues to grow. More renewable energy products enter the North American market every month. Among these are products intended to deliver heat from renewable energy sources, including solar thermal collectors, heat pumps, and wood-fired heating devices. The applications for such devices span the range of residential to large commercial systems. The two primary end uses for the thermal energy they deliver are space heating and domestic water heating.

The question of whether North Americans will embrace renewable energy as part of a *long-term* energy strategy remains unanswered. Market factors and human nature will continue to influence this issue as they have for decades. So too will the performance and long-term reliability of the renewable energy systems being installed at present.

The goal of this text is to help designers create efficient and reliable heating systems using common renewable energy sources. The foundation for this text is modern **hydronics** technology (e.g., water-based distribution of thermal energy). This technology provides an elegant, versatile, and highly energy-efficient means of controlling and distributing the energy gathered by the renewable energy heat source.

This text also emphasizes that any renewable energy heat source is only *part* of an overall system. For the renewable energy heat source to deliver optimum performance, it must be properly matched with other **balance-of-system** hardware and control strategies.

Designers need solid and unbiased technical information on how to create heating systems using renewable energy heat sources. Information that doesn't simply focus on the sociopolitical issues often surrounding use of renewable energy, but instead, focuses on the "nuts and bolts" of how to create systems that are practical, efficient, and reliable. Systems that maximize the **return on investment** in the renewable energy heat source while also providing consistent comfort for building occupants.

Many of the currently available technical publications covering solar thermal collectors, heat pumps, or wood-fired boilers do not adequately discuss *complete system design*. Instead, they focus on the heat source, and assume designers will access other sources of information when designing the remainder of the system. This often leads to compromised designs and mismatched hardware. Few existing publications emphasize the *commonality of design concept* that applies to many different renewable energy heat sources. Understanding these commonalities allows those who often design systems around one type of renewable energy heat source, such as solar thermal collectors, to quickly transpose their designs into systems supplied by heat pumps or wood-fired boilers.

This text provides a broad discussion of the hardware and procedures needed for complete system design. This chapter grounds the reader in the fundamental thermodynamics that pertain to all heating systems. It stresses concepts that are common to all heating systems. It provides simple tools for calculating the **unit cost of energy** from a wide variety of sources. In short, it provides information that is as important to those designing renewable energy heating systems as musical scales are to those learning to play an instrument. The foundation for successful system design and installation begins here.

1.2 Thermal Comfort

Contrary to common belief, heating systems are not created to heat buildings. Instead, *they are created for sustaining human thermal comfort within buildings.* Think about it: Does a window, concrete block, or insulation batt really "care" whether its temperature is 40 °F or 70 °F? Of course not. However, the selection and placement of these materials can significantly impact human thermal comfort, or lack thereof.

Providing comfort should be the primary objective of any heating system designer or installer. Unfortunately, this objective is too often compromised by other factors, the most common of which is installation cost. Even small residential heating systems affect the health, productivity, and general well-being of many people over several decades. It only makes sense to plan and install them with commensurate care.

The average North American homeowner spends little time thinking about the consequences of the heating system in his or her home. Most view such systems as a necessary but uninteresting part of a building. When construction budgets are tightened, it is often the heating system that is compromised to save money for other, more visible amenities.

Heating professionals should take the time to discuss comfort with their clients. Many people who have lived with uncomfortable heating systems simply don't realize what they have been missing. In retrospect, many would welcome the opportunity to live or work in truly comfortable buildings, and would willingly spend more money, if necessary, to do so.

Maintaining thermal comfort is not a matter of supplying heat to the body. Instead, it is a matter of *controlling how the body loses heat.* When interior conditions allow heat to leave a person's body at the same rate it is generated, that person feels thermally comfortable. If heat is released faster or slower than the rate it is produced, some degree of discomfort is experienced.

A normal adult engaged in light activity generates heat at a rate of approximately 400 **British thermal units (BTUs)** per hour. Figure 1-2 shows the various processes by which the body of a person at rest releases heat to a typical indoor environment.

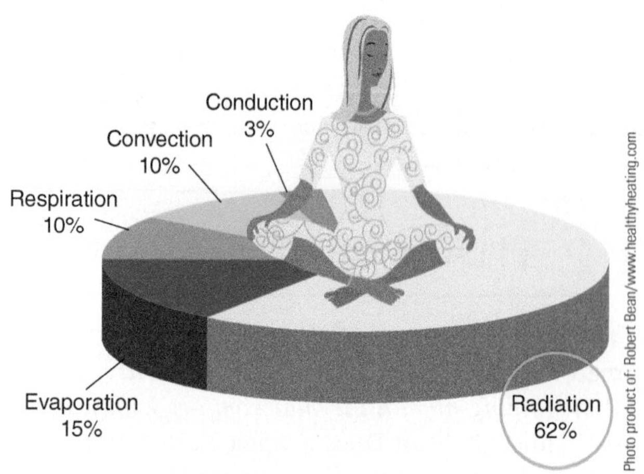

Figure 1-2 | Processes the body uses to release heat to a typical indoor environment.

Notice that a large percentage of the body's heat loss comes from **thermal radiation** to surrounding surfaces. Most people will not be comfortable in a room containing cool surfaces such as large windows or cold floors, even if the room's air temperature is 70 °F. The remaining heat loss occurs through a combination of convection, evaporation, respiration, and a small amount of conduction. The latter occurs through surfaces in direct contact with the body.

The body can adjust its heat loss mechanisms, within certain ranges, to adapt to different interior environments. For example, if air temperature around the body increases, **convection** heat loss will be suppressed. The body responds by increasing **evaporation** heat loss (perspiration), and increasing skin temperature to increase radiative heat loss.

Properly designed heating systems control both the air temperature and surface temperatures of rooms to maintain optimal comfort. Modern controls can maintain room air temperature to within ± 1 °F of the desired setpoint temperature. Hydronic heat emitters such as radiant floor or ceiling panels raise the average surface temperature of objects in the room. Since the human body is especially sensitive to radiant heat loss, these heat emitters significantly enhance thermal comfort. Comfortable humidity levels during the heating season are also easier to maintain in hydronically heated buildings.

Several factors, such as activity level, age, and general health, determine what is a comfortable environment for a given individual. When a group of people are living or working in a common environment, any one of them might feel too hot, too cold, or just right. Heating systems that allow various **zones** of a building to be maintained at different temperatures can adapt to the comfort needs of several individuals. This is called zoning. Although both forced-air and hydronic heating systems can be zoned, the latter is usually much simpler and easier to control.

1.3 Heat, Heat Transfer, and Thermodynamics

Before attempting to design any type of heating system, it is crucial to understand the entity being manipulated: **heat**.

What we commonly call heat can also be technically described as energy in thermal form. Other forms of energy, such as electrical, chemical, mechanical, and nuclear, can be converted into heat through various processes and devices. All heating systems consist of devices that convert one form of energy into another.

Heat is our perception of atomic vibrations within a material. Our means of expressing the intensity of these vibrations is called **temperature**. The more intense the vibrations, the greater the temperature of the material, and the greater its heat content. Any material above absolute zero temperature (–459.67 °F) contains some amount of heat.

There are several units for expressing a quantity of heat. In North America, the most commonly used unit of heat is the British thermal unit (Btu). A Btu is defined as the amount of heat required to raise 1 pound of water by 1 °F.

Heat always moves from an area of higher temperature to an area of lower temperature. In hydronic systems, this occurs at several locations, as illustrated in Figure 1-3.

In the liquid-based solar collectors, heat moves from the hot metal fins of the **absorber plate** into the cooler fluid passing through the absorber plate tubes. This heat then passes from the collector circuit fluid, through a **heat exchanger**, into slightly cooler water within a thermal storage tank. This water then flows through piping to one or more **heat emitters**, where it is released into the cooler surrounding air, or to cooler surfaces within the room. Finally, heat passes through the **thermal envelope** of the building, and into outside air or the soil beneath the building. In every instance, heat moved from an area of higher temperature to an area of lower temperature.

In this book, the rate of heat transfer is expressed in British thermal units per hour, abbreviated as Btu/hr. It is very important to distinguish between the *quantity* of heat present in an object (measured in Btu) and the *rate* at which heat moves in or out of the object (measured in Btu/hr). These terms are often misquoted by people, including heating professionals.

The rate of heat transfer from one location to another is governed by several factors. One is the temperature difference between where the heat is, and where it is going. Temperature difference is the "driving force" that causes heat to move. *Without a temperature difference between two locations, there can be no heat transfer.* The greater the temperature difference, the faster heat flows. In most instances, the rate of heat transfer through a material is directly proportional to the temperature difference across the material. Thus, if the temperature difference across a material were doubled, the rate of heat transfer through the material would also double.

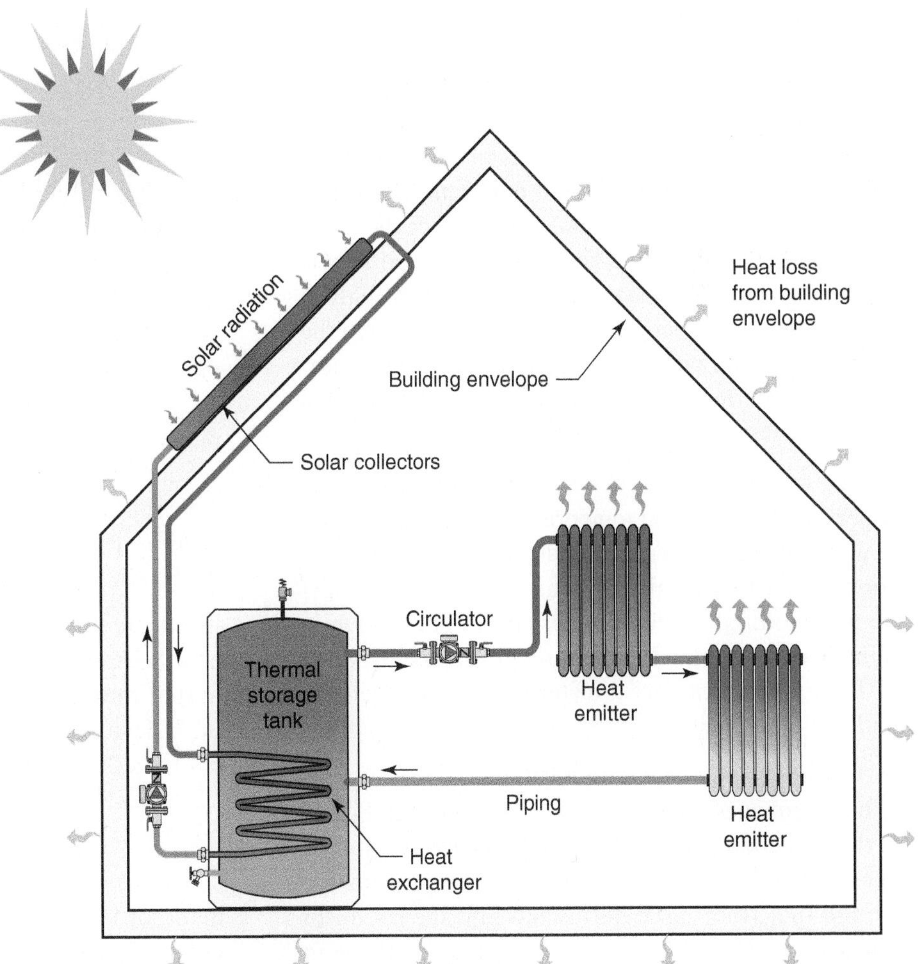

Figure 1-3 | Heat movement within a solar heating system and the building it serves.

Another factor affecting the rate of heat transfer is the type of material through which the heat moves. Some materials, such as copper and aluminum, allow heat to move through very quickly. Other materials, such as polyurethane foam, greatly inhibit the rate of heat transfer.

THREE MODES OF HEAT TRANSFER

Thus far, two important principles of heat transfer have been discussed. First, heat moves from an area of higher temperature to an area of lower temperature. Secondly, the rate of heat transfer depends on temperature difference and the type of material. To gain a more detailed understanding of how the thermal components of a heating system work, we need to classify heat transfer into three modes: **conduction**, convection, and thermal radiation.

Conduction is the type of heat transfer that occurs through *solid materials*. Recall that heat has already been described as atomic vibrations. Heat transfer by conduction is a dispersal of these vibrations from a source of heat, out across trillions of atoms that are bonded together to form a solid material. The index that denotes how well a material transfers heat is called its **thermal conductivity**. The higher a material's thermal conductivity, the faster heat can pass through it, all other conditions being equal. Heat moving from the inner surface of a pipe to its outer surface is an example of conduction heat transfer. Heat moving through a concrete basement wall is another.

The rate of heat transfer by conduction is directly proportional to both the temperature difference across the material and its thermal conductivity. It is inversely proportional to the thickness of a material. Thus, if one were to double the thickness of a material while maintaining the same temperature difference between its sides, the rate of heat transfer through the material would be cut in half.

In some locations within heating systems, designers try to enhance conduction. For example, the higher the thermal conductivity of a flooring material installed over a heated floor slab, the faster heat can pass upward through that flooring material and into the room. In contrast, the slower the rate of conduction through the insulation of a hot water storage tank, the better it retains heat. Equations for calculating heat flow by conduction are given in Chapter 2.

Convection heat transfer occurs when a *fluid* at some temperature moves along a surface at a different temperature. *The term* fluid *can refer to a gas or a liquid.*

Consider the example of water at 100 °F flowing through the absorber plate of a solar collector that has a temperature of 125 °F. The cooler water molecules contacting the warmer surface absorb heat from that surface. These molecules are being churned about as the water moves along. The heated molecules are constantly being swept away from the surface into the bulk of the water stream and replaced by cooler molecules. One can think of the heat as being "scrubbed" off the surface by the flowing water.

The speed of the fluid moving over the surface greatly affects the rate of convective heat transfer. The faster the fluid along the surface, the greater the rate of convective heat transfer. We've all experienced the increased "wind chill" effect as cool air is blown across our skin rather than lying stagnant against it. Although the air temperature may not be extremely cold, the speed it moves over our skin greatly increases the rate of convective heat loss. Although it may feel like the air is very cold, we are sensing the *rate of heat loss* from our skin rather than the air temperature. To achieve the same cooling sensation in calm air requires a much lower air temperature.

When fluid motion is caused by either a circulator (for water) or a blower (for air), the resulting heat transfer is more specifically called **forced convection**. When buoyancy differences within the fluid cause it to move along a surface, the heat transfer is more specifically called **natural convection**. Generally, heat moves much slower by natural convection than by forced convection. The warm air currents rising from the fin-tube element in Figure 1-4 are an example of natural convection. The heat transferred to the air pushed along by the blower is an example of forced convection.

Some hydronic heat emitters, such a fin-tube baseboard, are designed to release the majority of their heat output to the surrounding air by natural convection. Such devices are appropriately called **convectors**. Heat emitters that use fans or blowers to force air through a heat exchanger are usually called **fan coils** or **air handlers**.

Thermal radiation is probably the least understood mode of heat transfer. Just like visible light, thermal radiation is **electromagnetic energy**. It travels outward from its source in straight lines, at the speed of light (186,000 miles per second), and cannot bend around corners, although it can be reflected by some surfaces.

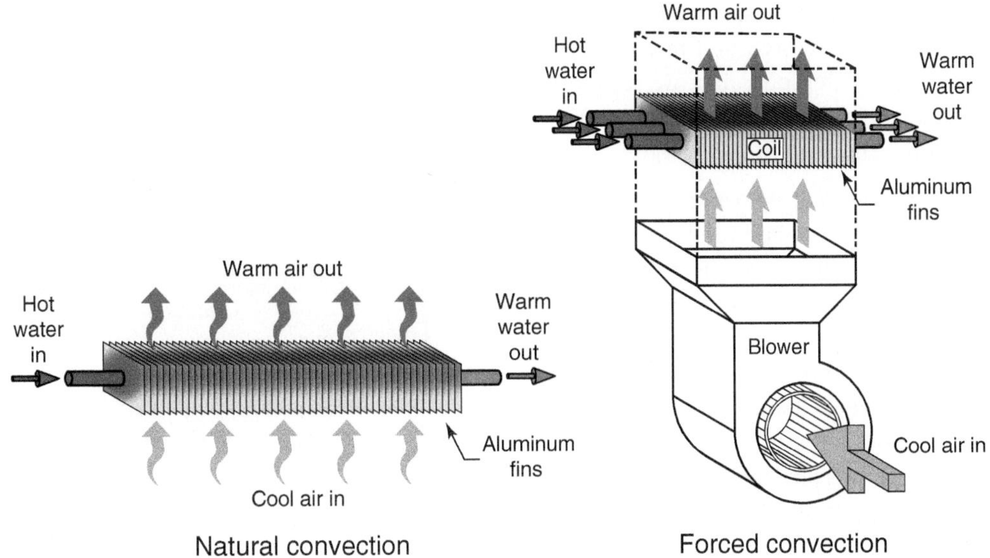

Figure 1-4 | Two types of convective heat transfer.

Unlike conduction or convection, thermal radiation does not require the presence of a material (e.g., solid, liquid, or gas) to transfer heat from one surface to another.

Consider a person sitting a few feet away from a campfire on a cold winter day. If pointed toward the fire his or her face probably feels warm, even though the air around the person is cold. This sensation is the result of thermal radiation, emitted from the burning wood, traveling through the air and being absorbed by the exposed skin. The air between the fire and the person is not heated as the thermal radiation passes through it. Likewise, thermal radiation emitted from the warm surface of a heat emitter passes through the air in a room *without directly heating that air*. When the thermal radiation strikes another surface in the room, most of it is absorbed. At that instant, a high percentage of the energy carried by the thermal radiation is absorbed by the atoms of that surface. This energy intensifies the vibration of those atoms, and thus raises the temperature of the surface. At that moment, the absorbed thermal radiation has been converted into heat.

Figure 1-5 shows the **electromagnetic spectrum**. This spectrum ranges from extremely short wavelength gamma radiation to much longer wavelength

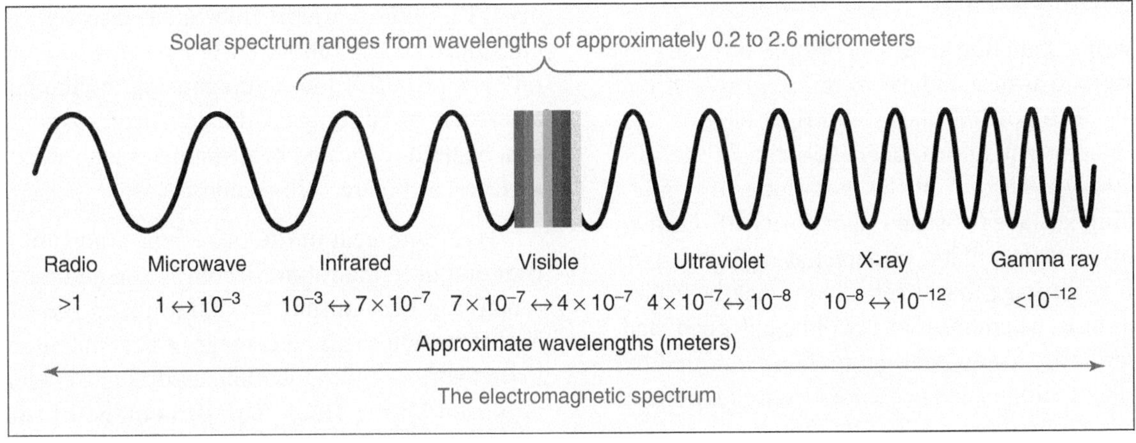

Figure 1-5 | The electromagnetic spectrum.

radio waves. Various regions of the spectrum are given names, such as X-rays, ultraviolet, and microwaves. The human eye can only detect a narrow range of wavelengths within the electromagnetic spectrum. This region is appropriately called the **visible range**.

The difference between *thermal* radiation and visible light is the wavelength of the radiation. Anyone who has watched molten metal cool has noticed how the bright orange color eventually fades to duller shades of red, until finally the metal's surface no longer glows. As the surface of the metal cools below approximately 970 °F, our eyes can no longer detect visible light from the surface. Our skin however, still senses that the surface is very hot. Though unseen, thermal radiation in the non-visible **infrared** portion of the electromagnetic spectrum is still being emitted by the metal's surface.

Any surface continually emits thermal radiation to any cooler surface within sight of it. The surface of a heat emitter that is warmer than our skin or clothing surfaces transfers heat to us by thermal radiation. Likewise, our skin and clothing give off thermal radiation to any surrounding surfaces at lower temperatures. Examples of the latter would be windows, or unheated walls, ceilings, or floors.

The term **mean radiant temperature** describes the area-weighted average temperature of all surfaces within a room. As the mean radiant temperature of a room increases, the air temperature required to maintain thermal comfort decreases. As the temperature of the room's surfaces increase, the heat released from the body by thermal radiation decreases, so the amount released by convection must increase to keep the total rate of heat release constant. The converse is also true. This is why a person in a room with an air temperature of 70 °F may still feel cool if he or she is close to cold surfaces, such as large, uncovered windows, on a cold winter night.

As thermal radiation strikes an opaque surface, part of it is absorbed as heat and part is reflected away from the surface. The portion that is absorbed or reflected is determined by the optical characteristics of the surface and the wavelength of the radiation. Most interior building surfaces absorb the majority of thermal radiation that strikes them, as depicted in Figure 1-6. The small percentage that is reflected typically strikes another surface, where most of it will be absorbed, and so on. Very little, if any, thermal radiation emitted by warm surfaces within a room escapes from the room.

Although the human eye cannot see thermal radiation, there are devices that can detect it and display an image that uses colors to represent different surface temperatures. Such an image is called an **infrared thermograph**. Figure 1-7a shows an infrared-detecting **thermographic camera** pointed at a tiled floor. The visible floor surface gives no evidence that it is being heated. Figure 1-7b is an infrared thermograph of the same floor area produced by the thermographic camera. The bright colors show areas of different surface temperatures.

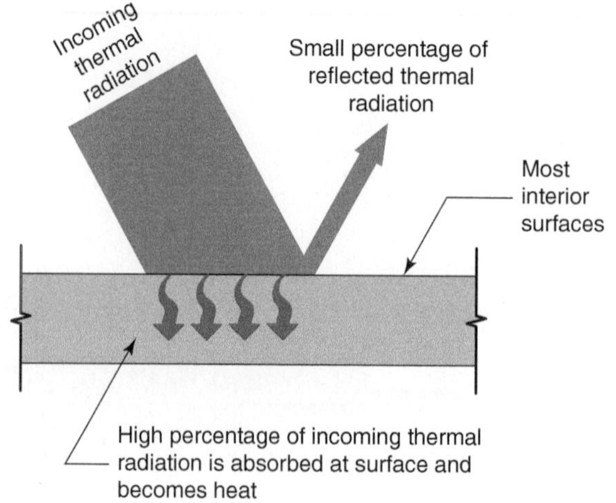

Figure 1-6 The majority of thermal radiation striking an interior building surface is absorbed at that surface, and instantly becomes heat.

The shape of the colored patterns indicates that some type of embedded element is heating the materials within the floor. In this case, the heat is coming from tubing carrying heated water. The color spectrum on the far right of the image gives the range of surface temperatures within the image. In this case, they range from the relatively cool aluminum sill of the patio door (about 42.0 °F), to the warmest floor areas directly above the embedded tubing (about 86.1 °F). Infrared thermography is a powerful tool in diagnosing the thermal characteristics of buildings and mechanical systems. It can also be used to located heated objects embedded behind surfaces, as Figure 1-7b demonstrates.

Hydronic heat emitters deliver a portion of their heat output to the room by convection and most of the remaining heat output by radiation. In the case of a heated floor, a small percentage of heat output may also leave the floor by conduction, assuming a cooler object is placed on the floor. The percentage of total heat output that is delivered by each mode of heat transfer depends on many factors, such as surface orientation, shape, type of finish, air-flow rate past the surface, and

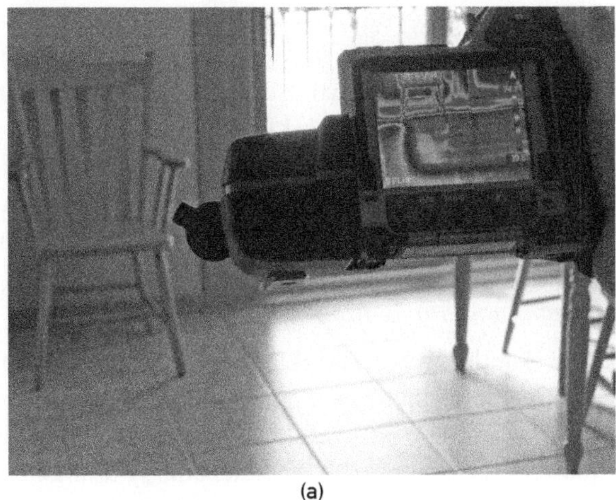
(a)

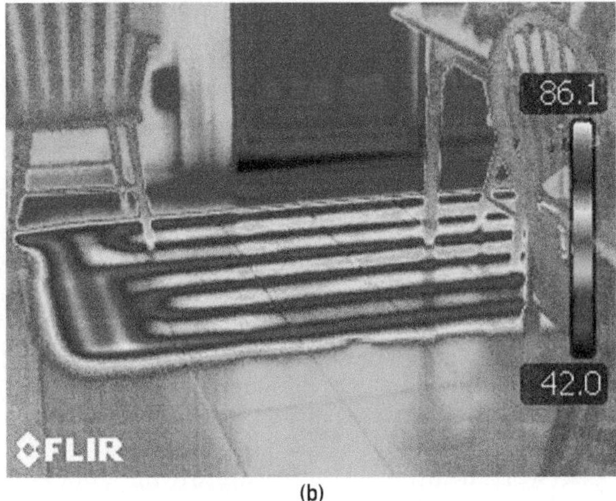

(b)

Figure 1-7 (a) An infrared camera aimed at a floor. (b) Infrared thermograph of the same floor reveals heat from tubing installed below the floor.

surface temperature. When a heat emitter transfers over 50 percent of its heat output by radiation, it is called a **radiant panel**. Heated floors, walls, and ceilings are all examples of radiant panels. They are discussed in more detail in Chapter 5.

THERMAL EQUILIBRIUM

If a material is not gaining or losing heat and remains in a single physical state (solid, liquid, or gas), its temperature does not change. This is also true if the material happens to be gaining heat from one source while simultaneously releasing heat to another material at the same rate. Under such conditions, the material(s) that remain at a stable temperature are said to be in **thermal equilibrium**.

Thermal equilibrium has many practical applications in the context of heating systems. For example, suppose you observed the temperature of water at the outlet of a continuously operating hydronic heat source flowing onward to a hydronic distribution system. Over the course of 20 to 30 minutes, you saw no change in the temperature of that water. What could you conclude? Answer: Since there was no change in the water's temperature, it did not undergo any net gain or loss of heat. Thus, the rate at which the heat source was adding heat to the flowing water was the same as the rate the heat emitters were extracting heat from that water (Figure 1-8).

Under these conditions, the system is in thermal equilibrium. *All hydronic heating systems will inherently try to find a point of thermal equilibrium and remain in operation at that point.* The goal of the system designer is to ensure that when thermal equilibrium is established, the system operates at conditions that maintain comfort within the building and do not adversely affect the operation, safety, or longevity of the system's components.

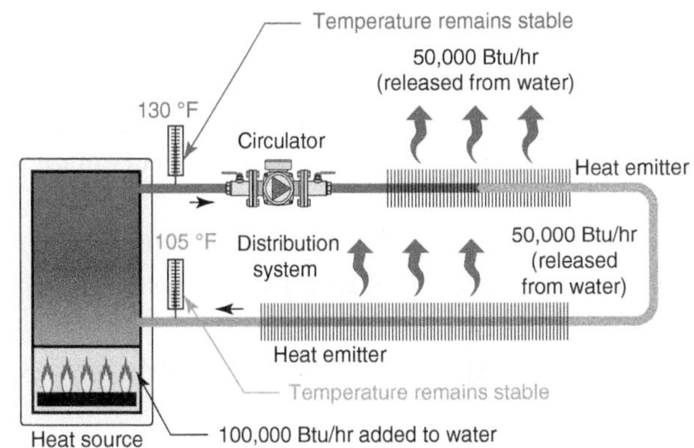

Figure 1-8 When the rates of heat input and heat release are equal, the system is in thermal equilibrium and fluid temperatures at all locations within the system are stable.

THE FIRST LAW OF THERMODYNAMICS

One of the most fundamental principles used in all areas of science and engineering is the **first law of thermodynamics**. It states that *energy cannot be created or destroyed, only changed in form.*

For example, when a gallon of propane, which contains about 91,800 Btus of energy in chemical form, is combusted within a boiler, the majority of the chemical energy is converted into heat and transferred to a stream of water flowing through the boiler. This is useful heat output. Another portion of the energy contained in the propane becomes heat that leaves through the boiler's exhaust system, and thus serves no useful purpose in heating the building. Yet another portion of the fuel energy drives chemical reactions during combustion, such as when hydrogen in the propane combines with oxygen in the air to produce water vapor.

If one were to precisely measure all the energy released from or stored within the boiler in this situation, in both useful and unusable forms, and add them up, the total would always equal the original chemical energy in the gallon of propane. Although the energy was converted to different forms within the boiler, none of it was ever destroyed.

This underlying principle of the first law of thermodynamics is sometimes stated as "energy in equals energy out." This concept is often used as the basis of **energy balance equations**. Such equations are mathematical statements that imply that, under steady state conditions, all the energy flowing into a process must equal the total energy flowing out of that process. **Heat balance equations** are useful in predicting energy flows that might not be easily measured or otherwise calculated. They are also used in establishing the **thermal efficiency** of devices such as solar collectors, heat pumps, and boilers.

The thermal efficiency of any device that converts one form of energy into heat is the ratio of the useful heat output from that device, divided by the energy input to the device. The greater this ratio, the higher the thermal efficiency of the device.

The principle of thermal efficiency also applies to a complete heating system. Here, the objective is to maximize the useful energy delivered to the heating load, per unit of energy contained in the original energy source. This objective applies to all facets of design, installation, and operation of the heating system. Any design details that improve the thermal efficiency of a heating system, while also being cost-effective, are generally desirable.

THE SECOND LAW OF THERMODYNAMICS

When designing heating systems, especially those with renewable energy heat sources, it's important to consider the **second law of thermodynamics**. Historically, this law has been stated in many ways by famous scientists such as Carnot, Kelvin, and Plank. Some of these statements are highly technical and beyond the scope of this text. However, there are several practical ways to consider the second law of thermodynamics. One of the simplest has already be mentioned in this chapter: *Heat always moves from an area of higher temperature to an area of lower temperature.*

Another way to think about the second law of thermodynamics involves a quantity called **entropy**, which is a measure of the usefulness of energy. Energy at low entropy is very useful, whereas energy at high entropy is less useful, or perhaps even useless. The second law of thermodynamics states that the **entropy** of any isolated system always increases with time. In other words, all energy in the universe is continually changing from lower entropy to higher entropy, and thus becoming less useful.

Examples of energy in *low entropy* form are electricity, and the heat available from a 2,500 °F flame. Both of these energy sources are highly useful. Electricity can be used to operate a wide range of devices such as motors or heat pumps or even converted into heat at very high temperatures. One example of the latter is the tungsten filament in an incandescent light bulb, which converts electrical energy into heat at a temperature of about 5400 °F. Heat from a 2,500 °F flame could heat several types of metal to their melting point. It could also be used to create steam, heat buildings, or operate a powerful steam engine. In short, both of these low entropy energy sources are highly useful in a wide range of potential applications.

Two examples of energy in *high entropy* form are the heat in air at 65 °F and the heat available when an ice melts into liquid water at 32 °F. There can be a large quantity of heat in a large volume of 65 °F air, but none of that energy can be *directly* transferred to maintain the temperature of a room at 70 °F. Although many people might disagree, there can also be a large quantity of heat in ice. Every pound of water at 32 °F must release 144 Btu of heat to change into a pound of ice at 32 °F. Imagine how much heat must be released from the water at the surface of a large lake when it freezes over in winter. Even though billions of Btus of heat energy are released from the water during this process, not one of them can be directly transferred to heat a room at 70 °F. That's because heat cannot directly move from an area of lower temperature to one at higher temperature. Thus, **high entropy energy**, although often plentiful in supply, is less useful than **low entropy energy**.

It's important not to confuse the entropy of energy with the quantity of energy. There can be vast amounts of energy present in a material, but in high entropy form. The energy released as a large lake freezes in winter is an example. Similarly, the entropy of heat available from the flame of a burning match makes that energy very useful, even though the quantity of heat available is very small. A completely burned wooden match only releases about 1 Btu of heat.

Consider the usefulness of heat (e.g., thermal energy) stored in a tank filled with water at 180 °F, versus the usefulness of the same quantity of heat after it has moved into the 65 °F air surrounding the tank in the basement of a house. The 180 °F water, when circulated through a hydronic distribution system, is hot enough to operate any hydronic heat emitter, and easily transfer heat into a 70 °F interior space. However, after the heat has moved from the 180 °F water into the basement air at 65 °F, it is no longer able to heat rooms at 70 °F. Even though the quantity of energy has not changed, it is now useless for directly heating rooms at 70 °F.

It's reasonable to argue that, from the standpoint of the second law of thermodynamics, using the heat from a 2,500 °F flame to directly heat a building at 70 °F is a poor use of that energy. After all, the low entropy energy in the 2,500 °F could be used to create steam, which could drive a turbine, which could spin a generator to produce electricity, which could power a heat pump or light a lamp. Such a process is routinely used in many gas- and coal-fired electrical generating facilities. However, this process only converts about one-third of the original energy in the fuel to electricity. The remaining two-thirds of the energy from the fuel becomes heat, which is often dissipated to the atmosphere using large cooling towers.

A good example of matching the entropy of energy with an application for that energy is using a heat pump to heat a building. Under certain conditions, heat pumps are capable of moving four to five units of low temperature heat from outside air or soil into a building with an interior air temperature of about 70 °F, using only one unit of electrical energy in this process. Thus, one unit of low entropy electrical energy has allowed the practical use of four or five units of high entropy energy. Such a process "respects" the second law of thermodynamics.

Another example of an application that respects the second law of thermodynamics is the use of a large surface area hydronic heat emitter, such as a floor with embedded tubing, used in combination with a renewable heat source such as solar thermal collectors or a heat pump. Both of these heat sources are more efficient at gathering renewable energy when operated at low water temperatures. Coupling either to a well-designed hydronic radiant panel could allow the average water temperature in the system to be in the range of 90 to 100 °F, even on a cold winter day. By contrast, using a heat emitter with a smaller surface area could force the system to operate at an average water temperature of 130 °F or higher. Although this may be possible, the efficiency of both heat sources would be significantly lower under such conditions. Good design practice seeks to match high entropy energy sources, such as solar heat stored in "cool" outside air, with low temperature heat emitters. The closer the temperature of the energy source is to the final temperature required for the process, the better the design from the standpoint of the second law of thermodynamics. From the standpoint of heating buildings, the lower the operating temperature of the hydronic distribution system, the better the system can use the relatively high entropy thermal energy available from the renewable energy heat sources discussed in this book.

1.4 The Need for Auxiliary Heating

Several renewable energy heat sources have the potential to *contribute* to the space heating and domestic water heating needs of buildings. However, in the context of meeting these loads, few renewable heat sources are practical as the *sole source* of heat. The reasons for this vary with the renewable heat source, the load(s) it serves, and the comfort tolerance of the building occupants.

For example, a solar thermal heating system installed in a New England home might be able to supply 100 percent of that home's heating needs *if* the occupants were willing to let the indoor temperature drop 15 °F to 20 °F *below* normal interior comfort temperatures, during a stretch of cold and cloudy midwinter days. Most North Americans are not accustomed to such a compromise and would not tolerate it as a necessary condition of living with a renewable energy heat source.

Similar reasoning applies to providing domestic hot water. For example, most people prefer to shower with water in the temperature range of 102 °F to 108 °F. Imagine taking a shower on a winter morning that followed a cloudy day. Domestic "hot" water for the shower is supplied by a solar-only water heating system. With the shower valve in the full hot position, the water temperature leaving the showerhead is 90 °F. Would this be acceptable to you? Grab a thermometer and try this for yourself, then answer the question. Next, ponder the implications of promoting this solar-only domestic water heating system, which would likely create this scenario several times each year, to North American consumers.

Another reason for providing an **auxiliary heat source** is that heat availability from renewable energy heat sources is often not synchronized with the load. Figure 1-9 illustrates this, on an annual basis, for a system that provides space heating using solar thermal collectors.

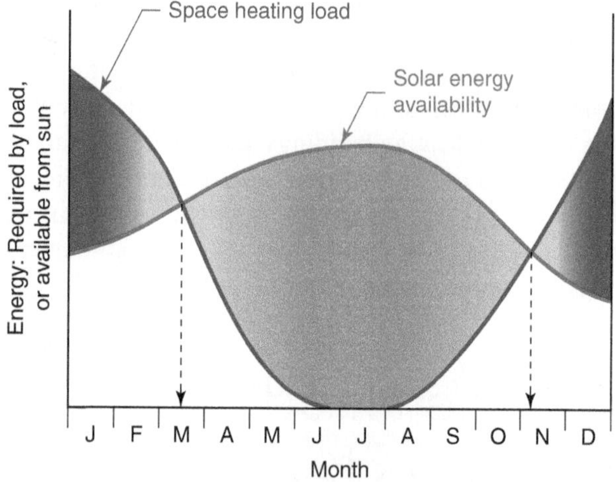

Figure 1-9 | Availability of solar thermal energy (orange curve) versus space heating load (blue curve) for a small, well-insulated house in Syracuse, NY.

On an average monthly basis, the space heating load exceeds the heat available from the solar collectors during January, February, and the first half of March as well as from early November through the end of December. During this time, the building must receive heat from some other source, or its interior temperature will drop, often to values that would be unacceptable to its occupants.

This "out-of-phase" relationship between heat availability and load also exits on a shorter term basis. For example, the occupants of a house heated by a wood-fired boiler system with minimal thermal storage may not be available to regulate the heat produced by that boiler 24 hours a day, 7 days a week. There will likely be times when the boiler's output exceeds the load as well as times when its output is significantly less than the load. Without an auxiliary heat source, it would be very difficult to maintain thermal comfort.

Another reason for auxiliary heating is convenience and life style. History has shown that it's possible to heat houses using nothing other than a woodstove. However, the choice to use a woodstove, which must be manually loaded with fuel, as the *sole means* of heating a house is as much a choice of lifestyle as it is a choice of fuel. Occupants who rely solely on a manually loaded wood-burning heater in a cold climate must be available to tend the fire, often several times during a typical winter day (Figure 1-10). If they want to be away from their house for a day or more during cold weather, they must make alternate arrangements to maintain the house at some minimum acceptable temperature. Although this is certainly possible, it's also a serious complication to modern life styles.

Figure 1-10 | Loading firewood into an outdoor wood-fired heater.

Thus, most of the systems discussed in this text include provisions for auxiliary heating that can supplement the heat produced by the renewable energy heat source when necessary.

The conditions under which the auxiliary heat source is operated can vary. In some cases, the building owner may choose to turn on the auxiliary heat source manually, such as by flipping a switch, or kindling a fire. However, in most cases, systems will be configured so that the auxiliary heating device turns on *automatically* when necessary, and thus allows for unattended operation. Such systems can transition from heat being supplied by the renewable heat source, to heat being supplied by the auxiliary heat source, with no change in interior comfort. In most cases the occupants would be totally unaware of this transition.

1.5 Energy Conservation Versus Energy Supply

According to the National Renewable Energy Laboratory (**NREL**), residential buildings use approximately 57 percent of the total energy consumed by *all* buildings in the United States. Single-family homes use 71 percent of that total. Heating and air-conditioning accounts for 43 percent of total residential energy use, appliances account for another 40 percent, and the rest is used for domestic water heating. Reducing space heating and domestic water heating energy requirements in residential buildings through energy efficiency and passive solar design can significantly lower the building's total energy consumption. It can also significantly reduce the size and installation cost of a renewable energy heating system for that building.

For most American homeowners, the primary reason to consider using renewable energy is *reduced operating cost* relative to conventional fuels. This statement is not meant to diminish the other benefits of using renewable energy, such as lower carbon emissions, reduced dependency on imported fuels, or even enhanced national security. It simply acknowledges a well-established market reality; that wide spread use of renewable energy will not occur without appealing economic justification. *The vast majority of North American homeowners will not make a significant investment into a renewable energy system without assurance of corresponding economic benefits.*

There are two fundamental ways to reduce the operating cost of systems that supply spacing heating and domestic hot water:

1. Find an energy source that costs less per unit of energy supplied.
2. Reduce the load so less energy is required.

When considering the use of a renewable energy heat source, many building owners initially assume that it can *completely replace* their building's need of conventional fuel. They might pose a question such as: "How many solar collectors would I need to heat my house?" They view the situation much like the option of converting their heating system from one fuel to another, based on the expectation that the renewable heat source will provide a lower operating cost.

Although it is possible to build renewable energy heating systems that can supply 100 percent of the thermal load, it is seldom cost-effective, nor does it generate the best long-term consumer satisfaction. *This is particularly true with solar thermal systems that supply space heating.*

Cost/Benefit Ratio

Most building owners have several available options for reducing their cost of providing space heating and domestic hot water. Some of these options pertain to energy conservation, while others apply to lowering the cost of the energy required. From the standpoint of determining the best way(s) to invest capital to reduce energy cost, it is prudent to compare the available options. There are several ways to do this. They range from very simple concepts, to sophisticated financial analysis. Chapter 16 provides a discussion of these options. However, for now, consider a simplified concept called **cost/benefit (c/b) ratio**. The c/b ratio, in its basic form, is found by dividing the initial cost of an improvement that lowers energy cost, by the annual savings generated by that improvement.

Example 1.1: A family is building a new home. They have the option of upgrading the standard insulation "package" offered by the builder to an "energy-efficient insulation package" that would reduce their current heating cost by an estimated $350 per year. The added cost of this insulation upgrade option is $5,000. They also have the option to add a solar thermal **combisystem** (i.e., one that provides space heating and domestic hot water) to the home. That system, on an average year, and based on the current cost of energy, will save them an estimated $650 per year. It will have a net installation cost, after deducting for any applicable tax credits and rebates, of $12,000. Which is the better option based on a simple c/b ratio?

Solution: The c/b ratio for the insulation upgrade is: $5,000/$350 = 14.3.

The c/b ratio for the solar combisystem is: $12,000/$650 = 18.5.

Discussion: Based on this simple analysis, the insulation upgrade option has the *lower* c/b ratio, and is therefore the better investment. Keep in mind, however, that this is a very simplistic comparison. It does not include financing costs, maintenance costs, or fuel price inflation over time. These factors will be introduced and discussed in more detail in Chapter 16.

The numbers used in Example 1.1 are simply made up and don't reflect any specific ratio or quotation. However, experience over many years and many installations has shown that *investments associated with reducing thermal loads through energy conservation usually provide more favorable economic returns than do investments that supply existing thermal loads using renewable energy heat sources.* This is especially true when thermal loads are relatively high due to wasteful usage, poorly insulated buildings, or inefficient mechanical systems. In short: *It seldom makes financial sense to add a renewable energy heating system to an energy wasteful building when the objective is to reduce the cost of space heating and domestic water heating.*

The best starting point when considering the use of any renewable energy heat source is what can be done to reduce the thermal load. This holds true for space heating as well as domestic water heating.

Reducing Space Heating Loads

When energy was relatively inexpensive, North American homes often had design space heating loads of 30 to 50+ Btu/hr/ft^2 of floor area. Thus, a 2,400 square foot house, on a cold winter day, could require 72,000 to 120,000 Btus of heat input, per hour, to maintain comfortable interior conditions. Providing even half of this heating requirement in a cold and marginally sunny location, would require several hundred square feet of solar thermal collectors and thousands of gallons of thermal storage.

Consider, for example, a 2,400-square-foot house in Milwaukee, Wisconsin, with a design space heating load of 40 Btu/hr/ft^2 and a domestic water heating load of 80 gallons per day heated to 120 °F. Using a solar thermal design software package called **f-chart**, it can be shown that 16 4-foot-by-8-foot flat plate solar collectors and a 1,000-gallon thermal storage tank would only provide about 29 percent of the annual energy required for space heating and domestic water heating. The installed cost of such a system would likely be in the range of $75,000 in 2013 dollars. Special structural and spatial modifications might have to be made just to accommodate the necessary equipment. Such a system would also suffer from significant overheating issues during warmer weather, when the heat generated by the solar collectors far exceeds the relatively small heating and domestic hot water loads of the home.

Today, there are many methods and materials that can be used to limit the design space heating load of houses and other small buildings to 10 to 20 Btu/hr/ft^2 of floor area. If the 2,400-square-foot house in Milwaukee was constructed so that its design heating load was 15 Btu/hr/ft^2, a system with *six* 4-foot-by-8-foot flat plate collectors and 380 gallons of storage could supply about 29 percent of the annual space heating energy.

The initial *cost* associated with reducing the space heating load of a new 2,400-square-foot house from 40 to 15 Btu/hr/ft^2 would likely be several thousand dollars. However, the installation cost savings associated with the smaller solar heating system would likely be several *tens of thousands* of dollars. In this situation, energy conservation easily trumps investing in a solar thermal system large enough to provide even 29 percent the existing load.

Energy-Efficient Construction Details

Heating professionals should be familiar with energy-efficient construction methods, especially if working with renewable energy heat sources. This knowledge can be used to advise clients who often don't understand the importance of energy conservation versus renewable energy supply options for buildings with marginal energy performance.

Without specific codes that require otherwise, the author recommends the following *minimum* insulation and air leakage rate values for new residential construction in cool and cold climates.

MINIMUM insulation and air leakage recommendations

Wall insulation: R-30 °F•hr•ft^2/Btu

Ceiling insulation: R-60 °F•hr•ft^2/Btu

Insulation for floors over partially heated spaces: R-30 °F•hr•ft^2/Btu

Windows and doors: R-3.5 °F•hr•ft^2/Btu

Basement wall insulation (heated basements): R-16

Underslab insulation: R-10 °F•hr•ft^2/Btu

Air leakage rate: 1.0 air changes per hour at 50 Pascal pressure differential

The insulation requirements are stated as **R-values**, which is a measure of *resistance* to heat flow through building panels, such as walls, ceilings, or windows, that separate heated space from unheated space. The air leakage requirement is stated in **air changes per hour** at a specific building pressurization (50 Pascal). Both

R-value and air changes per hour are discussed in more detail in Chapter 2.

Other construction techniques that help reduce space heating load include:

- Framing details that reduce or eliminates thermal bridging
- Raised heal roof trusses that allow full insulation thickness over exterior walls
- 24 inches on center inline framing (floors, walls, ceilings)
- Two-stud corners on exterior framed walls
- Use of structural insulated panels (SIPs) for walls and roofs
- Use of insulated concrete forms (ICFs) for below and above grade walls
- Spray foam insulation at the sill and rim joist areas of floor decks
- Spray foam insulation of stud cavities
- Fully taped building wrap over sheathing
- Insulated headers over windows and doors
- Complete interior air barrier with low-perm (vapor transmission) rating
- Dampers on all exhaust fans
- Avoidance of crawl spaces

Figure 1-11 shows a wall section containing several of these details. Keep in mind that these are only a small sampling of the construction details available to significantly reduce space heating loads using modern construction methods and materials.

PASSIVE SOLAR DESIGN

The intentional orientation of a building, and placement of its windows to capture sunlight, can significantly reduce the energy that must be supplied by that building's heating system. This approach to reducing space heating energy requirements is called **passive solar design**. The word "passive" refers to the architectural design of a building that enables it to capture and store solar energy, and maintain acceptable comfort, without the need of powered devices such as circulators, fans, or blowers to move heat within the building. This is a distinctly different approach from **"active" solar thermal technology**, which *does* require devices such as circulators, fans, or blowers, to move energy from where it is collected to where it is stored, and ultimately to where it is needed in the building. The roots of passive solar design can be traced to the ancient Greeks. Through trial and error, they learned that a building's orientation relative to the sun's path across the sky was not a trivial matter.

Passive solar design is most effective in new construction and on sites having good solar access. Those planning to use solar energy for domestic water heating, active space heating, and producing electricity using photovoltaic panels, should also consider some degree of passive solar design as a cost-effective way of reducing the building's net heating load.

Figure 1-12 shows an example of a "direct-gain" passive solar building with a generous south-facing window area.

Buildings that are intended to take advantage of passive solar gains should be oriented with their long axis facing within 30º of true south. Theoretically, the best orientation to maximize the annual solar radiation on a vertical wall is **true south**. However, some designers prefer to rotate the building 15º to 30º east of true south. This favors morning solar gain through south-easterly facing windows. Furthermore, this orientation reduces solar gain through these windows in afternoon, which helps reduce summer overheating.

The recommended amount of southerly exposed glazing for passive solar buildings varies with climate. Southerly glass areas of 8 to 12 percent of the floor area, that remain unshaded during most of the winter, are generally positive contributors to the heating load (e.g., their solar gains more than offset their heat losses). The lower end of this range pertains to well-insulated buildings in cold climates with marginal winter solar availability. The upper end applies to buildings with less insulation but in sunnier and warmer climates. For buildings in cold climates, the glazing used in southerly exposed windows should have a low **shading coefficient** to maximize solar energy transfer to the interior. All windows used in such climates should have low-E coatings that reduce heat loss by thermal radiation.

During midday hours on a sunny winter day, most passive solar buildings gain heat at rates that far exceed their rate of heat loss. Such a situation, if not properly addressed, can quickly lead to overheating. The most common solution is to incorporate **thermal mass** into the building. Thermal mass describes the ability of a material to store thermal energy. Common construction materials such as concrete and masonry, stone, ceramic floor tile, drywall, and even wood, all provide some

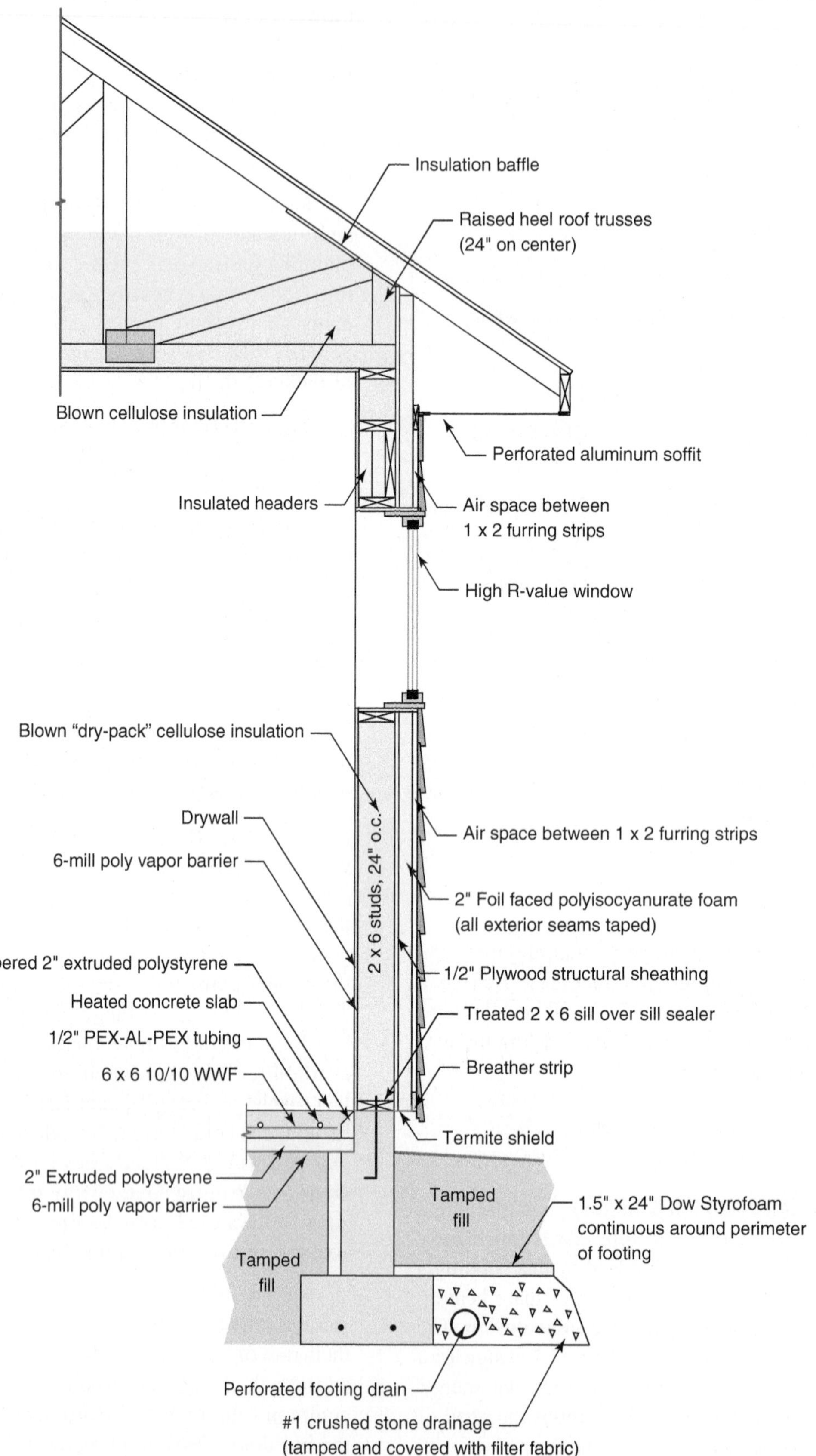

Figure 1-11 | Example of energy efficient construction details.

Copyright © Rachel Wagner, AIA

(a)

Copyright © Rachel Wagner, AIA

(b)

Figure 1-12 (a) South-facing exterior of a direct-gain passive solar structure in Minnesota. Note the overhang for summer shading. (b) Interior view with afternoon sunlight incident on tiled thermal mass floor and walls.

amount of thermal mass. Water can also be used as interior thermal mass, but it must be held in a container that provides acceptable aesthetics.

Thermal mass must be proportioned and placed within the structure so that it can absorb energy directly from the incoming sunlight, or indirectly by contact with heated room air (Figure 1-13a). *Absorbing energy directly from incoming sunlight is preferred because it helps reduce interior overheating.*

A common approach is to create a high thermal mass concrete floor directly behind southerly facing windows as shown in Figure 1-13b. To remain

Courtesy of Misha Rauchwerger

(a)

Courtesy of Misha Rauchwerger

(b)

Figure 1-13 (a) Exterior of a direct-gain passive solar home in California. Note how the low winter sun angle results in minimal shading of windows. (b) Interior showing polished concrete slab floor serving as thermal mass.

effective, such a floor cannot be covered with rugs or other objects that significantly cover the floor's surface. That surface should also have a color that encourages absorption of the incoming solar radiation. The floor slab must be placed over a minimum of 2 inches of extruded polystyrene underside insulation to limit downward heat loss. Additional thermal mass may be placed on the surface of interior walls or other architectural features exposed to direct or reflected sunlight.

The thermal mass in a passive solar building will interact with that building's auxiliary heating system. In some cases, this interaction can create undesirable, and often unanticipated effects. Consider the following scenario: While planning a direct-gain passive solar house, a designer contemplates a means of auxiliary heating for cloudy days when solar gains cannot maintain thermal comfort. Given that an insulated concrete slab floor is already planned, the designer decides to embed tubing within the slab and circulate warm water through this tubing when the building needs auxiliary heating. Many readers will recognize this approach as **radiant floor heating**.

Although this approach seems rational, it is likely to cause significant overheating whenever a cold night is followed by a sunny morning. The problem arises because the floor slab is often maintained at an elevated temperature (typically 75 °F to 85 °F at the upper surface) by the auxiliary heating system during the late night and early morning hours of a cold winter day. When solar gains become significant, at perhaps 9:00 AM the following morning, the slab is already "filled" with stored heat and thus very limited in its ability to absorb more heat from the impinging sunlight. This causes the building's air temperature to rise rapidly. It doesn't take long before the occupants start opening windows to purge the surplus heat in an attempt to maintain comfort. The net result is that much of the passive solar gains are lost through ventilation. Even worse, the building's cooling system may turn on, and thus create a significant electrical load. To avoid such situations, it is imperative to match the response characteristics of the supplemental heating system with the thermal mass characteristics of the building. This will be discussed in detail in Chapter 5.

REDUCING DOMESTIC WATER HEATING LOADS

The priority of energy conservation also applies to domestic water heating. Instead of trying to "overpower" a significant domestic water heating load using a renewable energy heat source, it's better to take steps that reduce that load. There are several ways to reduce domestic water heating loads that often yield favorable economic returns compared to increasing the size of a renewable energy heat source and its related hardware. Several of these measures reduce *both* the energy required to heat water and the quantity of water used.

One method for reducing the cost of supplying domestic hot water is to reduce the flow rate allowed through fixtures. A federal law enacted in 1992 set the maximum flow rate of an individual showerhead at 2.5 gallons per minute when supplied with water at 80 psi pressure. However, multiple showerheads are allowed within a single showering compartment, and thus the total hot water flow rate into the shower compartment could be much higher than 2.5 gallons per minute. In 2011, the **United States Department of Energy (DOE)** enacted new showerhead standards that would limit the *total* water flow rate into the *shower compartment* to 2.5 gallons per minute. However, as this text is being written, it remains unclear if the DOE will enforce these new regulations.

Politics aside, low-flow showerheads, combined with judicial "self-enforced" hot water usage, will obviously reduce energy use and likely increase contributions from solar water heating systems, as well as other types of renewable heat sources.

There are several other ways to reduce domestic water heating loads:

- *Install aerators on faucets.* Faucet **aerators** are inexpensive and easily installed devices that mix air into the water as it leaves the fixture. This reduces the flow rate from the fixture, improves the washing characteristics of the stream, and reduces splashing as seen in Figure 1-14. Faucet aerators that limit flows to 0.35 gallons per minute are now available.
- *Adding insulation blankets to storage-type water heaters.* The DOE estimates that adding insulation blankets to storage-type water heaters can reduce standby heat loss by 25 to 45 percent. This lowers water heating energy use by an estimated 4 to 9 percent. Such blankets should have a minimum insulating value of R-11. They are available at most home centers and are easily installed by the average homeowner. The insulating blanket should cover the sides and top of *electric* water heaters, as shown in Figure 1-15. Cutouts need to be made to access the thermostats and elements. Insulating blankets should

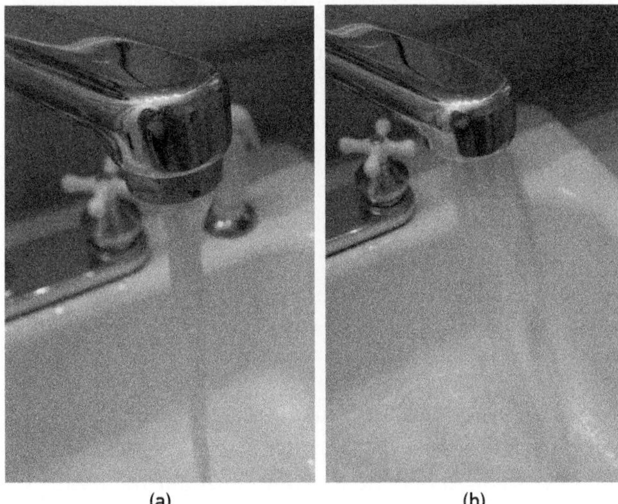

Figure 1-14 (a) Faucet with a 2.0 gallon-per-minute aerator installed. (b) Same faucet with aerator removed, operating with the same valve setting and the same water pressure.

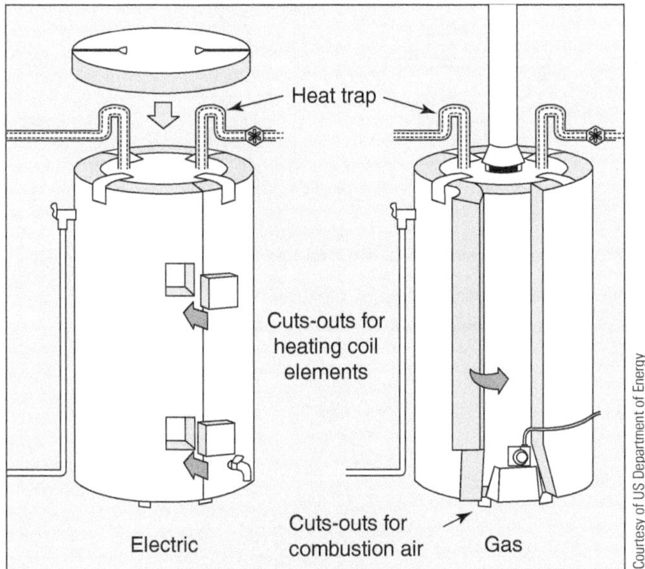

Figure 1-15 Insulation blankets installed on electric (left) and gas-fired (right) water heaters.

not be installed on the top of the gas-fired water heaters due to the proximity to the flue connector.

- *Placing insulation under* electric *storage water heaters.* Placing an electric storage-type water heater on a platform built from 2-inch-thick extruded polystyrene insulation board will reduce downward heat loss from electric storage water heaters. This detail is *not* recommended for gas-fired water heaters due to the proximity of the combustion chamber to the bottom of the heater.

- *Insulating hot water distribution piping. Adding ½-inch-thick foam rubber insulation to copper tubing carrying 120 °F water through a space at 55 °F air temperature reduces heat loss to about one-third that of uninsulated copper tubing. Such insulation is required by some state energy codes for any hot water piping passing through unheated or partially heated spaces. Pipe insulation, as shown in figure 1-16, is a relatively inexpensive and easily installed product that will pay for itself several times over during the life of the system.*

Figure 1-16 Piping insulated with foam rubber insulation.

- *Reducing hot water supply temperature.* The lower the temperature to which domestic water is heated, the less energy is required. From the standpoint of burn protection, hot water should never be delivered from fixtures such as lavatories, tubs, or showers at temperatures above 120 °F. However, from the standpoint of eliminating Legionella bacteria—the cause of **Legionnaire's disease**—many sources recommend heating water to a minimum of 140 °F and then reducing its delivery temperature at the fixtures using thermostatic mixing valves. There are several ways to do this in systems that combine renewable energy heat sources and auxiliary heating devices. They will be discussed in later chapters.

- *Installing a recirculating hot water system with fully insulated piping.* Hot water distribution piping in homes and small commercial buildings is often designed with a "main" pipe that supplies several branches leading to individual fixtures. When the distance between the hot water source and the farthest fixture is only a few feet, such a piping system provides acceptable performance. However, as the distance between the hot water source and the farthest fixture increases, so does the "wait time" for hot water delivery to the farthest fixture. This is especially noticeable when water in the hot water piping has cooled substantially due to lack of demand for several hours.

The wait for hot water, the loss of water that has cooled in piping between demands, and the energy required to heat additional water to make up for this wasted cool water can be largely eliminated by a **recirculating hot water distribution system** as illustrated in Figure 1-17.

This type of system uses a very small, low-wattage circulator to move hot water through insulated distribution piping. When there is no demand for hot water at the fixtures, the hot water circulates back to the storage-type water heater. With properly insulated pipes, the water loses very little heat as it travels through the circuit. For example, a ½-inch copper tube with ¾-inch-thick foam rubber insulation, routed through a space with a 65 °F air temperature, and carrying 120 °F water at a low-flow rate, only loses about 5.4 Btu/hr per foot of piping.

In some systems, the small recirculation circulator operates continuously. However, in most homes, it is not necessary to operate the circulator during nighttime hours or when the home is unoccupied. Recirculation

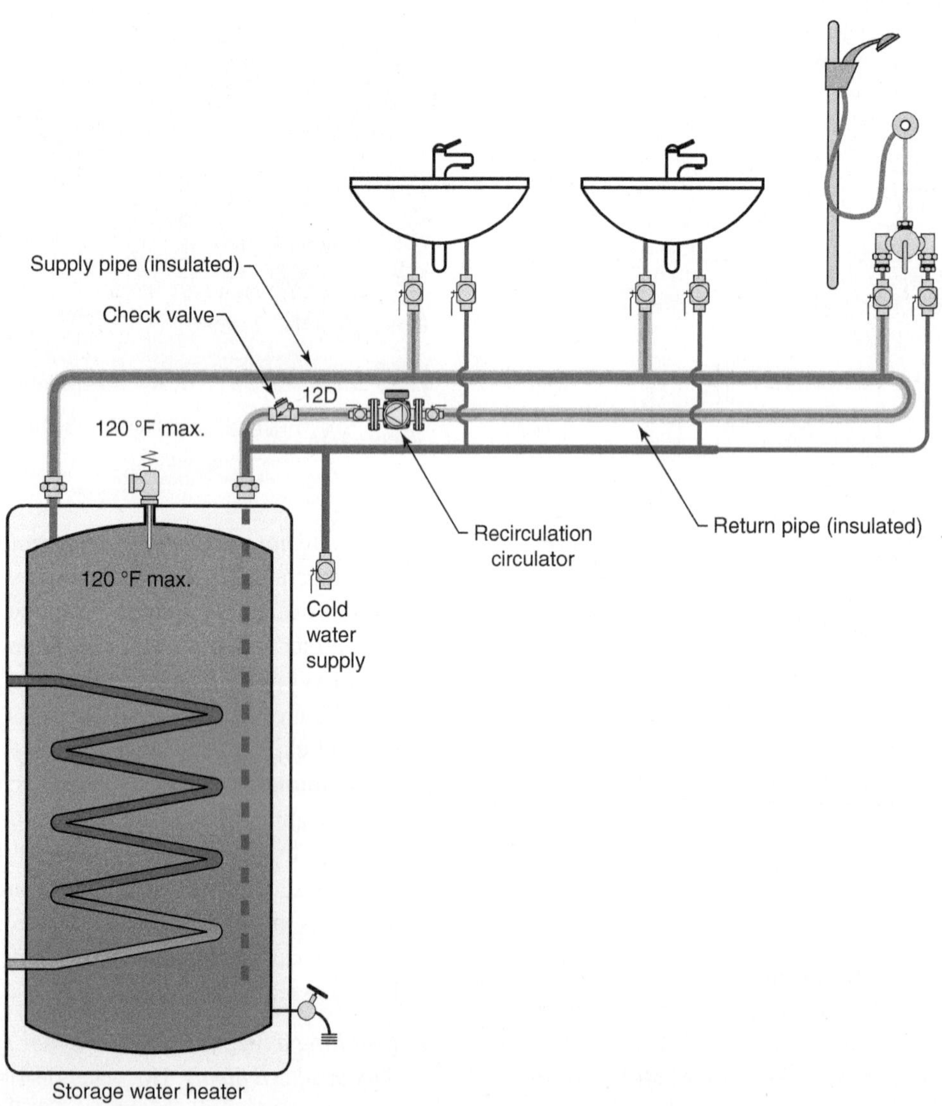

Figure 1-17 | A basic hot water recirculation system.

circulators are now available with built-in timers that allow their operating times to be matched to likely demands. An example of a small circulator for a recirculation hot water system is shown in Figure 1-18. When operating at full speed, this circulator draws only about 12 watts of electrical power.

Figure 1-18 | Example of a circulator for a recirculating domestic hot water system.

Because hot water is present in a recirculating domestic hot water system during much of the day (if not the entire day), it is imperative to insulate both the hot water supply and return piping. The minimum recommended R-value for such insulation is R-4.

It is also possible to create recirculation distribution systems in which the circulator is "demand operated." One approach is illustrated in Figure 1-19.

The hot water circulator is turned on by any of the pushbutton switches, each located in a bathroom or near other fixtures requiring hot water. Hot water moves quickly through the outgoing distribution piping. The circulator turns off when hot water warms a thermostat attached to the piping near the last fixture served by the recirculation loop. If the water temperature at the thermostat was already warm, based on previous hot water usage, the recirculation circulator does not turn on. This approach reduces the operating time of the circulator but still allows rapid availability of hot water at each fixture when needed.

- *Design distribution piping to minimize distance to fixtures requiring hot water.* It stands to reason that piping arrangements that minimize the distance from the hot water source to the fixtures reduce the volume of heated water in the piping. In a non-recirculating system, this will also reduce the wait time for hot water at the fixtures as well as the amount of cool water sent down the drain while waiting for hot water to arrive.
- *Repair any leaks or drips in hot water distribution piping or fixtures.* A leak or drip in a hot water piping system is a thermal load. It's also a waste of water. Although the flow is seemingly small in comparison to when the fixture is in use, a leaky faucet that drips 50 times per minute will waste about 340 gallons of water per year (Figure 1-20). This is needless waste that is usually easy to correct.
- *Add heat traps to the piping connections on storage-type water heaters.* **Heat traps** are designed to inhibit the tendency of hot water to rise from a storage-type water heater into piping above it. Heat traps can be fabricated by installing the piping carrying hot water out of a storage-type water heater so that it drops at least 12 inches lower than where it connects to the water heater, as shown in Figure 1-21.

If the dropped piping detail is not possible, heat traps can also be created using fittings installed at both the cold and hot water connections of storage water heaters. Some older heat trap fittings contain small balls that rise and fall to block convective flow from the heater when there is no demand. Newer heat trap fittings use flexible rubber check valves for the same purpose. An example of the latter is shown in Figure 1-22.

- *Use tankless water heaters to reduce standby losses.* Most North Americans associate domestic water heating with tank-type appliances that hold heated water ready for use. However, there are many water heating devices now available that do not require storage tanks. Instead, they heat water "instantaneously" as it flows through them and onward to fixtures. Such devices are called by different names, including "**tankless water heaters**," "on-demand water heaters," and "instantaneous water heaters." They are available in both gas-fired and electric models in a wide range of capacities for both residential and commercial applications. Because they

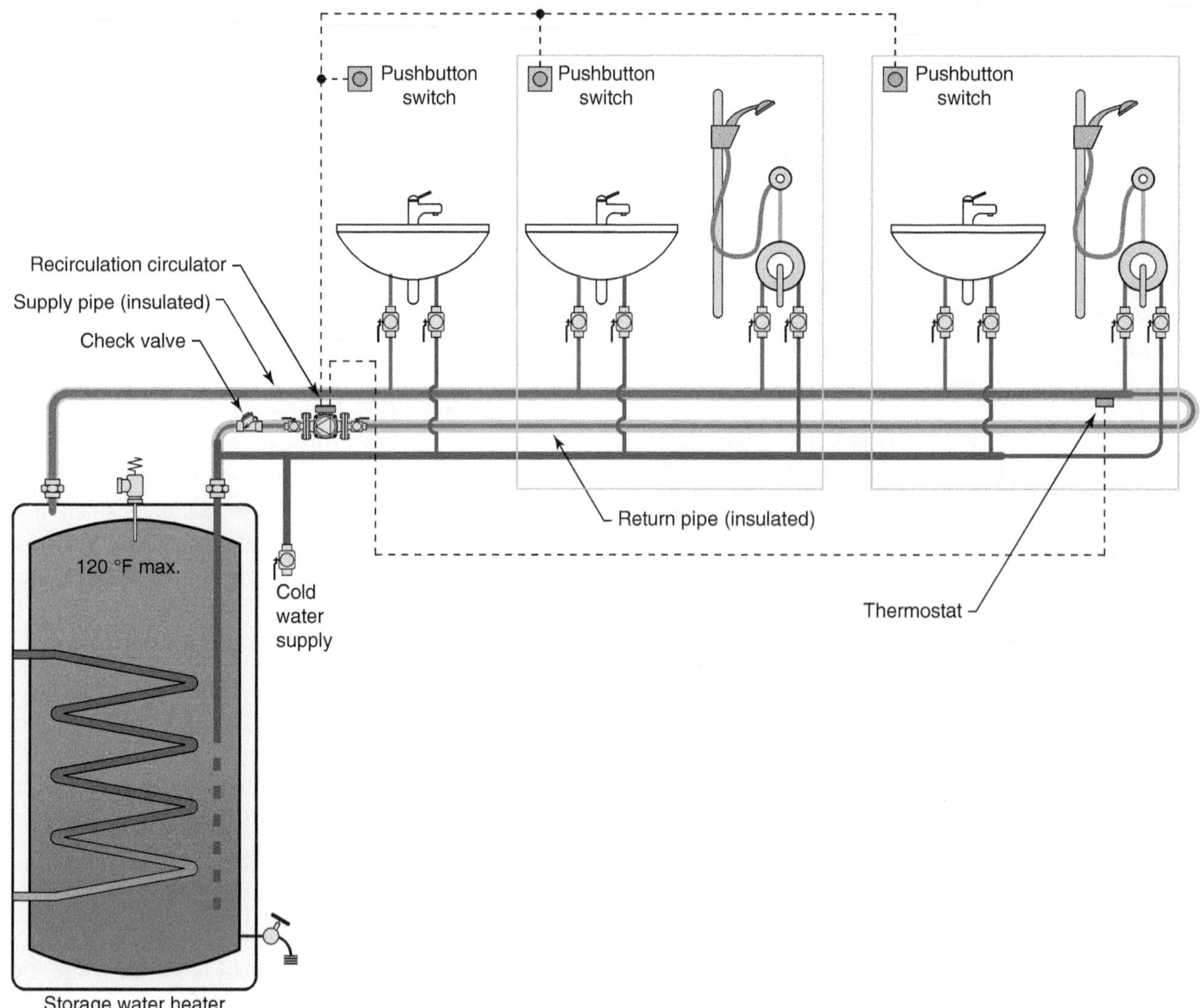

Figure 1-19 | Demand-operated recirculation hot water delivery system.

heat water only when it's needed, tankless water heaters eliminate the standby heat loss associated with storage-type water heaters. Figure 1-23 is an example of an electric tankless water heater designed for installation near a fixture such as a lavatory or shower.

Electric tankless water heaters have one or more resistance heating element housed in a pressure-tight chamber through which water flows. These elements are turned on by a flow switch that detects whenever flow through the unit reaches approximately 0.6 gallon per minute.

Some electric tankless water heaters are "on/off" devices. They operate at full heating capacity whenever the flow rate through them is above the turn-on threshold. Other electric tankless water heaters are thermostatically controlled, with delivered water temperature accuracies of +/−1 °F. Thermostatically controlled units monitor flow rate, incoming temperature, and leaving water temperature. They use this information to vary the input power to the heating elements. Thus, as flow through the unit increases, so does power input in an attempt to keep the outlet temperature very close to the setpoint. Thermostatically controlled electric tankless water heaters are preferred in

Figure 1-20 | A faucet that drips 50 times per minute will waste about 340 gallons of water per year.

Figure 1-22 | Heat trap fittings using rubber check valves.

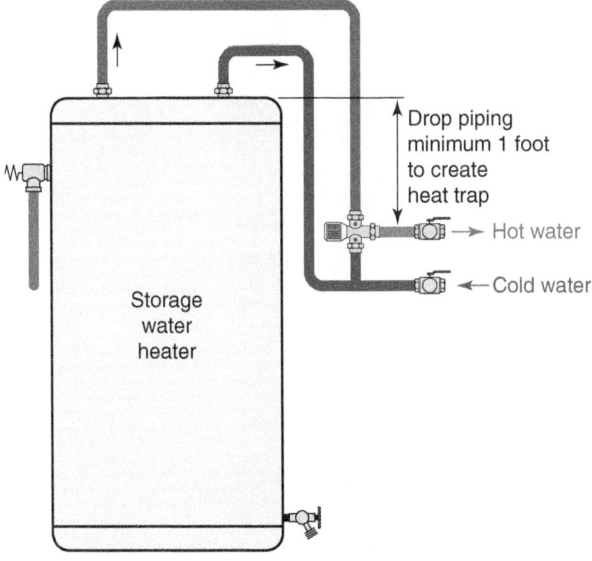

Figure 1-21 | A heat trap formed by dropping both the hot and cold water piping relative to where it connects to the heater.

Figure 1-23 | An electric tankless water heater.

any situation where preheated water is supplied from a renewable energy heat source. Later chapters show how such devices are used in systems where domestic water is typically preheated by a renewable energy heat source.

- *Use greywater heat recovery.* Much of the heat in hot water used for washing remains in that water as it goes down the drain. In most buildings, this heat is simply carried into the sewer. However, it is possible to recover up to 40 percent of this otherwise wasted heat using a **greywater heat exchanger**. An example of such a device is shown in Figure 1-24.

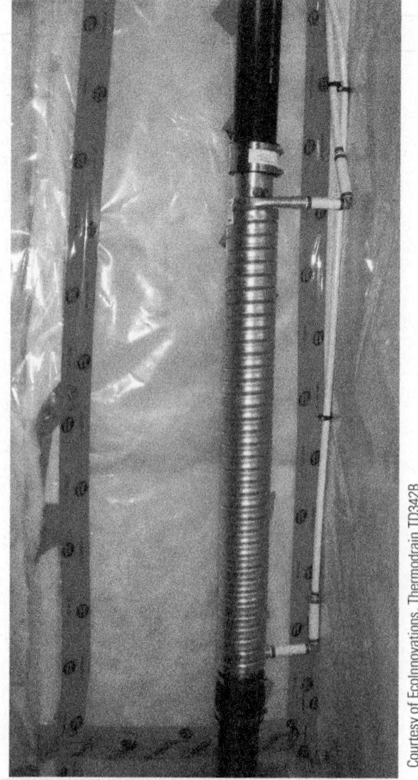

Figure 1-24 | An installed greywater heat exchanger.

Greywater heat exchangers are simple, passive devices. They consist of an inner copper pipe that's fitted inline with the main vertical greywater drainage pipe. This inner copper pipe is wrapped with one or more coils of tightly fitting, partially flattened copper tubing that is bonded to the outer surface of the inner pipe. The construction of a typical greywater heat exchanger is shown in Figure 1-25. A typical installation configuration is shown in Figure 1-26.

Greywater heat exchangers take advantage of frequent simultaneous flows of cold water and hot water as the latter flows to fixtures such as lavatories and showers. As hot water is being used, cold water enters the bottom connection of the coiled tubing and passes upward in a **counterflow** direction to the greywater. Heat from the greywater transfers through the copper tube walls and preheats cold domestic water. The two streams of water are always separated by *two* copper tube walls. A potential leak in either the inner pipe or outer coil would not cause contamination of domestic water.

In Figure 1-26, some of the preheated water leaving the greywater heat exchanger is piped directly to the "cold" water port of the shower valve. This reduces the required flow of fully heated domestic hot water to the shower valve.

Under typical operating conditions, the entering cold domestic water will be warmed 20 °F to 25 °F before it exits at the top of the coil. Thus, water entering the building at

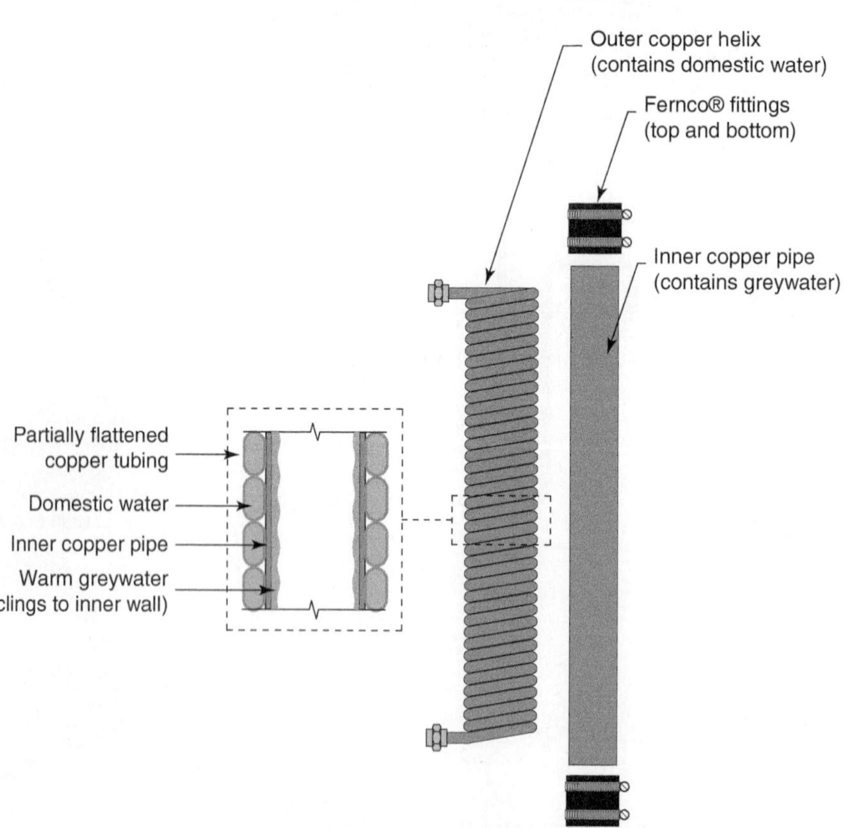

Figure 1-25 | Construction of a typical greywater heat exchanger.

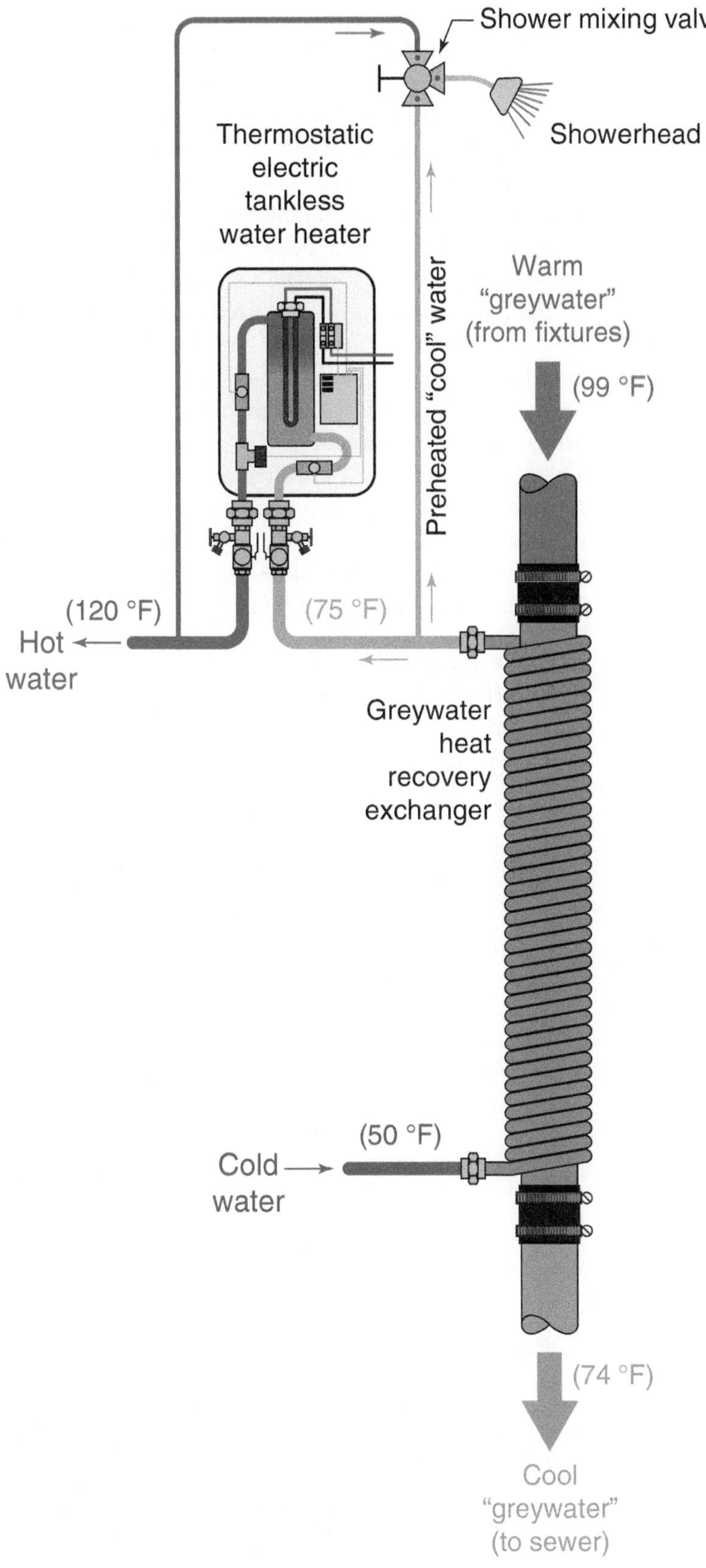

Figure 1-26 Installation of a greywater heat exchanger in combination with a thermostatically controlled electric tankless water heater.

50 °F would be preheated to a temperature of 70 °F to 75 °F before entering the water heater. This reduces the water-heating load by 29 to 36 percent, assuming a final desired delivery temperature of 120 °F. This is a significant contribution for a simple, unpowered and relatively inexpensive device.

Greywater heat exchangers should only be installed in vertical drainage pipes. They rely on the "film effect" of greywater passing through a *vertical* drainage pipe. Most of this water clings to the pipe wall rather than falling through the air space inside the pipe. This is ideal from the standpoint of extracting heat from the inner copper pipe.

1.6 Unit Pricing of Thermal Energy

Imagine someone who is about to build a new house, or have a new heating system installed in an existing house. It's likely they have several fuel options available to choose from. After doing some research with local suppliers, they have compiled the following list of fuel options and associated pricing:

- Natural gas priced at $1.10 per therm
- #2 fuel oil priced at $3.50 per gallon
- Propane priced at $2.75 per gallon
- Electricity priced at $0.12 per kilowatt•hour
- Wood pellets priced at $225 per ton

From the standpoint of energy cost, which of these options is the best buy? The answer is not obvious, because many of the fuels are sold in different units (e.g., therms, gallons, kwhr). Furthermore, these fuels will be converted into heat using devices that operate at different efficiencies.

To answer this question, the cost of each available fuel must be expressed on a **unit price** basis. The units commonly used are *dollars per million Btus delivered heat* (abbreviated **$/MMBtu**). "Delivered heat" means heat delivered to the distribution system by the heat source. The **seasonal efficiency** of each heat source is factored in when determining the unit price of delivered energy.

The equations given in Figure 1-27 can be used to convert the purchase price of several common fuels to a unit price in dollars per million Btus delivered.

Electric resistance heat	____cents/Kwhr x 2.93 = ____ $/MMBtu
Heat pump	(____cents/Kwhr x 2.93) / (____average COP) = ____ $/MMBtu
#2 fuel oil	(____ $/gallon x 7.14) / (____ AFUE (decimal)) = ____ $/MMBtu
Propane	(____ $/gallon x 10.9) / (____ AFUE (decimal)) = ____ $/MMBtu
Natural gas	(____ $/therm x 10) / (____ AFUE (decimal)) = ____ $/MMBtu
Firewood*	(____ $/face chord x 0.149) / (____ ave. efficiency (decimal)) = ____ $/MMBtu
Wood pellets	(____ $/ton x 0.06098) / (____ ave. efficiency (decimal)) = ____ $/MMBtu
Bituminous coal	(____ $/ton x 0.03268) / (____ ave. efficiency (decimal)) = ____ $/MMBtu
Shelled corn**	(____ $/bushel x 2.551) / (____ ave. efficiency (decimal)) = ____ $/MMBtu

NOTES:

1. $/MMBtu = Dollars per million Btu of heat delivered to building
2. Kwhr = Kilowatt • hr. = 3413 Btu
3. Average COP = Average Coefficient Of Performance during heating season (for geothermal heat pump with low temp. distribution system generally 2.5 to 4.0)
4. AFUE = Annual Fuel Utilization Efficiency of appliance (for typical oil or gas boiler use 0.75 to 0.80)
5. The average efficiency for biomass and coal fired boiler can vary widely depending on operating conditions.

* Assumes a 50/50 mix of maple and beech dried to 20% moisture content. Price is for 4 ft x 8 ft x 16 in. face chord split and delivered.

** Assumes 15% moisture content

Figure 1-27 Equations for determining the unit price of various fuels based on purchase price and efficiency.

Some fuels are priced on "sliding scales." The cost per unit of fuel decreases as the monthly usage increases. This is typical of natural gas pricing from many utilities. In such cases, use a weighted average price per unit of fuel based on usage.

In most cases, utilities include a **basic service charge** in their invoices for either electricity or natural gas. This charge can be thought of as "meter rental." It applies regardless of the amount of energy passing through the meter. The basic service charge for electrical service should be *subtracted* from the total invoice amount before calculating the unit cost of energy supplied by electric heating devices such as electric boilers, electric furnaces, electric space heaters, or heat pumps. This is because electrical service is almost always required for any new building, regardless of how it is heated. Thus, the basic service charge for electricity is not strictly assignable to the electrically driven heating equipment in that building. However, if a natural gas service is installed on the building *solely to supply space heating and domestic water heating*, the basic service charge for the gas meter *should be included* in the overall cost of operation. Example 1.3 will demonstrate how to do this.

Example 1.2: Estimate the unit price of the following fuel options and their associated thermal conversion efficiencies.

- Electric resistance heat at $0.12/kWhr (kilowatt-hour) and 100 percent efficiency
- Electric heat pump with seasonal average COP = 2.5*
- #2 fuel oil at $3.50 per gallon and 83 percent AFUE**
- Propane at $2.75 per gallon and 83 percent AFUE**
- Natural gas at $1.25 per therm and 83 percent AFUE**
- Firewood at $60 per face chord and 60 percent seasonal heat-source efficiency
- Wood pellets at $275 per ton and 83 percent seasonal heat-source efficiency
- Coal at $150 per ton and 83 percent seasonal heat-source efficiency
- Shelled corn at $9 per bushel and 80 percent seasonal heat-source efficiency

* **COP** stands for coefficient of performance. This is a performance index for heat pumps and is discussed in detail in Chapters 13 and 14.

** **AFUE** stands for annual fuel utilization efficiency. It represents the seasonal average conversion efficiency for oil and gas-fired heat sources.

Solution: Substituting these numbers into the equations in Figure 1-27 yields the results shown in Figure 1-28.

Electric resistance heat	12 cents/Kwhr x 2.93	= 35.16 $/MMBtu
Heat pump	12 cents/Kwhr x 2.93 / 2.5 average COP	= 14.06 $/MMBtu
#2 fuel oil	3.50 $/gallon x 7.14 / 0.83 AFUE (decimal)	= 30.11 $/MMBtu
Propane	2.75 $/gallon x 10.9 / 0.83 AFUE (decimal)	= 36.11 $/MMBtu
Natural gas	1.25 $/therm x 10 / 0.83 AFUE (decimal)	= 15.06 $/MMBtu
Firewood	60 $/face chord x 0.149 / 0.60 ave. efficiency (decimal)	= 14.90 $/MMBtu
Wood pellets	275 $/ton x 0.06098 / 0.83 ave. efficiency (decimal)	= 20.20 $/MMBtu
Bituminous coal	150 $/ton x 0.03268 / 0.83 ave. efficiency (decimal)	= 5.91 $/MMBtu
Shelled corn	9 $/bushel x 2.551 / 0.80 ave. efficiency (decimal)	= 28.70 $/MMBtu

Figure 1-28 | Completed fuel cost worksheet.

Discussion: The cost of each fuel and the conversion efficiencies shown in Figure 1-28 were not selected to skew opinions. Fuel costs vary widely based on location and market conditions. Likewise, the efficiency of converting various fuels to heat can vary over a wide range depending on the type of equipment, its age, and how well it is maintained. Thus, the relative rankings of unit prices can and often do change frequently. The results shown in Figure 1-28 are simply a "snapshot" based on specific assumptions.

The unit cost of a given energy source is independent of the device used to convert that energy into heat. Thus, a boiler, furnace, or small space heater operating with the same fuel and at the same conversion efficiency would produce the same estimated annual heating cost.

To determine the estimated annual cost of providing space heating and domestic water heating, one needs to know how much total energy, often expressed in units of **MMBtu** (e.g., millions of Btus) is needed by each load. Chapter 2 gives specific information on estimating these values.

Example 1.3 demonstrates how to estimate the annual cost of space heating and domestic water heating using the unit cost of energy, in combination with assumed loads.

Example 1.3: An 1,800-square-foot house, recently constructed in upstate New York, incorporates good insulation and air sealing details. The owner wants to select a single fuel to provide space heating and domestic water heating. Using methods that will be discussed in Chapter 2, the estimated annual energy needs for this house are as follows:

- Space heating: 37 MMBtu/year
- Domestic water heating: 12.6 MMBtu/year

The following fuel options are available to the homeowner:

1. Electric resistance heating at $0.13/kilowatt•hour
2. Electrically operated heat pump with a seasonal COP of 2.5
3. Fuel oil @ $3.50 per gallon used in a standard boiler with AFUE of 83 percent
4. Propane at $2.75 per gallon used in a high-efficiency boiler with AFUE of 93 percent
5. Natural gas @ $0.65 per therm, burned in a high-efficiency boiler with AFUE of 93 percent
6. Firewood at $70 per face cord, burned in a gasification boiler with average efficiency of 75 percent

These costs include the cost of the fuel, delivery charges, and applicable taxes. However, the following basic service charges also apply to utility-supplied electricity and natural gas:

Basic service charge for electricity: $16.50 per month

Basic service charge for natural gas: $17.85 per month

Determine which fuel option provides the lowest cost to the homeowner.

Solution: Start by using the applicable fuel types in Figure 1-27 to determine the unit cost of each available fuel option. These are given in Figure 1-29.

Based solely on these unit costs, natural gas appears to have a significant price advantage over the other options. However, a decision to use natural gas implies an additional $17.85 per month basic

Electric resistance heat	13 cents/Kwhr x 2.93	= 38.09 $/MMBtu
Heat pump	13 cents/Kwhr x 2.93 / 2.5 average COP	= 15.24 $/MMBtu
#2 fuel oil	3.50 $/gallon x 7.14 / 0.83 AFUE (decimal)	= 30.11 $/MMBtu
Propane	2.75 $/gallon x 10.9 / 0.93 AFUE (decimal)	= 32.23 $/MMBtu
Natural gas	0.65 $/therm x 10 / 0.93 AFUE (decimal)	= 6.99 $/MMBtu
Firewood	70 $/face chord x 0.149 / 0.75 ave. efficiency (decimal)	= 13.91 $/MMBtu

Figure 1-29 | Determining unit cost of each fuel option using Figure 1-27.

service charge. This charge must be considered since the natural gas would only be used for space heating and domestic water heating. In this example, no such charges apply to the other fuel options. Although there is a basic service charge for electricity, that charge will not be assigned to the cost of energy for space heating or domestic hot water because electrical service is necessary to operate many other devices and appliances in the house.

The total estimated energy required is that used for space heating, plus that required for domestic water heating.

$$37 \text{ MMBtu} + 12.6 \text{ MMBtu} = 49.6 \text{ MMBtu}$$

The estimated annual energy cost for space heating and domestic water heating is now easily calculated by multiplying the total energy requirement (in MMBtu), by the unit cost of each fuel (in $/MMBtu) and then adding in any applicable basic service charges:

1. Electric resistance heating:
   49.6 MMBtu × 38.09 $/MMBtu = $1889/year
2. Electric heat pump:
   49.6 MMBtu × 15.24 $/MMBtu = $756/year
3. Fuel oil:
   49.6 MMBtu × 30.11 $/MMBtu = $1493/year
4. Propane:
   49.6 MMBtu × 32.23 $/MMBtu = $1599/year
5. Natural gas:
   49.6 MMBtu × 6.99 $/MMBtu
   + ($17.85/month × 12months) = $561/year
6. Firewood:
   49.6 MMBtu × 13.91 $/MMBtu = $690/year

Discussion: Even with the basic service charge included, which in this case represents about 38 percent of the total, natural gas provides the lowest annual operating cost. Second place, based on the stated assumptions, goes to firewood burned in a high-efficiency wood gasification boiler. Electric resistance heating is the highest cost option, followed closely by fuel oil and propane.

It is imperative to view these results as a "snapshot" that only applies at a given time, in a given location, and with given assumptions on heating source performance. For example, as this text is being written:

- The cost of natural gas service in upstate New York is about *half* of what it was 4 years earlier.
- Fuel oil is relatively expensive due to worldwide demand for crude oil.

Changes in global markets can quickly alter fuel prices, in some cases drastically. Thus, the choice of fuel for space heating and water heating should include perspective on factors beyond current pricing. Issues such as local availability, environmental impact, safety, equipment life, maintenance requirements, on-site storage, any need for space cooling, and automatic versus manual feeding of the heat source should all be weighed into the decision. There is no universal answer to the question: "Which fuel choice is best?" However, the information presented in this section does help in making better-informed decisions.

1.7 Cooling and Ventilation Considerations

This text is focused on providing space heating and domestic hot water. Both are critically important in making homes and other buildings enjoyable, safe, healthy, and affordable. However, even the best space heating and domestic hot water systems can only provide part of what could be described as a high-quality interior environment.

Robert Bean (www.healthyheating.com) has done an excellent job of describing the "big picture" in what he refers to as **indoor environmental quality (IEQ)**. This concept encompasses heating, cooling, and ventilation, along with factors such as vibration, sound, lighting, and odors, to collectively determine the quality of the indoor environment, as illustrated in Figure 1-30.

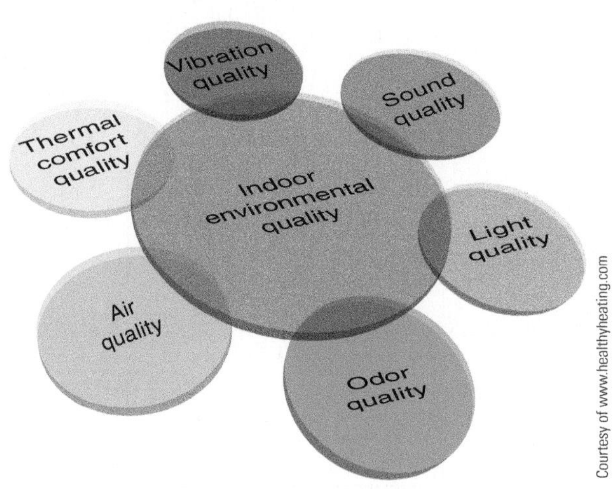

Figure 1-30 | Factors that collectively determine Indoor Environmental Quality (IEQ).

Each of these aspects of indoor environmental quality could fill a textbook and thus cannot be fully addressed here. However, in the more limited context of *thermal comfort* and *indoor air quality*, we will discuss methods that can be coordinated with systems using renewable energy heat sources for space heating and domestic hot water. Those who design such heating systems should have a basic familiarity with cooling and ventilation.

Cooling Options

Many discussions between comfort professionals and potential customers eventually move from heating to the inevitable question: "What do I do about cooling?"

The vast majority of homebuyers, in all but the coldest regions of the United States, expect their homes to be comfortable throughout the summer. Professionals who can integrate cooling, and ventilation, along with heating to provide year-round comfort and healthy indoor air are well positioned for projects where the alternative is to involve multiple designers and subcontractors for each aspect of the overall **HVAC (heating, ventilation, and cooling)** system.

Residential and light commercial buildings that use renewable energy heat sources have several options for cooling. These include dedicated systems that have very little, if any, interaction with the heating system as well as approaches that overlap with the heating system.

Perhaps the most obvious option is to install a separate central cooling system with traditional ducting. Although this has been done in many homes and light commercial buildings, it often involves the complications of routing ducts throughout the building. It should not be dismissed as a possibility, but neither should it be accepted as inevitable based on available alternatives.

It is possible to provide zoned cooling without ducting. Figure 1-31 shows an example of what is appropriately called a **"ductless" cooling system**. Some HVAC professionals also refer to this approach as a mini-split system.

Ductless cooling systems typically have a wall-mounted **evaporator unit** in each of the building's cooling zones. Within these units, heat is absorbed from interior air into a cold refrigerant that evaporates as it passes through a **coil** consisting of aluminum fins and copper tubing, as illustrated in Figure 1-32.

The incoming air is cooled as it passes across the coil. In many cases, water vapor in the incoming air condenses into droplets on the surfaces of the coil. These droplets fall from the coil into a drain pan and are routed outside the building or into the building's drainage system. The cooled and dehumidified air is discharged into the space by a blower within the unit.

A harness consisting of refrigerant tubes and electrical wiring runs from each indoor unit to an outdoor **condenser unit**. In some systems, a separate condenser unit is used for each indoor unit. In other systems, two, three, or even four independent indoor units are connected to a single outdoor condenser unit. The small diameter refrigeration piping, condensate drainage tube, and electrical wiring can be routed through partitions and around other building structures much easier than standard ducting.

A distinct advantage of ductless cooling is that each indoor unit can be operated as an independent zone, with its own interior temperature setting and operating schedule.

Some ductless systems also use **variable refrigerant flow (VRF)** technology, which allows the flow rate of refrigerant to vary among the indoor units in response to their individual cooling loads. These systems typically

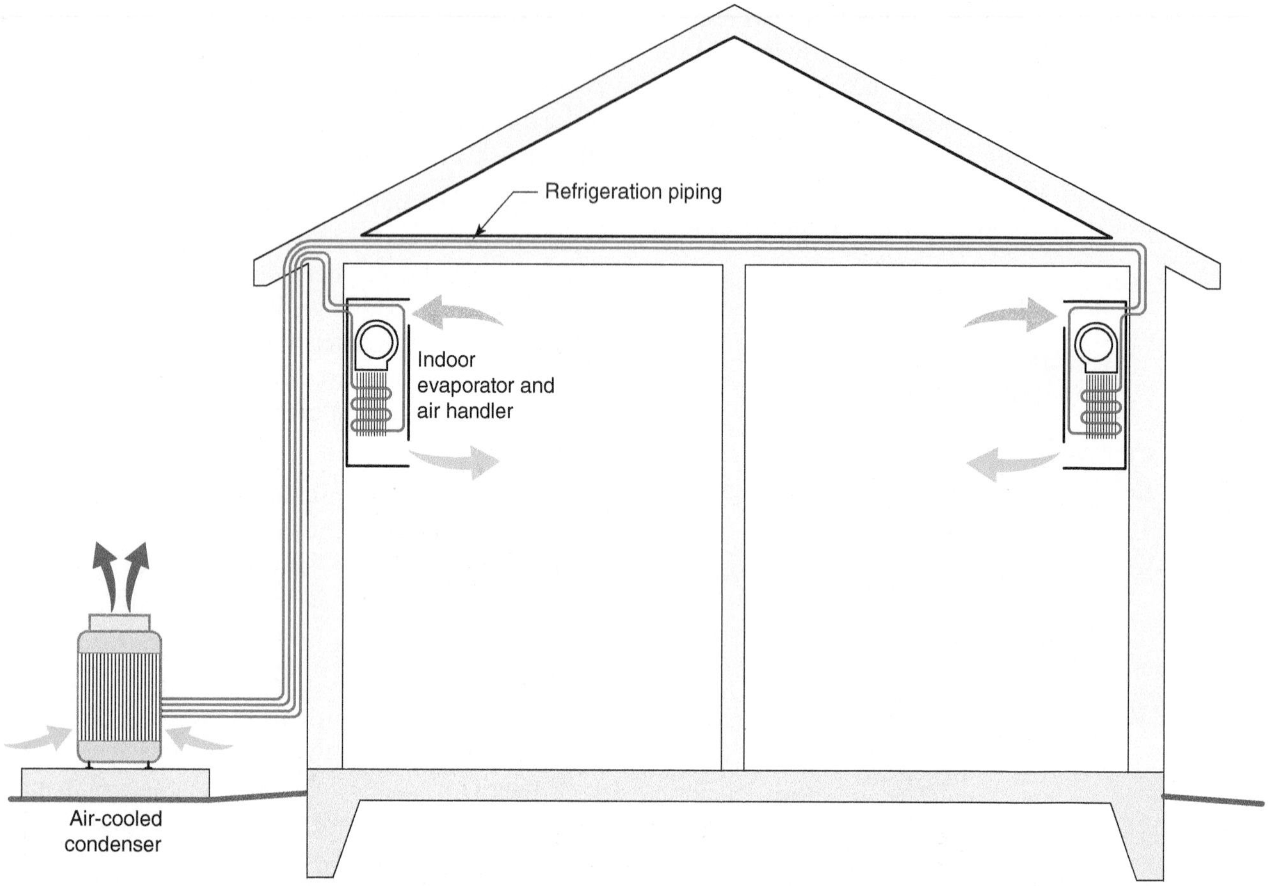

Refrigeration piping
Indoor evaporator and air handler
Air-cooled condenser

Figure 1-31 | A zoned ductless cooling system. Each indoor evaporator unit is connected to a separate outdoor condenser unit with refrigerant piping and wiring.

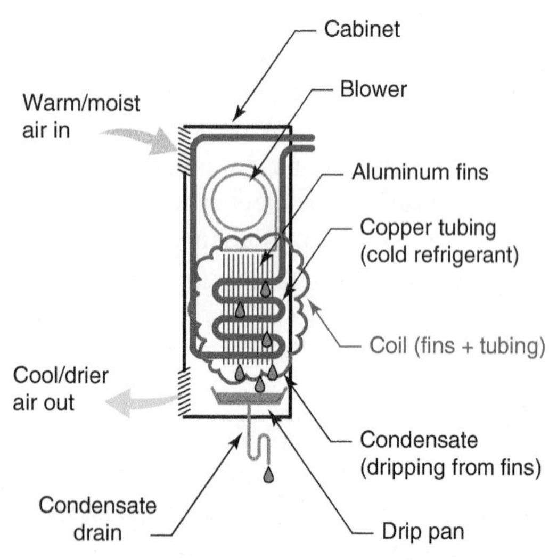

Figure 1-32 | Cooling and dehumidification of air passing through an indoor evaporator unit that is part of a ductless cooling system.

allow up to eight independent interior zones to be connected to a single outdoor condenser unit.

Some ductless systems can also operate as heat pumps, providing cooling in warmer weather and heating in cooler weather. Under certain conditions, such units provide a backup to the building's main heating system.

Mini-Duct Systems

Several North American companies now offer cooling systems that rely on small (e.g., 2-inch diameter) flexible ducting to distribute cool air to locations where conventional ducting won't fit. A schematic of this concept is shown in Figure 1-33.

The cooling for most mini-duct systems is supplied from a refrigerant-based evaporator coil within an air handler that is located in an attic or basement. However, air handlers are also available for chilled-water cooling. The latter is a form of hydronics technology using

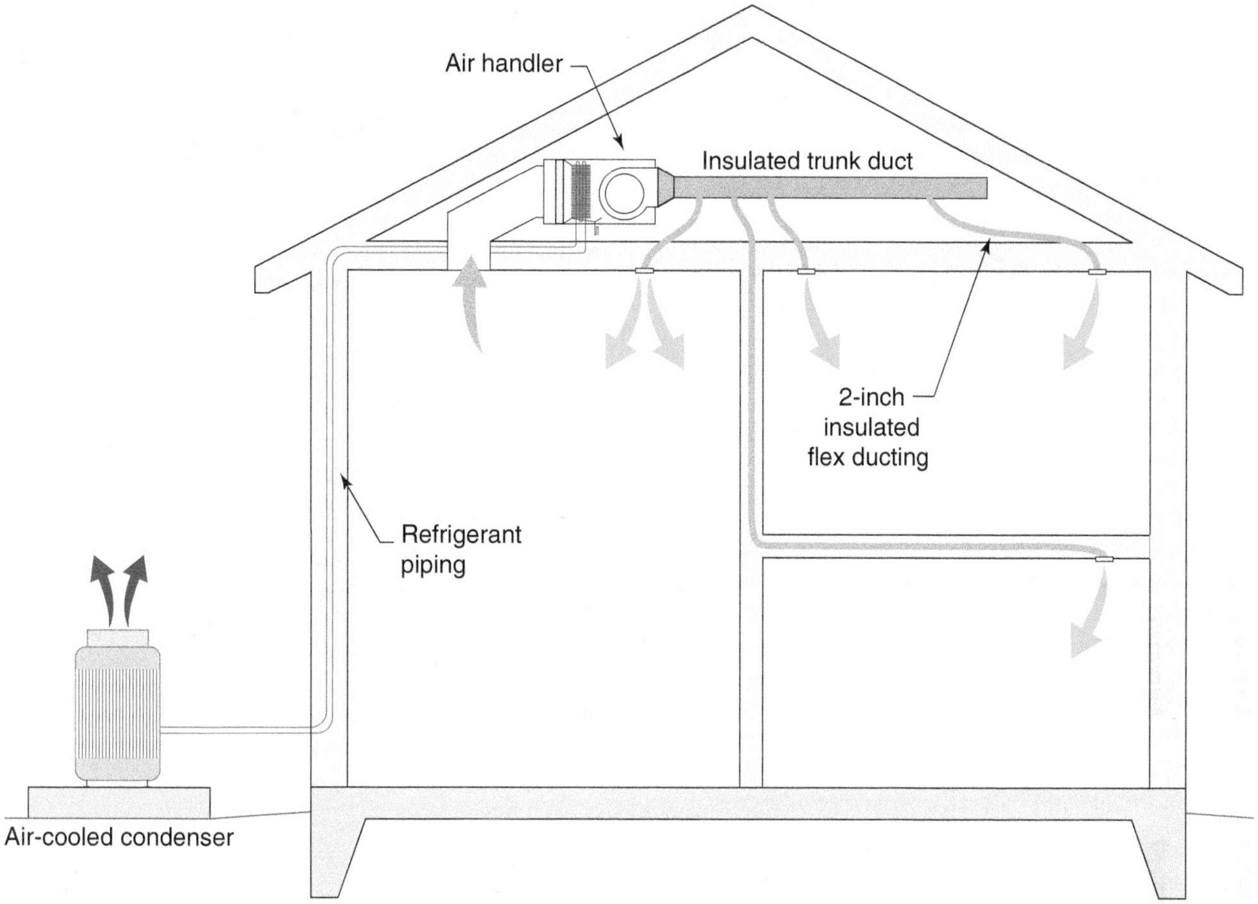

Figure 1-33 | A single indoor air handler delivers cooling through several insulated mini ducts.

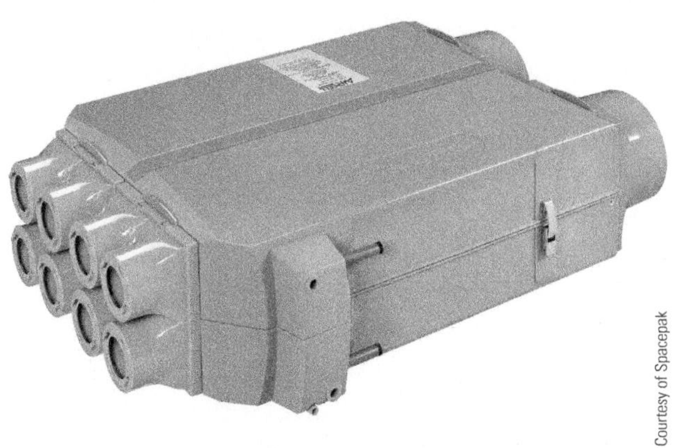

Courtesy of Spacepak

Figure 1-34 | Example of a small chilled water air handler designed to connect to mini ducts.

chilled water rather than heated water. Chilled water can be supplied from several types of heat pumps, and will be discussed later in the text. Figure 1-34 shows an example of a small chilled water air handler that is equipped to supply up to eight mini ducts.

INTERIOR POLLUTANTS

Before major efforts were made to reduce building air leakage as a means of lowering heating cost, the natural air leakage rate of most single-family homes provided acceptable indoor air quality. At times, when detectable pollutants were higher than normal (e.g., someone burned food on the stove), an exhaust fan was operated for a short time, or windows were opened. Beyond that, little attention was given to maintaining high-quality indoor air through mechanical ventilation.

As the rate of air leakage in smaller buildings has decreased, incidents of poor indoor air quality have correspondingly increased. Furthermore, some construction materials and interior furnishings contain chemicals known as **volatile organic compounds (VOCs)** that "**off-gas**" over time and reduce interior air quality.

Other possible interior pollutants include compounds associated with combustion, such as carbon dioxide, carbon monoxide, nitrogen oxides, and aldehydes. Such compounds are produced by fireplaces, gas ranges, stoves, furnaces, and boilers. Unvented combustion devices, such as some gas-fired fireplaces and portable kerosene space heaters, are especially high emitters of such compounds and, as such, are highly incompatible with good indoor air quality.

Over the last two decades, sealed combustion furnaces and boilers have become available. Such devices draw all air needed for safe and efficient combustion through ducting that leads outside the building. They also deliver all their exhaust products through sealed vent piping leading outside the building. These **sealed combustion** appliances minimize the chances that combustion gases will be inadvertently released inside the building due to **backdrafting**. This is caused when a relatively powerful fan or blower, such as might be used in a kitchen, lowers the air pressure inside a tightly sealed building to an extent that exhaust gases reverse their normal flow direction and spill into the interior air. Such conditions are potentially lethal and typically claim several lives within the United States each year.

Still other interior airborne pollutants include dust, mold spores, tobacco smoke, bacteria, viruses, radon, and both man-made and naturally occurring allergens.

Given the myriad of pollutants that can negatively affect interior air quality, it's readily apparent that an underventilated or improperly ventilated building has the potential to be a serious threat to the long-term health of its occupants.

There are many ways to reduce this threat. The most obvious approach is to *decrease the source* of the interior pollutants to the extent possible. Today, many manufacturers aggressively promote low-emission building products and furnishings, such as wood, concrete, stone, and ceramic tile. Source control also involves storing materials such as cleaning solutions, pool/spa chemicals, insecticides, herbicides, paints, and solvents away from occupied spaces or equipment used to ventilate those spaces.

VENTILATION

Once the source of the pollutants is minimized, whole-house ventilation can be used to maintain concentrations of pollutants within acceptable limits.

Whole-house ventilation can be categorized as follows:

- Powered exhaust systems
- Powered supply systems
- Balanced ventilation systems

Powered exhaust systems use fans or blowers to pull air out of buildings and rely on natural air leakage for incoming air. A kitchen or bathroom exhaust fan is a typical example. **Powered supply systems** do just the opposite. They force fresh air into the building and rely on natural air leakage to remove air. **Balanced ventilation systems** use fans or blowers to control both supply and exhaust air. Because they don't count on natural air leakage for proper operation, balanced systems provide the most consistent ventilation rates.

One of the most recognized standards for whole-house ventilation is ASHRAE Standard 62.2. This standard assumes that all ventilation air is supplied through mechanical means and does not include any contribution due to natural air leakage.

The ventilation rate recommended by this standard for residential buildings can be calculated using Equation 1-1:

(Equation 1.1)

$$Q = 0.01A + 7.5(n + 1)$$

Where:

Q = required air flow (ft^3/minute) or (**CFM**)

A = floor area of building (ft^2)

n = number of bedrooms in building

Example 1.4: A 2,400-square-foot house includes four bedrooms. What is the recommended air-flow rate for a mechanical ventilation system based on ASHRAE Standard 62.2?

Solution: The stated floor area and number of bedrooms are entered into Equation 1.1.

$$Q = 0.01A + 7.5(n + 1) = 0.01(2400) + 7.5(4 + 1) = 61.5CFM$$

Discussion: Equation 1.1 includes a background ventilation rate based on the size of the building (e.g., 0.01A). The rationale is that larger buildings contain more materials and thus have greater potential for pollutant emissions. The second portion of Equation 1.1 is meant to factor in the effect of occupants, which is often assumed to be the same as the number of bedrooms.

Balanced Heat Recovery Ventilation

Although powered exhaust and powered supply ventilation systems are simple, they can create conditions that are not ideal. For example, a powered supply system tends to slightly pressurize a building. This pressurization can force moisture-laden air outward through air leakage paths in exterior walls, ceilings, and other surfaces. During cold weather, the moisture driven into these building cavities can condense, leading to mold, and eventual rot of the wood framing. Outward moisture movement can also cause paint to peel away from exterior surfaces.

Powered exhaust systems tend to *depressurize* buildings. This helps prevent moisture-laden air from passing outward into wall and ceiling cavities. However, negative pressure within a building can increase infiltration of radon gas from surrounding soils. It might also interfere with devices, such as wood stoves, that rely on natural draft for venting.

Most powered supply and powered exhaust systems also rely on the building's natural air leakage. This leakage is hard to predict. It varies with construction quality, wind direction, building height, and outdoor temperature.

Balanced ventilation systems use blowers to control both supply air flow and exhaust air flow and thus do not rely upon natural air leakage. They can be adjusted to balance incoming and outgoing air-flow rates. Balanced ventilation systems also allow the use of heat recovery devices that can remove some heat from the outgoing (exhaust) air stream and use that heat to preheat the incoming air stream. This technique is called **heat recovery ventilation**. It has become increasingly popular in low-energy-use homes and is even code-mandated in some circumstances.

The key device in a heat recovery ventilation system is a **heat recovery ventilator (HRV)**. These devices contain blowers for both incoming and exhaust air and a compact heat exchanger core constructed of either polyethylene or aluminum sheets, as seen in Figure 1-35.

Figure 1-35 | Internal construction of a heat recovery ventilator.

The incoming and exhaust air streams flow through alternating channels within the core (e.g., warm exhaust air passes through channels 1, 3, 5, etc., while incoming cold air passes through channels 2, 4, 6, etc.). Heat from the warm exhaust air transfers through the thin polyethylene or aluminum sheets that separate the channels, to preheat incoming cold air. The two air streams remain separated, but up to 70 percent of heat that would otherwise be lost with the exhaust air stream is now recovered and transferred to the incoming air stream. Thus, proper rates of fresh air are supplied to the building, with significantly less energy loss compared to either exhaust- only or supply-only ventilation systems.

Figure 1-36 shows an HRV connected to small diameter ducting within a building. Exhaust air is typically collected from bathrooms and the kitchen, both areas of higher moisture concentration. Supply air is delivered to bedrooms and other living space.

Modern HRVs also contain controls that allow for periodic defrost during very cold weather. Furthermore, many HRVs now have variable speed blowers that can operate at reduced flow rate for **background ventilation** and, when necessary, operate at full speed to reduce relative humidity or concentrations of a specific pollutant such as carbon dioxide.

Some HRV systems use flexible polyethylene ducting with a nominal 2-inch diameter. This is relatively easy to install within framing cavities as well as under slabs when necessary. Such ducting can also be cleaned, when necessary, used flexible rotary brushes.

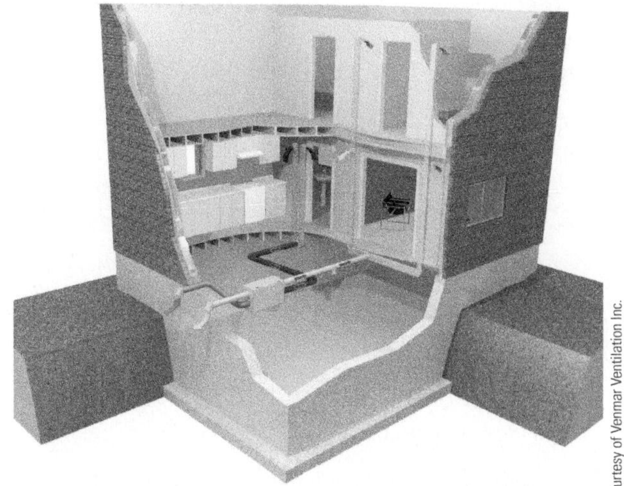

Figure 1-36 | Typical installation of a residential heat recovery ventilator.

SUMMARY

This chapter has presented considerations that serve as a starting point for decisions on using energy for space heating and domestic hot water. Later chapters will present analytical methods, hardware details, and practical system configurations that further support and refine such decisions. As you progress toward more detail, be retrospective on how your designs address the fundamentals of comfort, energy conservation, indoor air quality, and long-term reliability.

KEY TERMS

$/MMBtu
absorber plate
active solar thermal technology
active solar technology
aerator
AFUE
air handler
air change per hour
alternate energy source
Arab oil embargo
auxiliary heat source
backdrafting
background ventilation
balance-of-system
balanced ventilation system
basic service charge
biomass
British thermal unit (Btu)
CFM
combisystem
condenser unit
conduction
convection
convectors
COP
cost/benefit (c/b) ratio
counterflow
ductless cooling system
electromagnetic energy
electromagnetic spectrum
entropy
evaporation
evaporator unit
f-chart
fan coil
first law of thermodynamics
forced convection
greywater heat exchanger
heat
heat balance equation
heat emitter
heat exchanger
heat recovery ventilation
heat recovery ventilator (HRV)
heat trap
heating, ventilation, and cooling (HVAC)
high entropy energy
hydronics
hydropower
indoor environmental quality (IEQ)
infrared
infrared thermograph
Legionnaire's disease
low entropy energy
mean radiant temperature
MMBtu
natural convection
NREL
off-gas
passive solar design
powered exhaust system
powered supply system
R-value
radiant floor heating
radiant panel
recirculating hot water distribution system
renewable energy
return on investment
sealed combustion
seasonal efficiency
second law of thermodynamics
shading coefficient
tankless water heater
temperature
thermal conductivity
thermal efficiency
thermal envelope
thermal equilibrium
thermal mass
thermal radiation
thermographic camera
true south
United States Department of Energy (DOE)
unit cost of energy
variable refrigerant flow (VRF)
visible range
volatile organic compounds (VOCs)
zone

QUESTIONS AND EXERCISES

1. During the heating season, why is it better to surround a person with warm surfaces as opposed to just warm air?
2. What type of heat transfer creates the wind-chill effect we experience during winter?
3. At approximately what temperature does a surface that is cooling from a high temperature stop emitting *visible* radiation?
4. How does thermal radiation differ from visible light? In what way are they similar?
5. Describe thermal equilibrium within a hydronic system. What hydronic systems don't operate in thermal equilibrium?
6. Describe a situation in which conduction heat flow will affect thermal comfort.
7. Describe the difference between natural and forced convection.
8. What is the difference between ultraviolet and infrared radiation?

9. Why does very little infrared radiation emitted by a heat emitter within a room escape from that room?

10. A device that converts propane into heat is being monitored in a laboratory. The measured fuel input is 1,000 Btu/hr. The measured useful heat output is 852 Btu/hr. What is the thermal efficiency of this device? How would this efficiency change if the fuel being used were natural gas rather than propane?

11. From the standpoint of minimizing the increase in entropy, which is the best heat source for heating a building?
    a. A tank filled with 90 °F water
    b. A tank filled with 120 °F water
    c. A flame at 2000 °F
    d. An electric resistance-heating element

12. An advertisement appears for an electric space heater that uses a "monocrystalline titanium dioxide" heating element to convert 1 kwhr of electricity into 3,500 Btus of heat output. Refute this claim based on the first law of thermodynamics.

13. Consider a burning match and a 1,000-gallon tank filled 180 °F water. Which source provides the lowest entropy energy? Which source has the higher quantity of useful energy when used to maintain a building at 70 °F interior air temperature?

14. Describe at least two reasons for including auxiliary heating in systems using renewable energy heat sources.

15. Why is it important to examine energy conservation before designing a renewable energy heating system?

16. What is meant by the term "raised heel truss"? Why are they used?

17. Explain the difference between heat and temperature.

18. Based on preventing burns, what is the highest water temperature that should be delivered from fixtures such as showers and lavatories?

19. A homeowner is considering the following three possible energy upgrades and their associated costs:
    a. Replace boiler with higher efficiency unit: Cost $6,500, estimated annual savings: $300
    b. Add attic insulation: Cost $1,500, estimated annual savings: $175
    c. Replace all windows in house: Cost $13,000, estimated annual savings $550

    Determine the cost/benefit ratio of each of these. From the standpoint of c/b ratio only, which is the better investment?

20. Why is thermal mass necessary in a passive solar building?

21. Why do some designers prefer to rotate a passive solar building so that its southerly exposure is facing slightly to the east rather than due south?

22. Why should insulation blankets not be installed on the top of a gas-fired water heater?

23. Why must a greywater heat exchanger be installed in a vertical drainage pipe?

24. What does the term "counterflow" mean regarding greywater heat exchangers?

25. Why is insulation imperative on the piping of a recirculating hot water distribution system? What is the minimum recommended R-value of this pipe insulation?

26. Describe the operation of a "demand-operated" recirculating hot water system.

27. When should the basic service charge, assessed by utilities, be included when estimating the annual cost of space heating?

28. Determine the unit price of the following fuels when converted to heat at the stated conversion efficiencies:
    - Natural gas for $1.25 per therm, conversion efficiency = 90 percent
    - #2 fuel oil for $3.50 per gallon, conversion efficiency = 83 percent
    - Propane for $2.75 per gallon, conversion efficiency = 90 percent
    - Electricity for $0.12 per kilowatt•hour, conversion efficiency = 98 percent
    - Wood pellets for $225 per ton, conversion efficiency = 78 percent

29. Describe the advantages of a balanced ventilation system compared to both a powered supply system and a powered exhaust system.

30. A house is being planned that has 1,850 square feet of floor area. It will have three bedrooms. Determine the recommended ventilation rate based on ASHRAE Standard 62.2.

31. What steps should be taken before resorting to mechanical ventilation to dilute all interior pollutants to acceptable levels?

32. What is meant by the "core" of a heat recovery ventilator?

33. What are two advantages of ductless cooling systems relative to standard "ducted" systems? Explain each one.

R410A
R410A

CHAPTER 2

Space Heating and Domestic Water Heating Loads

OBJECTIVES

After studying this chapter, you should be able to:

- Describe what a design heating load is and why it is important to heating system design.
- Explain the difference between room heating loads and building heating loads.
- Determine the thermal envelope of a building.
- Calculate the effective total R-value of a building surface.
- Describe how the unit U-value for windows and doors is determined and how it is used.
- Estimate infiltration heat loss using the air change method.
- Determine the heat loss of foundations and slab floors.
- Explain what degree days are and how they are used.
- Estimate the annual space heating energy usage of a building.
- Estimate the daily energy used for domestic water heating.

2.1 Introduction

The design of any space heating or domestic water-heating system must start with an estimation of the **thermal load**. The care given to this step will directly affect system cost, efficiency, and most importantly, customer satisfaction.

It is natural for those learning about renewable energy heating systems to want to "dive into" a discussion of the hardware involved, especially the heat sources. However, this is like selecting the foundation for a building without knowing how much weight that foundation must support. It is pointless to start designing the heating system for a building without knowing the rate at which that building requires heat. To that end, this chapter lays out the basics for determining both space heating and domestic water-heating loads. Using the information in this chapter, designers can accurately estimate these loads and then optimize their system designs to meet them.

The majority of this chapter deals with estimating space heating loads. Simple mathematical methods for **conduction heat loss** and **air infiltration heat loss** are introduced, as are methods of estimating heat losses from slab on grade floors and basements. These methods are then applied to an example house on a room-by-room basis. Software-based approaches to heating load estimation are also discussed. The concept of **degree days** is introduced and used as a way to extrapolate the design heating load into an estimate of annual heating energy use for the building in a specific climate.

Information for estimating daily and peak hourly domestic water-heating loads is also presented. Domestic water-heating load information will be combined with space heating loads and used to design combisystems in later chapters.

2.2 Design Space Heating Load

It is important to understand what a design space heating load is before attempting to calculate it. *The **design heating load** of a building is an estimate of the rate at which a building loses heat during the near-minimum outdoor temperature.* This definition contains a number of key words that need further explanation.

First, it must be emphasized that a heating load is a calculated *estimate* of the rate of heat loss of a room or building. Because of the hundreds of construction details and thermal imperfections in an object as complex as a building, even a simple house, it is simply not possible to determine its design heating to the nearest Btu/hr. Other factors that add uncertainty to the calculations are:

- Imperfect installation of insulation materials
- Variability in R-value of insulation materials
- Shrinkage of building materials, leading to greater air leakage
- Complex heat transfer paths at wall corners and other intersecting surfaces
- Effect of wind direction on building air leakage
- Overall construction quality
- Traffic into and out of the building

Second, the heating load is a *rate* of heat flow from the building to the outside air. It is often misstated, even among heating professionals, as a number of Btus rather than a *rate of flow* in Btus per hour (Btu/hr). This is like stating the speed of a car in miles rather than miles per hour. Recall from the discussion of thermal equilibrium in Chapter 1 that when the rate of heat flow into a system matches the rate of heat flow out of a system, the temperature of that system remains constant. In the case of a building, *when the rate of heat input to the building equals the rate the building loses heat, the indoor temperature remains constant.* This is also true for each room within the building.

Finally, the design heating load is estimated assuming the outside temperature is *near* its minimum value. This temperature, called the **97.5 percent design dry bulb temperature**, is not the absolute minimum temperature for the location. It is the temperature that the outside air is at or above during 97.5 percent of the year. Although outside temperatures do occasionally drop below the 97.5 percent design dry bulb temperature, the duration of these low temperature excursions is short enough that most buildings can "coast" through them using heat stored in their thermal mass. Use of the 97.5 percent design temperature for heating system design helps prevent oversizing of the heat source.

BUILDING HEATING LOAD VERSUS ROOM HEATING LOAD

One method of calculating the heating load of a building yields a single number that represents the heating load of the entire building. This value is called the **building heating load**, and it is useful for selecting the heating output of the heat source as well as estimating annual heating energy use. However, this single number is not sufficient for designing the heat distribution system. It does not tell the designer how to properly proportion the heat output of the heat source to each room. Even when the building heating load is properly determined, failure to properly distribute the total heat output can lead to overheating in one room and underheating in another.

One should not assume that individual **room heating loads** are proportional to room floor area. For example, a small room with a large window area could have a greater heating load than a much larger room with smaller windows.

The proper approach is to perform a heat load estimate calculation for each room. Then, once these individual room heating loads are determined, they can be added together to obtain the building heating load.

WHY BOTHER WITH HEAT LOAD CALCULATIONS?

Would an electrician select wiring without knowing the amperage it must carry? Would a structural engineer select steel beams for a bridge without knowing the forces they must carry? Obviously the answer to these questions is no. Yet many so-called "heating professionals" routinely select equipment for heating systems with only a guess as to the building's heating load.

Statements attempting to justify this approach range from "I haven't got time for doing all those calculations," to "I'd rather be well oversized than get called back on the coldest day of the winter," to "I did a house like this one a couple of years ago, it can't be much different." Instead of listing excuses, let's look at some of the consequences of not properly estimating the building heating load.

- If the estimated heat loss is too low, the building will be uncomfortable during cold weather (a condition few homeowners will tolerate). An expensive callback will eventually result.
- Grossly *overestimated* heating loads lead to systems that needlessly increase installation cost. Oversized systems may also cost more to operate due to lower heat source efficiency. This reduced efficiency can waste *thousands* of dollars worth of fuel over the life of the system.

Most building owners never know if their heating systems are oversized and simply accept the fuel usage of these systems as normal. Most trust the heating professional they hire to properly size and select equipment in their best interest. Heating professionals who deserve that trust adhere to the following principle:

Before attempting to design a space heating system of any sort, always calculate the design heating load of each room in the building using credible methods and data.

The remainder of this chapter shows you how to do this.

2.3 Conduction Heat Losses

Conduction is the process by which heat moves through a solid material whenever a temperature difference exists across that material. The rate of conduction heat transfer depends on the **thermal conductivity** of the material as well as the temperature difference across it. The relationship between these quantities is given in Equation 2.1.

(Equation 2.1)

$$Q = A\left(\frac{k}{\Delta x}\right)(\Delta T)$$

where:

Q = rate of heat transfer through the material (Btu/hr)

k = thermal conductivity of the material (Btu/°F·hr·ft)

Δx = thickness of the material in the direction of heat flow (ft)

ΔT = temperature difference across the material (°F)

A = area across which heat flows (ft^2)

Example 2.1: Determine the rate of conduction heat transfer through the 6-inch-thick panel shown in Figure 2-1.

Solution: Substituting the data into Equation 2.1:

$$Q = A\left(\frac{k}{\Delta x}\right)(\Delta T)$$

$$Q = 200\text{ ft}^2\left(\frac{0.1\text{ Btu/°F·hr·ft}}{0.5\text{ ft}}\right)(70\text{°F} - 10\text{°F})$$

$$= 2{,}400\text{ Btu/hr}$$

Discussion: Notice how the units of the inserted data cancel out so that the resulting units are Btu/hr. Thermal conductivity data for various materials may be

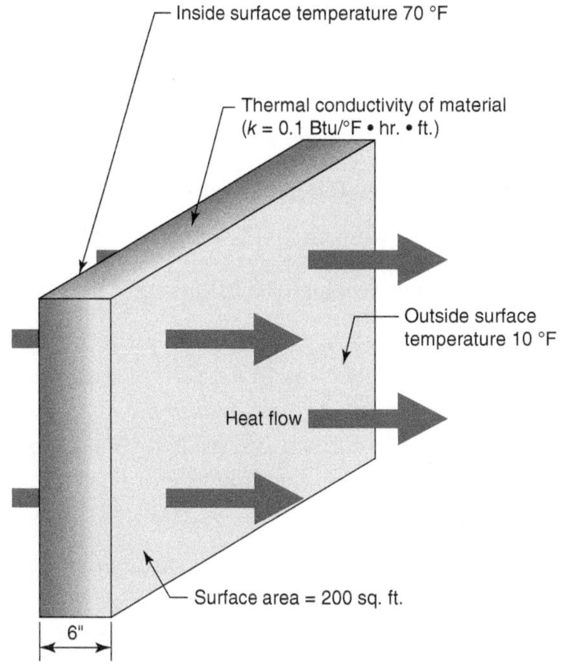

Figure 2-1 | Heat flow through a material by conduction.

stated units other than those give below Equation 2.1. If so, be sure to convert that conductivity value to the stated units, or the results of the calculation will be invalid.

THERMAL RESISTANCE OF A MATERIAL

Since building designers are usually interested in *reducing* the rate of heat flow from a building, it is more convenient to think of a material's *resistance* to heat flow rather than its thermal conductivity. An object's **thermal resistance** (also referred to as its **R-value**) can be defined as its thickness in the direction of heat flow divided by its thermal conductivity.

(Equation 2.2)

$$\text{R-value} = \frac{\text{thickness}}{\text{thermal conductivity}} = \frac{\Delta x}{k}$$

The greater the thermal resistance of an object, the slower heat passes through it when a given temperature differential is maintained across it. Equations 2.1 and 2.2 can be combined to yield an equation that is convenient for estimating heating loads:

(Equation 2.3)

$$Q = \left(\frac{A}{R}\right)(\Delta T)$$

where:

Q = rate of heat transfer through the material (Btu/hr)

ΔT = temperature difference across the material (°F)

R = R-value of the material (°F·ft²·hr/Btu)

A = area across which heat flows (ft²)

Some interesting facts can be demonstrated with this equation. First, *the rate of heat transfer through a given object is directly proportional to the temperature difference (ΔT) maintained across it.* If this temperature difference were doubled, the rate of heat transfer would also double.

Second, *the rate of heat transfer through an object is inversely proportional to its R-value.* For example, if its R-value were doubled, the rate of heat transfer through the object would be halved (assuming the temperature difference across the object remained constant).

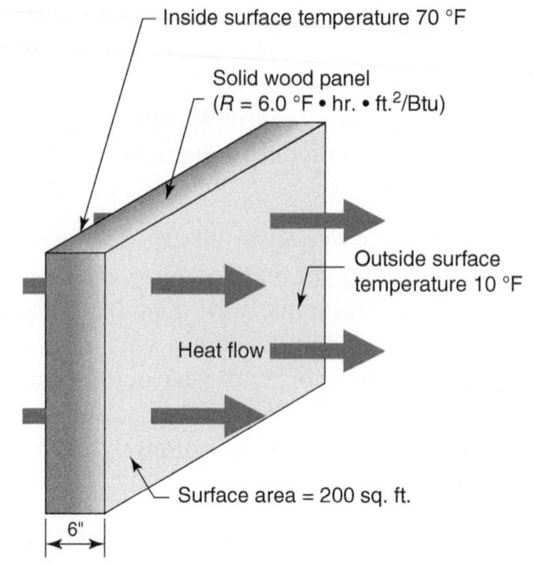

Figure 2-2 | Heat flow through a panel by conduction (using R-value).

Finally, the larger the object's surface area, the greater the rate of heat transfer through it. For example, a 4-foot by 4-foot window with an area of 16 ft² would transfer heat twice as fast as a 2-foot by 4-foot window with an area of 8 ft².

Example 2.2: Determine the rate of heat transfer through the 6-inch-thick wood panel shown in Figure 2-2.

Solution: Substituting the relevant data into Equation 2.3:

$$Q = \left(\frac{A}{R}\right)(\Delta T) = \left(\frac{200}{6}\right)(70 - 10) = 2{,}000 \text{ Btu/hr}$$

The R-value of a material is also directly proportional to its thickness. If a 1-inch-thick panel of extruded polystyrene insulation has an R-value of 5.4, a 2-inch-thick panel would have an R-value of 10.8, and a 1/2-inch-thick piece would have an R-value of 2.7. This relationship is very useful when the R-value of one thickness of a material is known, and the R-value of a different thickness needs to be determined.

The R-value of a material is slightly dependent on the material's temperature. As the temperature of the material is lowered, its R-value increases slightly. However, for most building materials, the change in R-value is small over the temperature ranges the material typically experiences and thus may be assumed to remain constant.

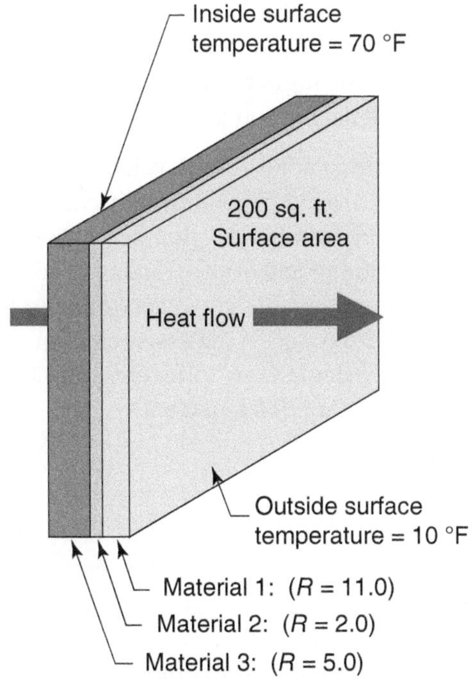

Figure 2-3 | Wall assembly consisting of three materials having different R-values.

TOTAL R-VALUE OF AN ASSEMBLY

Since the walls, floors, and ceilings of buildings are rarely constructed of a single material, it is often necessary to determine the total R-value of an assembly made up of several materials in contact with each other. This is done by adding up the R-values of the individual materials. For example, the total R-value of the assembly shown in Figure 2-3 is:

$$R_{total} = 11.0 + 2.0 + 5.0 = 18.0$$

The heat flow across the panel can now be calculated using the **total R-value** substituted into Equation 2.3:

$$Q = \left(\frac{A}{R}\right)(\Delta T) = \left(\frac{200}{18}\right)(70 - 10) = 667 \text{ Btu/hr}$$

The R-values of many common building materials can be found in Appendix B. To find the R-values of other materials, consult manufacturer's literature or the current edition of the *ASHRAE Handbook of Fundamentals*.

AIR FILM RESISTANCES

In addition to the thermal resistances of the solid materials, there are two thermal resistances, called the **inside air film** resistance and **outside air film** resistance, which affect the rate of heat transfer through a panel separating inside and outside spaces. These thermal resistances represent the insulating effects of thin layers of air that cling to all surfaces. The R-values of these air films are dependent on surface orientation, air movement along the surface, and reflective qualities of the surface. Values for these resistances can also be found in Appendix B. The R-values of air films include the combined effect of conduction, convection, and radiation heat transfer, but are stated as a conduction-type thermal resistance for simplicity.

Example 2.3 shows how the total R-value of the wall assembly shown in Figure 2-3 is affected by the inclusion of the air film resistances. Notice that the temperatures used to determine the ΔT in this example are the indoor and outdoor *air temperatures*, rather than the *surface temperatures* used in the previous examples.

Example 2.3: Determine the rate of heat transfer through the assembly shown in Figure 2-4. Assume still air on the inside of the wall and 15-mph wind outside. See Appendix B for values of the air film resistances.

Solution: The total R-value of the assembly is again found by adding the R-values of all materials, including the R-values of the inside and outside air films:

$$R_{total} = 0.17 + 11.0 + 2.0 + 5.0 + 0.68 = 18.85$$

The rate of heat transfer through the assembly can once again be found using Equation 2.3:

$$Q = \left(\frac{A}{R}\right)(\Delta T) = \left(\frac{200}{18.85}\right)(70 - 10) = 637 \text{ Btu/hr}$$

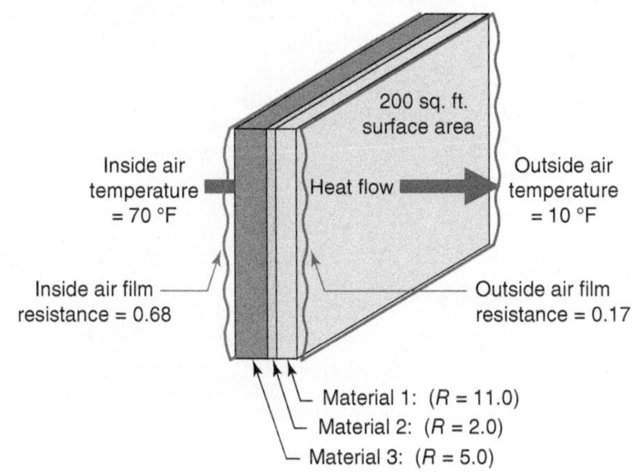

Figure 2-4 | Air film resistances add R-value to a wall assembly.

Discussion: Including the R-value of the inside and outside air films caused a slight decrease in the rate of heat transfer through the panel. For the panel construction shown in Figure 2-4, the decrease in the rate of heat transfer was about 4.5 percent. The greater the total R-value of the materials making up the panel, the smaller the effect of the air film resistances. However, the R-value of the air films should always be included when determining the total R-value of an assembly.

EFFECT OF FRAMING MEMBERS

The rate of heat flow through assemblies made of several materials stacked together like a sandwich can now be calculated. In such situations, and when edge effects are considered insignificant, the heat flow through the assembly is uniform over each square foot of surface area. However, few buildings are constructed with walls, ceilings, and other surfaces made of simple stacked layers of materials. This is particularly true of wood-framed buildings with wall studs, ceiling joists, window headers, and other framing members. Wooden framing members that span across the insulation cavity of a wall tend to reduce its **effective total R-value**. This is because the thermal resistance of wood, about 1.0 per inch of thickness, is usually less than the thermal resistance of insulation materials it displaces. Thus the rate of conduction through framing materials is usually faster than through the surrounding insulation materials. This effect can be detected with infrared thermography, as shown in Figure 2-5. In this case, the wall studs as well as top and bottom plates are readily visible to the infrared camera because they create "stripes" of higher surface temperature along the outside wall surface on a cold winter night.

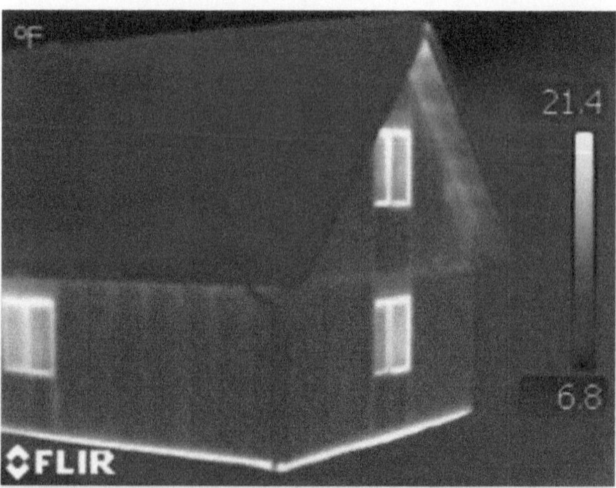

Figure 2-5 Infrared thermograph of framed walls on a cold winter night reveals "stripes" of higher outside surface temperatures due to lower R-value of wall framing relative to cavity insulation.

It is possible to adjust the R-value of an assembly to include the thermal effects of framing. This requires that the assembly's total R-value be calculated both between the framing and at the framing. The resulting total R-values are then weighted according to the percentage of solid framing in the assembly. Equation 2.4 can be used to calculate an "effective total R-value," which can then be applied to the entire panel area.

Equation 2.4:

$$R_{\text{effective}} = \frac{(R_i)(R_f)}{p(R_i - R_f) + R_f}$$

where:

$R_{\text{effective}}$ = effective total R-value of panel (°F·ft²·hr/Btu)

p = percentage of panel occupied by framing (decimal %)

R_f = R-value of panel at framing (°F·ft²·hr/Btu)

R_i = R-value of panel at insulation cavities (°F·ft²·hr/Btu)

Although the percentage of the panel occupied by framing could be calculated for every construction assembly, such calculations can be tedious and often make insignificant changes in the results. Typical wood-framed walls will have between 10 and 15 percent of the wall area as solid framing across the insulation cavity. The lower end of this range would be appropriate for 24 inches on-center framing, with insulated headers over windows and doors. The upper end of this range is typical of walls with 16 inches on-center framing and solid headers.

To illustrate these calculations, the effective total R-value of the wall assembly shown in Figure 2-6 will be calculated assuming that 15 percent of the wall is solid framing.

Notice that the effective R-value of this wall (R-22.9) is about 13.6 percent lower than the R-value through the insulation cavity (R-26.5). If the person estimating the heat loss simply ignored the effect of framing, the estimated heat loss for the wall would be 13.6 percent low. It follows that a heating system designed according to this underestimated heat loss might not be able to maintain comfort on a design load day.

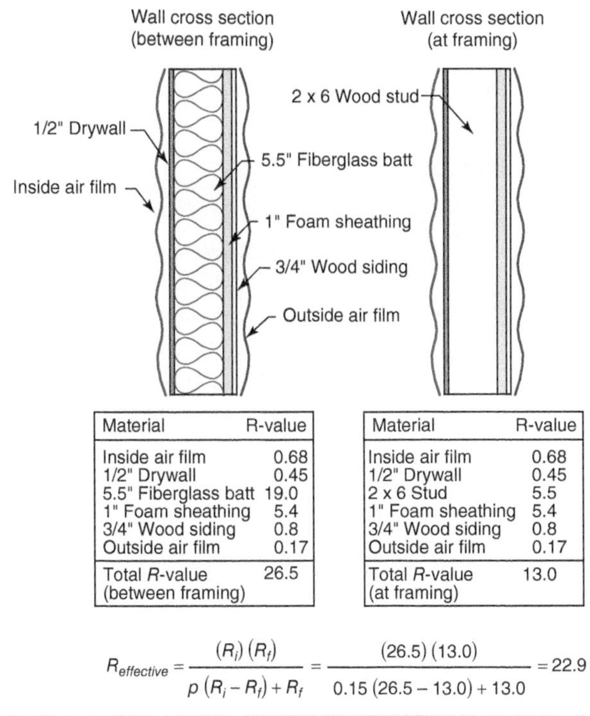

Material	R-value
Inside air film	0.68
1/2" Drywall	0.45
5.5" Fiberglass batt	19.0
1" Foam sheathing	5.4
3/4" Wood siding	0.8
Outside air film	0.17
Total *R*-value (between framing)	26.5

Material	R-value
Inside air film	0.68
1/2" Drywall	0.45
2 x 6 Stud	5.5
1" Foam sheathing	5.4
3/4" Wood siding	0.8
Outside air film	0.17
Total *R*-value (at framing)	13.0

$$R_{effective} = \frac{(R_i)(R_f)}{p(R_i - R_f) + R_f} = \frac{(26.5)(13.0)}{0.15(26.5 - 13.0) + 13.0} = 22.9$$

Figure 2-6 | Calculating the effective total R-value of a wall assembly.

To obtain the total heat loss of a room, the effective total R-value of each different **exposed surface** within the room must be determined using a procedure similar to that shown in Figure 2-6. Fortunately, many buildings have the same type of construction for most exposed walls, ceilings, and so on. In such cases, the effective R-value of the assembly need only be calculated once.

It is good practice to make a sketch or CAD drawing of the cross section of each exposed assembly (i.e., wall, ceiling, floor), similar to that shown in Figure 2-6. The R-value of each material can then be determined from Appendix B and the effective total R-value calculated. These sheets or CAD drawing files showing the cross sections of the assemblies and their effective total R-values can be saved and referenced on other buildings with the same construction.

2.4 Foundation Heat Loss

Heat flow from a basement or slab-on-grade foundation is determined by complex interactions between the building, the surrounding soil, insulation materials (if present), and the air temperature above grade.

Figure 2-7a shows a **finite element analysis** computer simulation of heat flow from a residential basement during February, in a cold winter climate. The basement has a width of approximately 26 feet. Only half of this width is shown because heat flow is assumed to be symmetrical about a vertical centerline through the basement. The floor slab is heated by embedded tubing and is assumed to be covered by a ¾-inch-thick hardwood flooring. *There is no underslab insulation.* The basement walls are insulated with 2 inches of expanded polystyrene.

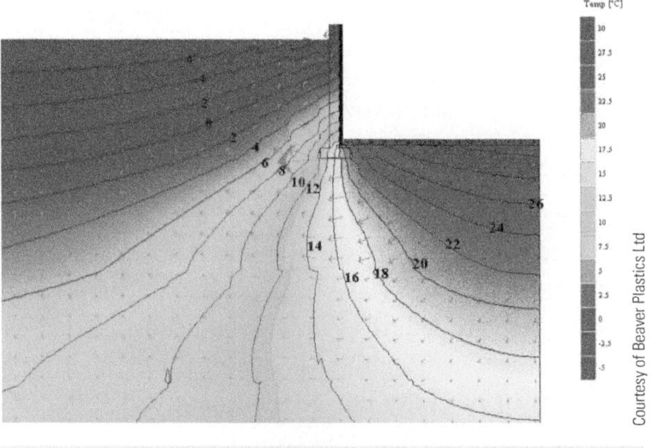

Courtesy of Beaver Plastics Ltd

Figure 2-7a | Computer-simulated heat flow through a basement wall and *uninsulated heated slab floor* during February.

The contour lines with temperatures labeled in °C are called **isotherms**. They represent locations having the same temperature. Heat flow is perpendicular to the isotherms in all locations. In this case, heat flows downward from the slab, passes under the footing, then passes outward and upward toward the cold exposed soil surface. In this case, almost half of the purchased energy used to heat the slab floor is being driven downward into the soil.

Figure 2-7b shows the same situation, but with 2 inches of expanded polystyrene insulation installed under the floor slab. Notice that soil temperatures below the slab are significantly lower, because less heat is being driven downward. It follows that heat loss from this basement will be significantly less. For the cold northern climate assumed in this simulation, which represented an average between the typical weather conditions in Toronto, Ontario, and Calgary, Alberta, the use of 2-inch expanded polystyrene insulation under the heated slab reduced heat loss by approximately 50 percent. The energy savings over a complete heating season were estimated at 8,945 Btu/ft^2 of floor area.

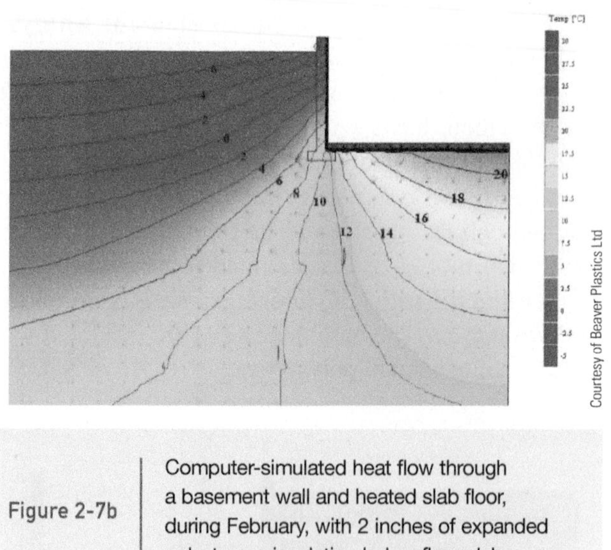

Figure 2-7b | Computer-simulated heat flow through a basement wall and heated slab floor, during February, with 2 inches of expanded polystyrene insulation below floor slab.

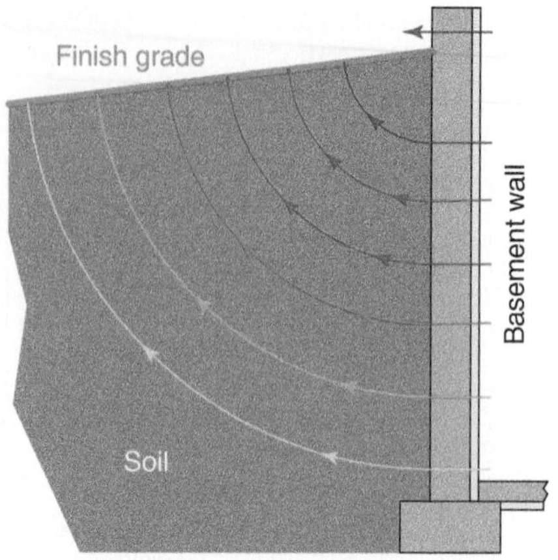

Figure 2-8 | Theoretical heat flow paths through a uniformly insulated basement wall.

The results of complex computer simulations of foundation heat loss have been used to develop simpler calculation procedures that can be used for load estimating. One such method is useful for estimating the heat loss of partially buried basement walls, the other for estimating the heat loss from floor slabs.

HEAT LOSS THROUGH BASEMENT WALLS

The heat loss of a basement wall is significantly affected by the height of soil against the outside of the wall. Soil adds thermal resistance between the wall and the outside air and thus helps reduce heat loss.

Figure 2-8 illustrates the theoretical heat flow paths through a uniformly insulated basement wall and adjacent soil. Notice that the heat flow paths from the inside basement air to the outside air are longer for the lower portions of the wall. As the path length increases, so does the total thermal resistance of the soil along that path.

One method of approximating the heat loss through basement walls is based on dividing the wall into horizontal strips based on the height of finish grade. The upper strip includes all wall area exposed above grade as well as the wall area to a depth of 2 feet below grade. The implicit assumption is that by mid-winter, shallow soils will be at approximately the same temperature as the outside air. The middle strip includes the wall area from 2 feet to 5 feet below grade. The lower strip includes all wall area deeper than 5 feet below grade. Figure 2-9 illustrates these areas.

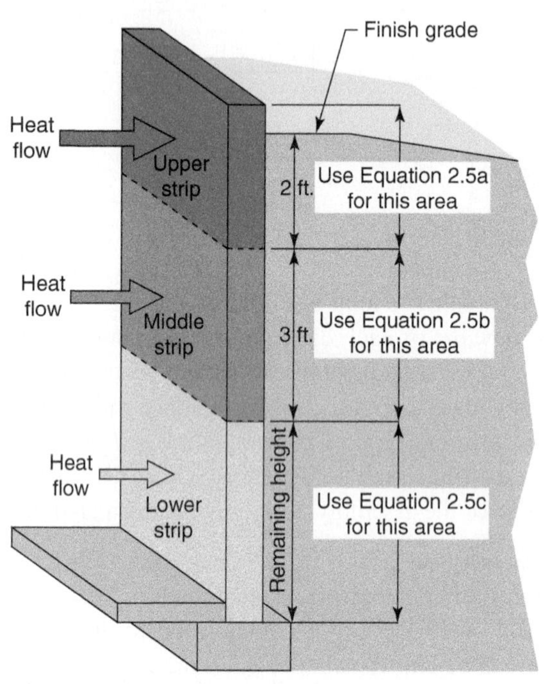

Figure 2-9 | Dividing the basement wall into three horizontal strips to estimate heat loss.

Use Equation 2.5a to calculate the effective R-value of basement walls for all above grade areas as well as areas down to 2 feet below grade.

(Equation 2.5a)

$$R_{effective} = R_{wall} + R_{insulation}$$

Use Equation 2.5b to calculate the effective R-value of basement walls for areas from 2 feet to 5 feet below grade.

(Equation 2.5b)

$$R_{\text{effective}} = 7.9 + 1.12(R_{\text{insulation}})$$

Use Equation 2.5c to calculate the effective R-value of basement walls for areas deeper than 5 feet below grade.

(Equation 2.5c)

$$R_{\text{effective}} = 11.3 + 1.13(R_{\text{insulation}})$$

The parameter ($R_{\text{insulation}}$) in Equations 2.5a through 2.5c represents the R-value of any insulation *added* to the foundation wall over the particular strip for which the effective R-value is being determined. The foundation wall itself is assumed to be a masonry or concrete wall between 8 and 12 inches thick.

Once the R-values of each wall strip have been determined, use Equation 2.3 to calculate the heat loss of each area. In each case, the ΔT in Equation 2.3 is the difference between the basement air temperature and the outside air temperature.

Example 2.4: Estimate the heat loss through a 10-inch-thick concrete basement wall 9 feet deep with 6 inches exposed above grade. The wall is 40 feet long and has 2 inches of extruded polystyrene insulation (R-11) added on the outside over its full depth. The basement air temperature is 70 °F. The outside air temperature is 10 °F. Assume the R-value of the 10-inch concrete wall is 1.0. A drawing of the wall is shown in Figure 2-10.

Solution: For the upper wall strip, use Equation 2.5a to determine the effective R-value. Then use Equation 2.3 to calculate the heat loss.

$$R = 1.0 + 11 = 12$$

$$A = 40 \text{ ft.} (2.5 \text{ ft.}) = 100 \text{ ft.}^2$$

$$\Delta T = (70\ °\text{F} - 10\ °\text{F}) = 60\ °\text{F}$$

$$Q = \left(\frac{A}{R}\right)(\Delta T) = \left(\frac{100}{12}\right)(60) = 500 \text{ Btu/hr.}$$

For the middle wall strip, use Equation 2.5b to determine the effective R-value. Then use Equation 2.3 to calculate the heat loss.

$$R = 7.9 + 1.12\,(R_{\text{added}}) = 7.9 + 1.12\,(11) = 20.2$$

$$A = 40 \text{ ft.} (3 \text{ ft.}) = 120 \text{ ft.}^2$$

$$\Delta T = (70\ °\text{F} - 10\ °\text{F}) = 60\ °\text{F}$$

$$Q = \left(\frac{A}{R}\right)(\Delta T) = \left(\frac{120}{20.2}\right)(60) = 356 \text{ Btu/hr.}$$

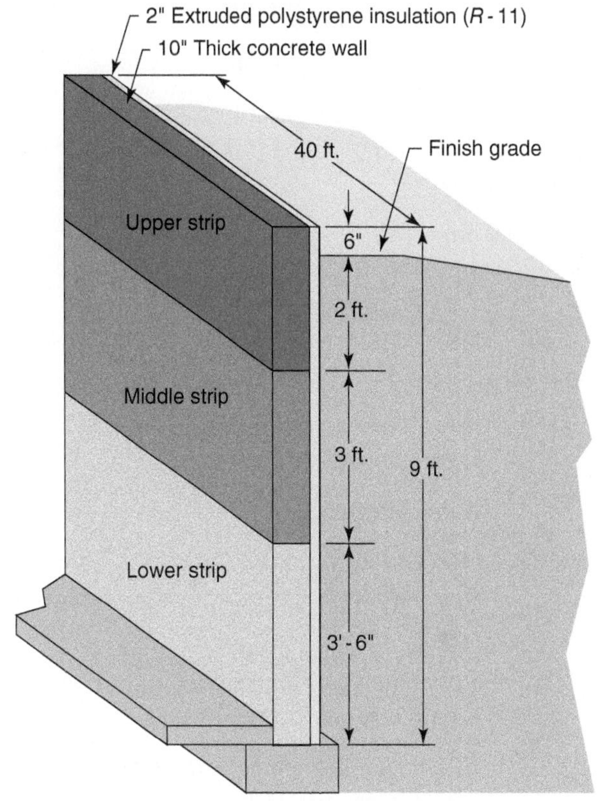

Figure 2-10 | Basement wall used in Example 2.4.

For the lower wall strip, use Equation 2.5c to calculate the R-value. Then use Equation 2.3 to calculate the heat loss.

$$R = 11.3 + 1.13\,(R_{\text{added}}) = 11.3 + 1.13\,(11) = 23.7$$

$$A = 40 \text{ ft.} (3.5 \text{ ft.}) = 140 \text{ ft.}^2$$

$$\Delta T = (70\ °\text{F} - 10\ °\text{F}) = 60\ °\text{F}$$

$$Q = \left(\frac{A}{R}\right)(\Delta T) = \left(\frac{140}{23.7}\right)(60) = 354 \text{ Btu/hr.}$$

The total heat loss for this wall is the sum of the heat losses from each wall strip:

$$Q_{\text{total}} = 500 + 356 + 354 = 1{,}210 \text{ Btu/hr.}$$

Discussion: Note that the same ΔT term (the difference between the basement air temperature and the outside air temperature) was used in all three calculations. It should also be noted that these results are based on a two-dimensional heat transfer model of the situation. Such a model does not account for the more complex three-dimensional heat transfer at wall corners, which will slightly increase basement heat loss. In general, narrow basements with several corners will have greater rates of heat loss compared to wider basements with minimal corners.

HEAT LOSS THROUGH BASEMENT FLOORS

Equation 2.6a can be used to estimate downward heat loss for basement floor slabs that are at least 2 feet below finish grade:

(Equation 2.6a)

$$Q_{\text{bslab}} = \left(\frac{A}{R_{\text{bslab}}}\right)\Delta T$$

where:

Q_{bslab} = rate of heat loss through the basement floor slab (Btu/hr)

A = floor area (ft^2)

R_{bslab} = effective R-value of slab *(designated as R_i in Equation 2.6b, or as R_u in Equation 2.6c)*

ΔT = a. (basement air temperature minus outside air temperature) (°F) *for unheated slabs*

b. (average slab temperature minus outside air temperature) (°F) *for heated slabs*

The effective R-value of the basement slab (R_{bslab}) is found using either Equation 2.6b or 2.6c. Use Equation 2.6b if the slab is *insulated with a least R-3 (°F·hr·ft²/Btu) underside insulation.* Use equation 2.6c if the slab is *uninsulated.* Equations 2.6b and 2.6c are based on the width of the shortest side of the floor slab.

(Equation 2.6b)

$$R_{\text{i}} = 20.625 + 1.6063(w)$$

(Equation 2.6c)

$$R_{\text{u}} = 14.31 + 1.1083(w)$$

where:

R_i = effective R-value for *uninsulated* slab (°F·hr·ft²/Btu)

R_u = effective R-value for *insulated* slab (°F·hr·ft²/Btu)

w = width of *shortest* side of floor slab (ft)

Example 2.5: Determine the rate of heat loss from an unheated basement floor slab measuring 30 feet by 50 feet, when the outside temperature is 10 °F, and the basement air temperature is 65 °F. Assume the basement floor has no underside insulation.

Solution: Since the slab has no underside insulation, Equation 2.6c will be used to calculate the effective total R-Value.

$$R_{\text{u}} = 14.31 + 1.1083(w)$$
$$= 14.31 + 1.1083(30) = 47.56$$

The heat loss from the slab can now be determined using Equation 2.6a.

$$Q_{\text{bslab}} = \frac{A}{R_{\text{bslab}}}(\Delta T) = \frac{30 \times 50}{47.56}(65 - 10) = 1{,}735 \text{ Btu/hr}$$

Discussion: The effective R-value of the slab determined using either Equation 2.6b or Equation 2.6c includes the effects of the slab, underside insulation (if present), and several feet of soil in the path of heat flow from the basement air to the outside air. Thus, the effective R-value in relatively high, even for an uninsulated slab.

HEAT LOSS THROUGH SLAB-ON-GRADE FLOORS

Many residential and commercial buildings have concrete floor slabs rather than crawl spaces or full basements. Heat flows from the floor slab downward to the surrounding earth and outward through any exposed edges of the slab. Of these two paths, outward heat flow through the slab edge tends to dominate. This is because the outer edge of the slab is exposed to outside air or to soil at approximately the same temperature as the outside air. Soil under the slab and several feet in from the perimeter tends to stabilize at somewhat higher temperatures once the building is maintained at normal comfort conditions. Hence, downward heat losses are relatively small from interior areas of the floor slab. Areas of the floor close to the exposed edges have higher rates of heat loss. This often justifies the use of higher R-value insulation near the perimeter of the slab compared to under interior areas.

As with basement walls, the heat flow paths from a slab to the soil and outside air are complex and vary with foundation geometry, soil conditions, and the insulation materials used. Figure 2-11 shows isotherms and arrows indicating the direction of heat flow for a heated slab on grade with continuous underside and edge insulation.

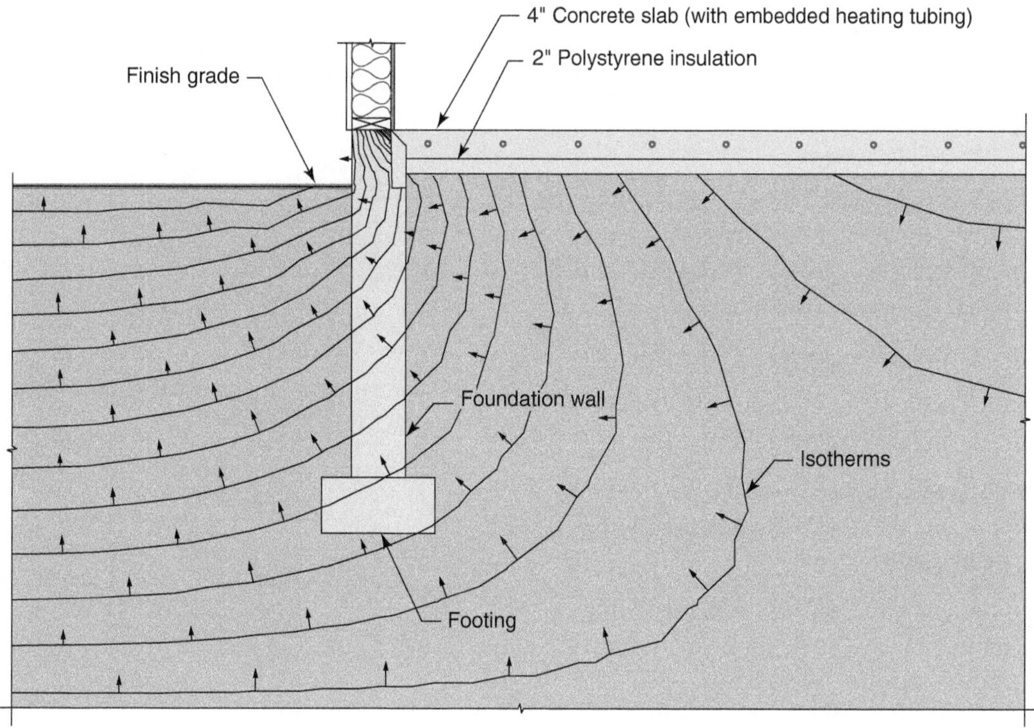

Figure 2-11 | Isotherms and arrows indicating the direction of heat flow for a heated slab-on-grade foundation.

The method used to estimate the heat loss from a slab-on-grade floor depends on the amount of insulation (if any) used near the perimeter of the slab, the orientation of that insulation, and the type of soil present at the perimeter of the slab. Dense soils with significant water content will conduct heat better than lighter/drier soils and thus increase rates of heat loss.

Two scenarios will be considered: One in which no underslab or edge insulation is used, and another where a rigid insulating panel having an R-value between 4 and 16 (°F·hr·ft²/Btu) is installed between the edge of the slab and the foundation as well as under the outer four feet of the slab. Both scenarios are shown in Figure 2-12.

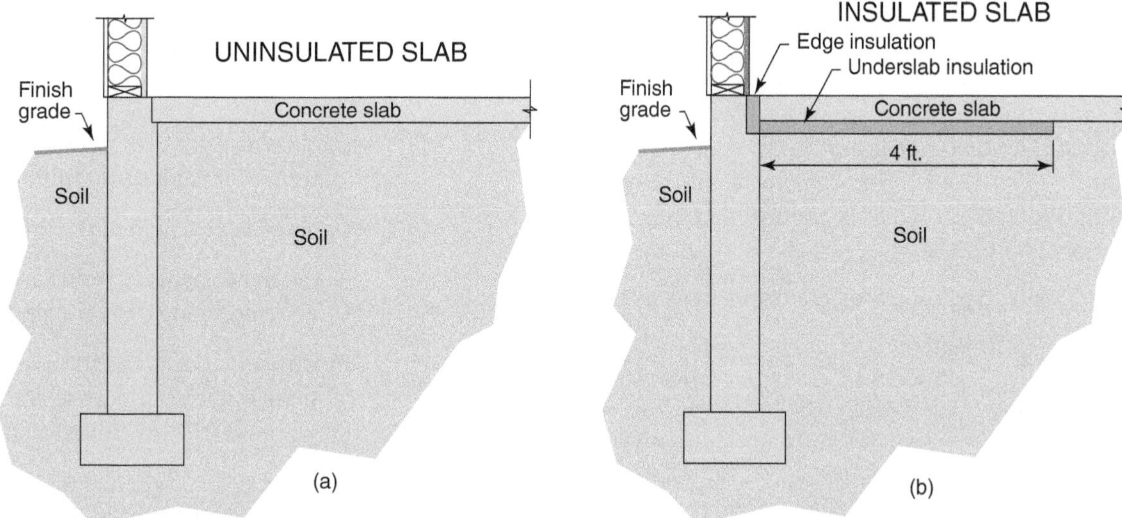

Figure 2-12 | (a) Uninsulated floor slab. (b) Insulated floor slab.

The heat loss from an *uninsulated* slab can be estimated using equation 2.7a.

(Equation 2.7a)

$$Q_{\text{slab}} = c_0 L(\Delta T)$$

where:

Q_{slab} = rate of heat loss through the slab edge *and interior floor area* (Btu/hr)

L = exposed edge length of the floor slab (ft)

ΔT = a. difference between inside and outside air temperature (°F) for unheated slabs

b. difference between *average slab temperature* and outside air temperature (°F) for heated slabs

c_0 = a number based on soil type (see table in Figure 2-13)

Heavy/damp soil	Heavy/dry soil (or) light/damp soil	Light/ dry soil
c_0 = 1.358	c_0 = 1.18	c_0 = 0.989

Figure 2-13 | Values of c_0 for Equation 2.7a (uninsulated slabs).

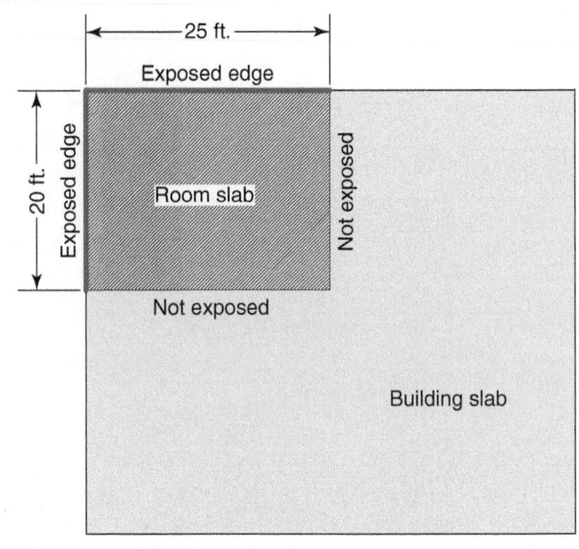

Figure 2-14 | Room with exposed slab edge used in Example 2.6.

Example 2.6: Determine the rate of heat loss from the shaded portion of an unheated and uninsulated floor slab shown in Figure 2-14. The soil outside the slab is classified as heavy and damp. The inside air temperature is 70 °F and the outside air temperature is 10 °F.

Solution: Only two sides of the room's slab are exposed to the outside. The total linear footage of exposed slab edge is 20 + 25 = 45 feet. The value of c_0 for heavy/damp soil in Figure 2-13 is 1.358. Substituting the data into Equation 2.7a yields:

$$Q_{\text{slab}} = c_0 L(\Delta T) = 1.358(45)(70 - 10)$$
$$= 3{,}670 \text{ Btu/hr}$$

Discussion: It should also be stressed that unlike many of the equations used for estimating heat loss, Equation 2.7a is based on the *length* of the exposed perimeter of the slab and not the slab's area.

If the slab contained embedded tubing for heating, it would typically be at temperatures higher than room air temperature during the heating season. This would increase the rate of downward and outward heat loss. *It is suggested that the air temperature in Equation 2.7a be replaced by a temperature that is 25 °F higher than room air temperature when estimating the heat loss of a heated floor slab under design load conditions.*

If the slab-on-grade floor is *insulated* as shown in Figure 2-12b, with insulation having an R-value between 4 and 16 (°F·hr·ft²/Btu), the heat loss can be determined using Equation 2.7b

(Equation 2.7b)

$$Q_{\text{slab}} = L(\Delta T)\left[b_0 + b_1 R_{\text{edge}} + b_2 (R_{\text{edge}})^2\right]$$

where:

Q_{slab} = rate of heat loss through the slab edge and interior floor area (Btu/hr)

L = exposed edge length of the floor slab (ft)

ΔT = a. difference between inside and outside air temperature (°F) for unheated slabs

b. difference between the average slab temperature and outside air temperature (°F) for heated slabs

R_{edge} = R-value of the edge and underslab insulation (°F·hr·ft²/Btu) ($4 \le R_{edge} < 16$)

b_0, b_1, b_2 = constants based on soil type (see Figure 2-15)

Heavy/damp soil	$b_0 = 0.755$	$b_1 = -0.0425$	$b_2 = 0.00126$
Heavy/dry soil (or) light/damp soil	$b_0 = 0.595$	$b_1 = -0.0360$	$b_2 = 0.00108$
Light/dry soil	$b_0 = 0.411$	$b_1 = -0.0293$	$b_2 = 0.0009$

Figure 2-15 | Values of b_0, b_1, and b_2 for Equation 2.7b (insulated slabs).

Example 2.7: Determine the rate of heat loss from the shaded portion of the unheated floor slab shown in Figure 2-14, *assuming 2 inches of extruded polystyrene insulation has been placed at the edge of the slab and under the outer 4 feet of the slab.* The soil outside the slab remains classified as heavy and damp. The inside air temperature is 70 °F and the outside air temperature is 10 °F.

Solution: Using the data in Appendix B, the R-value of 2 inches of extruded polystyrene insulation is 2 × (5.4) = 10.8 (°F·hr·ft²/Btu). Substituting the values of b_0, b_1, and b_2 for heavy/damp soil along with the remaining data into Equation 2.7b yields the following:

$$Q_{slab} = 45(70 - 10)[0.755 + (-0.0425)10.8 + (0.00126)(10.8)^2] = 1{,}196 \text{ Btu/hr}$$

Discussion: Comparing the results of examples 2.6 and 2.7 shows that there is a 67 percent decrease in heat loss from the slab when 2 inches of extruded polystyrene insulation is installed. Insulating slab floors in heated buildings is essential to good performance. The estimated heat loss includes loss through the exposed edge of the slab as well as downward heat flow from the shaded interior area. Equation 2.7b should only be used when the R-value of the underslab insulation is between 4 and 16 (°F·hr·ft²/Btu).

Designers and building owners should consider that there is most likely only one opportunity to properly insulate a concrete slab-on-grade floor. Although "retrofitting" under slab insulation is not impossible, it is extremely disruptive and very expensive. From a practical standpoint, such retrofitting is very unlikely to happen. Thus, the choice of underslab insulation will most likely have to suffice for the life of the building. It is imperative to "get it right" the first time. The author recommends a *minimum* of 2-inch-thick extruded polystyrene insulation (R-10.8 °F·hr·ft²/Btu) under *all* slab-on-grade floor areas in heated spaces. If the building program emphasizes high energy efficiency, even more underslab insulation (3 to 4 inches of extruded polystyrene) is likely justified, especially in cold northern climates. This insulation should be placed under all floor areas with exception of footings for load bearing columns or walls. Extruded polystyrene insulation is available with compressive stress ratings of 15 to 100 psi to accommodate slabs ranging from interior residential floors to heavy vehicle maintenance garages and aircraft hangers.

HEAT LOSS THROUGH WINDOWS, DOORS, AND SKYLIGHTS

The windows, doors, and skylights in a room can represent a large percentage of that room's total heat loss. The construction of windows and skylights can range from old, single glass assemblies with metal frames to modern multi-pane assemblies filled with Argon gas to suppress convection, low-E (low emissivity) coatings to suppress radiation heat transfer, and insulated polymer frames to reduce edge conduction losses. Doors can range from solid wood to foam-core insulated panels with steel or fiberglass claddings.

The thermal resistance of a typical window, door, or skylight unit varies depending upon *where* on the cross section of the unit the thermal resistance is evaluated. For example, a wooden window frame will provide higher thermal resistance than will two panes of glass with an air space in between. Similarly, the thermal resistance at the edge of a multi-pane glazing assembly will be less than that measured at the center of the glazing unit. This is the result of greater conduction through the edge spacers that separate the panes versus the air space at the center of the unit.

To provide a uniform and simplified approach to dealing with these effects, manufacturers, working cooperatively through the National Fenestration Rating Council (NFRC), have developed a standard for determining the **unit U-value** of a window, door, or skylight unit.

The R-value is the reciprocal of U-value. If either the U-value or R-value is known, the other can be easily determined.

$$R = \frac{1}{U}$$

The lower the unit U-value of a window, door, or skylight, the lower its heat loss, all other conditions being equal.

The methods used in the standard NFRC 100 *Procedures for Determining Fenestration Product U-Factor* account for the thermal resistance at the window frame as well as at the edge and center of the glazing unit. These thermal resistances are then "area weighted" based on the shape of the unit. An example of how this would apply to a simple rectangular window unit is shown in Figure 2-16.

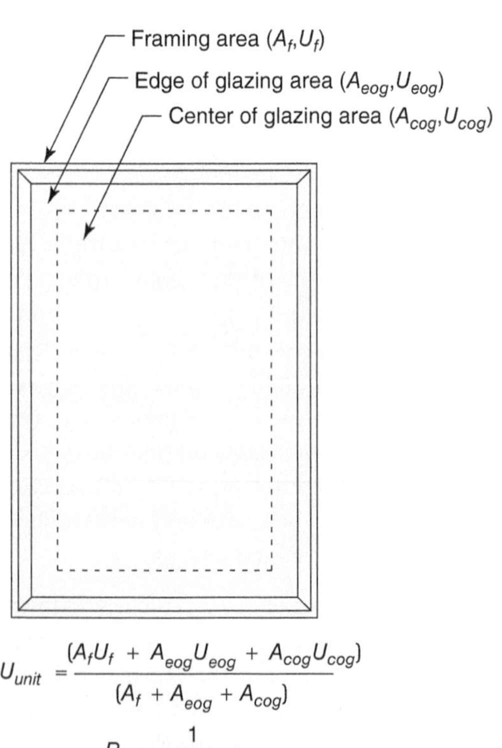

Figure 2-16 | Areas and formula for determining unit U-value of a rectangular window unit.

The unit U-value number can be thought of as the *effective* U-value of the unit. Values for it can be found in the specifications published by window and door manufacturers. Many building codes now require windows and doors used in new construction or remodeling to have unit U-values at, or below, a specified value.

Once the unit U-value is known, its reciprocal (unit R-value) can be used in the same manner at the effective R-value of a wall or ceiling when heat loads are calculated.

THERMAL ENVELOPE OF A BUILDING

Heat is lost through all building surfaces that separate heated space from unheated space. Together, these surfaces (walls, windows, ceilings, doors, foundation, etc.) are called the **thermal envelope** of a building. *It is only necessary to calculate heat flow though the surfaces that constitute this thermal envelope.* Walls, ceilings, floors, or other surfaces that separate one heated space from another should *not* be included in these calculations. Figure 2-17 illustrates a part of the thermal envelope for a house.

One can also envision each room in a building as a "compartment." The thermal envelope of that compartment would consist of all surfaces that *separate heated space from unheated space*. Heat loss calculations would only be performed for these surfaces.

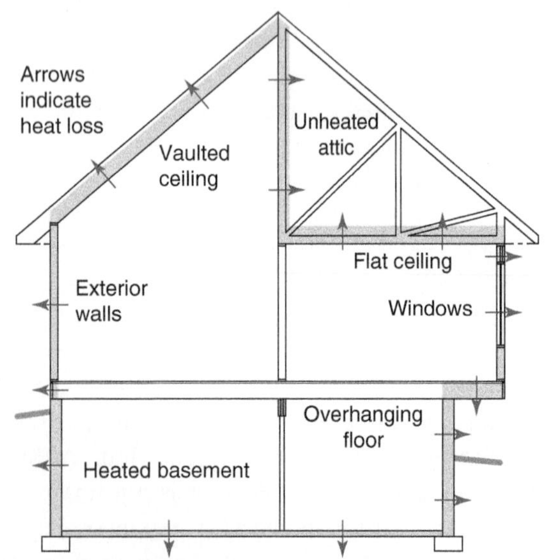

Figure 2-17 | A portion of the thermal envelope of a house (shown shaded).

2.5 Infiltration Heat Loss

In addition to conduction losses, heat is also carried out of buildings by uncontrolled air leakage. This is called **infiltration heat loss**. The faster air leaks into and out of a building, the faster heat is carried away by that air. The thermal envelope of an average building contains hundreds of imperfections through which air can pass. Some leakage points, such as fireplace flues, exhaust hoods, and visible cracks around windows and doors, are obvious. Others are small and out of sight, but when taken together represent significant amounts of leakage area. These include cracks where walls meet the floor deck, air leakage through electrical outlets, small gaps where pipes pass through floors or ceilings, and many other small imperfections in the thermal envelope. Some of these may be the result of structural or aesthetic detailing of the building.

It would be impossible to assess the location and magnitude of all the air leakage paths in a typical building. Instead, an estimate of air leakage can be made based on the air sealing quality of the building. This approach is known as the **air change method** of estimating infiltration heat loss. It relies on a somewhat subjective classification of air sealing quality as well as experience. The designer chooses a rate at which the interior air volume of the building (or an individual room) is exchanged with outside air. For example: 0.5 air change per hour means that half of the entire volume of heated air in a space is replaced with outside air once each hour.

The infiltration rates shown in Figures 2-18a and 2-18b are suggested in *ACCA Manual J* (8th edition). They are categorized based on single or two story constructions. Use Figure 2-18a for single-story construction and Figure 2-18b for two-story construction.

The descriptions for air sealing quality (e.g., tight, semi-loose) are as follows:

- **Tight:** All structural joints and cracks are sealed using *meticulous* workmanship along with air infiltration barriers, caulks, tapings, or packings. Window and doors are rated at less than 0.25 CFM per foot of crack at 25-mph wind speed. All exhaust fans are equipped with backdraft dampers. The building either does not have recessed light fixtures or has recessed lights with negligible air leakage. The house does not have powerful (150 CFM or greater) exhausting range hoods. The house either does not have combustion-type heat sources within conditioned space, or any combustion heat sources present are direct vented. If fireplaces are present

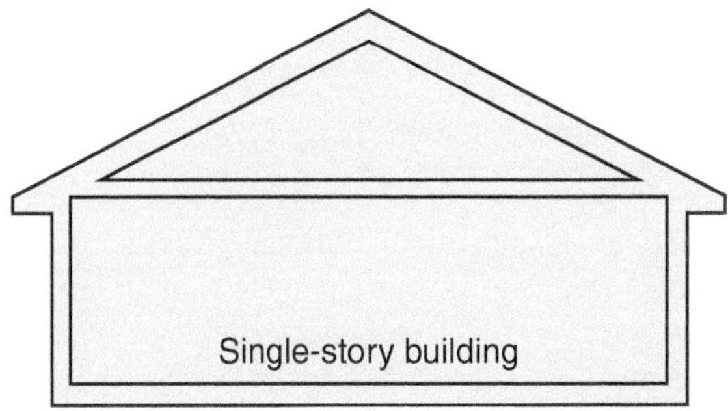

Single-story Floor area (ft.^{2})	900 or less	901–1,500	1,500–2,000	2,000–3,000	3,001 or more
Tight	0.21	0.16	0.14	0.11	0.10
Semi-tight	0.41	0.31	0.26	0.22	0.19
Average	0.61	0.45	0.38	0.32	0.28
Semi-loose	0.95	0.70	0.59	0.49	0.43
Loose	1.29	0.94	0.80	0.66	0.58

Figure 2-18a | Air change rates for single-story construction.

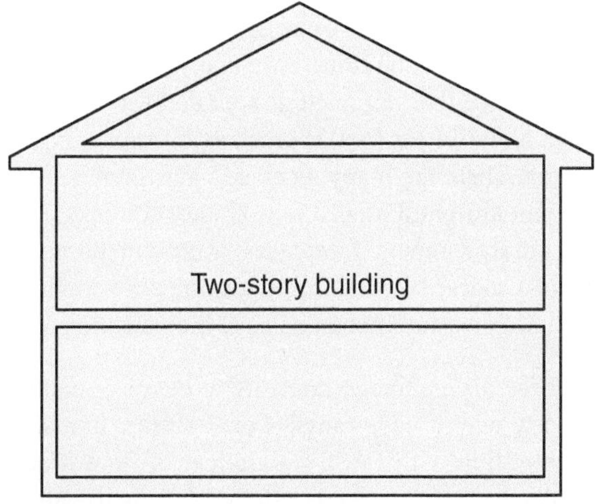

Two-story Floor area (ft.^{2})	900 or less	901–1,500	1,500–2,000	2,000–3,000	3,001 or more
Tight	0.27	0.20	0.18	0.15	0.13
Semi-tight	0.53	0.39	0.34	0.28	0.25
Average	0.79	0.58	0.50	0.41	0.37
Semi-loose	1.23	0.90	0.77	0.63	0.56
Loose	1.67	1.22	1.04	0.85	0.75

Figure 2-18b | Air change rates for two-story construction.

they are provided with their own combustion air and have tight fitting glass doors.

- **Semi-Tight:** Conditions range between tight and average.
- **Average:** All structural joints and cracks are sealed using *average* workmanship along with air infiltration barriers, caulks, tapings, or packings. Window and doors are rated at 0.25 to 0.50 CFM per foot of crack at 25-mph wind speed. All exhaust fans are equipped with backdraft dampers. The building either does not have recessed light fixtures or has recessed lights with negligible air leakage. The house does not have powerful (150 CFM or greater) exhausting range hoods. The house either does not have combustion-type heat sources within conditioned space or any combustion heat sources present are direct vented. If fireplaces are present, they are provided with their own combustion air and have tight fitting glass doors.
- **Semi-Loose:** Conditions range between average and loose.
- **Loose:** There is little or no effort to seals joints or cracks in the building envelope. Windows and doors are either not rated or are rated for more than 50 CFM per foot of crack at 25-mph wind speed. There are many recessed light fixtures that are not airtight. Exhaust fans are not equipped with backdraft dampers. Fireplaces, if present, do not have a source of outside combustion air or tight glass doors. Powerful kitchen exhaust blowers are present.

These air exchange rates are at best estimates and they vary over a wide range. For example, in a single-story building with 1,800 square feet of floor area, the air leakage rate associated with "loose" quality air sealing results in almost six times more heat loss due to air leakage compared to the same size building with "tight" air sealing.

Experience is also helpful in estimating infiltration rates. A visit to the building while it is under construction helps give the heating system designer a feel for the air sealing quality being achieved. An inspection visit to an existing building, though not as revealing, can still help the designer assess air sealing quality. Things to look for or consider include:

- Visible cracks around windows and doors
- Continuous spray foam insulation versus batt insulation
- Poorly installed or maintained weather stripping on doors and windows
- Slight air motion detected near electrical fixtures
- Smoke that disappears upward into ceiling lighting fixtures
- Deterioration of paint on the downwind side of building
- Poor-quality duct-system design (especially a lack of proper return air ducting)
- Interior humidity levels (drier buildings generally indicate more air leakage)
- Presence or lack of flue damper on fossil fuel heating system
- Windy versus sheltered building site
- Storm doors and entry vestibules help reduce air infiltration
- The presence of fireplaces (especially those of older masonry construction)
- Casement and awning windows generally have less air leakage than double hung or sliding windows
- Cold air leaking in low in the building generally indicates warm air leaking out higher in the building

BLOWER DOOR TESTING

The airtightness of an existing or partially constructed building can often be assessed through blower door testing. A **blower door** consists of an adjustable frame covered by an airtight fabric that expands to form an airtight seal around the perimeter of an exterior door opening, as shown in Figure 2-18c. A variable speed fan mounts through the fabric panel.

When the fan in the door panel is operated, a negative air pressure is created within the building. This causes outside air to move into the building (e.g., infiltrate) through any possible leakage paths. Infiltrating air can be detected by a **smoke pencil** held near suspected leakage paths such as the perimeter of doors and windows, electrical junction boxes or pipe penetrations, or seams and cracks in surfaces. The smoke pencil emits an inert smoke that responds to the slightest air movement. In many cases, the detected leakage paths can be repaired using sealants, tapes, or similar measures. This is especially true when the building is

Figure 2-18c | A blower door test being performed during house construction.

under construction. Ideally, a blower door test is performed when the thermal envelope is completed, but the wall, ceiling, and floor finishes have not yet been installed. This allows good access to detect and repair leakage paths. The blower door can then be used to assess the degree to which the leakage paths have been reduced or eliminated.

Some blower doors are equipped with instruments that precisely measure the differential pressure between outside and inside the building. The overall air tightness of the building is determined by increasing the fan speed until this differential pressure reaches 50 Pascals, which is approximately 0.2 inches of water column pressure. This creates a condition that approximates a 20-mph wind striking all surfaces of the building simultaneously. The air changes per hour leakage rate at 50 Pascals differential pressure can then be determined based on the fan speed and house volume. An average house without special air leakage detailing could have upward of 15 air changes per hour at a 50 Pascal differential pressure. An **EnergyStar-certified** house requires an air leakage rate less than or equal to 4 air changes per house at 50 Pascals. An ultra-tight house built to the German **PassivHaus standard** must have a tested air leakage rate not exceeding 0.6 air changes per hour at 50 Pascals differential pressure.

ESTIMATING INFILTRATION HEAT LOSS

Air leakage rates can be converted into rates of heat loss using Equation 2.8:

(Equation 2.8)

$$Q_i = 0.018(n)(v)(\Delta T)$$

where:

Q_i = estimated rate of heat loss due to air infiltration (Btu/hr)

n = number of air changes per hour, estimated based on air sealing quality (1/hr)

v = interior volume of the heated space (room or entire building) (ft^3) 0.018 = heat capacity of air (Btu/ft^3/°F)

ΔT = inside air temperature minus the outside air temperature (°F)

Example 2.7: Determine the rate of heat loss by air infiltration for a 30-foot by 56-foot single-story building with 8-foot ceiling height and average quality air sealing. The inside temperature is 70 °F. The outside temperature is 10 °F.

Solution: The floor area of this building is 30 × 56 = 1,680 ft^2. Referring to Figure 2-18a for a 1,680-square-foot *single-story* building with average air sealing quality, the suggested air leakage rate is 0.38 air changes per hour. Substituting this value along with the other data into Equation 2.8 yields:

$$\begin{aligned} Q_i &= 0.018(n)(v)(\Delta T) \\ &= 0.018(0.38)(30 \times 56 \times 8)(70 - 10) \\ &= 5{,}516 \text{ Btu/hr} \end{aligned}$$

As the conduction heat losses of a building are reduced, air infiltration becomes a larger percentage of total heat loss. When higher insulation levels are specified for a building, greater efforts at reducing air infiltration should also be undertaken. These include the use of high-quality windows and doors, close attention to caulking, sealed infiltration barriers, and so on.

2.6 Example of Complete Heating Load Estimate

Finding the total design heat loss of a room is simply a matter of calculating both the conduction heat loss and the infiltration heat loss and then adding them together. When this is done on a room-by-room basis, the resulting numbers can be used to size the heat emitters as the hydronic distribution system is designed. The sum of the room heat losses—the total building design heat loss—determines the required heat output of the heat source.

This section demonstrates a complete heat loss estimate for a small house having the floor plan shown in Figure 2-19. Simple tables show the results of using the equations present earlier in this chapter. Other assumed values for the calculations are as follows:

- Unit R-value for all windows: R-3.0
- Unit R-value for exterior door: R-5
- R-value of foundation edge insulation: R-10
- Rate of air infiltration in vestibule and utility room: 1.0 air change/hr
- Rate of air infiltration in all other rooms: 0.5 air change/hr
- Outdoor design temperature: −5 °F
- Desired indoor temperature: 70 °F

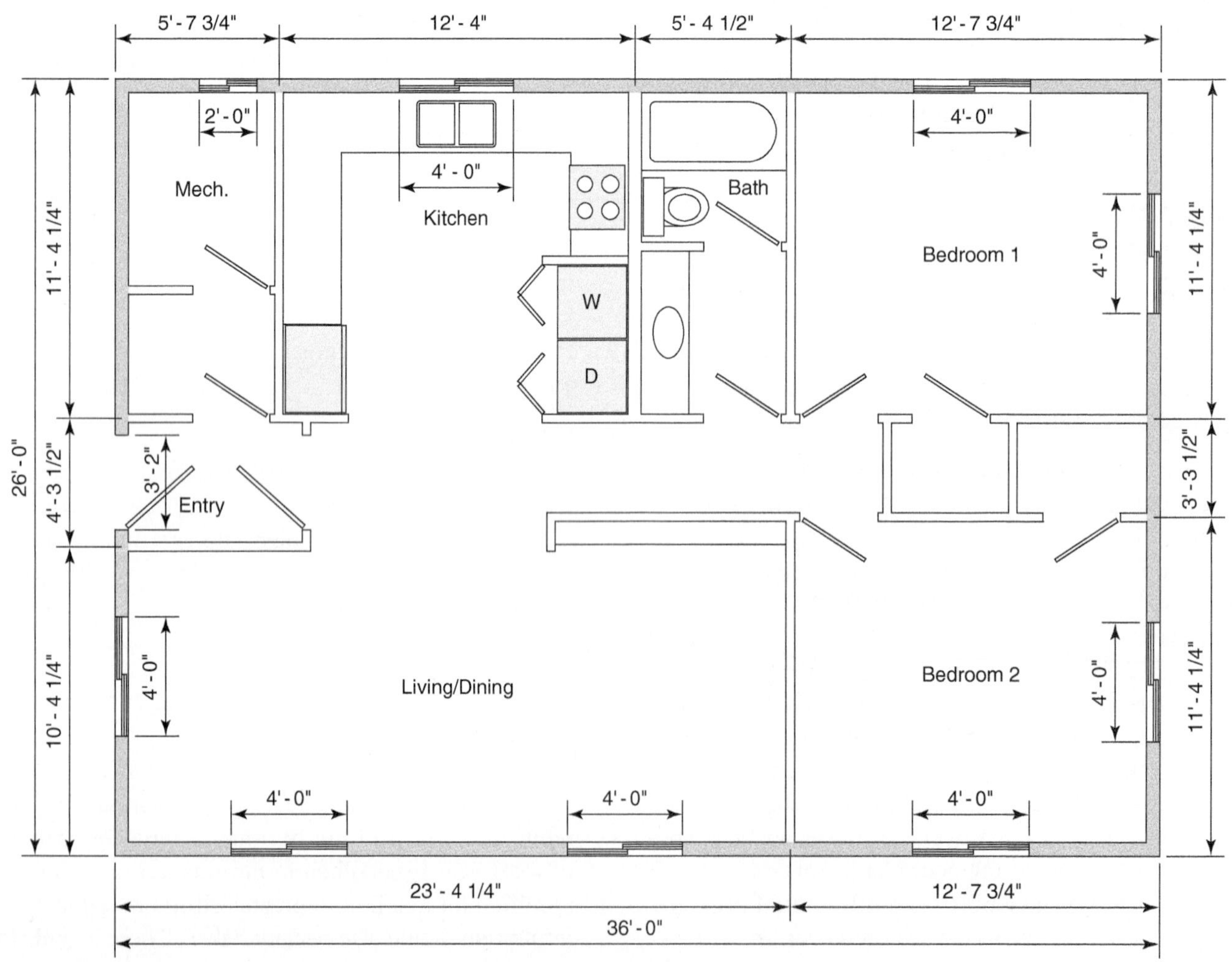

Figure 2-19 | Floor plan of example house for which heating load will be estimated.

- Window height: 4 feet
- Exterior door height: 6 feet 8 inches
- Wall height: 8 feet
- Light/dry soil under slab

DETERMINING THE TOTAL R-VALUE OF THERMAL ENVELOPE SURFACES

Walls: The wall cross section is shown in Figure 2-20. It is made up of 2 × 6 wood framing spaced 24 inches on center with R-21 fiberglass batt insulation in the stud cavities. The inside finish is 1/2-inch drywall. The outside of the wall is sheathed with 1/2-inch plywood, and covered with a Tyvek® infiltration barrier and vinyl siding.

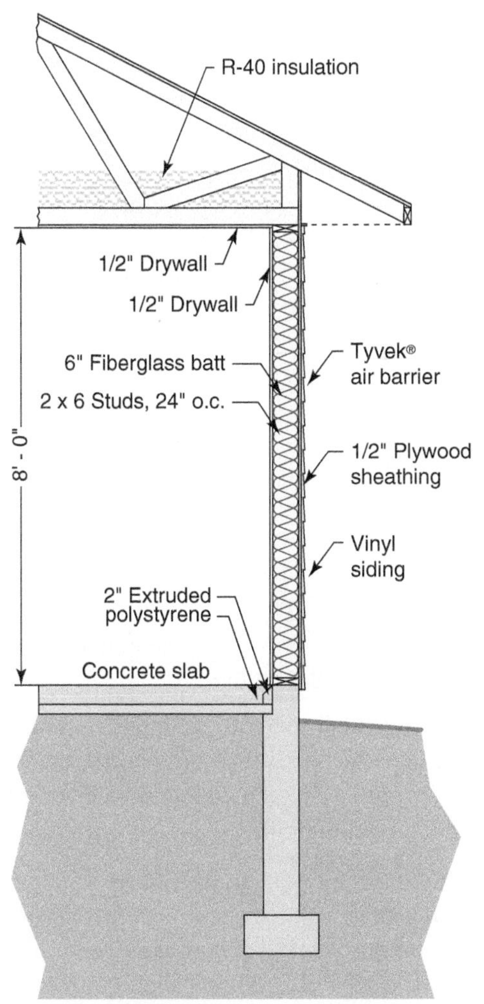

Figure 2-20 | Wall cross section for example house.

The R-values of the materials are as follows:

Material	R-value between framing	R-value at framing
Inside air film	0.68	0.68
1/2-inch drywall	0.45	0.45
Stud cavity	21.0	5.5
1/2-inch plywood sheathing	0.62	0.62
TYVEK® infiltration barrier	0.61	0.61
Vinyl siding	~0	~0
Outside air film	0.17	0.17
Total	23.5	8.03

Assuming 15 percent of wall is solid framing, the effective total R-value of the wall is determined using Equation 2.4:

$$R_{\text{effective}} = \frac{(R_i)(R_f)}{p(R - R_f) + R_f}$$

$$= \frac{(23.53)(8.03)}{0.15(23.53 - 8.03) + 8.03} = 18.3$$

Ceilings: The ceiling consists of 1/2-inch drywall covered with approximately 12 inches of blown fiberglass insulation. A cross section is shown in Figure 2-21. The roof trusses displace a minor amount of this insulation.

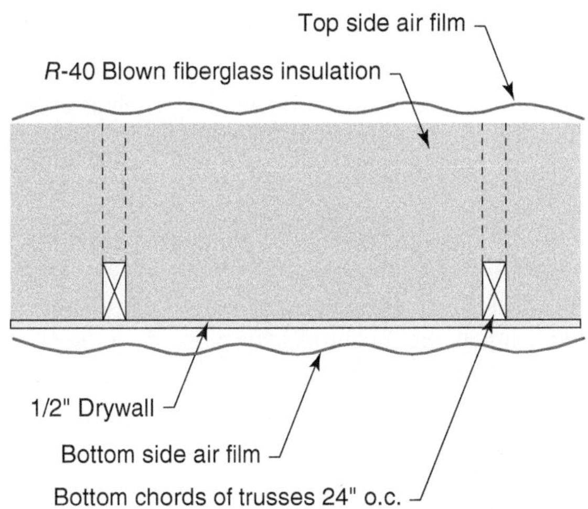

Figure 2-21 | Cross section of ceiling in example house.

The R-values of the materials are as follows:

Material	R-value between framing	R-value at framing
Bottom air film	0.61	0.61
1/2-inch drywall	0.45	0.45
Insulation	40	28.3
Framing	0	3.5
Top air film	0.61	0.61
Total	41.67	33.47

Assuming 10 percent of ceiling is solid framing, the effective total R-value of the ceiling is also determined using Equation 2.4:

$$R_{\text{effective}} = \frac{(R_i)(R_f)}{p(R_i - R_f) + R_f}$$

$$= \frac{(41.67)(33.47)}{0.10(41.67 - 33.47) + 33.47} = 40.7$$

This effective R-value is very close to the R-value between framing. The wooden truss chords have minimal effect on the effective total R-value.

ROOM-BY-ROOM CALCULATIONS

Using information from the floor plan and the total effective R-values of the various surfaces, the design heating load of each room can be calculated. This is done by breaking out each room from the building and determining all necessary information, such as areas of walls, and windows. This information will be entered into tables to keep it organized.

To expedite the process, many of the dimensions are shown on the floor plan of the individual rooms. If such were not the case, the dimensions would have to be estimated using an architectural scale along with the printed floor plan. If the floor plan were available as a CAD file, dimensions could also be determined using the dimensioning tool of the CAD system without having to print the drawing. As a matter of convenience, the room dimensions are taken to the centerline of the common walls.

When working with architectural dimensions, it is best to convert inches and fractions of inches into decimal feet since many dimensions will be added and multiplied together. Example 2.8 illustrates this.

Example 2.8: Convert the dimension 10 ft. 4 1/4 in to decimal feet.

Solution:

$$10\text{ ft. } 4\ 1/4\text{ in} = 10\text{ ft.} + \left(\frac{4.25}{12}\right)\text{ft.} = 10\text{ ft.} + 0.35\text{ ft.}$$

$$= 10.35\text{ ft.}$$

Discussion: The conversion from feet and inches into decimal feet has been performed before the calculation of areas and lengths in Figures 2-22 through Figure 2-26.

TOTAL BUILDING HEATING LOAD

The total building heating load is obtained by summing the heating loads of each room.

Living/dining room	4,384 Btu/hr
Bedroom 1	3,102 Btu/hr
Bedroom 2	3,102 Btu/hr
Kitchen	2,278 Btu/hr
Bathroom	688 Btu/hr
Mechanical room/entry	2,533 Btu/hr
Total	**16,087 Btu/hr**

A number of simplifying assumptions were used in this example.

- The rooms were divided at the centerline of the common partitions. This ensures that all the exterior wall area is assigned to individual rooms and not ignored as being between the rooms.
- Since the two bedrooms are essentially identical in size and construction, it was only necessary to find the heat load of one, then double it when determining the total building load.
- The exterior dimensions of the rooms were used to determine areas, volumes, and so forth. This makes the calculations somewhat conservative because the exterior wall area of a room is slightly greater than the interior wall area.

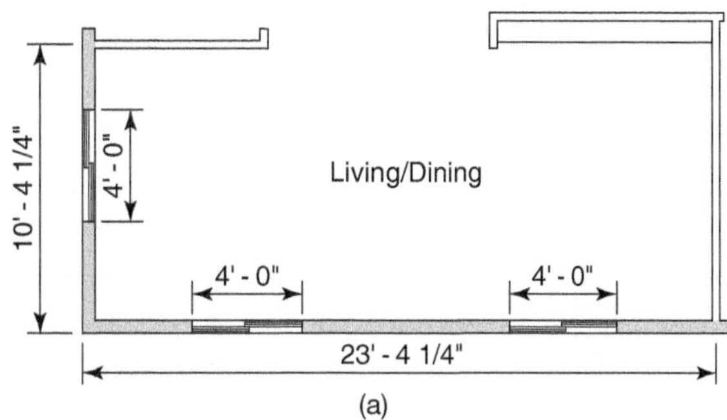

(a)

Gross exterior wall area = (10.35 ft. + 23.35 ft.)(8 ft.) = (33.708 ft.)(8 ft.) = 269.7 ft.^{2}

Window area = 3 (4 ft. × 4 ft.) = 48 ft.^{2}

Exterior door area = 0

Net exterior wall area = Gross exterior wall area − window and door area = 221.7 ft.^{2}

Ceiling area = (10.35 ft.)(23.35 ft.) = 241.7 ft.^{2}

Exposed slab edge length = (10.35 ft. + 23.35 ft.) = 33.7 linear ft.

Room volume = (10.35 ft.)(23.35 ft.)(8.0 ft.) = 1,933.4 ft.^{3}

		1	2	3	4	5	6	7	8	9	10
Room name		Exposed walls	Windows	Exposed doors	Exposed ceilings	Exposed floor	Exposed slab edge	Room volume	Air changes	Infiltration loss	Total room load
Living & dining	1	A = 221.7	A = 48	A = 0	A = 241.7	A = 0	L = 33.7	V = 1933	N = 0.5		
	2	R = 18.3	R = 3	R = n/a	R = 40.7	R = n/a	Re = 11				
	3	$\frac{A}{R}(\Delta T)$ = 909	$\frac{A}{R}(\Delta T)$ = 1200	$\frac{A}{R}(\Delta T)$ = 0	$\frac{A}{R}(\Delta T)$ = 445	$\frac{A}{R}(\Delta T)$ = 0	Q_{slab} = 525			Q_i = 1305	4384

(b)

Figure 2-22 | (a) Floor plan for living/dining room of example house. (b) Table with values for living/dining room of example house.

- The areas and volumes of the closets between the bedrooms were equally divided up between the two bedrooms. Since a separate heat emitter would not be supplied for each closet, this part of the load is simply assigned to the associated bedrooms.

2.7 Computer-Aided Heating Load Calculations

After gathering the data and performing all the calculations needed for a building heating load estimate, It becomes apparent that a significant amount of information goes into finding the results. If the designer then decides to go back and change one or more of the numbers, more time is required to unravel the calculations back to the point of the change and calculate the new result. Even for people used to handling this amount of information, these calculations are tedious and always subject to errors. The time required to perform accurate heating load estimates is arguably the chief reason why some "heating professionals" do not do them.

Like many routine design procedures, heating load calculations are now commonly done using computers, and sometimes even on smartphones. The advantages of using load estimating software are many:

- Rapid (almost instant) calculations to quickly study the effect of changes
- Automated referencing of R-values, and weather data, in some programs
- Significantly less chance of error due to number handling
- Ability to print professional reports for customer presentations
- Ability to store project files for possible use in similar future projects
- Ability to import area and R-value information directly from CAD-based architectural drawings

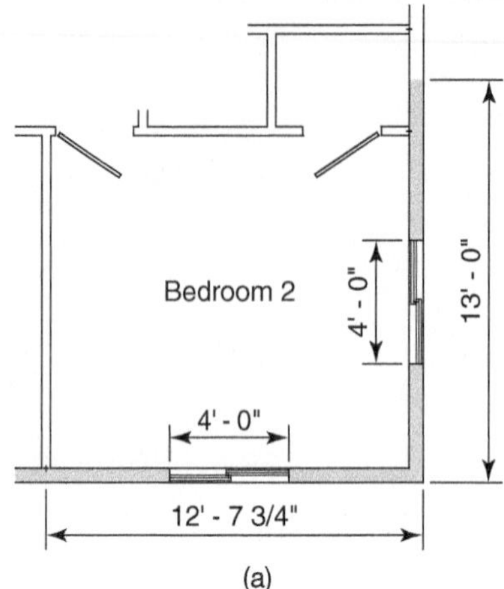

(a)

Gross exterior wall area = (12.65 ft. + 13.0 ft.)(8 ft.) = (25.65 ft.)(8 ft.) = 205.2 ft.^{2}
Window area = 2 (4 ft. · 4 ft.) = 32 ft.^{2}
Exterior door area = 0
Net exterior wall area = Gross exterior wall area – window and door area = 173.2 ft.^{2}
Ceiling area = (12.65 ft.)(13.0 ft.) = 164.5 ft.^{2}
Exposed slab edge length = (12.65 ft. + 13.0 ft.) = 25.65 linear ft.
Room volume = (12.65 ft.)(13.0 ft.)(8.0 ft.) = 1,315.6 ft.^{3}

		1	2	3	4	5	6	7	8	9	10
Room name		Exposed walls	Windows	Exposed doors	Exposed ceilings	Exposed floor	Exposed slab edge	Room volume	Air changes	Infiltration loss	Total room load ↓
Bedroom 2	1	A = 173.2	A = 32	A = 0	A = 164.5	A = 0	L = 25.7	V = 1316	N = 0.5		
	2	R = 18.3	R = 3	R = n/a	R = 40.7	R = n/a	Re = 11				
	3	$\frac{A}{R}(\Delta T)$ = 710	$\frac{A}{R}(\Delta T)$ = 800	$\frac{A}{R}(\Delta T)$ = 0	$\frac{A}{R}(\Delta T)$ = 303	$\frac{A}{R}(\Delta T)$ = 0	Q_{slab} = 401			Q_i = 888	3102*

(b)

Figure 2-23 | (a) Floor plan for Bedroom 2 of example house. (b) Table with values for Bedroom 2 of example house.

There are many software packages currently available for estimating the heating loads for residential and light commercial buildings. They vary in cost from free to over \$500. The higher-cost programs usually offer more features such as material reference files, automated areas determination from architectural drawings, and customized output reports. The lower-cost programs offer no frills output, but can nonetheless still yield accurate results.

Figure 2-27a shows the main screen of a program named ***Building Heat Load Estimator***, which is part of the more comprehensive software tool ***Hydronics Design Studio 2.0***, which was co-developed by the author. This software allows the user to build a list of rooms, each of which is defined based on it thermal envelope. The total heat loss of each room as well as the overall building is continually displayed as the user interacts with the software.

Building Heat Load Estimator allows the user to define the thermal envelope surfaces with each room, using pull-down menus for materials. It also provides a point-and-click selection of common wall, ceiling, floor, window and door assemblies. An example of how a framed wall is specified is shown in Figure 2-27b.

Building Heat Load Estimator also assists the user in determining surface areas and room volumes, as shown in Figure 2-27c.

A free demo version of this software can be downloaded at **www.hydronicpros.com**.

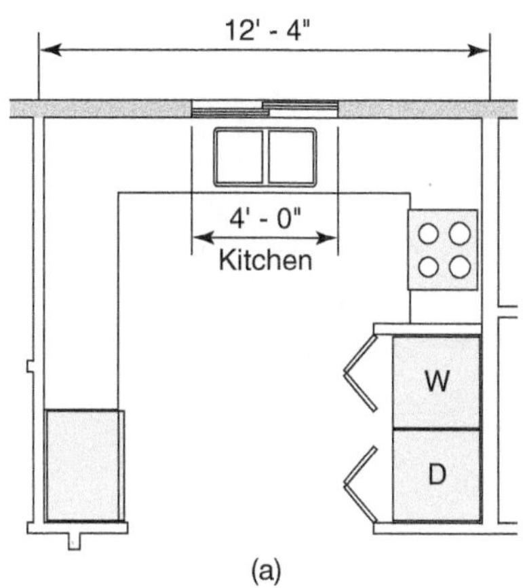

(a)

Gross exterior wall area = (12.33 ft.) (8 ft.) = 98.6 ft.^{2}
Window area = 1 (4 ft. × 4 ft.) = 16 ft.^{2}
Exterior door area = 0
Net exterior wall area = Gross exterior wall area – window and door area = 82.6 ft.^{2}
Ceiling area = (12.33 ft.) (11.0 ft.) = 135.6 ft.^{2}
Exposed slab edge length = 12.33 linear ft.
Room volume = (12.33 ft.) (11.0 ft.) (8.0 ft.) = 1,085 ft.^{3}

		1	2	3	4	5	6	7	8	9	10
Room name		Exposed walls	Windows	Exposed doors	Exposed ceilings	Exposed floor	Exposed slab edge	Room volume	Air changes	Infiltration loss	Total room load
	1	A = 82.6	A = 16	A = 0	A = 135.6	A = 0	L = 12.3	V = 1085	N = 0.5		
Kitchen	2	R = 18.3	R = 3	R = n/a	R = 40.7	R = n/a	Re = 11				↓
	3	$\frac{A}{R}(\Delta T)$ = 339	$\frac{A}{R}(\Delta T)$ = 400	$\frac{A}{R}(\Delta T)$ = 0	$\frac{A}{R}(\Delta T)$ = 250	$\frac{A}{R}(\Delta T)$ = 0	Q_{slab} = 401			Q_i = 888	2278

(b)

Figure 2-24 | (a) Floor plan for kitchen of example house. (b) Table with values for kitchen of example house.

2.8 Estimating Annual Heating Energy Use

Although heating system designers need to know the design heat loss of each room in a building before sizing a heating system, such information is often meaningless to building owners. Their interest usually lies in what it will cost to heat their building. Such estimates can be made once the building's design heating load and the seasonal efficiency of the heat source are established.

DEGREE DAY METHOD

The seasonal energy used for space heating depends upon the climate in which the building is located. A relatively simple method for factoring local weather conditions into estimates of heating energy use is the concept of heating degree days. *The number of heating degree days that accumulate in a 24-hour period is the difference between 65 °F and the average outdoor air temperature during that period* (see Equation 2.9). The average temperature is determined by averaging the high and low temperature for the 24-hour period.

(Equation 2.9)

$$DD_{daily} = (65 - T_{ave})$$

where:

DD_{daily} = daily degree days accumulated in a 24-hour period

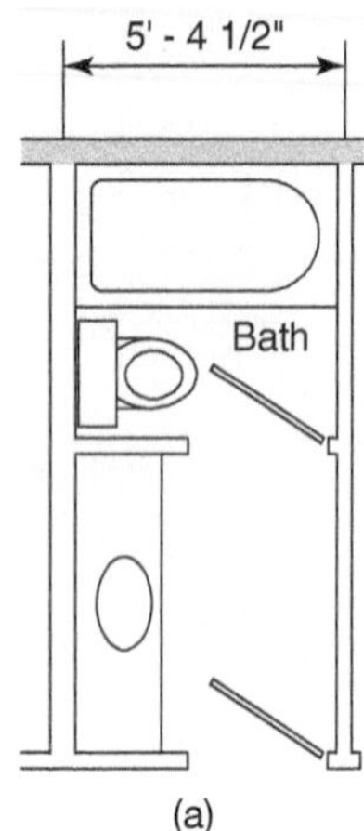

(a)

Gross exterior wall area = (5.38 ft.) (8 ft.) = 43.0 ft.^{2}
Window area = 0
Exterior door area = 0
Net exterior wall area = Gross exterior wall area – window and door area = 43.0 ft.^{2}
Ceiling area = (5.38 ft.) (11.0 ft.) = 59.2 ft.^{2}
Exposed slab edge length = 5.38 linear ft.
Room volume = (5.38 ft.) (11.0 ft.) (8.0 ft.) = 473 ft.^{3}

Room name		1 Exposed walls	2 Windows	3 Exposed doors	4 Exposed ceilings	5 Exposed floor	6 Exposed slab edge	7 Room volume	8 Air changes	9 Infiltration loss	10 Total room load
Bathroom	1	A = 43	A = 0	A = 0	A = 59.2	A = 0	L = 5.38	V = 473	N = 0.5		↓
	2	R = 18.3	R = 3	R = n/a	R = 40.7	R = n/a	Re = 11				
	3	$\frac{A}{R}(\Delta T)$ = 176	$\frac{A}{R}(\Delta T)$ = 0	$\frac{A}{R}(\Delta T)$ = 0	$\frac{A}{R}(\Delta T)$ = 109	$\frac{A}{R}(\Delta T)$ = 0	Q_{slab} = 84			Q_i = 319	688

(b)

Figure 2-25 | (a) Floor plan for bathroom of example house. (b) Table with values for bathroom of example house.

T_{ave} = average of the high and low outdoor temperature for that 24-hour period

The total heating degree days for a month, or an entire year, is found by adding up the *daily* heating degree days over the desired period. Many heating reference books contain tables of monthly and annual heating degree days for many major cities in the United States and Canada. A sample listing of annual degree days is given in Figure 2-28.

Heating degree day data has been used to estimate fuel usage for several decades. Such data is often recorded by fuel suppliers, utility companies, and local weather stations. Degree day statistics are also frequently listed in the weather section of newspapers as well as on Web-based weather reports.

When the heating degree day method was first developed, fuel was inexpensive and buildings were poorly insulated. Building heat loss was substantially greater than for comparably sized buildings using modern construction techniques.

Estimating seasonal energy consumption using heating degree days assumes that buildings need heat input from their heating systems whenever the outside temperature drops below 65 °F. While this may be true for many poorly insulated buildings, it is often incorrect for better-insulated buildings. Many energy efficient homes can maintain comfortable interior temperatures, without heat input from their space heating systems, even when outdoor temperatures drop into the 40 °F range. This is a result of greater internal heat

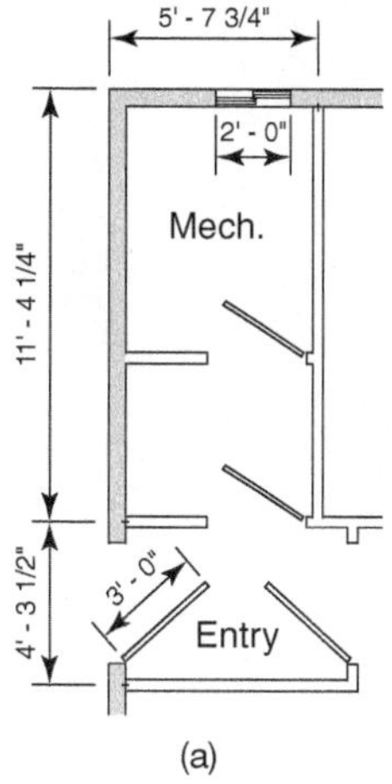

(a)

Gross exterior wall area = (4.29 ft. + 11.35 ft. + 5.65 ft.) (8 ft.) = (21.28 ft.) (8 ft.) = 170.2 ft.^{2}
Window area = (2 ft. × 4 ft.) = 8 ft.^{2}
Exterior door area = (6.66 ft. × 3 ft.) = 20 ft.^{2}
Net exterior wall area = Gross exterior wall area − window and door area = 142.2 ft.^{2}
Ceiling area = (4.29 ft. + 11.35 ft.) (5.65 ft.) = (15.64 ft. × 5.65 ft.) = 88.4 ft.^{2}
Exposed slab edge length = 21.28 linear ft.
Room volume = (15.64 ft.) (5.65 ft.) (8.0 ft.) = 707 ft.^{3}

		1	2	3	4	5	6	7	8	9	10
Room name		Exposed walls	Windows	Exposed doors	Exposed ceilings	Exposed floor	Exposed slab edge	Room volume	Air changes	Infiltration loss	Total room load ⬇
Mech. room & entry	1	$A = 142.2$	$A = 8$	$A = 20$	$A = 88.4$	$A = 0$	$L = 21.3$	$V = 707$	$N = 1.0$		
	2	$R = 18.3$	$R = 3$	$R = 5$	$R = 40.7$	$R =$ n/a	$Re = 11$				
	3	$\frac{A}{R}(\Delta T) =$ 583	$\frac{A}{R}(\Delta T) =$ 200	$\frac{A}{R}(\Delta T) =$ 300	$\frac{A}{R}(\Delta T) =$ 163	$\frac{A}{R}(\Delta T) =$ 0	$Q_{slab} =$ 332			$Q_i =$ 955	2533

(b)

Figure 2-26 | (a) Floor plan for mechanical room and entry of example house. (b) Table with values for mechanical room and entry of example house.

gains from appliances, lights, people, and sunlight combined with much lower rates of heat loss due to better insulation and reduced air leakage. Internal heat gains eliminate the need for a corresponding amount of heat from the building's heating system. Passive solar buildings are a good example. On sunny days, some of these buildings require no heat input from their "auxiliary" heating systems, even when outside temperatures are below 0 °F!

Another complicating factor is that some heating degree days occur in months when most heating systems are turned off. This is especially true in northern climates where some heating degree days are recorded in June, July, and even August. Obviously, if the heating system is turned off, there will be no fuel consumption associated with these degree days. Still, any degree days that occur in these summer months are included in the annual total.

To compensate for the effect of internal heat gains, and warm-weather degree days, ASHRAE devised a correction multiplier known as **Cd factor**. This factor is plotted against total heating degree days in Figure 2-29 and will be used to estimate seasonal heating energy use. The Cd value given by this graph is representative of average conditions over a large number of buildings. The specific Cd value that applies to a given building might be higher or lower than the value shown by the graph. For example, a passive solar building in a sunny climate such as Denver is likely to have a lower Cd value than given in Figure 2-29 because it may get a major part of its heating energy from the sun. On the other hand, a building with small windows, in a shaded

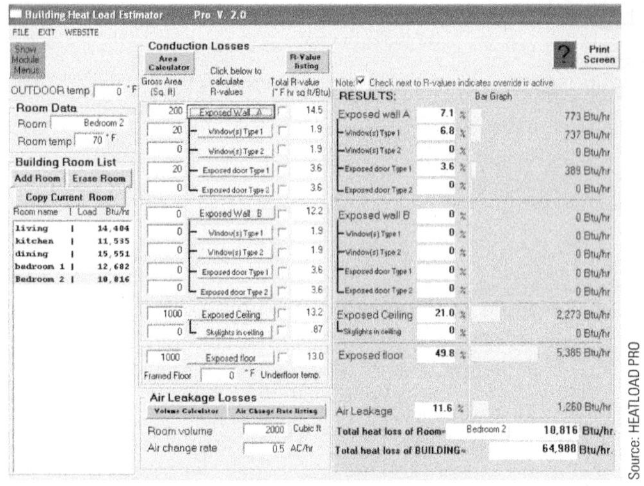

Source: HEATLOAD PRO

Figure 2-27a | Main screen of *Building Heat Load Estimator* software.

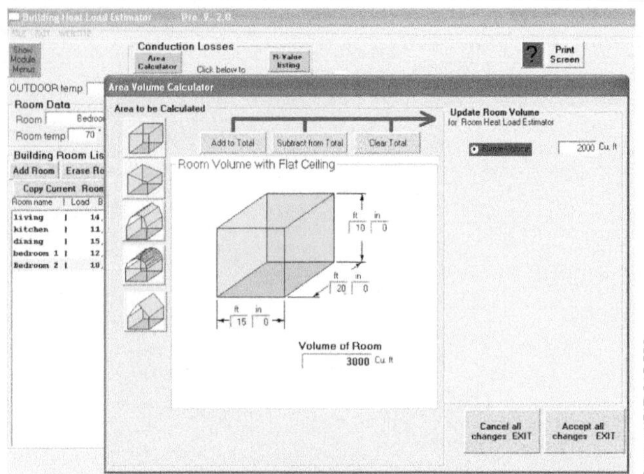

Source: HEATLOAD PRO

Figure 2-27c | *Building Heat Load Estimator* screen for assisting with room volume determination.

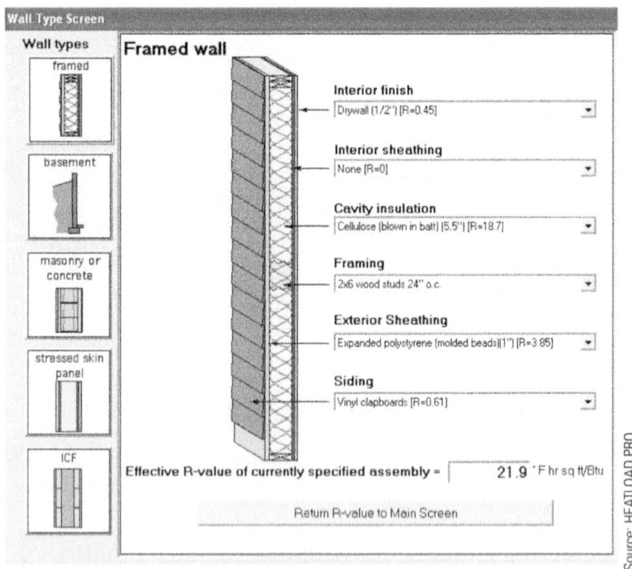

Source: HEATLOAD PRO

Figure 2-27b | HEAT LOAD PRO screen for defining and specifying wall construction.

location, or one with very little internal heat generation may have a larger value of Cd compared to the "typical" value listed in Figure 2-29. This variation will directly affect the estimate of annual fuel usage.

Equation 2.10 can be used to estimate annual heating energy required by the building:

(Equation 2.10)

$$E_{\text{annual}} = \frac{(Q_{\text{design}})(DD_{65})(24)(C_{\text{d}})}{1{,}000{,}000(\Delta T_{\text{design}})}$$

where:

E_{annual} = estimated annual heating energy required by the building (MMBtu)

NOTE: 1 MMBtu = 1,000,000 Btu

Q_{design} = design heating load of the building (Btu/hr)

DD_{65} = annual total heating degree days (base 65) at the building location (degree days)

C_{d} = correction factor from Figure 2-28 (unitless)

ΔT_{design} = design temperature difference at which design heating load was determined (°F)

Example 2.9: A house has a calculated design heating load of 55,000 Btu/hr when its interior temperature is 68 °F, and the outdoor temperature is –10 °F. The house is located in a climate having 7,000 annual heating degree days. Estimate the annual heating energy that must be delivered to this building for a typical heating season.

Solution: The Cd factor for a 7,000 heating degree day climate is estimated at 0.63 using Figure 2-29. This and the other given data are then substituted into Equation 2.10:

$$E_{\text{annual}} = \frac{(Q_{\text{design}})(DD_{65})(24)(C_{\text{d}})}{1{,}000{,}000(\Delta T_{\text{design}})}$$

$$= \frac{(55{,}000)(7{,}000)(24)(0.63)}{1{,}000{,}000(68 - (-10))} = 74.6 \text{ MMBtu}$$

CITY	°F-days (base 65)
Atlanta, GA	2,990
Baltimore, MD	4,680
Boston, MA	5,630
Burlington, VT	8,030
Chicago, IL	6,640
Denver, CO	6,150
Detroit, MI	6,290
Madison, WI	7,720
Newark, NJ	4,900
New York, NY	5,219
Philadelphia, PA	5,180
Pittsburgh, PA	5,950
Portland, ME	7,570
Providence, RI	5,950
Seattle, WA	4,424
Syracuse, NY	6,720
Toronto, ONT	6,827

Figure 2-28 | Annual heating degree days for some major cities.

Discussion: The building needs 74.6 MMBtu (or 74,600,000 Btus) delivered to it over a typical heating season. Keep in mind that *this calculation does not account for the inefficiencies of the heat source or the distribution system in producing and delivering this heat to the building*. These inefficiencies will make the *purchased energy requirement larger than the delivered energy requirement*. The lower the efficiency of the heat source, the higher the purchased energy requirement will be compared to the delivered energy requirement. These inefficiencies can be accounted for using the methods for unit pricing of energy discussed in Chapter 1.

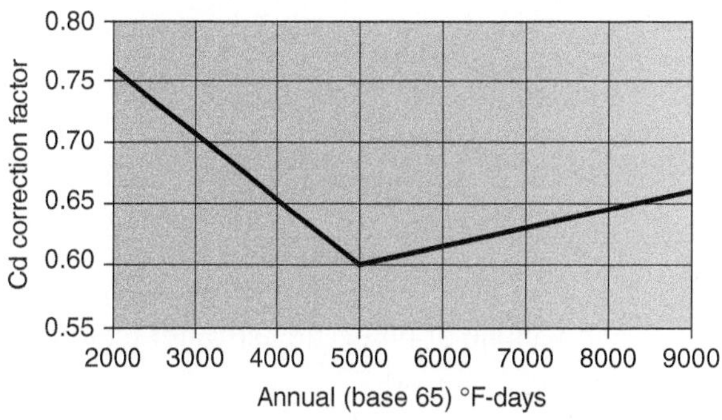

Figure 2-29 | Mean values of C_d versus annual heating degree days suggested by ASHRAE.

2.9 Estimating Domestic Water-Heating Loads

Domestic water heating is one of the largest energy consumers in a typical North American home. In northern climates, it is usually second only to the energy required for space heating. In southern climates, it is typically second behind space cooling energy use.

Because domestic water heating is required throughout the year, it is also one of the best practical applications for renewable energy heat sources, especially solar thermal collectors.

Most of the renewable energy based combisystems discussed in later chapters are configured to provide energy for both space heating and domestic water heating. The design of such systems must therefore factor in the thermal load for the latter. This section presents simple methods for estimating the total daily energy required for domestic water heating as well as the energy required during the hour of peak demand.

DAILY DOMESTIC HOT WATER (DHW) LOAD ESTIMATES

The daily energy requirement for heating domestic water can be estimated using Equation 2.11:

(Equation 2.11)

$$E_{daily} = (G)(8.33)(T_{hot} - T_{cold})$$

where:

E_{daily} = daily energy required for DHW production (Btu/day)

G = volume of domestic hot water required per day (gallons)

T_{hot} = hot water temperature supplied to the fixtures (°F)

T_{cold} = cold water temperature supplied to the water heater (°F)

The daily usage of domestic hot water is very dependent on occupancy, living habits, type of water fixtures used, water pressure, and the time of year. The following estimates are suggested as a guideline:

- House: 10 to 20 gallons per day per person
- Office building: 2 gallons per day per person
- Small motel: 35 gallons per day per unit
- Restaurant: 2.4 gallons per average number of meals per day

Example 2.10: Determine the energy used for domestic water heating for a family of four with average usage habits. The cold water temperature averages 55 °F. The supply temperature is set for 125 °F. The assumed "average" domestic hot water usage will be 15 gallons per day per person.

Solution: The daily domestic water-heating load is estimated using Equation 2.11:

$$E_{daily} = (4 \times 15)(8.33)(125 - 55) = 34{,}990 \text{ Btu/day}$$

The estimated annual usage would be the daily usage times 355 days of occupancy per year (assuming the family is away 10 days per year).

$$\left(34{,}990\frac{\text{Btu}}{\text{day}}\right)\left(355\frac{\text{occupied days}}{\text{year}}\right) = 12{,}421{,}000\frac{\text{Btu}}{\text{year}}$$

$$= 12.4\frac{\text{MMBtu}}{\text{year}}$$

Discussion: The cost of providing this energy can be estimated by multiplying the annual energy requirement by the cost of delivered energy in \$/MMBtu, as established using methods from Chapter 1. For example, at \$0.10/kwhr and 96 percent efficiency, electrical energy has a delivered cost of \$30.52/MMBtu. The annual energy cost of providing the domestic hot water in this example would be:

$$\left(12.4\frac{\text{MMBtu}}{\text{year}}\right)\left(30.52\frac{5}{\text{MMBtu}}\right) = \$378/\text{year}$$

The efficiency of 96 percent is typical for storage type electric water heater. It implies that 96 percent of the electrical energy used by the device leaves as heated water. The other 4 percent leaves as standby heat loss from the tank's jacket and attached piping.

DHW USAGE PATTERNS

The rate at which domestic hot water is used at the fixtures within a building varies considerably with the type of occupancy, and time of day. A typical residential **domestic hot water usage profile** is shown in Figure 2-30.

Notice the two distinct periods of high demand: one during the wake-up period, the other during the early evening. The greatest demand in this specific profile is from 7 P.M. to 8 P.M. However, this still represents only 11.6 percent of the total daily DHW demand. For a family using 60 gallons of domestic hot water per day, this would represent a peak hourly usage of just under 7 gallons.

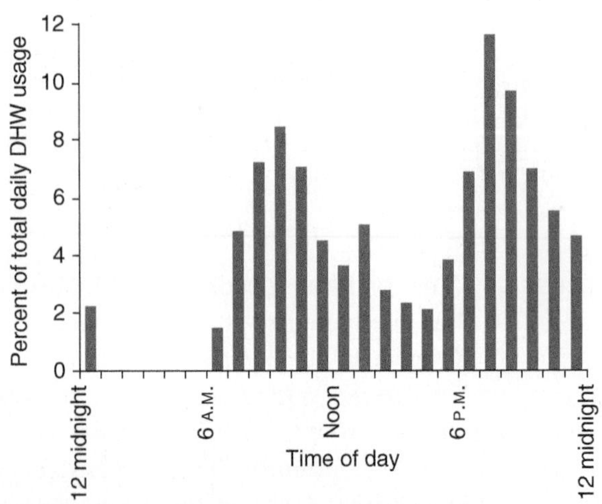

Figure 2-30 | Typical usage profile for residential domestic hot water.

The *average* rate at which energy is required for DHW production during this peak hour (assuming cold water and hot water temperatures of 55 °F and 125 °F, respectively) can be found using Equation 2.11 modified for a single peak hour:

$$E_{\text{peakhour}} = [(0.116)(60)](8.33)(125 - 55)$$
$$= 4{,}058 \text{ @ } 4{,}100 \text{ Btu/hr}$$

The first factor in this equation, (0.116), represents the peak demand of 11.6 percent of the total daily water volume.

The occurrence of two peak demand periods, one during the wake-up period, the other following the evening meal period, is common in residential buildings. Some households tend to be "high morning users" while others are "high evening users." Figure 2-31 compares these two categories.

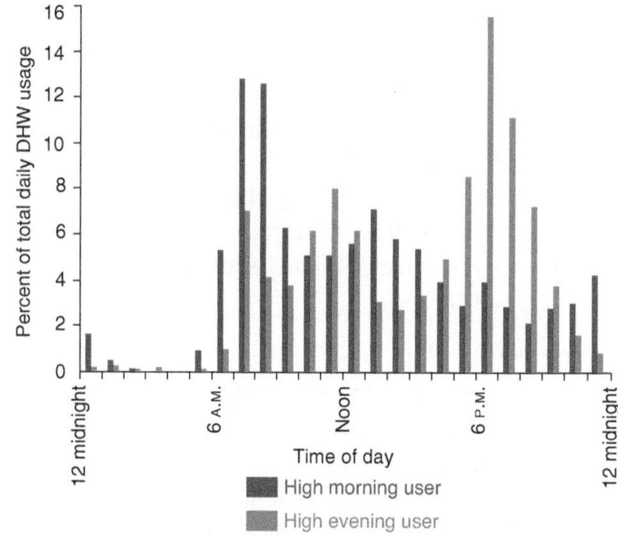

Figure 2-31 | Comparison of hourly usage of domestic hot water for a high morning user versus a high evening user.

PEAK HOURLY DEMAND FOR DHW

Another important aspect of domestic water-heating loads is the **peak hourly demand**. This is simply how much domestic hot water is required, in gallons, during the hour of highest demand.

Peak hourly demand can be estimated by first identifying the hour of probable highest demand for hot water based on expected usage habits. This varies considerably from one household or building to another. For design purposes, it can be established by discussing likely usage habits with occupants. In existing buildings, it can also be established by data logging domestic hot water flow versus time of day, over a period of several typical days.

If it is not possible to discuss or monitor domestic hot water usage habits with the occupants for

Use	Average gallons of hot water per use		Number of uses during peak hour		Gallons used during peak hour
Shower	12	×		=	
Bath	9	×		=	
Shaving	2	×		=	
Hands and face washing	4	×		=	
Hair shampoo	4	×		=	
Hand dishwashing	4	×		=	
Automatic dishwashing	14	×		=	
Food preparation	5	×		=	
Automatic clothes washer	32	×		=	
			TOTAL PEAK HOUR DEMAND	=	

Figure 2-32 | DOE table for estimating peak hour demand for domestic hot water.

whom the system is being designed, the table shown in Figure 2-32 can also be used to estimate peak hourly demand.

Figure 2-32 is based on statistics of *typical* usage as published by the U.S. Department of Energy. If the building has water-conserving showerheads and faucets, usage may be less. Likewise, if the building has high-flow fixtures, usage could be higher.

Figure 2-33 shows an example of how the table in Figure 2-32 is used.

Knowing the likely peak hourly demand assists in selecting a domestic water-heating device. All electric, gas, and oil-fired storage type water heaters sold in the United States have a **first-hour rating**. This rating, established by independent testing, is the number of gallons of domestic hot water the device can supply over one hour, starting from a normal heated standby condition. The first-hour rating includes the contribution of storage volume as well as the heating capacity of the burner or element. The gallons of hot water supplied as the first-hour rating are based on a 90 °F temperature rise between the incoming cold water and the leaving hot water.

Once the peak hourly demand for domestic hot water is established, common practice is to select a water-heating device with a first-hour rating that equals or slightly exceeds the peak hourly demand.

One might argue that this selection method only applies to "standard" methods of domestic water heating, without contributions from a renewable energy source. However, it is conservative practice to design renewable energy systems that include "auxiliary" heat sources that can assume 100 percent of the domestic water-heating load, if necessary. During normal operation, the auxiliary heat source should provide far less than 100 percent of the load. However, if the renewable heat source is *inoperable for any reason*, the entire domestic water-heating load must be assumed by the auxiliary heat source. In the author's experience, very few occupants will tolerate inadequate domestic hot water availability while waiting for the renewable energy heat source to come back online. Design for this possibility.

Use	Average gallons of hot water per use		Number of uses during peak hour		Gallons used during peak hour
Shower	12	×	3	=	36
Bath	9	×	0	=	0
Shaving	2	×	1	=	2
Hands and face washing	4	×	0	=	0
Hair shampoo	4	×	1	=	4
Hand dishwashing	4	×	1	=	4
Automatic dishwashing	14	×	0	=	0
Food preparation	5	×	0	=	0
Automatic clothes washer	32	×	0	=	0
			TOTAL PEAK HOUR DEMAND	=	46

Figure 2-33 | Example of how the table in Figure 2-32 is used to estimate peak hourly demand.

SUMMARY

This chapter has presented straightforward methods for estimating the design space heating load and domestic water-heating load of residential and light commercial buildings. Load estimating is an essential part of overall system design. It provides the foundation for all subsequent equipment selection.

KEY TERMS

97.5 percent design dry bulb temperature
air change method
air infiltration heat loss
blower door
Building Heat Load Estimator
building heating load
Cd factor
combisystems
conduction
conduction heat loss
degree days
design heating load
domestic hot water usage profile
effective total R-value
EnergyStar certified
exposed surface
finite element analysis
first-hour rating
Hydronics Design Studio 2.0
infiltration heat loss
inside air film
isotherms
outside air film
PassivHaus standard
peak hourly demand
R-value
room heating loads
smoke pencil
thermal conductivity
thermal envelope
thermal load
thermal resistance
total R-value
unit U-value

QUESTIONS AND EXERCISES

1. One wall of a room measures 14 feet long and 8 feet high. It contains a window 5 feet wide and 3.5 feet high. The wall has an effective total R-value of 15.5. Find the rate of heat flow through the wall when the inside air temperature is 68 °F and the outside temperature is 5 °F.

2. Using the data in Appendix B, determine the total effective R-value of the wall shown in Figure 2-34. Assume that 15 percent of the wall area is wood framing.

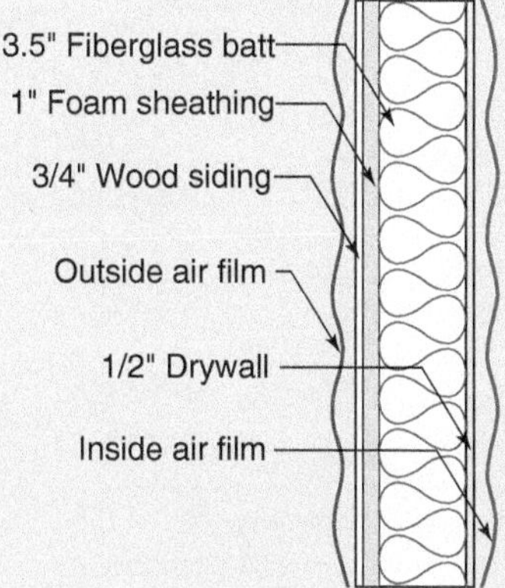

Figure 2-34 | Wall cross section for Exercise 2.

3. The rate of heat flow through a wall is 500 Btu/hr when the inside temperature is 70 °F and the outside temperature is 10 °F. What is the rate of heat flow through the same wall when the outside temperature drops to –10 °F and the inside temperature remains the same?

4. The perimeter of the basement of a house measures 120 feet. The masonry wall is 8 feet tall with 1 foot exposed above grade. The wall is insulated with 2 inches of extruded polystyrene insulation. The basement air temperature is maintained at 65 °F. Determine the rate of heat loss through the basement walls when the outside air temperature is 10 °F. Assume the R-value of the masonry wall is 2.0.

5. Determine the rate of heat loss through the slab-on-grade floor shown in Figure 2-35. The exposed edges are insulated with 1.5 inches of extruded polystyrene. The outside air temperature is 0 °F and the slab is maintained at an average temperature of 90 °F by embedded tubing.

6. Determine the design heat loss of the room shown in Figure 2-36. Assume the exterior walls are constructed as shown in Figure 2-34. Assume the outdoor design temperature is –10 °F and the desired indoor temperature is 70 °F.

7. Using the construction information given for the example house in Section 2.6, determine the design heat loss of the entire house shown in Figure 2-37. Assume all floors are slab on grade. Estimate any dimensions you need by proportioning it to the dimensions shown.

8. Estimate the cost of domestic water heating for a typical family of five assuming cold water enters the system at 45 °F and is supplied to the fixtures at 120 °F. Assume the water is heated by an electric water heater in an area where electricity costs $0.12/kwhr.

9. What is the difference in cost to heat 60 gallons of water from 50 °F to 70 °F compared to heating the same amount of water from 100 °F to 120 °F ?

10. It is determined that a family uses domestic hot water as shown in Figure 2-31. Assuming a total daily domestic hot water usage of 80 gallons, a cold water temperature of 50 °F, and a delivery temperature of 120 °F, how much energy (in Btus) is required to meet their peak hourly demand for domestic hot water?

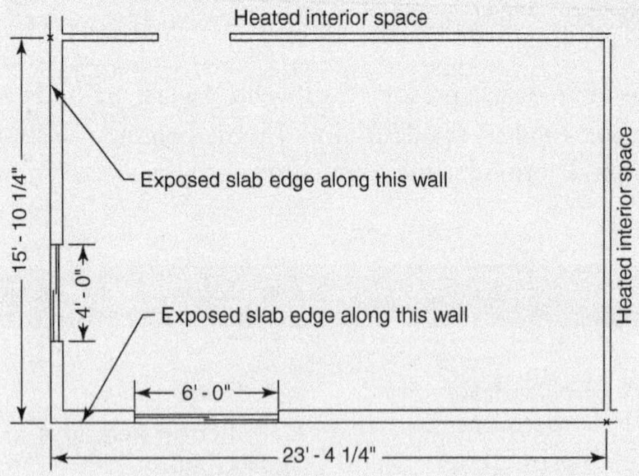

Figure 2-35 | Floor slab for Exercise 5.

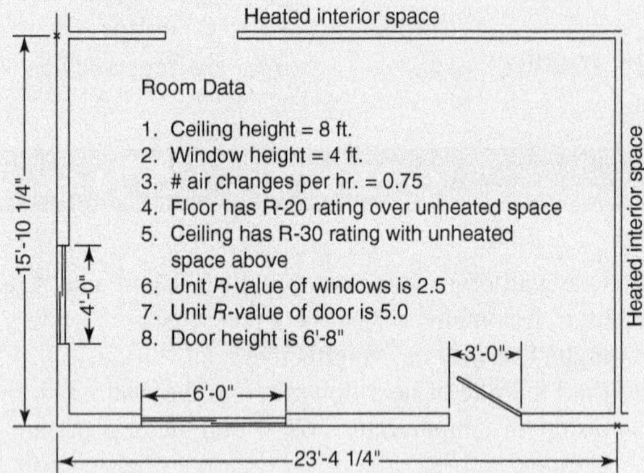

Figure 2-36 | Room floor plan for Exercise 6.

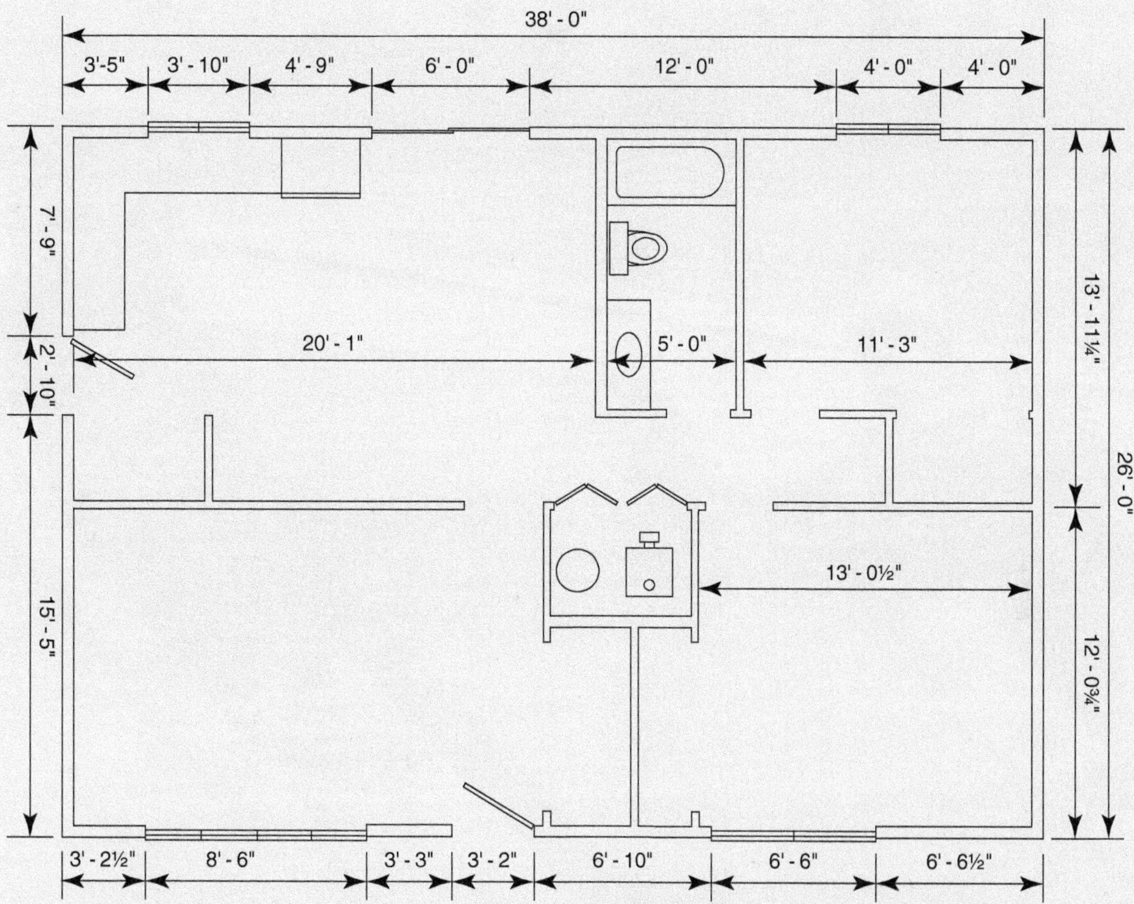

Figure 2-37 | House floor plan for Exercise 7.

R410A
R410A

CHAPTER 3

Universal Hydronic Concepts

OBJECTIVES

After studying this chapter, you should be able to:

- Describe the difference between closed-loop and open-loop piping system.
- List the advantages of closed-loop piping circuits.
- Describe the function of several common hydronic system components.
- Describe relevant physical properties of water including density, specific heat, heat capacity, viscosity, boiling point, and dissolved air content.
- Determine values for these physical properties.
- Calculate the heat stored in a given amount of water based on temperature change.
- Determine the rate at which flowing water transports heat.
- Compare the ability of water versus air as heat transport fluids.
- Calculate the static pressure at any location in a piping circuit.
- Describe "head energy," and how it is added or removed within a hydronic circuit.
- Calculate the equivalent length of a piping circuit.
- Construct the head loss curve of a piping circuit.

3.1 Introduction

Modern hydronics technology is the "glue" that holds together many thermally based renewable energy systems. Although these systems employ a variety of **heat sources**, such as solar collectors, heat pumps, or wood-fueled boilers, *none of them can operate to their fullest potential without good hydronic detailing.* Trying to design a great renewable-energy-based heating system without a solid understanding of basic hydronics is like trying to build a house without knowing how to saw lumber. Both will be frustrating and utterly pointless pursuits.

This chapter discusses some commonly used hydronic hardware. It also describes the physical properties of water that allow a hydronic system to deliver unsurpassed efficiency in conveying heat from where it is produced or stored to where it is needed. It introduces simple mathematical relationships that allow designers to quantify and predict system performance long before any hardware is assembled.

3.2 Fundamentals of Closed-Loop Hydronic Circuits

Nearly all hydronic systems, whether supplied by conventional heat sources or renewable energy heat sources, share common components. This section introduces those components and briefly describes their function. Later chapters describe this hardware in more detail and show how to size and assemble it into complete systems.

The vast majority of hydronic heating systems described in this text are **closed-loop systems**. This means that the piping components form an assembly that separates the fluid in the system from contact with the atmosphere. If the system's fluid is exposed to the atmosphere at any point, even through a tiny opening, that system is classified as an **open-loop system**.

Closed-loop circuits have several advantages over open loop systems. First, because they are sealed from the atmosphere, there is very little loss of fluid over time. Any slight fluid loss that occurs is usually the result of weeping valve packings or gaskets. These minor fluid losses typically do not create problems and can be automatically corrected with components discussed later in this section.

Another advantage of properly designed closed-loop systems is that the water within them contains a very small amount of **dissolved oxygen**. This greatly reduces the potential of corrosion, especially in systems containing iron or steel components.

Closed-loop systems can also operate under pressure. This helps in eliminating air from the system. It also helps in suppressing boiling within the system. The latter effect is especially useful in solar thermal systems where the fluid can sometimes reach temperatures well above the atmospheric boiling point of the fluids they contain.

THE BASIC CLOSED HYDRONIC CIRCUIT

Figure 3-1 shows a basic closed-loop hydronic circuit. The box representing the heat source could be a boiler, a heat pump, a thermal storage tank heated by solar collectors, or some other heat-producing device. In this section, the heat source is assumed to be a *closed* device, sealed from the atmosphere, and capable of operating under some pressure.

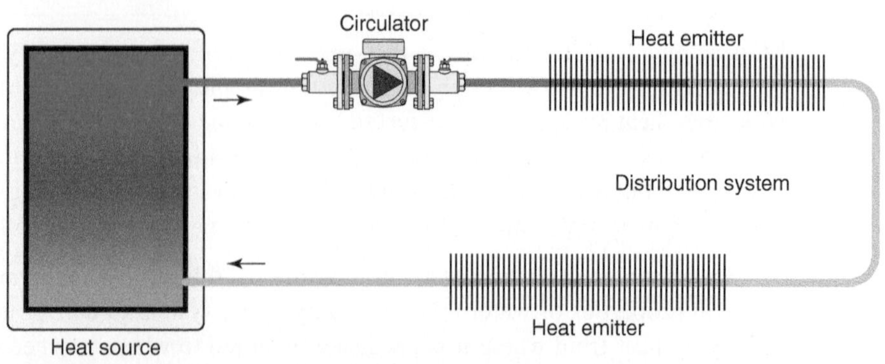

Figure 3-1 | Basic closed-loop hydronic circuit.

The fluid in a hydronic circuit serves as a *conveyor belt for heat*. That heat is absorbed into the fluid at the heat source, carried through the **distribution system** by the fluid, and dissipated from the fluid into the building at one or more **heat emitters**. After releasing heat, the fluid returns to the heat source to absorb more heat and repeat the process. The **circulator** maintains flow through the circuit. The same fluid remains in the closed-loop circuit, often for many years. It never loses its ability to absorb, transport, or dissipate heat.

BASIC HYDRONIC CONTROLLERS

In an ideal system, the rate at which the heat source produces heat would always match the rate at which the building, or domestic water, requires heat. Unfortunately, this is seldom the case with any real system, especially one supplied by an **intermittent heat source** such as solar collectors or a wood-fueled boiler. For this reason, it is necessary to provide **controllers** that manage both heat production and heat delivery.

One of the most basic controllers is a **thermostat**. In many systems, it determines when flow is needed within the hydronic circuit based on room air temperature. For the simple hydronic circuit being described in this section, the thermostat turns the circulator on and off to create flow or stop flow though the circuit. The thermostat's "goal" is to keep the room air temperature at, or very close to, its **setpoint temperature**. If the room's air temperature starts rising above the setpoint temperature, the thermostat turns off the circulator to stop further heat transport to the heat emitters. When the room temperature drops below the thermostat's setpoint, it turns on the circulator to start flow and resume heat transport.

Another common controller is called a **high limit controller**. Its function is to turn the heat source on and off so that the temperature of the fluid supplied to the distribution system remains within a useful range. High limit controllers are commonly used with heat sources such as boilers that generate heat by burning fuel. However, they may not be present in systems supplied by renewable heat sources. Instead, a **3-way mixing valve** may be used to blend cool water returning from the heat emitters with heated water from the heat source, or a thermal storage tank, to achieve the desired **supply temperature** to the heat emitters.

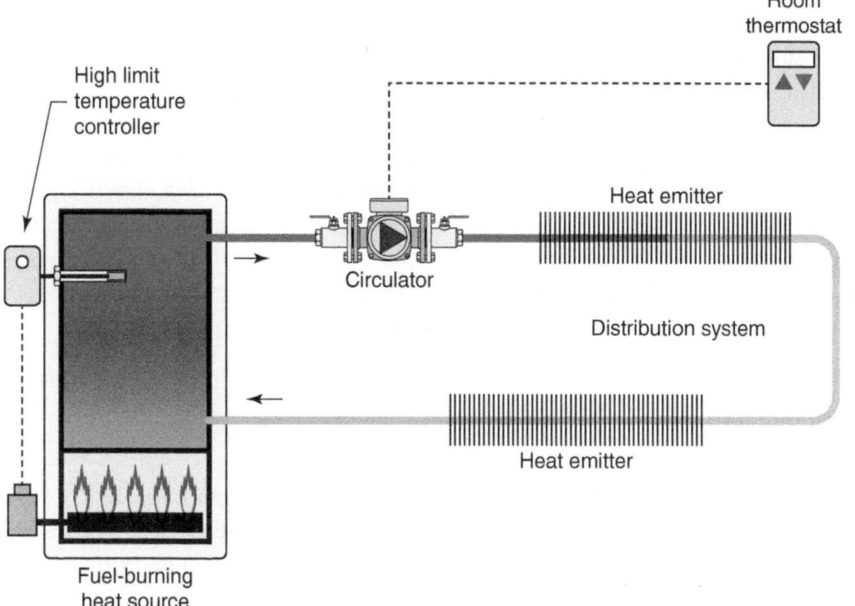

Figure 3-2 | Basic hydronic circuit supplied by fuel-burning heat source. Two controllers (e.g., a room thermostat and high limit controllers) have been added to regulate heat generation and heat delivery.

Figure 3-2 shows a basic hydronic system supplied by a fuel-burning heat source. The circulator is assumed to be controlled by a room thermostat. The water temperature within the heat source is controlled by the high limit controller.

Figure 3-3 shows a simple hydronic system where heat is supplied from a thermal storage tank. The water within this tank is assumed to be heated by a renewable energy heat source. A room thermostat turns the distribution circulator on and off based on room temperature. A 3-way motorized mixing valve regulates the water temperature supplied to the distribution system. The mixing valve prevents the potentially high temperature water within the storage tank from being

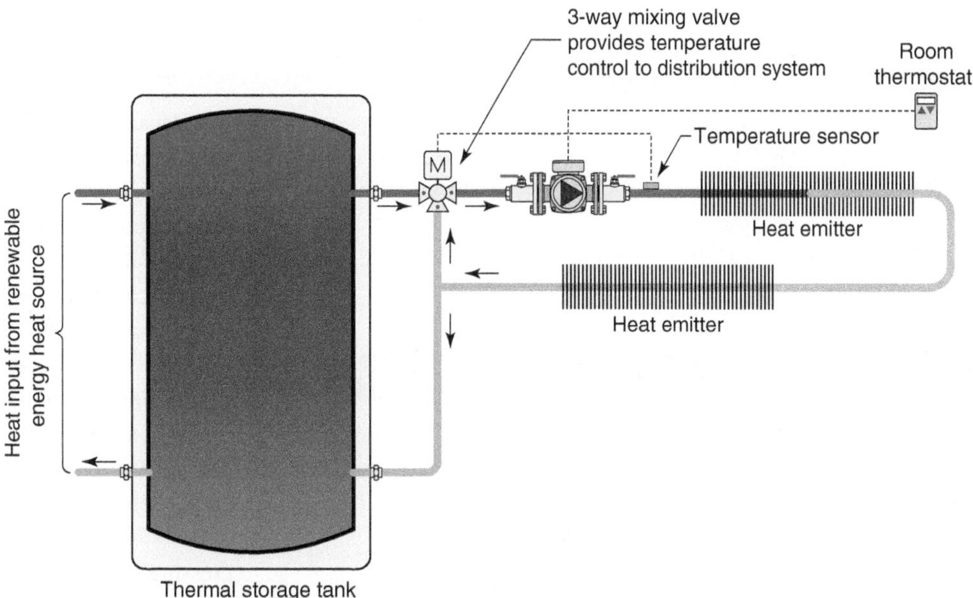

Figure 3-3 | Simple hydronic system, where heat is supplied from a thermal storage tank.

delivered directly to heat emitters that are intended to operate at lower water temperatures. Mixing valves are discussed in more detail in Chapter 4.

EXPANSION TANK

Water expands when it is heated. This increase in volume is an extremely powerful but predictable characteristic that must be accommodated in any type of closed-loop hydronic system. Figure 3-4 shows an **expansion tank** added to the basic system. The expansion tank contains a captive volume of air. As the water within the system expands, it pushes into the expansion tank and slightly compresses this air. As a result, the pressure within the system rises slightly. As the system's water cools, its volume decreases, allowing the compressed air to expand and system pressure to return to a lower value. This process repeats itself each time the system heats up and cools off.

Most modern hydronic systems use diaphragm-type expansion tanks. Such tanks contain air within a sealed flexible chamber. The sizing and placement of the system's expansion tank are crucial to proper system operation, and will be discussed in Chapter 4.

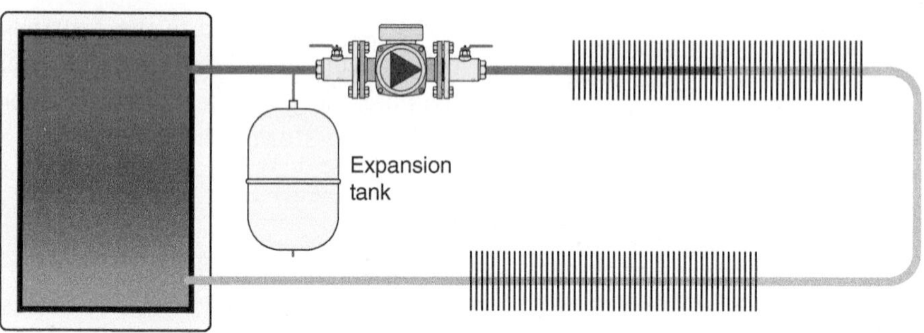

Figure 3-4 | Adding an expansion tank to the basic closed-loop circuit.

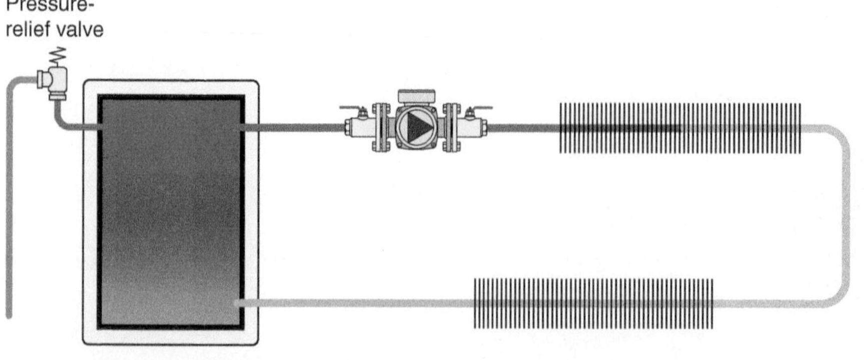

Figure 3-5 | Adding a pressure relief valve to the heat source.

PRESSURE-RELIEF VALVE

Consider the fate of a closed-loop hydronic system in which a defective controller fails to turn off the heat source after its upper temperature limit has been reached. As the water gets hotter and hotter, system pressure steadily increases due to the water's expansion. This pressure could eventually exceed the pressure rating of the weakest component in the system. The consequences of a system component bursting at high pressures and temperatures could be devastating. For this reason, all closed-loop hydronic systems must be protected by a **pressure-relief valve**. This is a universal requirement of mechanical codes in North America and other countries. Figure 3-5 shows a pressure-relief valve installed on the heat source. Because the heat source is part of the closed-loop system, this placement of the relief valve protects the entire system from excess pressure.

MAKE-UP WATER SYSTEM

Most closed-loop hydronic systems experience very minor water losses over time due to weepage from valve packings, pump seals, air vents, and other components. These losses are normal and must be replaced to maintain adequate system pressure. The common method for replacing this water is through an automatic **make-up water system** consisting of a **pressure-reducing valve**, **backflow preventer**, **pressure gauge**, and **shutoff valve**.

Because the pressure in a municipal water main or private water system is often higher than the pressure-relief valve setting in a hydronic system, such water sources cannot be directly connected to the loop. A pressure-reducing valve, also known as a **feed water valve**, is used to reduce and maintain a constant minimum pressure within the hydronic system. This valve allows water into the system whenever the pressure on at its outlet side drops below the valve's pressure setting. Most pressure-reducing valves have an adjustable pressure setting. Determining the proper setting is covered later in the text.

The backflow preventer stops any water that has entered the system from returning and possibly contaminating the potable water supply system. Most municipal codes require such a device on any heating system connected to a public water supply.

The shutoff valve is installed to allow the system to be isolated from its water source. An optional **"fast-fill" valve** is sometimes installed in parallel with the pressure-reducing valve so water can be rapidly added to the system. This is especially helpful when filling larger systems. The components that constitute a typical make-up water system are shown in Figure 3-6.

A **purging valve** is also shown in Figure 3-6. It allows most of the air within the circuit to exit as the system is filled with water. Purging valves consist of a ball valve, which is inline with the distribution piping, and a side-mounted drain port. To purge the system, the inline ball valve is closed and the drain port is opened. As water enters the system through the make-up water assembly, air is expelled through the drain port of the purge valve.

AIR SEPARATOR

An **air separator** is designed to separate air from water and eject the air from the system. Modern air separators create regions of reduced pressure as water passes through. The lowered pressure causes dissolved gases within the water to **coalesce** into bubbles. These bubbles are guided upward into a collection chamber, where an automatic air vent expels them from the system. For best results, the air separator should be located where fluid temperatures are highest, which is typically in the supply pipe from the heat source, as shown in Figure 3-7.

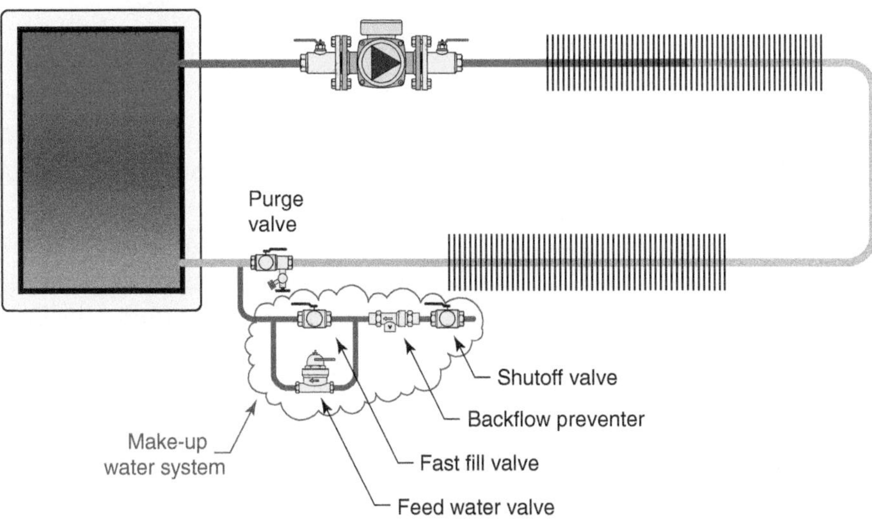

Figure 3-6 | Adding a make-up assembly to the basic circuit.

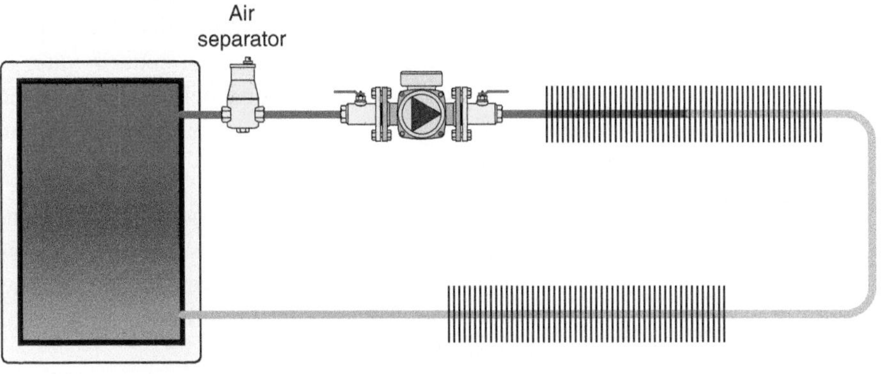

Figure 3-7 | Adding an air separator to the basic circuit.

PUTTING IT ALL TOGETHER

Figure 3-8 is a composite drawing showing all the components previously discussed in their proper positions within the simple hydronic circuit. By assembling these components, we have built a simple hydronic heating system. It must be emphasized, however, that just because all the components are present does not guarantee that the system will function properly. Combining these components is not a matter of choosing a favorite product for each and simply connecting them as shown. Major subsystems such as the heat emitters and heat source have temperature and flow requirements that must be properly matched if they are to function together as a system. The objective is to achieve a stable, dependable, affordable, and efficient overall system. Failure to respect the operating characteristics of all components will result in installations that underheat, overheat, waste energy, or otherwise disappoint their owners.

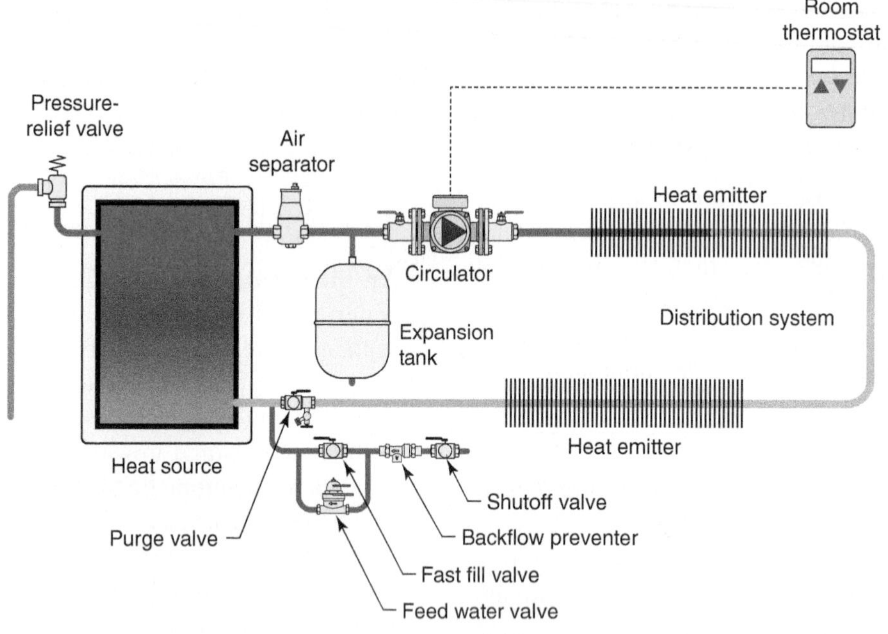

Figure 3-8 | Composite drawing showing all the basic hydronic circuit components.

3.3 Properties of Water and Water-based Fluids

Water is the essential material in any hydronic heating system. Its properties make it possible to deliver heat throughout a building using relatively small "conduits" (e.g., tubing), and minimal amounts of electrical energy.

This section discusses several properties of water that are relevant to hydronic heating using conventional heat sources and renewable energy heat sources. A solid understanding of these properties is essential in understanding how hydronic systems behave.

SENSIBLE HEAT VERSUS LATENT HEAT

The heat absorbed or released by a material while it remains in a *single phase* is called **sensible heat**. The word sensible means that the presence of the heat is able to be *sensed by a temperature change in the material.* As a material absorbs heat, its temperature increases. As the material releases heat, its temperature drops. The vast majority of heat transfer discussed in this text is sensible heat transfer (e.g., accompanied by a temperature change).

If the material *changes phase* while absorbing heat, its temperature does not change. The heat absorbed in such a process is called **latent heat**. For water to change from a solid (ice) to a liquid requires the absorption of 144 Btu/lb while its temperature remains constant at 32 °F. When water changes from a liquid to a vapor, it must absorb approximately 970 Btu/lb while its temperature remains constant.

SPECIFIC HEAT AND HEAT CAPACITY

One of the most important properties of water relevant to its use in hydronic systems is **specific heat**. The specific heat of any material is the number of Btus required to raise the temperature of 1 pound of that material by 1 degree Fahrenheit (°F). The specific heat of water is approximately 1.0 Btu per pound per °F. To raise the temperature of 1 pound (lb) of water by 1 °F will require the addition of 1 Btu of heat.

This relationship between temperature, material weight, and energy content also holds true when a substance is cooled. For example, to lower the temperature of 1 pound of water by 1 °F will require the removal of 1 Btu of energy.

The specific heat of any material varies slightly with its temperature. For water, this variation is very

Material	Specific heat (Btu/lb/°F)	Density* (lb/ft³)	Heat capacity (Btu/ft³/°F)
Water	1.00	62.4	62.4
Concrete	0.21	140	29.4
Steel	0.12	489	58.7
Wood (fir)	0.65	27	17.6
Ice	0.49	57.5	28.2
Air	0.24	0.074	0.018
Gypsum	0.26	78	20.3
Sand	0.1	94.6	9.5
Alcohol	0.68	49.3	33.5

Figure 3-9 | Specific heat, density, and heat capacity of some common substances.

*In this text, "density" is defined as weight per unit of volume, rather than mass per unit of volume, as used in classical physics. Its use in all subsequent equations in this text will be based on this definition, and the units of density expressed as (lb/ft³).

small over the temperature range used by the systems discussed in this text. Thus, to provide reasonable simplification, the specific heat of water will be considered to remain constant at 1.0 Btu/lb/°F.

The specific heat of water is high in comparison to other common materials. In fact, water has one of the highest specific heats of any known material. The table in Figure 3-9 compares the specific heats of several common materials. Also listed is the **density** of each material and its **heat capacity**.

The heat capacity of a material is the number of Btus required to raise the temperature of one *cubic foot* of the material by 1 °F. The heat capacity of a material can be found by multiplying the specific heat of the material by its density, as shown by Equation 3-1.

Equation 3-1

$$(\text{specific heat})(\text{density}) = \text{heat capacity}$$

$$\left(\frac{\text{Btu}}{\cancel{\text{lb}}\cdot{}^\circ\text{F}}\right)\left(\frac{\cancel{\text{lb}}}{\text{ft}^3}\right) = \left(\frac{\text{Btu}}{{}^\circ\text{F}\cdot\text{ft}^3}\right)$$

By comparing the heat capacity of water and air, it can be shown that *a given volume of water can hold almost 3,500 times as much heat as the same volume of air*, for the same temperature rise. This allows a given volume of water to transport vastly more heat than the same volume of air, assuming both undergo the same temperature change.

The difference in heat capacity of water versus that of air is the single greatest advantage of using water rather than air to convey heat. This is illustrated in Figure 3-10, where two heat transport systems of equal heat carrying ability are shown side by side. A 3/4-inch tube carrying water can convey the same amount of heat as a 14-inch by 8-inch duct conveying air when both systems are operating at typical temperature differentials and flow rates.

DENSITY

In this text, "density" refers to the number of pounds of the material needed to fill a volume of 1 cubic foot. For example, it would take 62.4 pounds of water at 50 °F to fill a 1 cubic foot container, so its density is said to be 62.4 pounds per cubic foot, (abbreviated as 62.4 lb/ft³).

The density of water depends on its temperature. Like most substances, as the temperature of water increases, its density decreases. As temperature increases, each molecule of water requires more space. This molecular expansion is extremely powerful and can easily burst metal pipes or tanks if not properly

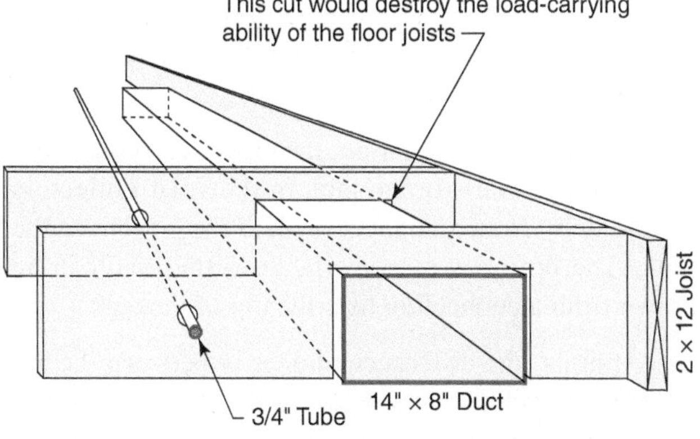

Figure 3-10 | The high heat capacity of water allows a 3/4-inch tube carrying water to convey the same amount of heat as a 14-inch by 8-inch duct carrying air when both systems are operating at normal temperature differentials and flow rates.

accommodated. A graph showing the relationship between temperature and the density of water is shown in Figure 3-11.

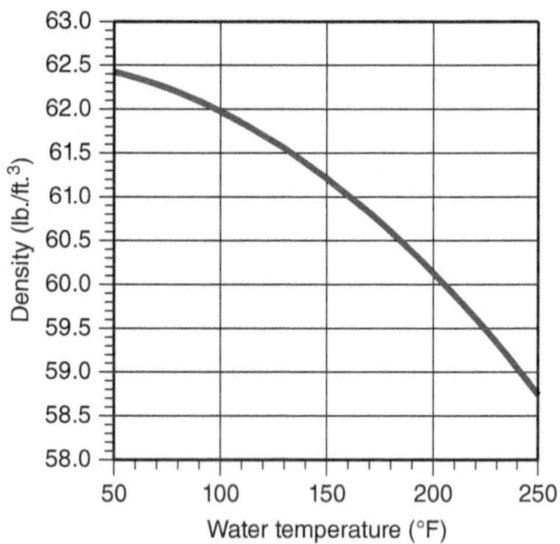

Figure 3-11 | Density of water as a function of temperature.

The change in water's density based on temperature has many implications in the design of hydronic systems. For example, the change in density of water or water-based antifreeze solutions directly affects the size of the expansion tank required on all closed-loop hydronic systems. Proper tank sizing requires data on the density of the system fluid both when the system is filled and when it reaches its maximum operating temperature.

Another effect caused by the change in density of system fluid is called **stratification**. Because hot water has a lower density than cool water, it will rise to the top of a thermal storage tank, as depicted in Figure 3-12. The coolest water will settle to the bottom of the storage tank.

Because of stratification, the piping carrying water from the thermal storage tank to the solar collectors should draw from a connection near the bottom of the tank. The hottest water available to the load can be drawn from a connection near the top of the tank.

At night, the difference in density between the hot water in the tank, and the cooler water in the collectors, can create a very *undesirable* condition: Hot water will rise up the piping to the top of the collectors, as cool water within the collectors and the piping supplying them, flows backward toward the storage tank. This **reverse thermosiphoning** can dissipate a significant amount of heat from the thermal storage tank during the night.

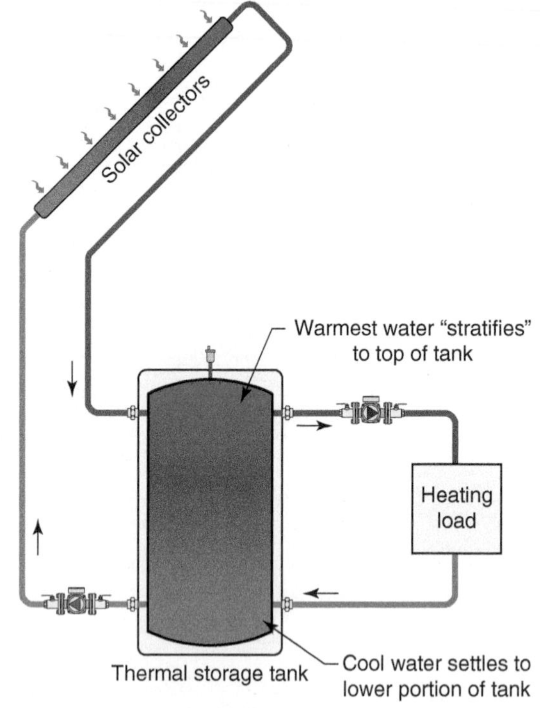

Figure 3-12 | The hottest (lower-density) water rises to the top of a thermal storage tank, while the coolest (higher-density) water settles near the bottom of the tank.

Fortunately, this effect can be prevented by installing a check valve in the piping leading to the collectors. Such detailing will be discussed at length in later chapters.

SENSIBLE HEAT-QUANTITY EQUATION

It is often necessary to determine the *quantity* of heat stored in a given volume of material that undergoes a specific temperature change. Equation 3.1a, known as the **sensible heat-quantity equation**, can be used for this purpose.

(Equation 3.1a)

$$h = 8.33v(\Delta T)$$

where:

h = quantity of heat absorbed or released from the water (Btu)

v = volume of water involved (gallons)

ΔT = temperature change of the material (°F)

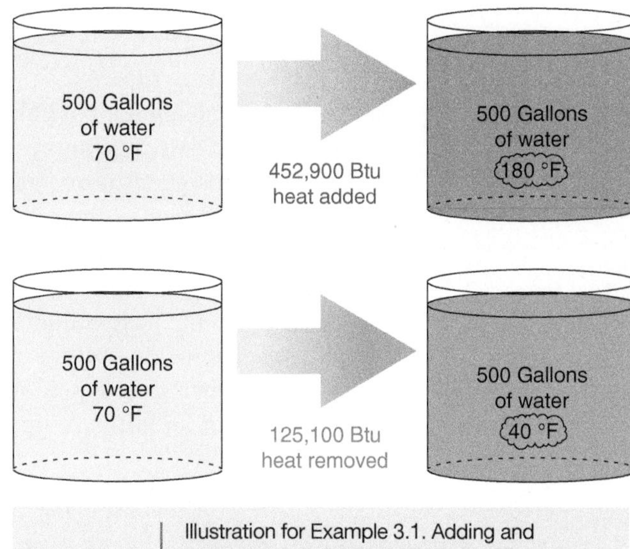

Figure 3-13 Illustration for Example 3.1. Adding and removing sensible heat causes a temperature change in water.

Example 3.1: The storage tank shown in Figure 3-13 contains 500 gallons of water initially at 70 °F. How much energy must be added or removed from the water to:

a. Raise the water temperature to 180 °F?

b. Cool the water temperature to 40 °F?

Solution:

a. Using the sensible heat-quantity equation for water (Equation 3.1a),

$$h = 8.33(500)(180-70) = 458{,}000 \text{ Btu}$$

b. Using the same sensible heat-quantity equation for water (Equation 3.1a),

$$h = 8.33(500)(70-40) = 125{,}000 \text{ Btu}$$

SENSIBLE HEAT-RATE EQUATION

Hydronic system designers often need to know the *rate* of heat transfer to or from a stream of water flowing through a device such as a heat source or heat emitter. This can be done using the **sensible heat-rate equation**:

(Equation 3.2)

$$Q = (8.01Dc)f(\Delta T)$$

where:

Q = rate of heat transfer into or out of the water stream (Btu/hr)

8.01 = a constant based on the units used

D = density of the fluid (lb/ft^3)

c = specific heat of the fluid (Btu/lb/°F)

f = flow rate of water through the device (gpm)

ΔT = temperature change of the water through the device (°F)

When using Equation 3.2, the density and specific heat should be based on the *average* temperature of the fluid during the process by which the fluid is gaining or losing heat.

For *cold water only,* this equation simplifies to:

(Equation 3.3)

$$Q = 500f(\Delta T)$$

where:

Q = rate of heat transfer into or out of the water stream (Btu/hr)

f = flow rate of water through the device (gpm)

500 = constant rounded off from 8.33 × 60

ΔT = temperature change of the water through the device (°F)

Equation 3.3 is valid for water at temperatures that are close to 60 °F. The factor 500 is based on the weight of 1 gallon of water at approximately 60 °F (8.33 pounds/gallon) multiplied by 60 minutes/hour. The factor 500 is easy to remember when making quick mental calculations for the rate of sensible heat transfer. However, at higher temperatures, the density of water decreases slightly. This, in turn, reduces its ability to transport heat.

Equation 3.2 is a more general form that allows for variations in both the density and specific heat of the fluid. It can be used for water as well as other fluids, such as water-based antifreezes, which are commonly used in solar thermal systems and other hydronic heating applications.

Example 3.2: Water flows into a heat emitter at 120 °F and leaves at 105 °F. The flow rate is 4.5 gpm, as shown in Figure 3-14. Calculate the rate of heat transfer from the water to the heat emitter using:

a. Equation 3.3

b. Equation 3.2

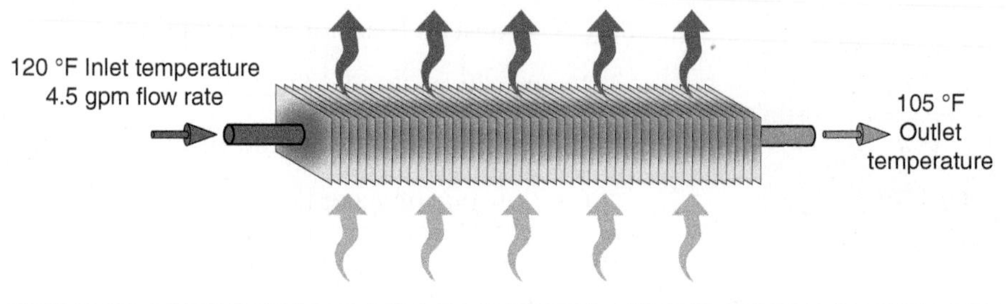

Figure 3-14 | Heat emitter for Example 4.2.

Solution: Using Equation 3.3, the rate of heat transfer is:

$$Q = 500f(\Delta T) = 500(4.5)(120 - 105)$$
$$= 33{,}750 \text{ Btu/hr}$$

To use Equation 3.2, the density of water at its average temperature of (120+105)/2=112.5 °F must first be estimated using Figure 3-11.

$$D = 61.81 \text{ lb/ft}^3$$

The specific heat of water can be assumed to remain 1.0 Btu/lb/°F.

Substituting these numbers into Equation 3.2 yields:

$$Q = (8.01Dc)f(\Delta T)$$
$$= (8.01 \times 61.81 \times 1.00)4.5(120-105)$$
$$= 33{,}419 \text{ Btu/hr}$$

Discussion: Equation 3.2 estimates the rate of heat transfer to be about 1 percent less than that estimated by Equation 3.3. This occurs because Equation 3.2 accounts for the slight decrease in the density of water at the elevated temperature, whereas Equation 3.3 does not. The variation between these calculations is arguably small. Equation 3.3 is generally accepted in the hydronics industry for quick estimates of heat transfer to or from a stream of water. However, Equation 3.2 yields slightly more accurate results when the variation in density and specific heat of the fluid can be factored into the calculation.

Example 3.3: A heating system is proposed that will deliver 50,000 Btu/hr to a load while circulating water at a flow rate of 2.0 gpm. What temperature drop will the water have to go through to accomplish this?

Solution: This situation again involves a *rate* of heat transfer from a water stream. Equation 3.3 will be used for the estimate. It will be rearranged to solve for the temperature drop term (ΔT).

$$Q = \left(\frac{Q}{500f}\right) = \left(\frac{50{,}000}{500(2.0)}\right) = 50 \text{ °F}$$

Discussion: Although it's possible that a hydronic distribution system could be designed to operate with a 50 °F temperature drop, this value is much larger than normal. Such a system would require very careful sizing of the heat emitters. This example shows the usefulness of the sensible heat-rate equation to quickly evaluate the feasibility of an initial design concept.

VAPOR PRESSURE AND BOILING POINT

*The **vapor pressure** of a liquid is the minimum pressure that must be applied at the liquid's surface to prevent it from boiling.* If the pressure on the liquid's surface drops below its vapor pressure, the liquid will instantly boil. If the pressure is maintained higher than the vapor pressure, the liquid will remain a liquid, and no boiling will occur.

The vapor pressure of a liquid is strongly dependent on its temperature. The higher the liquid's temperature, the higher its vapor pressure. Stated another way: The greater the temperature of a liquid, the more pressure must be exerted on its surface to prevent it from boiling.

Vapor pressure is stated as an **absolute pressure**, and has the units of *psia*. The "a" following psi stands for *absolute*. On the absolute pressure scale, zero pressure represents a complete vacuum. On Earth, the weight of the atmosphere exerts a pressure on any liquid surface open to the air. At sea level, this atmospheric pressure is about 14.7 psi. Stated another way, the absolute pressure exerted by the atmosphere at Earth's surface is approximately 14.7 psia. This implies that vapor pressures lower than 14.7 psia would represent partial vacuum from our perspective in Earth's atmosphere.

Did you ever wonder why water boils at 212 °F? What is special about this number? Actually, there's nothing special about it. It just happens that at 212 °F, the vapor pressure of water is 14.7 psia (equal to atmospheric pressure at sea level). Therefore, water boils at 212 °F at elevations near sea level. If a pot of water is carried to an elevation of 5,000 feet above sea level and then heated, its vapor pressure will rise until it equals the reduced atmospheric pressure at this elevation, and boiling will begin at about 202 °F.

Because water has been widely used in many heating and power production systems, its vapor pressure versus temperature characteristics have been well established for many decades. The vapor pressure of water between 50 °F and 250 °F can also be read from the graph in Figure 3-15.

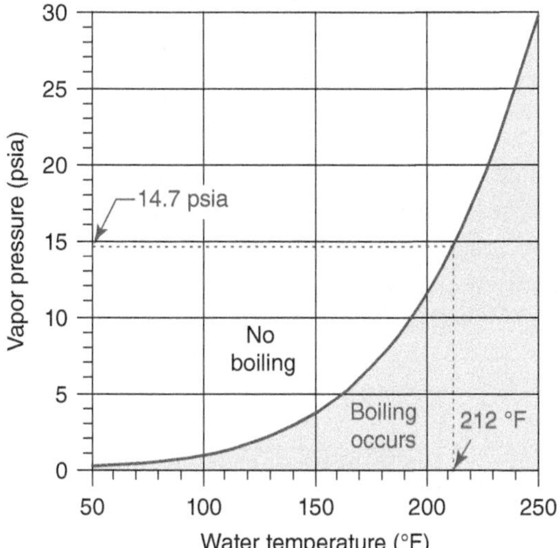

Figure 3-15 | Vapor pressure of water as a function of temperature.

Water will boil whenever its absolute pressure and temperature are within the pink shaded area under the blue curve in Figure 3-15. It will not boil if its absolute pressure and temperature are above the blue vapor barrier curve in Figure 3-15.

Closed-loop hydronic systems should be designed to ensure that the absolute pressure of the water at all points in the system remains safely above the water's vapor pressure under all operating condition. This prevents problems such as cavitation in circulators and valves, or "steam flash" in piping. The latter is a situation in which the pressure on the water drops below vapor pressure allowing the liquid water to instantly change to vapor (e.g., steam). This can cause loud banging noises in piping and wide pressure variations in the system.

VISCOSITY

The **viscosity** of a fluid is an indicator of its resistance to flow. The higher the viscosity of a fluid, the more drag it creates as it flows through pipes, fittings, valves, or any other component in a hydronic circuit. Higher viscosity fluids also require more pumping power to maintain a given flow rate compared to fluids with lower viscosities.

Like all other properties discussed in this chapter, the viscosity of water depends on its temperature. *The viscosity of water and water-based antifreeze solutions decreases as their temperature increases.*

In this text, a specifically defined quantity known as **dynamic viscosity** is used whenever viscosity is referenced. The relationship between the dynamic viscosity of water and its temperature is shown in Figure 3-16.

The physical units of dynamic viscosity can be confusing. They are a mathematical consequence of how dynamic viscosity is defined. The units used for dynamic viscosity within this book are (lb_{mass}/ft/sec). These units are compatible with all equations used in this book that require viscosity data. The reader is cautioned, however, that several other units for viscosity may be used in other references (such as manufacturer's literature). These other units must first be converted to (lb_{mass}/ft/sec) using appropriate conversion factors before being used in any equations in this book.

Solutions of water and glycol-based antifreeze have significantly higher dynamic viscosities than water alone. The greater the concentration of antifreeze, the higher the dynamic viscosity of the solution. This increase in dynamic viscosity can result in significantly lower flow rates within a hydronic system if it is not compensated for through use of more powerful circulators.

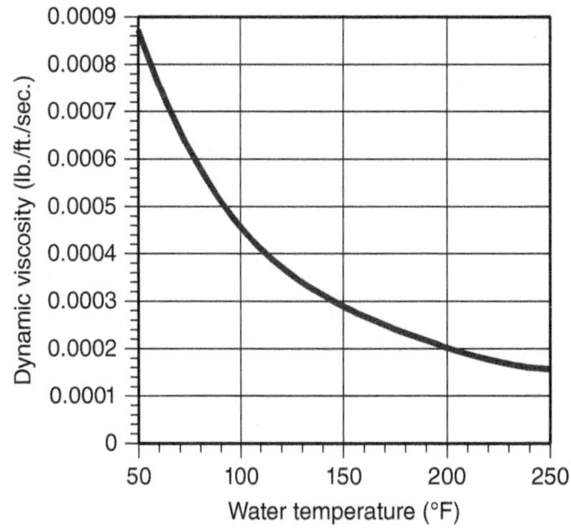

Figure 3-16 | Dynamic viscosity of water as a function of temperature.

The difference in dynamic viscosity between fluid commonly used in hydronic systems will be accounted for in equations presented later in this chapter.

DISSOLVED AIR IN WATER

Water can absorb air much like a sponge can contain a liquid. Molecules of the various gases that make up air, including oxygen and nitrogen, can exist in solution with water molecules. Even when water appears perfectly clear it can contain **dissolved air**. In other words, just because bubbles are not visible in the water does not mean air is not present.

When hydronic systems are first filled with water, dissolved air is present in the water throughout the system. If allowed to remain in the system, this air can create numerous problems including noise, cavitation, corrosion, and improper flows. A well-designed hydronic system must be able to efficiently and automatically capture dissolved air and expel it. A proper understanding of what affects the air content of water is crucial in designing a method to get rid of it.

The amount of air that can exist in solution with water is strongly dependent on the water's temperature. As water is heated, its ability to hold air in solution rapidly decreases. However, the opposite is also true. As heated water cools, it has a propensity to absorb air. Heating water to release dissolved air is like squeezing a sponge to expel a liquid. Allowing the water to cool is like letting the sponge expand to soak up more liquid. Together, these properties can be exploited to help capture and eliminate air from the system. The water's pressure also significantly affects its ability to hold dissolved air. When the pressure of the water is lowered, its ability to contain dissolved air decreases, and vice versa.

Figure 3-17 shows the *maximum* amount of air that can be contained (dissolved) in water as a percent of the water's volume. The effects of both temperature and pressure on dissolved air content can be determined from this graph.

To understand this graph, first consider a single curve such as the one labeled 30 (psi). This curve represents a constant pressure condition for the water. Notice that as the water temperature increases, the curve descends. This indicates a decrease in the water's ability to hold air in solution. The exact change in air content can be read from the vertical axis. This trend holds true for all other constant pressure curves.

This effect can be observed by heating a kettle of water on a stove. As the water heats up, bubbles can be seen at the bottom of the kettle (e.g., where the water is hottest). The bubbles are formed as gas molecules that were dissolved in the cool water, are forced out of solution by increasing temperature. The molecules join together to create the bubbles. This process is called **coalescing**. If the kettle of water is allowed to cool, these bubbles will eventually disappear as the gas molecules go back into solution.

One can think of the dissolved air initially contained in a hydronic system as being "cooked" out of solution as water temperature increases. This takes place inside the heat source when it is first operated after being filled with fresh cool water containing dissolved gases.

Figure 3-17 indicates the *maximum* possible dissolved air content of water at a given temperature and pressure. Properly deaerated hydronic systems will quickly reduce the dissolved air content of their water to a small fraction of 1 percent and maintain it at this condition.

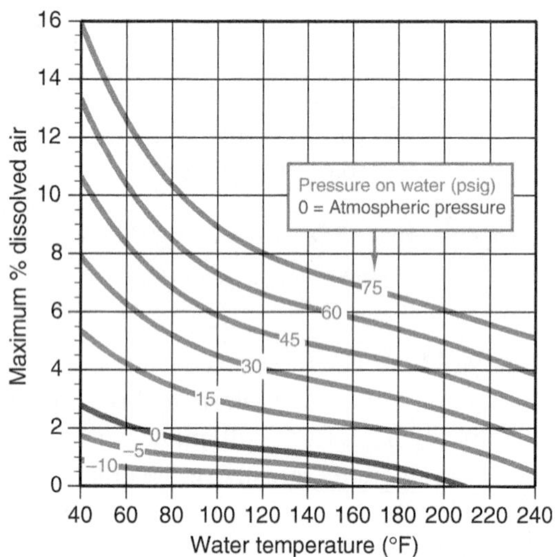

Figure 3-17 | Solubility of air in water at various temperatures and pressures.

INCOMPRESSIBILITY

When a quantity of water is put under pressure there is very little change in its volume. A pressure increase of 1 psi will cause a volume reduction of only 3.4 *millionths* of the original water volume. This change is so small that it can be ignored in the context of designing hydronic systems. For such purposes water, as well as most other liquids, can be treated as **incompressible**.

In practical terms, this means that liquids cannot be squeezed together or "stretched." It also implies that if a liquid's flow rate is known at any one point in a closed series piping circuit, it must be the same at all other points, regardless of the pipe size. If the piping circuit contains parallel branch circuits, the flow can divide up through them, but the sum of all branch flow rates must equal the total system flow rate.

Consider the arbitrary piping assembly shown in Figure 3-18. It is completely filled with water and contains several different branches, pipe sizes, circulators, and valves. The flows within the branches of such a complex maze of components can be difficult to determine and are certain to change depending on which circulators are operating and how the valves are set. However, the fact that water is incompressible guarantees that the flow rate at point (A) will always be the same as the flow rate at point (B).

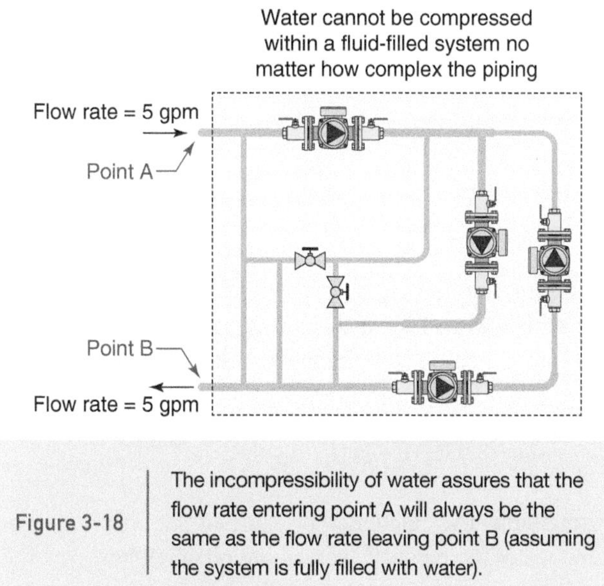

Figure 3-18 | The incompressibility of water assures that the flow rate entering point A will always be the same as the flow rate leaving point B (assuming the system is fully filled with water).

The total flow rate(s) entering any portion of a fluid-filled (non-leaking) piping assembly will always equal the total flow rate(s) exiting that assembly. This simple principle, based on incompressibility, forms the basis for analyzing fluid flow in complex piping systems.

Incompressibility also implies that liquids can exert tremendous pressure in any type of closed container when they expand due to heating. For example, a closed hydronic piping system completely filled with water, and not equipped with an expansion tank or pressure-relief valve, would quickly burst apart at its weakest component as the water temperature increased.

3.4 Static Pressure

Static pressure develops within a fluid due to its own weight (e.g., due to Earth's gravitational force). The static pressure that is present within a liquid depends on the *depth* of the liquid below (or above) some reference elevation.

Consider a pipe with a cross-sectional area of 1 square inch, filled with water to a height of 10 feet, as shown in Figure 3-19.

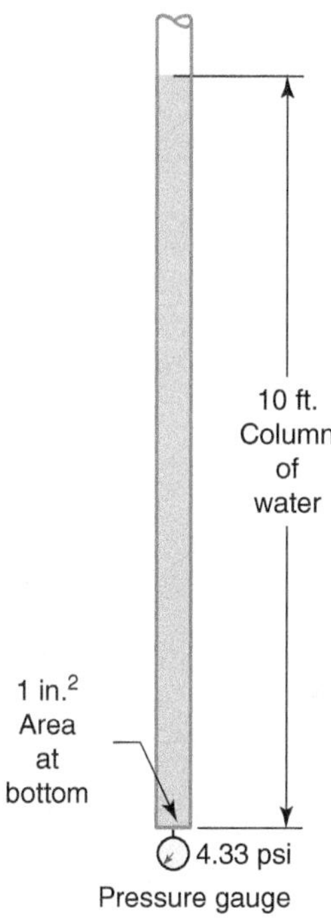

Figure 3-19 | Column of water 10 feet high in a pipe with a cross-sectional area of 1 square inch.

Assuming the water's temperature is 60 °F, its density is about 62.4 lb/ft³. The weight of the water in the pipe could be calculated by multiplying its volume by its density.

$$\begin{aligned}\text{Weight} &= (\text{volume})\,(\text{density})\\ &= (\text{cross-sectional area})\,(\text{height})\,(\text{density})\end{aligned}$$

$$\text{Weight} = (1\ \text{in}^2)(10\ \text{ft})\left(62.4\frac{\text{lb}}{\text{ft}^3}\right)\left(\frac{1\ \text{ft}^2}{144\ \text{in}^2}\right) = 4.33\ \text{lb}$$

The water exerts a pressure on the bottom of the pipe due to its weight. This pressure is equal to the weight of the water column divided by the area of the pipe. In this case:

$$\text{Static pressure} = \frac{\text{weight of water}}{\text{area of pipe}}$$

$$= \frac{4.33\ lb}{1\ in^2} = 4.33\frac{lb}{in^2} = 4.33\ psi$$

If a very accurate pressure gauge were mounted into the bottom of the pipe as shown in Figure 3-19, it would read exactly 4.33 psi.

The size or shape of the "container" holding the water has no effect on the static pressure at any given depth. Thus, an accurate pressure gauge placed 10 feet below the surface of a *lake* containing 60 °F water would also read exactly 4.33 psi.

The static pressure of a liquid increases proportionally from some value at the surface, to larger values at greater depths below the surface. The following equation can be used to determine static pressure of a liquid.

(Equation 3.4)

$$P_{static} = \left(\frac{D}{144}\right)h + P_{surface}$$

where:

P_{static} = static pressure at a given depth, h, below the surface of the liquid (psi)

h = depth below the liquid's surface (ft)

D = density of liquid in system (lb/ft³)

$P_{surface}$ = any pressure applied at the surface of the liquid (psi)

If the liquid is open to the atmosphere at the top of the container, even through a pinhole-size opening, then $P_{surface} = 0$. If the container is completely closed and filled with the liquid, $P_{surface}$ is the pressure of the liquid at the top of the container.

Knowing the static pressure at a given location in a hydronic system allows one to evaluate the potential for a circulator to cavitate. Static pressure can also help in determining if a piping loop of known elevation is completely filled with water. Static pressure at a given location is also important when sizing and pressurizing a diaphragm type expansion tank.

Example 3.4: Find the static pressure at the pressure gauge near the bottom of the solar collector circuit shown in Figure 3-20. The piping circuit has an overall height of 65 feet, is filled with 60 °F water, and has a static pressure of 10 psi at its uppermost point.

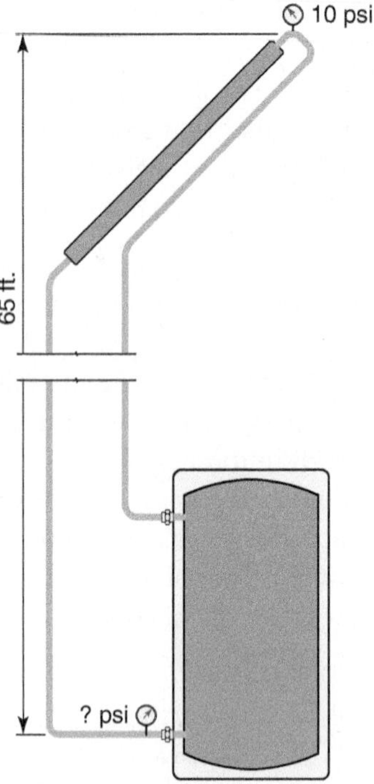

Figure 3-20 | Piping system for Example 3.4.

Solution: The density of water at a temperature of 60 °F can be found in Figure 3-11 (62.4 lb/ft³). Substituting this into Equation 3.4 yields the static pressure:

$$P_{static} = \left(\frac{62.4}{144}\right)65 + 10 = 38.2\ \text{psi}$$

Discussion: The calculated pressure of 38.2 psi is correct if the water temperature were about 60 °F. If, however, the same circuit contained water at 200 °F (having a density of 60.15 lb/ft³), Equation 3.4 would yield the following:

$$P_{static} = \left(\frac{60.15}{144}\right)65 + 10 = 37.2\ \text{psi}$$

The hot water would exert slightly less static pressure at the base of the system due to its lower density.

3.5 Flow Rate and Flow Velocity

BASIC FLUID MECHANICS

In a purely technical sense, the word "fluid" can mean either a **liquid**, such as water, or a **gas**, such as air. The science of fluid mechanics includes the study of both liquids and gases. In this book, the word fluid always refers to a liquid unless otherwise stated.

As discussed earlier in this chapter, most liquids, including water, are incompressible. This means that any given amount of liquid always occupies the same volume, with the exception of minor changes in volume due to thermal expansion and contraction. A simple example would be a long pipe completely filled with water. If another gallon of water were pushed into one end of that pipe, the exact same amount of water would have to flow out the other end. No additional liquid can be squeezed into a rigid container (such as a pipe) that is already filled with liquid.

By comparison, if the same pipe were filled with a gas such as air, it would be possible to push another gallon of air into the pipe without allowing any air out the other end. This action would increase in the pressure of the air within the pipe. Because of this, gases are said to be **compressible fluids**.

The fact that liquids are incompressible makes it easier to describe their motion though closed containers such as piping systems. Two basic terms associated with this motion are **flow rate** and **flow velocity**.

FLOW RATE

Flow rate is a measurement of the *volume of fluid that passes a given location in a pipe during a given time.* In North America, the customary units for flow rate are *U.S. gallons per minute* (abbreviated as **gpm**). Within this text, flow rate is represented in equations by the symbol f. The flow rate of a fluid through a hydronic system greatly influences that system's thermal performance.

FLOW VELOCITY

The motion of a fluid flowing through a pipe is more complex than that of a solid object moving in a straight path. The speed, or flow velocity of the fluid, is different at different points across the diameter of the pipe.

To visualize the flow velocity of a fluid, think about several small buoyancy-neutral particles that have been positioned along a line within the cross section of a pipe, and released at the same instant to be carried along with the flow. The particles would move faster near the center of the pipe compared to near the edge.

The speed of each particle could be represented by an arrow as shown in Figure 3-21a. A curve connecting the tips of all these arrows is called a **velocity profile**. The velocity profile shown in Figure 3-21a is two-dimensional. However, if one imagines this curve rotated around the centerline axis of the pipe, it forms a "nose cone" shape that represents a three-dimensional velocity profile as seen in Figure 3-21b.

In the context of hydronics, the term "flow velocity" is commonly understood to mean the *average* flow velocity of the fluid as it moves through a pipe. This average velocity would be the velocity that, if present across the entire cross section of the pipe, would result in the same flow rate as that created by the actual velocity profile. The concept of **average flow velocity** is illustrated in Figure 3-21c. In this text, the term "flow velocity" will always mean the average flow velocity in the pipe, unless otherwise stated. The common units for flow velocity in North America are feet per second, abbreviated as either ft/sec or FPS. Within equations, the average flow velocity is represented by the symbol, v.

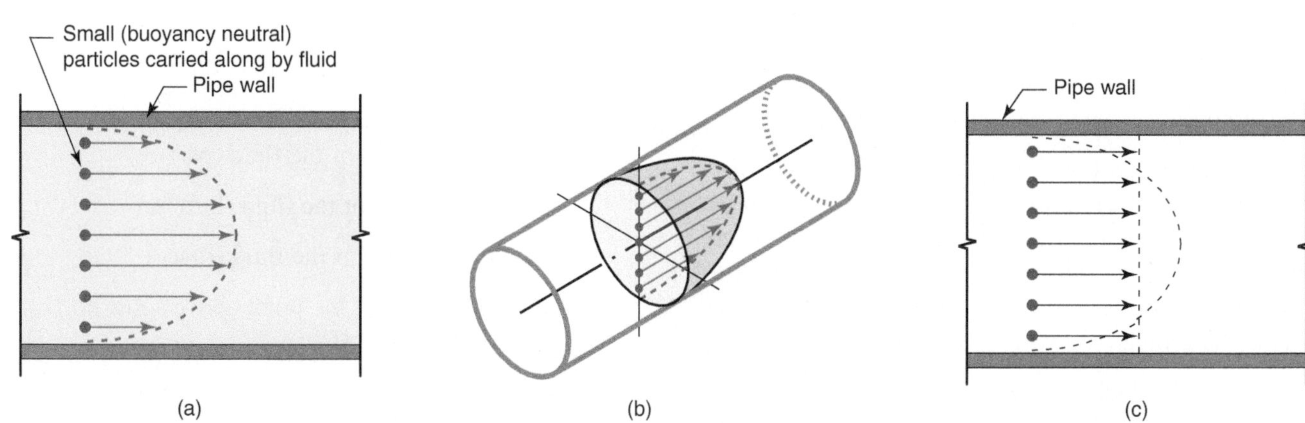

Figure 3-21 | (a) Representation of actual velocity profile. (b) Velocity profile in 3D. (c) Equivalent *average velocity* profile.

Equation 3.5 can be used to calculate the average flow velocity associated with a given flow rate in a round pipe.

(Equation 3.5)

$$v = \left(\frac{0.408}{d^2}\right)f$$

where:

v = average flow velocity in the pipe (ft/sec)

f = flow rate through the pipe (gpm)

d = exact inside diameter of pipe (inches)

Example 3.5: What is the average flow velocity of water moving at 6 gpm through a pipe with an inside diameter of 0.811 inches?

Solution: Substituting these numbers into Equation 3.5 yields:

$$v = \left(\frac{0.408}{d^2}\right)f = \left(\frac{0.408}{0.811^2}\right)6 = 3.72 \text{ ft/sec}$$

Discussion: Equation 3.5 can be used for any type of *round* pipe for which the internal diameter is known.

When selecting a pipe size to accommodate a given flow rate, the resulting average flow velocity should be calculated. *Pipes with a nominal diameter of 2 inches or less are typically selected so that the flow velocity through them remains between 2 and 4 feet per second. Some designers extend the range of this selection criteria to larger pipe sizes, while others allow the flow velocity to increase for larger piping.*

The lower end of this flow velocity range is based on the ability of flowing water to move air bubbles along a vertical pipe. Average flow velocities of 2 feet per second or higher can entrain air bubbles and move them in all directions including straight down. Entrainment of air bubbles is important when a system is filled and purged, as well as when air bubbles have to be captured and removed following maintenance of the system.

The upper end of this flow velocity range (4 feet per second) is based on minimizing noise generated by the flow. Average flow velocities above 4 feet per second, in piping with nominal inside diameters of 2 inches or less, can cause objectionable flow noise and should be avoided in hydronic systems.

Figure 3-22 tabulates the results of applying Equation 3.5 to selected sizes of type M copper, PEX, and PEX-AL-PEX tubing. Each equation can be used to calculate the flow velocity associated with a given flow rate. Also given are the flow rates corresponding to average flow velocities of 2 feet per second and 4 feet per second.

3.6 Head Energy

Fluids in a hydronic system contain both thermal and mechanical energy. We can sense the amount of thermal energy in a fluid by measuring its temperature. For example, the hot water leaving a heat source contains more thermal energy than cooler water returning from the distribution system. The increase in temperature of the fluid as it passes through the heat source is the "evidence" that thermal energy was added to it.

The mechanical energy contained in a fluid is called **head**. The amount of head present in a fluid depends on several factors including the fluid's pressure, density, elevation, and velocity at some point in the system. It is helpful to divide head energy into three categories:

- **Pressure head**: the mechanical energy a fluid contains due to its pressure
- **Velocity head**: the mechanical energy a fluid contains due to its velocity
- **Elevation head**: the mechanical energy the fluid contains due to its height in the system

Total head is the sum of the pressure head, velocity head, and elevation head. It can be calculated at any location within a hydronic system using **Bernoulli's equation**, stated as Equation 3.6:

Equation 3.6

$$\text{Total head} = \frac{P(144)}{D} + \frac{v^2}{64.4} + Z$$

where:

P = the pressure of the fluid (psi)

D = the density of the fluid (lb/ft^3)

v = the velocity of the fluid (ft/sec)

Z = the height of the point above some horizontal reference (feet)

144 = a number required for consistent units (144 in^2/ft^2)

Tubing size/type	Flow velocity	Minimum flow rate (based on 2 ft./sec.) (gpm)	Maximum flow rate (based on 4 ft./sec.) (gpm)
3/8″ copper	$v = 2.02\ f$	1.0	2.0
1/2″ copper	$v = 1.26\ f$	1.6	3.2
3/4″ copper	$v = 0.62\ f$	3.2	6.5
1″ copper	$v = 0.367\ f$	5.5	10.9
1.25″ copper	$v = 0.245\ f$	8.2	16.3
1.5″ copper	$v = 0.175\ f$	11.4	22.9
2″ copper	$v = 0.101\ f$	19.8	39.6
2.5″ copper	$v = 0.0655\ f$	30.5	61.1
3″ copper	$v = 0.0459\ f$	43.6	87.1
3/8″ PEX	$v = 3.15\ f$	0.6	1.3
1/2″ PEX	$v = 1.73\ f$	1.2	2.3
5/8″ PEX	$v = 1.20\ f$	1.7	3.3
3/4″ PEX	$v = 0.880\ f$	2.3	4.6
1″ PEX	$v = 0.533\ f$	3.8	7.5
1.25″ PEX	$v = 0.357\ f$	5.6	11.2
1.5″ PEX	$v = 0.256\ f$	7.8	15.6
2″ PEX	$v = 0.149\ f$	13.4	26.8
3/8″ PEX-AL-PEX	$v = 3.41\ f$	0.6	1.2
1/2″ PEX-AL-PEX	$v = 1.63\ f$	1.2	2.5
5/8″ PEX-AL-PEX	$v = 1.00\ f$	2	4.0
3/4″ PEX-AL-PEX	$v = 0.628\ f$	3.2	6.4
1″ PEX-AL-PEX	$v = 0.383\ f$	5.2	10.4

Figure 3-22 | Equations relating flow rate (f), and average flow velocity (v) for type M copper, PEX, and PEX-AL-PEX tubes. Flow rates corresponding to flow velocities of 2 feet/second and 4 feet/second are also given.

Head can be stated as the number of "foot pounds" of mechanical energy contained in each pound of fluid. This can be mathematically formatted as follows:

$$head = \frac{ft. \cdot lb.}{lb.}$$

Those who have studied physics will recognize the unit of ft. · lb. (pronounced "foot pound") as a unit of *energy*. As such, it can be converted to any other unit of energy, such as a Btu. However, engineers long ago chose to cancel the units of pounds (lb.) in the numerator and denominator of this ratio, and express head in the sole remaining unit of feet, as shown below.

$$\frac{ft. \cdot \cancel{lb.}}{\cancel{lb.}} = ft.$$

To distinguish between feet as a unit of *distance* and feet as a unit of *fluid energy*, the latter can be stated as feet *of head*.

Liquids can exchange total head energy back and forth between velocity head, pressure head, and elevation head. For example, when a liquid flows through a transition from a smaller to larger pipe size, its velocity

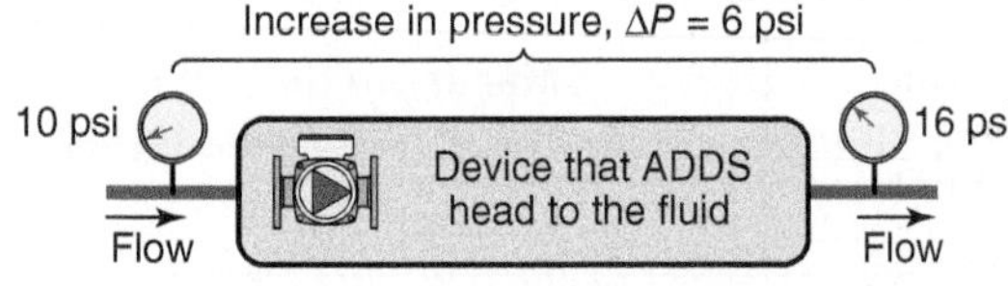

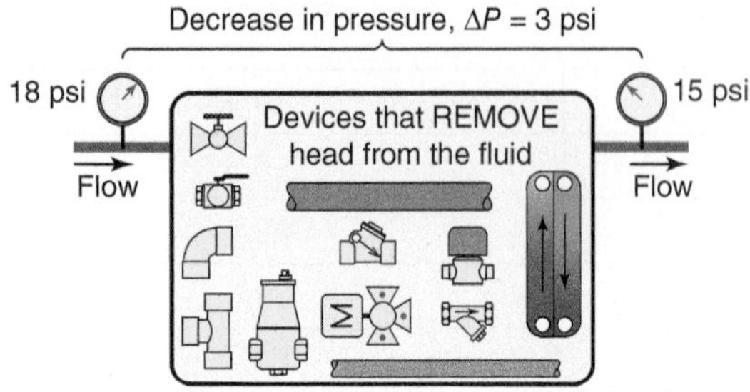

Figure 3-23 | Pressure changes associated with head that is added or removed from a flowing fluid.

decreases, and its pressure increases because some of the velocity head is instantly converted into pressure head.

When head energy is added to or removed from a liquid in a closed-loop piping system, there will always be an associated pressure change in the fluid. Just as a change in temperature is "evidence" of a gain or loss of thermal energy, a change in pressure is evidence of a gain or loss in head energy. When head is lost, pressure decreases. When head is added, pressure increases. This concept is illustrated in Figure 3-23.

Using pressure gauges to detect changes in the head of a liquid is like using thermometers to detect changes in the thermal energy content of that liquid.

Equation 3.7 can be used to convert an observed change in pressure across a component in a hydronic system to the associated gain or loss of head energy.

(Equation 3.7)

$$H = \frac{144(\Delta P)}{D}$$

where:

H = head energy added or lost from the liquid (feet of head)

ΔP = pressure change corresponding to the head added or lost (psi)

D = density of the liquid at its corresponding temperature (lb/ft^3)

This equation can be rearranged to calculate the pressure change associated with a given gain or loss in head:

(Equation 3.8)

$$\Delta P = \frac{HD}{144}$$

Example 3.5: Water at 140 °F flows through a heat emitter. A pressure gauge on the inlet side of the heat emitter reads 20 psi. A pressure gauge on the outlet side reads 18.5 psi. What is the change in the head energy of the water as it passes through?

Solution: The density of water at 140 °F is 61.4 lb/ft^3. Substituting this and the pressure data into Equation 3.7 yields:

$$H = \frac{144(\Delta P)}{D} = \frac{144(20 - 18.5)}{61.4} = 3.52 \text{ feet of head}$$

Discussion: Since the pressure decreases in the direction of flow, this is a head *loss*. If the fluid cools as it passes through the heat emitter, its density slightly increases. However, the variation in density is very small for the 10 °F to 30 °F temperature drop that occurs across a typical heat emitter and thus has negligible effect on the calculated change in head.

CALCULATING HEAD LOSS

Whenever a fluid flows, an energy-dissipating effect called viscous friction develops both within the fluid stream and along any surfaces the fluid contacts. This friction causes head energy to be converted to (or "dissipated" into) thermal energy. When designing a hydronic circuit, it is essential to know the head loss that occurs when a fluid passes through that circuit at a given flow rate. Equation 3.9 can be used to calculate this head loss *for piping circuits constructed of smooth tubing such as copper, PEX, or PEX-AL-PEX.*

(Equation 3.9)

$$H_L = (\alpha c L)(f)^{1.75}$$

where:

H_L = head loss of the circuit (feet of head)

α = fluid properties factor (see Equation 3.9 or Figure 3.23)

c = pipe size coefficient (see Figure 3.24)

L = total equivalent length of the circuit (ft)

f = flow rate through the circuit (gpm)

1.75 = an *exponent* applied to flow rate (f)

Equation 3.9 is a simplification of the **Darcy-Weisbach equation**, which is often used in the study of fluid mechanics. *Equation 3.9 should only be used for circuits constructed of smooth tubing.* This equation should also be limited to flows with Reynolds numbers between 2300 and 200,000. This range includes most flow situations that meet the pipe sizing criteria discussed earlier in this chapter.

Although the calculation of head loss using Equation 3.9 is straightforward mathematically, it requires the designer to gather information from other equations, graphs, or tables.

The value of the **fluid properties factor**, (α) alpha, can be calculated for various fluids using Equation 3.10.

(Equation 3.10)

$$\alpha = \left(\frac{D}{\mu}\right)^{-0.25}$$

where:

α = fluid properties factor

D = density of the fluid (lb/ft^3)

μ = dynamic viscosity of the fluid (lbm/ft/sec)

The α value is a complete mathematical description of the fluid's physical properties that are needed to determine head loss.

For water, as well as selected solutions of propylene glycol, the value of alpha (α) can also be read from the graph in Figure 3-24.

The value of the **pipe size factor (c)** in Equation 3.10 is a constant for a given tubing size. It incorporates dimensional information such as interior diameter, cross-sectional area, and appropriate unit conversion factors into a single number. A table of pipe size factors for several types and sizes of tubing is given in Figure 3-25.

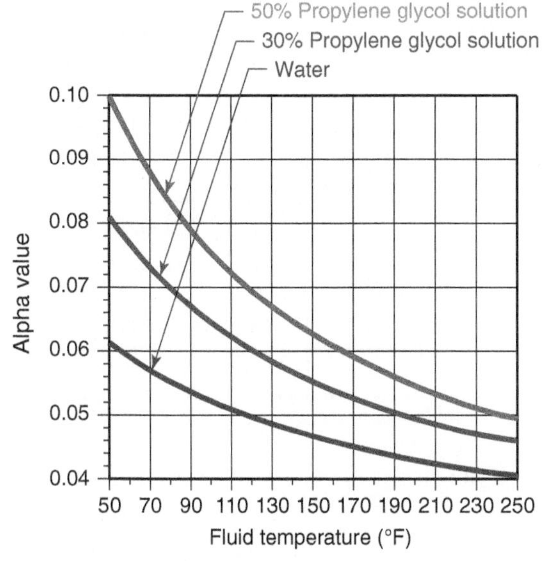

Figure 3-24 | Graph of alpha value of water and water-glycol solutions for temperatures of 50 °F to 250 °F.

Tube (size & type)	C value
3/8″ type M copper	1.0164
1/2″ type M copper	0.33352
3/4″ type M copper	0.061957
1″ type M copper	0.01776
1.25″ type M copper	0.0068082
1.5″ type M copper	0.0030667
2″ type M copper	0.0008331
2.5″ type M copper	0.0002977
3″ type M copper	0.0001278
3/8″ PEX	2.9336
1/2″ PEX	0.71213
5/8″ PEX	0.2947
3/4″ PEX	0.14203
1″ PEX	0.04318
1.25″ PEX	0.01668
1.5″ PEX	0.007554
2″ PEX	0.002104
3/8″ PEX-AL-PEX	3.35418
1/2″ PEX-AL-PEX	0.6162
5/8″ PEX-AL-PEX	0.19506
3/4″ PEX-AL-PEX	0.06379
1″ PEX-AL-PEX	0.019718

Figure 3-25 | Values of c (pipe size coefficient).

EQUIVALENT LENGTH OF FITTINGS AND VALVES

To use Equation 3.10 one must first determine the **total equivalent length** of the piping circuit. *The total equivalent length is the sum of the* **equivalent lengths** *of all fittings, valves, and other devices in the circuit, plus the total length of all tubing in the circuit.*

The equivalent length of a component is the amount of tubing, of the same size, that would produce the same head loss as the actual component, at the same flow rate. By replacing all components in the circuit with their

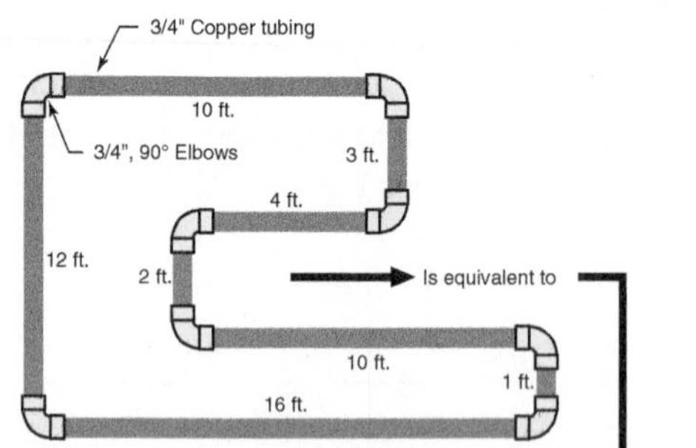

58 ft. of tubing + 8 × (2 equivalent ft. per elbow) = 74 ft. Total equivalent length

Figure 3-26 | Converting a piping circuit containing pipe and fittings to an equivalent length of straight pipe.

equivalent length of piping, the circuit can be treated as if it were a single piece of pipe having a length equal to the sum of the actual pipe length, plus the total equivalent lengths of all fittings, valves, or other devices.

The piping circuit shown in Figure 3-26 contains a total of 58 feet of 3/4-inch copper pipe and eight 90-degree elbows. Assuming that the equivalent length of each elbow is 2 feet of 3/4-inch pipe, the entire circuit could be thought of as if it were $58 + 8(2) = 74$ feet of 3/4-inch pipe.

Figure 3-27 gives the equivalent lengths of many commonly used fittings and valves.

The data in Figure 3-26 is representative of typical fittings and valves. The exact equivalent length of a specific fitting or valve will depend on its internal shape, surface roughness, and other factors.

Note that the values in the table in Figure 3-26 are for *soldered* fittings and valves. *These equivalent lengths should be doubled for threaded fittings.*

Example 3.6: Water at 140 °F and 6 gpm flows through the 3/4-inch piping circuit shown in Figure 3-28.

Copper tube sizes									
Fitting or valve[1]	3/8″	1/2″	3/4″	1″	1.25″	1.5″	2″	2 1/2″	3″
90-degree elbow	0.5	1.0	2.0	2.5	3.0	4.0	5.5	7.0	9
45-degree elbow	0.35	0.5	0.75	1.0	1.2	1.5	2.0	2.5	3.5
Tee (straight run)	0.2	0.3	0.4	0.45	0.6	0.8	1.0	0.5	1.0
Tee (side port)	2.5	2.0	3.0	4.5	5.5	7.0	9.0	12.0	15
B&G Monoflo® tee[2]	n/a	n/a	70	23.5	25	23	23	n/a	n/a
Reducer coupling	0.2	0.4	0.5	0.6	0.8	1.0	1.3	1.0	1.5
Gate valve	0.35	0.2	0.25	0.3	0.4	0.5	0.7	1.0	1.5
Globe valve	8.5	15.0	20	25	36	46	56	104	130
Angle valve	1.8	3.1	4.7	5.3	7.8	9.4	12.5	23	29
Ball valve[3]	1.8	1.9	2.2	4.3	7.0	6.6	14	0.5	1.0
Swing-check valve	0.95	2.0	3.0	4.5	5.5	6.5	9.0	11	13.0
Flow-check valve[4]	n/a	n/a	83	54	74	57	177	85	98
Butterfly valve	n/a	1.1	2.0	2.7	2.0	2.7	4.5	10	15.5

1. Data for soldered fittings and valves. For threaded fittings double the listed value.
2. Derived from C_v values based on no flow through side port of tee.
3. Based on a standard-port ball valve. Full-port valves would have lower equivalent lengths.
4. Based on B&G brand "flow control" valves.

Figure 3-27 | Representative equivalent lengths of common soldered fittings and valves (all values expressed as feet of copper tube of the same nominal size).

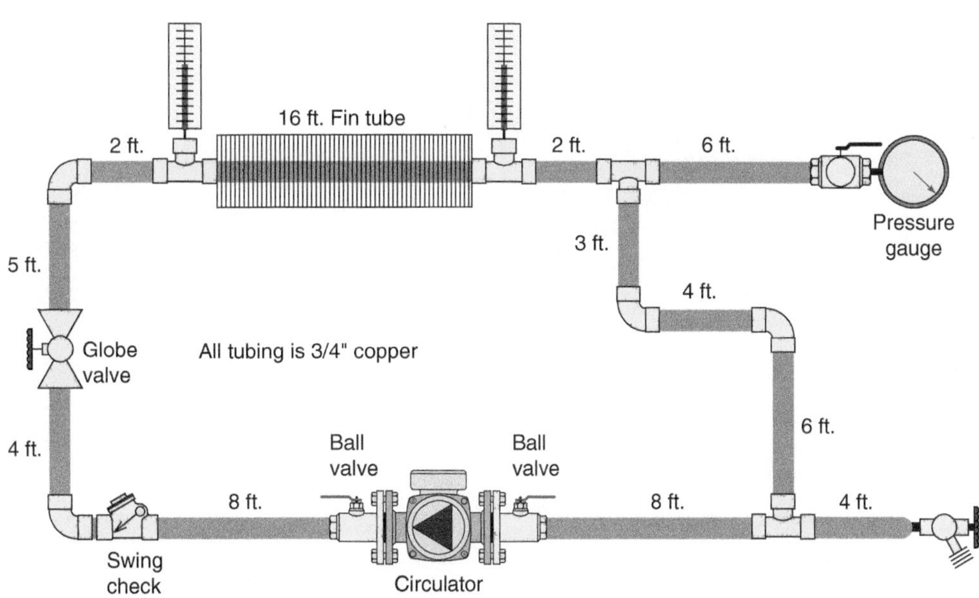

Figure 3-28 | Piping assembly for use in Example 3.6.

Determine the head loss and associated pressure drop around the circuit.

Solution: Start by finding the total equivalent length of the circuit by adding the equivalent lengths of the fittings and valves to the length of tubing as shown in Figure 3-29.

Notice that only tubing lengths *within the flow path* are included since flow does not occur in dead-end pipe branches. Figure 3-30 illustrates how the original piping assembly can now be treated at an equivalent length of 100.2 feet of ¾" type M copper tubing.

Components	Equivalent length
3/4″ straight tube	58 ft
3/4″ × 90° elbows	4 × 2 ft each = 8 ft
3/4″ straight run tees	2 × 0.4 ft each = 0.8 ft
3/4″ side port tees	2 × 3 ft each = 6 ft
3/4″ ball valves	2 × 2.2 ft each = 4.4 ft
3/4″ globe valves	1 × 20 ft each = 20 ft
3/4″ swing check	1 × 3 ft each = 3 ft
TOTAL EQUIVALENT LENGTH =	**100.2 ft**

Figure 3-29 | Adding the equivalent lengths of the fittings and valves in Figure 3-28.

Equation 3.9 can be used with this total equivalent length to determine the head loss at 6 gpm.

The α value for water at 140 °F can be determined from Figure 3-24 to be 0.0478.

The value of the pipe size factor c for 3/4-inch copper tube is found in Figure 3-25: $c = 0.061957$

This data can now be substituted into Equation 3.9 to determine the head loss:

$$\begin{aligned} H_L &= (\alpha c L)(f)^{1.75} \\ &= [(0.0478)(0.061957)(100.2)](6)^{1.75} \\ &= 6.83\,\textit{feet of head} \end{aligned}$$

The corresponding pressure drop around the circuit can be found using Equation 3.8. Note that this equation requires the density of water from Figure 3-11.

$$\Delta P = \frac{(6.83)(61.35)}{144} = 2.91\,psi$$

Discussion: Always be sure that the value 1.75 in Equation 3.9 is treated as an *exponent*, not a multiplying factor.

SYSTEM HEAD LOSS CURVE

An examination of Equation 3.9 shows that for any piping circuit, operating with a specific fluid, and at a constant temperature, the grouping of terms $(\alpha c L)$

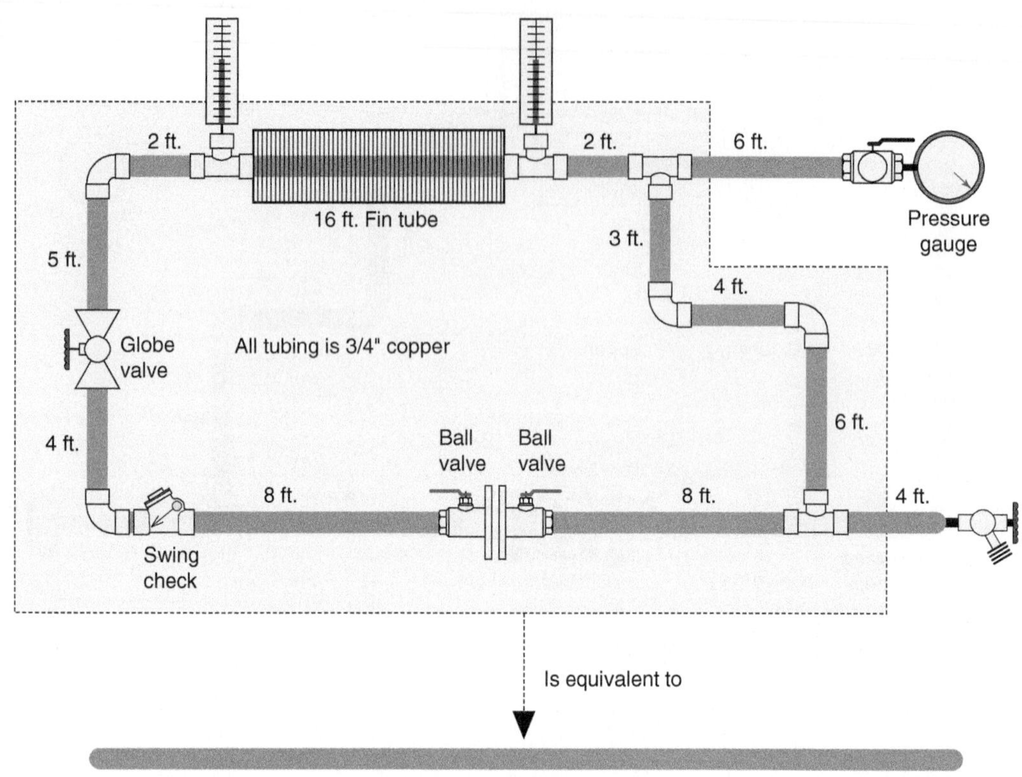

Figure 3-30 | Converting the piping assembly to an equivalent length of pipe.

remains constant. Thus Equation 3.8 can be treated as follows:

(Equation 3.11)

$$H_L = (\alpha c L)(f)^{1.75} = (number)(f)^{1.75}$$

Under these conditions, the head loss around the piping circuit depends only on flow rate. Equation 3.11 is a mathematical function that can be graphed by selecting several flow rates, calculating the head loss at each, and plotting the resulting points. Once the points are plotted, a smooth curve could be drawn through them.

Example 3.7: Using the piping circuit and data from Example 3.6, plot several points representing different flow rates and the associated head loss. Draw a smooth curve through these points.

Solution: Start by substituting the appropriate data into Equation 3.9:

$$H_L = (\alpha c L)(f)^{1.75} = [(0.0478)(0.061957)(100.2)](f)^{1.75}$$

Multiply the values of α, c, and L together to obtain a single number. In this case, it is 0.297. Thus, Equation 3.9 for this piping circuit simplifies to:

$$H_L = [0.297](f)^{1.75}$$

Flow rate (gpm)	Head loss (feet of head)
0	0
3	2.03
6	6.83
9	13.89
12	22.98
15	33.96

Figure 3-31 | Values of head loss calculated at several flow rates using Equation 3.11.

This equation can now be used to determine the head loss of the circuit at several (arbitrarily selected) flow rates, as given in Figure 3-31.

This data can now be plotted and a smooth curve drawn through the points, as shown in Figure 3-32.

Discussion: A graph similar to that shown in Figure 3-32 can be easily prepared using any spreadsheet or graphing utility program. The upward curvature indicates that head loss is increasing faster at higher flow rates. This is the result of the 1.75 exponent of the flow rate in Equation 3.11.

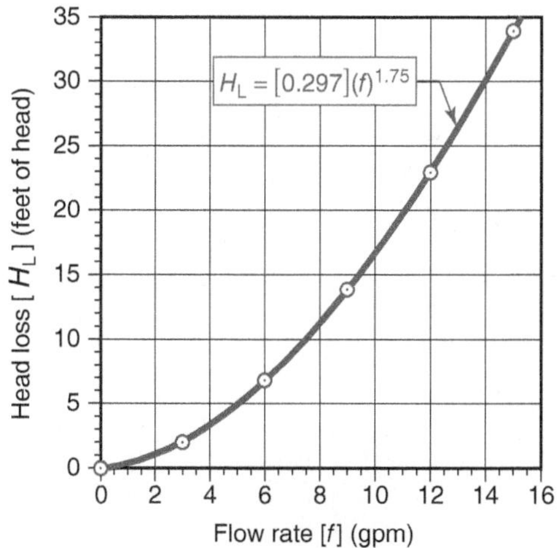

Figure 3-32 Example of a system head loss curve plotted using data from Figure 3-29.

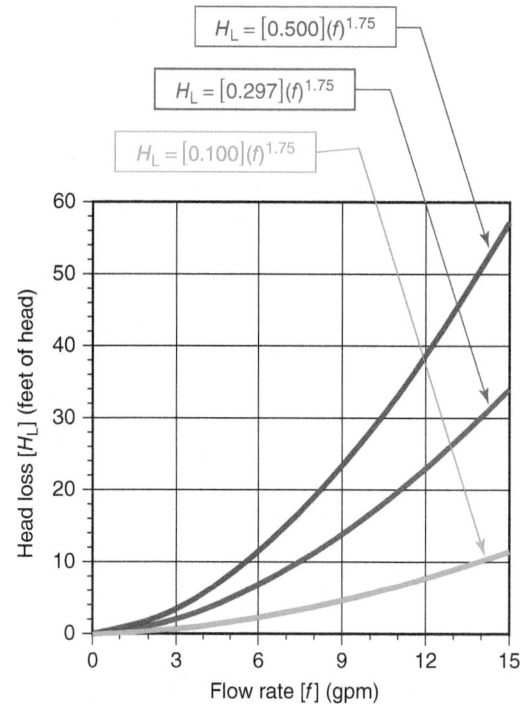

Figure 3-33 Examples of how the value of (αcL) effects the steepness of the system head loss curve.

A graph such as the one shown in Figure 3-32 is called a **system head loss curve**. It represents the relationship between flow rate and head loss for a given piping circuit using a specific fluid at a given temperature. *All piping circuits have a unique system head loss curve. It could even be thought of as the analytical "fingerprint" of that circuit. Determining the system head loss curve of a piping circuit is an essential step in properly selecting a circulator for that circuit.*

Any changes to the piping components, tubing, fluid, or average fluid temperature will change the value of (αcL) and thus effect the system curve, as shown in Figure 3-33.

If the value of (αcL) increases, the system curve gets steeper. If the value of (αcL) decreases, the system curve becomes shallower. The system curve for all piping circuits that are completely filled with fluid will always pass through the point (0,0) (e.g., zero head loss at zero flow rate).

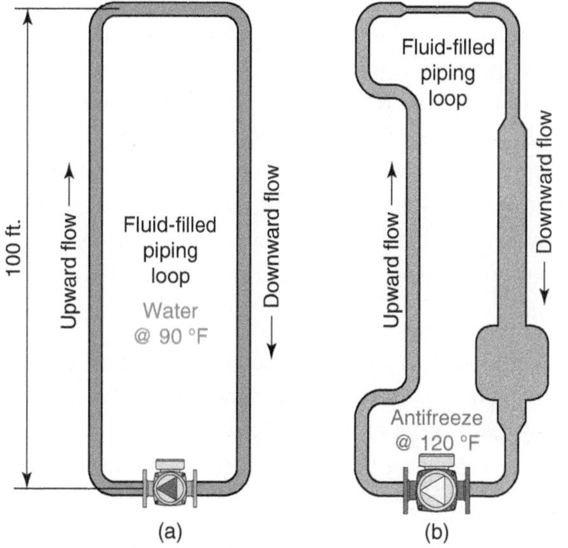

Figure 3-34 (a) A closed loop circuit filled with water at 90 °F. (b) Another closed loop circuit with different pipe sizes, fittings, circulator, and filled with an antifreeze solution at 120 °F. In either circuit, the associated circulator only has to replace the head loss due to viscous friction. *It does not have to "lift" liquid up the left side of the circuit.*

LIQUID-FILLED PIPING CIRCUITS

Most hydronic heating systems consist of *closed* piping systems. After all the piping work is complete, these circuits are completely water or some other liquid and purged of air. During normal operation, very little, if any, liquid enters or leaves the system.

Consider the **liquid-filled piping circuits** shown in Figure 3-34. Assume that circuit (a) is filled with water at some arbitrary temperature (i.e., 90 °F), while circuit (b) is filled with an antifreeze solution at some other arbitrary temperature (i.e., 120 °F). One might assume that the circulator in each circuit has to exert significant pressure to "push" the liquid 100 feet up the left side of the circuit, and that the liquid then "falls" back down the right side. Perhaps someone compares this situation to the physical effort required to lift a pail of liquid 100 feet above the ground, and then pour it out and let it fall back down 100 feet. This, however, is not the case. Because both liquids are incompressible, and because neither can escape from its closed piping circuit, any amount of liquid that moves up the left side of the circuit requires an equal amount of liquid to come down the right side of the circuit. Furthermore, this movement must occur over exactly the same time. This means that the weight of liquid moving up the left side of the circuit, over a given time, is always balanced by the weight of liquid moving down the right side during that time.

The following conclusion can thus be stated:

To maintain a constant flow rate in any liquid-filled piping circuit, the circulator need only replace the head loss due to viscous friction.

This principle remains true regardless of the size or shape of the piping path, the liquid within the circuit, or the circulator used, as illustrated in Figure 3-34. The flow rates and flow velocities will be different for circuit (a) versus circuit (b) due to differences in the fluids as well as differences in pipe size, length, and circulators. However, within each circuit, some stable flow rate will be quickly established when the circulator is turned on, and that flow rate will be constant throughout that circuit. Thus, the weight of fluid moving up left side of that circuit always equals the weight of fluid moving down the right side. In either circuit, the circulator only has to replace the head loss due to viscous friction. *It does not have to "lift" liquid up the left side of the circuit.*

This principle can be extended to multiple circuits and multiple circulators within the same system. For example, the two circulators shown in Figure 3-35 are only responsible for replacing the head energy lost due to friction through their respective flow paths within the system.

The liquid in a completely filled piping circuit acts like a Ferris wheel with the same weight in each seat, as illustrated in Figure 3-36.

The weight in the seats moving up exactly balances the weight in the seats moving down. If it were not for friction in the bearings and air resistance, this balanced Ferris wheel would continue to rotate indefinitely once started. Likewise, since the flow rate is always the same throughout a liquid-filled piping circuit, the weight of the liquid moving upward through the circuit in a given amount of time is always balanced by the weight of liquid moving downward through the circuit during that time. If it were not for the viscous friction of the liquid, it too would continue to circulate indefinitely within a piping circuit.

This principle is extremely important in hydronic system design. It explains why a small circulator can establish and maintain flow in a filled piping loop, even if the top of the loop is several stories above the circulator and contains thousands of gallons of liquid. Unfortunately, many circulators in hydronic systems are needlessly oversized because this principle is not understood.

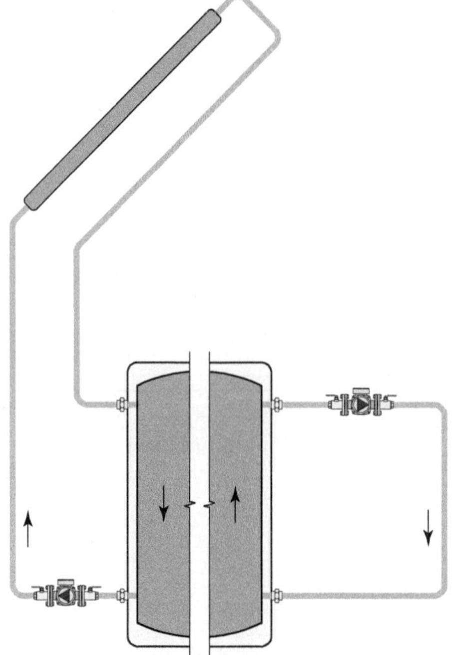

Figure 3-35 The weight of fluid moving up in each circuit always equals the weight of fluid moving down in that circuit. Each circulator only has to replace the head energy lost due to friction within its associated circuit path.

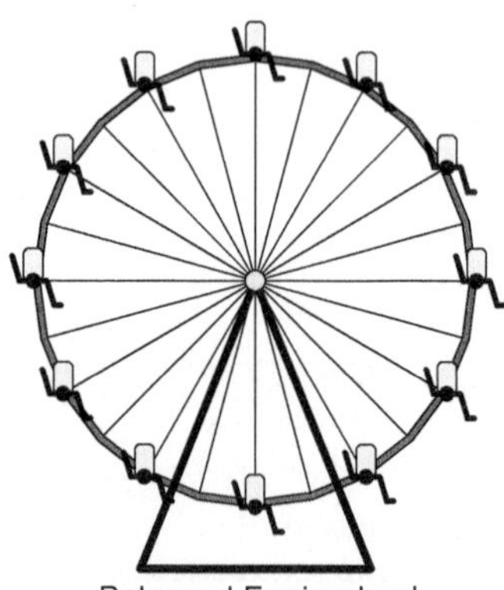

Figure 3-36 The liquid in a completely filled piping circuit acts like a Ferris wheel with the same weight in each seat.

SUMMARY

This chapter has presented several fundamental principles that collectively govern the operation of every hydronic system, regardless of whether it has a conventional heat source or a renewable energy heat source. If these principles are understood and respected, and if hardware devices are selected based on them, the resulting system will likely be successful. If these principles are ignored, and hardware devices selected without their regard, the likelihood of creating a truly successful system is very low. As in most technologies, the fundamentals provide the unchanging foundation upon which all systems are built.

KEY TERMS

absolute pressure
air separator
average flow velocity
backflow preventer
Bernoulli's equation
circulator
closed-loop system
coalesce
coalescing
compressible fluids
controllers
Darcy-Weisbach equation
density
dissolved air
dissolved oxygen
distribution system
dynamic viscosity
elevation head
equivalent lengths
expansion tank
fast-fill valve
feed water valve
flow rate
flow velocity
fluid properties factor (α)
gas (as one type of fluid)
gpm
head
heat capacity
heat emitters
heat sources
high limit controller
incompressible
intermittent heat source
latent heat
liquid(as one type of fluid)
liquid-filled piping circuits
make-up water system
open-loop system
pipe size factor (c)
pressure gauge
pressure head
pressure-reducing valve
pressure-relief valve
purging valve
reverse thermosiphoning
sensible heat
sensible heat-quantity equation
sensible heat-rate equation
setpoint temperature
shutoff valve
specific heat
static pressure
stratification
supply temperature
system head loss curve
thermostat
3-way mixing valve
total equivalent length
total head
vapor pressure
velocity head
velocity profile
viscosity

QUESTIONS AND EXERCISES

1. What component limits the pressure change in a hydronic system as the water heats up and cools down?

2. What device can be used to prevent very hot water in a storage tank from being circulated directly through low-temperature heat emitters?

3. Why is it necessary to have a pressure-reducing valve in the make-up water assembly supplying a hydronic system?

4. Explain at least two advantages of a closed-loop hydronic system in comparison to an open-loop system.

5. How many gallons of water would be required in a solar storage tank to absorb 250,000 Btu while undergoing a temperature change from 90 °F to 160 °F?

6. Water enters a panel radiator at 180 °F and at a flow rate of 1.5 gpm. Assuming the radiator releases heat into the room at a rate of 20,000 Btu/hr, what is the temperature of the water leaving the radiator?

7. How much pressure must be maintained on water to maintain it as a liquid when its temperature is 240 °F? State your answer in psia.

8. If you wanted to make water boil at 100 °F, how many psi of vacuum (below atmospheric pressure) would be required? Hint: psi vacuum = 14.7 – absolute pressure in psia.

9. Assume you need to design a hydronic system that can deliver 80,000 Btu/hr. What flow rate of water is required if the temperature drop of the distribution system is to be 10 °F? What flow rate is required if the temperature drop is to be 30 °F?

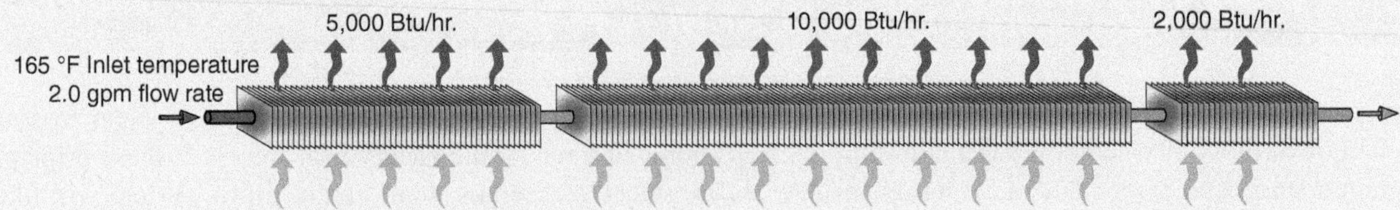

Figure 3-37 | Series-connected baseboard convectors for Exercise 10.

10. Water flows through a series-connected heat emitters system, as shown in Figure 3-37. The inlet temperature to the first baseboard is 165 °F. Determine the outlet temperature of the first baseboard and both the inlet and outlet temperatures for the second and third baseboards. Assume the heat output from the interconnecting piping is insignificant. Hint: Use the approximation that the inlet temperature for a baseboard equals the outlet temperature from the previous baseboard.

11. What physical property of water allows a ¾-inch tube to deliver the same rate of heat delivery as an 8-inch by 14-inch air duct?

12. You are designing a thermal storage system that must supply an average load of 30,000 Btu/hr for 18 consecutive hours. The minimum water temperature usable by the load is 110 °F. Determine the temperature the water must be heated to by the end of the charging period if:
    a. The tank volume is 500 gallons
    b. The tank volume is 1,000 gallons

13. A swimming pool contains 18,000 gallons of water. A boiler with an output of 50,000 Btu/hr is operated 8 hours per day to heat the pool. Assuming all the heat is absorbed by the pool water with no losses to the ground or air, how much higher is the temperature of the pool after the 8 hours of heating?

14. At what *gauge* pressure will water boil if its temperature is 250 °F?

15. Which combination of conditions is best for getting dissolved air to come out of solution with water?
    a. Low temperature and low pressure
    b. High temperature and low pressure
    c. Low temperature and high pressure
    d. High temperature and high pressure

16. A family of four uses 60 gallons of domestic hot water per day. The water enters the water heater at 50 °F and is heated to 140 °F by the time it leaves. Calculate the number of Btus per day used for this process.

17. Assume a hydronic system is filled with 50 gallons of water at 50 °F and 30 psi pressure. The water contains the maximum amount of dissolved air it can hold at these conditions. The water is then heated and maintained at 200 °F and 30 psi pressure. Using Figure 3-17, estimate how much air will be expelled from the water.

18. You are working on a system that will operate with water at an average temperature of 140 °F and a pressure of 30 psi. Write down the following properties of this water. Be sure to state the appropriate units in each answer
    a. Its density
    b. Its dynamic viscosity
    c. Its specific heat
    d. Its heat capacity
    e. The maximum amount of dissolved air it could contain

19. Examine the piping assembly shown in Figure 3-38.
    a. What is the flow rate exiting point B in the piping assembly?
    b. How would this change if only two of the four circulators were operating?
    c. Assume the flow rate exiting point B was 3.2 gallons per minute, and the other flow rates remain as shown. What would this imply is happening somewhere with the piping assembly?

20. What static water pressure would be required at street level to push water to the top of a 300-foot building? Express the answer in psi. If the static water pressure at the top of the building had to be 50 psi, how would this affect the static pressure at the street level?

21. Determine the head loss when 200 °F water flows at 10 gpm through 400 feet of 1-inch type M copper tube. Determine the pressure drop (in psi) associated with this head loss.

22. Find the total equivalent length of the piping path shown in Figure 3-39.

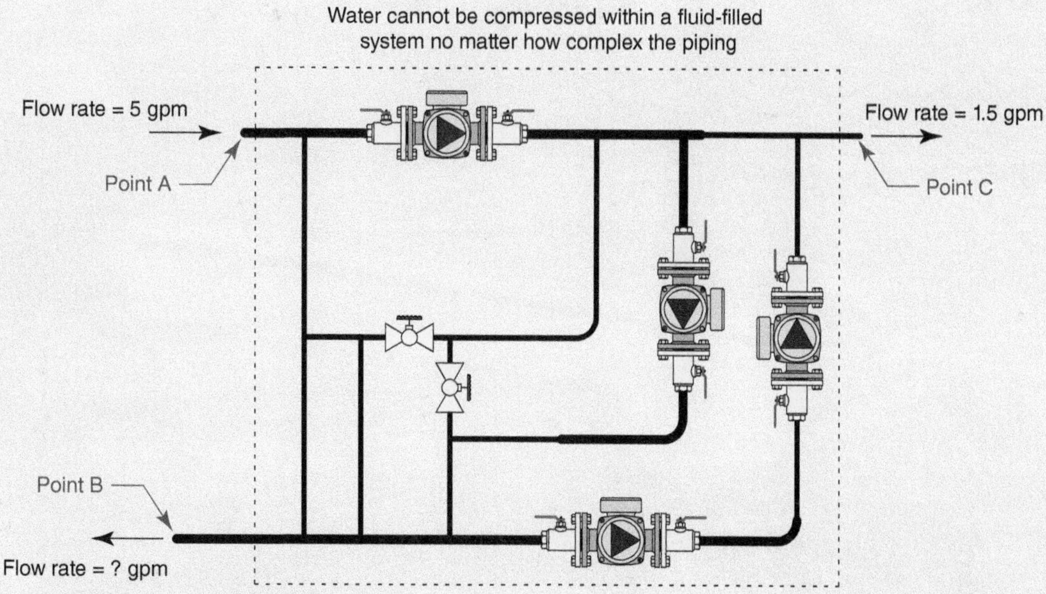

Figure 3-38 | Piping assembly for Exercise 19.

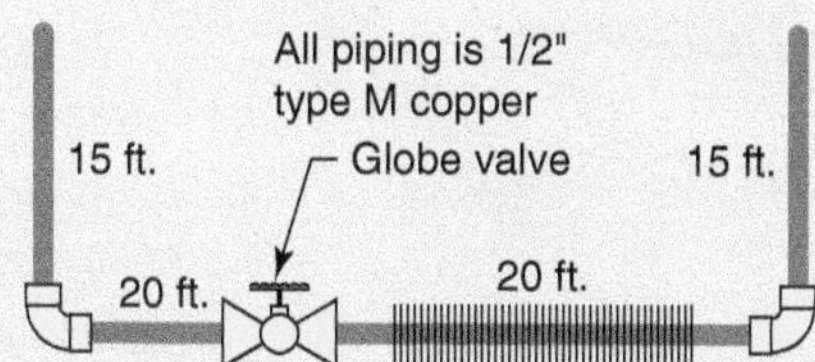

Figure 3-39 | Piping assembly for Exercise 22.

23. Assume another 25 feet of ½-inch copper tubing is added to the assembly shown in Figure 3-39 and that this completes a closed-loop circuit. Also assume that circuit operates with a 30 percent solution of propylene glycol antifreeze. Determine the value of (αcL) for that circuit.

24. Calculate five data points representing head loss versus flow rate through the closed-loop circuit described in Exercise 23. Plot this data and sketch a system head loss curve. Hint: Use Equation 3.9 to generate the necessary head loss versus flow rate data.

R410A
DAIKIN
R410A

CHAPTER 4

Essential Hydronic Hardware

OBJECTIVES

After studying this chapter, you should be able to:

- Understand the relationship between hydronics and renewable energy heat sources.
- Distinguish between different types of auxiliary hydronic heat sources.
- Describe the performance characteristics of conventional and condensing boilers as auxiliary heat sources.
- Describe where different types of tubing are best applied in hydronic systems.
- Describe several fittings used in hydronic systems.
- Explain the purpose and operation of several types of valves.
- Understand the types and proper application of 3-way mixing valves.
- Describe the proper placement of circulators within hydronic systems.
- Select an appropriate circulator for a specific hydronic circuit.
- Size a diaphragm-type expansion tank for a hydronic system.
- Explain methods for both air elimination and air management.
- Describe how flat plate heat exchangers are sized.

4.1 Introduction

Almost all systems using renewable energy heat sources, such as solar collectors, heat pumps, and wood-fueled boilers, share common hydronic components and subsystems. This will become increasingly evident as you progress through this text. These components and subsystems allow for efficient conveyance, storage, and delivery of heat, regardless of the heat source. A thorough understanding of these components and subsystems allows designers to apply them over a wide range of applications. In some cases, a system that uses solar collector as its primary heat source will use exactly the same component or subsystem as a system using a geothermal heat pump or pellet-fueled boiler as its primary heat source.

Many of the problems that have occurred in combisystems using renewable energy heat sources can be traced to poor selection or installation of components *other than the renewable energy heat source(s)*. For example, a poorly performing system having a properly sized and installed array of solar collectors might be traced to a poorly planned or incorrectly installed underfloor heating system, an inadequately sized circulator, or an improperly selected valve.

Therefore, before discussing the intricacies of renewable energy heat sources, it's imperative to gain a solid understanding of the essential hydronic hardware that becomes the "media" from which combisystems that rely on renewable energy heat sources are crafted.

4.2 Auxiliary Boilers

Most renewable energy heat sources are used in systems that are expected to provide *automatic* and *uninterrupted* heat delivery. Because most renewable energy heat sources provide intermittent heat, those systems must be equipped with an **auxiliary heat source**.

The auxiliary heat source could be entirely separate from the system that delivers heat generated by the renewable energy source. For example, a house that's partially heated by solar collectors that supply heat through a hydronic radiant panel delivery system could have electric baseboard heaters installed for auxiliary heating. However, it is often more cost-effective and aesthetically pleasing to use a *hydronic-based auxiliary heat source* to supply heat through the same distribution system and heat emitters that are connected to the renewable energy heat source.

The most common hydronic-based auxiliary heat source is a **boiler**. This term is a bit deceptive. None of the "boilers" discussed in this text are intended to boil water. Instead, they heat water to temperatures compatible with the distribution system, using fuels such as natural gas, propane, fuel oil, and electricity. Although the author prefers the term **hydronic heat source** to more accurately and generically describe such devices, most of the HVAC industry, as well as building owners, are familiar with the term *boiler*, and thus it will be used in this text.

BOILER CLASSIFICATIONS

Boilers can be classified in many ways, such as by the fuel they use, their heat output rating, the materials used for their heat exchanger, the shape of their heat exchanger, or how exhaust gases are removed from them. Chapter 3 in *Modern Hydronic Heating*, 3rd edition, provides extensive descriptions of a wide range of boilers.

For purposes of selecting a boiler as an auxiliary heat source, the following two classifications are useful:

- Conventional boilers
- Condensing boiler

Conventional boilers *are intended to operate so that the water vapor, produced during combustion, does not condense, on a sustained basis, within the boiler or its venting system.* **Condensing boilers** are specifically designed to operate with sustained flue gas condensation.

CONVENTIONAL BOILERS

Nearly all boilers with cast-iron, carbon steel, or copper tube heat exchangers fall into this category. Figure 4-1 shows an example of each of these boilers.

Given the right operating conditions, ANY boiler can operate with ***sustained flue gas condensation.*** This condensation is a mixture of water and chemical compounds that are based on the fuel and the efficiency at which that fuel is combusted. These compounds make the condensate acidic. As such, it is highly corrosive to carbon steel, cast iron, and copper. It can also quickly corrode galvanized steel vent connector piping, which could cause spillage of exhaust products into the building. Such situations must be avoided through proper design.

It is imperative that conventional boilers are applied and operated so that sustained flue gas condensation does NOT occur.

It's also important to realize that *all* boilers experience ***intermittent flue gas condensation*** when starting from a cool condition. However, when a

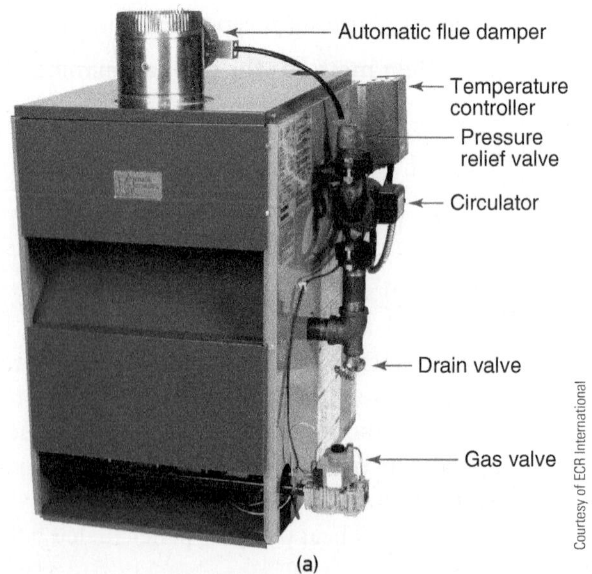

Figure 4-1 | (a) Example of a gas-fired sectional cast-iron boiler.

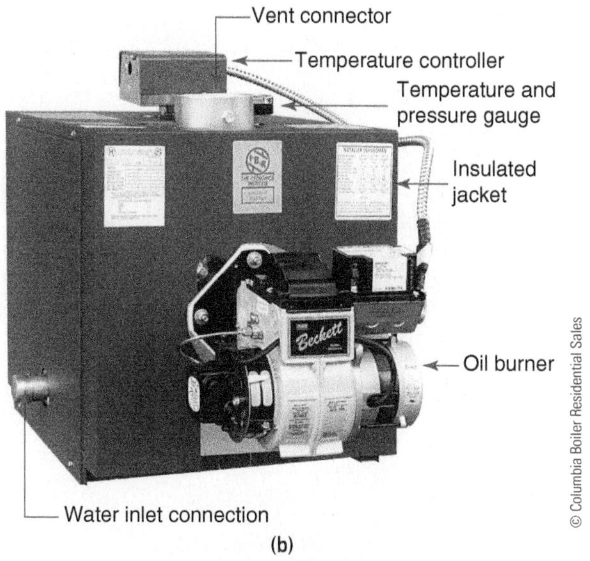

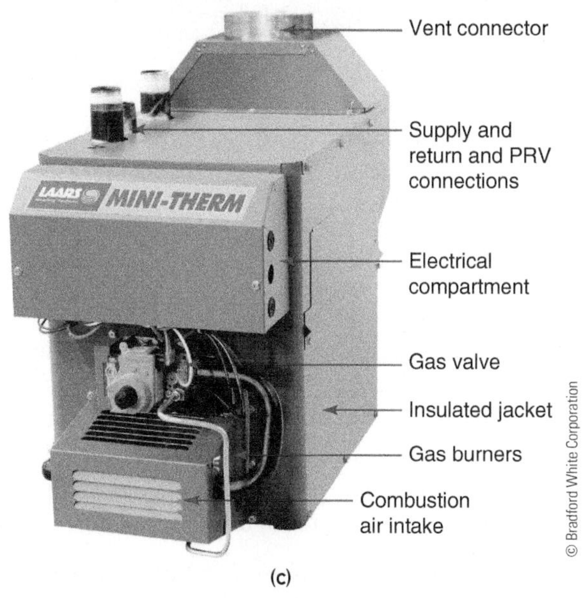

Figure 4-1 (b) Example of a steel fire-tube boiler. (c) Example of a copper fin-tube boiler.

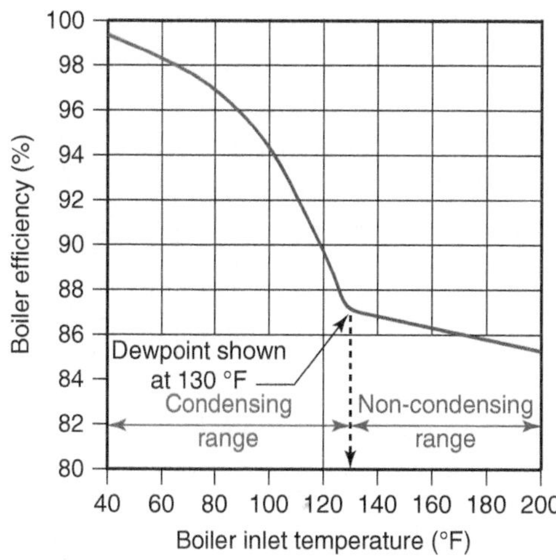

Figure 4-2 Relationship between water temperature entering boiler, its thermal efficiency, and its operation in either a condensing or non-condensing mode.

conventional boiler is properly applied, intermittent condensation quickly evaporates as the boiler warms. Intermittent flue gas condensation is generally not detrimental to the boiler.

The presence or absence of sustained flue gas condensation is determined by the temperature of the water entering the boiler and the air/fuel ratio at which combustion occurs. Of these, only the boiler inlet water temperature is easily controllable by how the boiler is applied.

Figure 4-2 shows a typical relationship between the temperature of the water entering a boiler and its associated thermal efficiency. The latter is the percentage of the fuel's chemical energy content that is converted to heat and transferred to the water in the boiler.

If the temperature of the water entering the boiler is relatively high (i.e., 200 °F), the combustion side surfaces of the boiler's heat exchanger are well above the **dewpoint temperature** of the water vapor in the exhaust stream. Under such conditions, no condensation occurs.

However, as the temperature of the water entering the boiler decreases, so does the temperature of the boiler's heat exchanger surfaces. At some entering water temperature, these surface temperatures drop to the dewpoint temperature of the water vapor in the exhaust stream. This temperature varies with the type of fuel burned and the air/fuel ratio supplied to the combustion chamber. For natural gas, flue gases begin condensing when the inlet water temperature drops to approximately 130 °F. The lower the entering water temperature, the greater the rate of flue gas condensation. If all the water vapor formed during combustion were condensed, approximately 1.15 gallons of liquid **condensate** would form for each therm (e.g., 100,000 Btu) of natural gas burned.

Figure 4-2 also shows that a boiler's thermal efficiency increases rapidly if the boiler operates within the condensing range of inlet water temperature. The lower the boiler's inlet water temperature, the greater the rate of condensation and the higher its thermal efficiency. For example, at an inlet water temperature of 100 °F, Figure 4-2 indicates a probable thermal efficiency of about 94%.

The increased thermal efficiency is the result of capturing the **latent heat** released as water vapor condenses into liquid. Each pound of water vapor converted to liquid releases about 970 Btu of heat to the boiler's heat exchanger. In a conventional boiler, this latent heat is carried away with the exhaust stream. When fuel was inexpensive, there was little motivation to capture this energy.

CONDENSING BOILERS

The energy crisis of the 1970s motivated manufacturers to create boilers that could capture the latent heat contained in the water vapor being carried up the chimney from conventional boilers. Early attempts at such boilers had limited success. However, over the past 35 years, great strides have been made in developing reliable boilers capable of operating with sustained flue gas condensation. These condensing boilers are constructed with large heat exchanger surfaces capable of extracting more heat from the exhaust gases compared to conventional boilers. When operated with suitably low inlet water temperatures, these boiler can easily cool the exhaust stream below the dewpoint temperature of the water vapor and thus allow condensation to occur.

Figure 4-3 | Example of a cast-iron boiler equipped with a stainless steel secondary heat exchanger, seen on left side.

Some condensing boilers use two separate internal heat exchangers. Combustion takes place with the **primary heat exchanger**, where temperatures are high enough that flue gases will *not* condense when the boiler is operated within its rated temperature range. The primary heat exchanger can therefore be constructed of steel, copper, or cast iron. After giving up some heat at higher temperatures, exhaust gases move from the primary heat exchanger to a **secondary heat exchanger**, which is typically made of high-grade stainless steel. Additional heat is extracted from flue gases as they move through the secondary heat exchanger, and condensation occurs. The secondary heat exchanger is designed to withstand the corrosive nature of the condensate. It is also equipped with a drain to remove condensate from the boiler. An example of this type boiler is shown in Figure 4-3.

MODULATING/CONDENSING BOILERS

Early generation condensing boilers operated at a fixed firing rate and thus a fixed rate of heat output. Today, nearly all condensing boilers can vary their heat output from some maximum heat output rate down to approximately 20 percent of that rate. These boilers are said to be **modulating**. The term **mod/con** is often used to describe boilers that can vary their firing rates and are capable of operating with sustained flue gas condensation. An example of such a boiler, in a wall-hung configuration, is shown in Figure 4-4.

Modulating the heat output of a boiler is similar to changing the throttle setting on an internal combustion engine. The rate of heat production is lowered by reducing the fuel and air sent into the combustion process. In mod/con boilers, a variable speed blower regulates the rate of air flow entering a sealed combustion chamber. Natural gas or propane is mixed into this air stream in proportion to the air-flow rate. The slightly pressurized mixture of gas and air is forced through a burner head and ignited. Burner heads are typically made of stainless steel or porous ceramic materials. The flame pattern they create remains close to the burner's surface, as seen in Figure 4-5.

The ability of a mod/con boiler to reduce its heat output (e.g., modulate) is expressed as its **turndown ratio**. This is the reciprocal of the lowest possible percentage of full heat output rate the boiler can maintain. For example: If a boiler can maintain stable operation down to 20 percent of its maximum heat output rate,

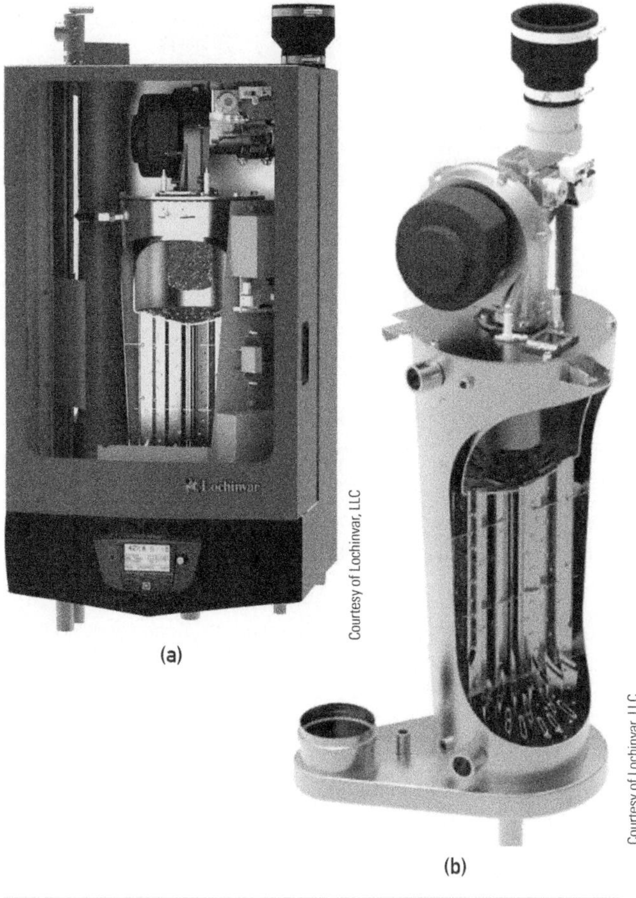

Figure 4-4 | (a) Example of a wall-hung mod/con boiler with vertical fire-tube heat exchanger. (b) Internal combustion system.

it would have a turndown ratio of 5:1 (e.g., 5 being the reciprocal of 20 percent expressed as a decimal number).

Figure 4-6 shows how the heat output of a modulating boiler with a turndown ratio of 5:1 will vary as the boiler attempts to match a changing heating load.

Figure 4-5 | Surface combustion on a stainless steel burner head within a modulating/condensing boiler.

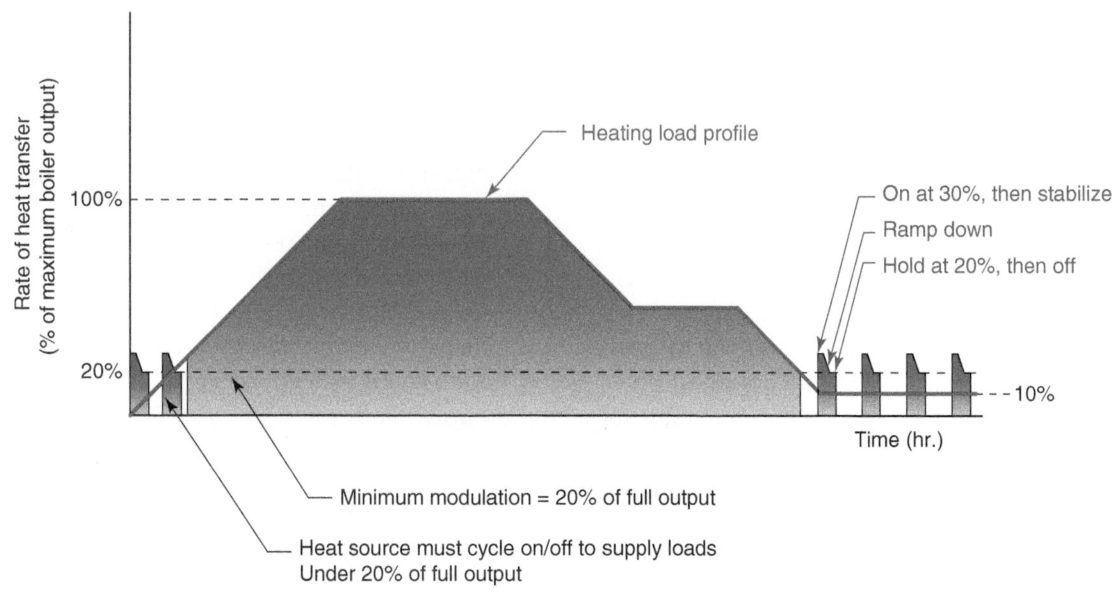

Figure 4-6 | Heat output rate of modulating/condensing boiler as it attempts to track a hypothetical heating load profile.

The blue line in Figure 4-6 is called a **heating load profile**. It represents the rate at which heat is required by the load over a period of several hours. In this case, the heating load begins at zero, steadily increases to a maximum value, and then reduces in various slopes and plateaus over time, eventually stabilizing at 10 percent of maximum load.

The red shaded areas represent the heat output from the modulating boiler. The pulses seen near the left and right extremes of the load profile result from the boiler turning on at approximately 30 percent output, then quickly reducing it heat output to the lowest stable output rate of 20 percent, and finally turning off for a short time. Such operation is typical of how current generation mod/con boilers respond to very low heating loads.

The large red shaded area shows how the mod/con boiler accurately tracks the heating load whenever it is 20 percent or more of the maximum load. This is ideal. The rate of heat output from the boiler is precisely matching the heating load.

When the heating load drops below the minimum modulation rate (in this case 20 percent), the burner must again cycle on and off to avoid supplying excess heat to the load. When these on/off cycles occur frequently they create a condition called **short cycling**. Such operation is always undesirable. It causes excessive wear on components such as the boiler's ignition system. Short cycling also results in higher emissions, and low thermal efficiency. Fortunately, it can be avoided through proper system design, which is discussed in later chapters.

DEALING WITH CONDENSATE

All condensing boilers must have means of capturing the condensate produced within them, and routing it to a suitable drain.

In some cases, local codes require the condensate to be neutralized from its initial acidic condition before it is released into a drain. This is done by passing the condensate through a **neutralizer** filled with an alkaline material such as crushed limestone chips or **magnesium hydrolite granules**. The reaction between the acidic condensate and alkaline material allows the condensate leaving the neutralizer to have a pH close to 7.0 (e.g., neutral). The level of alkaline fill in the neutralizer is checked annually and refilled as necessary. An example of a mod/con boiler equipped with a neutralizer is shown in Figure 4-7.

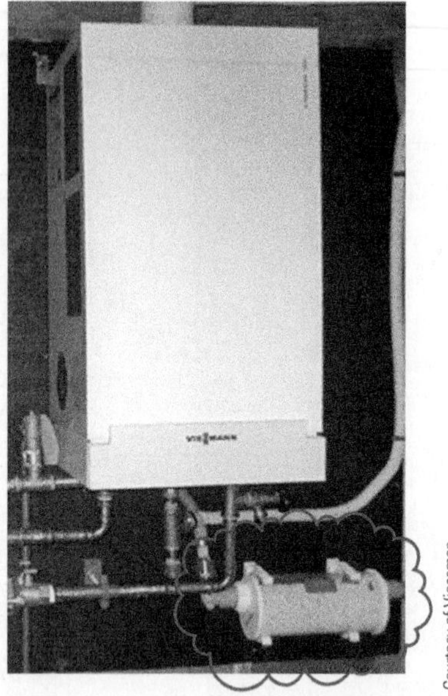

Figure 4-7 | Condensate neutralizer shown installed beneath mod/con boiler.

CONVENTIONAL VERSUS CONDENSING: WHICH SHOULD YOU USE?

There is no one answer to this question. In each situation, the answer depends on the type of hydronic distribution system the boiler will supply, the installation cost difference versus a conventional boiler, and energy savings that would eventually recoup any additional installation cost. The following issues should be weighed when considering a condensing boiler.

- Condensing boilers are significantly more expensive than conventional boilers of equal capacity. Their economic viability depends on recovering this higher installation cost through fuel savings, within a reasonable time. The best results are achieved when condensing boilers are used in situations where they can run at or near peak thermal efficiency (e.g., within condensing mode) most of their operating time. Low-temperature hydronic distribution systems consistently provide such operating conditions. *Furthermore, low water distribution systems are critical to achieving high thermal performance from any of the renewable energy heat sources discussed in this text.* Thus, mod/con boilers are well suited as auxiliary heat sources in system using renewable energy as the primary heat source. To ensure good performance, *the author recommends that any hydronic distribution system supplied by a renewable energy heat source, or a mod/con boiler, be configured to deliver full design load output using a maximum supply water temperature of 120 °F.* This recommendation will be emphasized throughout this text.

- If a condensing boiler will be used, the installation must provide for unattended condensate drainage. In a typical residential system, several gallons of condensate can be formed each day during cold weather. Never assume that it will simply evaporate or run down through cracks or holes in a concrete floor slab. In situations where no floor drain is available, a float-operated condensate pump may be required.
- Condensing boilers cannot be vented into existing masonry flues. The moist/acidic exhaust products can quickly deteriorate clay flues. Furthermore, the low-temperature exhaust stream may not create sufficient draft to properly carry exhaust products up a cold masonry flue. This could create leakage of toxic exhaust gases into the building.
- Only materials listed by the boiler manufacturer should be used for venting. Polypropylene, CPVC, and AL29-4C stainless steel vent pipe are generally acceptable. Galvanized steel flue pipe should never be used with condensing boilers. Solid-core schedule-40 PVC piping is currently listed as an acceptable venting material by some boiler manufacturers, but specifically excluded by others. Foam-core PVC should never be used for venting purposes. The author recommends against the use of *any* PVC pipe for venting boilers.
- Designers should also consider the head loss created as system fluid flows through the boiler. Some mod/con boilers, with compact heat exchangers, require a relatively powerful dedicated circulator. This increases installation cost as well as operating cost. Over the life of the system, the higher operating cost of an additional circulator, or a more powerful circulator, can often grow to several times the added installation cost. Boilers with low head loss characteristic are preferred whenever possible.

4.3 Hydronic Piping, Fittings, and Valves

A wide variety of piping materials are available for hydronic-based heating systems. They include metal, polymer, and composite pipes, fittings, and valves. This section gives a brief overview of the most common and appropriate piping materials. A more thorough description can be found in chapter 5 of *Modern Hydronic Heating,* 3rd edition.

COPPER TUBING

The type of tubing commonly used in residential and light commercial hydronic systems is called **copper water tube**. In the United States, copper water tube is manufactured according to the ASTM B88 standard. In this category, pipe size refers to the **nominal inside diameter** of the tube. The word "nominal" means that the measured inside diameter is approximately equal to the stated pipe size. For example, the actual inside diameter of a nominal 3/4" type M copper tube is 0.811 inch.

The outside diameter of copper water tube is always 1/8 inch (0.125 inch) larger than the nominal inside diameter. For example, the outside diameter (O.D.) of 3/4-inch type M copper tube is 0.875 inch. This is exactly 7/8 inch, or 1/8 inch larger than the nominal pipe size.

Copper water tube is available in three wall thicknesses designated as types K, L, and M in order of decreasing wall thickness. The outside diameters of K, L, and M tubing are identical. This allows all three types of tubing to be compatible with the same fittings and valves.

Because the operating pressures of residential and light commercial hydronic heating systems are relatively low, the thinnest-wall copper tubing (type M) is most often used. This wall thickness provides several times the pressure rating of other common hydronic system components, as can be seen in Figure 4-8.

Unless local codes require otherwise, type M is the standard for copper water tube used in hydronic heating systems. Some codes may require type L copper water tube for piping carrying **domestic water**. Type K is usually only required for directly buried underground piping.

Two common hardness grades of copper water tube are available. Hard drawn tubing is supplied in straight lengths of 10 and 20 feet. Because of its straightness and strength, hard drawn tubing is the most commonly used type of copper water tube in hydronic systems. So-called "soft temper" tubing is annealed during manufacturing to allow it to be formed with simple bending tools. It is useful in situations where awkward angles do

Nominal size	Inside diameter (inches)	Outside diameter (inches)	Rated working pressure (psi)*	Internal volume (gallons/foot)
3/8″	0.450	0.500	855 (456)	0.008272
1/2″	0.569	0.625	741 (395)	0.01319
3/4″	0.811	0.875	611 (326)	0.02685
1″	1.055	1.125	506 (270)	0.0454
1.25″	1.291	1.375	507 (271)	0.06804
1.5″	1.527	1.625	497 (265)	0.09505
2″	2.009	2.125	448 (239)	0.1647
2.5″	2.495	2.625	411 (219)	0.2543
3″	2.981	3.125	380 (203)	0.363

* Pressure ratings are based on operating temperatures not exceeding 200 °F. The first number is for drawn (hard) tubing. The number in ( ) is for annealed (soft temper) tubing.

Figure 4-8 | Physical data for selected sizes of type M copper water tube.

not allow proper tubing alignment with standard fittings. Soft temper tubing comes in flat coils having standard lengths of 60 and 100 feet. The minimum wall thickness available in soft temper copper tubing is type L.

The high temperature and pressure ratings of copper tubing make it one of the two most common piping for connecting solar thermal collectors. It is also commonly used for straight piping runs in mechanical rooms as well as for distribution piping through buildings.

Copper tubing has relatively good resistance to corrosion due to contact with water containing dissolved oxygen. However, it is not immune from all corrosion. The acids formed by glycol-based antifreeze that has thermally degraded due to extreme temperatures inside solar collectors can aggressively corrode copper tubing.

Copper water tube can be joined to fittings and valves by **soft soldering**, **brazing**, and **press fitting**. Soft solder currently remains the most common method. In most hydronic systems, soft soldering with 50/50 tin/lead solder is acceptable from a pressure and temperature standpoint. However, tin/lead solder cannot be used in any piping carrying domestic water. Only code-approved non-lead solders are suitable for such piping.

When copper piping is used to connect to solar collectors, or join piping within 5 feet of collectors, a higher temperature solder, such as 95/5 tin/antimony, should be used. Under stagnation conditions, it is possible for collector piping connections and nearby piping to reach temperatures that could soften 50/50 tin/lead solder.

Soft soldering requires the use of a **flux** to chemically clean the surfaces of the fittings and copper tube that will be metallurgical joined. Many fluxes contain acids. Such fluxes should be used sparingly and completely flushed out of the completed piping circuits before the system is filled with its final operating fluid. Residual soldering flux on the outside of tubing, fittings, or valves should be removed using detergent and water.

FITTINGS FOR COPPER TUBING

Many of the pipe fittings used to solder copper tubing together in hydronic systems are the same as used in domestic water supply systems. They include:

- Couplings (standard and reducing)
- Elbows (90° and 45°)
- Street elbows (90° and 45°)
- Tees (standard and reducing)
- Threaded adapters (male and female)
- Unions

The cross sections of most of these common pipe fittings are shown in Figure 4-9.

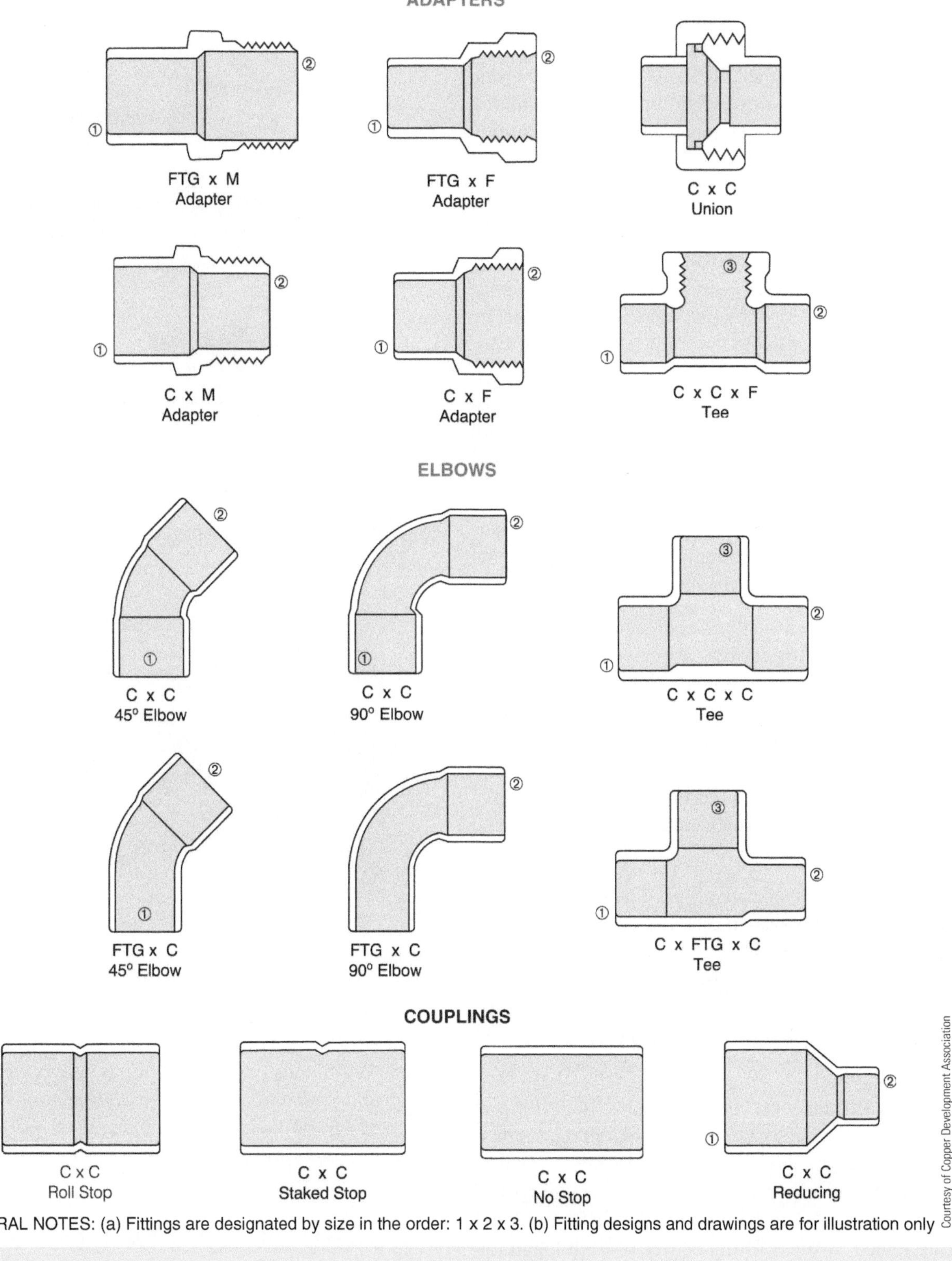

Figure 4-9 | Cross sections of several common fittings used with copper tubing.

Fittings for copper water tube are available in either wrought copper or cast brass. Both are suitable for use in hydronic systems. During soldering, wrought copper fittings heat up faster than cast brass fittings. They also have smoother internal surfaces. Of the two, wrought fittings are more typical in the smaller pipe sizes used in residential and light commercial systems.

Fittings are manufactured with solder-type socket ends, threaded ends, or a combination of the two. Standard sockets for solder-type joints create a gap of between 0.002 inch and 0.005 inch between the outside of the tube and the inside of the fitting. This gap allows solder to flow between the tube and fitting by capillary action.

The standard designation for solder-type sockets is the letter C. For example, a fitting designated as 3/4-inch C × C indicates that both ends of the fitting have solder-type sockets to match 3/4-inch copper water tube. All fitting manufacturers in North America use standardized dimensions and tolerances in constructing fittings to ensure compatibility with tubing. Because the outside diameter of types K, L, and M copper tubing are the same, the same fittings work with all three types.

In the United States, threaded pipe fittings use standardized national pipe thread (NPT). External threads are designated as male threads (**MPT** or M). Internal threads are designated as female threads (**FPT** or F).

Brazing is often used in the manufacturing of solar collectors and refrigeration systems built of copper. However, it is not commonly used for on-site joining of copper water tube. Brazing requires the use of an oxy-acetylene torch and specific brazing alloys. The temperature and pressure ratings of brazed joints are typically well above the operating requirements of piping circuits in residential and light commercial hydronic systems.

Press fitting is a relatively new method of joining copper tubing. This system uses special fittings containing elastomer (EPDM) O-rings that are mechanically compressed against the tube wall using an electromechanical tool to form pressure-tight joints. An example of a "press fit" joint is shown in Figure 4-10. Figure 4-11 shows the compression tool in use.

Press fitting does not require that the tubing or fitting be mechanically cleaned before joining. Furthermore, no heat or solder are required. However, the tube ends should still be reamed to remove any burrs due to cutting. It's also important that the surface of tubes being joined by press fitting be free of deep longitudinal scratches.

A wide variety of fittings and valves are now available for press fit joints in hydronic systems. Designers should always verify that the temperature and pressure ratings of press fit joints equal or exceed those that will be experienced in the system under the most severe conditions. *In particular, the stagnation temperature of solar collectors and the possible maximum temperature of wood-fueled boiler connections may exceed the temperature/pressure ratings of some press fit systems.*

Figure 4-10 | Example of a press fit joint in copper tubing using a mechanically compressed O-ring fitting.

Figure 4-11 | A press fit joint being made using a cordless compression tool.

CORRUGATED STAINLESS STEEL TUBING (CSST)

Many residential solar domestic water-heating systems benefit from hardware that allows fast and simple installation. One product that is especially well suited for such applications is **corrugated stainless steel tubing (CSST)**.

Constructed of type 316L stainless steel, CSST tubing can operate at temperatures as high as 350 °F and pressures up to 150 psi. It is typically sold as a pre-insulated twin-tube coil, as shown in Figure 4-12.

Pre-insulated coils of CSST can be supplied in lengths up to 75 feet. They are available in nominal tube sizes of 1/2 inch, 3/4 inch, 1 inch, 1.25 inches, 1.5 inches, and 2 inches.

The properties of CSST meeting the ASTM A240/A240M standard are listed in Figure 4-13.

Many manufacturers embed a light-gauge (18 AWG) shielded electrical cable in the pre-insulated harness that carries the CSST tube set. Its purpose is to connect the temperature sensor at the solar collector array to the system's controller, which is usually located near the other end of the tubing harness. Thus, a single flexible harness provides all the required piping and electrical connections between the collector array and the remainder of the system. This simplifies and speeds installation. The harness insulation is easily split with a knife when necessary to connect the supply and return tubes to the collector array, and other portions of the system.

Manufacturers of CSST use different methods for connecting it to standard threaded or soldered fittings. One supplier provides a system that requires a specialized tool to cut and shape the end of the CSST. This tool, seen in Figure 4-14, provides a square cut surface at the base of the corrugation and creates a flat sealing surface against which a fiber washer rests.

A hex nut is then slipped over the end of the tubing, and a hinged stainless steel retainer ring is placed into the corrugation directly behind the flattened end

Figure 4-12 | A 50-foot length of pre-insulated twin-tube CSST with embedded sensor cable.

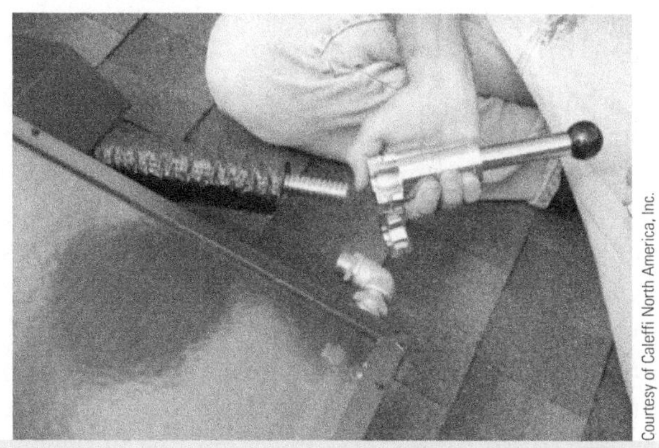

Figure 4-14 | End forming tool for CSST.

Nominal tube size (inch)	Outside diameter (inch)	Wall thickness (inch)	Inside diameter (inch)	Max. operating pressure (psi)	Volume (gallons/ft.)	Min. bend radius (uninsulated) (inches)
1/2	0.785	0.009	0.632	150	0.019	1
3/4	0.920	0.009	0.752	150	0.028	1
1	1.255	0.009	1.037	150	0.053	3
1.25	1.495	0.010	1.250	150	0.075	3
1.5	1.745	0.011	1.478	140	0.109	3
2	2.320	0.011	1.978	65	0.183	4

Figure 4-13 | Physical data for CSST conforming to ASTM A240/A240M.

surface. This ring prevents the nut from coming off the end of the tube as it is being tightened.

A fiber washer is placed between the flattened end of the tube and a brass or stainless steel fitting the tube connects to, as seen in Figure 4-15. A set of wrenches is used to tighten the nut to the fitting. This compresses the fiber washer, resulting in a pressure-tight joint.

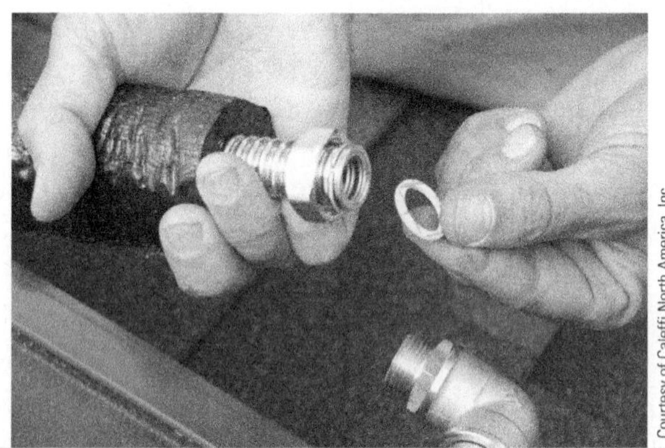

Figure 4-15 A fiber washer is inserted between the flattened surface of the CSST, and a brass fitting to which the CSST will be connected.

CSST has several advantages over rigid metal tubing when used to connect a small array of solar thermal collectors. First, the flexible pre-insulated twin tube set with embedded sensor wire allows for fast and simple installation, especially in retrofit applications. Second, all joints are created using a set of wrenches, eliminating any need for soldering. This is especially helpful in cramped attics, or on roofs during cold or windy weather. Eliminating soldering also eliminate the need for soldering flux, which must be flushed out of the tubing before the system is filled with its operating fluid. Third, CSST is supplied with all tubing ends capped. With reasonable care during installation, it provides a very clean piping system between the collector array and the balance of the system. This greatly reduces the potential of dirt entry into the circuit. Finally, CSST is capable of handling the potentially high temperatures and pressure encountered in some solar thermal systems, especially during **stagnation conditions**, when a collector is exposed to bright sunlight, but has no fluid flowing through it. CSST also provides good corrosion resistance if the antifreeze fluid passing through it becomes acidic due to thermal degradation.

One *undesirable* characteristic of CSST is its relatively high head loss characteristic relative to smooth tubing, as illustrated in Figure 4-16. The higher the head loss, the greater the circulator power required to maintain a given flow rate. The greater the circulator power requirement, the higher the operating cost of the system.

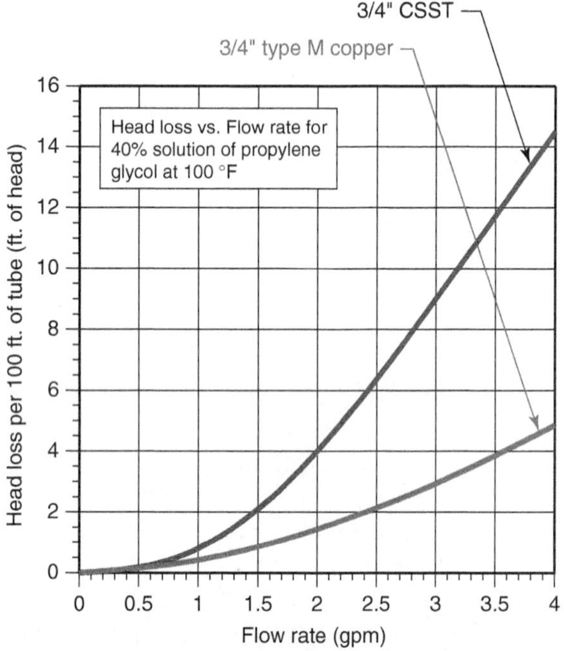

Figure 4-16 Comparison of head loss versus flow rate for 3/4-inch nominal size CSST and type M copper, when the circulating fluid is a 40 percent solution of propylene glycol and water at 100 °F.

Most experienced solar thermal designers are not comfortable specifying CSST in drainback solar applications. They prefer smooth tubing to ensure complete and expedient drainage of water from the collector array and its associated piping.

The analytical methods presented in Chapter 3 for determining the head loss of smooth tubing do not apply to CSST. Designers should use head loss versus flow data supplied by the CSST manufacturer.

CROSSLINKED POLYETHYLENE (PEX) TUBING

Cross-linked polyethylene tubing, commonly referred to as **PEX**, is a product that is now used extensively worldwide for a variety of hydronic heating applications. It is best known for its use in hydronic radiant panel heating systems as well as for hot and cold domestic water distribution in buildings. It is well suited for many applications involving renewable energy heat sources.

PEX has significantly higher pressure/temperature ratings than standard high-density polyethylene (**HDPE**) tubing. Figure 4-17 lists some physical data for PEX tubing that conforms to the widely accepted ASTM F876 standard.

With all plastic tubing, there is a trade-off between operating temperature and the corresponding maximum operating pressure. The ASTM F876 standard for PEX tubing establishes three simultaneous temperature/pressure ratings. The higher ratings for continuous service are 180 °F at 100 psi and 200 °F at 80 psi.

PEX tubing is sold in continuous coils ranging from 100 to more than 1,000 feet in length (depending on diameter and manufacturer). PEX is available in several diameters suitable for use in hydronic heating applications. The most commonly used nominal tube sizes range from 3/8 inch to 1 inch. Figure 4-18 shows samples of PEX tubing in sizes of 3/8 inch, 1/2 inch, 5/8 inch, and 3/4 inch.

The availability of long continuous coils, combined with the ability to bend around moderate curves without kinking, makes PEX tubing ideal for use in radiant panel heating systems. It also allows the tubing to run through confined or concealed spaces in buildings where working with rigid copper tubing would be difficult or impossible.

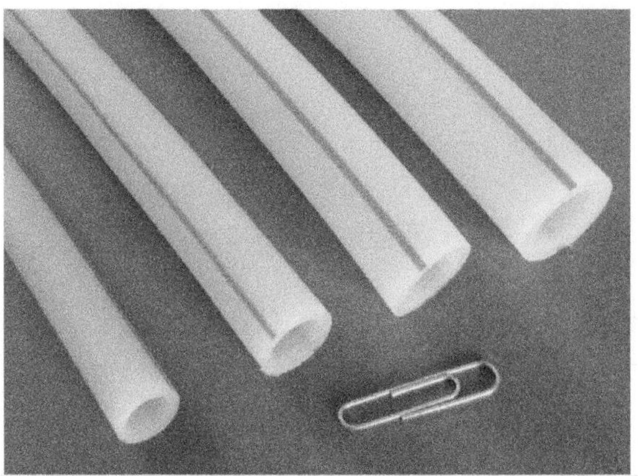

Figure 4-18 | PEX tubing in sizes of 3/8 inch, 1/2 inch, 5/8 inch, and 3/4 inch.

Specialized fittings are available from the tubing manufacturers to transition from PEX tubing to standard metal pipe fittings, both threaded and soldered. Examples of such fittings are shown in Figure 4-19. These fittings allow PEX tubing to be connected to heat emitters, or other system components, as well as used interchangeably with metal pipe.

It is the author's recommendation that all PEX tubing used in closed hydronic systems be specified with an **oxygen diffusion barrier**. This barrier is a thin layer of a special chemical that is laminated to the surface of the tubing, or in some cases incorporated as an interior layer. Its purpose is to reduce the diffusion of oxygen molecules through the tube wall and thus significantly reduce the potential for oxygen-based

Nominal tube size (inch)	Average outside diameter (inch)	Minimum wall thickness (inch)	Minimum burst pressure at 180 °F temperature (psi)	Internal volume (gallons/foot)
1/4	0.375	0.070	390	
3/8	0.500	0.070	275	0.005294
1/2	0.625	0.070	215	0.009609
5/8	0.750	0.083	210	0.01393
3/4	0.875	0.097	210	0.01894
1	1.125	0.125	210	0.03128
1.25	1.375	0.153	210	0.04668
1.5	1.625	0.181	210	0.06516
2	2.125	0.236	210	0.1116

Figure 4-17 | Physical data for several sizes of PEX tubing conforming to the ASTM F876 standard.

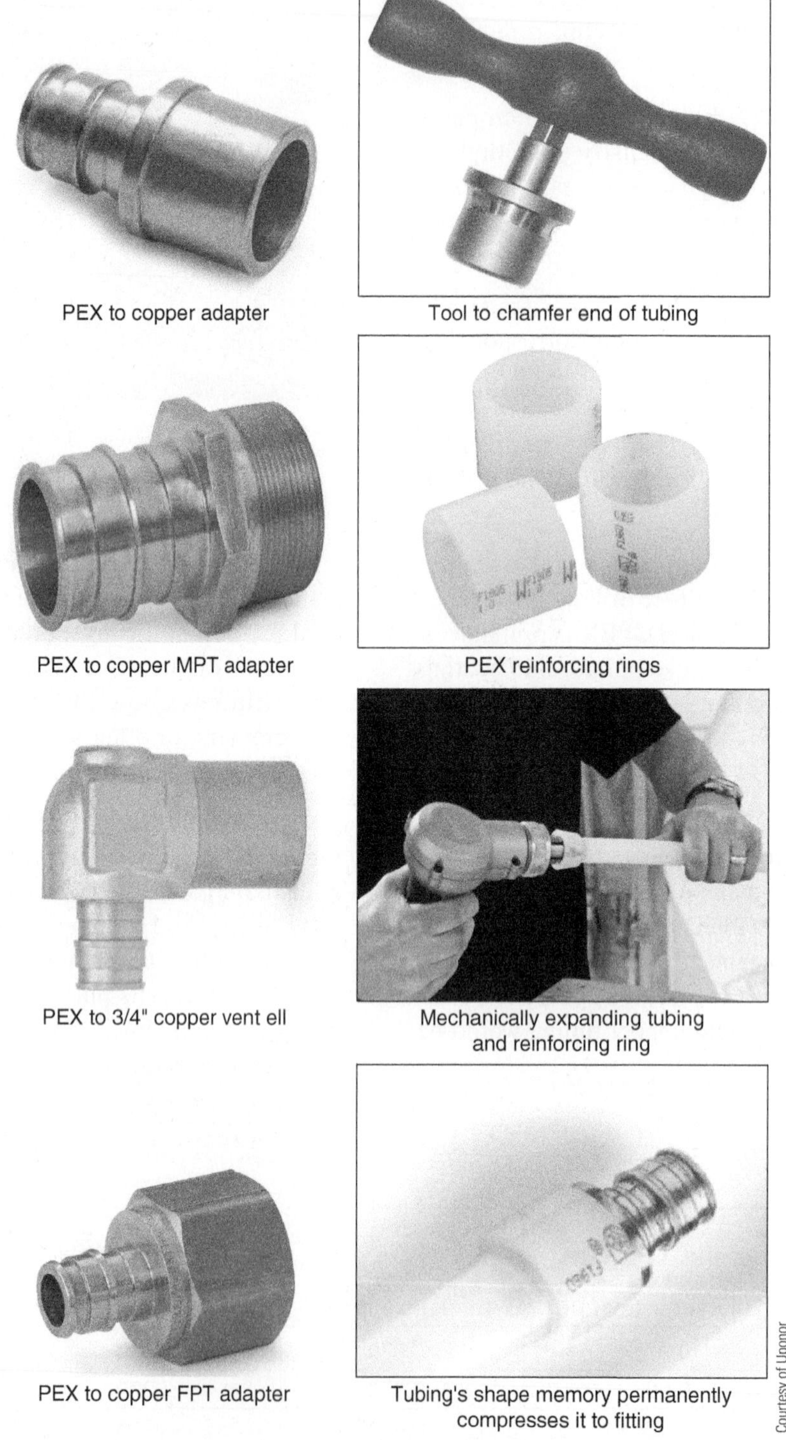

Figure 4-19 | Example of fittings and joining procedure used for PEX tubing.

corrosion within the system. All oxygen diffusion barriers should meet or exceed the requirements of the **DIN 4726 standard**. Use of non-barrier PEX tubing in hydronic systems containing ferrous metal components will result in corrosion and therefore should be avoided.

COMPOSITE (PEX-AL-PEX) TUBING

Another type of tubing that is well suited for hydronic heating applications is called composite **PEX-AL-PEX tubing**. It consists of three concentric layers bonded together with special adhesives. The inner and outer layers are PEX. The middle layer is longitudinally welded aluminum. A cross section of a PEX-AL-PEX tube is shown in Figure 4-20.

The PEX-AL-PEX tubing commonly used in hydronic heating systems conforms to the ASTM F1281 standard. Figure 4-21 lists some physical properties of PEX-AL-PEX tubing that meet this standard.

PEX-AL-PEX tubing has slightly higher temperature pressure ratings relative to PEX tubing. This is attributable to the added strength of the aluminum layer. At 180 °F, it is rated for pressures up to 125 psi. At 210 °F it is rated for pressures up to 115 psi. As with PEX, *the temperature ratings are not high enough to allow PEX-AL-PEX tubing to be used for piping circuits supplying solar collectors. For the same reason, PEX-AL-PEX tubing should not be connected directly to wood-fueled heat sources.*

Figure 4-20 | Cross section of PEX-AL-PEX tubing. Aluminum core is seen between inner and outer PEX layers.

The aluminum layer in PEX-AL-PEX tubing also significantly reduces its thermal expansion movement relative to other (all-polymer) tubes. This characteristic makes PEX-AL-PEX well suited for use in radiant panels where aluminum heat transfer plates are used as seen in Figure 4-22.

A wide variety of fittings are available for connecting PEX-AL-PEX tubing to itself as well as to copper tubing and other components such as valves and circulator flanges. Figure 4-23 shows some examples of these fittings as well as the steps used to join them to PEX-AL-PEX tubing.

THERMAL EXPANSION OF PIPING

All materials expand when heated and contract when cooled. In many hydronic systems, piping undergoes wide temperature swings as the system operates. This causes significant changes in the length of the piping

Courtesy of Harvey Youker

Figure 4-22 | Use of PEX-AL-PEX tubing in combination with aluminum heat transfer plates for radiant floor heating.

Nominal tube size (inch)	Minimum outside diameter	Minimum total wall thickness	Minimum aluminum layer thickness (inch)	Internal volume (gallons/foot)
3/8	12.00 mm/0.472″	1.6 mm/0.063″	0.007	0.00489
1/2	16.00 mm/0.630″	1.65 mm/0.065″	0.007	0.01038
5/8	20.00 mm/0.787″	1.90 mm/0.075″	0.010	0.01658
3/4	25.00 mm/0.984″	2.25 mm/0.089″	0.010	0.02654
1	32.00 mm/1.26″	2.90 mm/0.114″	0.011	0.04351

Figure 4-21 | Physical data for several sizes of PEX-AL-PEX tubing conforming to the ASTM F1281 standard.

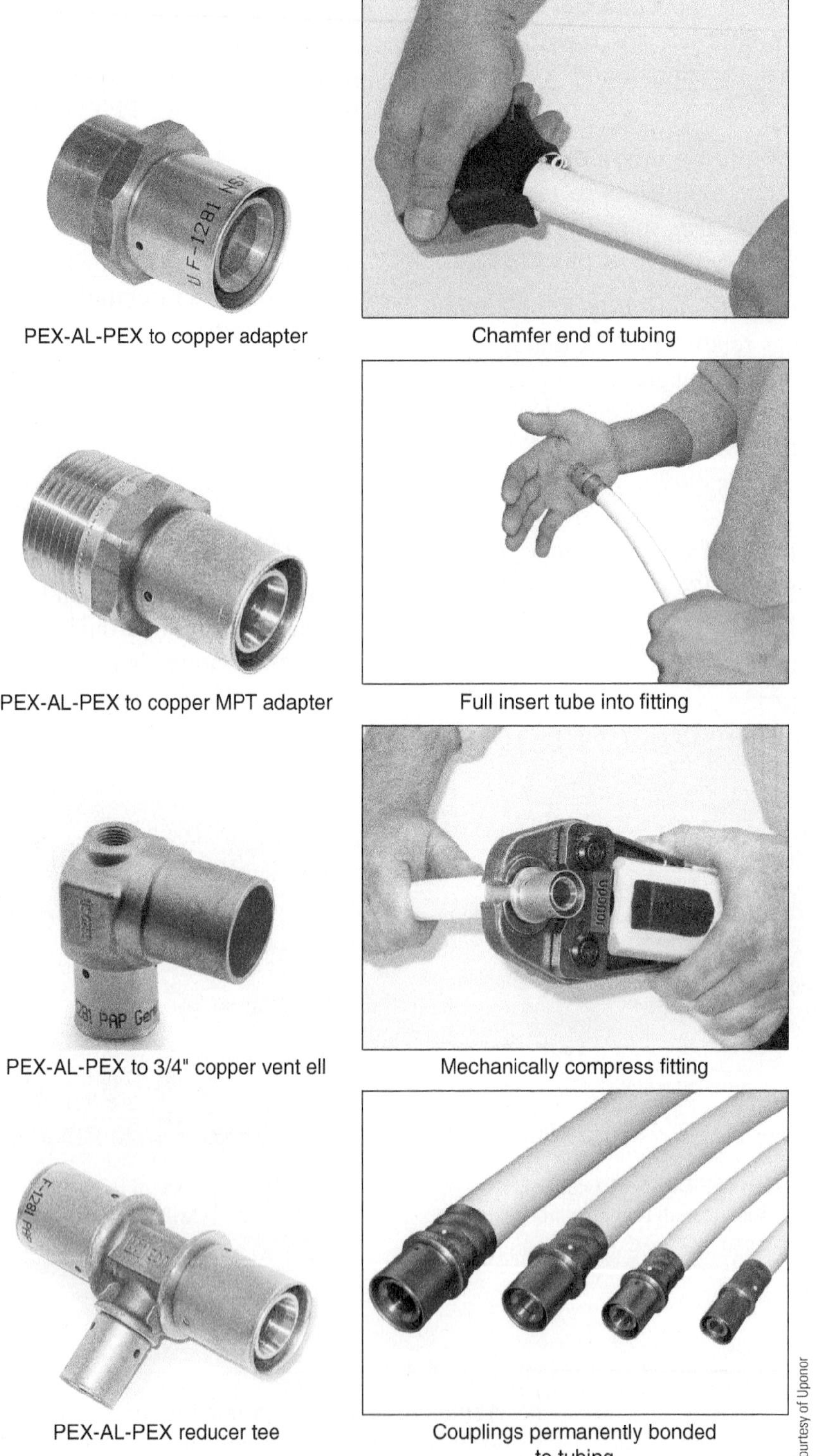

Figure 4-23 | Example of fittings and joining procedure used for PEX-AL-PEX tubing.

due to thermal expansion. The greater the length of the pipe, and the greater the temperature change, the greater the change in length. If the piping is rigidly mounted, this expansion can cause annoying popping or squeaking sounds each time the pipe heats up or cools down. In extreme cases, the piping can even buckle due to high compressive stresses that develop if the pipe is prevented from expanding.

The expansion or contraction movement of various types of tubing can be calculated using the following equation:

(Equation 4.1)

$$\Delta L = b(L)\Delta T$$

where:

ΔL = change in length due to heating or cooling (inches)

L = original length of the pipe before the temperature change (inches)

ΔT = change in temperature of the pipe (°F)

b = coefficient of linear expansion of tube material (see the following)

Tubing material	Coefficient of linear expansion (b) (inch/inch/°F)
Copper	0.0000094
PEX	0.000094
PEX-AL-PEX	0.000013

The smaller the coefficient of linear expansion, the less the tubing expands for a given change in temperature. The linear (lengthwise) expansion of the tube does not depend on the tube's diameter.

Example 4.1: Determine the change in length of a copper tube, 50 feet long, when heated from a room temperature of 65 °F to 200 °F. How does this compare to the change in length of a PEX and PEX-AL-PEX tube having the same length and undergoing the same temperature change?

Solution: The change in length of the copper tube is:

$$\Delta L = \left(0.0000094\frac{\text{inch}}{\text{inch}\cdot°\text{F}}\right)\times 50\text{ ft}\times\left(\frac{12\text{ inch}}{1\text{ foot}}\right)\times(200°\text{F}-65°\text{F}) = 0.76\text{ inch}$$

The change in length for PEX tube is:

$$\Delta L = \left(0.000094\frac{\text{inch}}{\text{inch}\cdot°\text{F}}\right)\times 50\text{ ft}\times\left(\frac{12\text{ inch}}{1\text{ foot}}\right)\times(200°\text{F}-65°\text{F}) = 7.60\text{ inch}$$

The change in length for PEX-AL-PEX tube is:

$$\Delta L = \left(0.0000013\frac{\text{inch}}{\text{inch}\cdot°\text{F}}\right)\times 50\text{ ft}\times\left(\frac{12\text{ inch}}{1\text{ foot}}\right)\times(200°\text{F}-65°\text{F}) = 1.05\text{ inch}$$

Discussion: The only difference in these calculations is the value of the coefficient of linear expansion for the different tubes. Notice that PEX tubing expands about 10 times as much as the copper tube for the same length and temperature change. PEX-AL-PEX tubing expands about 38 percent more than the copper, but significantly less than PEX. The expansion of the PEX-AL-PEX tubing is largely controlled by the aluminum core layer. Because the values for coefficient of expansion are very small, be careful to use the correct number of zeros when entering the number into calculations.

Regardless of the tubing used, expansion movement must be accommodated during installation, especially on long straight runs. In many small hydronic systems, the piping is not rigidly supported, or contains a sufficient number of elbows and tees to absorb the expansion movement without creating excessive stress in the pipe. Such situations only require piping supports that allow the tubing to expand and contract freely to prevent expansion noises.

When long straight runs of rigid tubing are required, it may be necessary to install an expansion compensator capable of absorbing the movement. A variety of such expansion compensator fittings are available. Those based on flexible tubing or reinforced hose require little if any maintenance. An example of a U-shaped compensator is shown in Figure 4-24. Similarly shaped compensators are available in other materials such as steel and PEX.

For any expansion compensator to work properly, the piping must be properly supported using a combination of fixed-point and sleeve or roller supports. Fixed-point supports are those that rigidly grab the pipe and do not allow movement between the support and pipe. Sleeve or roller supports support the weight of the pipe, but do not inhibit expansion movement. Figure 4-25 shows an example of how the expansion compensator must absorb all expansion movement between fixed-point supports.

The ability of an expansion compensator to absorb movement depends on the stress the piping is allowed to develop. Different materials and different pipe sizes affect this stress value. Manufacturers of expansion compensators list the amount of movement each compensator can safely accommodate.

Designers should keep in mind that cooling piping below the temperature at which it was installed will cause it to contract. Equation 4-1 can be used to determine the extent of this contraction based on the piping material and how much its temperature drops below that at which it was installed. Thermal contraction will occur in piping connected to solar thermal collectors, or within chilled water-cooling systems. It could also occur in a heating system application if the system was not operating during cold weather. Expansion compensators should be selected so they can accommodate pipe contraction due to cooling along with expansion due to heating.

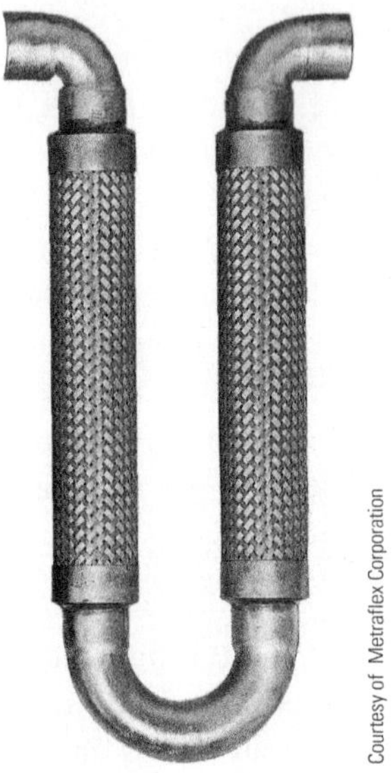

Figure 4-24 | U-type expansion compensator for copper tube.

COMMON VALVES

There are many types of valves used in hydronic-based renewable energy systems. Some are common designs used in all categories of piping, including hydronic, steam, plumbing, and pneumatics. Others are designed for a very specific function. The proper use of valves can make the difference between an efficient, quiet, and easily maintained system, and one that wastes energy, creates objectionable noise, or even poses a major safety threat. It is vital that hydronic system designers understand the proper selection and placement of several types of valves.

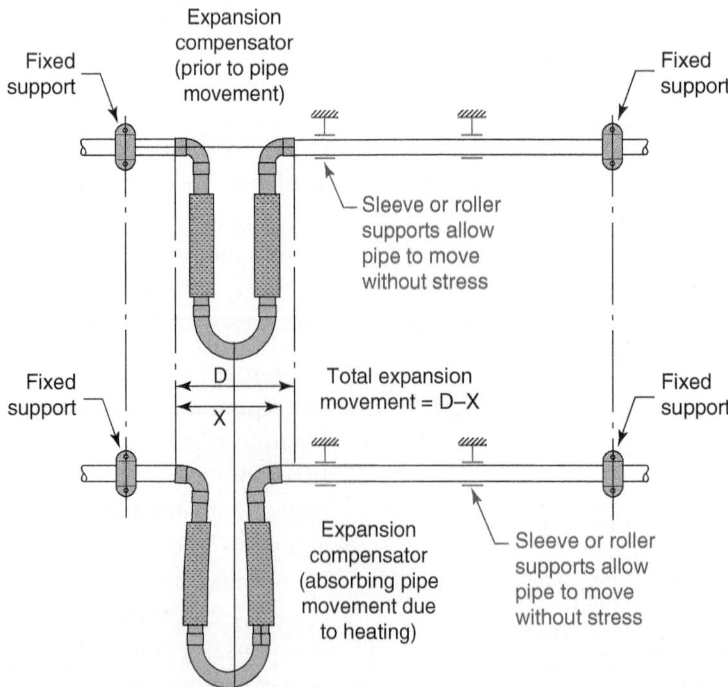

Figure 4-25 | Expansion movement absorbed by expansion compensator. Piping does not move at fixed support points.

Most of the common valve types are designed for one of the following duties:

- **Component isolation**
- **Flow regulation**

Component isolation refers to the use of valves to close off the piping connected to a device that may have to be removed or opened for maintenance. Examples include circulators, boilers, heat exchangers, and strainers. By placing valves on either side of such components, only minimal amounts of system fluid need to be drained or spilled when a component is removed for service. In many cases, other portions of the piping system can remain in operation during this time.

Flow regulation requires a valve that can be set to maintain a given flow rate within a piping system, or portion thereof. An example would be adjusting the flow rates in parallel piping branches. Valves used for flow regulation are specifically designed to remove mechanical energy (e.g., head) from the flowing fluid. This causes the fluid's pressure to drop as it passes through the valve. The greater the pressure drop, the slower the flow rate through the valve.

Figure 4-26 | A typical bronze gate valve.

GATE VALVES

Gate valves are designed specifically for component isolation. As such, *they should always be fully open or fully closed.* An example of a typical bronze gate valve is shown in Figure 4-26. When the handwheel is turned clockwise, an internal metal wedge moves downward from a chamber in the upper body until it totally closes off the fluid passage. Near the end of its travel, the wedge seats tightly against the lower body forming a pressure-tight seal.

In its fully open position, the valve's wedge is completely retracted into the body. In this position, the valve creates very little interference with the fluid stream moving through it. This results in minimal loss of pumping energy during normal system operation. This is very desirable considering that a typical isolation valve remains open most of its service life.

Gate valves should never be used to regulate flow. If the wedge is set to a partially open position, vibration and chattering are likely to occur. Eventually these conditions can erode the machined metal surfaces, possibly preventing a drip-tight seal when the valve is closed.

It may be necessary to occasionally tighten the packing gland of the valve to prevent minor leakage around the stem. A slight "snugging" of the packing nut is usually sufficient. Never over tighten the packing nut of any valve.

GLOBE VALVES

Globe valves are specifically designed for flow regulation. An example of a typical globe valve is shown in Figure 4-27.

Although their external appearance closely resembles that of a gate valve, there are major internal differences. The fluid's path through a globe valve contains several abrupt changes in direction. The fluid passes through a gap, the size of which is determined by the position of the valve stem. The smaller this gap, the greater the pressure drop through the valve. Always install globe valves so the fluid flows into the lower body chamber and upward toward the disc. Reverse flow through a globe valve can cause unstable flow regulation and noise. All globe valves have an arrow on their body that indicates the proper flow direction through the valve.

Courtesy of NIBCO, Inc.

Figure 4-27 | A typical bronze globe valve.

Globe valves should never be used for component isolation. The reason is that a fully open globe valve still removes considerably more head energy from the flowing fluid compared to a fully open gate valve. Therefore, during the thousands of operating hours when component isolation is *not* needed, the globe valve unnecessarily wastes head energy. Using a globe valve for component isolation is like driving a car with the brakes partially applied. It may be possible, but it is certainly not efficient.

BALL VALVES

A **ball valve** uses a machined spherical plug, known as the ball, as its flow control element. The ball has a large hole through its center. As the ball is rotated through 90 degrees of arc, the hole moves from being parallel with the valve ports (its fully open position), to being perpendicular to the ports (its fully closed position). Ball valves are equipped with lever-type handles rather than handwheels. The position of the lever relative to the centerline of the valve indicates the orientation of the hole through the ball. An example of a partially cut away ball valve is shown in Figure 4-28.

When the size of the hole in the ball is approximately the same as the pipe size, the valve is called a

Courtesy of Legend Hydronics

Figure 4-28 | External appearance of a full-port ball valve.

***full port* ball valve**. If the size of the hole is smaller than the pipe size, the valve is called a ***standard port* ball valve**.

Full port ball valves are ideal for component isolation. In their fully open position, they create very little flow interference or head loss.

CHECK VALVES

In many systems, it's necessary to prevent fluid from flowing backward through a pipe. A check valve, in one of its many forms, is the common solution.

Check valves differ in how they prevent reverse flow. The ***swing check***, shown in Figure 4-29, contains a disc hinged along its upper edge. When fluid moves through the valves in the allowed direction (as indicated by an arrow on the side of the valve's body), the disc swings up into a chamber out of the direct path of the fluid. The moving fluid stream holds it there. When the fluid stops, or attempts to reverse itself, the disc swings down due to its own weight and seals across the opening of the valve. The greater the back pressure, the tighter the seal.

Courtesy of NIBCO, Inc.

Figure 4-29 | A swing-check valve.

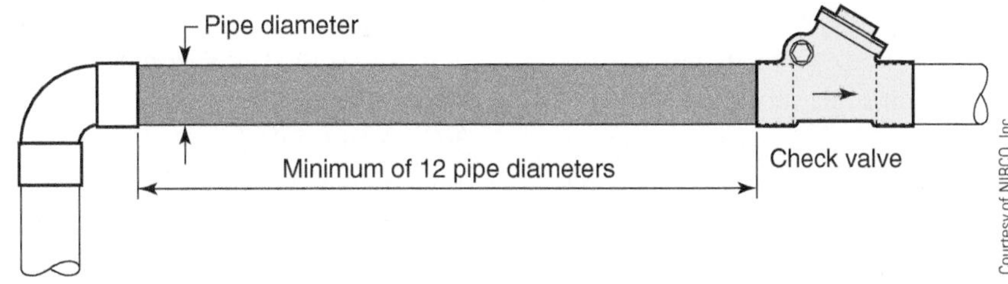

Figure 4-30 | Install at least 12 pipe diameters of straight pipe upstream of a check valve to prevent rattling sounds due to turbulence.

Figure 4-31 | Spring-loaded check valve.

For proper operation, swing check valves must be installed in horizontal piping with the bonnet of the valve in an upright position. Installation in other orientations can cause erratic operation, creating dangerous water hammer effects. It is also important to install swing check valves with a minimum of 12 pipe diameters of straight pipe upstream of the valve, as shown in Figure 4-30. This allows turbulence created by upstream components to partially dissipate before the flow enters the valve. Failure to do so can cause the valve's disk to rattle.

A spring-loaded check valve is designed to eliminate the orientation restrictions associated with a swing check. These valves rely on a small internal spring to close the valve's disc whenever fluid is not moving in the intended direction. *The spring action allows the valve to be installed in any orientation.* The force required to compress the spring does, however, create slightly more pressure drop compared to a swing check. As with a swing check, be sure the valve is installed with the arrow on the side of its body pointing in the desired flow direction. An example of a spring-loaded check valve is shown in Figure 4-31.

As with swing checks, be sure to install at least 12 pipe diameters of straight pipe upstream of all spring-loaded check valves to dissipate incoming turbulence.

FEED WATER VALVES

A **feed water valve**, also sometimes called a **pressure-reducing valve**, is used to lower the pressure of water from a domestic water distribution pipe before it enters a hydronic system. This valve is necessary because most buildings have domestic water pressure higher than the relief valve settings of small hydronic systems. The feed water valve allows water to pass through whenever the pressure at its outlet side drops below its pressure setting. In this way, small amounts of water are automatically fed into the system as air is vented out, or as small amounts of fluid are lost through evaporation at valve packings, pump flange gaskets, and other locations. An example of a feed water valve is shown in Figure 4-32.

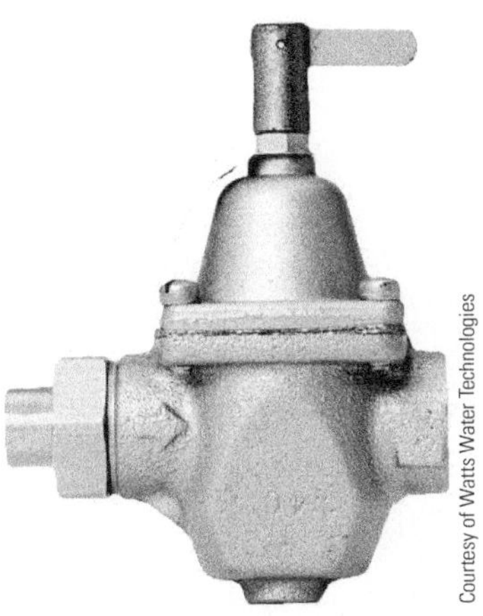

Figure 4-32 | Feed water valve. Notice flow direction arrow on body.

Over time, hydronic systems without feed water valves tend to lose pressure as a result of air venting and very small water losses through valves packing and pump flange gaskets. The resulting low-pressure operation can eventually cause noise and pump cavitation. In multistory buildings, the reduced pressure can also prevent circulation in upper parts of the system. The function of the feed water valve is especially important during the first few days of system operation as dissolved air is released from the fill water and vented out of the system.

The pressure setting of feed water valves can be varied by rotating a shaft or screw. Many feed water valves come preset for 12 psi system pressure. A later section in this chapter discusses how to properly adjust this pressure setting.

Some feed water valves also have a lever or knob at the top that allows the valve to be manually opened to quickly fill the system during start-up. This temporarily disables the pressure-regulating mechanism within the valve and allows water, driven by pressure in the building's plumbing system, to quickly enter the system and push air out of the piping. This action is called **purging** and is part of the commissioning procedure for nearly all closed hydronic circuits.

Most feed water valves are also equipped with removable strainers that prevent particulates in the supply water from entering the system. If the strainer becomes clogged, the valve may not supply water to the system. This is often evidenced by a slow drop in system pressure. *The strainer on a feed water valve should be checked during routine system maintenance.* Most feed water valves also have an internal check valve to prevent reverse flow. This check valve, in combination with an isolating ball valve on the supply side of the make up water assembly, allows the feed water valve to be isolated so the strainer can be removed, cleaned, and replaced with minimal water loss.

PURGING VALVES

Hydronic systems, once assembled, must be filled, purged of air, and filled with fluid. This is often accomplished at the same time by forcing fluid through the system at a high flow velocity. As this rapidly moving fluid progresses around the piping circuit it pushes much of the air ahead of it—like a piston moving through a cylinder. The water stream can also entrain air bubbles and drag them along. The air must exit the piping assembly as water goes in. A **purging valve** is specifically designed for this purpose.

Modern purging valves consist of two ball valve assemblies in a common body, as seen in Figure 4-33a. One ball valve is "inline" with the piping, while the other serves as a side outlet port. A purging valve is typically mounted within the piping system, as shown in Figure 4-33b.

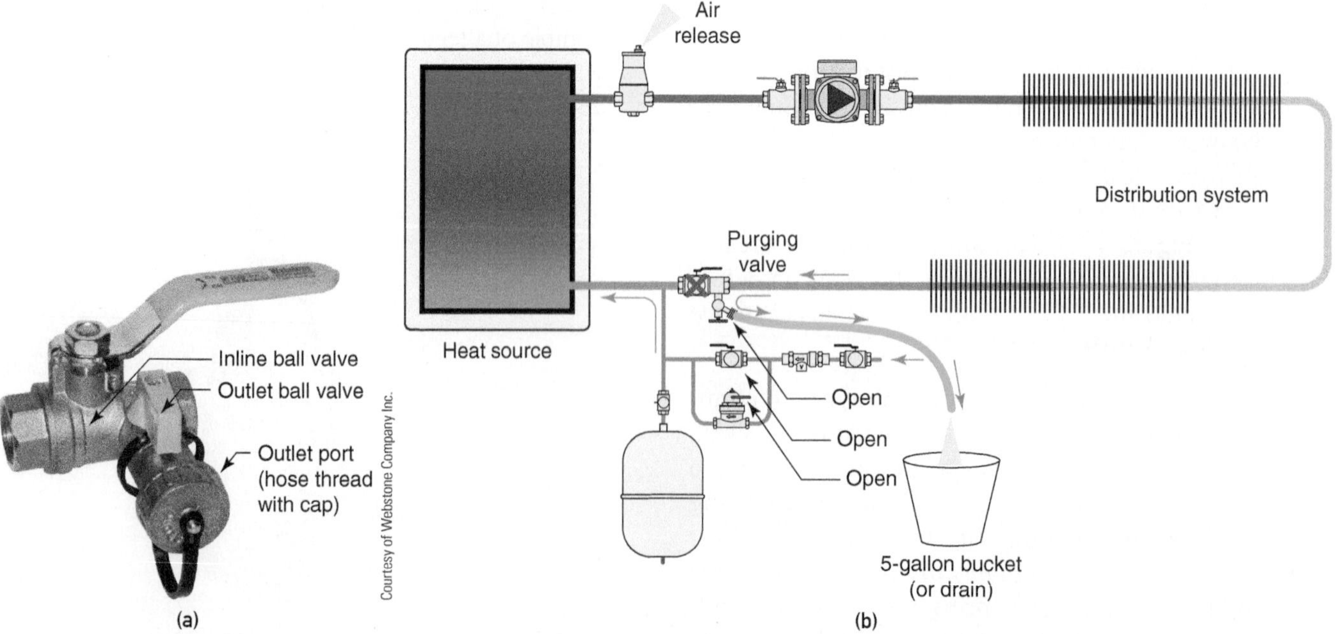

Figure 4-33 | (a) A modern purging valve. (b) Use of a purging valve to fill and flush air from a hydronic system.

During purging, the inline ball valve within the purging valve is closed and the outlet ball valve is opened. A hose connected to the outlet port of the purging valve routes exiting flow to either a pail or drain. The feed water valve is manually opened. In some systems, a separate ball valve mounted in parallel with the feed water valve is also opened. The objective is to force water through the system piping as fast as possible to efficiently capture and expel air in the piping and other system components. Once the fluid stream existing the outlet port of the purge valve is free of visible bubbles, the outlet ball valve is shut and the fast-fill settings of the make-up water assembly are turned off.

BACKFLOW PREVENTER

Consider a hydronic heating system filled with an antifreeze solution connected to a domestic water supply by way of a feed water valve. If the pressure of the domestic water system suddenly drops due to a rupture of the water main or other reason, the antifreeze solution could be pushed backward into the domestic water system by the pressurized air in the expansion tank. This could contaminate not only the water in the building, but neighboring buildings as well.

A **backflow preventer** eliminates the possibility. It consists of a pair of check valve assemblies in series, with an intermediate vent port that drains any backflow that migrates between the valve assemblies. An example of a small backflow preventer is shown in Figure 4-34.

Most plumbing codes require backflow preventers on all hydronic heating systems that are connected to domestic water piping. Even if local codes do not require this valve, it is still good practice to use it, especially if antifreeze or other chemical treatments might be added to the system in the future. A single check valve, or even two check valves in series, are *not* acceptable substitutes for a backflow preventer. In the latter case, if debris were to collect in both check valves, fluid could flow backward from the system into domestic water piping under certain conditions.

Figure 4-34 | Example of a small backflow preventer. Note flow direction arrow and vent port marking.

PRESSURE-RELIEF VALVE

A **pressure-relief valve** is a code requirement on any type of closed-loop hydronic heating system. It functions by opening at a preset pressure rating, allowing system fluid to be safely released from the system before higher pressures can develop. The pressure-relief valve is the final means of protection in a situation where all other controls fail to limit heat production. In its absence, some component in the system could potentially explode with devastating results.

Most mechanical codes require pressure-relief valves in any piping assembly that contains a heat source and is capable of being isolated by valves from the rest of the system. A pressure-relief valve is also required in any piping circuit supplied with heat through a heat exchanger.

Nearly all boilers sold in the United States are shipped with a factory installed ASME-rated pressure-relief valve. The typical pressure rating for relief valves on small boilers is 30 psi. Valves with higher-pressure ratings as well as adjustable pressure settings are also available.

The operation of a pressure-relief valve is simple. When the force exerted on the internal disc equals or exceeds the force generated by the internal spring, the disc lifts off its seat and allows fluid to pass through the valve. The spring force is calibrated for the desired opening pressure of the valve. This rated pressure, along with the maximum heating capacity of the equipment the valve is rated to protect, is stamped onto a permanent tag or plate attached to the valve as seen in Figure 4-35.

Figure 4-35 | A pressure-relief valve.

The lever at the top of a pressure-relief valve allows it to be manually opened. Such an opening may be required to remove air when the system is filled. Manually opening the valve to verify that it is not corroded or seized is also part of annual system maintenance.

All pressure-relief valves should be installed with their shaft in a vertical position. This minimizes the chance of sediment accumulation around the valve's disc. Such sediment, if allowed to accumulate, can interfere with proper seating of the valve's disc and cause the valve to leak.

All pressure-relief valves should also have a waste pipe attached to their outlet port. This pipe routes any expelled fluid safely to a drain, or at least down near floor level. The waste pipe must be the same size as the valve's outlet port, with a minimal number of turns and no valves or other means of shutoff. The waste pipe should end 6 inches above the floor or drain to allow for unrestricted flow if necessary. Be sure to check local codes for possible additional requirements on the installation of relief valves.

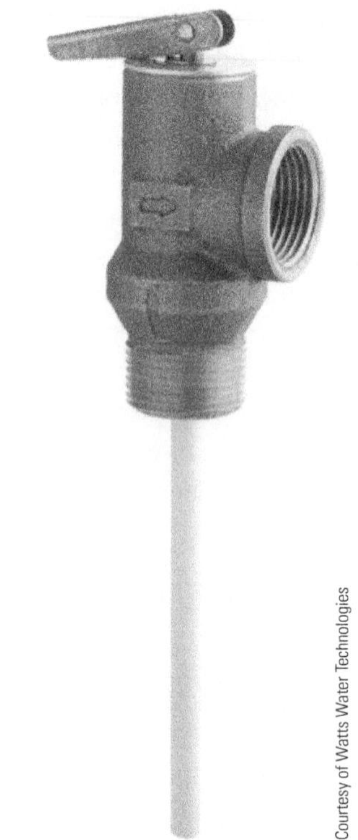

Courtesy of Watts Water Technologies

Figure 4-36 | Temperature and pressure (T&P) relief valve.

TEMPERATURE AND PRESSURE-RELIEF VALVE

Most tank-type domestic water heaters are required to be equipped with a **temperature and pressure (T&P) relief valve**. This valve has an internal spring that allows it to open at a specific pressure. The valve is also equipped with a temperature probe that is immersed in the heated water. A compound within this probe expands as the water is heated. This expansion is calibrated so that the valve opens at a temperature a few degrees below water's atmospheric boiling point. An example of a T&P relief valve is shown in Figure 4-36.

T&P relief valves are available in a variety of pressure ratings. The pressure rating must always be lower than the test pressure rating of the vessel the valve is protecting. In many cases, the pressure rating of a T&P valve is specified by local plumbing code.

The discharge port of any T&P relief valve should always be piped to a location where hot water discharge does not create a safety hazard. Ideally, the discharge piping would end about 6 inches above a floor drain. The discharge piping from a T&P relief valve must be the same size as the valve's port and should be routed to its final location using a minimum number of fittings. No other valves can be installed in this discharge piping. Other limitations may be imposed by local plumbing codes.

THERMOSTATIC RADIATOR VALVES

One benefit of hydronic heating is the ability to provide room-by-room temperature control. This can be done using electrically operated thermostats in each room, and wiring those thermostats to other flow-control hardware. However, "wireless" **thermostatic radiator valves (TRVs)** allow for room-by-room temperature control without the need for electrical power.

TRVs consist of a valve body matched with a thermostatic operator, as shown in Figure 4-37.

The nickel-plated brass valve body is available in pipe sizes from 1/2 inch to 1 inch, in either a straight or angle pattern. In the latter, the outlet port of the valve is rotated 90 degrees to the inlet port. Inside the valve is a plug mounted on a spring-loaded shaft. The plug is held in its fully open position by the force of the spring.

Figure 4-37 | A thermostatic operator ready to be attached to a radiator valve.

The fluid pathway through the valve is similar to that of a globe valve, thus making the valve well suited for accurate flow-rate control.

To close the valve, the shaft must be pushed inward against the spring force. No rotation is necessary. Shaft movement can be achieved manually using a knob that threads onto the valve body or automatically through use of a thermostatic operator.

Several types of thermostatic operators can be matched to radiator valve bodies. The most common configuration attaches the thermostatic operator directly to the valve body, as shown in Figure 4-37. The thermostatic operator body is aligned to the desired orientation and secured to the valve body by a threaded collar. It can be easily repositioned or removed if necessary.

The thermostatic operator contains a fluid in a sealed bellows chamber. As the air temperature surrounding the operator increases, the fluid expands inside the bellows. This expansion moves the shaft of the valve inward toward its closed position. This action decreases water flow rate through the heat emitter and thus reduces its heat output. As the room air temperature decreases, the fluid contracts, allowing the spring force to slowly reopen the valve and increase heat output from the heat emitter. Thus, TRVs provide fully modulating temperature control that continually adjusts flow through the heat emitter to maintain a constant (occupant-determined) room temperature.

TRVs are usually located on the *inlet* pipe supplying a heat emitter. The preferred mounting position is with the valve stem in a horizontal position, with the thermostatic operator facing away from the heat emitter, as shown in Figure 4-38. *TRVs should never be installed with the thermostatic operator above the heat emitter.* Doing so would cause the valve to close prematurely and greatly limit heat output from the heat emitter. Also be sure the valve is mounted so flow passes through it in the direction indicated by the arrow on the valve body.

Most manufacturers of thermostatic radiator valves also offer thermostat operators with a remote setpoint dial. This allows temperature adjustments to be made at normal thermostat height above the floor. A capillary tube runs from the setpoint dial assembly to the operator mounted on the radiator valve. Thermostatic operators can be ordered with capillary tubes up to 12 feet long. Figure 4-39 shows an example of a radiator valve fitted with a remote setpoint adjustment. Care must be taken not to kink or break the capillary tube during installation. Severing or severely kinking the capillary tube renders the operator useless.

MIXING VALVES

Many hydronic distribution systems, *especially those using intermittent heat sources such as solar collectors and wood-fueled boilers*, require water temperatures lower than what is available from the heat source. For example, a radiant ceiling heating system may require water at 110 °F, while the water in a thermal storage tank supplying this load is at 150 °F. In such cases, the lower water temperature can be created by blending cooler water returning from the distribution system with hot water from the heat source. The flow rate of each stream can be regulated by a mixing valve to attain the desired supply temperature.

There are several types of mixing valves used in hydronic heating systems. They include valves operated by electrically powered motorized **actuators** and valves operated by non-electric thermostatic operators. Both types of mixing valves have application in systems supplied by renewable energy heat sources.

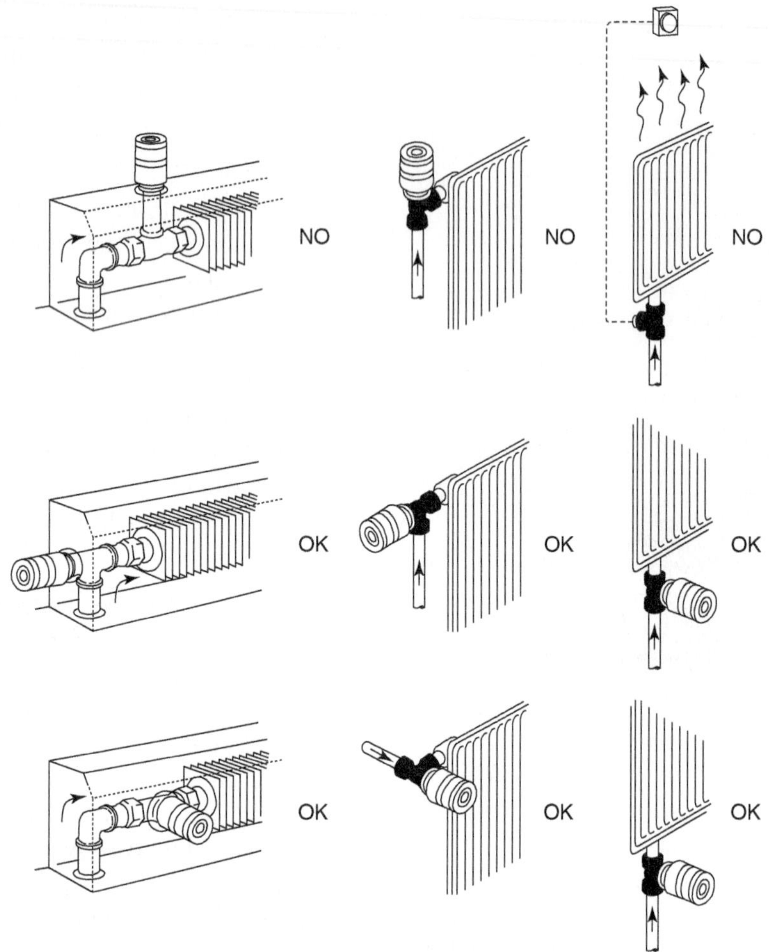

Figure 4-38 | Proper and improper mounting locations for a thermostatic operator. Do not mount the operator where it will be directly affected by convective air currents rising from the heat emitter.

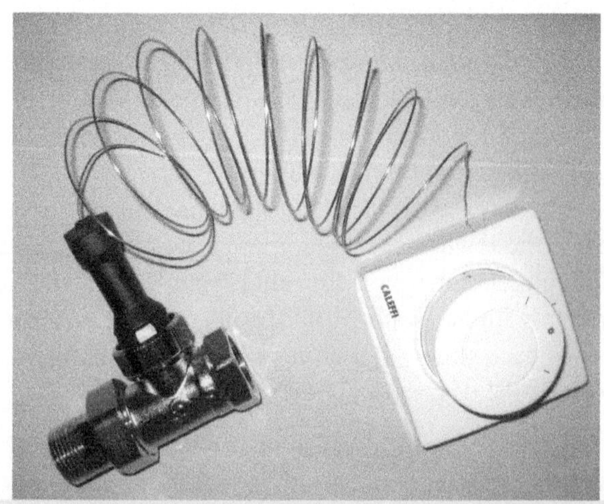

Figure 4-39 | A radiator valve fitted with a remote setpoint dial and operator assembly. Notice capillary tubing between setpoint dial and valve operator assembly.

3-WAY MOTORIZED MIXING VALVES

3-way motorized mixing valves have three ports. One is the inlet port for heated fluid, another is the inlet port for cooler fluid, and the third is the outlet port for the mixed fluid. The two inlet streams mix together inside the valve in proportions controlled by the position of the valve's flow element (or **spool**). The body of a typical **3-way rotary-type mixing valve** is shown in Figure 4-40a. An illustration showing how the position of the spool inside the valve varies the inlet flow proportions is shown in Figure 4-40b.

The position of the valve's shaft determines the amount of each incoming fluid stream entering the valve. The flow rate and temperature of each entering stream determines the mixed temperature leaving the valve. The mixed temperature will always

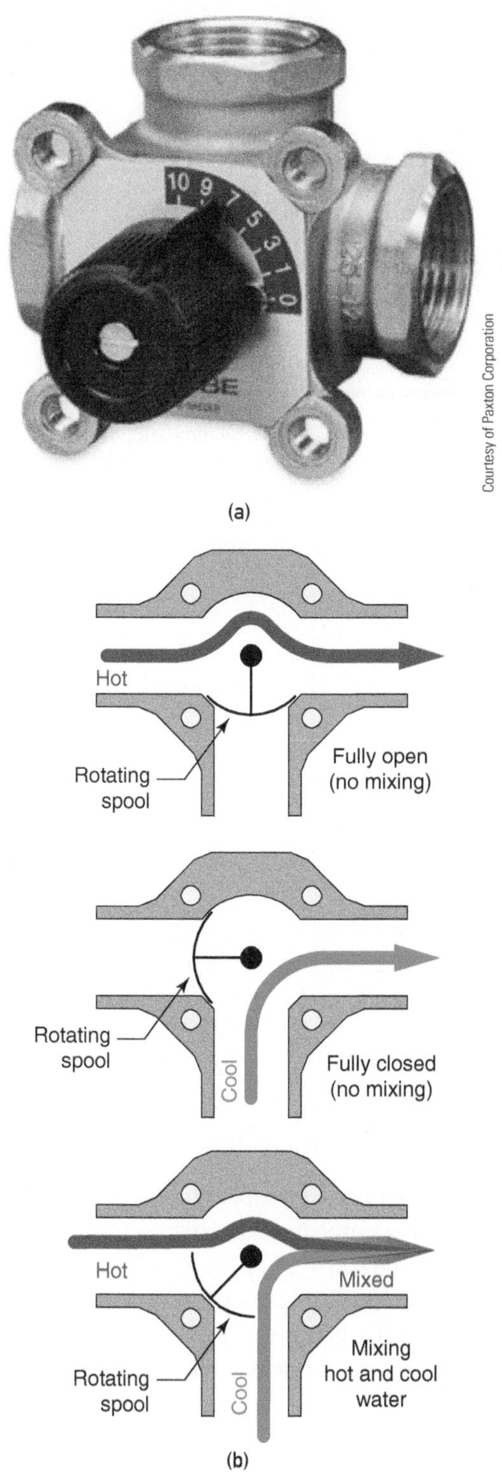

Figure 4-40 (a) 3-way rotary mixing valve with a manual adjustment knob. (b) Cross section through a 3-way rotary mixing valve with spool in different positions.

be between the temperature of the entering hot and entering cool streams. Most three-port valves have an indicator scale on the valve body. The pointer on the handle or knob indicates the setting of the valve between 0 and 10. The full rotating range of the shaft and internal spool is about 90 degrees.

Figure 4-41 Example of a 3-way rotary mixing valve with attached motorized actuator.

A 3-way rotary mixing valve, such as that shown in Figure 4-40a, does not adjust itself to compensate for changes in the temperature or flow rate of entering fluid streams. As such, it is called a "dumb" mixing valve. For automatic temperature control, the valve must be equipped with a motorized actuator operated by an electronic controller.

An example of 3-way rotary mixing valve with a motorized actuator mounted to it is shown in Figure 4-41. The actuator contains an electric motor, gear train, and the necessary electronics to operate the valve based on a signal sent from a temperature controller.

The controller that operates a 3-way motorized mixing valve may be integrated into the motorized actuator or consist of a separate device wired to the valve actuator. Figure 4-42 shows the latter.

When configured as shown in Figure 4-42, the mixing valve's controller uses a sensor to measure the water temperature supplied to the distribution system whenever that system is operating. The controller continually compares the measured temperature to a "target" supply water temperature that it is attempting

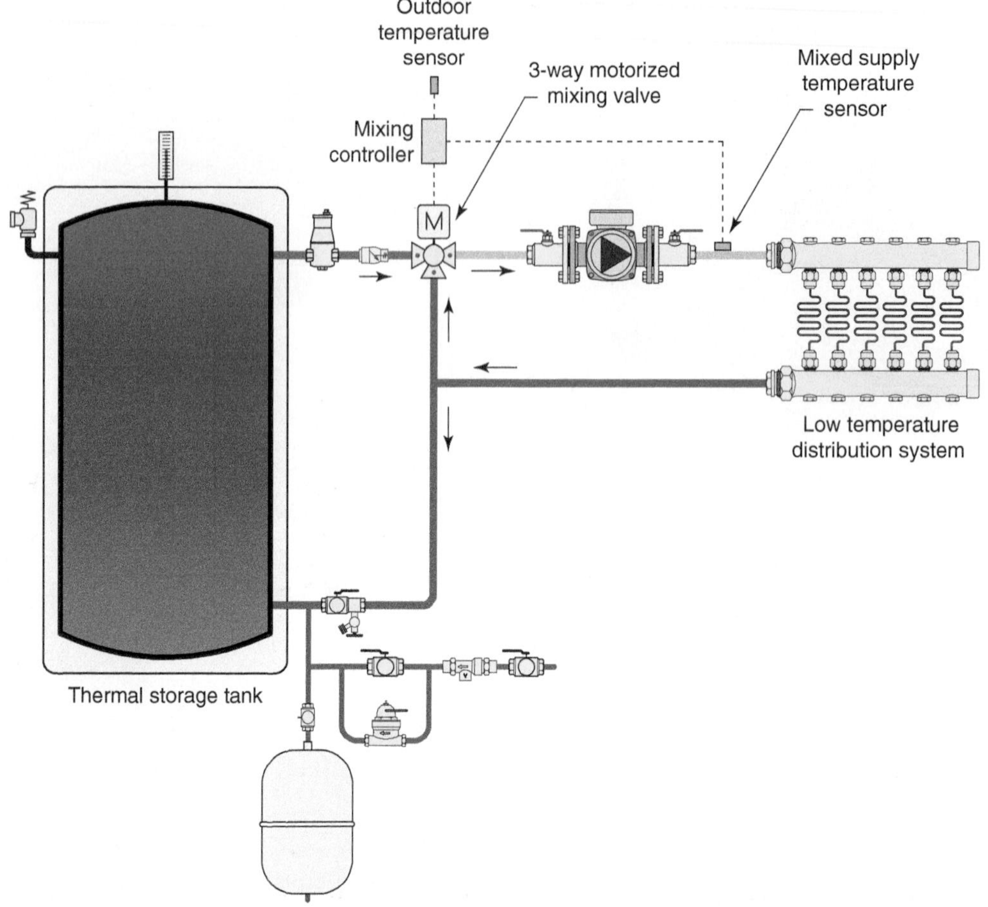

Figure 4-42 | A 3-way motorized mixing valve piped to reduce the water temperature supplied from a thermal storage tank to a low temperature distribution system.

to maintain. The target supply water temperature could be a fixed value set on the controller. It could also be a value that varies with outdoor temperature. The latter strategy is called **outdoor reset control** and is discussed later in this chapter. The controller operates the mixing valve's actuator to regulate the amounts of incoming hot and cool water. Its goal is to maintain the measured supply water temperature as close to the target water temperature as possible.

3-WAY THERMOSTATIC MIXING VALVES

Another type of mixing valve suitable for use in hydronic systems is called a **3-way thermostatic mixing valve**. Unlike a motorized 3-way mixing valve, a 3-way thermostatic mixing valve has an internal non-electric thermostatic cartridge that expands and contracts to adjust the flow of hot and cool water, with a goal of maintaining a set outlet temperature. An example of such a valve is shown in Figure 4-43.

The desired mixed temperature is set on the valve's knob. The two entering flow streams mix within the valve body and flow past the internal thermostatic cartridge. This cartridge creates movement of the valve's internal components based on the thermal expansion or contraction of a special wax compound sealed within it. This movement controls the openings that allow hot and cool water into the valve.

Although 3-way thermostatic valves can be used in certain low-flow hydronic mixing applications, most of them are not designed for a wide spectrum of such applications. Instead, they are designed to protect against scalding in domestic hot water delivery systems. One common configuration for this application is shown in Figure 4-44.

Figure 4-44 shows the 3-way thermostatic mixing valve in a "point of delivery" application. As such,

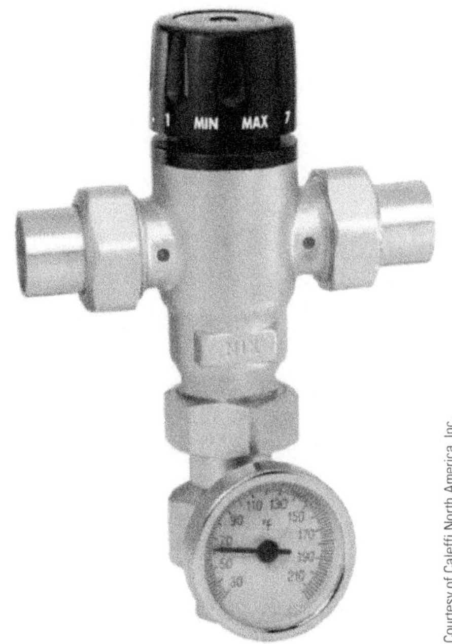

Figure 4-43 | Example of a 3-way thermostatic mixing valve.

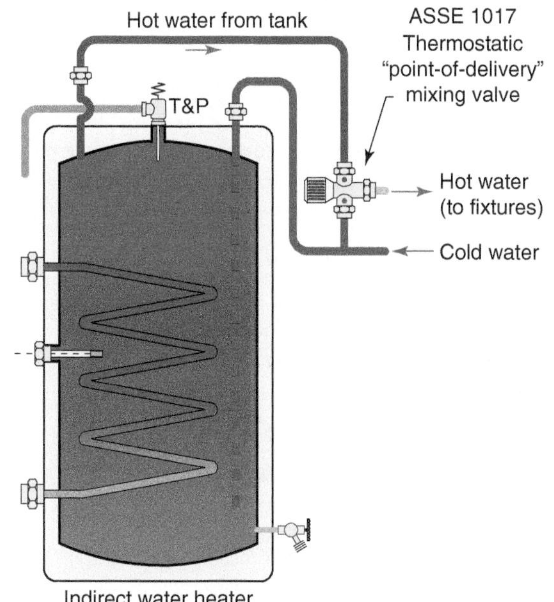

Figure 4-44 | Use of 3-way thermostatic mixing valve to provide scald protection in a domestic hot water delivery system.

it provides temperature protection for all domestic hot water piping downstream of the mixing valve. Another type of 3-way thermostatic mixing valve is specifically designed for "point of use" protection. Such valves are installed close to individual fixtures that require hot water. Figure 4-45 shows such a valve installed under a lavatory.

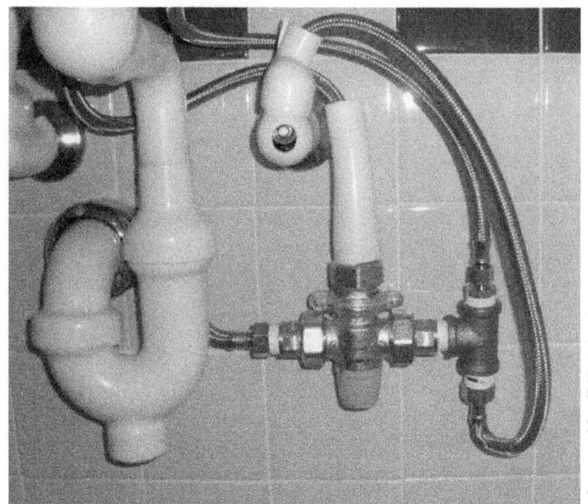

Figure 4-45 | 3-way thermostatic "point of use" mixing valve provides protection against scalding hot water being delivered to the lavatory above.

It is critically important to install a "point of delivery" thermostatic mixing valve, or multiple "point of use" thermostatic mixing valves, in *any* domestic hot water system supplied by an intermittent heat source such as solar collectors or a wood-fueled boiler. These sources often heat water to temperatures much higher than required at the fixtures. Without thermostatic mixing valves, there is a high probably of scalding hot water being delivered to the fixtures.

ZONE VALVES

Hydronic heating systems often use two or more piping circuits to supply heat to different zones of a building. One device that is often used to allow or prevent flow through a zone circuit is called a **zone valve**.

Several types of zone valves are available. They all consist of a valve body combined with an electrically powered actuator. The actuator is the device that produces movement of the valve shaft when supplied with an electrical signal. Some actuators use small electric motors and gears to produce a rotary motion of the valve shaft. Others use **heat motors** that expand and contract to produce a linear motion that moves the valve's shaft up and down. A zone valve equipped with a gear motor actuator is shown in Figure 4-46.

Zone valves are available that can operate with voltages ranging from 24 VAC to 240 VAC. The most common configuration for residential and light commercial hydronic systems operates on 24 VAC electrical power supplied through a transformer. A **"normally**

Figure 4-46 | Zone valve with detachable gear motor actuator

closed" zone valve will be closed until the actuator is supplied with power. A **"normally open" zone valve** will be open to flow until its actuator is supplied with power.

Zone valves operate in either a fully open or a fully closed position. *They do not modulate to regulate the flow rate in a piping circuit.* Those with gear motor actuators can move from fully closed to fully open within a few seconds of receiving power. Those with heat motor actuators usually require 2 to 3 minutes to open after receiving electrical power. Since the response time of a heating system is relatively slow, the slower opening of heat-motor-type zone valves does not make a noticeable difference in system performance.

A partial piping schematic for a four-zone system using zone valves and a variable speed circulator is shown in Figure 4-47. Notice that the zone valves are installed on the *supply* side of each zone circuit. In this location, a closed-zone valve prevents heat migration due to buoyancy-driven flow of hot water into zone circuits that are supposed to be off. Zone valves mounted on the return side of the zone circuits cannot provide this function.

Many zone valves are equipped with a built in **end switch**. This small, low-current switch closes its contacts whenever the zone valve reaches its fully open position. This contact closure is used to signal other electrical controls within the system that the zone is open and ready for flow, and therefore other control actions such as starting a circulator or firing a boiler need to occur. The most versatile zone valve configuration is a **4-wire zone valve**, which is schematically represented in Figure 4-48.

In a 4-wire zone valve, two of the wires are used to power the valve actuator's motor. The other two wires connect to an end switch that closes when the valve is fully open. Because the end switch is electrically isolated from the motor leads, it can be part of a different electrical circuit.

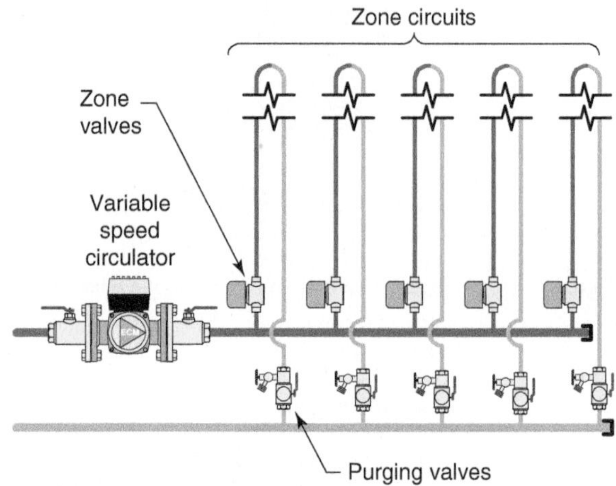

Figure 4-47 | Placement of zone valves on the supply side of zone circuits. This prevents heat migration into supply side of inactive zone circuits.

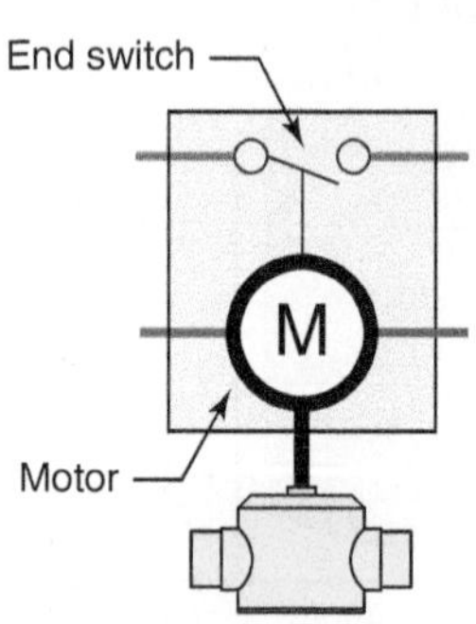

Figure 4-48 | Internal wiring of a 4-wire zone valve. Two wires operate the motor. The other two wires are connected to the isolated end switch.

3-WAY DIVERTER VALVES

Many hydronic systems require a means of routing flow through different portions of the system, depending on that system's current operating mode. A **3-way diverting valve** can be used for this purpose.

Figure 4-49a shows an example of a 3-way diverting valve. Figure 4-49b shows its internal construction.

This valve has one inlet port labeled "AB," and two outlet ports, one labeled "A" and the other labeled "B." When the valve's actuator is not powered, a flow path exists from port AB to port B. When the valve's actuator is powered, the internal ball or paddle moves to create a flow path from port AB to port A.

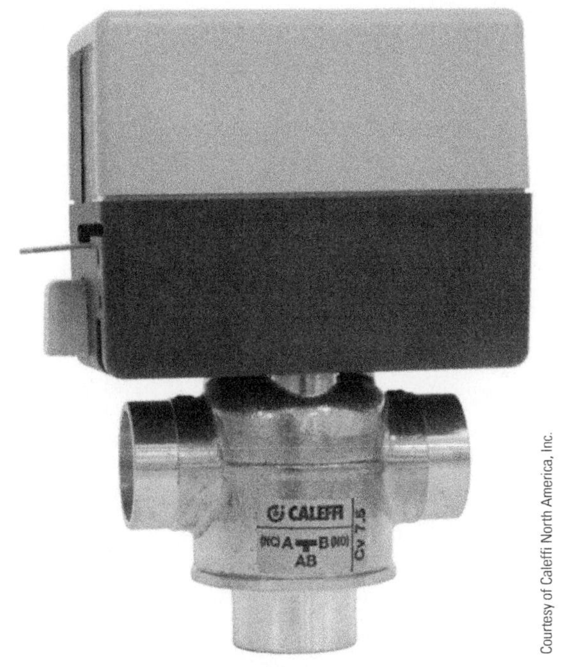

Figure 4-49a | Example of a 3-way diverter valve.

One common application for a 3-way diverting valve is in a system where the water supplying a distribution system can either come from a thermal storage tank heated by a renewable energy heat source, or from an auxiliary boiler. If the temperature of the thermal storage tank is high enough, it is the preferable heat source. However, if the tank's temperature is too low, the boiler needs to supply heat to the load. Figure 4-50 shows a typical set up for such a system.

When the diverting valve is *not* energized, flow passes from the AB port to the B port. This allows water returning from the distribution system to pass into the thermal storage tank, and not through the boiler. When the diverter valve is energized, flow through the tank is blocked. All flow must now pass through the auxiliary boiler.

Like zone valves, most diverting valves used in residential and light commercial systems operate on 24 VAC control voltage. This allows valves to be wired with standard thermostat cable. For certain applications, it may be desirable to operate diverting valves using line voltage. This option is available from some manufacturers.

When soldering a zone valve or diverting valve to copper tubing, the actuator should be removed from the

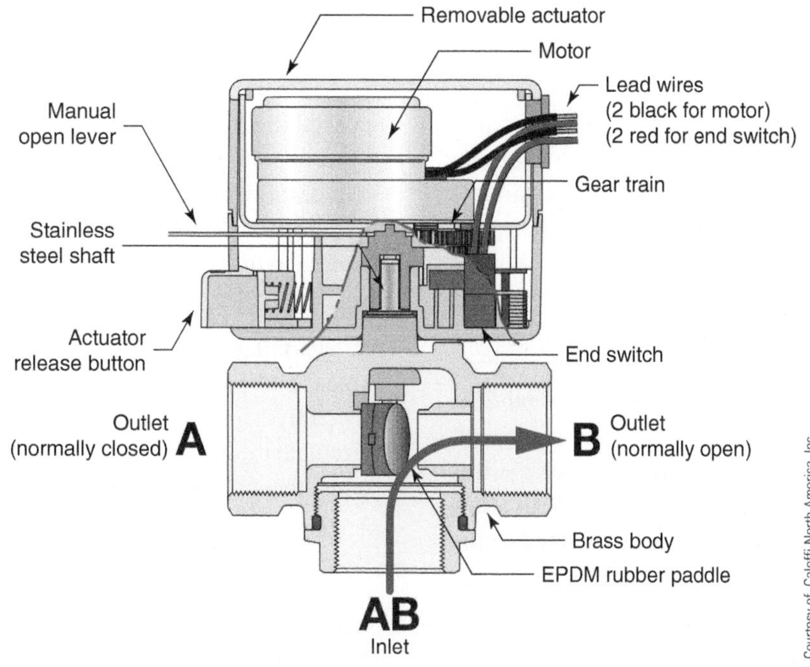

Figure 4-49b | Internal construction of a 3-way diverter valve.

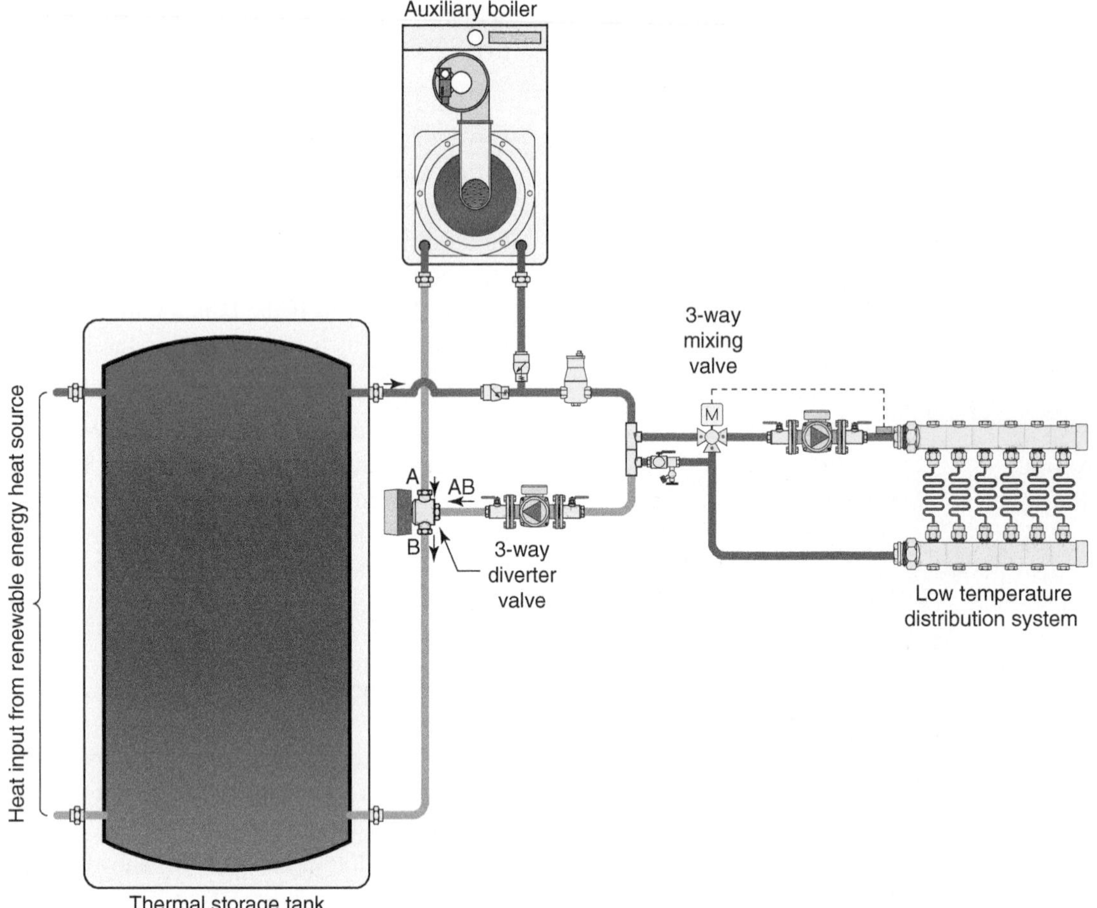

Figure 4-50 | A 3-way diverting valve used to route water returning from the distribution system to either a thermal storage tank or the auxiliary boiler.

valve body. The paddle that controls flow through the valve should be positioned away from the valve seats to minimize heating. Use a hot torch that can quickly heat the sockets of the valve body to minimize the potential for heat damage to other parts of the valve.

4.4 Circulators

All hydronic systems require flow. Almost all *modern* hydronic systems, including those supplied by renewable energy heat sources, use **circulators** to create this flow. A wide variety of circulators are available to allow designers to select models that are well matched to the flow requirements of a given hydronic circuit.

This section provides an overview of the type of circulators used in residential and light commercial hydronic systems. It describes what the **pump curve** of a circulator is and illustrates how to use it. This curve is then combined with the **system head loss curve** (discussed in Chapter 3) to find the flow rate at which the system "wants to" operate. The proper placement of one or more circulators within a system is also covered. Special attention is given to the subject of **cavitation** and its avoidance. The use of high-efficiency, variable-speed circulators is also discussed.

Circulators come in a wide variety of configuration, sizes, and performance ranges. In closed-loop, fluid-filled hydronic systems, the circulator's function is to circulate the system fluid around the piping. In most systems, the circulator does not "lift" fluid upward through a circuit, as discussed in Chapter 3. This is why the term *circulator* is a better descriptor than *pump* based on what this device does in a hydronic system. Still, the word "pump" has been around for many decades and is deeply ingrained in the vocabulary of hydronics. Therefore, in the context of hydronics, consider the words "circulator" and "pump" to be somewhat synonymous.

The type of circulator commonly used in residential and light commercial hydronic systems is known as a **wet-rotor circulator**. This design combines the **rotor**, shaft, and impeller into a single assembly that is housed in a chamber filled with system fluid. An example of a wet-rotor circulator is shown in Figure 4-51.

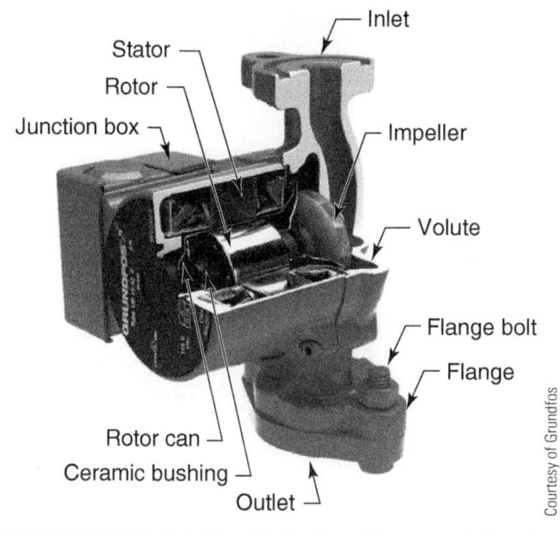

Figure 4-51 | Cut-away of a small wet rotor circulator.

The motor of a wet-rotor circulator is totally cooled and lubricated by the system's fluid. As such, it has no fan or oiling caps. The rotor assembly is supported on ceramic or graphite bushings within the **rotor can**. These bushings contain no oil, but ride on a thin film of system fluid. The rotor is surrounded by the **stator** assembly.

Wet-rotor circulators are available with cast-iron volutes for use in closed hydronic systems or with bronze or stainless steel volutes for direct contact with domestic water or open-loop hydronic systems. Most wet-rotor circulators have impellers constructed of stainless steel, bronze, or engineered polymers.

CIRCULATOR MOUNTING

Wet-rotor circulators are designed to be installed with their motor shafts in a horizontal position. This reduces the axial thrust load on the bushings due to the weight of the rotor and impeller.

The direction of flow through the pump is usually indicated by an arrow on the side of the volute. The installer should always check that the circulator is installed with the correct flow direction. As long as the shaft is horizontal, the circulator can be mounted with the flow arrow pointing upward, downward, or horizontally. Of these, upward flow is preferred when possible, because it allows the circulator to rapidly clear itself of air bubbles. Never mount a wet-rotor circulator with the motor facing down, as shown in Figure 4-52.

Figure 4-53 shows a circulator equipped with **isolation flanges**. These flanges each contain a ball

Figure 4-52 | Incorrectly installed circulators. The motor shaft should always be in a horizontal orientation.

Figure 4-53 | Circulator equipped with isolation flanges. The ball valve within each isolation flange is open when its handle is aligned with the piping.

valve that can be closed to isolate the circulator from the remainder of the system if it has to be serviced or replaced. The ball valves create very little head loss when in the open position. *The author recommends installing isolation flanges on all flanged circulators in every system.* If the circulator does not have flanged connections, individual full port ball valves, along with unions, should be installed in the piping on each side of the circulator to allow it to be isolated and easily removed if necessary.

Notice that the circulator in Figure 4-53 is also installed with the electrical cable coming from the bottom of the junction box. This **drip loop** in the cable protects against the possibility of water from a nearby leak traveling along the cable and entering the circulator's junction box.

All circulators should also be installed with a minimum of 12 diameters of straight pipe connected to their inlet port, as illustrated in Figure 4-54. Thus, if the circulator connects to 1-inch pipe, there should be a minimum of 12 inches of straight pipe leading to its inlet port. This straight pipe reduces turbulence into the impeller, which reduces sound output from the circulator.

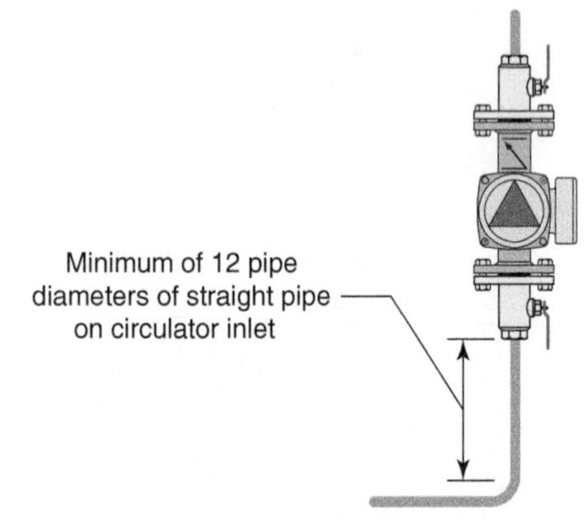

Figure 4-54 | Always install a minimum of 12 diameters of straight pipe on the inlet side of all circulators.

It is also important to properly support circulators so that their weight does not induce excessive stress on other components. Most of the wet-rotor circulators discussed in this text can be supported by the metal or rigid polymer piping to which they attach, provided that piping is itself supported within 6 inches of the circulator, as shown in Figure 4-55. If the circulator is connected to flexible tubing, it should be supported by a metal bracket attached to one or more of the flange bolts.

Figure 4-55 | Circulator supported by copper tubing that itself is supported by a channel strut secured to adjacent wall.

Although modern circulators can operate with very little sound, it is still preferable to avoid supporting them on brackets attached to hollow interior partitions that separate the mechanical room from adjacent living spaces. This is especially true if those adjacent spaces are sleeping rooms or used for other sound-sensitive purposes. One solution in these situations is to build a support structure using a steel channel strut that is not connected to the partition. The base of the support structure should be secured to a concrete floor. The upper portion of the support structure can be secured to the ceiling using **vibration dampening couplers** with rubber disks.

PLACEMENT OF CIRCULATORS IN A SYSTEM

The location of the circulator(s) within a piping circuit, relative to other components, can make the difference between quiet, reliable operation or constant problems. One guiding rule summarizes the situation: *Always install the circulator with its inlet close to the connection point of the system's expansion tank.*

To understand why this rule should be followed, one needs to consider the interaction between the circulator and system's expansion tank. In a closed-loop piping system, the amount of fluid, including that in the expansion tank, is fixed. It does not change regardless of whether the circulator is on or off. The expansion tank contains a captive volume of air at some pressure. The only way to change the pressure of this air is to either push more fluid into the tank to compress the air or to remove fluid from the tank to expand the air. This fluid would have to come from, or go to, some other location in the system. However, since the system's fluid is incompressible, and the amount of fluid in the system is fixed, this cannot happen regardless of whether the circulator is on or off. The expansion tank thus becomes the **point of no pressure change (PONPC)** within the system.

Consider a horizontal piping circuit filled with fluid and pressurized to some pressure, say 10 psi, as shown in Figure 4-56. When the circulator is off, the static pressure is the same (10 psi) throughout the piping circuit. This is indicated by the solid (red) horizontal line shown above the piping.

When the circulator is turned on, it immediately creates a pressure difference between its inlet and discharge ports. However, the expansion tank, being the point of no pressure change in the system, maintains the same (10 psi) fluid pressure at its point of attachment to the system. The combination of the pressure difference across the circulator, the pressure drop due to head loss in the piping, and the point of no pressure change results in a new pressure distribution, as shown by the dashed (green) lines in Figure 4-56.

Notice that the pressure *increases* in nearly all parts of the circuit when the circulator is turned on. This is desirable because it helps eject air from vents. It also reduces the chance of cavitation. The short segment of piping between the expansion tank and the inlet port of the circulator experiences a slight drop in pressure due to head loss in the piping. The numbers used for pressure in Figure 4-56 are illustrative only. The actual numbers will depend on flow rates, fluid properties, and pipe sizes.

Now consider the same system with the expansion tank *incorrectly* connected near the discharge port of the circulator. The dashed line in Figure 4-57 illustrates the new pressure distribution in the system when the circulator is on.

The point of no pressure change remains at the expansion tank connection. This causes the pressure in most of the system to *decrease* when the circulator is turned on. The pressure at the inlet port has dropped from 10 psi to 2 psi. This situation is not desirable because it reduces the system's ability to expel air. In some cases, it can also lead to cavitation.

To see how problems can develop, imagine the same system represented in Figure 4-57, but with a static pressurization of only 5 psi instead of the previous 10 psi. When the circulator starts, the same 9 psi differential will be established between its inlet and discharge ports. The pressure profile shown with dashed lines in Figure 4-57 will be shifted downward by 5 psi, as shown in Figure 4-58.

Notice that the pressure in the piping between the upper right-hand corner of the circuit and the inlet port of the circulator is now *below atmospheric pressure*.

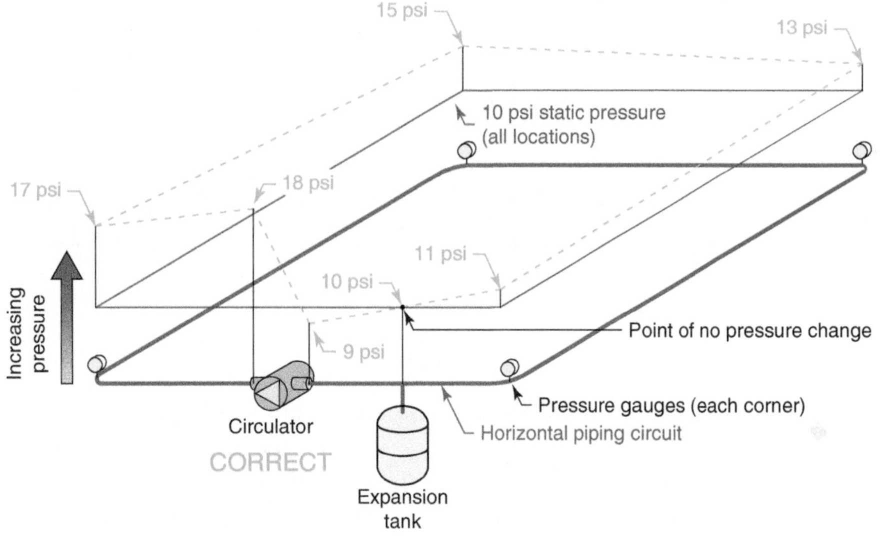

Figure 4-56 Pressure distribution in horizontal piping circuit. The solid red line is pressure distribution when circulator is off. The dashed green line is pressure distribution when circulator is on. The circulator's inlet is correctly located near the location where the expansion tank connects to the circuit.

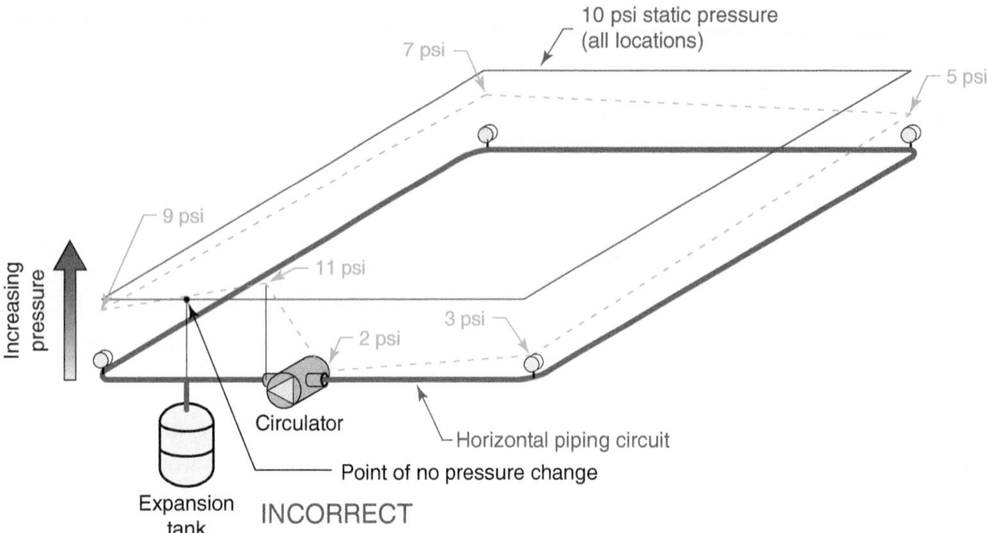

Figure 4-57 | Pressure distribution in horizontal piping circuit. The solid red line is pressure distribution when the circulator is off. The dashed green line is pressure distribution when the circulator is on. Note that the discharge port of the circulator is *incorrectly* located close to the expansion tank connection point.

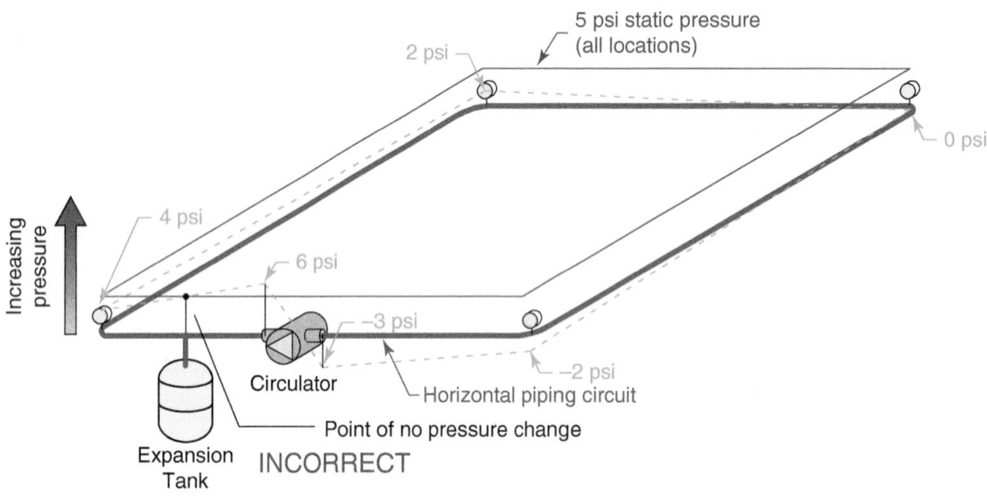

Figure 4-58 | Pressure distribution in horizontal piping circuit. The solid red line is pressure distribution when the circulator is off. The dashed green line is pressure distribution when the circulator is on. Note that the discharge port of the circulator is *incorrectly* located close to the expansion tank connection point. The pressure in some locations is sub-atmospheric when the circulator is on.

If air vents were located in this portion of the circuit, the sub-atmospheric pressure would suck air into the system every time the circulator turns on. The circulator is also much more likely to cavitate under these conditions.

This principle of locating the expansion tank close to the inlet side of a circulator also applies to multiple circulators mounted to a common header, as shown in Figure 4-59. The header supplying the circulators should be kept relatively short and sized for a maximum flow velocity of 2 feet per second when all the circulators are operating. This keeps the head loss and pressure drop along the header very low and minimizes interference between the circulators regardless of which circulators happen to be operating. Thus, this header would be described as a **low loss header**.

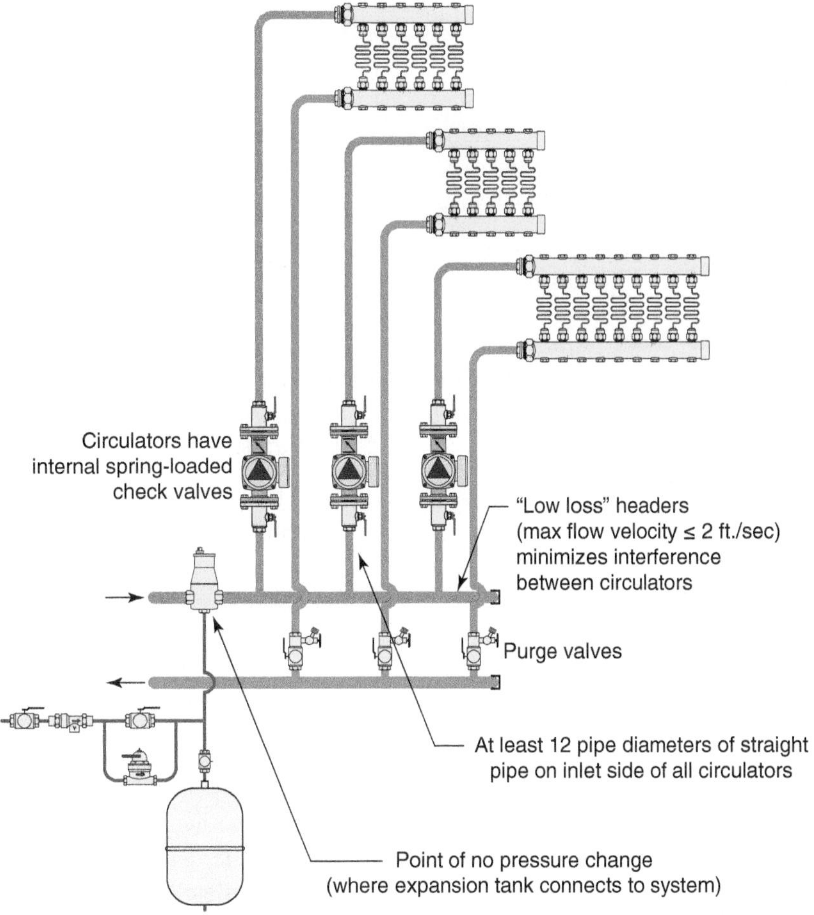

Figure 4-59 | Multiple circulators on a common low loss header. Inlet ports of circulators are correctly located close to the point where the expansion tank connects to the system.

CIRCULATOR HEAD AND DIFFERENTIAL PRESSURE

This section presents methods for describing the performance of circulators. These methods are vital in properly selecting a circulator to match the flow requirements of a hydronic circuit.

A circulator should be thought of as an energy conversion device. It converts electrical energy input into head energy output. Simply put: *An operating circulator adds head energy to a fluid as it passes through.* In Chapter 3, head was described as mechanical energy present in the fluid. For incompressible fluids such as the liquids used in hydronic systems, the added head (e.g., added mechanical energy) reveals itself as an increase in pressure between the inlet and discharge ports of the circulator. Thus, *an increase in pressure from the inlet to outlet of an operating circulator is the "evidence" that head energy has been added to the fluid.* This principle is illustrated in Figure 4-60.

The value of this pressure increase depends on the flow rate through the circulator and the density of the fluid being circulated. As the flow rate through a

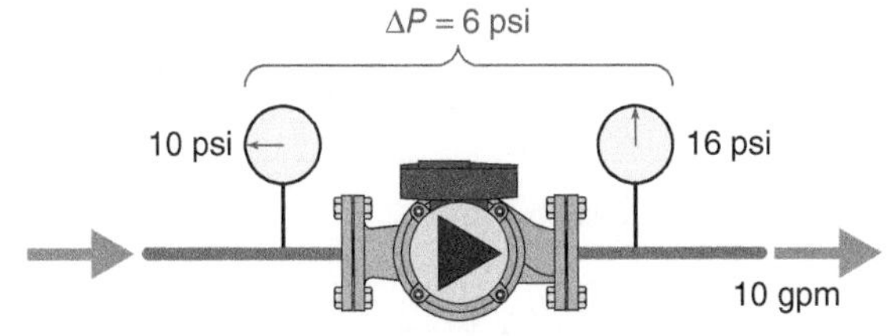

Figure 4-60 | An increase in pressure from the inlet to the outlet of an operating circulator is the evidence that head energy has been added to the fluid.

circulator increases, the pressure increase between the circulator's inlet and outlet is reduced, as shown in Figure 4-61.

In the United States, the head added to the fluid by a circulator is expressed in units of **feet of head**. This unit results from a simplification of the following units:

$$\text{Head} = \frac{\text{ft} \cdot \text{lb}}{\text{lb}} = \text{ft} = \frac{\text{energy}}{\text{lb}}$$

The unit ft·lb (pronounced "**foot pound**") is a unit of energy. Head may therefore be described as the *mechanical energy added (in foot·pounds) per pound of fluid passing through the circulator.*

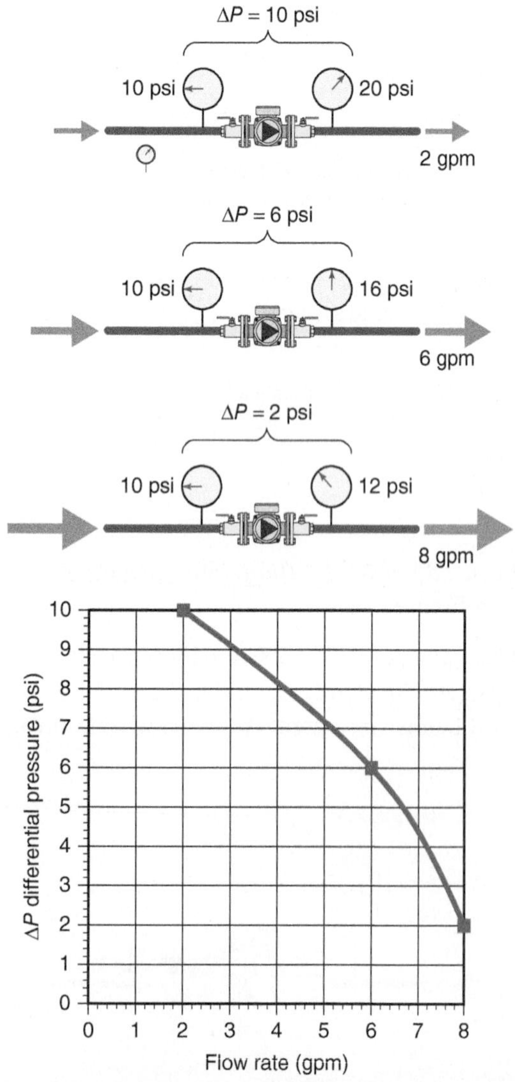

Figure 4-61 As the flow rate through a circulator increases, the pressure difference between the inlet and outlet ports decreases.

The head added to the fluid by an operating circulator is the same type of head that is removed from a flowing fluid due to friction, as discussed in Chapter 3.

The head added by a circulator at some given flow rate is essentially independent of the fluid being pumped. For example, a circulator adding a head of 20 feet while circulating 60 °F water at 5 gpm will also add 20 feet of head energy while circulating a 50 percent propylene glycol solution at 5 gpm. However, the pressure differential measured between the inlet and outlet ports of the operating circulator will be different because of the difference in the density of the fluids.

The head added to a fluid by a circulator can be determined by measuring the differential pressure across the circulator. Figure 4-62 shows four options for equipping a circulator with gauges to read differential pressure. On some circulators, threaded openings are provided in the volute flanges for direct attachment of pressure gauges.

Equation 4.2 can be used to convert the pressure differential measured across an operating circulator into the head added to the fluid. This calculation requires the determination of the density of the fluid being circulated.

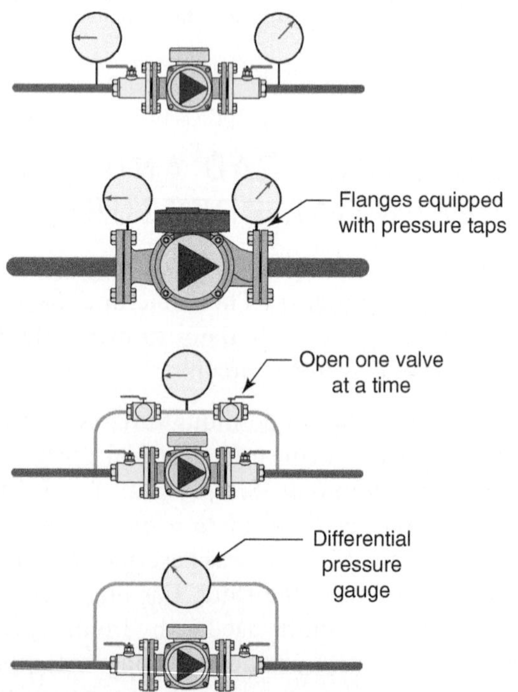

Figure 4-62 Four options for reading the pressure differential across a circulator.

(Equation 4.2)

$$H = \frac{144(\Delta P)}{D}$$

where:

H = head added by the circulator (feet of head)

ΔP = pressure differential between inlet and discharge ports of circulator (psi)

D = density of the fluid (lb/ft^3)

Example 4.2: Based on the pressure gauge readings, determine the head added to the fluid by the circulator in the two operating conditions shown in Figure 4-63.

Solution: In each case, the density of the water being circulated needs to be determined. The density of water at 60 °F and 180 °F can be found in Figure 3-11.

a. (60 °F water): density = 62.31 lb/ft^3

b. (180 °F water): density = 60.47 lb/ft^3

The head added by the circulator in each case can now be calculated using Equation 4.2:

a. $H = \frac{(20.8 - 10)(144)}{62.31} = 25.0$ feet of head

b. $H = \frac{(15.5 - 5)(144)}{60.47} = 25.0$ feet of head

Discussion: In both cases, the pressure rise across the circulator is different, yet the head energy imparted to each fluid is the same. This demonstrates that there is no "fixed" multiplier for converting differential pressure to head or vice versa. The relationship between head added and the corresponding increase in pressure will always depend on the density of the fluid. The density will depend on the fluid used as well as the temperature of that fluid.

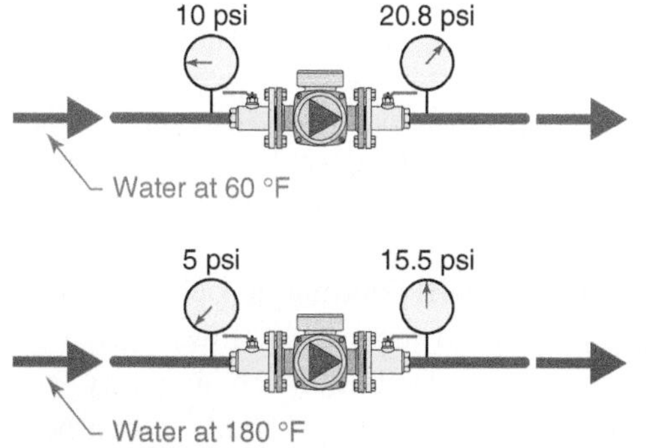

Figure 4-63 | Pressure differentials across circulators for Example 4.2.

PUMP CURVES

A graph of the head added to a fluid, versus the flow rate of that fluid through a circulator, is called a **pump curve**. An example of such a graph is shown in Figure 4-64.

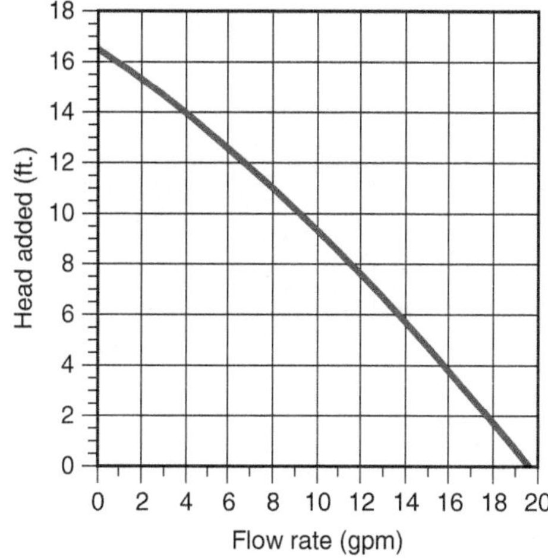

Figure 4-64 | Example of a pump curve for a small circulator.

Pump curves are developed from test data using water in the temperature range of 60 °F to 80 °F. For fluids with higher viscosities, such as glycol-based antifreeze solutions, there is a very small decrease in head and flow rate capacity of the circulator. However, for the fluids and temperature ranges commonly used in residential and light commercial hydronic heating systems, this variation is so small that it can be safely ignored. Thus, for the applications discussed in this text, *pump curves may be considered to be independent of the fluid being circulated.* Keep in mind, however, that the system head loss curve, which describes head *loss* versus flow rate in a specific hydronic circuit, is very dependent on fluid properties and temperature, as discussed in Chapter 3.

Pump curves are extremely important in matching the performance of a circulator to the flow requirements of a piping system. All circulator manufacturers publish these curves for the circulators they sell. In many cases, the pump curves for an entire series or **family of circulators** is plotted on the same set of axes so that performance comparisons can be made. Figure 4-65 shows an example of a "family" of pump curves.

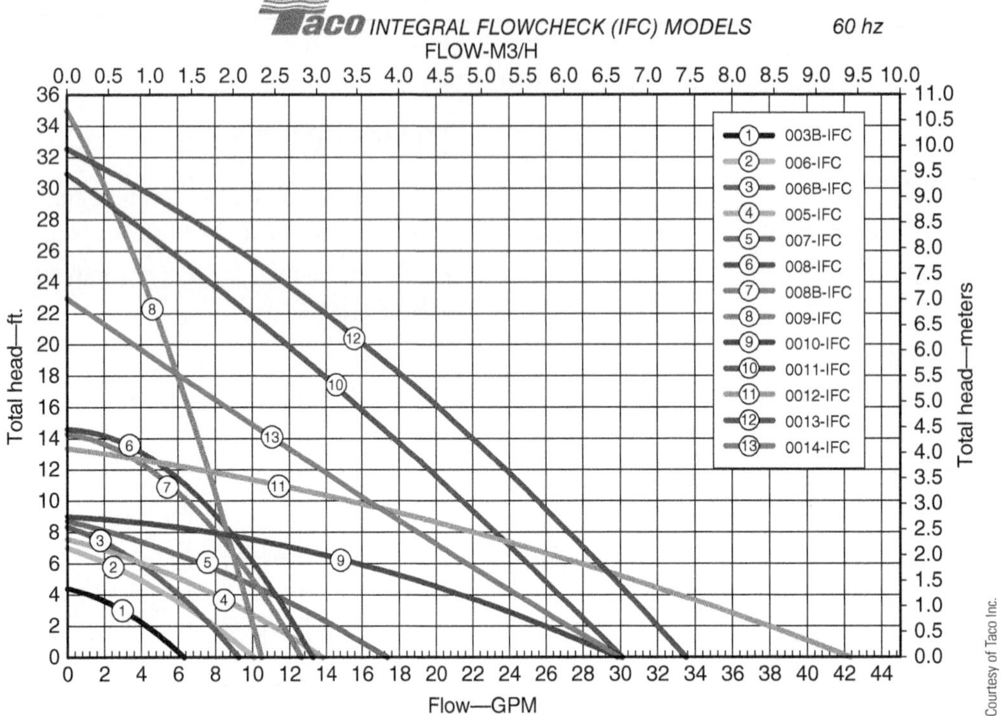

Figure 4-65 | Examples of several pump curves plotted on a common graph.

OPERATING POINT OF A HYDRONIC CIRCUIT

The system head loss curve of a piping system was discussed in Chapter 3. It shows the head loss of the hydronic circuit as a function of flow rate through it for a specific fluid at a specific temperature. It is a unique graphical description of the hydraulic characteristic of that circuit. An example of a system resistance curve is shown in Figure 4-66.

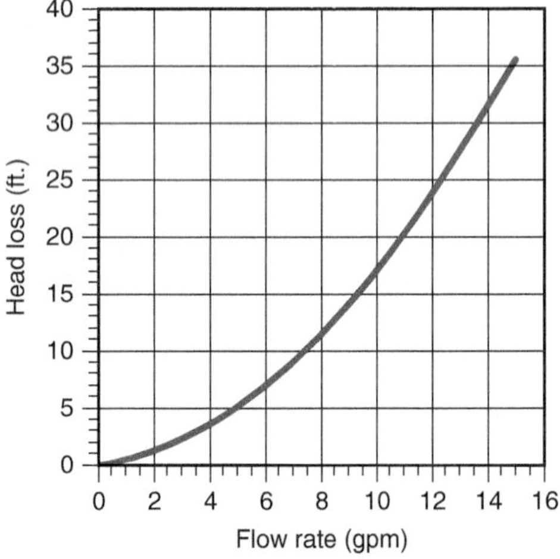

Figure 4-66 | Example of a system head loss curve for a specific circuit.

Notice that the pump curve in Figure 4-64 and the system head loss curve in Figure 4-66 have similar quantities on their vertical axes (e.g., head). The pump curve shows the head *added* on the vertical axis, whereas the system curve shows the head *loss* on its vertical axis. The pump curve shows the ability of the circulator to *add* mechanical energy to a fluid over a range of flow rates. The system head loss curve shows the ability of the piping system to *remove*, or *dissipate*, mechanical energy from the fluid over a range of flow rates. In both cases, the gain or loss of mechanical energy is expressed in feet of head.

The first law of thermodynamics states that when the rate of energy input to a system equals the rate of energy removal from the system, that system is in equilibrium and will continue to operate at those conditions until one of the energy flows is changed. This concept holds true for both thermal energy (e.g., heat) and mechanical energy (e.g., head). It can be summarized as follows:

When the rate at which a circulator adds head to the fluid in a piping circuit equals the rate at which the piping circuit dissipates head, that circuit is in ***hydraulic equilibrium****, and the flow rate through it will remain constant.*

The point at which hydraulic equilibrium occurs in a given piping system, with a given circulator, can be found by plotting the system head loss curve and pump curve on the same set of axes, as shown in Figure 4-67.

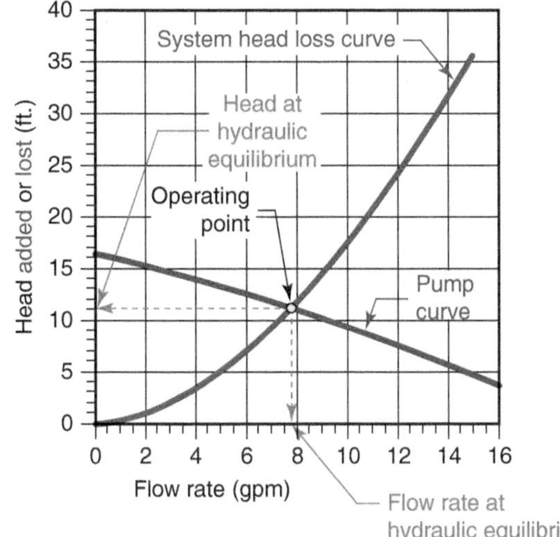

Figure 4-67 A pump curve and system head loss curve plotted on the same axes. The operating point is where the curves cross. This is where hydraulic equilibrium occurs.

The point where the curves cross represents hydraulic equilibrium and is called the **operating point** of the circuit. The flow rate through the circuit at hydraulic equilibrium is found by drawing a vertical line from the operating point down to the horizontal axis. The head input by the circulator (or head loss by the piping circuit) can be found by extending a horizontal line from the operating point to the vertical axis.

A performance comparison of several "candidate" circulators in a given piping circuit can be made by plotting their individual pump curves on the same set of axes as the circuit's head loss curve. The intersection of each circulator's pump curve with the circuit's head loss curve indicates the operating point for that particular circulator. By projecting vertical lines from the operating points down to the horizontal axis, the designer can determine the flow rate each circulator would produce within the circuit. This is illustrated in Figure 4-68.

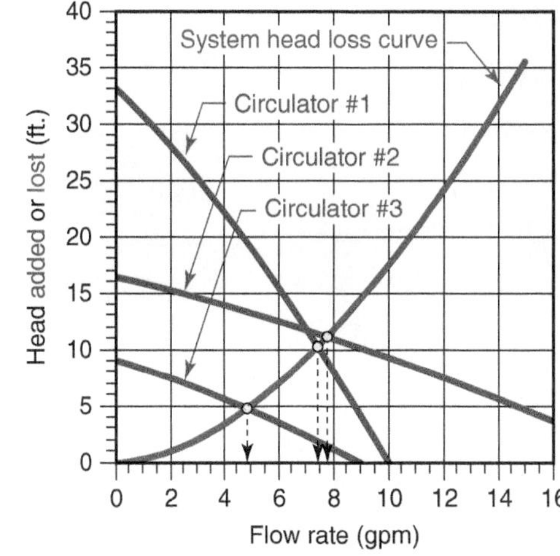

Figure 4-68 Three pump curves plotted along with a head loss curve for a specific piping circuit. The vertical lines descending from the operating points indicate flow rates produced by each circulator within the piping system.

Notice that even though the curves for circulators #1 and #2 are markedly different, they intersect the system curve at almost the same point. Therefore, these two circulators would yield very similar flow rates of about 7.5 gpm and 7.8 gpm in this piping system. The flow rate produced by circulator #3, about 5 gpm in this case, is considerably lower.

HIGH HEAD VERSUS LOW HEAD CIRCULATORS

Some circulators are designed to produce relatively high heads at lower flow rates. Others produce lower but relatively stable heads over a wide range of flow rates. These characteristics are fixed by the manufacturer's design of the circulator. Factors that influence the shape of the pump curve include the diameter and width of the impeller as well as the number and curvature of the impeller's vanes.

Figure 4-69 compares the internal construction of a circulator designed for high heads at low flow rates versus one designed for lower heads over a wider range of flow rates. Notice the impeller of the high head circulator has a relatively large diameter but a very small separation between its disks. The low head circulator, on the other hand, has a small diameter impeller with deeper vanes, and in this case, no lower disk.

The pump curves for the circulators shown in Figure 4-69 are plotted in Figure 4-70. The high head circulator is said to have a **"steep" pump curve**.

Figure 4-69 | Comparison between (a) impeller design for a steep pump curve and (b) impeller design for a flat pump curve.

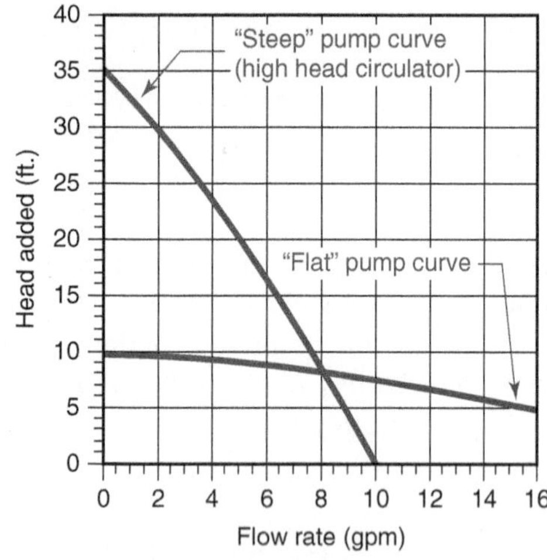

Figure 4-70 | Pump curves for circulators shown in Figure 4-68.

The other circulator would be described as having a relatively **"flat" pump curve**. The terms "steep" and "flat" are relative to each other.

Circulators with steep pump curves are intended for systems having high head losses at modest flow rates. Examples would include a drainback-protected solar collector circuit or an earth loop for a geothermal heat pump.

Circulators with flat pump curves should be used when the objective is to maintain a relatively steady differential pressure across the distribution system over a wide range of flow rates. An example of such an application is a multi-zone system in which zone valves are used to control flow through individual zone circuits. Depending on the number of zone valves open at a given time, and the shape of the pump curve, the differential pressure across the circulator could change considerably based on the pump curve of the circulator. This is illustrated in Figure 4-71.

A circulator with a flat pump curve will minimize changes in differential pressure across the headers in such systems and is therefore much preferred over a circulator with a steep pump curve.

The "ideal" pump curve for a circulator used in a system with valve-based zoning would be perfectly flat, as shown in Figure 4-72.

A circulator with a perfectly flat pump curve would produce no change in differential pressure between its inlet and outlet ports, regardless of the flow rate through it. Thus, the flow rate through any given zone circuit would remain essentially constant, regardless of what other zone circuits happened to be on or off at any given time. Unfortunately, it is not possible to create a centrifugal-type circulator with a perfectly flat pump curve, and thus any real circulator can only approximate this ideal conditions. However, it is possible to use a variable speed circulator to "mimic" the behavior of a circulator with a perfectly flat pump curve. This will be discussed shortly.

MULTI-SPEED CIRCULATORS

A single speed circulator has only one pump curve. This may be acceptable if that pump curve happens to match the flow and head requirements of a given piping circuit. If not, the designer needs to keep looking for a different circulator until a proper match is found.

Because piping circuits can be created in virtually unlimited configurations, many manufacturers offer a "family" of circulators with a wide range of pump curves. However, this requires wholesalers to stock an equally wide range of circulator models. To reduce the number of models needed and still cover a wide range of performance requirements, most circulator manufacturers offer wet-rotor circulators capable of operating at three different speeds. The desired speed is selected using an external switch on the circulator. Each operating speed has a corresponding pump curve, as shown in Figure 4-73.

Each pump curve will intersect the system's head loss curve at a different operating point and thus produce three different flow rates. This provides the same versatility as having three separate circulators, each with a single pump curve.

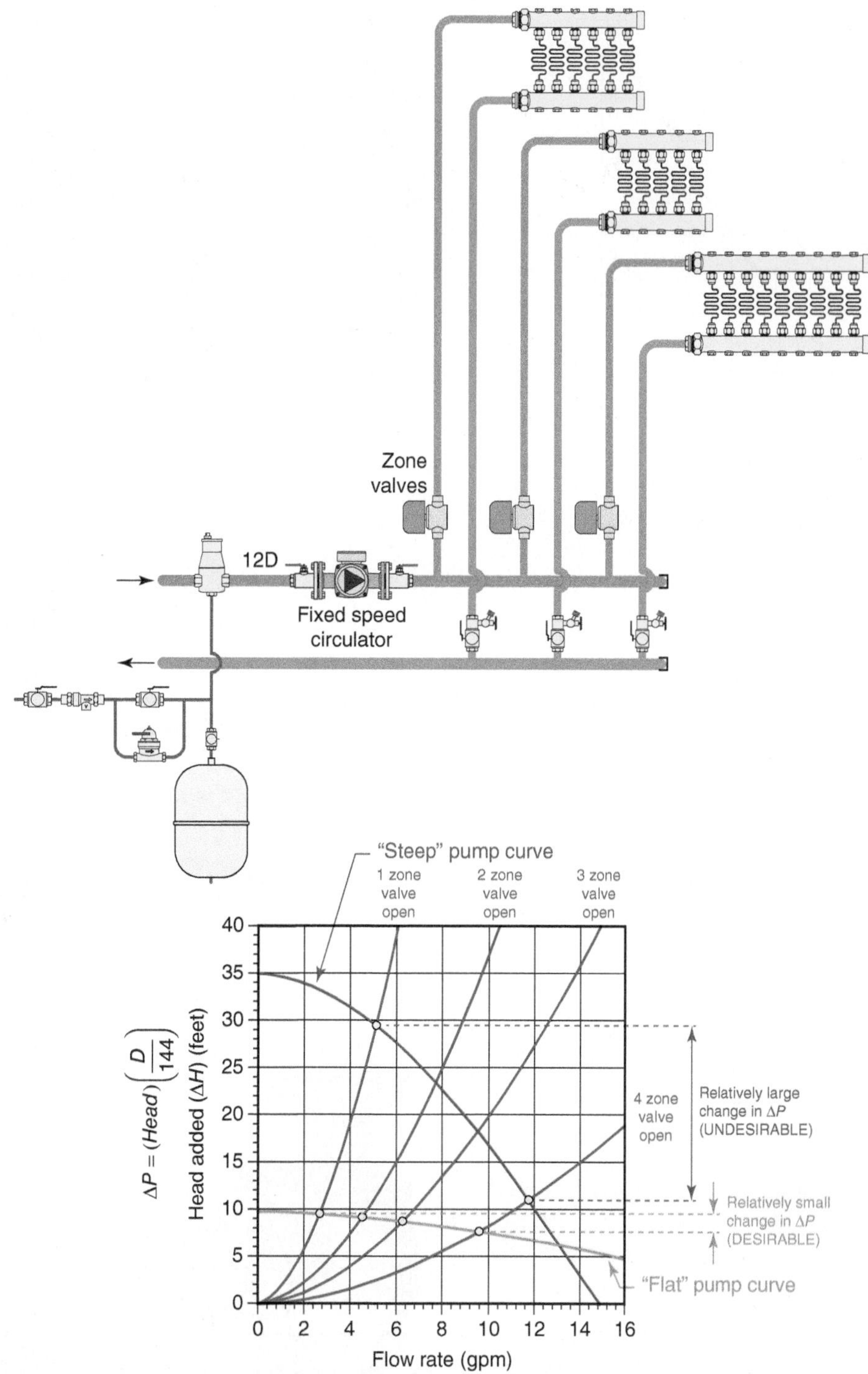

Figure 4-71 | Circulators with flat pump curves produce relatively small changes in differential pressure as zone circuits, controlled by zone valves, turn on and off. This is desirable.

SMART CIRCULATORS

One of the most important technical advances to have occurred in the hydronics industry over the past 20 years is the introduction of variable circulators with built in "intelligence." Such circulators can automatically vary their speed in response to attempted differential pressure changes in a hydronic distribution system. Some are capable of operating in several different modes, depending on how they are applied. These so-called **smart circulators** are quickly gaining market share in residential and light commercial hydronic systems. All indications are that

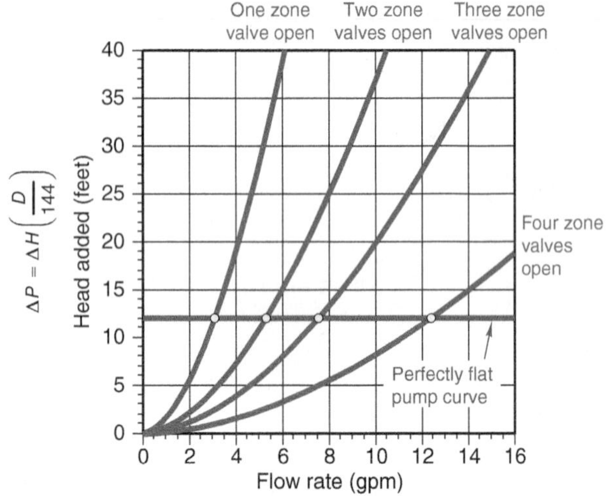

Figure 4-72 | Concept of a perfectly flat pump curve.

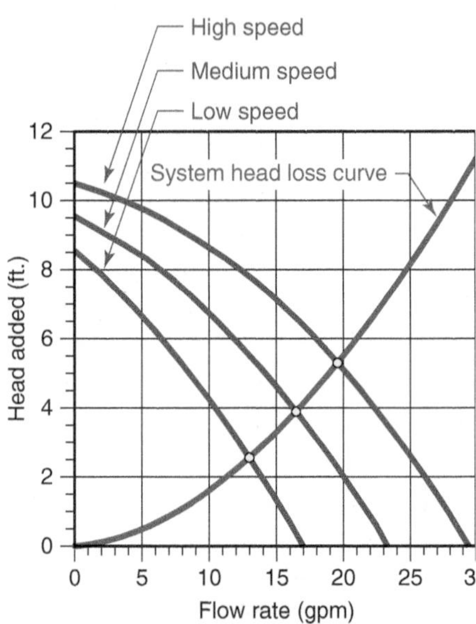

Figure 4-73 | A 3-speed circulator can operate on any of its three pump curves. Each curve would produce a different flow rate within a given piping circuit.

they will become the "new normal" in a wide variety of systems, including those using renewable energy heat sources.

The latest generation of smart circulators use **electronically commutated motors (ECM)**. These motors have very powerful **rare earth (neodymium) permanent magnets** sealed into a stainless steel **rotor**. There are no electrical windings in the rotor and no brushes between the rotor and stationary part of the motor. Because of this, ECM motors are sometimes also called **brushless DC motors**.

Figure 4-74 shows five examples of smaller ECM-powered circulators suitable for many residential and

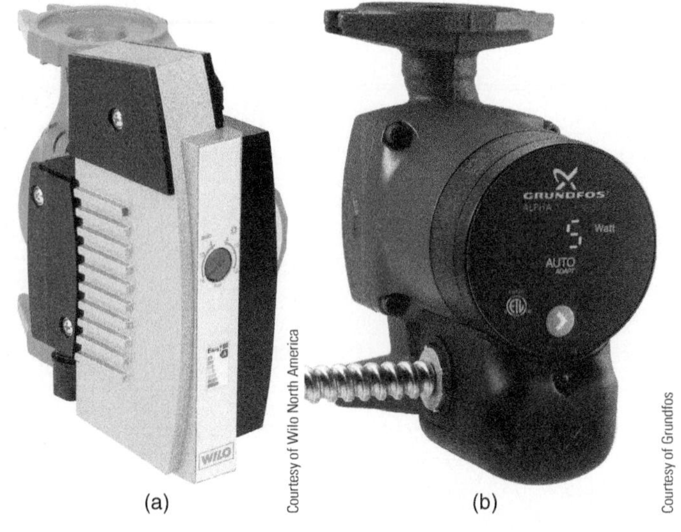

Figure 4-74 | Examples of smaller ECM-powered smart circulators.

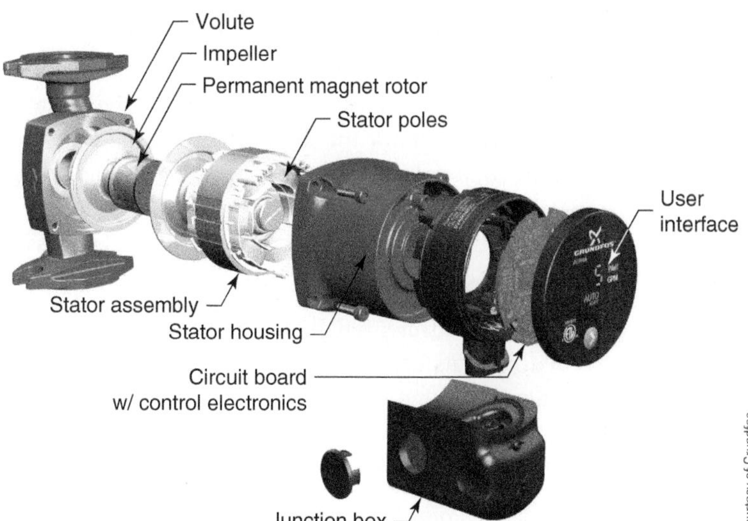

Figure 4-75 | Major components of an ECM-powered circulator.

light commercial applications. Larger models are also available for larger systems.

Figure 4-75 shows the internal construction and major components of an ECM-based circulator.

An ECM motor is regulated by control circuitry that changes the magnetic polarity of the **stator poles** in a way that causes the permanent magnet rotor to spin. This effect is depicted in Figure 4-76.

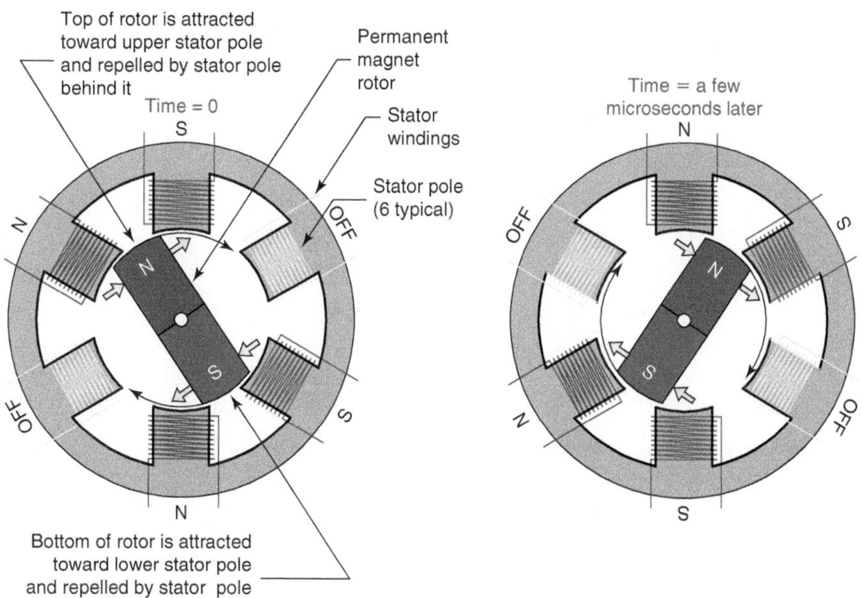

Figure 4-76 | Within an ECM circulator, the magnetic polarity of the stator poles is regulated by microprocessor-controlled switching circuitry. This creates torque on the permanent magnet rotor causing it to spin.

A microprocessor and associated solid-state switching circuitry control electrical current flow through each of the stator windings. This causes the magnetic polarity of each pole to change from north (N) to south (S) and vice versa, or remain off. The combined attraction and repulsion forces generated between the permanent magnet rotor and stator poles create a torque that causes the rotor to spin. The faster the stator poles change polarity, the faster the rotor spins.

Electronically commutated motors are ideal when accurate speed control is needed. They are also significantly more efficient at converting electrical energy into mechanical energy compared to the **permanent split capacitor (PSC) motors** used in most wet-rotor circulators. ECMs also create approximately four times higher starting torque than PSC motors. This greatly reduces the possibility of a "stuck rotor" condition after prolonged shut down periods.

The microprocessor-based speed control circuitry within an ECM powered circulator allows it to operate according to a coded instruction set that exists within the circulator's **EEPROM** (electrically erasable programmable read-only memory). This instruction set is also called the circulator's **firmware**. Firmware controls the circulator's microprocessor, similar to how software controls the microprocessor in a computer. However, the instructions contained in firmware are created and loaded into the circulator's solid-state components by its manufacturer and are typically not able to be changed without special equipment. Another difference between software and firmware it that instructions contained in firmware are not lost during power outages. They remain ready for operation even if the circulator is unpowered for several years. The ability to regulate the operation of an ECM-powered circulator based on such firmware allows tremendous versatility in customizing the circulator for specific tasks.

By changing the impeller speed, an ECM-powered circulator can generate a wide range of pump curves, as depicted in Figure 4-77. Whenever speed is decreased, the pump curve shifts down and to the left.

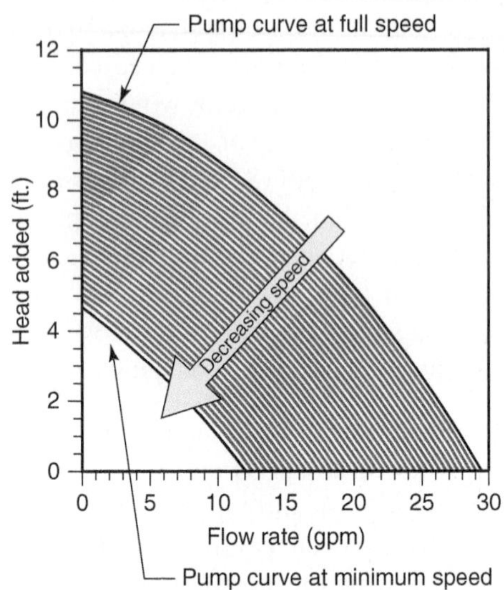

Figure 4-77 | When motor speed decreases, the pump curve shifts down and to the left.

CONSTANT DIFFERENTIAL PRESSURE CONTROL

One operating mode used in many ECM-powered smart circulators maintains a constant head (and constant differential pressure) over a wide range of flow rates. This **constant differential pressure control** mode creates the equivalent of a perfectly flat pump curve.

When the circulator is commissioned, the installer sets the circulator for the differential pressure desired when all zone valves are open (e.g., design load conditions). The circulator then automatically varies its speed to maintain this differential pressure setpoint regardless of which zone valves are open or closed at any given time.

Constant differential pressure control using a smart circulator is ideal for systems that meet *both* of the following criteria:

1. *They use valves for zoning.*
2. *They are configured so that most of the head loss of the distribution system occurs in the zone circuits and not in the common piping.*

The schematic in Figure 4-78 meets both of these criteria. It uses zone valves on each branch circuit and has relatively short headers sized for low head loss. The thermal storage tank also creates very little head loss.

When a zone valve closes, the system head loss curve steepens and the operating point begins shifting

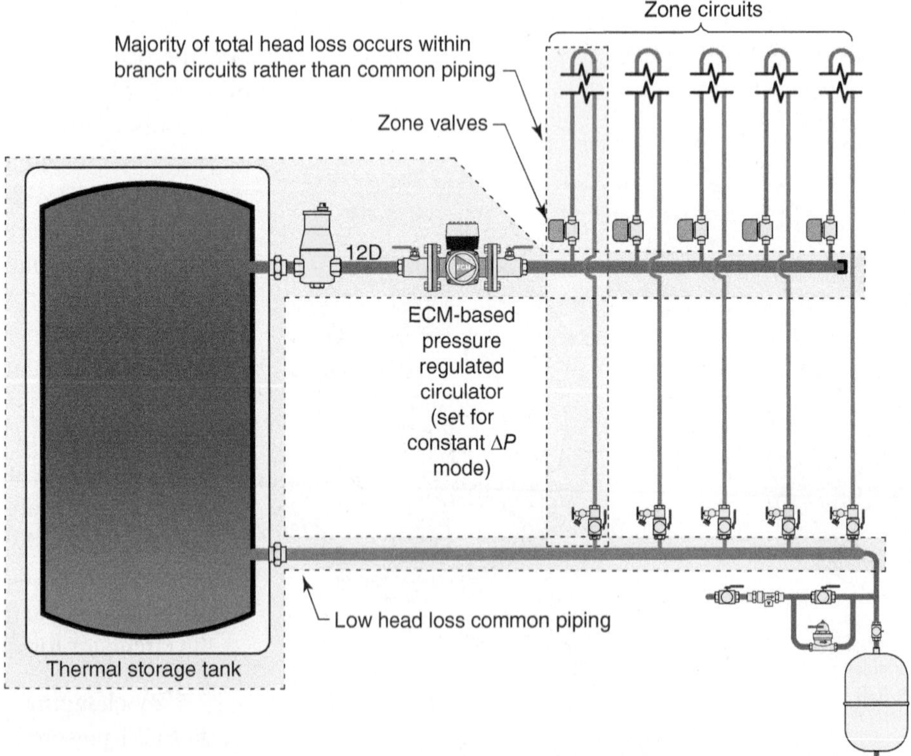

Figure 4-78 | Hydronic distribution system uses valves for zoning and has low head loss common piping. The smart circulator should be set for constant differential pressure control.

upward along the pump curve, as seen in Figure 4-79a. However, control circuitry within the smart circulator senses this departure from the set differential pressure setpoint and begins reducing motor speed. This causes the pump curve to shift down and to the left, as seen in Figure 4-79b. When the circulator detects that it is again operating at its original (set) differential pressure, it remains at that speed, as shown in Figure 4-79c. As other zone valves close, the process repeats itself. The net effect is that the operating point has shifted horizontally to the left, but remains at the same differential pressure and head on the vertical scale. This mimics what would happen with a fixed-speed circulator having a perfectly flat pump curve.

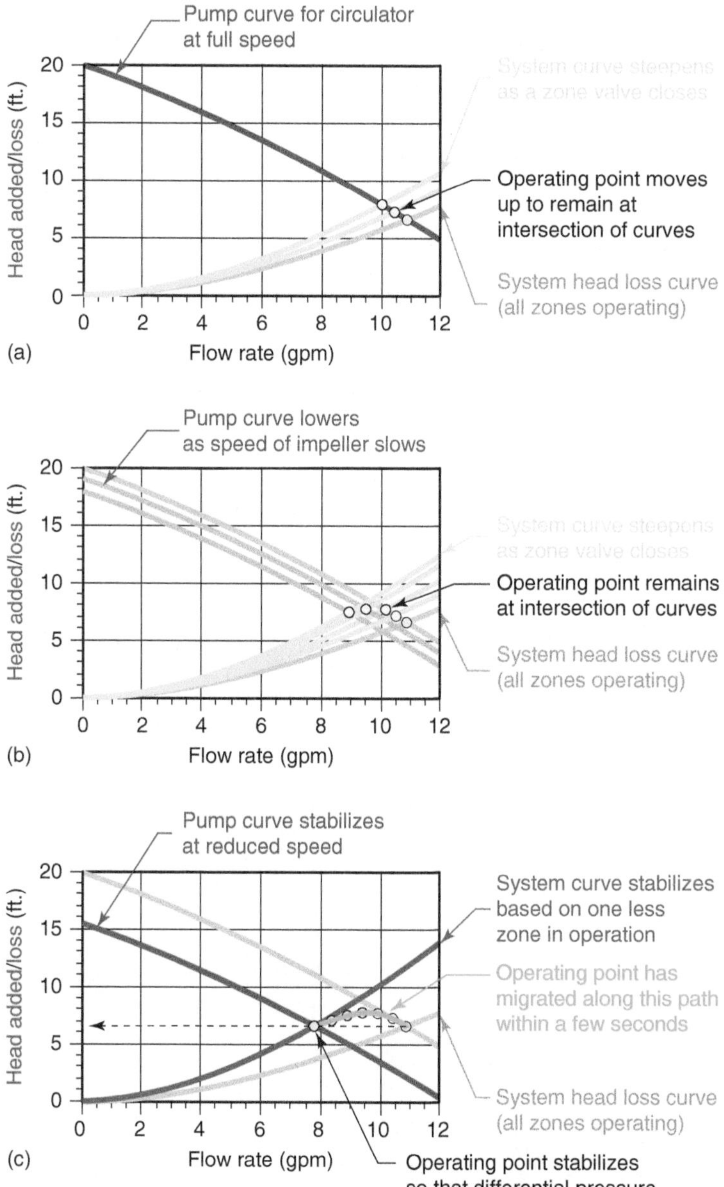

Figure 4-79 Sequence showing shifting of operating point as a zone valve closes in a system using an ECM-powered circulator set for constant differential pressure control. (a) Operating point climbs as zone valve closes. (b) Circulator detects operating point drift and reduces speed. (c) Circulator restores original differential pressure and remains operating at that speed.

When a zone valve opens, the process occurs in reverse. The initial drop in differential pressure is detected by the circulator. It responds by increasing speed until the original (set) differential pressure is restored.

These processes require only a few seconds and are completely regulated by the electronics and firmware within the circulator. No external pressure sensors are required. The desirable result is that differential pressure across the headers remains quasi-constant, regardless of the number of zones in operation at any time, as shown in Figure 4-80.

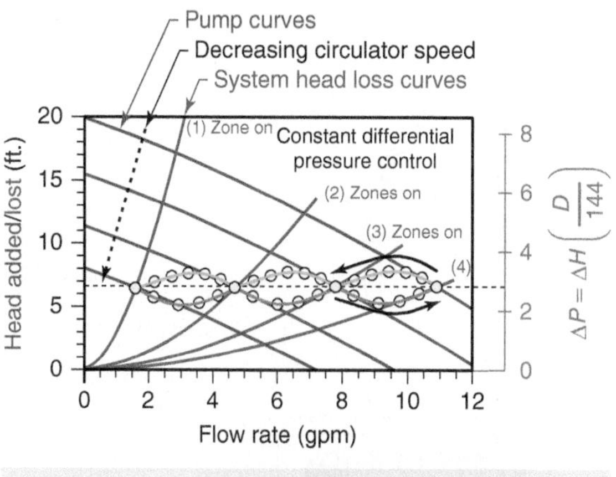

Figure 4-80 Constant differential pressure control allows the differential pressure across the headers to remain essentially constant regardless of the number of zones operating.

PROPORTIONAL DIFFERENTIAL PRESSURE CONTROL

Another operating mode built into many ECM-powered smart circulators is called **proportional differential pressure control**. This mode is best suited to hydronic systems where a significant amount of the total head supplied by the circulator is lost along the *mains* rather than the **crossover branches**. The piping arrangement shown in Figure 4-81 is an example of such a system.

Proportional differential pressure control also varies the circulator's speed in response to attempted changes in differential pressure detected by the circulator. However, in this control mode, the operating point moves downward as it moves to the left and upward as it moves to the right. The differential pressure at zero flow rate is always 50 percent of the set differential pressure at design (maximum) flow rate. As zone valves close, circulator speed is reduced in a way that causes the operating points to move downward along the sloping dashed line, as in Figure 4-82.

The objective of proportional differential pressure control is to minimize flow rate variations in the active crossover branches, regardless of how many crossover branches are active at any time. The movement of operating points shown in Figure 4-82 is the best way to approximate this objective in a 2-pipe direct return, or 2-pipe reverse return distribution system.

Many ECM-powered circulators can also detect when all zone valves in a system are closed. Under this condition, the circulator slows to a **sleep mode**, where it maintains just enough differential pressure to detect when one or more zone valves reopens. It then resumes normal operation. During sleep mode, the circulator requires about 5 watts of electrical power input. In systems with electrically controlled zone valves, the end

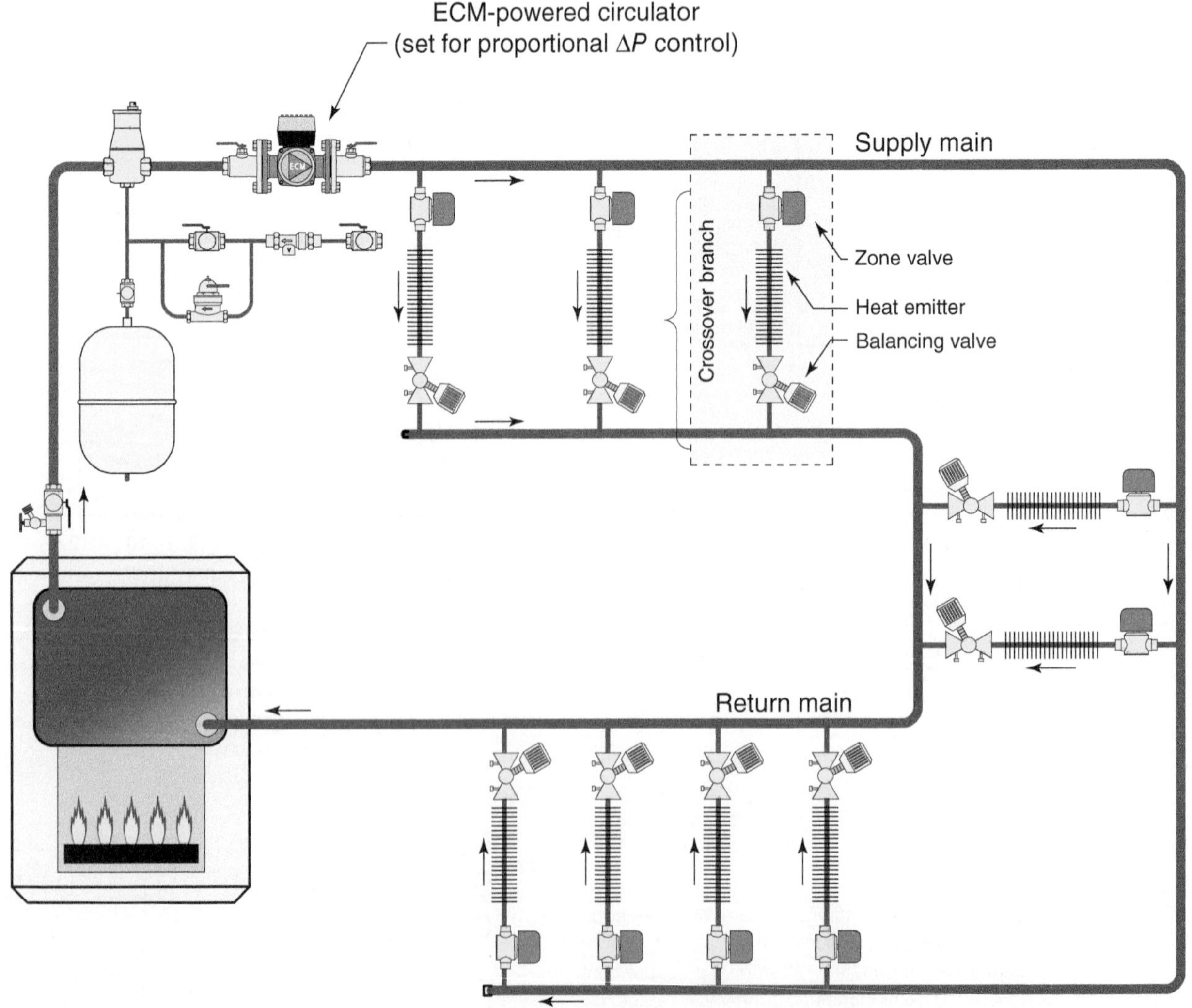

Figure 4-81 | Example of a 2-pipe reverse return distribution system using an ECM-powered circulator set for proportional differential pressure control.

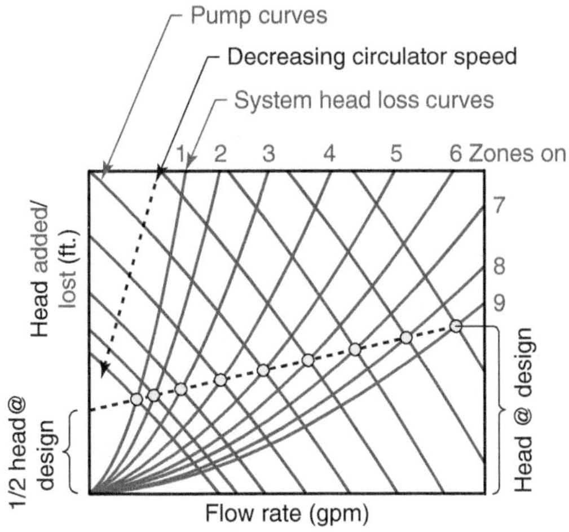

Figure 4-82 As zone valves close, the operating points move left and downward when an ECM-powered circulator operates in proportional differential pressure mode.

switches of the zone valves can be wired to turn the circulator off when no zones are calling for heat.

ENERGY SAVINGS USING SMART CIRCULATORS

Whenever the speed of a circulator decreases, so does its electrical power requirement. Every time the operating point shifts to the left (using constant differential pressure control), or to the left and downward (using proportional differential pressure control), the wattage supplied to the circulator's motor decreases.

Total savings depends on the accumulated time the circulator operates at various speeds (and associated input wattages) during a heating season. Circulator manufacturers use load simulation software, along with measured performance data, to compare operation of various circulators over an entire heating season. The results give estimated total electrical energy consumption over that season. These simulations indicate that electrical energy savings of 60 to 80 percent are possible using ECM-powered smart circulators when compared to the electrical energy consumption of standard wet-rotor circulators of equivalent peak hydraulic performance.

The implications of such energy savings are profound, considering that over 115 million small wet-rotor circulators are currently in use worldwide, with thousands more added each day. Energy savings also occur as existing circulators are eventually replaced by high-efficiency ECM-powered circulators. Most ECM-powered circulators sold in North America are built with the same flange dimensions as existing circulators and thus allow for easy replacement.

WIRE-TO-WATER EFFICIENCY OF A CIRCULATOR

Efficiency is typically defined as the ratio of some desired output divided by the necessary input. This concept can be applied to circulators. The combined efficiency of both the mechanical and electrical portions of a circulator can be accounted for by dividing the rate of mechanical energy transfer to the fluid (e.g., head), by the rate of electrical energy input to the circulator. This index is called the circulator's **wire-to-water efficiency**. It is a performance indicator that is especially relevant to small circulators that come with a non-interchangeable motor and a non-modifiable impeller.

Equation 4.3 can be used to calculate the wire-to-water efficiency of a circulator that is operating at a known flow rate and corresponding differential pressure.

(Equation 4.3)

$$\eta_{\text{wire-to-water}} = \frac{0.4344f(\Delta P)}{w}$$

where:

$\eta_{w/w}$ = wire-to-water efficiency of the circulator (decimal percent)

f = flow rate through the circulator (gpm)

ΔP = pressure differential measured across the circulator (psi)

w = input wattage required by the motor (watts)

0.4344 = units conversion factor

The input wattage required by a circulator is not a fixed value, but depends on where the circulator operates on its curve. Therefore, to make use of Equation 4.3, one needs to know the circulator's input wattage at the operating point on the pump curve. Data for input wattage as a function of flow rate may be available from the pump's manufacturer or can be measured using a simple wattmeter such as the **Kill-A-Watt meter** shown in Figure 4-83 used in combination with a flow meter.

Figure 4-84 shows the electrical power drawn by a small wet-rotor circulator with a permanent split capacitor (PSC) motor over a range of flow rates.

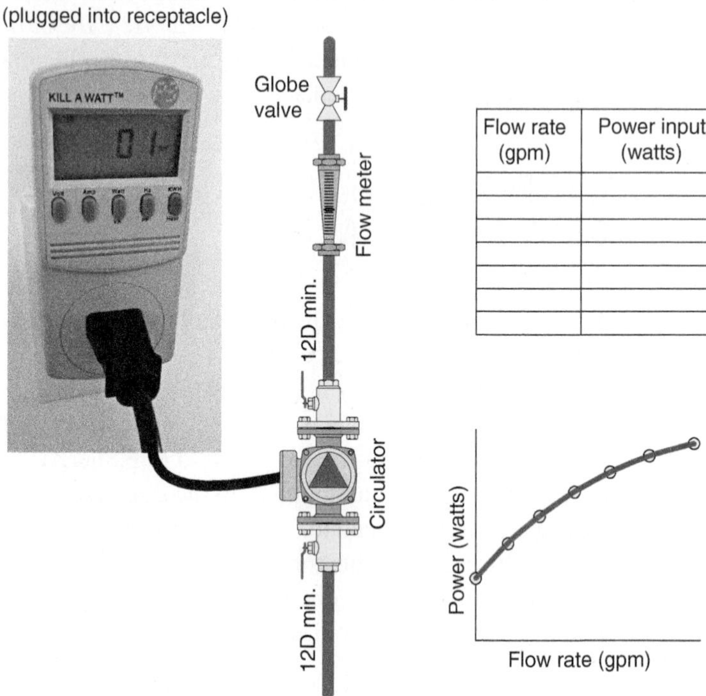

Figure 4-83 Example of a "Kill-A-Watt" meter, used in combination with a flow meter, to acquire performance data that allows a graph of power input versus flow rate to be constructed.

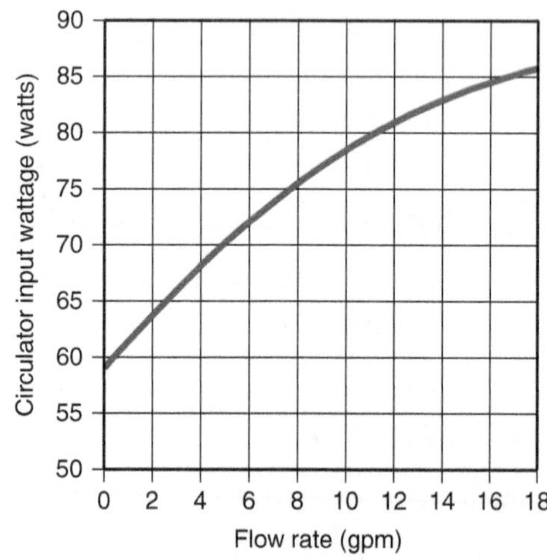

Figure 4-84 Input wattage required by a small wet-rotor circulator with PSC motor.

Figure 4-85 shows the pump curve of the same circulator used to prepare Figure 4-84, along with its calculated wire-to-water efficiency. Notice that peak wire-to-water efficiency occurs at a flow rate that is close to the center of the pump curve. For this circulator, peak wire-to-water efficiency occurs at a flow rate

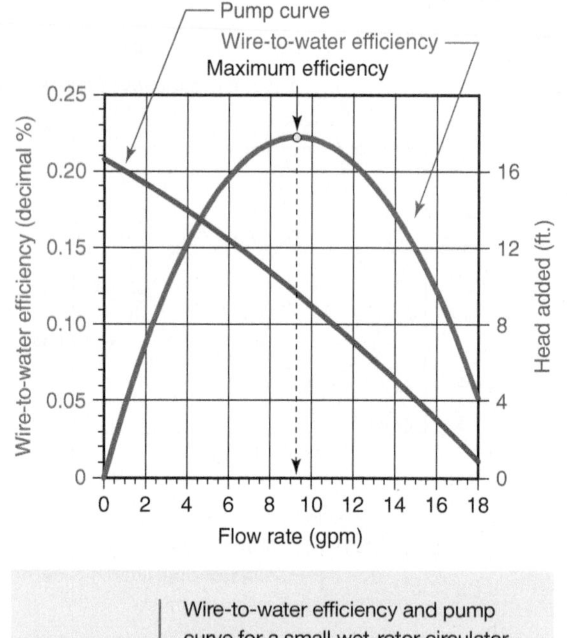

Figure 4-85 Wire-to-water efficiency and pump curve for a small wet-rotor circulator. Peak wire-to-water efficiency occurs at a flow rate near center of pump curve.

of about 9.3 gpm. Ideally, this circulator will be applied so that the intersection of the pump curve and system curve will be at, or near, this point of peak wire-to-water efficiency.

Example 4.3: A small wet-rotor circulator operates at a point where it produces a flow rate of 8.0 gpm of 140 °F water and a corresponding head of 9.7 feet. A wattmeter indicates the circulator is drawing 85 watts of input power. What is the wire-to-water efficiency of the circulator?

Solution: Equation 4.3 can be used, but it requires the pressure difference across the circulator, not the head. This differential pressure can be found using Equation 3.8. The fluid density is also required and is found from Figure 3-11:

$$\Delta P = \frac{(9.7)(61.35)}{144} = 4.13 \text{ psi}$$

This pressure difference can now be substituted into Equation 4.3 along with the other data:

$$\eta_{\text{pump \& motor}} = \frac{0.4344(8)(4.13)}{85}$$
$$= 0.169 \text{ or } 16.9\% \text{ efficient}$$

Discussion: This value seems surprisingly low but is typical of most current-generation, small, wet-rotor circulators using permanent split-capacitor (PSC) motors. Several factors contribute to this low efficiency. One is an unavoidable decrease in impeller efficiency as its diameter is reduced. Another is that the volute chambers of small circulators do not contain diffuser vanes to guide the fluid off the impeller and out of the volute with minimal turbulence.

Interestingly, the wire-to-water efficiencies of small circulators using electronically commutated motors (ECM) and operating at full speed are typically almost *double* those of small circulators using standard PSC motors. Thus, ECM-powered circulators, even if operated as fixed-speed devices, still hold significant operating cost advantages. Viewed another way: ECM circulators provide a given flow/head capability using about half the electrical power of a conventional circulator.

The hydraulic efficiency of most circulators is highest near the center of the pump curve, as shown in Figure 4-85. For the circulator to operate at or near its maximum efficiency, the system curve must pass through this region of reasonable efficiency. *For reasonable wire-to-water efficiency, the author recommends that the operating point (e.g., the intersection between the pump curve and system curve) should be within the middle third of the pump curve.* This recommended criteria is illustrated in Figure 4-86.

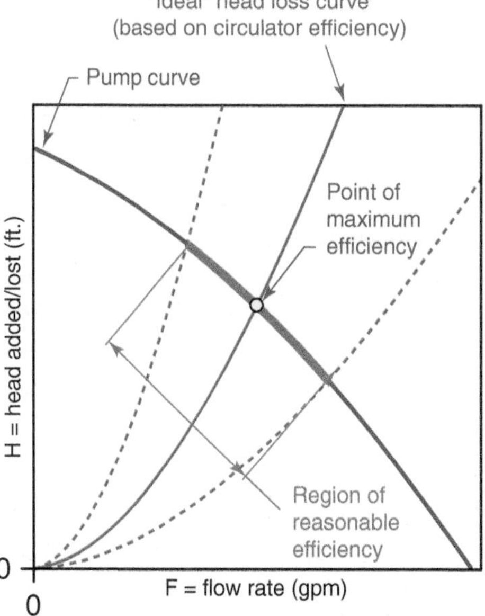

Figure 4-86 For reasonably good wire-to-water efficiency, the system head loss curve should cross over the middle third of the pump curve. The "ideal" situation is when the pump curve crosses the system head loss curve at the point of maximum wire-to-water efficiency.

CIRCULATOR CAVITATION

Chapter 3 discussed how liquids will begin boiling at specific combinations of pressure and temperature. The minimum pressure that must be maintained on a liquid to prevent it from boiling at some corresponding temperature is called its **vapor pressure**.

The process of boiling involves the formation of **vapor pockets** in the liquid. These pockets look like bubbles in the liquid, but should not be confused with air bubbles. They will form even in water that has been completely deaerated, instantly appearing whenever the liquid's pressure is lowered below the vapor pressure corresponding to its current temperature. This process is often described as the water "flashing" into vapor.

When vapor pockets form within liquid water, the density inside the vapor pocket is about 1,500 times lower than the surrounding liquid. This is comparable to a single kernel of popcorn expanding to the size of a baseball.

If the pressure on the liquid then increases above the vapor pressure, the vapor pockets instantly collapse inward in a process known as **implosion**. Although this sounds relatively harmless, on a microscopic level, vapor pocket implosion is a very violent effect, so violent that it can rip away molecules from surrounding surfaces, even hardened metal surfaces.

For a liquid at a given temperature, cavitation occurs anywhere the liquid's pressure drops below its vapor pressure. In hydronic systems, the most likely locations for cavitation are in partially closed valves, or at the eye of a circulator's impeller.

As liquid flows into the impeller, it experiences a rapid drop in pressure. If the pressure at the impeller's eye drops below the vapor pressure, thousands of vapor pockets instantly form. These vapor pockets are carried outward between the impeller vanes. As the liquid/vapor mix progresses outward toward the periphery of the impeller, it pressure is increasing. At some location, the fluid pressure rises above the vapor pressure. This causes the vapor pockets to instantly implode. This usually occurs near the outer edges of the impeller. The violent implosions create the churning or crackling sounds that are characteristic of a cavitating impeller. A noticeable drop in flow rate and head occurs because the fluid inside the circulator is now a partially compressible mix of liquid and vapor. If the circulator continues to operate in this **vapor cavitation** mode, the impeller will be severely eroded in as little as a few weeks of operation. Obviously, vapor cavitation must be avoided if the system is to have a long, reliable life.

There are analytical methods for predicting the onset of cavitation. They involve calculations for a quantity known as the **net positive suction head available** to the circulator within a given piping circuit. Interested readers should consult chapter 7 of *Modern Hydronics Heating*, 3rd edition, for an explanation of these methods.

Another type of cavitation that can develop within circulators is called **gaseous cavitation**. It occurs when the liquid pressure at the eye of the impeller drops below the saturation pressure of any dissolved gases—such as oxygen or nitrogen—that may be dissolved in the system fluid. These gases may also be present as small air bubbles entrained by the system fluid.

Gaseous cavitation allows bubbles to form on the lower pressure side of the impeller vanes. It is not as violent or damaging as vapor cavitation. It often occurs intermittently when systems are first put into operation and thus operating with fluids containing dissolved gases. In closed-loop systems, gaseous cavitation can be eliminated by proper purging methods along with use of high performance air separators.

AVOIDING CAVITATION

The following guidelines widen the safety margin against the onset of cavitation. Some strive to keep the state of the fluid entering the eye of the impeller non-turbulent. Others strive to prevent the fluid from approaching conditions under which boiling is about to occur.

- Keep the static pressure on the system as high as practical.
- Keep the fluid temperature as low as practical.
- Always place the expansion tank near the inlet side of the circulator.
- Keep the circulator low in the system to maximize static pressure at its inlet.
- Do not place any high-flow-resistance components (especially flow-regulating valves) near the inlet of the circulator.
- If the system has a static water level, such as a partially filled tank, keep the inlet of the circulator as far below this level as possible.
- Be especially careful in the placement of high head circulators or close-coupled series circulator combinations because they create greater pressure differentials.
- Provide a straight pipe that is at least 12 pipe diameters long upstream of the circulator's inlet.
- Install good deareating devices in the system.

CORRECTING EXISTING CAVITATION

Occasionally, a technician may need to correct an existing noisy circulator condition. If the circulator exhibits the classic signs of cavitation (i.e., churning or crackling sounds, vibration, and poor performance), the problem can probably be corrected by applying one or more of the previous guidelines.

The circulator should first be isolated and the impeller inspected for cavitation damage, especially if the problem has existed for some time. Look for signs of abrasion or erosion of metal from the surfaces of the impeller. If evidence of cavitation damage is present, the impeller (or rotor assembly) should be replaced.

Because every system is different, there is no definite order in which to apply these guidelines in correcting the cavitation problem. In some cases, simple adjustments such as increasing the system's pressure or lowering its operating temperature will solve the problem. In others, the layout may be so poor that extensive piping modifications may be the only solution. Obviously, corrections that involve simple setting changes are preferable to those requiring piping modifications. These should be attempted first.

In cases where cavitation is not severe, increasing the system's static pressure will often eliminate the cavitation. This can usually be done by increasing the pressure setting of the boiler feed water valve. Be sure the increased pressure does not cause the pressure-relief valve to open each time the fluid is heated to operating temperature.

If the system experiencing cavitation is operating at a relatively high water temperature (above 190 °F), try lowering the high limit controller setting by 20 °F. In this temperature range, the vapor pressure of water decreases rapidly as its temperature is lowered. This may be enough to prevent vapor pockets from forming (the cause of vaporous cavitation). If such an adjustment is made, be sure the system has adequate heat output at the lower operating temperature.

Always check the location of the expansion tank in relation to the circulator. If the circulator is pumping *toward* the expansion tank, consider moving the tank near the inlet of the circulator. This may not only eliminate cavitation, but also improve the ability of the system to rid itself of air.

Also look for flow restrictions near the circulator inlet. These include throttling valves, plugged strainers, or piping assemblies made up of many closely spaced fittings. If necessary, eliminate these components from the inlet side of the circulator.

SELECTING A CIRCULATOR (FOR A FLUID-FILLED CIRCUIT)

The ideal circulator for a given system would produce the exact flow rate desired while operating at maximum efficiency and would draw the least amount of electrical power. It would also be competitively priced and built to withstand many years of unattended service. These conditions, although possible, are not likely in all projects because of the following considerations:

- Only a finite number of circulators are available for any given range of flow rate and head requirements. This contrasts with the fact that an unlimited number of piping systems can be designed for which a circulator must be selected.
- Some small circulators are single-speed units that operate on a fixed pump curve. These circulators cannot be fine-tuned to the system requirements through impeller changes, motor changes, and so on, as is often possible with larger circulators. The exception is multi-speed (usually 3-speed) circulators where each speed represents a different pump curve.
- The choices may be further narrowed by material compatibility. For example, cast-iron circulators should never be used in open-loop hydronic applications because of corrosion.
- Purchasing channels and inventories of spare parts may limit the designer's choices to the product lines of one or two manufacturers.

The selection process must consider these limitations while also attempting to find the best match between the head and flow-rate requirements of the system and circulator. A suggested procedure for finding this match is as follows:

STEP 1. Determine the desired flow rate in the system based on the thermal and flow requirements of the heat emitters. The relationship between flow rate and heat transport is described by the sensible heat rate equation (Equation 3.2). The flow requirement of various heat emitters will be discussed in Chapter 5.

STEP 2. Use the system head loss curve to determine the head loss of the system at the required flow rate. Detailed methods for determining the system head loss curve were given in Chapter 3.

STEP 3. Plot a point representing the desired flow rate and associated head loss on the pump curve graph of the circulators being considered.

STEP 4. Select a circulator that delivers up to 10 percent more head than required at the desired flow rate. This provides a slight safety factor in case the system is installed with more piping, fittings, or valves than what was assumed.

STEP 5. Finally, the curve of the selected circulator should (ideally) intersect the system head loss curve within the middle third of the pump curve where the circulator's wire-to-water efficiency is relatively high.

The following example will pull together many of the previous information and procedures.

Example 4.4: A piping system is being planned to deliver 35,000 Btu/hr of heat using a design temperature drop of 20 °F. The fluid used will be water at an average temperature of 160 °F. Using the procedures in Chapter 3, the piping loop is estimated to have a total equivalent length of 325 feet of 3/4-inch type M copper tube. Determine the necessary flow rate in the system and select an appropriate circulator from the pump curves shown in Figure 4-73.

Solution: The "target" flow rate for delivering 35,000 Btu/hr of heat using water and a temperature drop of 20 °F is determined using Equation 3.2:

$$Q = (8.01Dc)f(\Delta T)$$

Q = rate of heat transfer into or out of the water stream (Btu/hr)

8.01 = a constant based on the units used

D = density of the fluid (lb/cubic ft)

c = specific heat of the fluid (Btu/lb/°F)

f = flow rate of water through the device (gpm)

ΔT = temperature change of the water (°F)

Using Figure 3-11, the density of 160 °F water is found to be 60.9 lb/ft³.

The specific heat of water at 160 °F is 1.00 Btu/lb/°F.

Equation 3.2 is now rearranged to solve for the flow rate:

$$f = \frac{Q}{8.01 \times c \times D \times \Delta T} = \frac{35{,}000}{8.01 \times 1.00 \times 60.9 \times 20}$$
$$= 3.59 \text{ gpm}$$

The head loss of the circuit can now be estimated using Equation 3.11. Doing so requires the value of the fluid properties factor (α) and the pipe size factor (c).

The value of α for water at an average temperature of 160 °F is found in Figure 3-24: $\alpha = 0.046$.

The pipe size factor for 3/4-inch copper tube is found in Figure 3.25: c = 0.061957.

These values are combined with the total equivalent length of the circuit and the target flow rate (using Equation 3.11) to estimate the head loss of the system:

$$H_L = [\alpha c L](f)^{1.75}$$
$$= [(0.046)(0.061957)(350)](3.59)^{1.75} = 9.34 \text{ ft}$$
$$H_L = [\alpha c L](f)^{1.75} = [0.9975](f)^{1.75}$$

The target operating point of 3.59 gpm and 9.34 feet of head loss is now plotted on the family of pump curves, as shown in Figure 4-87. Also plotted are several other random points along the system curve calculated using Equation 3.11 with different flow rates. These points allow the system curve to be sketched on the same graph with the pump curves.

Discussion: The pump curve for circulator #1 crosses over the system head loss curve at a flow rate (about 5.5 gpm) that is significantly higher than the target flow rate of 3.59 gpm. Because it is more powerful than needed in this application, this circulator would cost more to install and operate compared to other options. The pump curve for circulator #3 would only produce a flow rate of about 2.9 gpm, lower than required. The pump curve for circulator #2 passes slightly above the target operating point, which is acceptable to provide a slight safety factor. It also appears that the system head loss curve and pump curve #2 intersect within the middle third of the pump curve, and thus the circulator's wire-to-water efficiency would be relatively high. Thus, the circulator represented by pump curve #2 would be the best choice in this scenario.

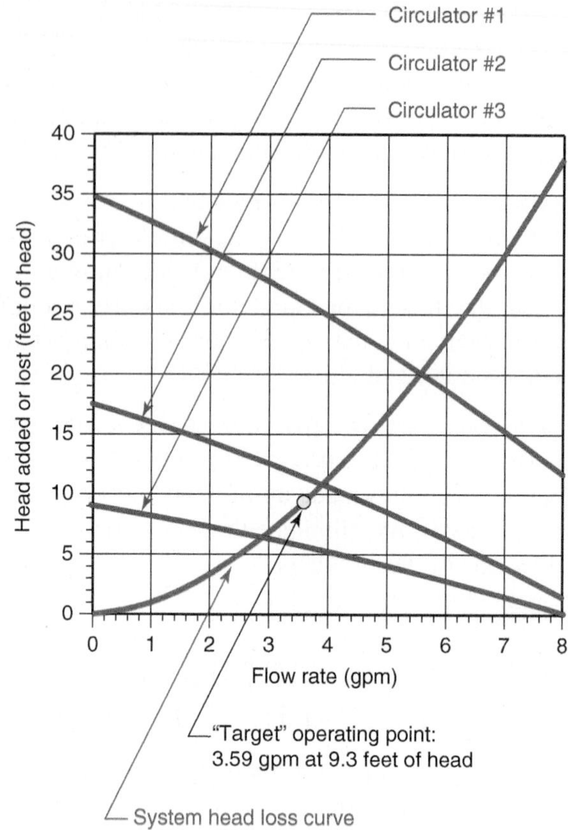

Figure 4-87 System head loss curve and pump curves for example 4.4.

SPECIAL SITUATIONS FOR CIRCULATORS

There are some special situations for circulators in systems using renewable energy heat sources or related hardware. Designers need to be aware of such situations and properly apply circulators for efficient performance and long life. These situations are introduced within the discussion of circulators in this chapter and then illustrated in several example systems in later chapters.

SPECIAL SITUATIONS: OPEN LOOPS

One situation that requires careful selection and placement of circulators is a circuit containing an unpressurized component such as an atmospherically vented thermal storage tank or outdoor wood-fired hydronic heater. This type of circuit is an "open loop." Figure 4-87a contrasts an open-loop circuit with a "closed-loop" circuit, which is completely closed to the atmosphere and usually operates under slight positive pressure.

The water in an open-loop system is directly exposed to the atmosphere, perhaps only through a small vent tube at the heat source or storage tank, but nonetheless exposed. As such, the water in the circuit will absorb oxygen molecules from the atmosphere each time it cools. This dissolved oxygen will be carried throughout the system every time the circulator operates. Oxygen is the "food" for corrosion. If ferrous metal components such as cast-iron circulators are present in open-loop systems, damage due to corrosion can be swift and serious. Therefore, *any circulators used in open loop systems must have volutes and other wetted surface components constructed of bronze, stainless steel, or other engineered polymers that will not corrode in the presence of dissolved oxygen.* Nearly all circulator manufacturers offer models constructed of these materials for use in open-loop systems.

Another issue with open systems is low static pressure. The water level in the system determines the zero static pressure line. Water *below* this level will be under slight positive pressure, which can be calculated using Equation 3.4. Water in the circuit *above* the zero

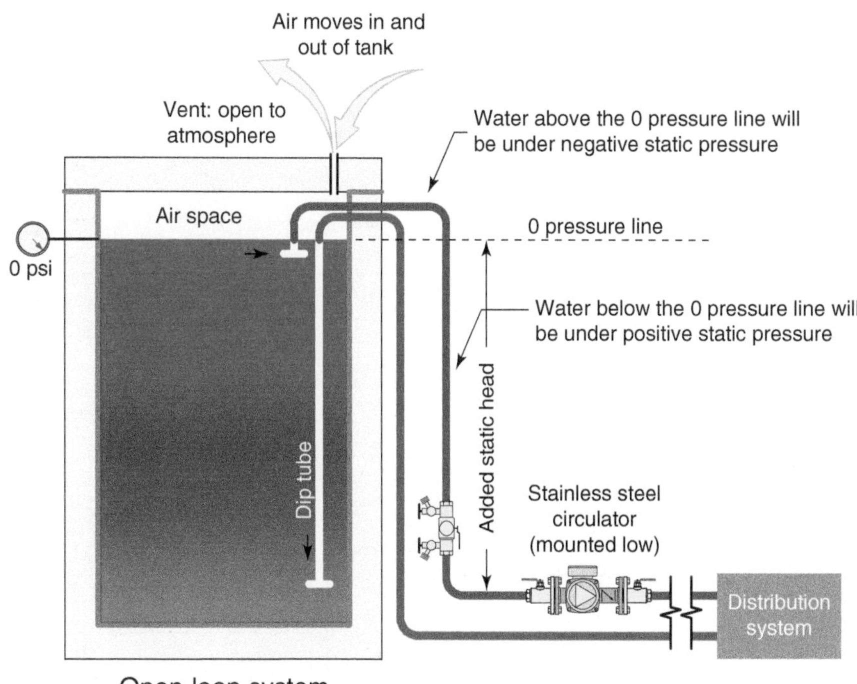

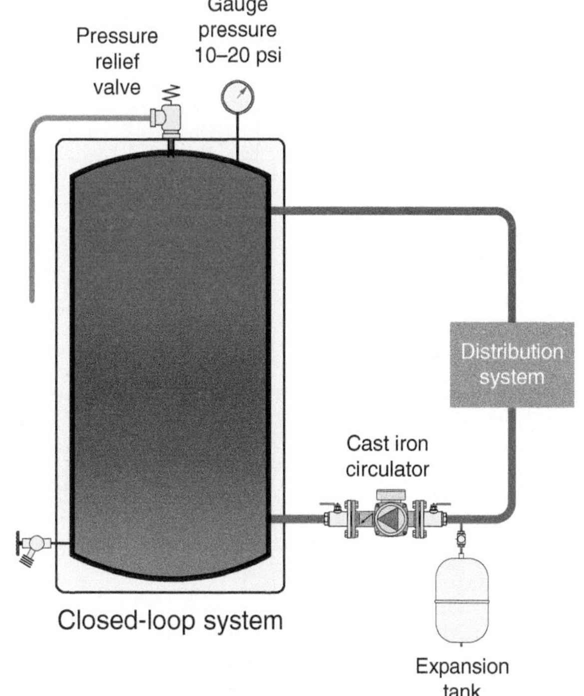

Figure 4-87a | Examples of open-loop and closed-loop circuits. In the open-loop system, the water in the storage tank is in direct contact with the atmosphere through the vent opening at the top of the tank.

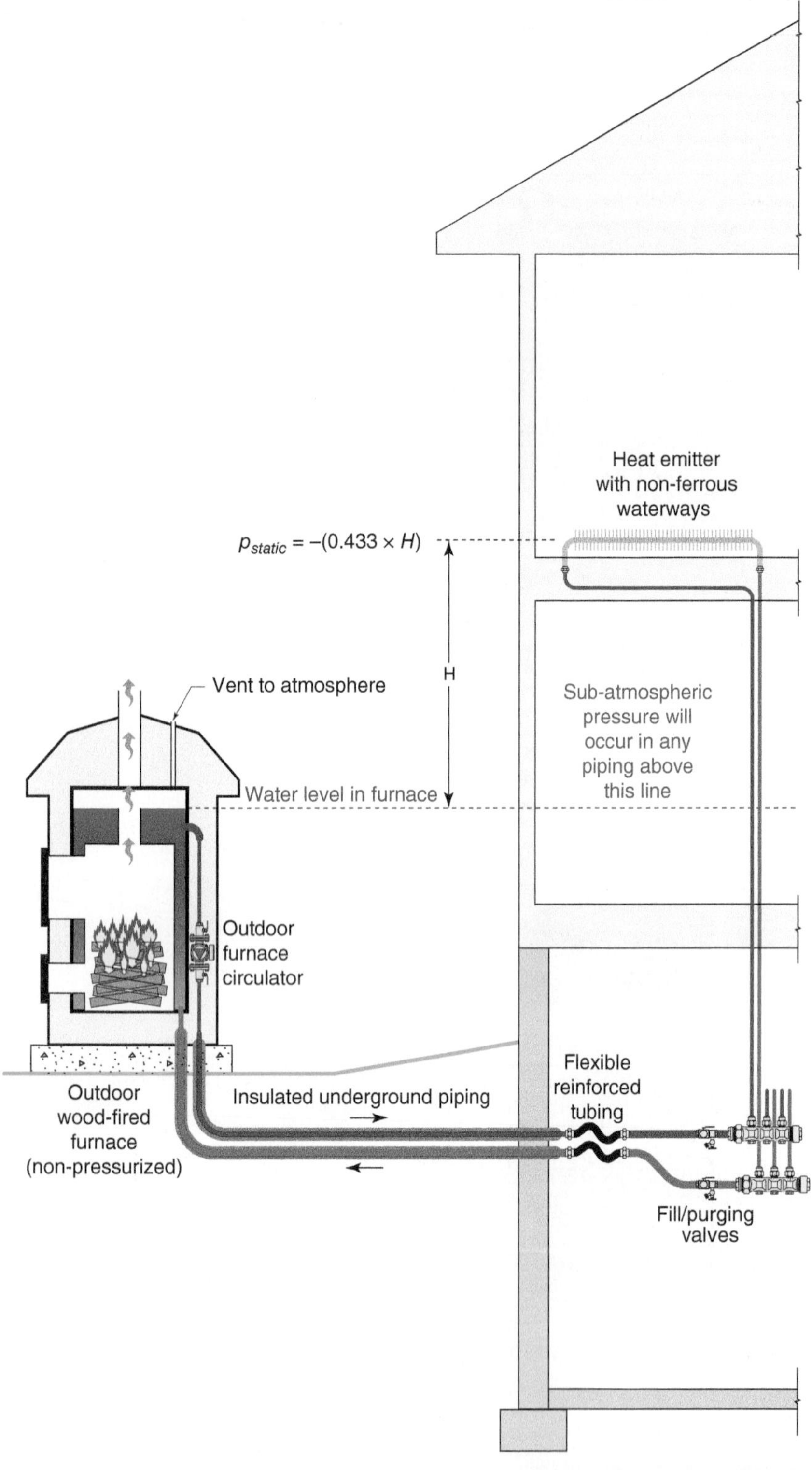

Figure 4-88 The zero static pressure line is at the top of the water in the outdoor furnace. Water below this level will be under slight positive pressure. Water above this line will be under negative pressure when the circulator is off.

pressure line will be under negative static pressure when the circulator is off. Figure 4-88 illustrates this situation for a system using an unpressurized outdoor wood-fired hydronic heater.

The circulator in open loop systems should be located as far below the water level as possible. This maximizes static pressure at the eye of the circulator and provides the best possible safety margin against circulator cavitation.

SPECIAL SITUATIONS: DRAINBACK SOLAR THERMAL SYSTEMS

Drainback solar thermal systems will be discussed at length in later chapters. However, the flow and head requirements of the collector circulator(s) in a drainback system are unique and appropriate to discuss in this section.

To prevent freezing, all the water in the solar collectors and exposed piping of a drainback-protected solar thermal system must flow back into heated space whenever the collector circulator is off. This action requires only gravity and properly sloped piping. Figure 4-89 shows a simplified drainback-protected solar collection subsystem in which the upper portion of the thermal storage tank serves as the drainback reservoir.

When turned on at the start of each energy collection cycle, the collector circulator in a drainback-protected system must push water upward through the collector supply piping and ensure that water flows over the uppermost piping in the collector circuit. This is very different from what a circulator does in a completely filled hydronic circuit.

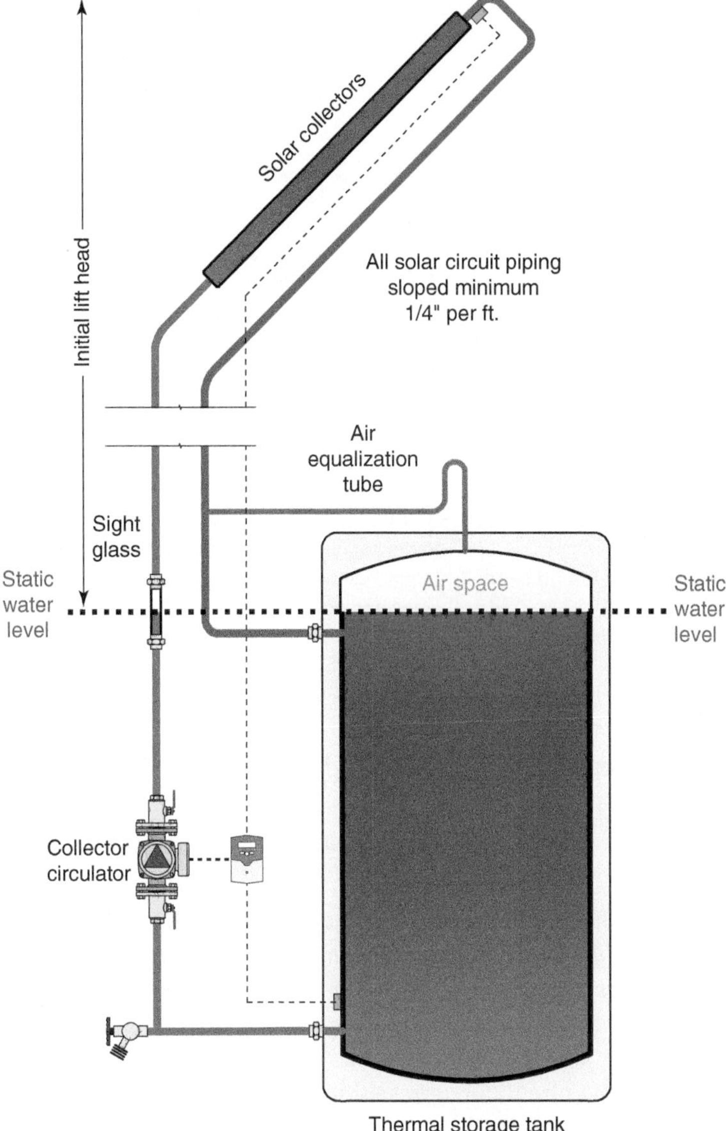

Figure 4-89 A drainback solar collection subsystem. Note the "lift head" from the water level in the tank to the top of the collector array.

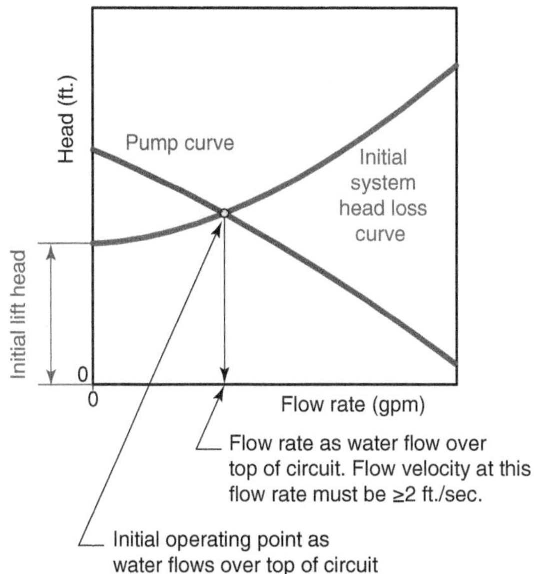

Figure 4-90 The initial system head loss curve is elevated by the lift head of the collector circuit. The curvature of the system head loss curve represents the frictional head loss of the collector supply piping up to the top of the circuit.

The relationship between the system head loss curve and pump curve shortly after the solar energy collection cycle begins is shown in Figure 4-90.

Notice that the *initial* system curve represents both the frictional head loss created by the collector supply piping up to the top of the circuit as well as the initial **lift head** representing the distance from the static water level in the storage tank to the top of the circuit.

The initial operating point where the initial system head loss curve intersects the pump curve is very important. The flow rate represented by this operating point must be able to produce a flow *velocity* in the collector return piping of at least 2 feet per second. This is the minimum flow velocity that can efficiently entrain air bubbles and move them downward in a vertical pipe. Thus, establishing a flow velocity of at least 2 feet per second in the collector return piping ensures that the air in this pipe will be forced back down into the storage tank.

Once the air is removed from the collector return piping, a **siphon** is established in the collector circuit. The collector return pipe is now filled with fluid. The weight of the fluid flowing down the return piping now counteracts the weight of the fluid moving up the supply piping. This effectively removes all, or nearly all, of the initial lift head, as illustrated in Figure 4-91.

Notice that the final system head loss curve is "steeper" than the initial system head loss curve. This happens because the frictional head loss now includes both the collector supply piping, the collectors, and the collector return piping.

However, even with great frictional head loss, the siphon formation and subsequent elimination of the initial

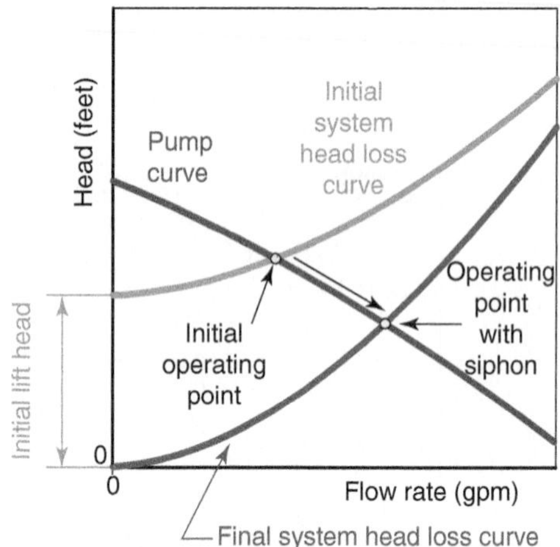

Figure 4-91 Transition of initial system head loss curve to final system head loss curve as the lift head is eliminated by formation of siphon in collector return piping.

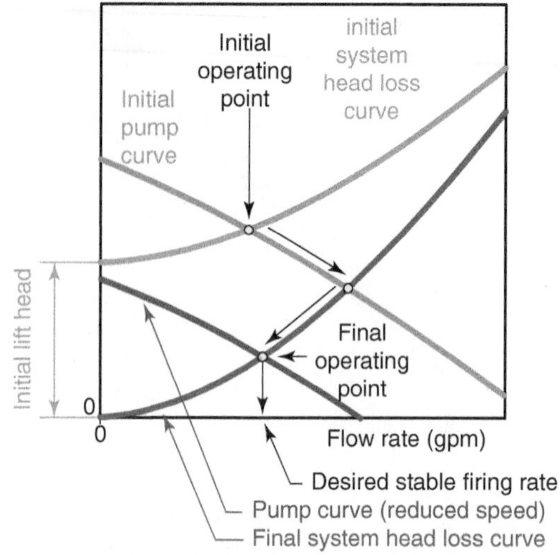

Figure 4-92 Reducing circulator speed after the siphon is formed allows desired flow rate at reduced circulator input power.

lift head cause the operating point to shift to the right and thus produce a higher flow rate through the collector array. Assuming the collector circulator continues to operate on the pump curve shown, this is the flow rate that will exist in the collector circuit for the remainder of the collection cycle. A typical fill time for a drainback system is only 30 seconds to perhaps 3 minutes. Thus, the higher flow rate would likely exist for a high percentage of the total solar collection cycle.

It is possible to take advantage of the siphon and resulting shift in the operating point to reduce circulator input power. This is done by reducing the speed of the collector circulator after the siphon is formed, as illustrated in Figure 4-92.

There are several controllers now available that can operate the collector circulator at full speed for a user-specified time to allow siphon formation and then reduce circulator speed to maintain an appropriate steady state flow rate. Reducing the circulator speed reduces its input power and thus lowers its operating cost.

4.5 Hydraulic Separation

Many hydronic systems, especially those using renewable heat sources, contain multiple, independently controlled circulators. These circulators can vary significantly in their flow and head characteristics. Some operate at fixed speeds while others operate at variable speeds.

When two or more circulators operate simultaneously in the same system, they each attempt to establish differential pressures based on their own pump curves. *Ideally, each circulator in a system will establish a differential pressure and flow rates that is* unaffected *by the presence of another operating circulator within the system.* When this desirable condition is established, the circulators are said to be **hydraulically separated** from each other.

Conversely, the lack of **hydraulic separation** can create very *undesirable* operating conditions in which circulators interfere with each other. The resulting flows and rates of heat transport within the system can be greatly affected by such interference, often to the detriment of proper heat delivery.

The degree to which two or more operating circulators interact with each other depends on the head loss of the piping path they have in common. This piping path is called the **common piping**, since it is common to, or shared, by both circuits. The lower the head loss of the common piping, the less the circulators will interfere with each other. Figure 4-93 illustrates this concept for a system with two independently operated circuits.

In this system, both circuits share common piping. The "spacious" geometry of this common piping creates very low flow velocity through it. As a result, very little head loss can occur across it.

Assume that circulator 1 is operating but circulator 2 is off. The lower system head loss curve (shown in blue) in Figure 4-93 applies to this situation. The point where this lower system head loss curve crosses the pump curve for circulator 1 establishes the flow rate in circuit 1.

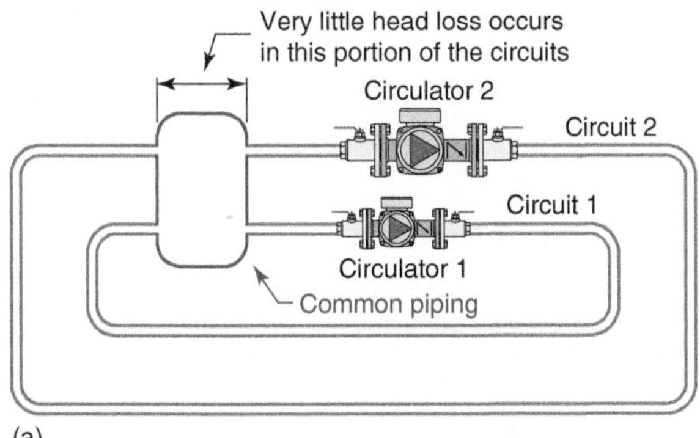

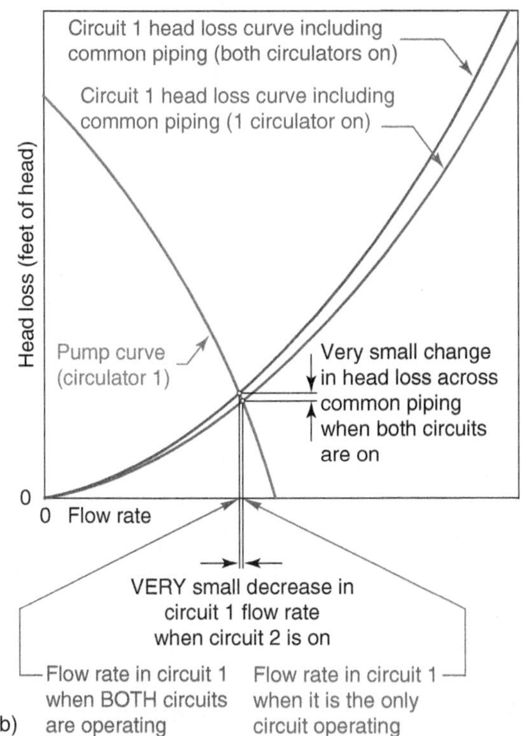

Figure 4-93 (a) Two independently controlled circuits that share a common piping path having very low head loss. (b) There is very little drop in the flow rate through circuit 1 when the circulator in circuit 2 is turned on.

Next, assume circulator 2 is turned on, and that circulator 1 continues to operate. The flow rate through the common piping will increase and so will the head loss across it. However, because of its spacious geometry, the increase in head loss across the common piping will be very low. The system head loss curve that is now "seen" by circulator 1 has steepened, but very slightly. It is shown as the green curve in Figure 4-93. The operating point of circuit 1 has moved very slightly to the left, meaning that the flow rate through circuit 1 has decreased very slightly.

Such a small change in circuit flow rate will have virtually no effect on the ability of circuit 1 to deliver heat. Thus, the slight interference created when circulator 2 was turned on is of no consequence. We could say that this situation represents almost perfect hydraulic separation between the two circulators.

One could imagine a hypothetical situation in which the head loss across the common piping was zero, even with both circuits operating. Because no head loss occurs across the common piping, it would be impossible for either circulator to have any effect on the other circulator. Such a condition would represent "perfect" hydraulic separation, and would be ideal.

Fortunately, perfect hydraulic separation is not required to ensure that the flow rates through independently operated circuits, each with their own circulator and each sharing the same low head loss common piping, remain reasonably stable and thus capable of delivering consist heat transfer. In animated terms—the simultaneously operating circulators cannot "feel" each other's presence within the system and thus operate as if they were essentially each in an independent circuit.

For all practical purposes, one can think of (and design) circuits that are hydraulically separated as if they were completely separate of each other, as illustrated in Figure 4-94.

Having stated that hydraulic separation is desirable, it is still worthwhile to consider a situation in which hydraulic separation is *not* present and observe the consequences.

Consider the system shown in Figure 4-95. The larger circulator is sized to move sufficient flow through the higher flow resistance circuit, including the high flow resistance heat source. When operating, the flow created by the large circulator creates a pressure drop of 5 psi between the supply and return headers connected to the heat source.

The smaller circulator can only produce 4 psi pressure differential between its inlet and outlet—even at zero flow. Thus, the check valve in the smaller circulator is held shut by the higher pressure at its

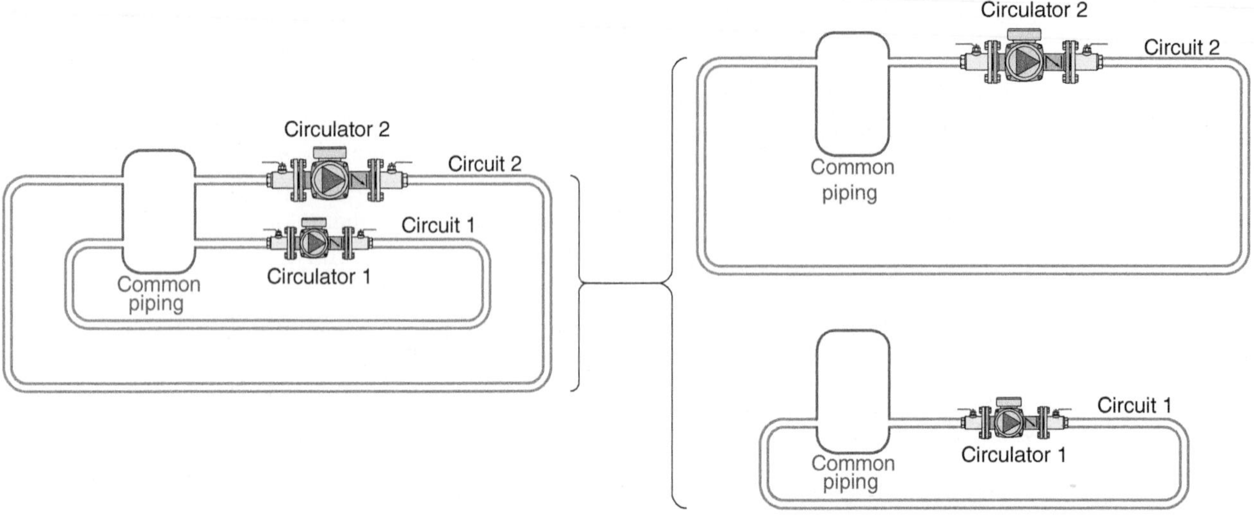

Figure 4-94 | For practical purposes, circuits that are hydraulically separated from each other can be visualized, and designed, as if they were totally independent of each other.

outlet relative to its inlet whenever the larger circulator is operating. Although the smaller circulator will continue to run, it is said to be "**deadheaded**" since there is no flow through it. Under such a condition, it will dissipate its input power as heat. This heat will be absorbed by the water in the circulator's volute as well as dissipated by the circulator's body. Small wet-rotor circulators can typically withstand this condition for a while, although it is not a condition that should be allowed by proper design. The solution is to create hydraulic separation between the circuits.

ACHIEVING HYDRAULIC SEPARATION

Any component, or combination of components, that has very low head loss and is common to two or more hydronic circuits can provide hydraulic separation between those circuits.

One way to create low head loss is to keep the flow path through the common piping very short. Another way is to significantly reduce the flow *velocity* through the common piping.

Examples of devices that use these principles include:

- closely spaced tees
- a tank (which might also serve other purposes in the system)
- a hydraulic separator

Each of these methods of achieving hydraulic separation will be illustrated.

The common piping in Figure 4-96 consists of the closely spaced tees and "generously sized" headers. These components are outlined by the dashed lines.

Because they are positioned as close to each other as possible, there is virtually no head loss between the tees. These tees form the common piping between the heat source circuit and the distribution circuits and thus provide hydraulic separation between these circuits.

The "generously sized" headers create low flow velocity, and low head loss, and thus provide hydraulic separation between the five distribution circulators. *These headers should be sized so the maximum flow velocity in the header, when all circulators served by the header are on, is no more than 2 feet per second.* Together, these details create hydraulic separation between all six circulators in the system.

The closely spaced tees allow the heat source circulator to only "see" the flow resistance of the heat source and the piping between the heat source and closely spaced tees. The heat source circulator does not assist in creating flow through the distribution circuits. Likewise, the five distribution circulators are only responsible for circulation through their respective circuits and do not assist in creating flow through the heat source.

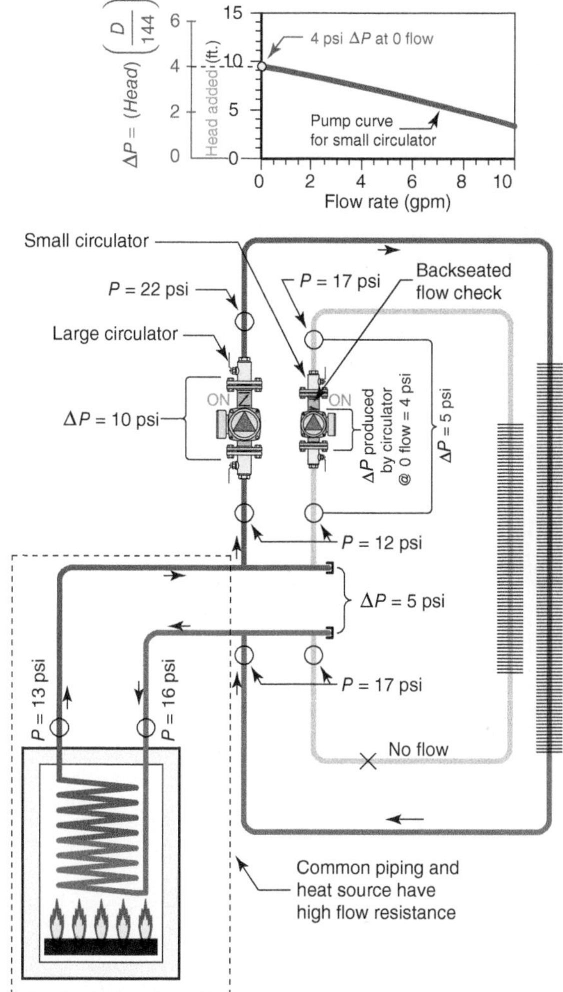

Figure 4-95 Two circuits, one with a larger circulator and the other with a smaller circulator, are jointed by a high-flow-resistance common piping path. The lack of hydraulic separation causes flow in the circuit with the smaller circulator to stop whenever the larger circulator operates.

Notice also that fixed speed and variable speed circulators, perhaps of different sizes, can be combined onto the same generously sized header system. Interaction between these circulators will be very minimal because of the generously sized (e.g., low head loss) headers.

Figure 4-97 shows a buffer tank and generously sized headers serving as the low head loss common piping that provides hydraulic separation between the heat source circulator and each of the distribution circulators. This demonstrates that hydraulic separation can sometimes be accomplished as an ancillary function to the main purpose of the device (e.g., hydraulic separation is not the main function of the buffer tank).

Still another method of providing hydraulic separation is using a device appropriately called a **hydraulic separator**. Although relatively new in North America, hydraulic separators have been used in Europe for many years. Figure 4-98 shows a hydraulic separator installed in place of the buffer tank of Figure 4-97. Note the similarity of the piping connections between the buffer tank and hydraulic separator.

The external appearance and internal construction of a hydraulic separator are shown in Figure 4-99.

Hydraulic separators, which are sometimes also called **low loss headers**, create a zone of low flow velocity within their vertical body. The diameter of the body is typically three times the diameter of the connected piping. This causes the vertical flow velocity within the vertical body to be no more than 1/9th of the flow velocity in the piping connected to either side of the hydraulic separator. Such low velocity creates very little head loss and very little dynamic pressure drop between the upper and lower connections. Thus, a hydraulic separator provided hydraulic separation in a manner similar to a buffer tank, only smaller.

The reduced flow velocity within a hydraulic separator allows it to perform two additional functions. First, air bubbles can rise upward within the vertical body and be captured in the upper chamber. When sufficient air collects at the top of the unit, the float-type air vent allows it to be ejected from the system. Thus, a hydraulic separator can replace the need for a high performance air separator.

Second, the reduced flow velocity inside the hydraulic separator allows dirt particle to drop into a collection chamber at the bottom of the vertical body. A valve at the bottom can be periodically opened to flush out the accumulated dirt. Thus, the hydraulic separator serves as a dirt removal device.

Modern hydraulic separators provide three functions: Hydraulic separation, air removal, and dirt removal. This makes them well suited for a variety of systems, especially those in which an older distribution system—one that may have some accumulated sludge from ferrous metal component—is connected to a new heat source.

Most hydraulic separators contain an internal **coalescing media** that enhances their ability to remove air and dirt. Coalescing media are discussed in more detail later in this chapter.

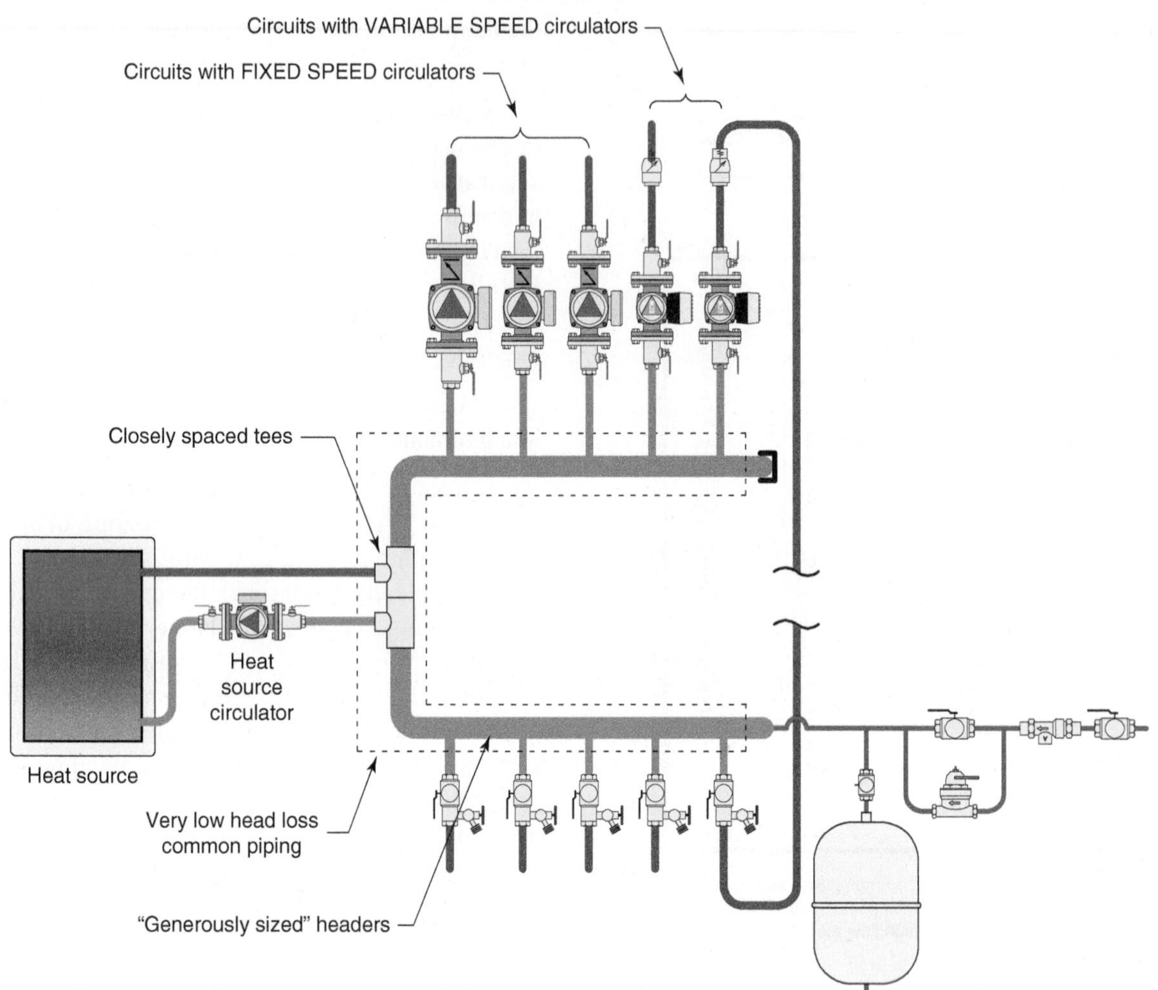

Figure 4-96 | Hydraulic separation is accomplished using closely spaced tees and generously sized headers.

Hydraulic separators are available in a several pipe sizes, from 1 inch to over 12 inches. Some contain removable magnets near their base that attract and hold iron particles to prevent them from circulating through the system. The magnets are periodically removed and cleaned.

Given the surface area of their bodies, *hydraulic separators should always be insulated* to minimize heat loss to their surroundings. This is especially true of larger hydraulic separators, which may have more surface area than a modestly sized radiator and without insulation would needlessly overheat the mechanical room.

MIXING AT THE POINT OF HYDRAULIC SEPARATION

Mixing can occur within any component, or group of components, that provides hydraulic separation. The most common scenario is when the flow rate on the secondary side of the hardware providing hydraulic separation is greater than the flow rate on the primary side. This situation is illustrated in Figure 4-100.

Equation 4.4 can be used to calculate the fluid temperature supplied to the distribution system from the component(s) providing hydraulic separation. It is based on a steady state energy balance across the separating component(s).

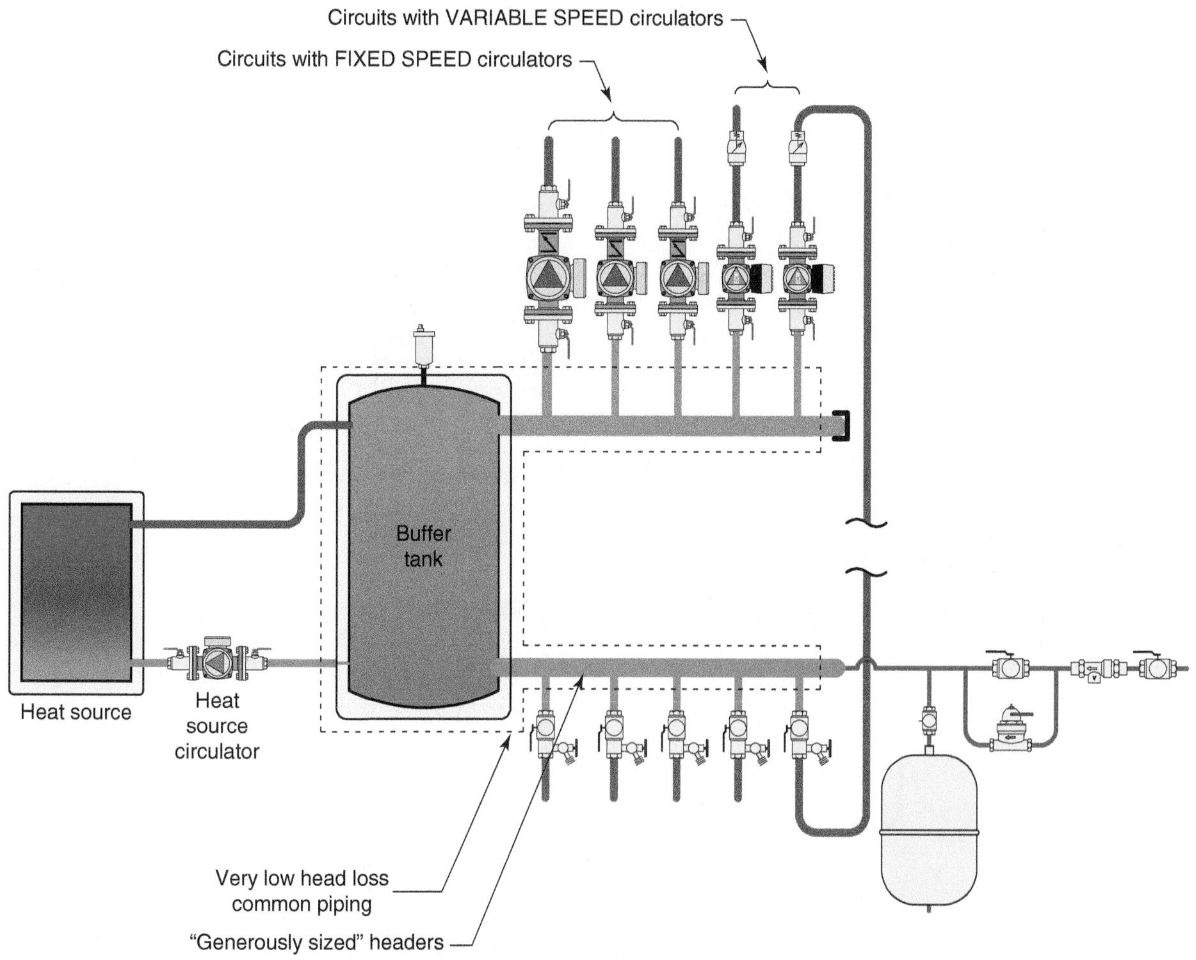

Figure 4-97 | Buffer tank and generously sized headers provide hydraulic separation between all circulators, both fixed and variable speed.

(Equation 4.4)

$$T_{mix} = \frac{(f_4 - f_1)T_4 + f_1 T_1}{f_4}$$

where:

f_4 = secondary flow rate returning from distribution system (gpm)

f_1 = primary flow rate entering from heat source (gpm)

T_4 = temperature of fluid returning from distribution system (°F)

T_1 = temperature of fluid entering from heat source (°F)

Example 4.5: At design load, the flow rate through a distribution system connected to the secondary side of a hydraulic separator is 25 gpm, as shown in Figure 4-101. Water returns from the distribution system at 100 °F and enters the lower right port of the hydraulic separator. At the same time, the heat source flow rate is 10 gallons per minute and the water temperature supplied to the upper left port of the hydraulic separator is 120 °F. What is the mixed water temperature supplied to the distribution system? And what is the water temperature returning to the heat source?

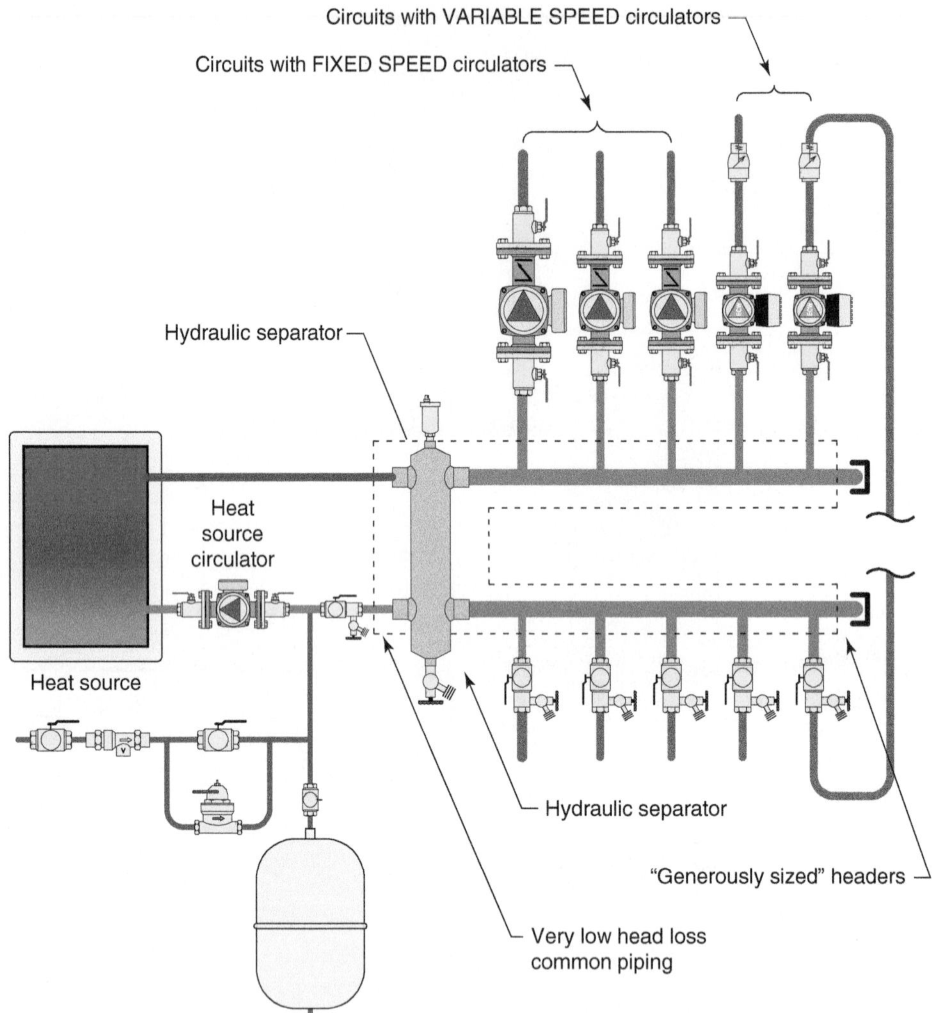

Figure 4-98 | Placement of a hydraulic separator between the heat source and multi-circulator distribution system.

Solution: The mixed water temperature is found using Equation 4.4:

$$T_{mix} = \left(\frac{(f_4 - f_1)T_4 + (f_1)T_1}{f_4}\right)$$

$$= \left(\frac{(25 - 10)100 + (10)120}{25}\right) = 108°F$$

The 25 gpm flow returning to the hydraulic separator splits up into 10 gpm flowing back to the heat source and the remaining 15 gpm flowing upward through the separator. The latter mixes with the 10 gpm of 120 °F water entering from the heat source to create the outgoing 25 gpm at the mixed temperature of 108 °F. The water temperature returning to the heat source is the same as that returning from the distribution system, 100 °F.

Discussion: In this case, the mixing resulting from the relatively high flow rate in the distribution system compared to the heat source causes a substantial difference between the water temperature at the heat source outlet and the temperature supplied to the distribution system. Any changes in flow rate on the secondary side of the hydraulic separator due to circulators turning on and off, or changing speed, or zone valves opening and closing, will affect the mixing proportions and therefore change the mixed water temperature supplied to the distribution system. *If the water temperature supplied to the distribution system is being controlled by modulating the heat source, the supply temperature sensor for that controller should be located on the outlet side of the hydraulic separator.* This allows the controller to adjust the heat source output based on the current supply temperature to the distribution system.

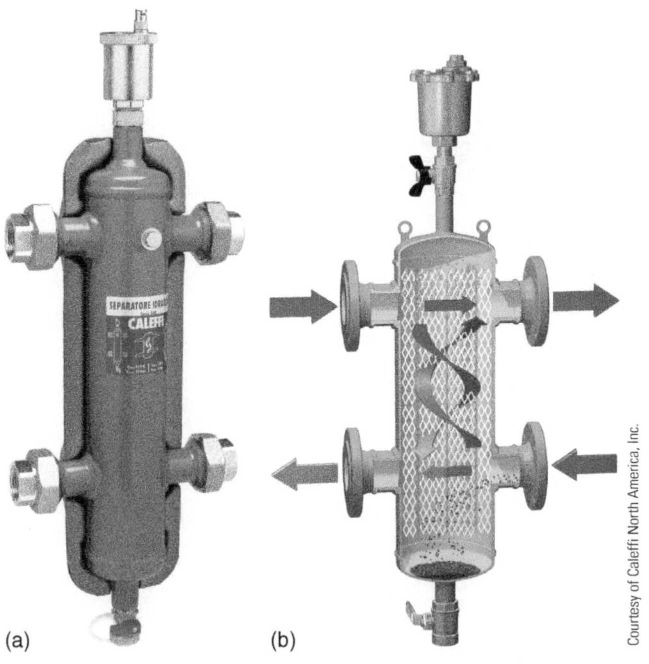

Figure 4-99 (a) A typical hydraulic separator with front portion of insulation shell removed. (b) Cut-away view of a hydraulic separator.

4.6 Air Elimination and Air Management

The terms **air elimination** and **air management** have different meanings. Air elimination applies to closed-loop, fluid-filled systems. It means that the air that is initially in the system should be eliminated, to the greatest extent possible, during commissioning and subsequent operation of the system. The goal of air elimination is to maintain the air content of the system as close as possible to zero. The only "desirable" air in a closed-loop, *fluid-filled* system is the captive air contained on the non-wetted side of the diaphragm-type expansion tank.

Air management applies to systems that are either open to the atmosphere at some point (i.e., a system with a non-pressurized thermal storage tank) or to systems that, under normal operation, are not completely filled with fluid. A solar drainback system is an example of the latter. Given the nature of such systems,

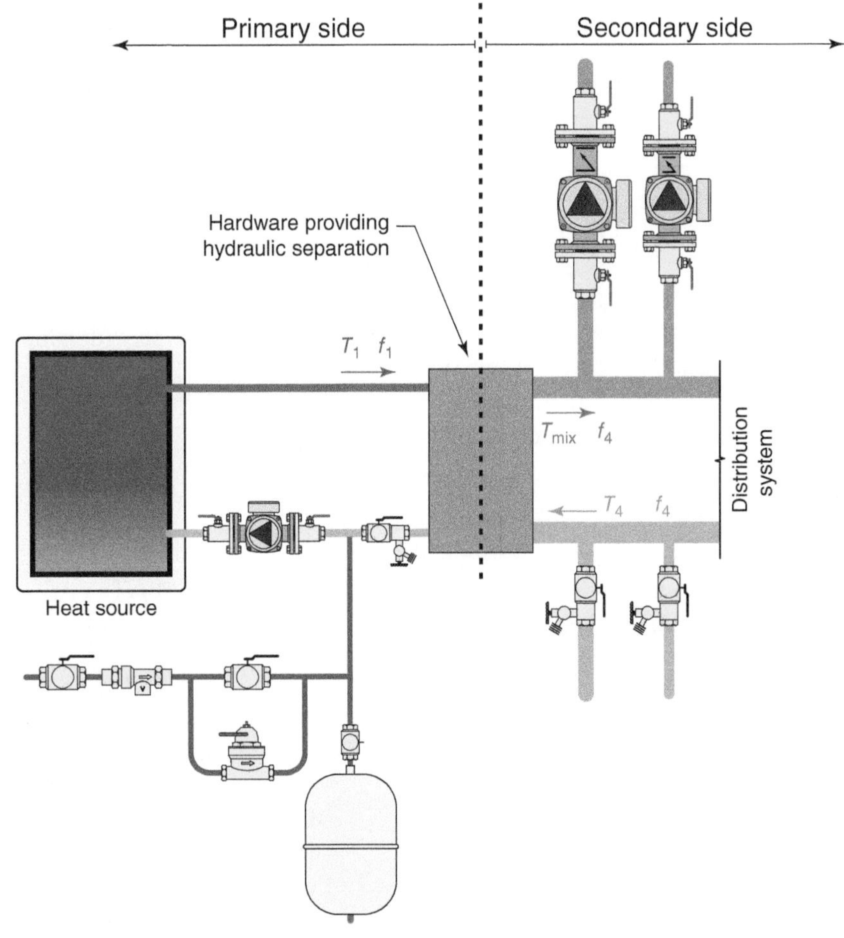

Figure 4-100 If the primary flow rate (f_1) is less than the secondary flow rate (f_4), mixing will occur within the hydraulic separation component(s). This will reduce the fluid temperature supplied to the distribution system.

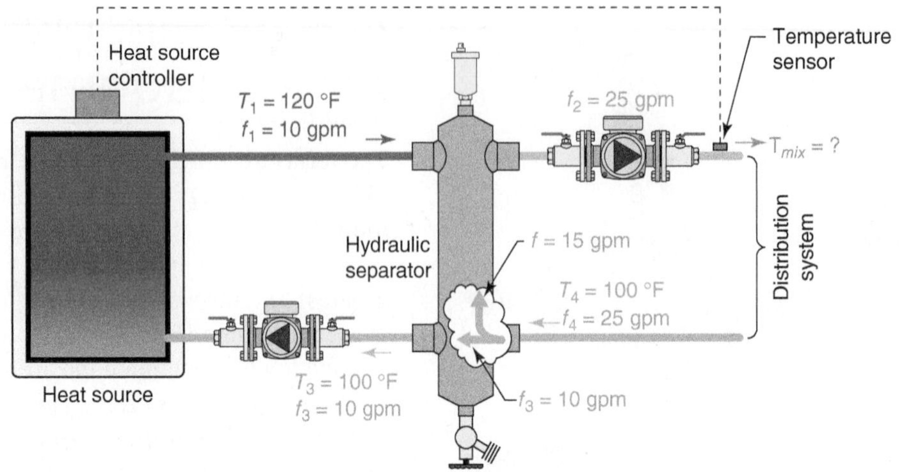

Figure 4-101 | Hydraulic separator for example 4.5.

there will always be some air within them. However, the goal of air management is to keep that air where it belongs in the system during all operating modes. For example, the air at the top of a thermal storage tank in a drainback solar combisystem should not be allowed to migrate into the heating distribution circuits of that system.

AIR ELIMINATION IN CLOSED-LOOP, FLUID FILLED CIRCUITS

For a closed loop, fluid-filled hydronic circuit to function properly, it must be essentially free of air. If this is not the case, problems ranging from occasional "gurgling" sounds in the pipes to complete loss of heat output can occur.

At some point, every hydronic system contains a mixture of water and air. This is especially true when the system is first filled and put into operation. *If properly designed, a closed-loop, fluid-filled system should rid itself of most air within a few days of initial startup.* The system should then maintain itself virtually air-free throughout its service life.

The problems associated with the lack of proper air removal include:

- Flow noise
- Accelerated corrosion due to presence of excess oxygen
- Inadequate flow
- Complete loss of flow
- Poor heat transfer
- Gaseous cavitation in circulators
- Circulator damage

Air can exist in three possible forms within hydronic systems:

- Stationary air pockets
- Entrained air bubbles
- Air dissolved within the fluid

All three of these forms can exist simultaneously, especially when the system is first put into operation. Each exhibits different symptoms in the system.

STATIONARY AIR POCKETS

Since air is lighter than water, it tends to migrate toward the high points of the system. These points are not necessarily at the top of the system. **Stationary air pockets** can form at the top of heat emitters, even those located low in the building. Air pockets also tend to form in horizontal piping runs that turn downward following a horizontal run. A good example is when a pipe is raised to cross over a structural beam or other obstruction and then drops to a lower elevation, as shown in Figure 4-102.

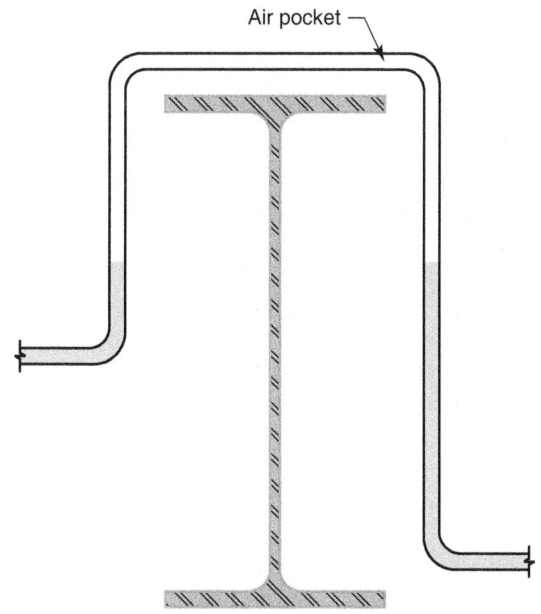

Figure 4-102 | A stationary air pocket can form at an intermediate high point in the circuit.

Figure 4-103 | Microbubbles slowly rising through a glass of water.

When a system is first filled with water, these high points are dead ends for air movement. In some cases, this trapped air can displace several quarts of fluid that eventually must be added to the system. Even after a system is initially **purged**, stationary air pockets can reform as residual air bubbles merge and migrate toward high points. This is especially likely in components with low flow velocities, where slow-moving fluid is unable to push or drag the air along with it. Examples of such components include large heat emitters, large diameter piping, and storage tanks.

ENTRAINED AIR BUBBLES

When air exists as bubbles, a moving fluid may be able to carry them along through the system. This **entrained air** can be both good and bad. It is good from the standpoint of transporting air from remote parts of the system back to a central air separating device; however, it is bad if the air cannot be separated from the fluid within that device. How well a fluid entrains air is best judged by its ability to move bubbles downward in a vertical pipe, against their natural tendency to rise. If the fluid moves downward faster than a bubble can rise, it will carry the bubble in its direction of flow. *A minimum flow velocity of 2 feet per second is recommended to effectively entrain air and carry in downward through a vertical pipe.*

One type of bubble often present in hydronic systems is called a **microbubble**. These bubbles are so small it is difficult to see a single microbubble. Dense groups make otherwise clear water appear cloudy. They can often be observed, momentarily, in a glass of water filled from a faucet with an aerator device, as shown in Figure 4-103.

Microbubbles can also be formed as dissolved air comes out of solution when a fluid is heated or when a component generates significant turbulence. *Microbubbles have very low rise velocities and are easily entrained by moving fluids. This characteristic makes it more difficult to separate microbubbles from the fluid.* Fortunately, there are several types of devices now available to efficiently capture and eject microbubbles from hydronic systems.

DISSOLVED AIR

Perhaps the least understood form in which air exists in a hydronic system is as **dissolved air**. Molecules of the gases that make up air including oxygen and nitrogen can exist "**in solution**" with water molecules, as discussed in Chapter 3. These molecules cannot be seen, even under a microscope. Although water may appear perfectly clear and free of bubbles, it can still contain a significant quantity of air in solution.

The amount of air that exists in solution with water is strongly dependent on the water's temperature and pressure. The curves shown in Figure 4-104 show the *maximum* dissolved air content of water as a percentage of its volume, over a range of temperature and pressure.

Notice that as the temperature of water increases, its ability to hold air in solution decreases. This explains why air bubbles appear on the lower surfaces of a pot of water being heated on a stove. It also explains the formation of microbubbles along the inside surfaces of heating devices such as boilers and solar collectors. As the water nearest these surfaces is heated, some of the air molecules come out of solution as microbubbles. The opposite is also true. *When heated water is cooled, it will absorb air molecules back into solution.*

The pressure of the water also markedly affects its ability to contain dissolved air. When the pressure of the water is lowered, its ability to contain dissolved air decreases, and vice versa.

It is always desirable to minimize the dissolved air content of the fluid in a closed-loop, fluid-filled system. This is accomplished by establishing conditions that "encourage" air to come out of solution (e.g., high temperatures and low pressures). When such conditions exist, the resulting microbubbles must be captured and ejected from the system. Devices for doing this will now be discussed.

HIGH-POINT AIR VENTS

The fact that air tends to collect at the high points of hydronic circuits makes such locations well suited to air removal. Any type of a device installed at a system high point, such as the top of a solar collector array, is called a **high-point air vent**.

The simplest type of high point venting device is a **manual air vent**. These components are small valves with a metal-to-metal seat. They thread into 1/8-inch or 1/4-inch FPT tappings and are operated with a screwdriver, square head key, or the edge of a dime. An example of a manual air vent, installed in the upper header side of a panel radiator, is shown in Figure 4-105.

Manual air vents are commonly located at the top of each heat emitter within a hydronic distribution system. They can also be mounted into special fittings called **baseboard tees,** as shown in Figure 4-106, or any fitting with the appropriate threaded tapping.

After a manual vent is opened, and any air below it is released, a small stream of fluid will continue to flow through the vent until it is closed. While it is operating, a person should stay near the vent with a can ready to catch any ejected fluid before it puddles on a wood floor, stains a carpet, or creates other undesirable results. This can be difficult (especially for one person), if all manual vents are left open as the system

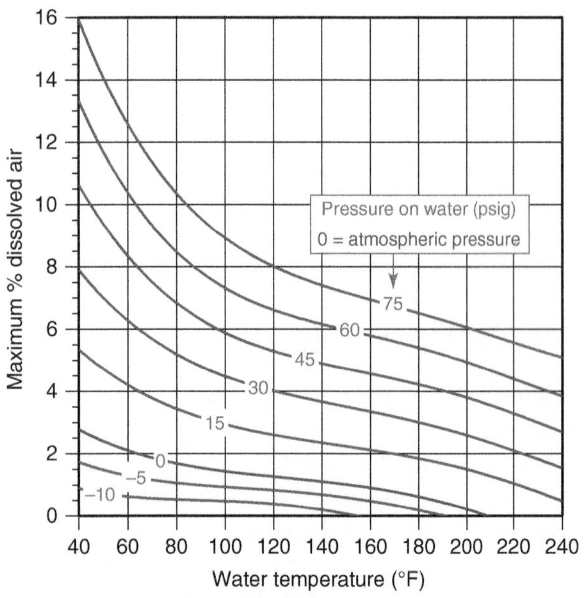

Figure 4-104 | Curves showing the *maximum* solubility of air in water as a percentage of its volume, over a range of temperatures and pressures.

Figure 4-105 | Example of a manual air vent.

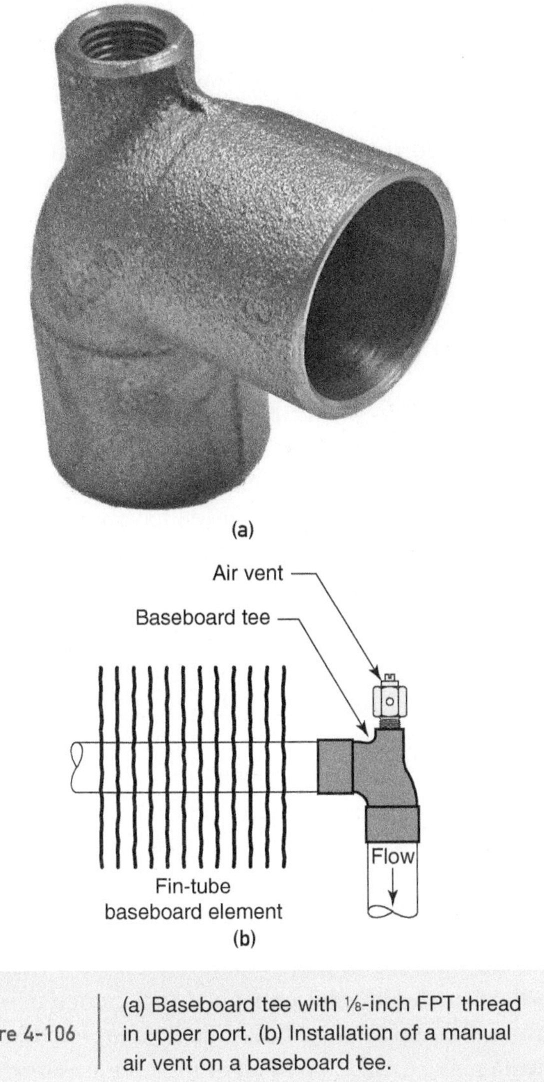

Figure 4-106 (a) Baseboard tee with ⅛-inch FPT thread in upper port. (b) Installation of a manual air vent on a baseboard tee.

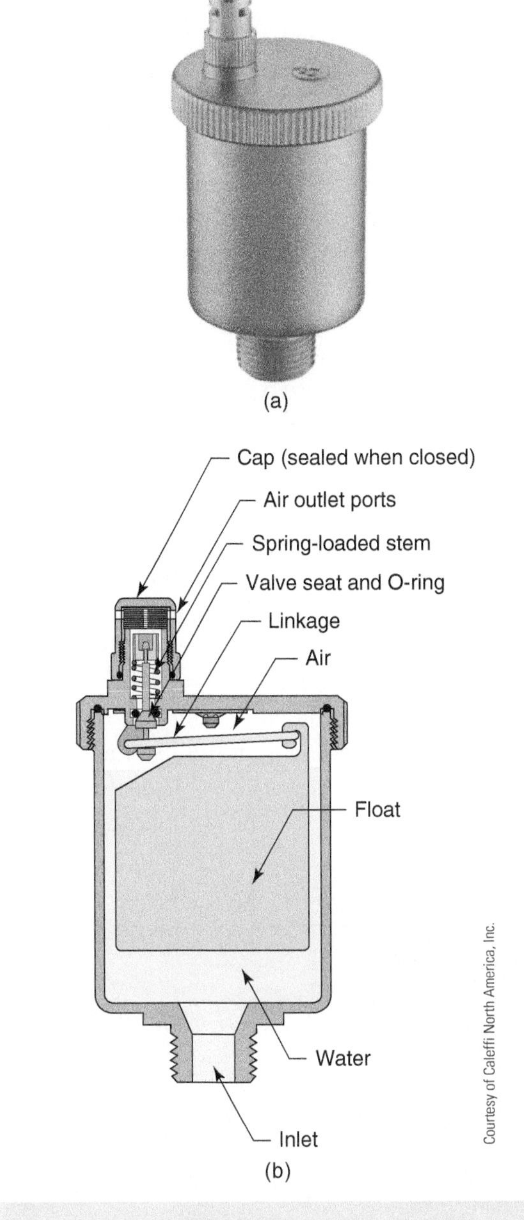

Figure 4-107 (a) Example of float type air vents. (b) Internal construction.

is filled. It is better to move from one vent to the next, operating them in sequence.

FLOAT-TYPE AIR VENTS

The need for fully unattended air venting requires a different type of device. A **float-type air vent** provides the solution. This device, shown in Figure 4-107, contains an air chamber, a float, and an air valve. When sufficient air accumulates in the chamber, the float drops down and opens the valve at the top of the unit. As air is vented, water rises into the chamber and lifts the float to close the air valve.

Most float-type air vents are equipped with a metal cap that protects the stem of the air valve. It is important that this cap remain loose when the vent is put into operation. If the cap is screwed down tight, air cannot be released.

Most float-type vents come with MPT threads and are designed to thread into threaded fittings such as baseboard tees or reducer tees in the same manner as manual air vents. They are available in different sizes and shapes that allow mounting both horizontal and vertical orientations. The large the piping connection, the more effective the air vent. *For the best performance, the author recommends a minimum ¼-inch piping connection, with a ½-inch size connection preferred whenever possible.*

Designers and technicians should remember that float-type air vents have a potential "Achilles heel." They can allow air to *enter* the system if the fluid pressure at their location falls below atmospheric pressure. This can result from a number of factors, most notably the improper placement of the expansion tank relative to the circulator. It is a deceptive problem because it often only occurs when the circulator is operating. The best way to prevent this from happening is to make sure there is always at least 5 psi of positive pressure at all locations where float-type vents are mounted.

Special versions of float-type air vents are available for venting the top of a solar collector circuit that is filled with an antifreeze solution. These vents use high-temperature-resistant elastomers for internal O-rings and other seals. They are typically installed above an isolation valve, as shown in Figure 4-108. This ball valve is closed a few days after the system is first put into operation, after which most of the air bubbles capable of being captured by the float vent have already passed through it. The closed isolation valve protects the air vent from potentially high temperatures and pressures that can occur when the solar collectors are stagnating (e.g., exposed to strong sunlight, but without any flow through them due to some malfunction in the system). The ball valve can be reopened if the system is serviced. It also allows the air vent to be replaced, if necessary, with very little fluid spillage.

Courtesy of Caleffi North America, Inc.

Figure 4-108 | A high temperature-rated float-type air vent at the top of a solar collector array. Note the isolating ball valve installed below the vent.

MICROBUBBLE AIR SEPARATORS

A **central air separator** is a device, mounted in the main system piping, that continually gathers any air passing through it and ejects that air from the system. Although hydronic systems often have several manual air vents or float-type air vents, they typically only require one central air separator.

Over the last several decades, many types of central air separating devices have been developed. Some perform better than others. Given the importance of proper air separation, it makes little sense to use anything other than the best available central air separators. Such devices can be described as **microbubble air separators**. Two examples of such devices, in sizes appropriate for residential and light commercial system, are shown in Figure 4-109.

All microbubble air separators create a region of reduced velocity and enhanced turbulence using an internal component called a coalescing media. This media may be made of metal or high-temperature polymer. As flow passes through it, the coalescing media creates thousands of tiny reduced pressure regions on the downstream side of its sharp facets. The reduced pressure causes dissolved gas molecules to "coalesce" (e.g., come together) into microbubbles. The coalescing media also encourage microbubbles to merge into larger bubbles. Eventually the bubbles reach a size where the surface tension forces holding them to the coalescing media are overcome by buoyancy forces. The bubbles then rise along the coalescing media toward the venting chamber at the top of the air separator. The venting chamber is equipped with a float-operated vent that eventually opens to eject the accumulated air from the system.

A microbubble air separator can maintain the system's fluid in an **unsaturated state of air solubility**. This means *the fluid is always ready to absorb additional air from any areas of the system where it may be present.* Once absorbed, the air is transported by system flow back through the microbubble air separator

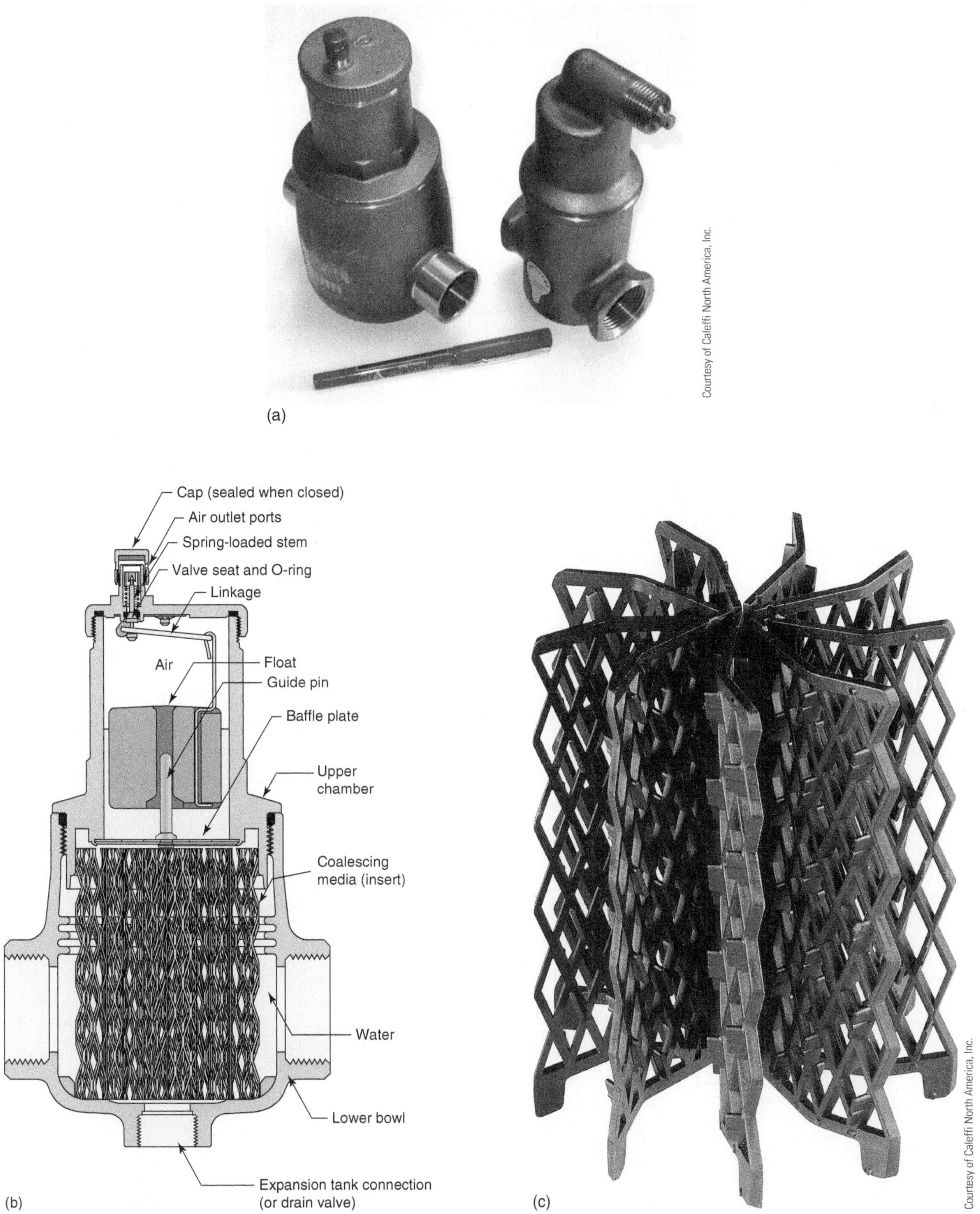

(a)

Courtesy of Caleffi North America, Inc.

Cap (sealed when closed)
Air outlet ports
Spring-loaded stem
Valve seat and O-ring
Linkage
Air
Float
Guide pin
Baffle plate
Upper chamber
Coalescing media (insert)
Water
Lower bowl
Expansion tank connection (or drain valve)

(b)

(c)

Courtesy of Caleffi North America, Inc.

Figure 4-109 | (a) Examples of microbubble separators. (b) Internal construction. (c) Coalescing media insert.

where it can be separated and ejected from the system. The fluid then returns to its unsaturated state and is ready to absorb more air, if available. This ability to continually seek out and collect air is very helpful for removing residual air pockets from all areas of the system, especially inaccessible areas that may not be equipped with vents.

Research has shown that microbubble air separators can reduce the dissolved air content of the system fluid below 0.4 percent. The amount of oxygen present in this small air content is of virtually no significance as far as corrosion is concerned.

To achieve the best air separation efficiency, the inlet flow velocity to microbubble air separators should be no higher than 4 feet per second. Microbubble air separators should also be placed where the entering flow is at the highest possible temperature and lowest possible pressure. This is typically near the outlet of the heat source, as shown in many schematics throughout this text.

FILLING AND PURGING CLOSED-LOOP, FLUID-FILLED SYSTEMS

Every closed-loop, fluid-filled hydronic system must be filled with water, or a water-based fluid, when it is commissioned. The goal is to get fluid into the system, while at the same time expelling as much bulk air from the system as possible. This process is called **purging**.

The amount of air that can be quickly expelled from a hydronic system during purging is substantially increased when the entering fluid has sufficient velocity to entrain air bubbles and carry them to an outlet valve. The greater the water pressure available from a well or water main, the faster forced-water purging can push air out of the system.

One method of purging uses a purging valve installed near the return side of the heat source, as shown in Figure 4-110, and numerous other schematics in this text.

To fill and purge the system:

1. Close the inline ball portion of the purging valve.
2. Open the outlet port of the purging valve.
3. Connect a hose to the outlet port of the purging valve and secure the end of it into a bucket or over a floor drain.
4. Lift the fast-fill lever on the feed water valve.
5. Fully open the fast fill ball valve, if present.

Water should now be flowing into the system at a high rate. Because the inline ball within the purging valve is closed, the entering water first fills the heat source. It then flows out through the remainder of the

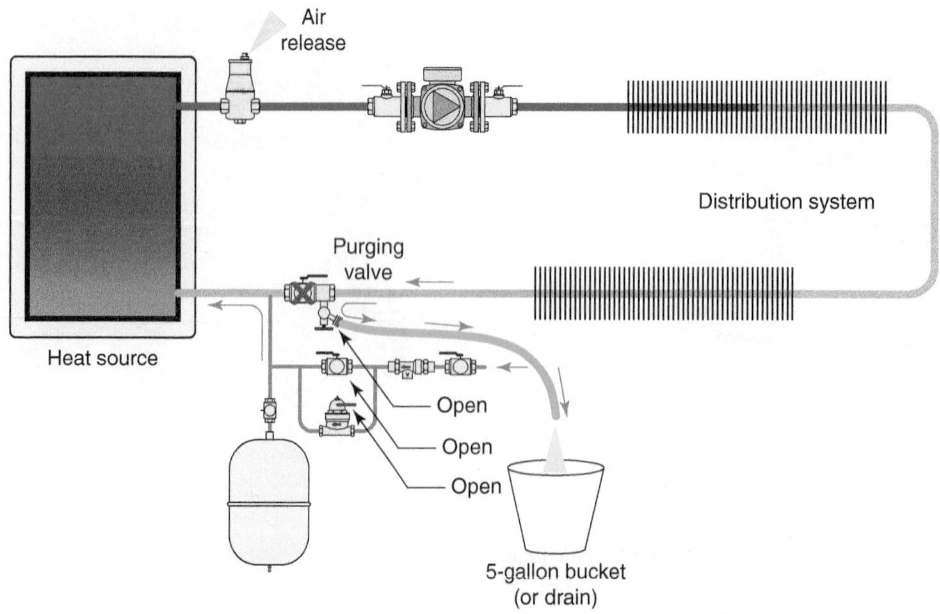

Figure 4-110 | Placement of purging valve near boiler inlet. The expansion tank could also be connected to the bottom of air separator.

distribution piping. Some of the bulk air in the system will be ejected through the central air separator; the remainder will be forced through the distribution system and exit through the outlet port of the purging valve.

When the water stream exiting the discharge hose is free of air bubbles for at least 30 seconds, most of the bulk air will have been purged from the system. At this point, the fast-fill lever is returned to normal position, and the fast fill ball valve, if present, is fully closed.

In systems with two or more branch circuits, it is best to purge each branch individually. This is done by closing valves in all but one branch while purging. When flow exiting the purge valve is free of air bubbles, the valve in the next branch circuit is opened and the valve in the current branch is closed. This process is repeated until each branch has been purged. This allows the maximum possible flow rate through each branch to dislodge and entrain as much air as possible.

After all branches have been purged individually, open all branch valves and continue purging. The reduced head loss of the fully open distribution system maximizes purging flow and helps dislodge any remaining air pockets in larger piping and components.

Placing the purging valve near the inlet connection of the heat source, as shown in Figure 4-110, helps ensure that debris such as small solder balls or dirt in the piping are flushed out of the system rather than into the heat source.

AIR MANAGEMENT

Consider the closed loop, drainback-protected solar combisystem shown in Figure 4-111.

For drainback freeze protection to work, there must be air in the collectors and any exposed piping whenever the collector circulator is off. Total air elimination, as previously discussed, would defeat the purpose of drainback freeze protection. However, the air in the collectors and upper portion of the storage tank should not be allowed to find its way into the distribution system where it could potentially cause problems such as noise, poor circulator performance, and trapped air pockets. Thus, the air in the system must be "managed."

Air management maintains the internal air volume in its proper location within the system. In the system shown in Figure 4-111, any air that is captured by the microbubble air separator is returned to the air-filled portion of the system rather than ejected from the system. This allows the pressurized closed-loop system to maintain its initial pressurization, since air is not being expelled from it. Likewise, when the collector circulator turns off, air from the top of the tank moves back through the air return tube and then up into the collector array. At the same time, water within the collector array and exposed piping flows back down to the tank. No air is ejected from the system.

Notice that there is no automatic make-up water assembly on this system. Such an assembly, if present, would eventually allow the system to fill with water should there be an air leak at any point. This would lead to freeze damage of the collectors and exposed piping.

Also note that there is no dedicated expansion tank in this system. The captive air volume at the top of the storage tank, if properly sized, provides the volume needed to accommodate the expansion volume of the system's water and serve as the drainback space.

4.7 Expansion Tanks

All closed loop hydronic systems require an appropriately sized expansion tank. Most modern systems use a **diaphragm-type expansion tank**, as shown in Figure 4-112.

On one side of the flexible **diaphragm** is a captive volume of air that has been pre-pressurized by the manufacturer. On the other side is a chamber for accommodating the expanded volume of system fluid. As more fluid enters the tank, the diaphragm flexes, allowing the air volume to be compressed.

Small, diaphragm-type expansion tanks are available in a variety of sizes and shapes. They range from 1 to 14 gallons in volume and are designed to hang from their 1/2-inch MPT piping connection.

Because of the potential fluid weight in an expansion tank, the piping supporting these tanks should itself be well supported. The preferred mounted orientation for such tanks is vertical, with the tank hung from the piping connection. The tank should be located so that it will not be stressed by accidental contact while maintaining nearby components. Brackets are available to support the weight of the tank in this orientation, as shown in Figure 4-113.

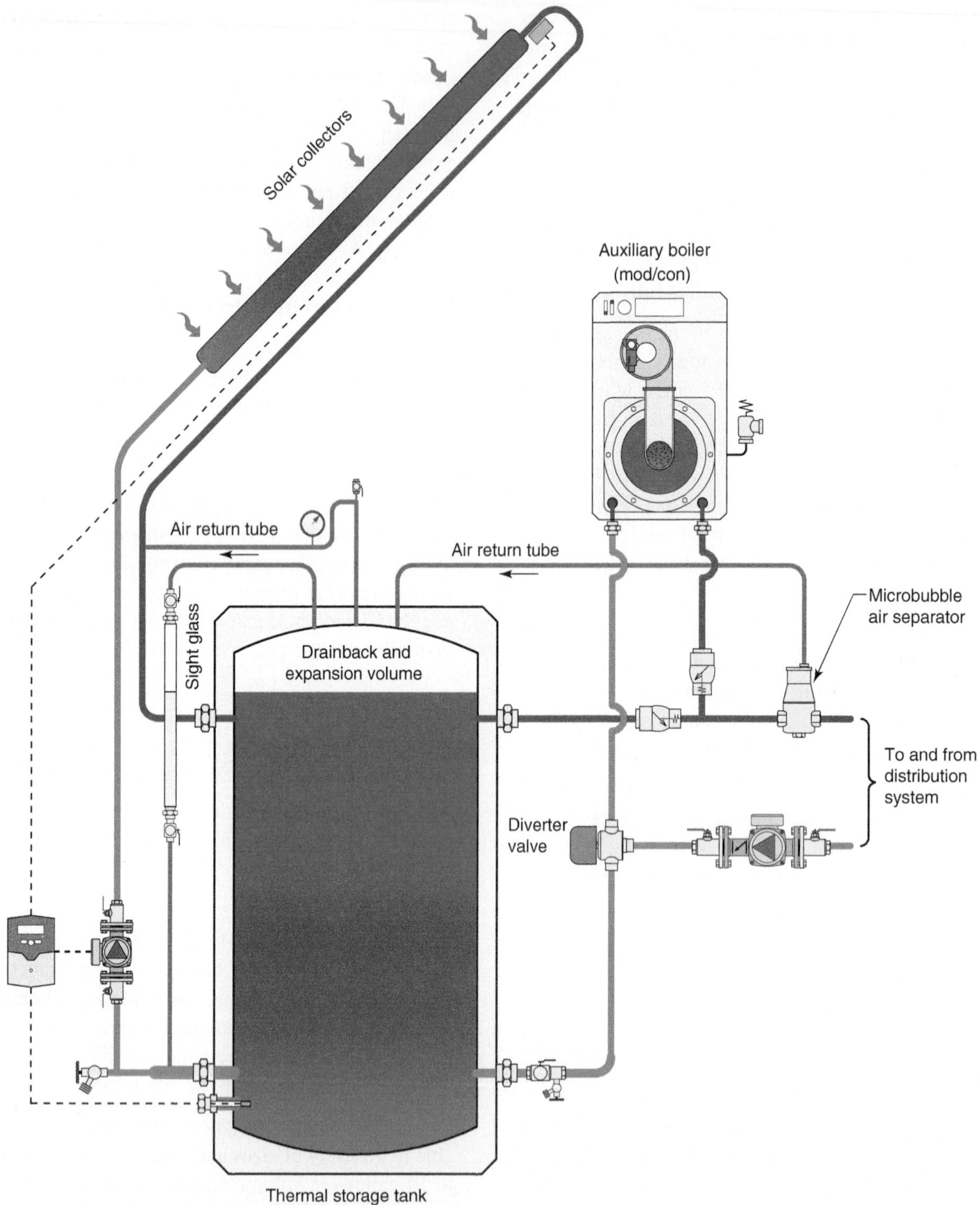

Figure 4-111 | A closed-loop drainback solar combisystem with captive total air volume.

Hydronic systems with large fluid volumes, such as those with large pressurized thermal storage tanks, often require expansion tanks larger than those illustrated in Figure 4-113. Due to their weight and volume, these larger expansion tanks must be supported by a floor rather than their piping connection. Figure 4-114 shows a floor-mounted diaphragm-type expansion tank in comparison to a small top-supported tank.

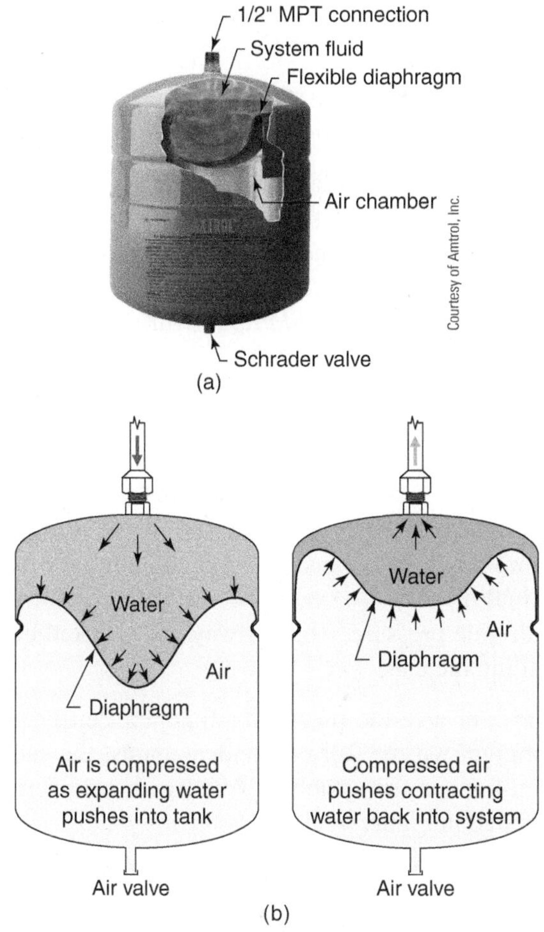

Figure 4-112 (a) Example of a small diaphragm-type expansion tank. (b) Flexing of the internal diaphragm as fluid enters or leaves the tank.

Figure 4-113 A small expansion tank supported by a wall bracket.

Figure 4-114 Example of a floor-mounted diaphragm-type expansion tank (in rear). Smaller diaphragm-type expansion tank that is designed to be hung from upper connection is shown in foreground.

Floor-mounted expansion tanks contain their captive air volume at the *top* of the pressure vessel. The fluid side piping connection is at the bottom of the tank and accessed through a hole in the lower support ring.

COMPATIBILITY OF TANK AND SYSTEM FLUID

It is very important that the diaphragm material used in the tank is chemically compatible with the fluid and/or any dissolved gases in the system. Incompatibilities can result in the diaphragm being slowly dissolved by the fluid. This can create sludge in the system that gums up system components. It will also cause the tank's diaphragm to eventually fail.

Most modern diaphragm-type tanks use a butyl rubber or **EPDM** diaphragm. Although such tanks are generally compatible with glycol-based antifreeze solutions, the installer should always verify this compatibility if it is not clearly stated in product specifications. Expansion tanks are also available with other, high-temperature/glycol-compatible diaphragms for specific use in solar collector circuits.

Another compatibility issue involves the presence of dissolved oxygen in the system water. Most expansion tanks sold for use in hydronic heating systems are intended for installation in closed-loop, oxygen-tight systems. System water is in direct contact with the steel shell of such tanks. If oxygen can migrate into the system, the shell will corrode and eventually fail. Dissolved oxygen could be present in any type of open-loop system or in systems using oxygen-permeable tubing. In such cases, an expansion tank rated for use in open-loop systems should be selected. Such tanks typically have a polypropylene liner that prevents contact between the system water and the steel tank shell.

Expansion tanks have maximum pressure and temperature ratings. This information is stamped on the label of the tank. Typical ratings are 60 psi and 240 °F. These rating are usually adequate for residential and light commercial systems.

SIZING DIAPHRAGM-TYPE EXPANSION TANKS

A properly sized diaphragm-type expansion tank allows the pressure at the pressure relief valve to reach a value approximately 5 psi lower than the pressure relief valve's rated opening pressure when the system reaches its maximum operating temperature. The 5 psi safety margin prevents the relief valve from leaking just below its rated opening pressure. It also allows for the inlet of the pressure relief valve to be slightly (up to 4 feet) below the connection point of the expansion tank and thus at a slightly higher static pressure.

The first step in sizing a diaphragm-type expansion tank is to determine the proper **air-side pressurization** of the tank, using Equation 4.5.

(Equation 4.5)

$$P_a = H\left(\frac{D_c}{144}\right) + 5$$

where:

P_a = air-side pressurization of the tank (psi)

H = distance from inlet of expansion tank to top of system (ft)

D_c = density of the fluid at its initial (cold) temperature (lb/ft^3)

The proper air-side pressurization is equal to the static fluid pressure at the inlet of the tank, plus an additional 5 psi allowance at the top of the system. *The pressure on the air side of the diaphragm should be adjusted to the calculated pressurization value before fluid is added to the system.* This is done by either adding or removing air through the Schrader valve on the shell of the tank. A small air compressor or bicycle tire pump can be used when air is needed. An accurate 0 to 30 psi pressure gauge should be used to check this pressure as it is being adjusted. Many smaller diaphragm-type expansion tanks are shipped with a nominal air-side pressurization of 12 psi. However, the air-side pressure should always be checked before installing the tank.

Proper air-side pressure adjustment ensures the diaphragm will be fully expanded against the shell of the tank when the system is filled with cold fluid, as illustrated in Figure 4-115.

Failure to properly pressurize the tank can result in the diaphragm being partially compressed by the fluid's static pressure before any heating occurs. If this occurs, the full volume of the expansion tank will not

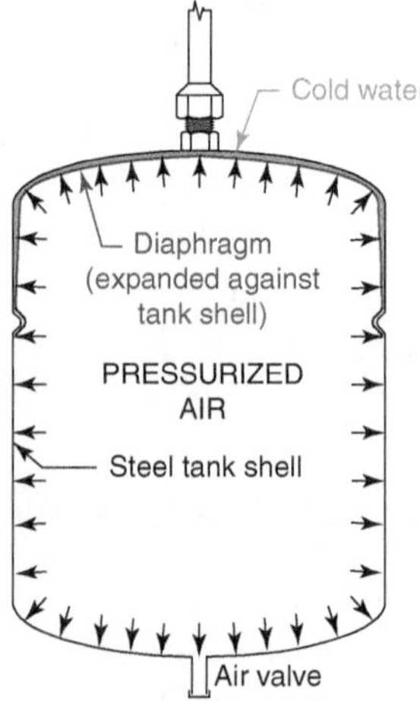

Figure 4-115 A properly prepressurized diaphragm-type expansion tank will have its diaphragm fully expanded against the tank's shell when the system is first filled with water but before the water is heated. This ensures the full working volume of the tank is available to accept the expansion volume of the heated water.

Copper (Type M)		PEX		PEX-AL-PEX	
	gal./ft.		gal./ft.		gal./ft.
3/8″ copper	0.008272	3/8″ PEX	0.005294	3/8″ PEX-AL-PEX	0.004890
1/2″ copper	0.01319	1/2″ PEX	0.009609	1/2″ PEX-AL-PEX	0.01038
		5/8″ PEX	0.01393	5/8″ PEX-AL-PEX	0.01658
3/4″ copper	0.02685	3/4″ PEX	0.01894	3/4″ PEX-AL-PEX	0.02654
1″ copper	0.0454	1″ PEX	0.03128	1″ PEX-AL-PEX	0.04351
1.25″ copper	0.06804	1.25″ PEX	0.04668		
1.5″ copper	0.09505	1.5″ PEX	0.06516		
2″ copper	0.1647	2″ PEX	0.1116		
2.5″ copper	0.2543				
3″ copper	0.3630				

Figure 4-116 | Internal volumes of tubing.

be available as the fluid heats up. An underpressurized tank will act as if undersized and possibly allow the system's pressure relief valve to open each time the system heats up. This situation must be avoided.

Once the air-side pressurization is determined, Equation 4.6 can be used to find the *minimum* required volume of the expansion tank:

(Equation 4.6)

$$V_t = V_s\left(\frac{D_c}{D_h} - 1\right)\left(\frac{P_{RV} + 9.7}{P_{RV} - P_a - 5}\right)$$

where:

V_t = minimum required tank volume (gallons) (not "**acceptance volume**")

V_s = fluid volume in the system (gallons) (see Figure 4-116)

D_c = density of the fluid at its initial (cold) temperature (lb/ft^3)

D_h = density of the fluid at the maximum operating temperature of the system (lb/ft^3)

P_a = air-side pressurization of the tank found using Equation 4.5 (psi)

P_{RV} = rated pressure of the system's pressure-relief valve (psi)

System volume can be estimated based on the total volume of the boiler, piping, and other components in the system. Figure 4-116 gives volumes for several common pipe types and sizes used in residential and light commercial systems.

The density of water and two solutions of propylene glycol antifreeze can be read from Figure 4-117.

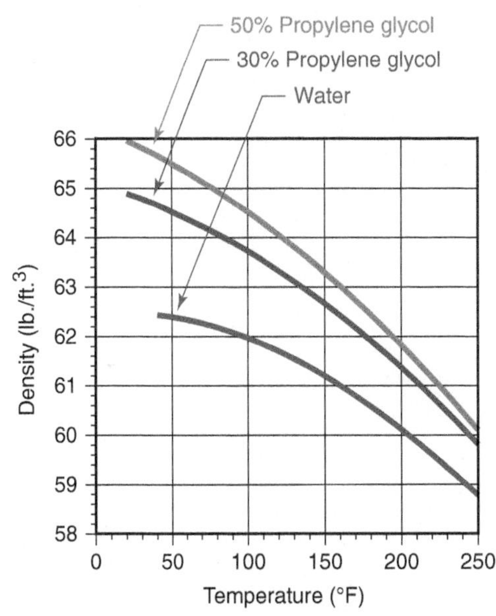

Figure 4-117 | Density of water and two solutions of propylene glycol antifreeze.

Example 4.6: Determine the minimum-size diaphragm-type expansion tank for the system shown and described in Figure 4-118.

Solution: Start by getting the density of water at both temperature extremes, using Figure 4-117 or other references:

$D_{50\,°F} = 62.4\ \text{lb/ft}^3$

$D_{200\,°F} = 60.15\ \text{lb/ft}^3$

The total estimated system volume is found by totaling up the volume in piping and other components:

$$V_s = (50)(0.09505) + (100)(0.0454) + (2000)(0.009609) + 23 + 238 = 290\ \text{gallons}$$

Equation 4.5 can now be used to calculate the cold-fill pressurization of the expansion tank:

$$P_a = H\left(\frac{D_c}{144}\right) + 5 = 12\left(\frac{62.4}{144}\right) + 5 = 10.2\ \text{psi}$$

Equation 4.6 can now be used to calculate the expansion tank volume:

$$V_t = 290\left(\frac{62.4}{60.15} - 1\right)\left(\frac{30 + 9.7}{30 - 10.2 - 5}\right) = 29.1\ \text{gallons}$$

Discussion: This expansion tank volume is "conservatively" large because it assumes that *all* water in the system reaches the upper temperature of 200 °F simultaneously. This is not the case for piping downstream of the mixing valve (e.g., the radiant panel circuits). Thus, it is not necessary to apply any additional safety factor to this volume. The expansion tank volume is also considerably larger than required in most residential hydronic systems. This is a direct result of the large buffer tank volume (238 gallons) out of the total system volume (290 gallons). In this case, the expansion tank volume was approximately 10 percent of the total system volume.

SPECIAL CIRCUMSTANCES

In many hydronic systems having only interior piping, the cold water used to fill the system will be at the lowest temperature it is likely to experience while in the system. This is assumed in the above discussion of proper air-side pressurization and tank sizing. However, an exception to this situation is a hydronic circuit serving outdoor equipment such as solar collectors or

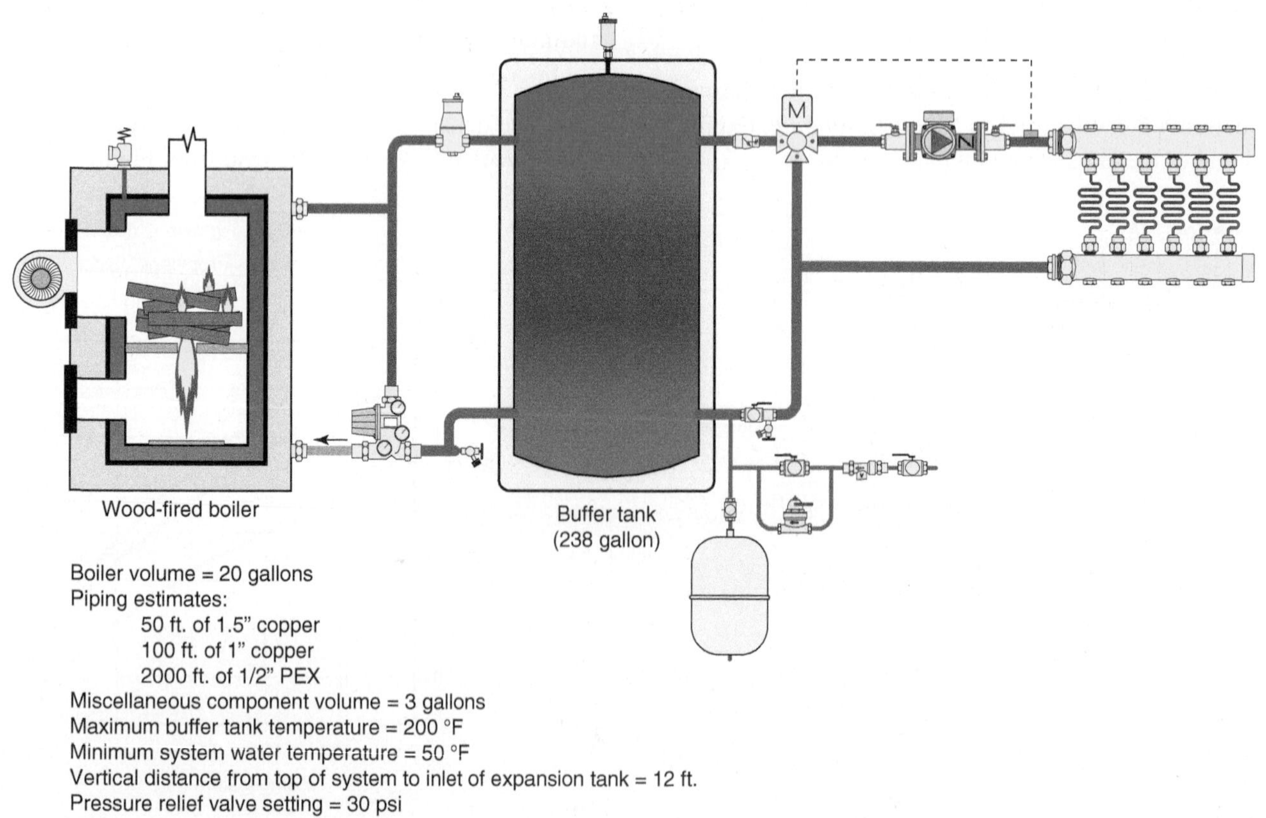

Figure 4-118 | System for example 4.6.

an air-source heat pump in a cold climate. These circuits are typically filled with an antifreeze solution. During cold weather, the antifreeze solution is likely to experience temperatures significantly lower than its temperature when the circuit was filled. This will cause a further decrease in the fluid's volume and could cause a negative pressure in the system if the expansion tank's diaphragm cannot further expand to allow more fluid into the circuit. Such situations require a slight reduction in the air-side pressurization compared to the pressure calculated using Equation 4.5. The reduced pressure allows a small amount of antifreeze fluid to remain in the tank at the conclusion of the cold fill procedure. A procedure to determine the reduced air-side pressurization in such circumstances is presented within the context of antifreeze-protected solar thermal systems in Chapter 10.

PLACEMENT OF THE EXPANSION TANK

The placement of the expansion tank relative to the circulator significantly affects the pressure distribution in the system when it operates. *The expansion tank should always be connected to the system near the inlet of the circulator.*

To understand why, one needs to consider the interaction between the circulator and expansion tank. In a closed-piping system, the amount of fluid (including that in the expansion tank) is fixed. It does not change regardless of whether the circulator is on or off. The expansion tank contains a captive volume of air at some pressure. The only way to change the pressure of this air is to either push more fluid into the tank compressing the air or to remove fluid from the tank expanding the air. This fluid would have to come from, or go to, some other location in the system. However, since the system's fluid is incompressible, and the amount of fluid in the system is fixed, this cannot happen regardless of whether the circulator is on or off. The expansion tank thus fixes the pressure of the system's fluid at its point of attachment to the piping. This is called the **point of no pressure change (PONPC)**.

Now consider a horizontal piping circuit filled with fluid and pressured to approximately 10 psi, as shown in Figure 4-119. When the circulator is off, the pressure is the same (10 psi) throughout the piping circuit. This is indicated by the solid horizontal pressure line shown above the piping.

When the circulator is turned on, it immediately creates a pressure differential between its inlet and outlet. However, the pressure at the point where the expansion tank is connected to the circuit remains the same. The combination of the pressure differential across the circulator, the pressure drop along the circuit, and location of the PONPC gives rise to the new **dynamic pressure distribution** shown by the dashed line in Figure 4-119.

Notice how the pressure increases in nearly all parts of the circuit when the circulator is operating. This is desirable because it helps eject air from vents. It also

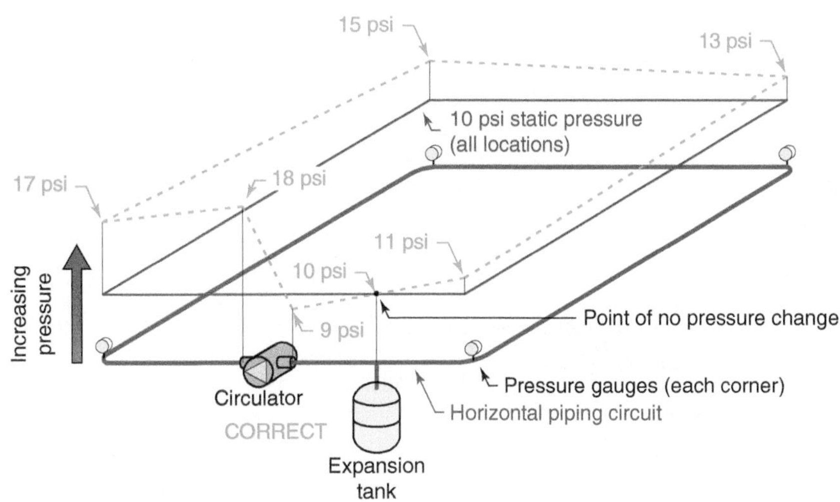

Figure 4-119 | Pressure distribution in horizontal piping circuit with circulator on and off. Note the expansion tank is (correctly) located near the inlet port of the circulator.

helps keep dissolved air in solution and minimizes the potential for cavitation at the circulator inlet. The short segment of piping between the expansion tank connection point and the inlet port of the circulator experiences a very slight drop in pressure due to the head loss in the piping. The numbers used for pressure in Figure 4-119 are illustrative only. The actual numbers will depend on flow rates, fluid properties, and pipe sizes.

Next, consider the same system with the expansion tank (incorrectly) located near the discharge port of the circulator. Figure 4-120 illustrates the pressure distribution that will develop when the circulator is operating.

The PONPC remains at the location where the expansion tank is attached to the system. This causes the pressure in most of the system to *decrease* when the circulator is turned on. The pressure at the inlet of the circulator has dropped from 10 psi to 2 psi. This situation is not desirable since it reduces the system's ability to expel air. It can also encourage circulator cavitation.

To see how problems can develop, imagine the same system with a static pressurization of only 5 psi. The same 9 psi differential will be established across the circulator when it operates, and the entire pressure profile shown with dashed lines in Figure 4-120 will shift downward by 5 psi as shown in Figure 4-121.

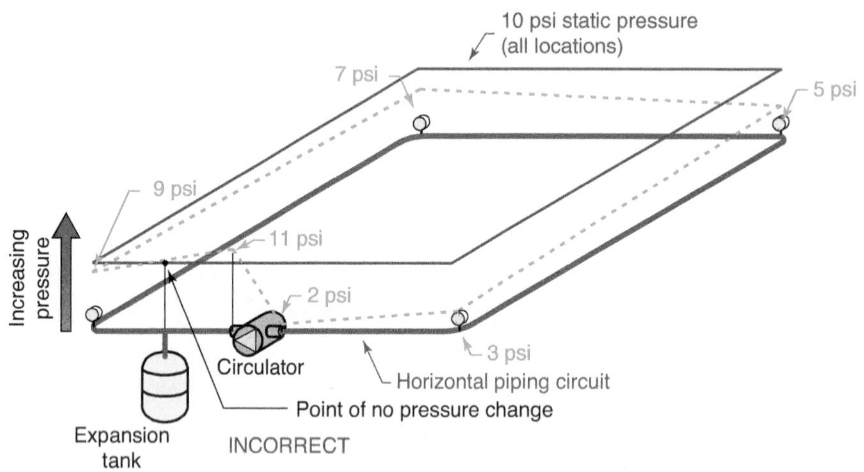

Figure 4-120 | Pressure distribution in horizontal piping circuit with circulator on and off. Note the expansion tank is (incorrectly) located near the discharge port of the circulator.

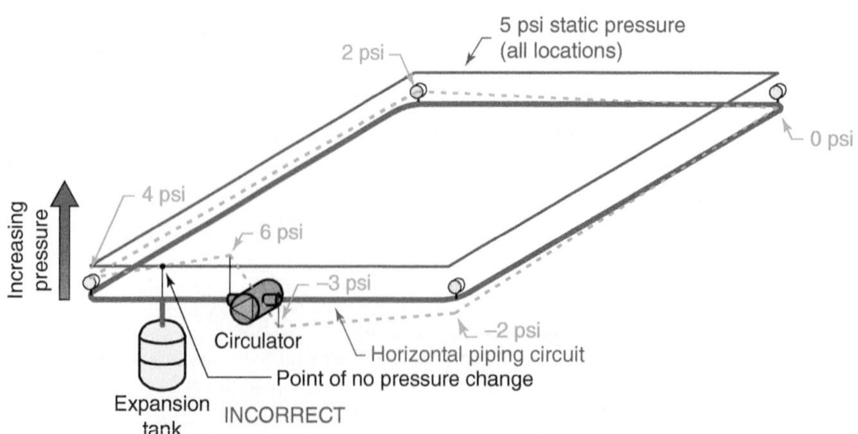

Figure 4-121 | Pressure distribution in horizontal piping circuit with low static pressure. Note the expansion tank is incorrectly located near the discharge port of the circulator. The pressure in a portion of the piping system becomes subatmospheric when the circulator operates.

Notice that the pressure in the piping between the upper-right corner and the circulator inlet is below atmospheric pressure when the circulator is operating. If there are air vents or slightly loose valve packings in this portion of the circuit, air will be sucked into the system every time the circulator operates. The absolute pressure at the inlet of the circulator is also lower; thus, the potential for vapor cavitation is increased, especially if the system operates at high fluid temperatures.

Some systems have heat sources or thermal storage tanks that create very little pressure drop as flow passes through them. In such cases, it is acceptable to locate the expansion tank on the upstream side of these components, even though the circulator may be on the downstream side and several feet away from the point where the expansion tank tees into the system. This is acceptable because there is very little pressure drop between the PONPC and the circulator inlet. This is illustrated in Figure 4-122.

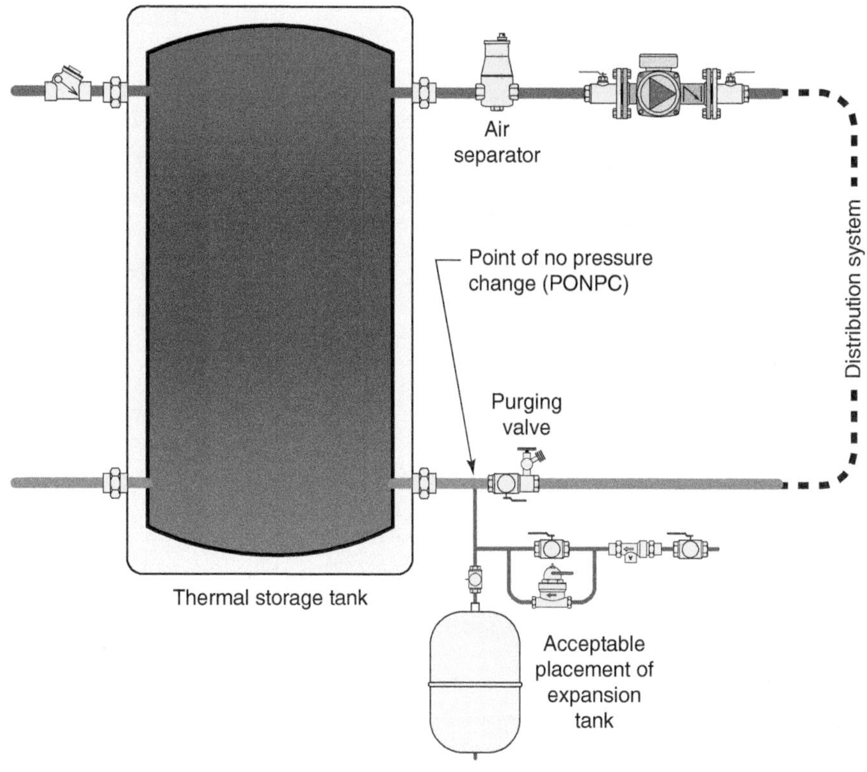

Figure 4-122 The circulator can be separated from the point where the expansion tank connects, provided that the flow path in between creates very low pressure drop.

Finally, *it is not advisable to install two or more expansion tanks at different locations within a system.* Doing so would create two points of no pressure change. When the circulator is operating, the dynamic pressures between such points could adjust in ways that create undesirable conditions such as air being drawn into the system through vents, circulator cavitation, or unnecessary opening of a pressure relief valve. This does not preclude the use of multiple expansion tanks, piped in parallel, and connected to the same point in the system.

4.8 Brazed-Plate Heat Exchangers

Many of the systems discussed in this text transfer heat between a fluid heated by the heat source and some other fluid within the system. An example would be transferring heat from the antifreeze solution circulating through solar collectors and water in that system's storage tank. Another example is transferring heat from the **system water** in a large thermal storage tank to **domestic water** used for washing and other purposes. These situations require the use of a **heat exchanger**, which prevents the two liquids from mixing yet allows heat to efficiently transfer from the higher temperature fluid to the lower temperature fluid.

There are several types of heat exchangers used in residential and light commercial hydronic systems. However, **brazed-plate heat exchangers** constructed of stainless steel are the most common in the type of systems discussed in this text, and we will thus limit our discussion to them. An example of a partially assembled brazed-plate heat exchanger is shown in Figure 4-123.

Brazed-plate heat exchangers are constructed by stacking several stainless steel plates that have been previously stamped to form alternating "herringbone" flow passageways. These plates are then pressed together, and their outer edges are brazed to form a pressure-tight assembly. The two liquids exchanging heat flow through narrow spaces between the formed plates. If these spaces were numbered, one liquid would flow through

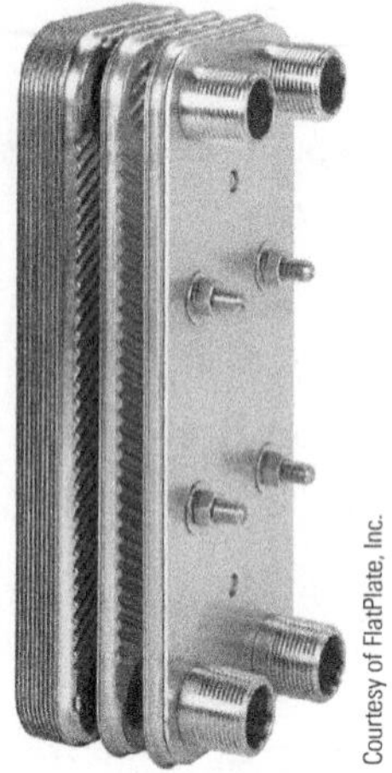

Figure 4-123 Example of a brazed plate heat exchanger. The plates at the rear (left side) of the heat exchanger are brazed together at their perimeter. The plates at the front of the heat exchanger are separated to show internal design.

spaces 1, 3, 5, and so on. The other fluid would pass through spaces 2, 4, 6, etc. Once brazed, the internal spaces between the plates are not accessible. Given the large internal surface area, the size of a brazed-plate heat exchanger is quite small relative to other types of heat exchangers of comparable performance.

The two fluids flowing through a heat exchanger should always move in opposite directions. This **counterflow** arrangement, seen in Figure 4-124, maximizes heat transfer between the two liquids for any given operating conditions. *When heat exchangers are installed, it is vitally important to pipe them for counterflow.* Be sure to check manufacturers piping drawings, as differences in piping connections do exist.

Whenever possible, brazed-plate heat exchanger should be mounted *vertically* as shown in Figure 4-125. This improves air elimination during commissioning. The heat exchanger, or the piping connected to the heat exchanger, should also be supported. Some brazed-plate heat exchangers are supplied with threaded mounting studs, as seen in Figure 4-123. These are for attaching a mounting bracket that is then fastened to a structural surface.

The spaces through which fluids pass in a brazed-plate heat exchanger are very narrow. When installed in a system, especially an older system, where dirt or other contaminants could be present, a dirt separator should be installed upstream of both fluid inlets. These separators can prevent debris such as solder balls or metal chips from entering and possibly lodging within the heat exchanger. In older systems, they can prevent precipitants or slug from adhering to the internal surfaces of the heat exchanger. If allowed to occur, such **fouling** can quickly reduce the heat transfer ability of any heat exchanger. Deposits can also cause pitting corrosion.

If one side of the heat exchanger operates with domestic water, lead-free combination isolation/flushing valves should be installed near both the inlet and outlet of the potable water connections. These allow the potable water side of the heat exchanger to be

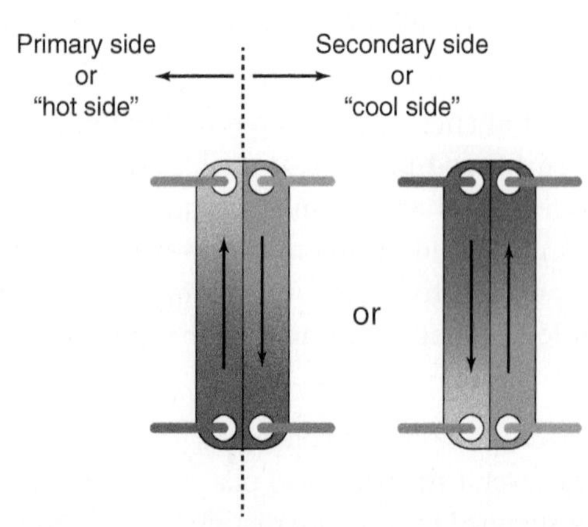

Figure 4-124 To maximize performance, heat exchangers should always be installed in "counterflow" (the two fluids moving in opposite directions through the heat exchanger).

Figure 4-125 A small BPHX with vertical mounting. This heat exchanger is supported by the connected piping. Larger brazed plate heat exchangers can be supported by a bracket attached to an adjacent structural surface.

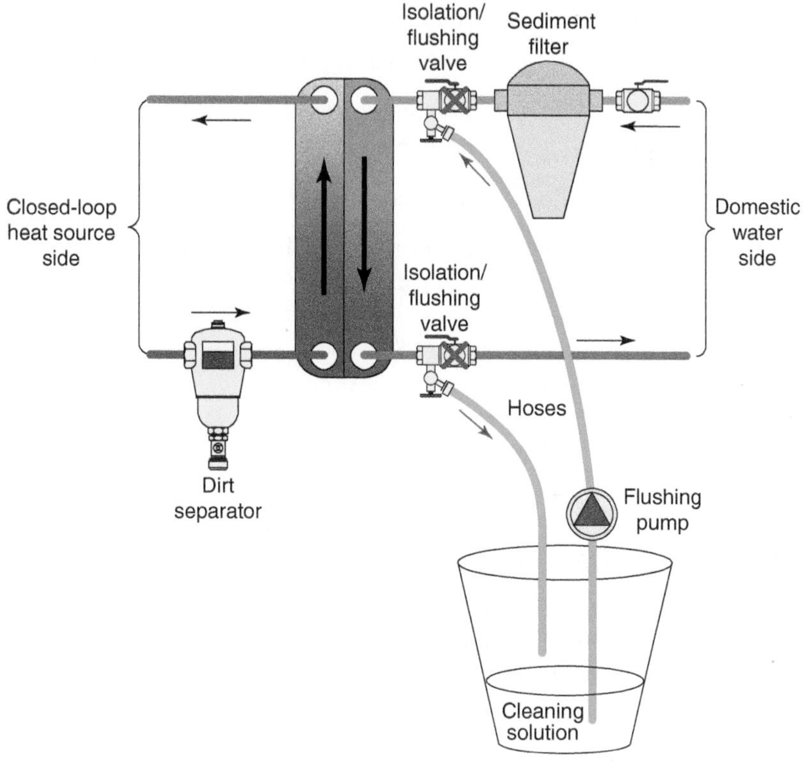

Figure 4-126 Brazed plate heat exchanger equipped with dirt separator on the primary side, and sediment filter on the domestic water side. No-lead combination isolation/flushing valves on the domestic water side allow for flushing/cleaning.

temporarily isolated from the remainder of the system and then flushed with a mild acid solution to remove lime scale and other deposits.

Figure 4-126 shows a brazed-plate heat exchanger equipped with appropriate dirt separators and isolation/flushing valves on the domestic water side.

During cleaning, the inline ball of the isolation/flushing valves are closed to isolate the domestic water side of the heat exchanger from the system. Hoses are shown connected to the flushing valves, as they would be during a flushing procedure. A small pump circulates a cleaning solution (typically a mild solution of acetic or muriatic acid) through the domestic water side of the heat exchanger to dissolve and flush out any accumulated scale.

APPROACH TEMPERATURE DIFFERENCE

The larger the internal surface area of a heat exchanger, compared to the rate of heat transfer, the smaller the required average temperature difference between the two sides of the heat exchanger. *Minimizing this temperature difference is especially important in systems using renewable energy heat sources such as solar collectors or heat pumps.* The lower the average temperature difference across the heat exchanger, the lower the operating temperature of the heat source relative to its load. Lower operating temperatures increase the thermal efficiency of most heat sources.

The **approach temperature difference** of a heat exchanger is the difference between the incoming hot fluid and the leaving heated fluid, as illustrated in Figure 4-127.

The lower the approach temperature difference, the lower the "**thermal penalty**" imposed by the presence of the heat exchanger. Thus, as the approach temperature difference nears zero, the thermal penalty associated with having the heat exchanger in the system also nears zero. The "perfect" heat exchanger with a zero approach temperature difference would simply transfer the heat at the desired rate, with no temperature drop between the fluid supplying the heat, and the fluid accepting the heat.

Unfortunately, it's impossible to create a real heat exchanger that operates with a zero approach temperature difference. However, it is possible to select heat exchangers with relatively low approach temperature differences in the range of 5 °F or less. This criteria is especially relevant to systems where heat source efficiency is markedly

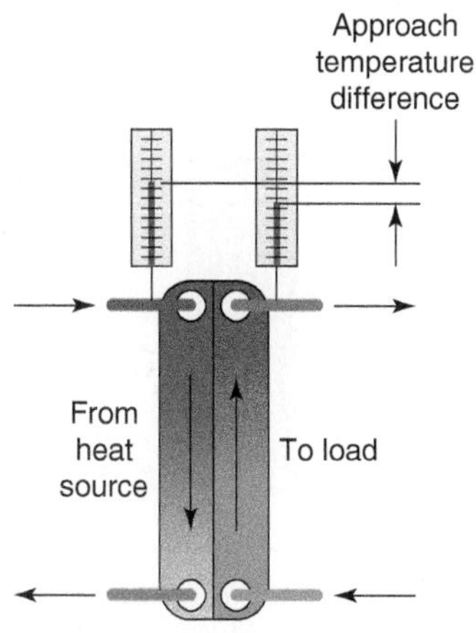

Figure 4-127 Approach temperature difference of a heat exchanger.

improved by operating at low fluid temperatures. Reducing the approach temperature difference usually requires a larger and more expensive heat exchanger.

EFFECTIVENESS

Another way to express the thermal performance of a heat exchanger is by its **effectiveness**, which is defined by Equation 4.7.

(Equation 4.7)

e = effectiveness =

$$e = \frac{\text{actual heat transfer rate}}{\text{maximum possible heat transfer rate}}$$

The actual rate of heat transfer can be determined based on the flow rate, specific heat, and temperature change of the fluid on either the hot side or cool side of the heat exchanger, as shown in Figure 4-128.

where:

Q_{actual} = actual rate of heat transfer across heat exchanger (Btu/hr)

8.01 = unit conversion factor

D_h = density of fluid through hot side of heat exchanger (lb/ft³)

D_c = density of fluid through cool side of heat exchanger (lb/ft³)

c_h = specific heat of fluid through hot side of heat exchanger (Btu/lb/°F)

c_c = specific heat of fluid through cool side of heat exchanger (Btu/lb/°F)

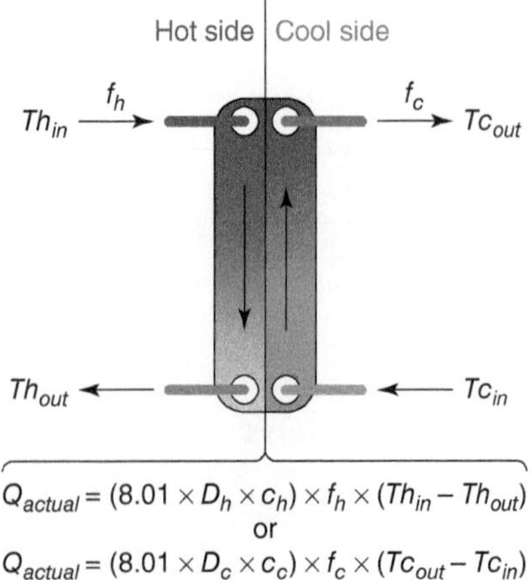

Figure 4-128 | Equations for determining the actual rate of heat transfer across a heat exchanger.

f_h = flow rate of fluid through hot side of heat exchanger (gpm)

f_c = flow rate of fluid through cool side of heat exchanger (gpm)

T = temperatures at locations shown in Figure (°F)

The maximum possible rate of heat transfer through the heat exchanger can be calculated using Equation 4.8.

(Equation 4.8)

$$Q_{max} = [8.01 \times D \times c \times f]_{min} \times (Th_{in} - Tc_{in})$$

Where:

$(8.01 \times D \times c \times f)_{min}$ = the smaller of the two **fluid capacitance rates**. This is found by calculating the product $(8.01 \times D \times c \times f)$ for both the hot and cool side of the heat exchanger and then selecting the smaller of the two.

Th_{in} = inlet temperature of the hot fluid (°F)

Tc_{in} = inlet temperature of the cool fluid (°F)

As the size of the heat exchanger increases compared to the required rate of heat transfer, its effectiveness approaches the theoretical limiting value of 1.0.

Example 4.7: A heat exchanger in a solar combisystem operates at the conditions shown in Figure 4-129. The fluid in the collector loop is a 40 percent solution of propylene glycol. The fluid on the cool side of the heat exchanger is water. Determine the rate of heat transfer across the heat exchanger and its effectiveness under these operating conditions.

Start by finding the fluid properties of both the 40 percent propylene glycol solution and water at the average temperature of each fluid as it passes through the heat exchanger. The density of the fluids at their

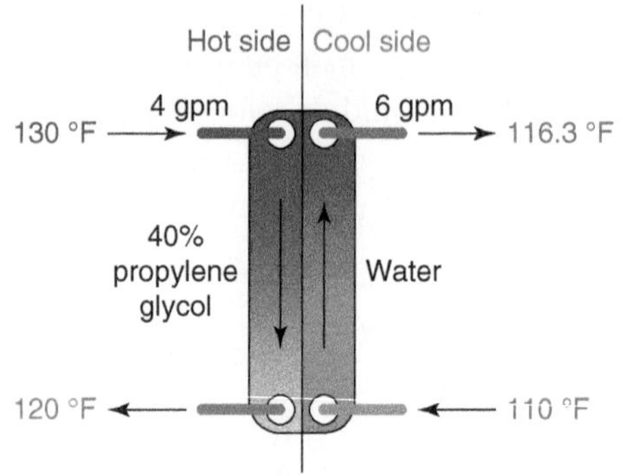

Figure 4-129 | Heat exchanger operating conditions for Example 4.7.

respective average temperatures within the heat exchanger can be determined from Figure 4-117. The specific heat of the water can be assumed to be 1.00 Btu/lb/°F. The specific heat of the 40 percent solution of propylene glycol can be found in manufacturer's literature.

For the 40 percent propylene glycol solution:

$D = 64.0$ lb/ft^3
$c = 0.91$ Btu/lb/°F

For water:

$D = 61.8$ lb/ft^3
$c = 1.00$ Btu/lb/°F

Next, calculate the actual rate of heat transfer across the heat exchanger. This can be done using data from either flow stream (e.g., the hot side or cool side of the heat exchanger). In this case, the data from the flow stream through the hot side of the heat exchanger (using the 40 percent propylene glycol solution) is used:

$$\begin{aligned} Q_{actual} &= (8.01 \times D_h \times c_h) \times f_h \times (Th_{in} - Th_{out}) \\ &= (8.01 \times 64.0 \times 0.91) \times 4 \times (130 - 120) \\ &= 18{,}600 \text{ Btu/hr} \end{aligned}$$

Next, determine which side of the heat exchanger has the *minimum* fluid capacitance rate (e.g., calculate the product $(8.01 \times D \times c \times f)$ for each flow stream and determine which is smaller).

For the hot side of the heat exchanger:

$$\begin{aligned} (8.01 \times D \times c \times f)_{40\%PG} &= (8.01 \times 64.0 \times 0.91 \times 4) \\ &= 1866 \frac{\text{Btu}}{\text{hr} \cdot {}^\circ\text{F}} \end{aligned}$$

For the cool side of the heat exchanger:

$$\begin{aligned} (8.01 \times D \times c \times f)_{water} &= (8.01 \times 61.8 \times 1.00 \times 6) \\ &= 2970 \frac{\text{Btu}}{\text{hr} \cdot {}^\circ\text{F}} \end{aligned}$$

The fluid capacitance rate on the hot side of the heat exchanger is the smallest.

Determine the maximum possible heat transfer across the heat exchanger, using Equation 4.8. This corresponds to a thermodynamic limit in which the outlet temperature of the fluid with the lower fluid capacitance rate approaches the inlet temperature of the other fluid stream. It is determined by multiplying the minimum fluid capacitance rate by the difference in temperature between the entering hot fluid and the entering cool fluid.

$$\begin{aligned} Q_{max} &= [8.01 \times D \times c \times f] \times (Th_{in} - Tc_{in}) \\ &= [8.01 \times 64.0 \times 0.91 \times 4] \times (130 - 110) \\ &= 37{,}320 \text{ Btu/hr} \end{aligned}$$

Finally, use Equation 4.7 to determine the effectiveness of the heat exchanger under these conditions.

$$e = \frac{Q_{actual}}{Q_{max}} = \frac{18{,}660}{37{,}320} = 0.50$$

The effectiveness of a heat exchanger is often used as an input by software simulation programs that estimate the performance of solar thermal systems. As a general guideline, the *minimum* effectiveness of any heat exchanger used between the collector array and thermal storage tank of a solar thermal system should be 0.55.

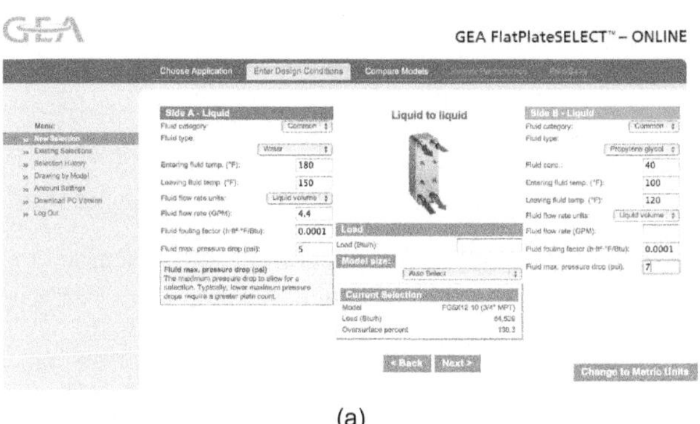

(a)

Note: To download a PDF file version of this report, navigate to the Print/Save page.

Model: FG5X12-120(1-1/4"MPT)			
Selection ID	NUD6E5E3Q	Model size	5x13
Application	Snow melt	Nominal surface (ft²)	45.2
Load (Btu/h)	260,000	Dimensions	5.1W x 13.3H x 11.0D
Log mean temp. diff. (°F)	12.3	Plate construction	Single wall
Overall HTC (Btu/h·ft²·°F)	469	Net weight (lb)	39.9
Oversurface percent	0.6		

Design Conditions	Side A - Liquid	Side B - Liquid
Fluid type	**Water**	**Propylene glycol**
Fluid conc.		50
Fluid mass flow rate (lb/min)	289.2	249.0
Entering fluid temp. (°F)	140.0	110.0
Leaving fluid temp. (°F)	125.0	130.0
Fluid flow rate (GPM)	35.2	29.1
Fluid fouling factor (h·ft²·°F/Btu)	0.00010	0.00010
Model Parameters		
Number of channels	59	60
Velocity (ft/s)	0.57	0.46
Pressure drop (psi)	1.7	1.4
Heat transfer coef. (Btu/h·ft²·°F)	1,942	706
Internal volume (ft³)	0.177	0.180

Ratings at Varying Conditions					
Percent difference	**-15%**	**-7½%**	**0%**	**7½%**	**15%**
Pressure drop (psi) (Side A)	1.3	1.5	1.7	2.0	2.2
Pressure drop (psi) (Side B)	1.0	1.2	1.4	1.5	1.8
Load (Btu/h)	**221,000**	**240,500**	**260,000**	**279,500**	**299,000**
Fluid flow rate (GPM) (Side A)	29.9	32.5	35.2	37.8	40.4
Fluid mass flow rate (lb/min) (Side A)	245.8	267.5	289.2	310.9	332.6
Fluid flow rate (GPM) (Side B)	24.7	26.9	29.1	31.3	33.5
Fluid mass flow rate (lb/min) (Side B)	211.6	230.3	249.0	267.6	286.3
Entering fluid temp. (°F) (Side A)	140.0	140.0	140.0	140.0	140.0
Entering fluid temp. (°F) (Side B)	110.0	110.0	110.0	110.0	110.0
Leaving fluid temp. (°F) (Side A)	125.0	125.0	125.0	125.0	125.0
Leaving fluid temp. (°F) (Side B)	130.0	130.0	130.0	130.0	130.0
Oversurface percent	7.3	3.8	0.6	-2.3	-4.9

Disclaimer

This software and the generated calculations provided herein are estimates only and should be treated as such. GEA PHE Systems North America, Inc. always strives to give complete and accurate information, but cannot provide any guarantees. This software and its output are provided "as is" and any express or implied warranties, including, but not limited to, the implied warranties of merchantability and fitness for a particular purpose are disclaimed. In no event shall GEA PHE Systems North America, Inc. be liable for any direct, indirect, incidental, special, exemplary, or consequential damages (including, but not limited to, procurement of substitute goods or services; loss of use, data, or profits; or business interruption) however caused and on any theory of liability, whether in contract, strict liability, or tort (including negligence or otherwise) arising in any way out of the use of this software, even if advised of the possibility of such damage.

(b)

Figure 4-130 (a) User interface for web-based sizing and selection software for brazed plate heat exchangers. (b) Detailed thermal and hydraulic operating conditions.

SIZING BRAZED-PLATE HEAT EXCHANGERS

Today, most manufacturers of brazed-plate heat exchangers offer software that can use the known or estimated values for some operating conditions to select an appropriate heat exchanger and calculate its performance. An example of the user interface from one such web-based software offering is shown in Figure 4-130.

As in other areas of system design, software-assisted analysis allows rapid comparisons of performance without time-consuming calculations. The software shown in Figure 4-130 also provided pressure drop information for the heat exchanger that can be converted to head loss and used with the methods presented in Chapter 3 to select an appropriate circulator for each side of the heat exchanger.

SUMMARY

This chapter has discussed a wide variety of hardware that is often used in hydronic-based renewable energy systems. Additional information on this hardware is available from manufacturers, much of it directly from manufacturer's websites.

The proper selection of the components discussed in this chapter is essential for proper system operation. Many of the schematics shown in later chapters will show this hardware in its proper relationship with renewable energy heat sources.

Appendix A provides a legend of the symbols used in these schematics.

KEY TERMS

3-way diverting valve
3-way motorized mixing valve
3-way rotary-type mixing valve
3-way thermostatic mixing valve
4-wire zone valve
acceptance volume
actuator
air elimination
air management
air-side prepressurization
approach temperature difference
auxiliary heat source
backflow preventer
ball valve
baseboard tees
boiler
brazed-plate heat exchanger
brazing
brushless DC motors
cavitation
central air separator
circulator
coalescing media
common piping
component isolation
condensate
condensing boilers
constant differential pressure control
conventional boilers
copper water tube
corrugated stainless steel tubing (CSST)
counterflow
cross-linked polyethylene tubing
crossover branches
deadheaded
dewpoint temperature
diaphragm
diaphragm-type expansion tank
DIN 4726 standard
dissolved air
domestic water
drainback solar thermal systems
drip loop
dynamic pressure distribution
EEPROM
effectiveness [of a heat exchanger]
electronically commutated motors (ECM)
end switch
entrained air
EPDM
family of circulators
feed water valve
feet of head
firmware
flat pump curve
float-type air vent
flow regulation
fluid capacitance rate
flux
foot pound
fouling
FPT
full port ball valve
gaseous cavitation
gate valve
globe valve
HDPE
heat exchanger
heat motor
heating load profile
high point air vent
hydraulic equilibrium
hydraulic separation
hydraulic separator
hydraulically separated
hydronic heat source
implosion
in solution
intermittent flue gas condensation
isolation flanges
Kill-A-Watt meter
latent heat
lift head
low loss header
magnesium hydrolite granules
manual air vent
microbubble
microbubble air separator

mod/con
modulating
MPT
net positive suction head available
neutralizer
nominal inside diameter
normally closed zone valve
normally open zone valve
operating point
outdoor reset control
oxygen diffusion barrier
permanent split capacitor (PSC) motor
PEX
PEX-AL-PEX tubing
point of no pressure change (PONPC)
press fitting
pressure-reducing valve
pressure-relief valve
primary heat exchanger
proportional differential pressure control
pump curve
purged
purging
purging valve
rare earth (neodymium) permanent magnets
rotor [of a circulator]
rotor can
secondary heat exchanger
short cycling
siphon
sleep mode
smart circulators
soft soldering
spool [of a valve]
stagnation conditions
standard port ball valve
stationary air pockets
stator
stator poles
steep pump curve
sustained flue gas condensation
swing check
system head loss curve
system water
temperature and pressure (T&P) relief valve
thermal penalty [of a heat exchanger]
thermostatic radiator valves (TRVs)
turndown ratio
unsaturated state of air solubility
vapor cavitation
vapor pockets
vapor pressure
vibration dampening coupler
wet rotor circulator
wire-to-water efficiency
zone valve

QUESTIONS AND EXERCISES

1. True or False: It is impossible for a conventional boiler to operate *with* sustained flue gas condensation? Justify your answer.
2. True or False: It is impossible for a condensing boiler to operate *without* sustained flue gas condensation? Justify your answer.
3. What are the differences between type K, L, and M copper tubing? Which is the most commonly used in hydronic heating systems?
4. A straight copper pipe 65 feet long changes temperature from 65 °F to 180 °F. What is the change in length of the tube? Describe what would happen if the tube were cooled from 65 °F to 40 °F.
5. What should be done before soldering a valve with a nonmetallic disc or washer?
6. Why should the thermostatic actuator of a radiator valve not be mounted above the heat emitter?
7. Why should a globe valve not be used for component isolation purposes? What type(s) of valves are better suited for this application?
8. What is an advantage of a spring-loaded check valve when compared to a swing check valve? What is a disadvantage?
9. Why is it important to install at least 12 diameters of straight tubing on the inlet side of a check valve?
10. Explain the difference between a 3-way motorized mixing valve and a 3-way thermostatic mixing valve. Be specific regarding the construction, operation, and flow characteristics of each.
11. What does an actuator do when fitted to a valve? Describe a common application in which an actuator is used on a valve in a hydronic system.
12. Why should a wet-rotor circulator always be mounted with its shaft horizontal?
13. Why should the expansion tank in a closed hydronic circuit be mounted close to the inlet side of the circulator in that circuit?
14. What is the evidence that head energy has been added to a fluid as it passes through an operating circulator?
15. What is meant by the term "hydraulic equilibrium"? When does it occur within a hydronic circuit?
16. What is the function of the end switch in a zone valve?
17. When can a swing check valve offer the same protection as a backflow preventer in the make-up water line of a hydronic system?
18. What creates the pressure tight seal when connecting CSST to a brass adapter fitting?
19. Which has greater head loss at a given flow rate: 3/4-inch CSST or 3/4-inch PEX tubing?
20. Assuming equal lengths of tubing and equal temperature increase, explain why a PEX-AL-PEX tube has less thermal expansion compared to a PEX tube.

21. A 100-foot length of 1-inch PEX tubing is installed within an insulated underground sleeve between an outdoor wood-fired hydronic heater and a mechanical room. When installed, the tubing is at a temperature of 50 ºF. As the boiler operates, this tubing conveys water at an average temperature of 180 ºF. Determine how much longer the tube becomes under this condition. Assume that it is free to expand within its insulated sleeve.

22. Describe the difference between a circulator with a steep pump curve and one with a flat pump curve. Which would be better for use in a multi-zone system involving zone valves? Why?

23. Describe the differences between "constant differential pressure mode" and "proportional differential pressure mode" as they relate to the operation of an ECM-based circulator. Which is appropriate for systems using a distribution manifold and low resistance common piping?

24. Describe the difference between the head of a circulator and the difference in pressure measured across the circulator using pressure gauges. How is this pressure difference converted to a head value?

25. A circulator with a pump curve, as shown in Figure 4-131, is installed in a system operating with 140 °F water. Pressure gauges on the inlet and discharge ports have readings as shown. What is the flow rate through the circulator?

26. A piping system has a system resistance curve described by the following equation:

$$H_L = (0.85)(f)^{1.75}$$

Where the head loss is in feet, and the flow rate (f) is in gpm. A circulator with the pump curve shown in Figure 4-131 is being considered for this system. Find the flow rate and head at the operating point of this piping system and circulator.

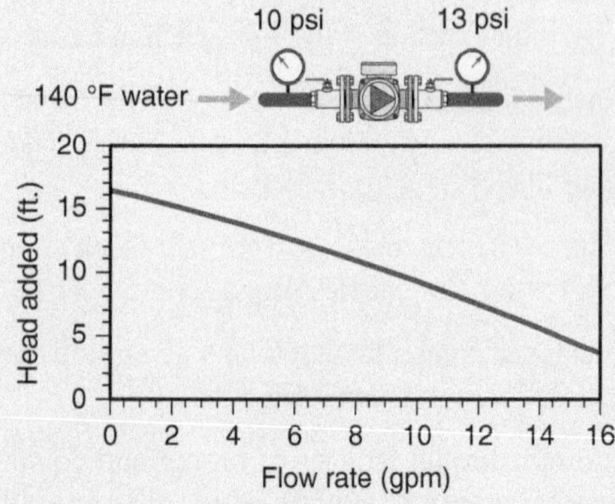

Figure 4-131 | Pressure gauge readings and pump curve for Exercise 25.

27. The static pressure within the circulator of a given system is 5 psig when the circulator is off. When the circulator operates, it adds 20 feet of head energy to the 140 ºF water flowing through it. Determine the readings of pressure gauges tapped into its inlet and discharge ports when it is running, assuming that:

a. The expansion tank is connected at the inlet port of the circulator.

b. The expansion tank is connected to the discharge port of the circulator.

28. A piping system and wet-rotor circulator have the head loss and head added curves shown in Figure 4-132. The measured power consumption of the circulator operating at these conditions is 75 watts. What is the wire-to-water efficiency of the circulator?

29. A circulator operates in a piping circuit filled with water at 100ºF. The head developed by the circulator is 18 feet. The flow rate through the circuit is 8 gpm. The wire-to-water efficiency of the circulator under these conditions is 20 percent. Determine the annual operating cost of this circulator assuming it operates for 3,000 hours per year in an area where the current cost of electricity is 15 cents per kilowatt-hour.

30. Describe three different hardware configurations that can produce hydraulic separation between the heat source circulator and a distribution system circulator.

31. True or false: To achieve hydraulic separation between circulators in two intersecting circuits, the piping that is common to both circuits should have high head loss.

32. Under what conditions does mixing occur within a device that provides hydraulic separation between two intersecting hydronic circuits?

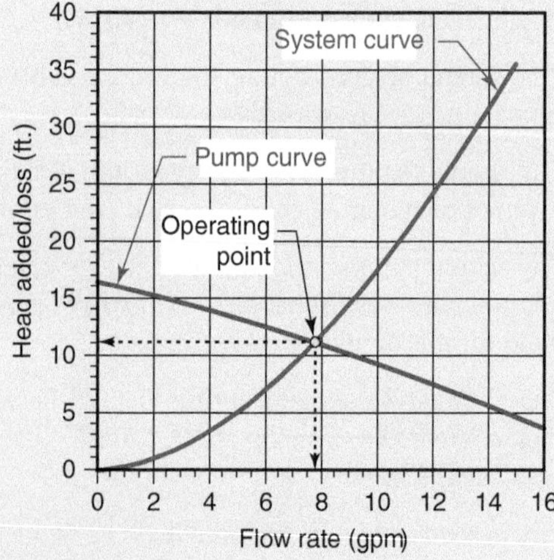

Figure 4-132 | Pump curve and system head loss curve for Exercise 28.

33. What is the purpose of a coalescing media within an air separator?

34. Why should an isolation valve be installed under a float-type air vent at the top of a solar collector array operating with an antifreeze solution?

35. Other than hydraulic separation, describe two functions provided by most modern hydraulic separators.

36. Why is it important to insulate the shell of a hydraulic separator operating within a heating system?

37. Explain why air management, rather than air separation, is appropriate for a drainback-protected solar thermal system.

38. A system contains the following hardware, all connected into the same circuit: 250 gallon storage tank, 200 feet of ¾-inch copper pipe, 50 feet of 1.5-inch copper pipe, and 2,000 feet of PEX tubing. The system is filled with water at 60 °F. The maximum operating temperature of the system is 140 °F. The top of the circuit is 35 feet above the location of the expansion tank and the 30-psi rated pressure-relief valve. Determine the following:

a. The proper air-side pressurization of the diaphragm-type expansion tank.

b. The minimum size of the diaphragm-type expansion tank.

39. Why is it important to use a small approach temperature difference when selecting a brazed plate heat exchanger for use between a solar collector array and thermal storage tank?

40. What would be the "effectiveness" value for a perfect heat exchanger? What would be the approach temperature difference for a perfect heat exchanger?

DAIKIN
R410A
R410A

CHAPTER 5

Low-Temperature Heat Emitters and Distribution Systems

OBJECTIVES

After studying this chapter, you should be able to:

- Explain the importance of low-water-temperature heat emitters in combination with renewable energy heat sources.
- Understand the advantages of parallel versus series distribution systems.
- Describe the differences between standard and low-temperature baseboard.
- Explain the construction and application of panel radiators.
- Determine the thermal output of various heat emitters over a range of water temperatures.
- Describe the heating characteristics of heated concrete slabs.
- Explain the sequence of installing materials to create heated slabs.
- Explain how the thermal mass of heat emitters will affect performance.
- Design and build low-mass radiant wall and ceiling panels.

5.1 Introduction

A **hydronic heat emitter** is any device designed to absorb heat from water flowing through it and release that heat into conditioned space. There are a wide variety of products available for this requirement. Some, such as fin-tube baseboard or panel radiators, are delivered to a construction site ready for installation. Others, such as a heated floor slab, are "site built." Both types have strengths and limitations.

Not all hydronic heat emitters are well suited for use with renewable energy heat sources. *Those that can deliver sufficient heat output to maintain comfort under design load conditions, while operating at supply water temperatures no higher than 120 °F, are greatly preferred.* Thus, the words "low temperature," as used in the title of this chapter, represent a criteria that must be observed if the overall system is to deliver good performance.

WHY LOW-WATER TEMPERATURE?

All the renewable energy heat sources discussed in this text operate at higher thermal efficiency when matched with low-water-temperature heat emitters and distribution systems. Although the thermal efficiency of devices such as solar thermal collectors, heat pumps, and wood-fired boilers is discussed in detail in later chapters, it's worth taking a preliminary look at how important low water temperature is.

Figure 5-1 shows how the instantaneous thermal efficiency of a flat-plate solar collector is affected by the temperature of water flowing into it.

If the outdoor air temperature and solar radiation intensity remain constant at the indicated values, which represent a sunny midwinter day in a cool climate, the thermal efficiency of the collector drops rapidly with increasing inlet fluid temperature.

For example, if the fluid temperature entering the collector is 80 °F, the outdoor air temperature is 30 °F, and the solar intensity is 250 Btu/hr/ft^2 (a bright midday winter sky), the collector is gathering about 59 percent of the solar radiation striking it. However, if the entering fluid temperature is 140 °F, with all the other conditions unchanged, the collector's thermal efficiency falls to about 40 percent. There is a significant penalty associated with operating the collector at the higher fluid inlet temperatures. Such operation increases the temperature difference between the collector's absorber plate and the outdoor air, which increases the rate of heat loss from the collector and thus lowers its thermal efficiency.

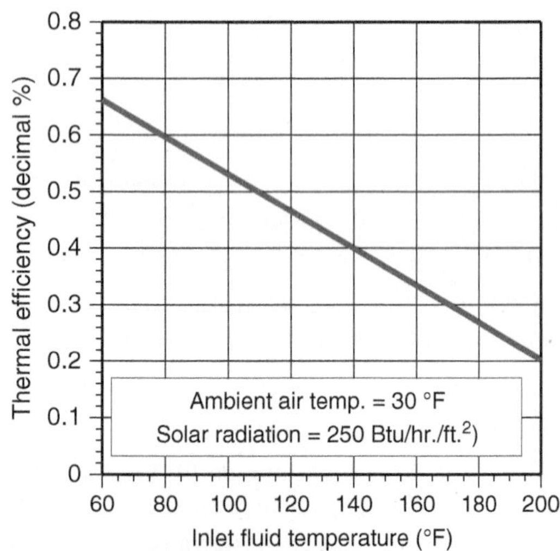

Figure 5-1 The instantaneous thermal efficiency of a flat-plate solar collector as a function of inlet water temperature. Higher inlet temperatures decrease efficiency.

Heat pumps also experience a drop in thermal efficiency when forced to operate at higher water delivery temperatures. Figure 5-2 shows the effect for a modern water-to-water geothermal heat pump being supplied with 45 °F water from an earth loop.

The **coefficient of performance (COP)**, which is plotted on the vertical axis of Figure 5-2, is an indicator of the heat pump's thermal efficiency when the heat pump operates in heating mode. Specifically, it is the ratio of the heat output rate (in Btu/hr) divided by the electrical input power (converted to Btu/hr) required to operate the heat pump. A COP of 4.0 means the heat pump is sending heat to its load at a rate four times greater than the rate of electrical energy input needed to operate it. The higher the COP, the lower the heat pump's operating cost.

Any design measures that decrease the temperature of water returning from the heating distribution system and into the heat pump will increase the heat pump's COP.

For wood-fired boilers, one could make the point that they are capable of producing relatively high water temperatures, even up to 200 °F. While this is true, it does not negate the benefits of matching them to low-temperature distribution systems. Most high-efficiency wood-fired boilers work best in combination with a thermal storage tank. When firing, they add heat to this tank. The heating distribution system then draws heat from this tank. The lower the tank temperature can drop, and still supply sufficient heat to the building, the less frequently the boiler needs to be fired. Less frequent and longer operating cycles increase the boiler's thermal efficiency and reduce its emissions.

Many auxiliary heat sources, such as gas-fired condensing boilers, also operate at higher efficiency when delivering heat to low-temperature distribution systems.

The author's recommendation is to *design every hydronic distribution system that is associated with a renewable energy heat source so that it can deliver the design heating load using a* ***design supply water temperature*** *no higher than 120 °F.*

The system designer must select and size heat emitters so that comfort and control are achieved in all areas of a building. Technical, architectural, and

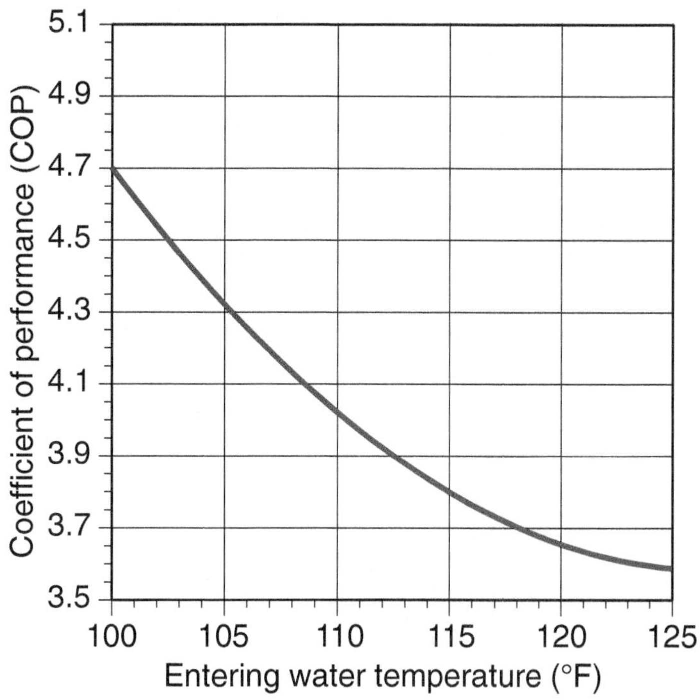

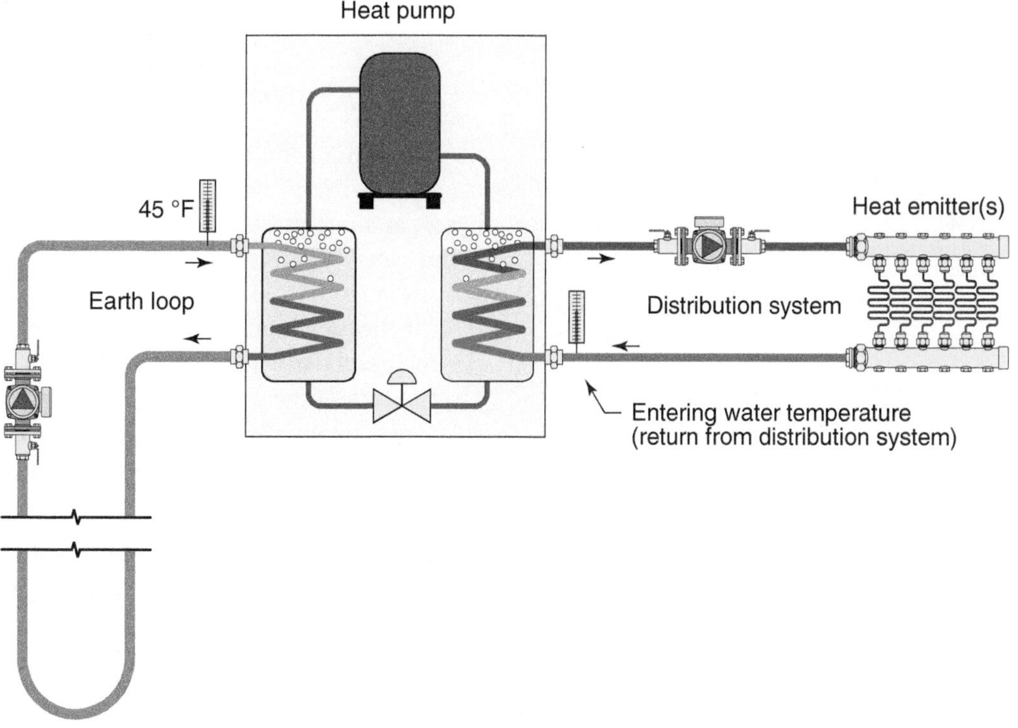

Figure 5-2 | The coefficient of performance (COP) of a water-to-water heat pump drops rapidly as the water temperature entering it, from the return side of the hydronic distribution system, increases.

economic issues must all be considered. The technical issues involve temperature, flow rate, and heat output characteristics of the heat emitter options. Architectural issues include the appearance of the heat emitters in the building as well as any interference they may create with furniture, door swings, and so on. Economic considerations can greatly expand or restrict heat emitter options. The designer must provide a workable, if not optimal, selection of heat emitters within a given budget. Since most heating systems are designed after the floor plan of the building is established, the heating system designer is often faced with the challenge of making the system fit the plan. A good knowledge of the available options can lead to creative and efficient solutions.

PARALLEL VERSUS SERIES DISTRIBUTION SYSTEMS

The piping configuration that connects the heat emitters into a **distribution system** is also critically important in achieving good performance. *When low supply water temperatures are used, that piping system should be configured so that heat emitters are supplied in parallel rather than series.* **Parallel piping systems** closely approximate the ideal scenario in which each heat emitter in the distribution system receives the same water temperature. Figure 5-3 shows two examples of distribution systems with parallel piping.

An example of a **series piping system** is shown in Figure 5-4.

In a series distribution system, flow passes from one heat emitter to the next. This creates a situation where the water temperature supplied to a given heat emitter is lower than that supplied to the heat emitter located upstream of it and higher than the heat emitter located downstream of it. This complicates the sizing procedure. It also leads to distribution systems that tend to have high head loss characteristics in comparison to those of parallel piping systems.

THERMAL EQUILIBRIUM

Before attempting to design a hydronic heating distribution system, it's critically important to understand what determines the water temperature in every hydronic heating system. Some designers, including many HVAC professionals, think that it's the heat source that determines the water temperature in a hydronic heating system. This is probably because many boilers come with a controller that has a dial (or perhaps digital interface) on which the installer "sets" a water temperature. Many think that by setting this temperature they are *guaranteeing* that the heat source will produce it. This simply isn't true. That temperature setting is only a *limit* on how high the water temperature leaving the heat source *might* climb. This controller is called a **high limit controller**, and its setting does not assure that the water temperature leaving the boiler will be at the set value.

As discussed in Chapter 1, *the water temperature in any hydronic distribution system only climbs as high as necessary for that system to achieve* **thermal equilibrium**—where the rate of heat release from the distribution system exactly balances the rate of heat input from the heat source. Once this condition is achieved, there is no "thermodynamic incentive" for the water temperature to climb higher, and it won't!

It's the design of the hydronic distribution system, rather than the setting of the heat source's high temperature limit controller, that determines the water temperature at which the system will operate.

Consider the two systems shown in Figure 5-5. Each system has an identical heat source that delivers heat at a rate of 20,000 Btu/hr to the circulating water and can operate over a wide range of temperature. System (a) has 31 feet of typical fin-tube baseboard for its heat emitter. System (b) contains 111 feet of the same baseboard. The controls on each heat source are set to turn off heat production if the water temperature leaving the heat source reaches 200 °F. Both systems are put into continuous operation at approximately the same water flow rate.

The water leaving the heat source in system (a) climbs to a temperature of 180 °F, and stabilizes, (e.g., reaches thermal equilibrium). At this temperature, the 20,000 Btu/hr being added to the water by the heat source is being dissipated from the water by the 31 feet of baseboard. *Notice that this supply water temperature is still 20 °F lower than the temperature at which the heat source will stop further heat production.* Even if the high limit controller on this heat source were set to 250 °F, the supply water temperature would not climb higher than 180°F. The water temperature doesn't need to climb any higher to create a balance between the rate of heat generation

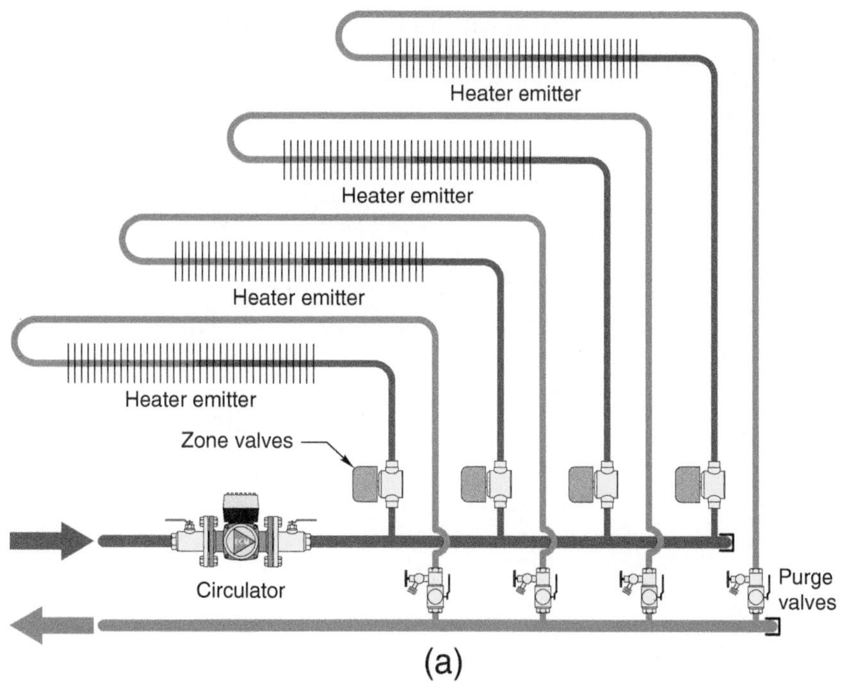

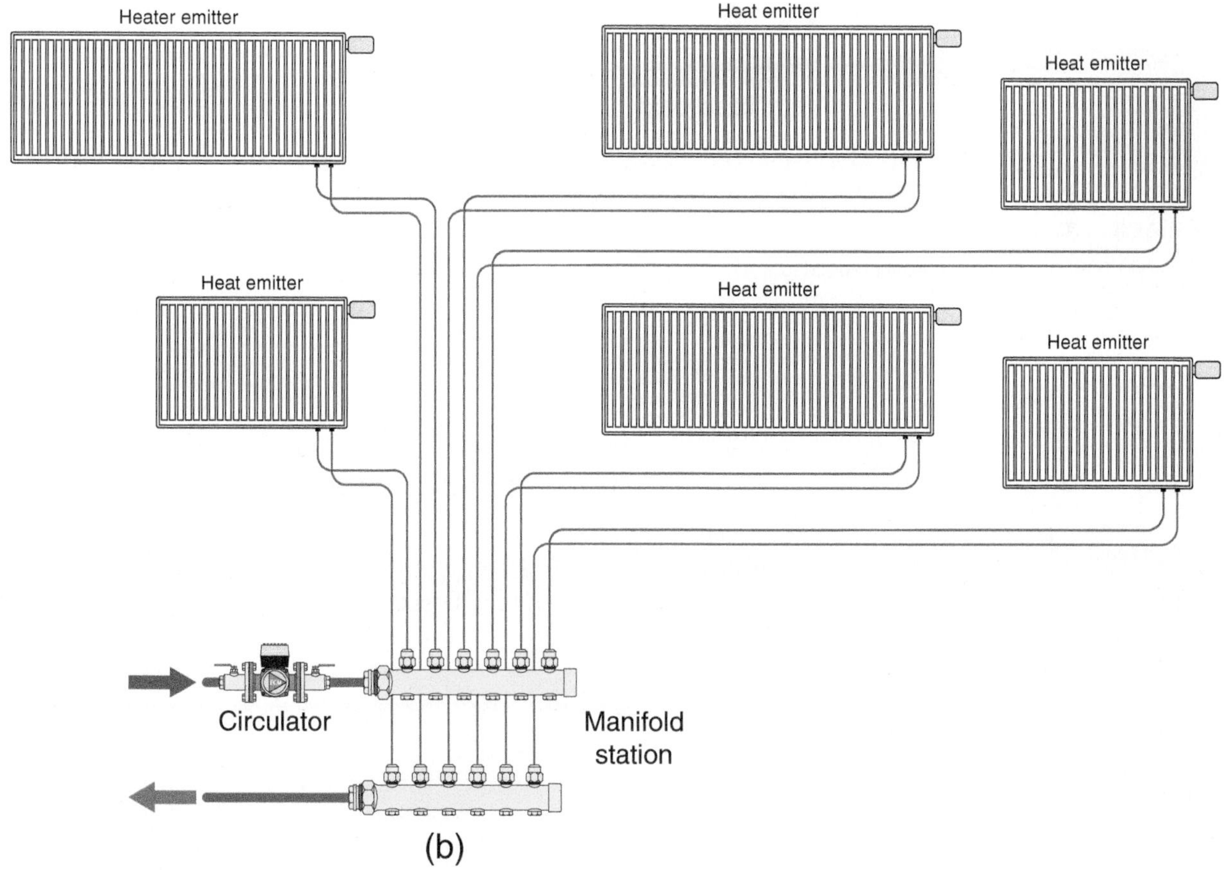

Figure 5-3 (a) Parallel piping of fin-tube baseboard using zone valves. (b) Parallel piping of panel radiators using a manifold station.

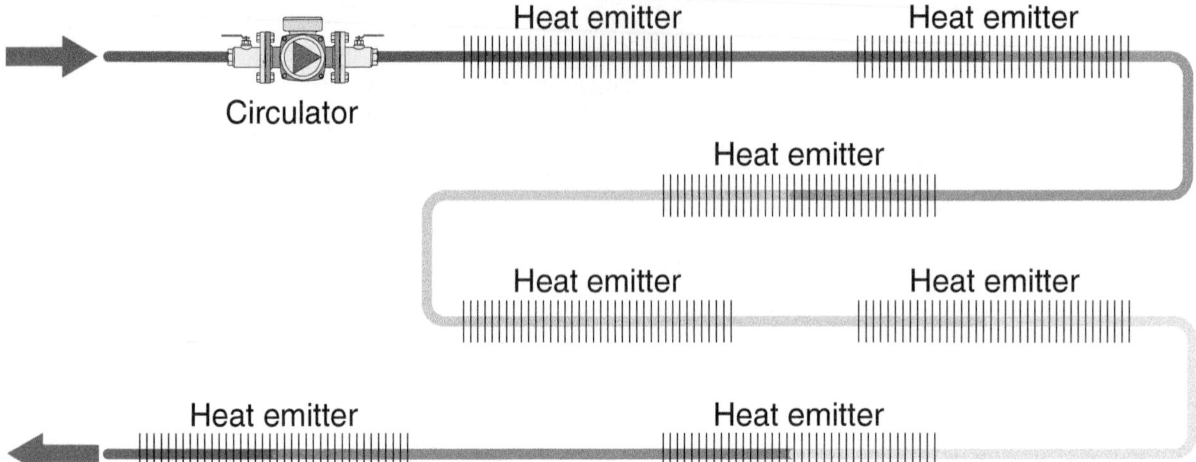

Figure 5-4 | Example of a series distribution system.

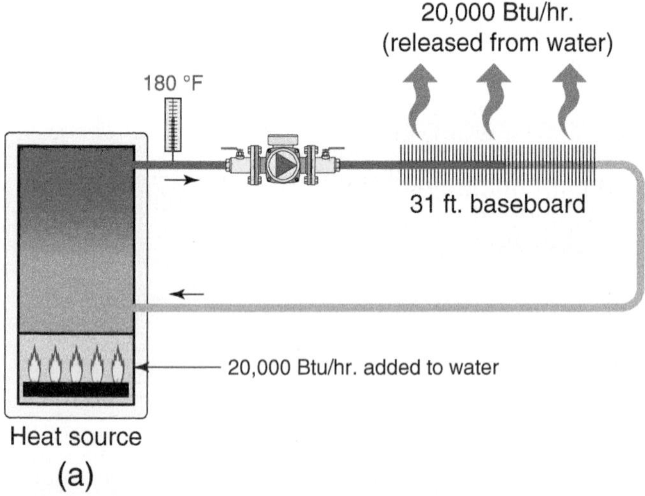

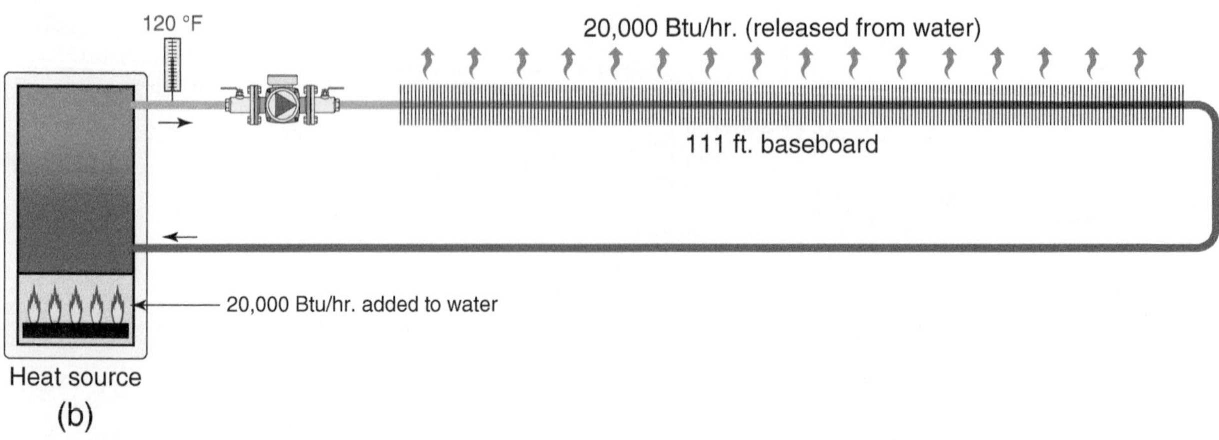

Figure 5-5 | (a) With 31 feet of baseboard as the heat emitter, thermal equilibrium is achieved at a supply water temperature of 180 °F. (b) With 111 feet of baseboard as the heat emitter, thermal equilibrium is achieved at a supply water temperature of 120 °F.

by the heat source, and the rate of heat dissipation by the distribution system.

If the length of baseboard were increased to 111 feet, and the circulator adjusted so that the circuit flow rate through the circuit remained the same, thermal equilibrium would occur at a supply water temperature of 120 °F. The fact that the heat source was set to allow water temperatures as high as 200 °F is irrelevant. In system (b), the supply water temperature has no reason to climb above 120 °F, since at that temperature, the 111 feet of baseboard can dissipate the full 20,000 Btu/hr of heat supplied by the heat source.

Assuming that the heat source's high limit controller is set high enough so as to not interfere with the process, all hydronic system *will inherently find a supply.water temperature at which thermal equilibrium occurs, and remain in operation at that condition.* The designer's goal is to ensure that when thermal equilibrium is established, the system operates at conditions that provide excellent comfort, and do not adversely affect the operation, safety, or longevity of the system's components.

5.2 Fin-Tube Baseboard

Standard fin-tube baseboard, such as shown in Figure 5-6, is commonly used in conventional hydronic systems. Originally developed in the 1930s as a substitute for cast-iron radiators, standard fin-tube baseboard has not changed much in several decades. When paired with a higher temperature boiler, it provides good performance based on the criteria of heat output per dollar of installed cost.

Standard fin-tube baseboard trades lower material content and lower installed cost for the availability of higher water temperature. It was designed in an era when high water temperature hydronic systems were common. Most systems using standard fin-tube baseboard are designed around water temperatures of 180°F or higher. Some manufacturer's literature still lists the heat output of standard fin-tube baseboard at water temperatures up to 240 °F.

Fin-tube baseboard releases most of its heat output through **natural convection**, especially when operated at high water temperatures. When mounted in the preferred location, at the base of exterior walls, this natural convection produces a gentle, virtually undetectable room air-circulation pattern, as shown in Figure 5-7.

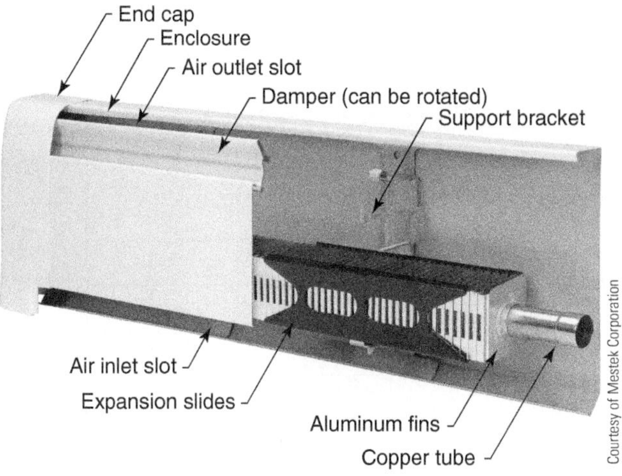

Figure 5-6 | Standard residential grade fin-tube baseboard.

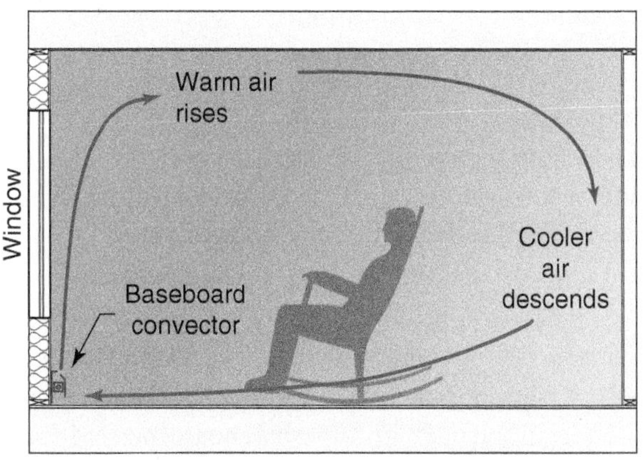

Figure 5-7 | Room air flow created by fin-tube baseboard mounted at the base of exterior walls.

This pattern of air flow counteracts the tendency for downward convection-driven air flow associated with cooler exterior surfaces, especially windows. Disrupting this downward flow helps eliminate cool drafts across the floor.

Figure 5-8 shows how the heat output of a typical residential-grade fin-tube baseboard varies based on the *average* water temperature in the baseboard.

While it is possible to lower the supply water temperature by adding more fin-tube baseboard to the system, lowering it from 180 °F to 120 °F requires about three times as much fin-tube length for the same heat output. Few rooms with design heat loads over 20 Btu/hr/ft^2 can physically accommodate this much baseboard, and few occupants would accept the aesthetics. Thus, the use of standard fin-tube baseboard

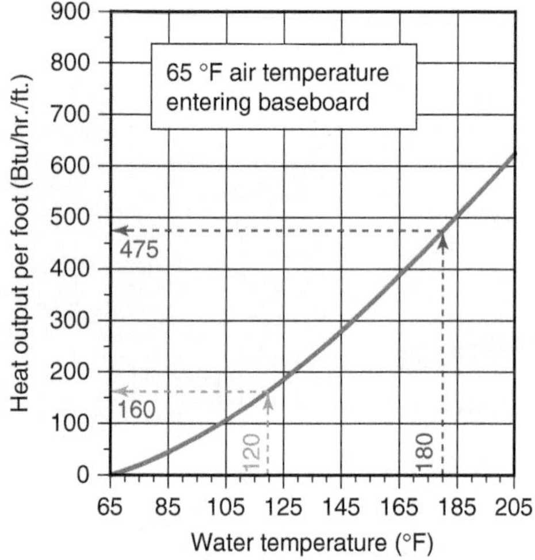

Figure 5-8 | Heat output versus average water temperature for standard fin-tube baseboard.

in a system with a supply water temperature no higher than the suggested 120 °F is rather impractical. It would only be possible in rooms with very low design heat loss and plenty of available wall space.

Example 5.1: A room measures 10 feet by 15 feet and has a design heat loss of 15 Btu/hr/ft². Determine the amount of standard fin-tube baseboard needed (based on the performance graph in Figure 5-8). Assume the supply water temperature at design load conditions is 120 °F and the flow rate through the circuit creates a temperature drop of 20 °F.

Solution: The room's design heat loss is easily calculated:

Design load = (10 ft)(15 ft)(15 Btu/hr/ft²) = 2,250 Btu/hr

The *average* water temperature in the baseboard is the supply water temperature minus half the circuit's temperature drop:

$$T_{ave} = 120 - \frac{20}{2} = 110\ °\mathrm{F}$$

Entering the horizontal axis of the graph in Figure 5-8 at 110 °F, reading up to the curve and over to the vertical axis, indicates a heat output of 125 Btu/hr/ft. Thus, the total length of baseboard needed is:

$$L = \frac{2{,}250\ \mathrm{Btu/hr}}{125\dfrac{\mathrm{Btu}}{\mathrm{hr}\cdot\mathrm{ft}}} = 18\ \mathrm{ft}$$

Discussion: The baseboard was sized based on the *average* water temperature rather than the supply water temperature. The first few feet of the 18 overall length of baseboard will release slightly more than 125 Btu/hr/ft because the water temperature is slightly higher than the average water temperature. Likewise, the last few feet of baseboard will have a heat output slightly less than 125 Btu/hr/ft because the water temperature will be less than average. The next section will present a more detailed mathematical model to track the temperature drop along the length of a baseboard.

LOW-TEMPERATURE FIN-TUBE BASEBOARD

Low-temperature heat sources such as condensing boilers, solar thermal collectors, and hydronic heat pumps will become increasingly common in future hydronic systems. This has motivated manufacturers to redesign traditional heat emitters in ways that make them compatible with these low-temperature heat sources.

Figure 5-9 shows one such heat emitter, known as **Heating Edge baseboard**. This product is available in North America, and provides substantially higher heat output at low average water temperatures, in comparison to standard fin-tube baseboard.

The fins used in Heating Edge baseboard have approximately three times the surface area of the fins used in standard residential fin-tube baseboard. The fin-tube element in Heating Edge baseboard also has two 3/4-inch copper tubes compared to the single 1/2-inch or 3/4-inch copper tube used in standard baseboard.

Figure 5-10 shows the heat output table supplied by the manufacturer of Heating Edge baseboard.

This rating table is based on test flow rates of 1 and 4 gpm. These reference flow rates are established by **Air-Conditioning, Heating, and Refrigeration Institute (AHRI)** rating standards. They do not imply that the baseboard *must* operate at these flows. Rather, they show that increasing the flow rate through the fin-tube element slightly increases its heat output. This is the result of increased convective heat transfer between the water and inner tube wall at the higher flow rates.

The heat output table also shows ratings for different flow arrangements through the dual-tube set. Note that the highest heat output for a given flow rate and average water temperature occurs with parallel flow through both tubes.

A footnote below the heat output table indicates that a 15 percent **heating effect factor** has been included in the thermal ratings. This means the listed values for heat output are actually *15 percent higher than the measured heat output established by laboratory testing.* The current rating standard for fin-tube baseboard allows manufacturers to add 15 percent to the tested performance, provided a note indicating such is listed along with the rating data. The origin of the heating effect factor goes back several decades. At that time, it was used to account for the higher output of baseboard convectors placed near floor level rather than at a height typical of standing cast-iron radiators. Unfortunately, it is a factor that has outlived its usefulness. Since the 15 percent higher heat output cannot be documented by laboratory testing, it is inaccurate to select baseboard lengths that assume its presence. The author recommends that baseboard heat output ratings that include a 15 percent heating effect factor be corrected by dividing these heat outputs by 1.15 before selecting baseboard lengths. This bases selection on actual tested performance.

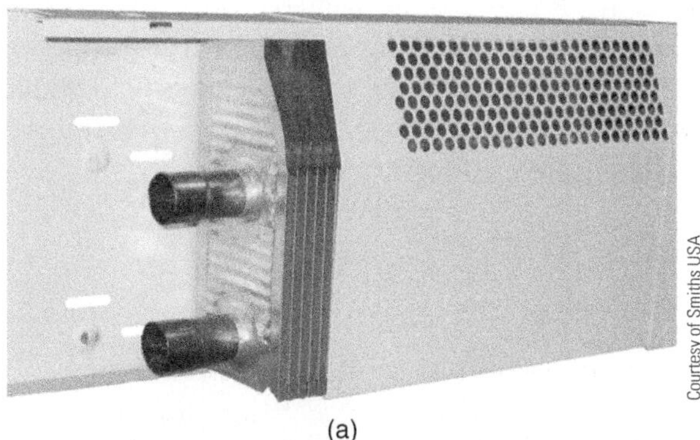

(a)

(b)

Figure 5-9 | (a) Low water temperature *Heating Edge* baseboard with large fins and dual water tubes. (b) Heating Edge baseboard installed.

Figure 5-11 compares the output of Heating Edge baseboard to standard residential baseboard. Both curves are based on a flow rate of 1 gpm and do not include the 15 percent heating effect factor. The output curve for the Heating Edge baseboard is based on parallel flow through both tubes (e.g., each tube flowing in the same direction at 0.5 gpm per tube flow rate).

One can view the higher output of the Heating Edge baseboard in two ways:

1. A given length of Heating Edge baseboard operating at a given condition provides significantly higher output. Thus, a shorter length of baseboard can provide the necessary heat output for a room compared to standard residential baseboard.

Heating Edge™ Hot Water Performance Ratings	Flow Rate GPM	PD in ft of H_2O	Average Water Temperature (BTU/hr/ft @AWT in °F) 90°F	100°F	110°F	120°F	130°F	140°F	150°F	160°F	170°F	180°F	190°F	200°F	210°F
TWO SUPPLIES PARALLEL	1	0.0044	130	205	290	385	460	546	637	718	813	911	1009	1113	1215
	4	0.0481	155	248	345	448	550	651	755	850	950	1040	1143	1249	1352
TOP SUPPLY BOTTOM RETURN	1	0.0088	105	169	235	305	370	423	498	570	655	745	836	924	1016
	4	0.0962	147	206	295	386	470	552	640	736	810	883	957	1034	1110
BOTTOM SUPPLY TOP RETURN	1	0.0088	103	166	230	299	363	415	488	559	642	730	819	906	996
	4	0.0962	140	212	283	350	435	524	623	722	792	865	937	1013	1093
BOTTOM SUPPLY NO RETURN	1	0.0044	75	127	169	208	260	311	362	408	470	524	576	629	685
	4	0.0481	85	140	203	265	334	410	472	536	599	662	723	788	850

Performance Notes: • All ratings include a 15% heating effect factor • Materials of construction include all aluminum "patented" fins at 47.3 per LF, mechanically bonded to two 3/4" (075) type L copper tubes ("Coil Block") covered by a 20 gauge perforated, painted cover all mounted to a backplate. Please see dimensional drawing for fin shape and dimensions • EAT=65°F • Pressure drop in feet of H_2O per LF.

Courtesy of Smiths USA

Figure 5-10 | Heat output table for Heating Edge baseboard.

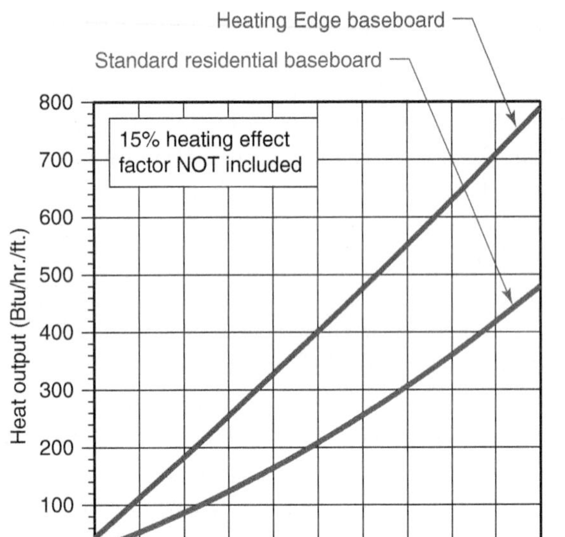

Figure 5-11 Comparative heat output versus average water temperature for standard residential fin-tube baseboard versus Heating Edge baseboard.

2. By using the same length of Heating Edge baseboard, the average water temperature required for a given heat output drops substantially.

Point 1 is useful when the maximum heat output needs to be "squeezed" into a given wall space. For example, suppose there is only one short wall space available within a room. At a given average water temperature, Heating Edge baseboard will yield approximately double the heat output of standard residential baseboard. This may allow a short length of baseboard to provide the necessary heat output.

Point 2 is of particular interest in the case of renewable energy heat sources or condensing boilers. The lower the average water temperature at which the baseboard can still provide design heating load, the higher the thermal efficiency of the heat source.

Example 5.2: The room described in Example 5.1 had a design heating load of 2,250 Btu/hr. At an average water temperature of 110 °F, it requires 18 feet of standard residential baseboard to meet this load. Determine the following:

a. If the average water temperature remained at 110 °F, how many feet of Heating Edge baseboard, with both tubes operating in parallel, and a total flow rate of 1 gpm, would be required?

b. If 18 feet of Heating Edge baseboard were used, with both tubes operating in parallel, and a total flow rate of 1 gpm, how much lower could the average water temperature be?

Solution: Based on the heat output graph shown in Figure 5-11, the Heating Edge baseboard at an average water temperature of 110 °F releases 250 Btu/h/ft. Thus, the required length of baseboard to meet the design heating load is:

$$L = \frac{2{,}250 \text{ Btu/hr}}{250\dfrac{\text{Btu}}{\text{hr} \cdot \text{ft}}} = 9 \text{ ft}$$

Assuming 18 feet of wall space were available and equipped with Heating Edge baseboard, the required heat output of each foot would be:

$$\frac{2{,}250 \text{ Btu/hr}}{18 \text{ ft}} = 125\frac{\text{Btu}}{\text{hr} \cdot \text{ft}}$$

Entering the vertical axis of Figure 5-11 at 125 Btu/hr/ft, draw a line to the heat output curve for Heating Edge baseboard and then down to the horizontal axis to find that an average water temperature of approximately 92 °F will be required to produce this heat output.

Discussion: The supply water temperature to the baseboard would be higher than 92 °F. How much higher depends on the flow rate. At low water temperatures, a drop of 10 °F to 20 °F across the heat emitter is typical under design load conditions. Thus, the supply water temperature to the Heating Edge baseboard in this example would be in the range of 97 °F to 102 °F (e.g., half the design temperature drop added to the average water temperature). Operating the balance of the system at a supply temperature of 102 °F rather than 120 °F at design load conditions would significantly boost the performance of solar thermal collectors, heat pumps, or a mod/con boiler. It would also extend the useful temperature range of a thermal storage tank supplied by a wood-fired boiler.

ANALYTICAL MODELS FOR BASEBOARD HEAT OUTPUT

It's often necessary to determine the heat output of fin-tube baseboard at operating conditions differing from those given in manufacturer's literature. These operating conditions include the water temperature at a given location along the baseboard, the air temperature entering the baseboard enclosure, and the flow rate through the fin-tube element. Equation 5.1 can be used to estimate the heat output per foot of *standard residential fin-tube baseboard.*

(Equation 5.1)

$$q = B(0.00096865)(f^{0.04})(T_w - T_{air})^{1.4172}$$

where:

q = heat output in Btu/hr/ft of fin-tube element length at a specific location on the element (Btu/hr/ft)

B = heat output of the baseboard at 200 °F average water temperature, 1 gpm from manufacturer's heat output table (Btu/hr/ft)*

f = flow rate of water through the baseboard (gpm)

T_w = water temperature at a specific location within the baseboard (°F)

T_{air} = air temperature entering the baseboard (°F)

Note: 0.04 and 1.4172 are exponents.

Equation 5.2 can be used to estimate the heat output of Heating Edge baseboard when the dual tubes are configured for parallel flow. This equation is based on performance that excludes the 15 percent heating effect factor from the manufacturer's heat output table.

(Equation 5.2)

$$q = 2.0063(f^{0.127})(T_w - T_{air})^{1.2643}$$

where:

q = heat output (Btu/hr/ft of fin-tube element length)

f = flow rate of water through the baseboard (gpm)

T_w = water temperature at a specific location within the baseboard element (°F)

T_{air} = air temperature entering the baseboard (°F)

Note: 0.127 and 1.2643 are exponents.

The coefficients and exponents in Equations 5.1 and 5.2 are the result of the curve fitting to data provided in the heat output tables.

*If the manufacturer's heat output table has a footnote indicating that the heat outputs include a 15 percent heating effect factor, the author recommends that the (B) value for heat output at 200 °F water temperature and 1 gpm flow rate be divided by 1.15. This removes the heating effect factor.

Example 5.3: Estimate the heat output of Heating Edge baseboard (in Btu/hr/ft) at a location where the water temperature in the fin-tube element is 115 °F and the total flow rate through the parallel piped tubes is 3 gpm. Assume the air temperature entering the bottom of the baseboard enclosure is 67 °F.

Solution: Substituting the stated operating conditions into Equation 5.2 yields the following:

$$q = 2.006(f^{0.127})(T_w - T_{air})^{1.2643}$$
$$= 2.006(3^{0.127})(115 - 67)^{1.2643} = 308\frac{\text{Btu}}{\text{hr} \cdot \text{ft}}$$

Discussion: The calculated heat output of 308 Btu/hr/ft only applies at the exact location along the fin-tube element where the water temperature is 115 °F. The rate of heat output would be slightly higher upstream of this location and slightly lower downstream of this location. This continual drop in water temperature must be accounted for when determining the total heat output of the baseboard.

TEMPERATURE DROP ALONG FIN-TUBE BASEBOARD

Consider the temperature of water passing through a fin-tube baseboard. Since heat is continually leaving the element, the water temperature must continually drop along the element, as illustrated in Figure 5-12.

If the drop in water temperature was linear (e.g., at same rate all along the entire element), then total heat output from the element could be accurately estimated by assuming the entire element operates at the *average* of the inlet and outlet water temperature. However, because the rate of heat transfer at every location on the element depends on the difference between the fluid temperature at that location and the surrounding air temperature, the rate of heat transfer is greater near the inlet of the element compared to near the end. This creates the curvature in the temperature profile seen in Figure 5-12. The slower the flow rate through the element, the more pronounced this curvature becomes.

When the inlet temperature and flow rate for a baseboard are known, the most accurate estimate of heat output comes from mathematically integrating the heat output functions given as Equations 5.1 and 5.2 along the length of the baseboard's element. This integration generates an equation that gives the outlet

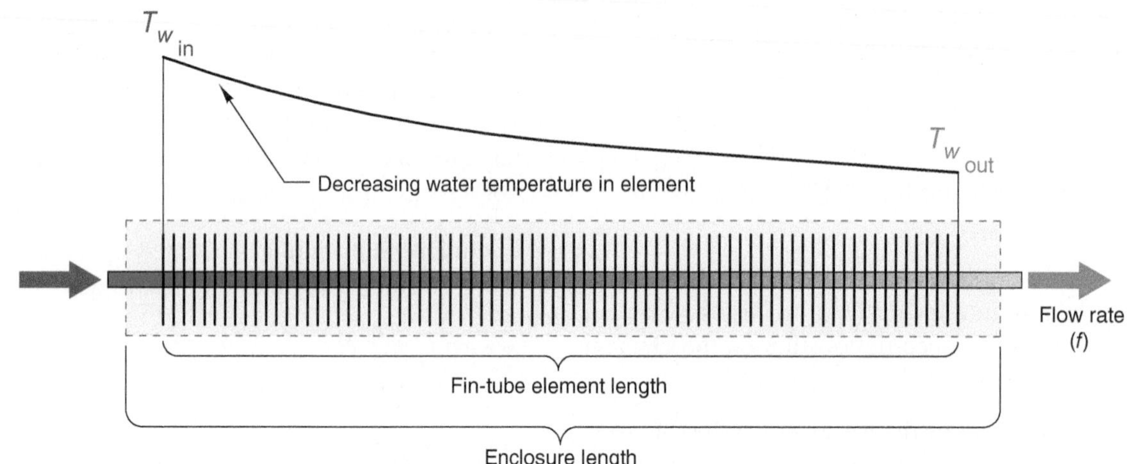

Figure 5-12 | The temperature of the water is continually decreasing as the water moves along a fin-tube element within a baseboard.

temperature of the fluid passing through the baseboard as a function of the other operating conditions. Equation 5.3 provides the result of this integration for a *standard residential fin-tube baseboard.*

(Equation 5.3)

$$T_{w_{out}} = T_{air} + \left[(T_{w_{in}} - T_{air})^{-0.4172} + \left(\frac{\beta(5.04 \times 10^{-5})}{Dc(f)^{0.96}}\right)l\right]^{-2.397}$$

where:

$T_{w_{out}}$ = fluid temperature leaving baseboard (°F)

$T_{w_{in}}$ = fluid temperature entering baseboard (°F)

T_{air} = air temperature entering baseboard (°F)

D = fluid density (lb/ft^3)

c = fluid specific heat (Btu/lb/°F)

f = fluid flow rate (gpm)

B = heat output rating of baseboard at 200 °F water/65°F air temp/1 gpm flow rate (Btu/hr/ft)*.

l = length of fin-tube element in baseboard (ft)

*If the manufacturer's heat output table has a footnote indicating that the heat outputs include a 15 percent heating effect factor, the author recommends that the (B) value for heat output at 200 °F water temperature and 1 gpm flow rate be divided by 1.15. This removes the heating effect factor.

−0.4172, 0.96, and −2.397 are all exponents (not multipliers)

Equation 5.4 provides the integration result specifically for the Heating Edge baseboard configured for parallel flow in its dual tube element.

(Equation 5.4)

$$T_{w_{out}} = T_{air} + \left[(T_{w_{in}} - T_{air})^{-0.2643} + \left(\frac{0.0662}{Dc(f)^{0.873}}\right)l\right]^{-3.7836}$$

where:

$T_{w_{out}}$ = fluid temperature leaving baseboard (°F)

$T_{w_{in}}$ = fluid temperature entering baseboard (°F)

T_{air} = air temperature entering baseboard (°F)

D = fluid density (lb/ft^3)

c = fluid specific heat (Btu/lb/°F)

f = fluid flow rate (gpm)

l = length of fin-tube element in baseboard (ft)

−0.2643, 0.873, and −3.7836 are all exponents (not multipliers)

Once the outlet temperature from the baseboard element is calculated, Equation 3.2 (the sensible heat rate equations) can be used to calculate the total heat output from the baseboard.

(Equation 3.2 repeated)

$$Q = (8.01Dc)f(\Delta T)$$

where:

Q = rate of heat transfer into or out of the water stream (Btu/hr)

8.01 = a constant based on the units used

D = density of the fluid (lb/ft^3)

c = specific heat of the fluid (Btu/lb/°F)

f = flow rate of water through the device (gpm)

ΔT = temperature change of the fluid as it passes through the device (°F)

Because they require the density and specific heat of the fluid passing through the element as inputs, Equations 5.3 and 5.4 allow heat output to be determined when fluids other than water are used. In the strictest sense, the density and specific heat of the fluid flowing through the fin-tube element are both functions of temperature and thus could be represented as temperature dependent when the integration is performed. However, for the fluids commonly used in hydronic heating systems, the variation in these properties is minor over the temperature drop range of most individual baseboards. To simplify the integration, these fluid properties can both be estimated at an estimated average fluid temperature within the fin-tube element.

Example 5.4: In Example 5.2, it was determined that 9 feet of Heating Edge baseboard, configured for parallel flow in the dual-tube element and operating at an average water temperature of 110 °F, and assumed water flow rate of 1 gpm, would provide the 2,250 Btu/hr required a given room at design load conditions. Use Equations 3.2 and 5.4 to assess the accuracy of this calculation.

Solution: If 9 feet of Heating Edge baseboard was releasing 2,250 Btu/hr at a flow rate of 1 gpm, the temperature drop across the element could be estimated using Equation 3.2. Water at the average temperature in the baseboard (e.g., 110 °F) has a density of 61.8 lb/ft^3. The specific heat of the water is assumed to remain at 1.00 Btu/lb/°F. Substituting these values along with the given data into a rearranged form of Equation 3.2 yields:

$$\Delta T = \frac{Q}{(8.01Dc)f} = \frac{2250}{(8.01 \times 61.8 \times 1.00) \times 1} = 4.55\ \text{°F}$$

If the average water temperature was 110 °F, the inlet water temperature can be estimated as follows:

$$110 + \left(\frac{4.55}{2}\right) = 112.3\ \text{°F}$$

This inlet water temperature along with the given flow rate is then entered into Equation 5.4 to calculate the outlet temperature from the fin-tube element.

$$T_{w_{out}} = T_{air} + \left[(T_{w_{in}} - T_{air})^{-0.2643} + \left(\frac{0.0662}{Dc(f)^{0.873}}\right)l\right]^{-3.7836}$$

$$= 65 + \left[(112.3 - 65)^{-0.2643} + \left(\frac{0.0662}{61.8 \times 1 \times (1)^{0.873}}\right)9\right]^{-3.7836} = 107.81\ \text{°F}$$

At this point, the inlet and outlet temperatures are determined and the flow rate is known. Thus, Equation 3.2 can be used to determine total heat output from the baseboard.

$$Q = (8.01Dc)f(\Delta T)$$
$$= (8.01 \times 61.8 \times 1.00) \times 1 \times (112.3 - 107.81)$$
$$= 2{,}223\ \text{Btu/hr}$$

Discussion: The total heat output calculated using Equations 5.4 and 3.2 (2,223 Btu/hr) is very close to the 2,250 Btu/hr estimated for 9 feet of Heating Edge baseboard operating at an average water temperature of 110 °F. A reasonable question then becomes: Why bother with the more complex calculations associated with Equation 5.4 (or Equation 5.3 in the case of standard baseboard), when they yield results very similar to those based on average water temperature? The answer is that the more complex relationships represented in Equations 5.4 and 5.3 are able to better model situations that deviate from the assumptions used in the manufacturer's heat output table. These include operating the baseboard at flow rates other than the 1 and 4 gpm values listed in the heat output tables as well as at different conditions for room air temperature and use of fluids other than water (such as a water/antifreeze solution).

DISTRIBUTION SYSTEM FOR FIN-TUBE BASEBOARD

Traditionally, residential fin-tube baseboard has been installed using *series* distribution systems, as shown in Figure 5-4. These piping systems are typically constructed of 1/2-inch or 3/4-inch copper tubing. Although this technique generally reduces the amount of straight pipe used relative to other piping options, it often requires installation of up to eight 90-degree copper elbows per baseboard. This happens because piping between baseboards is usually routed under a wood-frame floor, along the floor joists in basements. Such piping has to be offset to accommodate framed walls above the floor that are usually narrower than the foundation wall. Installing these fittings can be tedious and expensive. Series piping also causes a temperature drop from one baseboard to another. Although there are ways to calculate this temperature drop (see chapter 8 in *Modern Hydronic Heating*, 3rd edition), doing so increases design complexity.

Figure 5-13a shows a better way to pipe *Heating Edge* baseboards to create a parallel **"homerun" distribution system**.

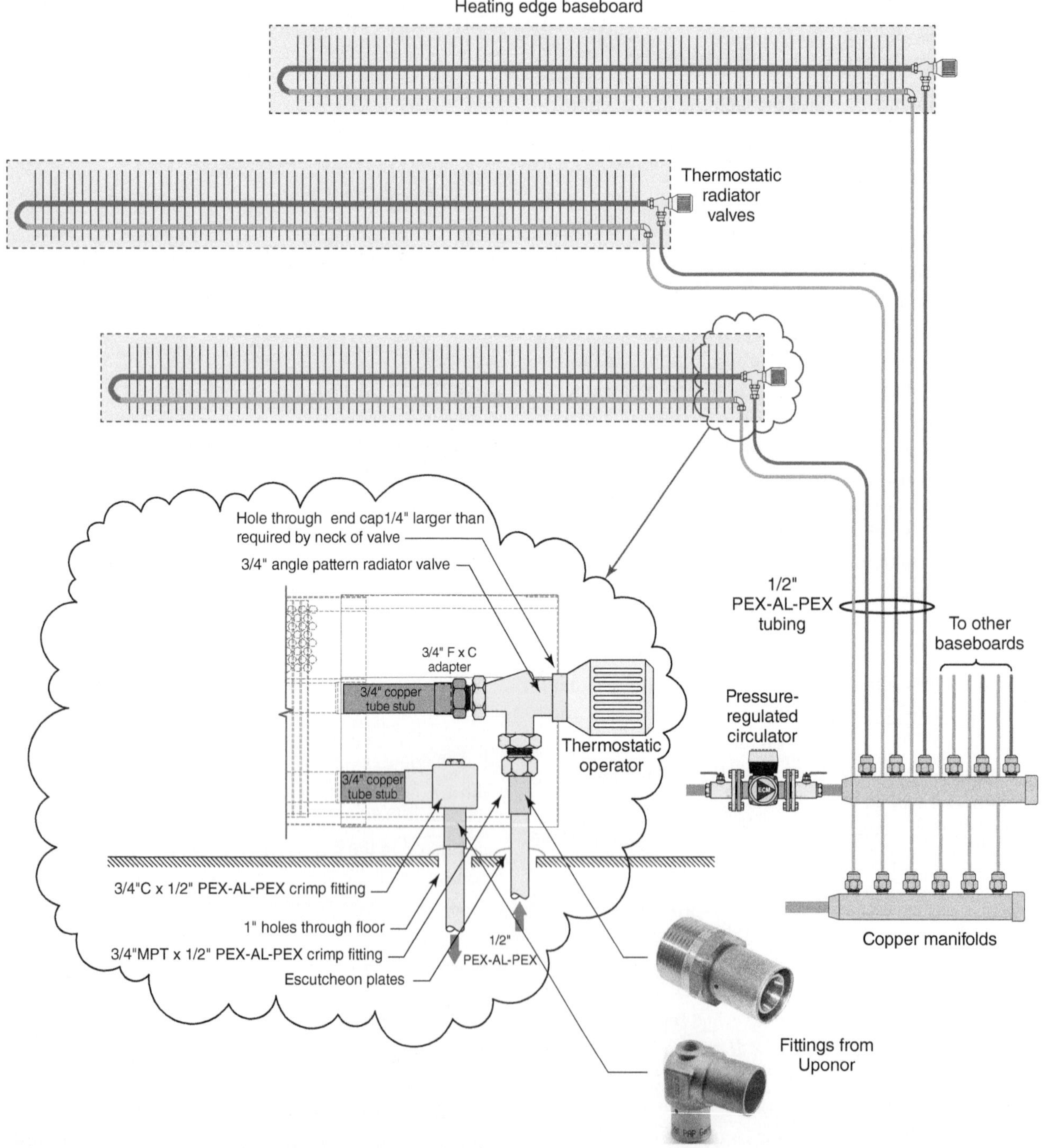

Figure 5-13a | Parallel "homerun" piping of Heating Edge baseboards. Thermostatic radiator valve installed on each baseboard for room temperature control. Fin-tube element piping connected in series.

In this system, each baseboard is piped to a set of 1/2-inch PEX-AL-PEX tubes for supply and return flow. The two tubes within the fin-tube element are connected in series using a U-bend fitting at the far end of each baseboard.

The small diameter flexible tubing is easily routed through framing cavities within the building and comes up through the floor directly beneath one end of the baseboard enclosure. Specialized fittings are used to adapt from the 1/2-inch PEX-AL-PEX tubing to 3/4-inch size copper piping as well as to a wireless thermostatic radiator valves. As discussed in Chapter 4, **thermostatic radiator valves** modulate flow through each baseboard to maintain a set room temperature. Thus, each baseboard represents a separate zone. This allows for easy room-by-room temperature control.

As the thermostatic valves vary the flow through each branch circuit of the system, the pressure-regulated circulator automatically adjusts speed to maintain a fixed differential pressure across the copper tube manifolds where each homerun circuit begins and ends. Whenever flow through the system is reduced, so is the wattage required to operate the pressure-regulated circulator.

This combination of hardware provides a simple, versatile, and highly energy-efficient approach to installing low-temperature fin-tube baseboard. It can also be adapted for use with other heat emitters discussed later in this chapter.

It is also possible to connect the fin-tube element piping in parallel, as shown in Figure 5-13b.

Connecting the two tubes within the fin-tube element in parallel increases heat output 22 to 25 percent compared to connecting the tubes in series, albeit it at the expense of more complex piping. If parallel piping is used, it is important to use the **bullhead tee** arrangement shown at each end of the fin-tube element in Figure 5-13b. This ensures even flow division between the upper and lower tubes.

SIZING PARALLEL PIPED BASEBOARDS

When baseboards are connected in parallel, and piping heat losses are minimal, each baseboard receives water at approximately the same temperature (e.g., the water temperature present in the supply manifold). However, the flow rate through each baseboard can vary depending on the length of homerun circuit, the length of the baseboard, and the head loss created by any specific fittings used in that circuit.

The flow rate through each baseboard circuit needs to be determined. This can be done using manual calculations based on the methods given in chapter 6 of *Modern Hydronic Heating*, 3rd edition. It can also be done using the **Hydronic Circuit Simulator** module in the **Hydronics Design Studio** software. Either method requires the equivalent length of each homerun circuit. This equivalent length includes the length of the ½-inch flexible tubing running to and from each baseboard, the equivalent length of the baseboard element, and the equivalent length of any fittings and valves in that circuit.

This creates a dilemma: How can one estimate the equivalent length of each circuit without first knowing the length of baseboard in that circuit? The solution to this dilemma is called **iteration**. It's a procedure that makes an initial estimate of a currently unknown quantity, or multiple quantities (in this case the lengths of baseboard in each room), and then analyzes the performance of the system based on those initial estimates to see how close the calculated results (in this case, the heat output of each baseboard at its initial estimated length) are to the ideal results (in this case, the design heat load of each room). After comparing the calculated results to the ideal results, changes might be made (in this case, to the length of one or more baseboards), that are likely to bring the next calculated results closer to the ideal results. The calculations are then redone to assess the effect of the changes. With reasonable skill, an iterative approach can typically produce acceptable results within one or two cycles of calculations.

Iterative calculations involving parallel piping paths can be time-consuming. A faster approach is to use software in combination with an initial estimated length for each baseboard. The following procedure can be used.

STEP 1: Determine the design heating load assigned to each baseboard.

STEP 2: Select a supply water temperature at design load conditions. For systems with renewable energy heat sources, the author recommends keeping this temperature at or below 120 °F.

STEP 3: Select a design *temperature drop* for the system under design load conditions. Common values for design load temperature drop are 10 °F to 20 °F.

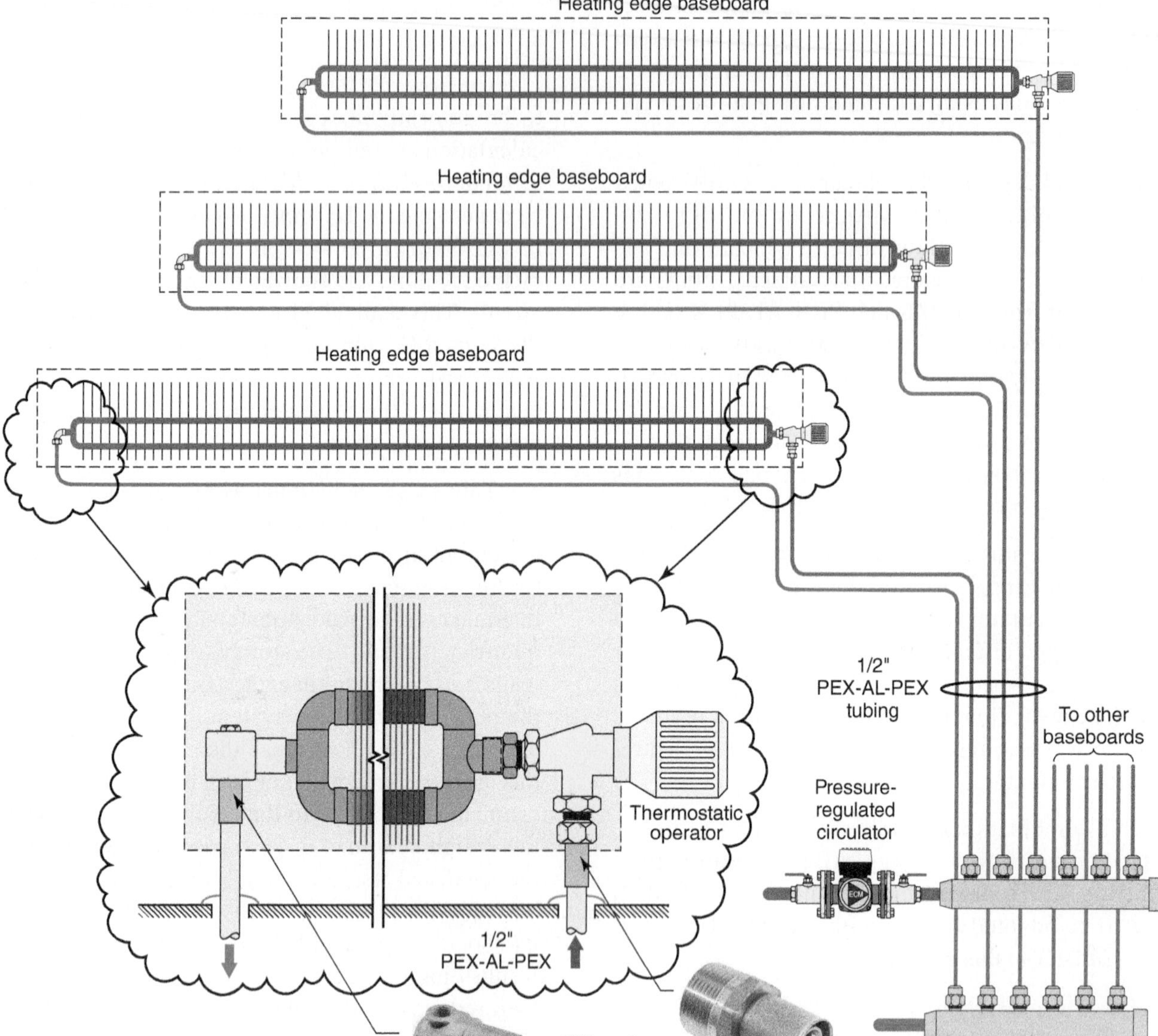

Figure 5-13b | Parallel "homerun" piping of Heating Edge baseboards. Thermostatic radiator valve installed on each baseboard for room temperature control. Fin-tube element piping connected in parallel.

STEP 4: Estimate the flow rate required in each baseboard circuit at design load conditions using Equation 5.5.

(Equation 5.5)

$$f_i = \frac{Q_i}{500(\Delta T)}$$

where:

f_i = flow rate required in a circuit serving baseboard "i" (gpm)

Q_i = design heating load of the space served by baseboard (i) (Btu/hr)

ΔT = target temperature drop of the circuit (°F)

500 = a constant for water*

STEP 5: Estimate the *average* water temperature in the baseboards using Equation 5.6.

* for other fluids, replace 500 with ($8.01 \times D \times c$) where:

D = fluid density at average circuit temperature (lb/ft^3)

c = the specific heat of fluid at average circuit temperature (Btu/lb/°F)

(Equation 5.6)

$$T_{ave} = T_{supply} - \left(\frac{Q_i}{1000 f_i}\right)$$

where:

T_{ave} = estimated average water temperature in the baseboard (°F)

T_{supply} = selected supply water temperature at design load conditions (°F)

Q_i = design heating load served by the baseboard (Btu/hr)

1,000 = 2 × 500*

STEP 6: Use the average temperature for each baseboard along with one of the following methods to estimate the heat output of the baseboard (in Btu/hr/ft).

a. Use data from the manufacturer's heat output table**

b. Use a graph, such as shown in Figure 5-11, prepared from data in the manufacturer's heat output table**

c. Use Equation 5.1 (for standard baseboard), or Equation 5.2 for Heating Edge baseboard.

Of these three options, (c) is preferred because it accounts for the variation in heat output based on the actual flow rate through the baseboard rather than assuming that the flow rate is either 1 or 4 gpm, as listed in the manufacturer's heat output ratings table.

STEP 7: Make an initial calculation for the length of baseboard needed for each room. Use the heat output (in Btu/hr/ft) established for each baseboard in step 6 along with the method shown in Example 5.2. For example, if the room's design heating load is 2,250 Btu/hr, and the heat output of the baseboard at the average water temperature is 230 Btu/hr/ft, then the initial calculated baseboard length is:

$$L = \frac{2250\dfrac{\text{Btu}}{\text{hr}}}{230\dfrac{\text{Btu}}{\text{hr}\cdot\text{ft}}} = 9.78 \text{ ft}$$

STEP 8: Round each initially calculated baseboard length *up* to the next whole foot (since baseboard is sold by the foot).

STEP 9: Place each of the initially calculated baseboard lengths in their selected positions on a scale floor plan of the building. Considering this placement, estimate the length of supply and return tubing from the distribution manifold to each baseboard.

STEP 10: You now have established a reasonably accurate estimate of the length of each baseboard, their placement within the building, and the length of tubing to connect each baseboard back to the manifold to form a homerun distribution system. The next step is to use either the manual calculation methods presented in chapter 6 of Modern Hydronic Heating, 3rd edition, or the Hydronic Circuit Simulator module in the Hydronics Design Studio software to determine the actual flow rates that will develop in each baseboard based on the circulator selected, and the system piping between the manifold and the heat source. The author recommends use of the software for much faster results. A free demo version of this software can be downloaded at www.hydronicpros.com. Figure 5-14 shows a screen shot of the software as it would be configured for a five-circuit homerun system.

Use the software to experiment with any of the parameters that govern the thermal and hydraulic performance of the system. These include supply fluid temperature, room air temperature, amount of tubing to and from each baseboard, circulator, size, system fluid, length of piping between manifold and heat source, or length of any baseboard. Within a few minutes you should be able to "fine tune" the system into a ready-to-install design.

*for fluids other than water, replace 1,000 with (16.02 × D × c), where:

D = fluid density at average circuit temperature (lb/ft^3)

c = the specific heat of fluid at average circuit temperature (Btu/lb/°F)

**If the manufacturer's heat output table indicates that a 15 percent heating effect factor is included in the ratings, the author's recommendation is to divide the rating by 1.15 to remove the heating effect factor.

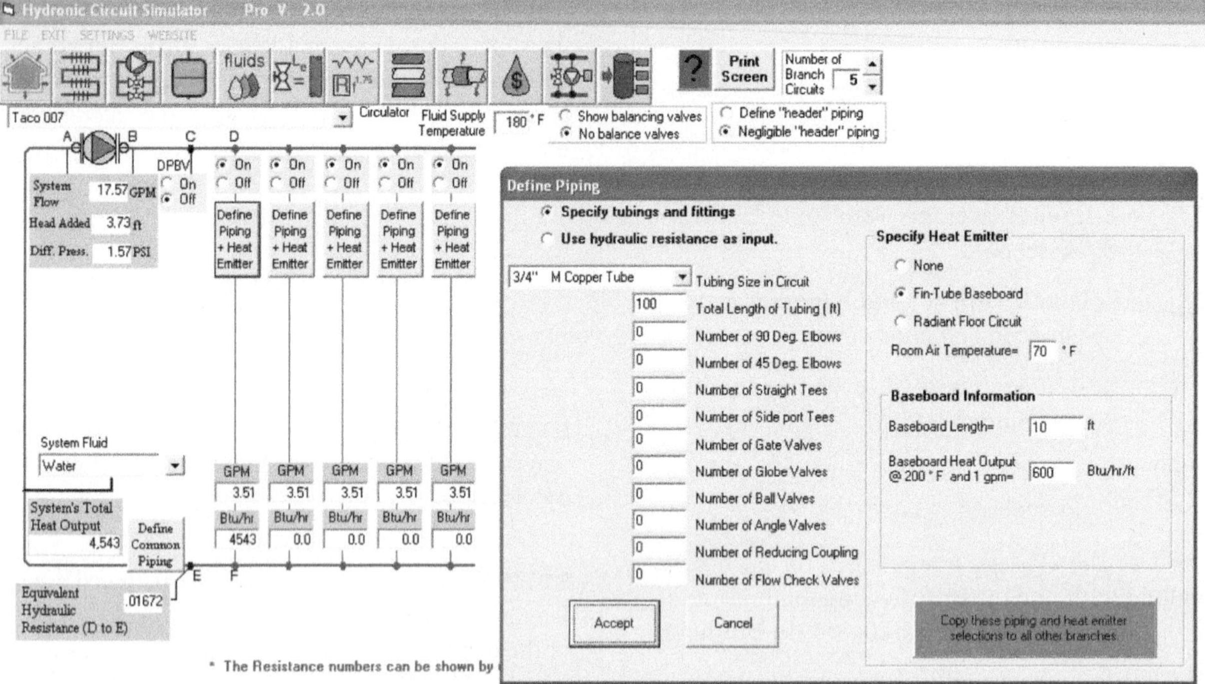

Figure 5-14 | The Circuit Simulator Module with the Hydronics Design Studio software, configured to model five parallel piping circuits.

5.3 Panel Radiators

Another type of hydronic heat emitter that can be used with renewable energy heat sources is called a **panel radiator**. Having evolved in Europe over several decades, modern panel radiators are now available in hundreds of sizes, shapes, colors, and heating capacities to fit different job requirements. They can be used throughout a building or in combination with other types of heat emitters.

Most panel radiators are made of steel. Some are built of preformed steel sheets welded together at their perimeter by highly automated machinery. Others are constructed of tubular steel components. *To prevent corrosion, steel panel radiators should only be used in closed-loop systems.*

Some panel radiators release a significant percentage of their heat as thermal radiation. Such panels typically have a relatively flat front and only project about 2.5 inches out from the wall surface to which they are mounted. An example of such a panel is shown in Figure 5-15.

The thermal radiation released from this type of panel tends to warm the objects in a room. The lower the water temperature at which panel radiators operates,

Courtesy of Myson, Inc.

Figure 5-15 | A thin panel radiator with a flat face produces both radiant and convective heat output.

the lower the total heat output. However, lower water temperatures increase the percentage of radiant versus convective heat output from the panel. This is often desirable because it improves comfort and reduces room air stratification relative to units that release the majority of their heat through convection (e.g., directly heating the room air).

Other panel radiators are designed to release a higher percentage of their heat output through convection. Such radiators are equipped with one or more rows of steel **fins** that dissipate heat into the surrounding air. This convective heat transfer is similar to that created by a fin-tube baseboard. These panel radiators tend to have deeper profiles that project 4 to 8 inches from the adjacent wall. This type of radiator is well suited for creating upward air currents to counteract downward drafts from large window areas. To produce strong upward convection, these panels need to operate at relatively high water temperature and thus are not well suited to lower temperature systems using renewable energy heat sources.

BENEFITS OF PANEL RADIATORS

There are several benefits to using panel radiators relative to other types of heat emitters. These include:

- Panel radiators typically require far less wall length than fin-tube baseboard sized for equivalent heat output and operating conditions. This reduces restrictions on furniture placement and usually improves aesthetics. In kitchens and bathrooms, the wall space required for properly sized fin-tube baseboard is often not available. However, because panel radiators come in a wide variety of widths, heights, and thicknesses, they can often be integrated into such limited wall space and still provide the necessary heat output.
- Most panel radiators contain very little water and relatively small amounts of metal. This results in low thermal mass and allows the panels to respond very quickly to variations in room air temperature or internal heat gains. The possibility of temperature overshoot in rooms with high internal heat gains from sunlight, lights, people, or heat-generating equipment gains is far less likely relative to systems that use high thermal mass heat emitters, such as heated floor slabs.

Figure 5-16 shows a sequence of **infrared thermographs** of a panel radiator taken over a period of approximately 4 minutes. The top image was taken 5 seconds after the panel's flow control valve was first opened. Before that, there was no flow through the panel and its temperature was approximately 60 °F. The sequence of images shows the surface temperatures resulting from the movement of 150 °F water through the

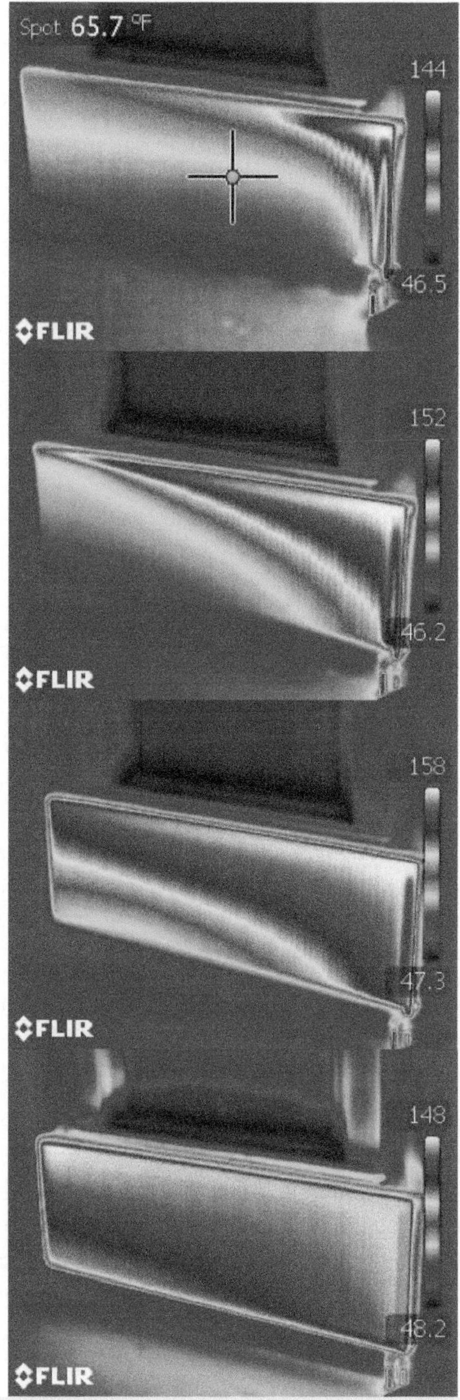

Figure 5-16 Sequence of infrared thermographs of a panel radiator warming to near steady state condition over approximately 4 minutes. Initial panel temperature equals 60 °F.

panel, beginning from the valve in its upper right corner. After 4 minutes, the panel is close to **steady-state conditions**, emitting both radiant and convective heat at nearly the maximum rate possible for its operating conditions.

- Panel radiators can operate with higher water temperature drops between their inlet and outlet, compared to some other heat emitters. In Europe, panel radiators are often sized to operate with a 20 °C (36 °F) temperature drop at design load. This is possible because occupants don't walk on wall-mounted panels, and thus wider variations in surface temperature are not as critical as they would be with floor heating. Larger temperature drops allow for lower flow rates, and lower flow rates allow for smaller tubing and reduced circulator power.
- Panel radiators can be sized to operate at relatively low water temperatures. This improves the efficiency of renewable energy heat sources such as solar thermal collectors and heat pumps.
- Panel radiators are well suited to new construction and in particular to remodeling. Their lightweight, easy-to-mount construction, in combination with modern flexible piping materials such as PEX or PEX-AL-PEX tubing, makes them easy to install with minimal disruption of existing surface finishes.
- Panel radiators are very durable. Their design and steel construction make them more resistant to physical damage than are most fin-tube baseboards. Most steel panel radiators have a high-quality powder-coat finish that provides excellent resistance to scratches or exterior corrosion. The latter is particularly important when heat emitters are located in humid spaces, such as bathrooms.
- Most panel radiators are wall-mounted and thus not effected by floor coverings.
- Many panel radiators release a significant portion of their heat output as radiant heat. This improves comfort and reduces room air stratification. In contrast, fin-tube baseboard releases almost all heat by convection. This can create room temperature stratification (e.g., warm air accumulating near the ceiling while cool air settles at floor level), especially in rooms with tall ceilings. Such stratification reduces comfort and increases heat loss from the room.
- The outward heat loss from the backside of panel radiators mounted on well-insulated exterior walls (e.g., R = 20 F·hr·ft^2/Btu) is less than 1 percent of their total annual heat output. By contrast, even a well-insulated (e.g., R = 10 F·hr·ft^2/Btu) heated floor slab can lose upward of 10 percent of its total heat output to the soil beneath.

FLAT-TUBE PANEL RADIATORS

Some panel radiators are built as an array of closely spaced, flat steel tubes. The end of each tube is welded shut to form a closed, pressure-tight chamber. Holes in the rear face, near the end of each tube, connect to headers. The headers distribute flow to and from each tube, as shown in Figure 5-17. The size and heating capacity of radiators constructed in this manner are determined by the length of the tubes and how many tubes are located side by side. The tubes may be oriented vertically, as seen in Figure 5-18, or horizontally, as seen in Figure 5-19.

Panel radiators are usually mounted with their bottom 3 to 6 inches above the floor. Most hang from rear-side mounting brackets fastened to the wall. These brackets should be fastened directly to wood-framing members or into solid masonry. In new construction with studded walls, wooden blocks should be positioned at the proper height and location to accept the fasteners for the brackets. It is important that the heating system designer specifies the position of these framing blocks so they are installed before the walls are closed up. Hanger locations are typically shown in manufacturer's technical literature.

Piping connections usually consist of 1/2-inch or 3/4-inch risers routed up through the floor and connecting to threaded openings at the bottom of the radiator. This is one location where steel pipe nipples provide better protection against physical damage and are preferred over copper tubing. Holes for the risers should be accurately located and neatly drilled about 1/4 inch larger than the outside diameter of the riser pipe. This prevents noise from thermal expansion. The holes can be covered with **escutcheon plates** for a neat appearance.

The detail shown in Figure 5-20 is useful when flexible tubing such as PEX or PEX-AL-PEX is used to supply a panel radiator. The transition fitting from the PEX or PEX-AL-PEX tubing to rigid steel pipe is concealed below the floor. However, the slack left in the flexible tubing allows this fitting to be pulled up through the floor if the panel ever needs to be removed for wall painting or other repairs. The steel pipe nipples can be painted to match the radiator. Escutcheon plates cover the oversized holes to provide a neat finish detail.

(a)

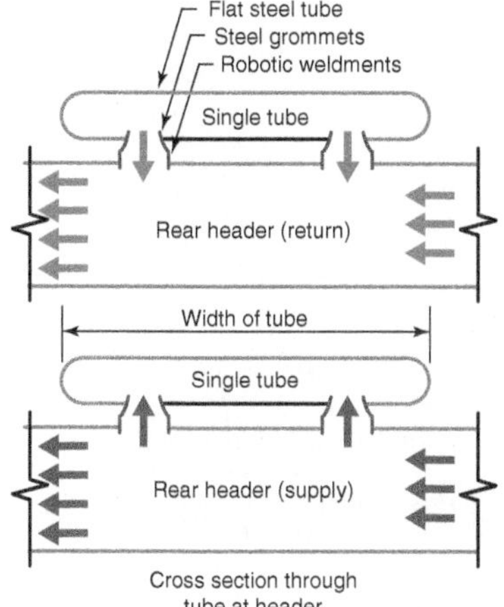

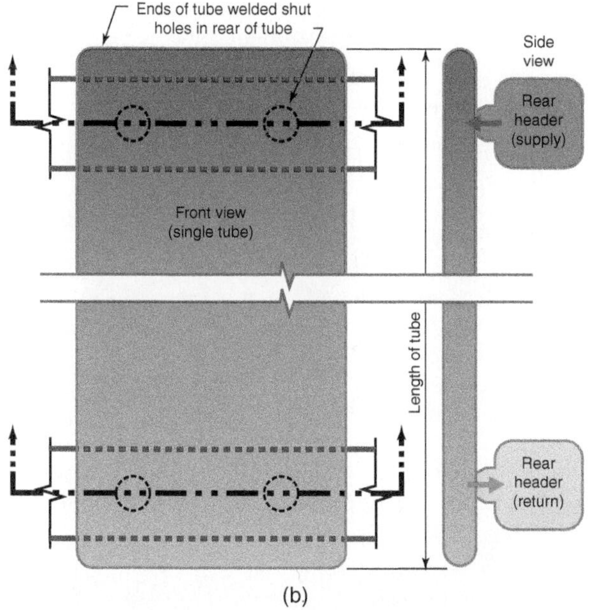

(b)

Figure 5-17 | Cross section through a typical flat tube panel radiator. The length and number of tubes determines the heat output of the radiator.

Figure 5-18 | Example of a vertical panel radiator.

Figure 5-19 | Example of a horizontal panel radiator.

To provide for individual heat output control, a wireless thermostatic radiator valve is often mounted in the *supply* pipe of the panel radiator. To facilitate removal of the radiator for wall painting, it is common to mount a shutoff valve in the return pipe. A **lockshield valve** with integral union is a good choice for this location. The combination of a thermostatic radiator valve with integral union in one riser and a lockshield valve with integral union in the other allows the panel to be isolated from the system, neatly drained through a hose, and easily

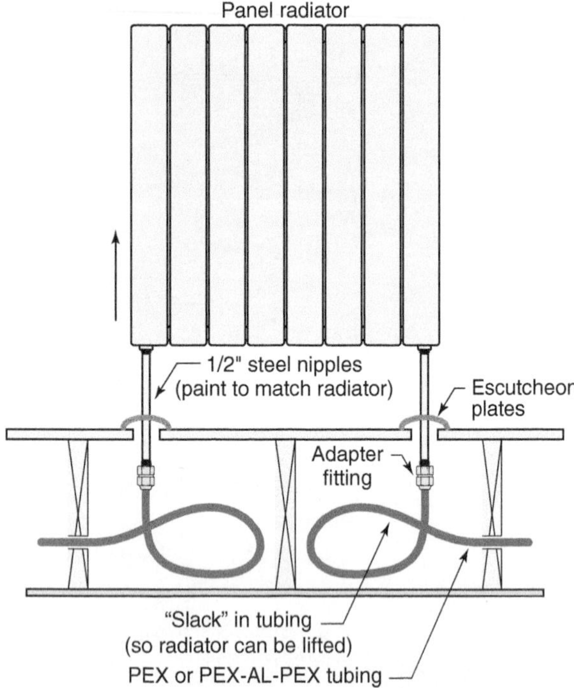

Figure 5-20 Panel radiator installation detail using steel pipe nipples above floor and PEX or PEX-AL-PEX tubing under floor. The tubing slack allows panel to be lifted off its mounting brackets if necessary.

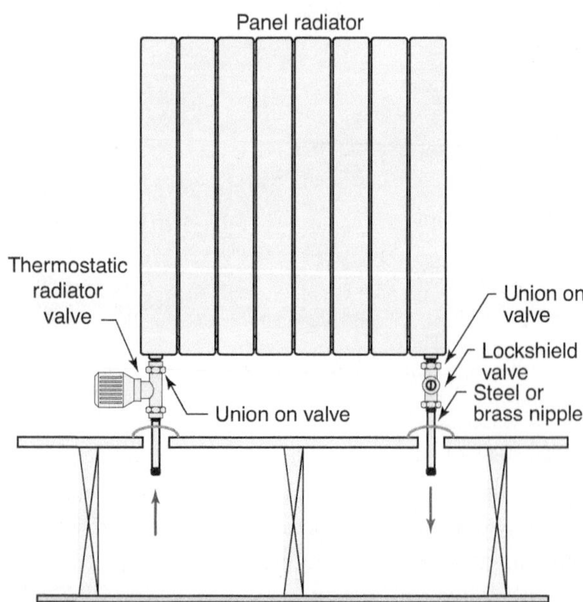

Figure 5-21 Use of a thermostatic radiator valve (TRV) on supply riser and lockshield valve on return riser of panel radiator. These valves allow heat output control as well as easy radiator removal, if necessary.

disconnected from the piping. It then can be lifted off its brackets for full access to the wall. A representation of these details is shown in Figure 5-21.

FLUTED STEEL PANEL RADIATORS

Another very common style for panel radiators is made of preformed steel sheets. Figure 5-22 shows an example of a **fluted steel panel radiator**.

(a)

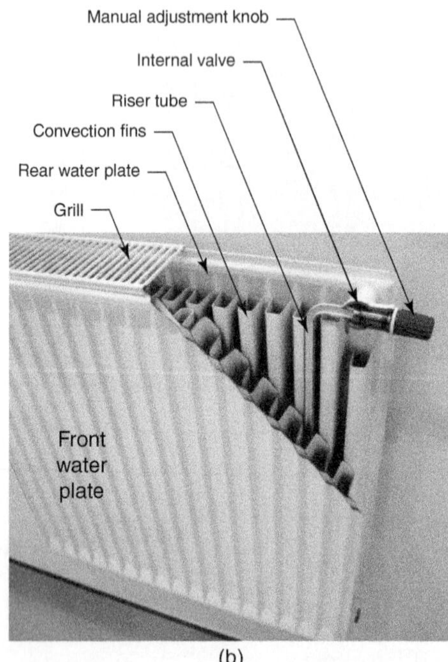

(b)

Figure 5-22 (a) An installed fluted steel panel radiator. Supply and return piping seen at bottom right. Wireless thermostatic radiator valve installed at top right. (b) Internal construction of a fluted steel panel radiator. This one has two water plates (front and rear).

The front of the radiator is called a **water plate**. It is made by pressing two steel sheets to form the vertical "flutes" and upper and lower manifold areas. These sheets are then welded together at their perimeters and several locations on the surface. The result is a thin sealed panel with relatively low water volume and low thermal mass. Folded steel fins are welded to the back of the water plate to enhance convective heat output. The resulting assembly provides a combination of radiant and convective heat output. The water plate and fins are fitted with side and top grills to form a complete radiator.

When the heat output requirement is high but wall space is limited, panel radiators with two and three water plates can be used. A comparison between panel radiators with one, two, and three water plates is shown in Figure 5-23.

Fluted steel panel radiators can have a variety of piping configurations. One of the most common is called a **compact-style panel radiator**. This style radiator has the supply and return piping connections at the bottom right side of the panel, as seen in Figure 5-24. The supply water enters at the left connection and passes up through an internal riser tube to the inlet of the panel's integral flow regulating valve. If this valve is closed, no flow can pass through the panel. If this valve is partially or fully open, flow can pass into the upper manifold, across the top of the panel, down through the flutes in the water plate(s), across the lower manifold, and eventually back to return connection at the bottom right of the panel. *It is important not to reverse the flow direction through the panel.* Doing so can create **cavitation** noise within the flow regulating valve.

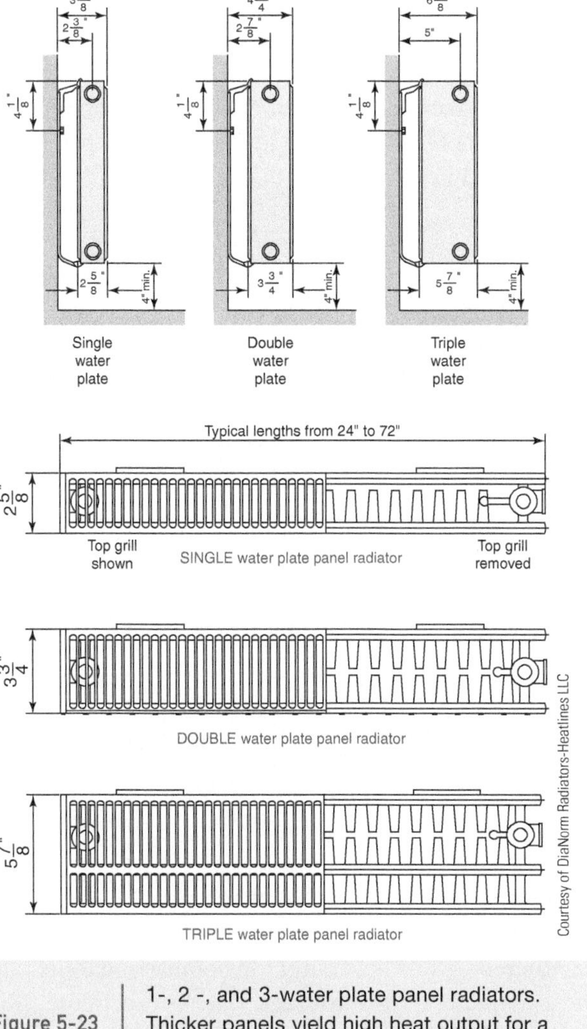

Figure 5-23 1-, 2 -, and 3-water plate panel radiators. Thicker panels yield high heat output for a given frontal area.

(a)

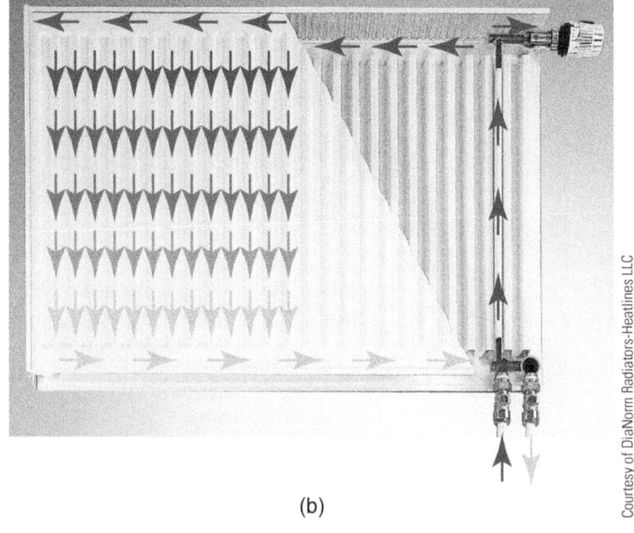

(b)

Figure 5-24 (a) Internal construction of a fluted panel radiator. (b) Water flow through panel radiator.

The location of the supply and return connections allow piping to enter from the floor or wall, as seen in Figure 5-25. It is common to install a dual isolation valve at the supply and return connections, also shown in Figure 5-25. This allows the panel to be isolated from the remainder of the system and disconnected from the piping if necessary.

THERMAL PERFORMANCE OF PANEL RADIATORS

Thermal ratings for panel radiators are usually given in chart form. For long vertical or horizontal flat-tube panels, thermal output ratings are usually expressed in Btu/hr per foot of panel length for various panel widths and average water temperatures. In most cases, the heat output ratings are based on a surrounding air temperature of 65 °F and a stated *average* water temperature within the panel (e.g., the average of the entering and leaving water temperature). Correction factors for other average water temperatures and room air temperatures usually accompany these ratings, as shown in Figure 5-26.

Figure 5-25 1/2-inch PEX-AL-PEX piping entering from floor through escutcheon plate. Dual isolation valve under radiator allows it to be isolated from remainder of system and removed if ever necessary.

Ratings for fluted steel panel radiators are usually listed by the manufacturer in Btu/hr for a given panel width, height, and thickness (e.g., number of water plates). These ratings are typically based on a stated *average* water temperature (e.g., the average of the entering and leaving water temperature) and room air temperature. Figure 5-27 shows a typical rating table.

The average water temperature assumed for the rating shown in Figure 5-27 is 180 °F. This is significantly higher than the water temperatures available from most renewable energy heat sources. **Correction factors** must be applied to "derate" the heat output for lower average water temperatures. Some manufacturers provide the derating factors as a table, graph, or equation, such as shown in Figure 5-28.

Example 5.6: A panel radiator has two water plates and measures 24 inches wide by 48 inches long. Its reference heat output ratings are given in Figure 5-27. Determine the heat output of this panel when operating at an inlet temperature of 120 °F, outlet temperature of 100 °F, and in a room maintained at 65 °F.

Solution: The reference heat output for this radiator is found in Figure 5-27 as 9,500 Btu/hr. This output is based on an average water temperature of 180 °F and room air temperature of 68 °F. A correction factor can be determined using the equation inset in the graph in Figure 5-28. The new ΔT for the panel is as follows:

$$\Delta T = (T_{ave} - T_{air}) = \left(\left(\frac{T_{in} + T_{out}}{2}\right) - T_{air}\right)$$

$$= \left(\left(\frac{120 + 100}{2}\right) - 65\right) = 45\ °\text{F}$$

The correction factor can now be calculated:

$$CF = 0.001882(\Delta T)^{1.33} = 0.001882(45)^{1.35} = 0.297$$

The corrected heat output is then calculated as:

$$Q_{corrected} = (CF)Q_{output} = (0.297)9{,}500 = 2{,}822\frac{\text{Btu}}{\text{hr}}$$

Discussion: Operating this panel radiator at the relatively low average water temperature of 110 °F significantly reduces its heat output in comparison to its rated heat output at 180 °F average water temperature. Still, even at the low average water temperature, the heat

	Btuh/ft. Ratings @ 65°F EAT					
	MODEL	HEIGHT in.	DEPTH in.	215°F	180°F	140°F
PERIMETER STYLE	R-1	2.8	1.6	230	**160**	90
	R-2	5.7	1.6	420	**300**	170
	R-3	8.6	1.6	620	**440**	250
	R-4	11.5	1.6	820	**580**	330
	R-5	14.4	1.6	1,020	**720**	410
	R-6	17.3	1.6	1,220	**860**	500
WALL PANEL	R-7	20.2	1.6	1,430	**1,010**	580
	R-8	23.1	1.6	1,640	**1,160**	660
	R-9	26.0	1.6	1,850	**1,300**	750
	R-10	29.0	1.6	2,060	**1,450**	830

(a)

		EAT										
		45°F	50°F	55°F	60°F	65°F	70°F	75°F	80°F	85°F	90°F	95°F
	240°F	1.365	1.350	1.304	1.266	1.220	1.171	1.124	1.086	1.039	1	0.953
	235°F	1.343	1.305	1.267	1.219	1.171	1.124	1.086	1.038	1	0.952	0.910
	230°F	1.305	1.267	1.219	1.171	1.124	1.086	1.038	1	0.952	0.910	0.868
	225°F	1.267	1.219	1.171	1.124	1.086	1.038	1	0.952	0.910	0.868	0.826
	220°F	1.219	1.171	1.124	1.086	1.038	1	0.952	0.910	0.868	0.826	0.785
	215°F	1.171	1.124	1.086	1.038	1	0.952	0.910	0.868	0.826	0.785	0.744
	210°F	1.124	1.086	1.038	1	0.952	0.910	0.868	0.826	0.785	0.744	0.704
	205°F	1.086	1.038	1	0.952	0.910	0.868	0.826	0.785	0.744	0.704	0.664
	200°F	1.038	1	0.952	0.910	0.868	0.826	0.785	0.744	0.704	0.664	0.625
	195°F	1	0.952	0.910	0.868	0.826	0.785	0.744	0.704	0.664	0.625	0.587
	190°F	0.952	0.910	0.868	0.826	0.785	0.744	0.704	0.664	0.625	0.587	0.549
	185°F	0.910	0.868	0.826	0.785	0.744	0.704	0.664	0.625	0.587	0.549	0.511
	180°F	0.868	0.826	0.785	0.744	0.704	0.664	0.625	0.587	0.549	0.511	0.474
	175°F	0.826	0.785	0.744	0.704	0.664	0.625	0.587	0.549	0.511	0.474	0.438
AWT	170°F	0.785	0.744	0.704	0.664	0.625	0.587	0.549	0.511	0.474	0.438	0.403
	165°F	0.744	0.704	0.664	0.625	0.587	0.549	0.511	0.474	0.438	0.403	0.369
	160°F	0.704	0.664	0.625	0.587	0.549	0.511	0.474	0.438	0.403	0.369	0.334
	155°F	0.664	0.625	0.587	0.549	0.511	0.474	0.438	0.403	0.369	0.334	0.301
	150°F	0.625	0.587	0.549	0.511	0.474	0.438	0.403	0.369	0.334	0.301	0.269
	145°F	0.587	0.549	0.511	0.474	0.438	0.403	0.369	0.334	0.301	0.269	0.237
	140°F	0.549	0.511	0.474	0.438	0.403	0.369	0.334	0.301	0.269	0.237	0.207
	135°F	0.511	0.474	0.438	0.403	0.369	0.334	0.301	0.269	0.237	0.207	0.177
	130°F	0.474	0.438	0.403	0.369	0.334	0.301	0.269	0.237	0.207	0.177	0.149
	125°F	0.438	0.403	0.369	0.334	0.301	0.269	0.237	0.207	0.177	0.149	0.122
	120°F	0.403	0.369	0.334	0.301	0.269	0.237	0.207	0.177	0.149	0.122	0.096
	115°F	0.369	0.334	0.301	0.269	0.237	0.207	0.177	0.149	0.122	0.096	0.071
	110°F	0.334	0.301	0.269	0.237	0.207	0.177	0.149	0.122	0.096	0.071	0.050
	105°F	0.301	0.269	0.237	0.207	0.177	0.149	0.122	0.096	0.071	0.050	0.030
	100°F	0.269	0.237	0.207	0.177	0.149	0.122	0.096	0.071	0.050	0.030	0.011

EXAMPLE: To find the Btuh/ft. Rating for an R-6 Panel at 155°F AWT and 65°F EAT, Multiply the Correction Factor (0.511) by the Btuh/ft. Rating at 215°F (1219), e.g., (0.511) × (1219) = 623 Btuh/ft.

(b)

Courtesy of Runtal Radiators

Figure 5-26 (a) Thermal performance ratings for flat tube panel radiators given as heat output per foot of panel length for several panel widths (R1 = 1 tube, R2 = 2 tubes, etc.). (b) Correction factors for heat output at different room air and average water temperatures.

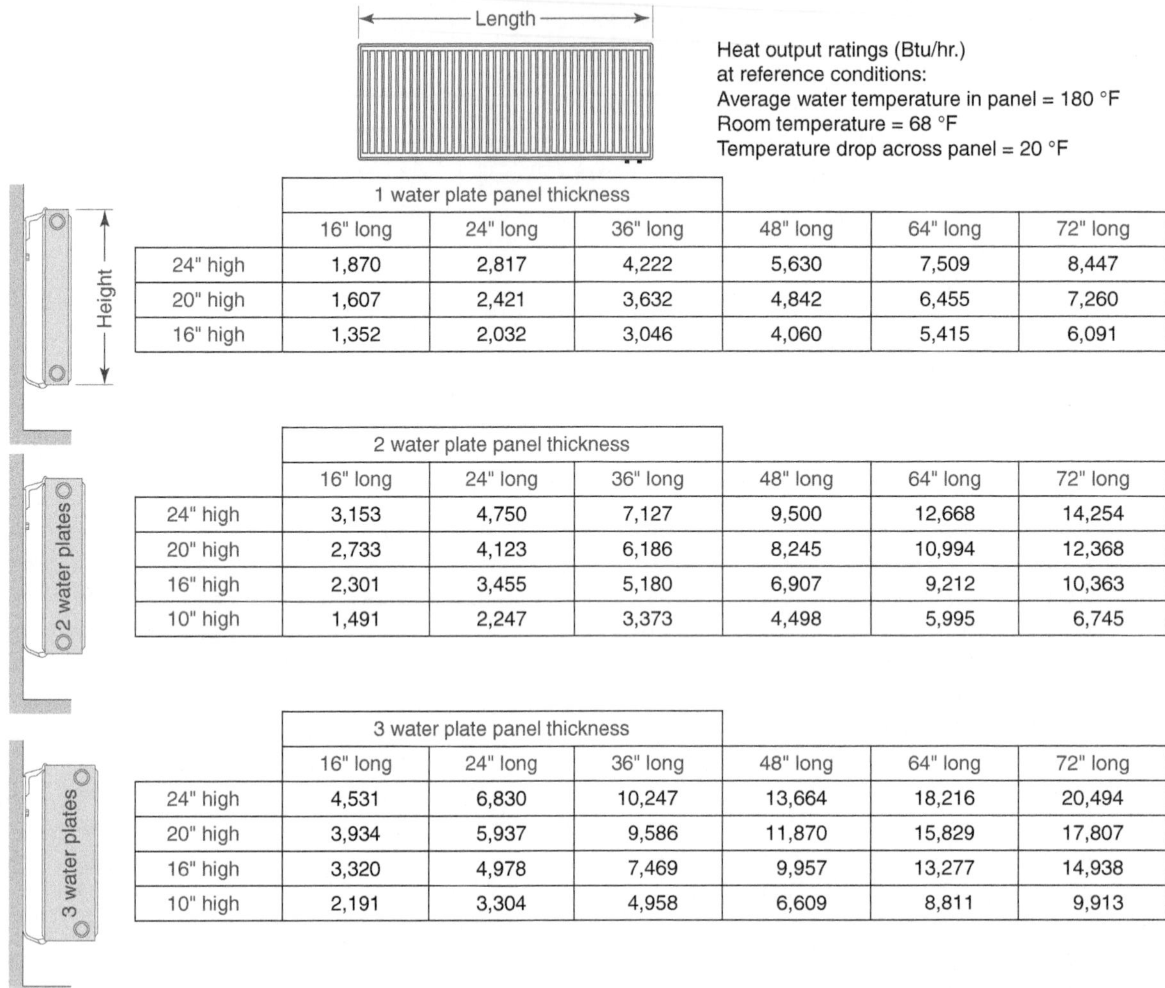

1 water plate panel thickness	16" long	24" long	36" long	48" long	64" long	72" long
24" high	1,870	2,817	4,222	5,630	7,509	8,447
20" high	1,607	2,421	3,632	4,842	6,455	7,260
16" high	1,352	2,032	3,046	4,060	5,415	6,091

2 water plate panel thickness	16" long	24" long	36" long	48" long	64" long	72" long
24" high	3,153	4,750	7,127	9,500	12,668	14,254
20" high	2,733	4,123	6,186	8,245	10,994	12,368
16" high	2,301	3,455	5,180	6,907	9,212	10,363
10" high	1,491	2,247	3,373	4,498	5,995	6,745

3 water plate panel thickness	16" long	24" long	36" long	48" long	64" long	72" long
24" high	4,531	6,830	10,247	13,664	18,216	20,494
20" high	3,934	5,937	9,586	11,870	15,829	17,807
16" high	3,320	4,978	7,469	9,957	13,277	14,938
10" high	2,191	3,304	4,958	6,609	8,811	9,913

Figure 5-27 Heat output rating table for fluted steel panel radiators based on panel dimensions and stated reference conditions. Note that these ratings are for an average water temperature of 180 °F and an assumed 20 °F temperature drop across the panel.

output of this panel could meet the design heating load of a 12 foot by 16 foot room with a design load of 15 Btu/hr/ft^2. This demonstrates the importance of low building heat loss in situations where reasonably sized heat emitters are expected to carry the load at low water temperatures.

GENERALIZED PERFORMANCE MODEL FOR PANEL RADIATORS

As this text is being written, there is no North American standard for testing the heat output of panel radiators. However, because panel radiators are widely used in Europe, there is a standardized method for converting rated heat output at specific conditions to expected heat output at other conditions. This procedure is part of a testing standard known as **EN442**. Under this standard, the *rated* heat output of the panel is based on a 50 °C (112 °F) *difference* between the average water temperature in the panel and the room air temperature.

The following procedure, based on EN442, can be used to convert rated heat outputs at specific conditions to heat outputs at other operating conditions.

STEP 1: Determine the value of u using Equation 5.7:

(Equation 5.7)

$$u = \frac{(T_{out} - T_{air})}{(T_{in} - T_{air})}$$

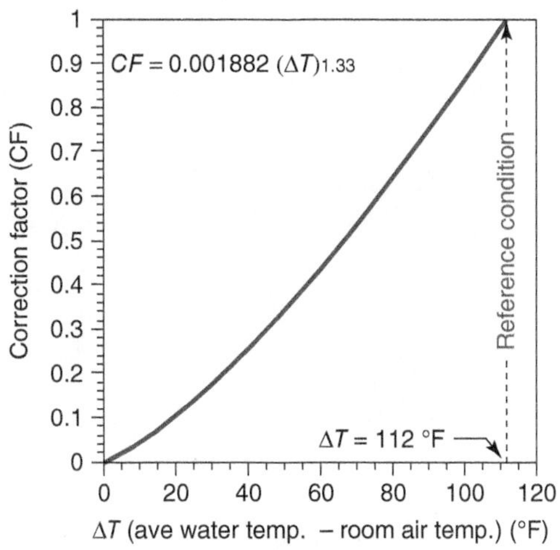

Figure 5-28 Correction factor equation and graph for heat output of fluted panel radiators at operating conditions that vary from the reference conditions listed in Figure 5-27.

where:

u = differential temperature ratio (unitless)

T_{out} = outlet fluid temperature from panel (°F)

T_{in} = inlet fluid temperature to panel (°F)

T_{air} = room air temperature (°F)

STEP 2: If $u < 0.7$, then determine the value of ΔT_d using Equation 5.8:

(Equation 5.8)

$$\Delta T_d = \frac{(T_{in} - T_{air})}{\ln\left(\frac{T_{in} - T_{air}}{T_{out} - T_{air}}\right)}$$

where:

ΔT_d = effective temperature difference (°F)

T_{in} = inlet water temperature to panel (°F)

T_{out} = outlet water temperature from panel (°F)

T_{air} = room air temperature (°F)

ln = natural logarithm function

If $u \geq 0.7$, then determine the value of ΔT_d using Equation 5.9:

(Equation 5.9)

$$\Delta T_d = \left[\left(\frac{T_{in} + T_{out}}{2}\right) - T_{air}\right]$$

where:

ΔT_d = effective temperature difference (°F)

T_{in} = inlet water temperature to panel (°F)

T_{out} = outlet water temperature from panel (°F)

T_{air} = room air temperature (°F)

STEP 3: After determining ΔT_d, the corrected heat output of the panel can be determined using Equation 5.10:

(Equation 5.10)

$$Q_e = Q_{112}\left(\frac{\Delta T_d}{112}\right)^{1.3}$$

where:

Q_e = expected heat output of the panel radiator (Btu/hr)

ΔT_d = temperature difference determined using either Equation 5.8 or 5.9 (°F)

Q_{112} = the output of the panel radiator when the difference between the *average* water temperature and room air temperature is 112 °F (or 50°C), in (Btu/hr)

1.3 = an exponent (not a multiplier)

Example 5.7: A panel radiator has a rated heat output of 8,500 Btu/hr when operated at an average water temperature of 180 °F and room air temperature of 68 °F. Using the methods from EN442, determine the panel's expected heat output rate when operated with a supply water temperature of 105 °F, a temperature drop of 15 °F, and in a room maintained at 65 °F.

Solution: If the supply temperature is 105 °F and the temperature drop is 15 °F, then the outlet temperature from the panel is 105 – 15 = 90 °F. The value of u can now be determined using Equation 5.7:

$$u = \frac{(T_{out} - T_{air})}{(T_{in} - T_{air})} = \frac{(90 - 65)}{(105 - 65)} = 0.625$$

Because u is less than 0.7, ΔT_d must be determined using Equation 5.8:

$$\Delta T_d = \frac{(T_{in} - T_{out})}{\ln\left(\frac{T_{in} - T_{air}}{T_{out} - T_{air}}\right)} = \frac{(105 - 90)}{\ln\left(\frac{105 - 65}{90 - 65}\right)} = \frac{15}{0.47} = 31.9\ °F$$

Equation 5.10 can now be used to determined the expected heat output:

$$Q_r = Q_{112}\left(\frac{\Delta T_d}{112}\right)^{1.3} = 8500\left(\frac{31.9}{112}\right)^{1.3} = 1{,}661\frac{\text{Btu}}{\text{hr}}$$

Discussion: Again, operating the panel at low water temperatures significantly reduces its heat output—in this case, down to 19.5 percent of its rated output. However, to keep things in perspective, the180 °F average water temperature at which the rated heat output is based is very high by today's design standards for hydronic systems, even those supplied by boilers. This temperature is "unrealistically" high for systems supplied by renewable energy heat sources. Thus, the derating to 19.5 percent of an unrealistically high heat output is not as bad as one might think. Larger panel radiators are necessary for low water temperature systems. It's simply a necessary trade-off between first cost and life-cycle operating cost.

Designers should use sizing procedures and data supplied by the panel radiator manufacturer—or, if no such information is available, the EN442 standard—to establish heat output correction factors. These factors can then be applied while using heat output tables to select appropriate panel radiators for each room.

DISTRIBUTION SYSTEMS FOR PANEL RADIATORS

As with fin-tube baseboard, a homerun distribution system is ideally suited for use with panel radiators. Figures 5-29 shows the concept of several panel radiators, each with their own wireless thermostatic radiator valve supplied from a common manifold station. The red and blue lines leading to each panel radiator are typically 1/2-inch or in some cases 3/8-inch size PEX or PEX-AL-PEX tubing. Such tubing is easily routed through framing, making this system well suited to both new installations and retrofits.

The piping schematic in Figure 5-30 expands on the distribution system shown in Figure 5-29. It adds a **pressure-regulated circulator**, thermal storage tank, and **3-way motorized mixing valve**.

The pressure-regulated circulator is set to operate in **constant differential pressure mode**. As each thermostatic radiator valve changes the flow through its associated panel radiator, the circulator automatically adjusts its speed to maintain a constant differential pressure across the manifold station.

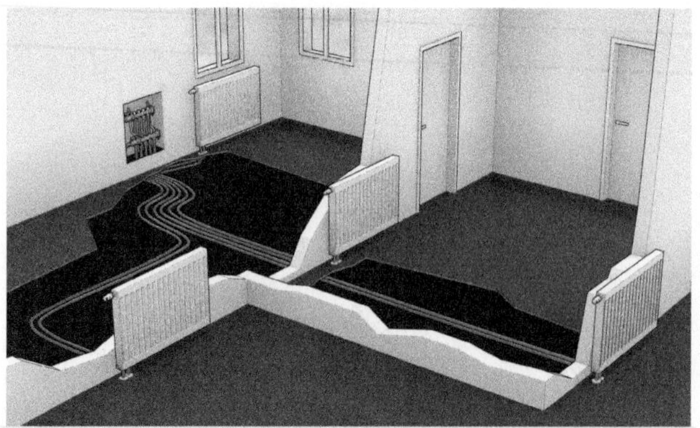

Courtesy of Caleffi North America, Inc.

Figure 5-29 Panel radiators, each with their own wireless thermostatic radiator valve, supplied from a common manifold station.

The 3-way motorized mixing valve installed between the thermal storage tank and the manifold station prevents what might be very hot water in the tank from flowing directly to the panel radiators. High water temperatures are very possible, even likely, with renewable energy heat sources such as solar thermal collectors and wood-fired boiler systems. Although the steel panel radiators can handle higher temperature water, providing a lower and controlled temperature reduces the potential for thermal expansion noise and provides more even heat delivery. It is also safer from the standpoint of panel surface temperature. The controller operating the 3-way mixing valve can also be configured to operate based on outdoor reset control, which is discussed in the next chapter.

FAN-ENHANCED PANEL RADIATORS

The heat output of standard panel radiators drops significantly with decreasing entering water temperature. Manufacturers who offer heat emitters specifically for low water temperature applications have found way to enhance heat output without having to use excessively large heat emitters, or without creating undesirable operating conditions. One approach is to enhance convective heat transfer at low water temperatures. Figure 5-31 illustrates the internal components in a state-of-the art product that combines the function of a panel radiator and convector.

Inside this heat emitter, and near its base, is a high-performance fin-tube element. Immediately above

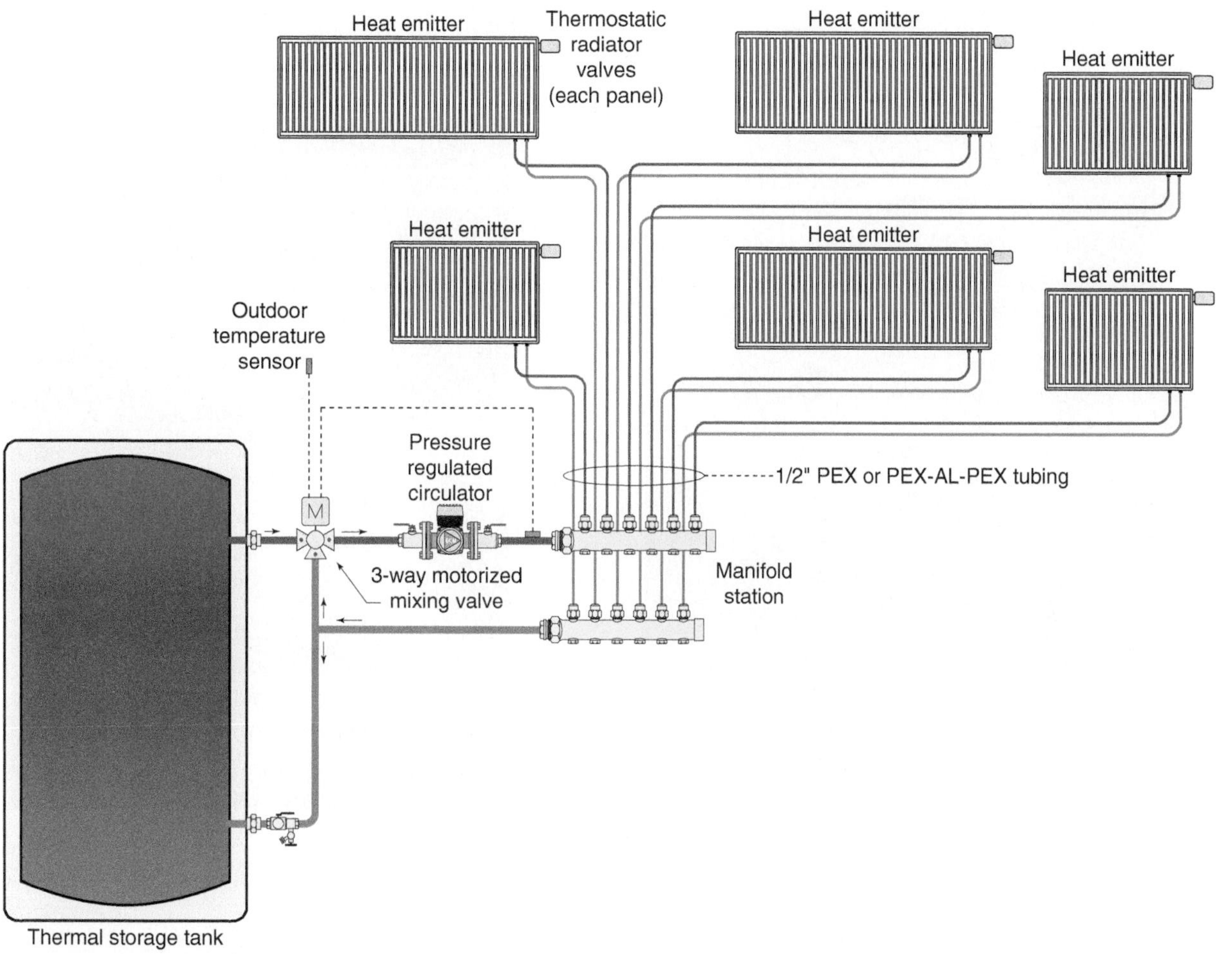

Figure 5-30 | Homerun distribution system supplying several panel radiators.

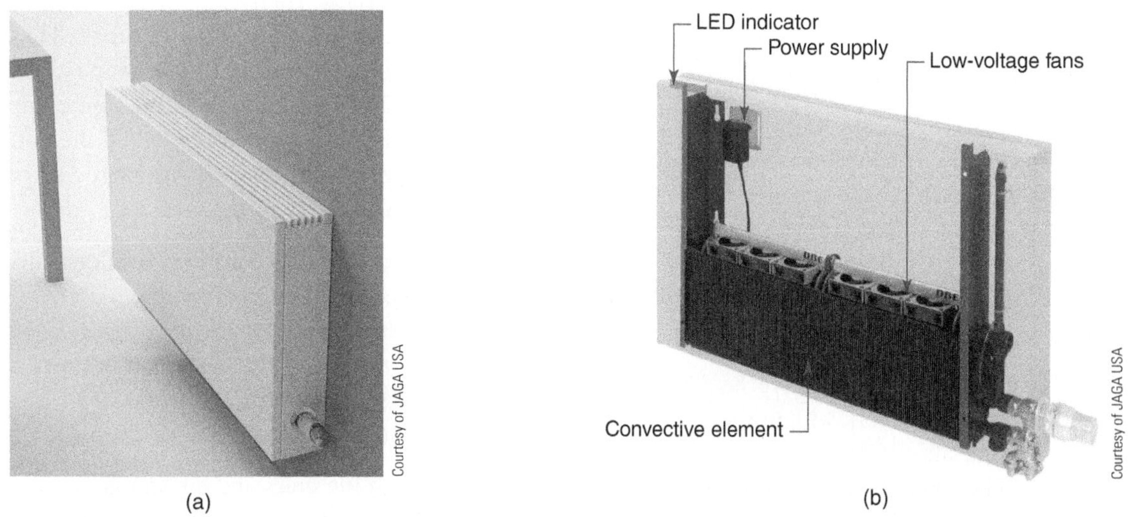

Figure 5-31 | (a) External appearance of a fan-enhanced panel radiator. (b) Internal design showing deep convective element with microfans at top.

(c)

(d)

Figure 5-31 | (c) Internals of unit being installed. (d) Finish installation.

that element is a rack of several low-voltage, variable-speed "**microfans**," similar to those used in personal computers. At full speed, each fan only requires about 1.5 watts of electrical power, but the enhanced air flow the fans create across the fin-tube element can boost heat output by 50 percent during normal comfort mode and by over 250 percent during recovery from a setback condition. The microprocessor-controlled fans vary their speed as necessary to hold the room at a setpoint condition. As the room temperature approaches setpoint, the fans slow to prevent temperature overshoot. The air currents within the enclosure also warm the front panel, which provides some radiant heat output.

The significant gain in convective heat transfer allows these heat emitters to work with low supply water temperatures, comparable to those required by bare heated floor slabs. The ability to operate at a supply water temperature as low as 95 °F can significantly increase the thermal efficiency of low-temperature hydronic heat sources, such as solar collectors, ground-source heat pumps, and condensing boilers.

5.4 Hydronic Fan Coils

Some rooms lack the wall space needed to mount fin-tube baseboard or panel radiators. One alternative is to use a hydronic **fan coil**, or **air handler**. These devices come in many different sizes, shapes, and heating capacities. They all have two common components: a fin-tube heat exchanger (often called the **coil**) through which system fluid flows, and a **blower** or **fan** that forces air to flow across the coil.

In some fan coils and air handlers, air is pulled through the coil by the blower. This arrangement is called a **draw-through fan coil**. In other products, air is blown through the coil. Such units are called **blow-through fan coils**. Figure 5-32 illustrates these arrangements between the coil and blower.

Although fan coils and air handlers share common characteristics, the term fan coil usually describes a smaller heat emitter that is mounted to the surface of wall or recessed into a wall cavity, as show in Figure 5-33. Fan coils typically deliver heat to a single room.

The term "air handler" describes a larger device that is usually designed for mounting in concealed spaces. Air handlers typically have larger coils and blowers compared to fan coils. These components are contained in a vertical or horizontal sheet metal cabinet that is designed for concealed mounting above or below the occupied space. They deliver an air stream to ducting that often distributes that air to more than one room. An example of a horizontal air handler is shown in Figure 5-34.

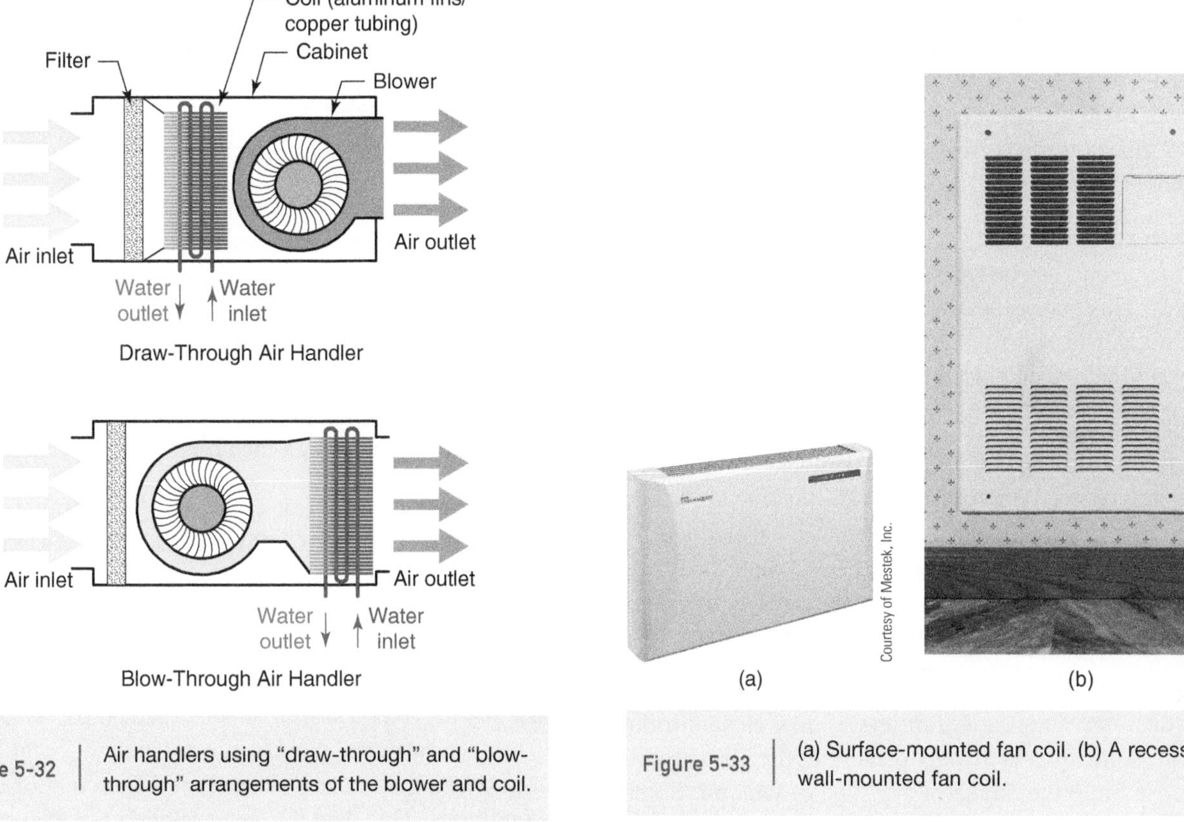

Figure 5-32 | Air handlers using "draw-through" and "blow-through" arrangements of the blower and coil.

Figure 5-33 | (a) Surface-mounted fan coil. (b) A recessed wall-mounted fan coil.

Figure 5-34 | Examples of a small air handler. Piping connections and condensate drain seen at right rear corner.

ADVANTAGE AND DISADVANTAGES OF FAN COILS

Unlike fin-tube baseboard and panel radiators, hydronic fan coils rely on **forced convection** to transfer heat from the coil surface to the room air. Forced convection is much more effective than natural convection in transferring heat from a surface to air. This allows the surface area of the coil to be significantly smaller than that of a fin-tube baseboard element of equivalent capacity. A fan-coil unit can often provide equivalent heating capacity using only a fraction of the wall space required by fin-tube baseboard.

Fan coils also have very little thermal mass and low water content. They can respond very quickly to a demand for heat. This makes them well suited for spaces that need to be quickly heated from setback temperature conditions or spaces that are only occasionally heated.

The compactness and fast-response characteristics of hydronic fan coils do not come without compromises. First, any type of fan coil will produce some noise due to the blower operation. Noises can be worsened by vibration, loose fitting grills, and trim. Manufacturers attempt to limit operating noise through design, but some operating noise will always be present with any fan coil.

Another factor associated with fan coils is dust (and dust movement). Regardless of how clean a room is kept, some dust is always present in the air and will be carried into the intake of any type of fan coil. Some fan coils have intake filters capable of removing most of the dust before it is deposited on the coil surfaces, others do not have such filters

Fan coils should be located so that air flow is not blocked by furniture. They should also be located so as not to blow air directly on room occupants, who, for example, may be seated at a work surface. Instead, they should be placed so that their discharge air stream will blend with room air before it contacts the room's occupants.

Fan coils and air handlers also require electrical power to be available where they are installed. A circuit must be provided to each location where a fan coil or air handler is installed. In some cases, it is also necessary to install low-voltage wiring from fan-coil units to thermostats or other controllers within the system.

Recessed fan-coil units need to be carefully coordinated with wall framing and other services such as plumbing, electrical work, and ducting. A "rough in" size for each recessed fan coil should be provided to those constructing the building to ensure that appropriately sized wall cavities are included.

THERMAL PERFORMANCE OF FAN COILS

The heat output of a hydronic fan coil is dependent on several factors. Some are fixed by the construction of the fan coil and cannot be changed. These include the:

- Surface area of the coil
- Size, spacing, and thickness of the fins
- Number of tube passes through the fins
- Air-moving ability of the blower

Other performance factors are dependent on the system into which the fan coil is installed. These include the:

- Entering water temperature
- Water flow rate through the coil
- Entering air temperature
- Air flow rate through the coil

When selecting a fan coil, designers must match the characteristics of the available units with the temperatures and flow rates the system can provide while also obtaining the required heating capacity. This is especially important when hydronic fan coils are to operate with low supply water temperatures. This usually involves a search of published performance data, from one or more manufacturers, in an attempt to find a good match among several simultaneous operating conditions. This data is often provided as tables that list the heat output at several combinations of temperature and flow rate for both the water side and air side of the unit. An example of a thermal performance table for a wall-mounted fan coil is shown in Figure 5-35.

Figure 5-35 lists the heating capacity of two different fan-coil models over a wide range of entering water temperatures and water flow rates. Each heat output value can be thought of as a "snapshot" of heating performance at specific conditions.

The heat output of the fan coil varies continuously over a range of temperatures and flow rates. There is no guarantee it will operate at one of the conditions listed in the table. Interpolation can be used when necessary to estimate the heat output of the unit at conditions other than those listed in the table. Another useful method, especially if the data will be used repeatedly, is to plot the data on a graph and create a smooth curve through the points. The estimated performance between data points is easily read from this curve. Some manufacturers also provide design-assistance software capable of quickly providing interpolated output ratings.

		HEAT OUTPUT (Btu/hr)				
		Entering water temperature				
Model	Flow rate (gpm)	100 °F	110 °F	120 °F	130 °F	140 °F
PSU10 Fan at LOW speed	0.5	720	896	1,440	1,852	2,299
	1.5	864	1,075	1,728	2,222	2,874
	2.0	**900**	**1,120**	**1,800**	**2,315**	**2,994**
	2.5	963	1,198	1,926	2,477	3,204
	3.0	1,008	1,254	2,016	2,593	3,353
	5.0	1,107	1,378	2,214	2,848	3,683
PSU10 Fan at HIGH speed	0.5	1,080	1,480	2,120	2,646	3,422
	1.5	1,296	1,776	2,544	3,175	4,107
	2.0	**1,350**	**1,850**	**2,650**	**3,307**	**4,278**
	2.5	1,445	1,980	2,836	3,539	4,578
	3.0	1,512	2,072	2,968	3,704	4,791
	5.0	1,661	2,276	3,260	4,068	5,262

Heat outputs based on 65 °F entering air temperature
Coil head loss at 2.0 gpm flow rate = 2.75 feet of head

Figure 5-35 | Thermal performance data for small wall-mounted fan-coil convector at different entering water temperatures, flow rates, and fan speeds.

Some manufacturers list the heat output of a specific fan-coil unit at a single reference condition consisting of a specified value for the entering temperatures and flow rates of both the water and air. Accompanying this will be one or more tables or graphs of correction factors that can be used to adjust the reference performance up or down for variation in temperature and flow rate of both the entering water and entering air. An example of a table listing correction factors for a fan coil is shown in Figure 5-36.

Notice that the correction factor is 1.000 at an entering water temperature of 200 °F and corresponding entering air temperature of 60 °F. This is the reference point for the table. The value of the correction factor decreases for any condition in which the entering water temperature decreases or the entering air temperature increases.

Figure 5-37 shows a graph of the correction factors given in Figure 5-36.

Notice that for a given entering water temperature, there is a linear relationship between the correction factor and the entering air temperature. This implies another linear relationship between the heat output from the fan coil and the temperature difference between the entering water and entering air temperature. This will now be discussed, along with other general principles that govern fan-coil performance.

GENERAL PRINCIPLES OF FAN-COIL PERFORMANCE

The heat output of fan coils can be described using several fundamental principles that can be verified by examining published ratings data. By varying one operating condition such as entering temperature or flow rate while all others remain fixed, the designer can gain an understanding of which factors have the greatest impact on heat output. This can be very helpful in evaluating the feasibility of various design options.

Principle #1: The heat output of a fan coil is approximately proportional to the temperature difference between the entering air and entering water.

	Entering water temperature (w/ 20 °F temp.s drop, and 6 0 °F entering air temp.)					
Entering air temp.	**100 °F**	**120 °F**	**140 °F**	**160 °F**	**180 °F**	**200 °F**
40 °F	0.439	0.585	0.731	0.878	1.025	1.172
50 °F	0.361	0.506	0.651	0.796	0.941	1.085
60 °F	0.286	0.429	0.571	0.715	0.857	**1.000**
70 °F	0.212	0.353	0.494	0.636	0.777	0.918
80 °F	0.140	0.279	0.419	0.558	0.698	0.837

Figure 5-36 | Correction factors, for a small fan coil, that are used to multiply the heat output rating at 200 °F entering water temperature, and 60 °F entering air temperature, to yield heat output at other entering water and entering air temperatures.

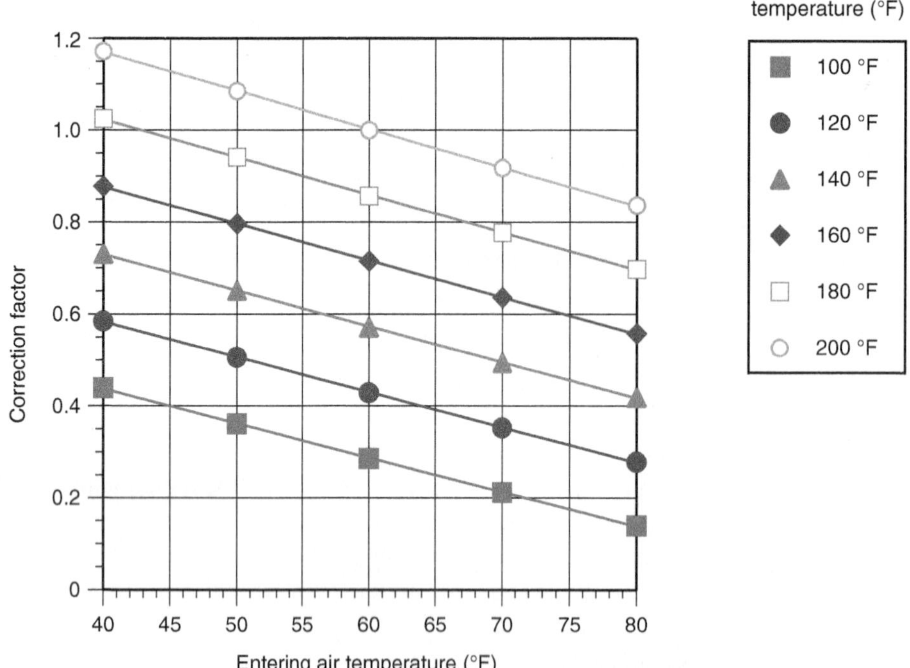

Figure 5-37 | Correction factors from Figure 5-36 plotted as a function of entering air temperature and six entering water temperatures.

If, for example, a particular fan coil can deliver 5,000 Btu/hr with a water temperature of 140 °F and room air entering at 65 °F, its output using 180 °F water can be estimated as follows:

$$\text{Estimated output at } 180° = \frac{(180\,°\text{F} - 65\,°\text{F})}{(140\,°\text{F} - 65\,°\text{F})} \times (5{,}000 \text{ Btu/hr}) = 7{,}670 \text{ Btu/hr}$$

A proportional relationship between any two quantities will result in a straight line that passes through or very close to the origin when the data is plotted on a graph. Figure 5-38 shows this to be the case when the heat output data from a manufacturer is plotted against the difference between the entering water and entering air temperatures.

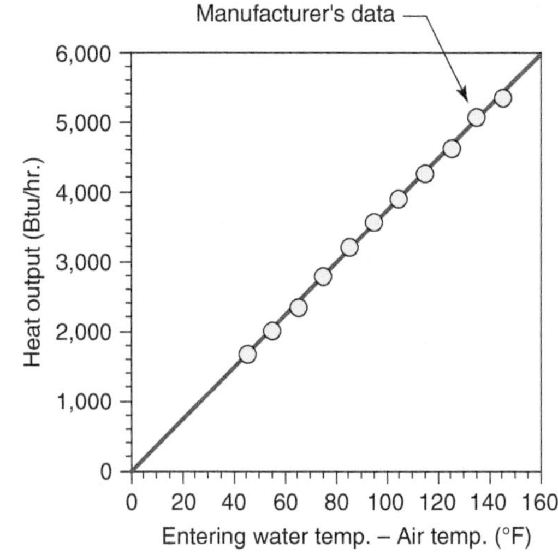

Figure 5-38 Plot of heat output versus temperature difference between entering water and entering air temperature based on manufacturer's data.

This relationship is useful for estimating heat output when a fan coil is used in a room kept above or below normal comfort temperatures. For example, if a fan coil can deliver 5,000 Btu/hr with an entering water temperature of 120 °F and entering air temperature of 70 °F, its output while heating a garage maintained at 50 °F could be estimated using the same proportionality method described above:

$$Q_{50°F\,av} = \left(\frac{120 - 50}{120 - 70}\right)5{,}000 = 7{,}000 \text{ Btu/hr}$$

Although this principle is helpful for performance estimates, *it should not replace the use of thermal rating data supplied by the manufacturer when available.*

Principle #2: Increasing the fluid flow rate through the coil will marginally increase the heating capacity of the fan coil.

Interestingly, many heating professionals disagree with this statement. They argue that because the water moves through the coil at a faster speed, it has less time in which to release its heat. However, the time a given molecule of water stays inside the fan coil is irrelevant in a system with continuous circulation. *The faster-moving fluid improves convective heat transfer between the fluid and the inner surface of the tubes, resulting in greater heating capacity.*

Another way of justifying this principle is to consider the *average water temperature* in the coil at various flow rates. As the flow rate through the coil is increased, this temperature difference decreases.

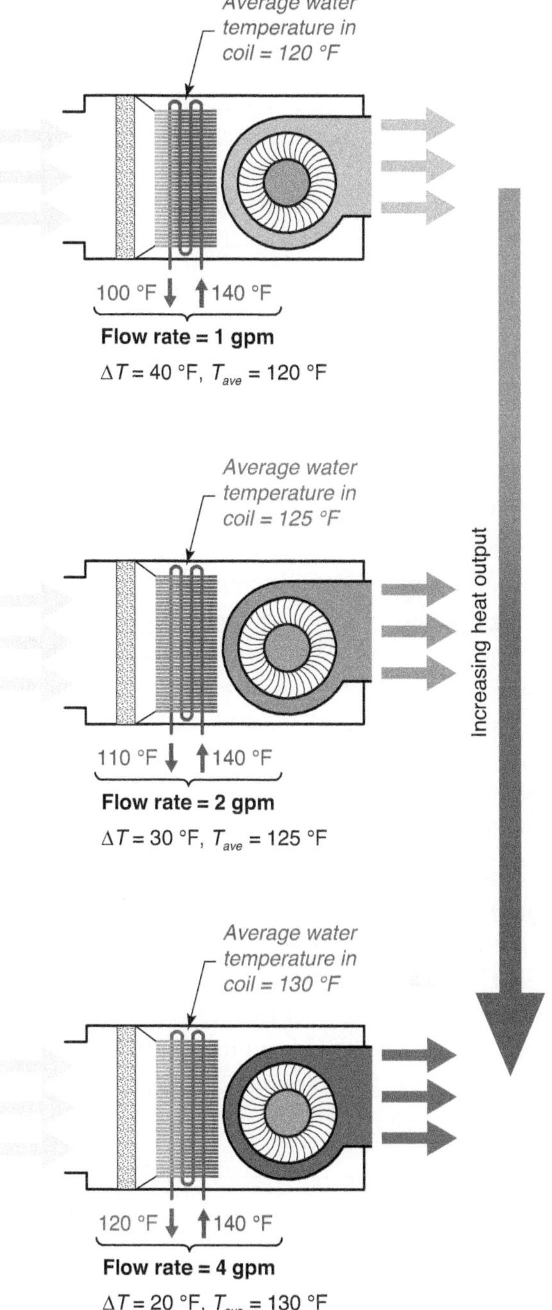

Figure 5-39 As the flow rate through the coil increases, the average water temperature in the coil also increases. This implies that heat output always increases with increased flow rate. This is true for all hydronic heat emitters.

This means that the average water temperature within the coil moves closer to the inlet temperature (i.e., it increases), and so does its heat output. This principle is illustrated in Figure 5-39. It also holds true for other heat emitters, including radiant panel circuits, panel radiators, and baseboard.

Those who claim that a smaller temperature difference between the ingoing and outgoing fluid implies less heat is being released are overlooking the fact that the rate of heat transfer depends on *both* the temperature difference *and* flow rate and the mathematical relationship between them. This relationship was described by the sensible heat rate equation introduced in Chapter 3.

The rate of gain in heat output is very fast at low flow rates but steadily decreases as the flow through the coil increases. This "nonlinear" characteristic can be seen in Figure 5-40, which plots manufacturer's data for heat output versus water flow rate for a small fan coil.

Notice that the fan coil has about 50 percent of its maximum heat output capacity at only about 10 percent of its maximum flow rate. The strong curvature of this relationship means that attempting to control the heat output of a fan coil by adjusting flow rate can be tricky. Very small valve adjustments can create large changes in heat output at low flow rates. However, the same amount of valve adjustment will create almost no change in heat output at higher flow rates. This is also true for other types of hydronic heat emitters and will be discussed in the context of control systems in Chapter 6.

The nonlinear relationship between flow rate and heat output also implies that attempting to boost heat output from a fan coil by operating it at an unusually high flow rate will yield very minor gains. The argument against this is further supported when one considers the increased head loss through the coil, and the associated higher circulator power requirements.

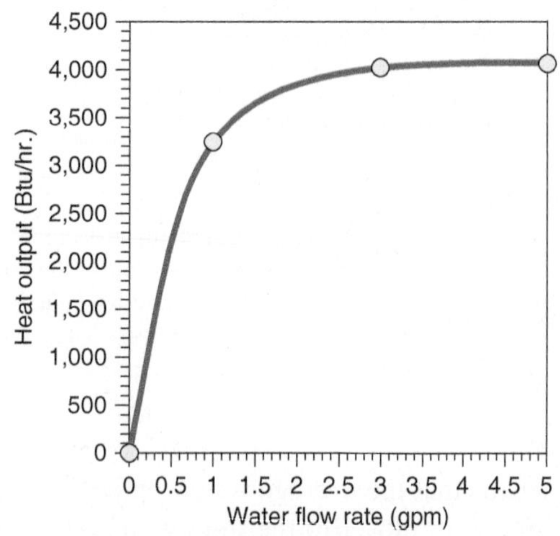

Figure 5-40 Heat output of a small fan coil at different flow rates and constant entering water and entering air temperatures.

Principle #3: Increasing the air-flow rate across the coil will marginally increase the heat output of the fan coil.

This principle is also based on forced-convection heat transfer between the exterior coil surfaces and the air stream. Faster-moving air "scrubs" heat off the coil surface at a greater rate. As with water flow rates, the gain in heat output is very minor above the nominal air-flow rate the unit is designed for.

Principle #4: Fan coils with large coil surfaces and/or multiple tubes passes through the coil can yield a given rate of heat output while operating at lower entering water temperatures.

Convective heat transfer is proportional to the contact area between the surface and the fluid. If this contact area is increased, the temperature difference between the fluid and the air streams can be reduced (for a given rate of heat output). This is illustrated in Figure 5-41.

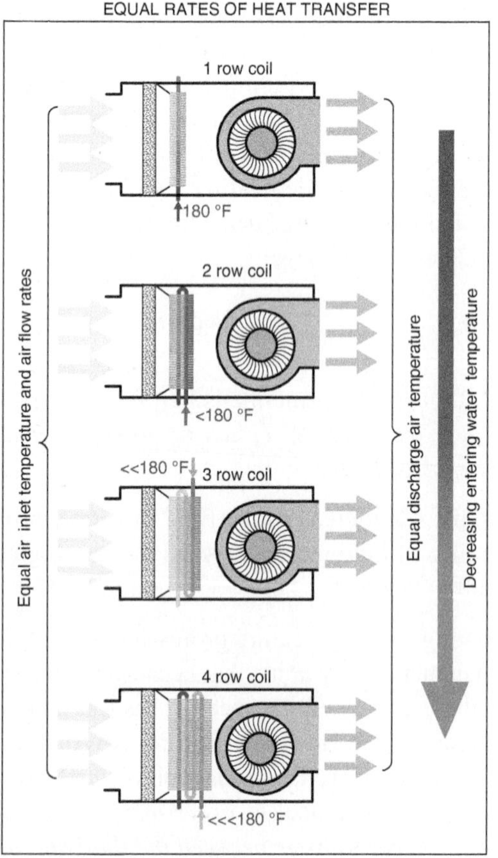

Figure 5-41 As the size and number of tube rows in the coil increases, the entering water temperature needed for a given rate of heat transfer decreases.

This principle is very important for hydronic systems that must operate with low-temperature heat sources such as solar thermal collectors or heat pumps. In these applications, the designer may need to use fan coils with larger coils and/or more **tube passes** through the fins of the coil, to drive heat from the coil at the required rate. On larger air handlers, manufacturers may offer higher performance coils with up to six rows of tubes as an option. This is usually not true on smaller residential fan coils. For these products, the only way to increase coil surface area is to install more fan coils.

Care must also be taken that comfort is not compromised when operating fan coils at reduced water temperatures. It is critically important to introduce the air stream from such fan coils into the space so that it mixes with room air before passing by occupants. Failure to do so will usually lead to complaints of "cool air" blowing from the fan coils, even though the room is maintaining the desired average temperature.

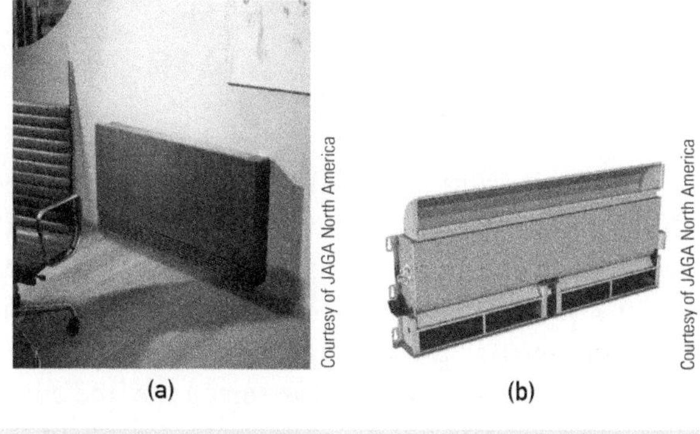

Figure 5-42 (a) Wall-mounting convector that can be used for low water temperature heating as well as chilled water cooling. (b) Internal construction.

USING FAN COILS FOR HYDRONIC COOLING

Many large buildings use hydronic distribution systems for supplying both heated and chilled water to fan coils or air-handler units. This can also be done in residential or light commercial systems if a water-chilling device such as a hydronic heat pump is used as the heating/cooling source.

Only fan coils or air handlers that are equipped with **condensate drip pans** *and drains should be used for chilled-water-cooling applications.* A drip pan acts as a catch basin for the water droplets that continually form on the coil during cooling operation. Drip pans are usually connected to a drainpipe or floor drain so that **condensate** can be disposed of at the same rate it is formed. If a floor drain is not available, a condensate removal pump is necessary. Many hydronic fan coils are designed only for heating. They do not have condensate pans and should never be used for chilled-water-cooling applications. Doing so would result in condensate running out the bottom of the unit, causing damage to the structure.

Figure 5-42 shows a slim (4.6-inch-deep) wall-mounted fan coil that's equipped with a condensate drip pan and thus can be used for heating and cooling applications.

Figure 5-43 shows a small air handler that is also equipped with a condensate drip pan and thus can be used for either heating or cooling. This unit is designed to discharge air to several 2-inch-diameter flexible "mini-ducts."

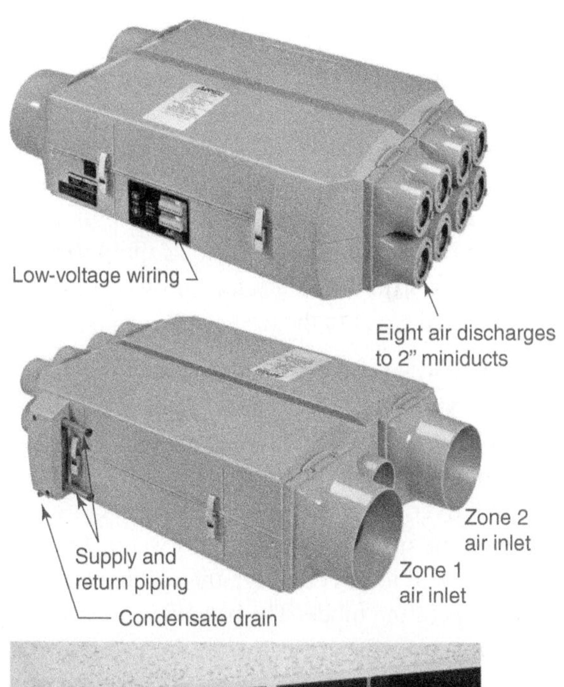

Figure 5-43 Small air handler equipped with condensate drip pan.

DISTRIBUTION SYSTEMS FOR FAN COILS AND AIR HANDLERS

Some fan coils and air handlers that are equipped with drip pans use the same internal coil for both heating and cooling operation. Such a unit is often called a **2-pipe fan coil** or **2-pipe air handler**. This name comes from the piping configuration. When combined with other units of the same type, the entire system must operate in either the heating mode or the cooling mode. Figure 5-44 shows a representation of this type of system, which is more often applied in residential and light commercial buildings. This type of system is well suited for climates where there are typically several days (or weeks) between outdoor conditions that require heating versus those that require cooling.

It is also possible to connect a fan coil or air handler to a **4-pipe distribution system,** as shown in Figure 5-45.

4-pipe distribution systems, which are more common in larger buildings, are capable of simultaneously conveying heated water and chilled water through buildings. Each air handler or fan coil connected to such a system can operate in either heating or cooling. This allows each fan coil or air handler to adapt to variations in internal heat gain or preferred comfort settings from one conditioned space to the next.

4-pipe distribution systems require additional valves. If piped, as shown in Figure 5-45, each fan coil or air handler requires two zone valves, two balancing valves, two purging valves, and one **diverter valve**. The zone valves and diverter valve are electrically operated by a control system such that only heated water, or chilled water, is allowed to pass through the coil depending on its current operating mode. The balancing valves are set to allow the proper flow rate through the coil, depending on it operating mode. The purging valves provide a way to quickly remove air from each fan coil or air handler. They also provide a positive shut off from the risers, if a fan coil or air hander has to be removed for maintenance.

INTERFACING TO A FORCED-AIR SYSTEM

Although most of the systems discussed in this text use hydronic distribution systems, there are situations where it makes sense to connect a hydronic-based renewable energy subsystem to a forced-air distribution system. This can reduce installation cost in retrofit situations such as when a solar combisystem is added to the home that has a properly designed and functioning forced-air duct system supplied by a furnace or heat pump. Figure 5-46 shows how a hot water coil can be mounted within the plenum of the air-handling device (furnace or heat-pump indoor unit).

The coil is mounted in the discharge plenum to avoid heating the components inside the air handler, or furnace. *Most manufacturers will not warrant a furnace or heat pump that receives preheated air above a specific temperature.*

The check valve installed in the circulator supplying the coil prevents reverse heat migration from the furnace to the thermal storage tank when the latter has cooled down and the furnace's burner or heat pump is producing heat.

An air temperature sensor is installed in the discharge plenum, downstream of the coil. This sensor is wired to the controller operating the 3-way motorized mixing valve and allows that valve to regulate the temperature of the supply air stream. This prevents the possibility of very hot water from a thermal storage tank heated by solar collectors or a wood-fired boiler from being sent directly to the plenum coil. Without this regulation, there could be bursts of very hot air delivered directly to the heated space.

The plenum coil should be specified with as many tube passes as possible. This lowers the required water temperature required to meet the design heating load. Coil manufacturers can size coils for the required rates of heat transfer at specified entering air and water temperatures and specified flow rates.

Figure 5-47 shows an example of a "10-row coil." This name implies that fluid makes 10 passes through copper tubing that's routed through the coil's aluminum fins. The coil shown in Figure 5-47 has 16 parallel circuits of small-diameter copper tubing between the supply header and the return header. This coil would yield excellent heat transfer at relatively low flow rates. However, it will also create more static pressure loss as air passes through the relatively "deep" fins. Designers need to assess this static pressure loss and be sure the blower in the air handler can accommodate it, while still providing the necessary air-flow rate.

This type of system may not create supply air temperatures as high as those created by a furnace. This is especially true if the storage tank that supplies

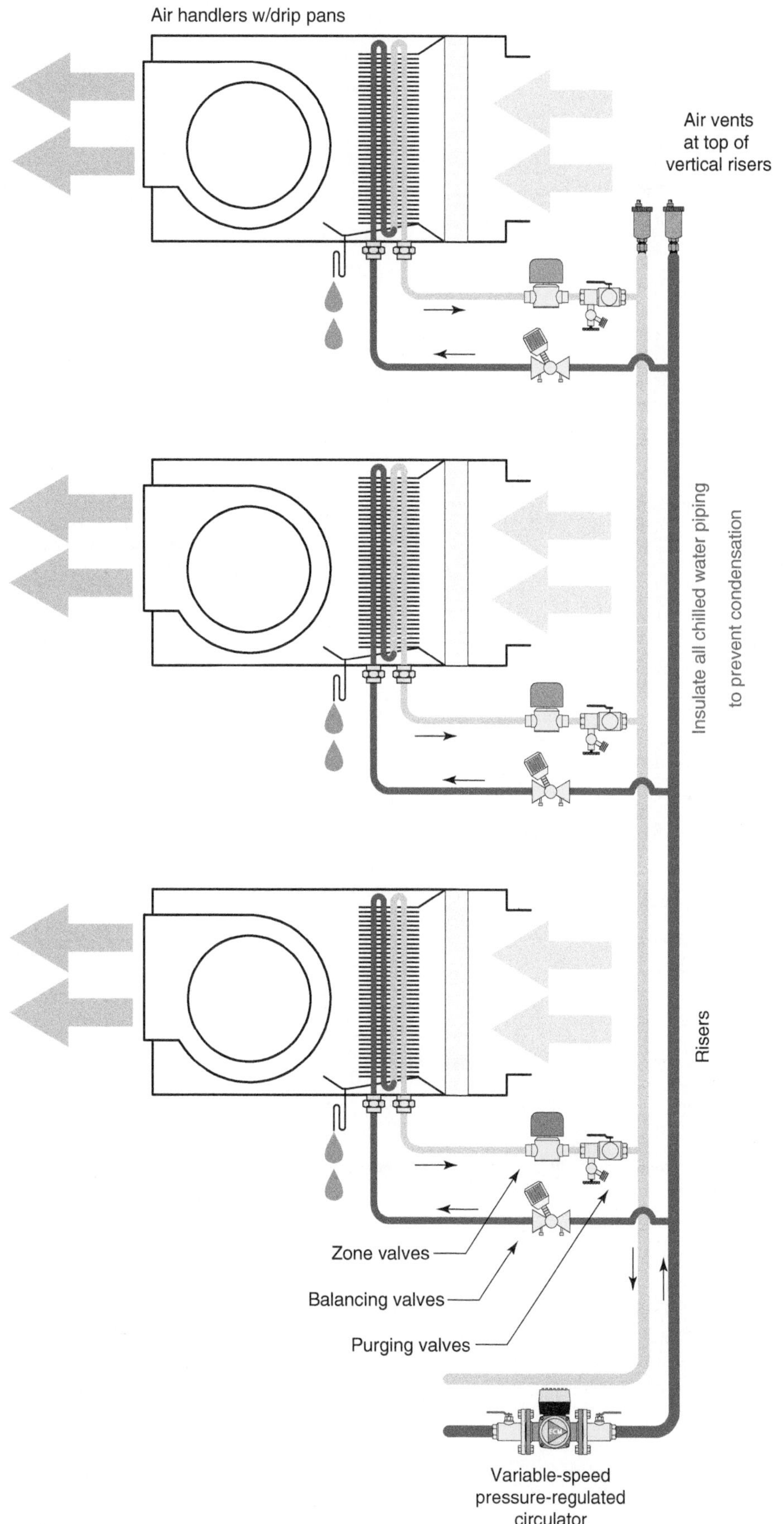

Figure 5-44 Example of a 2-pipe distribution system that operates all fan coils in either heating or cooling mode at the same time. Shown in cooling mode.

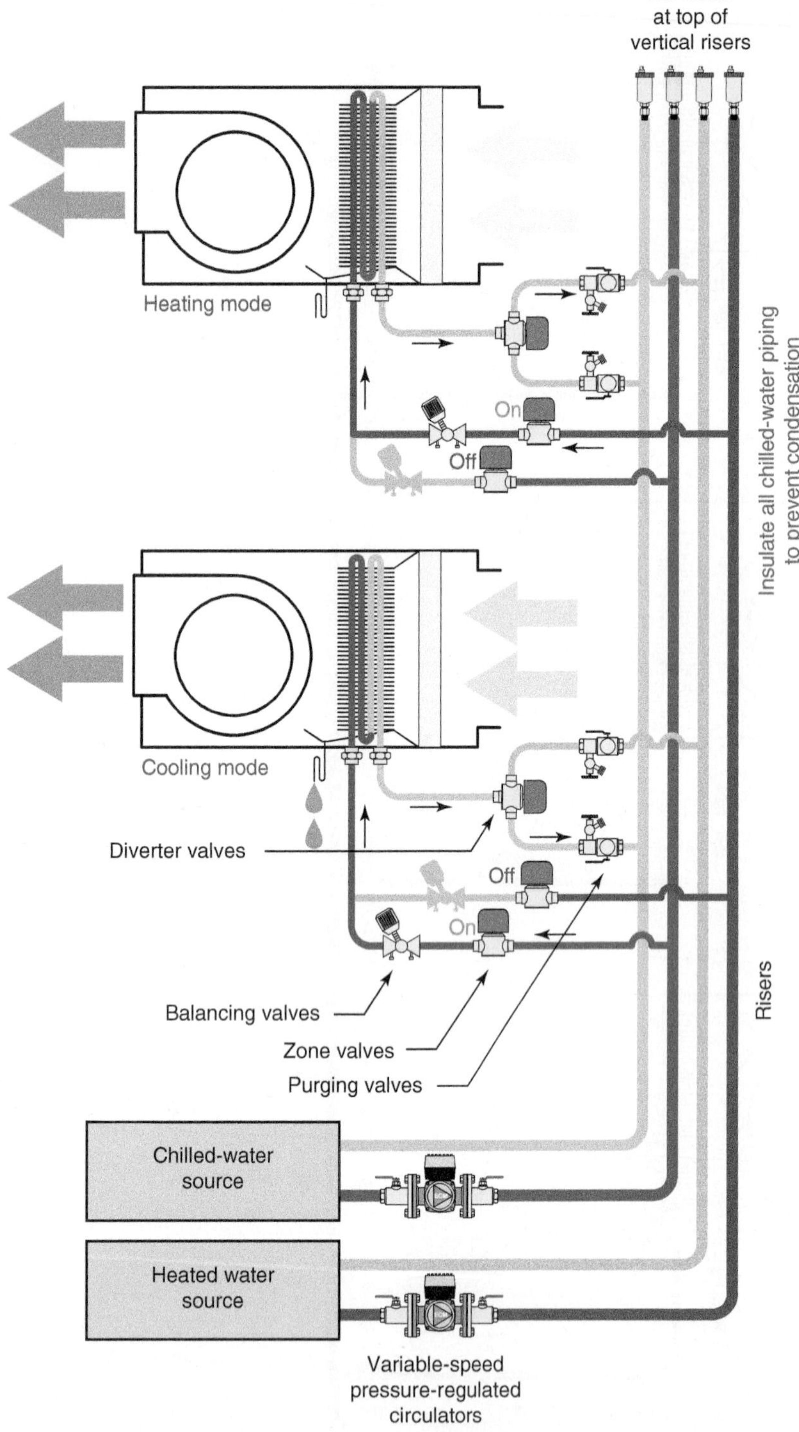

Figure 5-45 Example of a 4-pipe reverse return piping system in which any fan coil can operate in either heating or cooling mode independently of the others.

heated water to the plenum coil is allowed to operate down to temperatures in the range of 100 °F or less. To prevent the possibility of uncomfortable drafts, the **registers** and **diffusers** used for such a system should be carefully sized and placed so that the air they discharge will be mixed with room air above the **occupied zone**. Vertical air velocities into the occupied zones should be minimized to avoid drafts.

5.5 Hydronic Radiant Panel Heating

Hydronic **radiant panel** heating is becoming an increasingly popular option among consumers in North America. This form of heating has an established record of providing superior comfort as well as several other benefits.

Several types of hydronic radiant panels can produce excellent comfort while operating at low supply water temperatures. Such panels are well suited for use with renewable energy heat sources. This section provides an overview of **radiant *floor* panels**. The next section exams hydronic radiant *wall* and *ceiling* panels.

Because of the breadth of this topic, this section and the next cannot provide full design and installation details for all types of hydronic radiant panel options. However, these sections do give sufficient information to guide decision making regarding the application of radiant panels in combination with renewable energy heat sources. More extensive design and installation information is available in chapter 10 of *Modern Hydronic Heating,* 3rd edition.

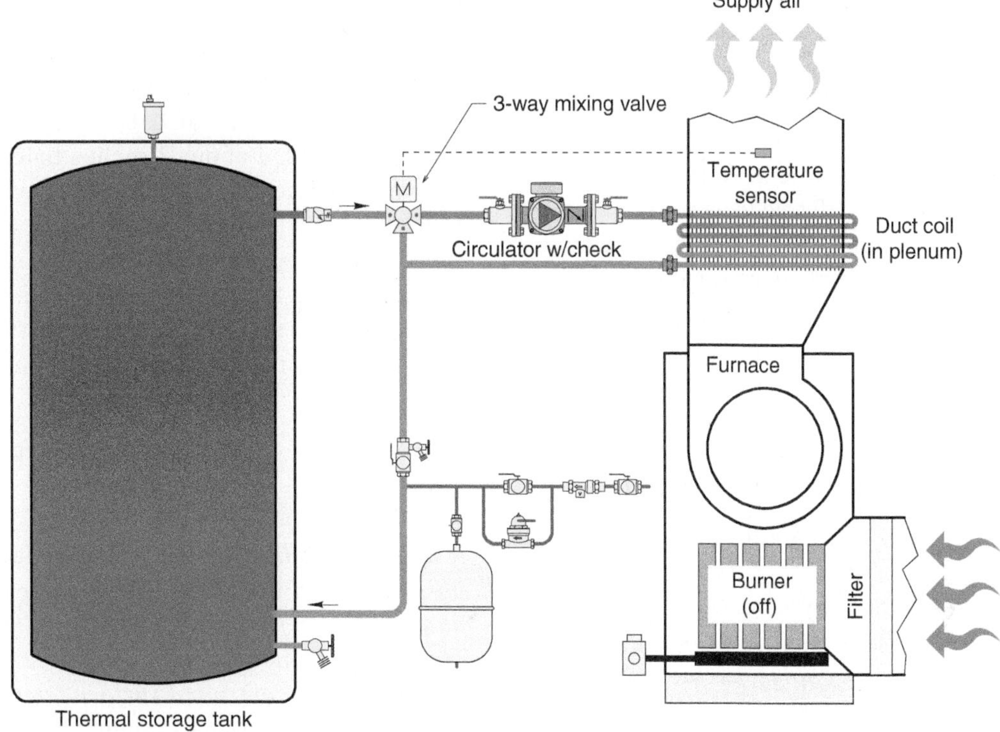

Figure 5-46 | Mounting a hydronic heating coil within the plenum of a forced air furnace.

Figure 5-47 | Example of a 10-row water-to-air heat exchanger coil. This coil has 16 parallel tubing circuits between its supply and return header. Note the reverse return connections on these headers.

WHAT IS RADIANT HEATING?

***Radiant heating** is the process of transferring thermal energy from one object to another by **thermal radiation**.* Chapter 1 briefly described thermal radiation as one of the three modes of heat transfer, along with conduction and convection. Whenever two surfaces are "within sight of each other" and are at different temperatures, thermal radiation travels from the warmer surface to the cooler surface. The rate of energy flow between these surfaces depends on the difference in their temperatures, the distance between the surfaces, the angle between the surfaces, and an optical property of the surfaces, known as emissivity.

All thermal radiation emitted by surfaces at temperatures lower than approximately 970 ºF will be in the **infrared** portion of the **electromagnetic spectrum** and as such cannot be seen by the human eye. All radiant panels discussed in this chapter operate well below this temperature and thus only emit infrared thermal radiation.

Other than the fact that it cannot be seen, infrared thermal radiation behaves similarly to visible light. It travels away from the emitting surface at the speed of light (about 186,000 miles per second) and can be partially reflected by some surfaces. *Like visible light, infrared thermal radiation travels equally well in any direction. This allows radiant energy to be delivered from heated walls and ceilings as well as from heated floors.*

Like other forms of electromagnetic radiation, thermal radiation does not require a media such as a solid, liquid, or gas to move energy from one location to another. Within a room, infrared thermal radiation emitted by a heated surface passes through the air with virtually no absorption of the energy it contains. Instead, the radiation is absorbed by the objects in the room. It is therefore accurate to state that thermal radiation (more commonly referred to as **radiant heat**) warms the objects in a room rather than directly heating the air. This difference is largely what separates radiant heating from convective heating. If the absorbed thermal radiation raises the surface temperature of an object above the room air temperature, some heat will be convected to the room air. Some heat may also be conducted deeper into the object if its interior temperature is lower than its surface temperature.

It is also important to understand that the thermal radiation emitter by a typical radiant panel is in no way unhealthy or harmful. *Thermal radiation is constantly emitted from our skin and clothing surfaces to any cooler surfaces around us.* A large portion of the heat generated through metabolism is released from our bodies by infrared thermal radiation. Evidence of this can be seen in **thermographic image** shown in Figure 5-48.

This image was taken by a thermographic camera that detects infrared radiation rather than visible light. The colors shown are based on the temperature scale seen at the right side of the images. Notice the relatively warm surfaces of exposed facial skin relative to body surfaces covered by clothing. Both skin and clothing surfaces are transferring thermal radiation to surrounding cooler surfaces. The body's ability to reject heat through thermal radiation greatly affects thermal comfort.

Figure 5-49 shows another thermographic image; in this case of a heated floor slab with tubing embedded about 2 inches below its surface. Thermographic cameras detect the invisible infrared radiation emitted by the surface and translate the detected wavelengths of this radiation into corresponding temperatures. They then "map" this temperature data into visible colors and display these colors as an output image.

WHAT IS A HYDRONIC RADIANT PANEL?

*A **hydronic radiant panel** is* any surface *warmed by passing heated water through tubing embedded in or attached to that surface that releases at least 50 percent of that heat to its surroundings as thermal radiation and that has a controlled surface temperature under 300 °F.* The heated water is simply the material used to deliver heat to a hydronic radiant panel.

Once heat has been transferred to the materials that make up a radiant panel, that panel's shape, orientation, surface temperature, surface properties, and surroundings determine its rate of heat output by both thermal radiation and convection.

The best-known type of hydronic radiant panel is a heated floor. Currently, floors account for over 90 percent of all hydronic radiant panel installations. However, *radiant panel heating is not limited to*

Figure 5-48 Thermal radiation is constantly emitted by skin and clothing surfaces to any surround surfaces. The colors correspond to the temperature scale at right of image.

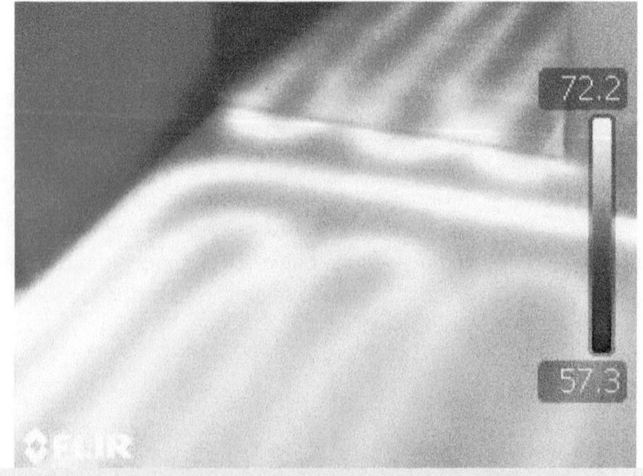

Figure 5-49 Thermographic image of a heated floor slab reveals tubing carrying heated water, and embedded about 2 inches below the top of the slab. The colors correspond to the temperature scale at right of image.

floors. There are several established methods of incorporating hydronic tubing into walls and ceilings. *Heating professionals should become familiar with the strengths and limitations of* all *radiant panels and apply each type where it holds advantages.*

The following radiant floor panel systems will be discussed in this chapter:

- Heated slab-on-grade floors
- Thin-slab floor systems
- Above-floor tube-and-plate system

These panels are best suited for use with low-temperature water and are thus most compatible with renewable energy heat sources.

5.6 Heated Slab-on-Grade Floors

Because a concrete slab is already part of many buildings, converting it into a radiant panel is straightforward and usually less expensive in comparison to other methods of floor heating. Heating concrete slab floors using embedded tubing is a technology that dates back well over a century. Early systems used steel, wrought iron, and copper pipe. An example of an early installation using metal tubing is shown in Figure 5-50.

Although many of these early installations were successful from the standpoint of comfort, issues developed with the metal piping's ability to survive the thermal stresses and potential chemical interactions within the concrete. Many early systems using metal piping eventually developed leaks. Although occupants enjoyed the comfort, they obviously didn't want to deal with leaks. Such leaks became the "Achilles heel" that led to a loss of confidence in such systems, both among consumers and design professionals. North American interest in radiant floor heating steadily declined through the 1970s and new installations of such systems, using metal piping, became very limited.

Figure 5-50 | An early application of slab-on-grade floor heating using metal tubing.

In the early 1980s, the availability of cross-linked polyethylene tubing (aka **PEX**) from Europe, began to reverse that trend. The reliability and installation ease offered by PEX was the catalyst for rekindled interest in radiant floor heating. Today, radiant floor heating is a widely recognized technology. Many North American companies offer products for creating radiant floor heating systems. Such systems, when properly design and installed, have earned back their reputation for superior comfort, energy efficiency, and reliability.

Heated floor slabs using modern materials and installation methods are suitable, under the right circumstances, for use with renewable energy heat sources. Successful applications require floors that can operate at low supply water temperatures and load characteristics that are compatible with the high thermal mass of concrete slabs.

ACHIEVING LOW SUPPLY WATER TEMPERATURE

Relatively close tube spacing and low-finish floor resistance are both vitally important when designing heated floor slabs that need to operate at low supply water temperatures. The graph in Figure 5-51 shows **upward heat output** (in Btu/hr/ft^2) from a heated slab based on tube spacing of 6 inches and 12 inches, and for finish floor resistances ranging from 0 to 1.0 (ºF·hr·ft^2/Btu). The steeper the line, the lower the supply water temperature will be for a given rate of heat output and given room temperature.

Figure 5-51 can be used to determine the **driving ΔT** required for a given rate of heat output from the slab. The driving ΔT is the difference between the average fluid temperature in the floor heating circuit and the room air temperature. For example, to achieve an upward heat output of 20 Btu/hr/ft^2 from a slab with no covering (e.g., Rff = 0) and 6-inch tube spacing, the driving ΔT has to be 17.5 ºF. Thus, in a room maintained at 70 ºF, the average water temperature in the circuit needs to be 87.5 ºF. The *supply* water temperature to this circuit would likely

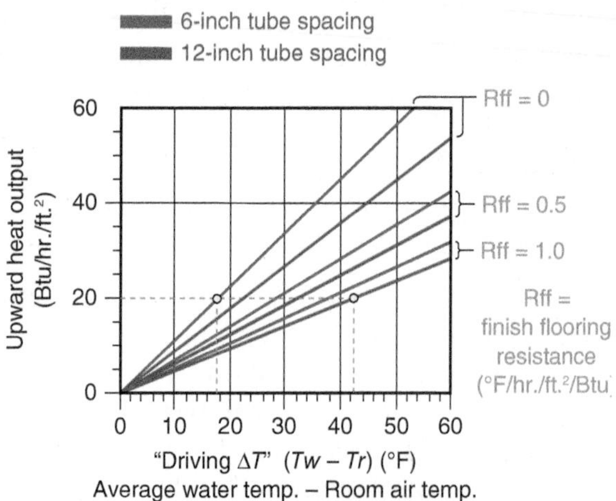

Figure 5-51 | Upward heat output of heated slab-on-grade for different tube spacing and finish flooring resistances.

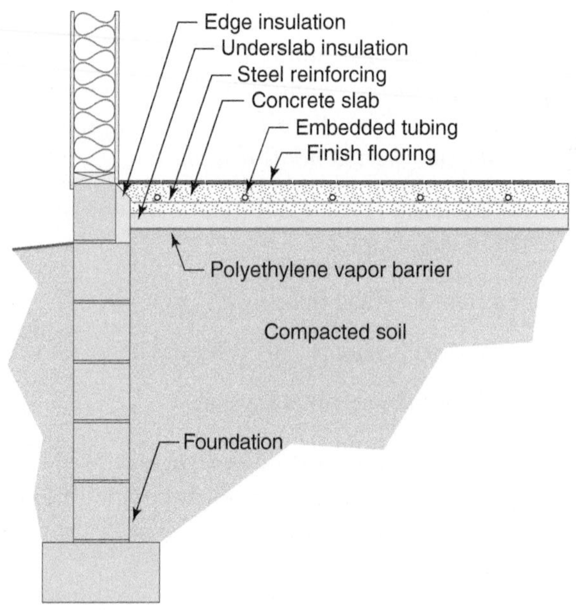

Figure 5-52 | A cross section of a heated slab-on-grade floor showing tubing at mid-depth of slab and good underside insulation.

be 7 to 10 °F *above* the average water temperature. In this case, the supply water temperature would be in the range of 95–98 °F. This is a relatively low temperature that could be supplied by all of the renewable energy heat sources discussed in this text.

For comparison, consider supplying the same 20 Btu/hr/ft² load using a heated floor slab with 12-inch tube spacing and a finish floor resistance of 1.0 (°F·hr·ft²/Btu). Figure 5-51 shows that the driving ΔT must now be 42.5 °F. The average circuit water temperature required to maintain a room temperature of 70 °F would be 70 + 42.5 = 112.5 °F, and the supply temperature likely in the range of 120–123 °F. These higher temperatures are at, and in some cases slightly above, the author's suggested upper limit of 120 °F for supply water temperature under design load conditions. This situation would be a "borderline" application. However, the situation could be improved by using closer tube spacing to achieve the same upward heat output at a lower supply water temperature or by decreasing the room's heat load so that the required upward heat output is reduced. Changing the floor covering to one having lower thermal resistance would also decrease the required supply water temperature.

Figure 5-52 shows a cross section for a heated floor slab. Notice that the tubing has been placed at approximately mid-depth within the slab and that the **underside insulation** and **edge insulation** of the slab are well insulated (R-10 °F·hr·ft²/Btu minimum). *Both of these details are imperative in achieving good low-temperature performance.*

THERMAL-MASS CONSIDERATIONS

Heated slab-on-grade floors have very high **thermal mass** in comparison to other hydronic heating emitters. For example, the thermal mass of a typical 4-inch thick concrete slab is over 200 times greater than that of the low-mass panel radiators discussed earlier in this chapter. The greater the thermal mass, the slower the response of heat emitter, both in warming up and in cooling down.

In some situations, high thermal mass is beneficial. For example, when a heated slab floor is used in a garage and the overhead doors of that garage are opened for a few minutes during cold weather, the heated slab can quickly increase its rate of heat output to help counteract the influx of cold air across the floor. Once the doors are closed, the slab's thermal mass helps quickly restore normal indoor comfort conditions. High thermal mass can also maintain reasonable comfort within a building for several hours without additional heat input, such as when a winter storm causes a power outage.

However, high thermal mass can also be a significant *hindrance* in buildings that are both well insulated and likely to experience significant and "unscheduled" internal heat gains from sunlight, occupants, or equipment.

Consider, for example, what could happen when hydronic tubing is installed in the floor slab of a passive solar house, with the intent of providing "auxiliary" heating on cold and cloudy days. Although it seems reasonable, this approach is likely to cause significant overheating whenever a cold night is followed by a sunny morning. The problem arises because the floor slab is often maintained at an elevated temperature (typically 75 °F to 85 °F at the upper surface) by the auxiliary heating system during the late night and early morning hours of a cold winter day. When solar gains become significant the following morning, the slab is already "filled" with stored heat, and thus very limited in its ability to accept further heat from the impinging sunlight. This causes the building's air temperature to rise rapidly. It doesn't take long before the occupants start opening windows to purge the excess heat in an attempt to maintain comfort. The result is that much of the passive solar gains are lost through ventilation. Even worse, the building's cooling system may turn on and thus create a significant electrical load. To avoid such situations, *heated floor slabs should not be used in buildings where significant internal heat gains can quickly arise. Heated floor slabs are also not recommended in buildings where frequent setbacks of desired interior temperature are likely.*

WHY DON'T THE FLOORS FEEL WARM?

Experience has shown that the primary reason consumers choose radiant floor heating is comfort. The experience of a putting bare feet on a warm floor is hard to beat, especially if those feet are cold to begin with.

When houses and other buildings had design heat losses in the range of 25 to 40 Btu/hr/ft^2, it was necessary for the average floor surface temperature to be in the range of 83 °F to 90 °F so that heat output from the floor could match the rate of building heat loss. These floor surface temperatures feel very comfortable to bare feet. However, the design heat loss of some well-insulated new homes may only be in the range of 10 to 15 Btu/hr/ft^2. This significantly reduces the average floor surface temperature necessary to meet the load.

Equation 5-11 can be used to estimate the **average floor surface temperature**, based on the rate at which it delivers heat and the room air temperature.

(Equation 5-11)

$$T_s = \frac{q}{2} + T_{air}$$

where:

T_s = average floor surface temperature (°F)

q = upward heat output (Btu/hr/ft^2)

T_{air} = room air temperature (°F)

Example 5.8: A room measures 12 feet by 18.5 feet. Its design heat loss is 2,000 Btu/hr. Assuming 90 percent of the room's floor area is heated, determine the average floor surface temperature required to maintain the room at 70 °F air temperature on a design day.

Solution: Start by calculating the required upward heat output from the floor:

$$q = \frac{\text{design load}}{\text{heated floor area}} = \frac{2000\frac{\text{Btu}}{\text{hr}}}{12\text{ ft} \times 18.5\text{ ft} \times 0.9}$$
$$= 10\frac{\text{Btu}}{\text{hr}\cdot\text{ft}^2}$$

Next, use Equation 5-11 to calculate the required average floor surface temperature:

$$T_2 = \frac{q}{2} + T_{air} = \frac{10}{2} + 70 = 75\text{ °F}$$

Discussion: This temperature is a few degrees *lower* than normal skin temperature for hands and feet. Thus, placing a thermally comfortable hand or foot on a surface with an average surface temperature of 75 °F is likely to result in conduction heat transfer *from* that hand or foot *to* the floor surface. Furthermore, the calculated average floor surface temperature of 75 °F is based on design load conditions. During partial load conditions, the average floor surface temperature will be lower. However, to its credit, the floor is still warmer than it would be if the room were heated by convective heat emitters, such as a fin-tube baseboard.

Figure 5-53 shows an infrared thermograph of a radiantly heated floor. This image shows embedded heating tubing within the slab. The footprint near the center of the image was created by placing a thermally comfortable foot against the floor for one minute and then removing it. The image clearly shows the residual heat absorbed *from the foot into the floor*. This implies that a floor surface that is only a few degrees above room air temperature will often be absorbing heat by

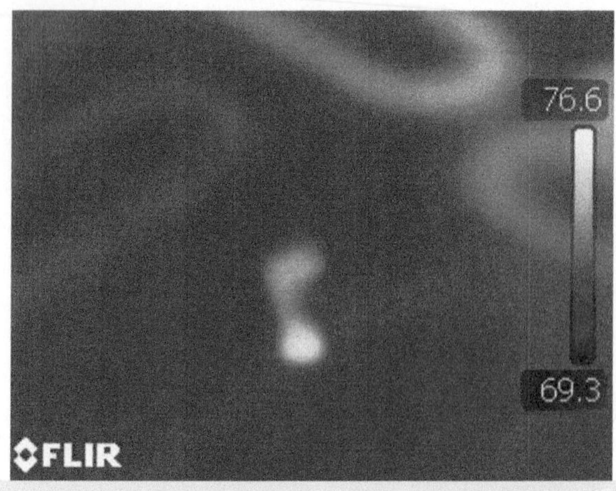

Figure 5-53 The footprint shows that heat was absorbed from a thermally comfortable foot into the heated floor slab.

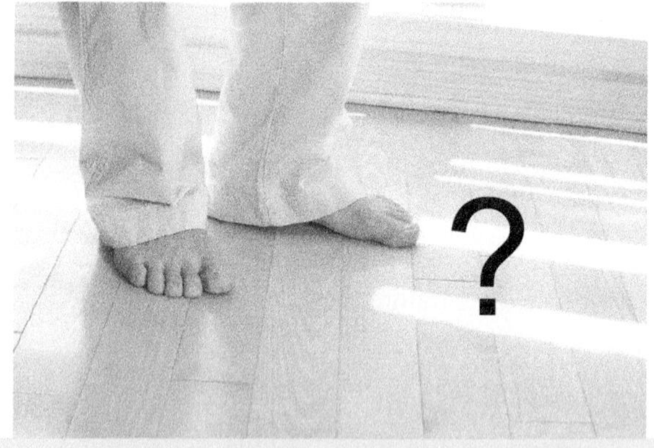

Figure 5-54 Be sure that occupants understand that heated floors in a low-energy-use building will usually *not* feel warm to the touch.

conduction from bare skin in contact with it. This may result in the floor feeling slightly cool to the touch, even though it is releasing sufficient heat to maintain the room at a normal comfort temperature.

Forcing the floor to operate at higher temperatures would quickly overheat the space. Windows would be opened and energy would be needlessly wasted.

This issue comes down to **occupant expectations**. If the occupants were informed that the floors would not necessarily feel warm when touched, even though interior air temperature would be maintained at the desired comfort level, and if they understood and agreed to this operating condition, there should not be any unfulfilled expectations. However, if the occupants expected their heated floors to feel warm to the touch, they are likely to be disappointed. The fact that the renewable energy heat source supplying this floor is operating at high thermal efficiency will probably not be much consolation. Therefore, it is essential to discuss occupant expectations regarding radiant floor heating, especially in homes with low heat loss. For some, the fact that comfort is being maintained and the heat source is operating at good efficiency may be sufficient justification for floor heating. For others, the absence of floors that feel warm to the touch may be reason enough to consider other types of heat emitters.

SIMPLIFIED DESIGN OF HEATED SLAB-ON-GRADE FLOORS

The following procedure is a simplified means of determining required tube spacing within each room of a heated slab-on-grade building as well as the supply water temperature required under design load conditions. This procedure combines the information presented in Figure 5-51 with the design heat loss of each room. It is also based on the preferred method of room-by-room circuit design.

Keep in mind that this is a *simplified procedure* and is specific to slab-on-grade radiant panels. Chapter 10 in *Modern Hydronic Heating*, 3rd edition, provides a more extensive discussion of sizing heated slab-on-grade floors.

STEP 1. Establish a **target supply water temperature** for design load conditions. The author recommends that this temperature not exceed 120 ºF in any system using a renewable energy heat source. Lower temperatures are always preferred when possible. For example, a target supply water temperature of 100 ºF would be appropriate for buildings with low heat loss. A temperature of 110 ºF would be reasonable for a building with average heat loss. Keep in mind that this temperature can be increased or decreased if necessary based on subsequent calculations.

STEP 2. Divide the design heat loss of each room by the floor area under which tubing can be placed within that room. That area is called **available floor area**. It should exclude areas covered by cabinets or other built-in objects. The resulting number is the **upward heat flux** required at design load conditions.

STEP 3. Determine the R-value of the finish floor covering (if any) used in each room. Figure 5-55 lists several common floor coverings with their associated R-value.

Floor Covering: R-values per Inch or per thickness given			
Description	R-value Per Inch h·ft²·°F/Btu	Typical Thickness inch	Typical R-value h·ft²·°F/Btu
Plywood/Wood Panels	1.08	3/4″	0.81
Plywood (Douglas Fir)	1.58	1/2″	0.79
Plywood (Douglas Fir)	1.58	5/8″	0.99
OSB	1.40	3/4″	1.05
Softwood	1.10	3/4″	0.83
Sheet Vinyl	1.60	1/8″	0.20
Vinyl Composition Tile (VCT)	1.60	1/8″	0.20
Linoleum	2.04	1/4″	0.51
Linoleum	2.04	1/8″	0.26
Dense Rubber Flooring	1.30	21/64″	0.25
Recycled Rubber Flooring	2.20	1/2″	1.10
Cement Board	0.14	1/4″	0.04
Concrete (40 lb/ft³ and k = 1.3)	0.78	1.5″	1.16
Concrete (120 lb/ft³ and k = 8)	0.13	4″	0.50
Concrete, (bare) no covering	0.00	0	0.00
Cork	3.00	3/8″	1.13
Cork/MDF/Laminate	2.35	1/2″	1.18
Brick	2.25	1/2″	3.38
Marble	0.90	1/2″	0.45
Ceramic Tile	1.00	1/4″	0.25
Thinset Mortar	0.80	1/8″	0.10
Terrazzo	0.08	3/8″	0.03
MDF/Plastic Laminate	1.00	1/2″	0.50
Laminate Floor Pad	1.92	5/32″	0.30
Engineered Wood	1.00	1/4″	0.25
Engineered Wood	1.00	3/8″	0.38
Engineered Wood	1.00	5/8″	0.63
Engineered Wood	1.00	3/4″	0.75
Engineered Wood Flooring Pad	1.60	1/8″	0.20
Engineered Bamboo	0.96	3/4″	0.72
Oak	0.85	3/4″	0.64
Ash	1.00	3/4″	0.75
Maple	1.00	3/4″	0.75
Pine	1.30	3/4″	0.98
Fir	1.20	3/4″	0.90
Carpet Pad/Slab Rubber 33 lb	1.28	1/4″	0.32
Carpet Pad/Slab Rubber 33 lb	1.28	3/8″	0.48
Carpet Pad/Slab Rubber 33 lb	1.28	1/2″	0.64
Carpet Pad/Waffle Rubber 25 lb	2.48	1/4″	0.62
Carpet Pad/Waffle Rubber 25 lb	2.48	1/2″	1.24
Carpet Pad/Frothed Polyurethane 16 lb	3.53	1/8″	0.53
Carpet Pad/Frothed Polyurethane 12.1b	3.48	1/4″	0.87
Carpet Pad/Frothed Polvurethane 10 lb	3.22	3/8″	1.20
Carpet Pad/Frothed Polvurethane 10 lb	3.22	1/2″	1.61
Hair Jute	3.88	1/2″	1.94
Hair Jute	3.88	21/64″	1.25
Synthetic Fibre Pad 20 oz	1.80	15/64″	0.42
Synthetic Fibre Pad 27 oz	1.98	18/64″	0.55
Synthetic Fibre Pad 32 oz	2.10	19/64″	0.63
Synthetic Fibre Pad 40 oz	2.20	11/32″	0.77
Prime Urethane	4.30	21/64″	1.40
Prime Urethane	4.30	1/2″	2.15
Bonded Urethane	4.20	21/64″	1.35
Bonded Urethane	4.20	1/2″	2.10
Carpet	2.80	1/4″	0.70
Carpet	2.80	3/8″	1.05
Carpet	2.80	1/2″	1.40
Carpet	2.80	5/8″	1.75
Carpet	2.80	3/4″	2.10
Wool Carpet	4.20	3/8″	1.58
Wool Carpet	4.20	1/2″	2.10

Courtesy of www.healthyheating.com

Figure 5-55 | R-values of common floor coverings.

The author recommends that finish floor coverings not exceed an R-value of 1.0 (°F·hr·ft²/Btu) whenever a renewable energy heat source is used in the system. This R-value determines which sloping lines should be used on the graph of Figure 5-51. If the R-value of the room's finish floor falls between the values listed for the sloping lines in Figure 5-48, it is acceptable to proportionally interpolate between the slopes and then draw in a new line.

STEP 4. Using the graph in Figure 5-51, draw a horizontal line from the upward heat flux required for the room, across the graph. This line will intersect the sloping line representing the finish floor covering R-value and a tube spacing (either 6 inches or 12 inches). From this intersection, drop a vertical line downward to read the required driving ΔT on the horizontal axis.

STEP 5. The required **average water temperature** for the circuit is found by adding the driving ΔT to the room's air temperature.

STEP 6. Repeat this procedure for each room.

STEP 7. Calculate the *average* of all the circuit average water temperatures. This is called the **system average water temperature**.

STEP 8. Ideally, the average water temperatures for each circuit are within +/− 5 °F of the system average water temperature. This will allow all circuits to be simultaneously supplied at the same supply water temperature. Doing so simplifies design for the remainder of the system. If this criteria is met, proceed to step 11. If this criteria is not met, identify all rooms where the average circuit water temperature is more than 5 °F above or below the system average water temperature. Certain details within these rooms will now be adjusted in an attempt to bring the average circuit water temperature of all circuits within +/− 5 °F of the system average water temperature.

STEP 9. If there are rooms where the circuit average water temperature is more than 5 °F *above* the system average water, the following steps can be taken to reduce circuit average water temperature so that it hopefully falls within 5 °F of the system average water temperature:

- Decrease tube spacing by 3 inches and recalculate the circuit average water temperature for the room. For 9-inch tube spacing, interpolate between the sloping lines representing 6-inch and 12-inch spacing in Figure 5-51.
- Consider changing the room's finish floor covering to one that has a lower R-value.
- Consider changes to the room's thermal envelope that would reduce the room's heat loss.

STEP 10. If there are rooms where the circuit average water temperature is more than 5 °F *below* the system average water, try increasing the tube spacing for that room by 3 inches and then recalculate the circuit average water temperature. See if that temperature now falls within 5 °F of the system average water temperature.

STEP 11. Recalculate a new system average water temperature each time a circuit is adjusted using steps 9 and 10. Repeat this procedure, as necessary, until the circuit average water temperature for each room falls within +/− 5 °F of the system average water temperature. If this cannot be achieved, the rooms with significantly higher or lower circuit average water temperatures will need to be supplied by separate subsystems.

STEP 12. Select a temperature drop for the system under design load conditions. The following are suggested values for this **design ΔT**:

- When the floor needs to be "barefoot friendly," the suggested design ΔT = 15 °F
- When the floor is in a location where "barefoot friendly" is not a major concern, the suggested design ΔT = 20 °F

STEP 13. Add one half of the selected design ΔT to the system average water temperature (see Equation 5-13). This will be the system's **design supply water temperature**:

(Equation 5-13)

$$T_s = T_{ave} + \frac{\Delta T}{2}$$

where:

T_s = design supply water temperature for the system (°F)

ΔT = design ΔT for the system (°F)

T_{ave} = system average water temperature (°F)

STEP 14. Determine the number of circuits in each room. In many cases, a single tubing circuit can be routed from a manifold station, across the floor of a room, and back to the manifold station. However, there are exceptions, which are based on circuit length limitations. Figure 5-56 lists the *approximate* length of tubing required per square foot of

Tube Spacing (inch)	Length of tubing per square foot of floor area (ft)
6	2
9	1.5
12	1

Figure 5-56 | Approximate length of tubing per square foot of heat floor area, based on tube spacing.

Nominal tube size	Maximum circuit length (ft)
3/8″	250
1/2″	300
5/8″	450

Figure 5-57 | Recommended maximum circuit lengths.

floor area for different tube spacings. Keep in mind that this is only approximate because it does not account for the exact lengths of return bends, offsets, or other "non-linear" placement of tubing.

Estimate the amount of tubing within each room by multiplying the heated floor area of the room (e.g., that floor area under which tubing is embedded) by the factor in the right-hand column of Figure 5-56. Add allowances for the tubing required between the manifold station and the room where the circuit is located. These lengths are called **leaders** and should be kept reasonably short wherever possible.

Figure 5-57 lists recommended maximum circuit length for tube sizes of 3/8 inch, 1/2 inch, and 5/8 inch. These lengths are based on limiting the head loss of the circuit and thus avoiding the need for higher power circulators. The lengths include the tubing on the floor within the room, plus any leader lengths required to route the circuit to a manifold station.

If the amount of tubing required within a room exceeds the maximum recommended circuit length in Figure 5-57, it will be necessary to use two or more circuits to supply that room. To determine the number of circuits required, divide the estimated total tube length for the room by the maximum recommend circuit length in Figure 5-57. Then, round up to the next whole number. This will keep all circuits within the room at approximately the same length.

SOFTWARE-ASSISTED DESIGN

It is also possible to use the Hydronic Circuit Simulator module in the Hydronics Design Studio V2.0 software to simulate the performance of radiant floor heating circuits. Figure 5-58 shows a screenshot of this software as it is being configured to simulate six radiant floor heating circuits on a common manifold station.

The Hydronic Circuit Simulator module can be configured with up to 12 circuits on the same manifold station. Each circuit can be configured independently of the others. It can have a specific length, tube size, floor covering, embedment method, and percentage of circuit in floor specification. The latter accounts for the fact that some percentage of a floor heating circuit will be routed as leaders between the floor being heated and the manifold station. As such, it will not add heat to the floor but *will* contribute to the head loss of the circuit. After the individual floor circuits are specified, the software can be configured with a specific circulator, common piping size and length, fluid, and supply water temperature. Figure 5-59 shows the piping schematic that is being modeled in Figure 5-58 along with inputs and selections.

The calculated results include the flow rate in each circuit and its heat output.

This software allows the designer to make modifications to the initially specified system and immediately see the results of those changes. Inputs such as supply water temperature, floor-covering resistance, tube spacing, circuit balancing valve settings, and circulator selection can be quickly altered. When the system has been configured so that the heat outputs of the circuits are reasonably close to the design requirements, that configuration represents an acceptable solution.

TUBING LAYOUT PLAN

The installation of heating tubing within a slab-on-grade floor is an integral part of the construction sequence of a building. It must be preplanned and closely coordinated with the other trades involved in the construction.

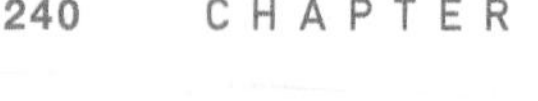

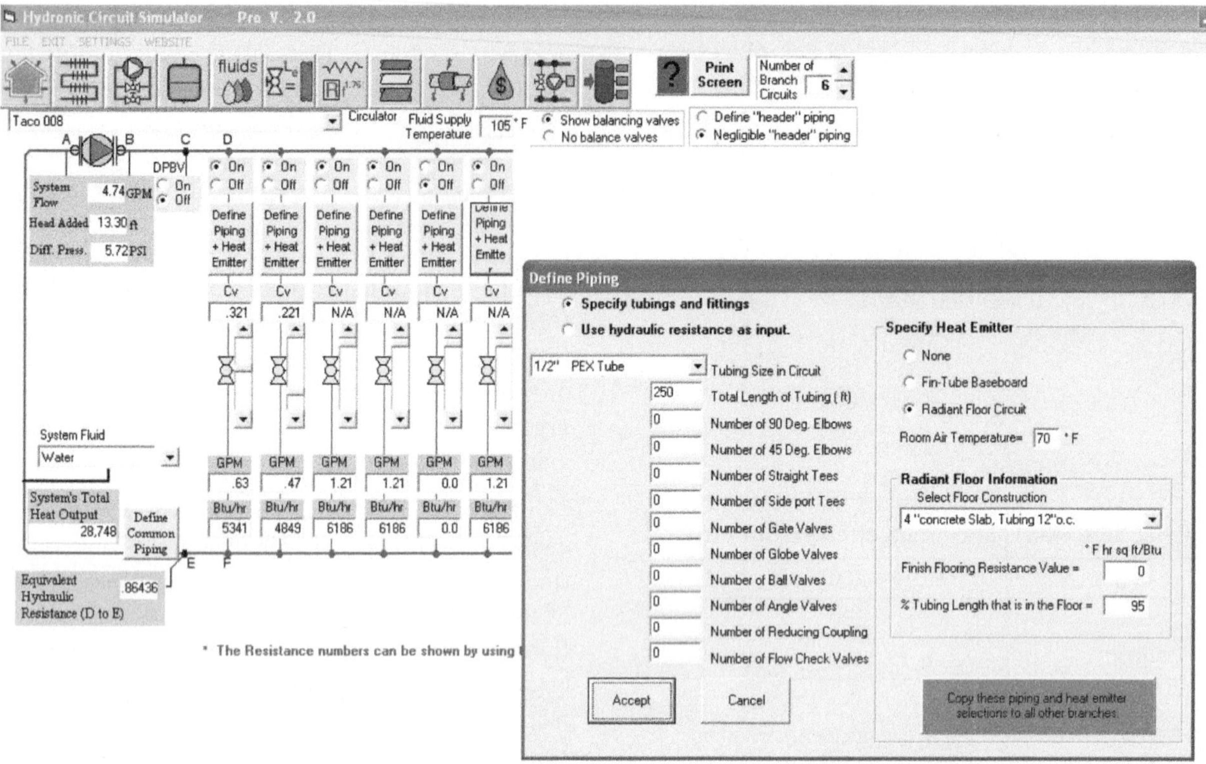

Figure 5-58 | Hydronics Design Studio software as individual radiant floor heating circuits are being specified.

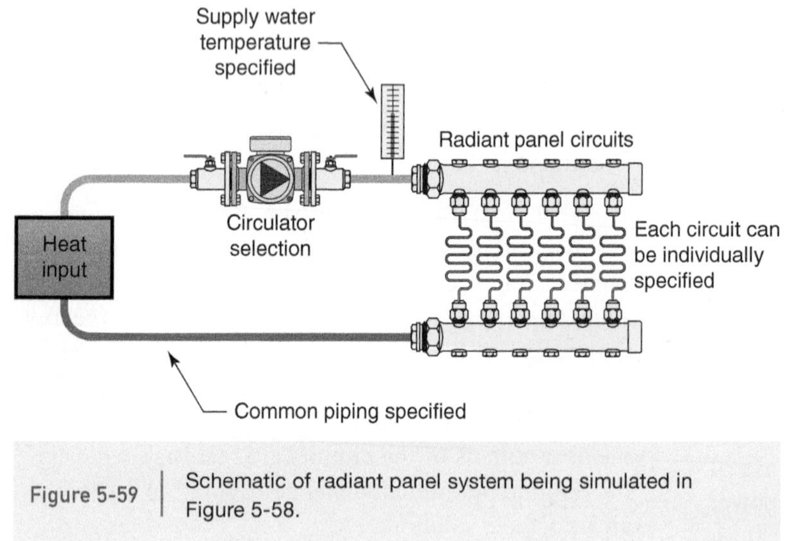

Figure 5-59 | Schematic of radiant panel system being simulated in Figure 5-58.

Part of this preplanning is to develop a **tubing layout plan**, which shows the placement of each floor heating circuit on a scale floor plan of the building. Without such a plan, even an experienced installer can spend hours trying to create properly sized and placed circuits by trial and error. The results are still likely to be less than ideal.

Figure 5-60 shows an example of a tubing layout plan for a slab-on-grade floor.

This type of drawing is best prepared using **CAD software**. Start with a scaled floor plan and add circuits at the appropriate tube spacings. Most CAD software allows the designer to join the various line segments, return bends, and arcs for a given circuit into a single entity often called a **polyline**. The software can then measure the length of the polyline to see if it falls within the recommended maximum circuit length given in Figure 5-57. If not, the circuit should be modified to reduce its length. When all circuits have been drawn, they should be labeled with a name and their length. Using separate colors for each circuit makes it easier to follow their placement on the floor plan. The author suggests adding 10 feet to the measured length of each circuit to account for the vertical risers needed to connect to wall-mounted manifolds and slight variations in tube placement compared to the layout plan. A final detail is to add flow direction arrows to all circuits.

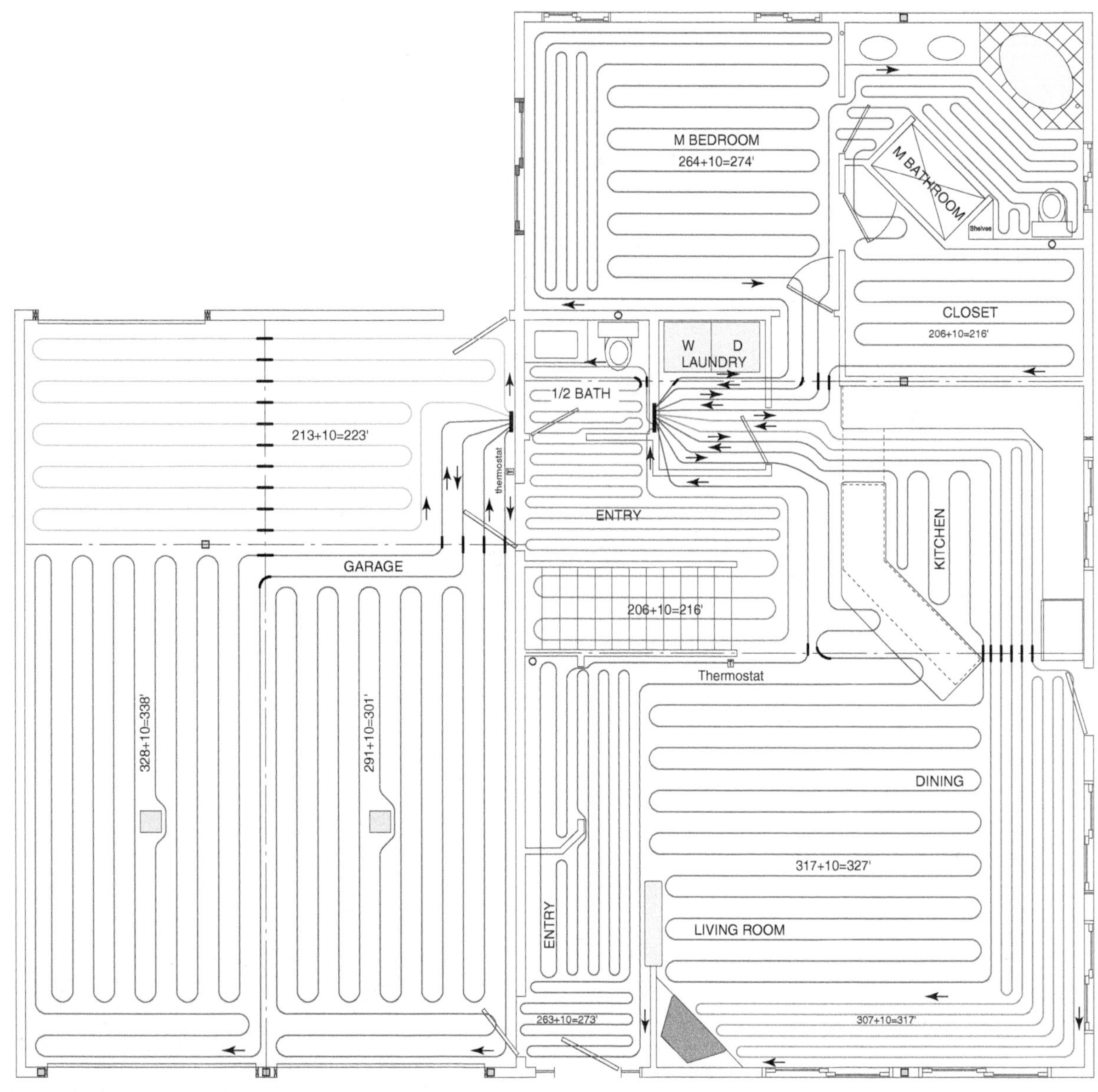

Figure 5-60 | A tubing layout plan prepared using CAD software.

INSTALLATION GUIDELINES FOR HEATED SLAB-ON-GRADE FLOORS

After proper design comes proper installation. Attention to details, such as sufficient **underslab insulation** and vertical positioning of tubing within the slab's thickness, are especially important for heated slabs supplied by renewable energy heat sources.

Figure 5-61 shows a cross section of a heated slab-on-grade floor.

Installation begins with a properly leveled and tamped **subgrade**. Tamping is essential in achieving a stable base for the slab. A level subgrade provides proper support for sheets of rigid foam insulation. A 6-mill polyethylene vapor barrier is placed directly on the tamped subgrade to inhibit upward migration of water vapor during warmer weather. Next comes installation of **extruded polystyrene** underslab and edge insulation. The author recommends a minimum of 2 inches of extruded polystyrene underside and edge insulation under all heated slabs. Greater thicknesses of the same material may be justified in very cold climates or when the goal is to achieve very low building heat loss. However, insulation should not be placed under load bearing footings.

As a minimum, the slab should be reinforced with sheets of **welded wire fabric**. For residential and

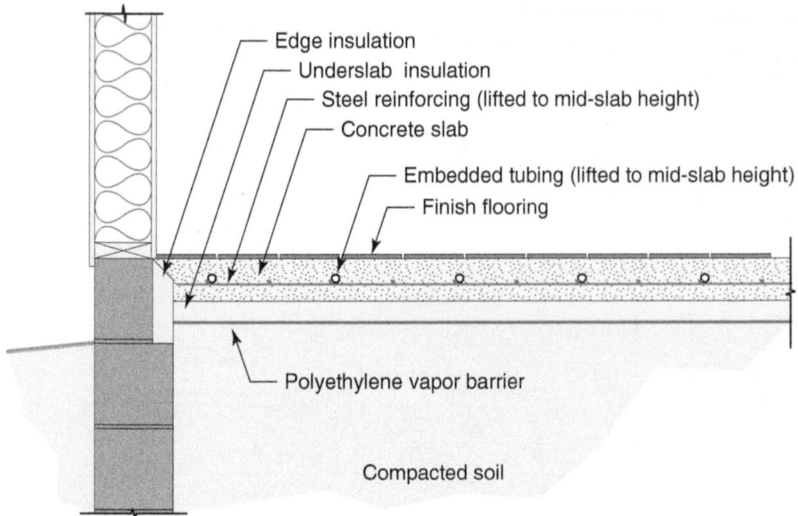

Figure 5-61 | Details for heated slab-on-grade floors.

light commercial buildings, welded wire fabric with a 6-inch grid spacing is common. This material should be laid directly on top of the underslab insulation. All sheets of welded wire fabric should be overlapped by 6 inches and tied together using wire ties. The tubing can then be placed on and secured to the welded wire fabric based on the tubing layout drawing. Several tube-fastening systems are available. The author prefers **wire bag ties** placed at intervals of 2 to 3 feet on straight runs. Corners and U-bends should have two to three fastenings each. All tubing circuits should begin and end at a manifold station, as shown in Figure 5-62.

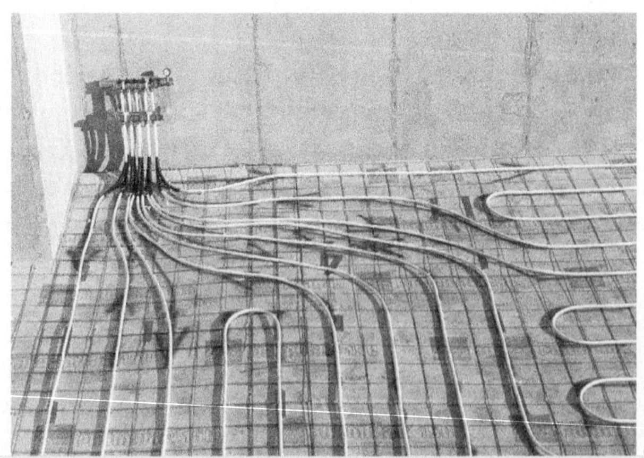

Courtesy of Harvey Youker

Figure 5-62 | All tubing circuits begin and end at a manifold station.

When all tubing circuits have been placed and connected to the manifold, the assembly should be pressure tested using compressed air. This is done by temporarily equipping the manifold station with a **Schrader valve** and pressure gauge, as shown in Figure 5-63. An air compressor is then used to bring the pressure in the assembly up to 75 psi. If any decrease in pressure is noticed, spray a 50 percent mixture of dish detergent and water on all the connections at the manifold station. It's very likely that's where the leak is located. Retighten or replace the fitting to stop the leak.

After all circuits have been placed, several photographs of the tubing installation should be taken. Photos should be taken around each manifold station as well as in any areas where tubing placement is tight or routed around objects such as foundation pads, toilet flanges, or other objects. Digital cameras allow such photos to be quickly gathered and stored on a Flash drive or CD for future reference.

The installation is now ready for concrete placement. There is no single recognized procedure

Figure 5-63 | Temporary installation of pressure gauge and Schrader valve for air pressure testing circuits prior to concrete placement.

for this. From the standpoint of protecting the tubing, the placement procedure used should minimize heavy traffic over the tubing and ensure that the tubing ends up at approximately mid-depth in the slab (other than where it passes beneath control joints).

Wheelbarrow traffic over tubing that is under pressure is generally not a problem; however, care should be taken not to pinch the tubing under the nose bar of the wheelbarrow as it is dumped. So-called power buggies are too heavy to be driven over the tubing and insulation. Likewise, concrete trucks should never be driven over these materials.

On small slabs, the concrete can often be placed directly from the chute of the delivery truck, as seen in Figure 5-64. For larger slabs, the use of a concrete pump truck equipped with an **aerial boom** is an ideal way to efficiently place the concrete with minimal heavy traffic over the tubing.

As the concrete is placed, lift the welded wire reinforcing and attached tubing to approximately mid-depth in the slab (other than at control joints). This is critically important from the standpoint of minimizing the supply water temperature at which the floor can operate. Leaving the tubing at the bottom of a 4-inch thick slab, which must drive heat upward at 15 Btu/hr/ft^2 through an R-0.4 °F·hr·ft^2/covering, will increase the required supply water temperature by 7 °F. This will create a measurable decrease in the efficiency of solar thermal collectors or heat pumps.

The welded-wire reinforcing with attached tubing is pulled upward to approximately one half the slab's eventual thickness using a **lift hook**, as shown in Figure 5-65. Once the coarse (stone) **aggregate** in the concrete has settled under the welded-wire reinforcing, it tends to support it quite well.

Figure 5-64 | Placement of concrete over tubing directly from truck chute.

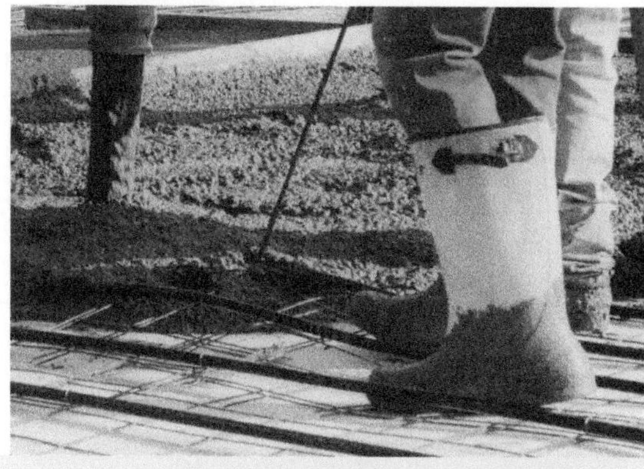

Figure 5-65 | Worker using a lifting hook to pull welded wire reinforcing and attached tubing to approximately mid-slab height. Note that hook is lifting reinforcing wire rather than directly lifting tubing.

After the concrete is placed, it is finished in the usual manner. Special care should be taken not to nick the tubing with trowels where it penetrates the slab under the manifold stations. It is also important to allow the concrete to properly cure, for a minimum of several days, before heated water is circulated through embedded tube circuits.

HEATED SLAB-ON-GRADE FLOORS: DESIGN AND INSTALLATION SUMMARY

The following guidelines summarize the design and installation of heated floor slabs that will be supplied by renewable energy heat sources:

- Do not use heated slab-on-grade floors in spaces subject to significant internal heat gain from sunlight, occupants, or equipment.
- Do not use heated slab-on-grade floors in spaces where frequent thermostat setbacks will be used or where rapid recovery from setback conditions is expected.
- Tube spacing within the slab should not exceed 12 inches. Closure tube spacings of 9 inches and 6 inches can be used to increase the heat output of the floor in areas with higher heat loss or where floors may be frequently wetted from tracked in snow or other activities.
- Try to adjust tube spacing as necessary so that all circuits within the system can operate at the

same supply water temperature under design load conditions. This simplifies system piping and control requirements.

- Always prepare a tubing layout plan based on the building's floor plan and room heat loads before placing tubing.
- All heated slabs should have *minimum* of R-10 rigid extruded polystyrene underside and edge insulation. Insulation should only be omitted under structural columns or footings.
- Tubing and welded wire reinforcing should be lifted to approximately half the slab depth below the surface.
- Bare, painted, or stained slab surfaces are ideal because the finish floor resistance is essentially zero.
- Other floor finishes should have a total R-value of 1.0 or less.

5.7 Heated Thin Slabs

There are many buildings where radiant floor heating is desirable but slab-on-grade construction is not used. One of the most common is a wood-framed floor deck in a residential or light commercial building.

One option is to install a **concrete thin slab** radiant panel over the **subfloor**. A cross-sectional drawing of this approach is shown in Figure 5-66.

Thin-slab systems use the same type of PEX or PEX-AL-PEX tubing as slab-on-grade systems. The tubing is fastened to the wood subfloor and then covered with a special concrete mix. Underside insulation is installed between the floor framing under the subfloor.

As with slab-on-grade systems, the concrete thin slab provides an effective "thermal wick" allowing heat

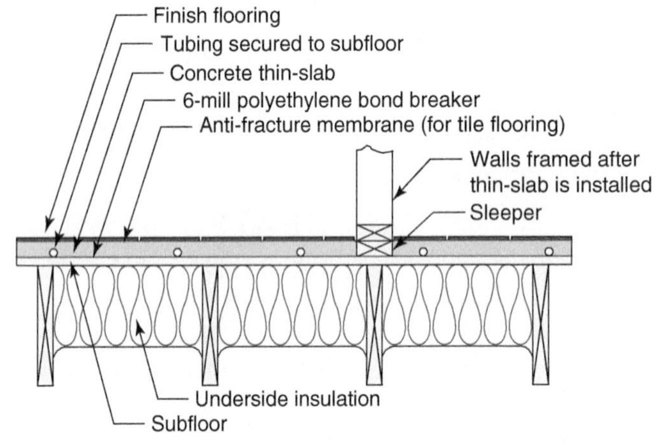

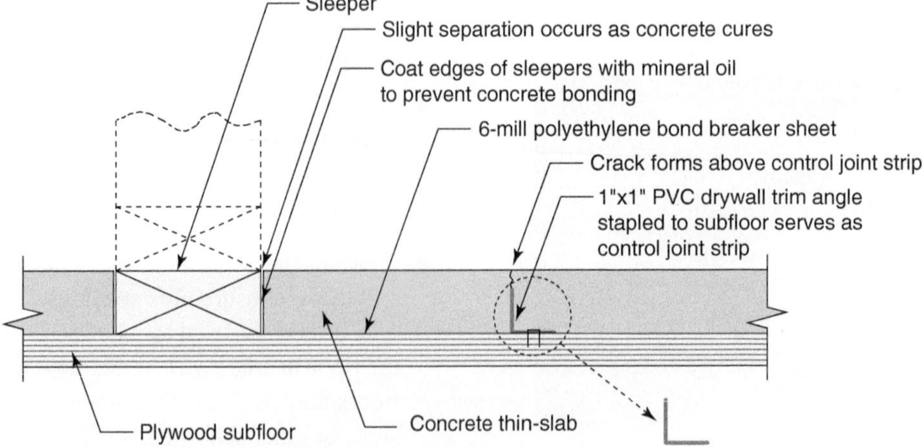

Figure 5-66 | Cross section of a concrete thin-slab radiant floor panel.

to diffuse laterally outward away from the tubing. The slab also provides moderate thermal mass to stabilize heat delivery.

Because the slab is thinner than a typical slab-on-grade floor, lateral heat diffusion is slightly less efficient. Slightly higher average water temperatures are therefore required for the same heat output relative to a heated slab-on-grade floor.

Concrete thin slab systems bring a number of issues into the planning process, not the least of which is the added weight of the slab on the floor deck. A typical 1.5-inch-thick concrete thin slab adds about 18 pounds per square foot to the **dead loading** of a floor. The floor-framing system must be capable of supporting this added load while remaining within code-mandated stress and deflection limits. Although significant, this extra weight can usually be accommodated by adjusting the spacing, depth, or width of floor framing. A competent structural designer or engineer should assess the necessary framing to support the added load. An often-cited **maximum deflection criteria** for such floors is 1/600th of the floor's clear span under full live loading.

Another consideration is the added height of the thin-slab. In most installations, thin slabs add 1.5 inches to the height of the floor deck. This affects the height of window and door **rough openings**, stair riser heights, and rough-in heights for closet flanges. Such adjustments are easily made if the thin-slab is planned for when the building is designed. However, these adjustments can be more difficult or even impossible if the decision to use a thin-slab system is made after the building is framed.

Figure 5-67 shows tubing fastened to a floor deck, awaiting placement of a concrete thin slab.

Figure 5-67 | Tubing installed over wooden floor deck awaiting placement of 1.5-inch thick concrete.

A concrete thin slab installation begins with a tubing layout plan, similar to that shown for a heated slab on grade system in Figure 5-60. Such a plan greatly speeds installation. It also ensures that circuit lengths, and proper circuit locations, are known before tubing gets stapled in place.

Onsite work begins with marking the location of all interior and exterior walls on the completed floor deck. The deck is then covered with 6-mill translucent polyethylene sheeting. This material acts as a **bond breaker** between the wooden floor deck, and the concrete slab. The absence of a bond between these materials reduces shrinkage cracking of the concrete. The next step is to install 2 × 4 or 2 × 6 lumber wherever there is an interior partition or exterior wall. This lumber establishes the thickness of the thin-slab. The tubing is then fastened in place using a specialized pneumatic stapler. Once all tubing circuits are installed and connected to their associated manifold, they are pressure tested with compressed air to verify that there are no leaks. **Control joint strips** are added where required. These details have all been completed in the photo shown in Figure 5-67. The final step is to place the concrete and screed it level with the top of the 2 × 4 or 2 × 6 lumber. The concrete is then troweled smooth. Framing can usually resume the following day.

The mix proportions suggested for a concrete thin slab are given in Figure 5-68.

This mix can be supplied by most batch plants. The small #1A "pea stone" aggregate along with the **superplasticizer** and **water-reducing agent** yield good flow characteristics and allow the concrete to fully encase the tubing. This is accomplished *without adding access water* that can cause shrinkage cracks, loss of strength, and dusty finished surfaces. The **Fibermesh®** additive provides tensile reinforcement without need for reinforcing steel.

THIN SLABS USING POURED GYPSUM UNDERLAYMENT

Thin slabs can also be constructed of **poured gypsum underlayment** rather than concrete. The difference in thermal performance, for slabs of the same thickness and tube spacing is very slight.

The installation sequence differs from that used for a thin concrete slab. First, there is no bond breaker layer required. Second, the slab is

Mix Design for 1 cubic yard of 3,000 psi thin slab concrete	
Type 1 portland cement	517 pounds
Concrete sand	1,639 pounds
#1A (1/4-inch maximum) crushed stone	1,485 pounds
Air entrainment agent	4.14 ounce
Hycol (water-reducing agent)	15.5 ounce
Fibermesh	1.5 pounds
Superplasticizer (WRDA-19)	51.7 ounce
Water	About 20 gallons

Figure 5-68 | Mix proportions for concrete thin slab.

typically poured after all walls have been installed and finished. The poured gypsum underlayment is mixed outside the building and pumped in through a hose. This material has a fluid consistency similar to pancake batter. As such, it provides a high degree of self-leveling as it is poured on the floor, as seen in Figure 5-69.

Poured gypsum underlayment is slightly lighter than concrete, adding about 14.5 pounds per square foot to the floor's dead loading, when installed at 1.5 inch thickness. During installation, poured gypsum underlayment brings hundreds of gallons of water into the building. Much of this water has to evaporate as the slab hardens. Buildings must be well ventilated during this time to avoid condensation on windows and other surfaces. Also keep in mind that poured gypsum underlayments are not waterproof, nor are they intended to serve as a finish floor surface. After curing, they must be protected against excessive water and covered with a finish floor surface.

Courtesy of Andrew Wormer

Figure 5-69 | Poured gypsum underlayment being poured over tubing to create a heated thin slab.

THERMAL PERFORMANCE OF THIN SLABS

Because they are thinner than a slab-on-grade floor, thin slabs have slightly lower heat dispersal characteristics. This means that slightly higher average water temperatures are required for a given rate of heat output compared to that required for a slab-on-grade.

As was true with heat slab-on-grade floors, *close tube spacing and low-finish floor resistance are both vitally important when designing heated thin slabs that need to operate at low supply water temperatures.*

The graph in Figure 5-70 shows upward heat output from a heated thin-slab based on tube spacing of 6 inches and 9 inches and for finish floor resistances ranging from 0 to 1.0 (°F·hr·ft^2/Btu). The steeper the line, the lower the supply water temperature will be for a given rate of heat output and given room temperature.

This graph is used in the same way as Figure 5-51. It gives the upward heat output from a thin-slab floor as a function of the difference between circuit's average water temperature and room air temperature. Lines are given for 6-inch and 9-inch tube spacing and finish floor resistances between 0 and 1.0 (°F·hr·ft^2/Btu).

For example, to achieve an upward heat output of 20 Btu/hr/ft^2 from a slab with an R = 0.5 (°F·hr·ft^2/Btu) floor covering, and 6-inch tube spacing, requires the "driving ΔT" (e.g., the difference between *average* water temperature in tubing and room air temperature) to be approximately 27 °F. Thus, in a room maintained at 70 °F, the average water temperature in the circuit needs to be 97 °F. The *supply* water temperature to this circuit would likely be 7 °F to 10 °F above the average water temperature. In this case, the supply

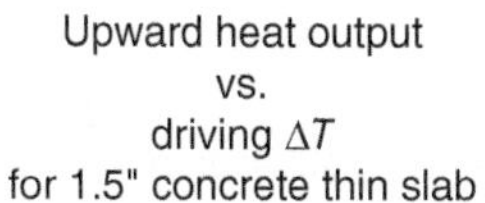

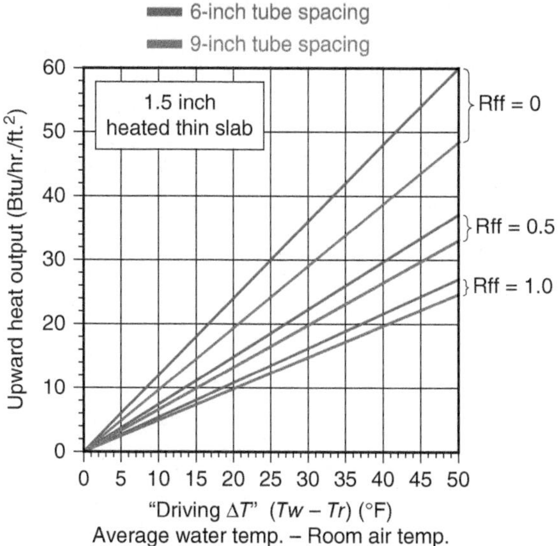

Figure 5-70 Upward heat output from a heated concrete thin slab as a function of the difference between circuit average water temperature and room air temperature, for 6-inch and 9-inch tube spacing and finish floor resistances between 0 and 1.0 (°F·hr·ft²/Btu).

water temperature would be in the range of 104–107 °F. This is a relatively low supply water temperature that is well within the range of what can be supplied by most renewable energy heat sources.

The design procedure for thin-slab systems would be essentially the same as that for a slab-on-grade system, but using data from Figure 5-70 rather than Figure 5-51.

HEATED THIN-SLAB DESIGN AND INSTALLATION SUMMARY

The following guidelines summarize the design and installation of heated thin slabs that will be supplied by renewable energy heat sources:

- Be sure that the floor on which the thin-slab will be placed provides adequate structure.
- Be sure the 1.5 inches added to the floor height is accounted for in other details such as door and window rough openings and stair heights.
- Tube spacing within the thin-slab should not exceed 9 inches when the system is supplied from a renewable energy heat source.
- Floors under thin slabs should have *minimum* of R-19 underside insulation (if the space under the floor is heated or semi-heated), or a *minimum* of R-30 underside insulation (if the space under the floor is unheated).
- Floor finishes installed over thin slabs should have a total R-value of 1.0 (°F·hr·ft²/Btu) or less.
- Never use **"lightweight" concrete** for heated thin slabs. It contains aggregates such as expanded shale or polystyrene beads, which significantly reduce the thermal conductivity of the slab and thus inhibit heat dispersal.
- Always install a suitable finish floor over poured gypsum underlayments.
- Remember that poured gypsum underlayments are not waterproof. They should only be installed when the building is fully enclosed.

5.8 Above-Floor Tube and Plate Systems

Although heated concrete floor slabs and thin slabs make excellent radiant panels, they are not suitable for all installations. For example, in some cases, the floor structure cannot support the added dead loading of a thin-slab. In other cases, a nailed down hardwood finish floor cannot be properly fastened to a slab.

In such situations, designers have the option of using an **above-floor tube and plate panel**. These panels use aluminum plates as "wicks" to pull heat away from the PEX-AL-PEX tubing and disperse it across much of the floor area. The thin layer of highly conductive aluminum becomes the substitute for a much thicker and heavier layer of concrete.

Figure 5-71 shows the cross section of an above-floor tube and plate radiant panel, along with different finish flooring options.

Figure 5-72 shows an above-floor tube and plate radiant panel during installation and before it is covered with finish flooring. The tubing in this installation is spaced 8 inches on center. This is the recommended spacing for good low-temperature performance, in combination with 6-inch-wide aluminum plates. Notice that this arrangement also leaves a nominal 2 inches of the wooden sleeper exposed for direct nailing of finish flooring materials, if needed.

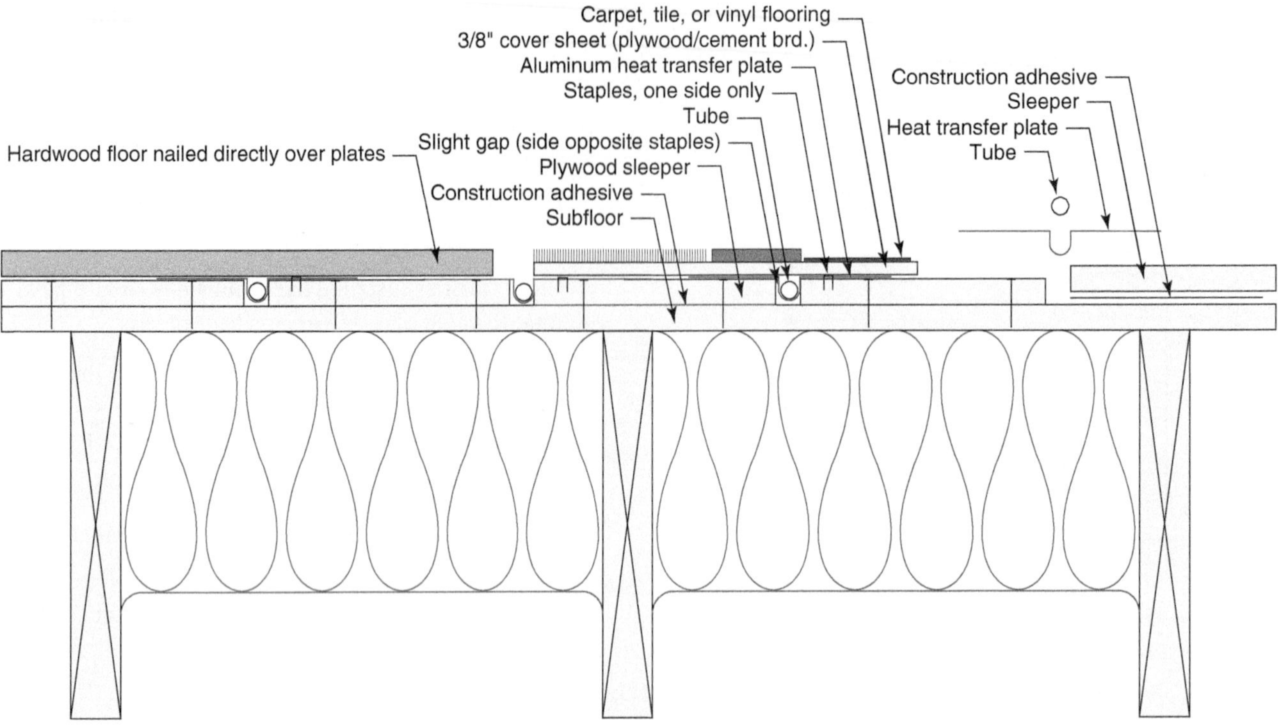

Figure 5-71 | Cross section of an above-floor tube and plate radiant panel showing different finish floor options.

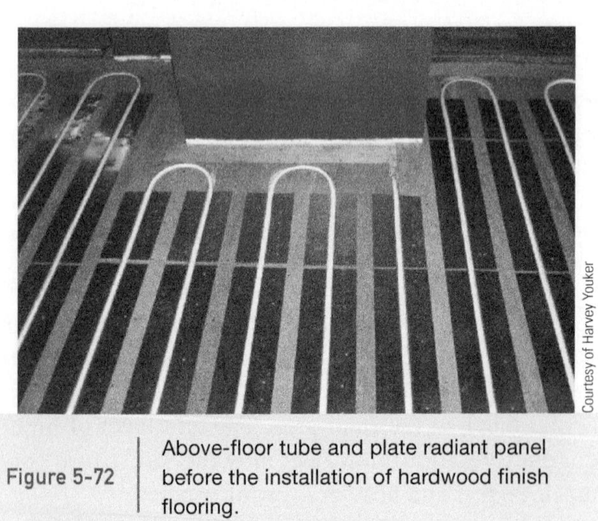

Figure 5-72 | Above-floor tube and plate radiant panel before the installation of hardwood finish flooring.

The installed PEX-AL-PEX tubing and aluminum plates seen in Figure 5-72 will eventually be covered by a nailed-down hardwood finish floor. In most cases, nailed-down hardwood flooring can be installed directly over, and at right angles to, the tubing. Flooring such as vinyl, ceramic tile, or thin carpet requires the installation of a 3/8-inch plywood **cover sheet** over the tubing and plates, which acts as a smooth substrate for the flooring.

Adequate insulation must be installed under all above-floor tube and plate radiant panels. If the space under the floor deck is heated or semi-heated, the insulation should have a *minimum* R-value of 19 (°F·hr·ft²/Btu). If the space under the floor is unheated, the insulation should have a *minimum* R-value of 30 (°F·hr·ft²/Btu).

The author recommends that only PEX-AL-PEX or PERT-AL-PERT tubing be used in combination with tube and plate radiant panels. The aluminum core of these tubes limits the thermal expansion of the tubing, keeping it close to that of the aluminum plates. This minimizes differential expansion and reduces the possibly of "ticking" sounds from the floor as the system is warming up.

THERMAL PERFORMANCE OF ABOVE-FLOOR TUBE AND PLATE PANELS

Above-floor tube and plate panels usually require higher water temperatures compared to slab systems for comparable heat output. Depending on the thermal resistance of the finish floor, average water temperatures can range as high as 145 °F under design load conditions. This is significantly higher than the 120 °F maximum supply water temperature recommended for any hydronic distribution system supplied by a renewable energy heat source. Thus, above-floor tube and plate systems must be used "judiciously" where low heating loads allow the supply water temperature, under design load conditions, to be 120 °F or less.

The graph in Figure 5-73 shows upward heat output (in Btu/hr/ft^2) from an above-floor tube and plate radiant panel based on tube spacing of 8 inches and **total top-side thermal resistance** ranging from 0.5 to 2.0 (°F·hr·ft^2/Btu). The total top-side resistance (R_{top}) includes the R-value of the finish flooring, plus the R-value of any cover sheet use.

This graph is used in the same way as Figures 5-51 and 5-70. It gives the upward heat output from the panel as a function of the difference between average circuit water temperature and room air temperature.

For example, to achieve an upward heat output of 20 Btu/hr/ft^2 from an above-floor tube and plate system with a total top-side R-value of 1.0 (°F·hr·ft^2/Btu) requires the "driving ΔT" (e.g., the difference between *average* water temperature in tubing and room air temperature) to be 41.5 °F. Thus, in a room maintained at 70 °F, the *average* water temperature in the circuit needs to be 41.5 + 70 = 111.5 °F. The *supply* water temperature to this circuit would likely be 7 °F to 10 °F above the average water temperature. In this case, the supply water temperature would be in the range of 118.5–121.5 °F. This is close to the recommended upper limit of 120 °F for use in systems with renewable energy heat sources. Thus, while top-side tube and plate radiant panels do have *possible* use in such systems, they are limited to situations of low building heat loss, and low-finish floor resistance.

Above-floor tube and plate tubing @ 8" spacing

R_{top} = 0.5

R_{top} = 1.0

R_{top} = 1.5

R_{top} = 2.0

Upward heat output (Btu/hr./ft.^{2})

0 5 10 15 20 25 30

0 5 10 15 20 25 30 35 40 45 50

"Driving ΔT" ($Tw - Tr$) (°F)
Average water temp. – Room air temp.

Figure 5-73 Upward heat output from an above-floor tube and plate radiant panel, with 8-inch tubing spacing and various total top side R-values for covering(s).

OTHER RADIANT FLOOR PANELS

There are several other methods and products available for constructing radiant floor panels. Several are covered in more detail in the text *Modern Hydronic Heating*, 3rd edition. There will surely be more products entering this market in the future. The author recommends the criteria of 120 °F maximum supply water temperature under design load conditions to evaluate any other radiant floor heating method for use in systems supplied by renewable energy heat sources.

A SITUATION TO AVOID

One "illegitimate" installation method, known as "**plateless staple-up**," has unfortunately been used in numerous North American homes over the last two decades. This approach relies on staples or clips to hold tubing against the underside of the subfloor, as shown in Figure 5-73. No aluminum heat transfer plates are installed. The author strongly recommends avoiding

Figure 5-73 Plateless staple-up tubing installation. This approach will not provide adequate heat transfer and should be avoided.

this approach, especially in systems that are intended to operate with low supply water temperatures.

Plateless staple-up tubing installation results in very poor heat transfer between the tubing and upper surface of the floor. At best there is "line contact" between the outer surface of the tubing and the bottom of the subfloor. Conduction heat transfer, which is critical in moving heat away from the tubing, is very limited. The wooden subfloor, by itself, provides poor lateral heat dispersion away from the tubing. Even when tubing is spaced 6 to 8 inches apart and the underside of the floor deck is insulated, the heat transfer path is extremely marginalized in comparison to the other methods of hydronic floor heating shown in this chapter. The origins of this method undoubtedly came from trying to reduce installation cost. While it may do so, it also results in severely compromised comfort, and poor system performance. It is pointless to sacrifice the potential performance of any hydronic heating system with such an approach.

5.9 Radiant Wall and Ceiling Panels

Radiant panels can also be integrated into walls and ceilings. In some cases, radiant wall or ceiling panels are even preferable over floor panels. Examples include situations where finish flooring with high thermal resistance will be used or when a large percentage of the floor area will be covered with objects that inhibit upward heat flow.

Radiant wall panels and **radiant ceiling panels** also have relatively low thermal mass in comparison to radiant floor panels, especially those involving slabs. This allows them to respond quickly to changes in load caused by internal heat gains or changes in thermostat settings. Radiant ceiling, in particular, are also ideally suited for radiant *cooling* applications.

Figure 5-74 shows one way to construct a low thermal mass radiant wall panel. This construction uses the same1/2-inch PEX-AL-PEX tubing and aluminum heat transfer plates as described for the above-floor tube and plate floor panel. The major difference is that these plates are bonded to strips of **foil-faced polyisocyanurate insulation** board, which in turn are bonded to a 7/16-inch **oriented strand board (OSB)** substrate fastened to the wall. All bonding is done use standard **contact adhesive**. After the tubing is installed and pressure tested, the assembly is covered by 1/2-inch drywall and finished in the usual manner.

Figure 5-75 shows a radiant wall panel being constructed using this method. Notice that the 2.5-inch drywall screws are being driven on lines

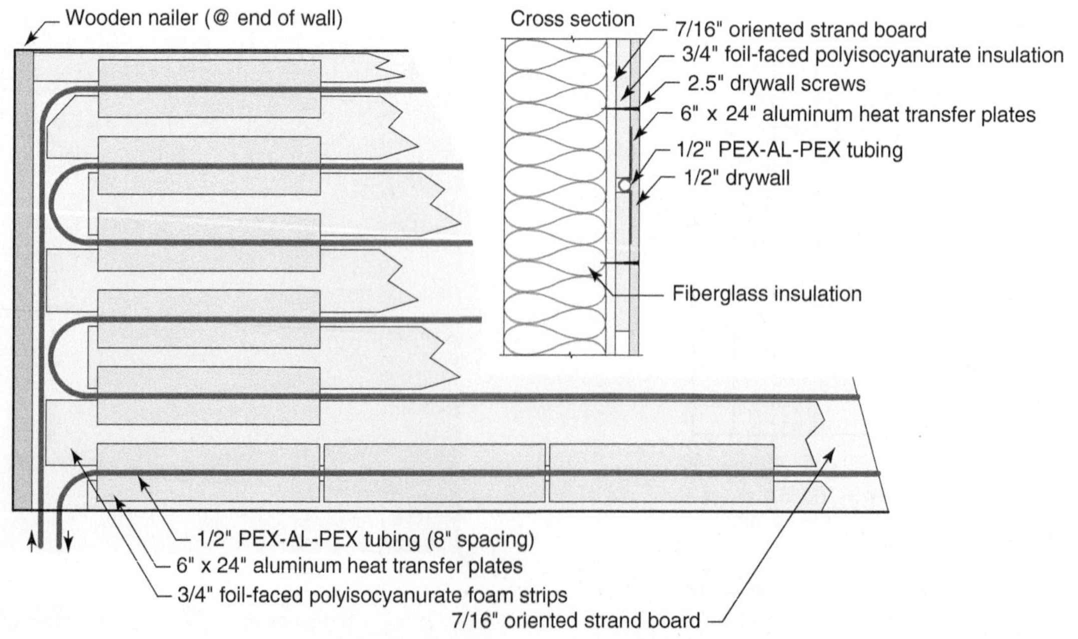

Figure 5-74 | Low-thermal-mass radiant wall construction.

Figure 5-75 Drywall being installed over the tube and plate assembly to complete the radiant wall. Note screw placement and how the plates are held back near the junction box.

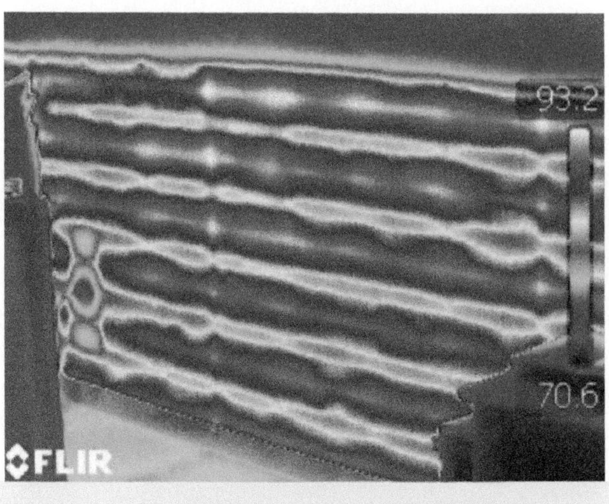

Figure 5-76 Infrared thermograph of the radiant wall panel shown in Figure 5-75 during operation. The temperatures represented by the colors are shown in the scale at the right side of the thermograph.

spaced halfway between the tubing, which itself is spaced 8 inches on center. This ensures that the screws don't penetrate the aluminum plates and are safely away from the tubing. These screws can be driven where the studs are located or between the studs (since they will secure to the oriented strand board).

When finished, this radiant wall panel is indistinguishable from a standard interior wall. Its low thermal mass allows it to respond quickly to changing internal load conditions or zone setback schedules.

It is best to limit the installed height of radiant wall panels to 4 feet. This concentrates radiant heat output into the occupied portion of the room. It also keeps the tubing below wall areas where pictures and shelves are typically fastened to the wall.

The author recommends that only PEX-AL-PEX or PERT-AL-PERT tubing be used in combination with the radiant wall panel construction shown. The aluminum core of these tubes limits the thermal expansion of the tubing, keeping it close to that of the aluminum plates. This minimizes differential expansion and reduces the possibly of "ticking" sounds from the wall panel as it is warming up.

Figure 5-76 shows an infrared thermograph of the finished radiant wall panel in operation. Notice the relatively consistent red stripes that indicate good lateral heat dispersion by the aluminum heat transfer plates in contact with the rear side of the drywall. The relatively low heat output where the plates were omitted near a receptacle is also evident.

RADIANT CEILING PANELS

Another way to incorporate radiant heating is by using heated ceilings. The construction details for one type of low thermal mass radiant ceiling panel are shown in Figure 5-77. This is essentially the same construction as used in the previously described radiant walls.

Figure 5-78 shows a sequence of installation photos for this radiant ceiling panel.

Figure 5-79 shows the final step of installing the 1/2-inch drywall over the tubing and plates.

Notice that the 2.5-inch drywall screws are being driven along a line spaced half way between adjacent tubes. This ensures that the screws do not penetrate the aluminum heat transfer plates, or the tubing. The 8-inch by 12-inch screw placement pattern also ensures that the 1/2-inch drywall is held tightly against the outer surfaces of the plates, to create good conduction heat transfer from the plates to the drywall.

As with the radiant wall, this radiant ceiling design has low thermal mass and responds quickly to interior temperature changes. Heated ceilings also have the advantage of not being covered or blocked by coverings or furniture and are thus likely to retain good performance over the life of the building.

The author recommends that only PEX-AL-PEX or PERT-AL-PERT tubing be used in combination with the radiant ceiling panel construction shown.

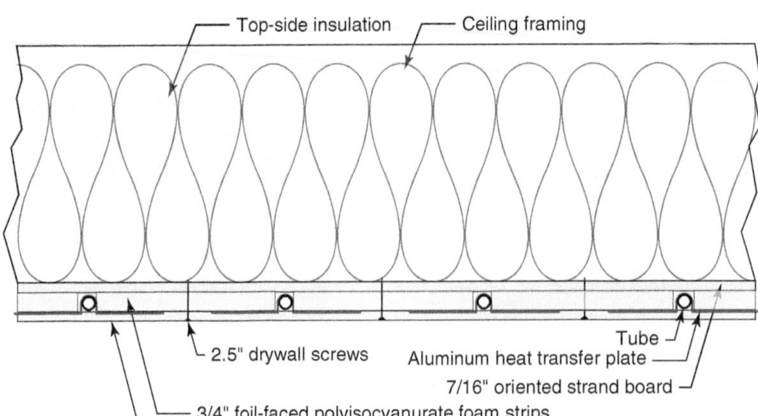

Figure 5-77 | Cross section of radiant ceiling assembly using aluminum heat transfer plates.

Figure 5-78 | Installation sequence for radiant ceiling panel. (a) Using roller to apply contact adhesive to foil-faced foam strips, on one side of 3/4-inch space. (b) Pressing 5-inch × 24-inch aluminum heat transfer plate into position. (c) Using wooden float to press 1/2-inch PEX-ALPEX tubing into groove of heat transfer plate. (d) Making a detour around a junction box. (e) Return bends of tubing circuits near partitions. (f) Tube routing from panel circuits to interior partition, then down to manifold station in basement.

The aluminum core of these tubes limits the thermal expansion of the tubing, keeping it close to that of the aluminum plates. This minimizes differential expansion and reduces the possibly of "ticking" sounds from the ceiling panel as it is warming up.

Figure 5-80 shows an infrared thermograph of the radiant ceiling panel as it is warming up. Notice that the aluminum plates are dispersing heat away from the tubing on the left side of the panel slightly better than on the right side. As the panel approaches steady-state operation, these differences are minimized.

THERMAL PERFORMANCE OF RADIANT WALLS AND CEILINGS

The graph in Figure 5-81 shows heat output from both the heated wall and heated ceiling panels discussed in this section. This performance assumes a standard

Figure 5-79 | Installation of a low-mass tube and plate radiant ceiling.

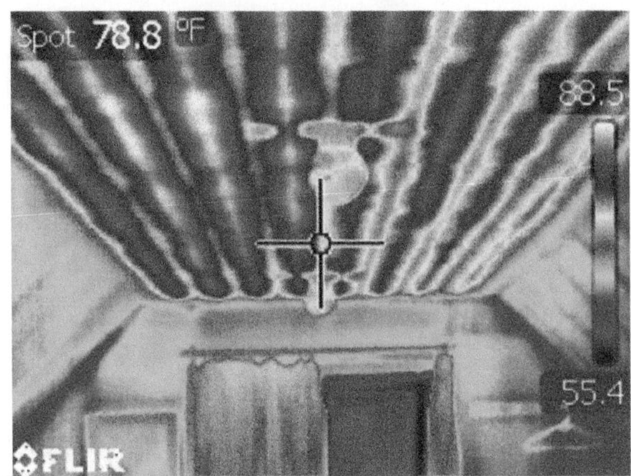

Figure 5-80 | Infrared thermograph of radiant ceiling panel as it warms up from left to right.

1/2-inch drywall finish on both the wall and ceiling panels.

A quick examination of Figure 5-80 shows that either of these panels can produce significant heat output with relatively low average water temperatures. For example, to produce an output of 20 Btu/hr/ft^2, the radiant *wall* panel requires a "driving ΔT" between the average water temperature and room air temperature of 25 °F. Thus, in a room maintained at 70 °F, the average water temperature needs to be 95 °F. The *supply* water temperature to this circuit would likely be 7 °F to 10 °F above the average water temperature. In this case, the supply water temperature would be in the range of 102 to 105 °F. For the same heat output, the supply water temperature for the radiant *ceiling* panel would be in the range of 106 °F to 109 °F. Both of these temperatures

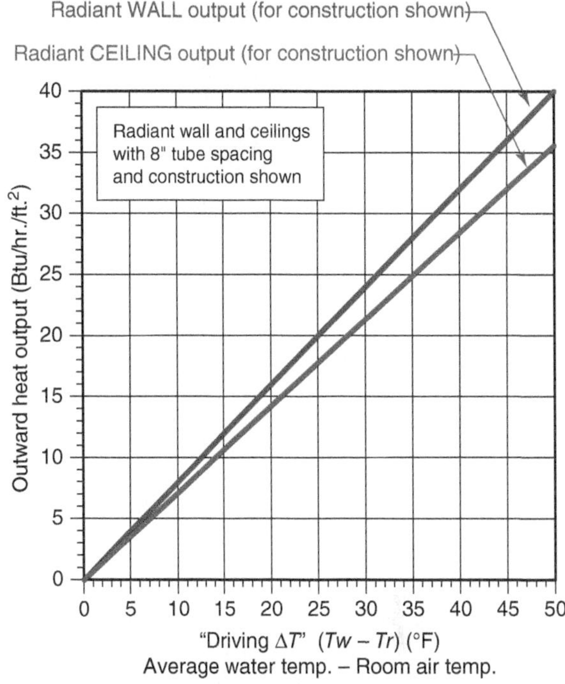

Figure 5-81 | Heat output of the radiant wall and radiant ceiling panels discussed in this section.

are well within the operating range of renewable energy heat sources.

5.10 Thermal Mass of Heat Emitters

The thermal mass of the heat emitters described in this chapter varies over a very wide range, as shown in Figure 5-82.

This graph compares the relative thermal mass of several hydronic heat emitters. It provide an "apples-to-apples" comparison by showing the thermal mass of each heat emitter based on the size of that heat emitter needed to release heat at 1,000 Btu/hr, when operating at an average water temperature of 110 °F, in a room maintained at 70 °F.

The low thermal mass panel radiator has less than 0.5 percent of the thermal mass of the 4-inch heated floor slab for equivalent output at the same operating conditions. It achieves this low mass by using a fin-tube element with very low water content. These panels can quickly warm up when necessary, such as during recovery from a setback condition. Just as importantly,

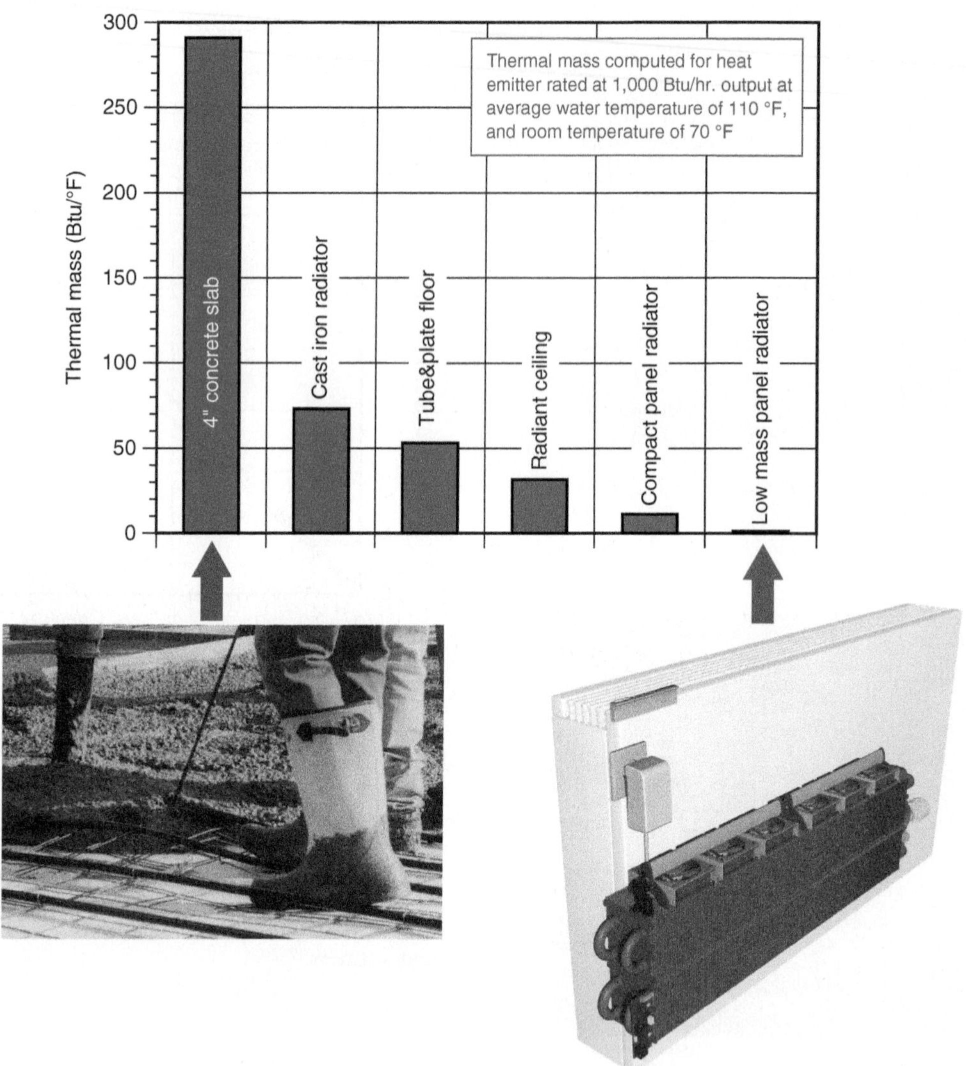

Figure 5-82 | Comparison of thermal mass of various heat emitters based on all emitters sized to provide an output of 1,000 Btu/hr, when operating at an average water temperature of 110 °F in a room maintained at 70 °F.

they can quickly stop emitting heat when solar energy or other internal heat gains present themselves.

Because of improvements in energy-efficient construction, new buildings with lower heating loads will be more effected by internal heat gains compared to older buildings. A given rate of internal heat gain will cause a greater rise in indoor air temperature in a building with a lower rate of heat loss.

It's also likely that more new buildings will incorporate passive solar design features and thus be more likely to experience quickly changing internal heat gains during a typical winter day.

The combination of low heat loss and passive solar gains creates the potential for faster and wider swings in interior temperature. This is especially true if the building's heating system has significant thermal mass and thus cannot quickly suspend heat output when an internal heat gain begins to raise the room air temperature.

For this reason, low thermal mass heat emitters will be preferred in many future buildings. They provide the ability to quickly turn on and—even more importantly—quickly turn *off* heat input to interior spaces.

SUMMARY

The chapter has discussed a wide range of hydronic heat emitters, many of which can be selected and sized to provide design load output at a supply water temperature not higher than 120 °F.

In some cases, a single type of heat emitter is well suited for use throughout a building. In other cases, it's possible to use two or more types of heat emitters, each best suited for a given location or heating requirement within the building. An example would be use of a heated floor slab in a basement in combination with radiant ceiling heating on the first floor.

When combining multiple heat emitters, the ideal scenario is to select and size them so that they can be supplied with a single water temperature. This simplifies system design and reduces installation cost. When this is not possible, multiple simultaneous supply water temperatures can be created using one of the mixing assemblies discussed in later chapters.

KEY TERMS

2-pipe air handler
2-pipe fan coil
3-way motorized mixing valve
4-pipe distribution system
above-floor tube and plate panel
aerial boom
aggregate
Air-Conditioning, Heating, and Refrigeration Institute (AHRI)
air handler
available floor area
average floor surface temperature
average water temperature [for a circuit]
blow-through fan coil
blower
bond breaker
bullhead tee
CAD software
cavitation
coefficient of performance (COP)
coil
compact-style panel radiator
concrete thin slab
condensate
condensate drip pan
constant differential pressure mode
contact adhesive
control joint strips
correction factors
cover sheet
dead loading
design ΔT
design supply water temperature
diffusers
distribution system
diverter valve
draw-through fan coil
driving ΔT
edge insulation
electromagnetic spectrum
EN442
escutcheon plates
extruded polystyrene
fan
fan coil
Fibermesh®
fin
fluted steel panel radiator
foil-faced polyisocyanurate insulation
forced convection
Heating Edge baseboard
heating effect factor
high limit controller
homerun distribution system
Hydronic Circuit Simulator
hydronic heat emitter
hydronic radiant panel
Hydronics Design Studio
infrared
infrared thermograph
iteration
leaders
lift hook
lightweight concrete
lockshield valve
maximum deflection criteria
microfan
natural convection
occupant expectations
occupied zone
oriented strand board (OSB)
panel radiator
parallel piping systems
PEX
plateless staple-up
plenum
polyline
poured gypsum underlayment
pressure-regulated circulator
radiant ceiling panel
radiant floor panel
radiant heat
radiant heating
radiant panel heating
radiant wall panel
registers
rough openings
Schrader valve
series piping system
standard fin-tube baseboard
steady-state condition
subfloor
subgrade
superplasticizer
system average water temperature
target supply water temperature
thermal equilibrium
thermal mass
thermal radiation
thermographic image
thermostatic radiator valves
total top-side thermal resistance
tube passes [in a coil]
tubing layout plan
underside insulation
underslab insulation
upward heat flux
upward heat output
water plate
water-reducing agent
welded wire fabric
wire bag ties

QUESTIONS AND EXERCISES

1. Why is the use of heat emitters that operate at low water temperatures important in systems using renewable energy heat sources?

2. Explain the relationship between the coefficient of performance of a water-to-water heat pump and the temperature of water entering that heat pump from the return side of the distribution system.

3. Describe the key advantage of parallel versus series distribution systems for use with renewable energy heat sources.

4. A boiler is connected to a distribution system. The high limit controller on the boiler is set to 200 °F. The boiler is turned on, but the water temperature leaving it will not climb above 145 °F, even after the boiler has operated continuously for several hours. Explain what is happening.

5. What is the dominant method of heat transfer from fin-tube baseboard: conduction, convection, or radiation? Explain why this is the case.

6. A fin-tube baseboard convector has the heat output characteristics given in Figure 5-8. Estimate the heat output of a 10-foot segment of this baseboard operating with an inlet temperature of 125 °F and water flow rate of 3 gpm in a room with 65 °F air near floor level. Use Equations 5.3 and 3.3.

7. Repeat exercise 6, assuming Heating Edge baseboard is used under the same operating conditions. Use Figure 5-10 and Equations 5.4 and 3.3.

8. Describe some advantages of homerun distribution systems.

9. What would be an advantage of using a "double water plate" versus a single water plate panel radiator, both of the same height and width, within a system supplied by solar thermal collectors?

10. As the average water temperature at which a panel radiator operates decreases, the percentage of its total heat output delivered by thermal radiation ________.

11. What is an escutcheon plate, and how is it used with panel radiators?

12. How often do the batteries in a thermostatic radiator valve have to be replaced?

13. What can happen when flow is allowed to move through a compact-style panel radiator with an integral valve in a direction opposite to that intended by the manufacturer?

14. A triple water plate panel radiator measuring 20 inches tall and 72 inches long, has been selected for use in a low-temperature distribution system. Water enters the panel at 120 °F and leaves at 100 °F. Determine the output of this panel in a room maintained at a 65 °F air temperature. Use information from Figures 5-27 and 5-28.

15. A panel radiator has a rated heat output of 8,500 Btu/hr when operated at an average water temperature of 180 °F and room air temperature of 68 °F. Using the methods from EN442, determine the panel's expected heat output rate when operated with a supply water temperature of 105 °F, a temperature drop of 20 °F, and in a room maintained at 68 °F.

16. Why is a 3-way mixing valve recommended between a thermal storage tank supplied by solar collectors and a low-temperature distribution system?

17. Describe the difference between a fan coil and an air handler.

18. Any fan coil or air handler that will be used for chilled water cooling must have a __________.

19. A fan coil can deliver heat at a rate of 5,000 Btu/hr with a water temperature of 140 °F and room air entering at 65 °F. Estimate its output using 120 °F water and the same room conditions.

20. What happens to the heat output of a fin-tube baseboard as the water flow rate through its element increases, but the inlet temperature remains constant?

21. A contractor is called in to correct the thermal performance of an air handler. He proposes adding a second circulator to the circuit supplying the air handler, stating that this will double the flow rate through the air handler and thus double its heat output. Discuss the validity of this claim in detail.

22. By using a coil with 4-tube passes versus 2-tube passes, and operating at the same flow rate, the required supply water temperature for a given rate of heat output will __________.

23. What is an advantage of a 4-pipe distribution system serving multiple air handlers in a commercial building?

24. Why should a hydronic heating coil, supplied by a wood-fired boiler, be installed on the discharge plenum of a furnace rather than on the air intake side?

25. Describe a situation in which a heated floor may not feel warm to a bare foot.

26. A room measures 12 feet by 16 feet, and has a design heat loss of 3,072 Btu/hr. It will be heated by a concrete slab-on-grade floor with a floor covering having a thermal resistance of 0.5 (°F·hr·ft^2/Btu) and tube spacing of 12 inches. Determine the circuit average water temperature required at design load if the room is to be maintained at 70 °F.

27. Why is it important to lift the tubing and welded wire fabric to about half the slab's thickness as the concrete is placed?

28. Describe several benefits of preparing a tubing layout plan before installing tubing for a heated slab-on-grade floor.

29. What is the purpose of a bond breaker layer under a concrete thin slab? Is this bond breaker used with poured gypsum underlayment?

30. A room measures 20 feet by 15 feet, and has a design heating load of 5,100 Btu/hr when maintained at 70 °F. It will be heated by a concrete thin slab covered with flooring having a resistance of 0.5 (°F·hr·ft^2/Btu) and with tubing spaced 9 inches on center. Determine the required circuit supply temperature, assuming the circuit will operate with a 15 °F temperature under design load conditions.

31. What is the advantage of using PEX-AL-PEX or PERT-AL-PERT tubing in the radiant wall and radiant ceiling panels described in this chapter?

32. What is the advantage of using a low-thermal-mass heat emitter in a house with low heat loss and significant passive solar gain?

R410A

CHAPTER 6

Control Principles and Hardware

OBJECTIVES

After studying this chapter, you should be able to:

- Understand several control outputs and algorithms used in hydronic heating systems.
- Describe the differences between controlling heat output using variable flow versus variable water temperature.
- Describe different methods of implementing outdoor reset control.
- Calculate the reset ratio for an outdoor reset controller used in a specific systems.
- Explain the operation of 3-way motorized mixing valves operated by floating control.
- Explain the operation of a differential temperature controller.
- Understand proper placement and wiring of temperature sensors.
- Explain two methods for determining the total thermal energy transferred by a process.
- Describe the terminology used for classifying switches and relays.
- Explain how electrical contacts are connected to create AND and OR logic.
- Create a ladder diagram and associated description of operation for a control system.

6.1 Introduction

The **control subsystem** is the "brain" of any hydronic system. It determines exactly when and for how long devices such as circulators, burners, compressors, and mixing valves will operate, and often what they do while they operate.

Many of the control devices and principles used in hydronic heating systems supplied by conventional heat sources, are also applicable to systems using renewable energy heat sources. This is especially true for fundamental actions, such as regulating the water temperature supplied to a hydronic distribution system, protecting the system against excessive temperature or pressure, or monitoring temperatures within a building to determine when and where heat needs to be supplied.

This chapter gives an overview of basics control principles and control used in hydronic systems. A complete description of principles and hardware for a broader range of applications is presented in chapter 9 of *Modern Hydronic Heating*, 3rd ed.

6.2 Closed-Loop Control Fundamentals

All heating systems rely on fundamental control principles to stabilize their operation and optimize their performance. It is imperative that those who design such systems understand these basic principles before learning about the hardware used to implement them. This section discusses these fundamentals and prepares the reader for a more detailed discussions of specific hardware later in the chapter.

Figure 6-1 depicts the elements of a simple **closed-loop control system** for hydronic heating. Some of the boxes represent information, while others represent physical devices.

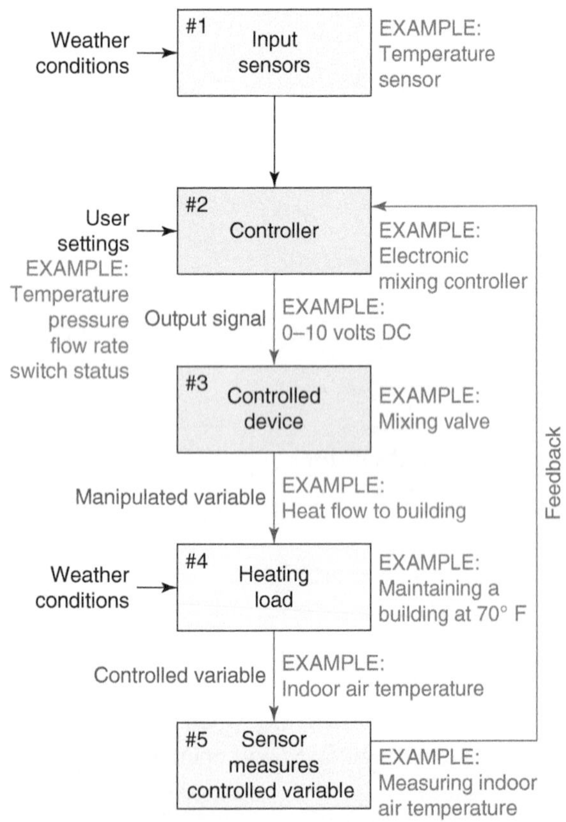

Figure 6-1 A block diagram of a feedback control system for space heating applications. Examples represent typical devices or measurements used for hydronic space heating.

Block #1 represents information that is sent to the control system by one or more sensors. In current generation hydronic systems, this information is usually outdoor air temperature. However, more sophisticated control system may expand this information to include measurements of windspeed, relative humidity, and solar radiation intensity. This information may be sent to the controller as an electrical signal such as a variable resistance, voltage, or current. It might also be sent as a **digitally encoded signal**.

Additional information comes into the controlled process as user **settings**. Examples include desired temperatures, pressures, flow rates, or switch status. The first three of these are examples of **analog inputs**. They all represent physical quantities that can be measured and that vary over a continuous range of numerical values. For example, the outdoor temperature may be represented by a number anywhere within the range of −50 °F to 120 °F, or a flow rate may be a value anywhere between 0 and 100 gpm. The last input—switch status—is an example of a **digital input**, which can only have one of two states: open or closed.

Block #2 represents a physical device called the **controller**. It accepts information from the input **sensor**(s). It also accepts and stores the user settings. The controller also receives information called **feedback**, sent from another sensor that measures the result of the controlled process. This result is represented by a **controlled variable**. For current generation hydronic space heating systems, the controlled variable is usually indoor air temperature.

The controller uses this information to compare the measured value of the controlled variable with the **target value**. The latter is the desired "ideal" status of the controlled variable. Any deviation between the target value and the measured value is called **error**. In this context, the word "error" does not imply that a malfunction has occurred. It means that there is some deviation between the target value and measured value of the controlled variable.

The controller uses a stored set of instructions called a **processing algorithm** to generate an output signal based on the error. This output may be a switch contact closure, a precise DC voltage, or even a variable frequency AC waveform.

The output signal is passed to a **controlled device** that responds by changing the **manipulated variable** (which in a heating system is usually heat flow). The change in the manipulated variable causes a change in the process being controlled (in this case, the heat released to the building). This causes a corresponding change in the controlled variable (in this case, the inside air temperature). The change in the controlled variable is sensed by the sensing element, which provides feedback to the controller. This entire process is continuous as long as the system is operating.

Closed-loop control systems can be tuned for very stable operation. The feedback inherent to their design makes them strive to eliminate any error between the target and measured value of the controlled variable.

6.3 Controller Outputs

A controller can interact several ways with a controlled device. In some cases, the controller produces a simple output action, such as opening or closing of an electrical contact. In other cases, the output action may be a constantly changing analog voltage or current signal from the controller to the controlled device. This section reviews the controller output types commonly used in hydronic heating systems.

ON/OFF OUTPUT

In **on/off control**, the controlled device can only have one of two possible control states at any time. If the controlled device is a valve, it can be either fully open or fully closed, but never at some partially open condition. If the controlled device is a heat source, it must be fully on or completely off. On/off output signals are usually generated by the closing and opening of electrical contacts within the controller.

When used to control a heating process, on/off control sends heat to the load in pulses rather than as a continuous process. In such applications, all on/off controllers must operate with a **differential**. This is the change in the controlled variable between the point where the controller output is on, to where it is off. In heating systems, the controlled variable is usually the temperature at some location within the system or the building it serves.

One of the most common examples of on/off controllers in heating systems is a **room thermostat**. Although the internal circuitry of modern electronic thermostats may be sophisticated, and even capable of communicating over the Internet, the output of most thermostats used in residential and light commercial buildings is an internal switch contact (e.g., relay contact), that is either open or closed. When this contact is closed, the thermostat is "calling" for heat from the remainder of the system. When its contact is open, the thermostat is said to be "satisfied."

Another device, called a **temperature setpoint controller**, operates in a similar manner. Like a room thermostat, a temperature setpoint controller monitors the temperature of its sensor. That sensor may be strapped to a pipe, mounted inside a solar collector, or located deep within a sensor well that has been threaded into the side of a thermal storage tank. The temperature setpoint controller has a user setting called the **setpoint**. The controller constantly compares the temperature of its sensor to its setpoint. When the difference between the measure temperature and setpoint reach a specific value, the temperature setpoint controller either opens or closes an electrical contact. This contact movement then initiates other control actions within the system, such as turning off the burner on a boiler, or turning on a heat dumping subsystem that's part of a solar combisystem.

All temperature setpoint controllers operate with a differential. On some, the differential is *below* the setpoint temperature, as shown in Figure 6-2a. In others, the differential is centered on the setpoint, as shown in Figure 6-2b.

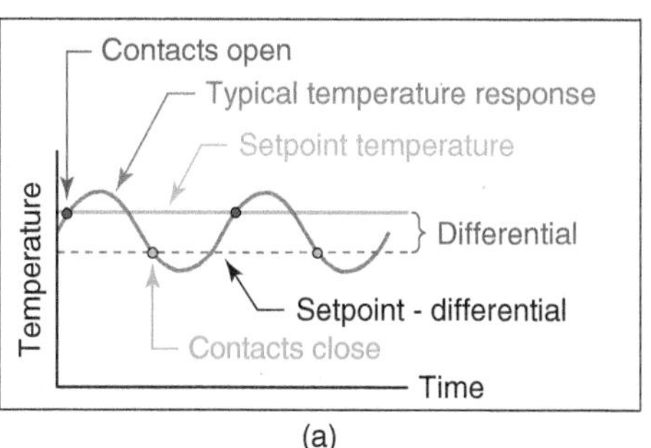

(a)

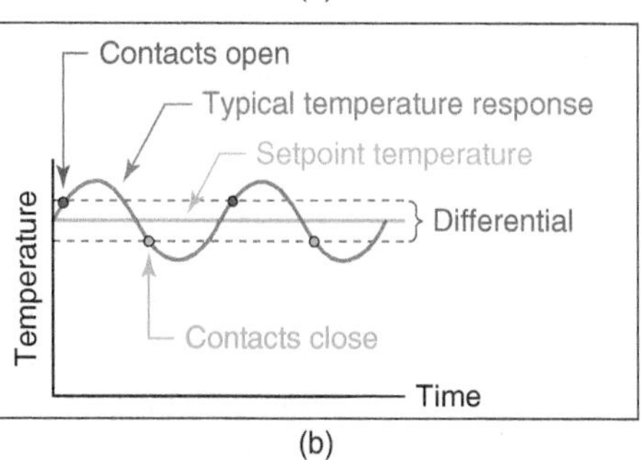

(b)

Figure 6-2 On/off temperature controllers operate with a differential that is either (a) entirely below the setpoint value or (b) centered on the setpoint value.

The smaller the differential, the smaller the deviation in temperature from the target value. However, if the differential is too small, the controller operates the controlled device in short but frequent cycles. This undesirable effect is called **short cycling**. In most cases, short cycling increases the wear on the equipment being controlled. This is especially true of devices such as burners in boilers or compressors in heat pumps.

If the differential is too wide, there can be large deviations between the target value and measured value of the controlled variable (usually temperature). In some situations this is acceptable, in others it is not. For example, the large thermal mass of a heated slab-on-grade floor can usually accept variations in water temperature of several degrees Fahrenheit without creating corresponding swings in room air temperature. However, a room thermostat with a differential of 5 °F, for example, would cause wide swings in room air temperature that would almost certainly lead to complaints.

The term **control differential** refers to the variation in temperature at which the controller changes the on/off status of the output. On some controllers, this is a factory set value and cannot be changed by the user. On other controllers, the control differential can be adjusted over a wide range.

The term **operating differential** refers to the *actual variation in the controlled variable as the system operates*. It depends on the combined effect of the controller and the system it is controlling. In heating systems, the operating differential is usually greater than the control differential because of a time lag caused by the thermal mass of the heating system components and the building. The term **overshoot** describes a situation where the controlled variable exceeds its intended upper limit. **Undershoot** occurs when the controlled variable drops lower than its intended lower limit. Figure 6-3 illustrates these terms for a temperature control system.

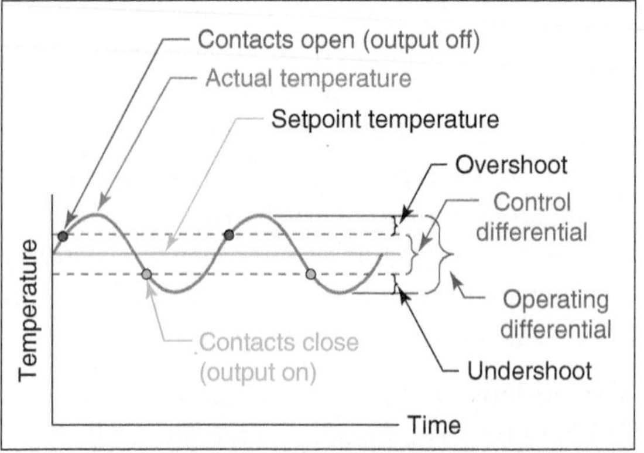

Figure 6-3 The *control differential* is the desired difference in the temperature between when controller output turns on and off. *Operating differential* is the actual temperature range experienced as the system operates. The operating differential is wider than the control differential due to overshoot and undershoot.

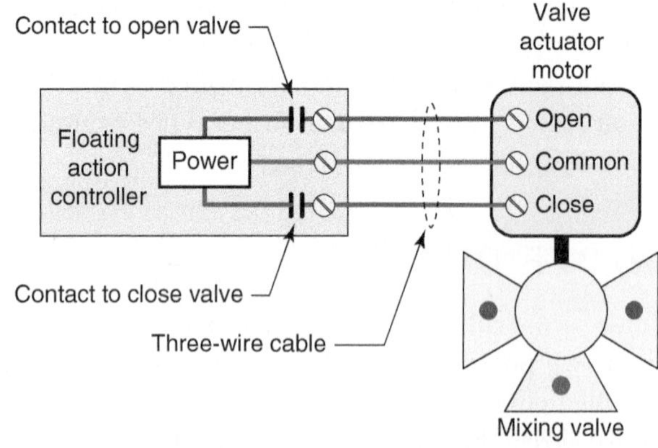

Figure 6-4 The operation of a 3-wire floating control system. Possible control states include opening valve, closing valve, or holding valve in present position.

FLOATING CONTROL

Another means of providing regulated heat transfer to a hydronic distribution system is called **floating control**. This control output was developed to operate motorized valves and dampers that need to be powered open as well as powered closed. *In hydronic heating systems, floating control is commonly used to drive 3-way motorized mixing valves.* Figure 6-4 shows the wiring between a controller, which operates with floating control, and a motorized 3-way mixing valve.

One electrical contact within the controller closes to drive the valve's actuating motor in a clockwise (CW) direction. The other electrical contact closes to drive the actuator motor counterclockwise (CCW). Only one contact can be closed at any time. The actuating motor driving the valve shaft turns very slowly. Some actuating motors can take up to 3 minutes to rotate the valve shaft over its full rotational range of 90 degrees. This slow operation is desirable since it allows the sensor that measures the mixed fluid temperature ample time to provide feedback to the controller for stable operation.

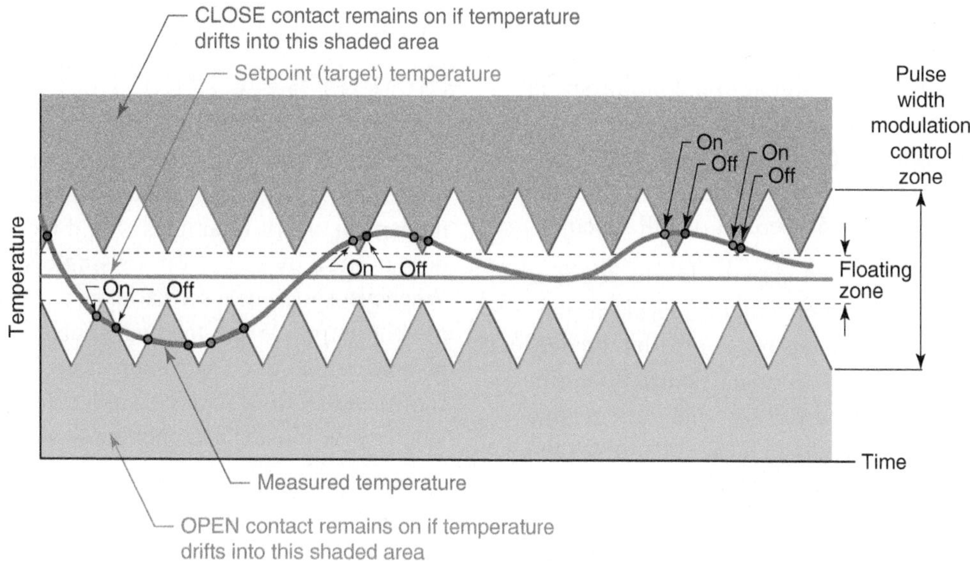

Figure 6-5 | A floating controller using pulse width modulation to determine the time and duration of the contact closures.

The floating controller determines the time at which one of the output contacts closes as well as the duration of that contact closure, as illustrated in Figure 6-5.

The green shaded area in Figure 6-5 represents conditions when the OPEN contact of the floating controller is active. Electrical power is flowing to the valve's actuator, and the valve's shaft is slowly rotating toward the open position. Likewise, the red shaded area in Figure 6-5 represents conditions when the CLOSE contact of the floating controller is active. The valve's shaft is now slowly rotating toward the closed position. The area between the red and green areas is called the **floating zone**. Whenever the measured temperature is within this zone, neither the OPEN or CLOSE contacts are active, and the valve's shaft does not move.

The "sawtooth" shape of the red and green shaded areas allows the duration of the contact closures to be proportional to the error between the target temperature and the measured temperature. The greater the error, the longer the duration of the contact closure as the controller attempts to counteract the error. The "width" (e.g., time duration) of the contact closure varies in response to the error. This is called **pulse width modulation (PWM)**.

If the measured temperature drifts above the red sawtooth shapes, it is significantly higher than the target temperature, and the controller keeps the CLOSE contact energized. Under this condition, the controller is providing its maximum corrective action to reduce the error. Likewise, if the measured temperature drops below the green sawtooth area, it is significantly lower than the target temperature, and the controller keeps the OPEN contact energized as it attempts to correct the error. The width and height of the triangles making up the pulse width modulation areas are determined by the controller's manufacturer to provide stable operation. Under normal operating conditions, the controller outputs gently "nudge" the valve's actuator to keep the measure temperature within the floating zone.

MODULATING DC OUTPUT

On/off control and floating control use the opening and closing of electrical contacts to send signals between the controller and controlled device. This type of output is not suitable for all controlled devices. For example, none of these outputs could use relay contacts to regulate the speed of a circulator or the heat output rate from a modulating boiler.

Some controlled devices, such a certain modulating valves, or variable speed circulators, require a continuous **analog signal** from a controller. The controller creates a variable voltage between 2 to 10 volts DC, or a variable current between 4 and 20 milliamps DC, and sends that signal to the controlled device. The controlled device responds by operating at a position or speed that is proportional to the input signal. For example, a control output signal of 2 volts DC fed to a motor speed controller means that the motor should be off. A 10-volt DC output signal to the motor speed controller means the motor should be operating at full speed. An output signal of 6 volts DC, which is 50 percent of the overall range of 2 to 10 volts, means that the motor should be running at 50 percent of full speed.

The reason these control signals do not begin at zero voltage or current is to prevent electrical interference or "noise" from affecting the controlled device. Wires running in proximity to other electrical equipment can experience induced voltages and currents due to electrical or magnetic fields. In most situations, raising the starting threshold to 2 volts DC, or 4 milliamps, prevents such interference. However, in some very noisy electrical environments, twisted pair wiring or shielded cable may be necessary between the controller and the controlled device. Consult the control manufacturer's recommendations regarding control wiring in such situations.

The 2 to 10 VDC or 4 to 20-milliamp signals generated by the controller are for control only, and do not supply the electrical power to drive the device. For example, the typical wiring of a 2–10 VDC modulating valve is shown in Figure 6-6. Notice that 24 volts AC power must also be supplied to the valve to power the motor and the actuator's circuitry. Likewise, a variable speed pump that is controlled by a 2 to 10 VDC or 4 to 20-milliamp signal requires line voltage (120 VAC) to supply operating power.

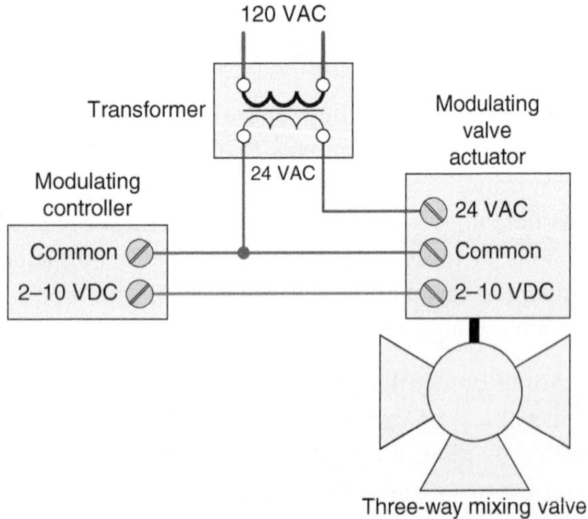

Figure 6-6 | Typical wiring for a valve actuator controlled by a 2–10 VDC modulating control signal.

6.4 On/Off Heat Source Control

Many residential and smaller commercial hydronic systems have a heat source that has only two operating modes: on or off. Examples would include conventional boilers fired by gas or oil, or single stage heat pumps. Whenever such a heat source is on, it produces a fixed rate of heat output. When the heat source is off, it produces no heat output. Such heat sources require a simple on/off control signal from their controller. This is typically provided by the closing and opening of an electrical contact.

The amount of heat transferred from the heat source to the distribution system over time is regulated by turning the heat source on and off. This situation is illustrated in Figure 6-7.

Part (a) of this illustration is a hypothetical **building heating load profile**. The load begins at zero, rises at a constant rate to its maximum design value, and then decreases at a steady rate back to half its design value. The time over which this occurs can be assumed to be a few hours.

The red shaded rectangles represent the heat generated by an on/off heat source. The height of all these pulses is the same, because the heat source, whenever it is on, creates a fixed rate of heat output.

In this scenario, the rated heat output of the heat source is assumed equal to the maximum (design) heating load of the building. Thus the height of the rectangles equals the maximum height of the load profile.

The width of the rectangles increases as the load increases. This means that the duration of the on cycle is increasing. When design load occurs, the rectangle remains uninterrupted. Since the heat source is sized to the design load of the building, it must operate continuously whenever design load conditions exist. As the load decreases, the rectangles become narrower. When the load stabilizes at 50 percent of design load, the width of the rectangles remain constant. Under this condition, the heat source is on 50 percent of the time. When on, its rate of heat output is twice the rate at which the building loses heat.

The red shaded area of each rectangle represents the quantity of heat delivered by the heat source during that on cycle.

Ideally, the blue shaded area under the load profile, out to some time, exactly equals the total area of

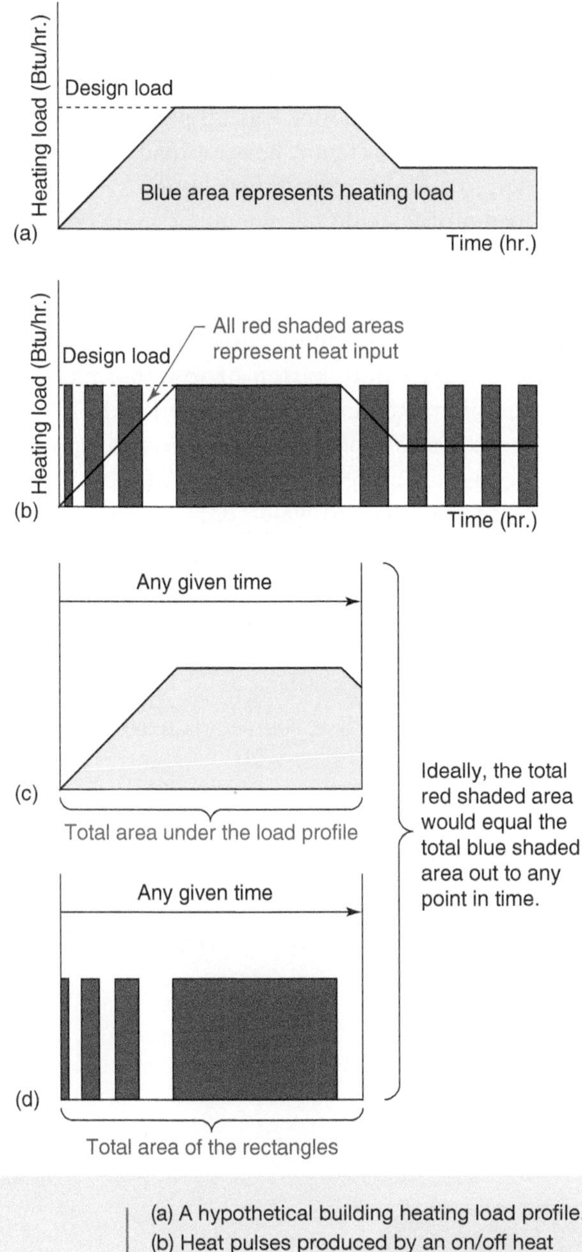

Figure 6-7 (a) A hypothetical building heating load profile. (b) Heat pulses produced by an on/off heat source. (c) Blue shaded area represents total heat added to building over a given period. (d) Ideally, the total area of the red shaded rectangles over the same period equals the blue shaded area under the load profile.

the red shaded rectangles, from the beginning of the graph to that time. This implies that the total heat produced by the heat source exactly matches the total heat required by the load over that period.

The match between heat output and heating load at low and medium load conditions is certainly not ideal. For example, consider the time when the load is very small. Under such conditions, the heat source sends short pulses of heat to the load. The height of these pulses corresponds to the full heating capacity of the heat source, which under partial load conditions may be several times higher than the load.

In heating systems with low thermal mass, on/off cycling of the heat source is often noticeable and usually undesirable. This condition is called "short cycling." It significantly increases wear on heat source components such as ignition systems, compressors, and contactors. In heat sources that combust fuel, short cycling increases emissions and decreases thermal efficiency.

Some hydronic systems, specifically those with thermal mass from significant water content (e.g., a thermal storage tank), or concrete (a heated slab), are better suited to accept pulses of heat input from an on/off heat source, without experiencing short cycling or causing noticeable variations in heat delivery to heated spaced. Many of the system discussed in later chapters leverage these forms of thermal mass to stabilize their operation.

6.5 Control of Modulating Heat Sources

In an ideal space heating system, the rate of heat output from the heat source would always match the rate of heat loss from the building. The heat output rate from the heat source could take on any value from zero to the full design load of the building. As the load changed due to changes in outdoor conditions or internal heat gain, the heat source would instantly readjust its output to match the load.

Figure 6-8 shows the same building heating load profile as Figure 6-7 as well as how a fully modulating heat source, ones that's perfectly matched to the load, would adjust its rate of heat output to that load.

This illustration shows what would be the *ideal* match between heat production and heating load. The area under the load profile is now fully shaded. Unlike the matching produced by an on/off heat source, and shown in Figure 6-7, there are no "white spaces" under the load profile, nor any instances when heat production exceeds heating load.

This type of heat production is described as **fully modulating**. The only currently available hydronic

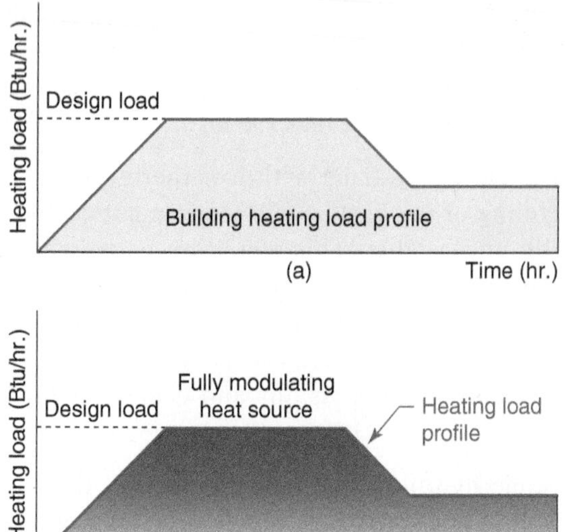

Figure 6-8 An "ideal" match between heat output of a *fully* modulating heat source (b) and a hypothetical load profile (a).

heat source that can produce this type of output is an electric resistance boiler, in which the wattage to the heating elements can be controlled from 0 to 100 percent using solid-state **triacs** to regulate current flow.

Many currently available, residential scale "mod/con" boilers, that operate on either natural gas or propane, can reduce (e.g., "**modulate**") their heat output down to approximately 20 percent of their full rated heat output. Thus, when the heating load they supply is 20 percent or more of their full output, they can accurately match their output to that load, as shown in Figure 6-9.

Notice that the mod/con boiler supplies pulses of heat to the load when that load is less than 20 percent of its rated output. The shape of these short pulses is based on the boiler initially starting itself at about 30 percent of rated output, and then attempting to modulate its heat output downward over the first few seconds of operation. When the boiler's internal controls find that the load is too low to be matched by downward modulation of the combustion system, they turn off the combustion system.

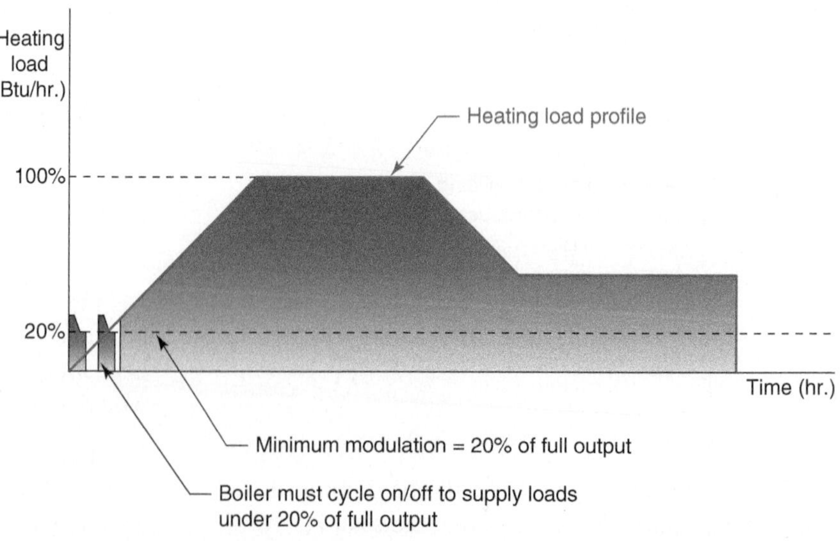

Figure 6-9 The match between the heat output of a mod/con boiler with a minimum modulation rate of 20 percent of full capacity and a hypothetical heating load profile. This graph assumes that the maximum boiler heat output equals the maximum heating load.

One might argue that the match between the building heating load profile and boiler heat output shown in Figure 6-9 is "not too bad." However, consider what would happen if this boiler was 50 percent oversized compared to the maximum heating load and directly connected to a highly zoned distribution system. The actual load profile could become much more complex, and thus harder to match, as shown in Figure 6-10.

The response of the mod/con boiler to this load profile produces significant short cycling. Over the last decade, the combination of low thermal mass mod/con boilers and extensively zoned hydronic distribution systems has produced many systems in which short cycling has been a problem. On/off heat pumps, both geothermal water-to-water, or air-to-water, would experience even worse short cycling as they attempted to match the heating load profile shown in Figure 6-10.

The following design guidelines will help reduce short cycling:

1. Don't oversize the heat source relative to the design heating load.
2. Use a heat source with high thermal mass and good insulation.
3. Use a **buffer tank** in systems that combine a low thermal mass heat source with a highly zoned distribution system.

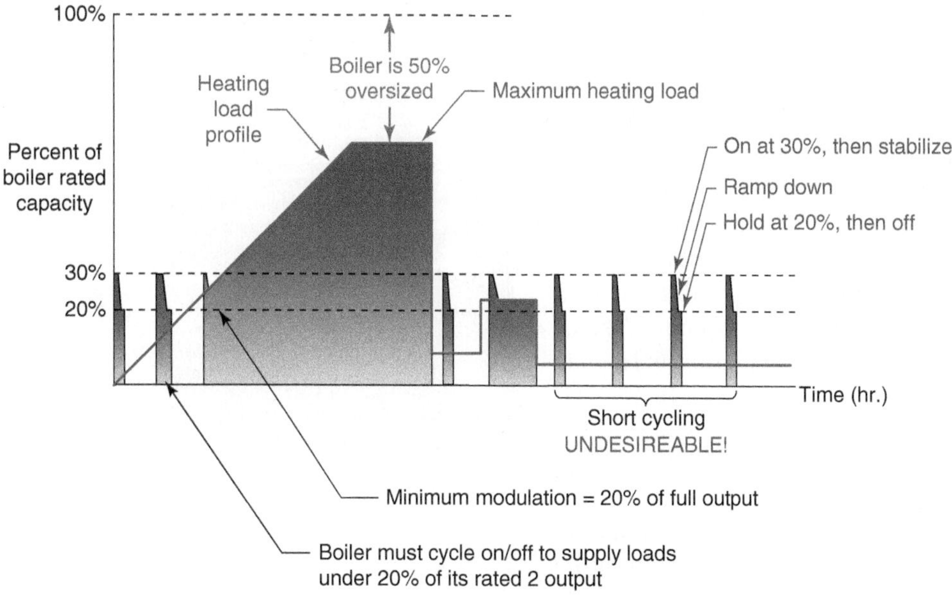

Figure 6-10 | A more complex heating load profile and how a mod/con boiler with a minimum modulation rate of 20 percent responds to that load.

6.6 Controlling the Heat Output of Heat Emitters

The primary goal of a heating control system is to maintain comfort by regulating the heat output of the system's heat emitters. This section describes two fundamental approaches and lays the groundwork for several types of control techniques and hardware that are discussed in later sections and chapters.

There are two fundamental methods of controlling the heat output of hydronic heat emitters:

1. *Vary the water temperature supplied to the heat emitter while maintaining a constant flow rate through the heat emitter.*
2. *Vary the flow rate through the heat emitter while maintaining a constant supply water temperature to the heat emitter.*

Both approaches have been successfully used in many types of hydronic heating applications over several decades. It is important for system designers to understand the differences as well as the strengths and weaknesses of each approach.

VARIABLE WATER TEMPERATURE CONTROL

Chapter 5 discussed how the heat output of a heat emitter increases in proportion to the difference between supply water temperature and room air temperature. This can be represented mathematically as follows:

(Equation 6.1a)

$$Q_O = k(T_s - T_r)$$

where:

Q_o = rate of heat output from heat emitter (Btu/hr)

k = a constant dependent on the heat emitter used

T_s = fluid temperature supplied to the heat emitter (°F)

T_r = air temperature of room where heat emitter is located (°F)

Every hydronic distribution system has its own value of k. For systems that have not yet been constructed, but for which the heat emitters and distribution system piping have been specified, the value of k is found by dividing the design load heat output of the distribution system by the difference between the supply water temperature at design load, minus the room air temperature.

It is also possible to determining the value of k for an *existing* hydronic distribution system. This requires instruments that can measure the rate of heat transfer to the distribution system. A later section in this chapter discusses these heat metering instruments. The value of k would be determined by measuring the rate of heat dissipation of the distribution system, under steady state conditions, and also recording the corresponding supply water temperature and room air temperature. Then, these numbers are used with Equation 6.1a and solve for k.

A graph of Formula 6-1a is a sloping straight line, as shown in Figure 6-11.

This graph could represent the heat output of a single heat emitter, such as a panel radiator, or it could represent the *total* heat output of a group of heat emitters, which are supplied by a parallel piping distribution system, as shown in Figure 6-12. Whichever is the case, the value of "k" in Formula 6-1a, that produces the graph in Figure 6-11, is 200 Btu/hr/°F.

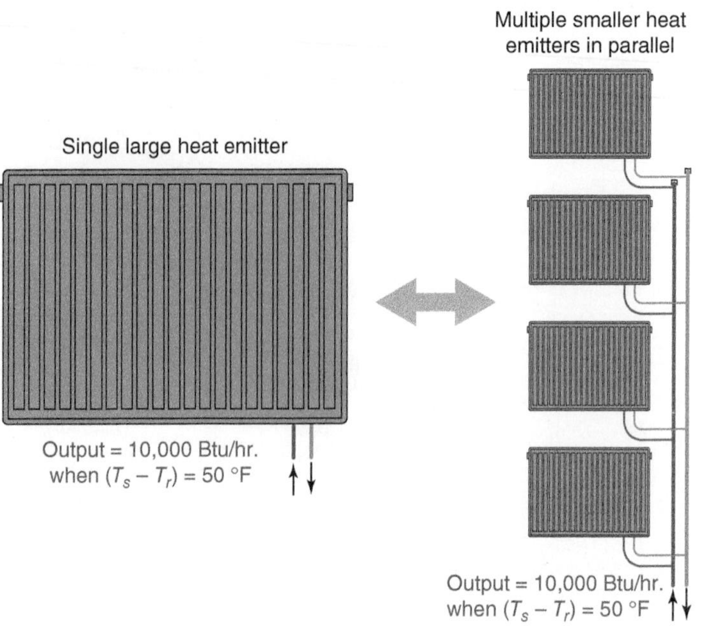

Figure 6-12 The proportional relationship between heat output versus (T_s–T_r) applies to either a single heat emitter or a group of heat emitters piped in parallel.

If water is supplied to the heat emitter(s) at room temperature (whatever that temperature happens to be), the difference (T_s–T_r) is zero, and so is the heat output from the heat emitter(s). As the temperature of the water supplied to the heat emitter(s) climbs above room air temperature, the heat output from the emitters also increases. The graph in Figure 6-11 shows that when the supply water temperature is 50 °F above the room air temperature, the heat emitter(s) will release 10,000 Btu/hr into the building. If the supply water temperature is only 25 °F above the room's air temperature, the heat emitter(s) releases 5,000 Btu/hr into the building. Thus, reducing the temperature difference by 50 percent also reduces the heat output from the heat emitter(s) by 50 percent.

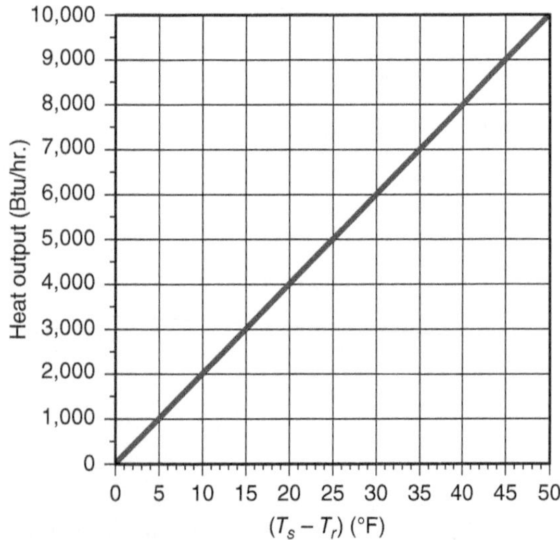

Figure 6-11 The proportional relationship between heat output and temperature difference between supply water temperature and room air temperature (°F).

For the heat emitter(s) represented in Figure 6-11, an increase of 1 °F in supply water temperature produces an increase of 200 Btu/hr in heat output, regardless of the starting temperature. Likewise, a decrease of 1 °F in supply water temperature reduces heat output by 200 Btu/hr.

If one assumes a desired room air temperature of 70°F, the graph shown in Figure 6-11 can be modified to that shown in Figure 6-13.

The only difference between Figures 6-11 and 6-13 is that the supply water temperature, rather than the difference (T_s–T_r), is shown on the horizontal axis. This proportional relationship between supply water temperature and heat output holds true, as a reasonable approximation, for most hydronic heat emitters. It is an important characteristic in the context of another control technique called **outdoor reset control**, which will be discussed later.

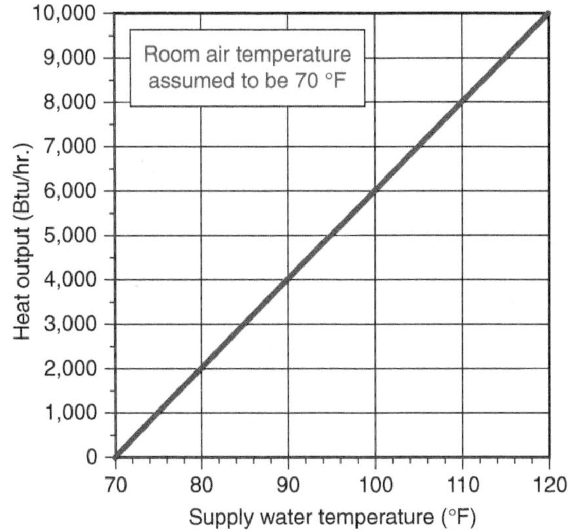

Figure 6-13 The proportional relationship between heat output and supply water temperature, assuming a room air temperature of 70 °F.

The slope of the line in a graph such as Figure 6-13 depends on the characteristics of the heat emitter. The greater the ability of the heat emitter to release heat, at a given supply water temperature, the steeper the slope of the line. For example, if a given size panel radiator produced the relationship shown in Figure 6-13, then a larger panel radiator would produce the same heat output at a lower supply water temperature and thus be represented by a steeper line. A smaller panel radiator would require a higher supply water temperature to produce this output and thus have a shallower line. This relationship is shown in Figure 6-14. The slope of the heat output versus supply water temperature line changes, depending on the heat dissipation ability of the heat emitter(s).

This type of linear relationship between the controlled variable (heat output) and the manipulated variable (supply water temperature) is preferable from a control standpoint. This relationship is relatively easy to simulate using electromechanical and electronic controls.

VARIABLE FLOW RATE CONTROL

The graph in Figure 6-15 shows how the heat output of a typical floor heating circuit varies as a function of the water flow rate through it, assuming the supply water temperature is held constant at 105 °F.

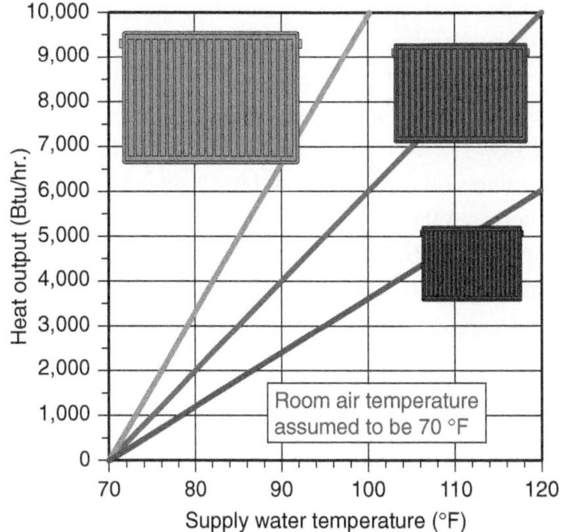

Figure 6-14 The greater the ability of a heat emitter to release heat, the steeper the line in the graph of its heat output versus supply water temperature.

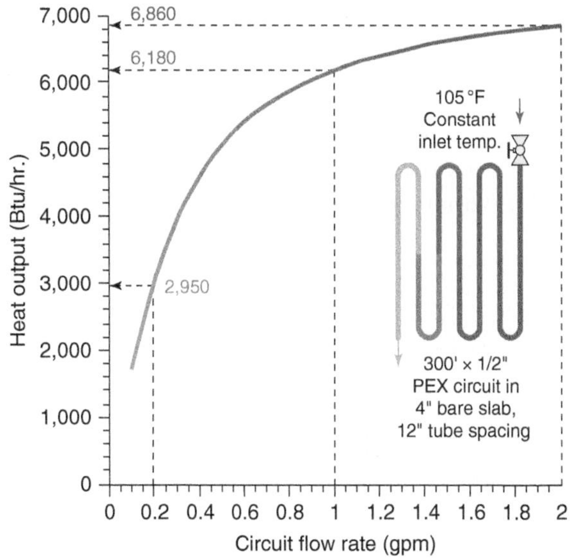

Figure 6-15 Heat output of a floor heating circuit supplied with water at 105 °F as a function of flow rate through it. The circuit is a 300 foot length of 1/2-inch PEX tubing embedded in a 4-inch bare concrete slab at 12-inch spacing. The room temperature is 70 °F.

This circuit's maximum heat output of 6,860 Btu/hr occurs at the maximum flow rate shown (2.0 gpm). Decreasing the circuit's flow rate by 50 percent to 1.0 gpm decreases its heat output to 6,180 Btu/hr, a drop of only about 10 percent. Reducing the flow rate to 10 percent of the maximum value (0.2 gpm) still allows the circuit to release 2,950 Btu/hr, about 43 percent of its maximum output. These numbers show that heat transfer decreases slowly at higher flow rates, but much faster at lower flow rates. This characteristic is typical of *all* hydronic heat emitters (and was discussed in Chapter 5). The relationship between the controlled variable (heat output) and the manipulated variable (flow rate) is very **nonlinear**. This relationship makes accurate control of heat output more challenging, especially under low-load conditions.

To provide accurate control under low-load conditions, the controlled device, which is typically a modulating valve, must compensate for the rapid rise in heat output at low flow rates as well as the much slower rise in heat output at higher flow rates. Special valves with an **equal percentage characteristic** have been developed for this purpose. The flow rate through such a valve as a function of its stem position is shown in Figure 6-16.

When a valve with an equal percentage characteristic regulates the flow rate through a hydronic heat emitter, the resulting relationship between the valve's stem position, and heat output from the heat emitter, is close to proportional, as shown in Figure 6-17.

The very slow increase in flow rate when the valve first begins to open compensates for the rapid rise in heat output from the heat emitter at low flow rates.

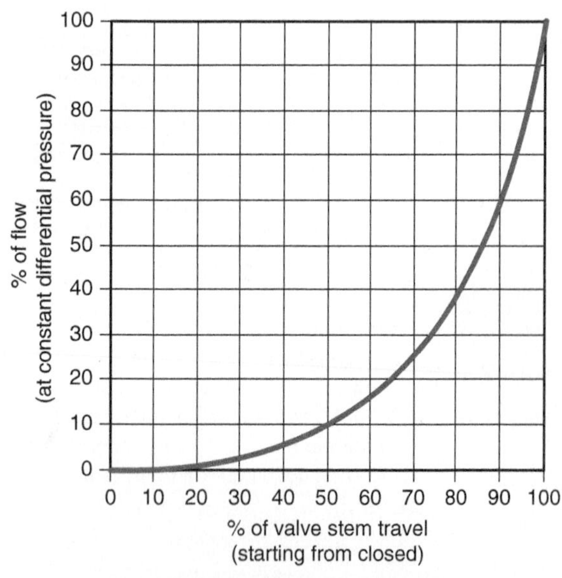

Figure 6-16 Percentage of flow through valve versus percentage of stem travel for a valve with an equal percentage characteristic operated at a constant differential pressure.

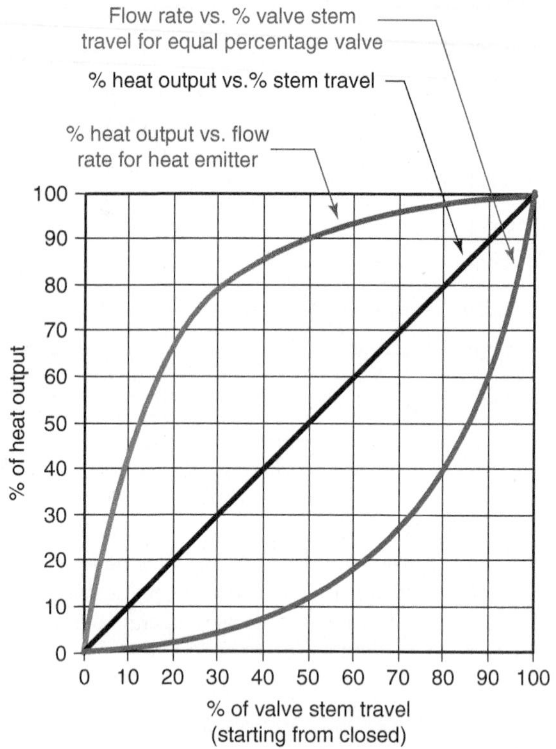

Figure 6-17 Heat output of heat emitter is approximately proportional to percentage of valve stem travel when flow through heat emitter is controlled by an equal percentage valve operated at a constant differential pressure. This is a desirable operating characteristic. Note: Supply water temperature to heat emitter is assumed constant.

Likewise, the rapid increase in flow at higher flow rates compensates for the relatively slow rise in heat output at high flow rates. The "net effect" is approximately linear, which is desirable from a control standpoint. It is especially important under low-load conditions where a quick opening valve (i.e., a standard ball valve) would make accurate and stable heat output control extremely difficult. *Valves with an equal percentage characteristic should be used in any situation where heat output is controlled by varying the flow rate through a heat emitter.*

6.7 Outdoor Reset Control

As outdoor temperatures change, so does the heating load of a building. Ideally, every heating system would constantly adjust its rate of heat delivery to match its building's current rate of heat loss. This would allow inside air temperature to remain constant regardless of outside conditions. Outdoor reset control was developed to do this by changing the temperature of the water supplied to the system in response to outdoor temperature.

BENEFITS OF OUTDOOR RESET CONTROL

There are several benefits associated with outdoor reset control, including:

- **Optimal use of renewable energy:** Outdoor reset control allows a renewable energy heat source to operate at the lowest possible water temperature that can satisfy the heating load of the building, based on current outdoor conditions. Reduced water temperatures improve the thermal efficiency and heating capacity of all the renewable energy heat sources discussed in this text. Outdoor reset control can also provide the "logic" necessary to turn on an auxiliary heat source when the ability to supply the heating load from a thermal storage tank heated by a renewable energy heat source is exhausted. With proper design, this transition occurs at the lowest possible water temperature and is "seamless" in terms of sustaining comfort within the building.
- **Stable indoor temperature:** Outdoor reset control reduces fluctuation of indoor temperature. When the reset control is properly adjusted, the water temperature supplied to the heat emitters is just high enough for the prevailing heating load. The rate of heat delivery is always maintained very close to the rate of building heat loss. This yields very stable indoor temperature compared to less sophisticated hydronic systems that deliver water to the heat emitters as if it were always the coldest day of winter. In the latter case, the flow of heated water to the heat emitters must be turned on and off to prevent overheating under partial load conditions. This often creates an easily detected and undesirable sensation that the heat delivery system is on versus off. With properly adjusted outdoor reset control, building occupants should have no sensation that heat delivery is on or off, just a sensation of continuous comfort.
- **Near-continuous circulation:** Because outdoor reset control supplies water just hot enough to meet the current heating load, the distribution circulator remains on most of the time. Flow through the distribution system stops only to prevent overheating when internal heat gains from sunlight, equipment, and interior lighting are present.

  Near-continuous flow can redistribute heat within heated floor slabs. Heat stored within interior portions of such slabs can be absorbed by the circulating water and carried to cooler perimeter areas of the slab. This can occur even when there is no heat input to the slab from the system's heat source. It is especially helpful when the slab has one or more **heat sinks**, such as the floor area just inside an overhead door where there is an insufficient **thermal break** between the interior slab and exterior pavement. The redistributed heat can help prevent freezing in such area during times when the heat source is off for several hours.
- **Reduced expansion noise:** The combination of near-continuous circulation and very gradual changes in water temperature minimizes expansion noises from the distribution piping and heat emitters. This is especially important if PEX tubing is used with metal heat transfer plates in radiant panel heating systems. During a typical heating season, the piping and heat emitters will experience thermal expansion movement similar to that in systems not using outdoor reset control. However, when outdoor reset control is used, the *expansion movement takes place over days, even weeks, compared to what might only be seconds in systems that simply turn the flow of hot water on and off.* Piping expansion noise is much more noticeable in systems where rapid changes in water temperature occur.
- **Indoor temperature limiting:** When water is supplied to the heat emitters at design temperature regardless of the load, occupants can choose to set the thermostat to a high temperature and open windows and doors to control overheating. Although this sounds like a foolish way to control comfort, it's often done in rental properties where tenants don't pay for their heat. However, if supply water temperature is regulated by outdoor reset control, it's just warm enough to meet the heating load with the windows and doors closed. Wasteful use of energy is discouraged.
- **Reduced energy consumption:** Outdoor reset control has demonstrated its ability to reduce fuel consumption in hydronic heating systems. The savings are a combination of reduced heat loss from boilers, reduced heat loss from distribution piping, and, in the case of condensing boilers, increased time in condensing mode operation. Exact savings will vary from one project to another. Conservative estimates of 10 to 15 percent are often cited.

ANALYTICS OF OUTDOOR RESET CONTROL

An **outdoor reset controller** continuously calculates the ideal **"target" temperature** of the fluid that should be supplied to a hydronic distribution system, at any time. This temperature depends on the type of heat emitters used in the system, as well as the current outdoor temperature. It therefore has the potential to change from one moment to the next.

Outdoor reset controllers use Formula 6-2 to determine the target water temperature.

Formula 6-2

$$T_{target} = T_{indoor} + (RR) \times (T_{indoor} - T_{outdoor})$$

where:

T_{target} = the "ideal" target supply water temperature to the system

T_{indoor} = desired indoor air temperature

RR = **reset ratio** (slope of reset line)

The graph in Figure 6-18 is a good way to visualize these relationships. In this case, the desired indoor temperature is assumed to be 70 °F, and the reset ratio (RR) to be 0.5.

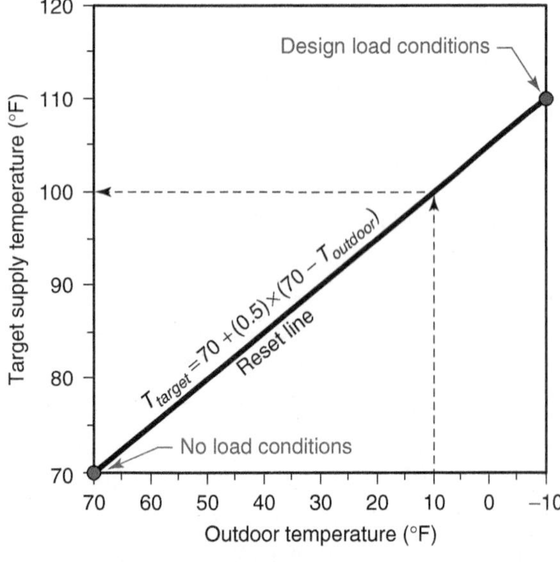

Figure 6-18 Calculated target supply water temperature as a function of outdoor temperature. The reset ratio = 0.5, and the room air temperature is assumed to be 70 °F.

The red dot in the upper right portion of the graph represents **design load conditions** (e.g., the coldest day of winter). For the graph as shown, the red dot indicates that the target supply water temperature should be 110 °F when the outdoor temperature is −10 °F.

The blue dot in the lower left corner represents **no load conditions** (e.g., where no heat output is needed from the heat emitters). Thus the target supply water temperature would be 70 °F when the outdoor temperature is 70 °F.

The sloping line that connects these two dots is called a **reset line**. Every hydronic distribution system can be thought of as having its own reset line. The slope of that line depends on the type of heat emitters used, how they are sized, and the range of outdoor temperature at the installation site.

The mathematical slope of a reset line is called the reset ratio. It can be calculated as the change in supply water temperature divided by the change in outdoor temperature between any two points on the reset line. This is represented by Equation 6-3. The end points of the reset line are typically used to make this calculation.

(Equation 6-3)

$$RR = \frac{\Delta T_{supply\ water}}{\Delta T_{outdoor}}$$

where:

RR = reset ratio

$\Delta T_{supply\ water}$ = change in supply water temperature between design load and no load conditions (°F)

$\Delta T_{outdoor}$ = change in outdoor temperature between design load and no load condition (°F)

Example 6-1: A building has a design heat loss of 80,000 Btu/hr when the indoor temperature is 70 °F and the outdoor temperature is −10 °F. The heat distribution system for this building has been designed so that it can release 80,000 Btu/hr when the supply water temperature is 110 °F and the inside air temperature is 70 °F. Determine the reset ratio for this system.

Solution: The change in water temperature between the no load and design load conditions is 110 − 70 = 40 °F. The corresponding change in outdoor temperature is

70 − (−10) = 80 °F. Thus, the necessary reset ratio when this heat distribution system is used in this building is:

$$RR = \frac{\Delta T_{\text{supply water}}}{\Delta T_{\text{outdoor}}} = \frac{40}{80} = 0.5$$

Discussion: *Every hydronic distribution system, in combination with the building in which it is installed, yields a unique reset line.* A building equipped with slab-type floor heating will have a different reset line compared to the same building using fin-tube baseboard convectors. This is due to the difference in supply water temperatures commonly used for these types of heat emitters.

Here are several other facts regarding the reset ratio:

- *The greater the value of the reset ratio, the steeper the reset line.*
- *The reset ratio can be interpreted as the necessary increase in supply water temperature per degree drop in outside temperature.*
- *The reset ratio is a unitless quantity and does not change when calculated using different temperature units (provided the same temperature units are used in the numerator and denominator).*

Figure 6-19 shows some representative reset lines and their associated reset ratios (RR) for several types of heat emitters that have been sized based on "customary" water temperatures. These water temperatures are not necessarily optimal or recommended, especially when the system is supplied by a renewable energy heat source.

According to this graph, the fin-tube baseboard distribution system has been sized so that it can provide design load heat output when the outdoor temperature is −10 °F, if supplied with water at 180 °F. Likewise, the panel radiator distribution system has been sized so that it can provide design load output when the outdoor temperature is −10 °F, if supplied with water at 150 °F. The lowest (e.g., "shallowest") reset line on the graph is for a radiant floor panel. It has been sized to provide design load output when the outdoor temperature is −10 °F, if supplied with water at 105 °F.

Keep in mind that the *sizing* of a given type of heat emitter, rather than choosing a different type of heat emitter, can have a significant effect on the reset ratio. For example, if a designer chooses to use larger panel radiators, which can meet the design heating load at lower supply water temperatures, the distribution system will have a lower the reset ratio, as illustrated in Figure 6-20.

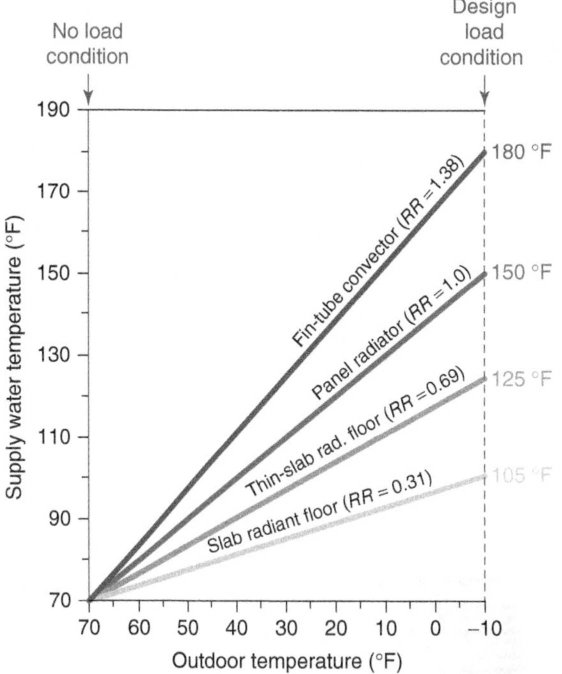

Figure 6-19 Representative reset lines for different types of heat emitters, based on "customary" selection of supply water temperature.

This is also true for other types of heat emitters such as fan coils, fin-tube baseboard, or site-built radiant panels. When the distribution system is supplied by renewable energy heat sources, the author recommends that heat emitters be selected, sized, and piped so that the distribution system can meet the building's design heating load with a supply water temperature no higher than 120 °F.

Readers with more interest in the theory underlying outdoor reset control, and the benefits it offers, are encouraged to read chapter 9 in *Modern Hydronic Heating,* 3rd ed.

IMPLEMENTING OUTDOOR RESET CONTROL

There are three ways to implement outdoor reset control in hydronic systems:

- **Heat source reset** (for on/off heat sources)
- Heat source reset (for modulating heat sources)
- **Mixing reset**

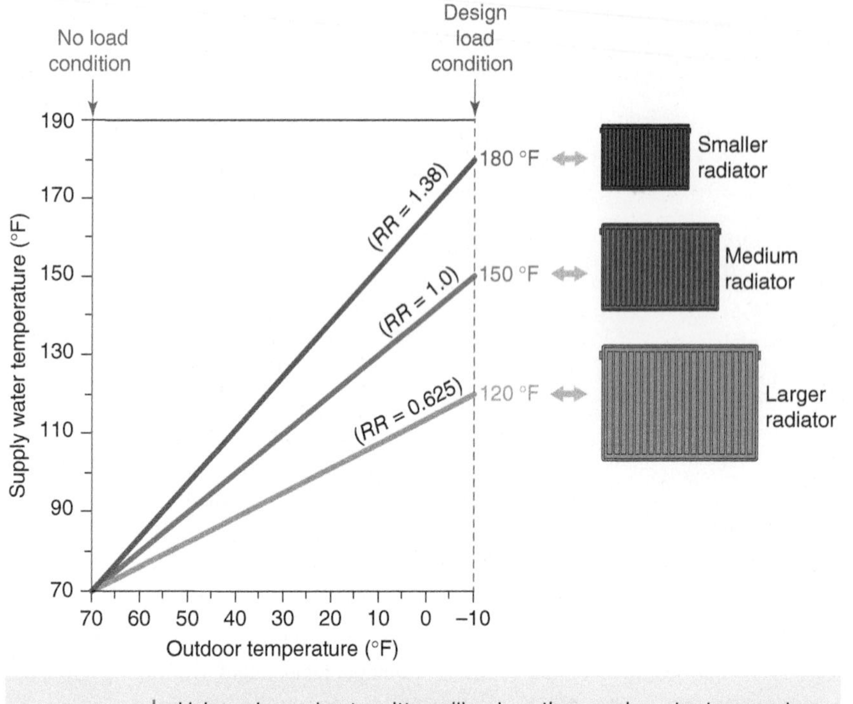

Figure 6-20 Using a larger heat emitter will reduce the supply water temperature required for a given rate of heat output, which will decrease the slope of the system's reset line.

These methods for implementing outdoor reset control require different controllers and can be used individually or combined when necessary.

HEAT SOURCE RESET (FOR ON/OFF HEAT SOURCES)

In some systems, the temperature of water supplied to a distribution system can be controlled by turning a heat source such as a boiler or heat pump on and off. This only requires the closure of an **electrical contact** within the outdoor reset controller. When this contact closes, the heat source is enabled to operate. When it opens, the heat source is turned off.

As described earlier in this chapter, on/off control action requires a differential between the temperature at which the electrical contacts open and when they close. Without this differential, the controller would be constantly switching between on and off, based on very minor changes in temperature. This short cycling is very undesirable because it shortens the life of components such as boiler ignition systems or heat pump compressors. It also increases emissions and reduces thermal efficiency.

The value of the differential can be adjusted on most outdoor reset controllers. Smaller control differentials reduce the variation in supply water temperature both above and below the target temperature. However, if the control differential is too small, the heat source will short cycle.

The differential of a typical outdoor reset controller is set during installation, based on the setting specified by the system's designer. Some outdoor reset controllers allow for differentials ranging from 4 °F to perhaps 40 °F. The wider the differential, the longer the operating time of the heat source. However, wider differentials also allow more deviation between the calculated target supply water temperature and the actual water temperature supplied to the distribution system.

Figure 6-21 illustrates the logic used by a typical reset controller that controls an on/off heat source. The sloping blue line is the reset line for the system, and as such, represents the ideal *target* supply water temperature over a range of outdoor temperatures. For example, if the outdoor temperature is 10 °F, the reset line indicates a target temperature of 105 °F, as indicated by the blue dot.

This graph also indicates that the differential of this outdoor reset controller has been set to 10 °F, and that

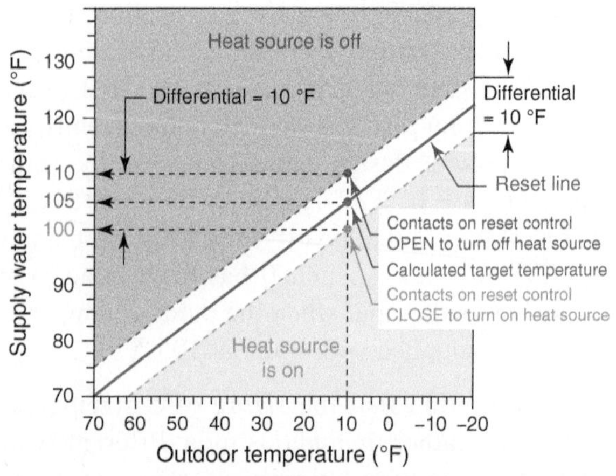

Figure 6-21 Outdoor reset control logic for an on/off heat source.

that differential is "centered" on the blue reset line (e.g., 5 °F above the reset line and 5 °F below the reset line).

When the outdoor reset controller is powered on, it immediately measures the outdoor temperature. It then uses this temperature, along with its settings, to calculate the target supply water temperature. Next, it compares the target temperature to the measured supply water temperature.

If the measured supply water temperature is equal or close to the target temperature, as represented by the sloping white band in Figure 6-21, no control action is taken. However, if there is sufficient deviation between the calculated target temperature and the measured supply water temperature, the outdoor reset controller takes action. To describe that action, assume that an outdoor reset controller is set as shown in Figure 6-21 and that the outdoor temperature is 10 °F:

- If the water temperature supplied to the distribution system is above 110 °F (e.g., 105 °F plus half the 10 °F differential), the contacts within the reset controller remain open and the heat source remains off.
- If the water temperature supplied to the distribution system is below 100 °F (e.g., 105 °F target minus half the 10 °F differential), the contacts within the reset controller close to turn on the heat source.

Once turned on, the heat source remains on until the water supplied to the distribution system reaches 110 °F (105 °F plus one half the control differential) or higher, or until the outdoor reset controller is turned off.

Figure 6-22 shows an example of a modern outdoor reset controller for an on/off heat source.

This particular controller is powered by 24 VAC, which is applied to the controller when switch (D1), as shown in Figure 6-22b, is closed. Sensors (S1) and (S2) measure the supply water temperature and outdoor temperature. The normally open relay contact seen between terminals 5 and 6 closes to turn on the heat source. When this contact opens, the heat source is turned off. Although the heat source is indicated as boiler, this controller could be adapted to any other type of on/off heat source that can be enable by a contact closure.

On/off outdoor reset controllers can also be used to determine if the water temperature in a thermal storage tank that's heated by a renewable energy heat source, is high enough to meet the current heating load. When the outdoor reset controller is powered on, it measures the outdoor temperature and calculates the target temperature. It also measures the temperature of a sensor in the upper portion of the storage tank. If the latter temperature is at or above the target temperature plus one half the differential, the storage tank will be used as the "heat source" for the distribution system. If the temperature in the upper portion of the tank is equal to or lower than the target temperature minus one half the differential, the contacts within the outdoor reset controller close. This contact closure enables the auxiliary heat source to operate. It may also change the operating status of other components such as diverter valves or circulators that are needed to deliver heat from the auxiliary heat source rather than the thermal storage tank. This type of control action is used in many of the systems designs shown in later chapters.

Figure 6-22 | (a) A typical outdoor reset controller for an on/off heat source.

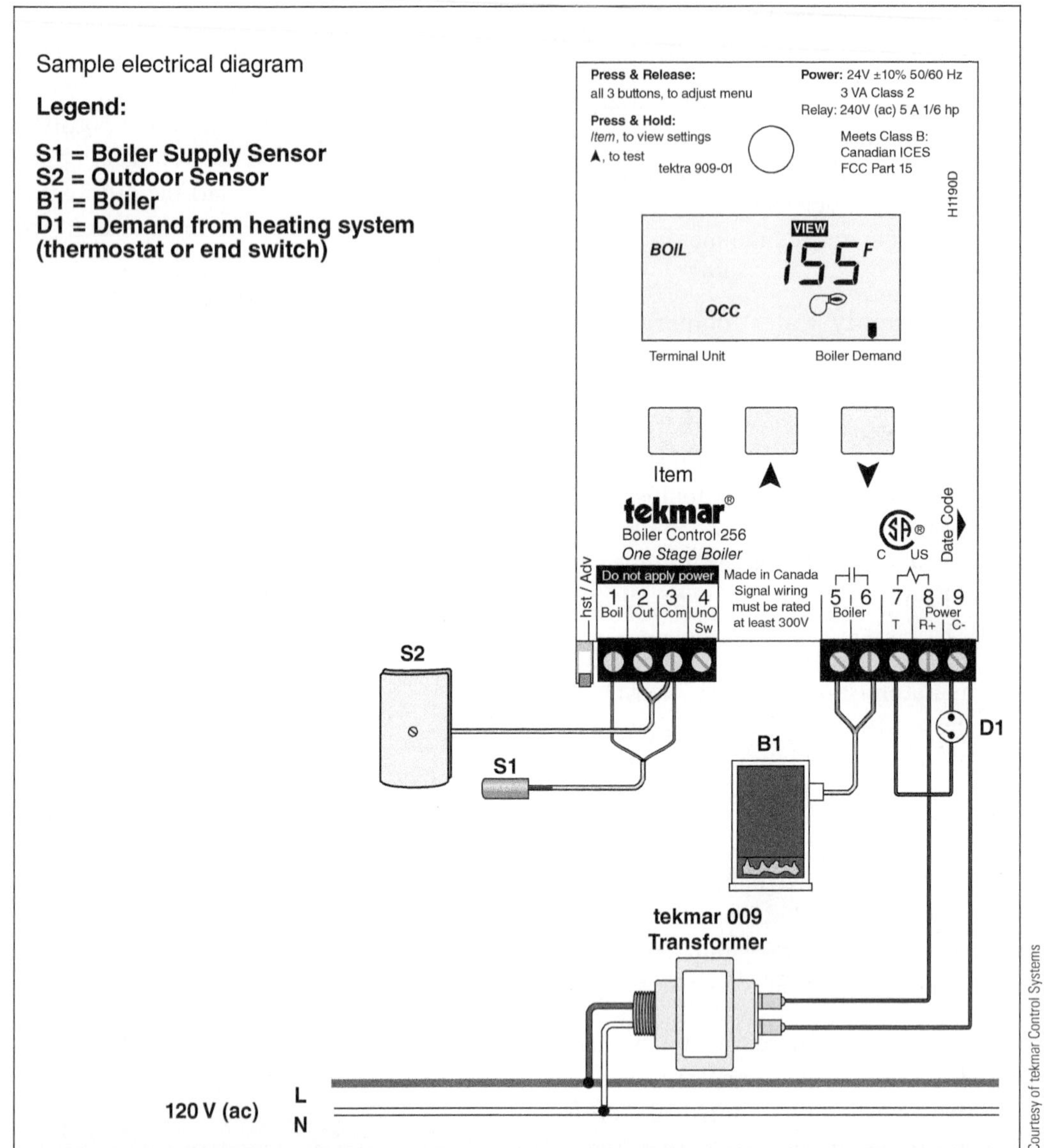

Figure 6-22 (b) Wiring of the controller shown in Figure 6-22a for an on/off heat source. The normally open relay contact shown between terminals 5 and 6 closes to turn on the heat source.

HEAT SOURCE RESET (FOR MODULATING HEAT SOURCES)

Many modern boilers, and even some heat pumps, can adjust their heat output over a relatively wide range. This is called modulation, and it allows the heat source to better match the heating load. Modulation reduces issues such as short cycling and temperature variations that are more common in systems using on/off heat sources.

Modulating heat sources typically have their own internal outdoor reset controllers. They continually measure outdoor temperature calculate the target supply water temperature, and compare it to the measured supply water temperature. Deviations between these temperatures cause the internal reset controller to regulate the speed of the combustion air blower on a modulating boiler or the compressor speed on a modulating heat pump. The goal is to keep the measured supply temperature very close to the calculated target temperature.

MIXING RESET

Outdoor reset control can also be implemented by any electronically controlled mixing assembly such as the motorized 3-way mixing valve shown in Figure 6-23. The **motorized actuator**, which creates rotation of the

Figure 6-23 | 3-way rotary type mixing valve with motorized actuator mounted to it.

valve's shaft, is mounted to the valve. Most of the actuators used for such valves are configured for **floating control**, which was discussed earlier in this chapter.

During normal operation, the motorized actuator slowly rotates the stem of the valve, as required to keep the mixed water temperature leaving the valve close to the calculated target value. The internal mechanism within the actuator can also be uncoupled from the valve's shaft if the valve needs to be manually adjusted, such as might be necessary during a power outage.

Figure 6-24 shows an example of a controller that creates a floating control output that can operate the motorized actuator shown in Figure 6-23.

The wires connected to terminals 9,10, and 11 are the three wires that connect the mixing controller to the motorized actuator on the valve. When the valve stem needs to rotate in the clockwise direction, 24 VAC is present between terminals 10 and 11. When the valve needs to rotate counterclockwise, 24 VAC is present between terminals 9 and 11. When the valve is to hold its present position, no voltage is sent to the motorized actuator.

Besides controlling the mixing valve, this particular controller can also operate the boiler and a system circulator. It has three temperature sensors. Sensor (S3) measure outdoor temperature. Sensor (S2) measures the water temperature supplied to the distribution system (e.g., downstream from the mixing valve). Sensor (S1) measures the boiler inlet temperature. This sensor is required on systems that use a **conventional boiler** (e.g., one that is not intended to operate with sustained flue gas condensation). It allows the mixing valve to limit hot water flow from the boiler into the mixing valve so that the boiler remains hot enough to avoid sustained flue gas condensation.

Some manufacturers offer 3-way motorized mixing valves that have the electronics necessary for outdoor reset logic integrated with the actuator. This eliminates the need for a separate controller. An example of such a valve is shown in Figure 6-25.

The logic used for mixing reset is similar to that already described for on/off and modulating heat sources. The outdoor temperature is measured and used along with the controller's settings, to calculate the target supply water temperature. The controller compares the target temperature to the measured supply water temperature to determine if any deviation (e.g., error) is present. The controller generates an output signal to operate the valve's actuator as necessary to eliminate any error.

Figure 6-26 shows how a motorized 3-way mixing valve would be piped in combination with a thermal storage tank that was heated by a renewable energy heat source such as solar collectors or a wood-fired boiler.

The outdoor reset controller operating the mixing valve only requires two sensors: outdoor temperature, and supply temperature to the distribution system. The third sensor is not required since the heat source is not a conventional boiler and doesn't require protection against sustained flue gas condensation.

Besides providing the optimal supply water temperature to the distribution system, the 3-way valve,

Courtesy of tekmar Control Systems

(a)

Sample electrical diagram

Legend

B1 = Boiler 1
D1 = Mix Demand
(From thermostat or end switch)
S1 = Boiler Supply Sensor
S2 = Mix Supply Sensor
S3 = Outdoor Sensor
P1 = Mix System pump
M1 = Floating Action Actuator

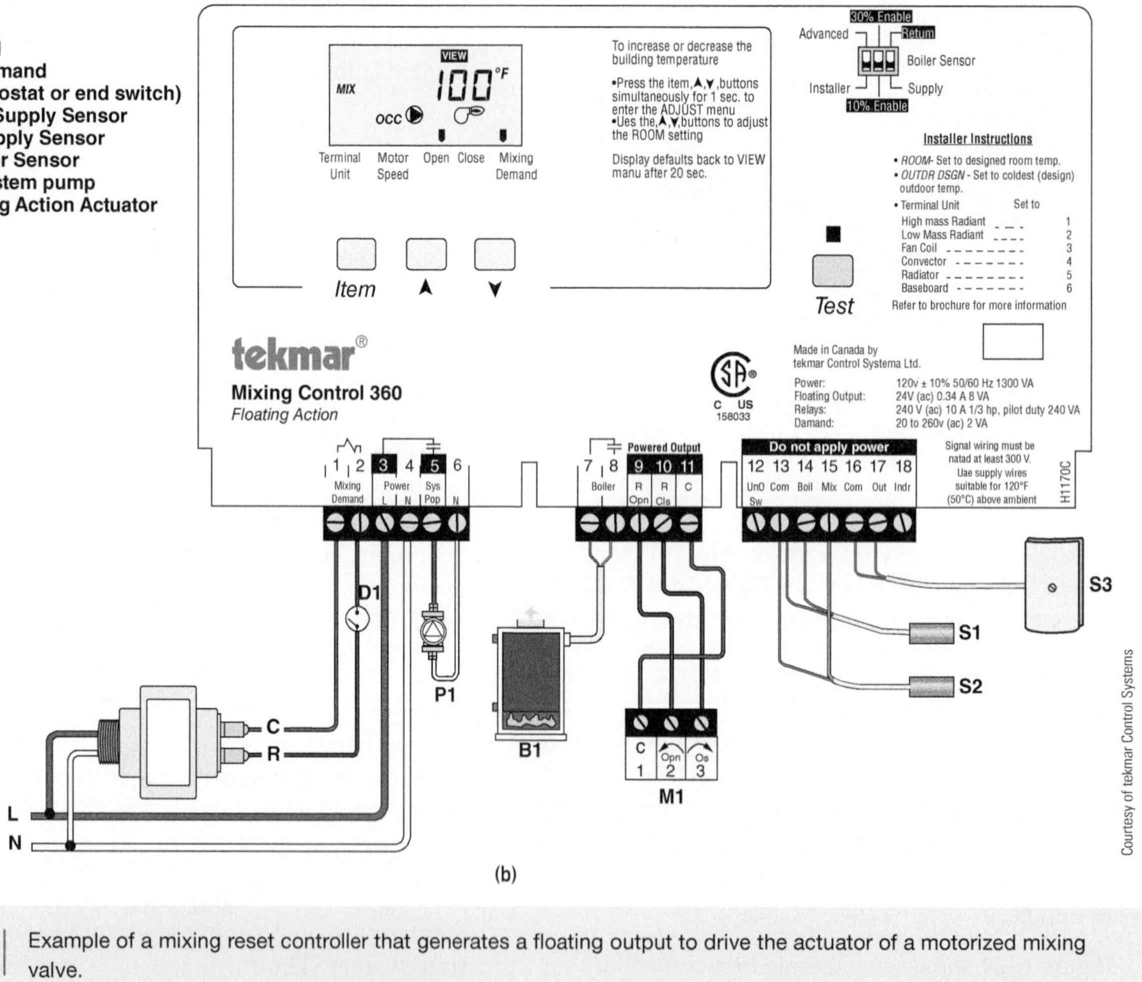

Courtesy of tekmar Control Systems

(b)

Figure 6-24 | Example of a mixing reset controller that generates a floating output to drive the actuator of a motorized mixing valve.

(a)

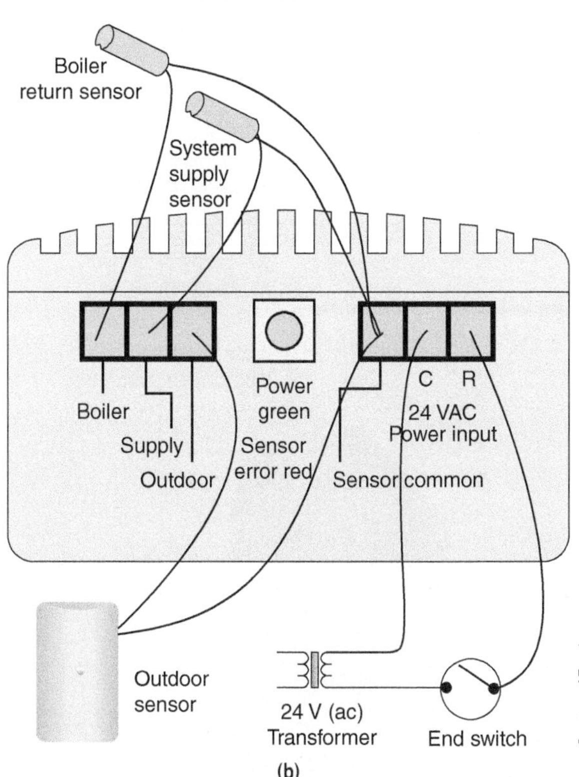

(b)

Figure 6-25 (a) Motorized 3-way mixing valve with self-contained logic for outdoor reset control. (b) Wiring diagram for this product.

operated by the outdoor reset controller, also protects low-temperature distribution systems from what could be very hot water in the thermal storage tank. Thus, this type of mixing assembly and controller is used on many systems that combine a renewable energy heat source with a low-temperature hydronic distribution system.

6.8 Differential Temperature Control

There are situations where a control action needs to be based on the *difference* between two temperatures rather than the value of either temperature. This is especially true in systems using renewable energy heat sources as well as systems that perform heat recovery. A special type of controller, appropriately named a **differential temperature controller**, is available for such applications.

A basic differential temperature controller has two temperature sensors: the **source temperature sensor** and the **storage temperature sensor**. The word "source" refers to any potential source of heat. Examples include a solar collector, a wood-fired boiler, or a tank containing heated water. The word "storage" refers to any potential destination for the heat from the source. The most common example is a thermal storage tank. However, "storage" could also be a swimming pool, a concrete slab, or any other media that could potentially accept heat from the source.

All basic differential temperature controllers also have two temperature differentials: the **on-differential** and the **off-differential**. The word "differential" refers to the difference in temperature between the source temperature sensor and the storage temperature sensor.

The controller closes its normally open contact whenever the temperature at the source temperature sensor is greater than or equal to the temperature at the storage temperature sensor, plus the *on*-differential. This logic can be represented as follows:

If $T_{source} \geq T_{storage} + \Delta T_{on}$, then relay contact CLOSES.

Once the relay contact has closed, the controller continues to monitor the temperatures at both sensors. If the temperature at the source sensor drops to, or below, the storage temperature + the *off*-differential, the relay contact opens. This logic can be represented as follow:

If $T_{source} \leq T_{storage} + \Delta T_{off}$, then relay contact OPENS.

The on-differential and off-differential of most modern differential temperature controllers are adjustable. Typical on-differentials are in the range of 9 °F to 15 °F. Typical off-differentials are in the range of 3 °F to 6 °F. The adjustment of these differentials must factor in the thermal mass of the heat source, as well as the heat loss of the piping between the source and

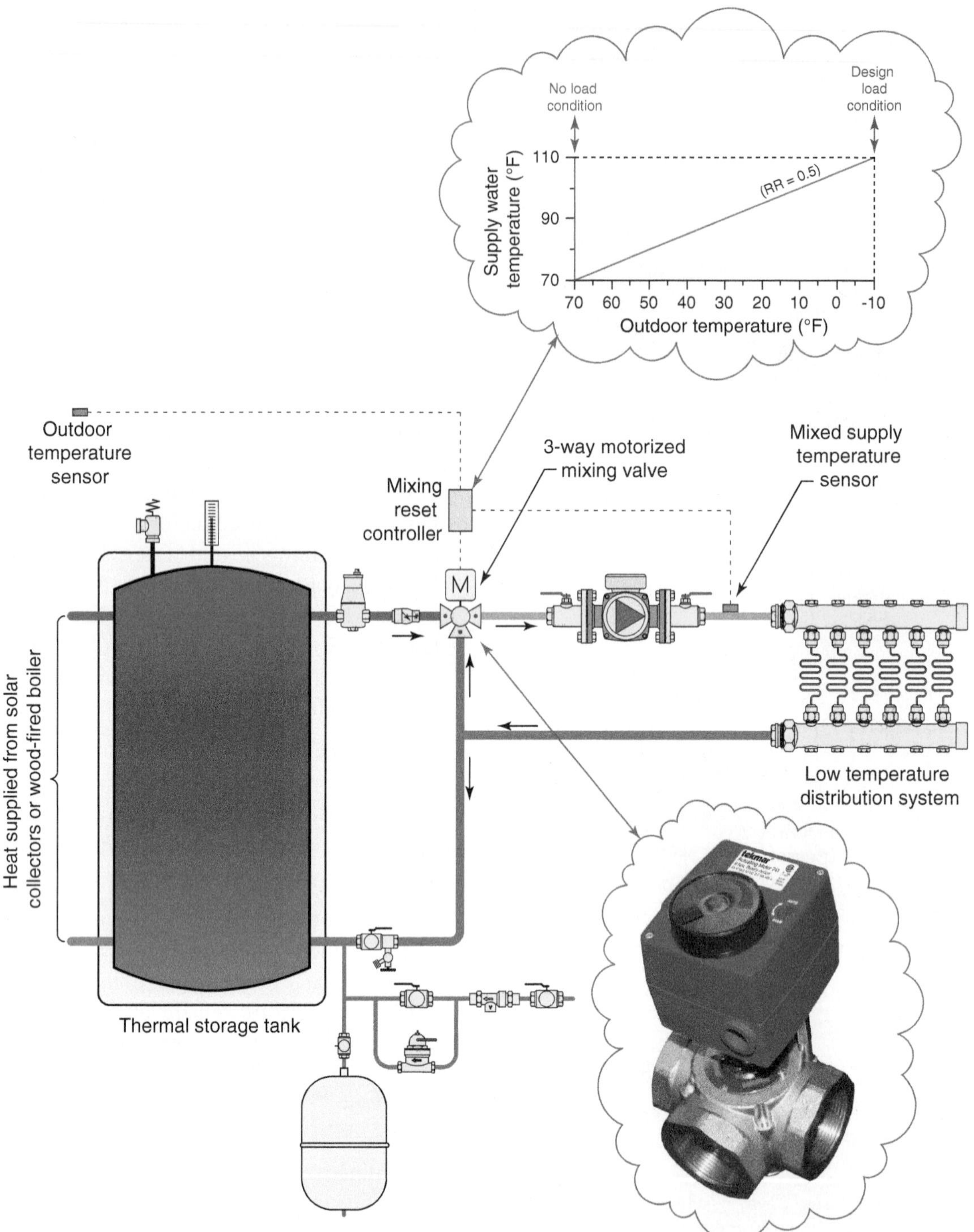

Figure 6-26 | 3-way motorized mixing valve, operated by an outdoor reset controller, regulates water temperature supplied from a thermal storage tank to a low temperature distribution system.

storage, and the flow rate through the circuit connecting the source to storage. The goal is to allow energy transfer from the source to storage whenever possible while also avoiding situations where the fluid may lose more heat than it gains if it continues to flow between the source and storage.

Figure 6-27 illustrates how the on-differential and off-differential relate to typical temperature changes within a solar collector and thermal storage tank during a sunny day.

Assuming a clear sky, the solar collector temperature sensor steadily warms during the morning and eventually

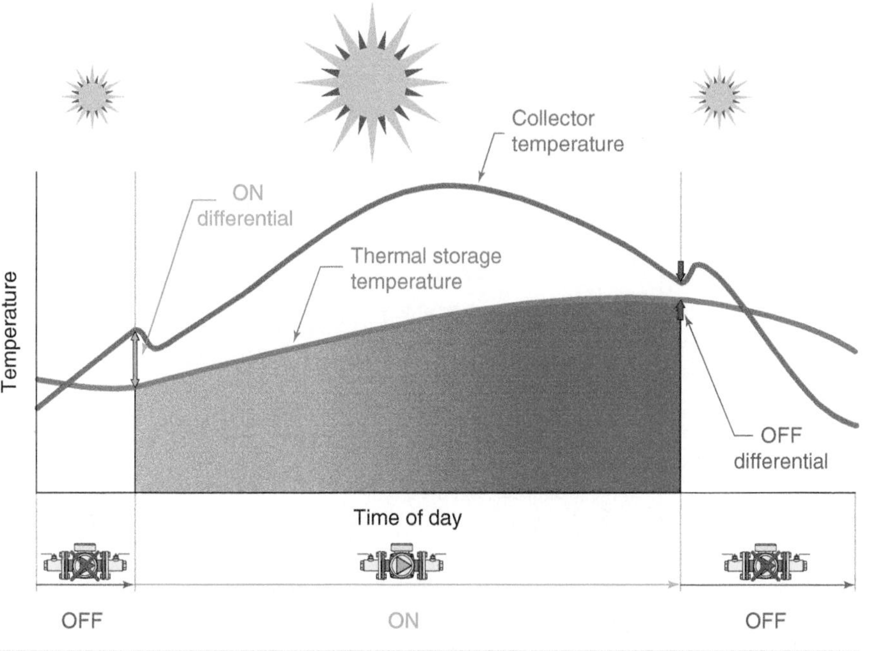

Figure 6-27 | Typical relationship between solar collector and storage temperature as well as on-differential and off-differential of a differential temperature controller during a sunny day.

reaches a temperature that equals the temperature of the storage tank sensor + the on-differential. At that point, the differential temperature controller turns on the collector circulator. Notice how the collector temperature dips sharply immediately after the collector circulator is turned on. This is caused by the rapid cooling effect of fluid moving through the collector. The setting for the on-differential and off-differential should be such that this dip doesn't cause the circulator to quickly turn off. If several such "short cycles" are observed during a sunny morning start up, the on-differential should be slightly increased and the off-differential slightly decreased.

Figure 6-27 shows a steady increase in thermal storage temperature during the midday period. This gain tends to level out as the afternoon continues and the collector temperature begins to drop. When the collector temperature minus the storage temperature drops to within the off-differential of the controller, the collector circulator is turned off. The collector temperature will usually rise immediately after this, since there is no fluid flow extracting heat from the absorber plate. Sometimes this temporary temperature rise causes the controller to restart the circulator. However, the collection cycle is likely to be very brief and will contribute very little if any further heat transfer to storage. Collector temperature will drop rapidly in late afternoon, and the collector circulator will remain off until the next time the sun sufficiently heats the collector.

Figure 6-28a shows an example of a modern differential temperature controller. Figure 6-28b show the wiring of this controller.

One of the most common applications for a differential temperature controller is operating the circulator that creates flow between an array of solar collectors and a thermal storage tank, as shown in Figure 6-29.

In this application, the source sensor is mounted to the upper portion of the collector's absorber plate or strapped to the outlet connection of the collector. The storage sensor is mounted within a **sensor well** that screws into the lower portion of the storage tank.

(a)

Figure 6-28 | (a) Example of a differential temperature controller.

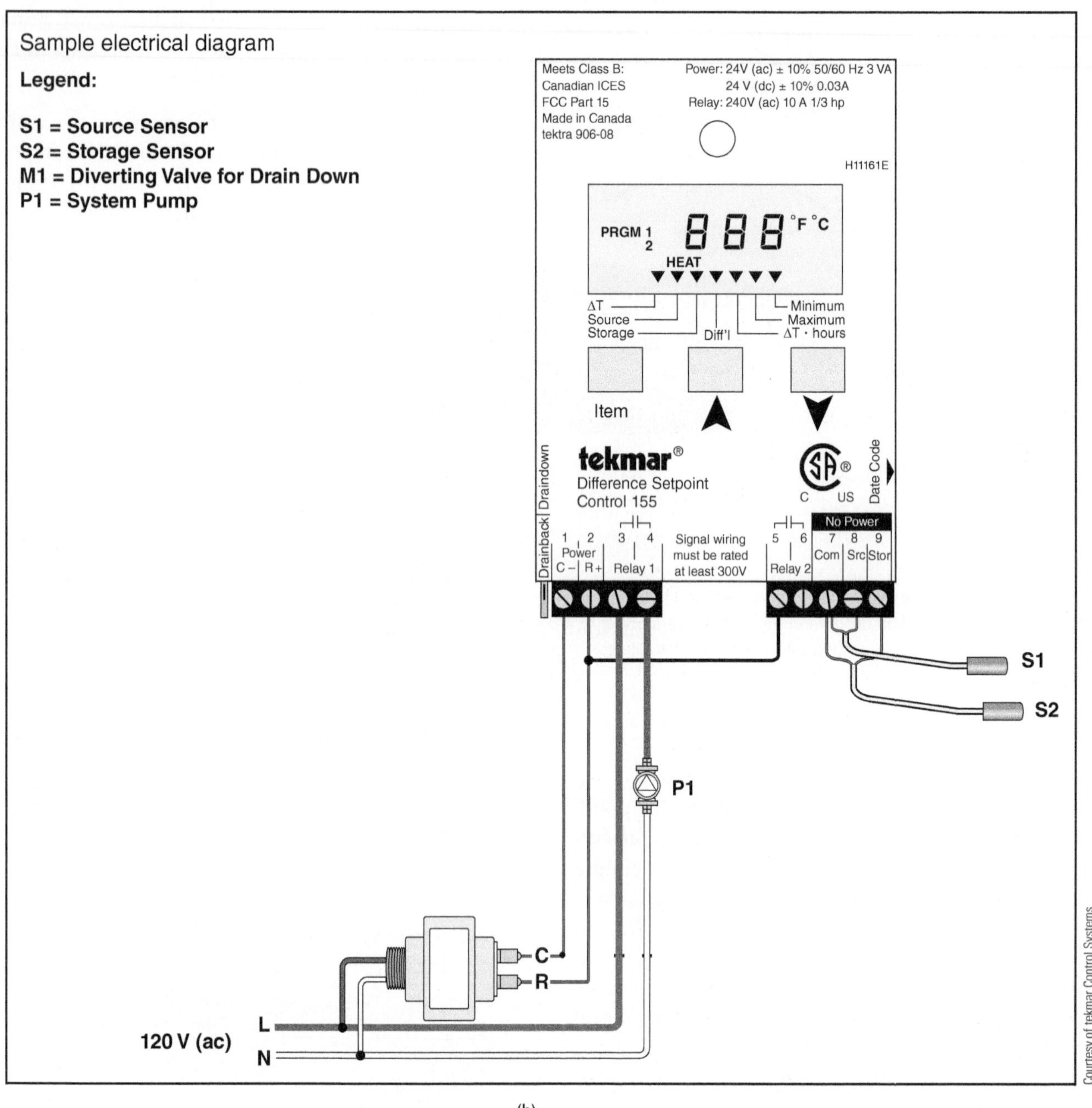

(b)

Figure 6-28 | (b) Typical wiring for this controller when operating a single circulator.

Assume that the on-differential of the differential temperature controller in Figure 6-29 is set for 8 °F and the off-differential is set for 3 °F. Also assume the temperature at the location of the storage sensor is currently 100 °F. The differential temperature controller would turn on the collector circulator when the collector temperature sensor climbed to 108 °F. It would remain on until the collector temperature dropped to 3 °F above the storage tank temperature. By late afternoon on a sunny day, the storage tank temperature may have climbed to 150 °F. Under this condition, the collector circulator would turn off when the collector temperature dropped to 153 °F. This action prevents the system from circulating heated fluid through the collector under conditions that don't allow the collector to transfer heat to that fluid. The off-differential of 3 °F also allows for a slight inaccuracy in sensor calibration as well as some error in the measured temperature because of how the sensor is mounted. For example, a sensor strapped to the outer surface of pipe cannot sense the temperature of the fluid within that pipe as accurately as it could if it were mounted within a sensor well immersed in the fluid stream.

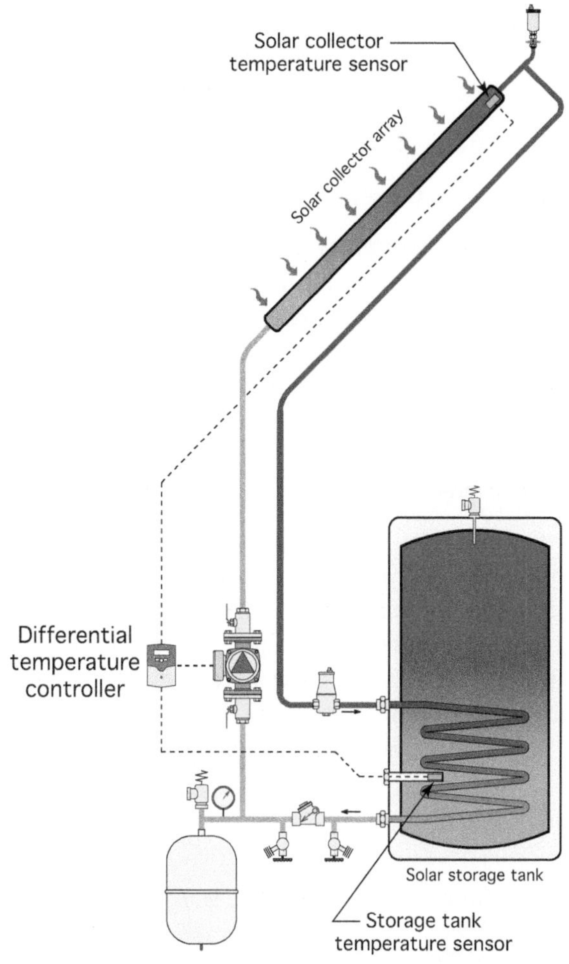

Figure 6-29 Use of a differential temperature controller to operate a circulator in a circuit between a solar collector array and an internal heat exchanger in a thermal storage tank.

ADVANCED FUNCTIONS OF DIFFERENTIAL TEMPERATURE CONTROLLERS

Many current generation differential temperature controllers offer capabilities beyond the basic switching of a single electrical contact. One of those capabilities is **variable speed control** of a wet-rotor circulator using a PSC-type motor. The speed can be controlled by varying the frequency and shape of the alternative current waveform supplied to the circulator. This requires a differential temperature controller with a **triac output** rather than a normally open relay output. A triac is a solid-state device that can vary the output waveform supplied to the circulator based on instructions from a microprocessor inside the controller. Like a relay contact, a triac can also operate the circulator at full-speed when necessary. Some manufactures refer to triac as "**solid-state relay**."

A typical variable speed output from an advanced differential temperature controller, operating a wet-rotor circulator in a solar thermal system, might be as follows:

1. *The controller monitors the temperature of the collector and storage sensors. When the collector temperature climbs above the storage temperature by the on-differential, the circulator is turned on, at full speed. It continues to operate at full speed for 10 seconds. This allows the collector temperature sensor to stabilize.*
2. *After 10 seconds at full speed, the controller reduces the circulator to minimum speed and continues to monitor the temperature difference between the collector and storage temperatures.*
3. *When (and if) the collector temperature climbs 20 °F above the storage temperature, the controller increases the speed of the circulator by 10 percent. With a higher flow rate through the collector, and the assumption of no change in sunlight intensity, the collector outlet temperature should drop slightly. However, if there is sufficient sunlight striking the collector, its outlet temperature may continue to rise, even at the slightly higher flow rate. If this is true, it implies that the circulator speed can be further increased. If the collector outlet temperature increases by a user set value (i.e., 4 °F) within a short time, the circulator speed is increased another 10 percent. This routine is repeated until the collector circulator either reaches full speed or there is no further rise in collector outlet temperature. This logic reduces the circulator power requirement under conditions of marginal sunlight, but can adapt quickly if sunlight intensity increases and higher circulator speed is needed.*
4. *If, during this variable speed operation, the difference between the collector temperature and storage temperature drops below the off-differential, the circulator is turned off.*

Some manufacturers offer differential temperature controllers that include features such as:

- Ability to provide a two independent differential temperature control functions
- Ability to operate motorized diverting valves within the system

- Ability to operate motorized mixing valves within the system
- Ability to turn an auxiliary heat source, such as a boiler, on and off

Figure 6-30 shows an example of a controller that can perform all of these functions. As such, it can operate a complete hydronic heating system that might include solar collectors, auxiliary boilers, provisions for heating domestic water, and space heating.

This type of controller can be configured to operate a wide variety of **schematic arrangements**, which represent specific combinations of hardware such as solar collector arrays, storage tanks, auxiliary boilers, and space heating distribution systems. Figure 6-31 shows some of the possible arrangements.

The large LCD display on these modern controllers can display all of these arrangements and indicate the operational status of devices such as circulators, valves, and auxiliary heat sources. Although these advanced controllers are sold as differential temperature controllers, the wide range of ancillary functions they provide allows them to function as **system controllers**.

Heat metering is another ancillary feature offered on some differential temperature controllers. This feature allows the controllers to record how much thermal energy has transferred from the source to the storage.

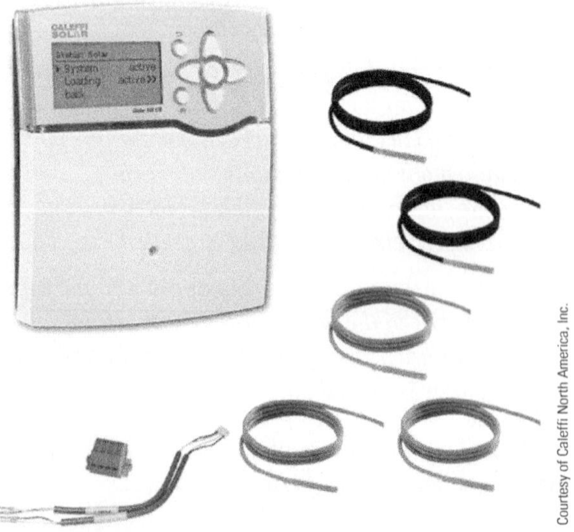

Figure 6-30 Example of an advanced differential temperature controller with several ancillary functions.

There are two methods for heat metering used in combination with differential temperature controllers. They are as follows:

1. **Accumulated ΔT hours**
2. **Fully instrumented heat metering**

These methods differ in accuracy and required hardware. The accumulated ΔT hours method doesn't require any additional hardware, but provides limited accuracy compared to controllers that use a flow rate transducer. The fully instrumented heat metering requires the installation of a flow transducer that works with the differential temperature controller. This approach is capable of greater accuracy, but also has a higher installation cost.

Controllers that use the accumulated ΔT hours method periodically record a number which is the instantaneous temperature difference between the source temperature and storage temperatures (e.g., ΔT), multiplied by a short time interval over which this ΔT is assumed to remain constant. This number is called **ΔT hours**. It can be represented as equation 6.4:

(Equation 6.4)

$$[\Delta T \cdot hours] = \sum_{time=0}^{time=now} (T_{source} - T_{storage})\Delta t$$

where:

$[\Delta T \cdot \text{hours}]$ = the TOTAL number of $\Delta T \cdot$hours to have accumulated between when measuring begins (e.g., time = 0) and when the meter is read (e.g., time = now) in units of °F·hours.

$(T_{source} - T_{storage})$ = instantaneous temperature difference between the source and storage sensors (°F)

Δt = a short time interval over which the value of $(T_{source} - T_{storage})$ is assumed to be constant (hours).

Example 6.2: A controller that logs ($\Delta T \cdot$hours) uses a time increment of 1 minute (e.g., 1/60th hour). Over 5 minutes of operation, it records the data shown in Figure 6-32. Determine the total number of ($\Delta T \cdot$hours) accumulated during these 5 minutes.

Solution: Each 1-minute time interval is converted to hours (e.g., 1 minute = 0.016667 hour). The difference between the source and storage temperatures is calculated for each of these time intervals. Next, each ΔT value is multiplied by the time interval to get the value of ($\Delta T \cdot$hours) for that time interval. Finally, all ($\Delta T \cdot$hours) values are added. These calculations are shown in Figure 6-33:

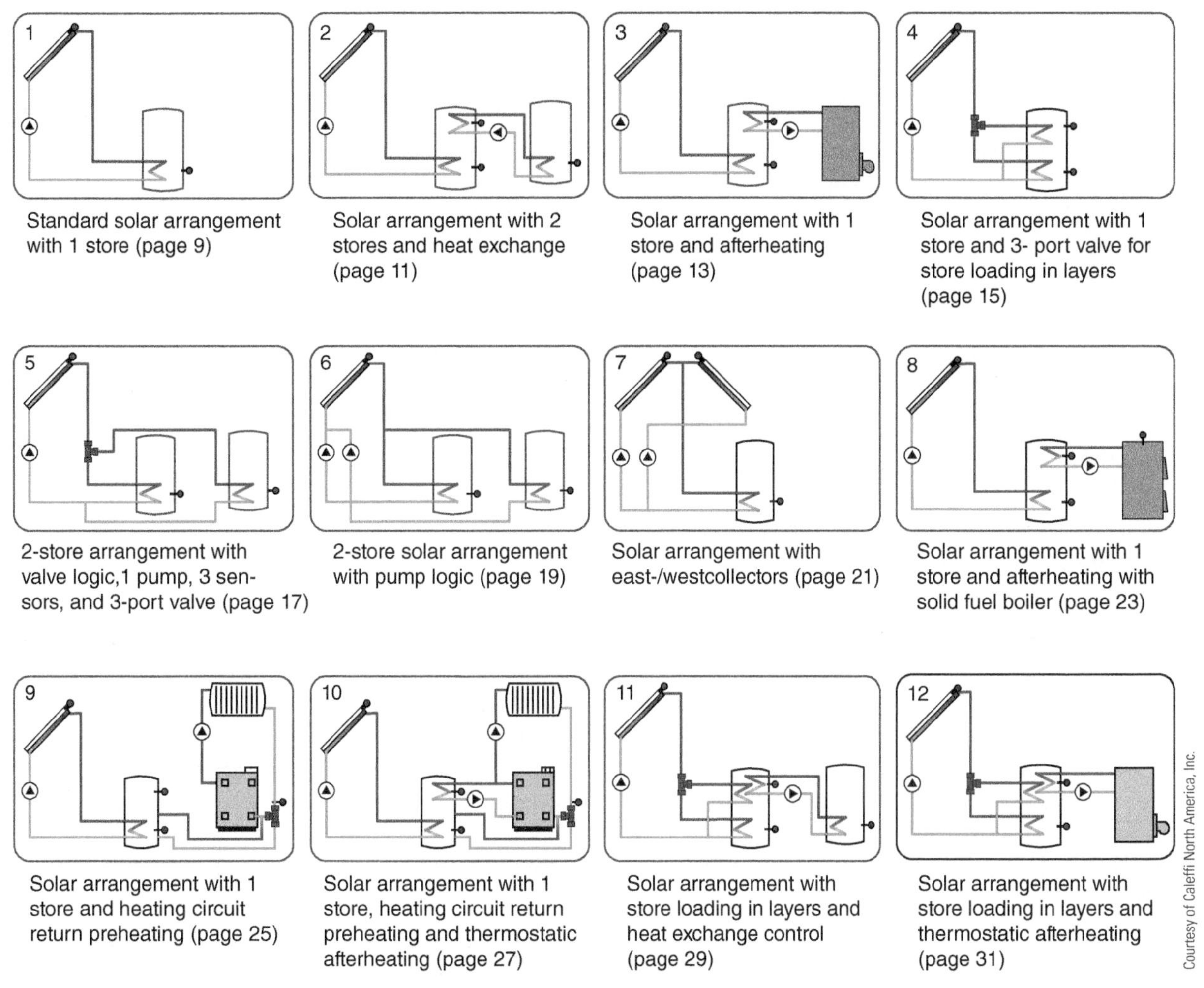

Figure 6-31 | Partial listing of the arrangements for which the controller shown in Figure 6-30 can be configured.

Time increment	T_{source}	$T_{storage}$
1st minute	100	75
2nd minute	100.5	75
3rd minute	101.5	75.1
4th minute	103	75.2
5th minute	105	75.2

Figure 6-33 | Data recorded during 5 minutes of operation.

Discussion: The total (ΔT·hours) is 2.24167. By itself, this number is of little use. However, when combined with a flow rate and the physical properties of the fluid, the total amount of heat transferred over a given time can be estimated.

Equation 6.5 can be used to convert (ΔT·hours) into an estimate of total heat transferred.

(Equation 6.5)

$$H = (8.01Dc)f\,[\Delta T \cdot \text{hours}]_{\text{total}}$$

where:

H = approximate total heat transferred from source to storage during a specific measurement period (Btu)

8.01 = a constant based on the units used

D = density of the fluid transferring heat (at the average temperature of the process (lb/ft^3)

c = specific heat of the fluid transferring heat (at the average temperature of the process (Btu/lb/°F)

f = flow rate of the fluid transferring heat (gpm)

$[\Delta T\text{·hours}]_{\text{total}}$ = total (ΔT·hours) accumulated over a specific measurement period (°F·hours)

Time increment	T_{source}	$T_{storage}$	ΔT	Hours	ΔT·hours
1st minute	100	75	25	0.016667	0.416667
2nd minute	100.5	75	25.5	0.016667	0.425
3rd minute	101.5	75.1	26.4	0.016667	0.44
4th minute	103	75.2	27.8	0.016667	0.46333
5th minute	105	75.2	29.8	0.016667	0.49667
				Total ΔT·hr =	2.24167

Figure 6-33 | Calculations to determine total (ΔT·hours) for Example 6.2.

The density and specific heat of the fluid conveying heat from the source to storage needs to be determined at the average temperature of that fluid while the heat is being conveyed. For example, assume a 40 percent solution of propylene glycol was flowing from a solar collector array to a storage heat exchanger. The fluid arrives at the storage heat exchanger at 140 °F and leaves at 128 °F. The average temperature of the fluid would be (140 + 128)/2 = 134 °F. Thus, to use Equation 6.5, the density and specific heat of this fluid should be determined at its average temperature of 134 °F.

Example 6.3: Assume a solar collector circuit operates with a 40 percent solution of propylene glycol and that its flow rate remains constant at 6.0 gpm over the same 5-minute interval used in Example 6.2. During this time, the average temperature of the glycol solution was 88.55 °F. Combine this information with the total (ΔT·hours) from Example 6.2 to determine the total heat transferred from the collectors to storage during this 5-minute interval.

Solution: The density and specific heat of a 40 percent solution of propylene glycol can be found in manufacturer's data or by using the **Hydronics Design Studio** software. They are as follows:

$D = 64.3\ \text{lb/ft}^3$

$c = 0.90\ \text{Btu/lb/°F}$

These numbers, along with the total (ΔT·hours) from Example 6.2, can now be combined using Equation 6.5.

$$H = (8.01Dc)f[\Delta T \cdot \text{hours}]_{\text{total}}$$

$$H = (8.01 \times 64.3 \times 0.90) \times 6.0 \times [2.24167] = 6234\ \text{Btu}$$

Discussion: This is an *estimate* of the total heat transferred during the 5-minute period. There are several assumptions embodied within this estimate. First, the flow rate is assumed to remain constant during this 5-minute interval. This a reasonable assumption. However, over a longer measurement period, such as an hour, there may be a slight increase in flow rate due to changes in viscosity and density as the fluid warms. Second, the average fluid temperature of 88.55 °F is also assumed to hold constant during the 5-minute interval. Again, this is a reasonable assumption based on the temperature data in Example 6.2. However, when the system is first turned on, or during a longer measurement period, there would likely be more variation in the average fluid temperature and thus more variation in its density and specific heat. The larger the variation, the more error there may be in the calculated result. Thus, *heat metering using the accumulated (ΔT·hours) method is an approximate method.* It's appeal lies in the fact that no additional hardware is required, beyond a basic differential temperature controller that records accumulated (ΔT·hours). Its usefulness is for *relative comparisons* of heat transferred over various time intervals rather than the measurement of the exact amount of heat transferred. The next section describes a more accurate approach to heat metering that accounts for variations in flow rate as well as variations in fluid properties.

The value of H in Equation 6.5 is an approximation of total heat transferred. The accuracy of this approximation depends on the time interval between the successive reading of the temperature difference. The shorter this time interval, the more accurate the approximation. The accuracy of the approximation is also partially determined by the type of sensors used for temperature readings and the type of flow meter used.

6.9 Heat Metering (fully instrumented)

The most accurate approach to heat metering is to mathematically integrate the *instantaneous rate* of heat flow, multiplied by the instantaneous temperature difference between the outgoing and return fluid streams, multiplied by the instantaneous fluid properties. It can be stated mathematically as follows:

(Equation 6.6)

$$H_{total} = \int_{t=0}^{t=low} [(8.01\,Dc) f \Delta T]\,dt$$

When the time increment during which the flow rate, fluid properties, and temperature difference is evaluated is kept very short, Equation 6.6 can be approximated as:

(Equation 6.7)

$$H_{total} \cong \sum_{time=0}^{time=low} (8.01\,Dc)\,f\,(T_{high} - T_{low})\Delta t$$

where:

H_{total} = total heat transferred during a specific measurement period (Btu)

8.01 = a constant based on the units used

D = density of the fluid transferring heat [at the average temperature of the process (lb/ft^3)]

c = specific heat of the fluid transferring heat [at the average temperature of the process (Btu/lb/°F)]

f = flow rate of the fluid transferring heat (gpm)

T_{high} = temperature of the hotter fluid stream (supplied from the heat source) (°F)

T_{low} = temperature of the cooler fluid stream (returning to the heat source) (°F)

Δt = short time increment (hours)

Figure 6-34 illustrates the difference between the exact amount of heat transferred, represented by the blue shaded area and Equation 6.6, and the *approximate* total heat transferred, based on the sum of the areas of the 15 red shaded rectangles and Equation 6.7.

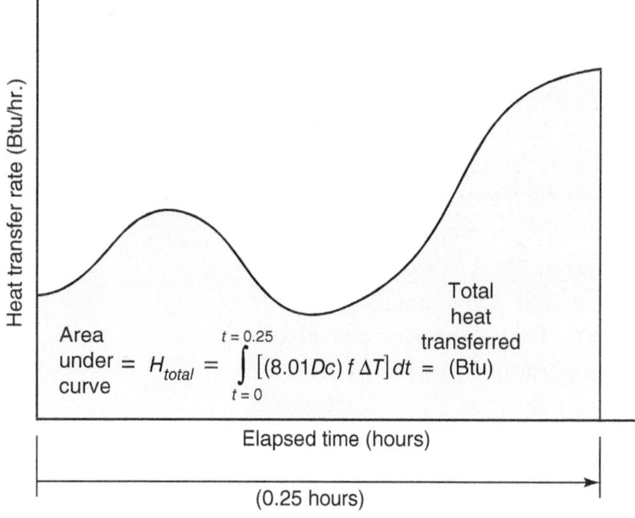

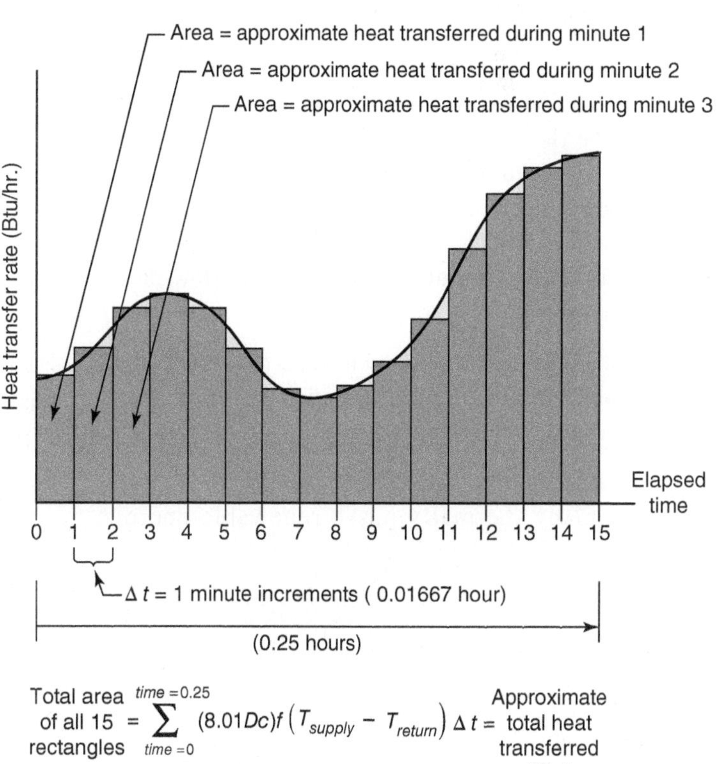

Figure 6-34 Comparison between (a) the exact heat transferred over a 15-minute time period versus (b) the approximation of the total heat transferred.

The shorter the time increment used in Equation 6.7, the narrower the rectangles and the greater the number of rectangles present over a given elapsed time. This increases the accuracy of the approximation relative to the actual value of total heat transferred. Time increments of 1 minute (e.g., Δt = 1/60th hour) are generally adequate for most heat metering applications.

HEAT METERING HARDWARE

The only way to get the accurate flow rate measurements required by Equation 6.7 is to have a **flow transducer** installed in the circuit for which heat is being metered. The flow transducer measures the flow rate and sends a signal to a digital circuit that interprets it as a mathematical value for flow rate. That digital circuit also receives temperature information from two temperature sensors. It combines all the incoming data, along with calculated values of fluid properties, and then performs the summation represented by Equation 6.7. An example of a small heat metering device and its associated sensors is shown in Figure 6-35.

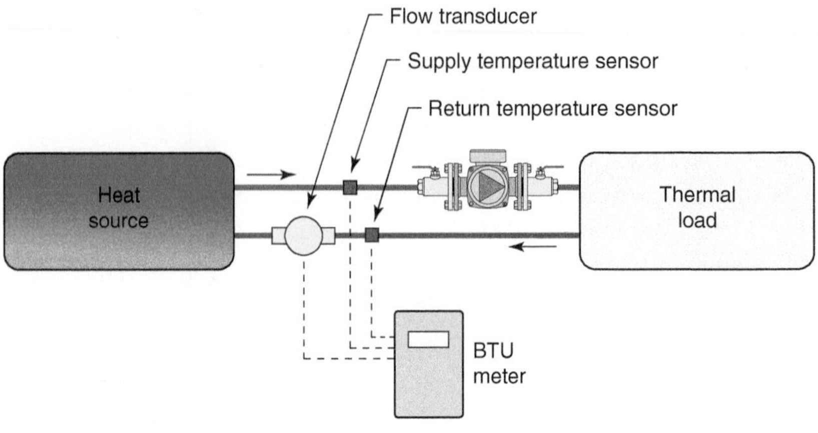

Figure 6-36 | Placement of the flow transducer and temperature sensors for a heat meter.

The brass bodied device in Figure 6-35 is the flow transducer. The electronics and digital display are mounted on top of the flow transducer. The temperature sensors are color coded: red for the higher temperature and blue for the lower temperature. The flow transducer should be installed with at least 12 diameters of straight upstream piping. This reduces turbulence and improves flow measurement accuracy.

Figure 6-36 shows how a heat meter is placed between a heat source and a load. This illustration shows the electronics as a separate module from the flow transducer.

Heat metering is now used to verify the performance of many thermally based renewable energy systems. Some government-sponsored incentive programs involving renewable energy heat sources even mandate the use of heat metering for performance verification. Heat metering is also used to record the thermal energy delivered to living quarters, such as apartments or condominiums as well as office or retail spaces, that are all supplied from a central heating plant. The occupants of those spaces are then billed for the thermal energy they consume for space heating and domestic water heating rather than being billed directly for the fuel used. This form of heat metering has been extensively used in Europe for several years and has proven to conserve energy relative to systems that don't charge occupants for the thermal energy they use.

Courtesy of Istec Corporation

Figure 6-35 | Example of a small, self-contained heat meter.

As this text is being written, **ASTM standard E44** is being developed as a basis for assessing the accuracy of heat metering devices in the United States. This standard is expected to be completed within 2015. Once finalized, this standard should allow for implementation of heat metering in a wide range of applications. More information on this type of heat metering is presented in chapter 14 of *Modern Hydronic Heating*, 3rd edition.

6.10 Temperature Sensors and Sensor Placement

All controllers discussed in this chapter use one or more temperature sensors. These sensors contain a small solid-state sensing element enclosed within a housing. Two wires leads, either detached or enclosed within a cable jacket, extend from the housing.

Different manufacturers use different solid-state sensing elements within these sensors. One of the most common is a **negative temperature coefficient (NTC) thermistor**. These sensing elements have a very repeatable correlation between their temperature and their electrical resistance. The controller "feels" the electrical resistance created by the sensor and the wiring connecting it to the controller. It uses this resistance to infer the sensor's temperature. Figure 6-37 shows the temperature/resistance curve for a typical 10K NTC thermistor sensor. The designation 10K means that the sensor has a resistance of 10,000 ohms at a reference temperature of 25 °C (77 °F).

To operate properly, all temperature-based controllers must be able to sense one or more temperatures that are at, or very close to, the actual temperature of the media or object for which the measured temperature is needed. Examples of common temperatures sensed by the controllers discussed in this text include:

- Water temperature within a storage tank
- Absorber plate temperature in a solar collector
- Outside air temperature
- Inside air temperature
- Fluid temperature entering a heat pump from a ground loop
- Fluid temperature within a pipe leaving a mixing valve

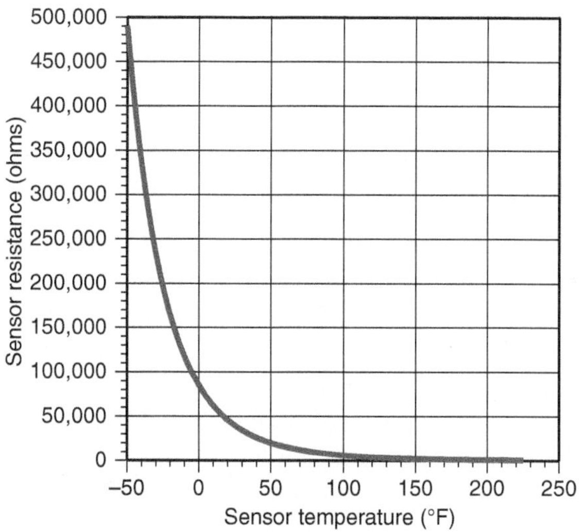

Figure 6-37 | Resistance versus temperature for a 10K @ 25 °C NTC thermistor sensor.

In some situations, the sensor can be directly immersed in the media. An example is a temperature sensor in the air stream leaving an air handler.

In most situations, the thermistor sensors commonly used with heating controllers cannot be directly immersed in liquids. When the temperature of a liquid needs to be sensed, the sensor is usually strapped to the outer surface of a metal pipe (preferably copper) or mounted within a sensor well that is itself immersed in the fluid. Both of these mountings allow the sensor to remain dry.

Figure 6-38 shows a thermistor temperature sensor designed to be strapped to the outside of a pipe. This type of sensor usually has a concave slot on its outer housing that reasonably matches the curvature of small pipes. The sensor is often strapped to the pipe with one or two nylon pull ties. These

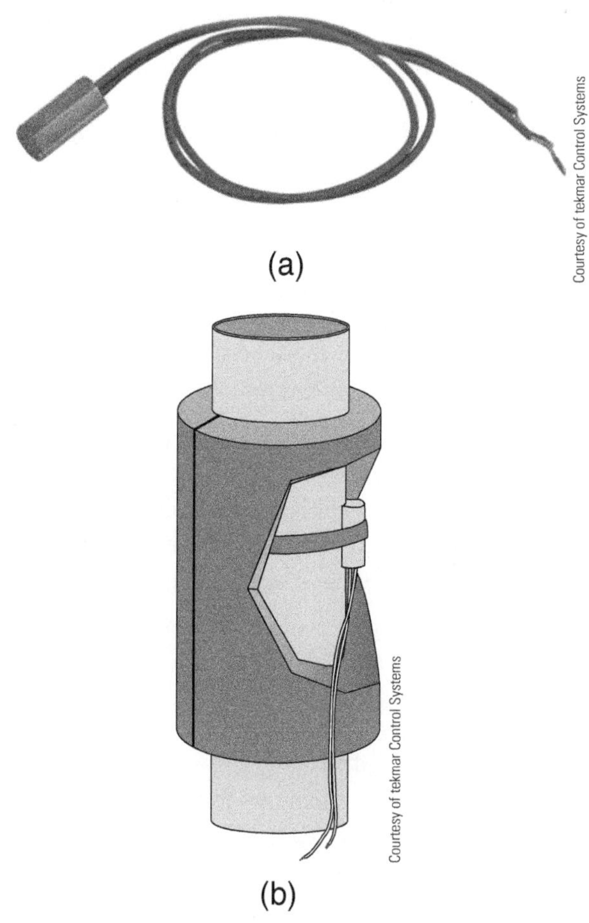

Figure 6-38 | (a) Temperature sensor designed to mount to the outer surface of a pipe. (b) Sensor secured to the pipe by a high temperature rated nylon pull tie. Note: This sensor needs to be covered with insulation.

pull ties should be rated to withstand the maximum temperature to which they may be exposed over several years of operation. For example, the absorber plate in a solar collector can reach temperatures over 350 °F during stagnation conditions. No nylon pull tie can withstand such temperatures. Under such conditions, the only acceptable means of attachment would be a stainless steel bolt and lock nut or a stainless steel hose clamp. If the latter is used, care must be taken not to overtighten the clamp, which can damage the sensor housing.

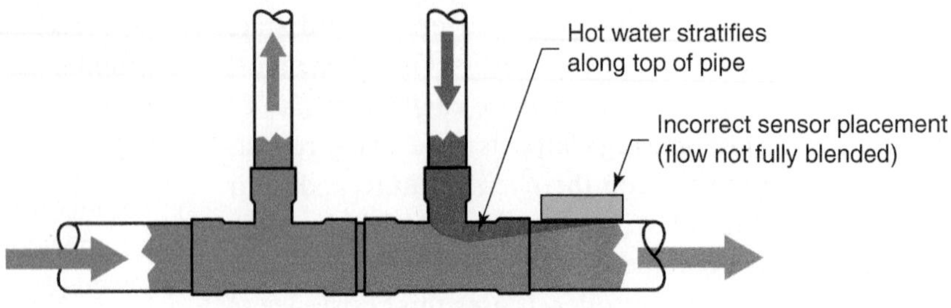

Figure 6-39 | Incorrect sensor placement for system using injection mixing.

All temperature sensors that are mounted to the outer surface of a pipe must be wrapped with insulation once they are mechanically secured. This helps ensure that the sensor is detecting the surface temperature of the pipe without being unduly influenced by the temperature of the surround air. Failure to install proper insulation as specified by the controller's manufacturer will lead to inaccurate control operation. Remember, the controller can only react to what its temperature sensors feel, not to what you think they should feel.

All electronic mixing controllers require a sensor that senses the blended temperature of two fluid streams that have been mixed together. That mixing could take place within a valve, a tank, a hydraulic separator, or a tee. Whatever the case, it is critically important that the sensor measuring the mixed fluid temperature is located so that it senses the *final* blended temperature.

One might assume that when two fluid streams flow together, they immediately mix to a final blended temperature. That is not necessarily true. For example, consider a case where hot water is being blended with cooler water in a tee, as shown in Figure 6-39.

Due to its buoyancy, the hot water entering the side port of the right hand tee will **stratify** (e.g., float along to upper part of the pipe). If a temperature sensor is placed immediately downstream of that tee, as shown in Figure 6-38, it will sense a temperature that could be much higher than the final blended temperature, which occurs farther downstream. This will lead to erratic controller response.

To avoid this situation, it is always preferable to locate the sensor for a mixing process downstream of the circulator, as shown in Figure 6-40. If this is not possible, there should be at least two 90-degree turns in the piping between were the mixing process begins and where the sensor that measures the blended temperature is located. These turns create turbulence that helps blend fluid molecules together, upstream of the sensor.

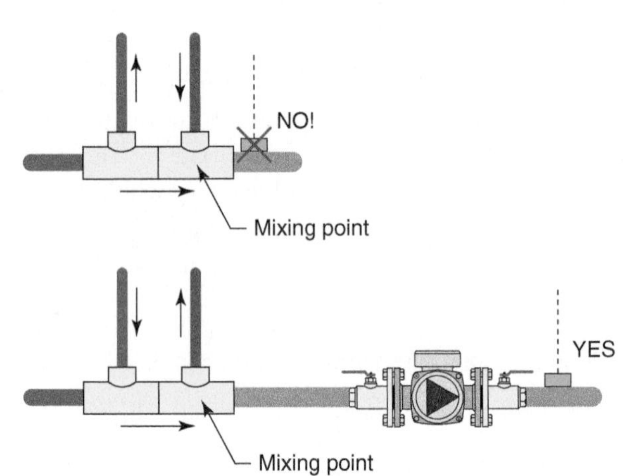

Figure 6-40 | Do not place temperature sensors immediately downstream of a mixing point. Instead, place them downstream of a circulator that itself is downstream from the mixing point.

SENSOR WELLS

One way to improve accuracy when a liquid temperature must be sensed is to mount the sensor within a well, as illustrated in Figure 6-41. Various types of sensor wells are available, depending on where the sensor needs to be located. These wells are usually made of copper or stainless steel. They allow the sensor to be surrounded by a highly conductive metal surface that is very close to the temperature of the liquid, but without being directly immersed in that liquid. They also allow the sensor to be removed and replaced, if necessary, without draining a liquid-filled component. For accurate sensing, any sensor inserted

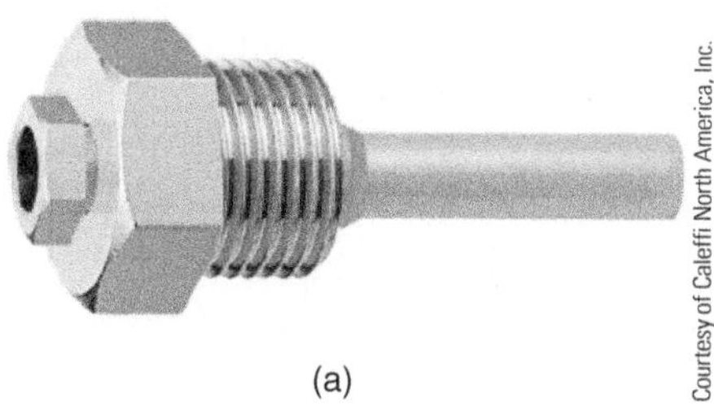

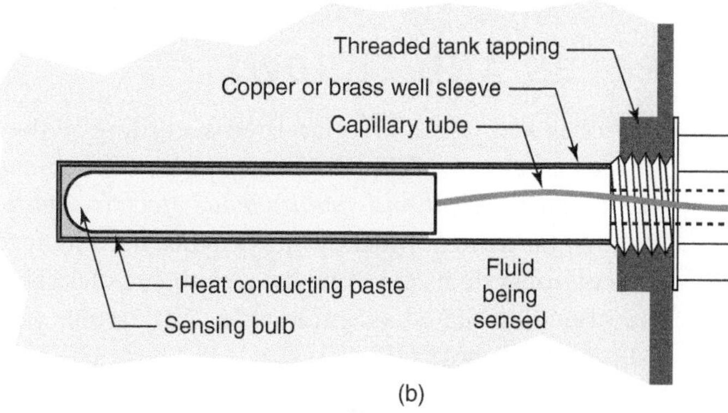

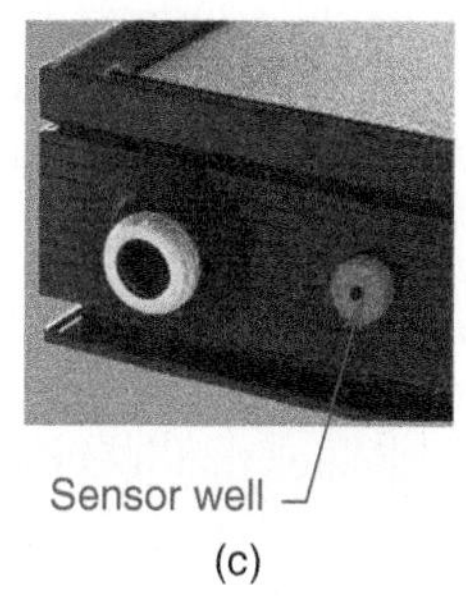

Figure 6-41 (a) A stainless steel sensor well that threads into an FPT connection. (b) Illustration of a sensor in a well. (c) A sensor well integrated into a solar collector.

into a well should first be coated with a heat conducting paste. This paste fills what would otherwise be air gaps between the sensor housing and the inside of the well.

Finally, be sure that the sensor, and the well it is intended to mount into, are compatible. Sensors do vary in diameter and length. The ideal well for a given sensor would have an inside diameter that is about 0.03 inches larger than the outside diameter of the sensor. This allows just enough gap for a thin coating of thermally conductive grease between the sensor and the inside surface of the well. It also allows the sensor to be pulled out of well by its leads, if necessary, without damaging the sensor.

SENSOR WIRING

Because the temperature sensors used by the controllers discussed in this text correlate temperature to electrical resistance, and because those sensors are often located significant distances from the controllers, it is imperative to limit the electrical resistance of the wiring connecting these sensors to their controllers. *The controller feels the combined electrical resistance of the sensor and the wiring between the sensor and controller*. If the resistance of the wiring is significant, it will affect the temperature the controller "believes" it is detecting at the sensor.

Most controller manufacturers specify the material, wire gauge, and maximum allowable length of the wiring between a resistance-type temperature sensor and the controller. In the absence of such specifications, use nothing smaller than 18 AWG copper conductors.

If the sensor wiring is routed close to devices such as motors, transformers, or florescent lighting, the sensor cable should contain either a **twisted pair of conductors** or be a **shielded cable**. Both types of cable minimize the tendency of strong inductive loads to impose **electrical noise** into the sensor circuits. If a shielded cable is used, the metallic shield within the cable jacket must be attached to an electrical ground *at only one end of the cable* (usually the end near the controller). The following additional guidelines also help minimize the potential for electrical noise in sensor wiring:

- Keep the sensor cable as short as practical and always at or under the manufacturer's maximum length specification.
- Never route sensor wiring next to AC wiring or other powerful electrical devices.
- Never route sensor wiring in the same conduit or wiring tray as AC wiring.

The splices between sensor leads and cable can also create undesirable electrical resistance, especially if oxidation occurs at the splice over time. *It is critically important to prevent moisture from entering a splice between a sensor lead and a cable conductor.* This is especially likely in exterior splices, such as where a collector temperature sensor, with short leads, splices to a cable. The best way to make splices in sensor wiring is with a **gel-filled compression connector**, an example of which is shown in Figure 6-42.

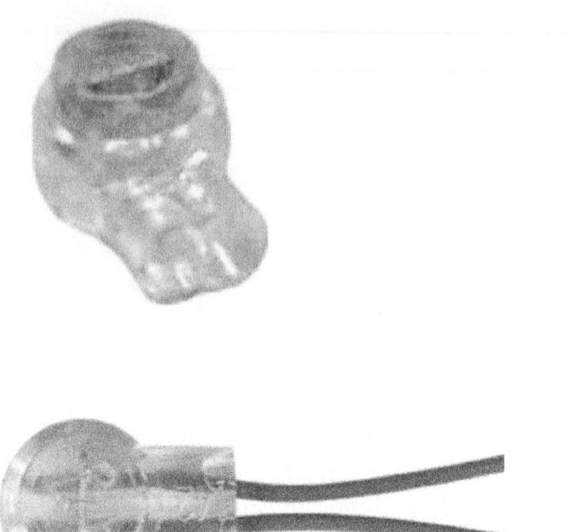

Figure 6-42 | Example of a gel-filled compression connector.

These connectors are widely used in the telecom industry. The two leads to be spliced are pushed, unstripped, into the barrels of this connector. The "button" at the top of the connector is then compressed with a common pliers, or a special crimping tool designed for use with these connectors. This forces a tiny metal connector through the insulation on both conductors, creating a permanent electrical bond. Furthermore, that bond is immersed within a moisture repelling gel inside the connector. This prevents corrosion of the electrical bond.

6.11 Switches, Relays, and Ladder Diagrams

This section describes many of the basic electrical switching devices used to build an overall control system. Almost every heating system, hydronic or otherwise, uses one or more of these switching devices within its control system. A working knowledge of these devices is essential in designing, installing, or troubleshooting hydronic systems. The section also presents ladder diagrams as a fundamental tool for documenting control system wiring.

SWITCHES

The operating modes of most heating and cooling systems are determined by a specific arrangement of electrical switch **contacts**. Such contacts can only have two possible states: open or closed. An open contact prevents an electrical signal (e.g., a voltage) from passing a given point in the circuit. A closed contact allows the signal to pass through. These contacts may be part of a manually operated switch or an electrically operated switch called a **relay** or **contactor**. When the settings of the various contacts allow a complete circuit to form, a current will flow and some predetermined control action will take place.

POLES AND THROWS

Switches and relays are classified according to their **poles** and **throws**. *The number of poles is the number of independent and simultaneous electrical paths through the switch.* Most of the switches used in heating control systems have one, two, or three poles. They are often designated as single pole (SP), double pole (DP), or triple pole (3P).

The number of throws is the number of position settings where a current can pass through the switch. Most switches and relays used in heating systems have either one or two throws, and are called single throw (ST) or double throw (DT) switches.

Figure 6-43a shows a typical double pole/double throw (DPDT) toggle switch. Notice the printing on the side of the switch indicates that it is rated for a maximum of 10 amps of current at 250 volts (AC) and 15 amps of current at 125 VAC. When selecting a switch for a given application, be sure its current and voltage ratings equal or exceed the current and voltage at which it will operate.

A schematic representation of this DPDT switch is shown in Figure 6-43b. The two lines on the left represent the independent electrical paths entering the switch (e.g., the poles). The two sloping lines connecting the circles on the left to the circles on the right represent the blades of the switch. These are internal metal conductors that move when the handle of the switch is moved. The dashed vertical line represents a nonconducting internal linkage that makes both blades move at the same time but does not allow any electrical interaction between the blades. The circles all represent terminals to which external wires can be connected.

In the upper portion of Figure 6-43b, the lever of the switch is set so that pole 1 on the left side conducts to terminal A on the right side. Similarly, pole 2 conducts to terminal C. Notice that the sloping lines representing the blades make contact from the center of the pole terminals to the edges of terminals A and C. Think of these blade lines as "pivoting" on the pole terminals

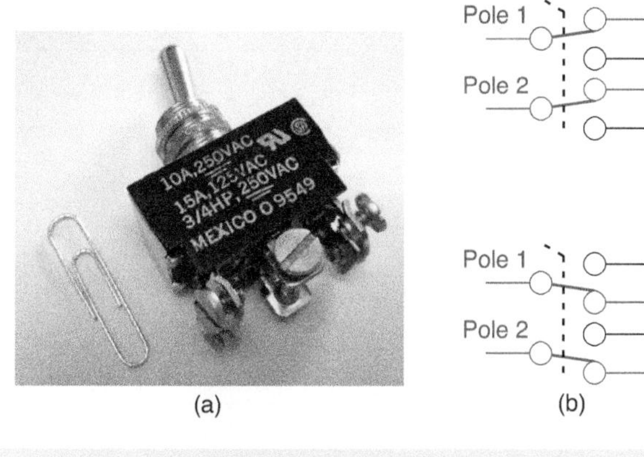

Figure 6-43 (a) Example of a double pole/double throw (DPDT) toggle switch. (b) Schematic representation of the DPDT switch in both possible positions.

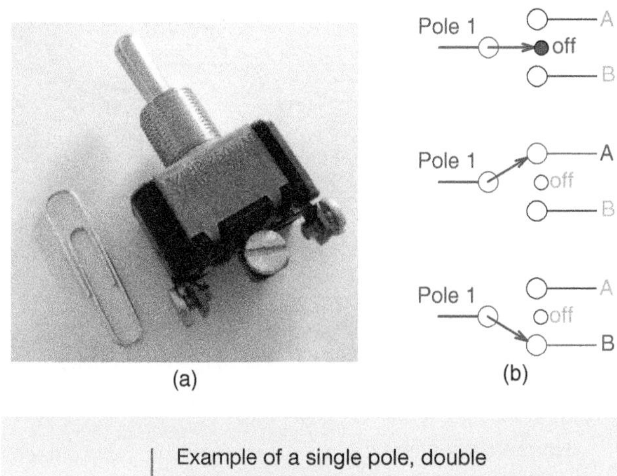

Figure 6-44 Example of a single pole, double throw, center off switch. (b) Schematic representation of the SPDT switch in all three possible positions.

and making contact with the periphery of terminals A and C. When the switch is in this position, terminals B and D are "floating." As such, they cannot pass any electrical signal through the switch.

When the lever of the switch is moved to its other position, pole 1 conducts to terminal B and pole 2 conducts to terminal D. Terminals A and C are now floating.

The two illustrations in Figure 6-43b show the two possible states of the DPDT switch. There is no other position in which the lever can be set to create a third possible state. Because the two terminals labeled poles are active in both states, they are often referred to as the "common" terminals.

CENTER OFF SWITCHES

Figure 6-44a shows a single pole, double throw switch with a **center off** setting (SPDT c/o). The switch is shown in its off setting, where the handle of the switch is straight up from the body. In this position, the middle contact is electrically isolated from both of the other contacts.

A typical application for a center off switch is within a control system that allows a mechanical system to operate in heating mode, cooling mode, or remain off. The latter occurs when the switch is in the center off position with its handle straight out.

The switches in Figures 6-43a and 6-44a have screw terminal connections. Similar switches are available with quick connect terminals. Switches of this type area also available with a wide range of current ratings and voltage ratings to suit different applications.

Designers need to specify switches that meet or exceed the maximum electrical current within the circuit the switch will be part of. In some cases there will be two current ratings: one for **resistive loads** and another for **inductive loads**. Typically the current rating for an inductive load such as a motor or transformer is lower than the current rating for a resistive load such as an incandescent light or heating element.

Switches are also rated for a maximum voltage. This is the maximum allowed voltage between any electrified portion of the switch and electrical ground. In most switches, the lever and any other metal portion of the body that contacts surrounding objects represent electrical ground. Applying a switch at voltages higher than its rating creates the possibility of dielectric breakdown and a ground fault current.

RELAYS

Relays are electrically operated switches. They can be operated from a remote location by a low-power electrical signal. They consist of two basic subassemblies: the **coil** and the contacts. Other parts include a spring, pendulum, and terminals. These components and basic operation of a relay are illustrated in Figure 6-45.

The spring holds the contacts in their "normal" position. When the proper voltage is applied to the coil, a magnetic force pulls the common contact attached to the pendulum to the other position, slightly extending the spring in the process. In most relays, this action takes only a few milliseconds. A click can be heard as the contacts move to their other position. When voltage

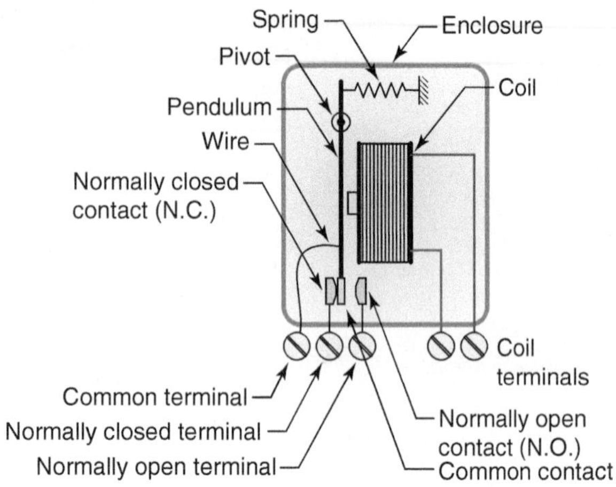

"Normal" mode: No voltage is applied to coil (terminals A and B). Spring holds pendulum so the common contact touches the normally closed contact. Current can flow from terminal 1 to terminal 2.

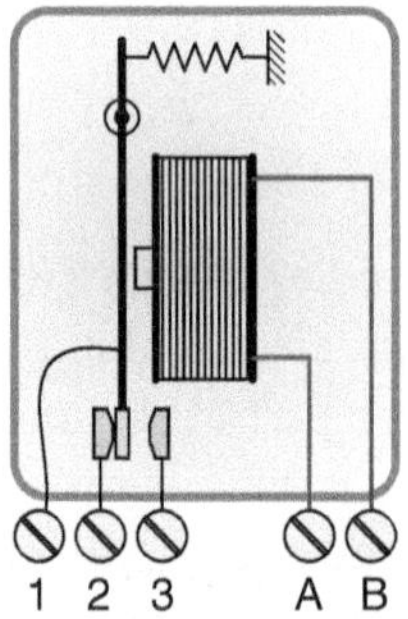

Energized mode: Voltage is applied to coil (terminals A and B). Pendulum pivots toward coil. The common contact touches the normally open contact. Current can flow from terminal 1 to terminal 3.

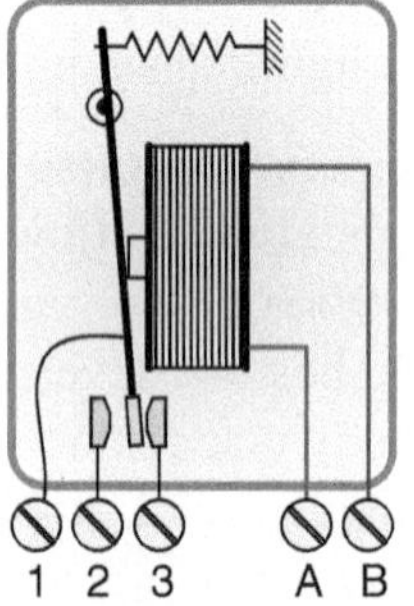

Figure 6-45 | Internal components and operating positions of a typical relay.

Contact designation	Switch contacts	Relay contacts
Single pole Single throw SPST	pole 1	pole 1
Double pole Single throw DPST	pole 1, pole 2	pole 1, pole 1
Triple pole Single throw 3PST	pole 1, pole 2, pole 3	pole 1, pole 2, pole 3
Single pole Double throw SPDT	pole 1	pole 1
Double pole Double throw DPDT	pole 1, pole 2	pole 1, pole 2
Triple pole Double throw 3PDT	pole 1, pole 2, pole 3	pole 1, pole 2, pole 3

Relay coil

Relay contact, normally open (N.O.) (closes when coil is energized)

Relay contact, normally closed (N.C.) (opens when coil is energized)

Figure 6-46 | Schematic symbols used to represent switches and relays having different numbers of poles and throws.

is removed from the coil, the spring pulls the contacts back to their deenergized (normal) position.

Relay contacts are designated as **normally open (N.O.)** or **normally closed (N.C.)**. In this context, the word "normally" means when the coil of the relay is not energized. A normally open contact will not allow an electrical signal to pass through while the coil of the relay is off. A normally closed contact *will* allow the signal to pass when the coil is off. Like switches, relays are also classified according to their poles and throws.

The coil is rated to operate at a specific voltage. In heating systems, the most common coil voltage is 24 VAC. Relays with coil voltages of 12, 120, and 240 VAC are also available when needed. Relays with coils designed to operate on DC voltages are also available, although less frequently used in heating control systems.

Figure 6-46 shows the schematic symbols for both switches and relays. On the double and triple pole switches, a dashed line indicates a nonconducting mechanical coupling between the metal blades of the switch. This coupling ensures that all contacts open or close at the same instant. Notice the similarity between the schematic symbols for switches and relays having the same number of poles and throws.

GENERAL PURPOSE RELAYS

Some control systems require one or more relays to be installed to perform switching or isolation functions. The type of relay often used is called a **general purpose relay**. They are built as plug-in modules with clear plastic enclosures. Their external terminals are designed to plug into a **relay socket**. These sockets have either screw terminal or quick-connect tabs for connecting to external wires. Relay sockets allow relays to be removed or replaced without having to disturb the wiring connections. Each terminal on the socket is numbered the same as the connecting pin on the relay that connects to that terminal. This numbering is very important because it allows the sockets to be wired into the control system without the relay being present. Relay manufacturers provide wiring diagrams that indicate which terminal numbers correspond to the normally closed, normally open, common, and coil terminals. In some cases, the wiring diagram is printed onto the relay enclosure.

Figure 6-47 shows a general purpose triple pole, double throw (3PDT) relay along with the socket it mounts.

Figure 6-48 shows the wiring schematic for the relay shown in Figure 6-47. It identifies the normally open and normally closed contacts, their corresponding terminal numbers, and the coil terminals.

Many relay sockets are designed to mount to a flat surface or to be snapped into an aluminum **DIN rail**. The latter mounting method allows relay sockets to be added or removed quickly and without fasteners. The DIN rail is a standard modular mounting system used in many types of control systems. Many small control devices are designed to mount to DIN rails. By mounting a generous length of DIN rail in a control cabinet, relays and other components can be easily added, removed, or moved to a different location. Figure 6-49 shows two different relay sockets mounted to a common DIN rail.

(a)

(b)

(c)

Figure 6-47 (a) 3PDT plug-in relay in plastic enclosure. (b) Relay socket for a 3PDT plug-in relay. Note numbers and letters designating terminals. (c) Relay mounted on socket.

TIME DELAY RELAYS

It is sometimes necessary to incorporate a time delay between two control events. One example is keeping a boiler circulator operating for a few minutes after the boiler has stopped firing to purge out the residual heat. Another is allowing both circulators in a drainback-protected solar thermal system to operate for a few minutes before turning one of those circulators off. These and other control functions can be accomplished using **time delay relays**.

Time delay relays are available with the following operating modes:

- Delay-on-make
- Delay-on-break
- Interval timing
- Repeat cycle

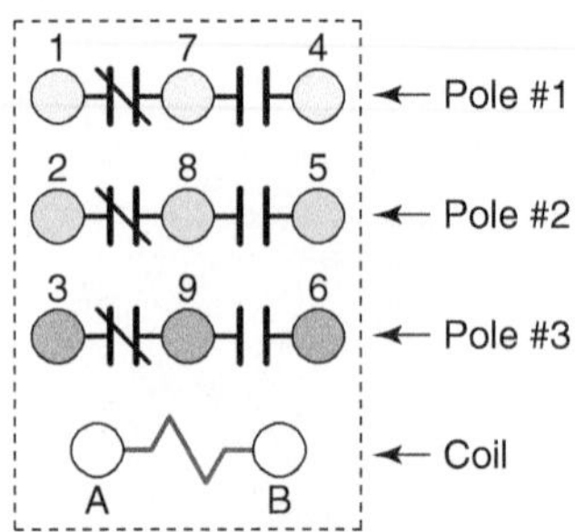

Figure 6-48 Schematic representation of relay associating contacts with terminal numbers. Coil connections are identified as A and B.

Figure 6-49 Different size relay sockets mounted to a common aluminum DIN rail.

These operating modes are described using timing charts in Figure 6-50. Of the four operating modes shown, the delay-on-make and delay-on-break functions are the most commonly used in hydronic heating applications.

The delay-on-make mode prevents the contacts from moving until a user-specified time delay period has elapsed from when the input signal is turned on. If the input signal is interrupted before the delay period has elapsed, the relay automatically resets the timing circuit to begin from zero the next time the input signal is energized.

One common use of a delay-on-make function is overriding priority control of a specific heating load in a multiple load system after a specified time has elapsed. For example, assume a designer wants domestic water heating to operate as the **priority load** for a maximum of 30 minutes. All other loads are to be temporarily turned off during this time. However, if the domestic water load is still

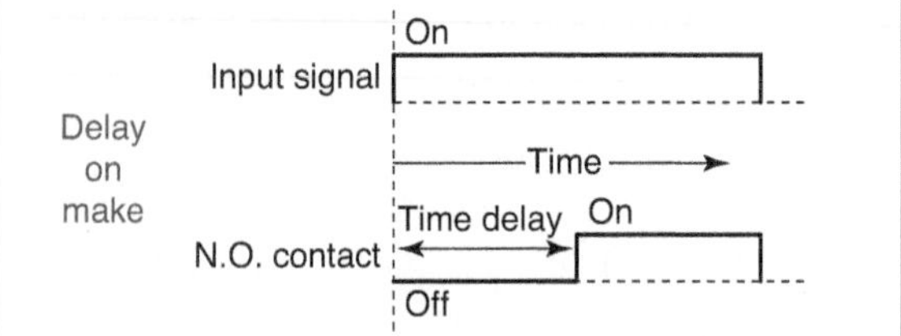

Upon application of voltage to the input terminals, the delay period begins. At the end of the time delay period, the contacts transfer and hold. When voltage to input terminals is interrupted the relay resets its timing circuit.

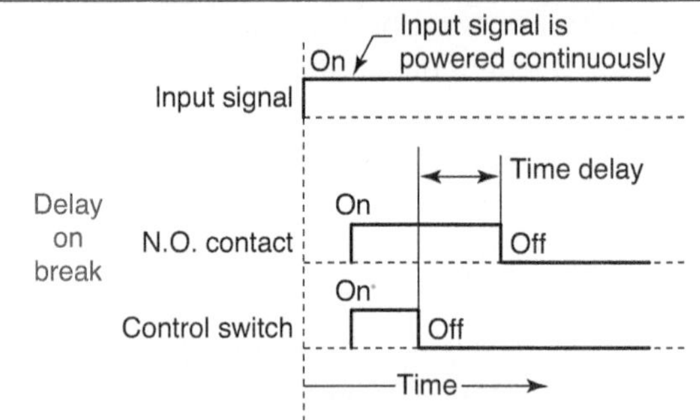

Voltage is applied to the input terminals at all times. Upon closure of the control switch the contacts transfer to on position and hold. When the control switch is opened, the time delay period begins. At the end of the time delay the contacts transfer to off position.

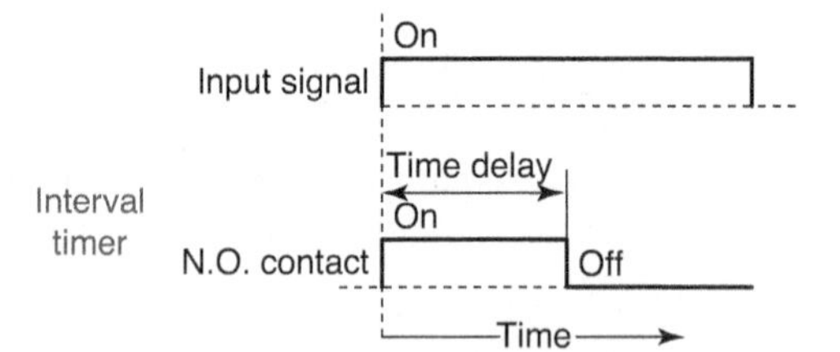

When voltage is applied to the input terminals the N.O. contacts close, and the off time delay begins. At the end of the delay period the contacts transfer back to off position.

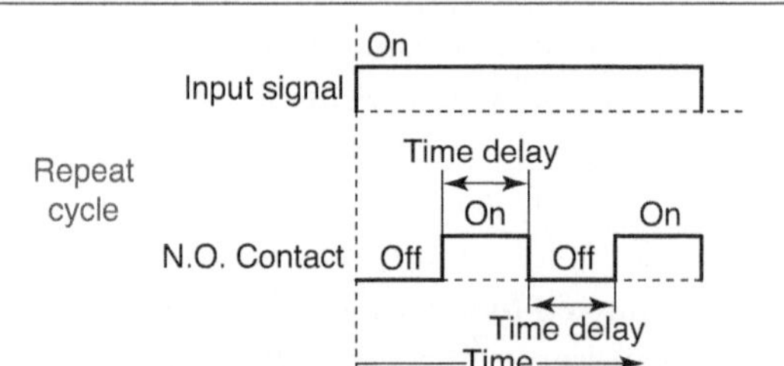

When voltage is applied to the input terminals the "off" time delay period begins. At the end of this time delay the contacts transfer, and the "on" delay period begins. At the end of the on time delay the cycle repeats. On some relays the on and off time delay periods are separately adjustable.

Figure 6-50 Operating modes for time delay relays. Some time delay can only provide one of these functions, while others can be set to provide any of these functions.

operating at the end of the 30-minute period, the other loads need to be turned back on to prevent potential freeze-ups.

The delay-on-break mode is the opposite of the delay-on-make mode. The contacts are held in their energized position until a user-specified time has elapsed. The input voltage to the relay is maintained at all times. When an external switch closure is detected across the control switch terminals of the relay, the contacts move to their energized position and remain there. When the external control switch opens, the time delay period begins. After the time delay period has expired, the relay contacts snap back to their "normal" position. If the control switch closes before the time delay period has elapsed, the time delay period is automatically reset to zero.

Time delay relays are available in several configurations of poles and throws. One of the most common is the double pole/double throw (DPDT) with a 120-VAC input signal.

Solid-state electronics have made it possible to combine several time delay relay functions into a single **multifunction time delay device**. An example is shown in Figure 6-51.

Notice the selector switches for both operating mode and time delay range on the multifunction time-delay device. These allow a single device to be configured for a wide range of functions and time delay periods.

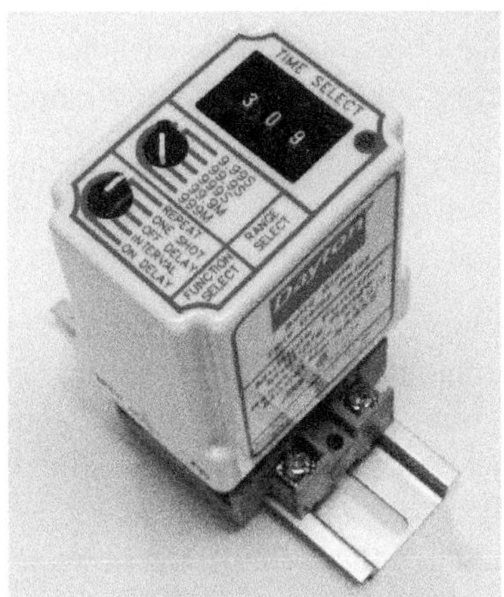

Figure 6-51 | Example of a multifunction time delay device.

HARD-WIRED LOGIC

One of the ways to create operating logic for a control system is by connecting switch and/or relay contacts in series, parallel, or combinations of series and parallel. Figure 6-52 shows basic series and parallel arrangements.

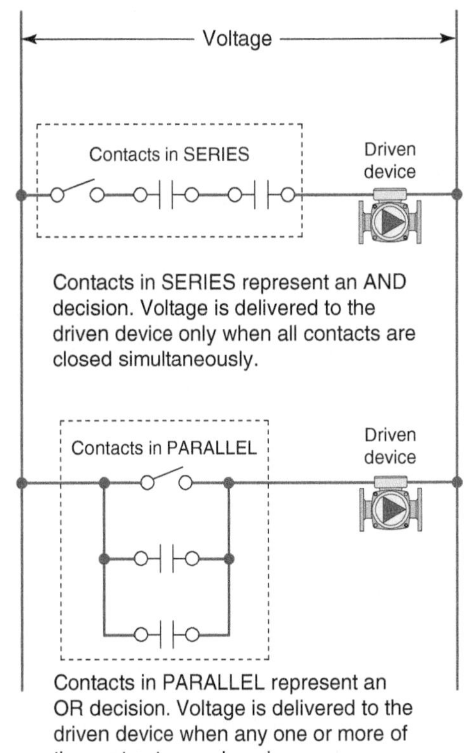

Figure 6-52 | Operating logic based on wiring of switches or relay contacts. Contacts in series provide "AND" logic. Contacts in parallel provide "OR" logic.

Contacts in series represent an **"AND" decision**. For an electrical signal to reach a driven device, all series-connected contacts must be closed simultaneously.

Contacts in parallel represent an **"OR" decision**. If any one or more of the contacts are closed, the electrical signal crosses the group of switches to operate the driven device.

When switch contacts, relay contacts, and contacts that are part of specialized hydronic controllers are physically wired together in a given manner, they create **hard-wired logic**. Such logic determines exactly how the control system functions in each of its **operating modes**. Some operating modes are planned occurrences while others may be fail-safe modes in the event a given component does not respond properly.

LADDER DIAGRAMS

It is often necessary to combine several control components to build an overall control system. The way these components are connected to each other as well as to driven devices such as circulators, zone valves, and boilers determines how the system operates.

A **ladder diagram** is a standard method for developing and documenting the electrical interconnections necessary to build a control system. The finished diagram can then be used for installation and troubleshooting. Such a diagram should always be part of the documentation of a hydronic heating system.

Ladder diagrams have two basic sections, the **line voltage section** at the top of the ladder, and the **low-voltage section** at the bottom of the ladder. A transformer separates the two sections. The primary side of the transformer is connected to line voltage. The secondary side of the transformer powers the low-voltage section. In North America, most heating and cooling secondary systems operate with a secondary voltage of 24 VAC.

An example of a simple ladder diagram is shown in Figure 6-53.

The vertical lines can be thought of as sides of an imaginary ladder. Any horizontal line connected between the two sides is called a rung of the ladder. A **rung** connected across the line voltage section is exposed to 120 VAC. A rung connected across the low-voltage section is typically exposed to 24 VAC. The

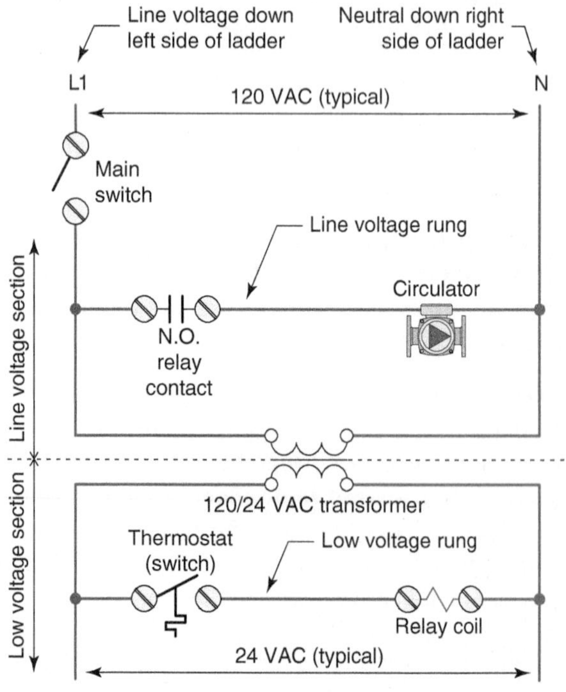

Figure 6-53 Example of a simple ladder diagram to operate a line voltage circulator using a low-voltage control circuit.

overall ladder diagram is constructed by adding the rungs necessary to create the desired operating modes of the system. The vertical length of the ladder can be extended as necessary to accommodate all required rungs.

Relays are often used to operate line voltage devices such as circulators, oil burners, or blowers based on the action of low-voltage components such as thermostats. Ladder diagrams are an ideal way to document how this is done. When a circuit path is completed through a relay coil in the low-voltage section of the ladder, one or more contacts of that relay may be used to operate devices in the line voltage section.

Consider a situation in which a line voltage circulator is to be operated by a low-voltage switch, such as a common thermostat. Since the circulator needs line voltage to run, it is connected across the line voltage section of the ladder diagram as shown in Figure 6-53. A normally open relay contact is wired in series with the circulator motor. When this contact is open, the motor is off. To close this contact, the coil of the relay must be energized. This requires a completed circuit path across the low-voltage section of the ladder. By wiring the relay coil in series with the thermostat, the coil is energized when the thermostat contacts are closed and deenergized when they are open. The overall operating sequence is as follows: The thermostat contacts close. Low voltage is then applied across the relay coil. The energized coil pulls the normally open contacts in the line voltage rung together. Line voltage is applied across the circulator motor to operate it.

Although this example is relatively simple, it illustrates the basic use of both the line voltage and low-voltage portions of the ladder diagram.

More complex ladder diagrams are developed by placing schematic symbols for additional components into the diagram. Some of these components might be simple switches or relays, others might be special purpose controllers that operate according to how the manufacturer designed them as well as how the adjustable parameters are set. When shown in a ladder diagram, the latter are called **embedded controllers** since they are part of a ladder diagram that documents the overall control system. An example of a ladder diagram with several rungs and an embedded controller is shown in Figure 6-54. The embedded controller is an on/off outdoor reset controllers designated as (ORC).

Although most of the electrical components in the ladder diagram shown in Figure 6-54 are shown within

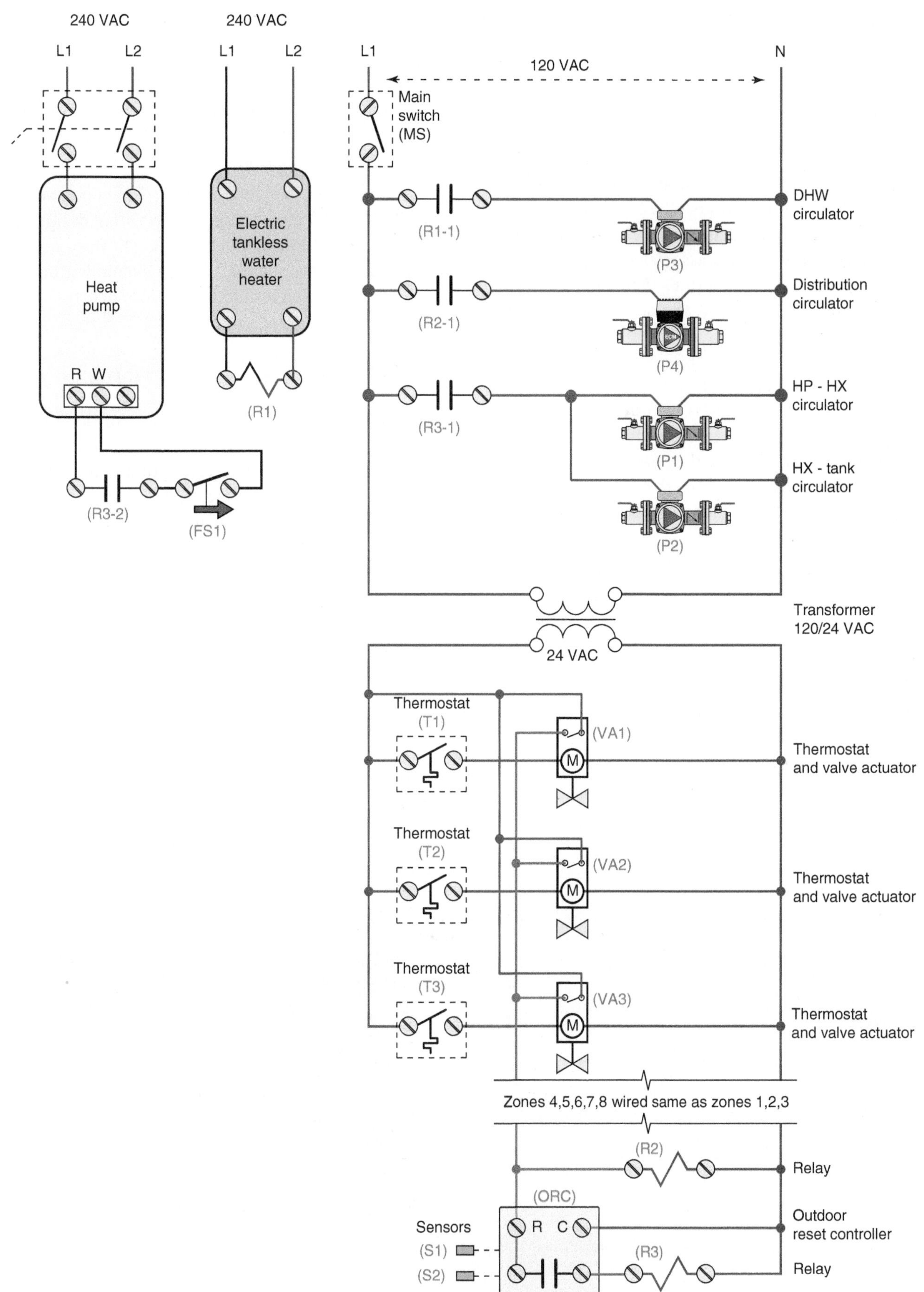

Figure 6-54 | Example of a ladder diagram with an embedded controller.

the main ladder, some other devices are shown outside the ladder. These include an electric tankless water heaters and an air-to-water heat pump. Both of these devices require their own dedicated electrical power circuits and thus do not draw power from the circuit supplying the ladder diagram.

Notice that all the devices are labeled (shown in green). A typical relay coil is labeled as (R2) or (R3) and its associated contacts are labeled as (R2-1) or (R3-2). These designations associate specific contacts with the relay coil that operates them. For example, relay coil (R2) operates relay contact (R2-1). The number after the dash indicates the pole number of the relay contact. For example, (R2-1) designates pole #1 on relay (R2), and (R2-2) designates pole #2 on the same relay. Such designations are critically important when interpreting the operating logic associated with the ladder diagram. Since all the relay coil symbols and contact symbols look the same, these designations are the only way to know which contacts are associated with a given relay coil.

Not all of the zone thermostats and zone valves are shown in Figure 6-54. A note in the lower portion of the ladder diagram indicates that zones 4 through 8 are wired identical to zones 1, 2, and 3. This allows the ladder diagram to be shortened when the detail is relatively simple and repeated. However, if space is available, all the repeated details can be shown.

Figure 6-55 shows the piping schematic of the system associated with the ladder diagram shown in Figure 6-54.

Notice that the electrically driven components on the piping schematic have corresponding designations on the ladder diagram. This makes it easy to cross-reference the two diagrams when examining the operating logic or troubleshooting the installed system.

DESCRIPTION OF OPERATION

Another critically important aspect of documenting a control system is to create a **description of operation**. This is a text-based description that narrates the sequence of events that must take place in each operating mode of the system. It is created by describing each component in the operating sequence, beginning with the component that "calls" for a specific mode of operation. Each operating mode should be described independently, while making frequent reference to the component designations in the ladder diagram. An example of a complete description of operation

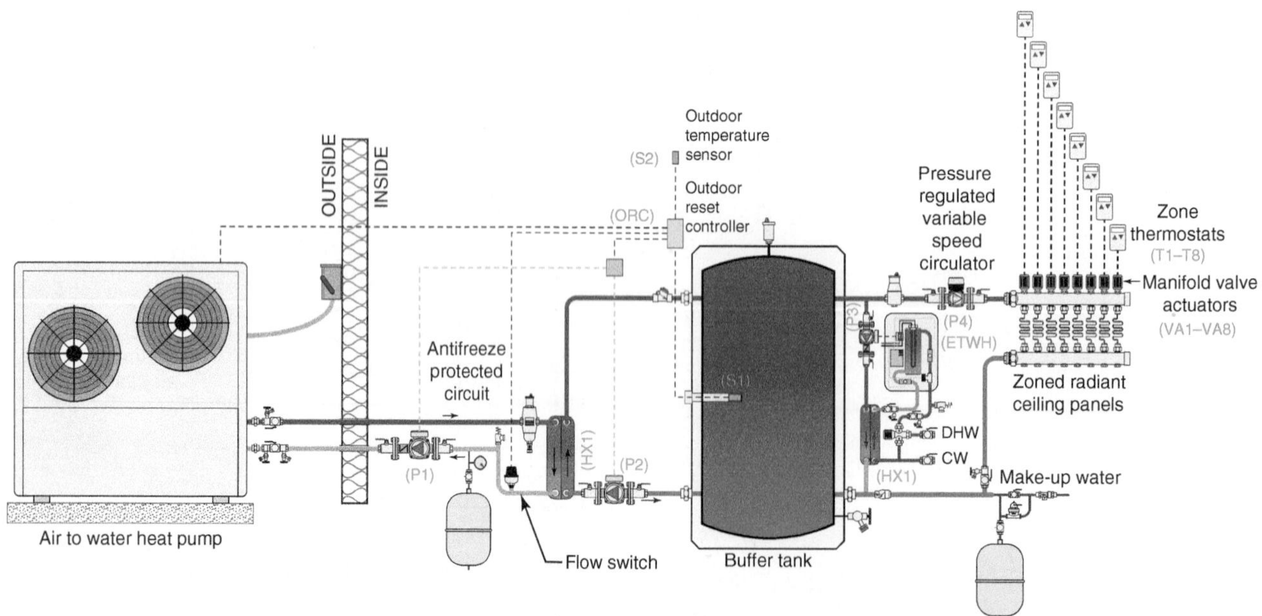

Figure 6-55 | The piping schematic of the system associated with the ladder diagram shown in Figure 6-54.

for the system shown in Figures 6-54 and 6-55 is as follows:

Description of Operation:

1. ***Space heating mode:*** *The distribution system has several zones, each equipped with a 24 VAC thermostat (T1, T2, T3, etc.). Upon a demand for heat from one or more of these thermostats, the associated 24 VAC manifold valve actuators (VA1, VA2, VA3, etc.) are powered on. When a valve actuator is fully open, its end switch closes. This supplies 24 VAC to the coil of relay (R2). This closes the normally open contacts (R2-1) that supply 120 VAC to circulator (P4). Circulator (P4) operates in constant differential pressure mode to supply the flow required by the distribution system. A call for heat also supplies 24 VAC to the outdoor reset controller (ORC). This controller measures outdoor temperature from sensor (S2). It uses this temperature, in combination with its settings, to calculate the target supply water temperature that needs to be maintained in the buffer tank. If the tank requires heating, the normally open relay contacts within the (ORC) close. This energizes the coil of relay (R3). One set of normally open contacts (R3-1) close to turn on circulators (P1) and (P2). This establishes flow through the heat pump and between the heat exchanger (HX1) and the buffer tank. Another normally open relay contact (R3-2) is wired in series with the contacts from the flow switch (FS1). When both of these contacts are closed, a 24 VAC circuit is completed between the R and Y terminals of the heat pump. This turns on the heat pump in heating mode and delivers heat to the buffer tank. The heat pump continues to run until: 1. The buffer tank reaches the target temperature of the outdoor reset controller (ORC) plus half of its differential setting, or 2. All zone thermostats are satisfied.*

2. ***Domestic water heating mode:*** *Whenever there is a requirement for domestic hot water flow of 0.6 gpm or higher, the flow switch inside the tankless electric water heater closes. This closure applies 240 VAC to the coil of relay (R1). The normally open contacts (R1-1) close to turn on circulator (P3), which circulates heated water from the upper portion of the buffer tank through the primary side of the domestic water heat exchanger (HX2). The domestic water leaving (HX2) is preheated to a temperature a few degrees lower than the buffer tank temperature. This water passes into the thermostatically controlled electric tankless water heater, which measures its inlet temperature. The electronics within this heater regulate current flow to the heating elements so that water leaving the heater is at the desired temperature. All heated water leaving the tankless heater flows into an ASSE 1017-rated mixing valve to ensure a safe delivery temperature to the fixtures. Whenever the demand for domestic hot water drops below 0.6 gpm, circulator (P3) and the tankless electric water heater are turned off.*

A full description of operation will be given for all the example systems discussed in later chapters.

DRAWING LADDER DIAGRAMS

Consistency is the key to drawing ladder diagrams that will be useful to many people. The following guidelines are recommended.

- Draw the ladder diagram with line voltage section at the top and the low-voltage section at the bottom.
- Specify the supply voltage and circuit **ampacity** required for the control system represented by the ladder diagram. Show this at the top of the line voltage section.
- Always show a main switch capable of completely isolating the ladder diagram from its power source when necessary.
- Use wider lines for line voltage conductors and thinner lines for low-voltage conductors to make the drawing easier to follow.
- When a component symbol appears in the system's piping schematic, and in the ladder diagram, be sure it has identical designations in both drawings (i.e., P1, T2, etc.). This allows for easy cross-referencing.
- Use separate abbreviations for each relay coil and its associated contact(s). Use a designation such as R1-2 to identify pole #2 of relay R1. R2-1 would identify pole #1 of relay R2, and so forth. Without such designations, it is impossible to know which contacts and coils are part of the same relay, especially in complex diagrams.
- It is customary to show switches and relay contacts in their off, or deenergized, positions.

- When several of the same components (i.e., relay contacts, circulators, zone valves, etc.) appear on several rungs, they should be vertically aligned. This not only makes the drawing look better, it also makes it easier to understand because of consistent symbol placement.
- Although ladder diagrams can be drawn by hand, there are many advantages to creating them using computer drawing software (e.g., CAD). For example, the vertical height of the ladder can be easily extended or shortened as needed to accommodate the necessary rungs. Component symbols and groups of symbols can be easily duplicated and moved to speed the design/drawing process.
- Always create a **description of operation** for the system represented by the ladder diagram. Describe each operating mode independently and make frequent reference to the component designations in the ladder diagram. This description greatly enhances the ability to understand what the system is supposed to do in each operating mode as well as the specific components involve in each operating mode.

The example systems shown in later chapters will have their control subsystems shown as ladder diagrams. Readers should look for the above-described common characteristics within these diagrams.

SUMMARY

This chapter has discussed several principals and hardware devices used to control modern hydronic systems, including those with renewable energy heat sources. Many of the example systems shown in later chapters will use these principles and devices.

As digital control technology advances, several manufacturers have begun offering advanced electronic controllers that, with proper programming, can handle many, and in some cases *all,* of the control functions needed in a residential combisystem using a renewable heat source. Some of these controllers are also Internet-enabled, allowing monitoring and adjustment through web-based devices such as laptops, tablets, and smartphones.

Although such controllers are increasingly available, many are based on preconfigured system templates for piping as well as control wiring. If a designer can "fit" one of these templates to the project requirements, these controllers often provide a convenient solution. However, none of them offer the ultimate flexibility that can be achieved using combinations of the basic principles and hardware devices discussed in this chapter.

Designers should also consider that an "all-in-one" controller that represents state-of-the-art technology when installed will likely not represent state-of-the art technology within 3 to 5 years. New products are constantly being developed to supersede older models. This raises legitimate concern over the availability of replacement components, or complete replacement controllers, over a typical system design life of 20+ years. A controller that constrains system design to a specific template but eventually becomes unserviceable could lead to costly but necessary modifications to keep the system in operation.

This scenario should be compared to one in which more rudimentary controllers such as thermostats, relays, outdoor reset controllers, motorized mixing valves, and differential temperature controllers are used. All of these devices have been used in the heating industry for several decades. Although more installation time may be required to configure an overall control system around such devices, versus and all-in-one controller, it is more likely that replacements for such devices will be readily available in the future. The ability to replace basic control components with new devices having identical or nearly identical capabilities helps ensure continued system operation without extensive or costly modifications.

KEY TERMS

ΔT·hours
accumulated ΔT·hours
ampacity
analog inputs
analog signal
AND decision
ASTM standard E44
building heating load profile
buffer tank
center off [switch]
closed-loop control system
coil
contactor
contacts
control differential
control subsystem
controlled device
controlled variable
controller
conventional boiler
delay-on-break relay
delay-on-make relay
description of operation
design load conditions
differential
differential temperature controller
digital input
digitally encoded signal
DIN rail
electrical contact
electrical noise
embedded controllers
equal percentage characteristic
error
feedback
floating control
floating zone
flow transducer
fully instrumented heat metering
fully modulating [heat source]
gel-filled compression connector
general purpose relay
hard-wired logic
heat metering
heat sink
heat source reset
inductive loads
ladder diagram
line voltage section
low-voltage section
manipulated variable
mixing reset
modulate
motorized actuator
multifunction time delay device
negative temperature coefficient (NTC) thermistor
no load conditions
nonlinear
normally closed (N.C.)
normally open (N.O.)
off-differential
on/off control
on-differential
operating differential
operating modes
OR decision
outdoor reset control
outdoor reset controller
overshoot
poles
priority load
processing algorithm
pulse width modulation (PWM)
relay
relay socket
reset line
reset ratio
resistive loads
room thermostat
rung [of a ladder diagram]
schematic arrangements
sensor
sensor well
setpoint
settings
shielded cable
short cycling
solid-state relay
source temperature sensor
storage temperature sensor
stratify
system controllers
target temperature
target value
temperature setpoint controller
thermal break
throws
time delay relays
triac
triac output
twisted pair of conductors
undershoot
variable speed control

QUESTIONS AND EXERCISES

1. Explain the difference between the *control differential* and the *operating differential* of an on /off temperature controller.
2. Explain the difference between the controlled variable and manipulated variable in a closed-loop control system for home heating.
3. Describe how floating control becomes more aggressive in correcting the error between a target temperature and measured temperature as the difference between these temperatures increases.
4. A heating system is designed to provide 120 °F water to the distribution system when the outside temperature is –5 °F. The building requires no heat input when the outside temperature is 65 °F. What would be the proper reset ratio for this system?
5. Explain why an analog output signal of 2–10 VDC or 4–20 milliamps doesn't start at zero voltage or zero current.
6. Describe a situation in which a modulating boiler that is capable of varying its firing rate from 100 percent down to 20 percent cannot match its heat output to the load. What happens under this condition? How can this condition be corrected?
7. Explain why a valve used to regulate heat output by varying the flow rate through a heat emitter should have an equal percentage characteristic.

8. When the total surface area of the heat emitters in a building increases, the supply water temperature required for a given rate of heat output__________, and the reset ratio of the outdoor reset controller operating the system ___________.

9. Why is it more difficult to regulate the heat output of a distribution system using variable flow rate control as compared to variable water temperature control?

10. Describe the difference in control output between an outdoor reset controller for a boiler versus an outdoor reset controller for a mixing valve.

11. The term *error* refers to the difference between the ________ temperature and the _________ temperature in a temperature control system.

12. Explain the difference between the number of poles and number of throws on a switch. Sketch a schematic symbol for each of the following switches:
    a. a DPST switch
    b. a DPDT c/o switch

13. Describe the capabilities of a triac output on a differential temperature controller that operates a standard wet-rotor circulator with a PSC motor.

14. What is the purpose of thermal grease when mounting a temperature sensor within a well?

15. Determine the total ΔT·hours represented by the measurements in Figure 6-56.

16. Assume water at a constant flow rate of 5 gpm was associated with the data in Exercise 15. Determine the total Btus transferred by the process over the 4-minute period.

17. Explain why fully instrumented heat metering with a flow transducer and two temperature sensors tends to be more accurate than the ΔT·hours method.

18. For a negative temperature coefficient thermistor, resistance goes up as the sensor's temperature goes __________.

19. Describe what can happen if the wiring between a thermistor temperature sensor and controller is either too long or uses a wire diameter that is too small.

Time Interval	T_{supply} (°F)	T_{return} (°F)
Minute 1	105.0	95.0
Minute 2	106.5	97.5
Minute 3	108.0	99.0
Minute 4	111.5	101.5

Figure 6-56 | Measurement data for Exercise 15.

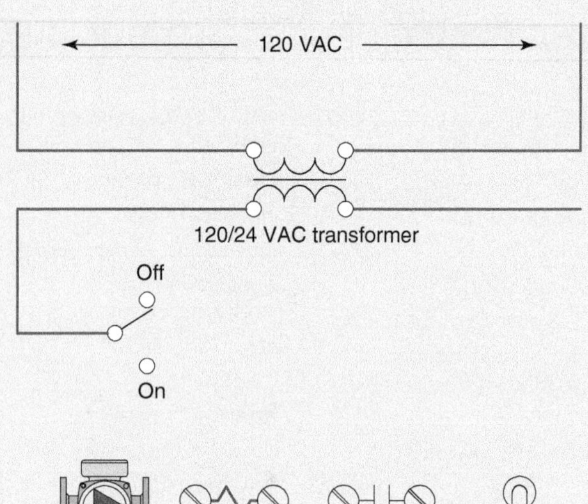

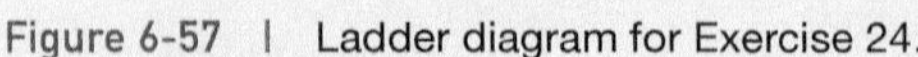

Figure 6-57 | Ladder diagram for Exercise 24.

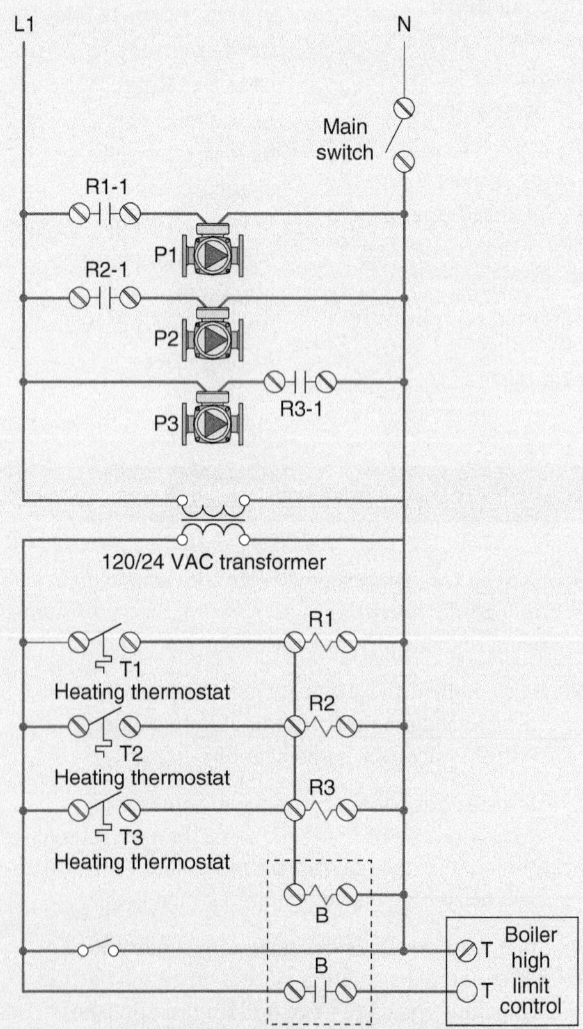

Figure 6-58 | Ladder diagram for Exercise 25.

20. Explain what can happen if moisture enters the electrical junction between a thermistor temperature sensor and the cable connecting it to a temperature controller.

21. Why is it preferable to install a temperature sensor for a mixing controller downstream of the circulator moving flow from the mixing device to the distribution system?

22. Describe a situation when either twisted pair cable or shielded cable should be used between a temperature controller and a thermistor temperature sensor.

23. Describe the difference between a delay-on-make and delay-on-break time delay relay. Which would be used to keep a circulator running for a few minutes after a boiler is turned off so that residual heat can be purged from that boiler?

24. Complete the ladder diagram shown in Figure 6-57 such that the following control action is achieved. When switch S1 is set in the off position, a red 24-VAC indicator light is on. When the switch is set to the on position, a green 24-VAC indicator light is on and a line voltage circulator is running. Label all components you sketch in the diagram.

25. A control system having the ladder diagram shown in Figure 6-58 is proposed to operate a three-zone hydronic system using a separate circulator for each zone. Identify any electrical errors in this diagram. Describe what would happen if the identified error were present when the system was turned on, or what is unsafe about the error. Sketch out how you would modify the schematic for proper and safe operation of the three-zone system.

R410A
R410A

CHAPTER 7

Solar Fundamentals

OBJECTIVES

After reading this chapter, you should be able to:

- Give a brief history of how solar energy has been used for heating.
- Describe the physical characteristics of solar radiation.
- Explain how the earth's atmosphere affects solar radiation.
- Describe solar altitude and azimuth angles.
- Determine solar altitude and azimuth angles for a given location.
- Explain the differences between direct and diffuse solar radiation.
- Use a sun path diagram to determine the sun's position in the sky.
- Convert between local clock time and local solar time.
- Determine true south at any location.
- Use a Solar Pathfinder® to estimate the solar energy available at a site.
- Access online resources for determining solar radiation available on various surfaces.

7.1 Introduction

When asked about heating with renewable energy, many people immediately think of "solar heating." Almost everyone understands that it's possible to derive heat from the sun. Most understand that this requires some sort of solar collecting device. Beyond that, the design of practical and reliable solar thermal systems remains largely unknown, even to the majority of North American heating professionals.

This is the first of several chapters that discuss using solar energy for heating purposes. It starts with a brief look at the history of solar energy usage for heating. It goes on to discuss the nature of solar radiation. It describes methods for locating the sun in the sky and assessing the potential solar energy available at a specific site, including the effects of shading. These are the basics of "solar energy science" that all systems designers and installers should be familiar with. The chapters that follow will introduce hardware, performance estimating, and system design.

Figure 7-1 | More solar energy falls on the earth in 1 hour than what is used by the entire world population in 1 year.

7.2 A Brief History of Solar Heating

Humans have used solar energy for heating for thousands of years. Archeologists assert that solar architecture flourished in Greece 2,500 years ago. This happened because the Greeks eventually consumed all their readily available firewood and were thus forced to find alternatives. In due course, they learned to construct buildings that allowed low winter sun to penetrate deeply into south-facing porticos while also blocking higher summer sun with overhangs. These methods remain fundamental to modern day **passive solar architecture**.

The ancient Romans followed a similar path in depleting readily available wood supplies. Necessity led them to adopt the solar design methods already used by the Greeks. They advanced these approaches through use of glass and even developed regional solar design variations based on climate.

Much of this skill vanished from Europe during the Dark Ages. Fortunately, Middle Easterners of that time preserved the mathematical knowledge that eventually led to the reappearance of solar devices. Leonardo da Vinci was well known for applying this knowledge to create solar mirrors.

As time progressed, devices that focused solar radiation were used to power water pumps and even printing presses. The first use of solar energy for domestic water heating can be traced to 1892, when Clarence Kemp of Baltimore patented the **Climax solar water heater**. The device consisted of multiple cylindrical tanks that were blackened and mounted in a glass-covered box. Figure 7-3 depicts an early advertisement for this system.

By the turn of the century, several companies had entered the solar water heating market with variants of this design. However, these devices lost much of the heat collected during a sunny day through subsequent nighttime cooling.

In 1909, a new approach that *separated the solar collector from the storage tank* was developed. It greatly reduced heat loss and made solar water heating more viable. It operated on the principle of **thermosyphoning**. Hot water within the collector rose through upward-sloped piping to an elevated storage tank. Cool water at the bottom of this tank dropped back to the collector. This approach was refined through the first half of the twentieth century, and the market for solar water heating systems flourished in warm, sunny locations such as Florida and Southern California.

Courtesy of US Department of Energy

Figure 7-2 | Anasazi cliff dwellings demonstrate early passive solar design; Mesa Verde, Colorado.

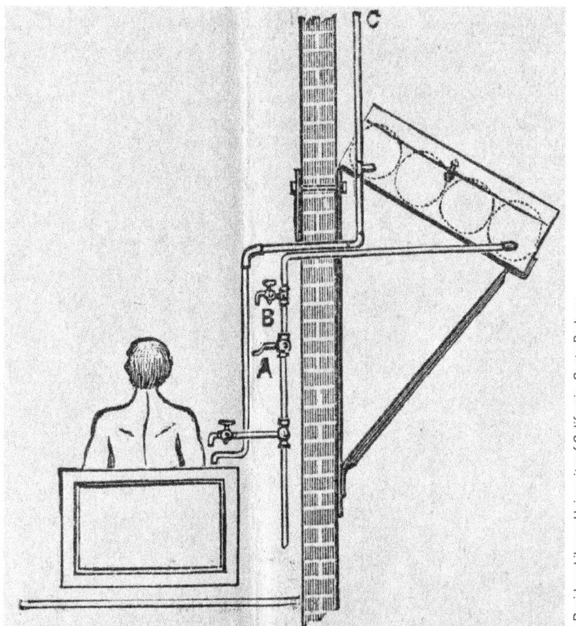

Fig. 5.—Shows a Climax Solar-Water Heater supported by a bracket on the wall.
A.—Is the cock to use when the hot water is wanted. This passes cold water into the heater, displacing the hot water and forcing it through a pipe to the bath tub.
B.—Is the drain cock which is used to prevent freezing.
C.—The air opening which prevents vacuum in the heater and siphonic action.

Courtesy of Department of Special Collections, Davidson Library, University of California, Santa Barbara

Figure 7-3 | Early device for solar water heating.

Courtesy of MIT Museum

Figure 7-5 | MIT Solar House #1, constructed in 1938.

In 1938, an ambitious solar heating demonstration project, **MIT Solar House #1**, got underway at the Massachusetts Institute of Technology. The small house, shown in Figure 7-5, used both liquid- and air-based solar collectors and separate thermal storage devices. By this point, electrically powered circulators and blowers were available to move fluids through systems. This allowed the storage vessel to be located lower than the collector—usually in the basement.

By 1949, MIT had constructed a solar demonstration house in the Boston area that gathered 75 percent of its heating energy requirement from the sun.

The 1950s and 1960s were a time when North America transitioned to an era of readily available and inexpensive energy. The notion that nuclear-supplied electricity would soon be "too cheap to meter" was put forth, and Americans were lulled into indifference when it came to energy and associated environmental issues. Electrical consumption within the United States increased by almost 500% during the 1950s and 1960s. Although the solar industry didn't completely disappear, interest quickly declined, as lower first-cost alternatives became readily available to supply a rapidly growing housing market. In short, Americans began enjoying an era of cheap energy and a collective belief that the reserves of such energy were virtually unlimited.

Copyright © John Perlin, Let It Shine: The 6000-Year Story of Solar Energy

Figure 7-4 | 1911 photo of a California home with solar collectors on roof supplying heat to storage tank in attic.

This situation changed quickly in the later 1970s following the **Arab oil embargo**. Scores of manufacturers started producing and delivering solar energy products to the North American market to capitalize on the fervor to replace conventional energy sources, many with highly volatile prices, with renewable energy. Federal **tax credit** legislation was passed in 1979. This, in combination with state tax credits, stimulated rapid growth in the solar energy industry.

Unfortunately, the rush to deploy solar energy technology by both the public and private sector resulted in less-than-ideal results. Many installations soon demonstrated their inability to provide long-term reliable performance. By the 1980s, changing political priorities, elimination of federal tax credits, and falling fuel prices caused the solar energy industry to all but collapse as North Americans returned to complacency about seemingly cheap and inexhaustible energy supplies.

Courtesy of Skip Fralick

Figure 7-6 | A poorly maintained solar thermal system from the 1970s.

Figure 7-7 | An array of modern flat-plate solar thermal collectors.

LIKE "DÉJÀ VU ALL OVER AGAIN"

Although he probably didn't have solar energy in mind, Yogi Berra's famous quotation does summarize the history of solar energy utilization.

Energy cost reduction has again become one of the highest priorities for homeowners and commercial building owners. The solar thermal market in North America is expanding, albeit at a relatively slow rate. Concerns over climate change and national security associated with oil importation are factoring into energy supply decisions. The United States and other industrialized countries are poised to enter an era where energy conservation and the use of renewable energy sources will play a major part in their future prosperity. This time around, the solar energy industry has better standards, better hardware, and better design tools.

Still, as this text is being written, much of the present North American solar thermal market remains heavily dependent on government subsidies offered as grants, rebates, and income tax credits. Those incentives were created to stimulate growth in the solar thermal market, hopefully to the point where it could sustain a market for solar thermal systems without further need of subsidies. The size of the U.S. debt, and efforts to reduce it, are likely to affect what, if any, subsidies remain within a few years (of the writing of this text). Thus, it is vitally important for the North American solar thermal industry to focus on developing systems that are technically sound, and **economically sustainable**, without subsidies. The information in this and other chapters is meant to inspire and facilitate that process.

7.3 Solar Radiation Fundamentals

The word sunlight, in a technical sense, refers to the total **electromagnetic radiation** given off by the sun. In this text, the term **solar radiation** is used synonymously with sunlight. Although solar radiation is produced by nuclear reactions at the sun, its transmission through space, and use within solar thermal systems, has nothing to do with **nuclear radiation**.

Because it is a form of electromagnetic radiation, solar radiation has some characteristics that are similar to other types of electromagnetic radiation, such as radio waves, x-rays, and even "radiant heat" emitted by a warm floor. It travels at the **speed of light** (186,000 miles per second) and can be changed in direction by reflection. The difference between solar radiation, and other types of electromagnetic radiation, is **wavelength**, as illustrated in Figure 7-8.

The wavelengths of solar radiation span from about 0.2 to 2.6 micrometers. A relatively narrow portion of this range, from about 0.4 to 0.7 micrometers, is visible to the human eye and thus called **visible light**.

Figure 7-9 shows the wavelengths of solar radiation and the relative intensity of energy carried by those wavelengths. This graph is called the **solar radiation spectrum**.

The composition of solar radiation that exists outside the earth's atmosphere is shown by the yellow areas in Figure 7-9. This radiation would be similar to what would be emitted by a theoretical **black body**, with a surface temperature of 5,250 ºC (9,482 ºF).

Just outside the earth's atmosphere, the intensity of solar radiation remains relatively constant throughout the year at an intensity of approximately 433 Btu/hr/ft^2. This value is called the **solar constant** and it represents the total energy content of the yellow shaded area in Figure 7-9. Although this value is often treated as constant, it does vary by approximately 1.5 percent over the course of a year due to variations in the distance between the earth and sun.

As solar radiation passes through the earth's atmosphere, gases such as water vapor, oxygen, and carbon dioxide as well as suspended dust absorb some of its

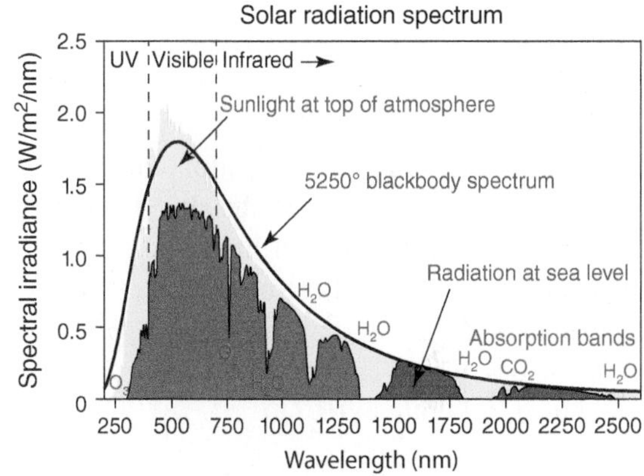

Figure 7-9 | The solar spectrum.

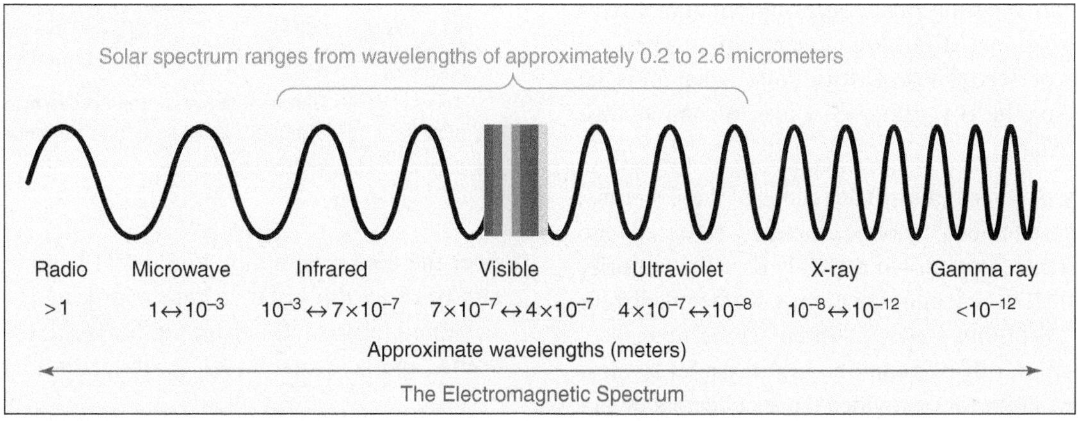

Figure 7-8 | Representation of the electromagnetic spectrum. Solar radiation falls within this spectrum, with wavelengths ranging from about 0.2 to 2.6 micrometers.

energy within specific, and relatively narrow ranges of wavelengths. These absorption bands result in the jagged red area shown in Figure 7-9. This reduces the intensity of the solar radiation reaching the earth's surface. That intensity varies with location, time of year, time of day, angle of the receiving surface, and clarity of the atmosphere. For example, under clear skies, at a location of 40° north latitude, on June 21, at solar noon, the solar radiation striking a horizontal surface will have an intensity of about 304 Btu/hr/ft^2. The same surface, under clear skies on January 21, at noon, will receive solar radiation at an intensity of about 164 Btu/hr/ft^2.

Just outside the earth's atmosphere, about 40 percent of the energy in solar radiation lies within wavelengths that are visible to humans. Fifty percent of the remaining energy lies within **infrared** wavelengths, and 10 percent within **ultraviolet** wavelengths. Although not seen by the human eye, both infrared and ultraviolet wavelengths are useable by solar collectors.

At the earth's surface, solar radiation can be qualitatively categorized as either **direct radiation** (also sometimes called **beam radiation**) and **diffuse radiation**. Direct solar radiation is radiation coming directly from the position of the sun in the sky. This radiation is traveling in straight parallel lines and, as such, can be easily reflected or concentrated onto a small target area by standard optical devices such as parabolically shaped mirrors, which are part of the solar thermal collector shown in Figure 7-10.

Courtesy of Pete Skinner

Figure 7-10 Parabolically shaped mirrors can reflect and concentrate direct solar radiation onto a small target area at the optical focal point.

If the earth had no atmosphere, almost all of the solar radiation reaching the earth's surface would be direct radiation. However, the gases and dust in the atmosphere scatter incoming solar radiation over the entire **sky dome**, as seen in Figure 7-11. This scattering creates diffuse solar radiation. Some sunlight arrives as diffuse solar radiation, even during times when the sky appears perfectly clear. Diffuse solar radiation is the reason we see the sky, rather than just the sun against black outer space.

Figure 7-11 Diffuse solar radiation comes from all directions of the sky due to atmospheric scattering.

Because it comes from all directions of the sky dome, diffuse solar radiation cannot be efficiently reflected onto a target or concentrated. On a cloudy day, the majority of solar radiation reaching the earth's surface is diffuse radiation. Although lower in intensity compared to direct radiation, diffuse radiation is still useful to most solar thermal collectors, provided those collectors do not rely heavily on reflectors.

The intensity of solar radiation striking a fixed surface, such as a typical solar collector, varies widely over the course of a day. Figure 7-12 shows the intensity of *clear day* solar radiation striking a surface that is sloped upward 40° from the horizontal, facing **true south**, and is located at 40° north **latitude**.

During a perfectly clear day, solar radiation is most intense at **solar noon** (e.g., that time of day when the sun is highest in the sky and directly above a true north/south line).

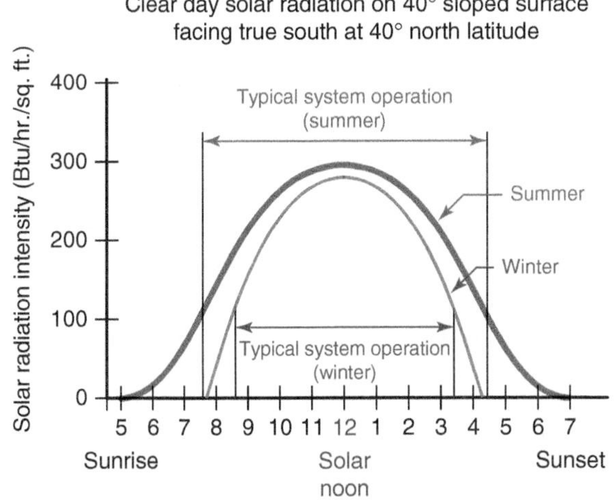

Figure 7-12 | Clear day solar radiation intensity striking a surface sloped 40° above horizontal and facing true south, at a 40° north latitude.

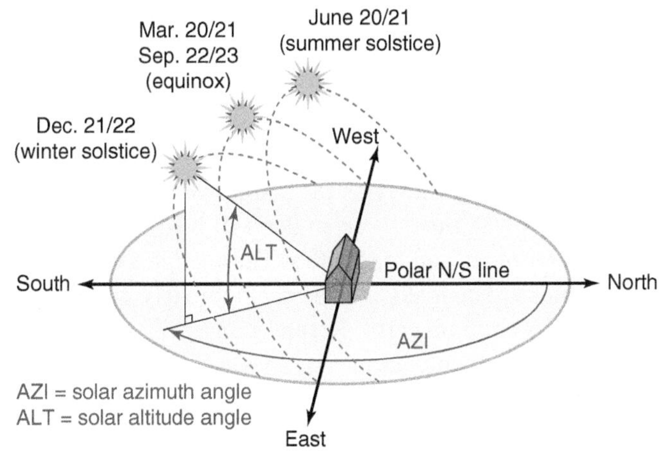

Figure 7-14 | The sun's path, as seen on earth, for specific days of the year.

7.4 Solar Geometry and Availability

The position of the sun in the sky has been very accurately measured and can be found using a variety of methods, both analytical and graphical. This section introduces the angles used to describe the sun's position in the sky for different locations and times of day.

The earth revolves once each day around an axis that passes through the North and South Poles. That axis is tilted 23.44° relative to the orbital plane of the earth around the sun, as shown in Figure 7-13.

The tilt of the earth's axis is called the **declination angle**. It's the reason that day length, between sunrise and sunset, changes as the earth makes its annual orbit around the sun. It also significantly affects the intensity of solar radiation striking a fixed surface at any geographic location. On earth, we observe this effect as a change in the sun's path across the sky, as seen in Figure 7-14.

The sun's position in the sky can be precisely described using two simultaneously measured angles.

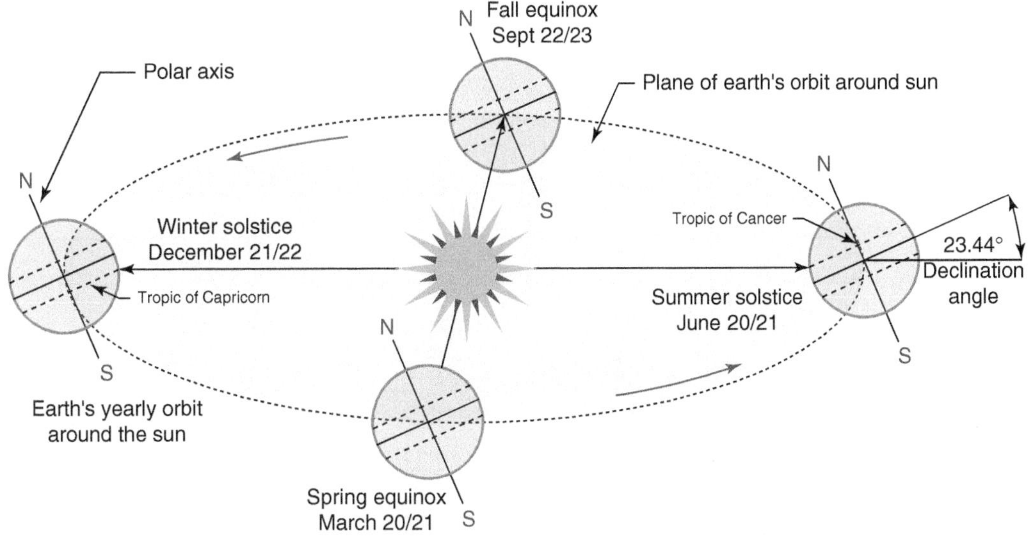

Figure 7-13 | Depiction of the earth's annual orbit around the sun. Note that the orientation of the earth's polar axis does not change during this orbit.

The **solar altitude angle (ALT)** is measured from a horizontal surface up to the center of the sun. The **solar azimuth angle (AZI)** is measured starting from true north (0°) in a clockwise direction (i.e., true south would have a solar azimuth of 180°). Both of these angles vary continuously as the sun moves across the sky. At any given time, these angles will be different at different latitudes and longitudes. These angles can be precisely calculated, albeit somewhat tediously, for any time and location on earth. Several references provide the necessary trigonometric equations needed to calculate the solar altitude and solar azimuth angles.

A much simpler method of determining the sun's position in the sky is to use a **sun path diagram**, such as shown in Figure 7-15.

This sun path diagram was provided from the website: http://solardata.uoregon.edu/SunChartProgram.html. This website allows one to specify the exact location, and time of year, for which the diagram is to be prepared. It then produces the diagram as a PDF file.

The curves on the sun path diagram shown in Figure 7-15 are for a specific day in each indicated month. These dates coincide with specific solar geometry events. The top arc represents June 21. This is the **summer solstice**, when the sun makes it highest arc across the sky. It is also the longest day of the year. The arc labeled December 21 presents the **winter solstice**, during which time the sun makes is lowest arc across the sky. This is also the shortest day of the year. The arc labeled March 20 represents the **vernal equinox**. On this day, the sun rises above the horizon at true east (e.g., 90° solar azimuth) and sets at true west (e.g., 270° solar azimuth). There are approximately 12 hours when the sun is above the horizon during both the vernal equinox, and its corresponding date, the **autumnal equinox**, when the sun follows the same path as on the vernal equinox.

Each blue arc on the sun path diagram can be used to find the solar altitude angle on the vertical axis as well as the solar azimuth angle, on the horizontal axis, based on the **solar time** indicated by the red curves.

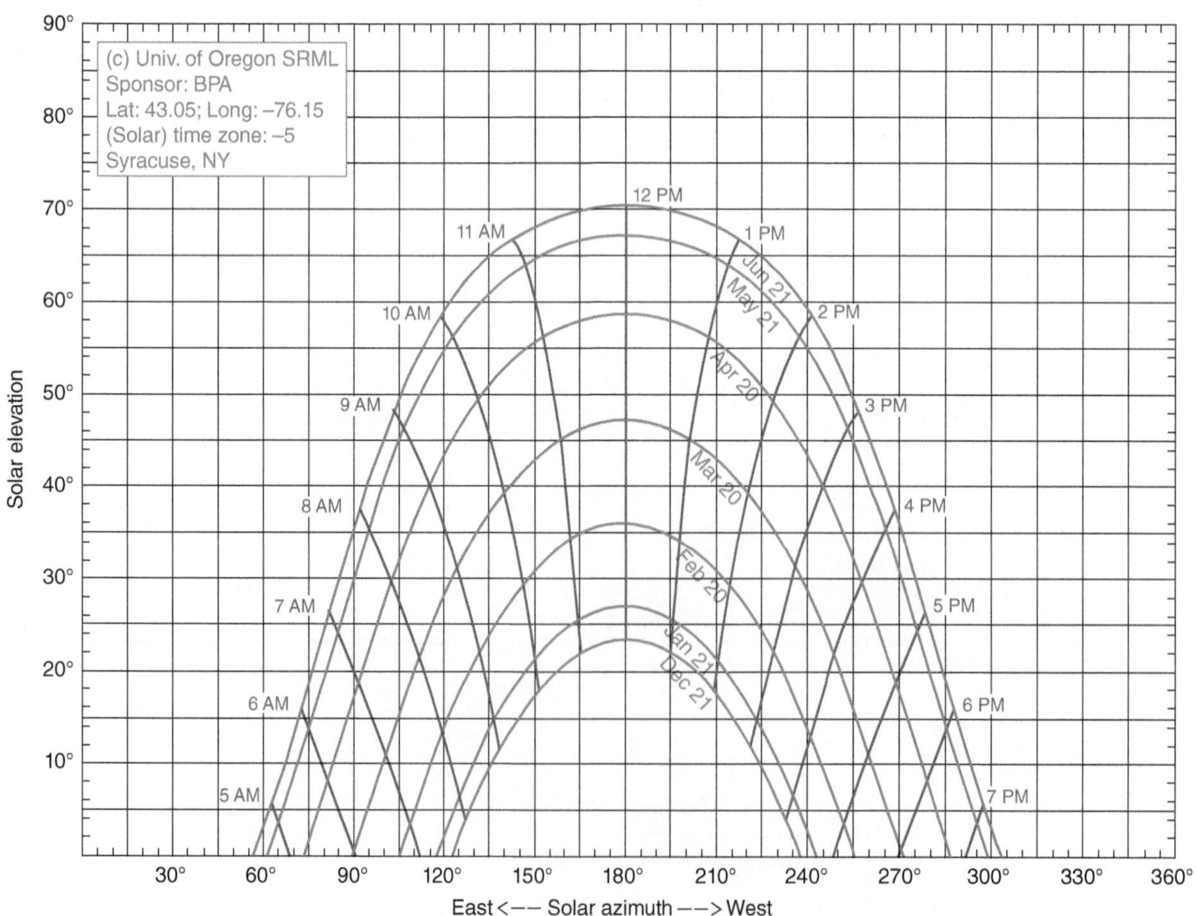

Figure 7-15 | Sun path diagram for Syracuse, NY.

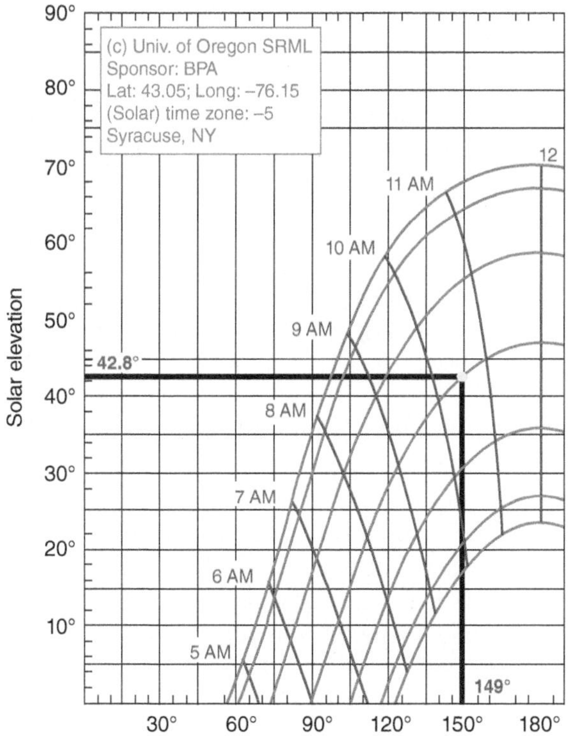

Figure 7-16 | Using the sun path diagram for Example 7.1.

Example 7.1: Find the solar altitude and solar azimuth angle on the sun path diagram for Syracuse, NY (Figure 7-15), for 10:30 AM solar time on March 20.

Solution: Make a mark on the blue arc representing March 20 and halfway between the red arc representing 10:00 AM and 11:00 AM. Draw a horizontal line from this mark to the vertical axis to read a solar altitude angle of 42.8°. Draw a vertical line down to the horizontal axis to read a solar azimuth angle of about 149°.

Discussion: These angles are approximate, as is any number read from a graph. However, with reasonable care, these angles can be estimated to within one 1°, which is adequate for any use in designing or installing solar thermal systems.

The motion of the sun across the sky dome can also be visualized using the online tool entitled **Motions of the Sun Simulator**, developed by the University of Nebraska–Lincoln. It can be accessed at the following URL: http://astro.unl.edu/naap/motion3/animations/sunmotions.html.

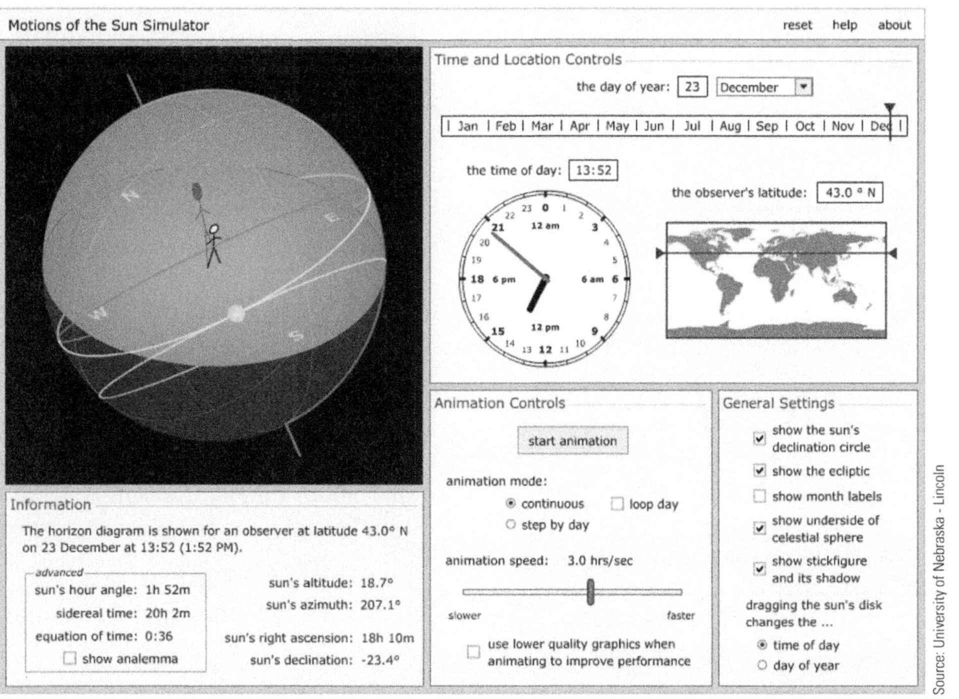

Figure 7-17 | Screen capture of web-based Motions of the Sun Simulator. Available at http://astro.unl.edu/naap/motion3/animations/sunmotions.html.

SOLAR TIME

All the times shown on sun path diagrams, or used in calculations for solar position, are **local solar times**. *Local solar time is usually not the same as local clock time*. Consider for example, that a time zone spans 15° of **longitude** (e.g., 360° divided into 24 time zones). Clock time is the same at the east and west extremes of each time zone, yet the sun's position cannot be the same at these extremes because of the significant differences in longitude (e.g., up to 15°). Thus, to know the exact position of the sun, *it is necessary to convert local clock time to solar time before using sun path diagrams or any equations that relate solar position to time*. This can be done using Equation 7-1 and Figure 7-18.

(Equation 7-1)

$$T_{solar} = T_{LS} + 4(L_M - L_{local}) + E$$

where:

T_{solar} = solar time at the location

T_{LS} = local *standard* time

L_M = longitude of the **standard meridian** for the time zone (°)

L_{local} = longitude at the location (°)

E = "equation of time" (read from Figure 7-18) (minutes)

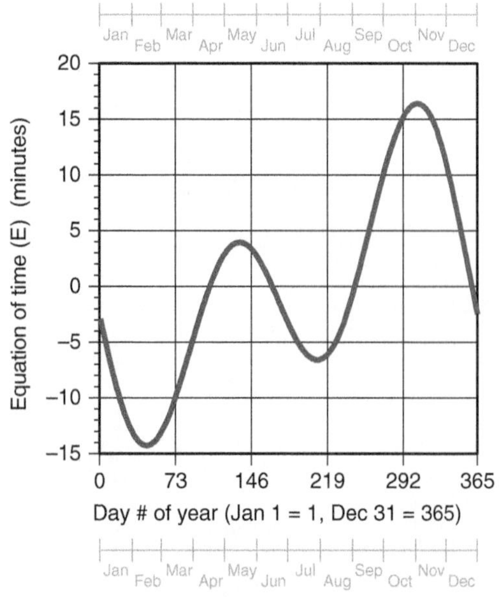

Figure 7-18 | Graph for equation of time.

The equation of time, as graphed in Figure 7-18, is a correction for slight variations in the speed at which the earth orbits the sun. Because that orbit is slightly elliptical, the earth's speed along its path is slightly faster in winter, when the earth is slightly closer to the sun, than in summer, when the earth is farther away from the sun.

Example 7-2: Determine the solar time that corresponds to 10:30 AM clock time on March 20 in Syracuse, NY.

Solution: The local longitude in Syracuse, NY is 76.1478° west. Syracuse, NY, is within the Eastern time zone, and the standard meridian for that time zone is 75° west. Daylight savings time *is* in effect on March 20. This means that the *standard time* corresponding to a clock time of 10:30 AM on March 20 is 9:30 AM. Finally, an estimate for the value of the equation of time (e.g., the value of E in figure 7-17) for March 20 is –8 minutes. Substituting this information into Equation 7.1 yields.

$$T_{solar} = 9{:}30 + 4(75 - 76.1478) + (-8)$$

$$T_{solar} = 9{:}30 - 12.59\ minutes = 9{:}17\ AM$$

Discussion: When the local clock indicates 10:30 AM on March 20, the solar time is 9:17 AM. The deviation between these two times is largely due to the difference between standard time and daylight savings time. To get local standard time, subtract 1 hour from the local clock time whenever daylight savings time is in effect. The remainder of the deviation is due to the location of Syracuse relative to the local time meridian (75° west longitude), and the slight variation in the earth's orbital speed accounted for by the equation of time.

All aspects of the sun's arc across the sky are symmetrical with respect to solar noon. The left half of a sun path diagram is symmetrical with the right half, about the vertical line representing solar noon. The sun is also directly above a true north/south line at exactly solar noon. Finally, the sun is always at its maximum altitude angle, for any given day, at exactly solar noon.

Equation 7-1 can be rearranged to calculate the clock time that corresponds to solar noon. This rearrangement is given as Equation 7-1a:

(Equation 7-1a)

$$T_{LS} = T_{solar} - 4(L_M - L_{local}) - E$$

where:

T_{LS} = local *standard* time

T_{solar} = solar time at the location

L_M = longitude of the standard meridian for the time zone (°)

L_{local} = longitude at the location (°)

E = "equation of time" (read from Figure 7-3) (minutes)

Example 7-3: Find the clock time corresponding to solar noon on March 20 in Syracuse, NY.

Solution: From Example 7.2, the longitude in Syracuse is 76.1478°. The longitude of the Eastern time meridian is 75°, and the value of E from Figure 7-17 is –8 minutes. Putting these values into Equation 7-1a yields:

$$T_{LS} = 12{:}00 + 4(75 - 76.1478) - (-8)$$

$$T_{LS} = 12{:}00 + 12.59 \text{ minutes} = 12{:}12{:}35$$

The time indicated as 12:12:35 means 12 minutes and 35 seconds past noon. Keep in mind that this calculated result is still *standard* time. On March 20, daylight savings time is in effect, so clock time is standard time +1 hour. Thus the *clock* time corresponding to solar noon is 1:12:35 PM, which can be rounded to 1:13 PM.

Discussion: Always use care when adding or subtracting times expressed in hours, minutes, and seconds.

The conversions between local clock time and local solar time can also be determined using online calculators, such as the one that is available at http://www.powerfromthesun.net/soltimecalc.html.

LOCATING TRUE SOUTH

Solar azimuth angles are always referenced to a **true north/south line**. This is a line at the earth's surface that's exactly parallel with the earth's polar axis. Because the earth's magnetic field is not always aligned parallel with its **polar axis**, there are locations within the continental United States where the needle of a compass can point as much as 20° east or west of a true polar north/south line.

The **isogonic chart** shown in Figure 7-19 indicates the deviation of a magnetic compass needle from true north. For example, in Boston, the isogonic chart indicates a deviation of approximately 15.5° west. *This*

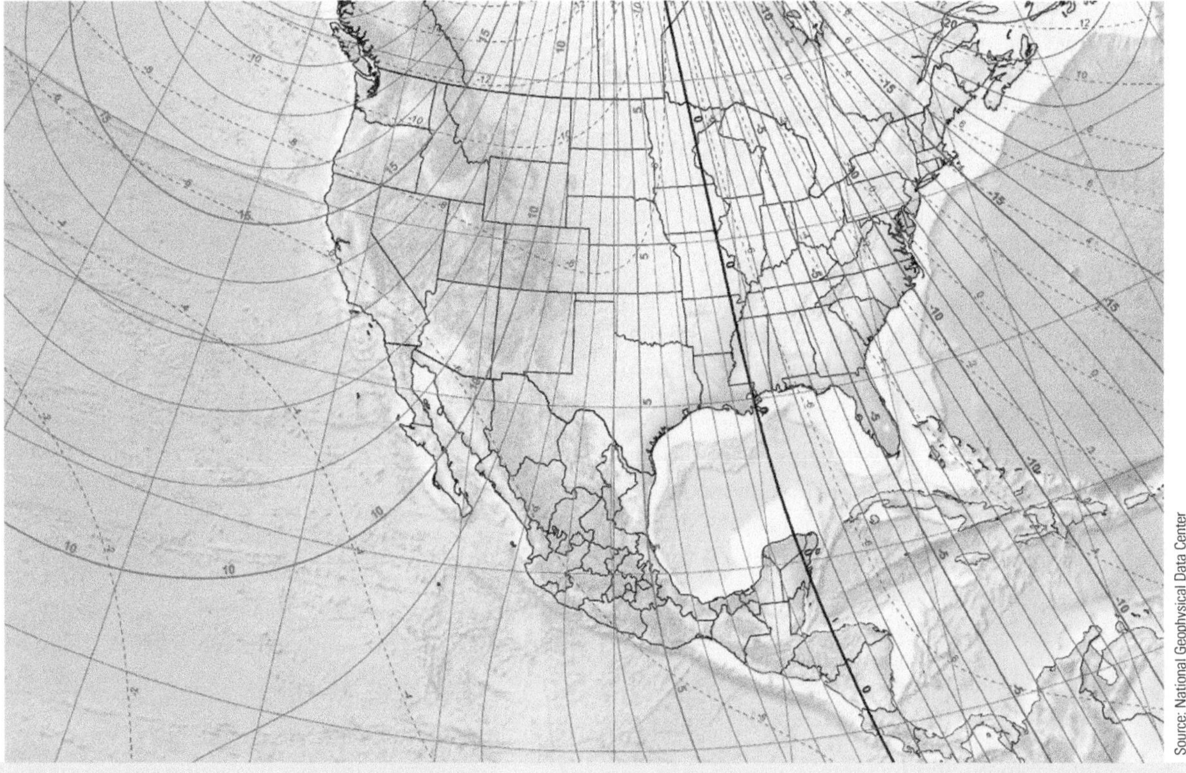

Figure 7-19 | An isogonic chart of the United States.

implies that true south is actually 15.5° west of the indicated compass south. In this case, solar collectors oriented to an uncorrected south compass direction would be facing approximately 15.5° east of true south.

The deviation between compass indicated north/south and "true" north/south is called **magnetic declination**. This declination changes slightly over time due to the magnetic characteristics of magma flowing deep beneath the earth's crust. Because of this, isogonic charts are dated. Fortunately, the deviations over time are very minor and will have no significant effect on solar collector installation and subsequent performance.

Magnetic declination can be precisely determined for any location based on latitude, longitude, and date using the calculator at the website http://www.ngdc.noaa.gov/geomag-web/#. This calculator generates a magnetic compass rose for the exact latitude and longitude of the site. The magnetic declination angle is also given allowing a compass reading to be converted to an angle relative to true north/south.

DETERMINING BUILDING ORIENTATION

Many solar thermal systems are placed on the roofs of existing buildings. It is important to know the orientation of a building's roof when assessing the feasibility of adding a solar thermal system to that building. Until recently, the only way to do this was to visit the site, align a compass with the side of the building, and read the deviation of the building's orientation relative to a magnetic north/south line. This angle could then be corrected based on the magnetic declination at the site.

Today, the availability of web-based satellite imaging software, such as **Google Earth**, **Apple Maps**, and **Bing™ Maps**, allows one to quickly display a satellite image of just about any building and use that image to establish the building's orientation relative to true north.

Start the process by entering the address of the site into one of the satellite imaging software tools. Zoom in on the image so that the building's major features are visible. Be sure the tilt setting of the image is set for zero, and then use the screen-capture feature available in Windows® or Mac OS to capture the image, as shown in Figure 7-20.

Use the following procedure to estimate the building's orientation relative to a true north/south line:

1. Import the captured building image into a CAD system. Most modern CAD systems can accept the image as an imported JPEG or PDF file.

Figure 7-20 | A screen capture of a house taken from Google Earth.

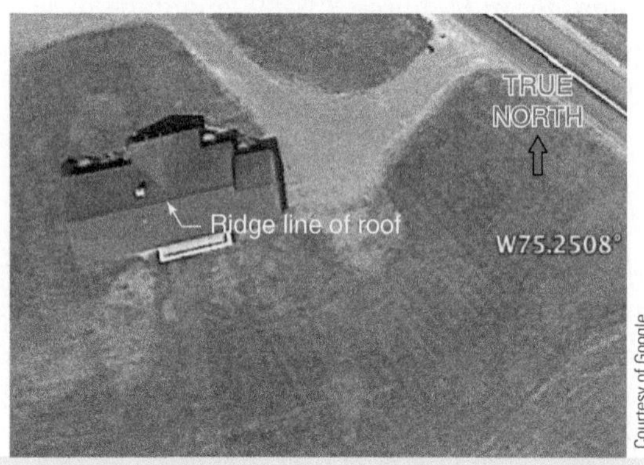

Figure 7-21 | Using the CAD system, draw a line over the image aligned with the horizontal feature on the building.

2. Select a feature on the building that represents a horizontal line. Examples include the horizontal ridge of a roof, the edge of a rectangular swimming pool, or the top of the railing along a deck. Longer features are preferred for better accuracy.

3. Using the CAD system, trace a straight line over the horizontal feature. Figure 7-21 shows such a line traced over the ridgeline of a roof.

4. Use the CAD system to create another line that is perpendicular to the first line and then center this line on the first line.

5. Use the CAD system to create a *vertical* line that passes through the intersection of the first two lines. This line, because it is vertical, represents *true north* on the satellite image.

6. Use the angular dimensioning tool of the CAD system to measure the angle between the true north line and the line perpendicular to the original reference line (i.e., the one along the roof ridgeline), as shown in Figure 7-22a.

Figure 7-22a A vertical line represents true north/south on the image. The angular dimensioning tool of the CAD system is used to measure the deviation between this line and a line representing the major axis of the building.

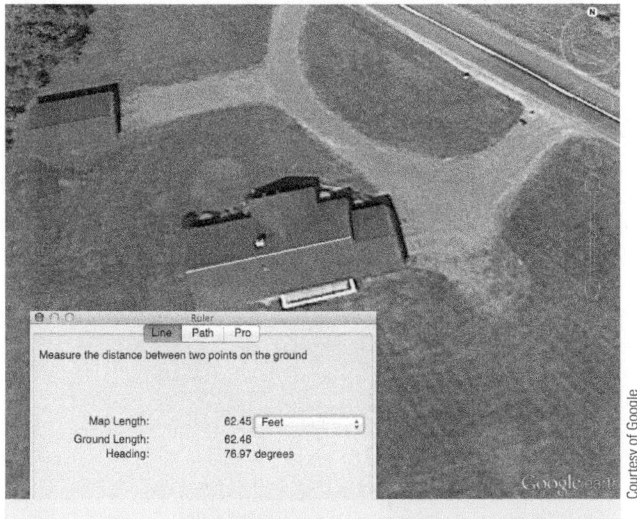

Figure 7-22b The line function under the ruler tool in Google Earth is used to draw the yellow line along the horizontal roof ridge. The heading of this line is given as 76.97° (or approximately 77°).

Figure 7-22a indicates that the long side of the building faces approximately 12.3° east of true south.

Care must be taken to select a *horizontal* line on the image. Selecting a sloping line, such as along the gable edge of a roof, a hip line on a roof, or the edge of a sloping sidewalk, will not provide a proper comparison with true north.

It is also possible to use the "line" function under the ruler tool within Google Earth to draw a line that overlies a horizontal roof ridge. The line is drawn by clicking on a starting and ending point using the computer's mouse. Once the line is drawn, its heading relative to a true north/south line will be displayed, along with its scaled length. Assuming the line was drawn from left to right along the roof ridge, as shown in Figure 7-22b, the azimuth angle of the southerly facing roof can be determined by adding 90° to this heading. An example is shown in Figure 7-22b.

The azimuth angle of the southerly facing roof on this house is approximately 77° + 90° = 167°. This is within 1° of the solar azimuth angle, as it would be determined from the angle shown in Figure 7-22a (e.g., 180° − 12.3° = 167.7°). These very slight variations in estimated azimuth angle are insignificant with regard to the estimated solar energy striking the roof surface.

SITE ASSESSMENT OF AVAILABLE SOLAR ENERGY

The ideal site for installing any type of solar energy collector would be free of obstacles that could cast shadows on the solar collectors. At this site, the sun's arc across the sky would be completely unobscured between sunrise and sunset, on every day of the year. Although some locations on the wide western planes of the North America might approach this condition, most sites at which solar collectors might be placed do not.

Various tools have been developed to help in assessing the shading potential of a given surface. They account for objects such as trees, adjacent buildings, and other visible terrain features, such as hills and mountains.

One of the "classic" instruments for assessing solar availability at a site is called the **Solar Pathfinder™**. This device was first created in 1978 and remains a relatively simple unpowered tool for solar heating professionals. It is shown in Figure 7-23.

The Solar Pathfinder™ has four main components:

- A clear plastic dome that creates a panoramic reflected image of the sky dome, including any visible objects above the horizon.
- A latitude-specific **sun path disc** that show the sun's arcing path for each month and solar times throughout the day.
- A magnetic compass for aligning the sun path disc relative to magnetic north. The disc can then be rotated to correct for any magnetic declination at the site.

Figure 7-23 | A Solar Pathfinder™ kit.

- A **bubble level** that is used to level the dome and sun path diagram and thus ensure the instrument is in a horizontal plane.

The sun path disc, magnetic compass, and bubble level are all located on the instrument section of the Solar Pathfinder™.

A sun path disc, as seen in Figure 7-24, has arcs that are oriented left to right. These arcs represent the sun's path for each month of the year. The disc also has vertically oriented curves that represent *solar time* during the day. Small numbers along each monthly arc indicate the percentage of total daily solar energy that occurs during each half hour segment of the sun's daily arc.

The Solar Pathfinder™ is set up by loading a sun path disc onto the instrument section, as shown in Figure 7-25. The notch in the sun path disc fits precisely over the blue triangle housing the bubble level. Sun path discs are available in increments of 6° of latitude.

To orient the Solar Pathfinder™, grip the blue triangle housing the bubble level and gently twist it so that the white dot on the perimeter of the instrument section aligns with the magnetic declination at the site. This is essential to properly orient the sun path disc to true south. Figure 7-26 shows the white dot properly positioned for a magnetic declination and of 13° west (represented as –13° on the sun path disc). With this setting, the 0° line of the sun path disc will be facing true south once the compass on the instrument section is aligned to magnetic north.

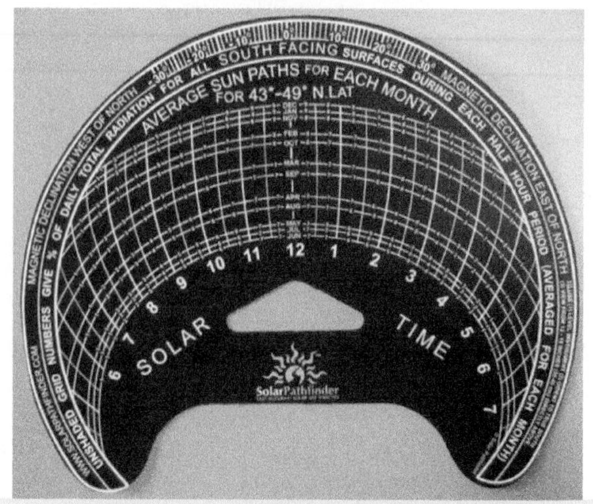

Figure 7-24 | A sun path disc (for latitudes of 43°–49°) ready to load onto instrument section of Solar Pathfinder™.

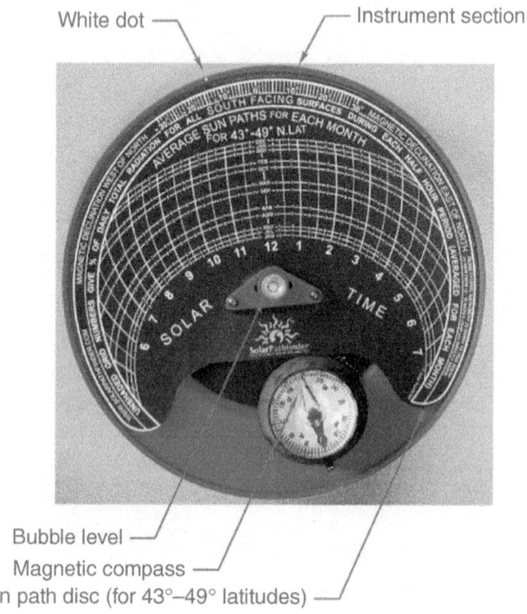

Figure 7-25 | A sun path disc loaded onto the instrument section of a Solar Pathfinder™.

The plastic dome is now placed over the instrument section and the assembly placed close to the anticipated location of the solar collectors. It can be placed on a small tripod, or temporarily propped up on one or more beanbags, as shown in Figure 7-27.

The instrument section must now be leveled by adjusting its supports (or beanbags) until the bubble level is centered. The instrument section can also be tilted, to a limited degree on the spherically shaped plastic base.

Figure 7-26 | The white dot must be aligned with the local magnetic declination.

Figure 7-27 | The instrument section of Solar Pathfinder™ is leveled over a sloping roof surface by adjusting it on a beanbag.

Next, the instrument section is rotated until the magnetic needle is directly over the north indicator. At this point, the sun path disc is symmetrically aligned with a true north/south line. The reflections of any surrounding objects are visible on the clear plastic dome. These reflections will likely cover a portion of the sun path disc. The areas of the sun path disc that are covered by the reflected image of surrounding objects indicate the solar time and month(s) that these objects will shade the location of the instrument.

To create a permanent record of the shading, one can trace the outline of the reflected images on the disc with a wax pencil. An alternate method is to stand directly over the Solar Pathfinder and take a digital photo, as shown in Figure 7-28.

Figure 7-29 shows an example of how surrounding objects appear on the dome of the Solar Pathfinder. The reflection of the person taking the photo is also seen. The camera should be held directly above the leveling bubble and pointed directly at that bubble.

Once the shading image has been created, it can be analyzed to determine the percentage of total daily sunshine available during each month. This can be done by manually adding the percentage of daily total sunshine information indicated by the small numbers on the monthly arcs of the disc. It can also be done using software, available for the Solar Pathfinder, that analyzes the digital camera image. This information can then be combined with other reference information on total daily solar radiation available for the site.

Another device for assessing on-site solar radiation availability is called a **SunEye 210**, as seen in Figure 7-30.

The SunEye 210 gathers data and performs shading analysis similar to the Solar Pathfinder™, but does so using digital electronics. It can gather and store shading information from multiple sites. When used with accompanying software, it can also estimate the total solar radiation that will be received on various surfaces based on the shading analysis.

Courtesy of The Solar Pathfinder Company

Figure 7-28 | Taking a digital photo of the reflected image on the plastic dome. Camera must be directly over the bubble level.

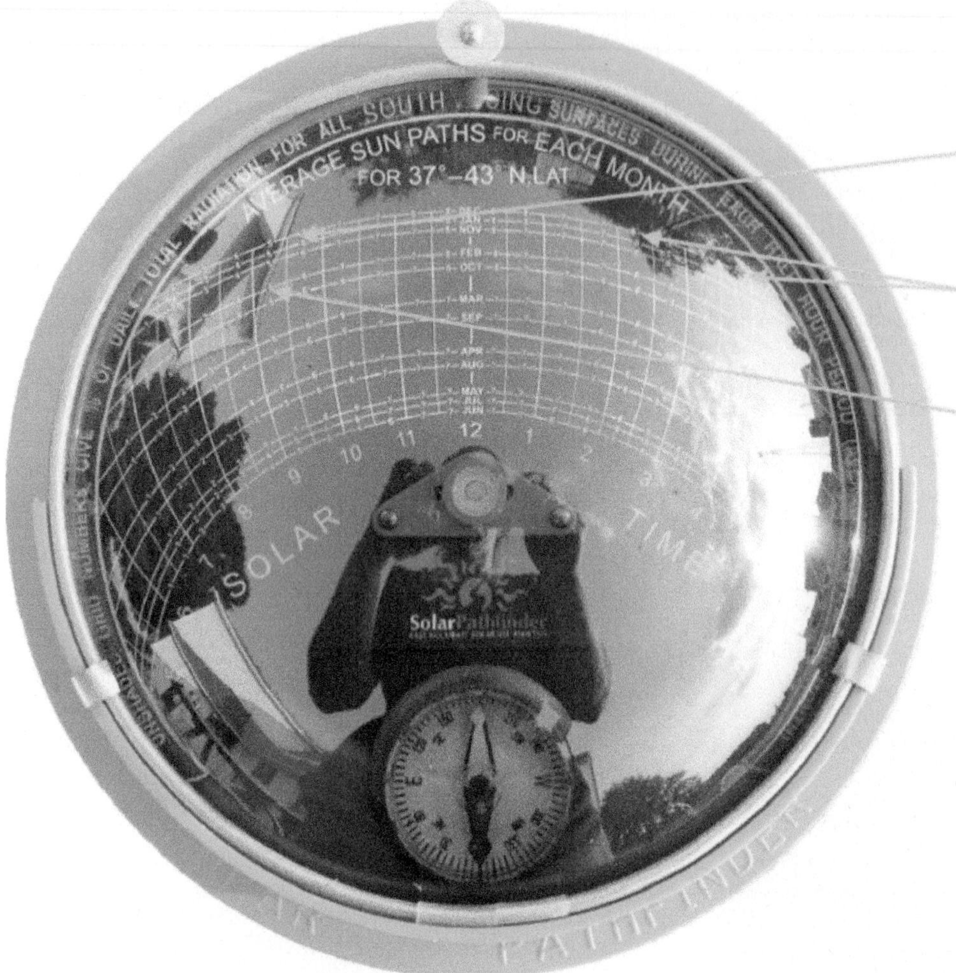

Courtesy of The Solar Pathfinder Company

Figure 7-29 | The areas of the sun path disc that are covered by the reflected images of surrounding objects indicate the solar time and month(s) during which the location of the instrument will be shaded.

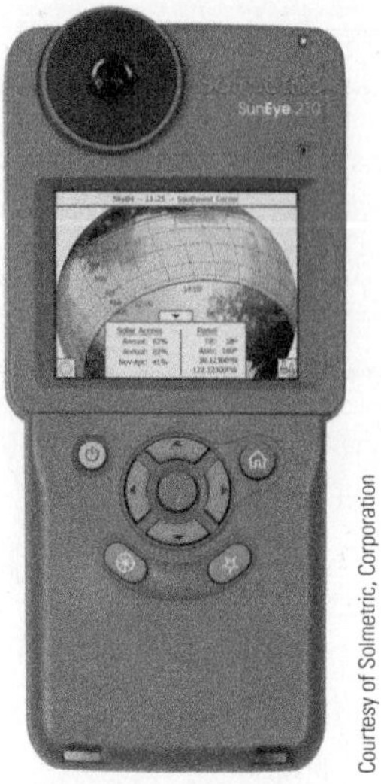

Courtesy of Solmetric, Corporation

Figure 7-30 | A SunEye 210 solar site assessment tool.

SOLAR RADIATION AVAILABILITY

A wide assortment of solar radiation data is available for North America. Within the United States, the **National Renewable Energy Laboratory (NREL)** provides easily accessed data for hundreds of locations.

One type of solar radiation data is provided as contour maps showing monthly averages of daily total solar radiation (e.g., direct plus diffuse) on surfaces at different orientations. This data is available for hundreds of locations within the United States, and for receiving surfaces including:

- horizontal surfaces
- south-facing vertical surfaces
- south-facing surfaces sloped at angle of local latitude –15°

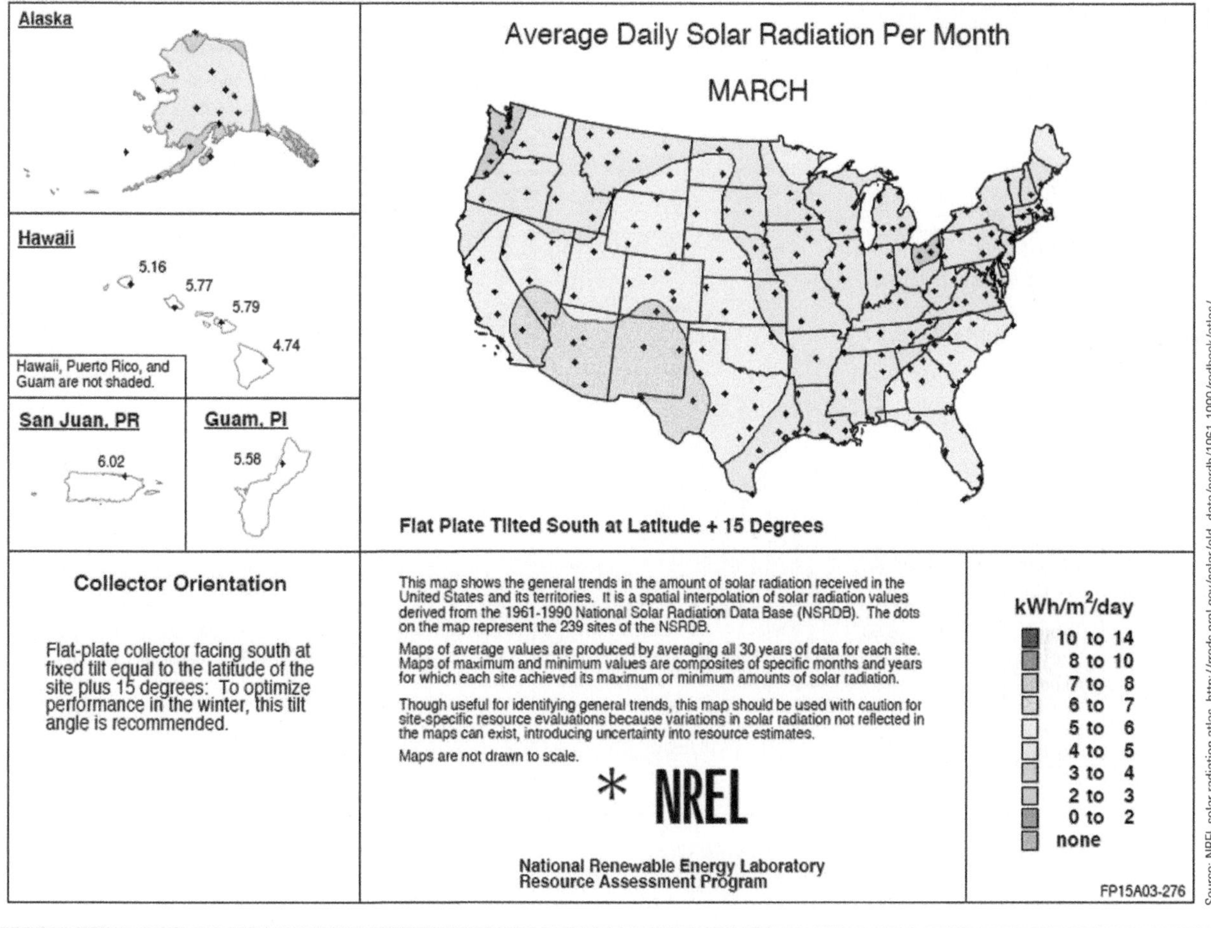

Figure 7-31 | Total daily solar radiation for March on a south-facing surface sloped at local latitude + 15°.

- south-facing surfaces sloped at angle of local latitude
- south-facing surfaces sloped at angle of local latitude +15°

An example of such a map, showing average daily total solar radiation for March on a south-facing surface sloped at local latitude +15°, is shown in Figure 7-31.

Note that the surface daily total radiation is given in units of kWh/m²/day. This can be converted to Btu/ft²/day using Equation 7.2.

(Equation 7.2)

$$\left(\frac{\text{kwh}}{\text{m}^2 \cdot \text{day}}\right) \times (317.078) = \left(\frac{\text{Btu}}{\text{ft}^2 \cdot \text{day}}\right)$$

Thus, 5 kWh/m²/day = 1,585 Btu/ft²/day.

NREL also published site-specific data for monthly average daily solar radiation on various surfaces, as shown in Figure 7-32.

Hourly solar radiation data for clear days on various south-facing surfaces is also available in ASHRAE Standard 93-2010. Figure 7-33 shows an example of this data.

Hourly solar radiation data can be used, in combination with a correction factor for the radiation's **incident angle** (e.g., the angle between the direction of beam solar radiation and a line perpendicular to the collector glazing), to estimate the daily solar energy collected by the system. Such calculations are tedious if done manually. Thus, hourly solar radiation data is most useful when incorporated in hour-by-hour software simulations.

Another web-based tool that is helpful in determining the *annual* total solar radiation striking a surface at a specific location, slope angle, and azimuth angle slope can be found at http://www.solmetric.com/annualinsolation-us.html. This online calculator uses **TMY3** (e.g., Typical Meteorological Year 3) data, published by NREL, for over 1,000 locations in the United States. It allows the user to select a state and city location for which several solar radiation values will be determined. These include optimal slope and azimuth

Syracuse, NY

WBAN NO. 14771

LATITUDE: 43.12° N
LONGITUDE: 76.12° W
ELEVATION: 124 meters
MEAN PRESSURE: 1001 millibars

STATION TYPE: Secondary

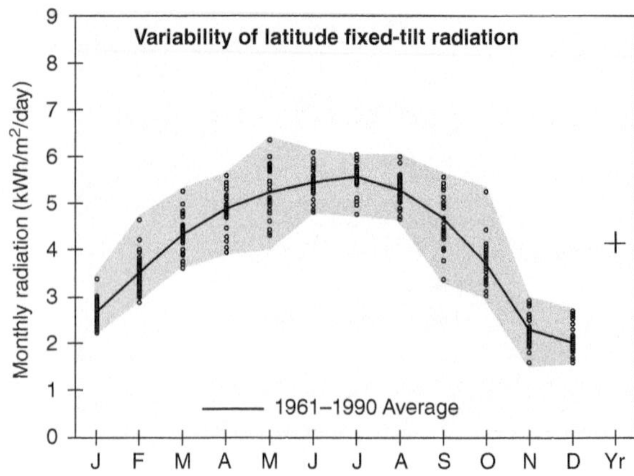

Solar radiation for flat-plate collectors facing South at a fixed tilt (kWh/m²/day), uncertainty ±9%

Tilt (°)		Jan	Feb	Mar	Apr	May	June	July	Aug	Sept	Oct	Nov	Dec	Year
0	Average	1.7	2.5	3.5	4.6	5.5	6.1	6.0	5.2	4.0	2.7	1.6	1.3	3.7
	Min/Max	1.5/1.9	2.1/2.9	3.1/4.0	3.9/5.1	4.4/6.6	5.4/6.8	5.3/6.5	4.7/6.7	3.1/4.5	2.3/3.4	1.3/1.8	1.1/1.5	3.6/3.9
Latitude −15	Average	2.4	3.3	4.2	5.0	5.6	5.9	6.0	5.5	4.7	3.5	2.1	1.8	4.2
	Min/Max	2.0/3.0	2.7/4.2	3.6/5.1	4.1/5.7	4.3/6.8	5.2/6.6	5.2/6.5	4.9/6.2	3.4/5.5	2.9/4.9	1.5/2.6	1.5/2.4	4.0/4.4
Latitude	Average	2.7	3.5	4.3	4.9	5.2	5.4	5.6	5.3	4.7	3.7	2.3	2.0	4.1
	Min/Max	2.2/3.4	2.9/4.7	3.6/5.3	3.9/5.6	4.1/6.3	4.8/6.1	4.8/6.0	4.7/6.0	3.4/5.6	3.0/5.3	1.6/2.9	1.6/2.7	3.9/4.4
Latitude +15	Average	2.8	3.6	4.2	4.5	4.6	4.7	4.9	4.8	4.5	3.7	2.3	2.1	3.9
	Min/Max	2.3/3.6	2.9/4.9	3.5/5.2	3.6/5.2	3.6/5.6	4.2/5.2	4.2/5.3	4.2/5.5	3.2/5.3	3.0/5.3	1.6/3.0	1.6/2.9	3.7/4.1
90	Average	2.7	3.3	3.5	3.1	2.7	2.6	2.8	3.1	3.2	3.1	2.0	2.0	2.8
	Min/Max	2.2/3.5	2.5/4.5	2.7/4.2	2.5/3.5	2.3/3.2	2.4/2.9	2.5/3.0	2.7/3.5	2.3/3.9	2.5/4.5	1.3/2.8	1.5/2.8	2.6/3.1

Solar radiation for 1-Axis tracking flat-plate collectors with a North-South axis (kWh/m²/day), uncertainty ±9%

Axis Tilt (°)		Jan	Feb	Mar	Apr	May	June	July	Aug	Sept	Oct	Nov	Dec	Year
0	Average	2.2	3.2	4.6	6.0	7.1	7.9	8.0	6.9	5.4	3.6	2.0	1.6	4.9
	Min/Max	1.9/2.8	2.7/4.4	3.8/5.8	4.6/7.2	5.0/8.9	6.6/9.1	6.7/8.8	6.1/7.9	3.8/6.4	3.0/5.2	1.4/2.5	1.3/2.1	4.6/5.2
Latitude −15	Average	2.8	9.1	5.2	6.4	7.3	7.9	8.0	7.2	5.9	4.3	2.4	2.0	5.3
	Min/Max	2.3/3.5	3.1/5.3	4.2/6.6	4.8/7.7	5.0/9.1	6.6/9.1	6.7/8.9	6.3/8.4	4.1/7.1	3.5/6.2	1.6/3.1	1.6/2.7	5.0/5.6
Latitude	Average	3.0	4.1	5.3	6.3	7.0	7.6	7.8	7.1	5.9	4.4	2.5	2.2	5.3
	Min/Max	2.5/3.9	3.3/5.7	4.3/6.8	4.7/7.6	4.8/8.9	6.3/8.7	6.4/8.6	6.2/8.2	4.1/7.2	3.6/6.5	1.7/3.3	1.7/3.0	5.0/5.7
Latitude +15	Average	3.1	4.1	5.2	6.1	6.6	7.1	7.3	6.7	5.8	4.4	2.6	2.3	5.1
	Min/Max	2.5/4.1	3.3/5.8	4.1/6.7	4.5/7.3	4.5/8.4	5.8/8.2	6.0/8.1	5.9/7.8	3.9/7.0	3.5/6.6	1.7/3.4	1.7/3.2	4.8/5.5

Solar radiation for 2-Axis tracking flat-plate collectors (kWh/m²/day), uncertainty ±9%

Tracker		Jan	Feb	Mar	Apr	May	June	July	Aug	Sept	Oct	Nov	Dec	Year
2-Axis	Average	3.1	4.1	5.3	6.4	7.4	8.1	8.2	7.2	6.0	4.5	2.6	2.3	5.4
	Min/Max	2.5/4.1	3.3/5.8	4.3/6.8	4.8/7.7	5.1/9.3	6.7/9.3	6.8/9.1	6.4/8.4	4.1/7.2	3.6/6.6	1.7/3.5	1.8/3.2	5.2/5.8

Direct beam solar radiation for concentrating collectors (kWh/m²/day), uncertainty ±8%

Tracker		Jan	Feb	Mar	Apr	May	June	July	Aug	Sept	Oct	Nov	Dec	Year
1-Axis, E-W Horiz Axis	Average	1.4	1.7	2.2	2.8	3.3	3.7	3.8	3.2	2.7	2.2	1.1	1.0	2.4
	Min/Max	0.6/2.0	0.9/3.0	1.5/3.6	1.7/3.7	1.4/4.8	2.5/4.7	2.6/4.6	2.5/4.1	1.4/3.7	1.5/3.9	0.5/1.9	0.4/1.8	2.1/2.8
1-Axis, N-S Horiz Axis	Average	0.9	1.4	2.4	3.6	4.3	4.9	5.0	4.2	3.2	2.1	0.8	0.6	2.8
	Min/Max	0.4/1.3	0.7/2.7	1.7/4.0	2.0/4.8	1.9/6.3	3.4/6.3	3.5/6.2	3.4/5.3	1.7/4.4	1.4/3.7	0.3/1.4	0.3/1.1	2.5/3.3
1-Axis, N-S Tilt = Latitude	Average	1.5	2.0	2.9	3.9	4.3	4.6	4.9	4.4	3.7	2.8	1.3	1.0	3.1
	Min/Max	0.7/2.1	1.0/3.7	2.1/4.8	2.2/5.2	1.9/6.2	3.2/5.9	3.4/6.0	3.5/5.6	1.9/5.1	1.8/4.9	0.5/2.1	0.4/1.9	2.7/3.6
2-Axis	Average	1.6	2.1	2.9	3.9	4.5	5.0	5.2	4.5	3.7	2.8	1.3	1.1	3.2
	Min/Max	0.7/2.3	1.0/3.8	2.1/4.8	2.2/5.2	2.0/6.5	3.5/6.5	3.7/6.4	3.6/5.7	1.9/5.1	1.9/4.9	0.6/2.2	0.5/2.0	2.8/3.7

Average climatic conditions

Element	Jan	Feb	Mar	Apr	May	June	July	Aug	Sept	Oct	Nov	Dec	Year
Temperature (°C)	−5.3	−4.4	1.1	7.6	13.9	18.5	21.3	20.2	16.4	10.4	4.7	−2.1	8.6
Daily Minimum Temp	−9.9	−9.2	−3.8	1.9	7.8	12.1	15.0	14.3	10.8	5.1	0.6	−6.1	3.2
Daily Maximum Temp	−0.8	−0.3	5.9	13.3	20.2	24.8	27.6	26.1	22.0	15.7	8.9	1.9	13.8
Record Minimum Temp	−32.2	−32.2	−26.7	−12.8	−3.9	1.7	7.2	4.4	−2.2	−7.2	−15.0	−30.0	−32.2
Record Maximum Temp	21.1	20.6	30.6	33.3	35.6	36.7	36.1	36.1	36.1	30.6	27.2	21.1	36.7
HDD, Base 18.3 °C	734	638	536	322	149	34	6	16	77	246	408	632	3797
CDD, Base 18.3 °C	0	0	0	0	13	39	99	74	19	0	0	0	243
Relative Humidity (%)	73	72	70	65	67	70	71	75	76	74	75	77	72
Wind Speed (m/s)	4.8	4.8	4.8	4.6	4.0	3.6	3.5	3.4	3.6	3.8	4.5	4.7	4.2

Source: http://rredc.nrel.gov/solar/pubs/redbook/

Figure 7-32 | Monthly average total solar radiation for various south-facing surfaces along with other average climatic conditions for a specific location (e.g., Syracuse, NY).

Solar position and insolation values for 40 degrees north latitude[a]											
Date	Solar time		Solar position		BTUH/sq. ft. total insolation on surface[b]						
	AM	PM	Alt	Azm	Normal[c]	Horiz.	South facing surface angle with horiz.				
							30	40	50	60	90
Jan 21	8	4	8.1	55.3	142	28	65	74	81	85	84
	9	3	16.8	44.0	239	83	155	171	182	187	171
	10	2	23.8	30.9	274	127	218	237	249	254	223
	11	1	28.4	16.0	289	154	257	277	290	293	253
	12		30.0	0.0	294	164	270	291	303	306	263
	Surface daily totals				2182	948	1660	1810	1906	1944	1726
Feb 21	7	5	4.8	72.7	69	10	19	21	23	24	22
	8	4	15.4	62.2	224	73	114	122	126	127	107
	9	3	25.0	50.2	274	132	195	205	209	208	167
	10	2	32.8	35.9	295	178	256	267	271	267	210
	11	1	38.1	18.9	305	206	293	306	310	304	236
	12		40.0	0.0	308	216	306	319	323	317	245
	Surface daily totals				2640	1414	2060	2162	2202	2176	1730
Mar 21	7	5	11.4	80.2	171	46	55	55	54	51	35
	8	4	22.5	69.6	250	114	140	141	138	131	89
	9	3	32.8	57.3	282	173	215	217	213	202	138
	10	2	41.6	41.9	297	218	273	276	271	258	176
	11	1	47.7	22.6	305	247	310	313	307	293	200
	12		50.0	0.0	307	257	322	326	320	305	208
	Surface daily totals				2916	1852	2308	2330	2284	2174	1484
Apr 21	6	6	7.4	98.9	89	20	11	8	7	7	4
	7	5	18.9	89.5	206	87	77	70	61	50	12
	8	4	30.3	79.3	252	152	153	145	133	117	53
	9	3	41.3	67.2	274	207	221	213	199	179	93
	10	2	51.2	51.4	286	250	275	267	252	229	126
	11	1	58.7	29.2	292	277	308	301	285	260	147
	12		61.6	0.0	293	287	320	313	296	271	154
	Surface daily totals				3092	2274	2412	2320	2168	1956	1022
May 21	5	7	1.9	114.7	1	0	0	0	0	0	0
	6	6	12.7	105.6	144	49	25	15	14	13	9
	7	5	24.0	96.6	216	214	89	76	60	44	13
	8	4	35.4	87.2	250	175	158	144	125	104	25
	9	3	46.8	76.0	267	227	221	206	186	160	60
	10	2	57.5	60.9	277	267	270	255	233	205	89
	11	1	66.2	37.1	283	293	301	287	264	234	108
	12		70.0	0.0	284	301	312	297	274	243	114
	Surface daily totals				3160	2552	2442	2264	2040	1760	724
Jun 21	5	7	4.2	117.3	22	4	3	3	2	2	1
	6	6	14.8	108.4	155	60	30	18	17	16	10
	7	5	26.0	99.7	216	123	92	77	59	41	14
	8	4	37.4	90.7	246	182	159	142	121	97	16
	9	3	48.8	80.2	263	233	219	202	179	151	47
	10	2	59.8	65.8	272	272	266	248	224	194	74
	11	1	69.2	41.9	277	296	296	278	253	221	92
	12		73.5	0.0	279	304	306	289	263	230	98
	Surface daily totals				3180	2648	2434	2224	1974	1670	610

Courtesy of ASHRAE

Figure 7-33 | Hourly, clear day solar radiation on various south-facing surfaces at 40° north latitude.

Solar position and insolation values for 40 degrees north latitude[a] *(continued)*											
Date	**Solar time**		**Solar position**		**BTUH/sq. ft. total insolation on surface[b]**						
	AM	**PM**	**Alt**	**Azm**			**South facing surface angle with horiz.**				
					Normal[c]	**Horiz.**	**30**	**40**	**50**	**60**	**90**
Jul 21	5	7	2.3	115.2	2	0	0	0	0	0	0
	6	6	13.1	106.1	138	50	26	17	15	14	9
	7	5	24.3	97.2	208	114	89	75	60	44	14
	8	4	35.8	87.8	241	174	157	142	124	102	24
	9	3	47.2	76.7	259	225	218	203	182	157	58
	10	2	57.9	61.7	269	265	266	251	229	200	86
	11	1	66.7	37.9	275	290	296	281	258	228	104
	12		70.6	0.0	276	298	307	292	269	238	111
	Surface daily totals				3062	2534	2409	2230	2006	1728	702
Aug 21	6	6	7.9	99.5	81	21	12	9	8	7	5
	7	5	19.3	90.9	191	87	76	69	60	49	12
	8	4	30.7	79.9	237	150	150	141	129	113	50
	9	3	41.8	67.9	260	205	216	207	193	173	89
	10	2	51.7	52.1	272	246	267	259	244	221	120
	11	1	59.3	29.7	278	273	300	292	276	252	140
	12		62.3	0.0	280	282	311	303	287	262	147
	Surface daily totals				2916	2244	2354	2258	2104	1894	978
Sep 21	7	5	11.4	80.2	149	43	51	51	49	47	32
	8	4	22.5	69.6	230	109	133	134	131	124	84
	9	3	32.8	57.3	263	167	206	208	203	193	132
	10	2	41.6	41.9	280	211	262	265	260	247	168
	11	1	47.7	22.6	287	239	298	301	295	281	192
	12		50.0	0.0	290	249	310	313	307	292	200
	Surface daily totals				2708	1788	2210	2228	2182	2074	1416
Oct 21	7	5	4.5	72.3	48	7	14	15	17	17	16
	8	4	15.0	61.9	204	68	106	113	117	118	100
	9	3	24.5	49.8	257	126	185	195	200	198	160
	10	2	32.4	35.6	280	170	245	257	261	257	203
	11	1	37.6	18.7	291	199	283	295	299	294	229
	12		39.5	0.0	294	208	295	308	312	306	238
	Surface daily totals				2454	1348	1962	2060	2098	2074	1654
Nov 21	8	4	8.2	55.4	136	28	63	72	78	82	81
	9	3	17.0	44.1	232	82	152	167	178	183	167
	10	2	24.0	31.0	268	126	215	233	245	249	219
	11	1	28.6	16.1	283	153	254	273	285	288	248
	12		30.2	0.0	288	163	267	287	298	301	258
	Surface daily totals				2128	942	1636	1778	1870	1908	1686
Dec 21	8	4	5.5	53.0	89	14	39	45	50	54	56
	9	3	14.0	41.9	217	65	135	152	164	171	163
	10	2	20.,	29.4	261	107	200	221	235	242	221
	11	1	25.0	15.2	280	134	239	262	276	283	252
	12		26.6	0.0	285	143	253	275	290	296	263
	Surface daily totals				1978	782	1480	1634	1740	1796	1646

[a] From Kreider, J. F, and F. Kreith, "Solar Heating and Cooling," revised 1st ed., Hemisphere Publ. Corp., 1977.

[b] 1 Btu/hr · ft² = 3.152 W/m². Ground reflection not included on normal or horizontal surfaces.

[c] Normal insolation does not include diffuse component.

Courtesy of ASHRAE

Figure 7-33 | Hourly, clear day solar radiation on various south-facing surfaces at 40° north latitude.

angles and total solar radiation striking a surface at those angles (measured in kWh/m^2). This calculator also displays a graphic image showing how the total annual solar radiation varies with different slope and azimuth angles, as seen in Figure 7-34.

The number shown below the graphic show total solar radiation striking a surface at a *user-specified slope and azimuth*. In this case, the slope was set to 45° and the azimuth to 160°. Roof pitches are also shown on the side vertical scale for cross-reference with slope angle. The TOF (tilt and orientation Factor) displayed beneath the graphic is the percentage of the maximum possible annual solar radiation striking an optimally oriented surface at that location. This tool can be used to quickly study the effect of surface slope and azimuth on annual total solar radiation. However, it cannot be used to divide this annual total into monthly totals.

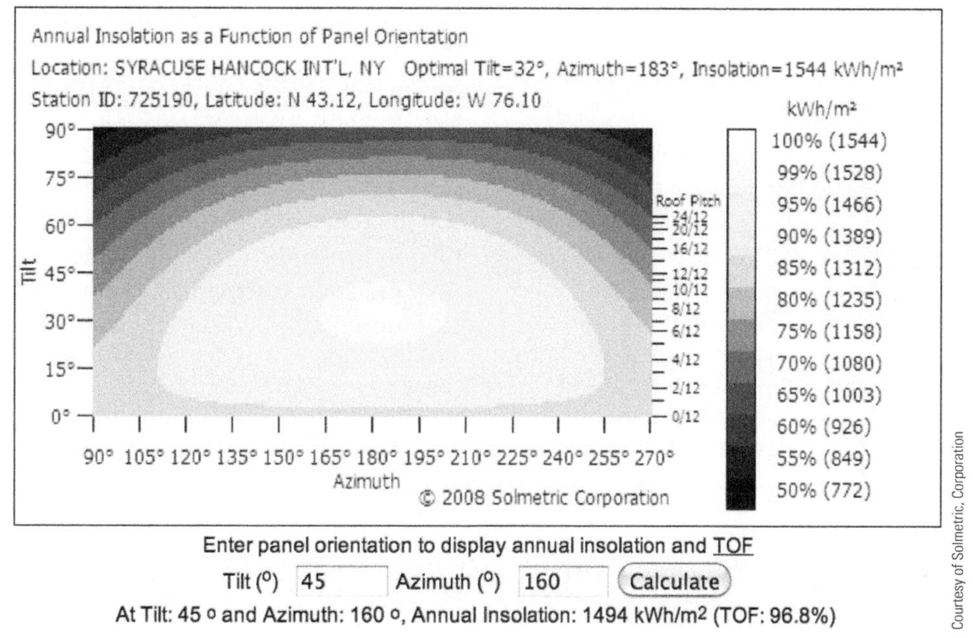

Figure 7-34 | Output of a web-based calculator showing total annual solar radiation based on surface slope and azimuth.

SUMMARY

This section has presented the basics of solar geometry and tools for estimating the probable solar radiation that will be incident on a given surface at a given time and location. It has also introduced tools that enable a technician to evaluate a potential installation site for the amount of solar radiation available to solar collecting devices. Although this text focuses on solar thermal applications, the fundamental information, tools, and resources discussed in this chapter is equally applicable to the siting of solar photovoltaic systems and used to evaluate solar gains potential in passive solar buildings.

ADDITIONAL RESOURCES

The following resources are recommended for additional reading on solar geometry and solar radiation availability:

1. Duffie, John. A., and William A. Beckman. 2013. *Solar Engineering of Thermal Processes*, 4th edition. John Wiley & Sons. ISBN 978-0-470-87366-3.

2. Goswami, D. Yogi, Frank Keith, and Jan F. Kreider. 2000. *Principles of Solar Engineering,* 2nd edition. Taylor & Francis Group. ISBN 978-1-56032-714-1.

3. Felix A. Peuser, Felix A., Karl-Heinz Remmers, and Martin Schnauss. 2002. *Solar Thermal Systems, Successful Planning and Construction*. James & James (Science Publishers) Ltd. ISBN 1-902916-39-5.

4. Perlin, John. 2013. *Let It Shine*. New World Library, ISBN 978-1-60868-132-7.

KEY TERMS

Apple Maps
Arab oil embargo
autumnal equinox
beam radiation
Bing™ Maps
black body
bubble level
Climax solar water heater
declination angle
diffuse radiation
direct radiation
economically sustainable
electromagnetic radiation
Google Earth
incident angle
infrared
isogonic chart
latitude
local solar time
longitude
magnetic declination
MIT Solar House #1
Motions of the Sun Simulator
National Renewable Energy Laboratory (NREL)
nuclear radiation
passive solar architecture
polar axis
sky dome
solar altitude angle (ALT)
solar azimuth angle (AZI)
solar constant
solar noon
Solar Pathfinder™
solar radiation
solar radiation spectrum
solar time
speed of light
standard meridian
summer solstice
SunEye 210
sun path diagram
sun path disc
TMY3
tax credit
thermosyphoning
true north/south line
true south
ultraviolet
vernal equinox
visible light
wavelength
winter solstice

QUESTIONS AND EXERCISES

1. Describe a trend that caused cyclical use of solar energy over the past two centuries in North America.

2. What is different about installing a solar thermal system today compared to 30 years ago?

3. Describe the difference between infrared and ultraviolet light. Which of these contributes more to the total solar radiation received on earth?

4. What causes the jagged nature of the solar spectrum on the surface of the earth in comparison to the solar spectrum outside the earth's atmosphere? Be specific.

5. Why is it difficult for a concentrating solar collector to make use of diffuse solar radiation?

6. Why is it more important to capture solar radiation between 11 AM and 1 PM than between 3 PM and 5 PM?

7. Assume you lived in Boston, and that the earth's axis was tilted 33.5° rather than its actual tilt of 23.5° relative to the plane of its orbit. Describe how the sun's path in the sky would be different on June 21 and December 21.

8. Two cities in the United States have the same longitude. Will the sun's altitude and azimuth angles be the same at the same solar time in both cities? Justify why or why not.

9. Go to the website http://solardata.uoregon.edu/SunChartProgram.html. Download and print out a sun path diagram for a city near you. Find the solar altitude and azimuth angle for 11 AM solar time on April 20.

10. For the same city used in Exercise 9, determine the local clock time corresponding to 11 AM solar time.

11. Determine the local clock time when the sun reaches its highest altitude in Chicago on June 21.

12. Use Google Earth to determine the azimuth angle of the roof on the right side of the ridgeline in Figure 7-35 relative to a true north/south line.

Figure 7-35 | Building for Exercise 12.

13. Why is it important to adjust the instrument section of the Solar Pathfinder® for the local magnetic declination? What happens if this is not done?

14. One side of a house is exactly aligned with the needle of a magnetic compass. The house is in Las Vegas, NV. Describe where true south is relative to that side of the house.

15. Why do isogonic charts have a date on them?

16. Go to the website http://rredc.nrel.gov/solar/old_data/nsrdb/1961-1990/redbook/atlas/. Find the location nearest to where you live and estimate the total daily solar radiation received in October on a surface tilted south at an angle of latitude $+15°$. Express this in units of Btu/ft^2/day.

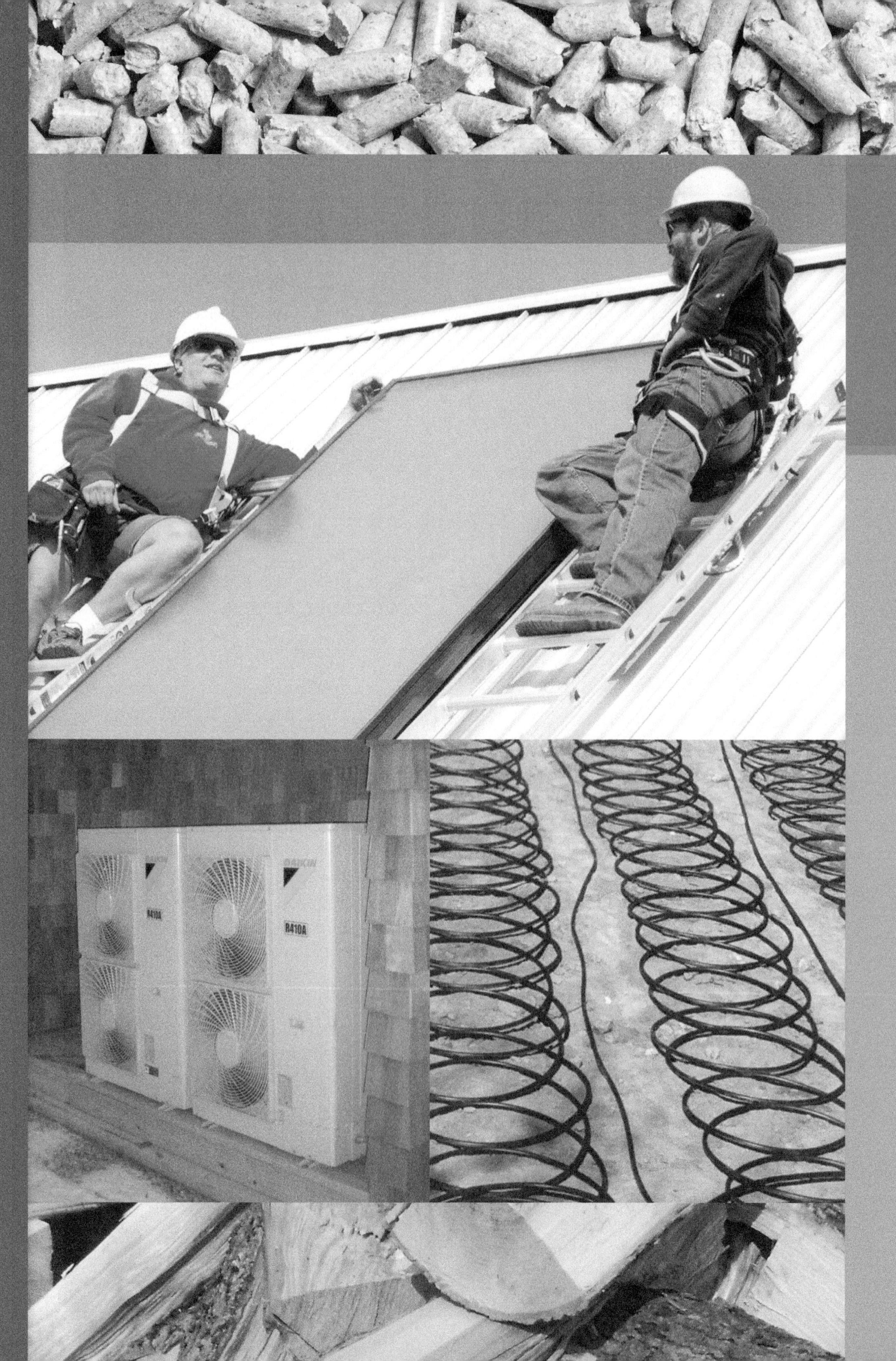
R410A
DAIKIN
R410A

CHAPTER 8

Solar Thermal Collectors

OBJECTIVES

After studying this chapter, you should be able to:

- Describe the construction and operation of flat-plate solar collectors.
- Describe different types of evacuated tube collectors.
- Discuss the strengths and limitations of both flat-plate and evacuated tube collectors.
- Define the instantaneous thermal efficiency of a solar collector.
- Explain how operating conditions affect the thermal efficiency of a solar collector.
- Calculate the instantaneous thermal efficiency of a collector based on operating conditions.
- Describe how a collector is tested for thermal efficiency.
- Understand what the incident angle modifier is, and how to use it.
- Access the information given on an SRCC collector rating sheet.
- Specify proper ways of mounting solar collectors to roofs.
- Explain the specifics of mounting collectors for use in drainback systems.
- Specify proper procedures for routing piping from collector arrays through roofs.

8.1 Introduction

Those who design hydronic heating systems using gas-fired boilers should know how those boilers operate, how they connect to the system, and how they perform from the standpoint of heat output and efficiency. All of these characteristics influence how the other portions of the overall system are designed. When it comes to designing a solar combisystem, the same rationale applies to **solar thermal collectors**. Solar combisystem designers must know how solar collectors operate, how they interact with other portions of the system, and how much heat they are expected to deliver.

The function of any solar thermal collector is to convert the highest possible percentage of solar radiation that shines on it into heat, and deliver that heat to a stream of fluid. Modern solar collectors should be able to do this for at least 20 years while exposed to wide temperature swings, intense solar radiation, high winds, hailstorms, and even potential lightning strikes. There are many solar thermal collectors now available that are up to this task. They are the results of refinements made over several decades of application experience, in combination with advances in materials and manufacturing methods.

This chapter discusses both **flat-plate collectors** and **evacuated tube collectors**. These are the type of collectors most commonly used in solar combisystems. It describes how they are made and how they perform. It also discusses online resources that can be used to access physical data and performance information on hundreds of solar thermal collectors that are now available in North America. It goes on to show a variety of mounting methods for solar collectors and how those collectors should be piped. The chapters that follow build upon this information to show how complete solar combisystems are constructed.

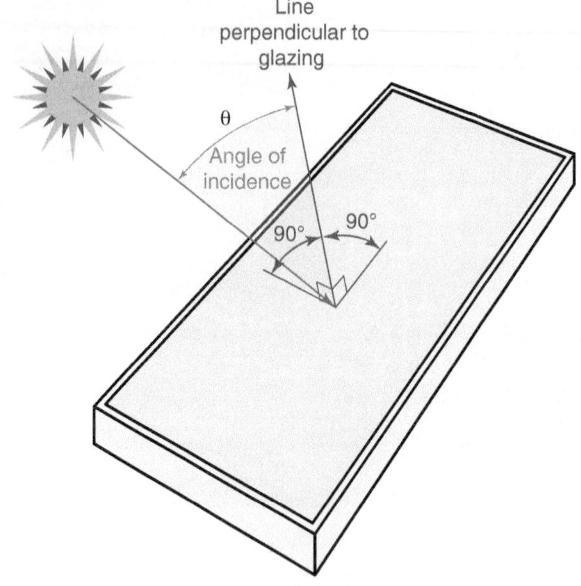

Figure 8-1 The angle of incidence is the angle between the sun's rays and a line perpendicular to the surface of the glazing.

8.2 Optical Properties of Solar Collector Materials

Several of the materials used to construct solar thermal collectors have optical properties that are crucial to good performance. The three optical properties of interest are:

- Transmissivity (τ)
- Absorptivity (α)
- Emissivity (ε)

Transmissivity is the percentage of the solar radiation striking a material that passes through the material. In the context of solar collectors, the **glazing** is the only material where transmissivity matters. Most solar collectors use special glass with **low iron oxide content** for their glazing. This glass typically allows 90 to 92 percent of the solar radiation that strikes perpendicular to its surface to pass through. Thus, the **"normal" transmissivity** of this glass is a value typically between 0.90 to 0.92. However, the transmissivity of glass decreases as the angle between the incoming solar radiation and a line perpendicular to (a.k.a., normal to) the glass surface increases. This angle is called the **angle of incidence**. It is illustrated in Figure 8-1.

The change in transmissivity is minimal for incident angles up to about 45°. However, at greater angles of incidence, transmissivity decreases rapidly, approaching zero at incident angles of 90°. Transmissivity is represented by the Greek letter *tau* (τ). The mathematical value of transmissivity for any material must be between zero (for any opaque material) and 1.0 (for a hypothetical material that would transmit all incident solar radiation).

Absorptivity is the percentage of solar radiation that is absorbed by the surface it strikes. For solar collectors, the absorptivity of the upper surface of the **absorber plate** should be a high as possible. Modern solar collectors use special **selective surfaces** such as **TiNOX®** on their absorber plates that absorb up to 95 percent of the solar radiation that strikes perpendicular to the plate's surface. As with transmissivity, the value of absorptivity decreases as the angle of incidence increases. Absorptivity is represented by the Greek letter *alpha* (α). The mathematical value of absorptivity for any material must be between zero (for a perfect reflector), and 1.0 (for a hypothetical perfect absorber).

Emissivity is the percentage of thermal radiation released by a surface compared to the thermal radiation that would be released from a hypothetical **black body** at the same surface temperature. The mathematical value of emissivity for any material must be between zero (for a hypothetical surface with no radiant heat release), and 1.0 (for a hypothetical black body). *For solar collectors, the emissivity of the upper surface of the absorber plate should be as low as possible.* The emissivity of the selective surface coatings used on modern solar collectors is typically in the range of 0.08 to 0.10. These low emissivities significantly reduce heat loss by thermal radiation between the upper surface of the absorber plate and the lower surface of the glazing. By comparison, the emissivity of a flat black-painted surface can be as high as 0.90. Emissivity is represented by the Greek letter *epsilon* (ε).

8.3 Flat-Plate Solar Collectors

Although there are differences from one manufacturer to the next, most flat-plate solar collectors share common materials and construction techniques. These include an aluminum frame, an absorber plate, tempered glass glazing, and insulation.

Figure 8-2 shows examples of modern flat-plate collectors.

The principal component in a flat-plate collector is the absorber plate. Its purpose is to "gather" the

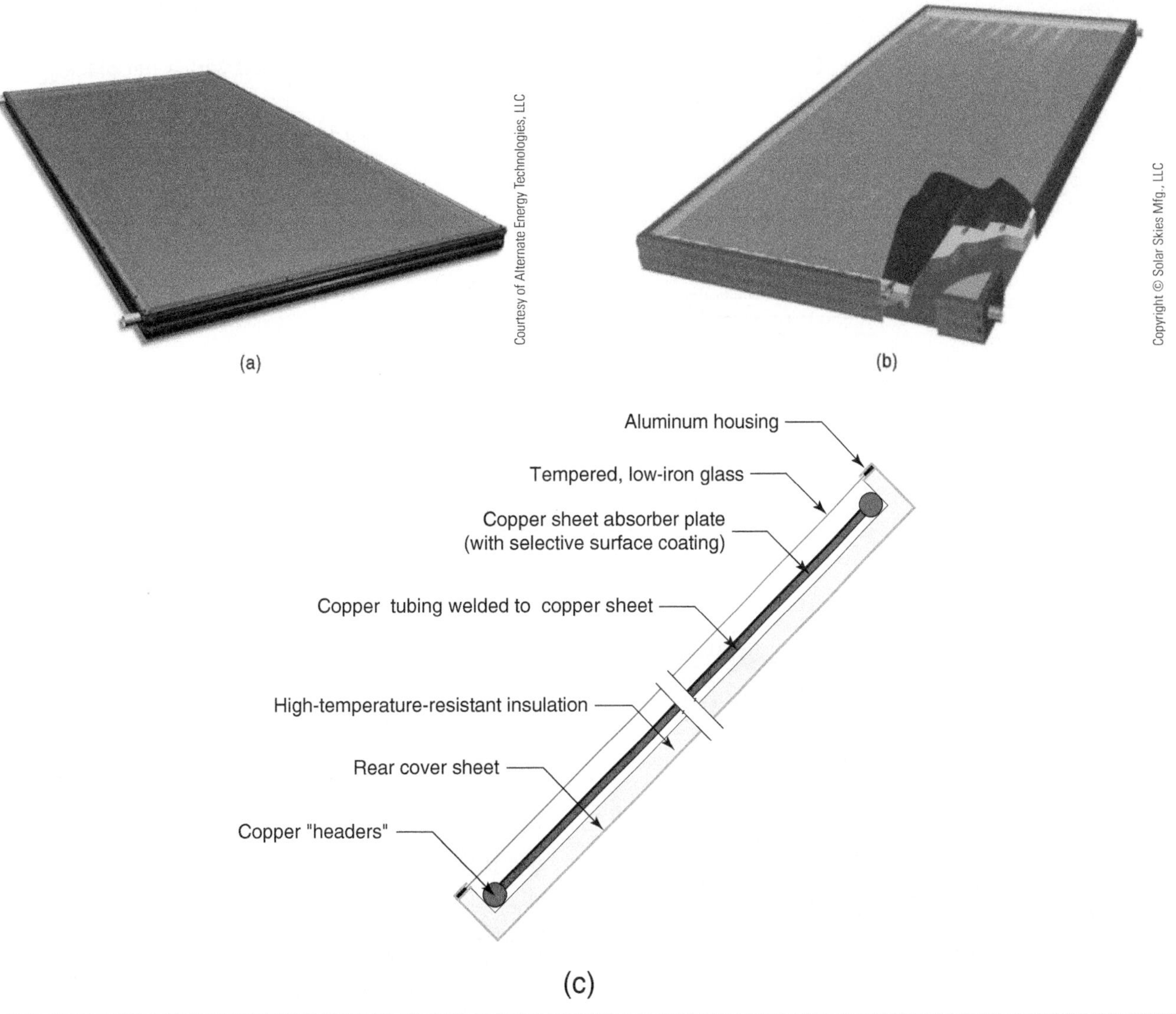

Figure 8-2 | (a) Example of a flat-plate collector. (b) Internal construction of a flat-plate collector. (c) Cross section of a typical flat-plate solar collector.

heat generated across its upper surface as it absorbs solar radiation and transfer that heat to a fluid circulated through the collector. The "ideal" absorber plate would have fluid flowing under every square inch of its surface. However, this is impractical from the standpoints of production cost and the ability to withstand potentially high fluid pressure. Instead, most modern absorber plates use a combination of copper fins, bonded to copper tubes, as illustrated in Figure 8-3.

Although there are different designs for absorber plates, one of the most common is an assembly of copper fin-tube **absorber strips**, arranged side by side, to form a **harp-style absorber plate**. The copper tubing used in each absorber strip is continuously bonded to the back side of the copper strip using high-frequency welding. Figure 8-4 shows an example of the resulting absorber strip.

The absorber strips are assembled, side by side, and brazed to copper headers at both ends. Figure 8-5 shows the resulting absorber plate.

After assembly and brazing, each absorber plate is pressurized with air and immersed in water to verify that it is free of leaks, as shown in Figure 8-6.

In modern collectors, the top surface of each fin-tube absorber strip is coated or electroplated with a selective surface that absorbs about 95 percent of the solar radiation that strikes it, but only emits about 8 percent of the infrared radiation that would be emitted by a black body at the same temperature. This ability to absorb a high percentage of the incoming solar radiation, while emitting very little thermal radiation, allows the collector to gather and retain a higher percentage of available solar energy compared to a collector in which the absorber plate was finished with a flat black paint.

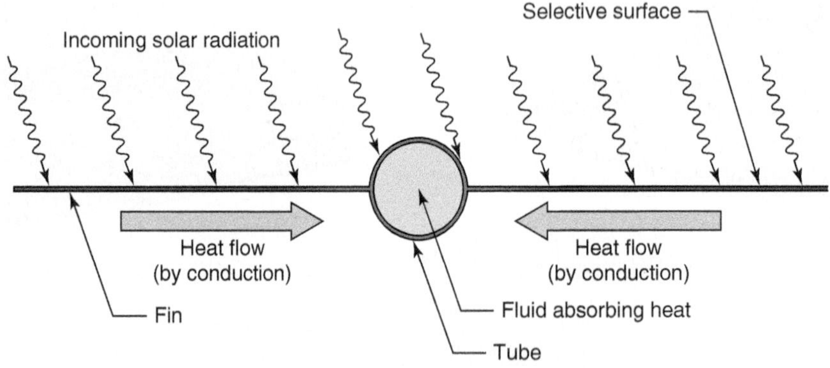

Figure 8-3 | Solar radiation strikes the upper surface of an absorber strip and is instantly converted to heat. The absorber strip conducts this heat to the tube, where it is transferred to a flowing fluid.

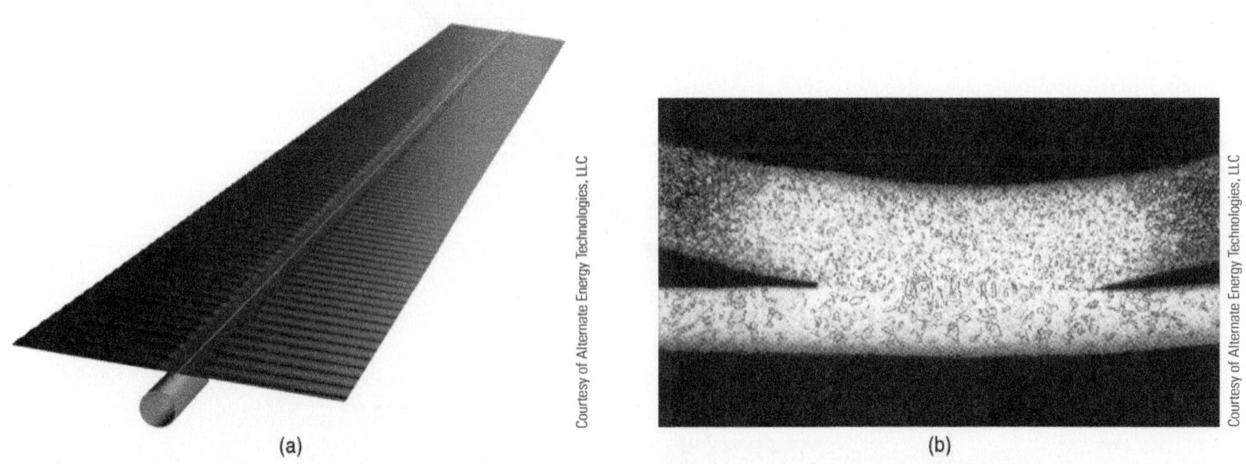

Figure 8-4 | (a) An absorber strip made by continuously welding a copper tube to the back of a copper sheet. The top of the copper sheet has a selective surface coating. (b) Magnified image of the metallurgical bond between the tube and sheet. This bond is essential for good thermal performance.

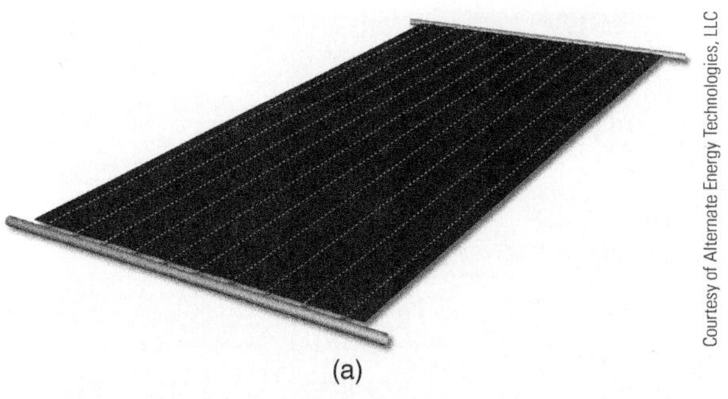

(a)

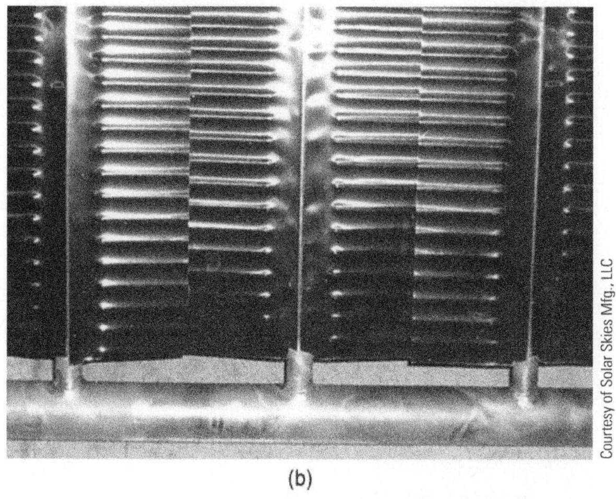

(b)

Figure 8-5 | (a) Copper absorber plate from a flat-plate collector. (b) Close-up of copper absorber strips joined side by side to form a harp-style absorber plate. The ends of each absorber strip are brazed to the copper header.

The instant solar radiation strikes the surface of the absorber plate it is converted from electromagnetic energy to heat. The copper absorber strips then act as "wicks" to conduct this heat toward the copper tubes. Heat moves toward the tubes because the fluid flowing through them is cooler than the copper strips. The fluid absorbs the heat and carries it to header tubes at the top of the absorber plate. The heated fluid then exits the collector through external piping connected to this header.

Other components use in a typical flat-plate collectors include:

- an aluminum frame, often with an **anodized finish**
- high-temperature-resistant insulation on the sides and behind the absorber plate
- tempered **"low-iron" glazing** that transmits 90+ percent of the solar radiation striking its outer surface

Figure 8-6 | An absorber plate being pressure tested to verify it is leak-free.

- a removable frame that holds the glazing to the collector enclosure
- a rear cover sheet made of aluminum, ABS, or PVC
- gaskets that seal the piping penetrations and the glazing to keep moisture out of the collector

Some flat-plate collectors also have **sensor mounting wells** that allow temperature sensors needed by controllers to be in direct contact with the absorber plate. Some also include **breather plugs** that allow air pressure to equalize between the inside and outside of the collector.

Many flat-plate collectors have four 1-inch copper piping connections located at the left and right ends of the upper and lower headers. Other collectors have male threaded connections at the ends of both headers. The latter are designed to be joined using special brass unions.

Figure 8-7 shows some of these features.

8.4 Evacuated Tube Solar Collectors

Another type of solar collector used in solar combisystems consists of several evacuated glass tubes, arranged side by side, and connected to a common header at the top. Figure 8-8 shows an example of an evacuated tube collector installed on a roof.

Each glass tube contains a narrow absorber strip that performs the same function as the absorber plate in a flat-plate collector.

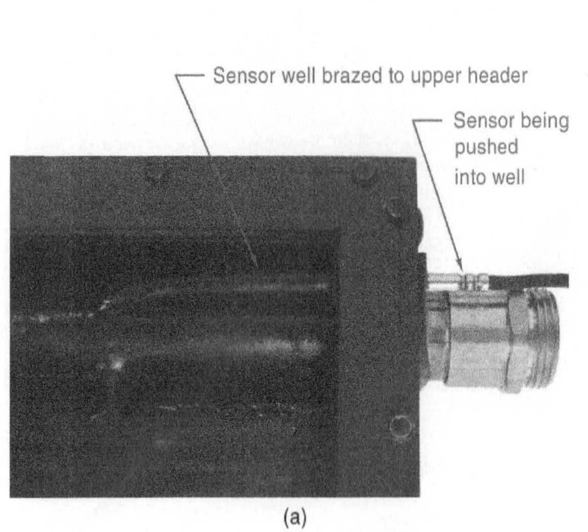

(a)

(b)

(c)

Figure 8-7 (a) A temperature sensor being inserted in a well that is brazed to the absorber plate during manufacturing. (b) A breather plug that equalizes air pressure between the inside and outside of the collector. (c) A double-union connection that allows collectors to be joined side by side using wrenches. The cable for a temperature sensor mounted into a sensor well within the collector is also visible.

The distinguishing feature of an evacuated tube collector is that the space between the glass tube and the absorber strip, or on some products, the space between two **concentric glass tubes**, is almost completely devoid of air (i.e., it is maintained under a high vacuum). *Removing air from this space reduces convective heat loss from the absorber strip to the glass to almost zero.* This, in combination with a low-emissivity selective surface on the absorber strip, significantly improves the ability of the collector to retain heat, even at low outside air temperature and under cloudy skies. In essence, the evacuated glass envelope acts like a transparent "Thermos® bottle" around the absorber strip.

As with flat-plate collectors, there are several variations on the details used in evacuated tube collectors. The evacuated tubes can be constructed with either a **single glass envelope** or **double glass envelope**.

Evacuated tubes with a single glass envelope minimize the optical losses associated with solar radiation passing through the glass. However, they also require a **glass-to-metal seal** that can sustain a high level of vacuum, for many years, over a wide range of temperatures. Achieving this seal is technically challenging. Figure 8-9 shows an example of single glass evacuated tubes mounted to a common collector frame.

Evacuated tubes with a *double* glass envelope do not require glass-to-metal seals to retain their vacuum. Instead, the upper ends of the concentric glass tubes are fused together as the vacuum between them is created.

One variation of the double glass tube design, sometimes called a **Sydney tube**, has the solar absorber coating applied to the full circumference of the inner glass tube. A thin, cylindrically shaped aluminum element, with spaces to accommodate copper tubing, is then pushed into the tube. Heat is transferred by conduction from the absorber coating on the inner tube, to the aluminum element, and then to the copper tubing. Figure 8-10 shows examples of this type of tube, with the glass envelope removed.

This type of evacuated tube is often paired up with a **parabolic reflector** that is fixed behind each tube within the collector's frame, as seen in Figure 8-11. This allows reflected sunlight to be absorbed by the back of the tube while incoming solar radiation is absorbed by the front of the tube.

Courtesy of David Yates, F.W.Behler, Inc.

Figure 8-8 | Evacuated tube collector installed on roof.

Courtesy of David Yates, F.W.Behler, Inc.

Figure 8-9 | Examples of single glass envelope evacuated tubes.

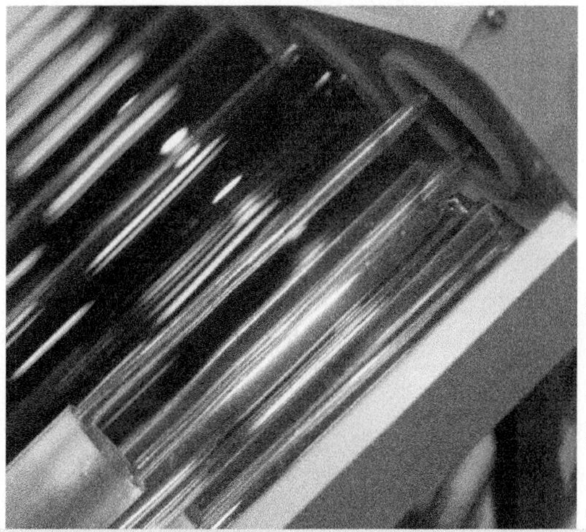

Courtesy of Solar Usage Now-OCH

Figure 8-10 | Double-glass tube with absorber coating on inner glass surface and form-fitting aluminum element that conducts heat to copper tubing.

Courtesy of Solar Usage Now-OCH

Figure 8-11 | Parabolic reflectors seen behind the Sydney tubes. Barium getter coating also visible at the end of tubes.

Some evacuated tubes also contain a silver-colored coating at the end of the tube. This is visible on the tubes in Figure 8-11. This coating is called a **barium getter**. Its function is to absorb any remaining gas molecules within the evacuated space as well as gas molecules that may diffuse through the glass-to-metal seal of the evacuated tube over its estimated 15- to 20-year life. If a tube loses its vacuum, the barium getter coating will turn white. This is an indication that the tube should be replaced.

There are two types of heat transport mechanisms used in evacuated tube collectors:

- heat pipes
- flow-through tubes

A **heat pipe** is a sealed copper tubing assembly that contains a **working fluid**. Most evacuated tubes use a solution of alcohol and hyper pure water as their working fluid. This mixture is completely sealed within the copper heat pipe at a pressure that allows it to begin vaporizing at a relatively low temperature. The alcohol content also serves as an antifreeze. The working fluid can exist as either a liquid or a gas, depending on its temperature and pressure. When a heat pipe is inclined, as shown in Figure 8-12, absorbed solar energy causes the working fluid to vaporize and rise upward within the copper tube attached to the absorber strip.

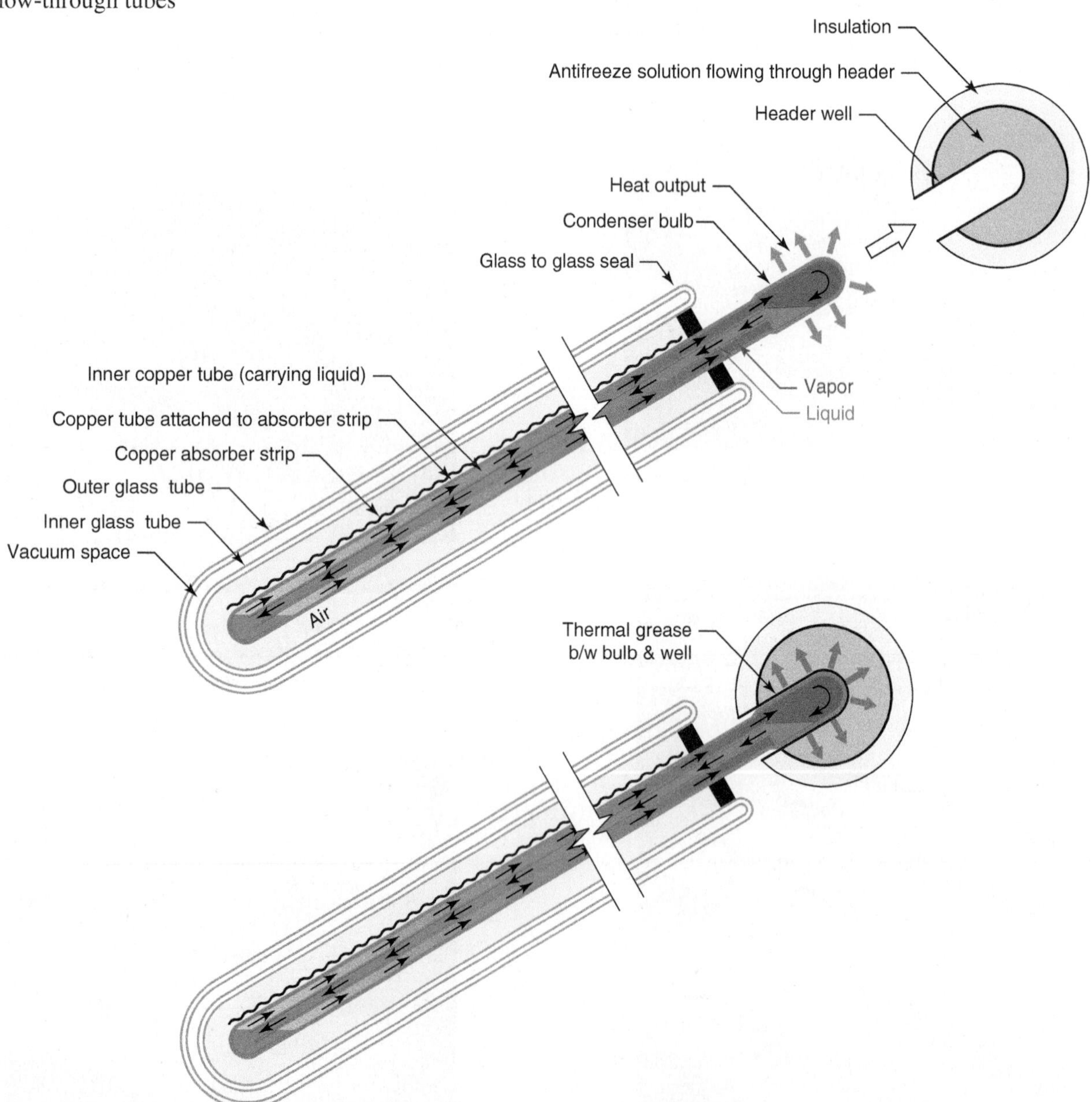

Figure 8-12 | Side view cross section of a double glass envelope evacuated tube collector using a heat pipe absorber.

Courtesy of Apricus, Inc.

Figure 8-13 | Copper condenser bulbs at the top of heat pipe evacuated tubes.

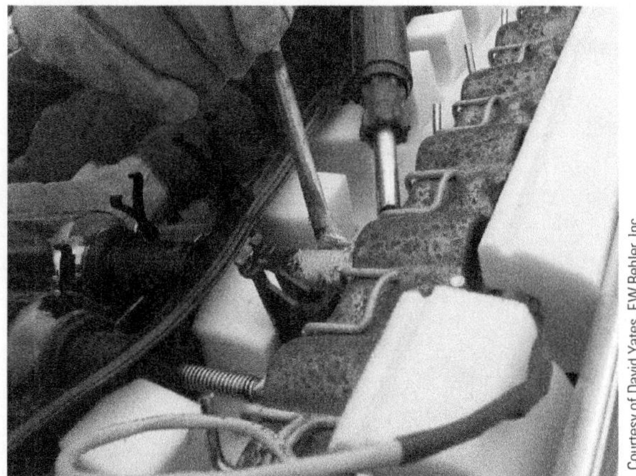
Courtesy of David Yates, F.W.Behler, Inc.

Figure 8-14 | Thermal grease being brushed onto condenser bulb of an evacuated tube as it is inserted into header well.

The upper end of the heat pipe is called the **condenser bulb**. When the collector is installed, the condenser bulb of each tube is inserted into a copper **header well**, as shown in the lower portion of Figure 8-12. The header well is an integral part of the header and is surrounded by the fluid passing through the header. Heat is transferred from the condenser bulb to the header well and finally into a fluid passing through the header. As it gives up heat, the working fluid within the tube condenses back to a liquid and flows downward through the inner tube of the heat pipe. Eventually, it reaches the bottom of the heat pipe, where it can again absorb heat, vaporize, and repeat the cycle. Figure 8-13 shows the copper condenser bulbs at the upper ends of several evacuated tubes.

Caution should always be used in handling evacuated tubes that are exposed to sunlight. Although the glass envelope will remain close to the surrounding air temperature, the condenser bulb can get very hot (up to 390 °F) depending on the intensity of the sunlight.

Most manufacturers require the condenser bulb to be coated with **thermal grease** before it is inserted into the header well, as shown in Figure 8-14. This grease is highly conductive to heat and thus ensures good heat transfer between the condenser bulb and header well.

After the mounting frame has been fastened to the roof or a ground mount, individual tubes are pushed into place and secured, as shown in Figure 8-15. If necessary, individual evacuated tubes can be replaced, without having to drain fluid from the system.

Courtesy of Apricus, Inc.

Figure 8-15 | Individual evacuated tubes being placed into header assembly.

Another type of evacuated tube is called a **flow-through evacuated tube**. Its construction is shown in Figure 8-16.

The copper absorber strip within each tube is metallurgically bonded to two small copper tubes. These tubes are connected with a U-bend at the lower end of the absorber strip. This type of tube is designed to have water or an antifreeze solution pumped through it, as is done with flat-plate collectors. Each evacuated tube requires two connections at the header assembly.

Evacuated tube collectors use high-temperature-resistant **borosilicate glass**. This glass is similar to that used for laboratory glassware, under brand names such as Pyrex™. Although not tempered, it is very resistant to wide temperature changes and thermal shock. It can

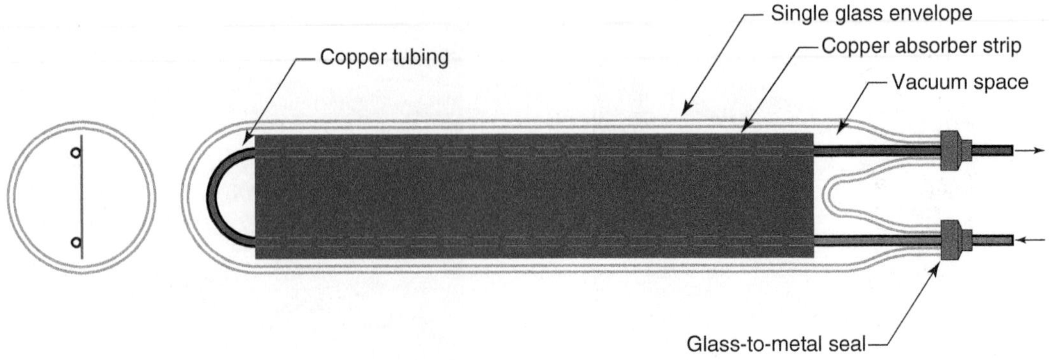

Figure 8-16 | Top view of a flow-through evacuated tube with single glass envelope.

also withstand the impact of smaller hailstones. However, large hailstones have been known to shatter the glass envelope of evacuated tube collectors.

COLLECTOR AREA

Many of the sizing calculations used in designing solar thermal systems require a numerical value for **collector area**. Some sizing calculations also determine the necessary collector area for a given application. Larger collectors, or several collectors within a **collector array**, obviously present more exposed area to the sun. However, there are different interpretations for what is meant by the term "collector area." The specific definitions of collector area are as follows:

- Gross collector area
- Net collector area (a.k.a., aperture area)
- Absorber area

Figure 8-17 illustrates the differences between these areas.

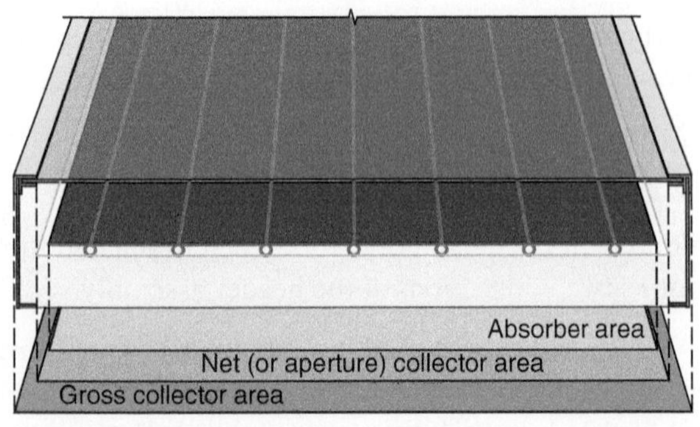

Figure 8-17 | Three specific definitions of collector area.

Gross collector area refers to the outer dimensions of the collector housing (excluding any piping connections). In the case of evacuated tubes, gross area would be the overall area of the mounting rack, tubes, and header assembly.

Net collector area, also sometimes called **aperture area**, is the area of the glazing exposed at the top of the collector. It does not include the area of the frame that holds the glazing to the collector housing. In the case of evacuated tube collectors, it would be the projected area of the outer glass tube surface onto the plane of the collector.

Absorber area is the total heat transfer area over which incoming solar radiation heats the fluid passing through the collector. It includes all the absorber plate and headers viewable through the collector glazing, regardless of the surface coating on these areas. It does not include the concealed areas of the headers where they pass through the housing or protrude outside the housing.

In an ideal collector, both the aperture area, and absorber area, would be equal to the gross area. However, in any real collector, the thickness of the housing and the need for a frame to hold the glazing in place require that the net area be smaller than the gross area. Similarly, the absorber area is typically slightly less than the net area.

In the case of some evacuated tube collectors, there can be significant differences between the gross area and the absorber area. The installation shown in Figure 8-18 bears out this point. Notice that a significant amount of white clapboard siding is visible behind the evacuated tubes. The area between the tubes, through which the siding shows, would not be considered part of the collector's net area. It would, however, be considered part of the collector's gross area.

Courtesy of ReVision Energy

Figure 8-18 The total absorber strip area, seen as dark purple, is significantly less than the gross area occupied by the overall collector assembly.

It is very important to know which of these collector areas is referenced when performing calculations that involve the size of the collector array. Some manufacturers base the thermal efficiency of their collectors on gross area, while others base it on net area or absorber area. This can lead to significant differences in stated efficiency.

Later sections in this chapter discuss the thermal efficiency of solar collectors. The definition of collector efficiency will be based on **ASHRAE standard 93-2010**, which references *gross collector area*. When comparing collectors, it is important that all efficiencies are based on the same definition of collector area.

FLAT-PLATE VERSUS EVACUATED TUBE COLLECTORS

The debate over the use of flat-plate versus evacuated tube solar collectors has gone on ever since the latter were first produced in the 1970s. There are strong opinions on both sides of this debate. Objectively, both types of collectors have strengths and limitations. The following is a qualitative discussion of these strengths and limitations.

Flat-Plate Collectors:

Strengths:

- Their tempered glass cover is very strong and resistant to hailstones
- They do not require long-lasting seals to maintain vacuum
- They have a higher ratio of absorber plate area to gross collector area
- They shed snow better than evacuated tube collectors when mounted at 40° or greater slope

Limitations:

- Large (4 feet × 10 feet) collectors weigh about 150 pounds, which requires suitable site planning for lifting and securing collectors, especially on high, steeply pitched roofs
- They are not as efficient as evacuated tube collectors when operating at higher fluid temperatures or in very cold air temperatures

Evacuated Tube Collectors:

Strengths:

- They are lighter during installation, since each tube is handled individually
- They retain heat better than flat-plate collectors when operating at higher temperatures

Limitations:

- Evacuated tubes can be broken by large hailstones or rough handling
- They are not good at shedding wet snow
- Heat pipe–type evacuated tubes must be mounted at a minimum slope of 25°
- Evacuated tubes must be replaced if vacuum seal fails
- They require more mounting space to attain equal absorber area

Comparisons between flat-plate and evacuated tube collectors should also address differences in thermal performance as well as life-cycle cost. Later sections in this chapter, and chapters to follow, present the information necessary to make such a comparison.

8.5 Solar Collectors for Pool Heating

Pool heating is often one of the most economically viable forms of solar energy utilization (assuming it displaces what would otherwise be conventional fuels used to heat the pool).

Most solar-pool-heating systems use unglazed and uninsulated flat-plate collectors. This is acceptable because in pool-heating applications, the absorber plate operates at temperatures that are often close to and sometimes even *lower* than ambient air temperature. Under such conditions, the absorber plate loses very little if any heat to the surrounding air and thus does not need an enclosure to prevent heat loss.

Unglazed pool-heating collectors are usually constructed of **UV-stabilized polymers** compatible with pool water chemistry. They consist of an upper and lower header with several polymer tube/plate assemblies thermally fused in between, as illustrated in Figure 8-19.

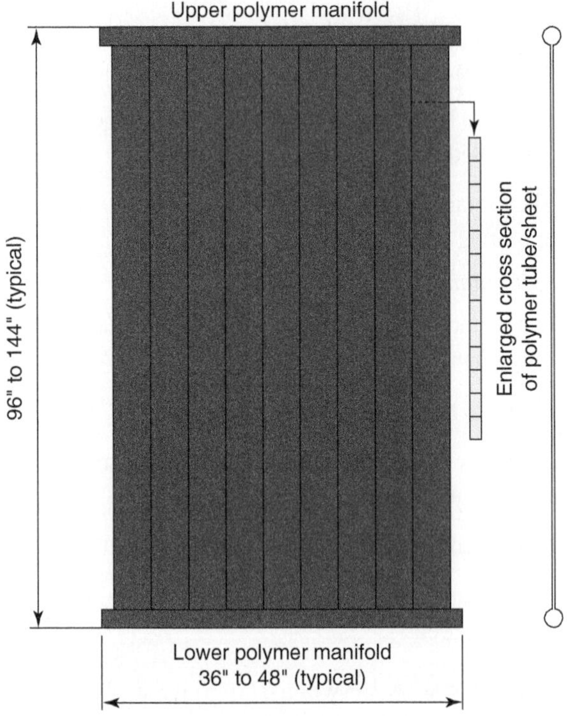

Figure 8-19 | Typical construction of an unglazed solar collector used for heating swimming pools.

Pool-heating collectors have a very high wetted surface area to compensate for the lower thermal conductivity of the polymer absorber plate compared to one made of copper. They are also designed to accommodate substantially higher flow rates than would be used with glazed flat-plate or evacuated tube collectors. These collectors are often mounted on roofs at relatively low slope angle to optimize summertime solar gain. Because they are unglazed and uninsulated, *solar collectors designed for pool heating are **not** suitable for domestic water-heating or space-heating applications.*

8.6 Thermal Performance of Solar Collectors

When designing solar thermal systems, it's necessary to know how much heat a given collector can deliver, as it operates over a wide range of sunlight intensities, ambient air temperatures, and fluid temperatures. This information is also helpful when comparing one type or model of collector to another in a specific application.

One method of expressing the thermal performance of a collector is a numerical value for **instantaneous thermal efficiency**. This efficiency is defined as the ratio of the instantaneous rate of heat output from the collector divided by the instantaneous rate of solar energy incident on the collector.

DERIVATION OF INSTANTANEOUS COLLECTOR EFFICIENCY

The useful heat output from a solar collector, at any instant, can be defined by Equation 8.1, which is simply a **heat balance** on the collector as it operates.

(Equation 8.1)

$$q_{out} = solar\ input - heat\ loss$$

where:

q_{out} = useful heat output from collector (Btu/hr.)

solar input = rate of energy delivered to the collector by solar radiation (Btu/hr.)

heat loss = rate of heat loss from the collector (Btu/hr.)

This heat balance can be expressed using known or measurable quantities, as given by Equation 8.2.

(Equation 8.2)

$$q_{out} = IA_a\tau\alpha - U_LA_a(T_p - T_{air})$$

where:

q_{out} = useful heat output from collector (Btu/hr.)

I = rate of energy input by solar radiation (Btu/hr./ft.^{2})

A_a = aperture area of the collector (ft.^{2})

τ = transmissivity of the glazing (decimal percent)

α = absorptivity of the absorber plate surface (decimal percent)

U_L = heat loss coefficient of the collector (Btu/hr./°F/ft.²)

T_p = average temperature of the absorber plate (°F)

T_{air} = ambient air temperature (°F)

The average temperature of the absorber plate is often difficult to determine; thus, a correction factor (F_R) has been derived that compensates for the difference between the average temperature of the absorber plates and the simplifying assumption that the absorber plate is at a *uniform temperature equal to the fluid inlet temperature*. The derivation of this factor is beyond the scope of this text, but can be found in the first reference listed at the end of this chapter.

By including this factor, Equation 8.2 can be modified into Equation 8.3.

(Equation 8.3)

$$q_{out} = F_R\left[IA_a\tau\alpha - U_LA_a(T_i - T_{air})\right]$$

where:

F_R = collector heat removal factor (unitless)

q_{out} = useful heat output from collector (Btu/hr.)

I = rate of energy input by solar radiation (Btu/hr./ft.²)

A_a = aperture area of the collector (ft.²)

τ = transmissivity of the glazing (decimal percent)

α = absorptivity of the absorber plate surface (decimal percent)

U_L = heat loss coefficient of the collector (Btu/hr./°F/ft.²)

T_i = fluid inlet temperature to the collector (°F)

T_{air} = ambient air temperature (°F)

The instantaneous thermal efficiency of the collector is defined as its rate of heat output divided by the rate at which solar radiation strikes the collector. Referencing ASHRAE standard 93-2010, the rate at which solar radiation strikes the collector is to be based on the *gross* collector area. Thus, Equation 8.4 can be written to represent instantaneous collector efficiency based on gross collector area.

(Equation 8.4)

$$\eta_g = \frac{q_{out}}{IA_g} = \frac{F_RIA_a\tau\alpha - F_RU_LA_a(T_i - T_{air})}{IA_g}$$

where:

η_g = instantaneous collector efficiency based on gross area (decimal percent)

A_g = gross collector area (ft.²)

F_R = collector heat removal factor (unitless)

q_{out} = useful heat output from collector (Btu/hr.)

I = rate of energy input by solar radiation (Btu/hr./ft.²)

A_a = aperture area of the collector (ft.²)

τ = transmissivity of the glazing (decimal percent)

α = absorptivity of the absorber plate surface (decimal percent)

U_L = heat loss coefficient of the collector (Btu/hr./°F/ft²)

T_i = fluid inlet temperature to the collector (°F)

T_{air} = ambient air temperature (°F)

The constants can be separated from the variables, which allows Equation 8.4 to be written as Equation 8.5:

(Equation 8.5)

$$\eta_g = \left(F_g\tau\alpha\frac{A_a}{A_g}\right) - \left(F_gU_L\frac{A_a}{A_g}\right)\left[\frac{T_i - T_{air}}{I}\right]$$

The parameters shown in parentheses are assumed to remain constant for a given collector. In the strictest sense, this is not true. For example, the heat loss coefficient (U_L) increases slightly with increasing collector temperature. However, this variation is relatively small for the temperature ranges of interest; thus, for the sake of simplicity, the quantities shown in parentheses in Equation 8.5 can be considered constant. As such, they can be reduced to two numbers, which remain constant for a given collector. Those numbers will be referred to as the **Y intercept** and **slope**. Thus, for any given collector, Equation 8.5 can be simplified as follows:

(Equation 8.6)

$$\eta_g = (Y\ intercept) - (slope)\left[\frac{T_i - T_{air}}{I}\right]$$

where:

η_g = instantaneous collector efficiency based on gross area (decimal percent)

Y intercept = a number for a specific collector (unitless)

slope = a number for a specific collector (Btu/hr./°F/ft.^{2})*

T_i = fluid inlet temperature to the collector (°F)

T_{air} = ambient air temperature (°F)

I = solar radiation intensity onto plane of collector (Btu/hr./ft.^{2})

The combination of terms $[(T_i - T_a)/I]$ is called the **inlet fluid parameter**. It represents the severity of the conditions under which the collector is operating, (e.g., the combined effect of the collector's inlet fluid temperature, sunlight intensity, and the ambient air temperature). *Anything that causes the inlet fluid parameter to increase will cause the instantaneous collector efficiency to decrease, and vice versa. Any change in operating conditions that lowers the inlet fluid parameter is desirable. Any change in operating conditions that increases the inlet fluid parameter is undesirable.*

If one were to graph the instantaneous collector efficiency as a function of the inlet fluid parameter, it would look similar to the graph shown in Figure 8-20.

Any straight line can be graphed based on a specified value for its Y intercept (e.g., where the line intersects the vertical axis of the graph) and another value for its slope. Thus, the values for Y intercept and slope directly relate the graph shown in Figure 8.21.

This graph shows that the instantaneous thermal efficiency of a solar collector can vary over a wide range, depending on its operating conditions. Those operating conditions include:

- inlet fluid temperature
- intensity of solar radiation striking the collector
- ambient air temperature

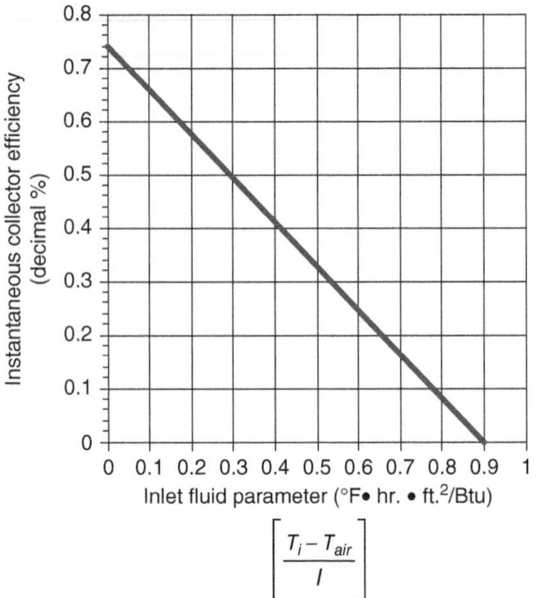

Figure 8-20 Representative graph of instantaneous collector efficiency versus inlet fluid parameter for a typical glazed flat-plate collector.

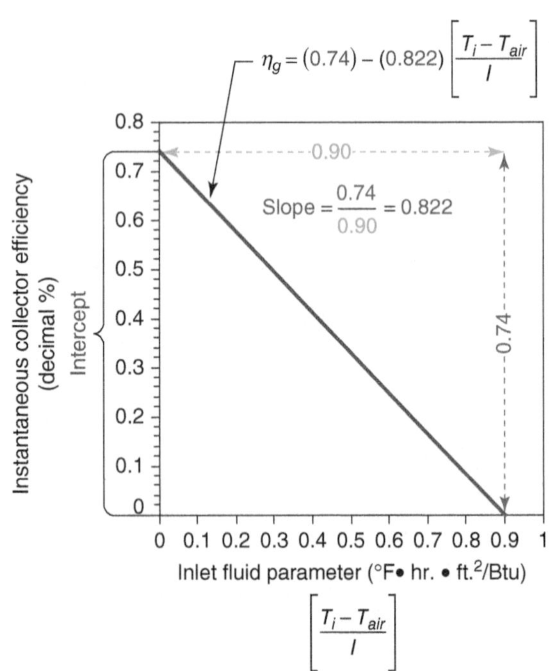

Figure 8-21 The red line indicates instantaneous collector efficiency. In this graph, the value of the Y intercept is 0.74 and the value of the slope is 0.822 (Btu/°F·hr.·ft.^{2}). The slope is calculated by dividing the vertical rise of the line by the horizontal run of the line—in this case, 0.74 divided by 0.90.

* From a mathematical perspective, the line representing instantaneous collector efficiency will always have a negative slope. A point moving along this line will always move down as it moves from left to right. This is represented by the negative sign in front of the word (*slope*) in Equation 8.6. Some technical literature may report the slope as a negative number, such as −0.8. In this case, the absolute value of the number (e.g., its positive value) should be used for the value of (*slope*) in Equation 8.6. This avoids the error of subtracting a negative number.

The "net effect" of these three conditions is represented by the inlet fluid parameter. A change in any one or more of these three operating conditions will *immediately* change the collector's thermal efficiency. For example, consider a day when passing clouds are causing relatively rapid and significant changes in the intensity of solar radiation striking a collector. This would cause the value of the inlet fluid parameter to change just as rapidly. A collector with an instantaneous efficiency of 0.60 (i.e., gathering 60 percent of the solar energy striking its gross area), at a moment when bright sunlight is striking the collector, might have its instantaneous efficiency drop to 0.20 (i.e., gathering 20 percent of the solar energy striking its gross area) a few seconds later, when the solar radiation is greatly reduced by a passing cloud. Thus, *a given value of instantaneous collector efficiency should never be used to estimate the amount of heat that a collector may produce over a period of hours, days, weeks, etc.*

One might wonder what happens if the collector operates at inlet fluid parameters that are higher than the value at which its efficiency line cross the horizontal axis. For example, what would happen if the collector represented by the red line in Figure 8-21 were operating at an inlet fluid parameter of 1.1? The answer is that it would be operating at **negative efficiency**. Under such conditions, the collector would be dissipating heat to the atmosphere. In most cases, this is undesirable. However, it can be useful as a means of **heat dumping** in situations where the system's thermal storage tank exceeds a maximum desired temperature. Heat dumping will be discussed in later chapters.

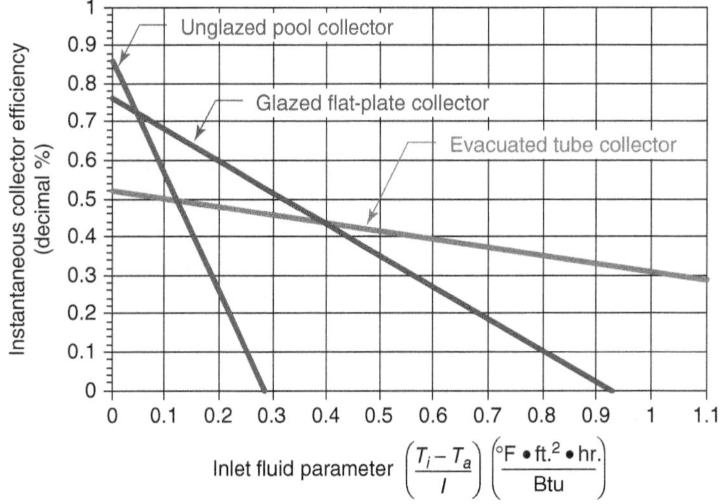

Figure 8-22 | Efficiency comparison for typical flat-plate, evacuated tube, and unglazed pool collectors.

COMPARING INSTANTANEOUS COLLECTOR EFFICIENCIES

It's possible to compare the thermal performance of different solar collectors by plotting their efficiency lines on a common graph. Figure 8-22 does this for a typical flat-plate, evacuated tube, and unglazed pool collector.

The unglazed pool collector has the highest Y intercept. Thus, when operating conditions allow very low values of the inlet fluid parameter (e.g., from 0 to about 0.05 °F·hr.·ft.²/Btu), the unglazed collector has a higher efficiency than that of the glazed flat-plate or the evacuated tube collector. This happens because there are no optical losses associated with glazing on an unglazed collector. However, the efficiency of the unglazed collector drops rapidly as the inlet fluid parameter increases. This is usually the result of decreasing air temperature around the collector. Without an insulated housing and glazing, such conditions quickly result in increased heat loss and thus lower efficiency. The very steep descent of the unglazed pool collector's efficiency line precludes any practical use other than for pool heating, or during times other than when the pool water is cooler than, equal to, or perhaps a few degrees above the ambient air temperature.

In comparison to the unglazed pool collector, the efficiency line for the glazed flat-plate collector begins at a lower Y intercept. This is due to optical losses through its glazing. However, the slope of the efficiency line for the glazed flat-plate collector is much shallower than that of the pool collector. This is the result of much better heat retention due to an insulated housing and glazing that maintain a warm environment for the absorber plate.

The efficiency line for the evacuated tube collector has a significantly lower Y intercept compared to that of the glazed flat-plate collector. This is the result of basing all efficiency calculations on *gross* collector area. Evacuated tube collectors have a significantly lower ratio of absorber area to gross area. However, the slope of the line for the evacuated tube collector is much shallower than that of the glazed flat-plate collector. This is the result of significantly better heat retention by the evacuated tube.

For the specific case shown in Figure 8-22, when the operating conditions result in an inlet fluid parameter of 0.4, the instantaneous thermal efficiency of the glazed flat-plate collector and evacuated tube collector are equal at 0.44, (e.g., each gathering 44 percent of the solar radiation striking their gross area and transferring that energy to the fluid passing through the collector).

If operating conditions result in an inlet fluid parameter lower than 0.4, the flat-plate collector will have a higher instantaneous efficiency than the evacuated tube collector. Likewise, if operating conditions result in an inlet fluid parameter greater than 0.4, the evacuated tube collector will have a higher instantaneous efficiency. Thus, at any particular moment, either collector may have the efficiency advantage. Ultimately, any quantitative performance comparison between a flat-plate versus evacuated tube collector should not be made based on instantaneous thermal efficiency. Instead, it should consider which collector gathers the greater amount of heat over an entire operating season. It should also factor in the total seasonal energy yield per dollar of collector array cost. This type of comparison requires simulation software, which will be described in later chapters.

Example 8.1: Two space-heating distribution systems are being considered for use with an array of flat-plate solar collectors. One distribution system is based on standard fin-tube baseboard. On a cold winter day, this system requires water at 160 °F to be supplied to the collector array. The other system is slab-type floor heating, which requires water at 95 °F to be supplied to the collector array. Assume that the efficiency of the collectors is represented by the red line (glazed flat-plate collector) in Figure 8-22. Compare the instantaneous efficiency of the collector in both situations, assuming it is operating at an outdoor air temperature of 20 °F and with a solar radiation intensity of 200 Btu/hr./ft.^{2}.

Solution: Calculate the inlet fluid parameter for both operating conditions.

For the higher temperature distribution system, the inlet fluid parameter is:

$$\left[\frac{T_i - T_{air}}{I}\right] = \left[\frac{160 - 20}{200}\right] = 0.7$$

For the lower temperature distribution system, the inlet fluid parameter is:

$$\left[\frac{T_i - T_{air}}{I}\right] = \left[\frac{95 - 20}{200}\right] = 0.375$$

Locating 0.7 on the horizontal axis of Figure 8-22, projecting up to the red line, then over to the vertical axis, indicates the collector's thermal efficiency is 0.16 (e.g., 16 percent). Thus, under these operating conditions, only 16 percent of the solar energy striking the gross area of the collector is being converted into useful heat.

Using the same procedure but entering the efficiency graph as an inlet fluid parameter of 0.375 shows the instantaneous collector efficiency to be 0.43, or 43 percent.

Discussion: The significant drop in the inlet fluid temperature, with other conditions remaining the same, results in much higher thermal efficiency. This comparison demonstrates that collector efficiency is extremely dependent on inlet fluid temperature. For the best performance, the inlet fluid temperature to any solar collector should be kept as low as possible. In space heating applications, this is done by designing the heat distribution system to operate at the lowest possible water temperature. As discussed in Chapter 5, *all space-heating distributions systems supplied by renewable energy heat sources should be designed to provide design load output without exceeding a supply water temperature of 120 °F.*

It should also be emphasized that the two efficiencies calculated in this example are "snapshots" under two specific sets of operating conditions. One cannot extrapolate the large difference in instantaneous collector efficiency over the course of an hour, a day, a month, etc. The total solar energy gathered by each system must be determined by software-based simulations of system performance versus a snapshot of instantaneous collector efficiency.

MEASURING INSTANTANEOUS COLLECTOR EFFICIENCY

There are two currently used standards for establishing instantaneous collector efficiency. They are entitled: ASHRAE Standard 93-2010 and **ISO 9806**. Both procedures require several simultaneous measurements of the solar radiation striking the collector, and the heat output from the collector. The testing can be done outside, on clear days with a high percentage of direct radiation, or inside using **solar simulators**. In either case, the collector needs to be oriented so that direct sunlight or simulated solar radiation strikes it perpendicular to the glazing, or in the case of evacuated tube collectors, perpendicular to the plane of the collector. Another way of stating this is that the angle of incidence, during testing, needs to be maintained as close to zero as possible, as shown in Figure 8-23. In some cases, this requires the collector to be mounted on a frame that can be adjusted for both slope and azimuth as the test is being conducted.

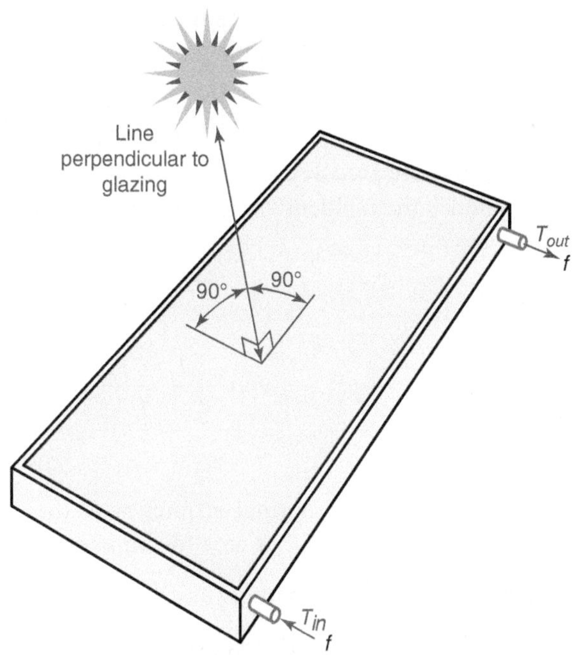

Figure 8-23 | During efficiency testing, the collector needs to be oriented so that the angle of incidence is maintained very close to zero (e.g., solar radiation strikes the collector perpendicular to its glazing).

The instantaneous intensity of solar radiation is measured by a **pyranometer**. An example of this instrument is shown in Figure 8-24.

Figure 8-24 | A pyranometer for measuring solar radiation.

During testing, the pyranometer is mounted in the plane of the collector and thus measures total radiation perpendicular to the collector. Several simultaneous measurements of solar radiation intensity, flow rate, inlet temperature, outlet temperature, and ambient air temperature are recorded. These measurements are taken over a wide range of inlet fluid parameters. The instantaneous collector efficiency is then calculated for each set of data using Equation 8.7.

(Equation 8.7)

$$\eta = \frac{(8.01\,Dc)f(T_{out} - T_{in})}{A_g(I)}$$

where:

η = instantaneous collector efficiency (decimal percent)

D = density of fluid at the average of T_{in} and T_{out} (lb./ft.^{3})

c = specific heat of fluid at the average of T_{in} and T_{out} (Btu/lb./°F)

f = flow rate through collector (gpm)

T_{in} = temperature of fluid at collector inlet (°F)

T_{out} = temperature of fluid at collector outlet (°F)

A_g = gross area of collector (ft.^{2})

I = solar radiation intensity (Btu/hr./ft.^{2})

Example 8.2: A flat-plate solar collector measures 47 inches width and 119 inches length. During a test, the following data was instantaneously recorded:

Water temperature at inlet = 100.2 °F

Water temperature at outlet = 108.8 °F

Water flow rate = 0.85 gpm

Solar radiation intensity (in collector plane) = 265 Btu/hr./ft.^{2}

Ambient air temperature = 36 °F

Determine the collector's instantaneous thermal efficiency and the inlet fluid parameter at which it is operating.

Solution: The specific heat and density of water need to be determined at its average temperature of (100.2 + 108.8)/2 = 104.5 °F. The specific heat of water can be assumed constant at 1.00 Btu/lb./°F. The density of water can be read from Figure 3-11: D = 61.8 lb./ft.^{3}.

The gross collector area is (47 inches × 119 inches)/144 in.^{2}/ft.^{2} = 38.84 ft.^{2}.

Substituting these values and the other data in Equation 8.7 yields:

$$\eta = \frac{(8.01\,Dc)f(T_{out} - T_{in})}{A_g(I)}$$

$$= \frac{(8.01 \times 61.8 \times 1.00)0.85(108.8 - 100.2)}{38.84(265)} = 0.35$$

The inlet fluid parameter at which the collector is operating is:

$$\left[\frac{T_i - T_{air}}{I}\right] = \left[\frac{100.2 - 36}{265}\right] = 0.24$$

Discussion: The pairing of an instantaneous efficiency of 0.35 (or 35 percent), at an inlet fluid parameter of 0.24, provides one point on the efficiency graph. Several other points would need to be established and plotted during the test.

Once these data points are plotted, a statistical method called **linear regression** is used to construct the **best straight line** through the points. The vertical intercept and slope of this line would then represent the Y intercept and slope values for this collector's efficiency equation. These values are provided on collector rating reports, which are discussed later in this chapter. These values can be used in combination with simulation software to predict the performance of systems based on use of a specific collector.

INCIDENT ANGLE MODIFIER

As previously described, collector efficiency testing is done with the collector oriented so that the angle of incidence is at or very close to zero (e.g., direct beam sunlight strikes the collector perpendicular to its glazing). Under this condition, the glazing allows the maximum percentage of incident solar radiation to pass through. The absorptivity of the absorber plate surface is also maximized under this condition.

The Y intercept of the collector's efficiency line depends on both the transmissivity of the glazing and the absorptivity of the absorber plate surface. Thus, the Y intercept value established during testing is technically only valid when beam solar radiation is perpendicular to the plane of the collector.

When installed in a typical fixed mounting, a solar collector spends *very little* time with an angle of incidence equal to zero. Instead, incoming solar radiation strikes the glazing and absorber plate at angles other than 90°, especially during early morning and late afternoon. The transmissivity of the glazing, and the absorptivity of the absorber plate coating, are both reduced under these conditions. Thus, the "effective" Y intercept of the collector's efficiency line is less than the Y intercept value established during testing.

The **incident angle modifier** was developed to correct the Y intercept values determined by the testing procedure for situations where the incident angle is not zero. Its use allows more accurate prediction of instantaneous thermal efficiency as well as the total energy collected over a simulated day.

The corrected collector efficiency equation, which includes the incident angle modifier, is given as Equation 8.8:

(Equation 8.8)

$$\eta_g = K(Y\,intercept) - (slope)\left[\frac{T_i - T_{air}}{I}\right]$$

where:

η_g = instantaneous thermal efficiency of the collector based on gross area (decimal percent)

K = incident angle modifier (unitless)

T_i = inlet fluid temperature (°F)

T_a = ambient air temperature (°F)

I = solar radiation intensity onto plane of collector (Btu/hr./ft.^{2})

Notice that K is a multiplier for the *Y intercept*, and not the slope. The maximum value of K is 1.0. This only occurs when the angle of incidence is zero.

Values for the incident angle modifier (K) are listed for incident angles between 10° and 70° on **SRCC collector rating reports**, which are discussed in the next section.

In theory, the value of K at angles of incidence between 0 and 90° can be calculated using Equation 8.9:

(Equation 8.9)

$$K = 1 - b_0\left(\frac{1}{\cos\theta} - 1\right)$$

where:

K = incident angle modifier (unitless)

θ = angle of incidence (°)

b_0 = a constant determined by testing (unitless)

The value of b_0 would be determined by plotting (K) versus [(1/cos θ) − 1] and then using linear regression to establish the best straight line through the data. The slope of the best straight line would be the value for b_0.

The smaller the value of bo, the lower the optical losses due to solar radiation striking the collector at incident angles other than 90°. Lower optical losses result in higher thermal efficiency and greater energy capture.

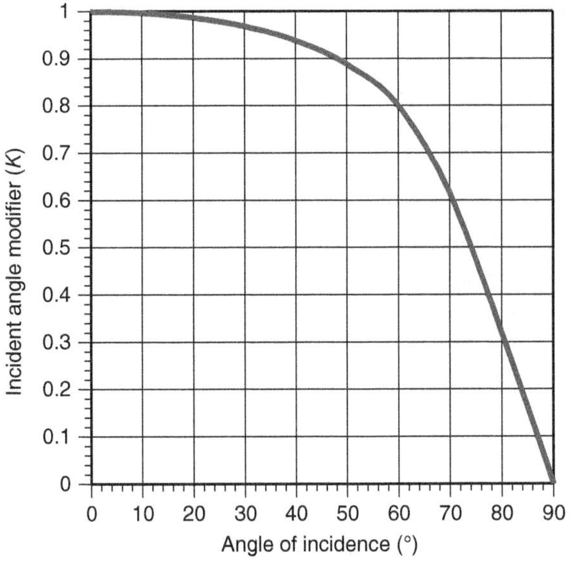

Figure 8-25 | Value of incident angle modifier versus angle of incidence of solar radiation for $b_0 = 0.20$.

The graph in Figure 8-25 shows the incident angle modifier (K) as a function of the angle of incidence based on a typical b_0 value of 0.20.

Notice that the value of K remains close to its maximum value of 1.0 until the angle of incidence reaches about 20°. K then decreases to about 0.9 when the angle of incidence reaches 47° and drops off rapidly at higher angles.

Most single-glazed flat-plate collectors lose very minor amounts of solar radiation during the **peak solar collection** window of 9:00 AM to 3:00 PM solar time due to the effects of the incident angle. For a south-facing collector sloped at 42°, and located at 32° latitude, the total energy losses between 9:00 AM to 3:00 PM solar time, due to the effects of the incident angle, are about 5.3 percent. *Relative* energy losses due to incident angle are more pronounced in earlier morning and later afternoon, when typical clear day solar radiation intensity is much lower.

The net effect of the incident angle modifier can be determined through calculations that use specific hourly values of solar radiation in the plane of the collector at a given location and orientation as well as specific hourly values for angle of incidence. The ASHRAE 93-2010 standard gives an example of such calculations.

SRCC SOLAR COLLECTOR RATINGS

Designers can access a wide range of performance data and physical data for solar thermal collectors through reports generated by the **Solar Rating & Certification Corporation (SRCC)**. This organization was established in 1980 to provide third-party evaluation and certification of solar thermal hardware. SRCC works in cooperation with accredited testing laboratories. It oversees the processes and procedures used to physically test solar thermal collectors as well as to certify those collectors that have successfully passed the requirements of a standard titled **OG-100 Operating Guideline for Certifying Solar Collectors**. Collectors that have achieved this certification have a label attached, an example of which is shown in Figure 8-26.

Within the United States, most of the current federal and state incentive programs that deal with solar thermal systems require collectors that are OG-100 certified. A listing of all currently certified collectors can be downloaded for free from http://www.solar-rating.org/.

Physical and performance data for each OG-100 certified collector is presented in a standardized report form and available as a free downloadable PDF file. An example of the SRCC rating sheet for a specific flat-plate collector is shown in Figure 8-27.

Courtesy of Caleffi North America, Inc.

Figure 8-26 | Example of a label for a collector that is certified to SRCC standard OG-100.

CERTIFIED SOLAR COLLECTOR

SUPPLIER:
Caleffi Solar
3383 W Milwaukee Road
Milwaukee, WI 53208 USA
www.caleffi.us

BRAND:	StarMax V
MODEL:	NAS15410
COLLECTOR TYPE:	Glazed Flat Plate
CERTIFICATION #:	10001745
Original Certification:	September 13, 2012
Expiration Date:	

The solar collector listed below has been evaluated by the Solar Rating & Certification Corporation™ (SRCC™) in accordance with SRCC OG-100, Operating Guidelines and Minimum Standards for Certifying Solar Collectors, and has been certified by the SRCC. This award of certification is subject to all terms and conditions of the Program Agreement and the documents incorporated therein by reference.

COLLECTOR THERMAL PERFORMANCE RATING							
Kilowatt-hours (thermal) Per Panel Per Day				Thousands of Btu Per Panel Per Day			
Climate -> Category (Ti-Ta)	High Radiation (6.3 kWh/m².day)	Medium Radiation (4.7 kWh/m².day)	Low Radiation (3.1 kWh/m².day)	Climate -> Category (Ti-Ta)	High Radiation (2000 Btu/ft².day)	Medium Radiation (1500 Btu/ft².day)	Low Radiation (1000 Btu/ft².day)
A (-5 °C)	15.0	11.3	7.6	A (-9 °F)	51.2	38.6	26.1
B (5 °C)	13.6	9.9	6.2	B (9 °F)	46.4	33.8	21.3
C (20 °C)	11.5	7.9	4.3	C (36 °F)	39.2	26.8	14.6
D (50 °C)	7.5	4.2	1.2	D (90 °F)	25.6	14.3	4.1
E (80 °C)	3.9	1.2	0.0	E (144 °F)	13.3	4.2	0.0

A- Pool Heating (Warm Climate) B- Pool Heating (Cool Climate) C- Water Heating (Warm Climate)
D- Space & Water Heating (Cool Climate) E- Commercial Hot Water & Cooling

COLLECTOR SPECIFICATIONS					
Gross Area:	3.554 m²	38.25 ft²	Dry Weight	63 kg	139 lb
Net Aperture Area:	3.484 m²	37.50 ft²	Fluid Capacity:	4.0 liter	1.1 gal
Absorber Area:	3.367 m²	36.24 ft²	Test Pressure:	1350 kPa	196 psi

TECHNICAL INFORMATION		Tested in accordance with: ISO 9806			
ISO Efficiency Equation [NOTE: Based on gross area and (P)=Ti-Ta]					
SI UNITS:	η= 0.735 - 3.58930(P/G) - 0.01010(P²/G)	Y Intercept:	0.740	Slope:	-4.198 W/m².°C
IP UNITS:	η= 0.735 - 0.63259(P/G) - 0.00099(P²/G)	Y Intercept:	0.740	Slope:	-0.740 Btu/hr.ft².°F

Incident Angle Modifier								Test Fluid:	Water	
θ	10	20	30	40	50	60	70	Test Mass Flow Rate:	0.0166 kg/(s m²)	12.24 lb/(hr ft²)
Kτα	0.99	0.98	0.95	0.90	0.81	0.66	0.34	Impact Safety Rating: 11		

REMARKS:

Jim Huggins
Technical Director

Print Date: November, 2012

www.solar-rating.org ♦ 400 High Point Drive, Suite 400 ♦ Cocoa, Florida 32926 ♦ (321) 213-6037 ♦ Fax (321) 821-0910

Page 1 of 3

Figure 8-27 | (a) First page of an OG-100 collector rating sheet showing thermal performance and physical data.

CERTIFIED SOLAR COLLECTOR

SUPPLIER:
Caleffi Solar
3383 W Milwaukee Road
Milwaukee, WI 53208 USA
www.caleffi.us

BRAND:	StarMax V
MODEL:	NAS15410
COLLECTOR TYPE:	Glazed Flat Plate
CERTIFICATION #:	10001745
Original Certification:	September 13, 2012
Expiration Date:	

The solar collector listed below has been evaluated by the Solar Rating & Certification Corporation™ (SRCC™) in accordance with SRCC OG-100, Operating Guidelines and Minimum Standards for Certifying Solar Collectors, and has been certified by the SRCC. This award of certification is subject to all terms and conditions of the Program Agreement and the documents incorporated therein by reference.

ADDITIONAL INFORMATION (click here to return to the rating page)			
Test Lab:	Bodycote	Test Report Date:	January 01, 0001
Test Report Number:		Test conducted:	indoors

SOLAR COLLECTOR CONSTRUCTION DETAILS					
Gross Length:	116.14 in	Gross Width:	47.44 in	Gross Depth:	3.90 in

COLLECTOR MATERIALS					
Outer Cover:	Glass sheet	Enclosure back:	Aluminum	Back Insulation:	Fiber, Foam
Inner Cover:	None	Enclosure side:	Aluminum	Side Insulation:	Foam, None
Absorber Description:	Tubes connected to Fins		Flow Pattern:	Parallel/Harp	
Riser Tube:	Copper		Fin:	Copper	
Absorber Coating:	Selective		Tube to fin connection	Solder	

Glazing	Outer Cover	Inner Cover
Material:	Glass sheet	None
Surface Characteristics:	Textured	
Thickness:	0.157 in	N/A
Transmissivity:	High (equal to or greater than 90%)	
Length:	114.61 in	
Width:	46.81 in	
Tube Glazing to Header Enclosure Seal:	EPDM gasket	

ABSORBER:				Absorber Coating:	Selective
Header Material:	Copper	Header OD:	1.102 in	Header Wall:	0.039 in
Riser Tube Material:	Copper	Riser Tube OD:	0.472 in	Riser Tube Wall Thickness:	0.039 in
Fin Material:	Copper	Fin Thickness:	0.005 in		

Page 2 of 3

Figure 8-27 | (b) Second page of an OG-100 collector rating sheet showing physical data.

Flow Pattern:	Parallel/Harp				
Number of Riser Tubes:	10	Tube Spacing:	4.5 in	Number of times each riser crosses the absorber:	10
Length of Flow Path:	9.580 ft	Riser to Fin/Plate Bond:	Solder		

INSULATION:					
Location	Type	Thickness	Location	Type	Thickness
Back – Top Layer:	Fiber	1.0 in	Sides – Inner Layer:	Foam	1.0 in
Back – Bottom Layer:	Foam	0.8 in	Sides – Outer Layer:	None	
Enclosure Fastening Methods:	Screws				

Power Output per Collector(W) [Ti-Ta, G = 1000 W/m²]				
0	10	30	50	70
2613	2482	2198	1885	1544

PRESSURE DROP				
Flow	ΔP		Flow	ΔP
ml/s	Pa		gpm	in H₂0
20	0.01		0.32	0.0
50	0.03		0.79	0.0
80	0.06		1.27	0.0

Page 3 of 3

Figure 8-27 | (c) Third page of an OG-100 collector rating sheet showing addition physical data.

These reports provide a wide range of information, which can be categorized as either physical data or performance data.

The physical data listed includes dimensions, weight, and fluid volume of the collector as well as its gross, net, and aperture areas. Additional data is given for the materials used to construct the collector and the size of its piping connections.

The performance data provided includes pressure drop versus flow rate, and a variety of thermal performance measurements. The latter can be further classified as follows:

- Simulated performance over a **standardized test day**, in several assumed applications
- Instantaneous efficiency data
- Incident angle modifier data

Simulated thermal performance is reported as the total amount of heat the collector gathers during three standardized test days. This information is presented as a table, an example of which is shown in Figure 8-28.

The simulated thermal performance is given in a table near the top of the first page of the OG-100 report. It is provided in both metric (left side of table) and standard U.S. units (right side of table). Performance is simulated, on an hour-by-hour basis, for three different standard days, based upon the total solar radiation received during that day. The standard days are defined as follows:

- High radiation (2,000 Btu/ft.^{2}/day)
- Medium radiation (1,500 Btu/ft.^{2}/day)
- Low radiation (1,000 Btu/ft.^{2}/day)

For each of these standardized test days, the collector performance is simulated for five predefined applications:

(A) Pool heating — warm climate

(B) Pool heating — cool climate

(C) Domestic water heating — warm climate

(D) Space heating and domestic water heating — cool climate

(E) Commercial hot water and cooling

Of these, category (D) is appropriate for the type of solar combisystems discussed in this text.

Example 8.3: Estimate the total heat gathered by the collector represented in Figure 8-28 in a combisystem application (space heating and domestic water heating) in a cool climate on a day with medium solar radiation intensity.

Solution: Enter the medium radiation intensity column and read down to line (D) to find the estimated total heat gain of 14.3 thousand Btu/day (e.g. 14,300 Btu/day).

Discussion: This estimated output is not a guarantee of performance. The listed collector might provide higher or lower total daily heat output based on factors such as actual weather conditions, temperature at which it operates, orientation, tilt, windspeed, and system control settings. However, because the SRCC ratings are based on the same predefined operating conditions, they are well suited for *relative comparisons* between collectors.

The first page of the SRCC OG-100 report lists two types of data for instantaneous collector efficiency.

COLLECTOR THERMAL PERFORMANCE RATING							
Kilowatt-hours (thermal) per Panel per Day				Thousands of Btu per Panel per Day			
Climate -> / Category ($T_i - T_a$)	High Radiation (6.3 kWh/ rrf.day)	Medium Radiation (4.7 kWh/ rrf.day)	Low Radiation (3.1 kWh/ rrf.day)	Climate -> / Category ($T_i - T_a$)	High Radiation (2,000 Btu/ ff.day)	Medium Radiation (1,500 Btu/ ff.day)	Low Radiation (1,000 Btu/ ff.day)
A (−5 °C)	15.0	11.3	7.6	A (−9 °C)	51.2	38.6	26.1
B (5 °C)	13.6	9.9	6.2	B (9 °C)	46.4	33.8	21.3
C (20 °C)	11.5	7.9	4.3	C (36 °C)	39.2	26.8	14.6
D (50 °C)	7.5	4.2	1.2	D (90 °C)	25.6	14.3	4.1
E (80 °C)	3.9	1.2	0.0	E (144 °C)	13.3	4.2	0.0

A- Pool Heating (Warm Climate) B- Pool Heating (Cool Climate) C- Water Heating (Warm Climate) D- Space & Water Heating (Cool Climate)
E- Commercial Hot Water & Cooling

Courtesy of Solar Rating & Certification Corp.

Figure 8-28 | Simulated collector performance for a flat-plate collector provided on an SRCC rating sheet.

TECHNICAL INFORMATION		Tested in accordance with: ISO 9806			
ISO Efficiency Equation [NOTE: Based on gross area and (P) = $T_i - T_a$]					
SI UNITS:	n = 0.735 − 3.58930(P/G) − 0.01010(P²/G)	Y Intercept:	0.740	Slope:	−4.198 W/mVC
IP UNITS:	rp 0.735 − 0.63259(P/G) − 0.00099(P²/G)	Y Intercept:	0.740	Slope:	−0.740 Btu/hr.ft.VF

Figure 8-29 | Portion of OG-100 report giving data and equations for instantaneous collector efficiency.

An example of this section of the report is shown in Figure 8-29.

On the right-hand side of this table, values are provided for the Y intercept and slope of the collector's instantaneous efficiency line in both metric (SI units) and standard U.S. units (IP units).

On the left side of the table, alternative equations of instantaneous collector efficiency are listed based on the ISO 9806 testing standard used by SRCC. These are **second-order equations**, which are slightly more accurate in tracking the changes in collector efficiency due to convective heat loss. For the collector represented in Figure 8-29, the second-order equation is as follows.

$$\eta_g = 0.735 - 0.63256\left(\frac{P}{I}\right) - 0.00099\left(\frac{P^2}{I}\right)$$

where:

$P = (T_i - T_a)$

T_i = inlet fluid temperature (°F)

T_a = ambient air temperature (°F)

I = solar radiation intensity onto plane of collector (Btu/hr./ft.²)

These two equations can be combined based on the standard format for the inlet fluid parameter, and the nomenclature used in this text, as Equation 8.10:

(Equation 8.10)

$$\eta_g = 0.735 - 0.63259\left[\frac{T_i - T_a}{I}\right] - 0.00099(I)\left[\frac{T_i - T_a}{I}\right]^2$$

where:

η_g = instantaneous thermal efficiency of the collector based on gross area (decimal percent)

T_i = inlet fluid temperature (°F)

T_a = ambient air temperature (°F)

I = solar radiation intensity onto plane of collector (Btu/hr./ft.²)

Example 8.4: Compare the efficiency "curve" generated by equation 8.10 to the instantaneous collector efficiency generated by a straight line for the flat-plate collector represented in Figure 8-29.

Solution: The straight line equation for the collector is as follows:

$$\eta_g = 0.74 - 0.74\left[\frac{T_i - T_a}{I}\right]$$

Figure 8-30 shows the straight line equation and two curves based on the second-order equation. One of the curves is plotted assuming a solar radiation intensity of 100 Btu/hr./ft.² and the other for an intensity of 250 Btu/hr./ft.².

Discussion: The second-order curve tends to lower instantaneous collector efficiency at higher values of inlet fluid parameter, and high solar-radiation intensity. Under lower solar-radiation intensity, the second-order equation tends to increase instantaneous collector efficiency compared to the straight line equation. Most (but not all) of the solar simulation software currently available uses the straight line model of instantaneous collector efficiency.

COLLECTOR STAGNATION TEMPERATURE

During its life, every solar collector will undergo periods when it is exposed to solar radiation, but no fluid is passing through its absorber plate. This condition is called **stagnation**. It can occur for many reasons, including:

- power outages
- collectors exposed to sunlight before they are piped and operating

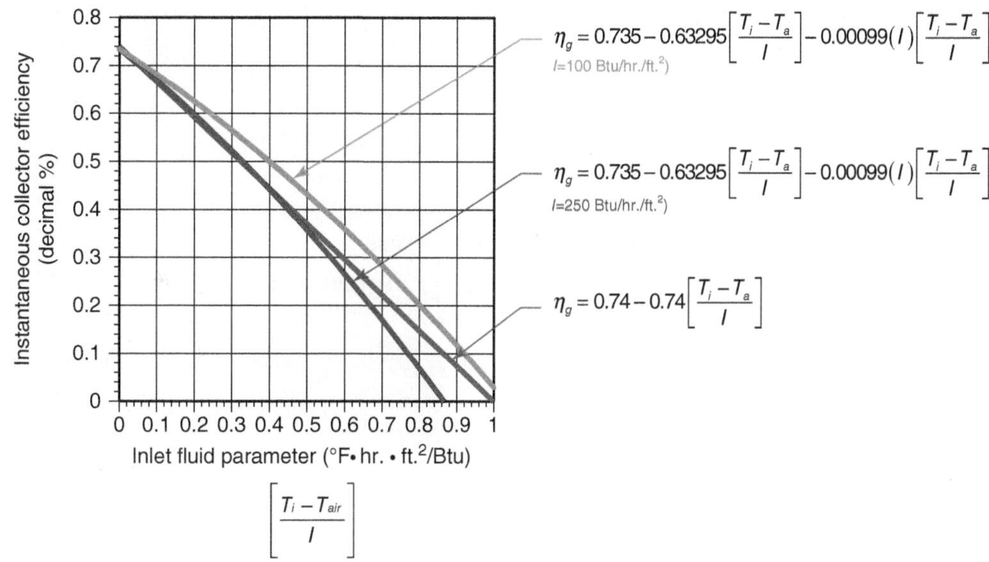

Figure 8-30 | Comparison between straight line equation and two curves based on the second-order equation. One curve is plotted for a solar radiation intensity of 100 Btu/hr/ft², the other for an intensity of 250 Btu/hr./ft.².

- failure of a controller or associated temperature sensor
- failure of the collector circulator
- a collector in a drainback-protected system when the circulator is off

During stagnation, the collector's internal temperature rises until the rate of heat loss from the collector's enclosure equals the rate of solar radiation absorber by the collector. The temperature of the absorber plate during stagnation can be estimated by setting the instantaneous collector efficiency equation equal to zero, since the useful heat output during stagnation is zero, and solving for the collector's inlet temperature. This results in Equation 8.11.

(Equation 8.11)

$$T_{stag} = \left(\frac{Y\ intercept}{slope}\right)I + T_{air}$$

where:

T_{stag} = estimated temperature of absorber plate during stagnation (°F)

Y Intercept = Y intercept value from collector's straight line instantaneous efficiency equation

slope = slope value from collector's straight line instantaneous efficiency equation (Btu/hr./ft.²/°F)

T_{air} = ambient air temperature (°F)

Example 8.5: A collector has the following straight line instantaneous efficiency equation:

$$\eta_g = 0.74 - 0.80\left[\frac{T_i - T_a}{I}\right]$$

Determine its stagnation temperature during a power failure, assuming the ambient air temperature is 85 °F and the solar radiation intensity is 300 Btu/hr./ft.².

Solution: The intercept value for this collector is 0.74. The slope value is 0.80 Btu/hr./ft.²/°F. Using Equation 8.11 along with the stated conditions yields:

$$T_{slag} = \left(\frac{intercept}{slope}\right)I + T_a = \left(\frac{0.74}{0.80}\right)300 + 85 = 362.5\ °F$$

Discussion: This is a very high temperature relative to normal operating conditions. The conditions used to calculate this stagnation temperature represent a bright and warm summer day, in which clear sky radiation was striking the collector at a low angle of incidence.

The materials used to construct solar collectors must be able to withstand the temperatures that can occur during stagnation. Likewise, any fluid or vapor that remains within the collector under such conditions must be able to survive without excessive thermal degradation. This can be a challenge for some glycol-based antifreeze solutions. Collectors that are used in drainback systems do not contain water or other fluids during stagnation. This so-called **dry stagnation** is preferred by many designers, including the author, as the best way for a collector to undergo extreme temperature excursions.

A portion of the OG-100 certification procedure requires collectors to undergo a minimum of 30 cumulative days of dry stagnation, where total solar radiation in the plane of the collector is at least 1,500 Btu/ft.^{2}/day. Following this exposure period, the collector is disassembled and inspected for any signs of material degradation.

Courtesy of Solar Skies Mfg., LLC

Figure 8-31 | Two flush-mounted flat-plate collectors.

8.7 Mounting Solar Collectors

Most solar collectors are mounted in a fixed position, either on a roof or a **ground mount**. It's vitally important that the mounting hardware used can secure the collector against the strongest possible winds that could occur at the site. Specific wind loading information is typically prescribed by local building code and can vary significantly between coastal areas and inland areas.

The durability of the collector mounting system is also critically important. All mounting systems should be free of corrosion or other deterioration for at least 25 years. Only stainless steel and aluminum, and in limited cases, galvanized steel, should be used for metallic portions of the mounting system. Stainless steel is most commonly used for fasteners, while aluminum tubing of various shapes is used for structural members. Treated lumber has been used for some ground mounting frames and other purposes, but is not as dimensionally stable as metal mounting hardware.

Courtesy of Solar Service Inc., Niles, IL

Figure 8-32 | Stand-off mounting used for steeply sloped collectors that favor winter solar angles.

FLUSH MOUNTING

The most common mounting location for solar collectors used in residential and light commercial buildings in on a roof. When the roof has suitable orientation and pitch, as assessed by design tools such as a Solar Pathfinder® and simulation software, solar collectors can be mounted parallel to the roof surface, as shown in Figure 8-31. This is often called **flush mounting**.

When a roof has a suitable orientation but its pitch is too shallow to achieve good performance, the common solution is to use **stand-off mounting**, an example of which is shown in Figure 8-32.

Most collector manufacturers provide mounting kits for their solar collectors. The hardware provided in such kits can usually be configured for either flush mounting or stand-off mounting.

Figure 8-33 shows the hardware supplied for a stand-off mount of a single flat-plate collector. This hardware includes hinged roof brackets that attach to the lower edge of the collector and allow its slope to vary over a wide range. The remaining hardware attaches the upper edge of the collector housing to square aluminum tubing that is cut so that the collector slopes at the desired angle.

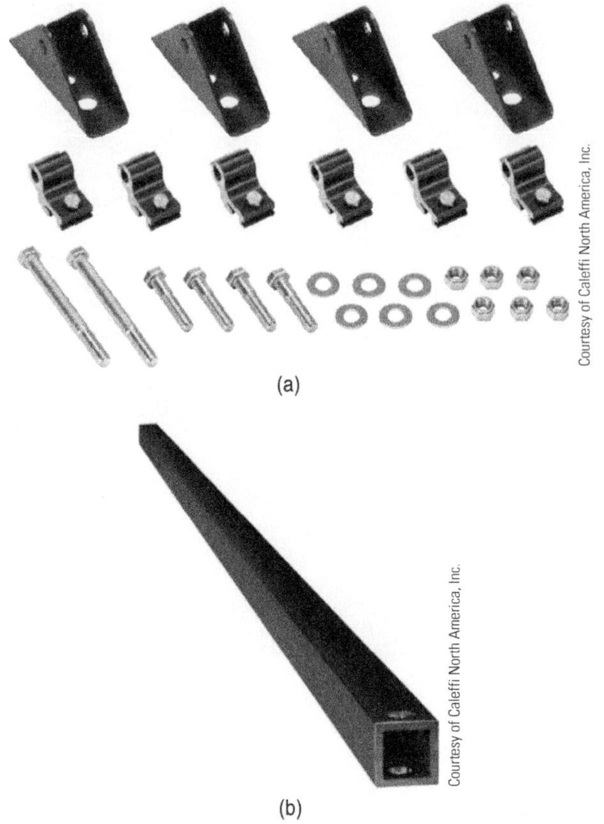

Figure 8-33 (a) Fastening and bracket hardware in typical collector mounting kit. (b) Square aluminum tubing that is cut to length on-site to support the read of the collector.

FASTENING TO ROOFS

Asphalt composite shingles are the most common roofing material used on North American residential and light commercial buildings. The structure supporting these roofs is most often wooden trusses or rafters. This section described options for fastening collectors to such roofs.

One of the most critical aspects of securing collector mounting hardware to roofs is that *all fasteners must penetrate into either the rafters or the top cord of wooden trusses or into solid wood blocking fastened between the rafters or trusses.* It is never acceptable to simply drive fasteners into roof decking such as oriented strand board or plywood. It is also important to pre-drill for all lag-screw fasteners to minimize the potential of splitting the rafter or truss cord.

Roof framing must be carefully located. A common technique is to make measurements from a known framing location, along with one or two very small

Figure 8-34 A collector mounting bracket secured to composite shingle roof. Chalk lines have been used to establish the exact location of the fasteners. Bracket is placed of roof sealing tape.

diameter "**probe holes**" drilled under the flap of an upturned shingle, that verify the exact location of the framing. These small probe holes are then filled with silicone caulk to prevent leakage.

Once the framing has been located, the exact position of the roof brackets can be determined. Ideally, a dimensioned drawing should be prepared that shows the location of all fasteners before any holes are drilled into the roof. When all locations have been determined and marked, the lower mounting brackets are secured to the roof using stainless steel lag-screws driven into the centerline of roof framing. A roof sealing tape, such as **EternaBond®**, should be applied to the underside of each bracket before it is screwed down. This tape is highly elastic and will reshape itself around the threads of the lag-screws and underside of the bracket to provide a watertight seal. Figure 8-34 shows a typical roof bracket mounted on roof sealing tape and secured to the roof framing by a stainless steel lag screw. A stainless steel washer is placed under the head of the lag screw to help distribute loading on the bracket.

After the roof mounting brackets are fastened in place, clips that will eventually form a hinge with the mounting brackets are loosely attached to the lower edge of each collector, as shown in Figure 8-35. Leaving these brackets slightly loose allows them to be moved sideways as the collector in placed and the piping connections are made. When all piping connections are completed, the mounting clips are tightened to the collector.

The upper edges of the collectors are secured in similar fashion. In systems where stand-off roof

(a)

(b)

Figure 8-35 | (a) Adjusting the collector mounting clips to align with the roof brackets. (b) Lower edge of collector secured to roof mounting brackets.

mounted is used, a length of square aluminum tubing needs to be cut so that the collector will eventually be sloped at the desired angle. The simplest way to do this is to temporarily support one collector at the desired angle and then cut the tube as necessary to fit

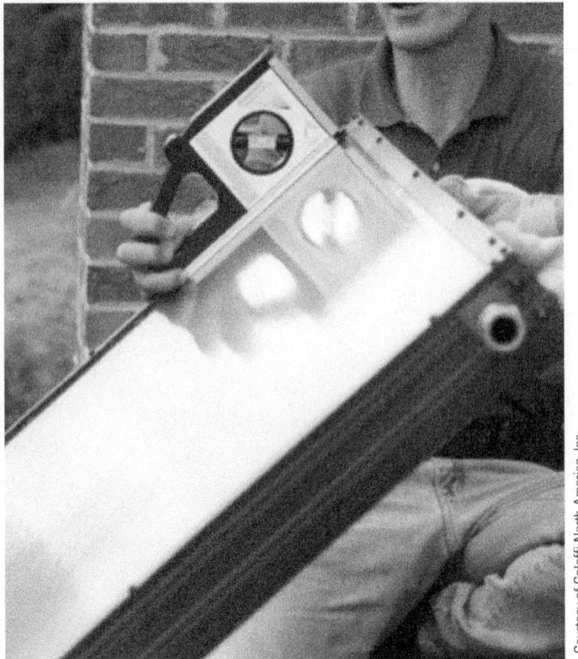

Figure 8-36 | The collector slope being measured using a protractor level.

between the roof bracket and collector mounting clip. A **protractor level**, as seen in Figure 8-36, can be used to verify the slope angle of the collector.

The upper end of square stand-off tube is bolted to two brackets that are fastened to the collector enclosure. The lower end is bolted to a roof mounting bracket. Both details are seen in Figure 8-37.

Figure 8-37 | A rear edge support bracket for a stand-off mount.

Depending on the length of the rear support tube, some manufacturers may require diagonal bracing to stiffen the mounting system against wind gusts.

Some installation codes and roofing warranties require fully flashed supports at all locations where the roof is penetrated. Figures 8-38 and 8-39 show examples of hardware available for this type of mounting. Several manufacturers now offer variations on fully flashed systems for creating **hard points** on shingled roofs. The hard points are typically used to support aluminum rails that in turn support the collectors.

Many of the fully flashed hard point systems are intended to support aluminum structural rails that then support the collectors. These systems allow the height of the rail to be adjusted at each hard point. This lets the installer compensate for any minor sags or other variations in the roof plane so that the mounting rails remain straight and level. Mounting rails also allow collectors to be fully supported as they are brought together at piping connections. They also provide temporary tie-offs for ladders and tool pails.

Figure 8-40 shows **aluminum UniStrut®** rails and associated fasteners that can be used to secure a

Courtesy of Wheat Ridge Solar

(a)

Courtesy of Wheat Ridge Solar

(b)

Figure 8-38 (a) Hardware for fully flashed structural "hard point" being installed on shingled roof. Structural stand-off bracket without flashing is seen at right. (b) Collectors mounted to structural rail supported on hard points.

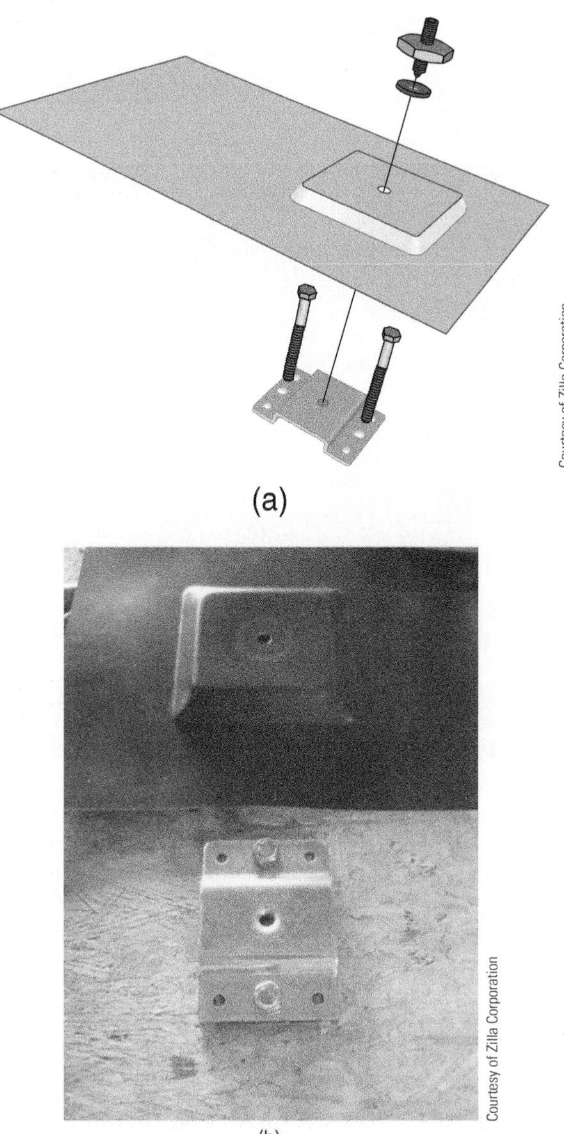

Courtesy of Zilla Corporation

(a)

Courtesy of Zilla Corporation

(b)

Figure 8-39 (a) Hardware assembly for a fully flashed "hard-point" support. (b) Structure cleat fastened into roof framing, with pre-shaped flashing about to be added.

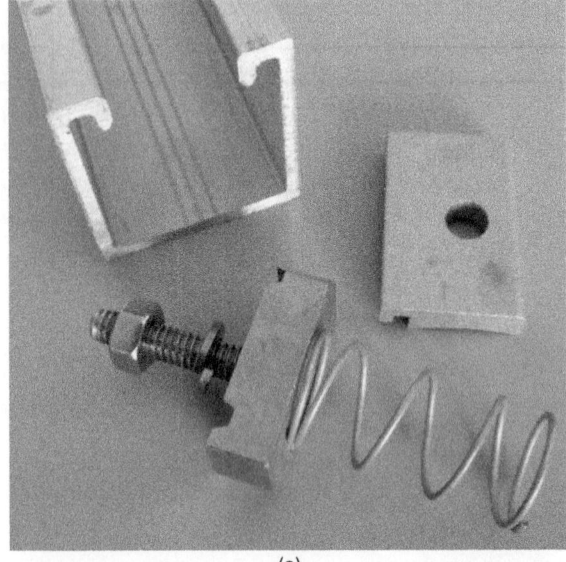

(a)

(b)

Figure 8-40 (a) Aluminum UniStrut®, strut bolt, and friction clip mounting hardware. (b) Collectors held to the rail by friction clip and stainless steel strut bolt. Note that the rail is elevated above shingles to allow rain water to pass underneath.

collector. A stainless steel **strut bolt** slides into the rail and twists in place under the rail's upper flanges. The spring keeps the strut bolt snug against the rail but still allows it to slide to the necessary location. A **friction clip** specifically designed for the collector is clamped down by the strut bolt to hold the collector securely to the rail.

Structural rails should always be elevated at least an 1/2 inch above the roof to prevent trapping water. If the rail is open at the top, several small holes should be drilled at the lower internal edge to allow water to drain out. All butt joints in rails should be made using splines inserts, as shown in Figure 8-41.

Figure 8-42 illustrates the use of an aluminum UniStrut® rail and friction clips in combination with the flashed double-stud hard-point mount shown in Figure 8-41.

Hardware is also available to fasten rails to standing seam metal roofs. A typical detail is shown in Figure 8-43. An **S-5 clip** is clamped to the standing seem. This seam is where a standing seem roof is fastened to the structural roof deck. This clip includes a threaded metal stud, which secures the aluminum rail.

Hardware is also available for fastening collectors to cedar shingle roofs as well as clay tile roofs.

Care should be taken to ensure that all rail mounting systems are installed so that the rail is perfectly flat under the collector array. *Mounting collectors to rails that are curved due to roof sag can create high stress in the collector housings and piping connections.* If necessary, stainless steel nuts and/or shim washers can be installed under the rail, at the mounting hard points, to ensure the rail is straight and flat.

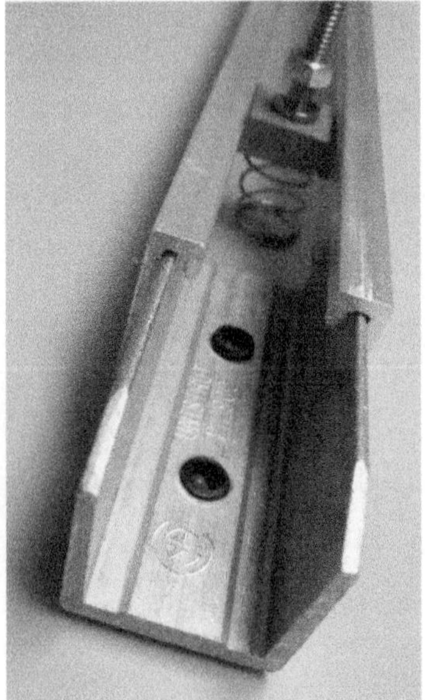

Figure 8-41 Stainless steel spline positioned into aluminum UniStrut® rail at the location of a butt joint. Spline is clamped to the rail with set screws.

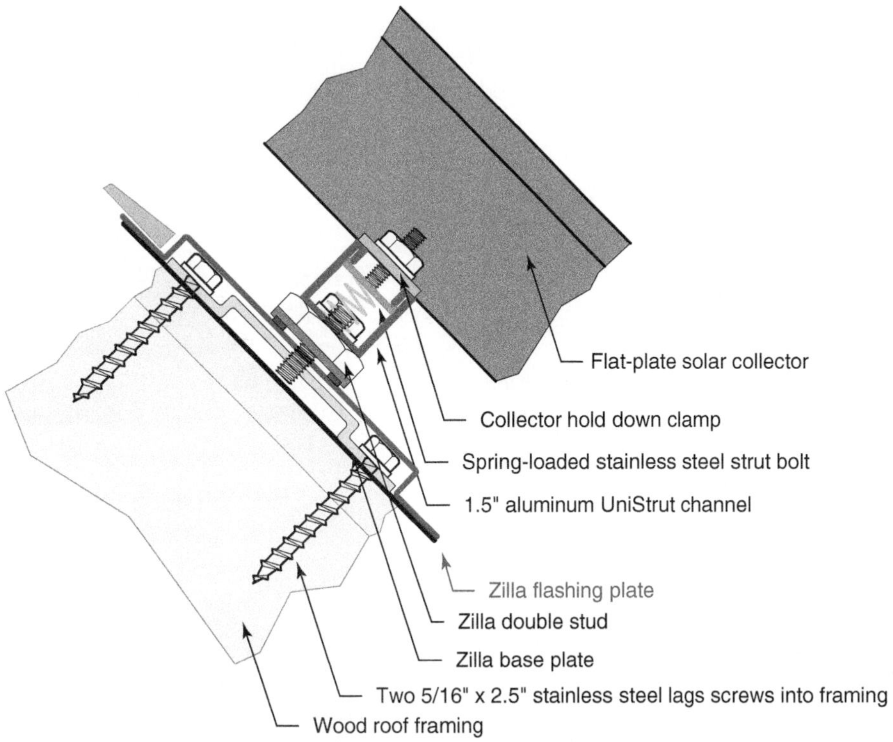

Figure 8-42 | Use of an aluminum rail and friction clamp, as shown in Figure 8-41, in combination with the flashed double-stud hard-point mount shown in Figure 8-39.

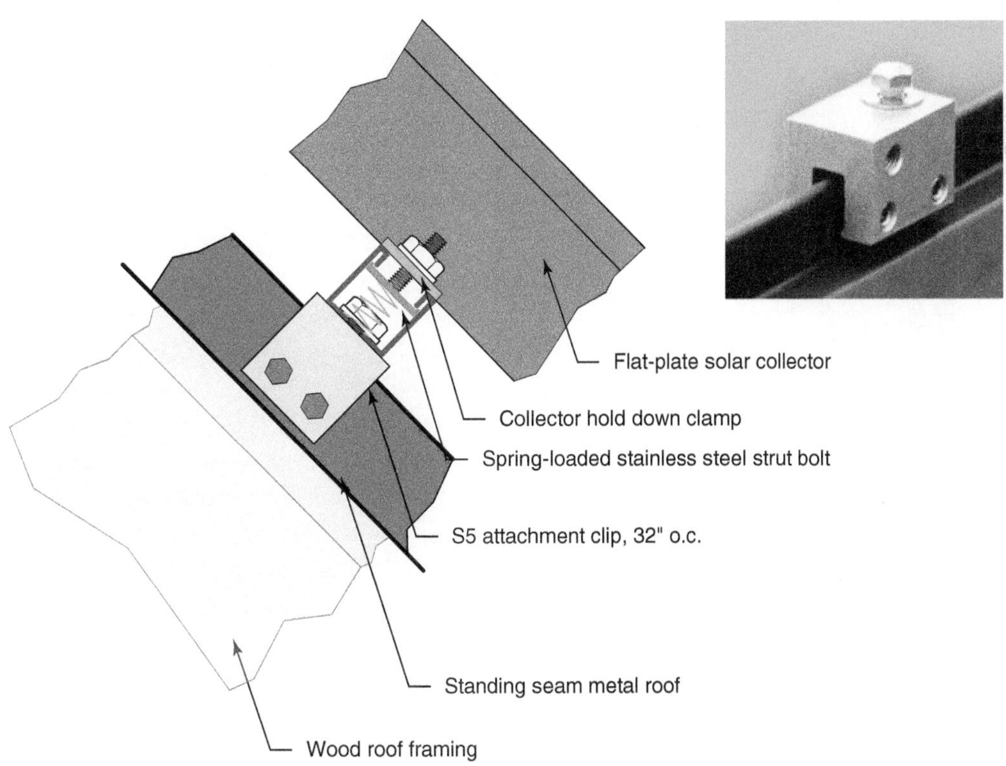

Figure 8-43 | Use of an S-5 clip to secure an aluminum UniStrut® rail to a standing-seam metal roof.

It's also important to make proper piping penetrations through the roof for the supply and return piping to a collector array. These penetrations should be weather tight, able to absorb expansion/contraction movement of the pipe, and thermally broken from the surrounding materials. Figure 8-44 shows an example of one fully flashed piping penetration system for use on shingled roofs.

GROUND MOUNTING

When roof mounting isn't practical due to building orientation, shading, or access, collectors can be mounted to ground frames. Figure 8-45 shows some examples of such mounting.

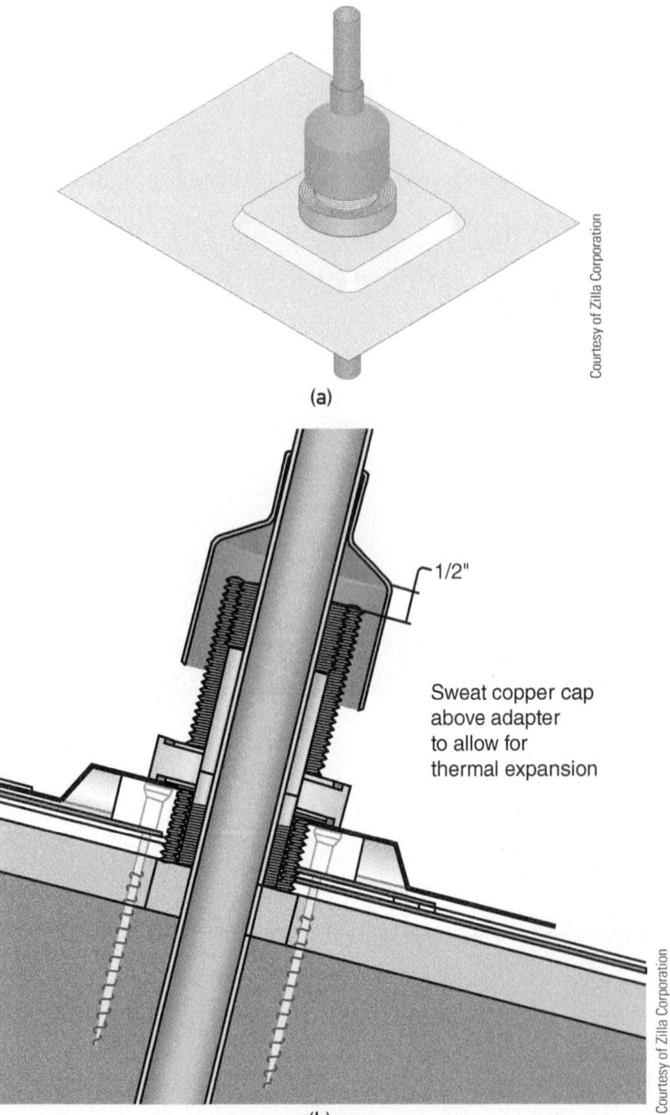

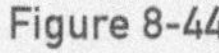

Figure 8-44 (a) Fully flashed pipe penetration system for shingled roofs. (b) Installation of copper cap to allow for piping expansion and contraction.

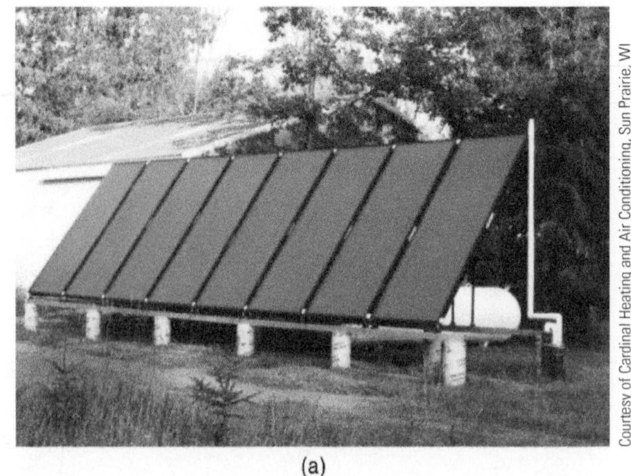

(a)

(b)

(c)

Figure 8-45 (a) Flat-plate collector array supported on galvanized strut rail secure to concrete piers. (b) Evacuated tube collectors supported on pressure-treated wood piers. (c) Specially fabricated collectors allow for "landscape" mounting.

The following design considerations pertain to all ground-mounted collector arrays:

- The frame should be designed and anchored to ensure structural integrity in the highest possible winds at the site. It should also be designed for the anticipated snow loading.
- Only noncorrodible fasteners, such as stainless steel, should be used.
- Metal frames made of galvanized steel or aluminum are generally preferred over treated wood frames for dimensional stability and low maintenance.
- Frames should be designed to keep the bottom of the collectors above the deepest expected snow as well as above areas where stones or other debris might be discharged from mowers.
- The area under the frame should be covered with a suitable weed barrier, and either stone ballast or mulch, to prevent growth of weeds.
- Ground frames should not be placed where snow and ice falling from nearby roofs could damage the collectors.
- If standard collectors will be used in a drainback system, the collector array, and possibly the ground frame, will need to have a minimum sideways pitch of 1/4 inch per foot.

There are currently many ground-mount frame systems on the market. Most are marketed for mounting solar photovoltaic modules. However, many of these frame systems can be slightly modified to support solar thermal collectors.

One metal racking system that can be used to support either solar photovoltaic modules or solar thermal collectors uses **helical augers**, as shown in Figure 8-46.

Helical augers are driven into the ground by rotation, much like a screw is driven into wood. Their size and depth can be varied to achieve a specific load-bearing requirement. The steel shaft of the auger is covered with a plastic sleeve that prevents the steel from bonding to frozen soil, and thus minimizes the potential for frost lift.

Once the helical augers are placed, and the tops are cut level, the remainder of the frame is attached to them using specially shaped aluminum joints and stainless steel fasteners. The upright pipes are also stabilized against side loading by diagonal bracing. A ground cover, such as landscape fabric covered with mulch or stones, prevents weeds from sprouting under the frame. When completed, this type of ground frame is solid and maintenance free.

OTHER COLLECTOR MOUNTING OPTIONS

Both flat-plate and evacuated tube collectors lend themselves to **awning mounting**, as shown in Figure 8-47. This allows the collector array to do double duty as a device that shades windows from high-angle summer sun. Awning mounting is typically done using the same mounting hardware used for stand-off roof mounting. It is critically important that all fasteners that secure mounting hardware to the building are anchored into structural framing or into solid concrete walls.

Collectors can also be mounted to walls using the same technique and hardware as awning mounting. An example is shown in Figure 8-48.

Collectors that are primarily used for space heating, especially in northern latitudes, are sometimes mounted to vertical walls, as shown in Figure 8-49.

For combisystem applications, vertical mounting typically decreases total annual solar energy yield compared with collectors mounted at more typical angles, such as local latitude +15°. However, in climates with snow cover, vertically mounted collectors will receive increased solar radiation due to reflection. Vertical mounting can also provide adequate solar gain in summer for domestic water heating, without excess overheating, as can occur when a collector array is sloped at latitude +15°. Vertical mounting generally simplifies installation and ensures that very little snow will accumulate on collectors. It also allows collectors to better coordinate with architectural features. Wall-mounted collectors also eliminate the requirement to temporarily remove collectors from roofs when shingles need replacing. Solar analysis software can be used to compare the difference in monthly as well as annual performance of vertically mounted collectors versus mounting at lower slopes.

COLLECTOR MOUNTING FOR DRAINBACK SYSTEMS

Drainback-protected solar thermal systems are designed to allow water to quickly drain from the collector array and exposed piping whenever the collector circulator

(a)

(d)

(b)

(e)

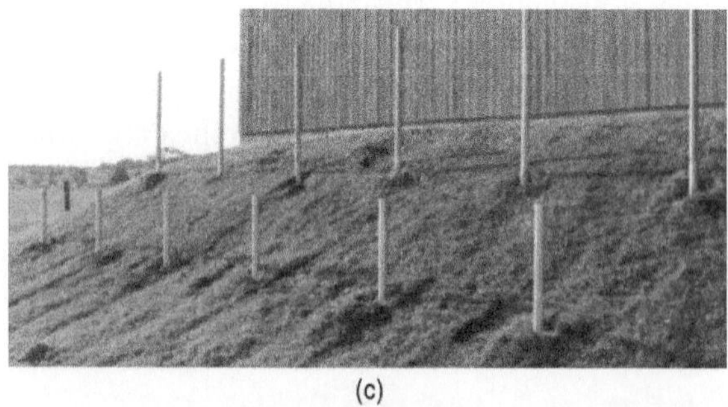
(c)

(f)

Figure 8-46 (a) 2-inch galvanized steel helical augers, with green plastic sleeves to prevent freezing soil from bonding to pipe. (b) Helical augers are twisted into the ground using a hydraulic driver. (c) Tops of 2-inch galvanized pipe are cut to level. (d) Aluminum joint pieces are installed to support horizontal 2-inch galvanized pipe and aluminum bracing tubes. (e) Close-up of aluminum joint pieces and stainless steel fasteners. (f) Completed ground frame with mulch ground cover to prevent weed growth.

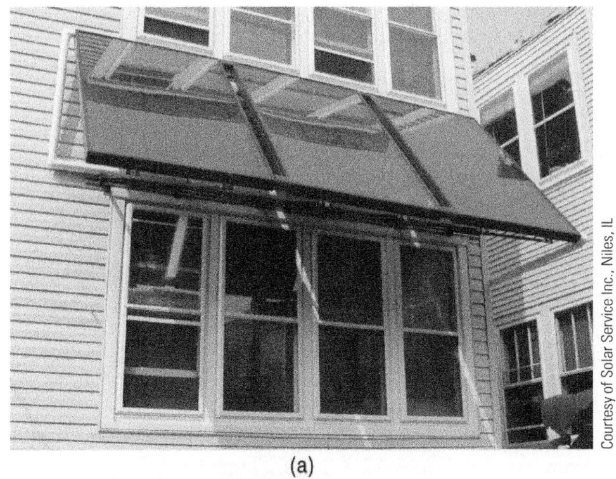
Courtesy of Solar Service Inc., Niles, IL

(a)

Courtesy of ReVision Energy

(b)

Figure 8-47 | (a) Flat-plate collectors installed in awning mount. (b) Evacuated tube collectors installed as awning.

Courtesy of SolarLogic, LLC

(a)

Courtesy of ReVision Energy

(b)

Figure 8-49 | (a) Vertically mounted flat-plate collectors. (b) Vertically mounted evacuated tube collectors.

Courtesy of Caleffi North America, Inc.

Figure 8-48 | Wall-mounted collector array.

is off. *Efficient drainage requires a minimum slope of 1/4-inch vertical drop per foot of horizontal run.* When mounted parallel to horizontal building lines, collectors with standard harp-style absorber plates do not provide sufficient slope of their internal headers to guarantee proper drainage. Thus, when this type of collector is used in a drainback system, it must have a sideways slope of at least 1/4 inch per foot, as shown in Figure 8-50.

Figure 8-51 shows an array of eight collectors, which are part of a system using drainback freeze protection. These collectors have standard harp-style absorber plates and thus are installed with

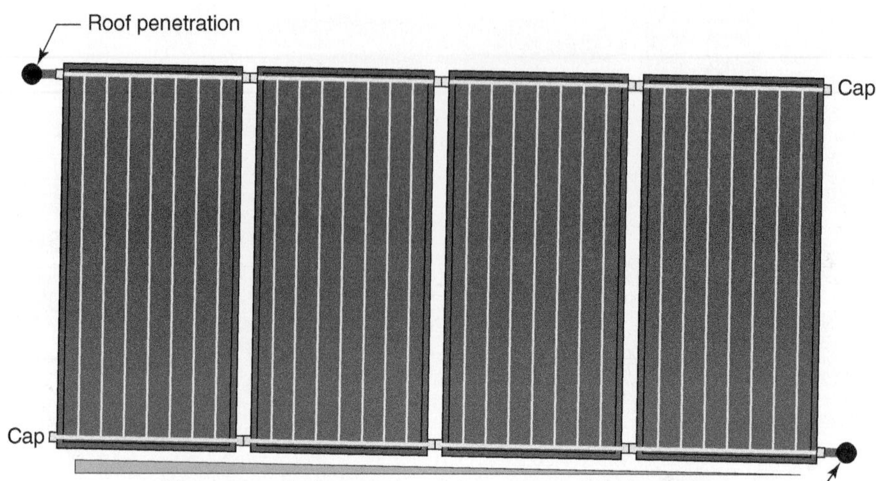

Figure 8-50 | Sideways slope of standard harp-style collectors to ensure proper drainage when used in a drainback system.

Figure 8-51 | Two banks of four collectors each, sloped to low-point roof penetration at center. These collectors are part of a drainback system.

side slopes. In this installation, each group of four collectors slopes toward a low point at the center. The outlet piping of each group leaves from the upper/outer corners. The supply piping for each group is connected by a tee at the lower center point, above a single roof penetration. The return piping for each group is also teed together, but inside the building.

Some specially designed collectors can be used in drainback applications, *without* side slope mounting. Figure 8-52 shows two examples.

The collectors using a "5th port" and sloping internal header require a sloping supply pipe along the bottom of the array, as seen in Figure 8-52a. The collectors with **serpentine absorber** tubing require a sloping supply header, and a sloping return header, as seen in Figure 8-52b.

ELECTRICAL GROUNDING

It is also recommended, and sometimes required by local code, that arrays of metal collectors be **bonded** to an electrical ground. This reduces the chances of

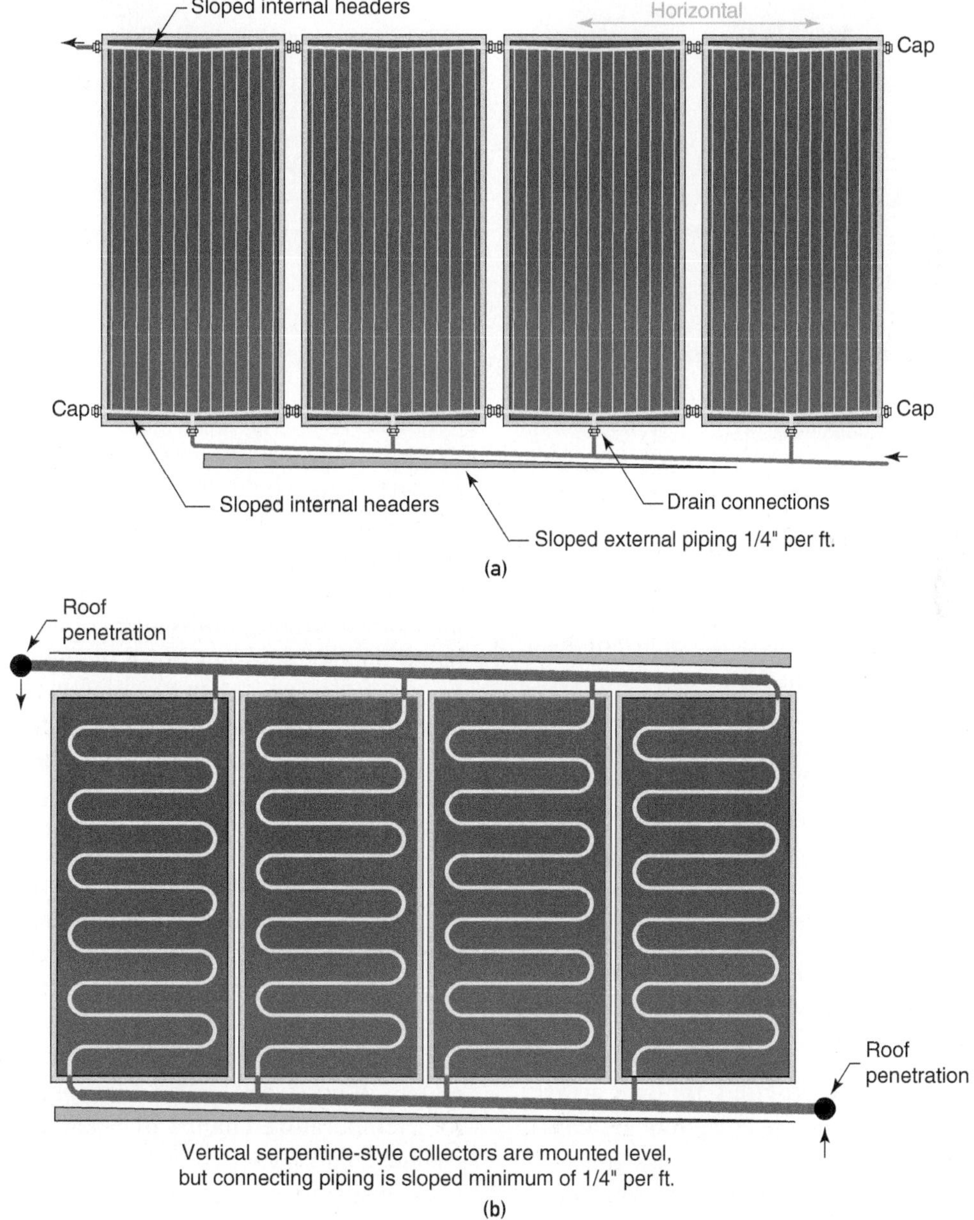

Figure 8-52 | (a) Collectors with bottom center drain connection, sloping internal headers, and sloping external supply piping. (b) Collectors with vertical serpentine absorber plate tubing and sloping external header piping.

damage due to a lightning strike. Grounding is typically done by routing a bare copper wire from a bonding terminal on the metal collector frame to a metal grounding rod that penetrates approximately 8 feet into the earth. Figure 8-53 shows an example of a bonding terminal. In cases where the collectors are not connected to a common metal frame, each collector should be individually bonded to the grounding conductor.

Figure 8-53 | Bonding a collector mounting frame to an electrical ground wire.

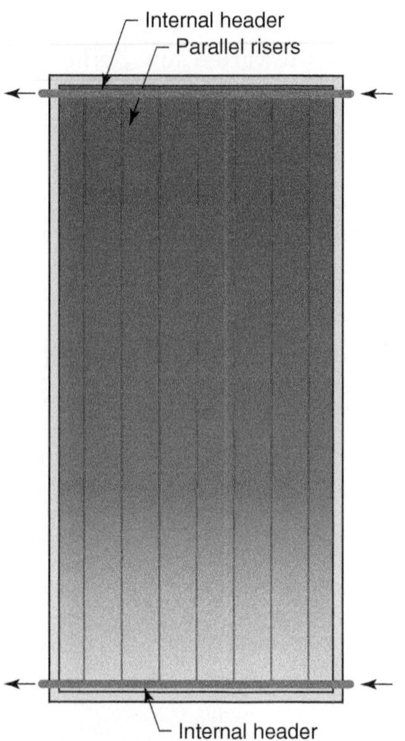

Figure 8-54 | Collector with typical harp-style absorber plate.

NUMBER OF COLLECTORS IN PARALLEL

Most residential and light commercial solar combisystems have no more than eight collectors. However, depending on the size of the collectors, and the available mounting space, there may be systems that require more than eight collectors. This raises the issue of how many collectors should be connected in a side-by-side arrangement. The answer depends on a number of factors, such as the flow design of the collectors and how they are connected to header piping.

The most common style of collector uses a harp-style absorber plate with four piping connections, as shown in Figure 8-54.

This style of collector provides its own array header, which is formed when the collectors are connected side by side. Most collectors of this design have nominal 1-inch copper piping connections. Depending on their size, each collector could have a flow rate up to approximately 1.5 gpm. At this flow rate, and with eight collectors arranged side by side, the flow rate entering the first collector of the array would be 12 gpm. This would also be the flow rate leaving the last collector in the array. The flow velocity corresponding to this flow rate in a 1-inch copper tube is about 4.4 feet per second. This is just above the recommended **maximum flow velocity** of 4.0 feet per second, based on avoiding flow noise in copper tubing. However, this flow rate is still low enough (e.g., $\leq$ 5 ft./sec.) to avoid **erosion corrosion** of the tubing. Thus, based on flow velocity, no more than eight harp-style collectors should be connected in a side-by-side parallel arrangement.

Other criteria further supports a limit of eight side-by-side collectors. One is limiting the **thermal expansion/contraction** movement of the copper headers across the array. In northern climates, the temperature of the header piping will vary from well below zero (perhaps -40 °F in some locations) to as high as stagnation allows, often $\geq$ 350 °F. A 4-foot-long header, exposed to a temperature change of -25 °F to 350 °F, will change length by 0.014 inches. Thus, eight collectors, side-by-side, would see a total change in the combined header length of about 0.113 inches. While most collectors can withstand this, it does impose mechanical stresses on connections and on seals where piping exists the collector housing.

Another reason to limit the number of side-by-side collectors with harp-style absorber plates is flow variations. Although side-by-side collectors are usually piped in reverse return, as shown in Figure 8-55, this does not ensure equal flow through each collectors. The reason is that the headers are not tapered in the direction of flow and thus are unable to create exactly equal head loss per unit length. The latter is required, along with reverse return piping, to provide equal flow through identical collectors.

The constant diameter of the overall array headers (e.g., the top and bottom headers formed by connecting harp-style collectors side-by-side) creates slightly higher flows in the outer collectors compared to the inner collectors. This is acceptable when arrays of collectors are limited to eight. However, longer arrays of collectors, especially collectors with low head loss in their risers compared to their headers, will worsen this undesirable effect.

This suggests that large collector arrays should be constructed of multiple subarrays connected in parallel reverse return. Each subarray should have no more than eight collectors.

COLLECTOR MOUNTING ANGLES

In the Northern Hemisphere, the ideal collector *azimuth angle* is 180° (e.g., collectors facing true south). This orientation maximizes total clear day solar radiation striking the array. However, existing building surfaces may not provide this orientation. There is also a possibility that surrounding objects, such as trees or buildings, will shade an array that faces directly south to a greater degree than if the array was rotated slightly east or west. Fortunately, the annual total solar energy captured by an unshaded collector array is not highly sensitive to the array's azimuth angle. Variations of 30° east or west of true south typically reduce annual solar energy collected by about 2.5 percent.

The ideal *slope angle* of a solar collector depends on latitude as well as the intended application.

For solar domestic water heating, the ideal collector slope angle is equal to local latitude. However, variations of +/−10° on this angle will have minimal effect on the annual total solar energy collected. Thus, it often makes sense to mount collectors parallel to existing roofs where the slope of the roof is within +/−10° of local latitude, and forego the need for bracketing that would only make minor adjustments to the collector slope. In climates with snow, a minimum tilt angle of 40° is suggested to encourage rapid shedding of snow when the sun reappears.

For solar combisystems that provide space heating and domestic water heating, somewhat steeper angles are preferred because of improved performance in winter. Slope angles equal to local latitude plus 10° to 20° are appropriate for such systems. These relatively steep slopes also reduce excess heat production by the collectors is summer. Even with this summer performance penalty, the larger collector arrays used in combisystems can usually provide a very high percentage of the domestic water-heating energy needed during warmer weather.

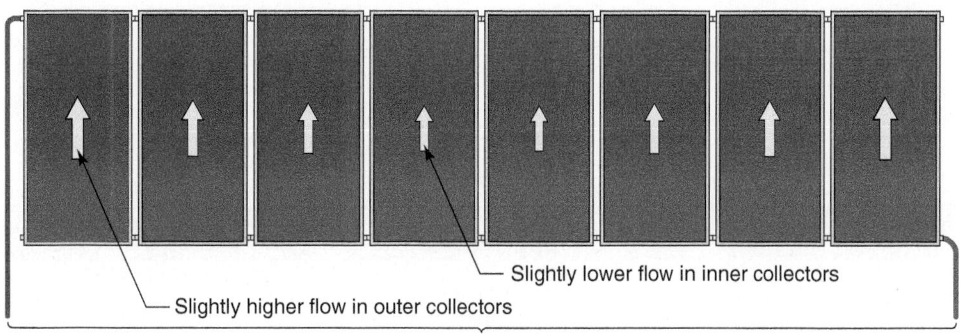

Figure 8-55 | Even though collectors are piped in reverse return, the outer collectors will receive slightly higher flow rates than the inner collectors.

OTHER FACTORS REGARDING COLLECTOR MOUNTING

It is the author's opinion that mounting collectors at readily visible compound angles, such as shown in Figure 8-56, should be avoided. Although the collectors can function as such angles, the aesthetics of the building are clearly compromised.

In climates where high-moisture snowfall is persistent throughout winter, and where collectors will be mounted at slopes of less than 60°, the author recommends using flat-plate collectors rather than evacuated tube collectors. As Figure 8-57 shows, wet snow will cling to evacuated tube collectors, and block sunlight. Furthermore, the low heat loss of evacuated tube collectors makes it difficult for them to melt any accumulated snow. It is often recommended that evacuated tube collectors not be flush mounted (e.g., mounted parallel to and just above the roof surface) in climates with significant snow accumulation.

It is also important that any roof surfaces below the bottom edge of the collector array do not inhibit snow sliding downward. Mounting the lower edge of the array within 2 or 3 feet of a lower slope roof, or a flat surface such as the ground, will interfere with snow sliding from the collector, as evidenced in Figure 8-58.

Courtesy of Huber House, Sanbornton, NH

Figure 8-57 | Heavy snow on evacuated tube collector mounted with less than 60° slope.

Figure 8-56 | Collectors mounted at compound angles.

Figure 8-58 | Snow sliding from collectors is limited by lower pitch roof under lower edge of array.

SUMMARY

This chapter has presented information on the design and performance of solar thermal collectors. Readers are encouraged to visit the SRCC website http://www.solar-rating.org and download further information about collector testing as well as ratings for collectors of interest. This information will be needed to assess overall system performance in later chapters. Readers can also consult the first reference under additional reading for more detailed derivations of thermal performance equations.

This chapter has also discussed several practical aspects of collector mounting. Designers and installers should also consult model codes such as **IAPMO/ANSI S1001.1-2013** as well as local construction codes for any specific structural or weatherproofing requirement associated with mounting solar collectors. Such mounting must also consider aesthetics, thermal performance, snow shedding, and long-term durability. Never compromise on proper mounting.

ADDITIONAL RESOURCES

The following references are recommended for additional reading on solar combisystems system design and installation:

1. Duffie, John. A., and William A. Beckman. (2013). *Solar Engineering of Thermal Processes*, 4th ed. John Wiley & Sons. ISBN 978-0-470-87366-3.
2. Peuser, Felix A., Karl-Heinz Remmers, and Martin Schnauss. *Solar Thermal Systems, Successful Planning and Construction* (2002). James & James (Science Publishers) Ltd. ISBN 1-902916-39-5.

KEY TERMS

absorber area
absorber plate
absorber strip
absorptivity
aluminum UniStrut®
angle of incidence
anodized finish
aperture area
ASHRAE standard 93-2010
awning mounting
barium getter
best straight line
black body
bonded [electrically]
borosilicate glass
breather plug
collector area
collector array
concentric glass tube
condenser bulb
double glass envelope
dry stagnation
emissivity
erosion corrosion
EternaBond®
evacuated tube collector
flat-plate collector
flow-through evacuated tube
flush mounting
friction clip
glass-to-metal seal
glazing
gross collector area
ground mount
hard point
harp style absorber plate
header well
heat balance
heat dumping
heat pipe
helical auger
incident angle modifier
inlet fluid parameter
instantaneous thermal efficiency
IAPMO/ANSI S1001.1-2013
ISO 9806
linear regression
low-iron glazing
low iron-oxide content
maximum flow velocity
negative efficiency
net collector area
normal transmissivity
OG-100 Operating Guideline for Certifying Solar Collectors
parabolic reflector
peak solar collection window
probe hole
protractor level
pyranometer
S-5 clip
second-order equation
selective surfaces
sensor mounting well
serpentine absorber
single glass envelope
slope [of efficiency line]
Solar Rating & Certification Corporation (SRCC)
solar simulator
solar thermal collector
SRCC collector rating reports
stagnation [collector]
stand-off mounting
standardized test day
strut bolt
Sydney tube
thermal expansion/contraction
thermal grease
TiNOX®
transmissivity
UV-stabilized polymer
working fluid [in a heat pipe]
Y intercept [of efficiency line]

QUESTIONS AND EXERCISES

1. Why should the emissivity of the coating on the upper surface of an absorber plate be as low as possible?

2. What is the significance of the bond between the absorber sheet and the tubes that carry fluid along the absorber plate?

3. Which location has a higher temperature when a collector is capturing solar energy?
   a. The edge of the absorber strip.
   b. Where the absorber strip contacts the tubing.

Justify your answer.

4. Why is low-iron glass used for the glazing on flat-plate solar collectors?

5. Give a detailed description of the function of the following collector details:
   a. a breather plug
   b. sensor well

6. What type of heat loss is reduced by eliminating the air inside an evacuated tube collector?

7. Describe both an advantage and disadvantage of a single glass envelope evacuated tube versus a double glass envelope evacuated tube.

8. What are two functions of the barium getter within an evacuated tube?

9. Arrange the following in order from smallest to largest, based on a typical flat-plate collector
   a. aperture area
   b. absorber area
   c. gross area

10. Which type of collector (flat plate or evacuated tube) typically has the largest ratio of absorber plate area to gross area?

11. Why is wet snow more likely to stick to an evacuated tube versus a flat-plate collector? Justify your answer.

12. Why can't collectors designed for pool heating be used to supply a low-temperature radiant-floor heating system?

13. As the angle of incidence of solar radiation increases, the transmissibility of the glazing __________ (increases or decreases) and the absorptivity of the absorber plate coating __________ (increases or decreases).

14. Explain why collectors being tested for thermal efficiency should face directly at the incoming sunlight.

15. Determine the Y intercept and slope of the evacuated tube collector efficiency line shown in Figure 8-22.

16. A collector has outside dimensions of 48 inches width by 96 inches length. It is being tested for thermal efficiency. The following data is recorded during the test.

Assume the following fluid properties are valid during the test:

- Specific heat of fluid = 1.00 Btu/lb./°F
- Density of fluid = 61.5 lb./ft.^{3}

   a. Using this data, determine the inlet fluid parameter and the instantaneous collector efficiency for each data set.
   b. Plot the calculated efficiency points versus the inlet fluid parameter and draw the best straight line through the data points.
   c. Determine the Y intercept and slope of the resulting efficiency line.

17. A collector has the following performance indices: (Y intercept = 0.78, slope = 0.85 Btu/hr./ft.^{2}/°F). It also has a b_0 value (for incident angle modifier) of 0.15. Determine the collector's instantaneous efficiency under the following conditions:
   a. T_i = 100 °F, T_a = 60 °F, I = 240 Btu/hr./ft.^{2}, incident angle = 0°
   b. T_i = 100 °F, T_a = 60 °F, I = 240 Btu/hr./ft.^{2}, incident angle = 30°
   c. T_i = 100 °F, T_a = 60 °F, I = 240 Btu/hr./ft.^{2}, incident angle = 60°

18. Use the data provided in the SRCC report shown in Figure 8-27 to make a graph of the incident angle modifier for incident angles between 0° and 70°.

19. Estimate the total heat output of an array of five collectors, having the SRCC rating report shown in Figure 8-27, for a combisystem in a cold climate location, and on a high solar radiation day.

20. Determine the collector stagnation temperature for the collector represented by Figure 8-27 under the following conditions: ambient air temperature = 80 °F, solar radiation intensity = 0.8 kWh/m^2.

21. Describe two advantages of mounting evacuated tube collectors against a vertical south facing, unshaded wall. Assume the collectors are part of a combisystem for space and domestic water heating.

22. What is an advantage of using a hard point collector mounting system on a roof with minor sags?

23. Describe three possible problems that could occur if 15 flat-plate collectors with standard harp-style absorber plates were mounted side-by-side.

Data set	T_{inlet} (°F)	T_{outlet} (°F)	$T_{ambient}$ (°F)	Solar radiation (I) (Btu/hr./°F)	Flow (gpm)
1	90.0	94.8	50	160	1.0
2	120.0	123.5	55	130	1.0
3	150.0	152.8	60	120	1.0

Figure 8-59 | Table for Question 16.

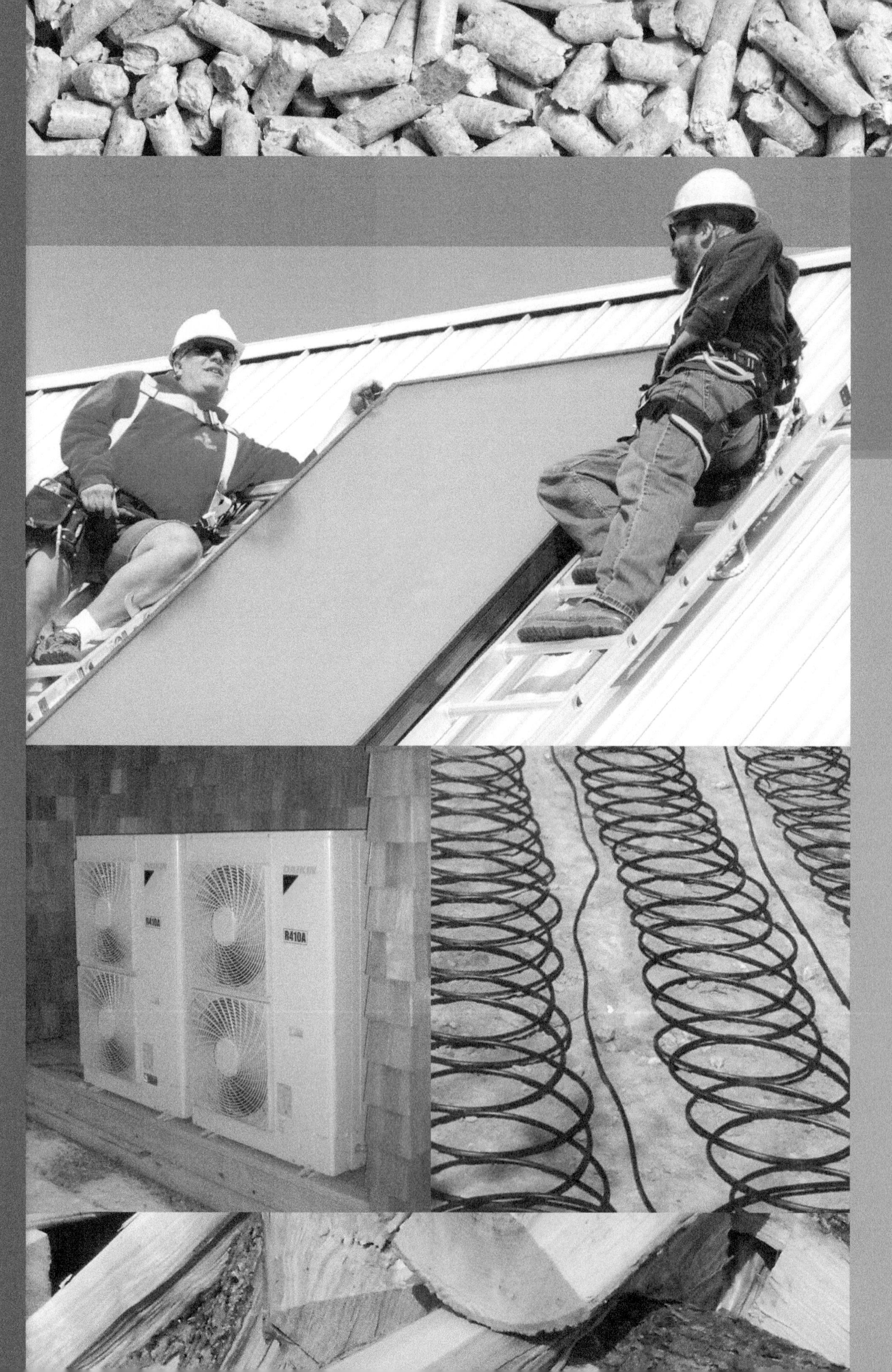
R410A
R410A

CHAPTER 9

Thermal Storage Tanks

OBJECTIVES

After studying this chapter, you should be able to:

- Compare strengths and limitations of pressurized versus unpressurized thermal storage tanks.
- Describe specific piping details for unpressurized tanks.
- Understand the importance of temperature stratification within thermal storage tanks.
- Calculate volumes of tanks based on dimensions.
- Describe several types of commercially available unpressurized tanks.
- Describe several types of commercially available pressurized tanks.
- Explain the differences between 4-pipe and 2-pipe storage tank configurations.
- Calculate the rate of heat loss from a thermal storage tank.
- Design a subsystem for producing domestic hot water from thermal storage.
- Explain methods for reversing flow through thermal storage tanks.

9.1 Introduction

The heat produced by many renewable energy heat sources is intermittent. This is very different from the quasi-consistent heat output of conventional heat sources such as boilers, which are capable of continuous operation at relatively steady conditions.

The energy demand of loads such as domestic water heating and space heating also tends to be intermittent. Although there may be times of continuous heat demand, such as when a building is recovering from a thermostat setback period, there can also be times when the load on the system lasts less than 1 minute!

Because of the intermittent nature of both the energy supply and the load(s), nearly all solar combisystems require thermal storage. So do most systems using wood gasification boilers or pellet-fired boilers. Other renewable energy heat sources, such as heat pumps, might also be coupled with thermal storage, depending on how they are operated. For example, a geothermal water-to-water heat pump may be operated during night-time hours to take advantage of lower **off-peak electrical rates**.

The systems discussed in this text use water in some type of insulated tank for thermal storage. The high heat capacity of water relative to other materials, combined with its relatively low cost and simple adaptability to hydronic based distribution systems, makes it a compelling choice for storing thermal energy.

Thermal storage is also essential when any type of on/off heat source is combined with a highly zoned hydronic distribution system. *Thermal storage allows the rate of heat generation to be significantly different from the rate of heat delivery to the load(s).* This helps prevent **short cycling** of the heat source. Thermal storage tanks used for this purpose are often called **buffer tanks**. With proper design and sizing, a single thermal storage tank can provide energy storage for a renewable energy heat source as well as buffering for an auxiliary heat source such as a propane-fueled boiler.

A wide variety of thermal storage tanks are currently available. Within this text, they will be categorized as follows:

- Unpressurized tanks
- Pressurized tanks

This chapter discusses both types of tanks. Later chapters show how they can be applied with a wide variety of renewable energy heat sources and hydronic distribution systems.

9.2 Unpressurized Thermal Storage Tanks

An **unpressurized tank** is any tank that has a direct connection to the atmosphere, even through a small tube. This connection prevents the tank from building internal pressure such as from the expansion of water as it is heated within the system. It also prevents a negative pressure from occurring as the water in the tank cools. The only pressure in such a tank is static water pressure against the sides and bottom of the tank based on the depth of water. This pressure can be determined using Equation 3.5.

Because of their connection to the atmosphere, unpressurized tanks are also described as "**open**" from the standpoint of hydronic system design. As such, they are more limited in how they can be applied. Only materials resistant to corrosion from oxygen should be used in any piping circuits connected to open tanks, or other open devices within the system. Materials such as copper, brass, bronze, PEX, PEX-AL-PEX, PERT, and stainless steel are acceptable. However, *components made of ferrous metals such as cast iron or carbon steel should not be used in piping circuits connected to open devices.* Doing so will cause rapid corrosion due to the constant availability of oxygen to "fuel" the corrosion reaction.

Most of the larger unpressurized thermal storage tanks on the market are shipped disassembled. They are designed to be assembled on-site. This allows these tanks to pass through standard doorways, which is a major advantage in existing buildings.

Figure 9-1 shows one example of an unpressurized thermal storage tank that is shipped in panels and assembled on-site. When assembled, this tank can hold several hundred gallons of water.

Figure 9-2 shows a collapsible cylindrical thermal storage tank. This tank is shipped on a narrow palette that can pass through a standard door. It is then expanded to form an insulated cylindrical tank.

Most large unpressurized tanks have a rectangular or cylindrical structural shell that supports a flexible inner waterproof **liner**. These liners are made of **EPDM rubber**, PVC, polypropylene, or other proprietary materials. They are hung within the structural shell and conform to its shape as water is added. All unpressurized thermal storage tanks must have an insulated cover that limits evaporation of water. *Designers should always verify that the maximum temperature limit of the liner is compatible with the intended application.*

The preferred method of connecting piping to unpressurized tanks is through the side of the enclosure, a few inches above the water line. This limits any possibility of leakage at the connections. These connections are typically made using a **bulkhead fitting** with compressible gasket, as shown in Figure 9-2d.

Another type of site-assembled unpressurized tank uses preformed expanded polystyrene "staves" in combination with circular bottom and top panels. The tank is assembled by interlocking the tongue and grooved staves around the circular rigid foam base, as shown in Figure 9-3. Each stave contains a preformed vertical steel "stud."

Once the staves are assembled, a preformed polypropylene liner is lowered into the foam shell. An outer aluminum jacket is then wrapped around the foam shell and fastened to the steel studs using stainless steel screws and aluminum rivets. This creates a very strong hoop, enabling the foam staves to withstand the static

(a)

(b)

Figure 9-1 Examples of a large unpressurized thermal storage tank. (a) Panels for the rectangular tank ready for shipment. (b) Panels assembled into a insulated shell that supports a flexible interior liner.

pressure of the water. It also protects the foam against physical or ultraviolet radiation damage over time.

The perimeter of the liner drapes over the top of the staves. It is gathered and fastened around the perimeter of the tank, as shown in Figure 9-3c. A preformed top is then placed on the tank to insulate and minimize

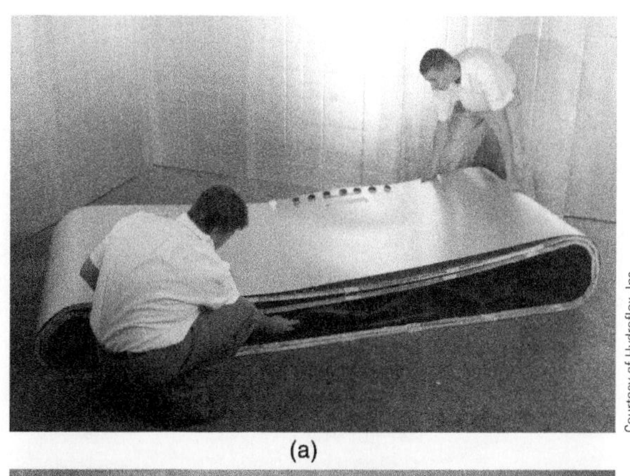

(a)

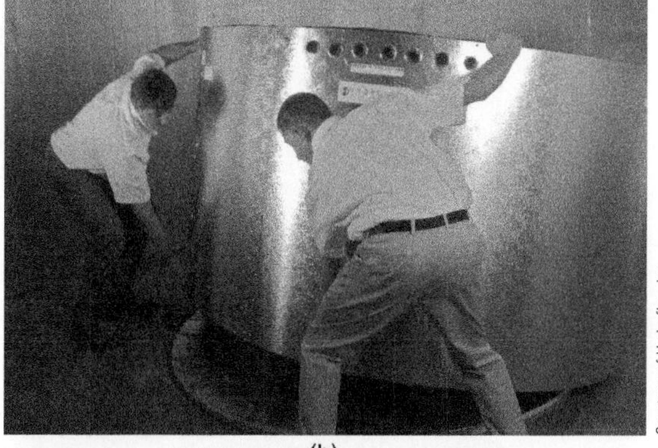

(b)

(c)

(d)

Figure 9-2 Examples of a large cylindrical unpressurized thermal storage tank. (a) Tank shipped in collapsed form for on-site assembly. (b) Tank shell is opened to fit over insulated base. (c) Piping connections into tank are above internal water level. (d) Close-up of piping connection.

Figure 9-3 (a) 6-inch thick preformed expanded polystyrene staves assembled around a circular foam base. (b) Preformed polypropylene liner inserted into foam shell. (c) Aluminum sheet jacket attached to steel studs in foam shell, with 6-inch foam lid. (d) Piping connections pass through side wall of tank above water line. Courtesy of Cocoon Tanks LLC.

any water loss from evaporation. Piping passes through sleeves that penetrate the side of the tank, and above the highest water level, as seen in Figure 9-3d.

This tank system is available in sizes from 165 to 2,014 gallons. It is designed to contain water at sustained temperatures up to 165 °F. The 6-inch-thick wall of the tank provides an R value of approximately 30 °F·hr.·ft.^{2}/Btu.

CONNECTING PIPING TO UNPRESSURIZED TANKS

In some cases, unpressurized tanks can connect directly to a hydronic circuit. In other situations, it is necessary to separate the water in the tank from the water in the piping circuit using a heat exchanger. Figure 9-4 shows some examples.

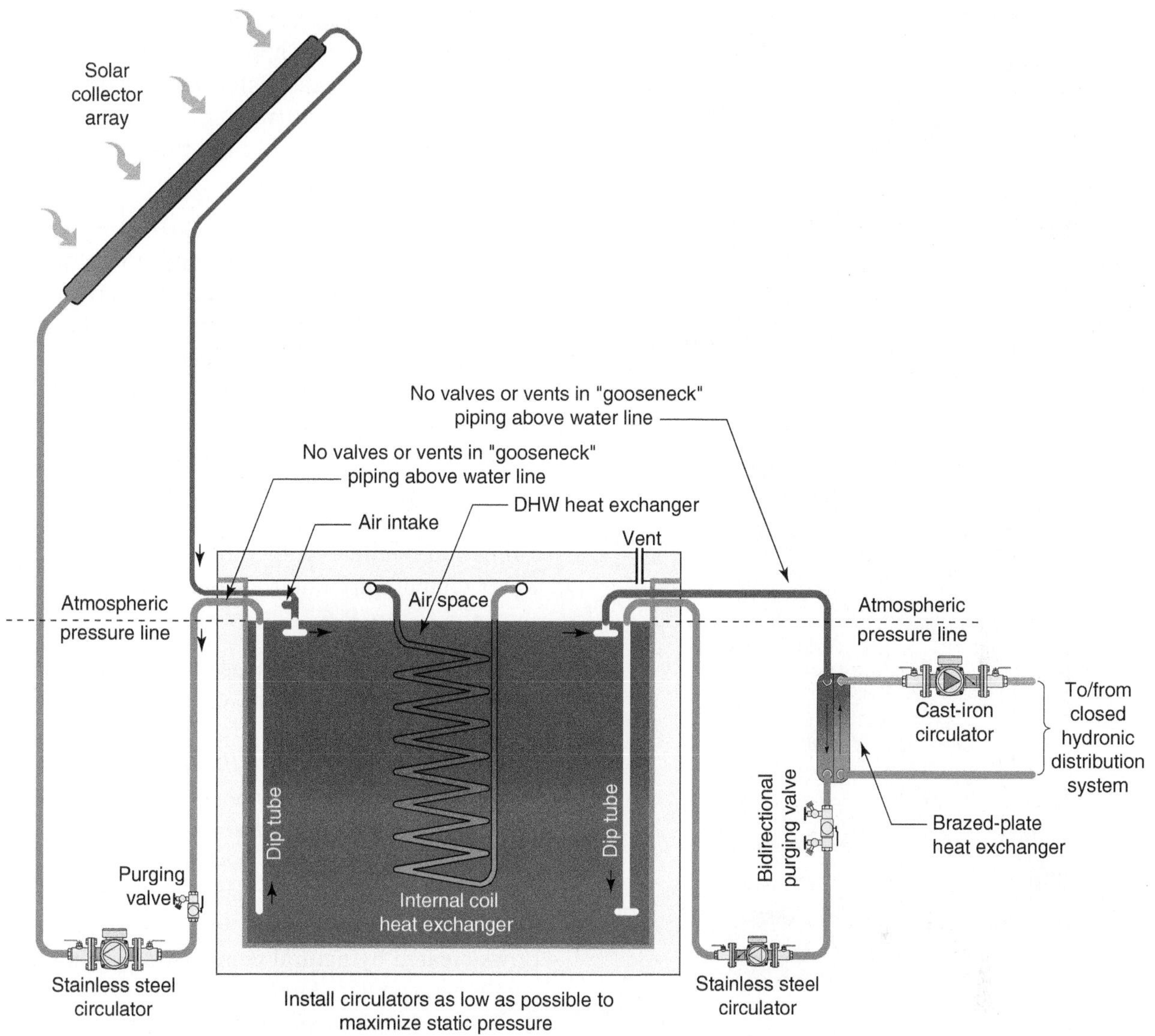

Figure 9-4 The drainback solar collector circuit, shown at the left of the tank, is connected directly to the tank water. The brazed-plate heat exchanger, shown at the right of the tank, provides complete isolation between the water in the "open" tank and a closed hydronic distribution system. The coil at the center of the tank can be used for domestic water heating.

When a circuit is connected directly to an unpressurized tank, all components within that circuit must be suitable for "open loop" operation. No carbon steel or cast-iron components can be used.

The solar collector circuit shown on the left in Figure 9-4 uses a **dip tube** that extracts the coolest water from near the bottom of the tank. A portion of the piping between the dip tube and circulator is slightly above the water level in the tank. This minimizes any chance of leakage where the pipes pass through the side wall of the tank. This is called the **gooseneck piping**. The pressure within this gooseneck will be slightly below atmospheric pressure. This is acceptable provided that no air can leak into the piping. *No valves, vents, or other piping devices with seals should be installed in the gooseneck piping.*

The air initially in the gooseneck piping must be blown downward and out the lower end of the dip tube when the system is commissioned. This is done with forced water flow through a **purging valve** seen near the inlet of the circulator. Once filled, and purged of air, the gooseneck will maintain its water content.

It is also important to install the circulator(s) as far below the tank's water level as possible in any piping circuit that directly connects to an unpressurized tank. This increases the static pressure in the circulator and reduces the possibility of cavitation.

(a)

Courtesy of American SolarTechnics

(b)

Courtesy of Hydroflex, Inc.

Figure 9-5 (a) Copper coil heat exchanger with four parallel circuits to reduce head loss. (b) Multiple copper coil heat exchangers suspended within unpressurized tank.

The other way of connecting piping circuits to unpressurized thermal storage tanks is through a heat exchanger. Both **internal heat exchangers** made of coiled copper tubing or corrugated stainless steel tubing and **external heat exchangers** made of stainless steel can be used. Figure 9-4 shows a coiled internal heat exchanger near the center of the tank. The right side of Figure 9-4 shows piping for an external stainless steel brazed plate heat exchanger. This heat exchanger provides complete isolation between the tank water and a closed loop hydronic distribution system.

Figure 9-5 shows examples of coiled copper internal heat exchangers that are intended to be suspended within unpressurized thermal storage tanks.

The greater the surface area of the coil heat exchanger, the lower its **approach temperature** will be for a given rate of heat transfer. The approach temperature is the difference between the hottest water in the tank and the temperature of the fluid leaving the coil heat exchanger.

Theoretical methods for estimating approach temperature and heat transfer rates through vertical coil heat exchangers are very complex. Heat transfer is effected by many factors, such as flow velocity inside the tube, natural convection coefficients on the outer coil surfaces, diameter of the coil, height of the coil, and overlapping versus staggering of vertically arranged coils. Manufacturers typically perform testing and use the results to generate tables that list the heat transfer rate of a specific coiler as a function of the average tank temperature and the water temperature entering the coil. An example is shown in Figure 9-6.

The flow direction through vertical copper coil heat exchangers should be complementary to temperature stratification within the thermal storage tank. *Coils adding heat to the tank should have hot water entering the top of the coil, and flowing downward. Coils extracting heat from the tank should have cooler water entering at the bottom of the coil, and flowing upward.* Both scenarios are shown in Figure 9-7. These flow directions, in combination with the flows created by natural convection within the tank, create **counterflow heat exchange**. This maximizes the rate of heat transfer for a given set of operating conditions.

Figure 9-8 shows an example of an unpressurized thermal storage tank made of **polypropylene**. It is well insulated and has a removable lid for full access

Heat transfer rate (Btu/hr) at coil flow rate = 10 gpm						
Coil inlet temp. (°F)	Average tank water temperature (°F)					
	80	100	120	140	160	180
130	43,730	23,690	6,340			
150	65,490	43,730	23,690	6,340		
170	88,540	65,490	43,730	23,690	6,340	
190	1,12,650	88,540	65,490	43,730	23,960	6,340
210	1,37,660	1,12,650	88,540	65,490	43,730	23,690

Figure 9-6 Example of heat output ratings of coiled copper heat exchangers.

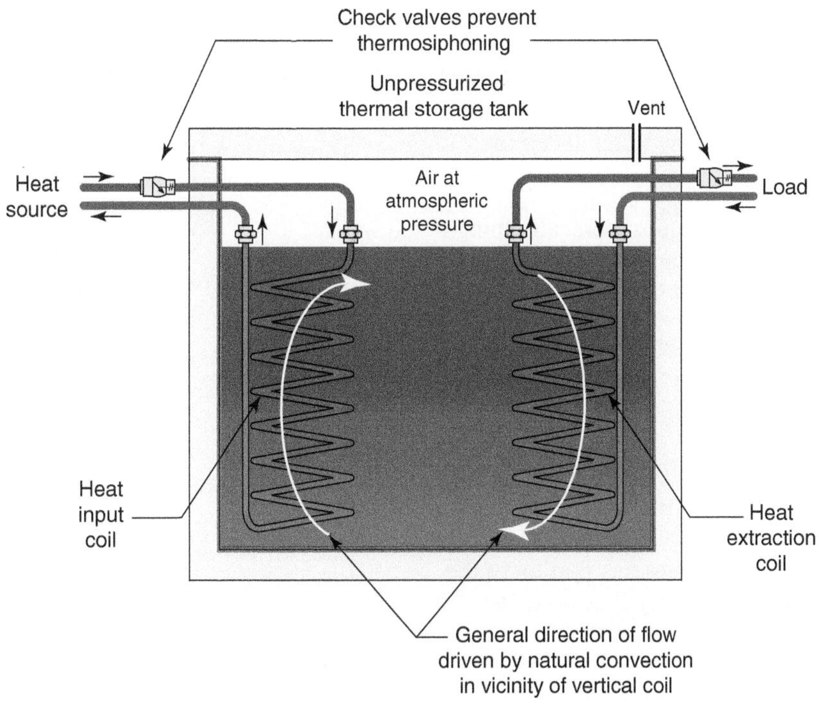

Figure 9-7 | Flow direction for coil heat exchangers adding and removing heat from an unpressurized tank.

Figure 9-8 | Unpressurized thermal storage tank with removal lid and two internal stainless steel coiled heat exchangers.

to either one or two internal stainless steel coil heat exchangers.

One of the internal coils absorbs heat from the tank water to heat domestic water. The other coil can be used to supply heat to a space heating load. It could also be used to add heat to the tank water from a second heat source. In small systems, these tanks can be used individually. In larger systems, two or more of these tanks can be manifolded together. This tank includes a float-type water level indicator seen at the top left of the cover.

Unpressurized thermal storage tanks have strengths and limitations:

Strengths:

- They are usually less expensive than pressurized tanks on a dollar per gallon basis.
- Because they are vented, unpressurized tanks are inherently protected from overpressure conditions.
- Most large unpressurized thermal storage tanks ship disassembled, or partially collapsed, and as such, can fit through standard passage doors.
- Unpressurized tanks are not subject to mechanical codes that require **ASME-certified welding** on **pressure vessels**.
- Provided that a suitable air space is maintained at the top of the tank, there is no need for a separate expansion tank.

Limitations:

- Because they are open to the atmosphere, some water loss occurs due to evaporation. The water level must be periodically checked and new water added to make up for losses.
- Because they are open to the atmosphere, all piping circuits connected directly to the tank must use nonferrous materials.
- Because they are open to the atmosphere, there is the possibility of biological growth in the tank. This can be controlled with **biocide additives**.
- Piping circuits connected directly to the tank and rising above the tank's water level will be below atmospheric pressure when circulators are off. Any component that allows air to enter this portion of the piping will cause water to drain back to the tank.
- Circulators in piping circuits connected directly to the tank water operate under relatively low static pressure and are thus more prone to cavitation.
- Most unpressurized tanks require piping connections to be above the water line.

THERMAL EXPANSION CONSIDERATIONS

Designers should allow sufficient space at the top of unpressurized thermal storage tanks to accept the expanded water volume when the system is operating at maximum temperature. This space should be such that the maximum height the water reaches is at least 1 inch below the bottom of sidewall piping connections.

Equation 9.1 can be used to estimate the increased water volume due to expansion.

(Equation 9.1)

$$V_e = V_s\left(\frac{D_c}{D_h} - 1\right)$$

where:

V_e = expansion volume of water due to temperature increase (gallons or ft^3)

V_s = volume of water at cold fill temperature (gallons or ft.^{3})

D_H = density of water at maximum temperature (lb./ft.^{3})

D_C = density of water at cold fill temperature (lb./ft.^{3})

Example 9.1: An unpressurized cylindrical tank measures 36 inches in diameter. Cold water is added to the tank until the water depth is 60 inches. The remaining portions of the system are estimated to contain 50 gallons of water and are directly connected to the tank. Determine the change in height of the water level in the tank assuming the water is heated from a cold fill temperature of 50 °F to a maximum temperature of 180 °F. The situation is illustrated in Figure 9-9.

Solution: Equation 9.1 can be used to determine the expansion volume in either cubic feet or gallons. The conversion factor between cubic feet and gallons is:

$$1 \text{ ft.}^3 = 7.49 \text{ gallons}$$

The solution requires the density of water at both 50 °F and 180 °F. This information can be found in Figure 3-11:

$$D_C = 62.4 \text{ lb./ft.}^3$$

$$D_H = 60.5 \text{ lb./ft.}^3$$

The volume of a cylinder can be determined using Equation 9.2:

(Equation 9.2)

$$V = \frac{\pi(d^2)h}{924}$$

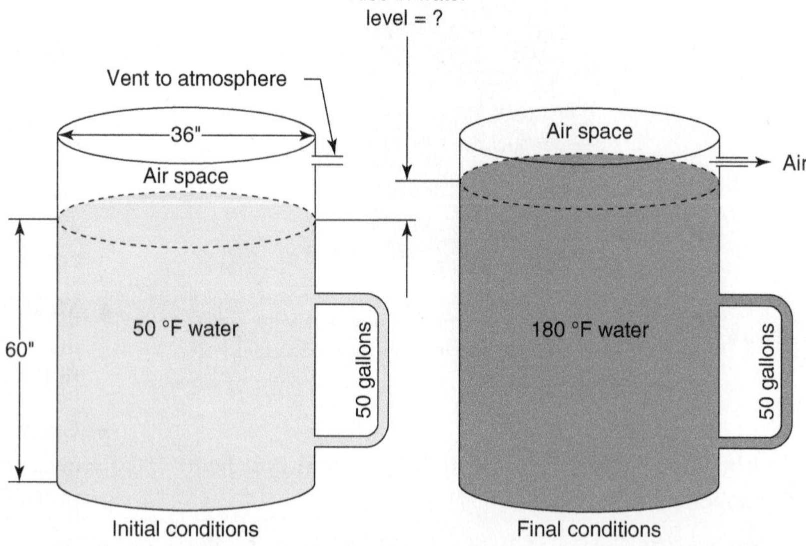

Figure 9-9 | Situation for Example 9.1.

where:

V = volume of a cylinder of diameter (d) and height (h) (in gallons)

d = diameter of cylinder (inches)

h = height of cylinder (inches)

Using Equation 9.2 to calculate the initial water volume in the cylinder yields:

$$V = \frac{\pi(d^2)h}{924} = \frac{\pi(36^2)60}{924} = 264.4 \text{ gallons}$$

Combining this with the volume contained in the remainder of the system yields the total cold water volume.

$$V_s = 264.4 + 50 = 314.4 \text{ gallons}$$

Assuming all of this water reaches a maximum temperature of 180 °F, the increased volume can be calculated using Equation 9.1.

$$V_c = V_s\left(\frac{D_c}{D_h} - 1\right) = 314.4\left(\frac{62.4}{60.5} - 1\right) = 9.87 \text{ gallons}$$

The increase in water level for this volume can be found by rearranging Equation 9.2 to solve for h:

(Equation 9.3)

$$h = \frac{924(V)}{\pi(d^2)} = \frac{924(9.87)}{\pi(36^2)} = 2.24 \text{ inches}$$

Discussion: This calculation is conservative because not all the water in the system will reach 180 °F at the same time. This calculation also does not account for the slight expansion of the materials making up the tank. This expansion would slightly increase the internal volume of the tank. It also assumes that the liner within the tank shell is completely conformed to the internal cylindrical shape of the shell. In reality, there are likely to be some minor folds in the liner that will slightly affect its internal volume. Another simplifying assumption is that there are no internal heat exchanger coils within the tank. If present, such coils would slightly decrease the tank's volume. This reduced volume could be accounted for if the length and diameter of any internal coils were known, and thus the volume of the water the coil displaces could be calculated. Finally, this calculation is useful in locating the height of any sidewall piping connections. The bottom of all such connections should be at least 1 inch above the highest water level in the tank to ensure no leakage occurs.

9.3 Pressurized Thermal Storage Tanks

Pressurized thermal storage tanks allow easier integration with closed hydronic subsystems that operate under slight positive pressure. Most piping circuits, other than those used for domestic water, can be connected directly to a pressurized tank.

Pressurized storage tanks are available in a wide range of sizes, pressure ratings, with or without insulation and equipped with one or more internal coil heat exchangers.

Several manufacturers offer tanks as fully assembled and pre-insulated products, in volumes up to 119 gallons. Figure 9-10 shows an example of one such product.

This tank has several 2-inch size piping connections on its sides, and some smaller connections on its top. This allows flexibility in how the tank is connected into solar thermal systems as well as systems with other types of heat sources. The larger pipe connections on the sides of this tank also minimize head loss, which, in some piping configurations, allows the tank to provide hydraulic separation of multiple circulators. A threaded brass or stainless steel bushing is used when smaller piping needs to be connected to the 2-inch tappings.

Figure 9-11 shows how this tank can be used for a drainback-protected solar thermal system.

Water that flows through the collectors also passes through the storage tank and out into the space heating distribution system. The absence of heat exchangers between these subsystems reduces cost and eliminates the **thermal penalty** associated with having a heat exchanger in the path of heat flow.

PRESSURIZED TANKS WITH INTERNAL HEAT EXCHANGERS

There are many tanks available in North America that have one or more internal heat exchangers. Many are designed for solar water heating systems. As such, the pressure vessel is made of either glass-lined steel or stainless steel, which are both suitable for direct contact with domestic water.

Tanks are available with a single coil heat exchanger near the bottom of the tank, a single coil

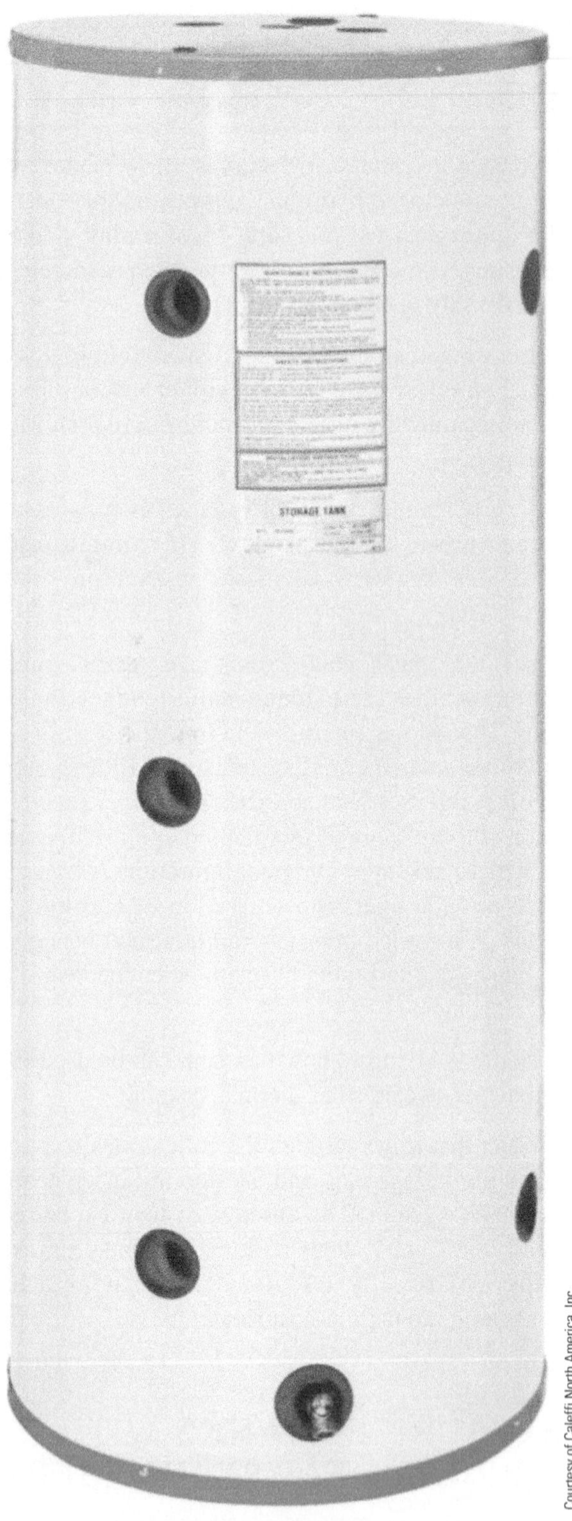

Figure 9-10 Pre-insulated thermal storage tank with multiple piping connections, in sizes up to 119 gallons.

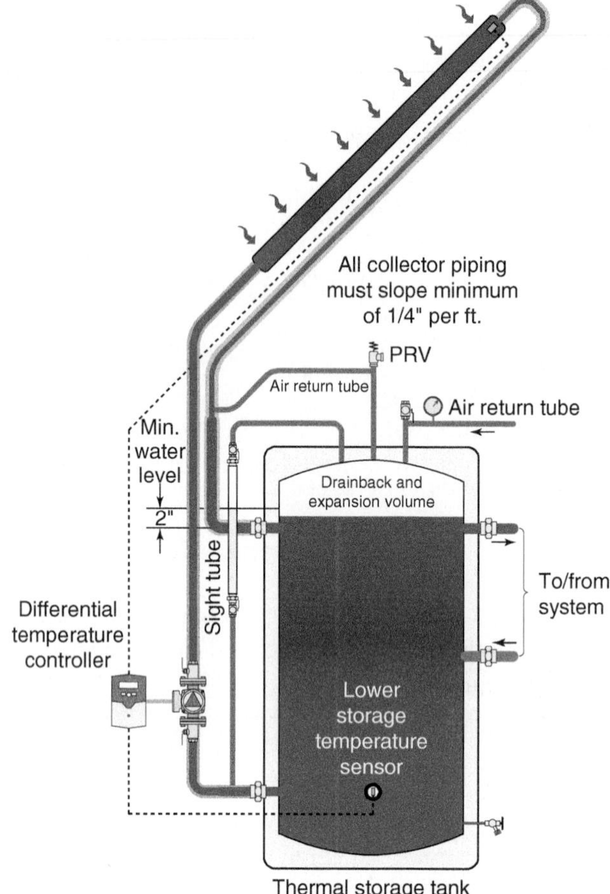

Figure 9-11 Pressurized thermal storage tank of Figure 9-9 configured for a drainback solar thermal combisystem.

heat exchanger near the top of the tank, or dual-coil heat exchangers, one near the bottom and the other near the top. Some of these tanks are adaptable to solar combisystems.

Figure 9-12 shows an example of a tank with a single internal heat exchanger mounted in the lower portion of the tank. This position places the heat exchanger in the coolest water and thus allows for a greater average temperature difference between the water in the tank and the fluid passing through the heat exchanger. Such tanks are commonly used for solar water heating systems, as shown in Figure 9-13. They are often equipped with a single electric resistance heating element in the upper portion of the tank. This element is controlled by a thermostat mounted close to it. The element supplies auxiliary heat, when needed, to ensure adequate domestic hot water delivery temperature.

Figure 9-12 Example of a pressure-rated tank with a single internal heat exchanger mounted in lower portion of the tank.

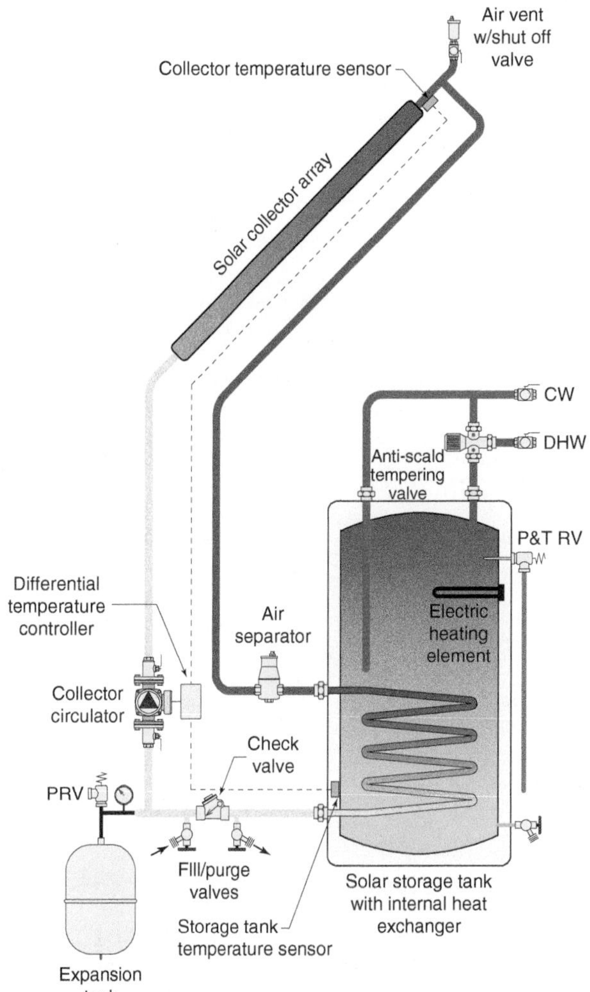

Figure 9-13 Typical solar water heating system using a tank with a single lower heat exchanger and an electric auxiliary heating element in the upper portion of the tank.

Figure 9-14 shows an example of a **dual-coil thermal storage tank**. Such tanks are also common for residential solar water heating systems. The upper coil is used, in combination with an auxiliary heat source, such as a boiler, to add heat to the water near the top of the tank. This allows the lower coil, which connects to the collector array, to operate in the favorable lower water temperatures near the bottom of the tank. A typical application is shown in Figure 9-15.

Thermal storage tanks intended for residential solar domestic water heating systems typically range from 60 gallons to 119 gallons.

Pressured-rated thermal storage tanks are also available in volumes larger than 119 gallons. However, some building codes, or mechanical installation codes, require any "pressure vessel" with a volume of 120 gallons or more to be **ASME certified**. Achieving this certification requires the tank to be manufactured according to Section VIII of the **ASME pressure vessel code**. Such tanks must also be labeled with a placard stating their maximum operating pressure limits. This adds significant cost to the tank. Designers contemplating the use of pressurized storage tanks in sizes over 119 gallons should check to see if the building and/or mechanical codes in effect at the installation site require ASME rating. If so, the designer must follow the code.

TANKS WITH DOUBLE-WALL HEAT EXCHANGERS

Some installation codes require thermal storage tanks with **double wall heat exchangers** in any system that involves an antifreeze solution and domestic water. A double-wall heat exchanger has two metal walls, with a **leakage path** in between. If a leak develops in either

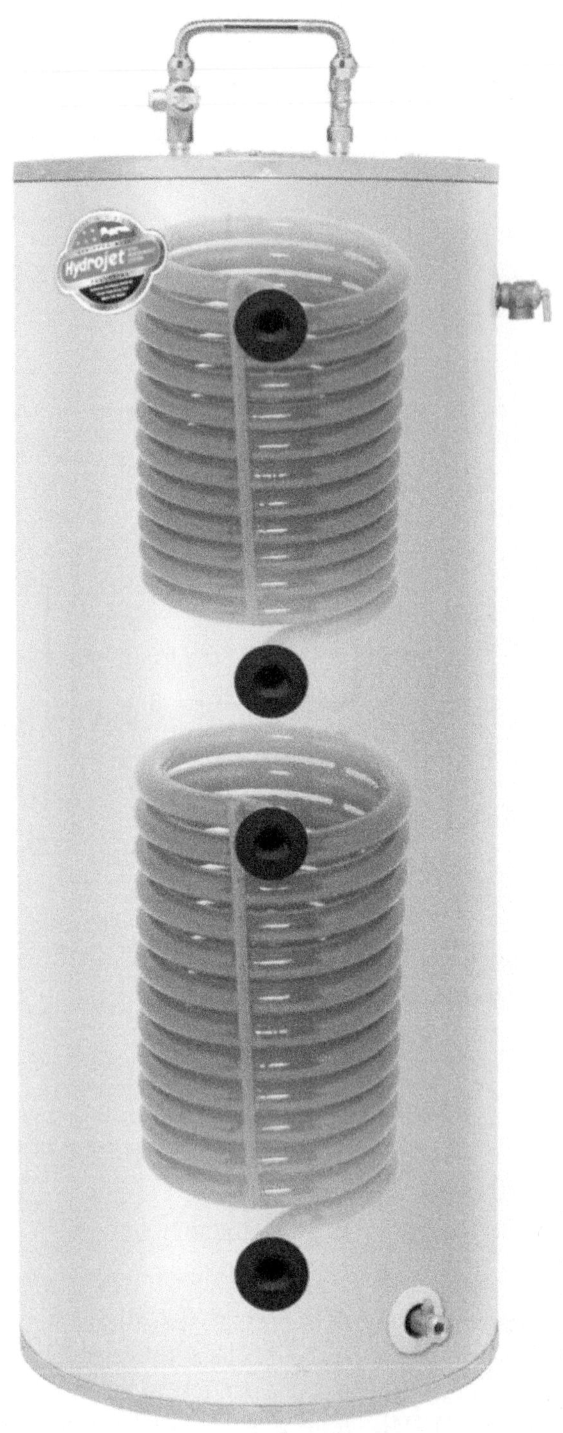

Figure 9-14 Thermal storage tanks with two internal coil heat exchangers. Components seen at top of tank provide thermostatic mixing to protect against scalds.

the inner or outer wall, the leaked fluid will pass along the leakage path until it reaches an opening outside the tank, where it will drip, and provide visual evidence of the leak.

Figure 9-16 shows a cross section of a double-wall copper tube heat exchanger. This tubing would be wound into a helical coil and braised inside the tank, similar to how a single-wall heat exchanger is mounted. The only difference is that the leakage path between the walls would end outside the tank jacket, at both the inlet and outlet of the heat exchanger.

Opinions vary on the need for double-wall heat exchangers, especially in systems that use nontoxic propylene glycol antifreeze. The thermal and hydraulic performance of double-walled heat exchangers is degraded compared to the same size heat exchanger with a single wall. Designers need to verify if local codes require double-wall heat exchangers, and plan accordingly.

HEAD LOSS OF HEAT EXCHANGER COILS

Any internal heat exchanger coil creates head loss as fluid passes through it. This head loss increases with flow rate. It also varies with the density and viscosity of the fluid being circulated through the heat exchanger. Manufacturers typically provide data or a graph showing the head loss of the coiled heat exchangers in their tanks as a function of flow rate. An example of such a graph is shown in Figure 9-17.

The head loss of the heat exchanger must be accounted for when determining the head loss curve for the solar collector circuit. This head loss of the heat exchanger coil would be combined with the head loss of the collector circuit piping, the collector array, and any other devices in the circuit.

MULTIPLE THERMAL STORAGE TANK ARRAYS

One alternative to a single, large pressurized tank is a group of smaller pressurized tanks arranged side by side, as illustrated in Figure 9-18.

One advantage of some **multiple tank arrays** is that each tank can pass through a standard 36-inch-wide doorway. This is important both during installation and if one or more tanks ever have to be replaced. Another advantage is that larger storage volumes can be attained without the need for ASME-rated pressure vessels, provided that individual tanks do not exceed a volume of 119 gallons.

A disadvantage of multiple tank arrays is that they have significantly higher combined surface area than a

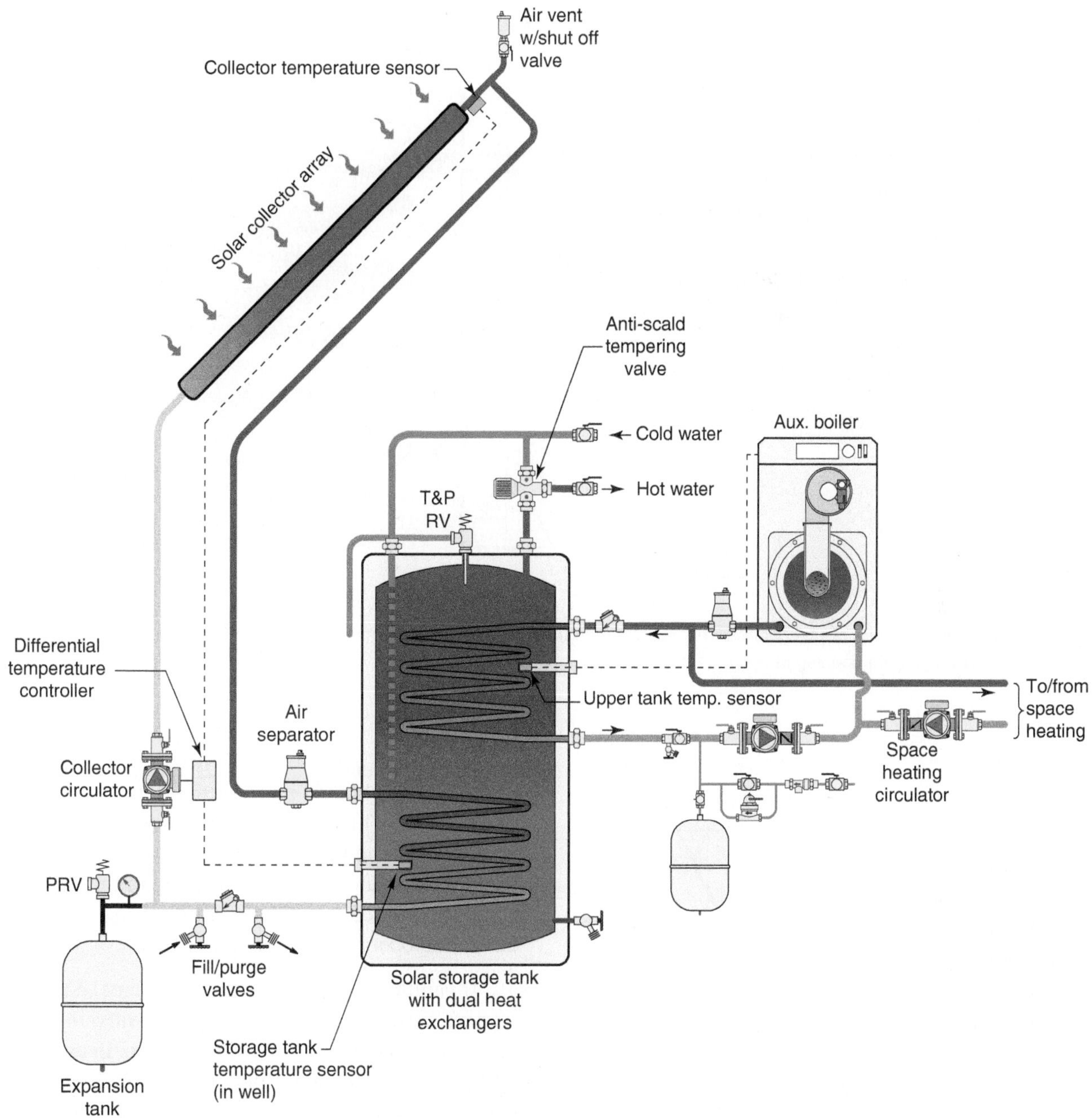

Figure 9-15 | Typical piping for a dual coil tank, where a boiler provides auxiliary heat to the upper portion of the tank.

single larger tank of equal volume. This creates greater standby heat loss, unless additional insulation is added to compensate for the greater surface area.

Another disadvantage is that significantly more piping, fittings, and valves are needed to properly connect multiple tanks.

Multiple tank arrays can be piped in **parallel reverse return**, as illustrated in Figure 9-18. This helps in achieving equal flow through each tank. The piping shown between points A and B is parallel reverse return, as is the piping between points C and D. To reduce total piping heat loss, it's best to use the longer piping assemblies on the piping supplying cooler water to the lower connections on the tank, rather than for the piping carrying higher temperature water to or from the upper portion of the tanks.

In any type of reverse return piping, the pipe sizes should be stepped down in the downstream flow

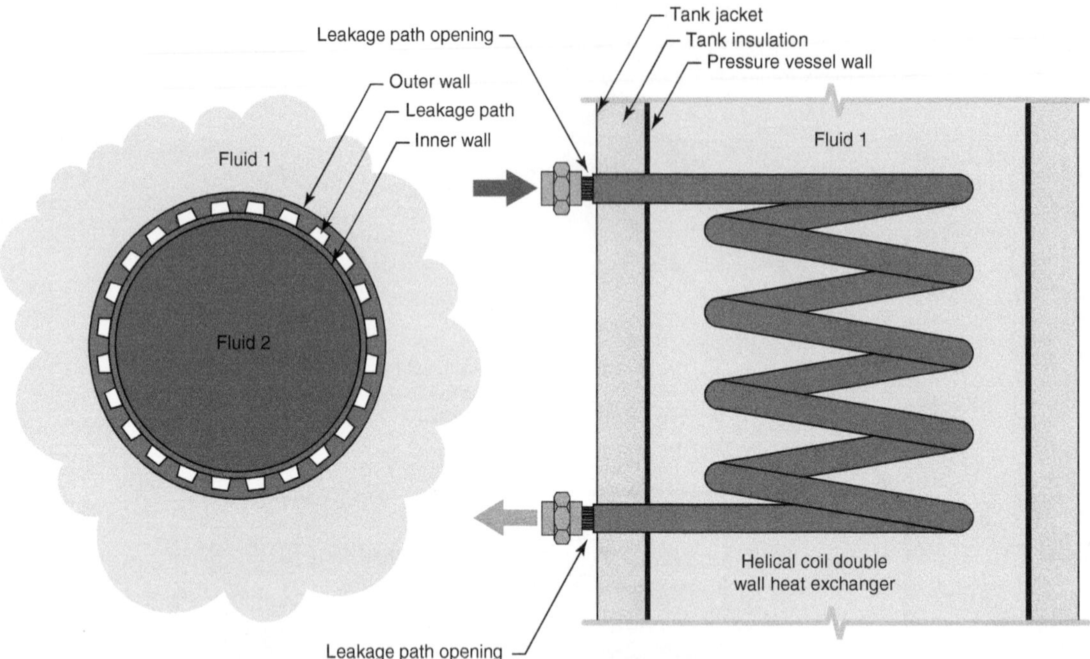

Figure 9-16 | Illustration of a double-wall internal heat exchanger. A leakage of either fluid will result in that fluid dripping from the leakage path opening outside the tank jacket.

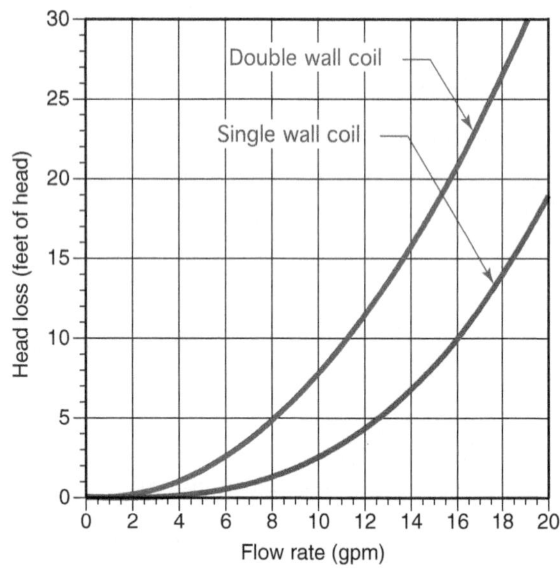

Figure 9-17 | Example of head loss versus flow rate for internal heat exchanger in storage tank. Note the significantly higher head loss associated with the double-wall heat exchanger.

direction to approximate equal head loss per unit length of piping. This detailing, along with symmetrical piping between the headers and tanks, helps produce approximately equal flow rates through each tank.

Note that isolation ball valves and unions are installed in all branch piping leading from the headers to the tanks. This, in combination with proper space planning around the tanks, and pipe placement above or behind the tanks, should allow any of the tanks to be isolated, drained, and removed, if necessary, without disrupting operation of the other tanks.

If parallel reverse return piping cannot be provided, a combination flow meter/balancing valve should be installed in lieu of a standard ball valve, in each parallel flow path. This detailing is shown in Figure 9-19. Each balancing valve would then be adjusted to provide equal flow rates through each tank.

The advantage of the piping arrangements shown in Figures 9-18 and 9-19 is that each tank can be isolated from the array. With proper space planning and pipe placement, it is possible to remove any of the tanks without affecting the other tanks or disrupting system operation.

Although this is an "advantage," the likelihood of having to replace a high quality pressure-rated tank that has been properly applied in a closed loop system is very small. Given this low probably of failure, some designers opt for potentially simpler and less costly ways to install multiple tanks, while foregoing the ability to individually isolate each tank. One such approach is shown in Figure 9-20.

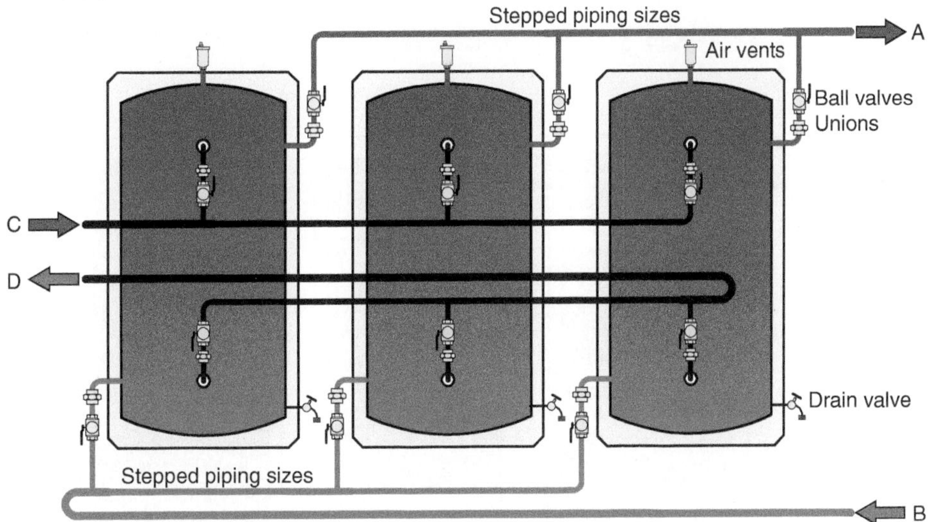

Figure 9-18 | Three thermal storage tanks connected in parallel using reverse return piping. Connections C and D are in parallel reverse return. So are connections A and B.

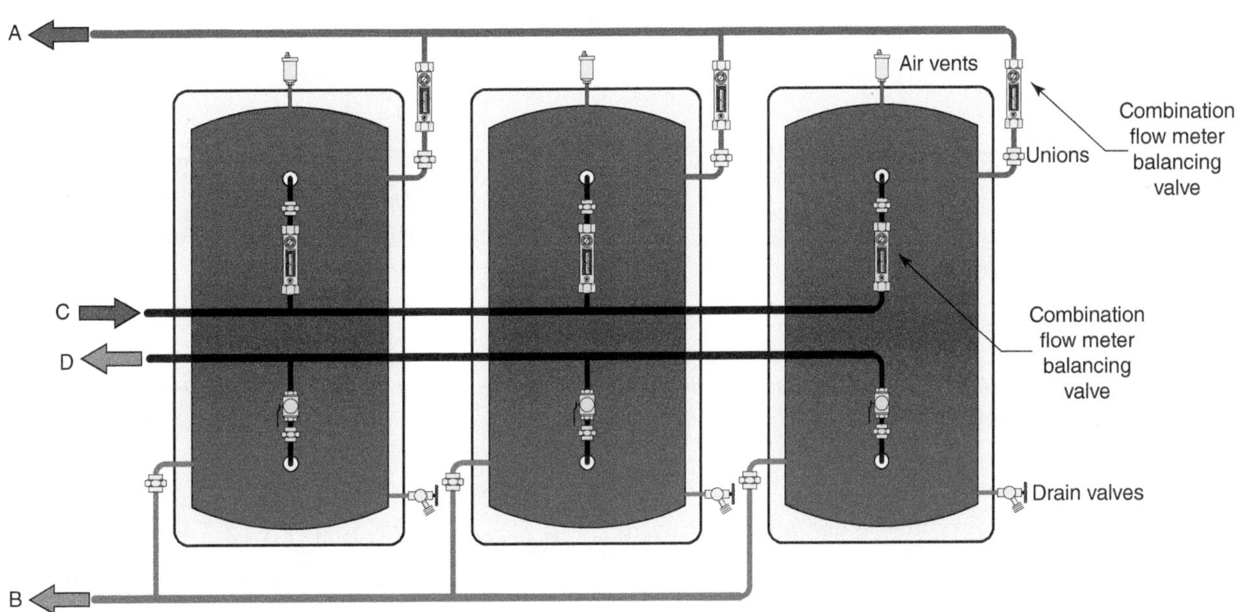

Figure 9-19 | When parallel reverse return piping is not used, a combination flow meter/balancing valve should be installed in each parallel flow path and then adjusted to allow equal flow through each tank.

This hybrid arrangement uses short piping connectors between adjacent tanks. The head loss through these connectors must be kept very low. This helps evenly balance flow between the tanks. It also allows the tank array to provide hydraulic separation between points A and B, which are the common points where the heat source circuit and load circuit come together.

The piping connectors between tanks should also be slightly flexible. This is necessary because of small variations in how the tanks are fabricated. A very slight misalignment from one tank connection to another, in combination with short and relatively large diameter metal piping connectors, would make it impossible to install rigid metal piping between adjacent tanks.

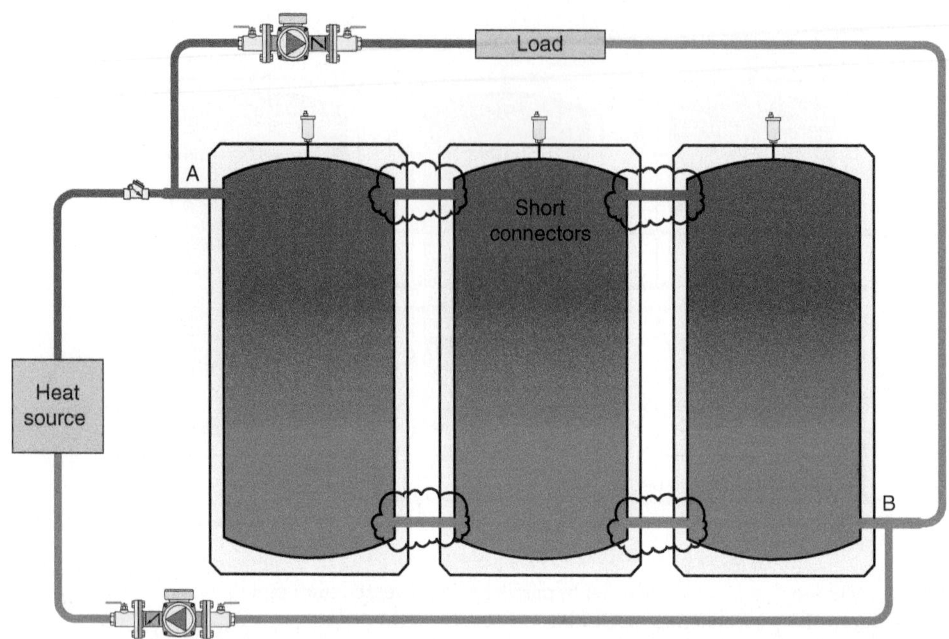

Figure 9-20 | A hybrid method of connecting multiple storage tanks in a "close coupled" arrangement.

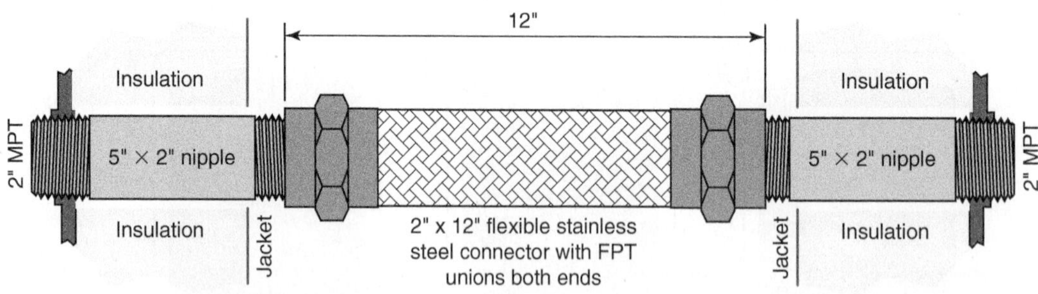

Figure 9-21 | Use of a flexible stainless steel connector to closely couple two identical thermal storage tanks while allowing for very slightly misaligned connections.

Flexible connectors also help compensate for floors that might not be perfectly level under the tanks.

When the tanks have threaded connectors, it is possible to closely couple them with a braid-reinforced corrugated stainless steel connector, as shown in Figure 9-21.

This connection requires a steel pipe nipple threaded into each tank connection. The nipple needs to be long enough to have its threads fully exposed outside the tank's insulation and jacket. A semi-flexible connector with FPT unions on each end connects between these nipples. When the pipe size of the connection is 2-inch, the shortest practical connector length is about 12 inches. If the thermal storage tanks have flange connections it is possible to use short expansion compensators with matching flanges between the tanks.

LARGE PRESSURIZED THERMAL STORAGE TANKS

Some solar combisystems, as well as systems using wood gasification boilers, require several hundred gallons of thermal storage. In these situations, the designer may prefer to use a single pressurized tank, rather than an array of small tanks. Larger pressure-rated tanks are available both as standard products as well as custom welded products. Some are supplied with insulation and an outer jacket. Others are supplied as a bare pressure vessel that must be insulated on-site, usually after all piping connections are made.

Many of the "stock" pressure-rated tanks currently available in North America, in sizes of 200 to 1,000 gallons, are intended for large-scale domestic hot water storage. As such, they are usually **glass lined**. Although such a lining does not preclude their use in systems where the tank contains nonpotable water, it does add cost compared to a standard carbon steel tank, which is adequate for use in closed hydronic systems.

A greater concern is that tanks designed for large-scale domestic water heating often do not have piping connections that are optimally placed or sized for hydronic system applications. If such tanks are used, the system's performance can be reduced due to factors such as disruption of temperature stratification, or less than optimal flow patterns caused by insufficient connections. It is the author's opinion that spending several thousands of dollars on a tank that is not well suited for an application is poor design practice.

Figure 9-22 shows an example of a 210-gallon ASME-certified thermal storage tank that has connections specifically designed for hydronic system applications.

This tank is available without insulation, as shown in Figure 9-22, or fitted with an insulating jacket. It has flat top and bottom plates that are supported by a grid of internal steel stay rods. Flat top and bottom plates are easier to insulate in comparison to semi-elliptical top and bottom domes. Four 2-inch piping connections are provided at upper and lower sidewall locations. Additional connections are provided on the sidewall for mounting temperature sensor wells, thermometers, or other piping. This tank is supported on four legs, which allows for good insulation placement under the tank. Another connection is provided in the top plate to accommodate an air vent.

Figure 9-23a shows an example of a large commercially available thermal storage tank that is designed to enhance temperature stratification when used in solar thermal systems.

Fluid from the solar collector array circulates through a helically shaped copper heat exchanger contained within a small vertical chamber seen on the side of the main tank. This chamber has several connections between it and the main storage tank. Because higher temperature water is less dense, it rises within the small chamber, and is guided by the upper piping from the small chamber to the appropriate level or "**strata**" within the main storage tank. Similarly, water at other temperatures and densities is routed to the appropriate strata by other piping connections between the chamber and tank. The effectiveness of this design is demonstrated by the thermal images seen in Figure 9-23b.

Figure 9-24 shows a common piping arrangement for this type of tank.

The solar thermal subsystem on the left side of the tank circulates an antifreeze solution between the collector array and the coiled copper heat exchanger. This provides solar-derived heat to the tank, when available.

Courtesy of Hydronic Specialty Supply

Figure 9-22 A 210-gallon ASME-certified thermal storage tank with connections sized and placed for hydronic system applications. The smaller sidewall connections for temperature sensors (if needed).

(a)

Courtesy of Lochinvar, LLC

Figure 9-23a A large thermal storage tank with side arm heat exchanger. The piping connections between the heat exchanger chamber and tank promote temperature stratification within the tank.

(b)
3) After 30 mins.
Tank cold, approx. +30 °C
4) After 75 mins.
Stratified charging at 2 levels
5) After 125 mins.
Upper tank at +40 °C
Centre of tank at +30 °C
Lower tank at +15 °C
6) After 170 mins.
Copyright © Lochinvar, LLC

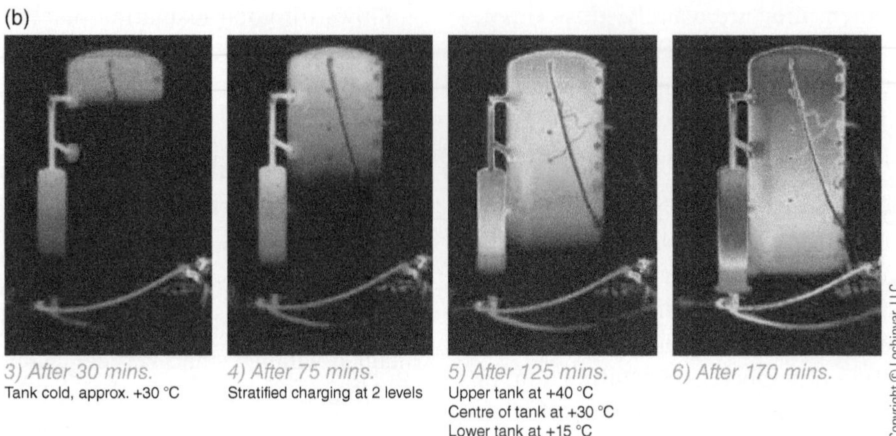

Figure 9-23b | "Stacking" of hot water due to temperature stratification induced by side arm heat exchanger on solar storage tank.

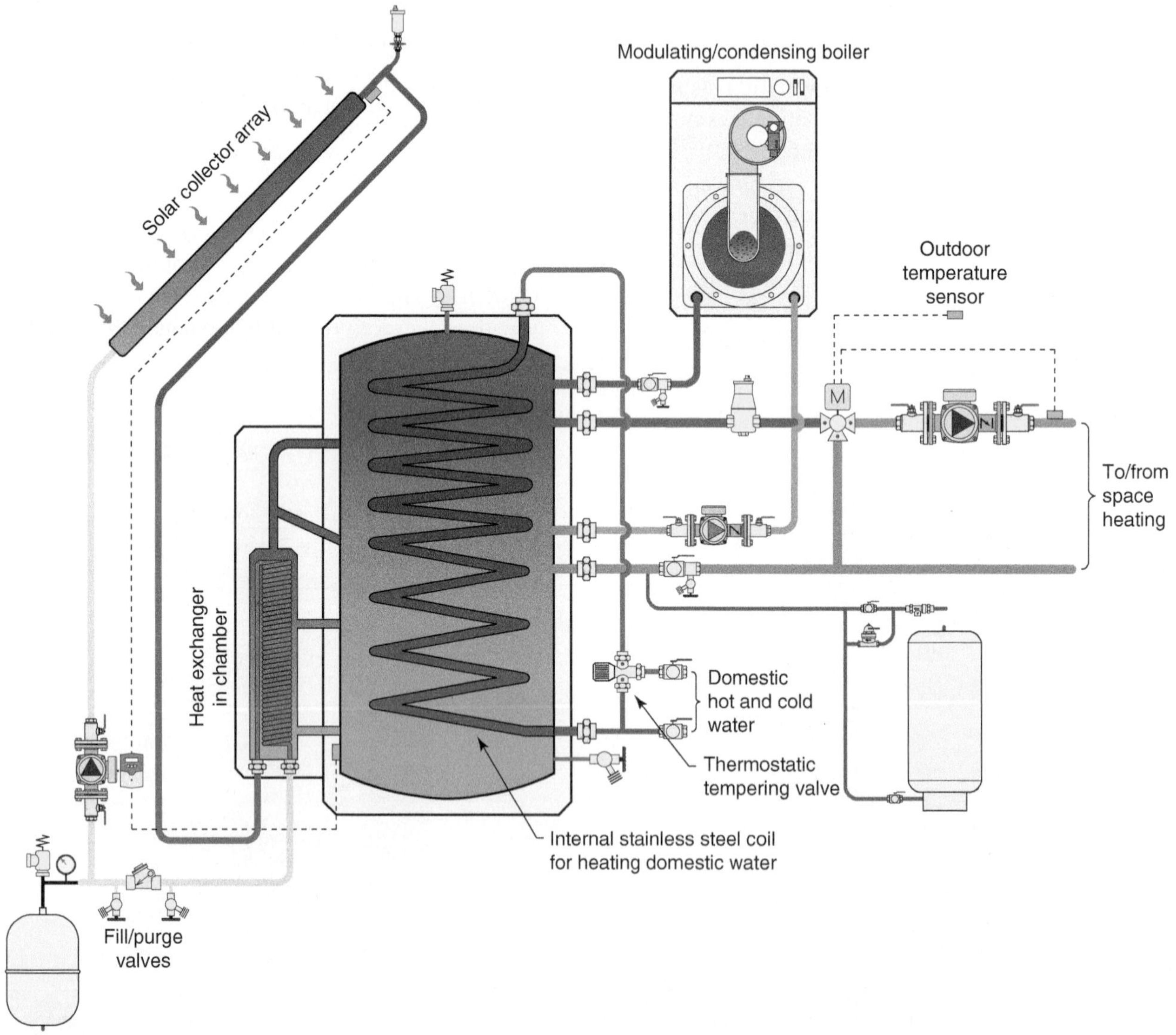

Modulating/condensing boiler
Solar collector array
Outdoor temperature sensor
M
To/from space heating
Heat exchanger in chamber
Domestic hot and cold water
Thermostatic tempering valve
Internal stainless steel coil for heating domestic water
Fill/purge valves

Figure 9-24 | Example of a highly stratified thermal storage tank used for a solar combisystems application.

The auxiliary boiler is piped and controlled so that it maintains a minimum setpoint temperature in the upper portion of the tank. This provides domestic hot water without disturbing the temperature stratification that allows the solar subsystem to contribute heat to the cooler water in the lower portion of the tank. The thermal mass of the hot water at the top of the tank also provides excellent buffering for what may be a highly zoned space heating distribution system. Hot water for space heating is drawn from the upper portion of the tank and blended with cooler water returning from the distribution system, using a 3-way motorized mixing valve. The water temperature supplied to the distribution system is adjusted by this mixing valve based on outdoor reset control. This valve is essential for preventing what could be very hot water in the tank from being supplied directly to a low temperature spacing heating subsystem.

REPURPOSED PROPANE STORAGE TANKS

Another option for large volume pressurized storage is a **repurposed propane tank**. These tanks are available in sizes of 250, 500, and 1,000 gallons. They are typically constructed to ASME specifications, and are pressure-rated to 250 psi, which is considerably higher than the nominal 30 psi relief valves used on most residential and light commercial hydronic systems. They are constructed of carbon steel and thus should only be used in closed loop systems.

Reconditioned propane tanks, *without* ASME certification, typically cost about $2.80 to $3.00 per gallon. These prices are often lower than those of other tank options.

The piping connections on most propane storage tanks are not ideal for thermal storage applications. Different piping connections can be cut into the tank shell, and a threaded steel coupling welded to the location. However, this usually voids any warranty on the tank and may or may not allow the tank to retain its ASME section VIII pressure vessel designation, depending on who does the welding. Again, designers should check on local code requirements regarding ASME certification of tanks and plan accordingly.

Figure 9-25 shows examples of a single 500-gallon storage tank and a stacked arrangement of two such tanks. The stacked arrangement allows for a smaller footprint compared to a single 1,000-gallon horizontally oriented propane storage tank, which is about 16 feet long.

Courtesy of AHONA.com

(a)

Courtesy of AHONA.com

(b)

Figure 9-25 Examples of propane storage tanks that have been repurposed for thermal storage. (a) Single 500-gallon tank. (b) Stacked 500-gallon tanks.

If the tank has contained propane, it's very likely that it also contains some residual of **mercaptan** (methanethiol), which is a strong odorant added to propane so that its presence can be detected by smell. Mercaptan smells like rotten cabbage. This odorant needs to be removed from the tank before it is used for thermal storage. This can be done by adding strong detergents, such as Oxiclean®, and water to the tank and then mechanically agitating it. The cleaning solution is then drained and the tank internally pressure washed with fresh water.

Propane storage tanks are supplied without insulation. They must be insulated on-site. The preferred insulating material is 2 lb./ft.^{3} density spray urethane, which is applied as a **monolithic coating** after all piping connections have been made and pressure tested. A minimum of 4 inches of spray foam is recommended; this should be applied in multiple layers to allow for proper curing. The foam should have a temperature rating equal to or above the maximum water temperature that might occur inside the tank. Some building codes may also require a fire retardant **intumescent coating**, over the urethane insulation.

An alternative is to build wooden framed enclosure around the tank and fill all space between the tank and the enclosure filled with a blown insulating material such as fiberglass or rock wool.

The following design considerations apply to repurposed propane tanks:

- They are made of carbon steel and as such only suitable for use in closed hydronic systems.
- These tanks are heavy. A typical 500-gallon tank weighs about 950 pounds, and a 1,000-gallon tank weighs about 1,800 pounds. Designers should always consider the logistics of placing the tank within a building. They should also have a planned exit strategy in case the tank needs to be removed from the building.
- The horizontal orientation of propane storage tanks does not allow good temperature stratification. However, stacked tanks, as shown in Figure 9-25b, if piped in series, will provide better stratification.
- Always provide a drain connection at the bottom and a vent connection at the top of the tank.
- Add connections for temperature sensor wells. Provide a well that ends near the top of the tank and another well that ends near the bottom of the tank.
- To help preserve stratification, these tanks can be provided with interior fittings that direct incoming flows horizontally. These fittings can be welded in place as part of the tank modifications.

Figure 9-26 shows a schematic in which a reconditioned propane tank is used as part of a drainback-protected solar combisystem.

This system also includes a wood gasification boiler. A well is provided for a sensor that measures the temperature near the bottom of the tank. That sensor connects to a differential temperature controller, which also monitors the temperature of the absorber plate in one of the solar collectors. When a suitable temperature differential is reached, the collector circulator operates to extract heat from the collectors and eventually add it to the main thermal storage tank. The wood gasification boiler, when operating, also adds heat to the thermal storage tank.

Another sensor well is provided to measure the water temperature near the top of the storage tank. This temperature will likely be needed by other system controllers to determine if the temperature of the tank is sufficient to meet the heating load(s).

Notice that the main piping connections leading in the tank are equipped with internal elbows that direct flow horizontally within the tank. This helps preserve some stratification. It also helps direct heated water across the upper strata of the tank from the inlet connection supplied by both heat sources to the outlet that supplies the space heating distribution system. This arrangement leverages the lower density of hot water, keeping it within the upper portion of the tank and thus expediting its delivery to the load. Similarly, horizontal flow near the bottom of the tank helps in returning cooler water from the load to the heat sources without fully mixing the tank.

The water level within the combination drainback/expansion tank varies based on operation of the solar subsystem as well as the temperature of the water in the main storage tank.

Designers should also consider that any system with a large pressurized thermal storage tank will require a relatively large expansion tank. This tank can be sized using the methods discussed in Chapter 4. A *starting estimate* for the volume of the expansion tank when used in systems that can generate relative high water temperatures is 10 percent of the storage tank volume.

9.4 Stratification In Thermal Storage Tanks

When hot water is carefully introduced into the upper portion of a storage tank, it tends to "float" above cooler water already in the tank. This phenomenon is called temperature stratification; it is a desirable effect in nearly all thermal storage applications.

Figure 9-27 shows two identical thermal storage tanks. Both are assumed to be used in a heating application. The tank on the left is stratified. The water temperature at the top is 120 °F. The water temperature at the bottom is 100 °F. Assuming a linear temperature gradient from top to bottom, the average water temperature in this tank is 110 °F. The tank on the right is fully mixed. The water temperature at the top, middle, and bottom of this tank is a uniform 110 °F.

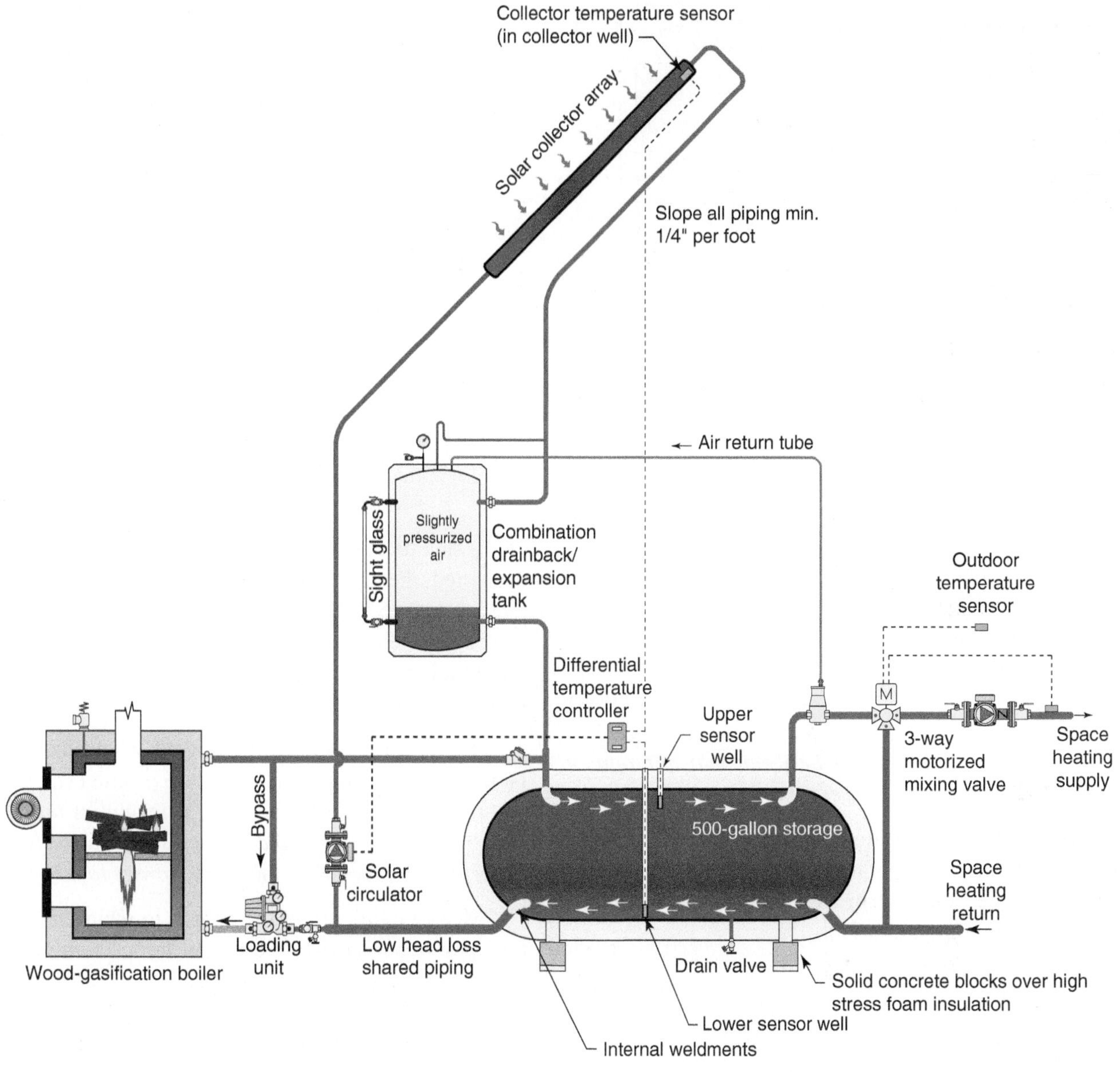

Figure 9-26 | Schematic showing use of 500-gallon repurposed propane tank for thermal storage.

Both tanks contain the same *amount* of heat. However, the temperature stratification present in the left tank gives it a "thermodynamic advantage." To understand why, one needs to consider the thermodynamic principle of **exergy**, which is based on the second law of thermodynamics.

Exergy is a number that determines the maximum ability of the energy contained in a material to affect change to its surroundings. In simpler terms, *exergy* can be thought of as the "usefulness" of the energy in a material, rather just the *amount* of energy in that material. *Energy that has higher exergy is always preferred because it is more useful.*

Although both tanks in Figure 9-27 contain the same *amount* of thermal energy, it can be shown that the *exergy* of the energy in the stratified tank is greater than the *exergy* of the fully mixed tank.

Here is one simple rationale that supports this claim: The water in the upper half of the stratified tank has a temperature from just over 115 to 120 ºF. This water could transfer heat to another material that has a temperature of 115 ºF. However, *none* of the 110 ºF water in the fully mixed tank could directly transfer heat to another material at 115 ºF. Thus, the water in the stratified tank is more useful and provides greater potential for how the energy can be used to supply a load.

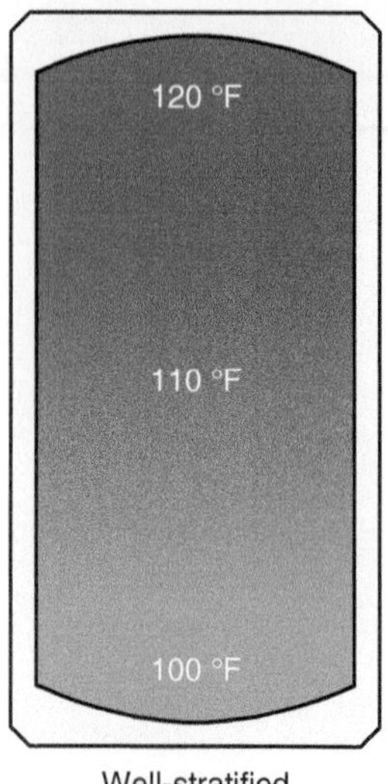

Well-stratified thermal storage tank

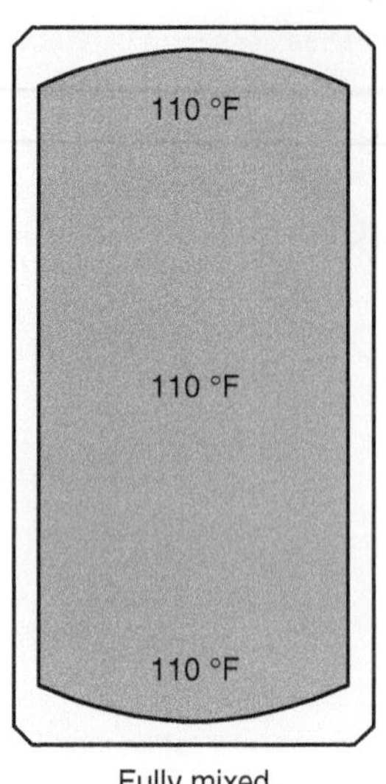

Fully mixed thermal storage tank

Figure 9-27 Two identical tanks. The left tank is well stratified. The right tank is fully mixed. Both tanks hold the same amount of heat.

Whenever mixing occurs between warmer and cooler fluids there is a loss of exergy.

One way to demonstrate this principle is to calculate the "equivalent temperature" (T_e) of a fully mixed tank that has the same *exergy* as a stratified tank. This can be done using Equation 9.3.

(Equation 9.3)

$$T_e = 2(T_{mixed}) - e^{\left[\frac{1}{H}\int_{h=0}^{H} ln[T(h)]dh\right]}$$

where:

T_e = equivalent temperature of a fully mixed tank that would have the same *exergy* as a stratified tank (°F)

T_{mixed} = actual temperature of fully mixed tank (°F)

h = vertical position above bottom of tank (ft.)

ln = natural logarithm

$T(h)$ = a function that gives the temperature at some height (h) within the tank (°F)

e = 2.718281828

To evaluate this equation, the temperature profile from the bottom to the top of the tank needs to be expressed as a function of tank height (h).

For example, consider a thermal storage tank that is 6 feet tall. Assume that the water temperature at the bottom of the tank is 100 °F and the water temperature at the top is 120 °F. If a linear temperature profile is assumed from bottom to top, the water temperature as a function of tank height can be expressed as Equation 9.4:

(Equation 9.4)

$$T(h) = 100 + \left[\frac{20}{6}\right]h$$

where:

$T(h)$ = temperature at some height (h) above the bottom of the tank (°F)

h = vertical position above bottom of tank (ft.)

If this function is substituted into Equation 9.3 and evaluated, the equivalent temperature (T_e) can be calculated as 110.15 °F. This is slightly greater than the actual mixed tank temperature of 110 °F. Since the equivalent temperature (T_e) is higher than the mixed temperature, there would be a loss of *exergy* if mixing occurs.

It can be mathematically proven that this will always be true for any possible temperature profile established by stratification within the tank. *This implies that mixing should always be avoided to preserve the usefulness of the heat contained in a thermal storage tank.*

Many factors affect the degree to which temperature stratification exists in a storage tank. These include:

- The temperature of water being added to the tank
- The position and orientation of the piping inlets to the tank
- The presence of specific devices called inlet flow diffusers in the tank
- The thermal conductivity of the tank walls
- The insulation on the tank
- The timing and rate at which water is added and removed from the tank

Thermal storage tanks used in hydronic systems should be designed and operated to encourage thermal stratification. The following guidelines help in this regard:

- Use vertically oriented rather than horizontally oriented tanks.
- Introduce heated water into the upper portion of a thermal storage tank.

- Extract cooler water from the lower portion of the tank.
- Provide piping connections, and possibly internal details, that allow incoming flow to enter at low velocities and in a horizontal direction.
- Avoid piping connections that create vertical flow jets within the tank.
- Use good insulation on the tank (R-18 °F·hr.·ft.^{2}/Btu is a suggested minimum).
- When possible, use tanks built of materials with low thermal conductivity

4-PIPE VERSUS 2-PIPE TANK CONFIGURATIONS

Figure 9-28 shows a common tank piping arrangement that embodies many of the stratification encouraging details just listed.

This is called a **4-pipe buffer tank configuration** (based on the four primary connections that connect the tank to the heat source (at points A and D), and to the load (at points B and C).

This configuration allows heated water entering the tank at point (A) to remain in the upper portion of the tank due to its lower density relative that of cooler water lower in the tank.

Still, there will be some mixing between the entering water and the water already in the upper portion of the tank. This will create a delay between the time that hot water enters the tank at point (A) and when water *at that same temperature* leaves the tank for the load at point (B).

For systems that maintain a relatively consistent load, this is not a problem. However, for systems that need to transfer heat to the load as quickly as possible after the heat source turns on, or following a recovery from a temperature setback, any interaction with the thermal mass of the tank is undesirable. In these situations, the **2-pipe buffer tank configuration** shown in Figure 9-29 provides advantages.

In a 2-pipe configuration, some of the flow from the heat source *may* be diverted to the load before reaching the thermal storage tank. This situation, when it occurs, provides two significant benefits:

- When necessary, it allows flow from the heat source to pass directly to the load without first "interacting" with the thermal mass in the upper portion of the tank.
- When there is flow through the heat source, and flow to the load, the flow rate entering the upper portion of the thermal storage tank is reduced. This is illustrated in Figure 9-30 for a heat source flow rate of 10 gpm, and a corresponding load flow rate of 8 gpm.

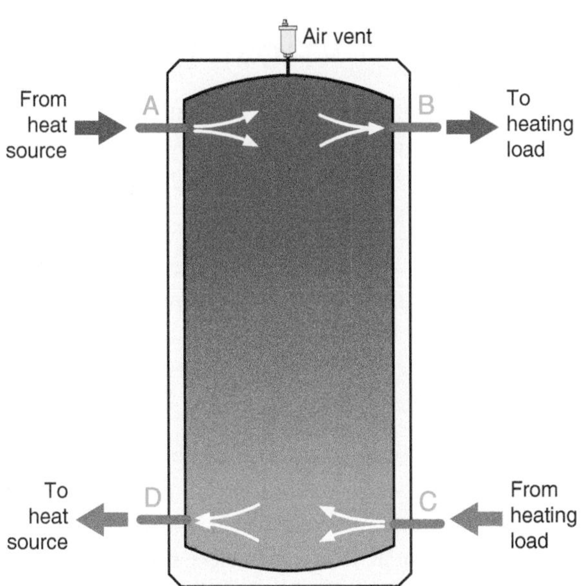

Figure 9-28 | 4-pipe buffer tank configuration.

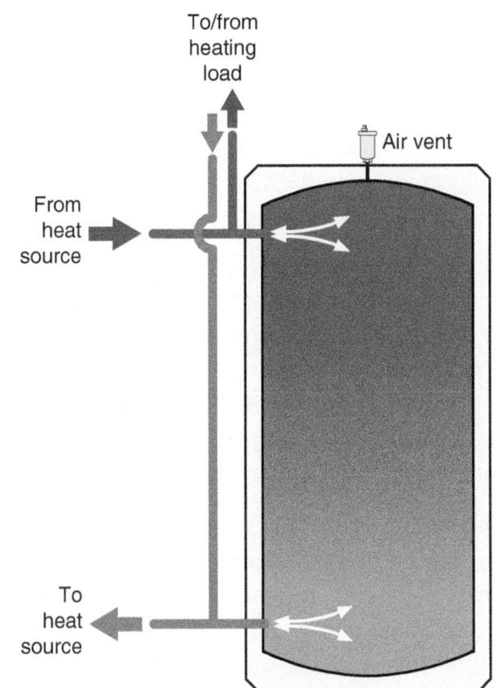

Figure 9-29 | 2-pipe buffer tank configuration.

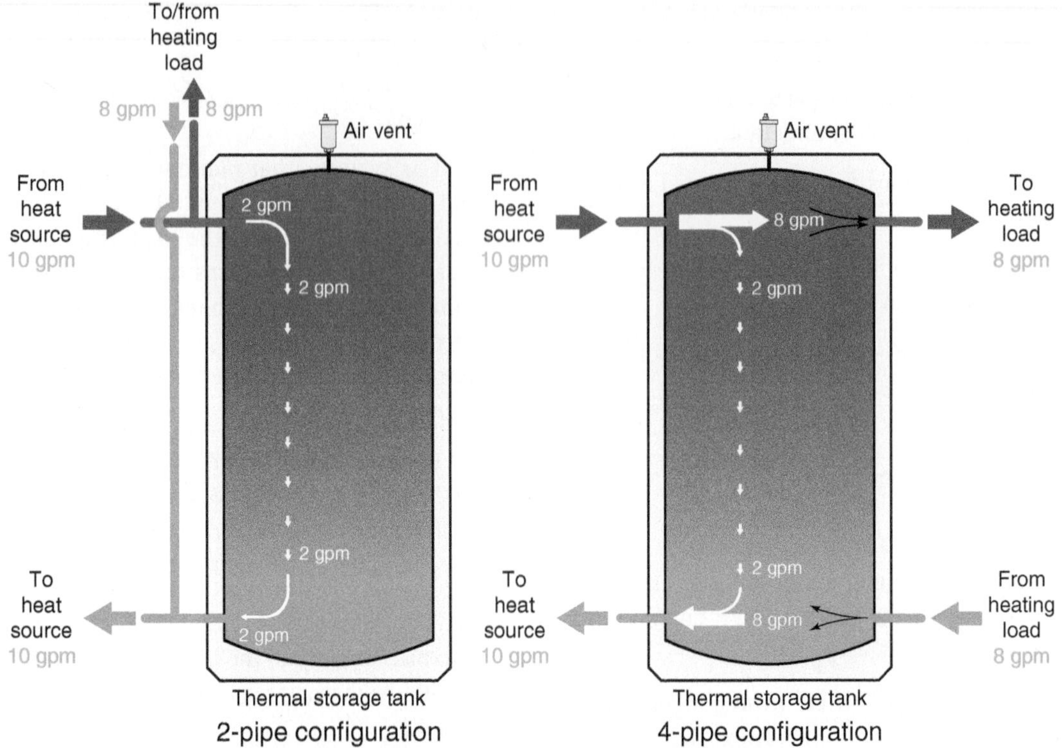

Figure 9-30 | A comparison of internal flow rates between a 4-pipe buffer tank configuration and a 2-pipe configuration. Note that in this scenario, internal flow rates are higher in the 4-pipe configuration.

With the 2-pipe configuration, the flow rate entering the upper portion of the tank is 2 gpm, as is the downward flow rate through the tank. However, if the same tank were set up with a 4-pipe configuration, the flow rate entering the tank is 10 gpm, which divides up between 8 gpm flowing across the tank toward the upper right outlet, and 2 gpm vertically downward. The same flow proportions also occur at the bottom of the tank.

The higher the flow rate entering the tank, the higher the flow velocity and thus the greater the mixing action within the tank. Mixing tends to degrade temperature stratification and should be avoided. Thus, a 2-pipe configuration, under this flow scenario, is preferred from the standpoint of preserving temperature stratification.

Thermal storage tanks set up as either 4-pipe or 2-pipe configurations can provide excellent hydraulic separation between circulators used for heat source(s) or loads. *When using a 2-pipe configuration, it is important to keep the tees connecting to the load piping as close to the tank as possible.* The short header piping between these tees and the tank should also be generously sized to reduce its head loss to almost zero. Thus, the piping that the heat source circuit and the load circuit have in common has very low head loss. This provides hydraulic separation between the circulators in these circuits.

AVOIDING VERTICAL FLOW JETS

Some pressurized tanks are made with semi-elliptical heads joined to cylindrical shells. When this type of tank is installed vertically, it is often equipped with top and bottom center connections.

If flow from the return side of the distribution system is routed to the bottom center connection, a **vertical flow jet** will be established that causes mixing in the tank. Experience has shown that this will significantly reduce stratification.

This type of connection should be avoided *unless the tank is equipped with an internal flow diffuser* that can reorient the flow from vertical to horizontal and slow the velocity at which this flow mixes with water in the lower portion of the tank. Figure 9-31 illustrates two piping details to avoid and one concept for a flow diffuser that significantly reduces internal mixing.

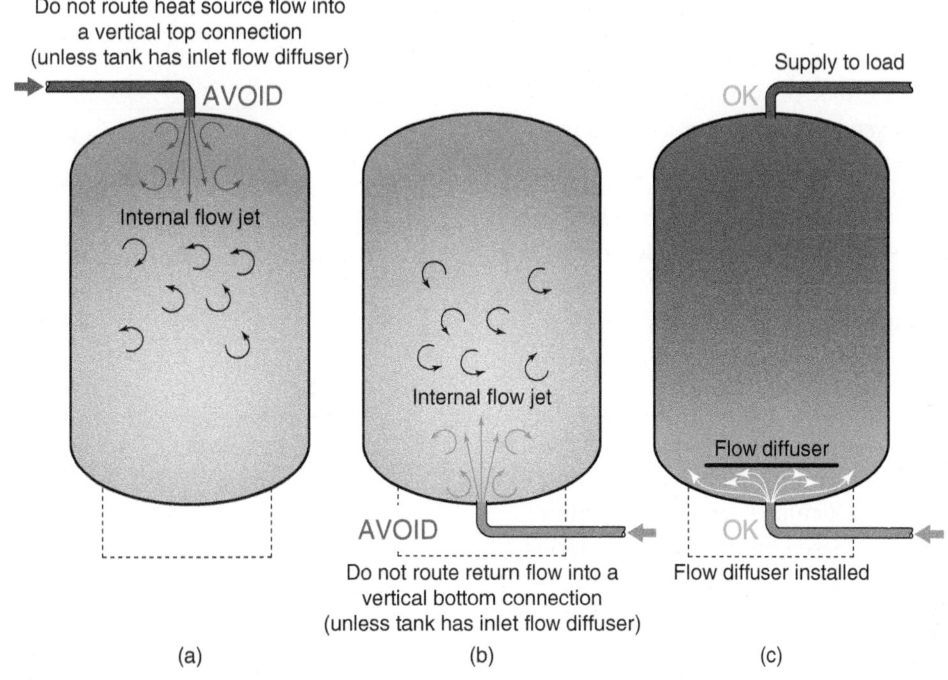

Figure 9-31 | (a) Avoid downward flow of entering hot water. (b) Avoid upward flow of entering cooler water. (c) Acceptable piping connections, provided the tank is equipped with a flow diffuser at or near the lower inlet connection.

9.5 Heat Capacity and Heat Loss of Thermal Storage Tanks

The amount of heat any thermal storage tank can hold is determined by its volume, along with the temperature change of the water within the tank. Equation 9.5 can be used to calculate the heat added to or removed from a water-filled thermal storage tank based on these factors:

(Equation 9.5)

$$Q = 8.33v(\Delta T)$$

where:

Q = amount of heat added or removed from tank (Btu)

v = volume of tank (gallons)

ΔT = temperature change of water in tank (ºF).

8.33 = a quasi-constant for water (Btu/gal./ºF)

Example 9.2: A 500-gallon thermal storage tank is heated from a uniform average initial water temperature of 60 ºF to a uniform final temperature of 140 ºF. How much heat was added to the tank in this process?

Solution: This is a very straightforward calculation:

$$Q = 8.33v(\Delta T) = 8.33(500)(140 - 60)$$
$$= 333{,}200 \text{ Btu}$$

Discussion: This calculation assumes that *all* water in the tank is 60 ºF when heat input begins. It also assumes that *all* water in the tank is 140 ºF when heat input stops. Although the starting assumption of a uniform 60 ºF temperature is reasonable for cold water, a well-designed thermal storage tank would develop temperature stratification as heat is added. Thus, it's likely that the water temperature within the tank will not be uniform, from top to bottom, when heat input stops. If the upper portion of the tank was at 145 ºF, and the lower portion at 135 ºF, and one assumed a linear temperature gradient from top to bottom within the tank, the calculation for total heat added would be accurate. However, if one were to just measure the water temperature at the top of the tank at 145 ºF and base the calculation using Equation 9.4 on that temperature, it would overestimate the total heat added.

USEABLE HEAT FROM STORAGE

Although Equation 9.5 is thermodynamically correct, it implies that the tank is either fully mixed to a uniform temperature or that reasonably accurate average temperatures are used for the calculation.

The amount of heat from storage that is *useable* by the space heating distribution system is determined by the *average* water temperature of the storage tank at the start of the heat extraction cycle and the *lowest temperature at which the tank can still supply heat to the load.*

In many systems, the latter criteria is determined by an outdoor reset controller that monitors the temperature in the upper portion of the thermal storage tank and compares it to a calculated target temperature. The target temperature is based on the current outdoor temperature and the settings of the reset controller. If the temperature in the upper portion of the tank is at or above the calculated target temperature, the reset controller "allows" the tank to serve as the heat source. The following scenario will illustrate this principle.

Consider a hydronic distribution system that can deliver design load heat output at a supply water temperature of 120 °F and a temperature drop of 20 °F from supply to return. The design load is based on an outdoor temperature of 0 °F and an indoor temperature of 70 °F.

Figure 9-32 shows the theoretical supply and return water temperatures for this hydronic system as a function of outdoor temperature, assuming the supply water temperature was maintained exactly at it calculated "target" value by an outdoor reset controller and that the flow rate through the distribution system remained constant. This graph also assumes that the desired indoor air temperature is 70 °F and that there are no internal heat gains in the building.

Notice that the temperature drop from supply to return is 20 °F under design load conditions, but decreases proportionally as the outdoor temperature increases. Assuming the flow rate through the system remains constant, this decreasing temperature difference is the only way that heat output decreases. In theory, when the outdoor temperature reaches 70 °F, both the supply and return water temperatures are 70 °F. Thus, no heat would be transferred into a building space at 70 °F.

An on/off type outdoor reset controller must operate with a temperature differential between where its output relay contacts close and open. When this type of controller is used to determine if the thermal storage tank temperature can supply the space heating load, that differential should be kept small. The author suggests 5 °F. Figure 9-33 shows this differential superimposed on the target supply temperature line from Figure 9-32. The lower (green) dashed line shows the temperature at which the reset controller would call for operation of the auxiliary heat source.

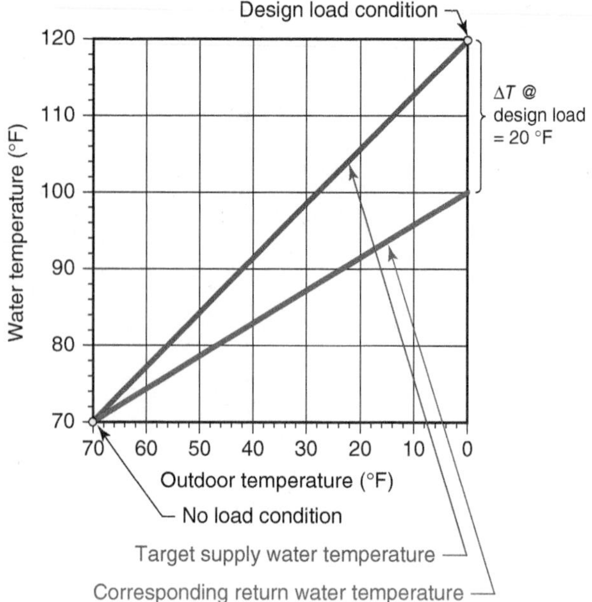

Figure 9-32 Theoretical (target) supply water temperature and corresponding return water temperature for a hydronic distribution system that requires 120 °F supply water at design load conditions.

At design load, the reset controller will allow water temperatures as low as 117.5 °F (e.g., one half the 5 °F control differential below the target water temperature), to be supplied from the thermal storage tank to the distribution system. This supply water temperature will drive slightly less heat from the distribution system, compared to the 120 °F target supply water temperature.

However, when the auxiliary heat source is turned on by the reset controller, it remains on until the supply water temperature reached 122.5 °F (e.g., one half the 5 °F differential above the target water temperature). This would slightly increase heat output from the distribution system compared to its output when supplied with 120 °F water. Over time, these two effects compensate for each other and keep the total heat output from the distribution system about the same as if it were supplied with a constant 120 °F water temperature.

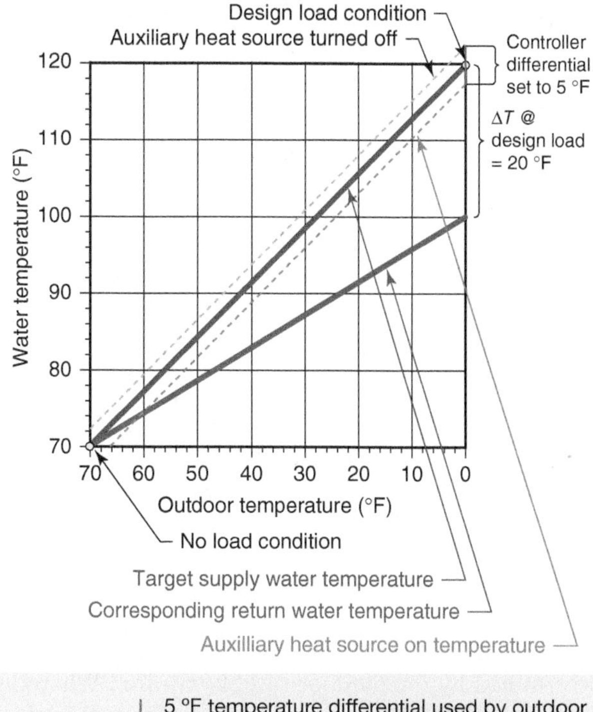

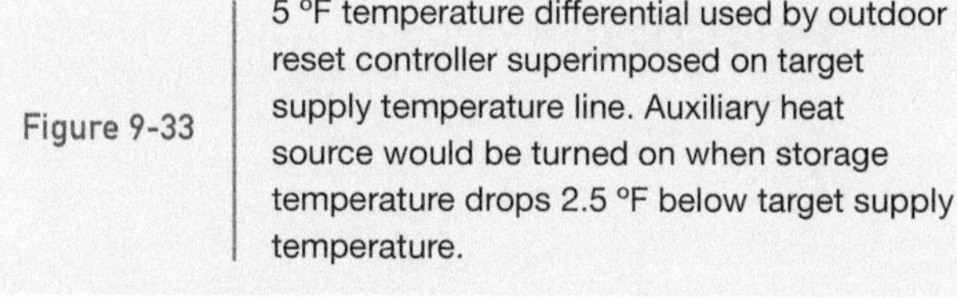

Figure 9-33 5 °F temperature differential used by outdoor reset controller superimposed on target supply temperature line. Auxiliary heat source would be turned on when storage temperature drops 2.5 °F below target supply temperature.

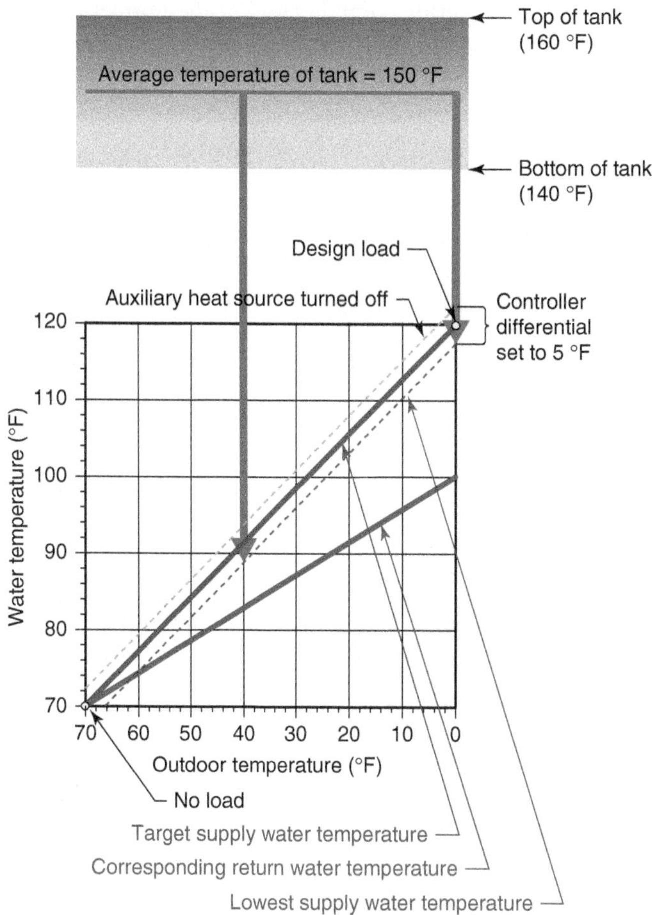

Figure 9-34 Temperature drop of thermal storage tank that is usable by distribution system, assuming an average tank temperature of 150 °F at start of discharge cycle. The longer the vertical arrow, the greater the amount of heat extracted from the tank.

These slight deviations above and below the target supply water temperature are also "smoothened out" by the thermal mass of the distribution system. It's very unlikely that building occupants will even notice these subtle changes in heat output, especially in comparison to standard on/off heat delivery systems.

When the outdoor temperature is 60 °F, the target supply water temperature is about 77 °F, as indicated by the sloping red line in Figure 9-33. However, the lowest available water temperature from storage (e.g., before the auxiliary heat source is turned on) is 2.5 °F lower (e.g., 74.5 °F). At this supply water temperature, the heat output is only 9.3 percent of the design load heat output. In theory, the required heat output should be 14.3 percent of design load. One could view this as a significant difference, and cause for concern. However, there are several factors that suggest it will not create a problem. First, it's very likely that there *will be* internal heat gains in the building. At these relatively high outdoor temperatures, those gains may even be greater than the heating load. In such a case, there would be no heat demand and the difference between theoretical and available heat output of no concern. Second, the lower rate of heat output will only exist for a short time. Any further drop in the thermal storage tank temperature will cause the outdoor reset controller to turn on the auxiliary heat source. Once on, it will bring the water temperature delivered to the distribution system up to about 79.6 °F before it turns off, assuming the heat demand still exists.

Based on this type of control logic, one could view the **usable temperature drop** of the thermal storage tank, as shown in Figure 9-34.

The heat extracted from the tank is represented by the large vertical arrows. The longer the arrow, the greater the amount of heat extracted from the tank.

This illustration is based on the assumption that the tank is at an *average* temperature of 150 °F at the start of the heat extraction period. Thus, if the temperature at the top of the tank were 160 °F and the temperature at the bottom were 140 °F, and assuming the temperature gradient was linear from top to bottom, the initial thermal energy content of the tank would be the same as if the entire tank were at 150 °F.

Heat is available from storage until the temperature at the sensor location of the outdoor reset controller drops to 2.5 °F (half of the controller's 5 °F differential) below the target temperature. At design load conditions, that will be 117.5 °F. However, at an outdoor temperature of 40 °F, the target temperature is lower, and heat can be delivered from storage to the distribution system until the temperature in the upper portion of the tank drops to about 88 °F. *This shows the advantage of using an outdoor reset controller to make the determination of whether heat is supplied from storage or by the auxiliary heat source.* It also demonstrates the advantage of designing the distribution system for low supply water temperatures. As previously stated, the author recommends a maximum supply water temperature of 120 °F, supplied to the distribution system under design load conditions.

ADDITIONAL HEAT EXTRACTION

In some systems, it is possible to simultaneously supply heat from thermal storage along with supplemental heat from the auxiliary heat source. The goal is to extract additional heat from storage provided its temperature remains above the *return* temperature from an active distribution system.

Consider the system shown in Figure 9-35. It's a space heating system supplied from a thermal storage tank, and when necessary, an auxiliary boiler.

An outdoor reset controller determines if the space heating load is sourced from the thermal storage tank or the auxiliary boiler. The tank serves as the sole heat source until its temperature drops below a value determined by the outdoor reset controller. When this occurs, the outdoor reset controller prevents flow from

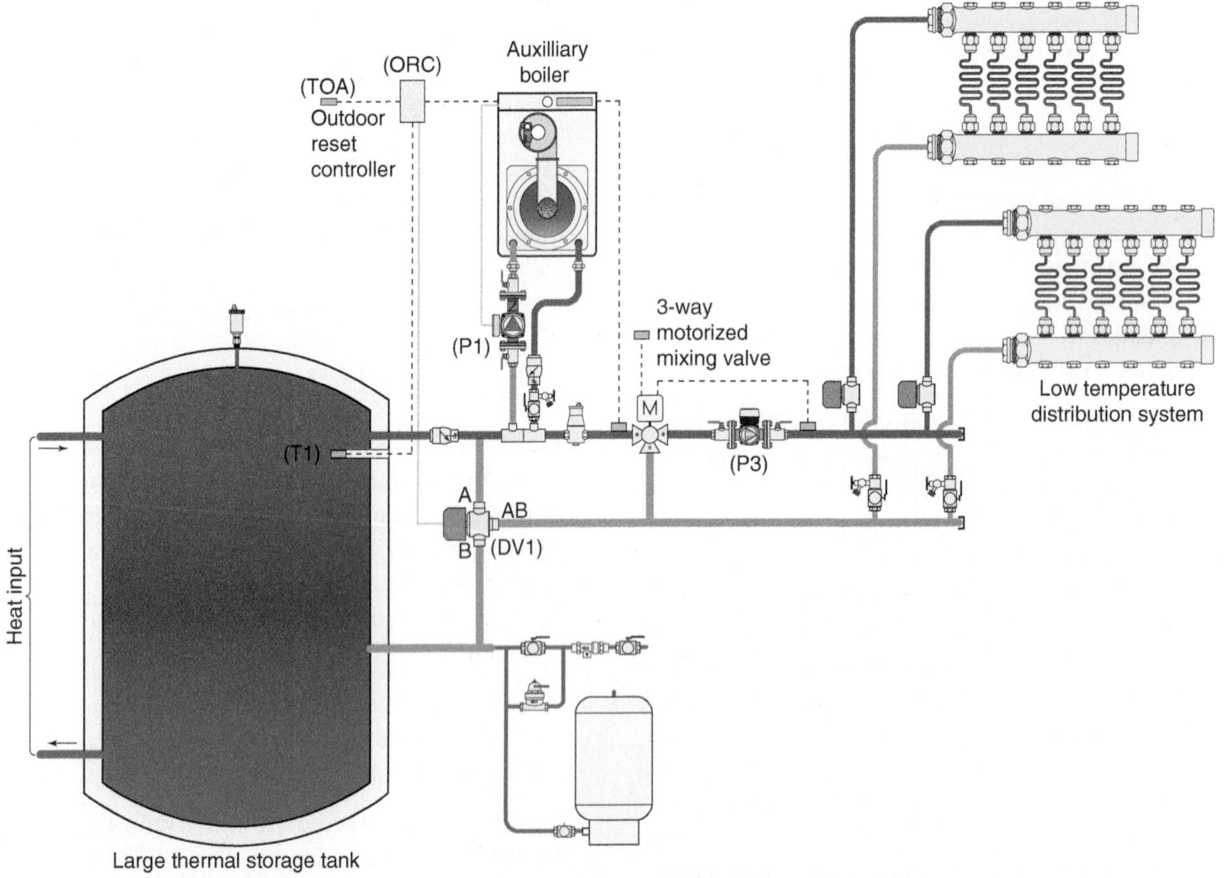

Figure 9-35 | Thermal storage tank supplying a low temperature distribution system from a thermal storage tank, or an auxiliary boiler.

returning to the thermal storage tank by turning on diverter valve (DV1), which routes flow returning from the distribution system around, rather than through, the thermal storage tank. The outdoor reset controller also enables operation on the auxiliary boiler and its associated circulator (P1). Under this control scenario, there is no mode where both the thermal storage tank *and* the auxiliary boiler could both supply heat at the same time.

However, consider a situation in which the temperature at the top of the storage tank is slightly less than the lower limit established by the outdoor reset controller, *but still warmer than the return temperature of the distribution system.* Under these conditions, the tank can still contribute some heat to the space heating load. The remaining heat would be supplied by the auxiliary boiler.

This additional heat can be extracted by adding a **differential temperature controller**, as shown in Figure 9-36.

The differential temperature controller measures the temperature difference between the top of the thermal storage tank (T3) and the return side of the distribution system (T2). As long as the top of the tank is warmer than the return side of the distribution system, the tank can contribute some heat to the load, with the remainder added by the auxiliary boiler.

Figure 9-37 shows a ladder diagram that contains the operating logic for a system piped as shown in Figure 9-36.

If the outdoor reset controller (ORC) determines that the tank is too cool to supply the space heating load, its normally open contact closes to power up the differential temperature controller (DTC) as well as a relay (R1) that enables the boiler to operate.

The differential temperature controller (DTC) measures the temperature difference between water at the top of the storage tank (at sensor T3) and that returning from the distribution system (at sensor T2). If the top of tank temperature (T3) is at least 4 ºF above the distribution return temperature (T2), the normally open contact in the differential temperature controller

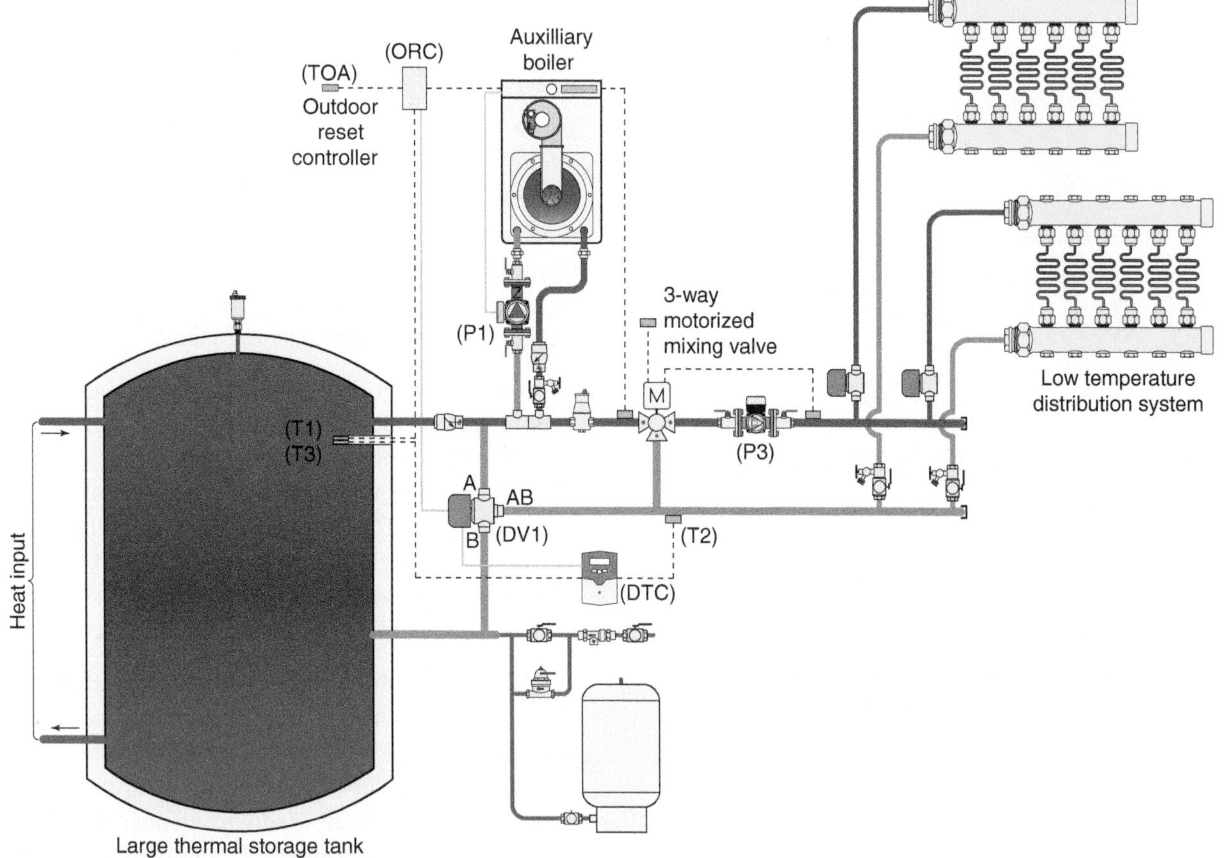

Figure 9-36 Adding a differential temperature controller that measures the temperature difference between the top of the thermal storage tank (T1) and the return side of the distribution system (T2).

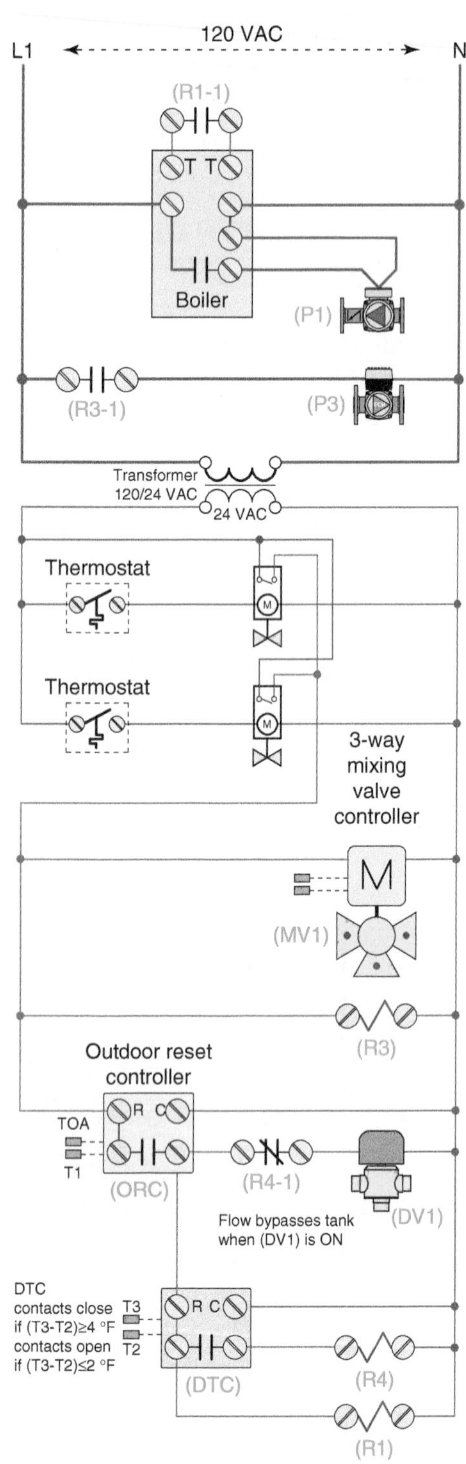

Figure 9-37 Ladder diagram with control logic for extracting additional heat from storage using a differential temperature controller.

(DTC) closes. This energizes the coil of relay (R4). The normally closed contact (R4-1) opens, which turns the diverter valve (DV1) off. This allows flow returning from the distribution system to continuing flowing *through* the thermal storage tank (e.g., from the AB port of the diverting valve to the B port).

The boiler and circulator (P1) are also operating at this time. The boiler is monitoring the temperature downstream of the closely spaced tees. Ideally it modulates to add just enough heat to maintain this temperature close to the target temperature required by the distribution system based on the settings of the boiler's internal controller.

The mixing valve (MV1), operating under its own outdoor reset control logic, and with the same settings as the boiler's internal reset controller, should be at, or very close to, its fully open position, and thus providing little if any mixing.

If the temperature difference between the top of the storage tank and the return side of the distribution system drops to 2 °F or less, there is very little useful heat remaining in the thermal storage tank. Under this condition, the differential temperature controller turns off relay (R4). The normally closed contact (R4-1) closes allowing 24 VAC to energize diverter valve (DV1). This stops flow returning from the distribution system from passing into the thermal storage tank. The auxiliary boiler and circulator (P1) remain on to supply the space heating load as required.

When the storage tank temperature again rises to where the temperature differential between the top of the tank and the return side of the distribution system reaches 4 °F or more, and there is a demand for space heating, the diverter valve is turned off and flow returning from the distribution system is again allowed to pass through the thermal storage tank.

This strategy makes sense when a *large* storage tank is used in the system. The author suggests that it is appropriate when the storage tank volume is at least 500 gallons. The ability to lower 500 gallons of water by an additional 10 °F implies a release of almost 42,000 additional Btus from the tank. Smaller tanks would contribute proportionally smaller amounts of heat and thus provide less justification for the additional controls.

When using this strategy it is very important that the temperature sensors for the differential temperature controller are identical in their temperature versus resistance characteristic curves. It is also important that both sensors are mounted in an *identical* manner. Ideally, both sensors should be mounted in identical sensor wells, immersed in the system water, and with ample coatings of thermal grease. If mounted to the surface of copper tubing, be sure the sensor surface makes good contact, is well secured, and is fully wrapped with insulation. Remember, *this strategy requires the differential temperature controller to detect differences in temperature as low as 2 °F.* Such low temperature differences are difficult for most people to detect with their bare fingers.

Figure 9-38 shows how this additional heat extraction strategy, using a differential temperature controller in combination with the reset controller, increases the useful heat available from storage.

The additional heat extraction potential is greatest at design load conditions and decreases as the temperature drop of the distribution system decreases.

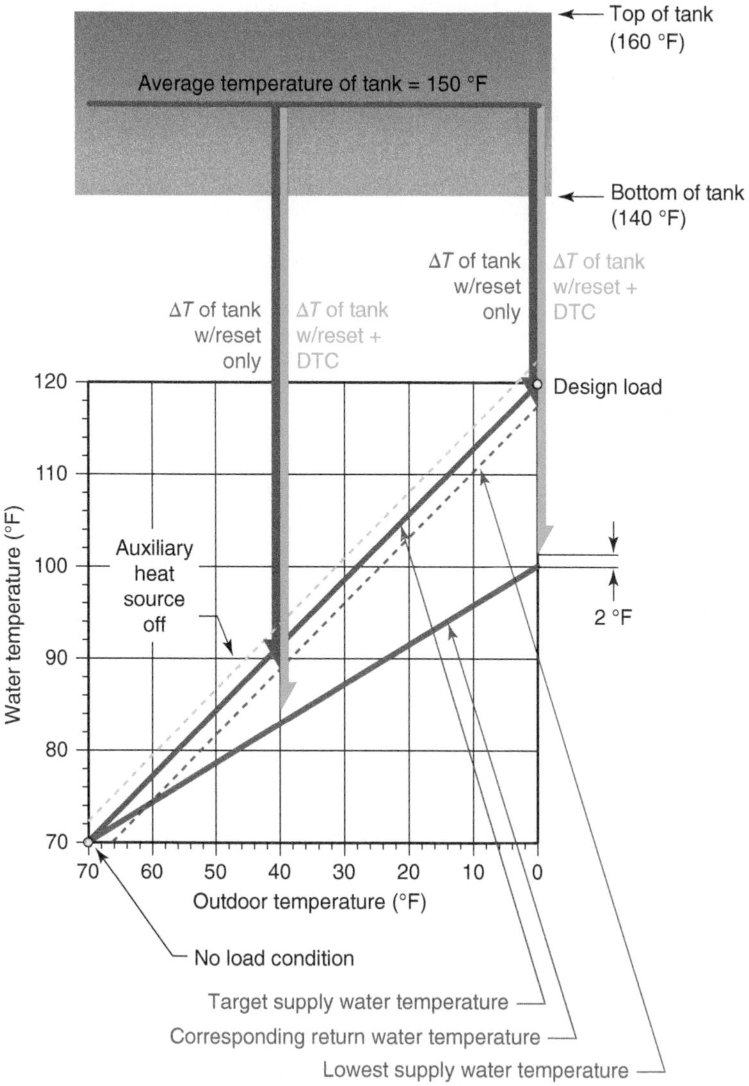

Figure 9-38 Vertical green arrows represent heat extraction potential when a differential temperature controller is used in combination with an outdoor reset controller. Vertical pink arrows represent heat extraction using only outdoor reset control.

The ability to heat domestic water would be required to qualify the system shown in Figure 9-36 as a combisystem. A method for doing this will be shown later in this chapter.

HEAT LOSS FROM STORAGE

The instantaneous rate of heat loss from a vertical *cylindrical* tank can be *approximated* using Equation 9.6:

(Equation 9.6)

$$Q = \left[\left(\frac{2\pi kL}{\ln\left(\frac{d_o}{d_i}\right) + \frac{1.36k}{d_o}}\right) + \frac{\pi d_o^2}{2R_{tb}}\right](T_w - T_a)$$

where:

Q = instantaneous rate of heat loss from tank (Btu/hr.)

d_o = outer diameter of tank insulation (ft.)

d_i = inner diameter of tank insulation (ft.)

L = height of cylindrical tank (ft.)

k = thermal conductivity of tank sidewall insulation (Btu/°F·hr.·ft.)

R_{tb} = R-value of insulation on top and bottom of tank (°F·hr.·ft.²/Btu)

p = 3.141592654

T_w = *average* temperature of water in tank (°F)

T_a = temperature of air surrounding tank in tank (°F)

The relevant dimensions can be seen in Figure 9-39.

Example 9.3: A thermal storage tank measure 35 inches in diameter, and 80 inches tall. Assume it has flat heads, as shown in Figure 9-39. Also assume that all surfaces of the tank have 3-inch thick polyurethane insulation (R= 6.0 per inch, k = 0.01389 Btu/hr.·ft.·°F). The average water temperature in the tank is 150 °F. The air temperature surrounding the tank is 70 °F as is the temperature of the floor slab under the tank. Use Equation 9.6 to estimate the rate of heat loss from the tank under these conditions.

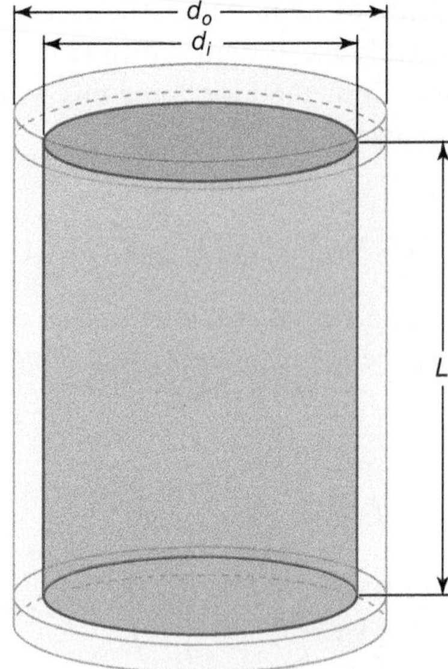

Figure 9-39 | Dimensions for vertical cylindrical tank for use with Equation 9.6.

Solution: It's important to check that all quantities have the units indicated for equation 9.6:

d_i = 35 inch = 2.9167 ft.

d_o = (35+3+3) inch = 41 inch = 3.4167 ft.

R_{tb} = 3 × (6.0/inch) = 18.0 (°F·hr.·ft.²/Btu)

Putting the numbers into Equation 9.6, and *carefully* processing them, yields.

$$Q = \left[\left(\frac{2\pi kL}{\ln\left(\frac{d_o}{d_i}\right) + \frac{1.36k}{d_o}}\right) + \frac{\pi d_o^2}{2R_{tb}}\right](T_w - T_a)$$

$$= \left[\left(\frac{2\pi(0.01389)(6.667)}{\ln\left(\frac{3.4167}{2.9167}\right) + \frac{1.36(0.01389)}{3.4167}}\right) + \frac{\pi(3.4167)^2}{2(18.0)}\right]$$

$$(150 - 170) = 366\frac{\text{Btu}}{\text{hr}}$$

Discussion: It should be emphasized that this result is an approximation. For mathematical simplicity, it assumes that the tank has a flat top and bottom. It also assumes that the floor temperature under the tank is the same as the air temperature surrounding the tank. It does not deduct for any heat loss associated with heat transfer to piping connected to the tank. It also assumes a linear temperature gradient from the top to the bottom of the tank and is thus based on the *average* water temperature in the tank. Finally, it assumes no thermal resistance for the metal walls that form the tank. A final point is that this rate of heat loss is only valid at these conditions. As the tank cools, the water temperature decreases and so does the rate of heat loss from the tank.

The time required for a vertical cylindrical storage tank to cool, on its own (e.g., without heat extraction by the loads it serves), can be estimated using Equation 9.7.

(Equation 9.7)

$$T_H = T_a + (T_{wi} - T_a)e^{-\left(\frac{c}{8.33v}\right)t}$$

where:

T_a = room air temperature (°F)

T_{wi} = initial average temperature of tank water (°F)

t = time (hours)

v = volume of water in tank (gallons)

e = 2.718281828

$$c = \left[\left(\frac{2\pi kL}{\ln\left(\frac{d_o}{d_i}\right) + \frac{1.36k}{d_o}}\right) + \frac{\pi d_o^2}{2R_{tb}}\right]$$

Readers will recognize the expression for "c" as the portion of Equation 9.6 that appears in [ ], and before the expression ($T_w - T_a$). The value of "c" is entirely defined by the dimensions and insulation of the storage tank. It does not depend on the temperatures involved.

Example 9.4: Assuming the storage tank from Example 9.3 starts at an initial average water temperature of 150 °F, and that no heat is removed by the loads. The tank is surrounded by 70 °F air. Estimate the tank's average temperature 24 hours later.

Solution: The value of "c" from Example 9.3 is 4.57.

It is also necessary to determine the tank's water volume. This can be determined using Equation 9.2:

$$v = \frac{\pi d_i^2 h}{924} = \frac{\pi(35)^2 80}{924} = 333 \text{ gallons}$$

Note that Equation 9.2 requires the diameter and height of the tank to be entered in inches.

This and the other values can now be substituted into Equation 9.6:

$$T_H = T_a + (T_{wi} - T_a)e^{-\left(\frac{c}{8.33v}\right)t} = 70 + (150 - 70)e^{-\left(\frac{4.57}{8.33(333)}\right)24}$$
$$= 70 + 80e^{-0.03954} = 146.9\ °F$$

Discussion: This calculation estimates that the average water temperature in the tank drops by just over 3 ºF during the 24-hour period. This is a very low rate of heat loss. However, this calculation does not account for heat loss from piping—especially uninsulated piping—that may be connected to the tank. The equation also doesn't include the thermal mass effect of the metal tank walls. It only factors in the thermal mass of the water in the tank. Interestingly, if this tank were constructed of 0.25-inch thick steel on all surfaces, the thermal mass of that steel would only equal an extra 1.46 gallons of water in storage. Thus, excluding the thermal mass of the steel from the calculation has very little effect on the results.

The **exponential decay** of the tank's water temperature can be illustrated by plotting its temperature at several time intervals. An example is shown in Figure 9-40.

The upper curve labeled (c = 4.57) represents the situation described in Examples 9.3 and 9.4. This situation was based on a well-insulated thermal storage tank. The curve labeled (c = 22.85) represents a situation in which the value of c in Examples 9.3 and 9.4 was increased by a factor of 5. This would represent a tank with much less insulation and thus significantly higher rates of heat loss. The curvature resulting from the exponential decay is much more pronounced in the lower curve. So is the faster decline in average temperature.

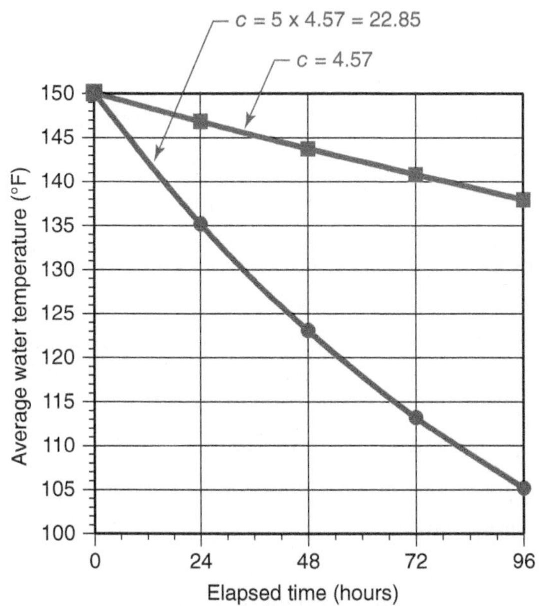

Figure 9-40 | Temperature drop versus time for the tank modeled in Example 9.4 and starting at an average water temperature of 150 ºF.

Equations 9.6, 9.7, and 9.2 can be combined to simulate the performance of a wide variety of vertical cylindrical thermal storage tanks. Similar expressions can be developed for tanks with other geometries, such as rectangular tanks.

9.6 Storage-Enhanced On-Demand Domestic Water Heating

Many combisystems supplied by one or more renewable energy heat sources have a thermal storage tank. The thermal mass of this storage can be leveraged, in combination with other hardware, to create a highly stable subsystem for heating domestic water. An example of that subsystem is shown in Figure 9-41.

The components within the light blue shaded area of Figure 9-41 form the subsystem. When a fixture or appliance demands hot water, and the flow rate reaches 0.6 gallons per minute, the **flow switch** closes its electrical contacts. This contact closure is used either directly, or indirectly through a relay, to turn on the small circulator. The warmest water at the top of the thermal storage tank is immediately circulated through the primary side of a stainless steel brazed plate heat exchanger, and back into the thermal storage tank. At the same time, cold domestic water enters the secondary side of this heat exchanger and absorbs heat from the primary side.

Depending on how the heat exchanger is sized, the water temperature leaving its secondary side is typically 5 to 10 ºF cooler than the water at the top of the thermal storage tank. Thus, if the upper portion of the storage tank is at a temperature of 85 ºF, the domestic water leaving the heat exchanger will be preheated to approximately 75 ºF to 80 ºF. Supplemental heat is needed to provide an acceptable delivery temperature.

That supplemental heat is provided by a ***thermostatically controlled*** **tankless electric water heater**

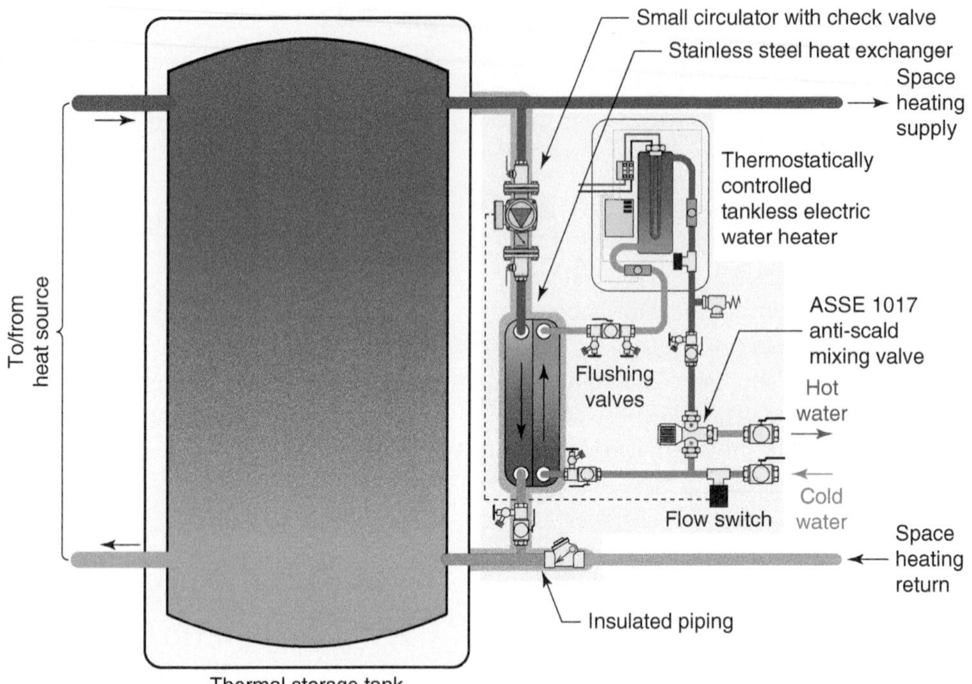

Figure 9-41 | Subsystem (shown in light blue background) for storage-enhanced on-demand domestic water heating.

that is piped inline with and downstream from the heat exchanger. This type of water heater is also turned on by its own internal flow switch that closes at a flow rate of about 0.6 gpm. When the heater's flow switch closes, an electrical contactor within the heater is energized to supply 240 VAC to **triacs**, which regulate electrical current flow through the heating elements. The electrical power delivered to the elements is regulated by electronics within the heater that measure incoming and outgoing water temperature. The power delivered to the elements is limited to that required to provide the desired outgoing water temperature. Thus, if preheated water is delivered to this heater at 115 °F, and the tankless heater's setpoint temperature is 120 °F, the power supplied to the elements will only be that required for the 5 °F temperature boost. If the water entering the tankless heater is already at or above the setpoint temperature, the elements remain off. This proportioning of wattage based on the difference between inlet and outlet water temperature is ideal for "topping off" preheated water from the heat exchanger.

Designers should be aware that not all tankless electric water heaters are *thermostatically controlled* as described above. Instead, some are simple on/off device that apply their full rated wattage to the water stream whenever flow is detected. *Such units are not recommended for this application.*

If the water temperature at the top of the thermal storage tank is at 140 °F after a sunny spring day, the domestic water leaving the heat exchanger is likely to be 130 °F to 135 °F. This temperature is higher than necessary, and thus no auxiliary energy is required. The hot water leaving the heat exchanger simply passes through the electric tankless water heater without further heat input.

To ensure a safe delivery temperature under all conditions, the hot water leaving the heater passes through an ASSE 1017–rated thermostatic mixing valve. This valve, which was described in Chapter 5, automatically blends the proper flow rate of cold domestic water to produce a delivery temperature no higher than its setting. The maximum delivery temperature allowed to fixtures should not exceed 120 °F.

There are several benefits to this method of domestic water heating:

- It takes advantage of the low temperatures at which domestic water heating begins. Some heat can be transferred from the thermal storage tank to domestic water under just about any condition. For example, even if the water within the thermal storage tank has cooled to 75 °F, it can still deliver heat to cold domestic water entering the heat exchanger at 40 to 50 °F. Remember: heating a given amount of water from 50 to 60 °F requires the same amount of energy as heating that water from 110 to 120 °F.

- It eliminates the need of a coil heat exchanger within the thermal storage tank. Although pressurized tanks with such coiled heat exchangers are available, they are usually limited to volumes of 119 gallons or less—as discussed earlier in this chapter. Furthermore, coils within tanks are generally not serviceable. If a leak occurs in such a coil, the entire tank usually has to be replaced. This is a very expensive scenario.
- The response time of a stainless steel brazed plate heat exchanger is extremely fast. This is the result of a very high ratio of heat transfer area per unit of volume. This typically allows the heat exchanger to achieve **steady-state heat transfer** conditions 1 to 2 seconds after steady inlet flows are present on both sides of the heat exchanger.
- Very little heated domestic water remains in the assembly at the end of a hot water draw. This reduces the potential for **Legionella** growth.
- This assembly is adaptable to a wide range of systems that include thermal storage. Examples of it will be shown in many of the combisystems discussed in later chapters.
- Stainless steel brazed plate heat exchangers are available in a wide range of sizes. Thus, this approach is reasonably scalable based on load requirements.
- The thermal mass provided by water in the thermal storage tank helps to stabilize hot water delivery in high demand situations. It also allows hot water delivery rates that are higher than those which could be provided by the tankless electric water heater alone.
- Very little electrical energy is required to operate the tank-to-heat exchanger circulator. Estimates have put the total operating cost of a properly sized ECM-based circulator, in this application for a single-family house, at *under one dollar per year*.
- This subsystem keeps the domestic water heat exchanger completely external to the tank. As such, it can easily be cleaned or replaced if necessary. The flushing valves shown in Figure 9-41 allow both the heat exchanger and the tankless electric water heater to be isolated and flushed with a **descaling solution**, if necessary, to remove deposits caused by hard water.

Figure 9-42 shows a brazed plate heat exchanger fitted with **combination isolation/flushing valves** that allow the domestic water side of the heat exchanger to be completely isolated from the remainder of the system for flushing and descaling. Flow for these combination valves enters and leaves through the side port ball valves (with red and blue handles), while the inline ball valves (with yellow handles) are closed to isolate the heat exchanger from domestic water piping. Notice that the upper valve includes a tapping for mounting a pressure-relief valve on the domestic water side of the heat exchanger.

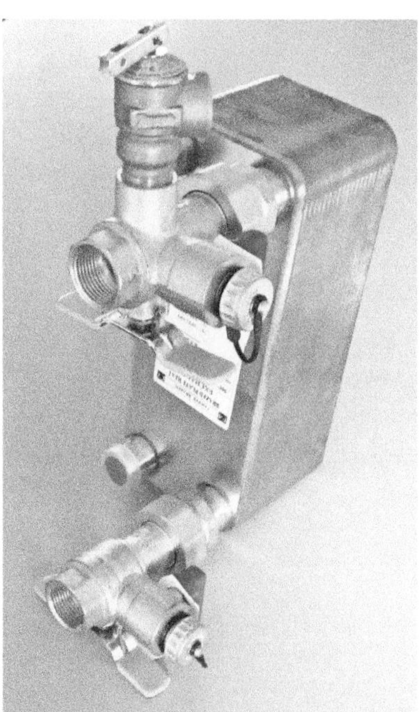

Figure 9-42 | Combination isolation/flushing valves mounted to a 5-inch wide by 12-inch tall by 40 plates deep stainless steel brazed-plate heat exchanger.

It is important to understand the function of each component in this domestic water heating subsystem.

Figure 9-43 shows examples of **domestic water flow switches**. This type of switch is commonly used in commercially available tankless water heaters. It has a brass or PVC body that connects to an assembly containing a **flow paddle** and **reed switch**. The flow

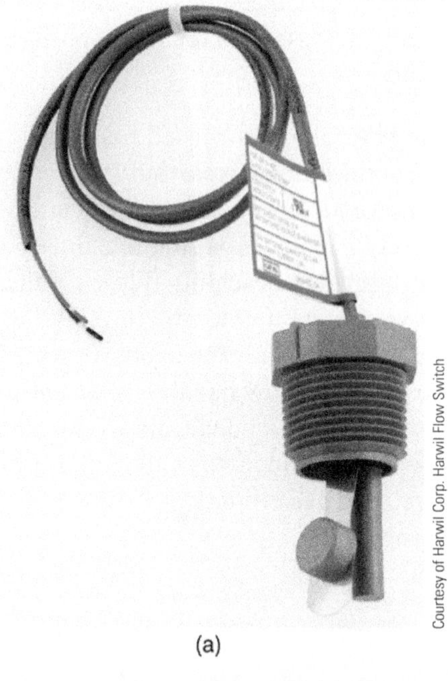

(a)

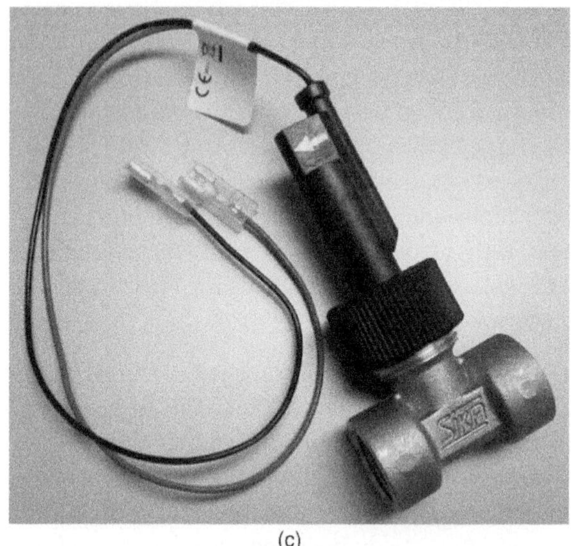

(c)

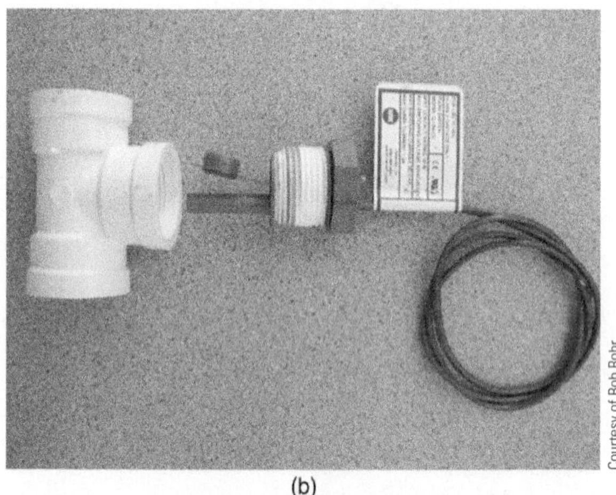

(b)

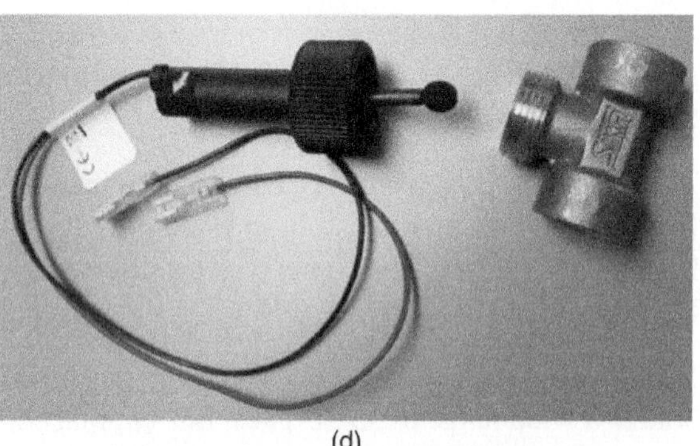

(d)

Figure 9-43 (a) Harwil Q-12N domestic water flow switch. Courtesy of Harwil. (b) Harwil flow switch ready to thread into a 3/4-inch PVC tee. Courtesy of Bob Rohr. (c) SIKA domestic water flow switch. (d) Disassembled SIKA flow switch showing internal paddle that moves when flow reaches a specific value.

paddle is oriented so that its flat side faces incoming cold water flow. A small spring holds the paddle in its forward position within the body. The paddle is sized so that it will move when the flow rate passing around it reaches a specified value. At that point, the force of the water against the paddle exceeds the spring force that normally holds the paddle in its forward position. This movement creates a corresponding movement of a small magnet, which in turn closes a set of light duty electrical contacts. These contacts are completely sealed from the water passing through the switch. This contact closure provides the "signal" that sufficient flow is present for the remainder of the system to operate.

When selecting flow switches, be sure they are rated for use with domestic water. All such switches used in the United States must now comply with a national "no-lead" law.

Flow switches with 1/2-inch or 3/4-inch piping connection are sufficient for most residential applications. Larger-size flow switches are available if needed; however, the minimum flow rate at which the switch contacts close typically increases with the pipe size of the flow switch. If a flow switch larger than 3/4-inch pipe size is needed, its higher activation flow rate may present a problem when only minimal flow of hot water is needed. An alternative wiring arrangement, presented later in this section, is the best alternative for such a situation.

Stainless steel brazed plate heat exchangers were discussed in Section 4.9. They are readily available in a variety of standard plate sizes and thermal capacities. Of those plate sizes, 5-inch by 12-inch heat exchangers are good for residential on-demand domestic water heating applications. Figure 9-44 shows an example of this size heat exchanger.

Figure 9-44 | Example of a 5-inch wide by 12-inch tall stainless steel brazed-plate heat exchanger.

For a given set of operating conditions, the rate of heat transfer across a 5 × 12 heat exchanger is determined by the number of plates that are stacked and brazed together along their perimeter. This is also the true for heat exchangers of other standard plate sizes. For on-demand domestic water heating, the heat exchanger should be selected so that it can supply the full design domestic water heating load, while operating at an approach temperature difference no higher than 10 °F, as illustrated in Figure 9-45.

Software available from heat exchanger manufacturers allows for quick and selection based on this criteria. An example of such software is shown in Figure 9-46. In this case, the heat exchanger is 5-inches by 12-inches

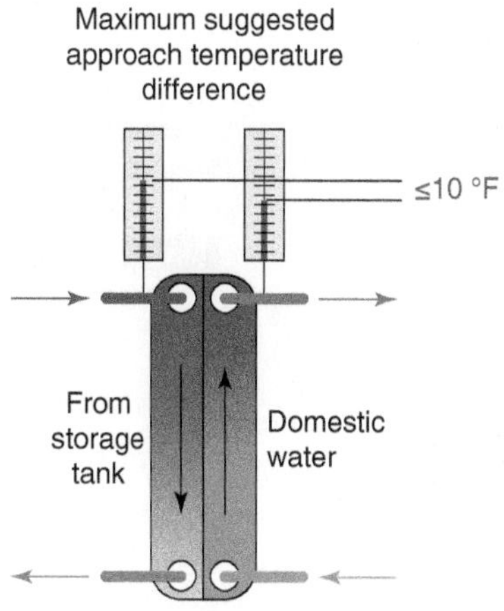

Figure 9-45 | Select a heat exchanger so that it can provide the design domestic water heating load with an approach temperature difference not exceeding 10 °F.

by 36 plates thick. It is capable of delivering 4 gpm of domestic water, heated from 65 °F to 115 °F, when its primary side is supplied with 120 °F water at 10 gpm. The rate of heat transfer across the heat exchanger is approximately 99,600 Btu/hr.

Any temperature boost needed after the domestic water has exited the heat exchanger is provided by a thermostatically controlled tankless electric water heater. Figure 9-47 shows the external appearance of one such heater. Figure 9-48 shows the interior details.

These heaters are available in a wide range of wattages depending on the load they are expected to meet. The smallest units have rated power inputs up to 3 kW and are intended for supplying domestic hot water to a single lavatory. The larger (single-phase residential) units are rated up to 40 kW and designed to supply the full hot water demand of a typical home.

All tankless electric water heaters have limits on the rate of domestic hot water production. This rate is based on their maximum electrical input power, which results in a relationship between flow rate through the heater and the maximum possible temperature rise that water flow can attain. Figure 9-49 shows an example of this relationship for a 20 kW–rated heater.

The heater represented by Figure 9-49 can raise the temperature of the water passing through it at 2 gpm by up to 68 °F. However, at a flow rate of 6 gpm, it can only raise the water temperature up to 23 °F. For the application being discussed, it is important to select a heater that can provide the design domestic hot water flow rate at a temperature rise corresponding to the

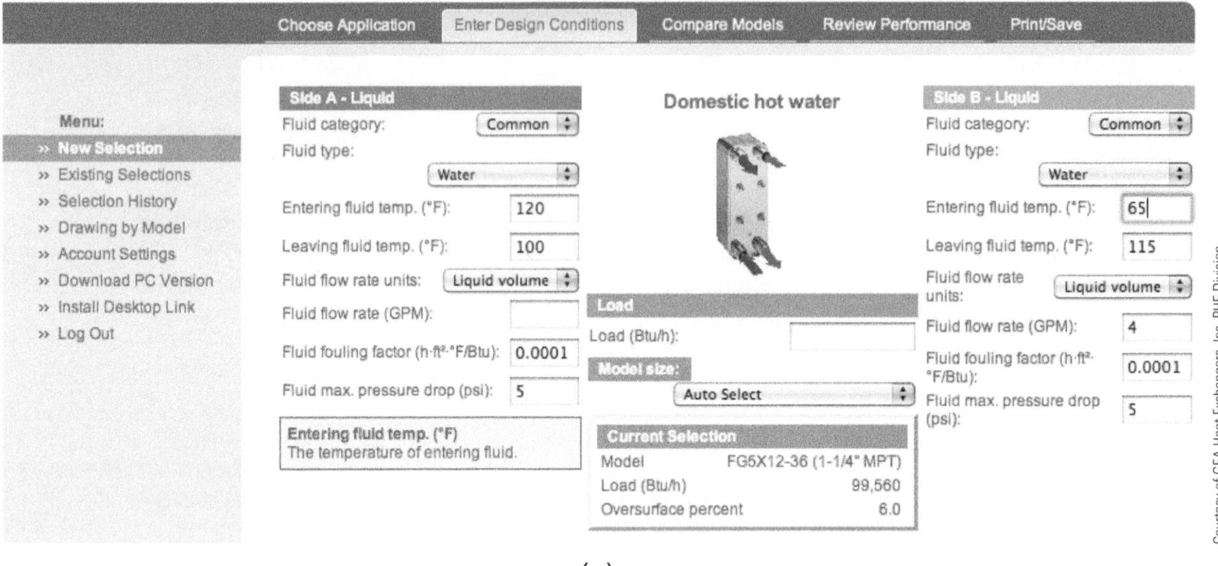

(a)

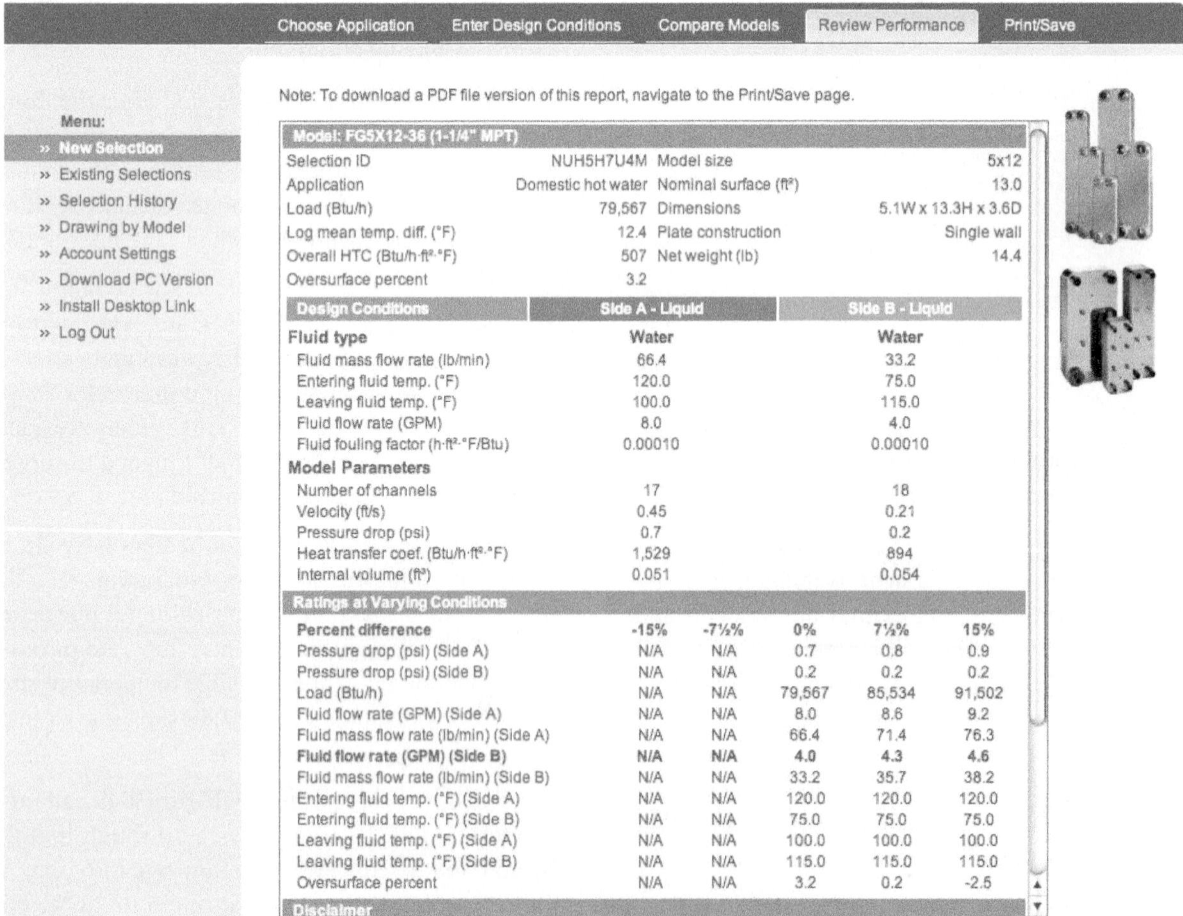

(b)

Figure 9-46 | (a) Using online software to select a heat exchanger. In this case, the selected unit will be a 5-inch wide by 12-inch tall heat exchanger with 36 plates. (b) Other operating conditions for the selected heat exchanger.

Figure 9-47 | Example of a thermostatically controlled tankless electric water heater.

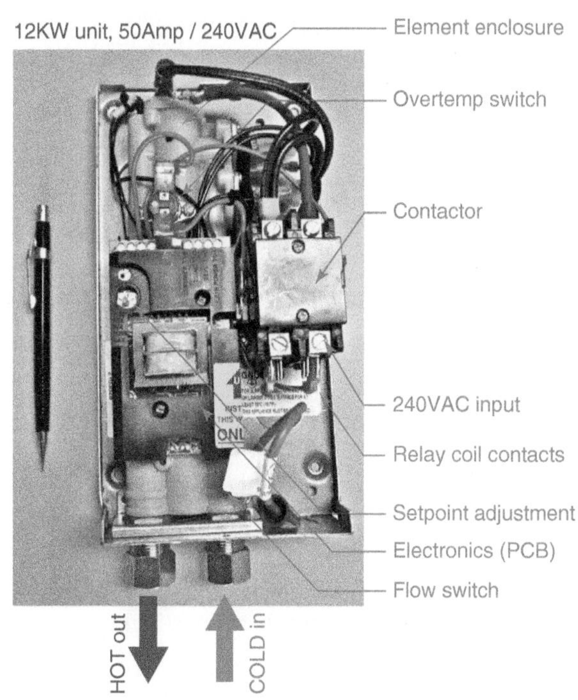

Figure 9-48 | Internal construction of a similar tankless electric water heater.

minimum expected preheat temperature from the heat exchanger and the expected delivery temperature. The author suggests a minimum preheat temperature of 65 °F, and a minimum acceptable supply temperature of 115 °F. With these temperatures, the maximum temperature rise would be (115 – 65) = 50 °F. The heater represented by Figure 9-49 could supply about 2.8 gpm under this condition.

Tankless electric water heaters, especially those intended to supply an entire house, have relatively high amperage requirements relative to tank-type electric water heaters. The amperage required for any tankless heater connected to a single phase 240 VAC electrical circuit is easily determined from the kW rating of the heater using Equation 9.8.

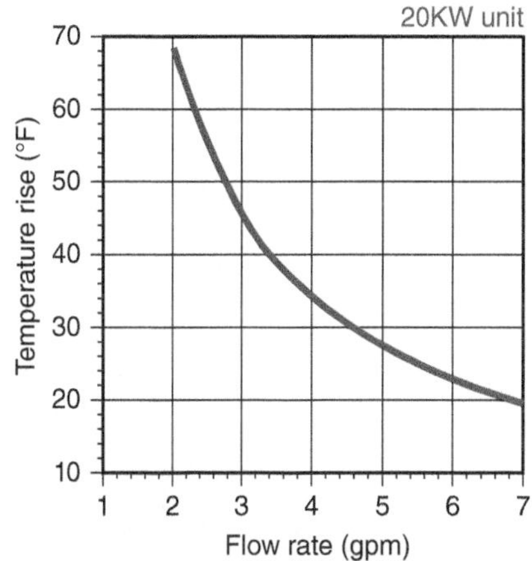

Figure 9-49 | Relationship between water flow rate and maximum temperature rise for a 20 kW–rated tankless electric water heater.

(Equation 9.8)

$$A = \frac{P}{0.24}$$

where:

A = required amperage (amps)

P = rated power (kilowatts)

0.24 = a constant based on a 240 VAC single-phase circuit

Example 9.5: Determine the necessary amperage required for a 3 kW–rated tankless water heater as well as that required for a 40 kW tankless heater.

Solution: For the 3 kW–rated heater.

$$A = \frac{3}{0.24} = 12.5 \text{ amp}$$

For the 40 kW–rated heater:

$$A = \frac{40}{0.24} = 167 \text{ amp}$$

Discussion: The amperage demand of the 3 kW heater is easily supplied by a single 20 amp/240 VAC circuit. However, the very high amperage demand of the 40 kW heater will require multiple circuits of lower **ampacity**. A 40 kW–rated heater would typically be supplied by *three* separate 60 amp/240VAC circuits. Given such high amperage requirement, it is critical that designers assess the electrical service entrance that is present, or planned for a building, before committing to the use of tankless water heaters. A *minimum* 200-amp electrical service entrance is suggested in such applications.

In some cases, it is possible to eliminate the domestic water flow switch shown in Figure 9-41. The minimum activation flow rate is instead detected by the flow switch built into the tankless water heater. When this internal flow switch detects sufficient flow, the main contactor in the heater is turned on to provide voltage to the heating elements. If the coil of this contactor has extra terminals, it may be possible to wire those terminals to the coil of another relay having the same coil voltage rating. This secondary relay can be used to supply line voltage to the tank-to-heat exchanger circulator as shown in Figure 9-50.

It is also important to determine if the tankless water heater has a **maximum temperature safety switch**. This switch, if present, is designed to stop operation of the heating elements if the water within the heater reaches a factory set maximum temperature. A common factory setting is 140 °F. Although this temperature is appropriate for normal (non-preheat) applications, there are likely to be times when the entering preheated water temperature from the heat exchanger is above this temperature. This is caused by high temperature water in the storage tanks and is likely when the tank is supplied by solar thermal collectors, or wood-fired boilers. Such a situation will cause the safety switch to immediately turn off the heating elements. This action usually requires the heater to be manually reset before it resumes operation. Some tankless heaters can be ordered with safety switches that open at 180 °F, which should be high enough to avoid nuisance trips.

If a higher temperature safety switch is not available, another option is to install a thermostatic mixing valve, as shown in Figure 9-51.

This additional thermostatic mixing valve reduces the water temperature entering the tankless heater to

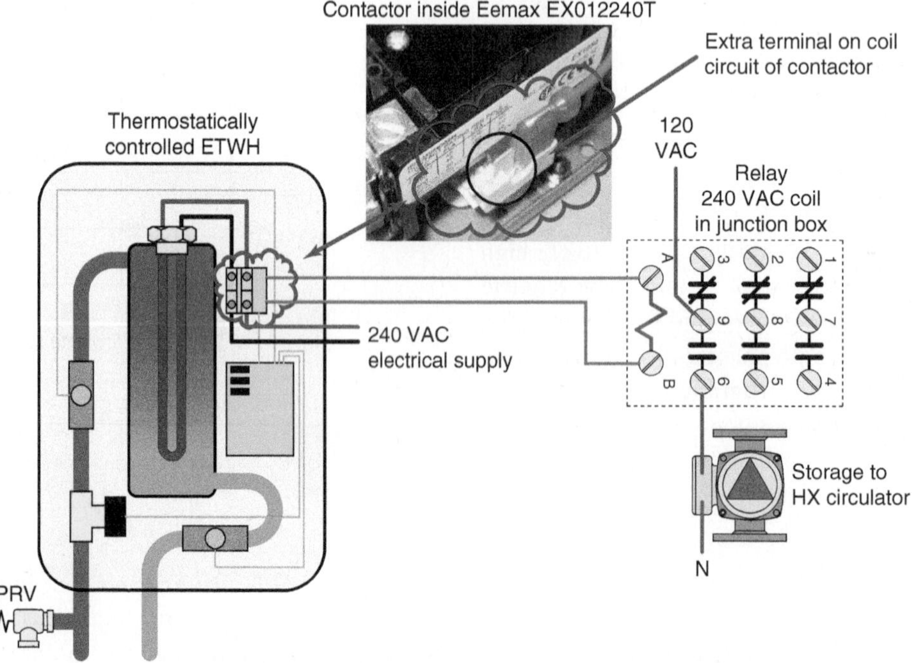

Figure 9-50 Using contactor in tankless water heater to turn on tank-to-heat exchanger circulator, via secondary relay. ***Caution: This wiring should only be used after verifying that it is acceptable to the manufacturer of the tankless water heater.***

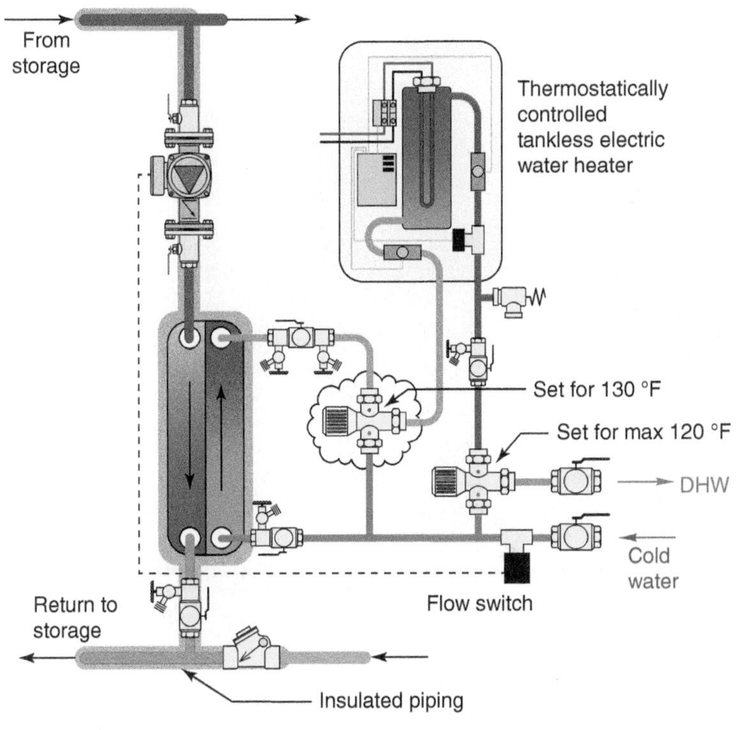

Figure 9-51 Installation of a second thermostatic mixing valve to reduce inlet water temperature to tankless heater to 130 °F.

not more than 130 °F, which is low enough to prevent tripping of the heater's internal safety switch. It does not, however, eliminate the need for the ASSE 1017–rated mixing valve on the outlet side of the heater.

9.7 Flow Reversal Through Storage

Some thermal storage tanks need to store heated water during the heating season and chilled water during the cooling season. This can force a compromise between the ideal piping and flow directions that encourage stratification in heating mode versus those for cooling mode. However, there are ways to eliminate this compromise.

One approach uses a single 4-way valve operated by a 2-position motorized actuator. An example of such a valve with the actuator attached is shown in Figure 9-52.

The 2-position motorized actuator needs to rotate the valve shaft 90° when changing from heating to cooling mode, or vice versa. One simple option is to use a **spring-return actuator** that is powered on to rotate the shaft 90° and then unpowered to allow the shaft to rotate in the opposite direction by 90°. 2-position spring return actuators are widely available and can be configured for either 24 VAC or 120 VAC operation.

Figure 9-53 shows how the 4-way valve would be mounted between a heat pump and thermal storage tank.

When the internal vane in the valve is rotated 90° the flow direction through the thermal storage tank is reversed. This allows optimal temperature stratification within the tank in both heating and cooling modes.

A motorized 4-way valve can also be used, when necessary, to reverse the flow direction through piping on the *load* side of the thermal storage tank. Figure 9-54 shows how this is done.

Be sure there are no check valves installed in any piping that will carry flow in both directions.

Flow reversal on the load side of a thermal storage tank maintains the same flow direction through terminal units that operate in heating as well as cooling mode and in combination with a single distribution circulator that is also used in both modes.

If separate terminal units are used for heating and cooling, and each subsystem has its own circulator, flow reversal on the load side of the thermal storage tank is not needed.

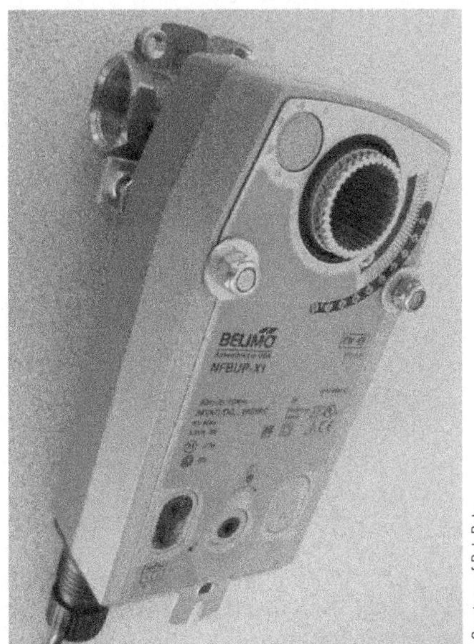

Courtesy of Bob Rohr

Figure 9-52 A spring-return motorized actuator mounted to a 4-way valve.

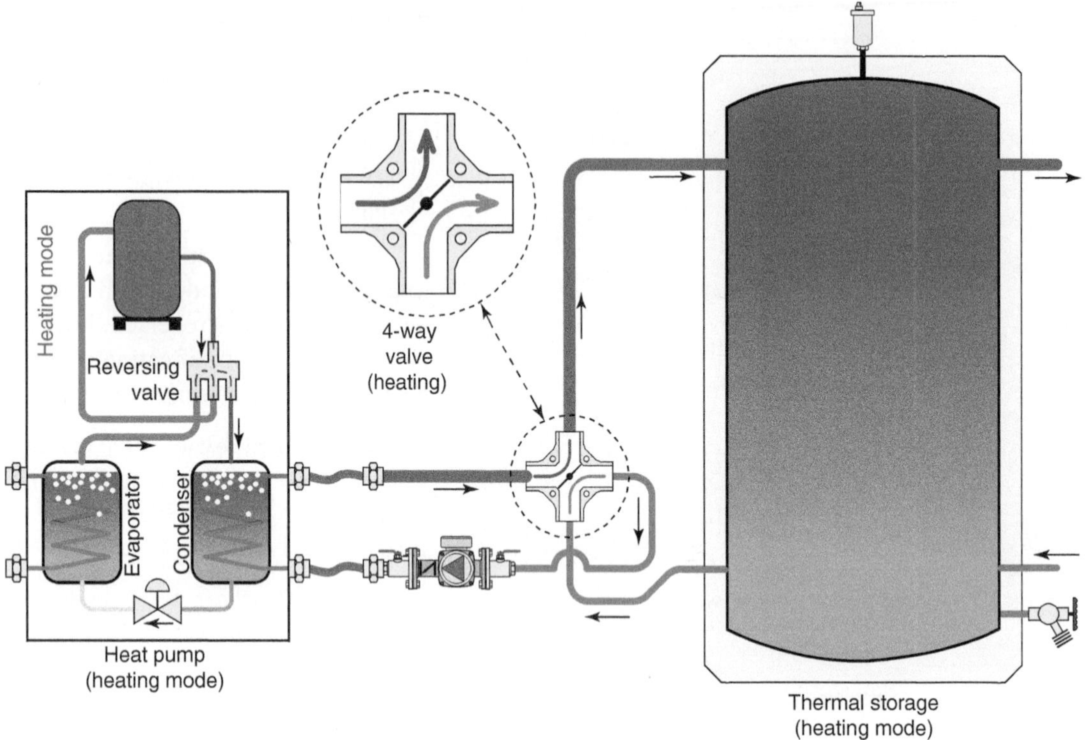

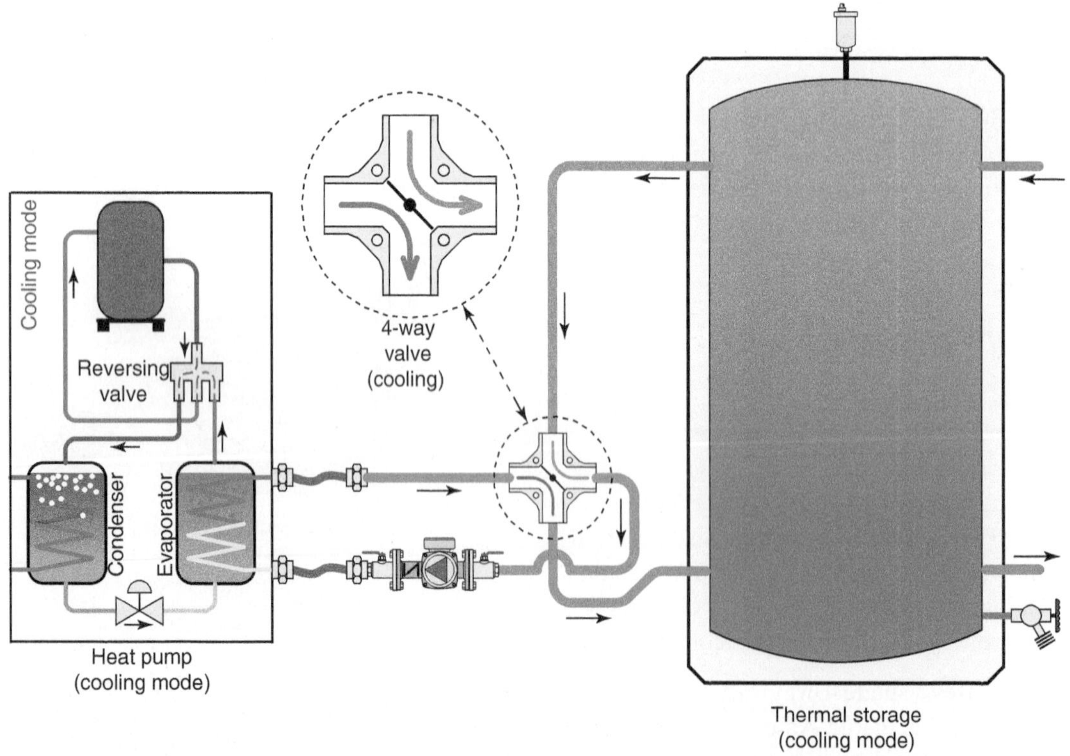

Figure 9-53 Use of a 4-way motorized valve, operated by a motorized actuator, to reverse flows in and out of thermal storage based on heating or cooling mode operation. Note the direction of the vane inside the 4-way valve in heating versus cooling mode.

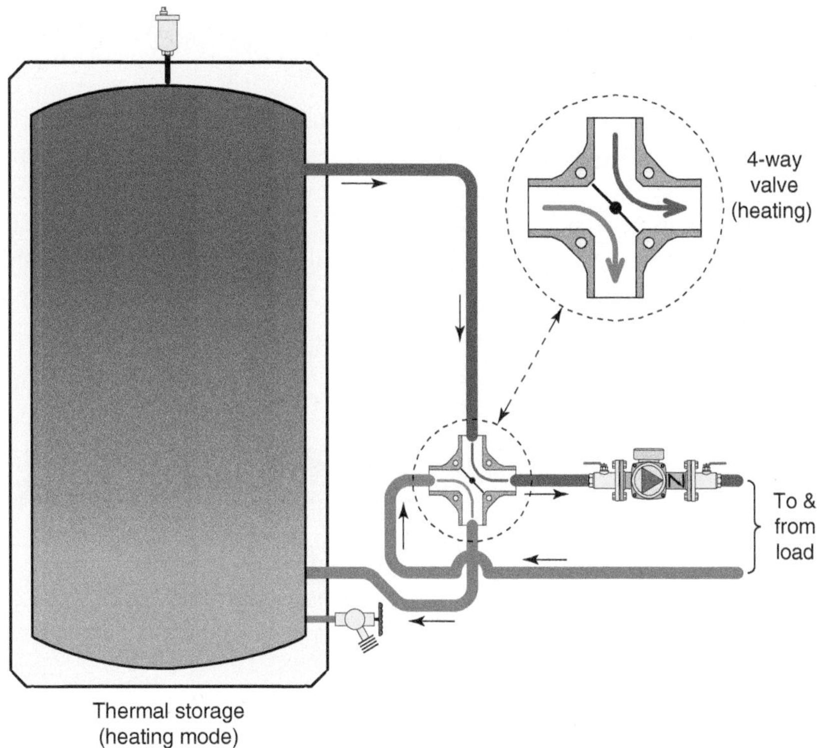

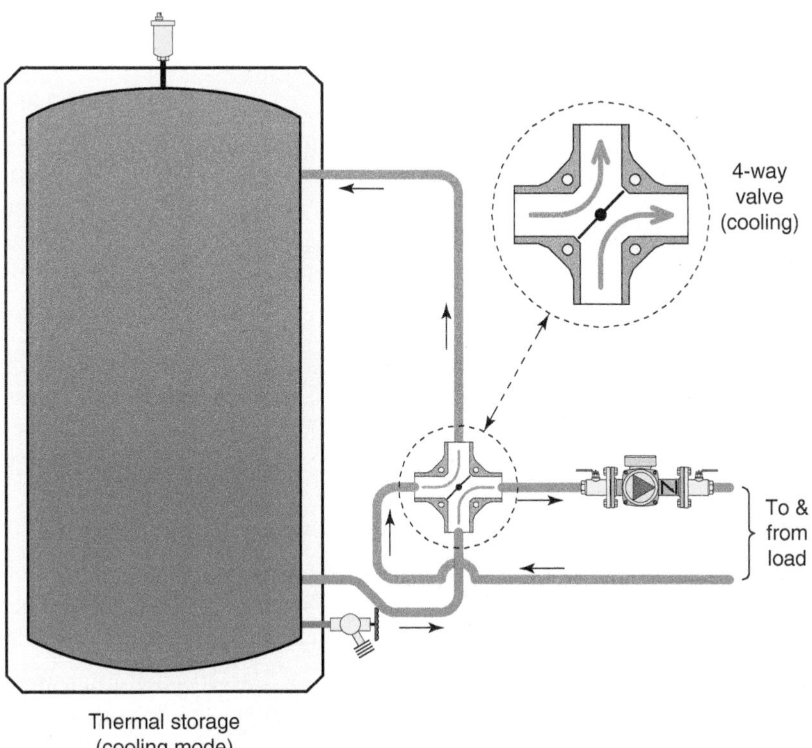

Figure 9-54 Use of a 4-way motorized valve, operated by a motorized actuator, to reverse flows in and out of the load side of a thermal storage tank based on heating or cooling mode operation. Note the direction of the vane inside the 4-way valve in heating versus cooling mode.

SUMMARY

The thermal storage tanks discussed in this chapter are an integral part of nearly every combisystem discussed in this text. Designers of such systems should be familiar with several tank options and match these options to the application requirements. Designers are encouraged to use the equations presented within this chapter to model the thermal performance of various tank sizes and geometries, especially how they relate to standby energy loss. Designers are also encouraged to play close attention to details that encourage temperature stratification within thermal storage tanks. The "on-demand" domestic water heating subassembly described in this chapter provides a way to extract heat from a variety of thermal storage tanks. The system examples given in later chapters will make extensive use of this subassembly.

KEY TERMS

2-pipe buffer tank configuration
4-pipe buffer tank configuration
ampacity
approach temperature
ASME certified
ASME-certified welding
ASME pressure vessel code
biocide additives
buffer tank
bulkhead fitting
combination isolation/flushing valves
counterflow heat exchange
descaling solution
differential temperature controller
dip tube
domestic water flow switches
double-wall heat exchangers
dual-coil thermal storage tank
EPDM rubber
exergy
exponential decay
external heat exchangers
flow paddle
flow switch
glass-lined
gooseneck piping
internal heat exchangers
intumescent coating
leakage path [in heat exchanger]
legionella
liner [of a tank]
maximum temperature safety switch
mercaptan
monolithic coating
multiple tank array
off-peak electrical rate
open [hydronic system]
parallel reverse return
polypropylene
pressure vessel
purging valve
reed switch
repurposed propane tank
short cycling
spring-return actuator
stainless steel brazed-plate heat exchangers
steady-state heat transfer
strata
thermostatically controlled tankless electric water heater
triacs
unpressurized tank
usable temperature drop
vertical flow jet

QUESTIONS AND EXERCISES

1. Describe a situation where an unpressurized thermal storage tank would be preferred over a pressurized tank.

2. Why are piping connections on unpressurized tank typically made above the tank's water line?

3. A vertical coil copper heat exchanger is used to extract heat from an unpressurized thermal storage tank. Should the water flowing through this coil flow from top to bottom or bottom to top? Justify your answer.

4. What kind of circulator is required in any piping circuit that connects directly to an unpressurized "open" thermal storage tank?

5. Why should circulators in circuits carrying water out of unpressurized thermal storage tanks be located as low as possible below the tank's water line?

6. Internal coiled heat exchangers rely on ___________ convection on the inside of the tube and ___________ convection on the outside of the tube.

7. An unpressurized 500-gallon tank is initially filled with 60 °F water. Determine the expanded volume if all this water is heated to 160 °F.

8. A vertical cylindrical tank needs to have a volume of 400 gallons. Assuming it has a diameter of 35 inches, determine its height. Assume the tank has a flat top and bottom.

9. Assume the tank in the previous exercise contains 350 gallons of water initially at 60 °F. Determine the rise in water level within the tank if all the water is heated to 180 °F.

10. What is the top coil in a dual-coil vertical cylindrical thermal storage tank typically used for?

11. Describe the purpose of a double-wall internal coil heat exchanger.

12. Explain why multiple storage tanks used in parallel need to have reverse return piping. Suppose that only direct return piping is possible. What details need to be added to the piping connecting the tanks?

13. What is a disadvantage of using three 100-gallon tanks, connected in parallel, compared to a single 300-gallon tank?

14. Explain the function of an inlet flow diffuser in a thermal storage tank.

15. Explain why fluid flowing from a thermal storage tank to a solar heating array should be drawn from the lower connection on the tank.

16. Why is more of the surface area of an internal coil heat exchanger used for heating domestic water typically located in the upper portion of the storage tank?

17. Describe two advantages and two disadvantage of using a repurposed propane storage tank for thermal storage applications compared to a custom fabricated tank.

18. A thermal storage tank contains 600 gallons of 60 °F water. How much heat is required to raise the temperature of this water to an average of 160 °F? How much heat would have to be removed from the tank to chill the water down to 40 °F?

19. Explain how the use of an outdoor reset control to invoke the auxiliary heat source can extend the usable temperature drop of a thermal storage tank compared to use of a simple temperature setpoint controller.

20. At design load conditions, an outdoor reset controller is programmed to stop flow from a 300-gallon thermal storage tank when the water temperature at the top of the tank drops to 115 °F. Assuming the distribution system operates with a 22 °F temperature drop when supplied with 115 °F water, and that an auxiliary boiler supplies supplemental heat as needed, determine how much additional heat could be extracted from the tank and transferred to the load if a differential temperature controller with a 2 °F off differential is used, as shown in Figure 9-37.

21. A vertical cylindrical thermal storage tank measures 24 inches in diameter and is 84 inches tall. It has a flat top and bottom. Determine the rate of heat loss from this tank when the average water temperature in it is 160 °F, the surrounding air and floor temperature is 70 °F, and the tank is insulated as follows:
a. 2" inches of spray polyurethane foam on all surfaces ($R = 6.0$/inch, $k = 0.1389$ Btu/hr.·ft.·°F)
b. 3" inches of spray polyurethane foam on all surfaces ($R = 6.0$/inch, $k = 0.1389$ Btu/hr.·ft.·°F)
c. 4" inches of spray polyurethane foam on all surfaces ($R = 6.0$/inch, $k = 0.1389$ Btu/hr.·ft.·°F)

22. Assuming the thermal storage tank described in Exercise 21 has 2 inches of spray polyurethane foam on all surfaces, and is initially filled with 140 °F water, determine the average water temperature after 48 hours of no heat input or extraction by loads.

23. Describe three advantages of using an external stainless steel brazed plate heat exchanger, compared to an internal coil heat exchanger, for domestic water heating.

24. Describe the operation of a thermostatically controlled electric tankless water heater versus a standard electric tankless water heater in the following scenario: Setpoint temperature of heater = 115 °F, and the temperature of preheated water entering heater is 110 °F.

25. What is meant by selecting a heat exchanger with an approach temperature difference of 5 °F?

26. An electric tankless water heater is rated at 12 kW.
a. Determine the maximum amperage required by the heater when supplied with a 240 VAC circuit.
b. Determine the maximum outlet temperature of water entering at 3.0 gpm and 65 °F.

DAIKIN
R410A
R410A

CHAPTER 10

Antifreeze-Protected Solar Combisystems

OBJECTIVES

After studying this chapter, you should be able to:

- Describe why freeze protection is necessary.
- Understand the difference between propylene glycol and ethylene glycol.
- Determine if the on-site water is suitable for use with glycol-based antifreeze.
- Describe the difference between freeze protection and burst protection.
- Explain how a collector loop is filled with fluid and purged of air.
- Describe how a passive heat dump subsystem works.
- Describe several methods of over-temperature protection and heat dumping.
- Size a heat emitter for heat dumping from a solar collector array.
- Size and select expansion tanks for antifreeze-based collector circuits.
- Properly place the expansion tank within the collector circuit.
- Discuss antifreeze-based solar domestic water heating systems.
- Describe the use of F-CHART software for analyzing solar DHW system performance.
- Describe the operation of several antifreeze-based combisystems.

10.1 Introduction

Several previous chapters have discussed the "building blocks" of solar thermal systems including solar collectors, storage options, and control devices. This chapter will show how to assemble those building blocks into complete solar combisystems that provide domestic hot water and space heating.

Almost every solar thermal system installed in North America, and solar combisystems in particular, require protection against freezing. Even climates such as Florida, Arizona, and Southern California may, at times, produce conditions that can cause water within solar collectors and exposed piping to freeze. This could occur on a night when the air temperature drops to 32 °F or less. It can also occur when the air temperature is several degrees above 32 °F, due to **radiative cooling** under clear nighttime skies.

In solar combisystems, there are two common methods of protecting the collectors and the piping connected to them from freezing:

- Using an antifreeze solutions in a closed collector circuit (e.g., **antifreeze-based freeze protection**)
- Draining water from the collector array and associated piping whenever the solar collectors are not collecting energy (e.g., **drainback freeze protection**)

This chapter discusses systems that use antifreeze-based freeze protection. The next chapter will discuss systems using drainback freeze protection.

10.2 Antifreeze Fluids

There are several antifreeze products currently available for use in solar thermal systems. Most are based on **propylene glycol** or **ethylene glycol**.

Propylene glycol is a nontoxic chemical. It is a widely used in food, pharmaceutical, and personal care products. It is also used for aircraft deicing. In solar thermal systems, as well as other HVAC applications, mixtures of propylene glycol and water can provide freeze protection to temperatures as low as −60 °F.

Ethylene glycol is a chemical with moderate oral toxicity. It is the base chemical used in most automotive antifreezes. Although it has several engineering properties that are preferable to those of propylene glycol, its toxicity is a major concern in any system that heats domestic water, because that water might be ingested by humans or animals. For this reason, *the author recommends against the use of ethylene glycol antifreeze in any heating system that provides domestic hot water.*

Most commercially available propylene glycol fluids for use in HVAC systems are not 100 percent propylene glycol. Instead, they are mixtures of propylene glycol and specific chemicals called **inhibitors**. These chemicals extend the useable life of the fluid by reducing its tendency to become acidic with age, especially when used at high temperatures. Solar collectors can reach very high temperatures during **stagnation**. Because of this, only properly inhibited propylene glycol fluids should be considered for such system. Many manufacturers offer specific formulations of inhibited propylene glycol for use in solar thermal applications.

It is also very important to use water with suitable properties when preparing antifreeze solutions. Various impurities in water can cause pitting in cast iron or steel, reduce the protection of corrosion inhibitors, and create scale within the system. Dow Chemical recommends the following minimum specifications for the dilution water that will be mixed with its propylene glycol-based fluid.

If on-site water does not meet these specifications, **distilled water** or **deionized water** should be used. Alternatively, some manufacturers offer prediluted solutions of inhibited propylene glycol, which are ready to add to the system without further mixing.

The temperature at which ice crystals begin forming in solutions of inhibited propylene glycol depends on the percentage of propylene glycol in the solution with water. Figure 10-2 shows the **freeze point**

Impurity	Maximum level
Chlorides	25 ppm
Sulfates	25 ppm
Total hardness as $CaCO_3$	100 ppm

Figure 10-1 Minimum water quality specified for use with Dowfrost inhibited propylene glycol antifreeze. Note: 17.1 ppm = 1 grain hardness.

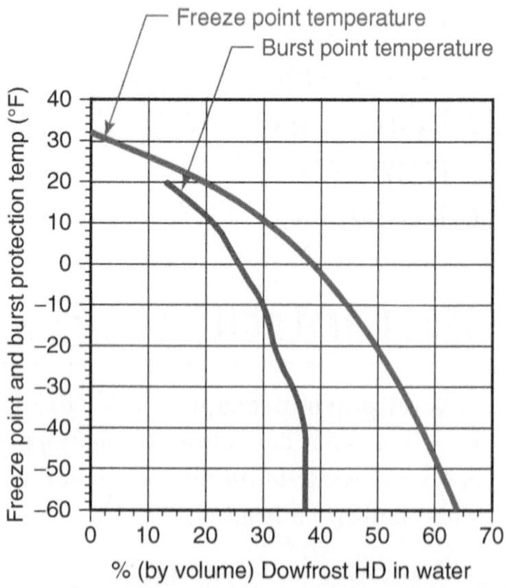

Figure 10-2 Freeze point temperature and burst point temperature of various concentrations of Dowfrost HD inhibited propylene glycol.

temperature and **burst point temperature** of a specific formulation of inhibited propylene glycol known as **Dowfrost HD**, based on the antifreeze's percent (by volume) concentration in water.

The freeze point temperature is where ice crystals begin forming in the solution. The fluid is still able to flow at this temperature. If the temperature drops lower, the solution transitions from a liquid to a **slush** and its volume increases due to the continued formation of ice crystals. Assuming the system has an expansion tank that can absorb this increase in volume, the slush will not cause piping components to burst until it reaches the burst point temperature, which is typically well below the freeze point temperature. However, circulators and other piping components in the system are not designed to handle a mixture of liquid and ice crystals (e.g., slush). Thus, the freeze point temperature is the lowest temperature at which flow should be maintained, whereas *the burst point temperature is only relevant to preventing damage in systems that are not operating.*

Concentrations of less than 25 percent propylene glycol by volume are not recommended. The concentration of the inhibitor may be too low to be effective. Certain microorganisms may also be able to grow in solutions with less than 25 percent inhibited propylene glycol. In most northern U.S. climates, concentrations of 50 percent are adequate. This concentration provides freeze protection to approximately −23 °F and burst protection to below −60 °F.

Courtesy of Alpha-Fernox

Figure 10-3 Fernox F3 system cleaner for pre-commissioning piping systems before adding antifreeze solutions.

PREPARING A SYSTEM FOR AN ANTIFREEZE SOLUTION

It is very important that the piping system be properly cleaned before it is filled with a propylene glycol solution. Impurities such as solder flux, cutting oil, scale, and dirt can react with propylene glycol to form sludge that inhibits heat transfer and evenly interferes with the smooth operation of internal mechanisms such as those contained in automatic air vents and certain types of spring-loaded valves. Various chemical products, such as the one shown in Figure 10-3, are available for cleaning the inside of piping before the final operating fluid is added to the system.

These cleaning compounds are typically injected into a system that already contains water. They are then circulated for several hours, while maintained at specific minimum temperatures based on the manufacturer's specifications. After the cleaning cycle, all the fluid in the system should be drained and the system flushed with fresh water to dilute and remove any residual cleaning fluid. Follow the manufacturer's instructions for proper disposal of the cleaning solution. If ample drain valves are not present, compressed air can sometimes be used to dislodge pockets of fluid from low points in piping circuits.

The preferred method of adding antifreeze to a system is by premixing it with water and pumping the required volume into a properly cleaned, empty piping

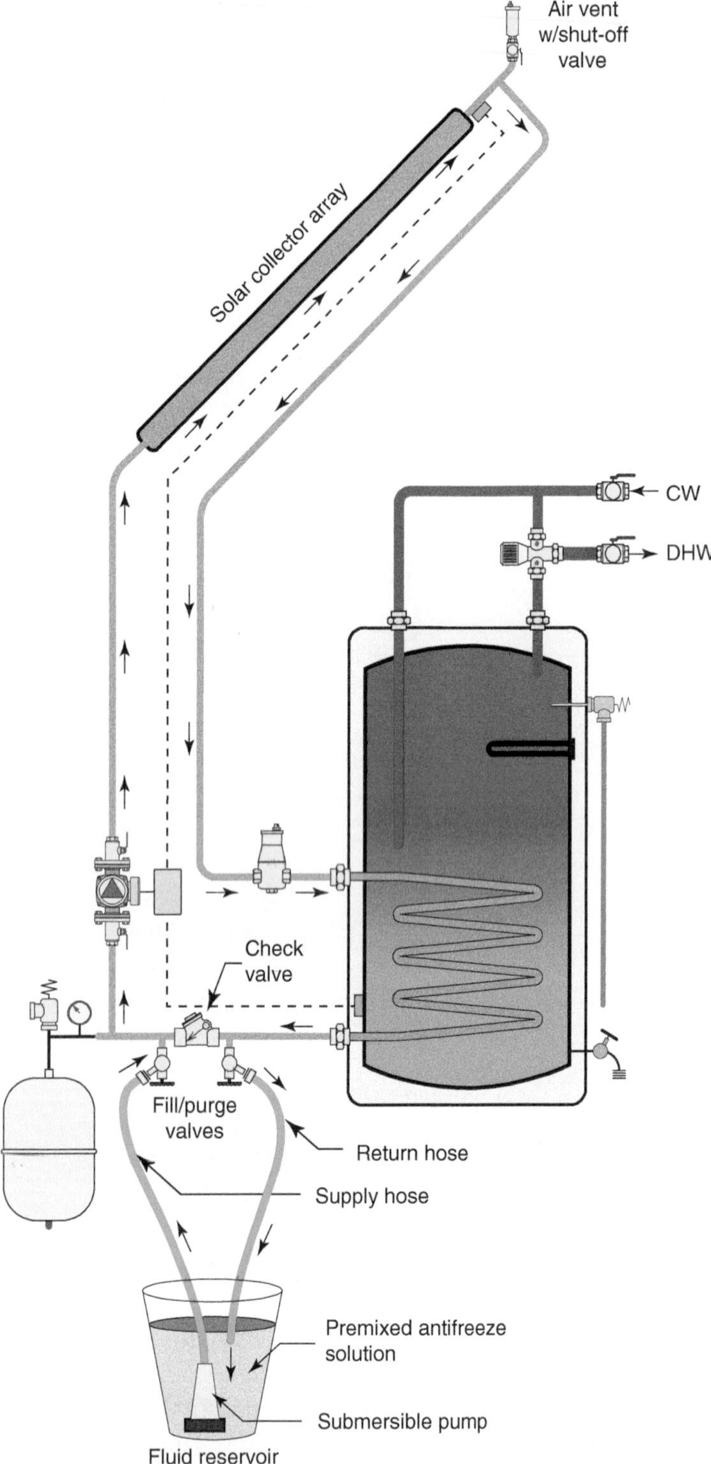

Figure 10-4 | Setup for filling a solar collector circuit with antifreeze solution and purging it of bulk air.

Courtesy of Bob Rohr

Figure 10-5 | Example of a small hand pump for increasing system pressure.

circuit. Figure 10-4 shows a setup for adding antifreeze solution to a typical solar collector circuit while simultaneously allowing air to exit that circuit. A small submersible pump is usually adequate to fill a small solar collection circuit.

The two valves that connect the solar collector circuit to the purging hardware are located on opposite sides of a spring-loaded check valve. The primary purpose of this check valve is to prevent **reverse thermosyphoning** of the antifreeze solution when the collector array is cooler than the storage tank. However, when this check valve is placed as shown in Figure 10-4, it backseats in the closed position during purging and thus forces the incoming antifreeze solution to flow around the collector circuit before exiting at the hose drain valve to the right of the check valve.

Once the fluid returning from the collector circuit to the fluid reservoir has been free of visible bubbles for several minutes, the outlet valve is closed and the system is slightly pressurized, using the purging pump. Any additional fluid needed to increase system pressure can be added using a small hand pump such as the one shown in Figure 10-5.

Larger **purging carts** are also available when higher purging flow rates are needed. Figure 10-6

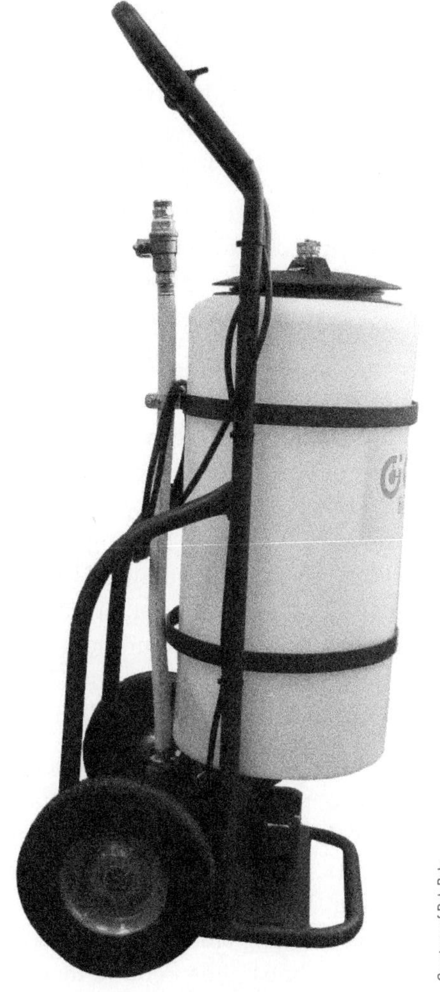

Courtesy of Bob Rohr

Figure 10-6 | Example of a purging cart designed for filling and purging solar collector circuits.

shows an example of such a cart with a fluid reservoir, pump (beneath the fluid reservoir), pipe connections, and wiring.

10.3 Design Details for Antifreeze-Based Systems

There are several design details that pertain to solar thermal systems using an antifreeze solution in the collector circuit. These details deal with the nuances of protecting both the antifreeze solution and the components within the collector circuit under extreme conditions.

HEAT DUMP SUBSYSTEMS

All solar thermal systems will experience periods during which solar radiation is incident on the collectors, but no flow is passing through the collectors. This condition is called stagnation. All solar collectors certified to the SRCC OG-100 standard have passed a test to verify that the materials they contain will not be adversely affected by stagnation. However, glycol-based antifreeze fluids are not part of this test. The nominal 350 °F absorber-plate temperatures that are possible within flat-plate collectors under warm weather stagnation conditions can quickly cause glycol-based antifreeze fluids to become acidic. This in turn will lead to corrosion within the collectors and other parts of the collector circuit. Thus, it is prudent to include a means of protecting glycol-based antifreeze solutions against stagnation temperatures.

Solar-specific propylene glycol fluids such as Dowfrost HD and **Tyfocor LS** contain inhibitors and **pH buffers** that stabilize them to temperatures of approximately 325 °F to 338 °F. *The level of these inhibitors and buffers should be checked on an annual basis.* Manufacturers of solar-specific glycols typically provide test kits for this purpose. To protect against corrosion, inhibitor chemicals may need to be replenished, especially if the collectors experience many hours of stagnation.

To complement the fluid's ability to withstand high temperatures, many designers of antifreeze-based systems prefer to include a **heat dump** within the collector circuit. The term *heat dump* applies to any practical and automatic means of dissipating heat from the collectors under excessively high temperature conditions. The objective of a heat dump is to prevent the glycol-based antifreeze solution from thermally degrading under such conditions.

The following devices can be used as heat dumps:

- Thermosiphon cooling loop
- Photovoltaic powered circulator
- Swimming pools
- Fan coils/fluid coolers
- External pavements with embedded tubing
- Earth loops for geothermal heat pumps
- Automatic dumping of hot domestic water from the tank

A thermosiphon cooling loop is built by suspending one or more rows of **fin tube** (e.g., copper tubing with aluminum fins) along a sloping piping path behind the collectors, as shown in Figure 10-7.

Commercial fin tube is available in copper tube sizes from 3/4 inch to 1.25 inch and with aluminum fins in sizes from 3.25 inches by 3.25 inches to 4.25 inches by 4.25 inches. The number of fins per foot of tube length is typically 32, 40, or 48. Tube lengths from 2 to 12 feet are available. Fins are available in either steel or aluminum. The latter are preferred in heat dump applications due to corrosion resistance when exposed to the elements.

Figure 10-7 | Example of passive fin-tube convector behind a solar collector array.

The fin tube is sloped downward in the direction of flow. This allows the cooling antifreeze solution to flow without the need of a circulator.

The fin tube can be connected, in combination with other components, to form a high-temperature activated circuit between the outlet and inlet of the collector array, as shown in Figure 10-8.

The **thermosiphon cooling loop** is run in parallel with the collector array's supply and return piping. It is filled with the same antifreeze solution as the collector array when the system is commissioned. The **thermostatic valve** seen in the upper right corner of the array may have a manual opening lever to allow flow through the valve during purging. If not, the valve should be gently heated with a hot air gun to open its thermostatic element and allow purging flow through. Purging flow exits through a drain valve located near the lower left of the collector array in Figure 10-8.

Flow through the heat dump piping occurs whenever the temperature of a thermostatic valve at the outlet of the collector array reaches a specific temperature (such as 220 °F), which infers that stagnation is occurring. When the thermostatic element in the valve senses this temperature, it opens a port leading through the thermosiphon cooling loop. Hot antifreeze fluid can now pass into the downward sloping fin tube, dissipating heat as it flows. As the fluid cools

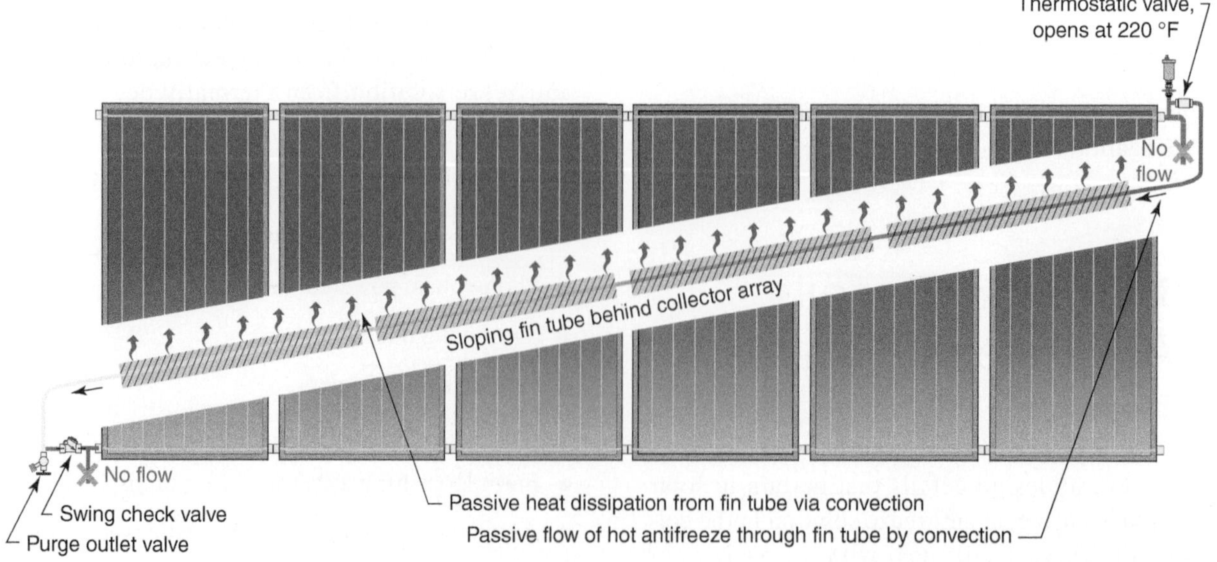

Figure 10-8 | A thermosiphon cooling loop. Note placement of thermostatic valve at upper right as well as swing check valve and purging outlet valve in lower left.

along the fin tube, its density increases compared to that of the hot fluid in the collectors. This provides sufficient differential pressure for thermosiphon flow and keeps the antifreeze solution from reaching temperatures where rapid thermal breakdown occurs. As the solar intensity and air temperature decrease, so will the temperature of the fluid within the collectors. When the thermostatic valve drops a few degrees below its established activation temperature, it closes off the thermosiphon cooling loop and allows normal operation to resume.

A typical residential fin-tube element mounted with a slight slope can dissipate about 250 Btu/hr. per foot of length, assuming an inlet fluid temperature 85 °F higher than the outdoor air temperature. Commercial fin-tube elements can dissipate heat at higher rates due to their larger fins.

One downside of such elements is that the spaces between fins are subject to clogging from wind-borne debris and insects. They should be periodically inspected and cleaned with a shop vacuum or pressure washer.

There are several other methods for heat dumping that require some type of electrical power. That power could be line voltage AC, supplied from a backup generator, or low-voltage DC, supplied from a **solar photovoltaic module**. The latter would function during a utility power outage and without a backup generator.

Figure 10-9 shows an example of a heat dump subsystem that can maintain flow between the collector array and the fin-tube heat dump during a loss of utility power and without a backup generator.

When the collector circulator is operating normally, 120 VAC is applied to the 3-way diverter valve. This routes flow from the AB port to the A port. The electrical circuit between the photovoltaic panel and circulator is open under this condition. During a utility power outage, the diverter valve is off and the flow passage opens between port AB and port B. Port A closes. The electrical circuit between the photovoltaic panel and the small DC circulator is closed, allowing the circulator to operate. Flow passes from the collector array, through the diverter valve, onward through the fin-tube heat dissipator, and then through the remainder of the collector circuit. The higher the intensity of solar radiation, the faster the DC circulator will operate. When utility power is restored, the DC circulator is disconnected from the photovoltaic panel and the system resumes normal circulation.

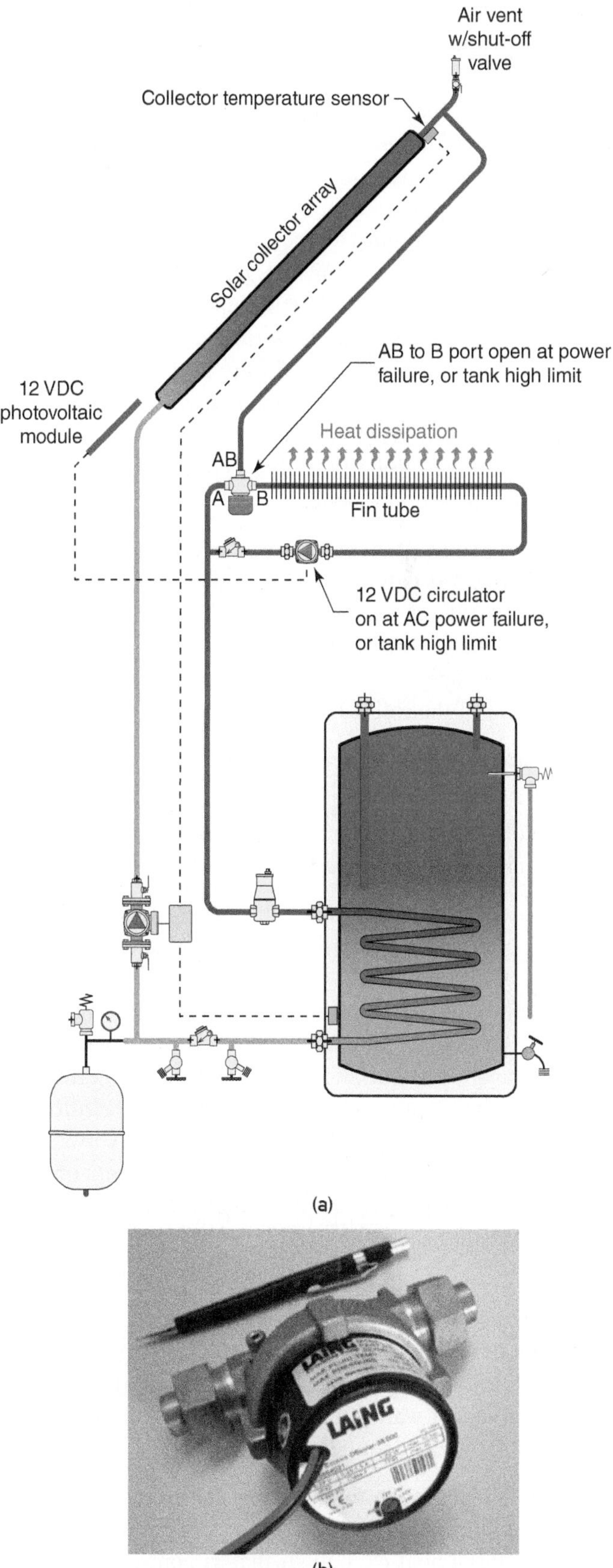

Figure 10-9 (a) Heat dump using small DC circulator power by photovoltaic module. (b) Small DC circulator.

OVER-TEMPERATURE PROTECTION

There will be times when every solar combisystems is able to heat the thermal storage tank to very high temperatures. Since most temperature and pressure relieve valves automatically open at 210 °F, it makes sense to limit the storage tank temperature so that the relief valve will only operate in an emergency. A temperature of 180 °F is often selected as the upper limit for the storage tank. If the storage tank reaches this temperature, most current-generation differential temperature controllers initiate an **over-temperature protection mode**. This differs from heat dumping, because utility power is assumed to be available to operate the over-temperature protection mode.

One method of over-temperature protection is called **nocturnal cooling**. In this mode, the collector array serves as a heat *emitter* rather than a heat collector. The collector circulator turns on when the differential temperature controller operating the system detects *both* of the following conditions:

a. The tank temperature is at or above a high limit setting (typically about 180 °F)

b. The collector temperature is a few degrees lower than the tank high limit temperature

Once this mode is initiated, the controller operates the collector circulator until the tank cools to a lower setpoint of about 140 °F to 150 °F, at which point the over-temperature protection mode ends. Although it may not seem to be the best way to conserve energy (thermal and electrical), nocturnal cooling is an effective way to prevent accelerated chemical degradation of glycol-based collector fluids. It is particularly useful during warm sunny weather when the occupants of a home may be away on vacation, and thus there is no domestic water heating demand.

Nocturnal cooling only works in systems using flat-plate collectors. The vacuum envelope of evacuated tube collectors—the same detail that retains heat under desirable operating conditions—significantly limits heat loss if warm fluid is intentionally circulated through the evacuated tubes.

Another approach to over-temperature protection is to circulate hot antifreeze fluid from the collector array though a heat exchanger that is connected to a swimming pool, as shown in Figure 10-10.

A **flow switch** is used to verify that pool water is flowing through the heat exchanger during this mode. It obviously makes no sense to circulate hot antifreeze through one side of the heat exchanger without pool water flowing through the other side.

Other over-temperature protection methods include heat diversion to any of the following:

- Fan coil
- PEX or PEX-AL-PEX tubing embedded in exterior pavement for snowmelting
- Geothermal earth loops

Of these, the fan-coil option is arguably the least complicated. The fan-coil unit could be mounted inside or outside. If mounted inside, air would be ducted from outside and discharged outside. If outside mounting is planned, the fan coil should be constructed of materials that will not corrode due to exposure to the elements. Preference should be given to mounting the fan coil in the shade, on the north side of the building, and preferably under the cover of an overhang, as shown in Figure 10-11.

Exterior pavements with embedded tubing can also absorb surplus heat. Pavement areas in shade will provide better heat dissipation than those in direct sun. A system that supplies snowmelting could likely be set up as a heat dump using the piping shown in Figure 10-12.

Under normal operation, the diverter valve is unpowered. Flow returning from the collector array passes from the valve's AB port to its B port. If the controller determines that the storage tank has reached its high temperature limit, it applies power to the diverter valve. This opens the flow path between the AB port and the A port and closes the B port. The end switch in the diverter valve also closes when the A port is fully open. This switch closure, in combination with a relay, turns on the circulator for the embedded pavement circuits. The hot antifreeze solution is routed through the brazed-plate heat exchanger, which transfers heat to the fluid in the embedded tubing circuits. The bypass valve across the heat exchanger allows some blending of the potentially hot fluid exiting the load side of the heat exchanger, with the cooler fluid returning from the embedded circuits. This helps prevent thermal shock to the pavement. The use of a heat exchanger between the solar subsystem and the embedded pavement circuits allows for separated fluids, expansion tanks, pressure relief valves, and the like.

The question of using the earth loop of a ground-source heat pump as a means of shedding excess solar heat often arises. This is a possibility when the heat

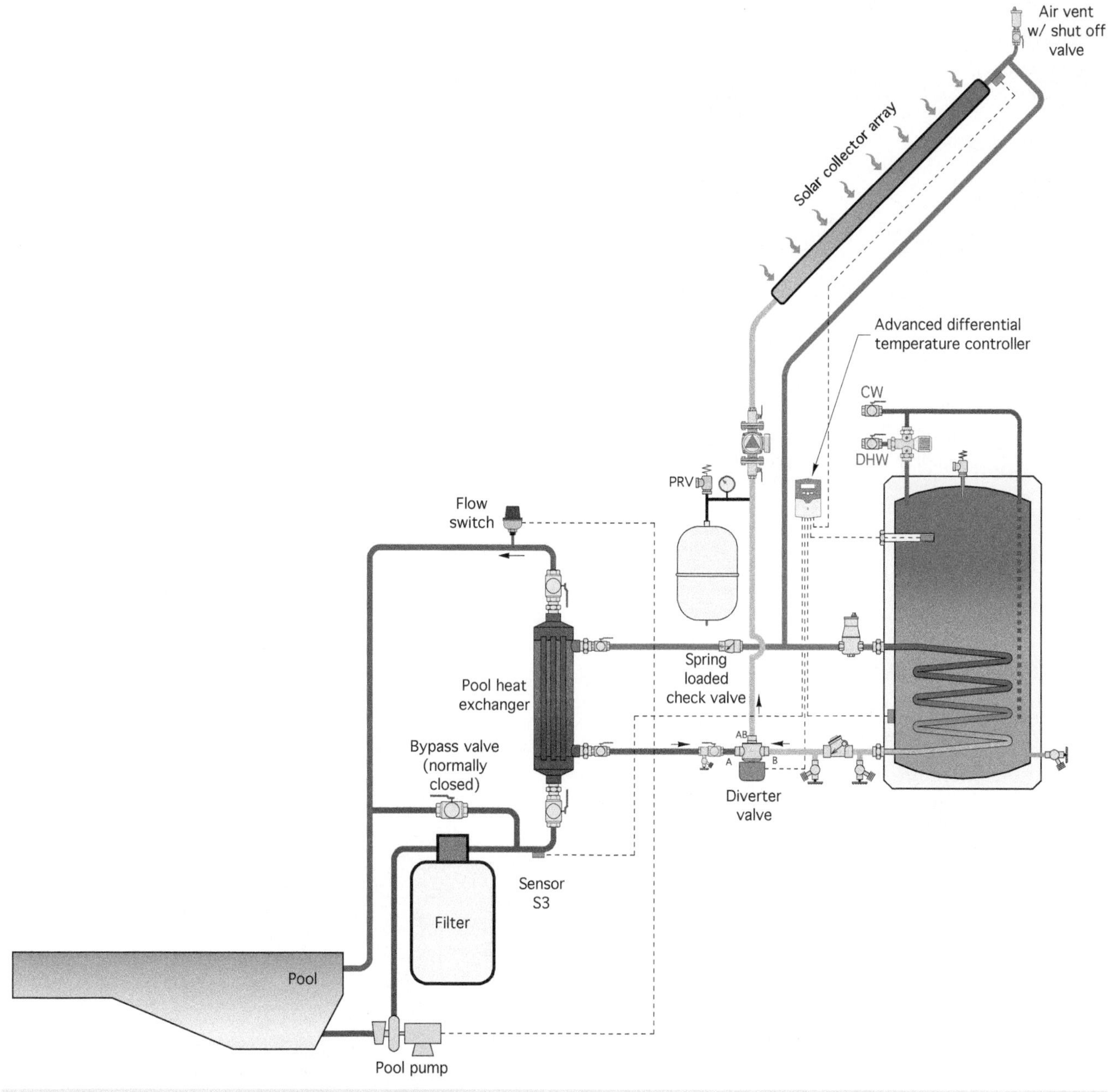

Figure 10-10 | Using a swimming pool as a heat dump.

input to the earth loop does not negatively affect heat dissipation from the heat pump operating in its cooling mode. This is a serious limitation in warm climates, where heat pump systems are often operating in cooling mode at the same time solar thermal systems may be reaching peak temperatures. In the author's opinion, this approach only makes sense in northern climates with minimal cooling loads or in heating-only heat pump systems. In the latter case, any heat retained by the soil around the earth loop would improve the capacity and COP of the heat pump when it operates in heating mode. Unfortunately, there is no simple way to predict how much of the heat dissipated into the soil remains available for later use by the heat pump. This would vary from one system to another depending on soil characteristics, subsurface water movement, and time between when heat is dissipated into the soil and when it is required for subsequent heating.

Figure 10-11 Two large exterior fan-coil units used for over-temperature protection. They are mounted under an overhang on the north side of a building and dissipate heat from an 1,800-square-foot array of flat-plate collectors.

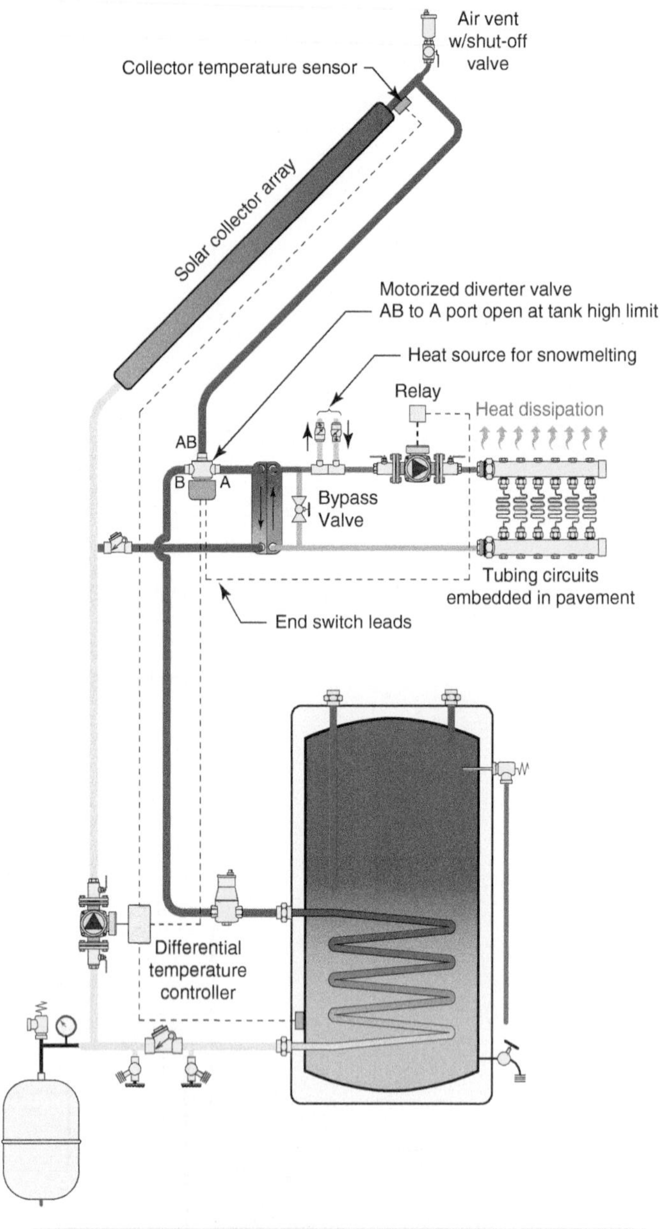

Figure 10-12 Using tubing embedded in pavement for over-temperature protection of the solar thermal system.

SIZING A HEAT DUMP

To be effective, a heat dump subsystem must be sized for a "design" level of heat dissipation at a selected inlet fluid temperature and, in the case of convectors, a corresponding outdoor temperature. The required rate of heat dissipation will depend on the total collector area as well as the efficiency of the collectors at design conditions.

The best way to present the sizing procedure is through an example.

Assume that a solar combisystem has 10 4-foot × 8-foot flat-plate collectors. The straight-line efficiency equation for these collectors is as follows:

$$\eta = 0.706 - 0.865\left(\frac{T_i - T_a}{I}\right)$$

where:

η = collector efficiency

T_i = collector inlet temperature (°F)

T_a = ambient air temperature (°F)

I = solar radiation intensity (Btu/hr./ft.^{2})

The collectors are operating with a 40 percent solution of propylene glycol and at a flow rate of 1 gpm per collector. On a hot summer afternoon, when heat dumping is most likely, assume a strong solar radiation intensity of 317 Btu/hr./ft.^{2} (equal to 1 kw/m^2) and a corresponding outdoor temperature of 90 °F.

The system's controls are set to operate the heat dumping subsystem when the storage tank reaches a temperature of 180 °F, and deactivate it when the tank drops to 160 °F.

The least favorable operating conditions for the heat dump subsystem will be when the tank is at the *lower* end of the heat dumping temperature range. Because there is a heat exchanger between the collector fluid and tank fluid, the collector temperature will be estimated at 165 °F when the tank is at 160 °F.

The collector efficiency under these conditions is:

$$\eta = 0.706 - 0.865\left(\frac{165 - 90}{317}\right) = 0.501(50\%)$$

The total heat output of the collector array under these conditions is found by multiplying the gross area of the collector array by the solar radiation intensity and efficiency:

$$Q = (A_{gross})(I)\eta = (320\text{ ft.}^2)\left(317\frac{\text{Btu}}{\text{hr.}\cdot\text{ft.}^2}\right)(0.501)$$

$$= 50{,}820\text{ Btu/hr.}$$

The outlet temperature from the collector array can now be determined based on the heat output, flow rate, and fluid properties:

$$T_{out} = T_{in} + \frac{Q}{[8.01Dc]f}$$

$$= 165 + \frac{50820\frac{\text{Btu}}{\text{hr.}}}{\left[\left(8.01\frac{\text{ft.}^3\cdot\text{min.}}{\text{gal}\cdot\text{hr.}}\right)\left(64.0\frac{\text{lb.}}{\text{ft.}^3}\right)\left(0.91\frac{\text{Btu}}{\text{lb.}\cdot{}^\circ\text{F}}\right)\right]10\frac{\text{gal}}{\text{min.}}}$$

$$= 175.9\ {}^\circ\text{F}$$

The values for density (D) and specific heat (c) of the 40 percent propylene glycol solution used in this equation were obtained from manufacturer's literature, assuming an average fluid temperature of 170 °F.

The criteria for sizing the heat dump device is now determined as:

- required rate of heat dissipation: 50,820 Btu/hr.
- corresponding inlet temperature to heat dump device = 175.9 °F
- corresponding outdoor temperature = 90 °F

The remaining task is to search for a device capable of operating at or close to these conditions. One example discussed earlier was a passive fin-tube element mounted outside in a shaded area. With an output rating of 250 Btu/hr./ft. under the above conditions, this system would require just over 200 linear feet of fin-tube element. A fan coil or fluid cooler is a more likely choice for a collector array of this size.

FINAL THOUGHTS ON HEAT DUMPS AND OVER-TEMPERATURE PROTECTION

Drainback-type solar thermal systems do not require any additional hardware for heat dumping or over-temperature protection. During a power failure, or when the storage tank reaches a preset upper temperature limit, the collector circulator turns off and the water quickly empties from the collector array and goes back into the tank. Any modern collector with an OG-100 rating from the SRCC (Solar Rating and Certification Corporation) has been tested for stagnation survival and should be capable of withstanding "**dry stagnation**" conditions without damage. Given the added complexity and cost of heat dumping and over-temperature protection, drainback freeze protection, especially for solar combisystems, makes good sense.

10.4 Expansion Tanks for Antifreeze-Protected Solar Collector Circuits

If stagnation occurs on a hot and bright summer afternoon, the temperature of the collector's absorber plate could exceed 350 °F. Along with protecting the antifreeze fluid from thermal degradation, designers need to deal with the thermal expansion of this fluid under extreme conditions. Specifically, they need to select a diaphragm-type expansion tank that prevents the system's pressure relief valve from opening under stagnation conditions.

Figure 10-13 shows the relationship between **absolute pressure** and boiling point for a 40 percent solution of propylene glycol, a common mix percentage in solar combisystems. Similar curves could be prepared for other concentrations of propylene glycol using manufacturer's data.

To maintain this solution as a liquid at a stagnation temperature of 350 °F would require an *absolute pressure* of approximately 118 psia (corresponding to a gauge pressure of about 103 psi) in the solar collectors. This is not practical, and may even violate some mechanical codes that require the pressure in the collector circuit to be no higher than the pressure of the domestic water.

Thus, because designers can't count on pressurization to prevent boiling at maximum stagnation

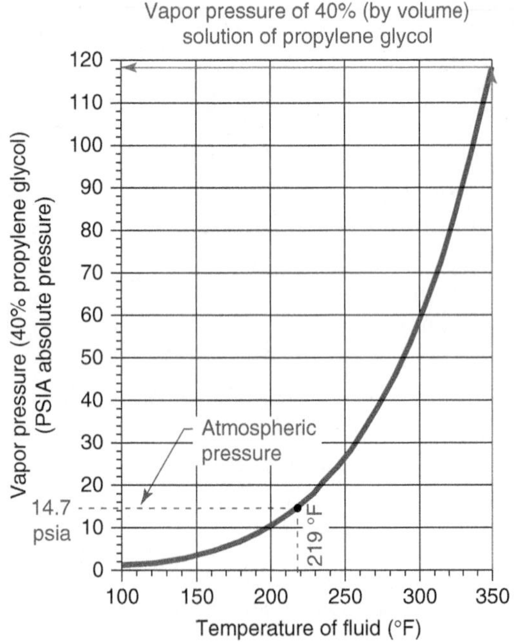

Figure 10-13 | Vapor pressure versus temperature for 40 percent solution of propylene glycol.

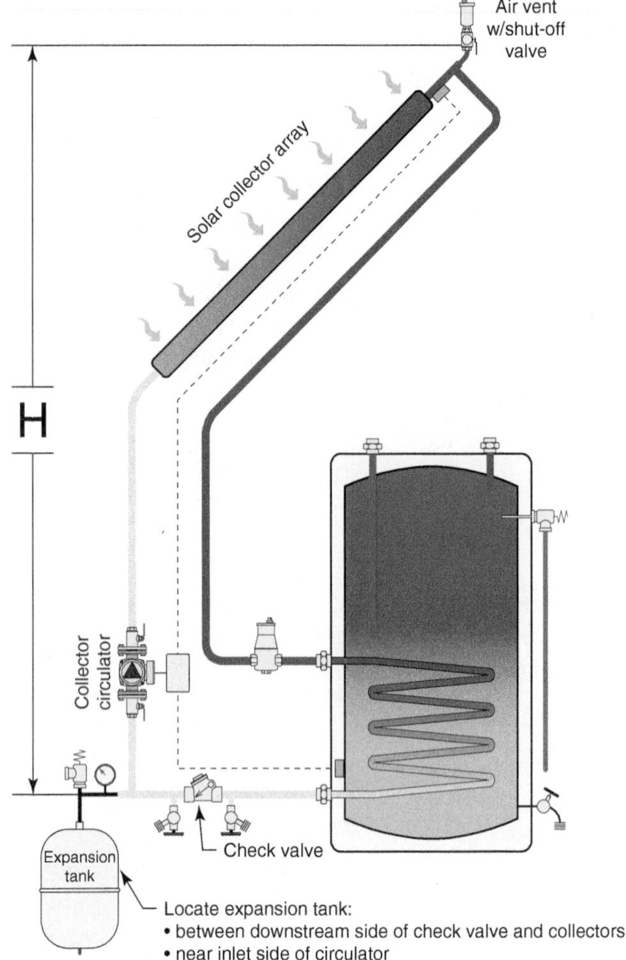

Figure 10-14 | Height (H) for use in Equation 10.1.

conditions, there will be times when the fluid in the collector will vaporize. Under these conditions, the liquid in the piping leading to and from the collector array could also be very hot.

To prevent the relief valve from opening under these conditions, the expansion tank must absorb both the fluid displaced by vaporization in the collectors as well as the expanding volume of the remaining liquid in the circuit.

Before calculating the volume of the expansion tank, it's necessary to know the rated opening pressure of the pressure relief valve in the collector circuit. Start by checking local codes to see if they mandate a maximum pressure relief valve setting in solar thermal systems. If this is the case, the code mandated rating is the benchmark.

If there are no code restrictions on the rating of the collector circuit pressure relief valve, the author suggests it be determined by assuming a cold fill static pressure of 25 psi at the top of the collector array and calculating the corresponding pressure at the relief valve location. Then, pick a relief valve that is rated to open 15 to 20 psi above this pressure.

Use Equation 10.1 to determine the static pressure at the relief valve location. See Figure 10-14 for the corresponding terms and dimensions.

(Equation 10.1)

$$P_{@PRV} = P_{@top} + \left(\frac{64.9}{144}\right)H$$

where:

$P_{@PRV}$ = static cold fill pressure at pressure relief valve location (psi)

$P_{@top}$ = static cold fill pressure at top of collector array (psi)

H = vertical distance from pressure relief valve to top of collectors (feet)

64.9 = density of 40 percent propylene glycol solution at 50 °F

144 = units conversion factor

Example 10.1: The top of a solar collector array is located 20 feet above the location of the system's pressure relief valve. The solar collector circuit operates with a 40 percent solution of propylene glycol. Assuming a 25 psi static cold fill pressurization at the top of the array, determine the static pressure at the relief valve when the fluid is at 50 °F.

Solution: The density of 40 percent propylene glycol at 50 °F is 64.9 lb./ft.^{3}. This value and the height of the system are used in Equation 10.1 to calculate the static pressure at the relief valve:

$$P_{@PRV} = P_{@top} + \left(\frac{64.9}{144}\right)H = 25 + \left(\frac{64.9}{144}\right)20 = 34 \text{ psi}$$

Discussion: The designer should now select a pressure relief valve that has a rated opening pressure 15 to 20 psi higher than this cold fill static pressure. Assume that a 50-psi relief valve is selected.

Equation 10.1 can be rearranged to determine the pressure at the top of the collector array just as the selected pressure relief valve is about to open—a condition that should be avoided at stagnation conditions.

(Equation 10.2)

$$P_{@top} = P_{PRVrated} - \left(\frac{64.9}{144}\right)H$$

where:

$P_{@top}$ = pressure at top of collector array as relief valve reaches rated opening pressure (psi)

$P_{PRVrated}$ = rating of pressure relief valve (psi)

H = vertical distance from pressure relief valve to top of collectors (feet)

64.9 = density of 40 percent propylene glycol solution at 50 °F

144 = units conversion factor

Assuming the selection of a 50-psi relief valve in Example 10.1, the pressure at the top of the system would be:

$$P_{@top} = P_{PRVrated} - \left(\frac{64.9}{144}\right)H = 50 - \left(\frac{64.9}{144}\right)20 = 41 \text{ psi}$$

The gauge pressure of 41 psi corresponds to an absolute pressure of 41 + 14.7 = 55.7 psia at the top of the collectors. According to Figure 10-13, this pressure would maintain the 40 percent propylene glycol solution in the collectors as a liquid to a temperature of about 297 °F. This is definitely higher than the fluid would see in normal operation, but not high enough to prevent vapor formation under summer stagnation conditions. Thus, it is very likely the fluid will boil if stagnation occurs on a warm and sunny afternoon. The expansion tank for the collector circuit needs to be selected to accommodate this possibility.

SIZING THE EXPANSION TANK

Sizing an expansion tank for an antifreeze-protected solar collector circuit is different from sizing an expansion tank for a "typical" hydronic system. The key difference is that under stagnation conditions, the antifreeze solution in the solar collectors could be heated to the point where it flashes into vapor. The volume required to contain this vapor is much greater than the volume required for the liquid that boiled to form the vapor. Thus, a small amount of antifreeze solution flashing to vapor in a collector is sufficient to force the remaining liquid within the absorber plate out of the collector. If the pressure relief valve within the collector circuit is to remain closed under this condition, the liquid forced out of the collectors has to be "parked" somewhere within the circuit. The only practical location is within the expansion tank. The procedure that follows sizes the expansion tank to receive this liquid volume, without creating pressures that would otherwise force the pressure relief valve in the collector circuit to open.

STEP 1: Using Equation 10.3, calculate the fluid volume that the expansion tank must absorb at stagnation. This equation assumes that boiling will occur and, as a result, all liquid will be forced out of the collectors at stagnation. It also assumes that the liquid temperature in the remainder of the system will reach 200 °F above the temperature at which the system was filled. The latter is a conservatively safe assumption.

(Equation 10.3)

$$V_a = 1.1[(V_c + V_p)0.08 + V_c]$$

where:

V_a = expansion volume to be accommodated (gallons)

V_c = total volume of collector array (gallons)

V_p = total volume of collector piping (excluding the collectors) (gallons)

0.08 = expansion factor for 40 percent propylene glycol solution for 200 °F temperature rise

1.1 = 10 percent added safety factor to allow for system volume estimates

Tube type/size	Gallons/foot
3/8″ type M copper	0.008272
1/2″ type M copper	0.0132
3/4″ type M copper	0.0269
1″ type M copper	0.0454
1.25″ type M copper	0.068
1.5″ type M copper	0.095
2″ type M copper	0.165
2.5″ type M copper	0.2543
3″ type M copper	0.3630

Figure 10-15 | Volume of type M copper tubing in gallons/foot.

The volume of a solar collector is usually listed in the manufacturer's specifications as is the volume of the tank's internal heat exchanger. Collector volume can also be found on the **SRCC OG-100 certification report**.

The volume of copper tubing can be estimated using data from Figure 10-15. If other types of tubing are used for the collector circuit, obtain volume data from the tubing manufacturer.

STEP 2: Calculate the cold fill static pressure at the location of the pressure relief valve. This is the pressure caused by the weight of fluid in the collector circuit above the pressure relief valve location, plus the static pressure maintained at the top of the system. It can be calculated using Equation 10.4.

(Equation 10.4)

$$P_{static} = P_{@topcold} + \left(\frac{64.9}{144}\right)H$$

where:

P_{static} = static pressure at the relief valve location (psi)

$P_{@topcold}$ = pressure at top of collector array at cold fill (psi)

H = height of collector circuit above location of pressure relief valve (feet)

64.9 = density of 40 percent propylene glycol solution at 50 °F

144 = units conversion factor

The air chamber in the diaphragm expansion tank must be pressurized to this calculated static pressure *before* fluid is added to the collector circuit. This ensures the diaphragm is fully expanded against the tank shell before the fluid begins to warm and expand.

STEP 3: Calculate the minimum required expansion tank volume using Equation 10.5, which is derived from **Boyle's law**.

(Equation 10.5)

$$V_T = V_a\left(\frac{P_{RV} + 14.7}{P_{RV} - P_{static}}\right)$$

where:

V_T = *minimum* required expansion tank volume (gallons)

V_a = expansion volume to be accommodated (from Step 1) (gallons)

P_{static} = cold fill static pressure at the relief valve location (from Step 2) (psi)

P_{RV} = maximum allowed pressure at the relief valve location (psig). Recommended value is pressure relief valve rating minus 3 psi. This allows for a slight safety factor against relief-valve leakage as the pressure approaches the valve's rating.

Example 10.2: A residential solar water heating system has the following components:

- Four collectors, each having a volume of 1.5 gallons
- Total of 120 feet of 1-inch copper tubing between heat exchanger and collector array
- Heat exchanger volume = 2.5 gallons
- Pressure relief valve rating = 50 psi
- Collector circuit fluid = 40 percent solution of propylene glycol

The distance from the top of the collector array down to the relief valve is 20 feet. Determine the minimum size of a diaphragm-type expansion tank for the system such that the relief valve doesn't open under stagnation conditions. The cold fill pressure at the top of the system is 25 psi.

Solution: The total collector array volume is 4 × 1.5 = 6 gallons. The total volume of the piping and heat exchanger is 120 ft. × (0.0454 gallons/foot) + 2.5 = 7.95 gallons.

STEP 1:

$$V_a = 1.1[(V_c + V_p)0.08 + V_c]$$
$$= 1.1[(6 + 7.95)0.08 + 6] = 7.83 \text{ gallons}$$

STEP 2:

$$P_{static} = P_{@topcold} + \left(\frac{64.9}{144}\right)H = 25 + \left(\frac{64.9}{144}\right)20$$
$$= 34 \text{ psi}$$

STEP 3:

$$V_T = V_a \times \left(\frac{P_{RV} + 14.7}{P_{RV} - P_{static}}\right) = 7.83 \times \left(\frac{47 + 14.7}{47 - 34}\right)$$
$$= 37.2 \text{ gallons}$$

The PRV value in Step 3 was set to 50 – 3 = 47 psi to guard against premature leakage from the valve as it approaches it rated pressure.

Discussion: As you can see, the expansion tank for the collector circuit is significantly larger that what is typical in hydronic heating systems of similar volume. This is partially the result of very conservative assumptions. It is also heavily influenced by the likelihood that some of the antifreeze solution in the collectors will vaporize during stagnation and push the remaining liquid from the collectors into the expansion tank.

The volume of the expansion tank could be reduced by reducing system volume (e.g., smaller tubing, smaller heat exchanger). It could also be reduced by lowering the static pressure at the top of the collector array and/or increasing the pressure rating of the relief valve. These changes all have consequences, but they may be possible in some applications.

In cases where the minimum required expansion tank volume exceeds the volume of available tanks, it is acceptable to use multiple tanks connected in parallel. However, be sure the air-side pressure in each tank is set to the calculated static pressure before filling the collector circuit.

FINAL PRESSURE ADJUSTMENT OF EXPANSION TANK

Unlike most hydronic circuits, the temperature of a solar collector circuit filled with an antifreeze solution can drop well below the "cold fill" temperature. If the diaphragm in the circuit's expansion tank is fully expanded against the tank's shell when the circuit's fluid is at the cold fill temperature, it cannot expand further to compensate for additional fluid contraction as the fluid temperature drops, in many cases well below 32 °F. This can create sub-atmospheric pressure in the collector circuit, which could allow air to be sucked in through float-type air vents or other components.

To prevent this from occurring, a small amount of antifreeze solution should be added to the circuit to compensate for **cold night fluid shrinkage**. When added, this extra fluid slightly compresses the diaphragm in the expansion tank. This slightly increases the air-side pressure above the static fill pressure calculated using Equation 10.4. As the collectors and exposed piping approach their lowest temperatures, the diaphragm can now expand as the extra fluid moves out of the tank. This prevents negative pressures within the collector circuit.

Equation 10.6 can be used to determine the final air-side pressure in the expansion tank after sufficient extra fluid has been added to compensate for cold night fluid shrinkage.

(Equation 10.6)

$$P_{final} = \frac{(P_{static} + 14.7)V_T}{V_T - \left[\frac{D_{vc}}{D_c} - 1\right](V_c + V_{ep})} - 14.7$$

where:

P_{final} = final air-side pressure in expansion tank, achieved by adding more antifreeze solution to the circuit (psi)

P_{static} = calculated static pressure at tank location (from Equation 10.4) (psi)

V_T = volume of expansion tank (gallons)

V_c = volume of collector array (gallons)

V_{ep} = volume of all collector circuit piping exposed to cold temperatures (gallons)

D_{vc} = density of fluid at lowest possible outdoor temperature (lb./ft.^{3})

D_c = density of fluid at cold fill conditions (typically 60 °F) (lb./ft.^{3})

Example 10.3: A solar system has four solar collectors, each with a volume of 1.5 gallon. It also has a 38 gallon expansion tank. The collector array as well as 60 feet of 3/4" type M copper tubing in the collector circuit will be exposed to -20 °F on the coldest night of the winter. The system operates with a 50 percent solution of Dowfrost HD–inhibited propylene glycol antifreeze. The static pressure at the expansion tank will be 34 psi when the system is filled with fluid at 60 °F. Determine the final adjusted air-side pressure in the expansion tank so that sufficient fluid is present to compensate for cold night fluid shrinkage in the collector array and exposed piping on the −20 °F night.

Solution: The fluid properties of 50 percent Dowfrost HD can be found in the manufacturer's literature:

$$D_c = D_{60F} = 65.33 \text{ lb./ft.}^3$$

$$D_{vc} = D_{-20F} = 66.46 \text{ lb./ft.}^3$$

The volume of the collector array is 4 × 1.5 gallons per collector = 6 gallons

The volume of 60 feet of 3/4-inch copper tubing is found using data from Figure 10-15:

60 feet × 0.0269 gallons/foot = 1.61 gallons

Substituting into Equation 10.6 yields the final air-side pressure in the expansion tank:

$$P_{final} = \frac{(P_{static} + 14.7)V_T}{V_T - \left[\frac{D_{vc}}{D_c} - 1\right](V_c + V_{ep})} - 14.7$$

$$= \left\{\frac{(34 + 14.7)38}{38 - \left[\frac{66.46}{65.33} - 1\right](6 + 1.61)} - 14.7\right\}$$

$$= 34.17 \text{ psi}$$

Discussion: The added air-side pressure (0.17 psi) is too small to measure with standard pressure gauges. Thus, the installer is likely to add fluid until the pressure goes up by perhaps 1 psi. This is conservative because it adds slightly more fluid to the expansion tank than needed to compensate for the cold night fluid shrinkage. At the same time, there is minimal loss of acceptance volume in the expansion tank to handle the opposite temperature extreme during summer stagnation. Even if the expansion tank volume in this system were 10 gallons, the increase in air pressure would only be about 0.65 psi. Thus, adding fluid to the purged system until the expansion tank's air side pressure is nominally 1 psi above the calculated static air pressure is generally adequate.

The volume of fluid needed to compensate for the fluid shrinkage can be directly calculated using Equation 10.7, which is part of Equation 10.6.

(Equation 10-7)

$$V_{added} = \left[\frac{D_{vc}}{D_c} - 1\right](V_c + V_{ep})$$

where:

V_{added} = volume of antifreeze solution to be added to compensate for cold night shrinkage (gallons)

V_c = volume of collector array (gallons)

V_{ep} = volume of all collector circuit piping exposed to cold temperatures (gallons)

D_{vc} = density of fluid at lowest possible outdoor temperature (lb./ft.^{3})

D_c = density of fluid at cold fill conditions (typically 60 °F) (lb./ft.^{3})

For the conditions assumed in Example 10.3, this works out as follows:

$$V_{added} = \left[\frac{D_{vc}}{D_c} - 1\right](V_c + V_{ep}) = \left[\frac{66.46}{65.33} - 1\right](6 + 1.61)$$

$$= 0.132 \text{ gallons}$$

The added volume, at just over 1 pint, is small compared to the size of the expansion tank.

As an alternative to measuring the pressure increase on the air side of the expansion tank, the installer could also measure out this calculated amount of fluid and add it to the system using a small hand pump.

PLACEMENT OF THE EXPANSION TANK

The expansion tank should be located so that it is close the inlet of the collector circulator. It should also be installed between the collector array and the downstream side of the circuit's check valve, as shown in

Figure 10-14 and in many other schematics in this chapter. If the circulator is supplied with an internal check valve, it should be removed. These details allow the vaporizing fluid to exit from both the top and bottom of the collector. *This is important in order to avoid "**slugging**" of the liquid if it can only exit at the top of the collector array when stagnation temperatures cause it to boil.*

10.5 Antifreeze-Protected Solar Domestic Water Heating Systems

Domestic water heating has proven to be one of the most practical applications for solar thermal systems. This is true for both residential and commercial applications. There are two fundamental reasons for this:

1. Domestic water heating is a year-round load. Thus, solar domestic water heating systems can take advantage of the high solar availability and coincident higher ambient temperatures available in summer as well as portions of the spring and fall.
2. Domestic water heating starts at relatively low temperatures, typically 40 °F to 60 °F. It takes just as much energy to rise a gallon of water from 40 °F to 50 °F, as it does to raise that gallon of water from 110 °F to 120 °F. Solar energy is best used at the lower temperature end of the domestic water heating process because of significantly higher collector efficiency at lower fluid inlet temperatures.

The favorable economics of solar domestic water heating must be respected when extending the system to provide some contribution to space heating. All practical solar combisystems will provide a significant portion of a building's domestic water heating requirement. In addition, they will provide a smaller fraction of the building's space heating load. The author prefers to describe practical solar combisystems as "**domestic hot water plus**." The "plus" refers to the system's ability to contribute some energy toward space heating, primarily in spring and fall and to a much lesser extent in midwinter.

Before exploring the design of solar combisystems, designers should be familiar with solar domestic water heating systems. This knowledge better prepares one to understand the "domestic hot water plus" concept when it's time to include space heating in the overall system design.

One of the most common solar domestic water heating systems uses a single thermal storage tank with an internal coil heat exchanger mounted in the lower portion of the tank, as shown in Figure 10-16.

A closed piping circuit is created between the internal heat exchanger and the collector array. This circuit is filled with a propylene glycol antifreeze solution. A differential temperature controller constantly compares the temperature of a sensor mounted at the top of one of the collectors to the temperature of a sensor mounted near the bottom of the storage tank. When the collector temperature is sufficiently above the tank

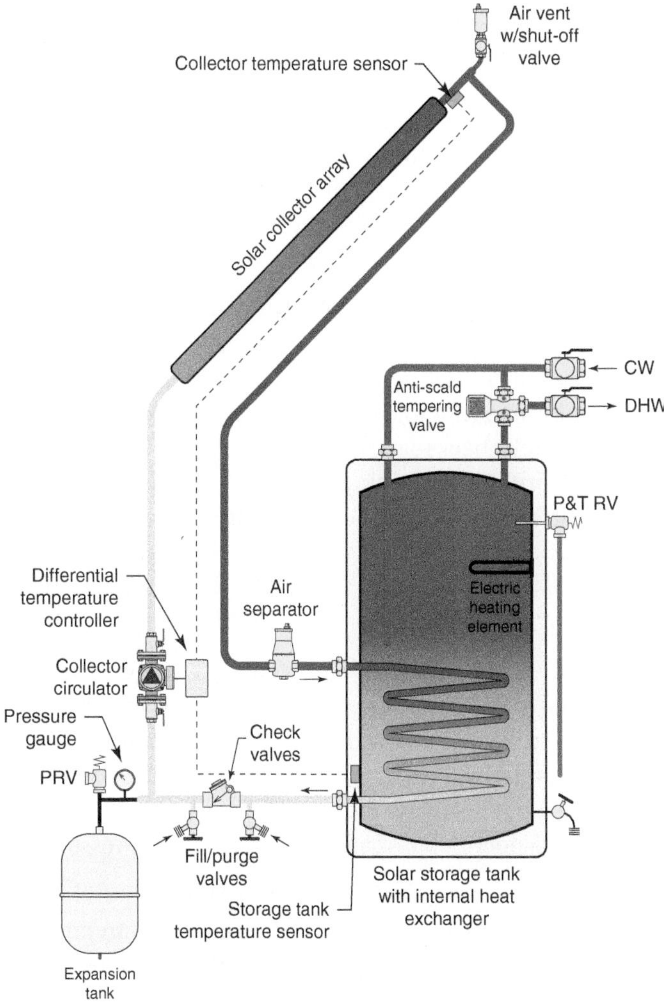

Figure 10-16 Closed-loop antifreeze-protected solar domestic water heating system with an electric element in the upper portion of the thermal storage tank for auxiliary heating.

temperature, this controller turns on the collector circulator. The antifreeze solution absorbs heat as it passes through the collectors and releases heat as it passes through the tank's internal heat exchanger.

The collector circuit contains several basic hydronic components, including:

- check valve
- fill/purging valves
- expansion tank
- pressure relief valve
- pressure gauges
- air separator

The check valve prevents reverse thermosiphoning, which would otherwise occur when the tank is warmer than the collectors. Without it, much of the heat added to the storage tank would be dissipated back through the cooler collectors at night. The check valve should be located upstream of the point where the expansion tank connects to the circuit. This important detail allows fluid to come down the supply pipe and into the expansion tank if boiling occur in the collectors under stagnation conditions.

Positioning the check valve between the fill/purging valves also allows the premixed antifreeze solution to be pumped into the downstream valve and flow up through the collector array. The developed pressure immediately backseats the check valve. As the fluid moves through the piping, it pushes air ahead of it. This air exits through the outlet-purging valve. Any residual air that was dissolved in the circuit's fluid will eventually be separated and ejected by the air separator.

The collector circuit also contains an expansion tank that compensates for the changes in volume of the antifreeze solution as its temperature changes. The pressure relief valve provides protection against excessively high pressure in the circuit. The pressure gauge provides a continuous reading of the circuit's pressure.

All these common components can be individually purchased and assembled; however, the trend in the solar thermal industry is to sell them as a preassembled **solar circulation station**, such as shown in Figure 10-17. This assembly contains the necessary hardware (other than the tubing) for an antifreeze-based solar collection circuit and significantly reduces installation time.

Courtesy of Caleffi North America, Inc.

Figure 10-17 A solar circulation station for a typical antifreeze-protected solar water heating system.

Another significant difference between current systems and their predecessors is the tubing used for the collector circuit. First-generation solar water heating systems typically used copper tubing for the collector circuit. Although it generally worked fine, assembling and soldering copper piping in an attic, or up on a roof on a windy day, can be challenging. Today, most solar water heating systems are installed with flexible **corrugated stainless steel tubing (CSST)**, as described in Chapter 4. All connections are made with wrenches, and no soldering is required. A harness of preinsulated stainless steel tubing (supply and return tube plus a cable for the collector temperature sensor) is simply uncoiled and routed from the location of the storage tank to the location of the collector array. This tubing is "factory clean" and shipped with end caps to keep it that way. As such, it is ready to be filled with antifreeze solution when the collector circuit is completed. When soldered copper tubing is used, all residual soldering flux must be flushed out of the circuit to prevent chemical interaction with the propylene glycol antifreeze.

There will certainly be times when the solar energy captured by the system will not be able to supply adequate domestic hot water. Therefore, it is necessary to provide auxiliary heat. This can be done in several ways. One of the simplest is a single electric resistance heating element mounted in the upper portion of the solar storage tank. This placement ensures that water leaving the tank is adequately heated, while minimizing heat migration to water in the lower portion of the tank. Keeping the lower portion of the tank cool improves the performance of the heat exchanger and solar collectors.

Every solar water heating system should include an **ASSE 1017–rated thermostatic mixing valve** at the point where it delivers hot water to the building's plumbing. The purpose of this valve is to reduce the water temperature delivered to hot water distribution piping to not more than 120 °F. It is not uncommon for domestic water temperatures in solar thermal systems to reach 180 °F, especially after two or three sequential sunny days when there is little or no demand for hot water because the occupants are away on a summer vacation. If water at such temperatures were delivered directly to fixtures, it could cause instant third-degree burns. As of January 4, 2014, this valve, as well as any other components in newly installed or renovated domestic water piping, must be in compliance with **EPA Public Law 111-380**, commonly known as the "no-lead law," if installed within the United States.

Another variation on solar domestic water heating uses an external heat exchanger in place of the internal heat exchanger. Figure 10-18 shows a typical arrangement of components.

With exception of the heat exchanger, the components in the solar collector circuit are the same as in Figure 10-16.

Modern systems of this design typically use a stainless steel brazed-plate heat exchanger, as discussed in Chapter 5. The heated antifreeze solution from the collector array circulates through the primary side of this heat exchanger while domestic water circulates through the secondary side. It is very important that these two flow streams pass through the heat exchanger in opposite directions. This creates **counterflow heat exchange**, which increases the rate of heat transfer relative to heat exchangers where both fluids flow in the same direction. It is also important to verify if local plumbing codes require the use of a double-wall heat exchanger in this type of system. Stainless steel brazed-plate heat exchangers are available in double-wall configurations when required.

The system is piped so that cooler water near the bottom of the storage tank is supplied to the heat exchanger, and heated water leaving the heat exchanger passes into the tank's dip tube. The latter conveys the entering water into the lower portion of the tank and thus helps preserve temperature stratification within the tank. A swing check valve is placed where the pipe carrying heated domestic water from the heat exchanger tees into the pipe leading to the dip tube. Its purpose is to prevent reverse thermosyphoning of heated tank water through the heat exchanger when the circulators are off.

A single electric heating element in the upper portion of the tank provides auxiliary heating as needed.

The circulator that conveys domestic water from the lower portion of the tank through the heat exchanger

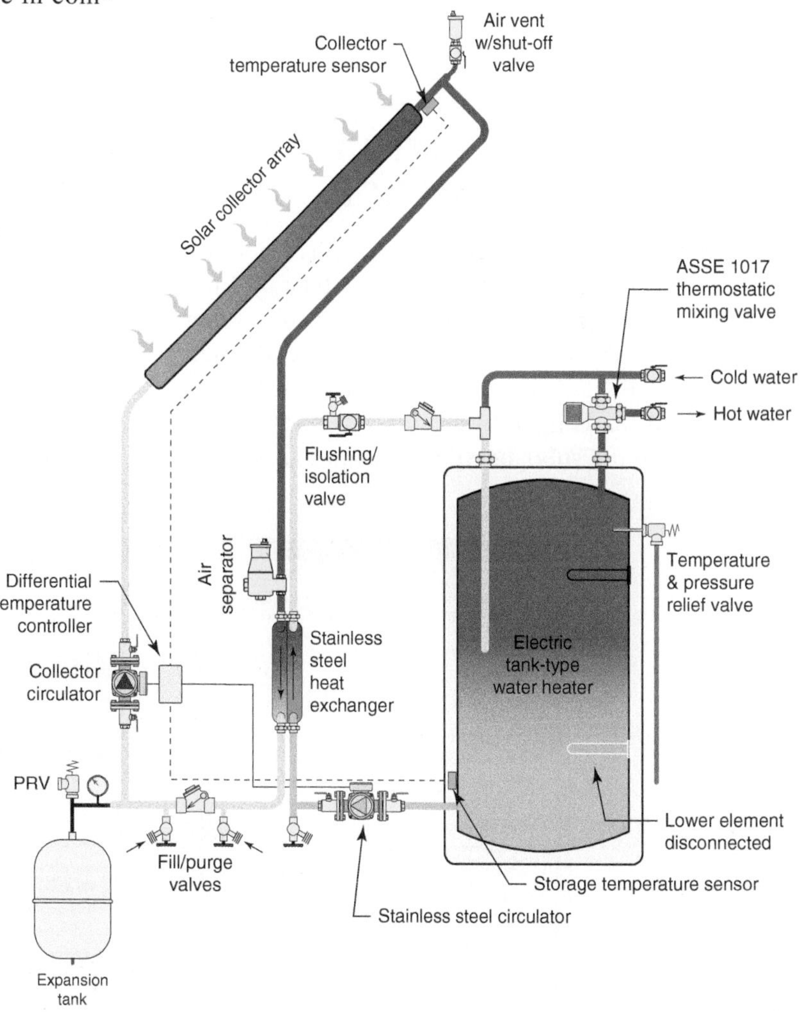

Figure 10-18 | Typical arrangement for an antifreeze-protected solar water heating system using an external heat exchanger.

must be of either stainless steel or bronze construction. It is typically wired in parallel with the collector circulator and thus runs at the same time.

Systems with external heat exchangers may be able to use an existing tank-type water heater for thermal storage. To do so, the tank should have a volume of at least 1 gallon per square foot of collector area, with volumes of 1.25 to 2.0 gallons per square foot of collector preferred for optimal performance. These higher storage volumes are preferred for systems in sunny/warm climates. If the existing tank has two electric heating elements, it may require rewiring so that only the upper heating element can operate.

Systems using external heat exchangers also have the advantage that the heat exchanger can be easily cleaned or replaced, if necessary, due to aggressive water. As an option, combination isolation/flushing valves can be installed near the inlet and outlet of the domestic side of the heat exchanger. These valves are shown in Figure 9-36.

SOLAR WATER HEATING SYSTEM PERFORMANCE

The performance of solar domestic water heating systems can be simulated using a variety of software. In this text, the performance of both solar water heating, and solar combisystems, will be estimated using **F-CHART** software. The f-Chart method was developed at the University of Wisconsin, Madison, for estimating the **solar heating fraction** that a given system configuration could produce for either domestic water heating or combisystem applications. The "f" in f-Chart stands for fraction.

The f-Chart method was developed based on correlating the results of thousands of **TRNSYS** (Transient System Simulation program) software simulations. This complex software, now in its 17th version, performs hour-by-hour simulations based on detailed component models and weather data. Because of its complexity, TRNSYS is seldom used for routine system design. However, the accuracy it provides is largely embedded in the much simpler f-Chart method.

Fundamentally, the f-Chart method is simple enough for manual calculations. However, its availability as a Windows®-based software tool has all but eliminated such calculations.

Figure 10-19 shows an example of the F-CHART input screens.

One window on the input screen accepts values for the solar collectors being used. The designer specifies collector performance by entering values for the Y intercept and slope of the collector's instantaneous efficiency line. The inputs to F-CHART also allow for specification of the collector array, with inputs for the gross collector area, number of collectors, slope angle, and azimuth angle. Inputs are also provided for the incident angle modifier, which can be modeled by either entering a value for the constant b_o, as represented by Equation 8.9, or by selecting the number of glazings on the collector. Another option for the incident angle modifier is to enter the value for each 10°

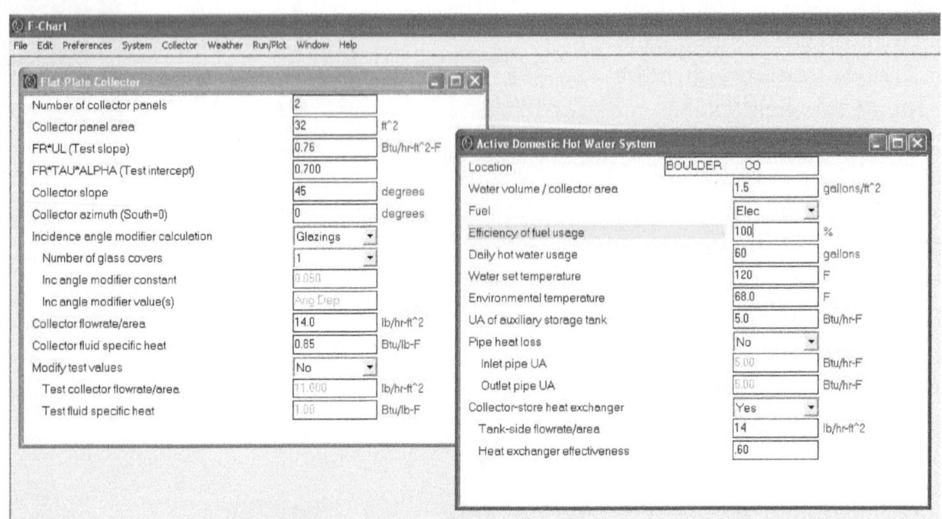

Figure 10-19 | F-CHART input screen configured for a domestic water heating system.

increment of the incident angle, as found on an SRCC OG-100 test report. Other F-CHART inputs include the collector's flow rate and the specific heat of the fluid used in the collector circuit.

Another window on the input screen accepts information about the load, storage tank size (specified as gallons of storage per square foot of collector), the effectiveness of the collector heat exchanger, and the system's location (specified by selecting from the built in database).

A window is also available for entering information about the cost of the system, the cost of auxiliary energy, and how the system will be financed. This window is not shown in Figure 10-19 but will be discussed in Chapter 12.

After the inputs are configured to describe the system, the software is run by pressing the F2 key. The results are instantly displayed, as shown in Figure 10-20.

The output window lists several results on a monthly basis. The first column ("Solar") lists the total monthly solar radiation incident on the collector array. The second column ("Dhw") lists the total domestic water heating load, including standby heat loss from the tank. The third column ("Aux") lists the auxiliary energy necessary to satisfy the Dhw load. The final column ("f") is the fraction of the load supplied by solar energy, expressed as a decimal percentage. For example, an *f* value of 0.405 indicates that 40.5 percent of the load was supplied by solar energy. The last row of this window provides an annual total of the energy quantities and the annual average solar fraction. For the system modeled in Figure 10-19, F-CHART is estimating that 84.4 percent of the total domestic water heating load, including losses from the storage tank, would be supplied by solar energy.

The user can now make changes to any of the inputs and quickly see the resulting changes in results. They can also perform **parametric studies** on a specific system be varying just one of the inputs and generating results that show the effect of this one parameter on overall system performance. This is a very helpful technique for gaining an understanding of which changes have significant versus minor effects on overall system performance.

To demonstrate this, multiple F-CHART runs were made for a solar water heating system described by the inputs in Figure 10-19. Only the slope of the collector array was changed from one run to the next. The resulting annual solar fractions were generated and plotted on the graph shown in Figure 10-21.

This parametric study shows that the annual solar fraction reaches a peak when the collectors are sloped at about 44°, which is about 4° greater than the local latitude in Boulder, CO. The solar fraction drops very

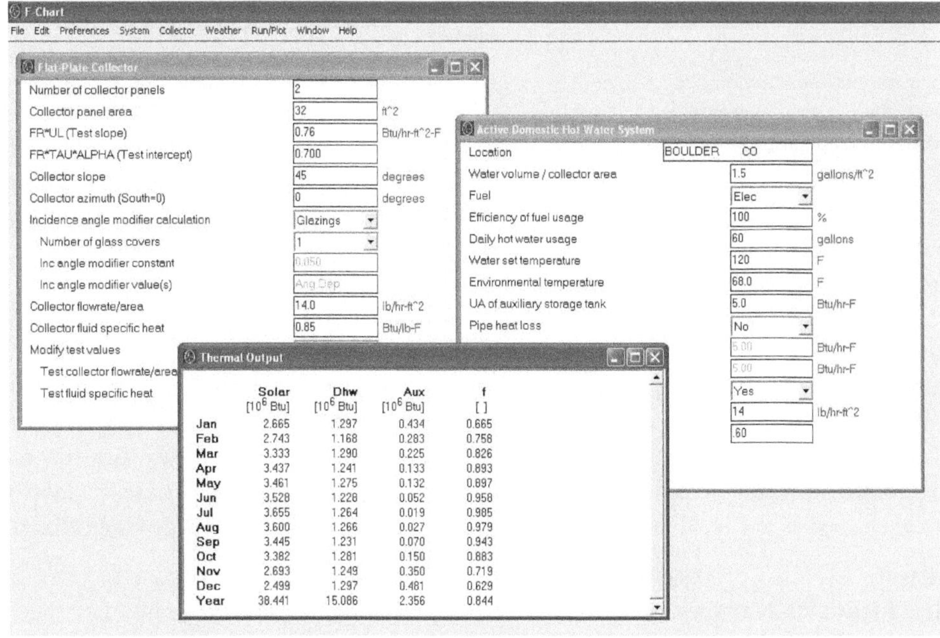

	Solar [10^6 Btu]	Dhw [10^6 Btu]	Aux [10^6 Btu]	f [ ]
Jan	2.665	1.297	0.434	0.665
Feb	2.743	1.168	0.283	0.758
Mar	3.333	1.290	0.225	0.826
Apr	3.437	1.241	0.133	0.893
May	3.461	1.275	0.132	0.897
Jun	3.528	1.228	0.052	0.958
Jul	3.655	1.264	0.019	0.985
Aug	3.600	1.266	0.027	0.979
Sep	3.445	1.231	0.070	0.943
Oct	3.382	1.281	0.150	0.883
Nov	2.693	1.249	0.350	0.719
Dec	2.499	1.297	0.481	0.629
Year	38.441	15.086	2.356	0.844

Figure 10-20 | F-CHART window showing thermal output for the system in foreground and input information in background.

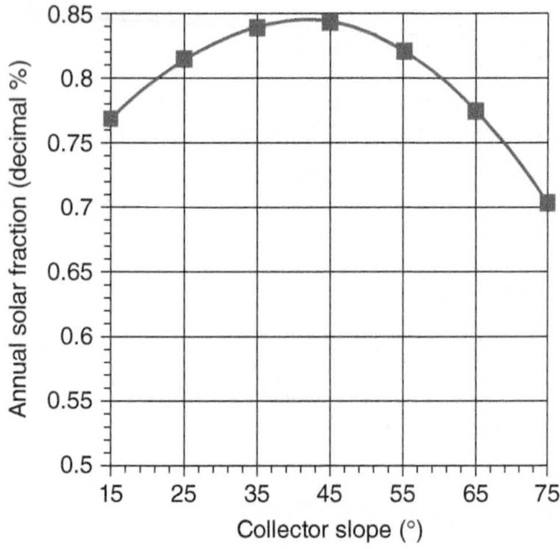

Figure 10-21 The effect of varying collector slope for the solar water heating system described by the numerical inputs in Figure 10-19.

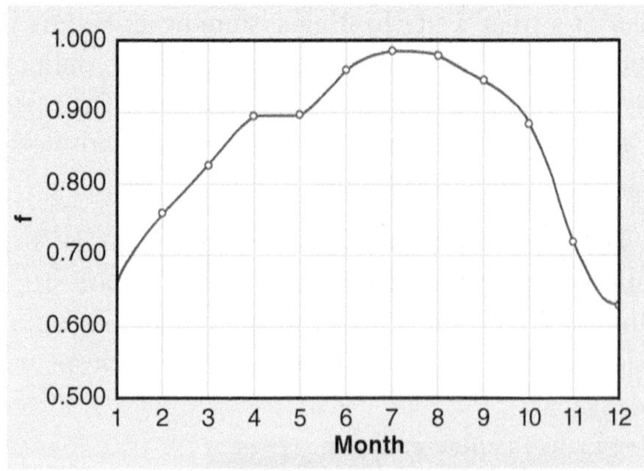

Figure 10-22 Solar fraction versus month for a solar water heating system in Boulder, CO, and described by the inputs in Figure 10-19.

slightly from about 84.4 percent to about 84 percent when the slope drops from 45° to 35°. Similarly, the solar fraction drops to about 82 percent if the collector slope is increased from 45° to 55°. Both of these changes show that the annual solar fraction provided by this specific system is relatively insensitive to collector slope angles that are within +/−10° of the optimal collector slope for domestic water heating systems, which is generally considered as the local latitude.

F-CHART software can also prepare graphs for the output quantities. Figure 10-22 shows a graph generated by F-CHART that plots solar fraction versus month for the system described by the inputs listed in Figure 10-19. Figure 10-23 shows a plot for incident

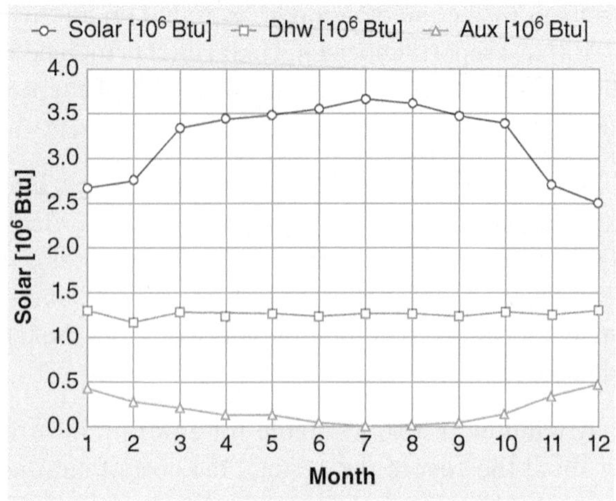

Figure 10-23 Plot of incident solar radiation, domestic water heating load (Dhw), and auxiliary energy for a solar water heating system in Boulder, CO, and described by the inputs in Figure 10-19.

solar radiation, domestic water heating load, and auxiliary energy, all on a monthly basis, for the same system.

Readers interested in more information on F-CHART software can find it at the website www.fchart.com.

10.6 Antifreeze-Protected Solar Combisystems

Practical solar combisystems are designed to supply a high percentage of a home's domestic hot water load and a smaller percentage of the home's space heating load. There are many possible configurations for such systems, largely due to the wide variation in space heating subsystems. The systems shown in this chapter were selected to be representative of modern hydronic design. They include low-temperature heat emitters, high-efficiency circulators, and other details that together allow systems to be affordable, efficient, and reliable. These systems make extensive use of the details and devices discussed in earlier chapters.

Designers should remember that the schematics presented in this chapter, and throughout the text, are conceptual drawings rather than final designs for a specific installation. These drawings may not include all detailing required by local codes. For example, some codes require auxiliary boilers to be protected by a low water cut-off switch or a manual reset high-limit controller. Other code requirements that vary from one location to another include the use of double-wall heat exchangers, collector array grounding, or use of temperature and pressure-relief valves vs. pressure-relief valves on tankless water heaters. Such details can all be incorporated into the conceptual drawings given in this text, where required. That is the responsibility of the system's designer.

ANTIFREEZE-PROTECTED COMBISYSTEM #1

The system shown in Figure 10-24 combines several previous discussed subsystems.

The solar subsystem consists of a collector array supplying a brazed-plate heat exchanger. The latter has been generously sized, using software available from the heat exchanger manufacturer, to provide a minimum effectiveness of 0.70. This minimizes both the **approach temperature** difference and the **thermal penalty**

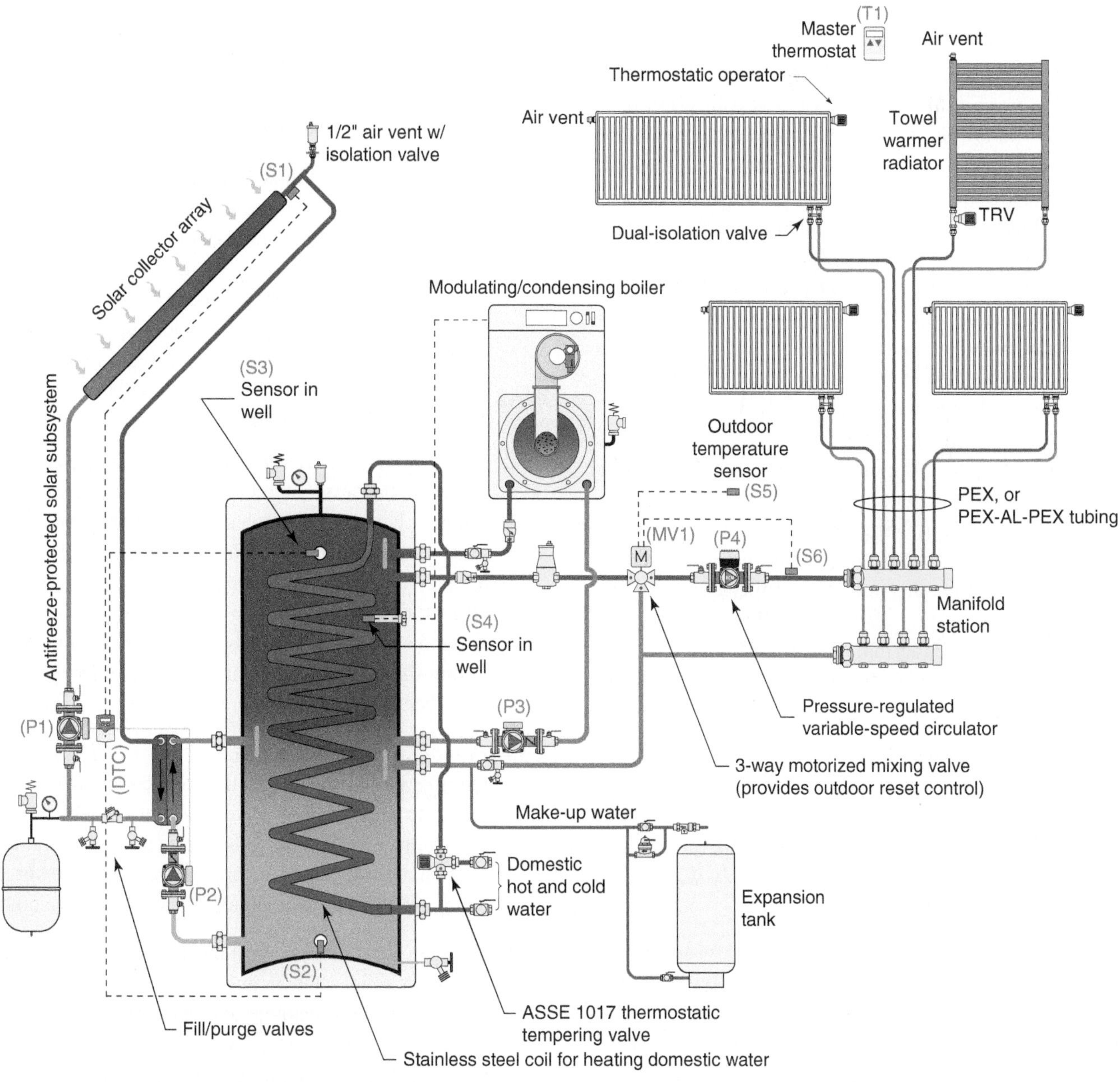

Figure 10-24 | Piping schematic for solar combisystem #1.

associated with having a heat exchanger between the collectors and thermal storage tank.

Operation of the collector circuit is managed by a differential temperature controller. The collector circulator and tank-to-heat exchanger circulator are both turned on when the collector sensor temperature reaches 15 °F above the temperature in the lower portion of the storage tank. The circulators continue to operate until the collector sensor temperature drops to within 8 °F above that of the lower storage tank sensor. Thus, solar derived heat is added to the storage tank whenever it is available based on these controller settings. This heat can contribute to either domestic water heating, through a large internal stainless steel heat exchanger, or to space heating.

The water temperature in the *upper portion* of the thermal storage tank is maintained at a minimum of 120 °F by the boiler. This allows domestic water to be heated to approximately 115 °F in a single pass through the internal stainless steel heat exchanger. Cold domestic water enters the bottom of this heat exchanger and passes upward to create counterflow heat exchange between the domestic water in the coil and the system water in the tank. The coil is shaped so that a higher percentage of its surface area is located near the top of the tank. This allows the hottest water in the tank to create heat exchange rates higher than would be possible if the coil surface area were evenly distributed from bottom to top within the tank. Heated domestic water passes from the coil to the hot port of an ASSE 1017–rated thermostatic mixing valve. If necessary, this valve mixes cold domestic water into the hot domestic water to prevent unsafe delivery temperatures to fixtures. This is especially important following a hot and sunny day, when the thermal storage tank is likely to be significantly hotter than the maximum safe hot water delivery temperature of 120 °F.

Water for space heating is drawn from near the upper portion of the tank. It passes through a 3-way motorized mixing valve that protects the low temperature distribution system from excessively high temperatures and provides outdoor reset control of the supply water temperature to the panel radiators.

The radiators have been sized to provide design heat output to each room when supplied with 120 °F water. They are supplied by a homerun distribution system driven by an ECM-based variable-speed pressure-regulated circulator. That circulator is operating in **constant differential pressure mode**. The output of each radiator is controlled by a wireless thermostatic radiator valve. This provides room-by-room temperature control. The thermal mass of the upper portion of the thermal storage tank stabilizes the highly zoned hydronic distribution system and prevents the boiler from short cycling under partial load conditions.

Because there are separate connections at the thermal storage tank for the supply and return end of the boiler circuit, the space heating distribution subsystem, and the heat exchanger circuit, the tank provides **hydraulic separation** between the circulators in all of these circuits. This is especially important because of the variable-speed pressure-regulated circulator in the space heating distribution system. At times, this circulator may be running at very low power levels (e.g., when only one or two panel radiator are active). If not for the hydraulic separation provided by the tank, the other circulators could interfere with its operation under such conditions.

Other important details on this system include:

1. *Air vents:* A 1/2-inch NPT float-type air vent is installed above a high-temperature-resistant isolation valve at the top of the collector array. This vent is helpful when the system is filled with fluid. After a few days of operation, the isolation valve below the vent is closed to protect the vent against high-temperature fluid. The isolation valve also allows the vent to be serviced or replaced without draining the collector circuit fluid.
2. *Check valves:* There are several check valves in this system. They are located to stop reverse thermosiphoning heat loss through circuits where it would otherwise develop when circulators are off. They are installed in the following locations: the solar collector circuit, the piping near the tank outlet serving space heating loads, the boiler circuit, and within the circulator that supplies tank water to the solar heat exchanger.
3. *Purging valves:* Valves that allow both filling and purging are located within the solar collector circuit (e.g., the two-hose bib valves on either side of the swing check valve). Purging valves are also installed near the return side of the boiler circuit and the space heating distribution subsystem, near the storage tank.
4. *Pressure relief valves:* Most mechanical codes require a pressure relief valve in any circuit that contains a heat source and is capable of being isolated from the remainder of the system. Thus, a pressure relief valve is installed on the boiler. Since it is possible to isolate the boiler from the remainder of the system, another pressure-relief valve must be installed to protect

the remainder of the system, which contains a solar heat source. In this system, that valve is installed at the top of the thermal storage tank, alongside a pressure gauge and air vent. A third pressure relief valve is also installed in the solar collector circuit, which is isolated from the remainder of the system by the solar heat exchanger.

5. *Expansion tanks:* One expansion tank limits pressure variation within the solar collector circuit. It should be sized based on methods discussed earlier in this chapter. A second expansion tank is required for the remainder of the system. It should be sized based on methods discussed in Chapter 5. Depending on its size, one or both of these tanks may have to be floor mounted. The piping connecting the expansion tank to the system tees into the return side of the space heating distribution system, near where it enters the thermal storage tank. This placement keeps the **point of no pressure change** relatively close (from the standpoint of pressure drop along the circuits) to the inlet side of both the boiler circulator and the distribution system circulator. It also allows for efficient filling and purging in combination with the purging valve located just upstream of the connection tee.

6. *Make-up water assembly:* The make-up water assembly functions just as it would in a nonsolar hydronic system, automatically adding water into the system, as necessary, to make up for minor losses over time.

7. *Piping and component insulation:* Although not shown in the schematic, all piping in a solar combisystem should be insulated to a minimum of R-3.0 °F·hr.·ft.^{2}/Btu, or as otherwise required by local code. The author recommends a minimum thermal storage tank insulation R-value of 24 °F·hr.·ft.^{2}/Btu. It makes little sense to detail a combisystem for optimal collection of solar energy and subsequently allow anything more than a "trickle" of that energy to be lost in an uncontrolled manner. Keep in mind that a thermal storage tank is losing heat to the building on the *hottest day of summer* as well as the coldest day of winter. Don't scrimp on proper insulation.

Readers are encouraged to study all these details and consider how they fulfill the functions and placement consideration of hydronic components discussed in previous chapters.

The electrical wiring diagram for antifreeze-based combisystem #1 is shown in Figure 10-25.

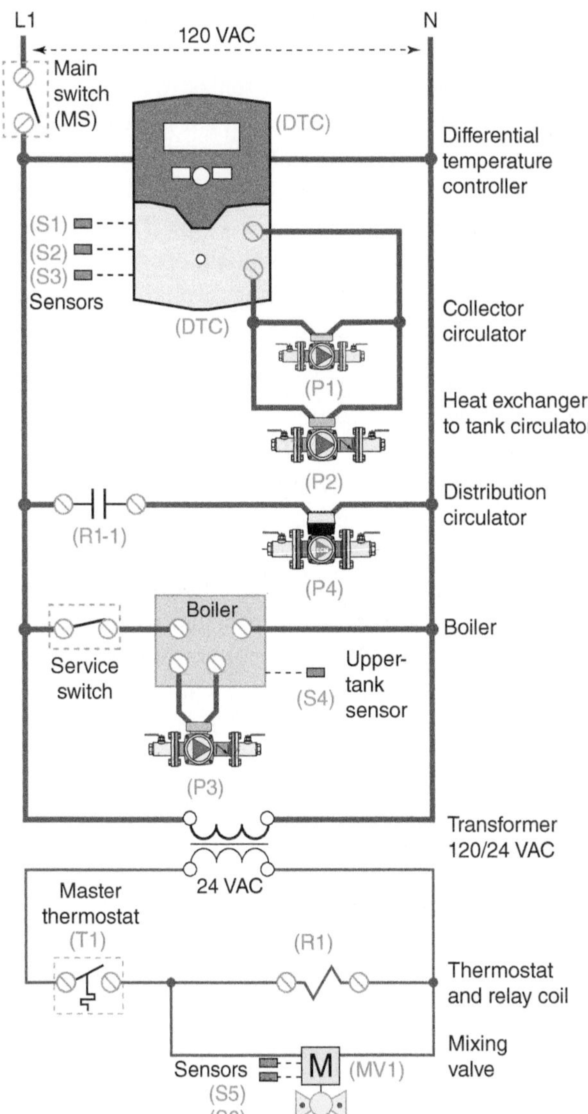

Figure 10-25 | Electrical schematic for antifreeze-based combisystem #1.

The following is a description of operation of the system shown in Figures 10-24 and 10-25. Readers are encouraged to cross-reference the component designations, shown in parentheses, within both the piping and electrical schematics.

Description of Operation:

1. *Solar energy collection mode:* Whenever the main switch (MS) is closed, line voltage is applied to the differential temperature controller (DTC). This controller continuously monitors the temperature of the collector sensor (S1) and the lower sensor in the thermal storage tank (S2). Whenever the temperature of the collector sensor (S1) is 15 °F or more above the

temperature of the lower storage temperature sensor (S2), the (DTC) turns on circulators (P1) and (P2). Whenever the temperature of sensor (S1) is 8 °F or less above the temperature of sensor (S2), the (DTC) turns of circulators (P1) and (P2). The (DTC) also monitors sensor (S3) in the upper portion of the thermal storage tank. If the sensor reaches 180 °F, the (DTC) initiates a nocturnal cooling mode in which it operates circulators (P1) and (P2) to release heat through the flat-plate collector array until the tank temperature drops to 160 °F.

2. *Domestic water heating mode:* Domestic water is fully heated in a single pass through the stainless steel heat exchanger within the thermal storage tank. This heat exchanger is always surrounded by heated tank water. Whenever the temperature in the upper portion of the storage tank, as detected by boiler sensor (S4), drops below 115 °F, the boiler is turned on in setpoint mode. The boiler turns on circulator (P3). This operation adds heat to the upper portion of the thermal storage tank until sensor (S4) reaches 125 °F, at which point the boiler and circulator (P3) are turned off.

3. *Space heating mode:* A master thermostat (T1) provides a call for space heating. Whenever its contacts close, 24 VAC is applied to the coil of relay (R1). This closes the normally open contacts (R1-1), which supplied 120 VAC to circulator (P4). Circulator (P4) operates in constant differential pressure mode to supply the flow required by the homerun distribution system, based on the status of all wireless thermostatic radiator valves. The master thermostat's call for heat also supplies 24 VAC to the motorized mixing valve (MV1). This valve is equipped with a combined actuator/outdoor reset controller. It measures outdoor temperature from sensor (S5). It uses this temperature, in combination with its settings, to calculate the target supply water temperature to the distribution system. It then adjusts the valve's shaft to steer the temperature measured by sensor (S6) as close as possible to the calculated target temperature. When the master thermostat (T1) is satisfied, power is removed from mixing valve (MV1) and circulator (P4). A spring-load check valve near the storage tank prevents heat migration from the thermal storage tank into the space heating distribution system.

4. *Auxiliary boiler:* The auxiliary boiler monitors the temperature of sensor (S4) at the top of the thermal storage tank. When this temperature drops below 115 °F, the boiler fires. Circulator (P3) is turned on by the boiler. Water flows from the mid-height connection on the thermal storage tank, through the boiler, and back to the upper connection on the thermal storage tank. When the temperature at sensor (S4) reaches 125 °F, the boiler and circulator (P3) are turned off.

Keep in mind that there are several potential variations on this wiring depending on the exact equipment selected and the available operating modes of that equipment. For example, some differential temperature controllers would allow for variable speed operation of the collector circulator (P1) and possibly circulator (P2). Some boilers may be supplied with an internal circulator, which would replace circulator (P3). Some motorized mixing valves may require an external controller and actuator, rather than the integral controller/actuator shown in Figures 10-24 and 10-25. Some designers might also choose methods other than a master thermostat for operating circulator (P4) and mixing valve (MV1). These variations all rely on a basic knowledge of the control fundamental discussed in Chapter 6.

ANTIFREEZE-PROTECTED COMBISYSTEM #2

The system shown in Figure 10-26 is a variation on the system shown in Figure 10-24. The main difference is how domestic water is heated. The thermal storage tank in Figure 10-26 does not have an internal coil heat exchanger. Instead, domestic water is heated "on demand" using an external stainless steel brazed-plate heat exchanger. The details for this approach were discussed in Chapter 9.

When there is a flow demand for domestic hot water of 0.6 gpm or higher, the flow switch inside the tankless electric water heater closes. This closure is used, in combination with a relay, to turn on the circulator (P5), which routes water from the upper portion of the storage tank through the primary side of heat exchanger (HX2). An electrical contactor within the tankless heater closes to supply 240 VAC to triacs within the heater, which regulate electrical current flow through the heating elements. That current flow, and hence the power delivered to the elements, is regulated by electronics within the tankless heater that measure the incoming and outgoing water temperatures. The electrical power delivered is limited to that required to provide the desired outgoing water temperature. Thus, if solar preheated water is delivered to this heater at 115 °F, and its setpoint temperature is 120 °F, the electrical current supplied to the elements will only be that required for the 5 °F temperature increase. If the water

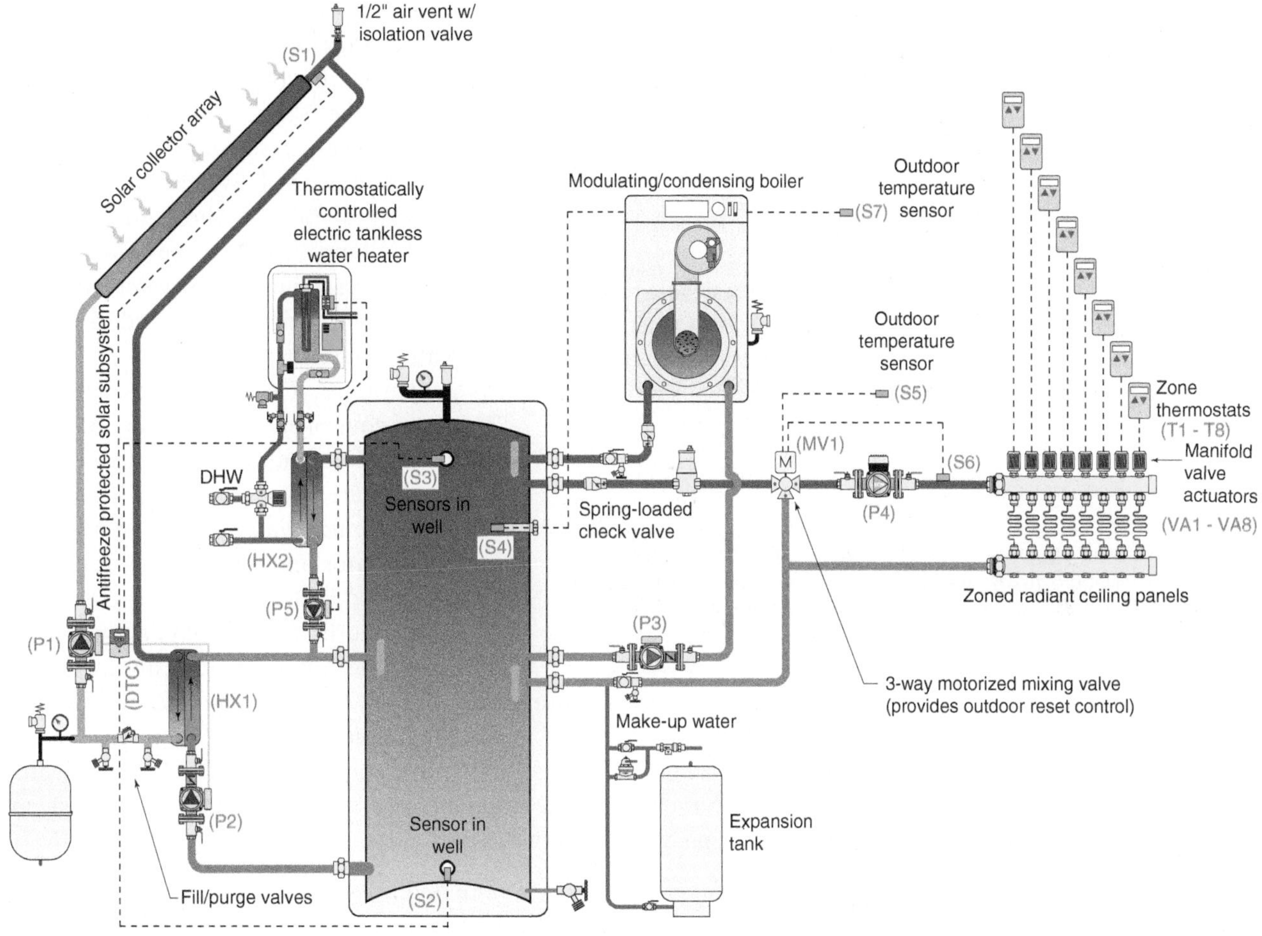

Figure 10-26 | Piping schematic for solar combisystem #2.

entering the tankless heater is already at or above the setpoint temperature, the elements are not turned on and the heated water simply passes through the tankless water heater. All heated water leaving the tankless heater flows into an ASSE 1017–rated mixing valve to ensure a safe delivery temperature to the fixtures.

Because of the way domestic water is heated, the thermal storage tank in this system does not have to be maintained at a temperature suitable to provide the full temperature rise of the domestic hot water. Instead, its temperature can be maintained by the boiler, based on outdoor reset control. If space heating is provided by low-temperature radiant ceiling panels, the maximum water temperature at the top of this tank may only have to be 120 °F. On partial load days, the boiler's internal outdoor reset controller would allow for even lower tank temperatures, perhaps in the range of 90 °F to 95 °F. The lower the tank temperature, the better the chances of heat contribution from the solar collectors. Figure 10-27 shows how the outdoor reset control within the boiler could be configured for this system.

As with the previous system, the thermal mass of the heated water in the upper portion of the thermal storage tank provides excellent buffering for the highly zoned heating distribution system. This highly desirable trait greatly reduces the possibility of boiler short cycling under low partial load conditions. The thermal storage tank also provides hydraulic separation between circulators (P2), (P3), (P4), and (P5).

This design allows the boiler to be completely turned off during times of the year when space heating is not required. Under this condition, the use of electricity is limited to "topping off" the temperature of domestic water, which will almost always be preheated by solar-derived heat within the thermal storage tank.

The electrical wiring diagram for antifreeze-protected combisystem #2 is shown in Figure 10-28. The zone thermostats and zone valves for zones 3, 4, 5, 6, and 7 are and not shown on the schematic to conserve space, but are wired identical to those in zones 1, 2, and 8, which are shown on the schematic.

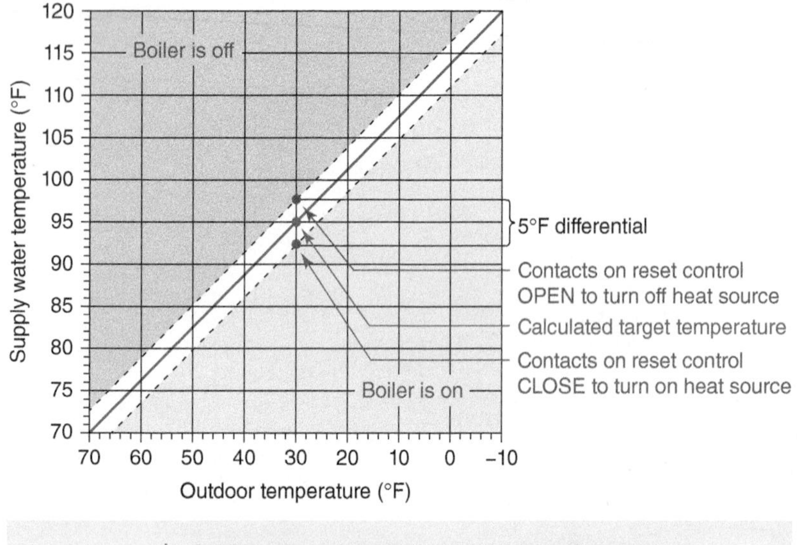

Figure 10-27 Configuration of outdoor reset for the boiler shown in Figure 10-26, assuming that all heat emitters are sized for 120 °F supply water temperature at design load.

The following is a description of operation of the system shown in Figures 10-26 and 10-28. Readers are encouraged to cross-reference the component designations, shown in parentheses, within both the piping and electrical schematics.

Description of Operation:

1. *Solar energy collection mode:* Whenever the main switch (MS) is closed, line voltage is applied to the differential temperature controller (DTC). This controller continuously monitors the temperature of the collector sensor (S1) and the lower sensor in the thermal storage tank (S2). Whenever the temperature of the collector sensor (S1) is 15 °F or more above the temperature of the lower storage temperature sensor (S2), the (DTC) turns on circulators (P1) and (P2). Whenever the temperature of sensor (S1) is 8 °F or less above the temperature of sensor (S2), the (DTC) turns of circulators (P1) and (P2). The (DTC) also monitors sensor (S3) in the upper portion of the thermal storage tank. If the sensor reaches 180 °F, the (DTC) initiates a nocturnal cooling mode in which it operates circulators (P1) and (P2) to release heat through the flat-plate collector array, until the tank temperature drops to 160 °F.

2. *Domestic water heating mode:* Whenever there is a demand for domestic hot water of 0.6 gpm or higher, the flow switch inside the tankless electric water heater closes. This closure applies 240 VAC to the coil of relay (R2). The normally open contacts (R2-1) closes to turn on circulator (P5), which circulates heated water from the upper portion of the storage tank through the primary side of the domestic water heat exchanger (HX2). The domestic water leaving (HX2) is either preheated, or fully heated, depending on the temperature in the upper portion of the storage tank. This water passes into the thermostatically controlled tankless water heater, which measures its inlet temperature. The electronics within this heater control electrical current flow to the heating elements based on the necessary temperature rise (if any) to achieve the set domestic hot water supply temperature. If the water entering the tankless heater is already at or above the setpoint temperature, the elements are not turned on. All heated water leaving the tankless heater flows into an ASSE1017–rated mixing valve to ensure a safe delivery temperature to the fixtures. Whenever the demand for domestic hot water drops below 0.4 gpm, circulator (P5) and the tankless electric water heater are turned off.

3. *Space heating mode:* The distribution system has eight zones, each equipped with a 24 VAC thermostat (T1, T2, T3, etc.). Upon a demand for heat from one or more of these thermostats, the associated 24 VAC manifold valve actuators (VA1, VA2, VA3, etc.) are powered on. When a valve actuator is fully open, its end switch closes. This supplies 24 VAC to the coil of relay (R1). The normally open contacts (R1-1) close to supply 120 VAC to circulator (P4). Circulator (P4) operates in constant differential pressure mode to supply the flow required by the distribution system, based on how many zones are calling for heat. A call for heat also supplies 24 VAC to the controller operating mixing valve (MV1). The valve's controller measures outdoor temperature from sensor (S5). It uses this temperature, in combination with its settings, to calculate the target supply water temperature to the distribution system. It then adjusts the mixing valve's stem to steer the temperature detected by sensor (S6) toward the target temperature. When all thermostats are satisfied, power is removed from mixing valve (MV1) and circulator (P4). A spring-load check valve near the storage tank prevents heat migration from the thermal storage tank into the space heating distribution system when no zones are calling for heat.

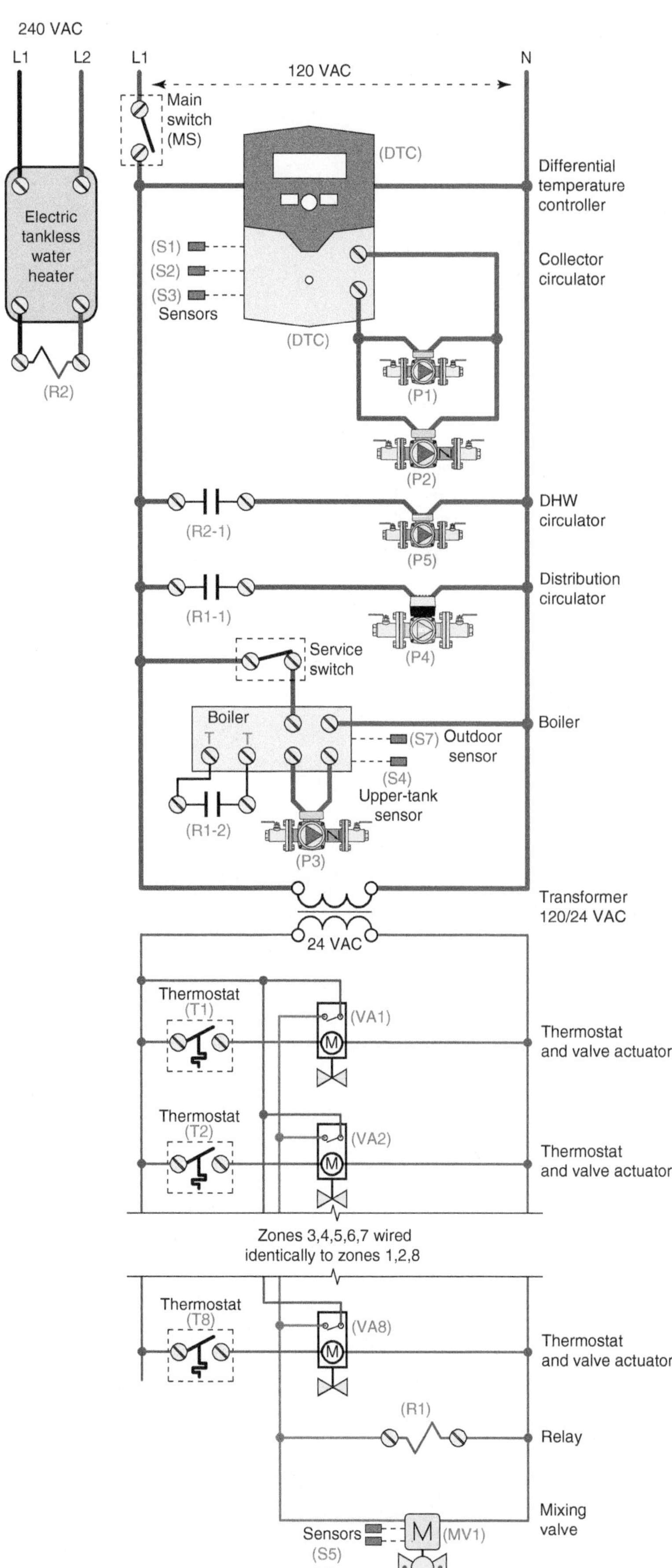

Figure 10-28 | Electrical wiring for solar combisystem #2.

4. *Auxiliary boiler:* The auxiliary boiler is enabled to operate by the closures of relay contact (R1-2) whenever one or more zones are calling for heat. Once enabled, the boiler continuously monitors the outdoor temperature detected by sensor (S7). The boiler's internal reset controller uses this temperature, along with its settings, to calculate the target temperature required by the distribution system. It then compares the temperature in the upper portion of the thermal storage tank, as detected by sensor (S4), with the calculated target temperature. When necessary, the boiler turns on circulator (P3) and operates its burner to maintain the temperature in the upper portion of the thermal storage tank close to the target temperature.

ANTIFREEZE-PROTECTED COMBISYSTEM #3

In systems where the minimum load of the distribution system is comparable to the minimum modulation rate of the boiler, short cycling should not occur. This eliminates the need to use the upper portion of the thermal storage tank to buffer the space heating load. This allows the system shown in Figure 10-26 to be reconfigured, as shown in Figure 10-29.

The solar collector subsystem and domestic water subsystem are the same, as shown in Figure 10-26.

The piping on the load side of the storage tank includes a 3-way diverter valve that directs flow returning from the space heating distribution system into either the midpoint of the thermal storage tank or to the inlet of the auxiliary boiler. This valve is controlled by an outdoor reset controller. When this controller determines that the water temperature in the upper portion of thermal storage tank is high enough to supply the space heating load, the diverter valve remains off and flow returning from the distribution system is routed back into the thermal storage tank. If the outdoor reset controller determines this is not the case, it turns on the diverter valve and return flow is directed to the boiler. The end switch in the diverter valve closes to enable boiler operation.

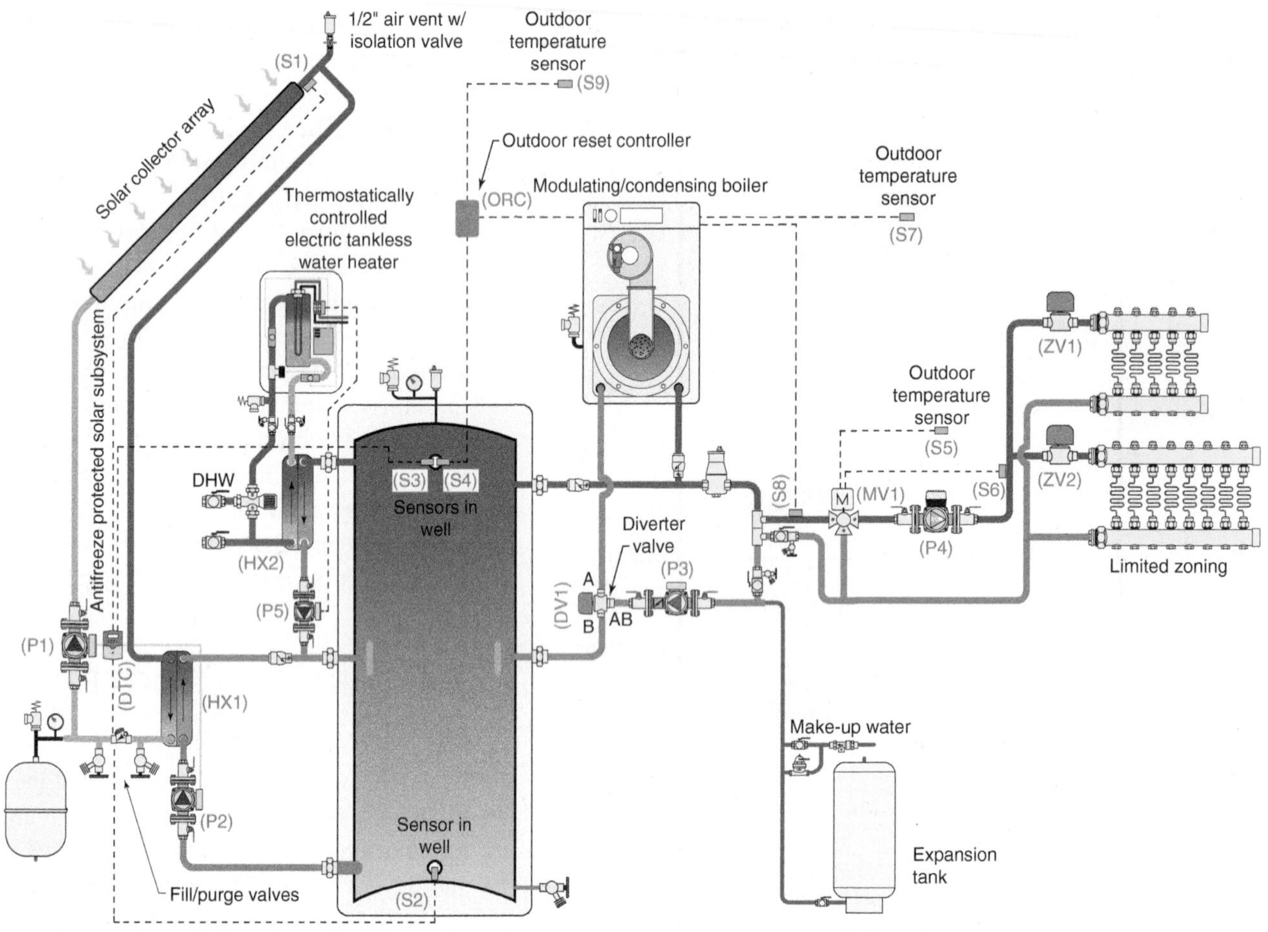

Figure 10-29 | Piping schematic for solar combisystem #3 with minimal zoning. Boiler does not maintain any minimum temperature in the thermal storage tank.

This approach allows the thermal storage tank to cool all the way down to room temperature during extended periods of cloudy weather. If the tank is located in conditioned space, the heat is gives up contributes to the space heating load. When the sun returns, the cool water in the tank allows the collectors to initially operate at low inlet temperatures, and thus high efficiency.

When there is a demand for domestic hot water, the cold domestic water entering heat exchanger (HX2) is still be preheated by the water in the thermal storage tank. The thermostatically controlled tankless electric water heater provides the necessary temperature boost to ensure acceptable domestic hot water delivery temperature.

The electrical wiring diagram for antifreeze-protected combisystem #3 is shown in Figure 10-30.

The following is a description of operation of the system shown in Figures 10-29 and 10-30. Readers are encouraged to cross-reference the component designations, shown in parentheses, within both the piping and electrical schematics.

Description of Operation:

1. *Solar energy collection mode:* Whenever the main switch (MS) is closed, line voltage is applied to the differential temperature controller (DTC). This controller continuously monitors the temperature of the collector sensor (S1), and the lower sensor in the thermal storage tank (S2). Whenever the temperature of the collector is 15 °F or more above the temperature of the lower storage temperature, the (DTC) turns on circulators (P1) and (P2). Whenever the temperature of sensor (S1) is 8 °F or less above the temperature of sensor (S2), the (DTC) turns of circulators (P1) and (P2). The (DTC) also monitors sensor (S3) in the upper portion of the thermal storage tank. If the sensor reaches 180 °F, the (DTC) initiates a nocturnal cooling mode in which it operates circulators (P1) and (P2) to release heat

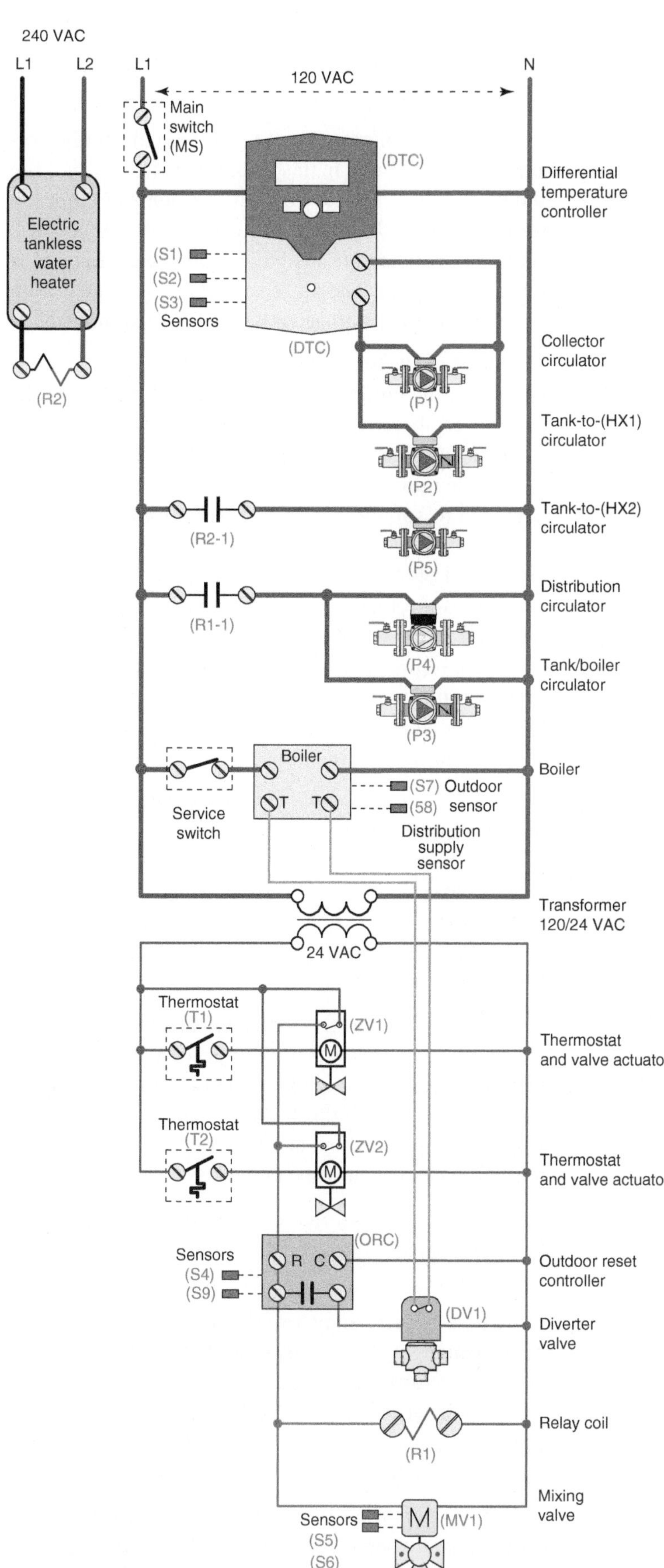

Figure 10-30 | Electrical wiring for solar combisystem #3.

through the flat-plate collector array, until the tank temperature drops to 160 °F.

2. *Domestic water heating mode:* Whenever there is a demand for domestic hot water of 0.6 gpm or higher, the flow switch inside the tankless electric water heater closes. This closure applies 240 VAC to the coil of relay (R2). The normally open contacts (R2-1) closes to turn on circulator (P5), which circulates heated water from the upper portion of the storage tank through the primary side of the domestic water heat exchanger (HX2). The domestic water leaving (HX2) is either preheated, or fully heated, depending on the temperature in the upper portion of the storage tank. This water passes into the thermostatically controlled tankless water heater, which measures its inlet temperature. The electronics within this heater control electrical current flow to the heating elements based on the necessary temperature rise (if any) to achieve the set domestic hot water supply temperature. If the water entering the tankless heater is already at or above the setpoint temperature, the elements are not turned on. All heated water leaving the tankless heater flows into an ASSE1017–rated mixing valve to ensure a safe delivery temperature to the fixtures. Whenever the demand for domestic hot water drops below 0.4 gpm, circulator (P5) and the tankless electric water heater are turned off.

3. *Space heating mode:* The distribution system has two zones, each equipped with a 24 VAC thermostat (T1, T2). Upon a demand for heat from either thermostat, the associated 24 VAC zone valves (ZV1, ZV2) are powered on. When either zone valve reaches its fully open position, its internal end switch closes. This supplies 24 VAC to the coil of relay (R1). The normally open contacts (R1-1) close to supply 120 VAC to circulators (P3) and (P4). Circulator (P4) operates in constant differential pressure mode to supply the flow required by the distribution system, based on how many zones are active. Circulator (P3) provides flow to either the thermal storage tank or the boiler, depending on which is currently serving as the heat source for the distribution system.

A call for heat also supplies 24 VAC to the controller operating mixing valve (MV1). The valve's controller measures outdoor temperature from sensor (S5). It uses this temperature, in combination with its settings, to calculate the target supply water temperature to the distribution system. It then adjusts the valve's stem to steer the temperature detected by sensor (S6) toward the target temperature. When all thermostats are satisfied, power is removed from mixing valve (MV1) and circulator (P4). A spring-load check valve near the storage tank prevents heat migration from the thermal storage tank into the space heating distribution system when neither zone needs heat.

Whenever there is a call for heating, 24 VAC is also supplied to the outdoor reset controller (ORC). This controller then measures the outdoor temperature using sensor (S9). It uses this temperature, in combination with its settings, to calculate the target supply temperature for the distribution system. It then compares this target temperature to the temperature in the upper portion of the storage tank, measured by sensor (S4). If the temperature measured by (S4) is high enough to supply the space heating load, the relay contact within the (ORC) remains open and the diverter valve remains off. Flow returning from the distribution system is directed back to the mid-height connection on the thermal storage tank. If the tank's temperature is too low to supply space heating, the relay contact within the (ORC) closes, supplying 24 VAC to the diverter valve. This valve then directs flow returning from the distribution system to the boiler. When the diverter valve is fully open, its isolated end switch closes. This provides a contact closure across the boiler's (T T) terminals, enabling it to operate. The boiler then monitors its outdoor temperature sensor (S7). It uses this temperature, in combination with its settings, to calculate the target supply water temperature to the distribution system. It compares the temperature at sensor (S8) to this target temperature and adjusts its firing rate as necessary to keep the temperature at (S8) close to the target supply temperature. There is no heat transfer from the boiler to the thermal storage tank.

ANTIFREEZE-PROTECTED COMBISYSTEM #4

Nearly all the space heating distribution systems shown in this text are hydronic. This is not to imply that only hydronic distribution systems are suited for use with renewable energy heat sources. The system shown in Figure 10-31 is a hybrid that combines hydronics for solar energy capture and storage, with forced air for heat distribution.

The solar collector circuit and thermal energy storage components used in this system are identical to those shown in Figure 10-29. So is the on-demand domestic water heating subsystem. The difference is that space heating is delivered to the building using a single-zone forced air system. The latter is supplied with heat from either the thermal storage tank or a gas-fired furnace.

Upon a call for heat from the thermostat, power is applied to the outdoor reset controller. It measures the outdoor temperature and uses this temperature, along with its settings, to determine if the water temperature in the upper portion of the thermal storage tank is adequate to supply the space heating load. If it is, the 3-way motorized mixing valve and load circulator (P4) are turned on. Heated water from the thermal storage tank flows through a **water-to-air heat exchanger coil** mounted in the **discharge plenum** of the furnace. The furnace's blower is also turned on, but its combustion system remains off. Air is moved through the system by the blower and is heated as it passes across the water-to-air heat exchanger coil.

The 3-way mixing valve monitors the discharge air temperature from the coil and adjusts the proportions of water from the thermal storage tank, and the return side of the coil, to maintain a comfortable discharge air temperature. That temperature can be as low as 100 °F, provided the duct system is carefully designed to diffuse air into heat spaces without creating drafts.

If the outdoor reset controller determines that the storage tank is not warm enough to supply the load, the mixing valve and load circulator are turned off and control is passed to the furnace's internal controller, which operates the blower and combustion system in a normal manner to heat the building.

A check valve in the load circulator prevents heat generated by the furnace from migrating back toward the storage tank.

Some readers may wonder why the water-to-air heat exchanger coil is not mounted on the air *intake* side of the furnace. The reason is that such mounting would preheat the air passing through the blower, voiding the warranty on most furnaces. The manufacturer's concern is potential overheating of the blower motor.

The water-to-air coil should be generously sized, with *at least* four tube rows. The larger the coil, the lower the water temperature at which it can deliver a

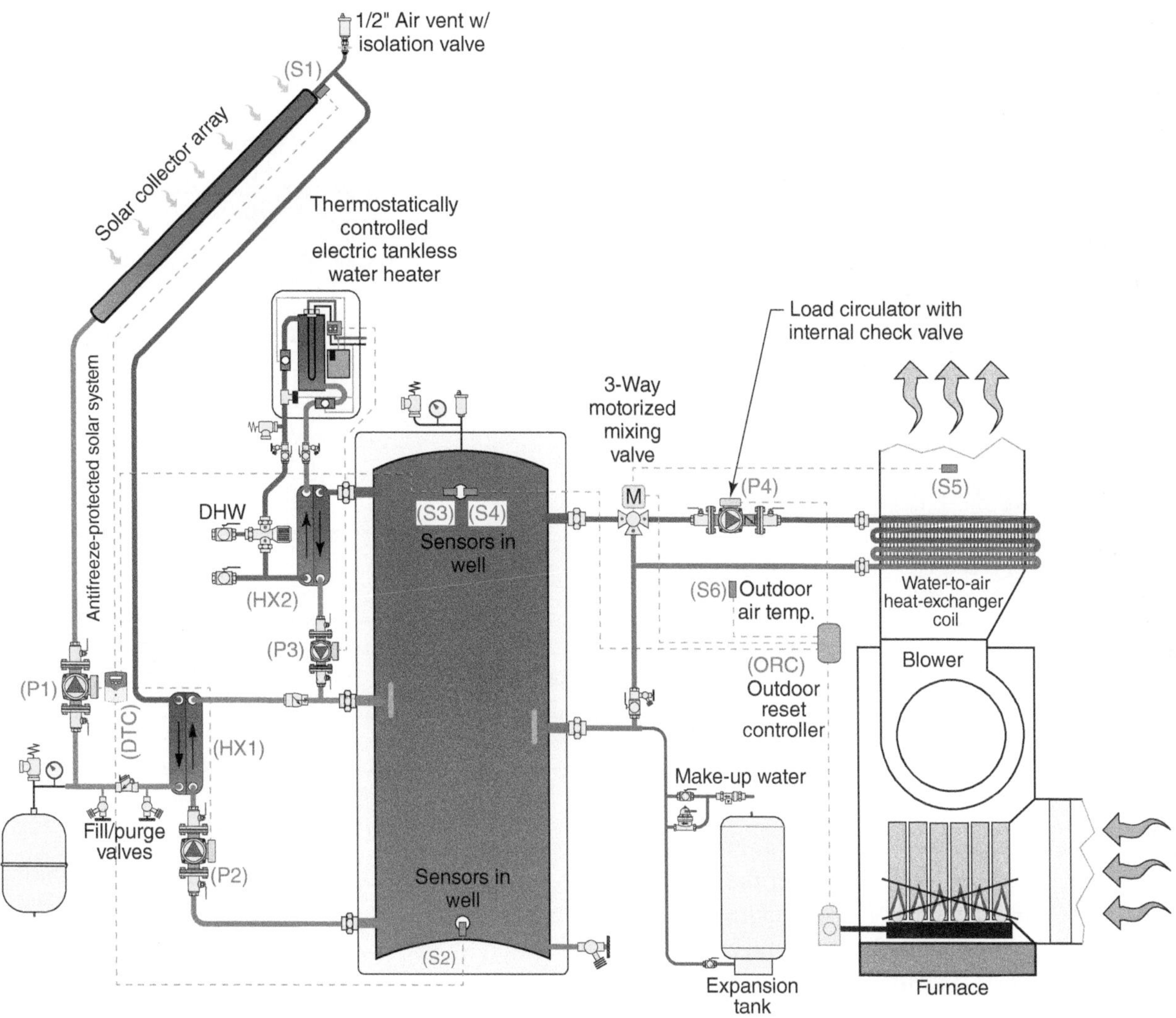

Figure 10-31 | Piping schematic for solar combisystem #4. This system uses forced air heat delivery.

given rate of heat to the building. Figure 10-32 shows an example of a 10-tube row coil.

When selecting the coil, remember that discharge air temperatures lower than 100 °F are seldom recommended. The reset line of the outdoor reset controller should be adjusted to prevent water temperatures lower than approximately 105 °F from supplying the coil. Likewise, the control logic operating the 3-way mixing valve should be configured to maintain discharge air temperatures in the range of 100 °F to 120 °F.

Designers should always check that static pressure drop imposed by the coil to ensure that it will not lower the air flow rate through the ducting to the point of causing comfort problems or inadequate rates of air delivery. The coil should also allow the manufacturer's minimum recommended air flow rate through the furnace.

This system design could also be used to supply solar-sourced heat to the air handler of an air-to-air or water-to-air heat pump. Auxiliary heat would then be provided by the heat pump's refrigeration system.

The electrical schematic for this system is shown in Figure 10-33.

The following is a description of operation of the system shown in Figures 10-31 and 10-33. Readers are encouraged to cross-reference the component designations, shown in parentheses, within both the piping and electrical schematics.

Description of Operation:

1. *Solar energy collection mode:* Whenever the main switch (MS) is closed, line voltage is applied to the differential temperature controller (DTC). This controller continuously monitors the temperature of the

Figure 10-32 | Ten-row water-to-air heat exchanger coil.

Courtesy of J. Allan Antcliffe, P.Eng.

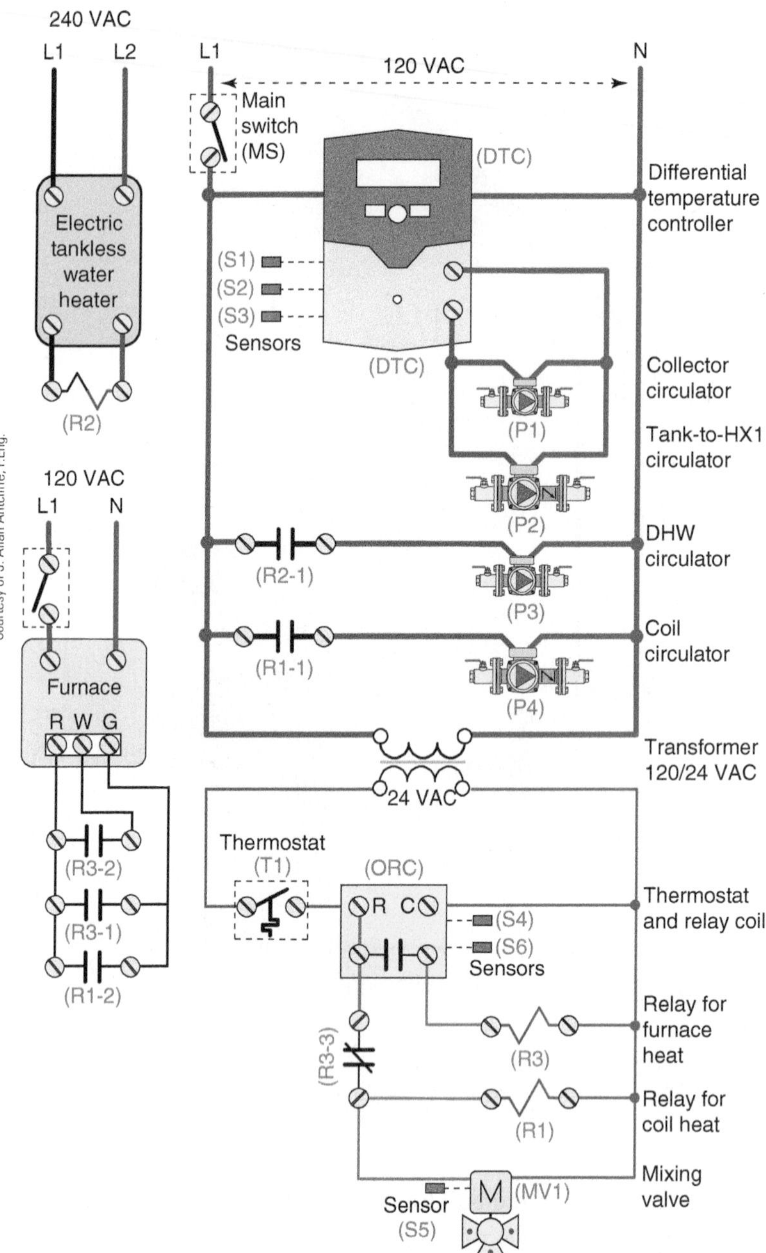

Figure 10-33 | Electrical wiring for solar combisystem #4.

collector sensor (S1) and the lower sensor in the thermal storage tank (S2). Whenever the temperature of the collector is 15 °F or more above the temperature of the lower storage temperature, the (DTC) turns on circulators (P1) and (P2). Whenever the temperature of sensor (S1) is 8 °F or less above the temperature of sensor (S2), the (DTC) turns of circulators (P1) and (P2). The (DTC) also monitors sensor (S3) in the upper portion of the thermal storage tank. If the sensor reaches 180 °F, the (DTC) initiates a nocturnal cooling mode in which it operates circulators (P1) and (P2) to release heat through the flat-plate collector array until the tank temperature drops to 160 °F.

2. *Domestic water heating mode:* Whenever there is a demand for domestic hot water of 0.6 gpm or higher, the flow switch inside the tankless electric water heater closes. This closure applies 240 VAC to the coil of relay (R2). The normally open contacts (R2-1) closes to turn on circulator (P5), which circulates heated water from the upper portion of the storage tank through the primary side of the domestic water heat exchanger (HX2). The domestic water leaving (HX2) is either preheated or fully heated, depending on the temperature in the upper portion of the storage tank. This water passes into the thermostatically controlled tankless water heater, which measures its inlet temperature. The electronics within this heater control electrical current flow to the heating elements based on the necessary temperature rise (if any) to achieve the set domestic hot water supply temperature. If the water entering the tankless heater is already at or above the setpoint temperature, the elements are not turned on. All heated water leaving the tankless heater flows into an ASSE 1017–rated mixing valve to ensure a safe delivery temperature to

the fixtures. Whenever the demand for domestic hot water drops below 0.4 gpm, circulator (P5) and the tankless electric water heater are turned off.

3. *Space heating mode:* Whenever there is a call for heating, 24 VAC is supplied to the outdoor reset controller (ORC). This controller then measures the outdoor temperature using sensor (S6). It uses this temperature, in combination with its settings, to calculate the target supply temperature for the distribution system. It then compares this target temperature to the temperature in the upper portion of the storage tank, measured by sensor (S4). If the temperature measured by (S4) is high enough to supply the space heating load, the relay in the (ORC) remains open and the combustion system in the furnace (R to W) remains off. 24 VAC also passes through the normally closed relay contact (R3-3) and powers up relay coil (R1) and mixing valve (MV1). The normally open contacts (R1-1) close to turn on circulator (P4), which circulates water from the upper portion of the thermal storage tank through the plenum coil. Another set of contact (R1-2) close between the furnace terminals (R and G) to turn on the blower. Mixing valve (MV1) measures the temperature of the air leaving the plenum coil using sensor (S5), and adjusts water proportions to maintain this temperature close to its target setting of 105 °F.

If the water temperature in the upper portion of the thermal stage is too cool to supply the load, the normally open contacts in the outdoor reset controller (ORC) close. This applies 24 VAC to the coil of relay R3. The normally open contacts (R3-3) open to interrupt 24 VAC to both relay (R1) and mixing valve (MV1). Another set of contacts (R3-1) close to complete the furnace control circuit (R to G), which turns on the blower. Another set of contacts (R3-2) close to complete the furnace control circuit (R to W), which turns on the furnace's burner system. There is no mode where the furnace's burner is operating simultaneously with the plenum coil.

ANTIFREEZE-PROTECTED COMBISYSTEM #5

The piping schematic for combisystem #5 is shown in Figure 10-34. It is based on a nonpressurized thermal storage tank.

The thermal storage tank is coupled to a closed-loop solar collector circuit through heat exchanger (HX1). It is also coupled to a closed-loop space heating distribution system through heat exchanger (HX2). An internal copper-coil heat exchanger is used for domestic water preheating.

The solar collector circuit shown in Figure 10-34 is similar to those shown in previous systems. However, the circuit between the thermal storage tank and collector heat exchanger (HX1) is now part of an "open" system. This circuit contains **"gooseneck" piping**, which is routed through the tank's sidewall above the water level. When the circulator in this circuit is off, the water in the piping above the water level will be under slight sub-atmospheric pressure. This is not a problem provided that proper design details are followed. First, there should not be any valves, vents, or other devices in the piping assembly that is above the water level. If present, such devices could allow air to enter the piping, and thus break the siphon. Second, the top of this piping should be maintained as low as possible to minimize the sub-atmospheric pressure. Third, the circulator should be placed as low as possible relative to the water level in the tank. This maximizes the static pressure at the circulator inlet and helps prevent cavitation, especially when the water is at a high temperature.

When the system is commissioned, the inline ball of the bidirectional purging valve located upstream of circulator (P2) is closed and water is forced into the upstream side port of this valve. The entering water forces air through the gooseneck piping and eventually out the bottom of the dip tube. After this air has been purged, water is forced into the other side port of the bidirectional purging valve. This water flows through circulator (P2), heat exchanger (HX1), and the return gooseneck piping, again pushing air ahead of it and releasing that air into the tank. At this point, both gooseneck piping assemblies should be filled with water and purged of air. Water flow for purging can now be turned off, and both goosenecks will retain the water within them. The circuit is now primed and ready to operate.

The heat exchanger subassembly supplying space heating uses the same details. Using a heat exchanger between the tank water and space heating distribution system allows the latter to be designed as closed system, which has several advantages, especially if the piping within it needs to be routed several feet above the water level in the thermal storage tank.

Circulators (P2) and (P3) are both part of the open system, and as such, must be made of either stainless steel, bronze, or a polymer. A cast-iron circulator in these locations would experience constant corrosion due to the dissolved oxygen in the water. These circulators should be installed as low as possible relative to

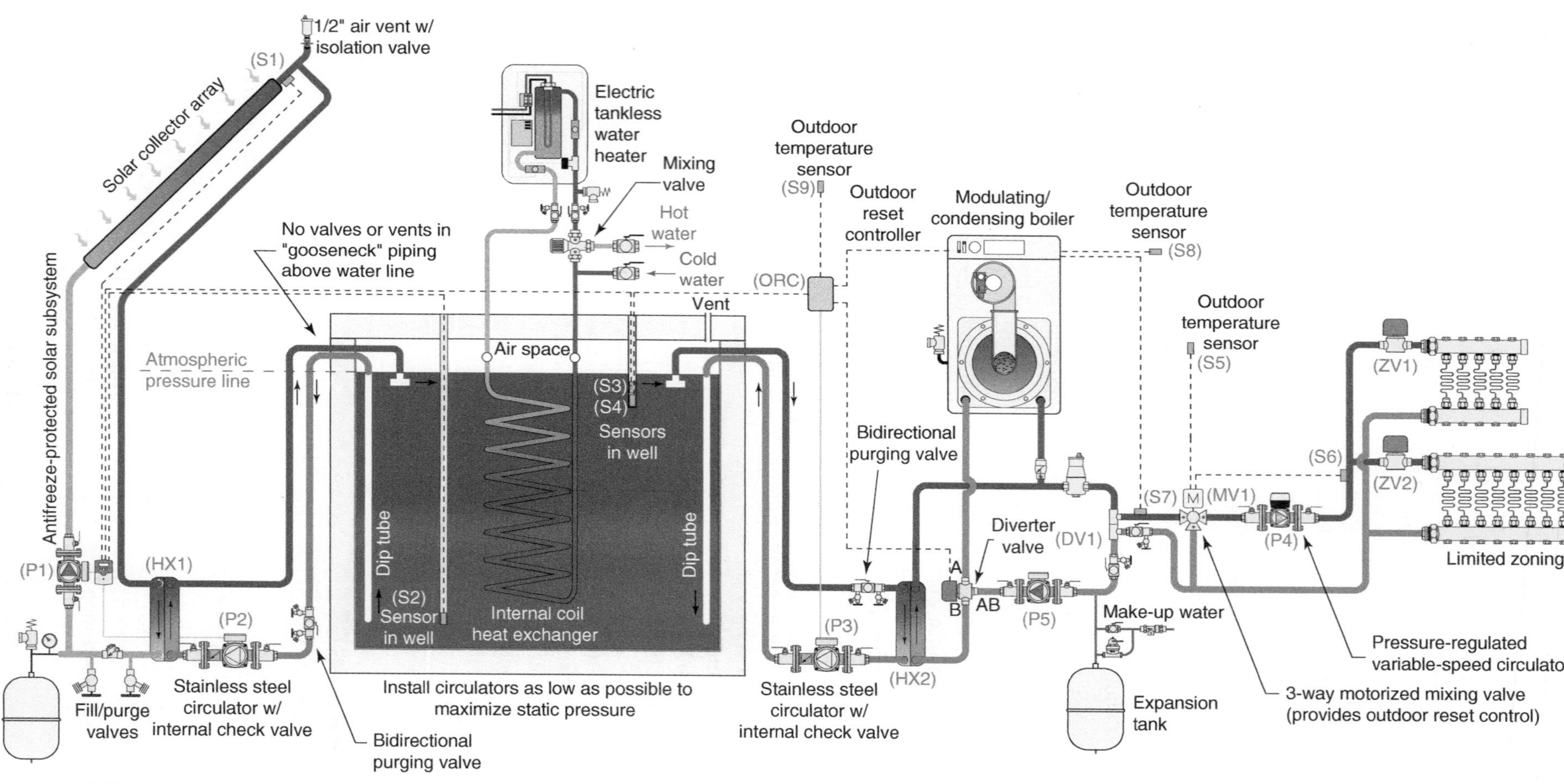

Figure 10-34 | Piping schematic for solar combisystem #5 using a nonpressurized thermal storage tank.

the water level in the tank. Both circulators contain internal spring-load check valves to prevent reverse thermosiphoning when the circulators are off.

The author recommends sizing the solar circuit heat exchanger, as well as the heat exchanger supplying space heating, for a maximum approach temperature difference of 5 °F. This reduces the thermal penalty associated with having a heat exchanger in these paths of heat transfer. The solar heat exchanger should also provide a minimum effectiveness of 0.70. Refer to Chapter 4 for methods of calculating the effectiveness of a heat exchanger.

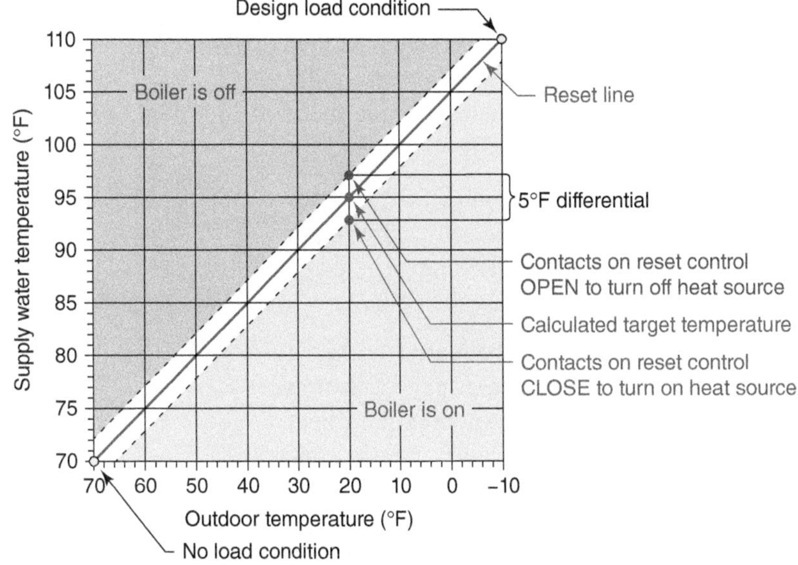

Figure 10-35 Outdoor reset control logic used to select which heat source supplies the space heating load for the system shown in Figure 10-34.

Upon a call for space heating, an outdoor reset controller is powered on. Circulators (P4) and (P5) are also turned on. This outdoor reset controller measures the current outdoor temperature as well as the temperature near the top of the storage tank. Based on these temperatures, and its settings, it determines if the storage tank temperature is sufficient to supply the current space heating load.

Figure 10-35 shows the control logic used by the outdoor reset controller when determining if the thermal storage tank or the auxiliary boiler will serve as the heat source for space heating. uses which heat source (e.g., the thermal storage tank, or the auxiliary boiler) will supply the space heating load.

This graph assumes that the heat emitters require a supply water temperature of 110 °F at an outdoor design temperature of –10 °F. It also shows that the supply water temperature will be 70 °F at an outdoor temperature of 70 °F, which represents a no-load conditions. The blue line between the no-load condition and design load condition is the reset line. It represents the ideal target supply water temperature that should be supplied to the heat emitters, based on the current outdoor condition. The green shaded area represents conditions in which the thermal storage tank is too cool to supply the load, and thus the auxiliary boiler needs to operate. The red shaded area represents conditions in which the tank is warm enough to supply the load, and hence the auxiliary boiler remains off. If the water temperature in the tank falls within the unshaded area of the graph, the boiler will remain off until the tank cools to the temperature represented by the upper edge of the green shaded area, at which point the boiler will operate. It will continue operating until the water temperature reaches the lower edge of the red shaded area. The colored dots along the vertical line representing an outdoor temperature of 20 °F, show that the auxiliary boiler is off until the water temperature at the top of the tank drops to 92.5 °F. The boiler is then turned on, and remains on until either the heating load is satisfied or solar heat input has warmed the water at the top of the thermal storage tank to 97.5 °F.

If the thermal storage tank is warm enough to supply the load, the 3-way diverter valve remains off and flow passes from its (AB) port to its (B) port. This flow passes through the heat exchanger (HX2) and eventually into the set of closely spaced tees that provide hydraulic separation between circulator (P4) and the distribution circulator (P5). Heated water then passes into the 3-way motorized mixing valve (MV1), where its temperature is adjusted, as necessary, to properly supply the low temperature heat emitters. Circulator (P3) is also operating during this mode to provide flow between the thermal storage tank and heat exchanger (HX2).

The system continues to operate in this mode as long as there is a call for heat, and the storage tank temperature is high enough to meet the load. However, when the storage tank temperature drops to the point where it cannot supply the load, the outdoor reset controller performs three functions. First, it turns off circulator (P3) to stop flow from the thermal storage tank to heat exchanger (HX2). Second, it powers on the diverter valve (DV1), which now directs flow from its (AB) port to its (A) port and thus through the boiler. Third, it enables the boiler to fire to provide auxiliary energy to the space heating distribution system.

In the latter operating mode, the thermal storage tank does not provide any buffering of the space heating load. Thus, zoning should be limited to avoid short cycling the boiler.

Domestic water is preheated within a copper coil heat exchanger suspended within the thermal storage tank. Any auxiliary heat needed is added by a thermostatically controlled electric tankless water heater. All domestic hot water must pass through an ASSE 1017–rated thermostatic mixing valve, which limits delivery temperature to the fixtures.

The electrical schematic for this system is shown in Figure 10-36.

The following is a description of operation of the system shown in Figures 10-34 and 10-36. Readers are encouraged to cross-reference the component designations, shown in parentheses, within both the piping and electrical schematics.

Description of Operation:

1. *Solar energy collection mode:* Whenever the main switch (MS) is closed, line voltage is applied to the differential temperature controller (DTC). This controller continuously monitors the temperature of the collector sensor (S1) and the lower sensor in the thermal storage tank (S2). Whenever the temperature of the collector sensor (S1) is 15 °F or more above the temperature of the lower storage sensor (S2), the (DTC) turns on circulators (P1) and (P2). Whenever the temperature of sensor (S1) is 8 °F or less above the temperature of sensor (S2), the (DTC) turns off circulators (P1) and (P2). The (DTC) also monitors sensor (S3) in the upper portion of the thermal storage tank. If the sensor reaches 180 °F, the (DTC) initiates a nocturnal cooling mode in which it operates circulators (P1) and (P2) to release heat through the flat-plate collector array, until the tank temperature drops to 160 °F.

2. *Domestic water heating mode:* Whenever there is a demand for domestic hot

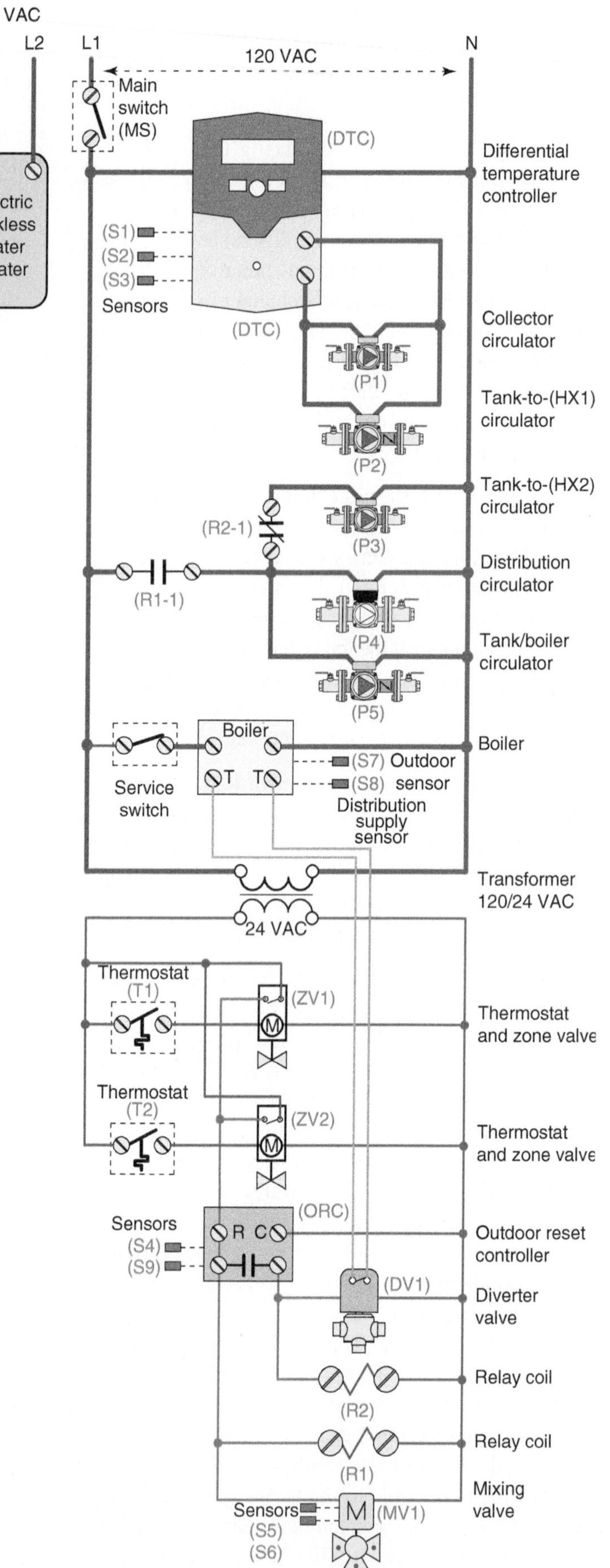

Figure 10-36 | Electrical wiring for solar combisystem #5.

water, cold domestic water passes through the copper-coil heat exchanger suspended within the thermal storage tank. The temperature at which the water leaves this coil depends on the temperature of the thermal storage tank. In some cases, it will be heated to or above the desired delivery temperature. In other cases, it will only be preheated. All water exiting the coil passes into the thermostatically controlled tankless electric water heater. When the domestic water flow rate reaches 0.6 gpm or higher, the flow switch inside the tankless electric water heater closes. The electronics within this heater control electrical current flow to the heating elements based on the necessary temperature rise (if any) to achieve the set domestic hot water supply temperature. If the water entering the tankless heater is already at or above the setpoint temperature, the elements are not turned on. All heated water leaving the tankless heater flows into an ASSE 1017–rated mixing valve to ensure a safe delivery temperature. Whenever the demand for domestic hot water drops below 0.4 gpm, the elements in the tankless electric water heater are turned off.

3. *Space heating mode:* The distribution system has two zones, each equipped with a 24 VAC thermostat (T1, T2). Upon a demand for heat from either thermostat, the associated 24 VAC zone valve (ZV1, ZV2) is powered on. When the zone valve is fully open, its end switch closes. This supplies 24 VAC to the coil of relay (R1). The normally open contacts (R1-1) close to supply 120 VAC to circulators (P4) and (P5), and, through a normally closed contact (R2-1) to circulator (P3). Circulator (P4) operates in constant differential pressure mode to supply the flow required by the distribution system, based on how many zones are active.

A call for heat also supplies 24 VAC to the controller operating mixing valve (MV1). The valve's controller measures outdoor temperature from sensor (S5). It uses this temperature, in combination with its settings, to calculate the target supply water temperature to the distribution system. It then adjusts the valve's stem to steer the temperature detected by sensor (S6) toward the target temperature. When all thermostats are satisfied, power is removed from mixing valve (MV1) and circulator (P4).

Whenever one of the zone valve end switches is closed, 24 VAC is also supplied to the outdoor reset controller (ORC). This controller measures the outdoor temperature using sensor (S9). It uses this temperature, in combination with its settings, to calculate the target supply temperature for the distribution system. It then compares this target temperature to the temperature in the upper portion of the storage tank, measured by sensor (S4). If the temperature measured by (S4) is high enough to supply the space heating load, the relay in the (ORC) remains open, and the diverter valve remains off. Flow returning from the distribution system is directed back through heat exchanger (HX2). If the tank's temperature is too low to supply space heating, the relay in the (ORC) closes, supplying 24 VAC to the diverter valve. This valve then routes flow returning from the distribution system to the boiler. When the diverter valve is fully open, its isolated end switch closes. This provides a switch closure across the boiler's (T T) terminals, enabling it to fire. The boiler then monitors its outdoor temperature sensor (S8). It uses this temperature, in combination with its settings, to calculate the target supply water temperature to the distribution system. It compares the temperature at sensor (S7) to this target temperature and adjusts its firing rate as necessary to keep the temperature at (S7) close to the target supply temperature. The relay in the (ORC) also powers on relay coil (R2). This opens the normally closed relay contact (R2-1) and turns off circulator (P3) while the boiler is supplying heat.

SUMMARY

This chapter has presented information on designing and detailing antifreeze-protected solar combisystems. Readers are encouraged to compare the piping schematic and electrical control schematics for the five systems given. There are several common details among these systems. In many cases, these details come from the hydronic fundamentals discussed in previous chapters. Designers are also reminded to verify any local code requirements if adapting one of these conceptual designs to a specific installation drawing or specification.

ADDITIONAL RESOURCES

The following references are recommended for additional reading on solar combisystems system design and installation:

1. Duffie, John A., and William Beckman. (2013). *Solar Engineering of Thermal Processes*, 4th ed. John Wiley & Sons. ISBN 978-0-470-87366-3.
2. Goswami, Yogi D., Frank Keith, and Jan F. Kreider. *Principles of Solar Engineering*, 2nd ed. (2000). Taylor & Francis Group. ISBN 978-1-56032-714-1.
3. Peuser, Felix A., Karl-Heinz Remmers, Martin Schnauss. *Solar Thermal Systems, Successful Planning and Construction*, 2nd ed. (2002). James & James (Science Publishers) Ltd. ISBN 1-902916-310-5.
4. *Solar Hot Water Systems Lessons Learned 1977 to Today.* (2004). Tom Lane, Energy Conservation Services Gainesville, Florida.
5. Ramlow, B., with Benjamim Nusz. *Solar Water Heating: A Comprehensive Guide to Solar Water and Space Heating Systems* (2010). New Society Publishers. ISBN 978-0-86571-668-1.
6. Skinner, Peter, P. E., et al. (2011). *Solar Hot Water Fundamentals, Siting, Design, and Installation.* E2G Solar, LLC. ISBN 978-098335520-5.

KEY TERMS

absolute pressure
antifreeze-based freeze protection
approach temperature
ASSE 1017–rated thermostatic mixing valve
Boyle's law
burst point temperature
cold night fluid shrinkage
commercial fin tube
constant differential pressure mode
corrugated stainless steel tubing (CSST)
counterflow heat exchange
deionized water
discharge plenum
distilled water
domestic hot water plus
Dowfrost HD
drainback freeze protection
dry stagnation
EPA Public Law 111-380
ethylene glycol
F-chart
fin tube
flow switch
freeze point temperature
gooseneck piping
heat dump
hydraulic separation
Inhibitors
nocturnal cooling
over-temperature protection mode
parametric studies
pH buffers
point of not pressure change
propylene glycol
purging carts
radiative cooling
reverse thermosyphoning
slugging
slush
solar circulation station
solar heating fraction
solar photovoltaic module
SRCC OG-100 certification report
stagnation
thermal penalty [due to heat exchanger]
thermostatic valve
thermosiphon cooling loop
TRNSYS
Tyfocor LS
water-to-air heat exchanger coil

QUESTIONS AND EXERCISES

1. Explain the difference between ethylene glycol and propylene glycol. Which is preferred for use in solar combisystems that heat domestic water? Explain why.
2. Describe the difference between the freeze point temperature and burst point temperature of an antifreeze solution. Which temperature should be used when preparing an antifreeze solution for use in an operating solar combisystem?
3. What is the function of an inhibitor in a propylene glycol antifreeze solution?
4. If the on-site water is tested and found unsuitable for mixing with glycol-based antifreeze, what other types of water should be used?
5. Why is it necessary to install an isolation valve at the top of the collector piping circuit in combination with and under a float type air vent?
6. After a submersible pump is used to add the antifreeze solution to the collector circuit, what is the proper tool for increasing the pressure in that circuit?
7. What happens to solutions of propylene glycol when maintained at high temperatures within stagnating collectors?
8. Explain the operation of a passive heat dump system. Include discussion of the thermostatic valve, fin-tube, and swing check valve.

9. Explain the operation of a photovoltaic-powered heat dump system.

10. Describe the operation of an antifreeze-protected solar thermal system during nocturnal cooling mode.

11. Describe a situation in which it is not wise to use the earth loop of a geothermal heat pump system as a heat dump.

12. A heat dump device needs to be sized for an array of 10 collectors. Each collector is 4 feet wide by 8 feet long. The collectors have the following performance indices $(F_R\tau\alpha) = 0.75$, $(F_RU_L) = 0.10$. The array operates at a total flow rate of 8 gpm, with a 40 percent solution of propylene glycol antifreeze. Determine the heat dissipation rating of a heat dump device that must maintain this array at an inlet temperature of 160 °F, on a day when the solar radiation in the plane of the collectors is 300 Btu/hr./ft.^{2} and the ambient air temperature is 85 °F.

13. Determine the minimum expansion tank volume for the following solar circuit. Assume that the static pressure at the top of the collector circuit is 25 psi at cold fill conditions, and that the 50 psi relief valve, as well as the inlet of the expansion tank, is located 20 feet below to the top of the circuit:

- size of collectors, 4 feet × 10 feet
- volume of each collector, 1.25 gallons
- number of collectors: four
- estimated length of 3/4-inch copper tube in circuit = 75 feet
- volume of heat exchanger coil in thermal storage tank = 3 gallons
- fluid in circuit: 40 percent solution of propylene glycol
- cold fill temperature = 50 °F
- maximum fluid temperature rise = 200 °F

14. Assume that the collector array, and half of the collector circuit tubing, cool from their cold fill temperature of 50 °F to –25 °F on a bitterly cold night. Determine the "shrinkage volume" of the 40 percent propylene glycol solution in the collector circuit under these conditions.

15. Why should the expansion tank in a solar collector circuit be located downstream of the check valve in that circuit? Why should it be located close to the inlet side of the circulator?

16. Explain the concept of "domestic hot water plus" as it pertains to practically sized solar combisystems.

17. Why should the hot antifreeze solution flow through a brazed-plate heat exchanger in the opposite direction from the domestic water being heated?

18. Explain what the "f" in f-Chart stands for. What does an annual "f" value of 0.458 mean for a solar domestic water heating system?

19. Describe what is meant by a "parametric study" when using a solar simulation software tool such as F-CHART.

20. The following questions will reference the schematic shown in Figure 10-24:

a. Why is there a higher percentage of the total surface area of the internal stainless steel heat exchanger within the upper portion of the storage tank?

b. Why is the temperature of the upper portion of the thermal storage tank, and not the entire tank, maintained at some minimum value by the boiler?

c. Explain how the piping connections on the tank provide hydraulic separation between the boiler circulator (P3), variable speed distribution circulator (P4), and heat exchanger circulator (P2).

d. Why is it necessary to include an ASSE 1017–rated mixing valve on the domestic hot water piping leaving this system?

e. Describe two functions of the motorized 3-way mixing valve (MV1) in this system.

f. Why is there a spring loaded check valve on the hot water pipe leaving the upper right side of the tank leading to the 3-way motorized mixing valve?

g. Explain the relationship between the heated water in the upper portion of the buffer tank, the highly zoned distribution system, and the mod/con boiler.

21. Explain why the water temperature in the upper portion of the thermal storage tank in Figure 10-24 has to be maintained, whereas the temperature in the thermal storage tank of Figure 10-29 does not. Which of these systems will likely yield higher-efficiency operation of the solar collectors? Why?

22. Why should the discharge air temperature from the plenum coil in Figure 10-31 be no less than 100 °F?

23. Why is it important to limit the number of heating zones when using either combisystem #3 or #5?

24. Why is it important not to have any valves, vents, or other devices within the gooseneck piping used with unpressurized thermal storage tanks?

25. Explain the logic used by an outdoor reset controller for determining if the thermal storage tank in a combisystem can serve as the heat source for space heating.

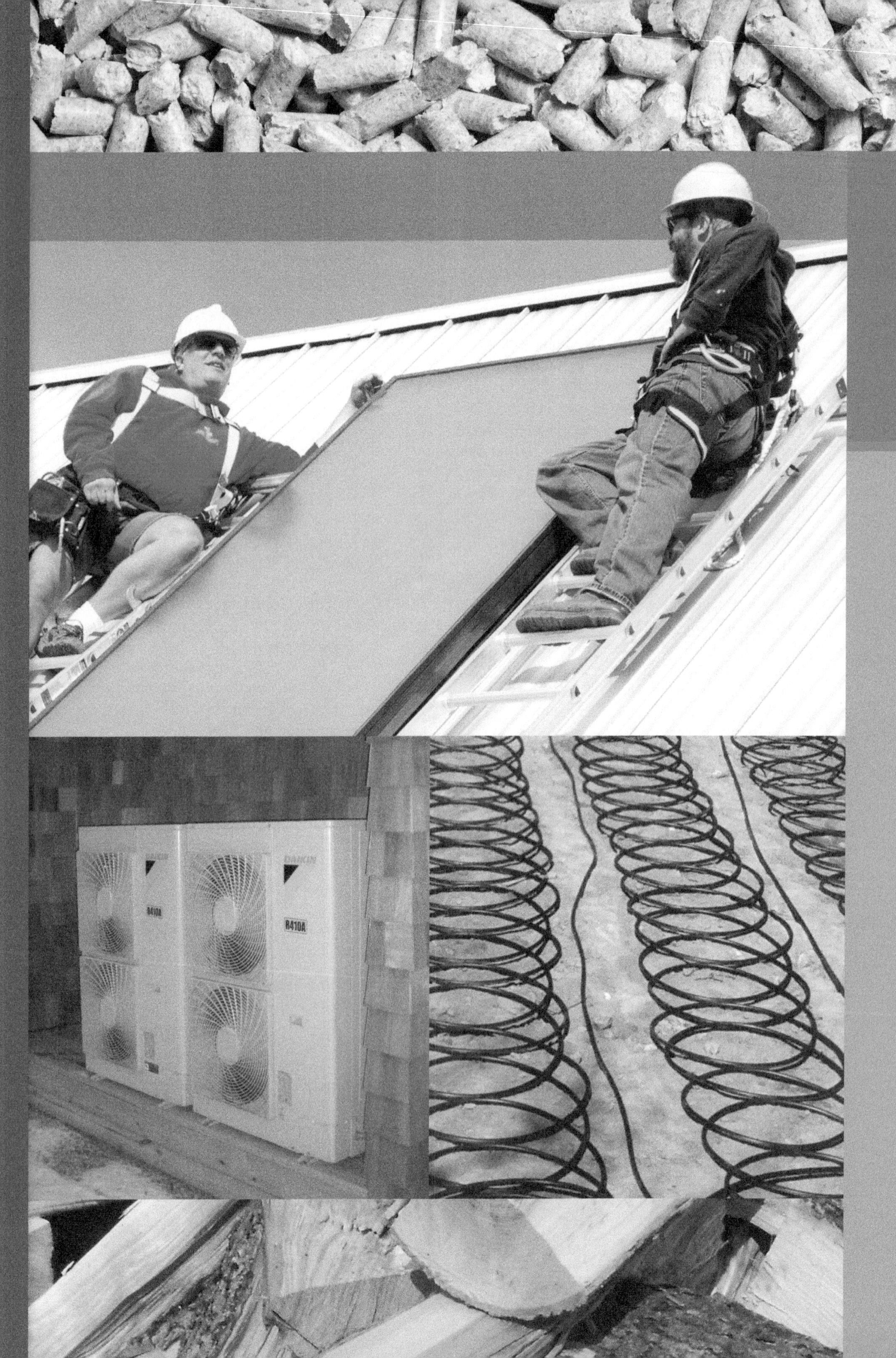
R410A
R410A

CHAPTER 11

Drainback-Protected Solar Combisystems

OBJECTIVES

After studying this chapter, you should be able to:

- Explain several advantages of drainback-protected solar combisystems.
- Learn advantages and limitations of pressurized drainback-protected systems.
- Learn advantages and limitations of unpressurized drainback-protected systems.
- Calculate the max height of a siphon in a drainback-protected system.
- Describe details for using drainback-protected systems in tall buildings.
- Explain the necessary collector mounting for drainback-protected systems.
- Size the airspace at the top of a drainback-protected tank.
- Design collector return piping to ensure that a siphon forms.
- Explain the difference between air control and air elimination in a drainback system.
- Describe how a double-pumped drainback-protected system works.
- Size the circulator(s) for a drainback-protected system.
- Describe the piping and controls used in several drainback-protected systems.

11.1 Introduction

The previous chapter described solar combisystems that rely on antifreeze fluids within the collector circuit to protect it against freezing. There are tens of thousands of antifreeze-protected solar thermal systems now in operation around the world. However, as mentioned in the previous chapter, the use of antifreeze does bring some "baggage" into the design process.

- Antifreeze, especially the high-temperature-resistant propylene glycols now commonly used for solar thermal applications, adds cost to the system. As of this writing, these formulations of propylene glycol solutions sell for \$18 to \$36 per gallon. The use of deionized water to dilute this antifreeze, where required, adds an additional cost.
- All glycol-based fluids must be protected against degradation due to high temperature stagnation conditions in solar collectors. This usually requires additional hardware for heat dissipation, which also increases cost.

- The glycol-based antifreeze solutions used in solar thermal systems should be tested, on a yearly basis, to ensure that their pH and levels of corrosion inhibitors are correct. If they are not, the collector loop circuit will require additives to restore the proper fluid chemistry.
- The use of antifreeze solutions requires a heat exchanger between the solar collector array and thermal storage tank. This increases installation cost. It also creates a **thermal penalty** by forcing the collectors to operate at higher temperatures than would be necessary if no heat exchanger were present. It is not uncommon for such heat exchangers to reduce annual solar energy yield by 3 to 5 percent.
- Cold antifreeze solutions increase the thermal mass of collectors, which slows the rate at which the absorber plate temperature can increase as solar radiation reaches useable levels. This delays the start of solar energy collection relative to "dry" collectors.
- The use of antifreeze also requires hardware such as a high-point air vent and an associated isolation valve, purging valves, additional pressure-relief valve, and a dedicated air separator within the collector circuit. Some of this hardware can be eliminated by drainback freeze protection.

Drainback-protected solar thermal systems *are designed so that the water, or other fluid within the collector circuit, drains back to an interior storage tank whenever the collector circulator is not operating.* A film of water, or tiny droplets of water, may remain in the collectors and piping components after drainback. However, this very small amount of water will not cause damage to the collectors or piping when it freezes.

To ensure proper drainage, all water tubes that are part of the collector's absorber plate must have a ***minimum downward pitch*** *of 1/4 inch per foot of horizontal travel.*

11.2 Drainback-Protected Systems Using Unpressurized Thermal Storage

Drainback systems can be designed using either unpressurized or pressured thermal storage tanks. Figure 11-1 shows a basic drainback system using an unpressurized thermal storage tank.

Because the storage tank is vented to the atmosphere, the **gauge pressure** at the water surface is

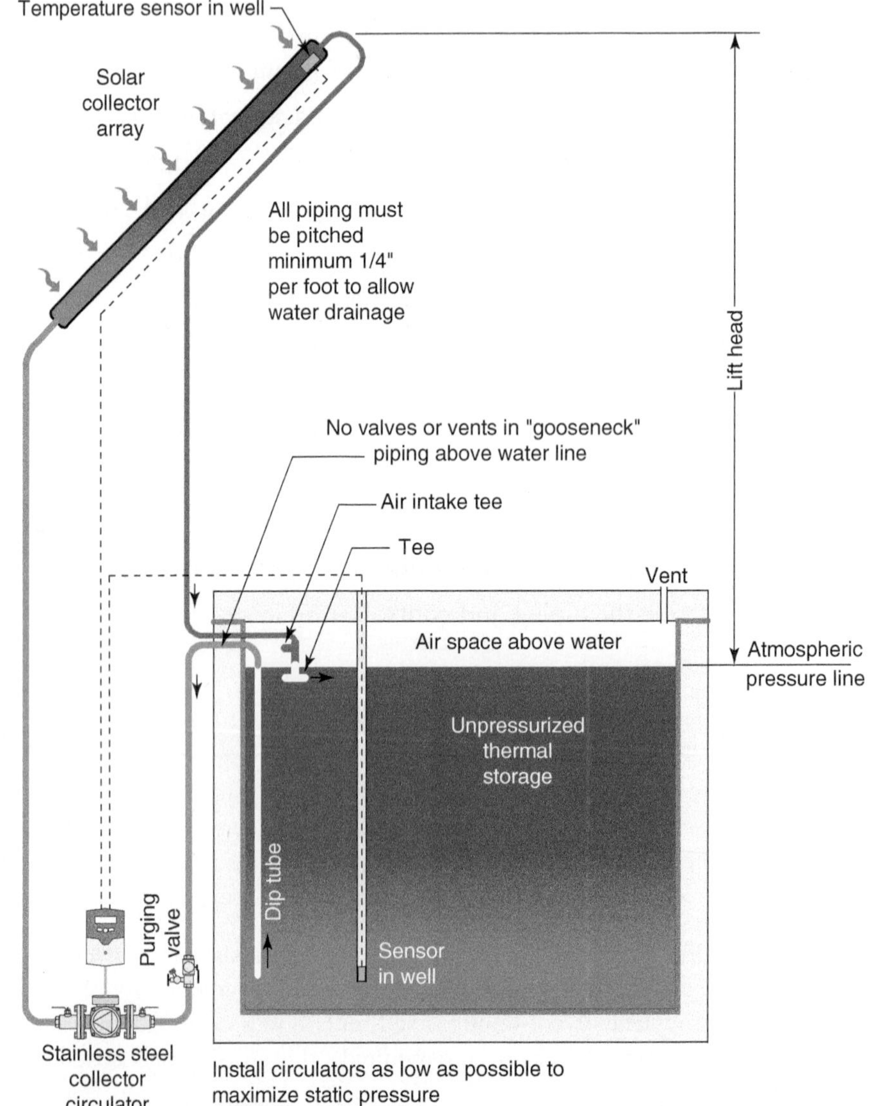

Figure 11-1 | A drainback-protected system using unpressurized thermal storage.

always zero. The air space within the tank, and above the water, serves two purposes. First, it provides volume to accommodate the water that drains back from the collector array and its associated piping. Second, it provides volume to accommodate the thermal expansion of the water in the tank, and piping connected directly to the tank, as its temperature increases. This eliminates the need for an expansion tank in this portion of the system. However, a smaller expansion tank will still be required for any closed-loop circuits that connect to the tank using a heat exchanger.

As with antifreeze-protected systems, the piping connections to most unpressurized tanks are made above the water level. This minimizes the potential for leaks at penetrations through the side of the tank.

The coolest water from the lower portion of the thermal storage tank needs to be routed to the collectors. Getting this water from the bottom of the tank, and through a piping connection above the water line, requires the use of a **dip tube**, as shown in Figure 11-1. When the collector circulator is operating, flow moves up the dip tube, through the **gooseneck piping** above the water level, and then downward to the circulator inlet. The collector circulator should be located as low as possible relative to the water level in the tank. This increases the static pressure at the circulator's inlet and reduces the potential for cavitation. The head loss of the dip tube, gooseneck piping, and the piping leading to the inlet of the circulator should also be kept low. Sizing these pipes for flow velocities in the range of 3 to 4 feet per second will accomplish this, while still allowing for air entrainment.

Because it conveys water directly from an "open" tank, the collector circulator must be constructed of stainless steel, bronze, or a polymer to prevent it from corroding.

At the start of each solar energy collection cycle, the collector circulator must lift water from the static water level in the tank to the top of the collector array. It must also produce sufficient **flow velocity** within the collector return piping to **entrain** air bubbles, and return them to the storage tank. This operating condition causes the return piping to quickly fill with water. This forms a **siphon**, which greatly reduces the **initial lift head** requirement of the collector circulator. This is very desirable because it can, with suitable controls, reduce the electrical power required by the collector circulator for the remainder of the energy collection cycle.

When the differential temperature controller determines that the collector has cooled to within a few degrees of the temperature in the lower portion of the storage tank, it turns off the collector circulator. Air immediately enters an **air intake tee** in the collector return piping located within the air space of the tank. Air quickly moves up the return piping as water drains back into the tank. At some point, the water remaining in the collectors will flow backward and down through the collector supply piping. Within a minute or two, all but a few droplets of water will have drained out of the collector array and exposed piping and into the thermal storage tank. The collector array and exposed piping can now drop to subfreezing temperatures without being damaged.

11.3 Drainback-Protected Systems Using Pressurized Thermal Storage

Figure 11-2 shows a **closed-loop drainback-protected system** designed around a pressure-rated thermal storage tank.

In this system, the air space for holding the collector drainback water, and for accommodating thermal expansion of system water, is at the top of the thermal storage tank. Closed-loop drainback systems can also use a separate drainback/expansion tank installed above a completely filled thermal storage tank. Examples of the latter are shown later in this chapter.

Because this system is completely closed from the atmosphere, it can be operated at a slight positive pressure. This reduces the potential for circulator cavitation. In certain circumstances, it also allows for greater heights between the water level in the tank and the top of the collector array. One of the biggest benefits of a closed-loop drainback system is that it can be connected *directly* to the space heating distribution system, without the use of a heat exchanger. This eliminates the thermal penalty associated with a heat exchanger and reduces the operating temperature of the collectors relative to the temperatures required by the space heating distribution system. The thermal efficiency of the collectors is improved, which ultimately leads to more collected energy. Finally, the circulators used in a closed-loop system can be of lower-cost cast-iron construction.

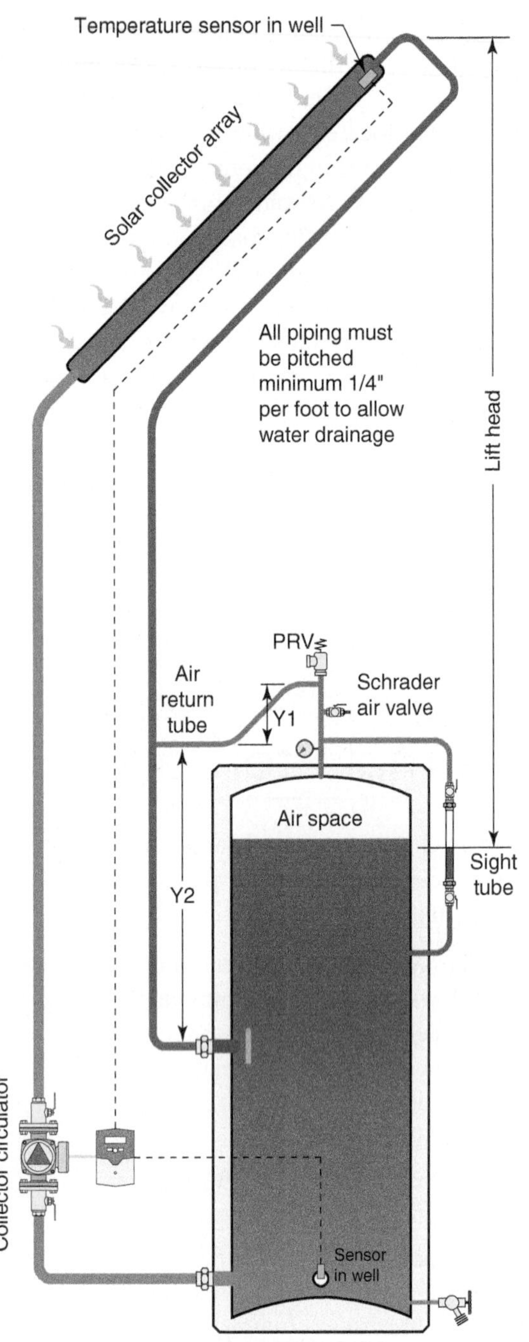

Figure 11-2 | Closed-loop drainback-protected system using a pressurized thermal storage tank.

When the collector circulator is off, the water in the collector supply and return piping will be at the same level as the water in the thermal storage tank. This water level can be monitored by a transparent **sight tube**, installed as shown in Figure 11-2. This sight tube can be made of Pyrex® glass or translucent PEX tubing. It should be several inches longer than the anticipated water level change in the system due to drainback and thermal expansion. It should be installed with part of this additional length above the maximum anticipated water level and some below the lowest anticipated water level. The sight tube assembly should have unions and isolating ball valves at the top and bottom. These allow the sight tube to be isolated, removed, and replaced, if necessary, with very little water loss from the system.

When the collector circuit is operating, water returning from the collector array will flow into the side connection of the thermal storage tank. Some water might also flow into the air return tube connected to the top of the tank. The latter can be minimized by raising the air return tube several inches vertically before routing it to the top of the tank, as shown in Figure 11-2. This vertical rise, shown as dimension "Y1" in Figure 11-2, should be at least 25 percent greater than the *head loss* of the collector returning piping, between the connection point of the air return tube and the point where the return piping enters the thermal storage tank. The latter length of tubing is identified as "Y2" in Figure 11-2. The point where the air return tube connects to the collector return piping should also be at least 2 inches above the highest static water level in the tank. Be sure that there are no sags in the air return tube from its high point to where it connects to the collector return piping. Half-inch tubing is usually sufficient for the air return tube in residential systems.

When the collector circulator turns off, air will immediately flow through the air return tube and up into the collector piping and array. Water drains down from both sides of the collector array. Within 2 or 3 minutes, the collector array and piping should be drained of essentially all water.

The pressure of the air space can be monitored by a pressure gauge, installed as shown in Figure 11-2. This pressure can be adjusted up or down by adding or releasing air from the **Schrader fitting** in the air return tube. A nominal air pressure in the range of 10 to 15 psi is typical.

It's also important to incorporate a pressure-relief valve in any closed drainback system, keeping in mind that it might, at times, be isolated from the pressure-relief valve on the auxiliary boiler, or other heat source.

SIPHON LIMITATIONS

Once a siphon is established within a drainback system, it's important to maintain it for the duration of the solar energy collection cycle.

Along with adequate flow velocity in the collector return piping, siphon stability depends on a relationship between the water's temperature, its corresponding vapor pressure, and the vertical distance between the top of the collector circuit and the water level in the storage tank.

A conservative estimate for the **maximum siphon height** that can exist is found using Equation 11.1:

(Equation 11.1)

$$H_{max} = \left(\frac{144}{D}\right)(P_a + P_{top} - P_v)$$

where:

H_{max} = maximum siphon height (ft.)

D = density of water at maximum anticipated temperature (lb./ft.^{3})

P_a = atmospheric pressure (**psia absolute pressure**)

P_{top} = pressurization (above atmospheric pressure) at the top of the collector circuit (psig)

P_v = vapor pressure of water at maximum anticipated temperature (psia absolute pressure)

The vapor pressure and density of water needed for Equation 11.1 can be calculated using Equations 11.2 and 11.3.

(Equation 11.2)

$$P_v = 0.771 - (0.0326)T + (5.75 \times 10^{-4})T^2 - (3.9 \times 10^{-6})T^3 + (1.59 \times 10^{-8})T^4$$

(Equation 11.3)

$$D = 62.56 + (3.413 \times 10^{-4})T - (6.255 \times 10^{-5})T^2$$

where:

P_v = vapor pressure of water (psia)

D = density of water (lb./ft.^{3})

T = temperature of water (°F)

Example 11.1: Determine the maximum siphon height for water at 200 °F in a system at sea level (where P_a = 14.7 psia), where the air pressure at the top of the collector circuit is 10 psi above atmospheric pressure.

Solution: The vapor pressure of water at 200 °F is found using Equation 11.2:

$$P_v = 0.771 - 0.0326 \times 200 + (5.75 \times 10^{-4})200^2 - (3.9 \times 10^{-6})200^3 + (1.59 \times 10^{-8})200^4 = 11.49 \text{ psia}$$

The density of water at 200 °F is found using Equation 11.3:

$$D = 62.56 + (3.413 \times 10^{-4})200 - (6.255 \times 10^{-5})200^2 = 60.1 \text{ lb./ft.}^3$$

The maximum siphon height is then calculated using Equation 11.1:

$$H_{max} = \left(\frac{144}{D}\right)(P_a + P_{top} - P_v) = \left(\frac{144}{60.1}\right)(14.7 + 10 - 11.49) = 31.7 \text{ feet}$$

Discussion: If the system were installed with a greater vertical distance from the top of the collector circuit to the water level in the storage tank, the water in the return piping would flash to vapor (e.g., boil) and break the siphon. The maximum siphon height decreases with increasing water temperature because as temperature increases, the water moves closer to its vapor flash point.

The maximum siphon height can be increased by raising the pressure within a closed-loop drainback system. Increased pressure helps suppress water from boiling. This is a significant advantage of a closed-loop pressurized drainback system compared to an open-loop system where pressurization is not possible.

Equation 11.1 does not include the effect of frictional pressure drop in the return piping or the potential effect of adding a flow-restricting device near the end of the return piping to increase pressure in that pipe and thus further suppress vapor flash. Thus the siphon height is conservatively estimated by Equation 11.1.

If the height of the building is such that the collector circuit must be taller than the maximum siphon height, a separate drainback tank should be used. It should be installed as high as possible within a *non-freezing area* of building to limit the lift head, as shown in Figure 11-3.

The initial lift head, and siphon height, is now limited to the distance from the water level in the upper (drainback/expansion) tank to the top of the collector array. This approach allows drainback systems to be used in buildings that are several stories high, without concern over siphon breakage.

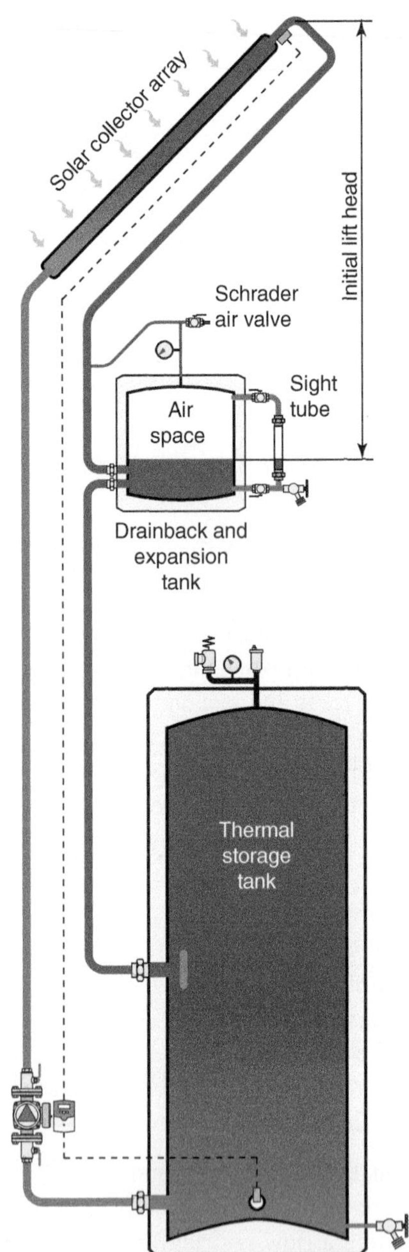

Figure 11-3 | Use of an elevated drainback tank to reduce initial lift head in a drainback system.

11.4 Design Details for Drainback Systems

It is critically important that all piping components within the collector circuit of a drainback-protected system, which could be exposed to freezing temperatures, have sufficient slope for efficient drainage. This drainage begins immediately after the collector circulator turns off.

There are those who suggest that a "dead level" pipe, or collector, should drain to an extent that freezing damage will not occur. While this is true, hypothetically, it fails to recognize potential imperfections such as a sagging length of horizontal tubing between supports. The sag may be caused by the tube's own weight when filled with water or by objects that are inadvertently placed or hung on the tubing. A collector array could also contain inadvertent low points created by installers who fail to compensate for roof sags. Whatever the cause, *if water at a low point fills the tubing to a depth greater than its radius, freezing damage is likely.* Arguments such as "You can't count on installers to ensure proper pitch" have been made as supposed justification for not installing drainback-protected systems. A plausible counterargument is that the drainage, waste, and ventilation (DWV) piping in *every building* must be adequately sloped for proper effluent flow. *Anyone who is not competent enough to provide this slope should not be installing DWV piping or drainback-protected solar thermal systems.* Plan the tube routing with slope in mind, keep an accurate level handy and use it often, and always properly support the pipe to maintain its installed pitch.

There are also those who advocate using a solution of propylene glycol antifreeze in drainback systems. They cite that this as a "belt and suspenders" approach to freeze protection. Thus, if the collector array or its piping are not properly sloped, the antifreeze solution will still prevent freezing. While this is true, the use of antifreeze in a drainback system introduces further compromises and concerns. One is that there will be a residual film of antifreeze solution left in the collectors after every drainage cycle. While this is of no concern during normal operation, there are times when it becomes an issue. Under very high, 350+ °F stagnation temperatures, this residual fluid film could harden into a permanent semi-solid layer inside the absorber plate tubes. This creates a "**fouling factor**" that reduces heat transfer and thus reduces collector performance. Since it's very impractical to fill an entire system with a 40 to 50 percent solution of propylene glycol, a heat exchanger will be required between the collector array and the thermal storage tank. This adds significant cost and reduces thermal performance.

COLLECTOR MOUNTING IN DRAINBACK SYSTEMS

If collectors with standard **"harp-style" absorber plates** are use in a drainback system, they must have a minimum sideways slope of 1/4 inch per foot to provide proper drainage, as shown in Figure 11-4.

When larger arrays of harp-style collectors are used, they can be subdivided, sloped, and piped, as shown in Figure 11-5.

Some collectors have been specially designed with sloping internal headers. They can be mounted *without side slope,* as shown in Figure 11-6.

These collectors, which use a "fifth" port and sloping internal headers, require a sloping external (supply) header along the bottom of the array, as seen in Figure 11-6.

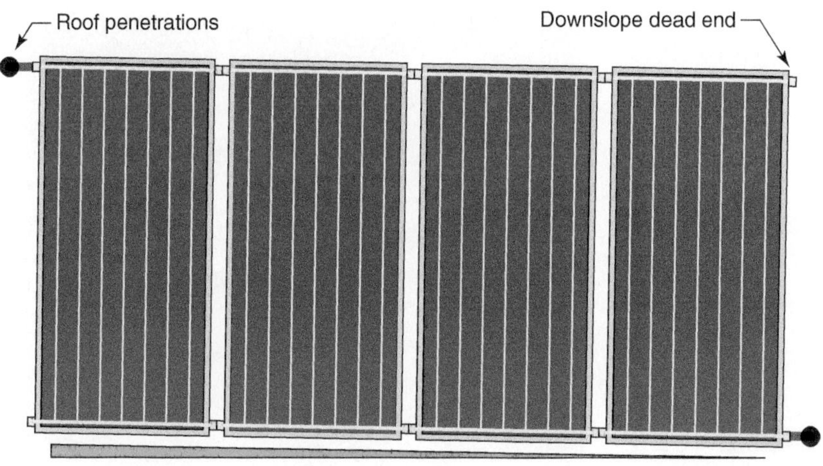

Figure 11-4 | Sideways slope of standard harp-style collectors to ensure proper drainage when used in a drainback-protected system.

CIRCULATOR SELECTION FOR DRAINBACK-PROTECTED SYSTEMS

The collector circulator in a drainback-protected system must perform a task that is very different from that performed by most circulators used in a closed-loop (fluid-filled) hydronic systems. It must lift water from the surface level in the system to the top of the collector array every time a solar collection cycle begins. This is illustrated in Figure 11-7.

Overcoming the initial lift head typically requires a circulator with a steeper pump curve that is capable of providing more head compared to a circulator that just overcomes friction head loss in a fluid-filled hydronic circuit. Methods for selecting this circulator will be discussed shortly.

What happens when the water column reaches the top of the collector array is also very important to the functioning of the system. There are two possibilities:

1. If the water has sufficient velocity, it will entrain air bubbles and quickly move them back to the storage tank. This allows the return piping to fill with water, and create a siphon that eliminates most of the initial lift head.
2. If the water's velocity is too low, it will not entrain all the air and there will be a mixture of water and air "gurgling" down the return piping. A siphon will not form under these conditions.

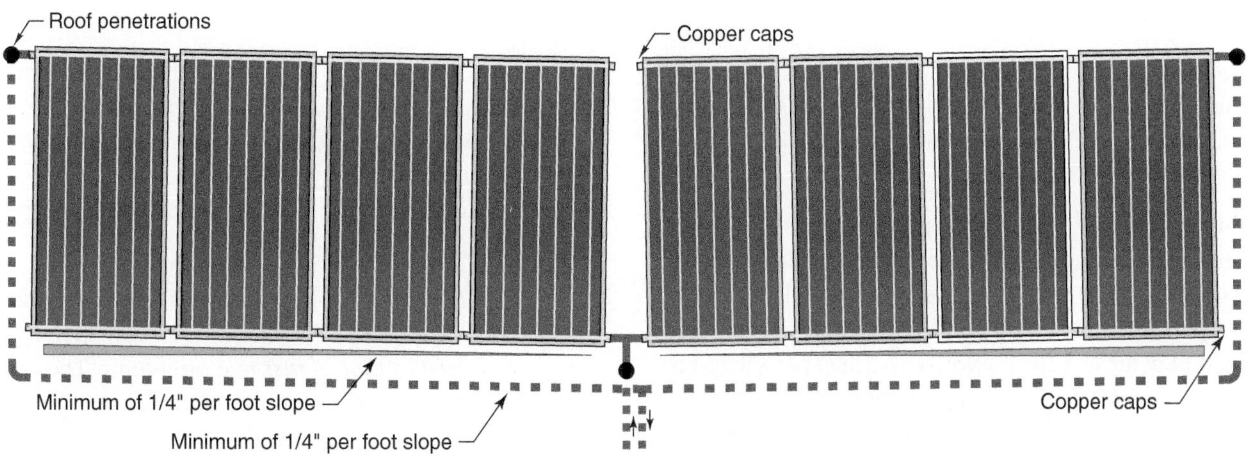

Figure 11-5 | Two banks of four collectors each, sloped to low-point roof penetration at center.

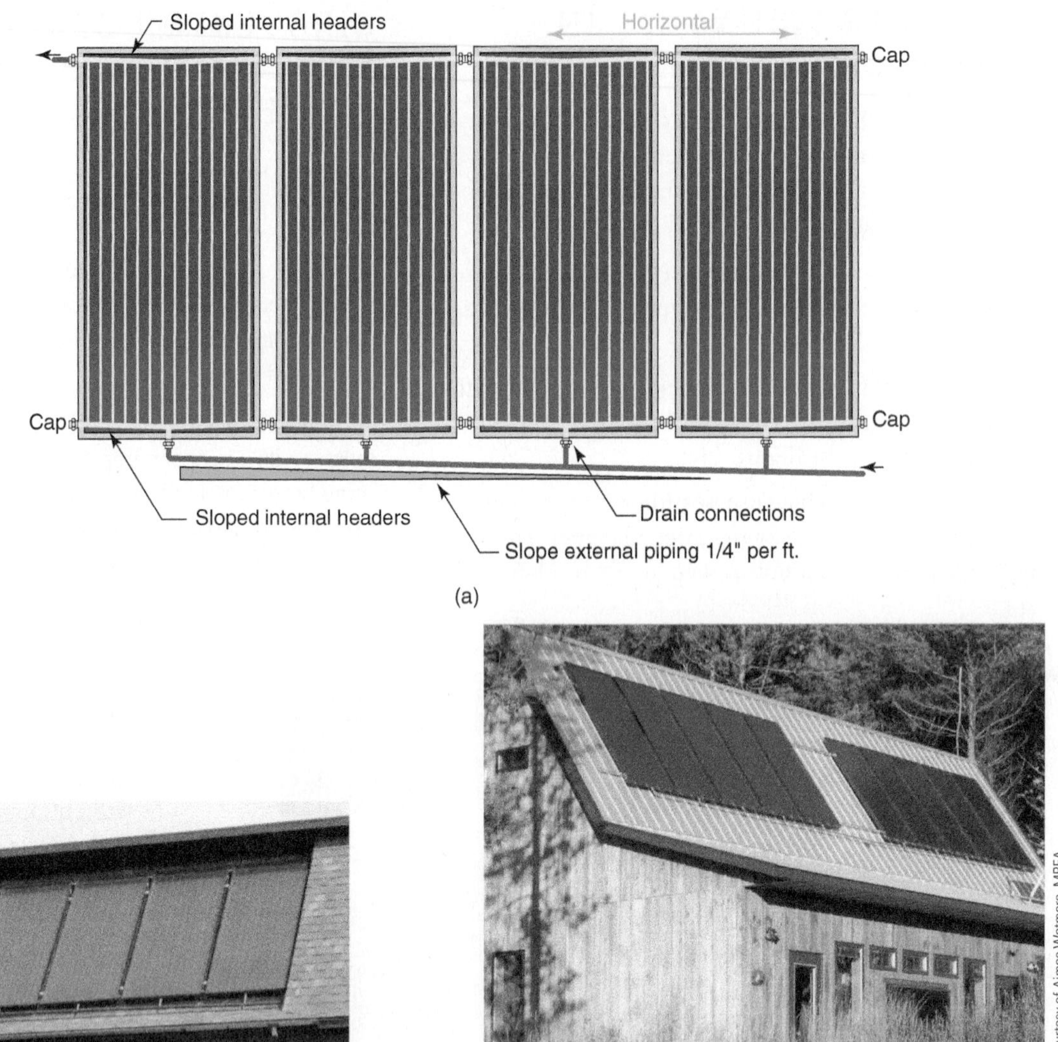

Figure 11-6 (a) Collectors with internal sloped headers and a "fifth port" for use in drainback-protected systems without side slope. However, lower header must be sloped. (b) Sloping lower header. (c) Two banks of five drainback collectors with center bottom supply and return at upper outer corners.

Although both situations will move heat from the collector array to the storage tank, the first is much preferred because it can significantly reduce pumping energy and operating noise.

The flow velocity within the piping as the water reaches the top of the collector circuit is critically important to siphon formation. That velocity should be *at least* 2 feet per second. The flow rate (in gallons per minute) needed to achieve a flow velocity of 2 feet per second is listed for several sizes of copper tubing in Figure 11-8. Alternatively, the flow rate corresponding to a flow velocity of 2 feet per second can be calculated, for any round pipe, using Equation 3.5.

One way to create the flow and head needed to provide the initial lift, and establish the siphon, is to install two identical circulators in a **series-coupled arrangement**, as shown in Figure 11-9. When this strategy is used, the system is called a **"double-pumped" drainback system**.

When two identical circulators are installed in series, they produce twice the head at any given flow rate, compared to that of just one circulator. The pump curve for the two series-connected circulators can thus be drawn by doubling the height of the single circulator curve at each flow rate. An example of these curves is shown in Figure 11-10.

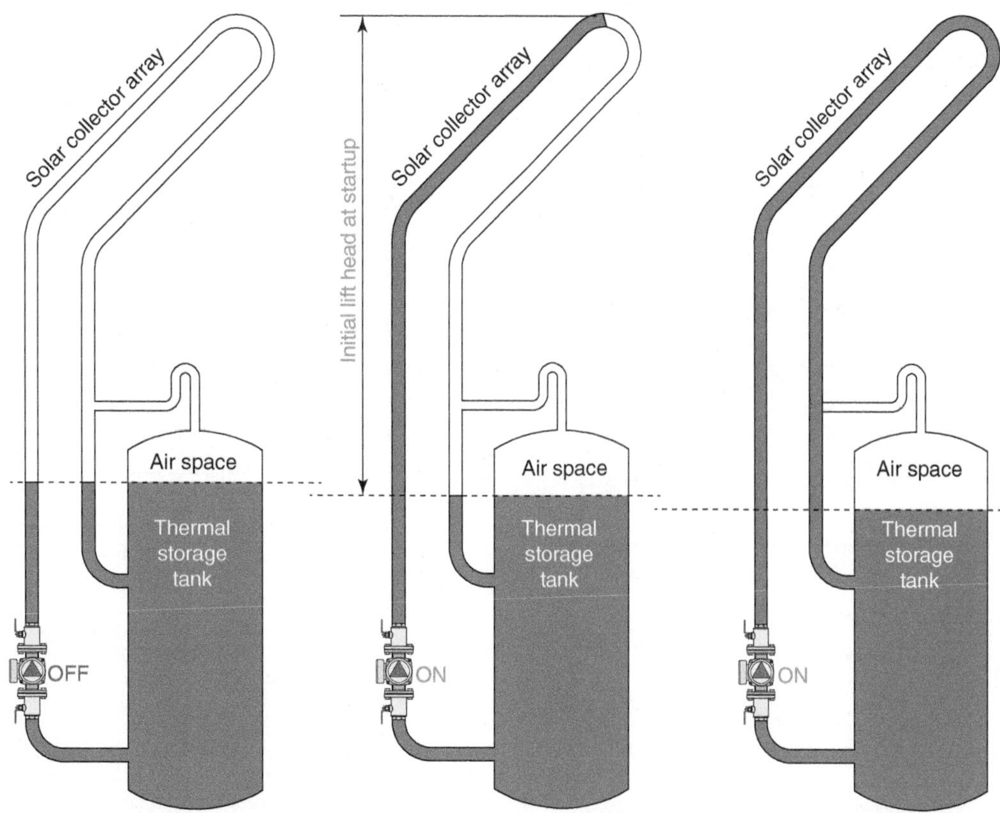

Figure 11-7 | At the start of each collection cycle, the circulator in a drainback-protected system must lift water from the water level in the thermal storage tank to the top of the collector array.

Tubing	Flow rate to establish 2 ft./sec. flow velocity
1/2″ type M copper	1.6 gpm
3/4″ type M copper	3.2 gpm
1″ type M copper	5.5 gpm
1.25″ type M copper	8.2 gpm
1.5″ type M copper	11.4 gpm
2″ type M copper	19.8 gpm
2.5″ type M copper	30.5 gpm
3″ type M copper	43.6 gpm

Figure 11-8 | Flow rate corresponding to a flow velocity of 2 feet per second in several sizes of copper tubing.

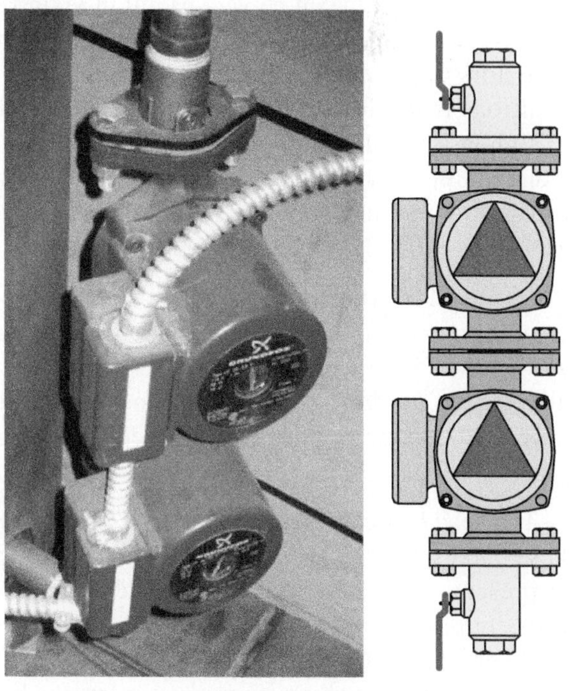

Figure 11-9 | Two identical circulators installed in close-coupled series arrangement.

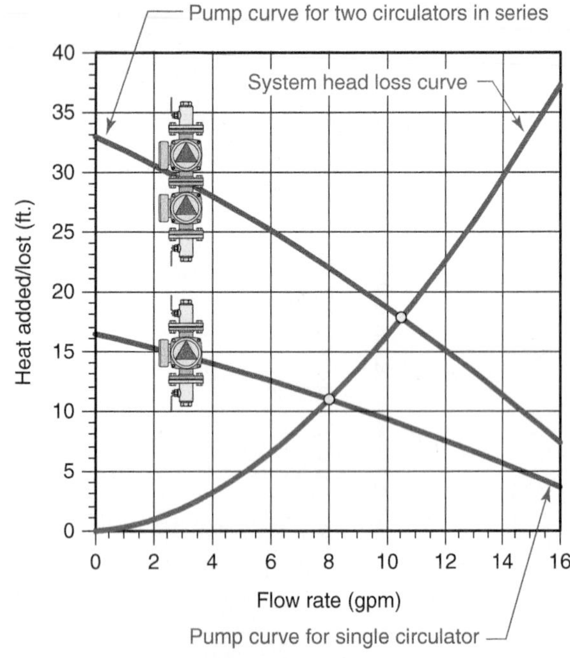

Figure 11-10 Two identical circulators installed in close-coupled series arrangement produce twice the head at all flow rates compared to that of a single circulator.

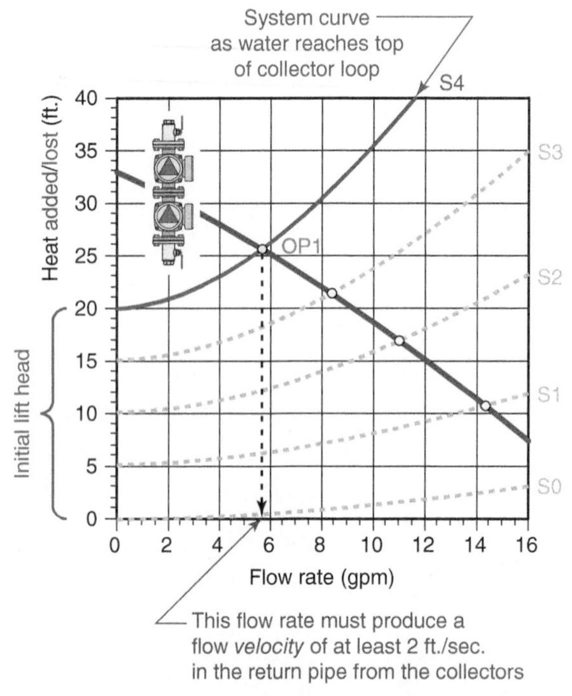

Figure 11-11 Pump curve for series-coupled circulators and changing system head loss curves, as water fills the collector array.

This relationship can be very useful in drainback systems. It allows the possibility of operating both circulators to provide the initial lift head as well as conditions that allow siphon formation. Then, after the siphon is stable, the downstream circulator can be turned off and the remaining circulator can sustain proper flow in what is now a water-filled collector circuit.

Figure 11-11 illustrates the relationship between the pump curve for the series-coupled circulators and the constantly changing head loss curve for the collector circuit during the first seconds of circulator operation at the beginning of an energy collection cycle.

The head loss curves are labeled (S0, S1, S2, etc.). These curves represent a sequence of conditions that are present during the first few seconds after both series-coupled circulators are turned on. Notice that the left end of each head loss curve gets higher on the head axis. The point where each curve meets the head axis represents the difference in height between the water in the collector supply pipe and the water level in the thermal storage tank (or elevated drainback tank). Also notice that each head loss curve is slightly steeper than the one below it. This is caused by the increased frictional head loss as the rising water column moves farther along the piping path on its way to the top of the collector array.

The curve labeled (S4) represents the collector circuit head loss curve at the moment the water reaches the top of the collector array. It is called the **maximum head loss curve**. The point where this curve crosses the pump curve for the series-coupled circulators is called the operating point (OP1). The flow rate through the piping at this moment is found by dropping a vertical line down to the flow rate axis. *To establish a siphon in the return pipe, the flow velocity corresponding to this flow rate must be at least 2 feet per second.* If the circulators, operating in series, cannot produce this minimum flow velocity, the siphon may not form or may become unstable if it does form. Slightly higher flow velocities in the range of 3 to 4 feet per second are acceptable and provide a better safety factor, ensuring rapid siphon formation.

The maximum head loss curve for a specific system can be constructed by combining two effects:

1. The initial lift head, which is the distance from the water level in the tank to the top of the collector array.
2. The frictional head loss associated with the collector supply piping, the collector array, and 50 percent of the head loss of the collector return piping. The latter is added as a safety factor to ensure conditions that will form a stable siphon.

Example 11.2: A closed-loop drainback-protected system is constructed as shown in Figure 11-12. Develop the maximum head loss curve for this system.

Solution: To determine the maximum head loss curve of this circuit, the designer needs to divide it up into segments based on flow rates. The 55 feet of 3/4-inch copper tubing on the supply side of the collector array will operate at some flow rate (f). That flow will divide up when it reaches the collector array. Since the collectors are piped in reverse return, the flow should divide in approximately equal portions through each collector. Thus, the flow rates along the lower header will vary for each segment of the header. These flows, from right to left along the header, are represented as f, $0.8f$, $0.6f$, $0.4f$, and $0.2f$. To maintain accuracy, it is necessary to determine the head loss of each of these segments and then add them together for the total head loss along the lower header.

Collector manufacturers typically provide a head loss curve for their collectors. This data can also be found in the OG-100 test report from SRCC. This data can be plotted, and a curve having the form given as Equation 11.4 can be fit to the data.

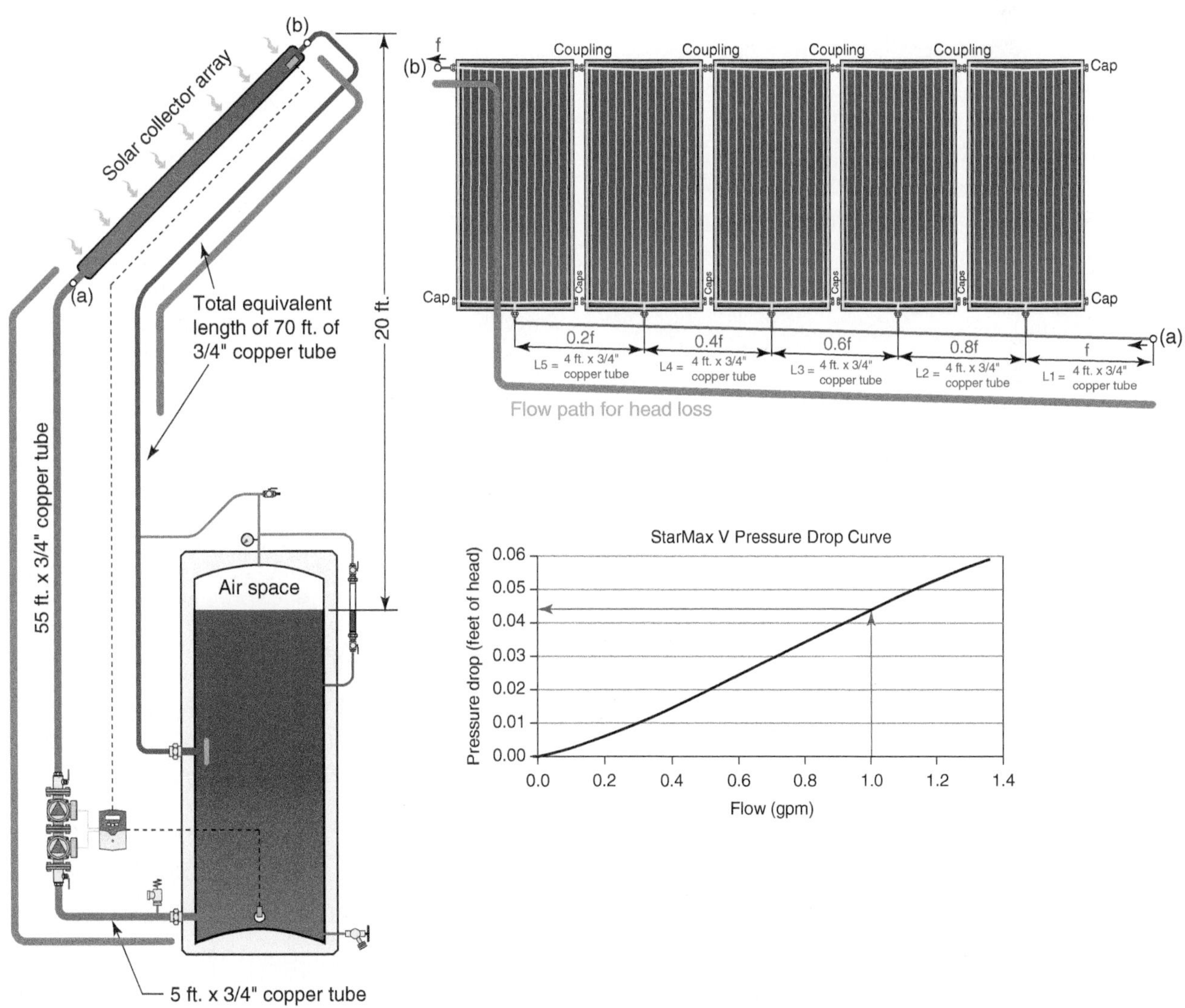

Figure 11-12 | System for Example 11.2. The heavy green lines indicate the piping path to be analyzed for head loss.

(Equation 11.4)

$$H_L = R(f)^{1.75}$$

where:

H_L = head loss (feet of head)

R = hydraulic resistance

f = flow rate (gpm)

This equation has the same form as the equation used to model the head loss of smooth tubing, as discussed in Chapter 3. The value for "R" in Equation 11.4 is found by substituting in the head loss at a flow rate comparable to that at which the collector is expected to operate. In this case, that flow rate will be 1.0 gpm. The manufacturer of the collector shown in Figure 11-12 lists a head loss of 0.044 feet at a flow rate of 1.0 gpm. Thus, the value of R in Equation 11.4 is 0.044.

$$H_{L(collector)} = 0.044(f)^{1.75}$$

The flow path to be analyzed for head loss is shown by the heavy green line in Figure 11-12. It includes all the collector supply piping, the various header segments under the collector array, the path through the far left collector, and, as a safety factor, 50 percent of the return piping.

The head loss of the copper piping segments can be analyzed using the methods from Chapter 3. Expressed mathematically, the relationship between head loss and flow rate is represented by Equation 3.9.

Equation 3.9 (repeated)

$$H_L = (\alpha c L)(f)^{1.75}$$

where:

H_L = head loss of the circuit (feet of head)

α = fluid properties factor (see Equation 3.9 or Figure 3.24)

c = pipe size coefficient (see Figure 3.25)

L = total equivalent length of the circuit (ft.)

f = flow rate through the circuit (gpm)

1.75 = an *exponent* applied to flow rate (f)

In this system, all the tubing in the collector circuit is 3/4-inch copper. The value of (c) for this type and size of tubing is found in Figure 3-25 ($c = 0.061957$).

The fluid is water at 80 °F. The value of (α) for this fluid is found in Figure 3-24 ($\alpha = 0.0553$).

The head loss along any of the piping segments can thus be represented as:

$$H_L = (\alpha c L)(f)^{1.75} = (0.0553 \times 0.061957 \times L)(f)^{1.75}$$
$$= (0.003426L)(f)^{1.75}$$

where:

L = the length of the piping segment (ft.)

f = the flow rate in the piping segment (gpm)

The head loss of the collector supply piping is:

$$H_{LS} = (0.003426 \times 55)(f)^{1.75}$$

The head loss along each of the five header segments can be represented as follows:

$$H_{L1} = (0.003426 \times 4)(f)^{1.75}$$
$$H_{L2} = (0.003426 \times 4)(0.8f)^{1.75}$$
$$H_{L3} = (0.003426 \times 4)(0.6f)^{1.75}$$
$$H_{L4} = (0.003426 \times 4)(0.4f)^{1.75}$$
$$H_{L5} = (0.003426 \times 4)(0.2f)^{1.75}$$

Notice that the flow rate in each segment is represented as a decimal percentage of the full circuit flow rate (f). For example, the flow rate in the second header segment, (L2), is shown as $0.8f$ because 20 percent of the total circuit flow has already passed up through the first collector on the right hand side of the array.

Finally, to make the calculation conservative, the head loss of 50 percent of the collector return piping will be included. This head loss can be represented as:

$$H_{LR(50\%)} = (0.003426 \times 35)(f)^{1.75}$$

It is now possible to factor and add all the coefficients of the flow rate, keeping in mind the following algebraic relationship:

$$(0.8f)^{1.75} = 0.8^{1.75} \times f^{1.75} = 0.6767(f)^{1.75}$$

When the constant 0.003426 is factored out of the piping segment terms, the head loss of the collector is added and the remaining coefficients are combined:

$$H_{L(total)} = 0.003426[55 + 4 + 4(0.8)^{1.75} + 4(0.6)^{1.75} + 4(0.4)^{1.75} + 4(0.2)^{1.75} + 35](f)^{1.75} + 0.044(f)^{1.75}$$
$$H_{L(total)} = 0.003426[99.39](f)^{1.75} + 0.044(f)^{1.75}$$
$$H_{L(total)} = 0.3845(f)^{1.75}$$

The maximum head loss curve can now be represented as the sum of the lift head (20 ft.), plus the total frictional head loss:

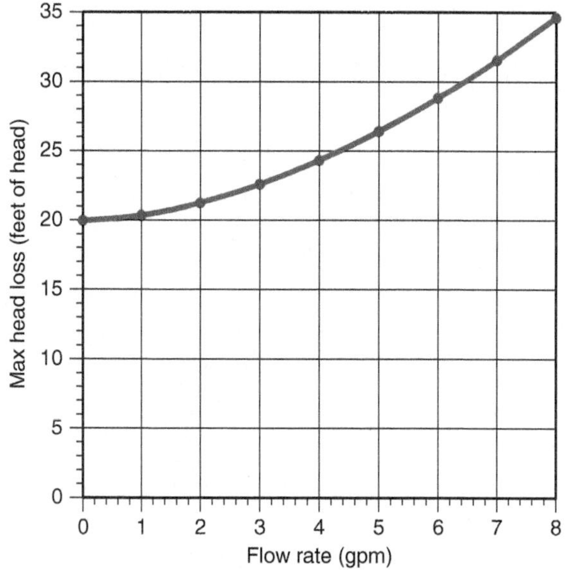

Figure 11-13 | Circuit maximum head loss curve at siphon formation for Example 11.2.

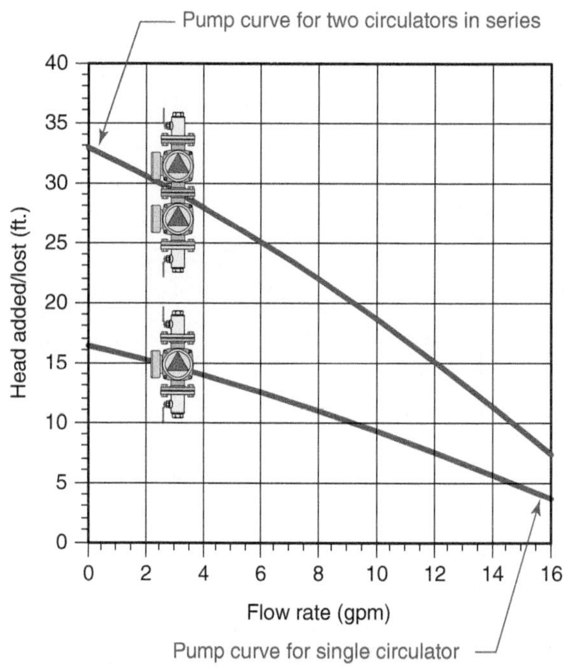

Figure 11-14 | Pump curves for tentative circulators used in drainback-protected system with maximum head loss curve shown in Figure 11-13.

(Equation 11.5)

$$H_{L(max)} = 20 + 0.3845(f)^{1.75}$$

This can be plotted by substituting in several values for flow rate (f) and calculating the resulting head loss. Such a plot is shown in Figure 11-13.

Discussion: The preparation of this curve draws upon the hydronic system design fundamentals presented in Chapter 3; specifically, how to calculate the head loss of a piping circuit or portion of a circuit based on the flow rate through it. This example is somewhat more complicated based on how the flow divides up among the supply header segments. It also requires that the initial lift head of the system be added to the frictional head loss.

Example 11.3: The designer is contemplating using two series-coupled circulators for the drainback system discussed in Example 11.2. Each circulator has the pump curve shown in Figure 11-14 (lower curve). Determine if these circulators, operating simultaneously, will be able to establish a siphon in the 3/4-inch return pipe from the collector array.

Solution: The pump curve for two of these circulators, in a series-coupled arrangement, is also shown, as the upper curve, in Figure 11-14. This curve is easily constructed by doubling the height of the single circulator pump curve at each flow rate.

The pump curve for the series-coupled circulators is now plotted on the same graph with the maximum head loss curve found in the previous example, as shown in Figure 11-15.

The point where the pump curve crosses the head loss curve represents the operating point just as a siphon would begin to form. Dropping a line from this point to the horizontal axis shows a corresponding flow rate of 5.0 gpm. Referencing Figure 11-8 for 3/4-inch copper tubing shows that this flow rate is well above that required to produce a flow velocity of 2 feet per second in the tube, and thus the siphon will form.

Discussion: The actual flow velocity corresponding to a flow rate of 5.0 gpm through a 3/4-inch copper tube can be calculated using Equation 3.5:

Equation 3.5 (repeated)

$$v = \left(\frac{0.408}{d^2}\right)f = \left(\frac{0.408}{0.811^2}\right)5.0 = 3.1 \text{ ft./sec.}$$

where:

v = average flow velocity in the pipe (ft./sec.)

f = flow rate through the pipe (gpm)

d = exact inside diameter of pipe (in.)

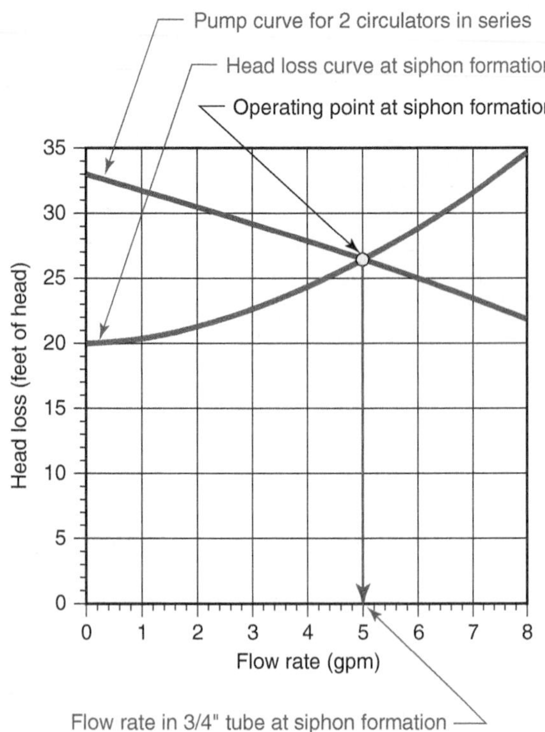

Figure 11-15 Plotting the pump curve for the series-coupled circulators along with the maximum head loss curve.

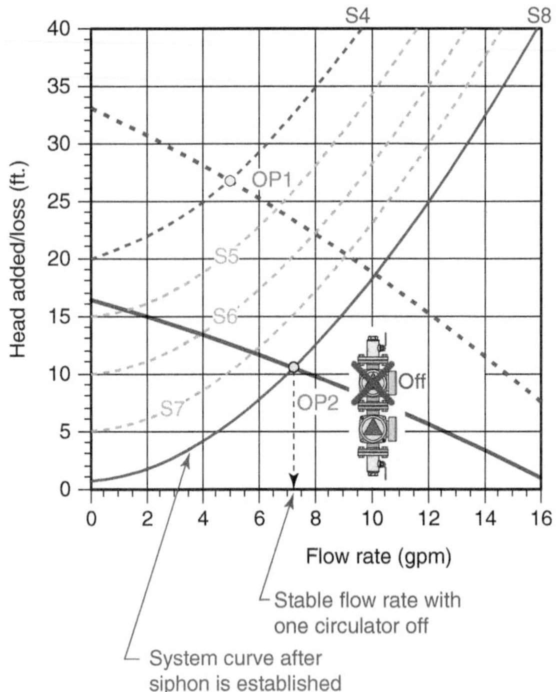

Figure 11-16 Head loss curve drops as siphon is formed in return piping.

A flow velocity of 3.1 feet per second through a 3/4″ copper tube is more than adequate to form a siphon. It is also under the maximum recommended flow velocity of 4 feet per second for copper tubing. Thus, this series-coupled circulator pair is a good choice for this system.

As the siphon forms in the return piping, the circuit's head loss curve quickly drops, as shown in Figure 11-16.

This happens because the weight of the water, which is starting to fill the return piping is "counterbalancing" the weight of the water in the collector supply piping, and thus the initial lift head is quickly disappearing. This causes the left side of the head loss curve to move down the vertical axis of the graph. This movement is represented by the curves (S5, S6, S7, and S8) in Figure 11-16. When the siphon is fully formed, the head loss curve will settle to some position for the remainder of the energy collection cycle. This is represented by head loss curve (S8). This curve has a very minor amount of lift head (about 1 ft.), due to possible siphon breakage where the air return tube tees into the return piping. This location is usually slightly above the surface water level in the thermal storage tank, and hence the slight "residual" lift head is accounted for.

It's also evident that head loss curve (S8) is steeper than curve (S4). This is the result of frictional head loss through the *entire* collector circuit, including the full length of the return piping, after the siphon has formed.

The significant drop in the head loss curve suggests that just one of the series-coupled circulators could maintain flow through the collector circuit once the siphon is fully established. If this is indeed the case, the electrical power required to operate the collector circuit could be about half that required when both circulators are operating. This condition is represented by operating point (OP2) in Figure 11-16. Notice that the flow rate at (OP2) is slightly higher than the flow rate when both circulators are operating, and the siphon is just beginning to form, as represented by operating point (OP1).

Example 11.4: Assume that the downstream circulator in the previous example is turned off after the siphon has formed. Estimate the flow rate through the collector circuit with just the remaining circulator operating.

Solution: This calculation requires information about the head loss through a circulator that is not operating.

This is typically not published by circulator manufacturers. However, based on some test data obtained by the author, the following equation can be used to approximate this head loss for a small wet-rotor circulator that is off:

(Equation 11.6a)

$$H_L = 0.045(f)^{1.75}$$

where:

H_L = head loss through the non-operating circulator (feet of head)

f = flow rate through the circulator (gpm)

The total friction head for the circuit can now be determined by adding the friction head loss from the previous example, along with the friction head loss of the remaining 35 feet of return piping, and the head loss estimate of the non-operating circulator. These terms are combined below.

$$H_{L(fullcircuit)} = 0.3845(f)^{1.75} + (0.003426 \times 35)(f)^{1.75} + 0.045(f)^{1.75}$$

$$H_{L(fullcircuit)} = 0.5494(f)^{1.75}$$

To be conservative, it is prudent to leave 2 feet of lift head in the final head loss curve to cover the possibility that the siphon may break up at the air return tube, which is typically a short distance above the water level in the tank. Thus, the head loss curve for the collector circuit after the siphon has formed can be represented as follows:

(Equation 11.6b)

$$H_{L(fullcircuit)} = 2 + 0.5494(f)^{1.75}$$

where:

H_L = head loss (feet of head)

f = flow rate (gpm)

Equation 11.6b can be plotted, along with the pump curve for a single circulator, as shown in Figure 11-17.

The flow rate at this condition can now be found by dropping a line from the operating point to the horizontal axis. In this case, the circuit will stabilize at a flow rate of about 5.4 gpm, with just one circulator operating.

Discussion: The eventual steady flow rate, with a single circulator operating, is estimated at 5.4 gpm. This is slightly higher than the flow rate at the point where

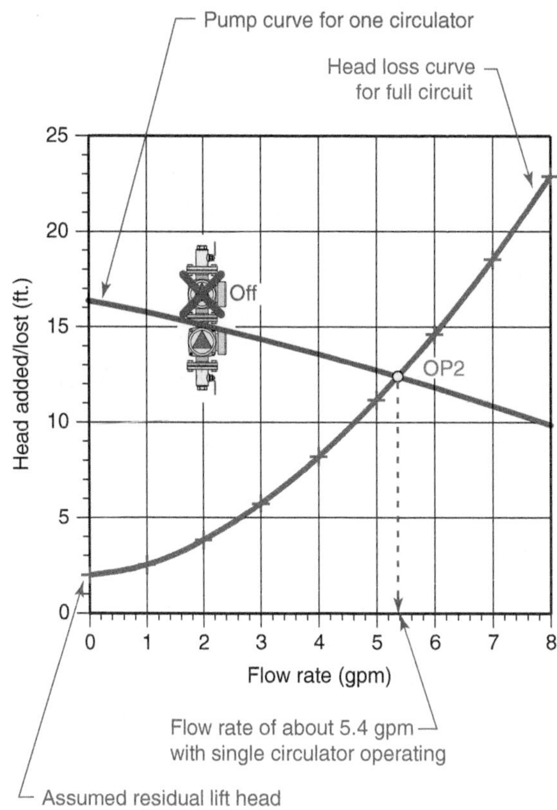

Figure 11-17 Head loss curve for circuit after siphon has formed, plotted with pump curve for a single circulator. The head loss of the nonoperating circulator is accounted for in the head loss curve.

both circulators were operating and the siphon was about to form. This is the result of a marked reduction in lift head that occurs as the siphon is formed. In most residential drainback-protected systems, the siphon will be fully formed and deareated within 3 minutes of starting the circulators. Thus, the circuit can operate with a single circulator for the vast majority of its total operating time. This significantly reduces the electrical energy consumed for pumping compared to a system where both circulators remain in operation during the full energy collection cycle.

Several differential temperature controllers are now available that can coordinate this type of double-pumped drainback system. They provide separate output relays to drive each of the two circulators. Both relays are closed whenever the solar collection cycle begins. The relay powering the upper circulator turns off after a preset time delay period, which is typically about 3 minutes. This action can also be accomplished using an external time delay relay that breaks line voltage to the upper circulator after a preset time.

Although double-pumped drainback systems have been successfully used for several decades, the recent availability of ECM-based variable-speed circulators is likely to set a new standard for drainback systems. That standard will be a single variable-speed circulator with the "intelligence" to operate at full speed for a specified time and then reduce speed to another preset pump curve that can maintain adequate flow through the collector array after the siphon has formed. This approach will simplify installation and eliminate the need to pump water through an inoperable upper circulator, as is required in a double-pumped drainback system. The high wire-to-water efficiency of an EMC circulator will further reduce the electrical energy needed to operate the collector circuit.

SIZING THE AIR SPACE IN AN *UNPRESSURIZED* DRAINBACK-PROTECTED SYSTEM

All drainback systems require an air space to accommodate the water volume that drains from the collector array. This space is also needed to absorb the expanded volume of water as it is heated within the system. When the system uses an unpressurized thermal storage tank, an air space above the water level in the tank can handle both functions. The required depth of this space depends upon the dimensions of the thermal storage tank, the total water volume in the system, the maximum overall temperature change of that water, the volume of the collector array, and the volume of the piping that drains along with the collectors. Equation 11.7 can be used to estimate the change in height of water within an open storage tank based on these factors.

(Equation 11.7)

$$\Delta H = \frac{230.7\left[(V_c + V_{ep}) + \left(\frac{D_c}{D_H} - 1\right)(V_{wt} + V_d)\right]}{(A_t)}$$

where:

ΔH = change in water level in tank (in.)

V_c = volume of the collector array (gallons)

V_{ep} = volume of all piping to and from the collector array that drains (gallons)

V_{wt} = volume of water in thermal storage at cold fill conditions (gallons)

V_d = volume of water is distribution system (gallons)

D_c = density of water at cold fill conditions (lb./ft.^{3})

D_H = density of water at maximum tank temperature (lb/ft^3)

A_t = cross sectional area of storage tank (in.^{2})

Example 11.5: An unpressurized cylindrical thermal storage tank has a liner shell measuring 60 inches tall and 48 inches in diameter. It is filled with cold (60 °F) water to within 6 inches of the top of the liner. The space heating distribution system is directly connected to this tank and holds an additional 50 gallons of water. The solar collector array contains five collectors with a volume of 1.2 gallons each. There is 100 feet of 3/4-inch copper tubing that must drain back to the thermal storage tank at the end of each energy collection cycle. The system controls are configured so that the maximum temperature the thermal storage tank can reach is 180 °F. Determine the change in height of the water in the tank when the collector array is drained and the water is at its maximum temperature.

Solution: The density of water at both 60 °F and 180 °F is required. These densities can be estimated from Figure 3-11. The density of water could also be calculated using Equation 11.3.

$$D_{60} = 62.3 \text{ lb./ft.}^3$$

$$D_{180} = 60.5 \text{ lb./ft.}^3$$

Equation 11.7 also requires the cross-sectional area of the tank. In this case, the tank is a vertical cylindrical and thus has a cross-sectional area of:

$$A_t = \frac{\pi d^2}{4} = \frac{\pi(48)}{4} = 1{,}810 \text{ in.}^2$$

This data and the remaining values can now be substituted into Equation 11.7:

$$\Delta H = \frac{230.7\left[(V_c + V_{ep}) + \left(\frac{D_c}{D_H} - 1\right)(V_{wt} + V_d)\right]}{(A_t)}$$

$$= \frac{230.7\left[(6 + 2.685) + \left(\frac{62.3}{60.5} - 1\right)(423 + 50)\right]}{(1{,}810)}$$

$$= 2.9 \text{ in.}$$

Discussion: The estimated change in water level of 2.9 inches is conservative. It assumes that *all* the water in the system reaches the same maximum temperature of 180 °F simultaneously. This is virtually impossible because of stratification within the tank and the temperature drop that must occur in any distribution system that is dissipating heat. Still, this example gives the designer a good estimate of water level changes. The result can be used to determine where the cold-fill water level should be set to maintain the risen water level safely below any sidewall piping penetrations.

In systems that link the space heating distribution system to the storage tank using a heat exchanger, the volume of water in the distribution system, represented as (v_d) in Equation 11.7, would be close to zero. This will reduce the change in water level.

SIZING THE AIR SPACE IN A *PRESSURIZED* DRAINBACK-PROTECTED SYSTEM

One way to configure a pressurized drainback system is with the air space at the top of the thermal storage tank. This space must accommodate the water that drains back from the collector array and associated piping. It must also accommodate the increased volume of system water as it is heated, without raising system pressure to a point where the pressure-relief valve opens. This section describes how to determine that volume based on a vertically oriented cylindrical tank. For simplicity, the top of that tank will be assumed to be flat.

To determine the required space at the top of the tank, one must establish a criteria for the changes in temperature, pressure, and volume of the captive air that will occur between two limiting conditions.

One of those limiting conditions is when the system is filled with cold water and the collector circulator is not operating. *All* water in the system (e.g., that which will eventually be in the collectors as well as that in the tank and distribution system) is assumed to be at some initial, relatively cool temperature, such as 60 °F. The air at the top of the tank and in the collectors is also assumed to be at this same temperature. The pressure of the air at the top of the tank and in the empty collector array and piping is assumed to be at some initial pressure. It could be atmospheric pressure, or a slightly higher pressure.

The other limiting condition occurs when all water and air in the system is at some maximum allowable temperature (180 °F, for example). Under this condition, the air pressure at the top of the tank cannot exceed some limiting value. The limiting pressure assumed in the equations to follow is the rated opening pressure of the pressure-relief valve at the top of the tank.

These two limiting conditions are shown in Figure 11-18.

The *minimum* height (h) of the cylindrical air space at the top of a tank with flat ends can be calculated using Equations 11.8a, 11.8b, and 11.8c.

(Equation 11.8a)

$$h = \left(\frac{924}{\pi d^2}\right)\left[\frac{R(v_T + v_d) + v_c(S - 1)}{1 + R - s}\right]$$

(Equation 11.8b)

$$S = \left[\frac{(p_i + 14.7)(T_{max} + 460)}{(P_{RV} + 14.7)(T_i + 460)}\right]$$

(Equation 11.8c)

$$R = \left[\frac{D_c}{D_h} - 1\right]$$

where:

h = MINIMUM height of cylindrically shaped air space required at top of tank (in.)

d = tank diameter (in.)

v_T = volume of entire tank (water and air space at top) (gallons)

v_d = volume of water in distribution system (gallons)

v_c = volume of collector array plus collector piping above water level in tank (gallons)

D_c = density of "cold" water when system is filled and pressurized (lb./ft.^{3})

D_h = density of water in system at maximum temperature (lb./ft.^{3})

p_i = initial air pressure in tank when water is cool (psi gauge)

p_{RV} = rated opening pressure of pressure-relief valve (psi gauge)

T_{max} = maximum temperature of water and air in system (°F)

T_i = initial temperature of water and air in system (°F)

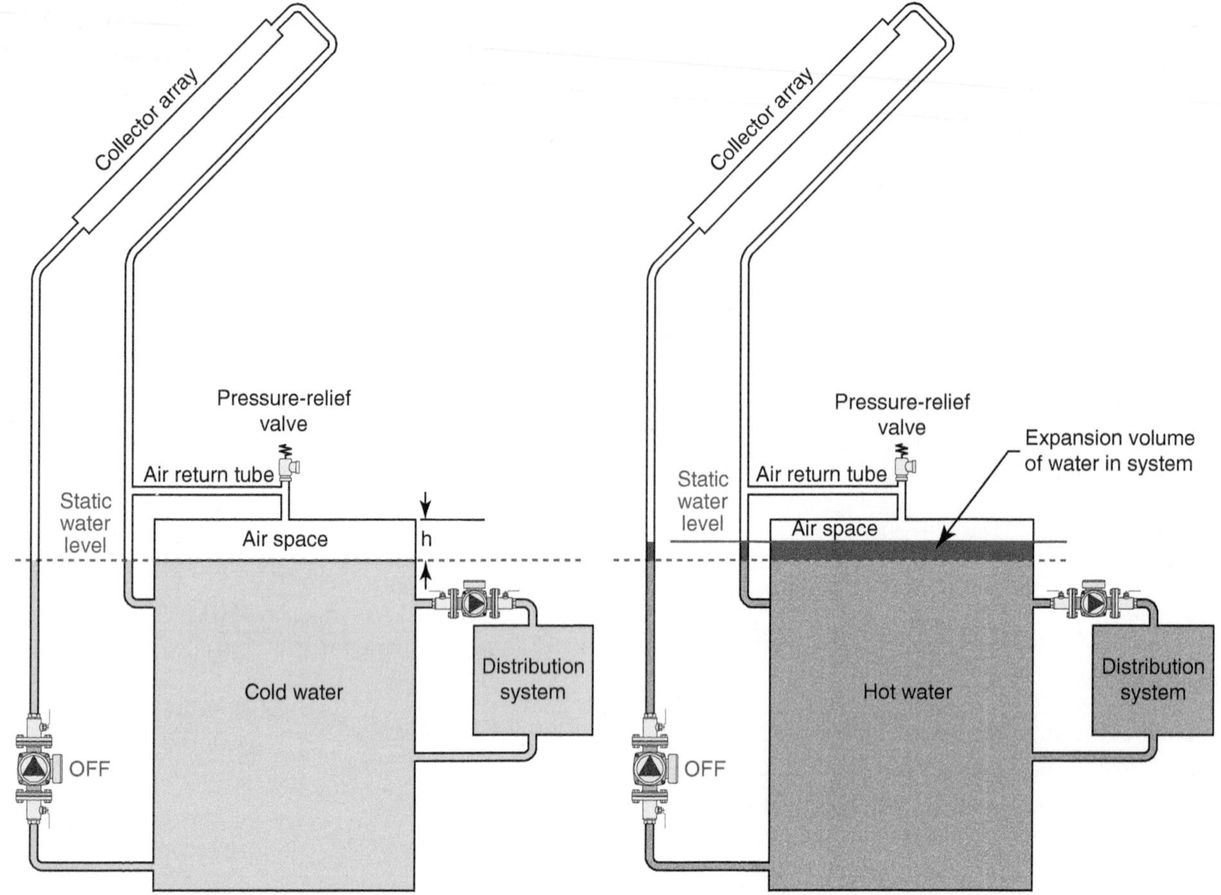

Figure 11-18 | Two limiting conditions for the air space volume.

The relationships embodied in Equations 11-8a, b, and c are based on the ideal gas law, which can be stated as Equation 11.9.

(Equation 11.9)

$$\frac{(p_{absolute})(v)}{T_{absolute}} = \text{constant}$$

The equation states that the absolute pressure of a quantity of air multiplied by its volume and divided by its absolute temperature remains a constant.

Example 11.6: Assume that you have selected a tank for a drainback system. It has flat ends, a volume of 250 gallons, and an internal diameter of 30 inches. The volume of the collector array and piping that will be drained is 10 gallons. The system is filled with water at 60 °F, and the air space at the top is left at atmospheric pressure (e.g., gauge pressure = 0). The distribution system (exclusive of the storage tank and collector subsystem) contains 50 gallons of water. The maximum temperature condition assumes that *all* water and captive air in the system reaches 180 °F. A 30-psi rated pressure-relief valve is installed at the top of the tank. What is the minimum vertical dimension of the air space at the top of the tank to accommodate expansion and drainback?

Solution: There is sufficient information given to immediately evaluate Equation 11.8b:

$$s = \left[\frac{(p_i + 14.7)(T_{max} + 460)}{(P_{RV} + 14.7)(T_i + 460)}\right]$$

$$= \left[\frac{(0 + 14.7)(180 + 460)}{(30 + 14.7)(60 + 460)}\right] = 0.4047$$

Before using Equation 11.8a, we need to know both S and R. To find R, we need the density of water at 60 °F and 180 °F. These values can be found in Figure 3-11 or Equation 11.3.

At 60 °F, the density of water is: $D_c = 62.4$ lb./ft.^{3}

At 180 °F, the density of water is: $D_h = 60.5$ lb./ft.^{3}

Equation 11.8c can now be used to find R:

$$R = \left[\frac{D_c}{D_h} - 1\right] = \left[\frac{62.4}{60.5} - 1\right] = 0.0314$$

The values of R and S, along with the other given data can now be used in Equation 11.8a:

$$h = \left(\frac{924}{\pi d^2}\right)\left[\frac{R(v_T + v_d) + v_c(S - 1)}{1 + R - s}\right]$$

$$= \left(\frac{924}{\pi(30)^2}\right)\left[\frac{0.0314(250 + 50) + 10(0.4047 - 1)}{1 + 0.0314 - 0.4047}\right]$$

$$= 1.8 \text{ in.}$$

Discussion: Just under 2 inches of vertical space at the top of the storage tank, in combination with the volume of air in the collectors and collector piping (above the static water level), is sufficient to keep the pressure from exceeding the pressure-relief valve rating when all water and air in the system are at a temperature of 180 °F.

The actual volume of this expansion space within the tank can be determined using Equation 11.10:

(Equation 11.10)

$$v = \frac{\pi d^2 h}{924} = \frac{\pi(30)^2 1.8}{924} = 5.5 \text{ gallons}$$

where:

v = volume of the expansion space within the tank (gallons)

d = tank diameter (in.)

h = minimum height of cylindrical air space (in.)

Those familiar with sizing non-diaphragm-type expansion tanks will likely see this volume as comparatively small for a system containing approximately 300 gallons of water that is heated from 60 °F to 180 °F. However, keep in mind that the volume of the collectors and the piping to and from them is also filled will air and thus acts as an expansion space. If the collector circulator is running, this air volume is just moved from the collector array and its associated piping into the storage tank. In Example 11.6, the collector array and associated piping add another 10 gallons to the system's air volume, making the total expansion air volume about 15.5 gallons.

Keep in mind that the dimension (h) in Equation 11.8a is the *minimum* air space height required. The design can be made more conservative by including more air space volume or by changing the limiting constraints. For example, the author often suggests subtracting 5 psi from the rated pressure of the relief valve as the maximum pressure the valve should reach with assurance that it will not "dribble." Subtracting 5 psi from the rated opening pressure of the relief valve is also appropriate if that valve is mounted low in the system, such as near the bottom of the thermal storage tank. Another conservative assumption built into this procedure is that the tank has a flat top surface. Although this is possible, it is seldom the case for pressure-rated tanks. The additional air space within the **dished head** at the top of such a tank is extra expansion space, as shown in Figure 11-19.

Anything that increases the air volume relative to water volume will reduce the pressure fluctuation as the system heats up and cools down.

It's also important to consider the placement of the tank inlet connection from the collector array. It should enter the tank horizontally and at least 2 inches below the water level in the tank when the circulator is on. This helps reduce splashing noise as water enters the tank. Also be sure that piping leading to the distribution system is at least 2 inches below the water level in the tank when the collector circulator is operating.

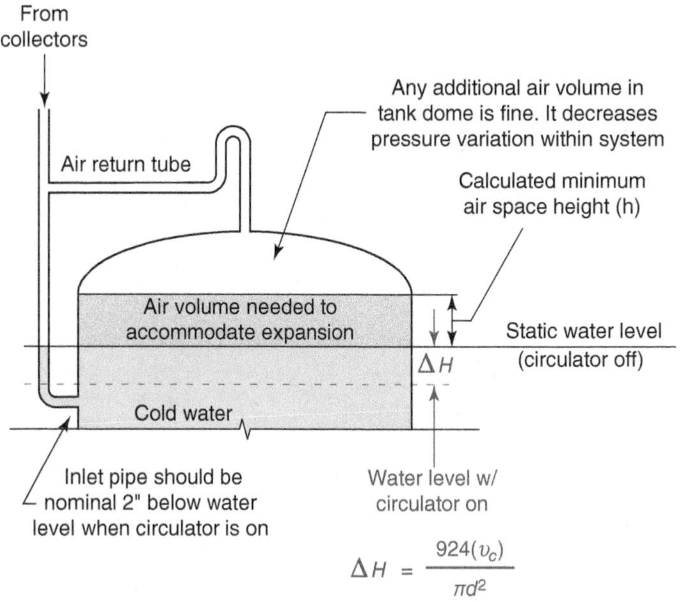

Figure 11-19 The air space within a dished head at top of tank increases the available expansion volume and thus lowers the pressure increase due to water expansion.

Equation 11.11 can be used to calculate the drop in water level (ΔH) within the cylindrical tank when the collector circulator is operating.

(Equation 11.11)

$$\Delta H = \frac{924(v_c)}{\pi d^2}$$

where:

ΔH = change in water level within tank when collector is on versus off (in.)

d = diameter of tank (in.)

v_c = volume of collector array plus collector piping above water level in tank (gallons)

In Example 11.6, where the collector array and associated "drainable" piping had a total volume of 10 gallons, the drop in water level when the collector circulator is running would be:

$$\Delta H = \frac{924(v_c)}{\pi d^2} = \frac{924(10)}{\pi(30)^2} = 3.27 \text{ in.}$$

Example 11.7: Assume the thermal storage tank from Example 11.6 is 22 inches in diameter but still has a total volume of 250 gallons. The collector array and associated piping above the water level in the tank have a total volume of 5 gallons. The distribution system is directly connected to the tank and has a volume of 50 gallons, as in Example 11.6. Assume the pressure-relief valve is rated at 15 psi, and that the minimum and maximum water temperatures in the system are 60 °F and 180 °F. The initial air pressure in the tank is 0 psi. Determine the minimum height of the air space at the top of the tank.

Solution: The value of S is now:

$$S = \left[\frac{(p_i + 14.7)(T_{max} + 460)}{(P_{RV} + 14.7)(T_i + 460)}\right]$$
$$= \left[\frac{(0 + 14.7)(180 + 460)}{(15 + 14.7)(60 + 460)}\right] = 0.6092$$

Since the minimum and maximum water temperatures are the same as in the previous example, the value of R will also be the same:

$$R = \left[\frac{D_c}{D_h} - 1\right] = \left[\frac{60.4}{60.5} - 1\right] = 0.0314$$

The minimum air space height can now be calculated using Equation 11.8a:

$$h = \left(\frac{924}{\pi d^2}\right)\left[\frac{R(v_T + v_d) + v_c(S - 1)}{1 + R - s}\right]$$
$$= \left(\frac{924}{\pi(22)^2}\right)\left[\frac{0.0314(250 + 50) + 5(0.6092 - 1)}{1 + 0.0314 - 0.6092}\right]$$
$$= 10.75 \text{ in.}$$

Discussion: This is a substantial change in the minimum air space height compared to that calculated in the previous example. It is due to a decrease in tank diameter, a decrease in the pressure-relief valve setting, and a smaller collector volume.

These calculations can also be used to determine the required air space in drainback systems that have a separate drainback/expansion tank, as shown in Figure 11-3.

ADDITIONAL DESIGN DETAILS FOR DRAINBACK SYSTEMS

The following details apply to *all* drainback-protect solar combisystems:

1. *Properly sloped piping and collectors:* All piping in the collector circuit should be sloped a minimum of 1/4-inch per foot to ensure proper drainage. Furthermore, all piping must be properly supported to ensure there are no sags that could trap water and thus lead to a freeze. If the collectors are not specifically designed for drainback applications, they must also be **side sloped** a minimum of 1/4-inch per foot. Although theory suggests that a "dead-level" collector array should drain, in practice, it is highly risky. Any sags that might develop with time, or inattentive installation, could create a low point that eventually causes a freeze. A single hard freeze that ruptures a pipe within an unheated attic could cause thousands of dollars in damages.
2. *Collector sensor mounting:* The sensor used to measure collector temperature in a drainback-protected system should always be mounted to the absorber plate or within a copper well that is brazed to the upper portion of the absorber plate. *Never mount the collector sensor on the piping outside the collector housing, as is sometimes done in antifreeze-protected systems.* If the collector does not have a sensor well, a sensor with a flat copper "tongue" can be riveted directly to the upper portion

of the absorber plate. This requires the glazing frame to be temporarily removed on one collector.

3. *Piping and component insulation:* Although not shown in the schematic, all piping in a solar combisystem should be insulated to a minimum of R-3 (°F·hr.·ft.²/Btu), or higher if required by code. The author suggests a *minimum* of R-24 (°F·hr.·ft.²/Btu) insulation for the thermal storage tank. It makes no sense to detail a solar combisystem for optimal collection of thermal energy and allow anything more than a "trickle" of that energy to be lost in an uncontrolled manner. It's also worth remembering that a thermal storage tank will be losing heat to the building on the hottest day of summer as well as the coldest day of winter. *Don't scrimp on proper insulation.*

4. *Avoiding control faults:* Older differential temperature controllers could create situations where water circulates through collectors under freezing conditions. This can happen if the collector temperature sensor fails in a **shorted** or near shorted condition with very low electrical resistance. Older differential temperature controllers do not detect this condition. Instead, they interpret the very low resistance of the failed sensor as a very *high* temperature at the collector and respond by turning on the collector circulator. Initially, the warm water from the storage tank will not freeze as it passes through the piping and collectors. However, given sufficient time and subfreezing temperatures, the water will cool, and ice crystals will form, eventually blocking further flow. This could create a hard freeze that bursts the piping or ruptures the absorber plates in the collectors. Modern digital differential temperature controllers avoid this scenario by reporting a fault condition if any temperature sensor fails as either an electrical short, or an open circuit. The controller keeps the circulator off if there is a sensor fault. Thus, only digital differential temperature controllers, which are now widely available, should be used in drainback systems. Many of the digital controllers also have a minimum collector temperature setting, below which the collector circulator will not be turned on.

5. *Air return provision:* All drainback-protected solar thermal systems must have provisions for allowing air to enter the return piping from the collector array and thus allow the drainback process to occur. In systems using an unpressurized tank, a tee with an open side port can be installed in the return piping from the collector array, inside the tank, and above the water line, as shown in Figure 11-1. Alternatively, a hole can be drilled into the side of the return pipe, again above the water line and inside the tank.

The following details are specific to drainback systems using a closed pressurized thermal storage tank and where there are no heat exchangers between the collector array, thermal storage tank, and space heating distribution system.

a. ***Air return provision:*** Figure 11-20 shows several hardware configuration that allow air at the top of a pressurized thermal storage tank to enter the return piping from the collector array, and thus allow drainback.

The most common air return provision is an air return tube, as shown in Figure 11-20(a) and several other schematics within this chapter and text. This tube exits the high point of the storage tank and rises several inches above the point where it connects to the collector return piping. This forms an inverted trap that minimizes water flowing from the collector return piping into the top of the drainback tank. Although such flow doesn't create a serious problem, it can result in dripping or gurgling sounds from the tank. The vertical rise of the inverted trap, indicated as (Y1) in Figure 11-20(a), must be greater than the *head loss due to friction* in the piping from the tee where the air return tube connects to the collector return piping down to the point where the collector return pipe enters the tank. The latter dimension is labeled as (Y2) in Figure 11-20(a). There should not be any sags in the air return tube that could trap water.

Another possibility is shown in Figure 11-20(b). The air return connection should be a high as possible on the tank's sidewall and well above the highest water level in the tank.

Figure 11-20(c) shows another possibility. The collector return piping connects to a prefabricated piping assembly that can be lowered into the top of the tank and sealed with a bushing. An air inlet hole is drilled into the tube under the bushing. The submerged tee at the bottom of this tube reduces flow velocity and diffuses incoming flow horizontally across the upper portion of the tank to minimize vertical jetting and preserve temperature stratification.

The details shown in Figure 11-20(b) and 11-20(c) may allow some of the flow returning from the collector array to cascade out of the air inlet provision and fall to the water surface below.

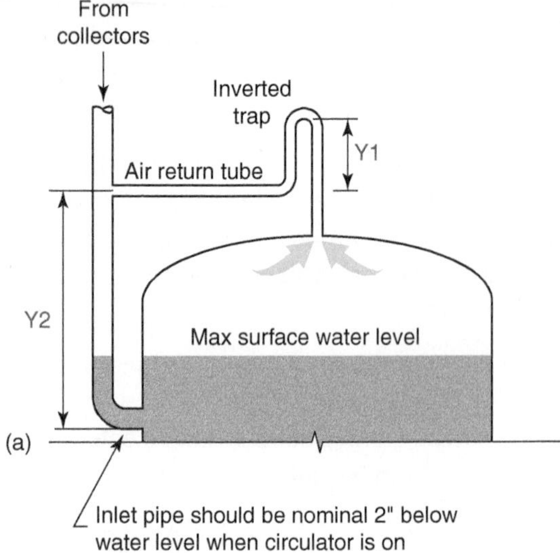

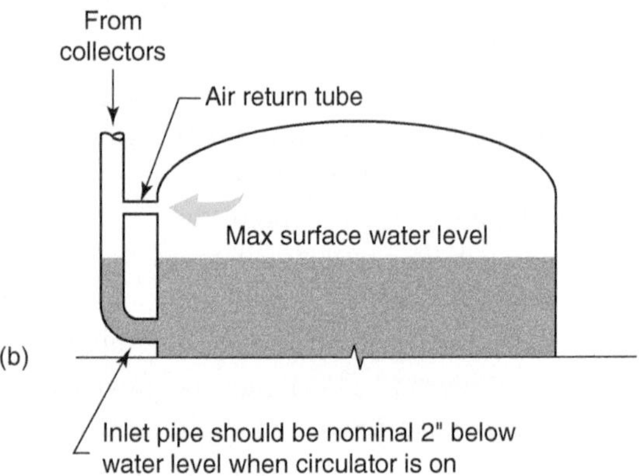

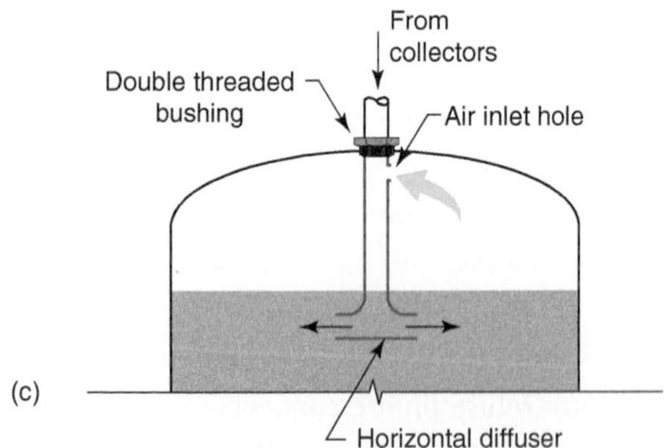

Figure 11-20 Options for a return air provision at top of pressurized storage tank used in a drainback-protected system.

Although this doesn't impair energy collection or drainback, it can create noticeable sound. The tank's insulation will muffle this sound to some extent. If the tank is located in a mechanical room away from living space, this sound is unlikely to be an annoyance. However, if the tank is mounted in a sound sensitive area, the air return tube shown in Figure 11-20(a) is the preferred option.

A second air return tube should connect from the outlet of a microbubble air separator near the beginning of the distribution system to the top of the drainback tank. This routes air that has been separated from water entering the distribution system back to the air space in the drainback tank. This is "**air control**" rather than **air elimination**. Without this detail, any air released by the air separator would cause an eventual drop in system pressure. Special fittings are available from air separator manufacturers that enable this tube to be properly connected to the air separator, as shown in Figure 11-21. A length of 3/8-inch or 1/4-inch PEX or copper tubing, with air-tight connections at both ends, works well for this detail. This detail is shown on several system schematics later in this chapter.

b. ***No make-up water system:*** *No automatic make-up water assembly should ever be used in a closed pressurized drainback-protected system in which the distribution system is directly connected to the thermal storage tank and collector circuit.* Having such a system invites the possibility of a **"waterlogged" system** in which any air leakage allows more water into the system. Eventually, the air space needed for drainback and expansion would become filled with water. The collectors and their associated piping could eventually freeze. The water level in the system should be periodically checked at the sight tube. If more water is needed, it can be added through a hose while the water level is monitored. Once the proper water level is restored, the hose should be disconnected from the system.

c. ***No automatic air vents:*** Automatic air vents are commonly used in hydronic systems to capture and eject air. While this action is useful in many hydronic systems, in combination with an automatic make-up water assembly, it is undesirable in a pressurized drainback-protected system. Any ejected air will cause a drop in the system's air pressure. Components such as manifold assemblies or boilers are often supplied with air vents. In such cases, these automatic air vents should be replaced by a manual air vent that can be completely closed, other than when air venting may be needed during system commissioning.

(a)

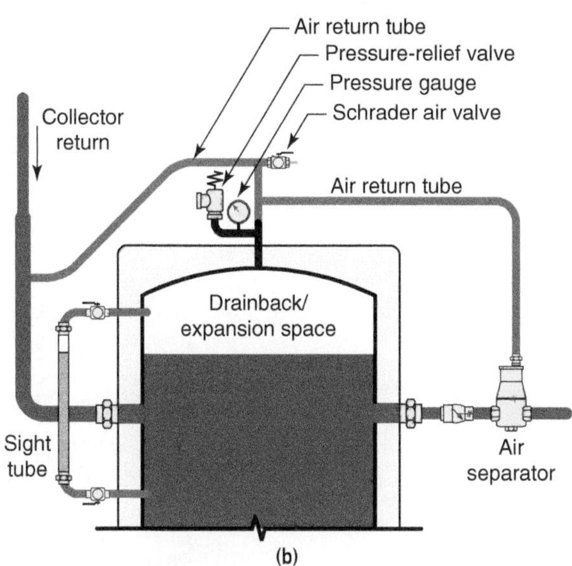

(b)

Figure 11-21 (a) Special adapter fittings allow an air return tube to be connected from the discharge port of the microbubble air separator back to the air space in the drainback/storage tank. This allows the discharged air to be contained within the system (e.g., air control) rather than ejected from the system (e.g., air elimination). (b) Routing of air return tube from air separator to tank.

d. ***No separate expansion tank:*** The drainback tank should be sized, using the procedure described earlier in this chapter, to accommodate the increased water volume at higher temperatures. This eliminates the need for a separate expansion tank in the system.

e. ***Pressure-relief valves:*** Most mechanical codes require a pressure-relief valve in any circuit that contains a heat source and is capable of being isolated from the remainder of the system. Thus, a pressure-relief valve is commonly installed on the auxiliary boiler in a combisystem. However, in most systems, that boiler can be isolated from the remainder of the system, if necessary for repair. Thus, another pressure-relief valve should be installed that protects the system against high pressure potentially created by the solar collectors. This valve is usually installed on piping attached to the thermal storage tank, as seen on many schematics in this chapter.

11.5 Drainback-Protected Solar Domestic Water Heating Systems

Domestic water heating has proven itself to be one of the best applications for solar thermal systems. This is true for both residential and commercial applications. There are two fundamental reasons for this:

1. Domestic water heating is a year-round load. Thus, solar domestic water heating systems can take advantage of the high solar availability and coincident higher ambient temperatures available in summer as well as portions of the spring and fall.
2. Domestic water heating starts at relatively low temperatures, typically 40 °F to 60 °F. It takes just as much energy to raise a gallon of water from 40 °F to 50 °F as it does to raise that gallon of water from 110 °F to 120 °F. The lower end of the temperature rise is where solar energy is best applied because of increased collector efficiency.

The favorable economics of solar domestic water heating must be respected when extending the system to provide some contribution to space heating. All practical solar combisystems will provide a significant portion of a building's domestic water heating requirement. In addition, they will provide a smaller fraction of the building's space heating load. The author prefers to describe practical solar combisystems as "**domestic hot water plus**." The "plus" refers to the system's ability to contribute some energy toward space heating, primarily in spring and fall, and to a much lesser extent during midwinter.

Many of the "packaged" solar domestic water heating systems now on the market use antifreeze-based freeze protection. Such systems were discussed in Chapter 10. However, solar water heating systems can also be designed to use simple drainback freeze protection. One example of such a system is shown in Figure 11-22.

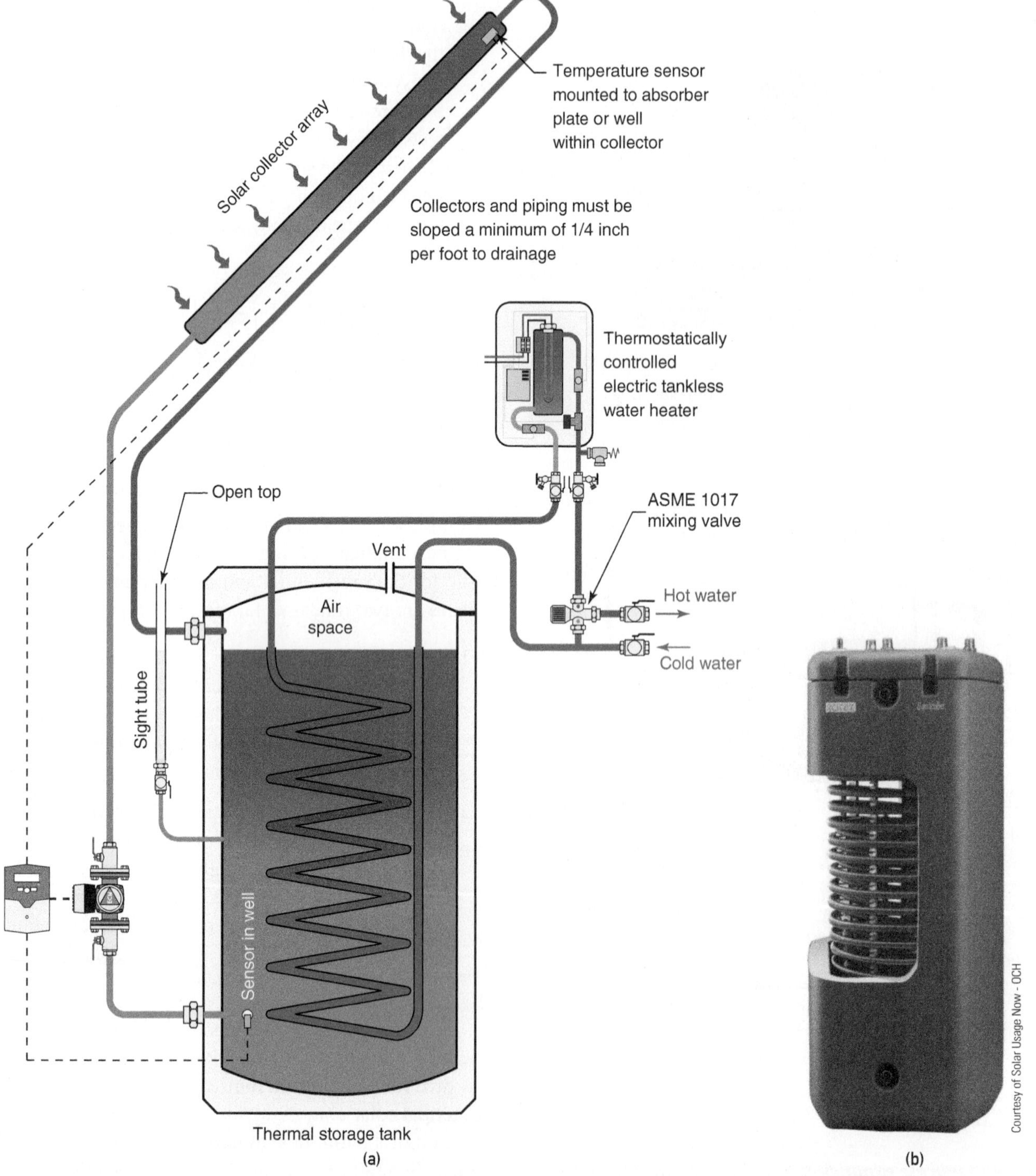

Figure 11-22 | (a) Schematic for a solar domestic water heating system using drainback freeze protection and an unpressurized thermal storage tank. (b) Unpressurized thermal storage tank with internal stainless steel coil.

This system uses an unpressurized thermal storage tank with polypropylene inner and outer shells separated by a generous amount of foam insulation. The tank has an air space at the top that accommodates the thermal expansion of water within the tank as well as the volume of water that drains back from the collector array and associated piping. When a differential temperature controller determines that solar energy is available for collection, it turns on the collector circulator(s) to establish flow through the collector array. In some systems, a single variable-speed circulator can provide the necessary lift head and flow velocity to ensure that a siphon quickly forms in the collector return piping. The circulator then reduces speed to maintain adequate flow through the collector array for the duration of the energy collection cycle. In other systems, two series-coupled circulators are used. Both circulators operate to fill the collector array and establish the siphon. The downstream circulator eventually turns off, and the remaining circulator maintains adequate flow for the duration of the energy collection cycle. When this circulator is turned off, all water in the collector array and associated piping drains back into the tank. Because the tank is vented to the atmosphere, the water level needs to be periodically checked.

As it makes a single pass through the 144-foot long corrugated stainless steel heat exchanger inside this tank, domestic water is heated to within a few degree of the temperature at the top of the storage tank. This water exits the coil and passes through a thermostatically controlled electric tankless water heater. This heater provides any necessary auxiliary energy to boost the water to the desired delivery temperature.

Another design approach for a drainback-protected domestic water heating system uses a pressurized thermal storage tank in combination with an external heat exchanger. An example is shown in Figure 11-23.

Operation of the collector circulator is managed by a differential temperature controller. The variable-speed circulator is turned on when the collector sensor temperature reaches 8 °F above the temperature in the lower portion of the storage tank. It operates at full speed for 3 minutes to ensure that a siphon is formed in the piping returning from the collector array. The circulator then reduces speed to some user-selected level that is adequate to maintain flow in the now-filled collector circuit. The circulators continue to operate at this speed until the collector sensor temperature drops to 4 °F above that of the lower storage tank sensor, at which point it turns off.

The system shown in Figure 11-23a uses the on-demand domestic water heating subsystem described in Chapter 9. Water from the thermal storage tank is circulated through the primary side of a stainless steel brazed-plate heat exchanger whenever there is a minimum demand for hot water. Domestic water is preheated as it passes through the secondary side of this heat exchanger. The circulator creating flow through the primary side of the heat exchanger is controlled by a flow switch within the electric tankless water heater. This flow switch closes whenever the domestic hot water flow reaches 0.6 gpm. At that point, an electrical contactor within the heater closes to supply 240 VAC to triacs, which regulate current flow through the heating elements. The electrical current, and hence the power delivered to the elements, is regulated by electronics within the heater that measure incoming and outgoing water temperature. The power delivered is limited to that required to provided the desired outgoing water temperature. Thus, if solar preheated water is delivered to this heater at 110 °F, and the tankless heater's setpoint temperature is 120 °F, the power supplied to the elements will only be that required for the final 10 °F temperature boost. If the water entering the tankless heater is already at or above the setpoint temperature, the elements are not turned on. All heated water leaving the tankless heater flows into an ASSE 1017–rated thermostatic mixing valve to ensure a safe delivery temperature to the fixtures.

11.6 Examples of Drainback-Protected Combisystems

Practical solar combisystems are designed to supply a significant portion of a home's domestic hot water load as well as a modest percentage of the home's space heating load. There are many possible ways to configure such systems, largely due to the wide variation in space heating subsystems. The systems shown in this chapter were selected to illustrate a variety of drainback-protected applications using both unpressurized and pressurized thermal storage tanks. Each system includes a piping schematic and associated electrical wiring schematic. Each system also includes a description of operation. Readers are encouraged to cross-reference between these drawings and the description of operation to ensure a complete understanding of how each system operates.

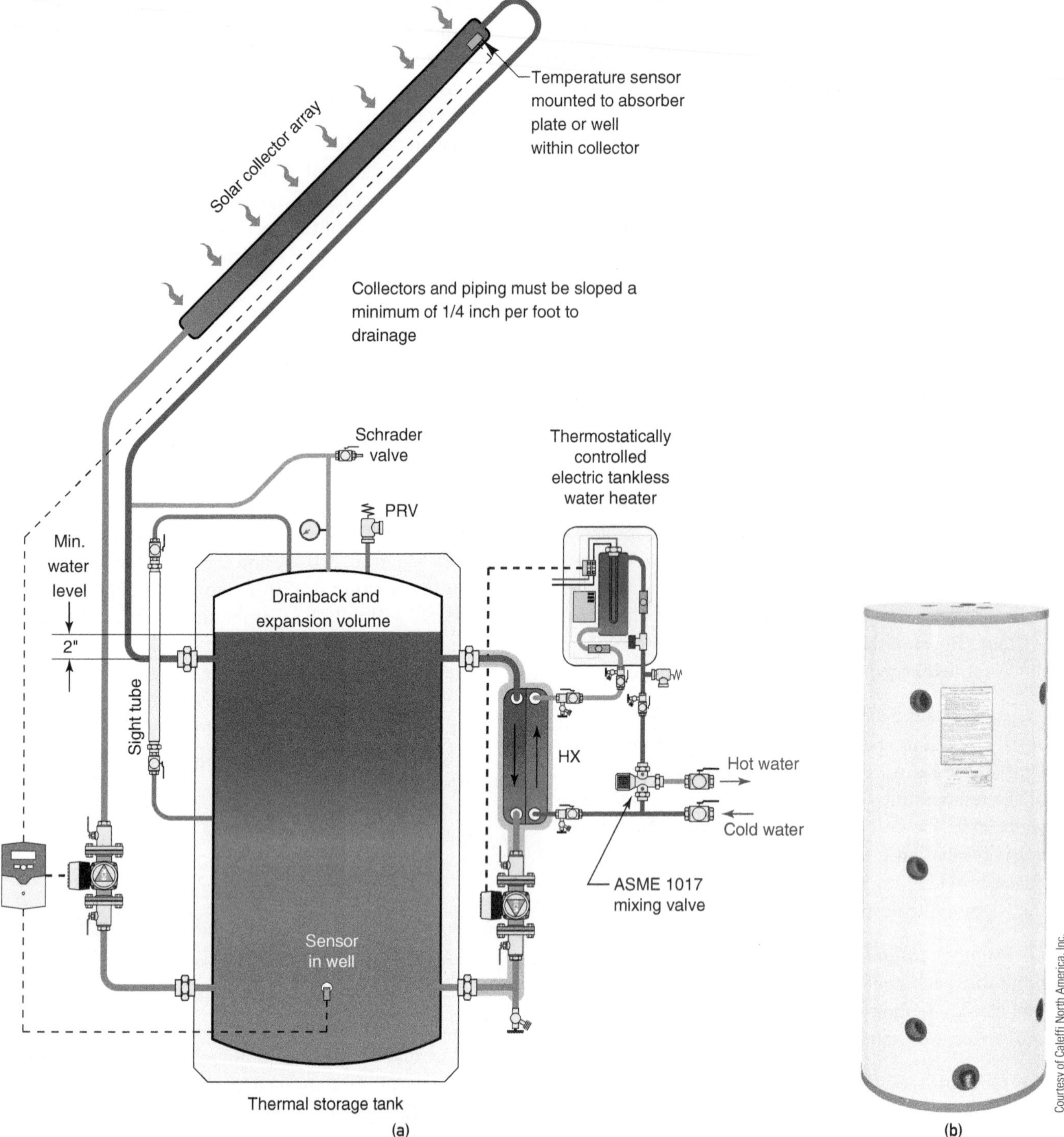

Figure 11-23 (a) Solar domestic water heating system using drainback freeze protection and pressurized thermal storage. (b) Pressure-rated thermal storage tank with multiple piping connections.

The schematics shown in this chapter, and throughout the text, are conceptual drawings rather than final designs for a specific installation. These drawings may not include all detailing required by local codes. For example, some codes require auxiliary boilers to be protected by a low-water cut off and a manual reset high limit controller. Other code requirements that vary from one location to another include use of double-wall heat exchangers, collector array grounding, and T&P versus PRV valves on tankless water heaters. Such details can all be incorporated into the conceptual drawings, if required. That is the responsibility of the system's designer.

DRAINBACK-PROTECTED COMBISYSTEM #1

The system shown in Figure 11-24 is a modification of the antifreeze-based system shown in Figure 10-24. It uses drainback freeze protection rather than antifreeze-based freeze protection.

This design eliminates the collector-to-storage heat exchanger and other hardware associated with the use of antifreeze-based freeze protection. It allows the water that flows through the collector array to also flow through the thermal storage tank and the space heating distribution system. The only heat exchanger present is the external stainless steel brazed-plate heat exchanger used for domestic water heating.

The collector circulator is controlled by a differential temperature controller. The variable-speed circulator is turned on when the collector sensor temperature reaches 8 °F above the temperature in the lower portion of the storage tank. It operates at full speed for 3 minutes to ensure that a siphon is formed in the collector return piping. The circulator then slows

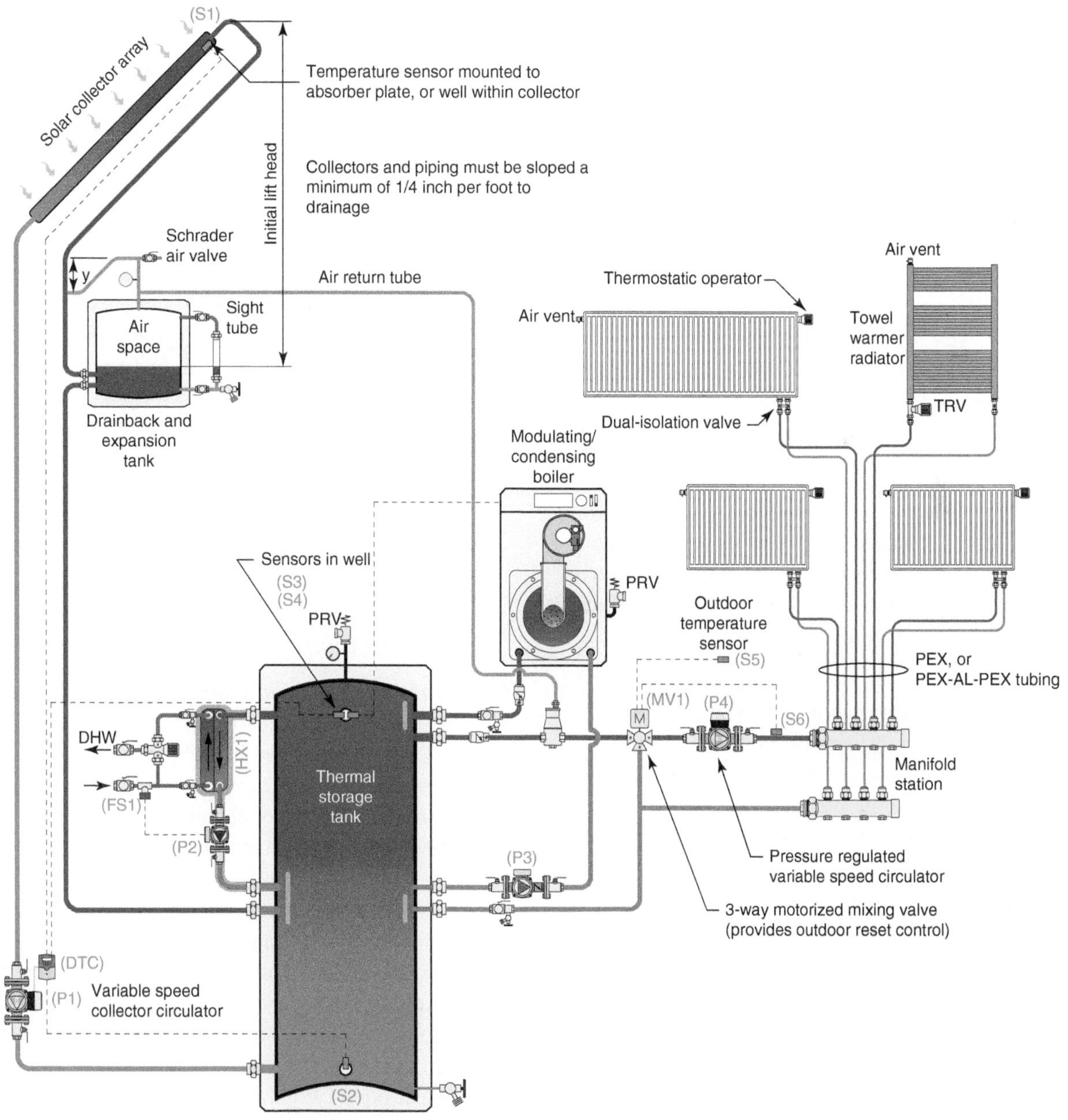

Figure 11-24 | Piping schematic for drainback-protected combisystem #1, using elevated drainback/expansion tank.

to approximately 50 percent of full speed to maintain flow in the now-filled collector circuit. It continues to operate at this speed until the collector sensor temperature drops to 4 °F above that of the lower storage tank sensor, at which point it turns off. The solar-derived heat can contribute to either domestic water heating or space heating.

The water temperature in the *upper portion* of the thermal storage tank is maintained at a minimum temperature of 120 °F by the boiler.

Domestic water is heated "on demand" using an external stainless steel brazed-plate heat exchanger. A flow switch detects whenever domestic hot water demand reaches 0.6 gpm or higher. This turns on circulator (P2) to circulate water from the top of the thermal storage tank through the primary side of the heat exchanger. This heat exchanger has been sized for a 5 °F approach temperature difference and thus can heat domestic water, flowing at design rate, to 115 °F on a single pass. The heated domestic water passes from the heat exchanger to the hot port of an ASSE 1017–rated thermostatic mixing valve. If necessary, this valve mixes cold domestic water into the hot domestic water to prevent excessively high delivery temperatures to fixtures. This is especially important following a hot and sunny day, when the thermal storage tank is likely to be much hotter than the maximum safe hot water delivery temperature of 120 °F.

Water for space heating is drawn from near the top of the tank. It passes through a 3-way motorized mixing valve that protects the low temperature distribution system from excessively high temperatures and provides outdoor reset control of the supply water temperature to the panel radiators.

The panel radiators have been sized to provide design load heat output to each room when supplied with 120 °F water. Flow through the homerun distribution system serving the panel radiators is provided by an ECM-based variable-speed pressure-regulated circulator operating in constant differential pressure mode. The output of each radiator is controlled by a wireless thermostatic radiator valve, and room-by-room temperature control is thus possible. The thermal mass of the upper portion of the thermal storage tank stabilizes the highly zoned hydronic distribution system and prevents the boiler from short cycling under partial load conditions.

Because there are separate connections at the thermal storage tank for the supply and return end of the boiler circuit, the space heating distribution subsystem, and the heat exchanger circuit, the thermal storage tank also provides **hydraulic separation** between the circulators in these circuits. This is especially important because of the variable-speed pressure-regulated circulator in the space heating distribution system. At times, this circulator may be running at very low power levels (e.g., when only one or two panel radiators are active). If not for the hydraulic separation provided, the other circulators could interfere with its operation under such conditions.

Other important details on this system include:

1. ***Properly sloped piping and collectors:*** All piping in the collector circuit must be sloped a minimum of 1/4-inch per foot to ensure efficient drainage. Furthermore, all piping must be properly supported to ensure there are no sags that could trap water and thus lead to a freeze.
2. ***Air return tubes:*** There is an air return tube from the top of the drainback/expansion tank to the collector return piping. It allows air to move back into the collector return piping to initiate drainage at the end of an energy collection cycle. This tube rises several inches above the point where it connects to the collector return piping to form an inverted trap. This stops water in the collector return pipe from flowing into the air space at the top of the tank. A second air return tube runs from the microbubble air separator near the beginning of the distribution system to the top of the drainback tank. This routes any air that is separated at the beginning of the distribution system back to the air space in the drainback tank.
3. ***Drainback/expansion tank details:*** The size of the drainback/expansion tank is determined using procedures described in Section 11.4. The "ideal" connections on the drainback tank allow the water returning from the collectors to enter the tank below the lowest water level. This minimizes splashing noise. It also allows air bubbles to separate within the tank and not be entrained in the flow leaving the drainback tank. The drainback tank is equipped with a sight tube to monitor the water level in the system. The drainback tank is located as high as possible within the heated portion of the building but still low enough to provide proper piping slope from the collector array to the tank. This minimizes the initial lift head, which reduces the required pumping power.

4. ***No make-up water system:*** No automatic make-up water assembly is used in this system for reasons discussed earlier in this chapter.

5. ***No automatic air vents:*** No automatic air vents are used in this system for reasons discussed earlier in this chapter.

6. ***Properly placed check valves:*** There are two spring-loaded check valves in this system. They are located in the piping supplying the space heat load and in the boiler circuit. Both check valves reduce tank heat loss by preventing thermosyphoning. The boiler circulator is also shown with an integral check valve to provide further protection against thermosyphoning. There are no check valves in the solar collector circuit. *Installers should always verify that the collector circulator(s) in a drainback-protected system do not contain check valves. Such valves would prevent collector drainage.*

7. ***Purging valves:*** One purging valve is installed near the return side of the boiler circuit. Another is installed on the return side of the space heating distribution subsystem near the storage tank. These valves allow efficient forced water purging of the system when it is filled and commissioned.

8. ***Pressure-relief valves:*** Most mechanical codes require a pressure-relief valve in any circuit that contains a heat source and is capable of being isolated from the remainder of the system. Thus, a pressure-relief valve is shown on the boiler. Since it is possible to isolate the boiler and its pressure-relief valve from the remainder of the system, another pressure-relief valve is installed at the top of the thermal storage tank.

9. ***Collector sensor in well:*** The sensor used to measure collector temperature is mounted into a well on the absorber plate of one collector.

10. ***Piping and component insulation:*** Although not shown in the schematic, all piping in a solar combisystem should be insulated to a minimum of R-3 (or higher if required by local code).

The electrical wiring diagram for drainback-protected combisystem #1 is shown in Figure 11-25.

The following is a description of operation of the system shown in Figures 11-24 and 11-25. Readers are encouraged to cross-reference the component designations, shown in parentheses, within both the piping and electrical schematics.

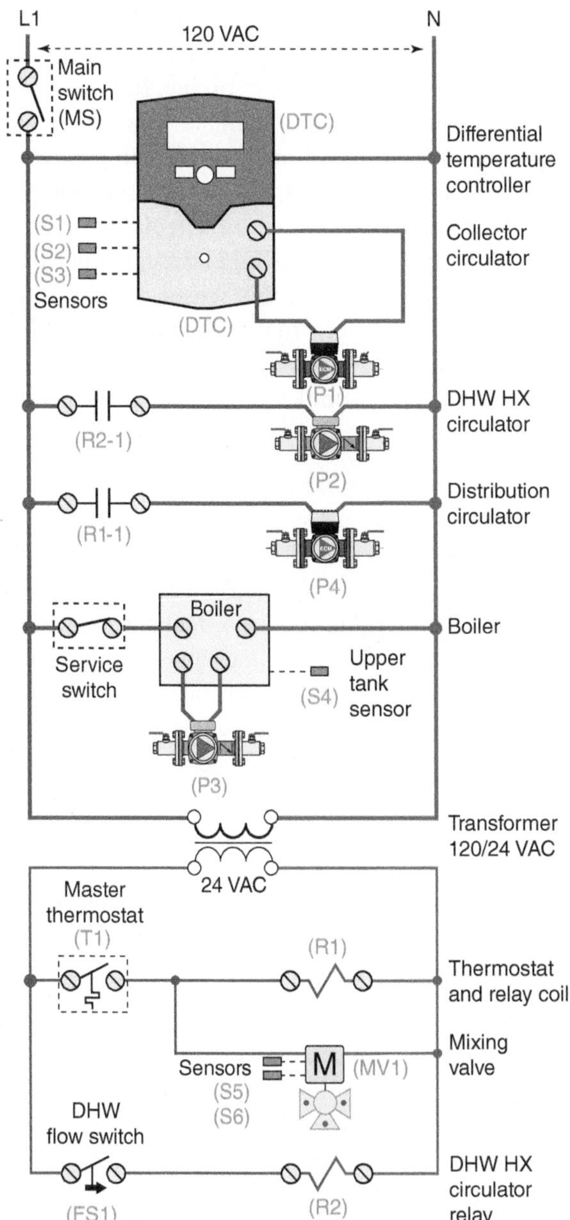

Figure 11-25 | Electrical schematic for drainback-protected combisystem #1.

Description of Operation:

1. *Solar energy collection mode:* Whenever the main switch (MS) is closed, line voltage is applied to the differential temperature controller (DTC). This controller continuously monitors the temperature of the collector sensor (S1) and the lower sensor in the thermal storage tank (S2). Whenever the temperature of the collector sensor is 8 °F or more above that of the lower storage temperature sensor, the (DTC) turns

on circulator (P1) at full speed. This circulator continues to run at full speed for 3 minutes to ensure that a siphon is formed in the collector return piping. It then slows to 50 percent of full speed to maintain flow in the now-filled collector circuit. Whenever the temperature of sensor (S1) is 4 °F or less above the temperature of sensor (S2), the (DTC) turns off circulator (P1). The (DTC) also monitors sensor (S3) in the upper portion of the thermal storage tank. If this sensor reaches 180 °F, the (DTC) initiates a nocturnal cooling mode in which it operates circulator (P1) to release heat through the flat-plate collector array, until the upper tank temperature drops to 160 °F.

2. *Domestic water heating mode:* Whenever there is a demand for domestic hot water of 0.6 gpm or higher, the flow switch (FS1) closes. This closure applies 24 VAC to the coil of relay (R2). The normally open contacts (R2-1) close to turn on circulator (P2), which circulates heated water from the upper portion of the storage tank through the primary side of the domestic water heat exchanger (HX1). The domestic water leaving (HX1) is fully heated and may be hotter than the desired domestic hot water delivery temperature if the thermal storage tank is at a high temperature. All heated water leaving heat exchanger (HX1) flows into an ASSE 1017–rated mixing valve to ensure a safe delivery temperature to the fixtures. Whenever the demand for domestic hot water drop below 0.4 gpm, flow switch (FS1) opens, and circulator (P2) is turned off.

3. *Space heating mode:* A master thermostat (T1) provides a call for space heating. Whenever its contacts close, 24 VAC is applied to the coil of relay (R1). This closes the normally open contacts (R1-1), which supplied 120 VAC to circulator (P4). Circulator (P4) operates in constant differential pressure mode to supply the flow required by the distribution system, based on the status of all thermostatic radiator valves. The thermostat's call for heat also supplies 24 VAC to the controller operating mixing valve (MV1). This controller measures outdoor temperature from sensor (S5). It uses this temperature, in combination with its settings, to calculate the target supply water temperature to the distribution system. The controller operates the valve's actuator, which adjusts the valve's stem to maintain the temperature detected by sensor (S6) close to the target temperature. When the master thermostat (T1) is satisfied, power is removed from mixing valve (MV1) and circulator (P4). A spring-load check valve near the storage tank prevents heat migration into the space heating distribution system when there is no demand.

4. *Boiler:* The boiler continuously monitors the temperature of sensor (S4) in the upper portion of the thermal storage tank. When this temperature drops below 117 °F, the boiler fires. Circulator (P3) is also turned on by the boiler. Water flows from the mid-height connection on the thermal storage tank, through the boiler, and back to the upper connection on the thermal storage tank. When the temperature at sensor (S4) reaches 123 °F, the boiler and circulator (P3) are turned off.

There are several potential variations on the wiring and control logic given for this system. These variations depend on the exact equipment selected and the capabilities of that equipment. For example, some differential temperature controllers may not allow variable-speed operation of the collector circulator (P1). Some boilers may be supplied with an internal circulator that would replace circulator (P3). Some motorized mixing valves may require an external controller and actuator, rather than the integral controller/actuator assumed in this system. Some designers might also choose methods other than a master thermostat for turning circulator (P4) and mixing valve (MV1) on and off. These variations are often based on personal preferences and should be scrutinized against the fundamentals discussed in previous chapters.

DRAINBACK-PROTECTED COMBISYSTEM #2

The system shown in Figure 11-26 is another variation on the system shown in Figure 10-24.

This system uses a pressurized thermal storage tank with sufficient volume for the air space needed for drainback and thermal expansion. All piping connections carrying water into or out of the thermal storage tank are a minimum of 2 inches below the lowest water level in the tank. This prevents splashing sounds when the system is operating.

A sight tube has been connected between the pipe entering the top of the tank and the pipe leading from the tank to the boiler. The connections allow the sight tube to be mounted without additional tappings into the tank. Isolation valves have been provided on both ends of the sight tube to allow it to be isolated and replaced if necessary. The sight glass can be used to verify the proper water level in the tank. The water level should be checked when all circulators are off.

Other details seen near the top of the tank include a pressure-relief valve, pressure gauge, Schrader air

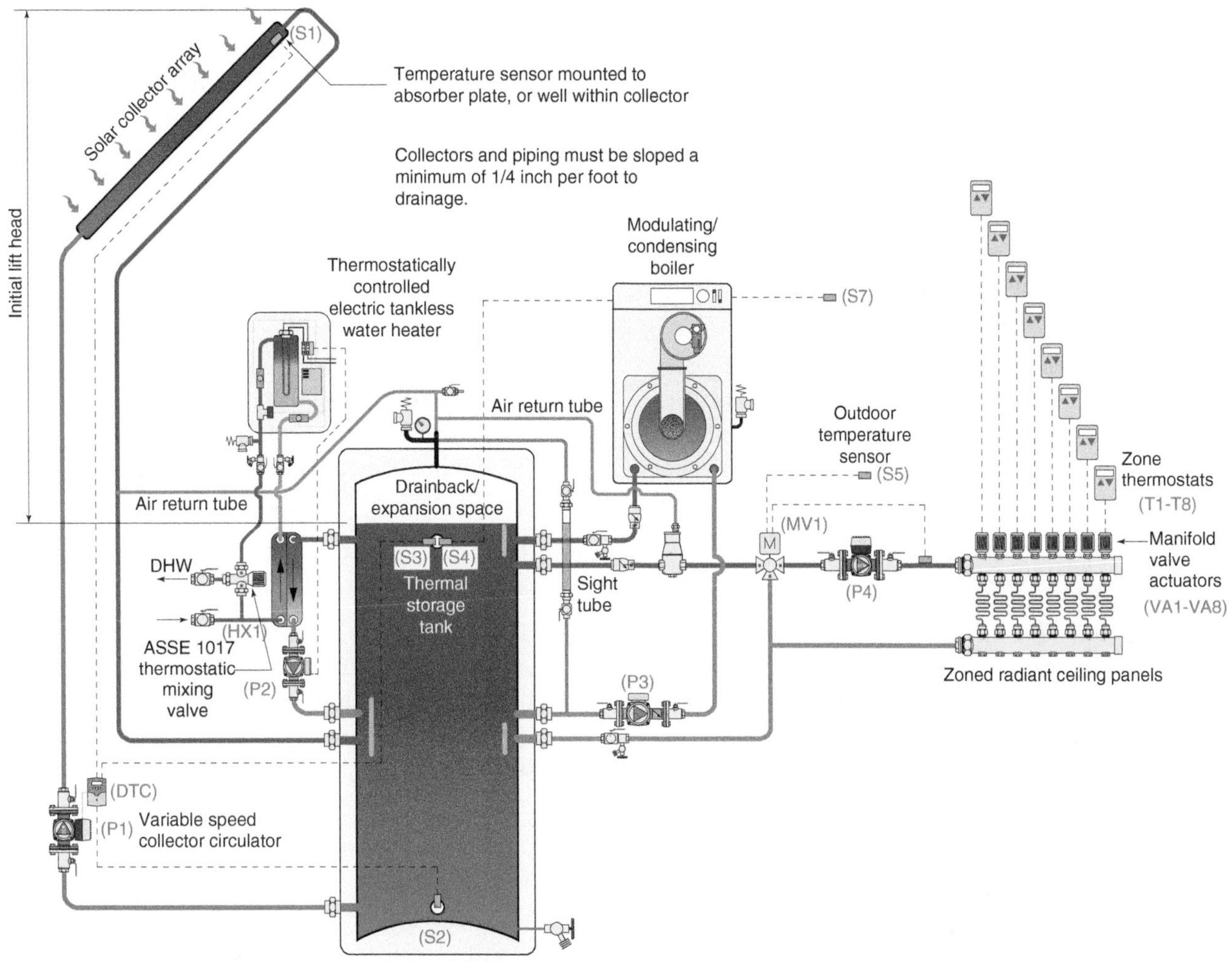

Figure 11-26 | Piping schematic for drainback-protected combisystem #2, using air space at top of thermal storage tank.

valve, inverted trap for the air return tube, and a separate air return tube from the microbubble air separator in the distribution system.

The collector circulator is operated by a differential temperature controller with a variable-speed output. That output could be a variable frequency AC wave used to operate a standard wet-rotor circulator with a PSC motor. It could also be a **digital pulse width modulation** output intended to operate an ECM-type circulator equipped with a PWM input. In either case, the circulator is turned on when the collector sensor temperature reaches 8 °F above the temperature in the lower portion of the storage tank. It operates at full speed for 3 minutes to ensure that a siphon is formed in the collector return piping. The circulator then slows to approximately 50 percent of full speed to maintain flow in the now-filled collector circuit. It continues to operate at this speed until the collector sensor temperature drops to 4 °F above the lower storage tank sensor, at which point it turns off.

Domestic water is heated instantaneously using an external stainless steel brazed-plate heat exchanger. Water from the thermal storage tank is circulated through the primary side of this heat exchanger whenever a hot water fixture is opened. The circulator creating this flow is under the control of a flow switch within the electric tankless water heater shown above the heat exchanger. This flow switch closes whenever domestic hot water flow reaches 0.6 gpm. At that point, an electrical contactor within the tankless heater closes to supply 240 VAC to triacs, which regulate the electrical power delivered to the elements based on measurements of incoming and outgoing water temperature. The power delivered is limited to that required to provide the desired delivery water temperature. Thus, if solar preheated water is delivered to this heater at 110 °F, and the tankless heater's setpoint temperature is 120 °F, the power supplied to the element will only be that required for the final 10 °F temperature boost. If the water entering the tankless heater is already at or above the setpoint temperature, the elements are not turned on. All heated water leaving the tankless heater flows into an ASSE 1017–rated mixing valve to ensure a safe delivery temperature to the fixtures.

Because of the method used for domestic water heating, *the thermal storage tank in this system does not have to be maintained at a temperature suitable to provide the full temperature rise of the domestic hot water.* Instead, its temperature can be maintained by the boiler, when necessary, based on outdoor reset control. If space heating is provided by low-temperature heat emitters, the target water temperature at the top of this tank should be no higher than 120 °F under design load conditions. On partial load days, the outdoor reset controller would allow for lower tank temperatures, perhaps in the range of 90 °F to 95 °F. The lower the tank temperature, the better the chances of solar heat gain from the collectors.

Figure 11-27 shows how the outdoor reset control within the boiler could be configured for this system. This configuration assumes that the radiant ceiling panels used in the system require a supply water temperature of 120 °F under design load conditions.

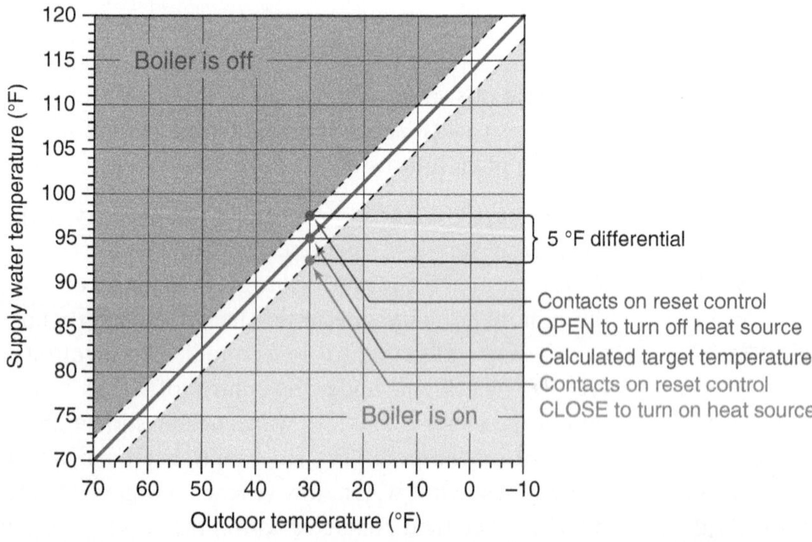

Figure 11-27 Configuration of the boiler's outdoor reset control for the system shown in Figure 11-26. The heat emitters are sized for 120 °F supply water temperature at design load.

As with the previous system, the thermal mass of the heated water in the upper portion of the thermal storage tank provides excellent buffering for the highly zoned distribution system. This greatly reduces the possibility of boiler short cycling.

This design allows the boiler to be completely turned off during times when space heating is not required. Under this mode of operation, the use of electricity is limited to "topping off" the temperature of domestic water, which will always be partially heated by solar-derived heat in the thermal storage tank. Given that a high percentage of the domestic water heating load should be supplied by solar energy, minimal amounts of electric heating should be required.

The electrical wiring for drainback-protected combisystem #2 is shown in Figure 11-28.

The following is a description of operation of the system shown in Figures 10-26 and 10-28. Readers are encouraged to cross-reference the component designations, shown in parentheses, within both the piping and electrical schematics.

Description of Operation:

1. *Solar energy collection mode:* Whenever the main switch (MS) is closed, line voltage is applied to the differential temperature controller (DTC). This controller continuously monitors the temperature of the collector sensor (S1) and the lower sensor in the thermal storage tank (S2). Whenever the temperature of the collector sensor is 8 °F or more above the temperature of the lower storage temperature sensor, the (DTC) turns on circulator (P1) at full speed. This circulator continues to run at full speed for 3 minutes, to ensure that a siphon is formed in the collector return piping. It then slows to 50 percent of full speed to maintain flow in the now-filled collector circuit. Whenever the temperature of sensor (S1) is 4 °F or less above the temperature of sensor (S2), the (DTC) turns off circulator (P1). The (DTC) also monitors sensor (S3) in the upper portion of the thermal storage tank. If this sensor reaches 180 °F, the (DTC) initiates a nocturnal cooling mode in which it operates circulator (P1) to release heat through the flat-plate collector array until the tank temperature drops to 160 °F.

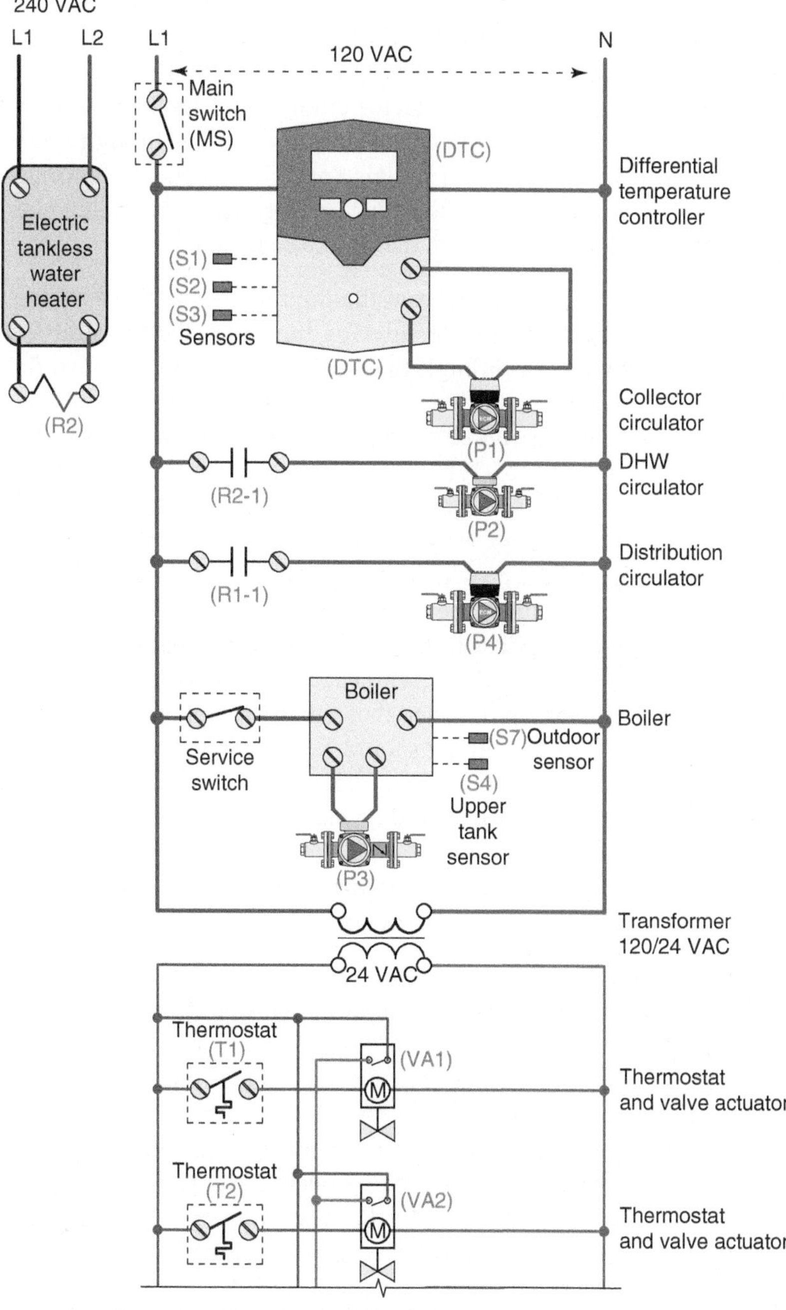

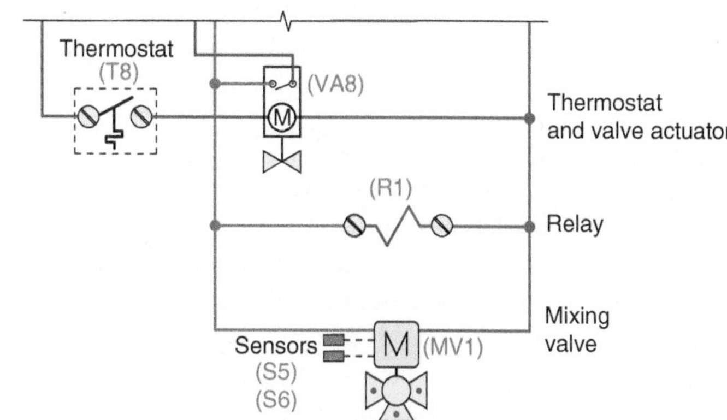

Figure 11-28 | Electrical wiring for drainback-protected combisystem #2.

2. *Domestic water heating mode:* When there is a demand for domestic hot water of 0.6 gpm or higher, the flow switch inside the tankless electric water heater closes. This closure applied 240 VAC to the coil of relay (R2). The normally open contacts (R2-1) close to turn on circulator (P2), which circulates heated water from the upper portion of the storage tank through the primary side of the domestic water heat exchanger (HX1). The domestic water leaving (HX1) is either preheated or fully heated, depending on the temperature in the upper portion of the storage tank. This water passes into the thermostatically controlled tankless water heater, which measures its inlet temperature. The electronics within this heater control the electrical power supplied to the heating elements based on the necessary temperature rise to achieve the set domestic hot water supply temperature. If the water entering the tankless heater is already at or above the setpoint temperature, the elements are not turned on. All heated water leaving the tankless heater flows into an ASSE 1017–rated mixing valve to ensure a safe delivery temperature to the fixtures. When the demand for domestic hot water drops below 0.4 gpm, circulator (P2) and the tankless electric water heater are turned off.

3. *Space heating mode:* The distribution system has several zones, each equipped with a 24 VAC thermostat (T1, T2, T3, etc.). Upon a demand for heat from one or more of these thermostats, the associated 24 VAC manifold valve actuators (VA1, VA2, VA3, etc.) are powered on. When a valve actuator is fully open, its end switch closes. This supplies 24 VAC to the coil of relay (R1), which closes its normally open contacts (R1-1) to supply 120 VAC to circulator (P4). Circulator (P4) operates in constant differential pressure mode to supply the flow required by the distribution system, based on how many zones are active. A call for heat also supplies 24 VAC to the controller operating mixing valve (MV1). The valve's controller measures outdoor temperature from sensor (S5). It uses this temperature, in combination with its

settings, to calculate the target supply water temperature to the distribution system. The controller operates the valve's motorized actuator, which rotates the valve's stem to steer the temperature detected by sensor (S6) toward the target temperature. When all thermostats are satisfied, power is removed from mixing valve (MV1), and circulator (P4). A spring-loaded check valve near the storage tank prevents heat migration into the space heating distribution system.

4. *Boiler:* During the heating season, the service switch to the boiler is closed. The boiler continuously monitors the outdoor temperature detected by sensor (S7). The boiler's internal reset control logic uses this temperature, and its settings, to calculate the target temperature required by the distribution system. It then compares the temperature in the upper portion of the thermal storage tank, as detected by sensor (S4), with the calculated target temperature. If the temperature in storage is more than 2.5 °F below the target temperature, the boiler turns on, and so does circulator (P3). Heat is added to the tank until the temperature detected by sensor (S4) is 2.5 °F above the target temperature. The boiler and circulator (P3) are then turned off. When the heating season ends, the service switch to the boiler is turned off. Under this condition, the only heat input to the storage tank is from the solar collector array

DRAINBACK-PROTECTED COMBISYSTEM #3

The system shown in Figure 11-29 is based on an unpressurized thermal storage tank. It is very similar to the system shown in Figure 10-34. The difference is drainback freeze protection of the collector array.

This system uses a "double-pumped" collector circuit. When the collector temperature climbs 8 °F above the temperature at the bottom of the thermal storage tank, both stainless steel collector circulators (P1a and P1b) are turned on. They continue to operate for 3 minutes, at which time the siphon is fully established in the collector return piping. The downstream circulator (P1a) is then turned off by the differential temperature controller. Circulator (P1b) maintains adequate flow through the collector array for the duration of the operating cycle. Heated water enters the thermal storage tank through a tee, which directs it horizontally. This detail helps preserve temperature stratification.

Flow continues through the collector circuit until the collector temperature drops to within 4 °F of the temperature at the bottom of the thermal storage tank. At that point, collector circulator (P1b) is turned off. Air enters the collector return piping through a tee, which opens to the air space in the thermal storage tank. Water in the collector array and associated piping drains back to the thermal storage tank.

The piping from the dip tube in the thermal storage tank, to the inlet of the collector circulators, is routed through the sidewall of the tank above the water level. This circuit contains "gooseneck" piping, which is routed through the tank's sidewall above the water level. When the circulator in this circuit is off, the water in the piping above the water level will be under slight subatmospheric pressure. This is not a problem, provided that proper design details are followed. First, there should not be any valves, vents, or other devices in the piping assembly that is above the water level. If present, such devices could allow air to enter the piping and thus break the siphon. Second, the top of this piping should be maintained as low as possible to minimize the subatmospheric pressure. Third, the circulator should be placed as low as possible relative to the water level in the tank. This maximizes the static pressure at the circulator inlet and helps prevent cavitation, especially when the water is at a high temperature.

When the system is commissioned, one of the isolation flanges on the collector circulators is closed and the inline ball of the purging valve is opened. Water is forced into the side port of the purging valve located just upstream of these circulators. The entering water forces air through the gooseneck piping and eventually out the bottom of the dip tube. After this, the inline ball of the purging valve is closed and the isolation flange on the circulator is opened, and water is momentarily forced through the collector circulators to displace air within them. Water flow for purging can then be turned off, and the inline ball of the purging valve reopened. The gooseneck piping will retain the water within it. The circuit is now primed and ready to operate.

The heat exchanger circuit supplying space heating has similar details. Using a heat exchanger between the tank water and space heating distribution system allows the latter to be designed as closed system, which has several advantages, especially if the piping within it needs to be routed several feet above the water level in the thermal storage tank.

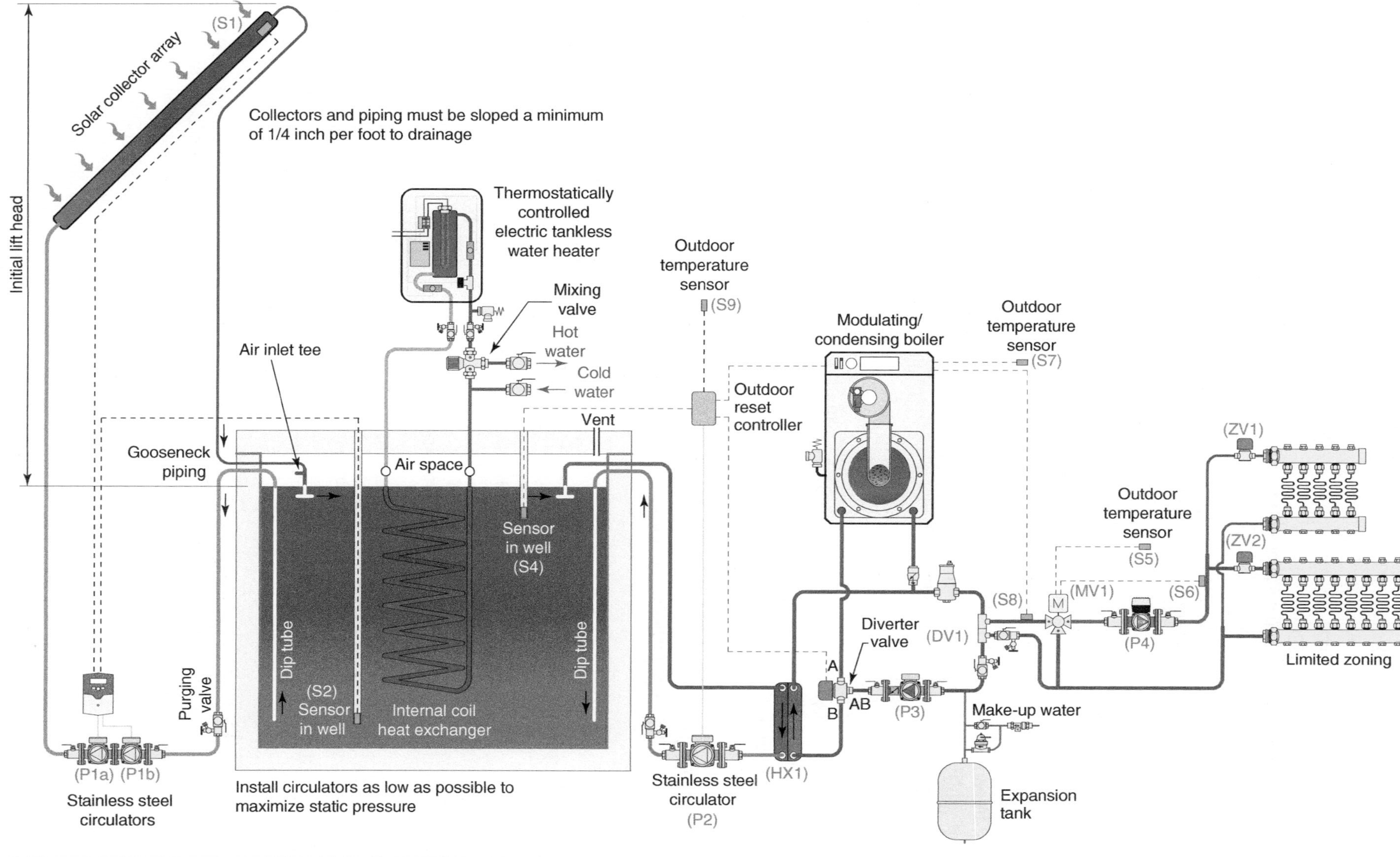

Figure 11-29 | Piping schematic for drainback-protected combisystem #3, using unpressurized thermal storage tank and minimal zoning.

Upon a call for space heating, an outdoor reset controller is powered on. Circulators (P2, P3, and P4) are also turned on. The outdoor reset controller measures the current outdoor temperature as well as the temperature near the top of the storage tank. Using these temperatures, and its settings, it determines if the storage tank temperature is high enough to supply the current space heating load.

Figure 11-30 shows the logic that the outdoor reset controller uses to determine which heat source (e.g., the thermal storage tank or the auxiliary boiler) will supply the space heating load.

This graph assumes that the heat emitters require a supply water temperature of 110 °F at an outdoor design temperature of –10 °F. It also shows that the supply water temperature will be 70 °F at an outdoor temperature of 70 °F, which represent no-load conditions. The blue line between the no-load condition and design load condition is the reset line. It represents the ideal target supply water temperature that should be supplied to the heat emitters, based on the current outdoor condition. The green shaded area represents conditions in which the thermal storage tank is too cool to supply the load, and thus the boiler needs to operate. The red shaded area represents condition where the tank is warm enough to supply the load, and hence the auxiliary boiler remains off. If the water temperature in the tank falls within the unshaded area of the graph, the boiler remains off until the tank cools to the temperature represented by the upper edge of the green shaded area, at which point the boiler will operate. It will continue operating until the water temperature reaches the lower edge of the red shaded area. The colored dots along the vertical line representing an outdoor temperature of 20 °F show that the auxiliary boiler is off until the water temperature at the top of the tank drops to 92.5 °F. The boiler is then turned on and remains on until either the heating load is satisfied, or solar heat input has warmed the water at the top of the thermal storage tank to 97.5 °F.

If the thermal storage tank is warm enough to supply the load, the 3-way diverter valve (DV1) remains off, and flow passes from the valve's (AB) port to its (B) port. This flow passes through the heat exchanger (HX1) and eventually into the set of closely spaced tees that provide hydraulic separation between circulator (P3) and the distribution circulator (P4). Heated water then passes into the 3-way motorized mixing valve (MV1), where its temperature is adjusted, as necessary, to properly supply the low temperature heat emitters. Circulator (P2) is also operating during this mode to provide flow between the thermal storage tank and heat exchanger.

The system continues to operate in this mode as long as there is a call for heat, and the storage tank temperature is high enough to meet that load. However, when the storage tank temperature drops to where it cannot supply the load, the outdoor reset controller performs three functions. First, it turns off circulator (P2) to stop flow from the thermal storage tank to heat exchanger (HX2). Second, it powers on the diverter valve (DV1), which now directs flow from its (AB) port to its (A) port and thus through the boiler. Third, it enables the boiler to fire to provide heat to the space heating distribution system.

In the latter operating mode, the thermal storage tank does not provide any buffering of the space heating load. Thus, zoning should be limited to avoid short cycling the boiler.

Domestic water is preheated within a copper coil heat exchanger suspended within the thermal storage tank. Any auxiliary heat needed is added by a

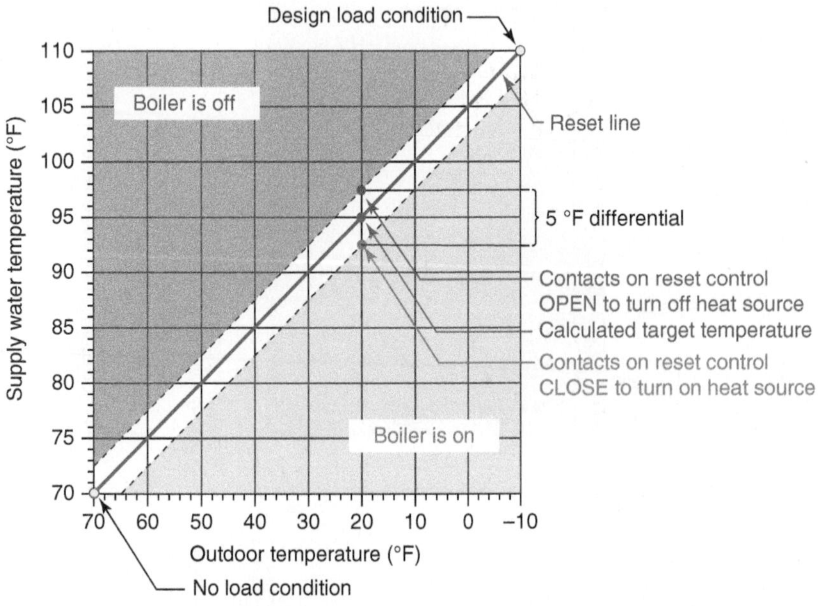

Figure 11-30 | Outdoor reset logic used to select which heat source supplies the space heating load for the system shown in Figure 11-27.

thermostatically controlled electric tankless water heater. All domestic hot water must pass through an ASSE 1017–rated thermostatic mixing valve, which limits delivery temperature to the fixtures.

The electrical wiring diagram for drainback combisystem #3 is shown in Figure 11-31.

The following is a description of operation of the system shown in Figures 11-29 and 11-31. Readers are encouraged to cross-reference the component designations, shown in parentheses, within both the piping and electrical schematics.

Description of Operation:

1. *Solar energy collection mode:* Whenever the main switch (MS) is closed, line voltage is applied to the differential temperature controller (DTC). This controller continuously monitors the temperature of the collector temperature sensor (S1) and the lower sensor temperature in the thermal storage tank (S2). Whenever the temperature of the collector is 8 °F or more above the temperature of the lower storage temperature sensor, the (DTC) turns on circulators (P1a) and (P1b). Both circulators operate for 3 minutes to ensure that a siphon forms in the collector return piping. Circulator (P1a) then turns off. Circulator (P1b) continues to provide flow through the now-filled collector circuit. When the temperature of sensor (S1) drops to 4 °F above the temperature of sensor (S2), the (DTC) turns of circulators (P1a) and (P1b).

2. *Domestic water heating mode:* Whenever there is a demand for domestic hot water, cold domestic water passes through the copper coil heat exchanger suspended within the thermal storage tank. The temperature at which it leaves this coil depends on the temperature of the thermal storage tank. In some cases, it will be heated to or above the desired delivery temperature. In other cases it will only be preheated. All water exiting the coil passes into the thermostatically controlled tankless electric water heater. When the domestic water flow rate reaches 0.6 gpm or higher, the flow switch inside the tankless electric water heater closes. The electronics within this heater control electrical current flow to the heating elements based on the necessary temperature rise (if any) to achieve the set domestic hot water supply temperature. If the water entering the tankless heater is already at or above the setpoint temperature, the elements are not turned on. All heated water leaving the tankless heater flows into an ASSE 1017–rated mixing valve to ensure a safe delivery

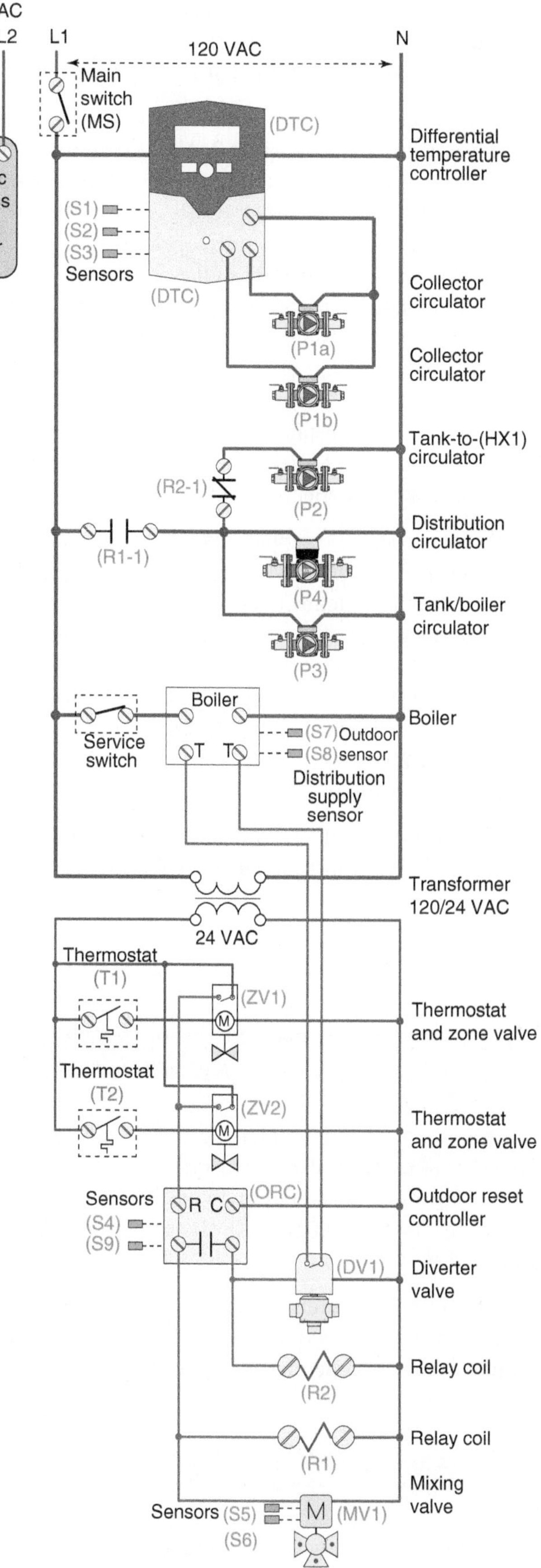

Figure 11-31 | Electrical wiring diagram for drainback-protected combisystem #3.

temperature. Whenever the demand for domestic hot water drops below 0.4 gpm, the elements in the tankless electric water heater are turned off.

3. *Space heating mode:* The distribution system has two zones, each equipped with a 24 VAC thermostat (T1, T2). Upon a demand for heat from either thermostat, the associated 24 VAC zone valve (ZV1, ZV2) is powered on. When the zone valve is fully open, its end switch closes. This supplies 24 VAC to the coil of relay (R1) (wired through the outdoor reset controller [ORC]). The normally open contacts (R1-1) close to supply 120 VAC to circulators (P3) and (P4) and, through a normally closed contact (R2-1), to circulator (P2). Circulator (P4) operates in constant differential pressure mode to supply the flow required by the distribution system, based on how many zones are active.

A call for heat also supplies 24 VAC to the controller operating mixing valve (MV1). The valve's controller measures outdoor temperature from sensor (S5). It uses this temperature, in combination with its settings, to calculate the target supply water temperature to the distribution system. It then adjusts the valve's stem to steer the temperature detected by sensor (S6) toward the target temperature. When both thermostats are satisfied, power is removed from mixing valve (MV1) and circulator (P4).

Whenever one of the zone valve end switches is closed, 24 VAC is also supplied to the outdoor reset controller (ORC). This controller measures the outdoor temperature at sensor (S9). It uses this temperature, in combination with its settings, to calculate the target supply temperature for the distribution system. It then compares this target temperature to the temperature in the upper portion of the storage tank, measured by sensor (S4). If the temperature measured by (S4) is high enough to supply the current space heating load, the relay in the (ORC) remains open and the diverter valve remains off. Flow returning from the distribution system is directed back through heat exchanger (HX1). If the tank's temperature is too low to supply space heating, the relay in the (ORC) closes, supplying 24 VAC to the diverter valve. This valve then routes flow returning from the distribution system to the boiler. When the diverter valve is fully open, its isolated end switch closes. This provides a switch closure across the boiler's (T T) terminals, enabling it to fire. The boiler then monitors its outdoor temperature sensor (S7). It uses this temperature, in combination with its settings, to calculate the target supply water temperature to the distribution system. It compares the temperature at sensor (S8) to this target temperature and adjusts its firing rate as necessary to keep the temperature at (S8) close to the target supply temperature. The relay in the (ORC) also powers on relay coil (R2). This opens the normally closed relay contact (R2-1) and turns off circulator (P2) while the boiler is supplying heat.

SUMMARY

This chapter has presented the design principals and required details for drainback-protected solar combisystems. These systems hold several advantages relative to antifreeze-protected systems. However, they also require carefully detailing and installation to ensure adequate pitch in the collector circuit piping as well as proper pipe sizing to ensure siphon formation. Designers should carefully note such details on all plans associated with drainback-protected systems.

The author's preference is for pressurized drainback-protected systems that allow the same water that flows through the collectors to also flow through the thermal storage tank and the space heating distribution system. This eliminates the cost and thermal penalties associated with heat exchangers.

ADDITIONAL RESOURCES

The following references are recommended for additional reading on solar combisystems system design and installation:

1. Duffie, John A., and William A. Beckman. (2013). *Solar Engineering of Thermal Processes*, 4th edition. John Wiley & Sons. ISBN 978-0-470-87366-3.

2. Goswami, D. Yogi, Frank Keith, and Jan F. Kreider. (2000). *Principles of Solar Engineering*, 2nd edition. Taylor & Francis Group. ISBN 978-1-56032-714-1.

3. Peuser, Felix A., Karl-Heinz Remmers, and Martin Schnauss. (2002). *Solar Thermal Systems, Successful Planning and Construction*. James & James (Science Publishers) Ltd. ISBN 1-902916-310-5.

4. (2004). *Solar Hot Water Systems Lessons Learned 1977 to Today.* Tom Lane, Energy Conservation Services Gainesville, Florida.

5. Ramlow Bob, with Benjamin Nusz. (2010). *Solar Water Heating: A Comprehensive Guide to Solar Water and Space Heating Systems*. New Society Publishers. ISBN 978-0-86571-668-1.

6. Skinner, Peter, P. E., et al. (2011). *Solar Hot Water Fundamentals, Siting, Design, and Installation*. E2G Solar, LLC. ISBN 978-098335520-5.

KEY TERMS

air control
air elimination
air intake tee
closed-loop drainback-protected system
digital pulse width modulation
dip tube
dished head
domestic hot water plus
double-pumped drainback system
drainback-protected solar thermal system
entrain
fouling factor
flow velocity
gauge pressure
gooseneck piping
harp-style absorber plate
hydraulic separation
initial lift head
maximum head loss curve
maximum siphon height
minimum downward pitch
psia absolute pressure
Schrader fitting
series-coupled arrangement
shorted
side sloped
sight tube
siphon
thermal penalty
waterlogged system

QUESTIONS AND EXERCISES

1. Name three advantages of unpressurized drainback-protected solar combisystems compared to pressurized drainback systems.

2. Name three *disadvantages* of unpressurized drainback-protected solar combisystem compared to pressurized drainback systems.

3. Describe the conditions necessary for siphon formation as water arrives at the top of the collector array piping during the start of an energy collection cycle.

4. Why should automatic air vents not be installed at the top of drainback systems?

5. What is the minimum piping slope necessary for collector circuit piping in a drainback system?

6. Describe how to adapt a drainback system to a five-story tall building, where the collector array will be on the roof and the storage tank in the basement.

7. Describe a specific advantage of a single-variable-speed collector circulator versus a "double-pumped" drainback system using 2 fixed-speed circulators.

8. What happens at the onset of a power failure in a drainback-protected system?

9. Explain why the water pressure in the gooseneck piping of a drainback collector circuit is slightly below atmospheric pressure.

10. Explain why an automatic make-up water subassembly should not be used in a pressurized drainback-protected system. What might happen if this subassembly were installed and operated as normal?

11. What is the purpose of the slope in the air return tube shown in Figure 11-2?

12. Explain the difference between air elimination and air control in the context of a drainback-protected system.

13. Determine the maximum siphon height in a drainback-protected system that has the following operating conditions:
    a. Air pressure at top of system = 0 psi, maximum water temperature = 190 °F
    b. Air pressure at top of system = 15 psi, maximum water temperature = 175 °F.

    Explain any difference in these siphon heights.

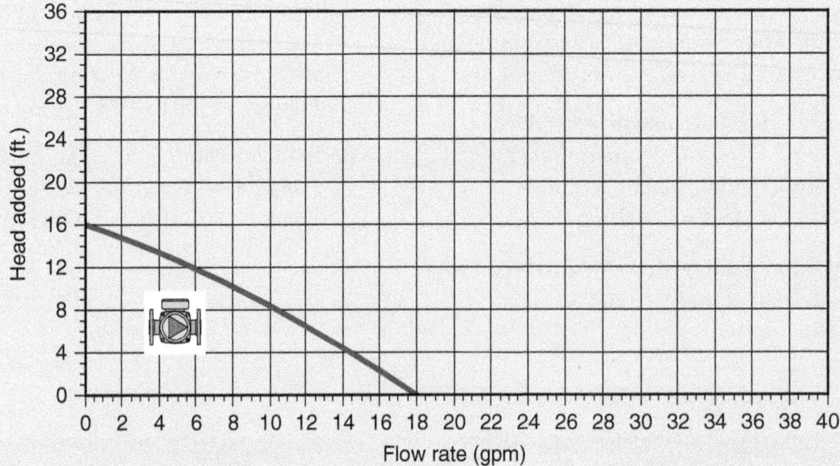

Figure 11-32 | Pump curve for Exercise 15.

14. Describe some situations in which "dead level" collectors or piping could freeze.

15. A circulator with the pump curve shown in Figure 11-32 is being considered for a double-pumped drainback-protected system. Accurately sketch the pump curve for two of these circulators connected in a close-coupled series arrangement.

16. In a double-pumped drainback system, the downstream circulator is turned off after the siphon has formed. Assume that the head loss of this inoperable circulator is represented as follows.

$$H_L = 0.045(f)^{1.75}$$

Use this head loss curve, in combination with the pump curve for a single circulator shown in Figure 11-32, to accurately sketch the "net" pump curve of this upstream circulator while it pushes flow through the inoperable circulator.

17. Calculate and accurately sketch the *maximum head loss curve* for the following drainback system:

Initial lift head = 15 feet

Collector supply piping = 50 feet (total equivalent length) of 3/4-inch type M copper

Collector return piping: 50 feet (total equivalent length) 1/2-inch type M copper

Collectors: 3 collectors piped in reverse return parallel, with 1-inch internal headers

Collector head loss relationship

$$H_{L(collector)} = 0.044(f)^{1.75}$$

18. Determine the change in water level within a cylindrical tank having a diameter of 42 inches, filled with 60 °F water to an initial depth of 60 inches, when the water is heated to 170 °F.

19. Assume that you have selected a pressurized tank for a drainback-protected system. It has flat ends, a volume of 300 gallons, and an internal diameter of 35 inches. The volume of the collector array and piping that will be drained is 8 gallons. The system is filled with water at 60 °F, and the air space at the top is left at atmospheric pressure (e.g., gauge pressure = 0). The distribution system (exclusive of the storage tank and collector subsystem) contains 40 gallons of water. The maximum temperature condition assumes that *all* water and captive air in the system reaches 180 °F. A 30-psi rated pressure-relief valve is installed at the top of the tank. What is the minimum vertical dimension of the air space at the top of the tank to accommodate expansion?

20. Describe the difference in how the boiler operates in drainback-protected combisystem #1, (Figure 11-24) versus how it operates in drainback-protected combisystems #2 and #3 (Figures 11-26 and 11-29).

21. Explain the pros and cons of using a heat exchanger between an open (unpressurized) thermal storage tank and the heating distribution system, as shown in Figure 11-29.

22. Describe how the boiler would operate based on Figure 11-27, when the outdoor temperature is 10 °F.

23. Explain the purpose of the air return tube, which shown as an orange line between the air separator and drainback tank in Figure 11-24. Why is this tube necessary? What happens if it is omitted?

24. Why is it necessary for the boiler in Figure 11-24 to maintain the upper portion of the storage tank at about 120 °F, even when there is no space heating load?

25. Why are there separate piping connections on the storage tank shown in Figure 11-26 for the boiler circuit and the space heating distribution circuit? Why not just tee these circuits together outside the tank and eliminate two of the piping connections on the tank?

26. Assume the differential of the outdoor reset controller represented in Figure 11-30 were set to 15 °F, rather than as shown at 5 °F. Describe how system operation and comfort would be affected at an outdoor temperature of 55 °F.

DAIKIN
R410A
R410A

CHAPTER 12

Estimating Solar Combisystem Performance Using The *f-Chart* Method

OBJECTIVES

After studying this chapter, you should be able to:

- Describe the types of analysis used to predict the performance of solar thermal systems.
- Describe the origin and intent of the f-Chart method.
- Explain what a dimensionless quantity is.
- Understand the quantities used to calculate the dimensionless X and Y parameters used in the f-Chart method.
- Use a simplified version of the f-Chart method for calculating the performance of a solar combisystem.
- Describe the various inputs and outputs used by F-CHART software.
- Convert units on input data to those required by F-CHART software.
- Use F-CHART software to model the performance of an antifreeze-based combisystem.
- Use F-CHART software to model the performance of a drainback-protected combisystem.
- Describe what a parametric study is.
- Use F-CHART software to perform parametric studies regarding-combisystem performance.
- Use F-CHART software to prepare graphs of system performance.

12.1 Introduction

Preceding chapters have shown that there are many ways to design a solar combisystem. To ensure proper operation, proposed system designs must be analyzed for both hydraulic and thermal performance.

Hydraulic performance analysis involves tasks such as:

- Determining flow rates based on loads
- Determining pipe sizes based on the flow rates
- Determining head losses based on flow rates, piping, fittings, and the fluids being conveyed
- Selecting circulators based on flow rates and associated head losses

All of these tasks are fundamentally important when designing any hydronic system. Methods for performing these tasks were presented in previous chapters.

Thermal performance analysis involves tasks such as:

- Estimating design heating loads
- Sizing heat emitters based on supply water temperature at design load
- Determining reset ratios based on building characteristics and heat emitter selections
- Sizing an auxiliary heat source
- Determining the contribution of the solar collector array to both space heating and domestic water heating

The first four of these tasks are also fundamentally important when designing hydronic heating systems and were covered in previous chapters.

The final task, however, is very different. Assessing the monthly and seasonal contribution of a specific array of solar collectors involves much more complex calculations. For example, while the efficiency of some heat sources, such as an electric boiler, is relatively stable, the thermal efficiency of solar collectors can vary continuously based on changing air temperature, incoming fluid temperature, and solar radiation. Furthermore, variations in collector array area, slope, azimuth, shading, collector flow rate, storage tank volume, and heat exchanger performance will all affect how much thermal energy is gathered and used by the system. Given all these variables, the only practical way to make complete and often iterative analysis of solar thermal systems is through computer simulation.

Methods for computer simulation of solar thermal systems date back to the 1970s and **mainframe computers**. One of the most widely used software programs for simulating solar thermal systems was developing during this time at the University of Wisconsin–Madison. It is called **TRNSYS**, which stands for Transient System Simulation program, and is pronounced "TRAN-sis." It is currently available in its 17th version. More detailed information on TRNSYS is available at http://sel.me.wisc.edu/trnsys/index.html.

TRNSYS is a high-level mathematical modeling tool that performs hour-by-hour simulations of user-defined systems based on detailed component models and weather data. It uses its own system description language and has a large library of component models. It is widely used by researchers and engineers. However, because of its complexity, learning curve, and cost, TRNSYS is seldom used for routine design of solar thermal systems.

During the late 1970s, researchers at the University of Wisconsin–Madison sought to create a simpler method for predicting the performance of solar thermal systems, one that would be useful to designers who did not have access to TRNSYS or mainframe computers. Personal computers were not widely available at the time, so the method had to be executable using a scientific calculator. This effort led to the development of the **f-Chart method**.

The f-Chart method is based on **empirical correlations** of the results of several hundred detailed computer simulations of solar thermal systems using TRNSYS software. These simulations cover a wide variety of system configurations, including different collector areas, collector slope angles, and collector azimuth angles. They also cover a range of other hardware configurations for storage volume, collector-to-storage heat-exchanger effectiveness, flow rates, and the size of the heat exchanger delivering heat to the space heating load. The configuration of the system used for simulating liquid-based solar combisystems is shown in Figure 12-1.

The f-Chart method empirically correlates the **monthly solar fraction (f)** of the thermal load to two **dimensionless quantities** called X and Y. The term "dimensionless" means that the calculated values of X and Y are simply two numbers. These numbers do not have units. This happens because the units of the **physical quantities** used to calculate the dimensionless quantities mathematically cancel each other out. Dimensionless quantities also have the same value in any consistent set of units. The values of X and Y calculated using inputs expressed in metric units will be the same numbers as when calculated using inputs expressed in standard U.S. units.

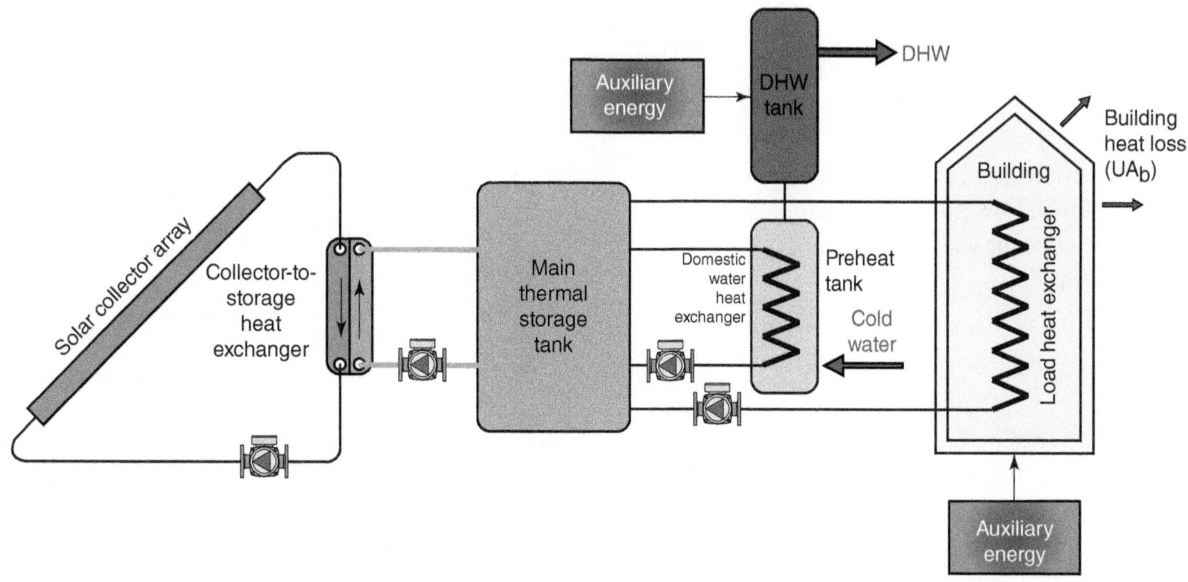

Figure 12-1 | Functional diagram for liquid based combisystem used in development of the f-Chart method.

The dimensionless quantity X is defined by Equation 12.1.

(Equation 12.1)

$$X = (F_R U_L)\left(\frac{F'_R}{F_R}\right)(212 - \bar{T}_a)(t_m)\left(\frac{A_c}{L}\right)$$

The dimensionless quantity Y is defined by Equation 12.2:

(Equation 12.2)

$$Y = (F_R(\tau\alpha)_n)\left(\frac{F'_R}{F_R}\right)G(\bar{H}_\tau)N\left(\frac{A_c}{L}\right)$$

where $(F_R U_L)$ is the **slope** of the collector's **instantaneous efficiency line** discussed in Chapter 8 and listed on **SRCC rating certificates**. The standard U.S. units of $F_R U_L$ are (Btu/hr./ft.^{2}/°F).

$(F_R(\tau\alpha)_n)$ is the **Y intercept** of the collector's instantaneous efficiency line discussed in Chapter 8 and available on SRCC rating certificates. $F_R(\tau\alpha)_n$ is a unitless quantity.

(F'_R/F_R) = **heat exchanger penalty factor** defined by Equation 12.5 (no units)

212 = 212 °F, a reference temperature used in the development of the f-Chart method

T_a = monthly average outdoor air temperature (°F)

t_m = hours in month (hr.)

A_c = gross area of the solar collector array (ft.^{2})

L = monthly total load for both space heating and domestic water heating (Btu)

G = a unitless quantity defined by Equation 12.6

H_T = monthly average value of the total daily solar radiation incident on the plane of the collector array (Btu/ft.^{2}/day)

N = days in month

When working with dimensionless quantities such as X and Y it is important to verify that the units of the physical quantities that combine to make the dimensionless quantity cancel each other out. If they don't, one or more of the physical quantities used in the calculation are incorrect.

Substituting the common U.S. units for each of the physical quantities used to define X in Equation 12.1 and Y in Equation 12.2 proves that the units do mathematically cancel to yield a dimensionless quantity (e.g., a number).

$$X = \left(\frac{\text{Btu}}{\text{hr.}\cdot\text{ft.}^2\cdot{}^\circ\text{F}}\right)\left(\frac{\text{no units}}{\text{no units}}\right)({}^\circ\text{F})(\text{hr.})\left(\frac{\text{ft.}^2}{\text{Btu}}\right)$$

$$= \text{unitless number}$$

$$Y = (\text{no units})\left(\frac{\text{no units}}{\text{no units}}\right)(\text{no units})\left(\frac{\text{Btu}}{\text{ft.}^2 \cdot \text{day}}\right)(\text{day})\left(\frac{\text{ft.}^2}{\text{Btu}}\right)$$

$$= \text{unitless number}$$

Once the values of X and Y are determined, the monthly solar faction for the liquid-based combisystems discussed in this text can be calculated using Equation 12.3:

(Equation 12.3)

$$f = 1.029Y - 0.065X - 0.245Y^2 + 0.0018X^2 + 0.0215Y^3$$

where:

f = fraction of the monthly heating load supplied by solar energy (decimal %)

X = dimensionless quantity determined from Equation 12.1

Y = dimensionless quantity determined from Equation 12.2

An alternate method of finding the monthly solar fraction is to use the values of X and Y in combination with the graph in Figure 12-2. This graph is the basis of the word "chart" in the f-Chart method.

This chapter discusses the f-Chart method in two ways:

1. As a simplified version suitable for manual calculations, provided that certain assumptions apply to the system being analyzed.

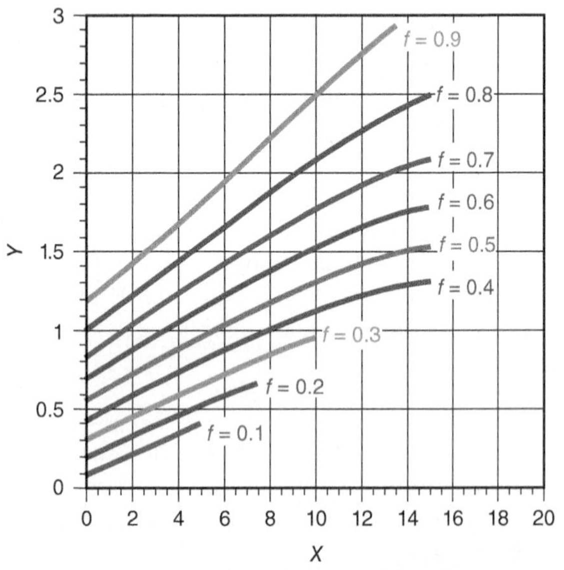

Figure 12-2 Graph called an "f-Chart" used to find monthly solar fraction based on the values of dimensionless quantities X and Y.

2. As an overview of F-CHART software that is currently available for use on Windows-based computers.

12.2 Simplified Version of the f-Chart Method for Manual Calculations

This section presents the f-Chart method in a form that can be used *without* software. To do this while maintaining reasonable expectations on the amount of calculations required, this section presents a *simplified version* of the full f-Chart method. The simplifications employed greatly reduce the number and complexity of calculations required. However, simplifications also involve assumptions that must be respected if the result is to remain reasonably accurate and useful. These assumptions are stated where they apply. If the stated assumptions do not apply to the system being analyzed, the reader can find detailed information about a more generalized version of the f-Chart method in the first reference listed at the end of the chapter. Another alternative is to use F-CHART software, which is discussed later in the chapter.

CALCULATING *X*, *Y*, AND *f*

The calculations to determine X, Y, and f using Equations 12.1, 12.2, and either Equation 12.3 or Figure 12-2 are straightforward. However, several inputs to Equations 12.1 and 12.2 can require a significant amount of referencing or additional calculations before X, Y, and finally f can be calculated.

Some of the numerical inputs are relatively easy to find, as they are defined in Equations 12.1 and 12.2. For example, $F_R(\tau\alpha)_n$ is the Y intercept of the collector's instantaneous efficiency line and can be determined from the collector's SRCC rating certificate, discussed in Chapter 8. The quantity $F_R U_L$ is the slope of the collector's instantaneous efficiency line and can also be determined from the collector's SRCC rating certificate. The monthly average outside air temperature can be found in many climate references. It can also be estimated using Equation 12.4.

(Equation 12.4)

$$T_a = 65 - \frac{(DD_{65})_{65}}{N}$$

where:

T_a = average outdoor air temperature for the month (°F)

65 = base of °F·day data

DD_{65} = # of base 65 °F days in the month (°F·days)

N = number of days in month (days)

The number of hours in a month is simply the number of days in that month multiplied by 24.

The value for L is the total estimated energy used for space heating and domestic water heating for each month. The space heating energy use can be estimated using the degree day method discussed in Chapter 2. The energy used for domestic water heating can also be estimated, on a monthly basis, using information from Chapter 2.

The remaining inputs to Equations 12.1 and 12.2 require additional calculations and will be discussed separately.

The heat exchanger penalty factor (F'_R/F_R) is a correction factor that accounts for the loss in system performance associated with having a heat exchanger between the solar collector array and thermal storage tank. Such a heat exchanger forces the collectors to operate at a higher fluid temperature, compared to a system in which no heat exchanger is used between the collector array and thermal storage tank. The larger the heat exchanger, and the higher its **effectiveness**, the closer this factor approaches the "ideal" value of 1.0. *In a drainback-protected system that has no heat exchanger between the collector array and thermal storage tank, the value of this factor is 1.0.*

For systems with collector-to-storage heat exchangers, the heat exchanger penalty factor can be calculated using Equation 12.5.

(Equation 12.5)

$$\left(\frac{F'_R}{F_R}\right) = \left[1 + \left(\frac{(F_R U_L)A_c}{(8.01D_c c_c)f_c}\right)\left(\frac{(8.01D_c c_c)f_c}{e(8.01Dcf)_{min}} - 1\right)\right]^{-1}$$

where:

(F'_R/F_R) = the heat exchanger penalty factor (no units)

$(F_R U_L)$ = the slope of the collector efficiency line discussed in Chapter 8 and available on SRCC rating certificates (Btu/hr./ft.²/°F)

A_c = gross collector area (ft.²)

8.01 = units factor

D_c = density of fluid in collector circuit (lb./ft.³)

c_c = specific heat of fluid in collector circuit (Btu/lb./°F)

f_c = flow rate through the collector array (gpm)

e = effectiveness of collector-to-storage heat exchanger (no units)

$(8.01Dcf)_{min}$ = the *lower* of the two **fluid capacitance rates** for the two fluids flowing through the collector-to-storage heat exchanger. Where D = density of the fluid (lb./ft.³), c = specific heat of the fluid (Btu/lb./°F) and f = flow rate of fluid (gpm). The quantity ($8.01Dcf$) will have units of (Btu/hr./°F).

The quantity $(8.01Dcf)_{min}$ is found by first calculating the product ($8.01Dcf$) for each fluid stream flowing through the collector-to-storage heat exchanger and then selecting the lower of the two values. In most systems, the lower value of ($8.01Dcf$) will be for the collector circuit side of the heat exchanger. Examples of these calculations, and methods for determining the effectiveness of a heat exchanger, are given in Chapter 4 (see Section 4.8).

For the simplified version of the f-Chart method, the heat exchanger penalty factor (F'_R/F_R) can be estimated from Figure 12-3, *when the following assumptions apply:*

- Assumption #1: Collector fluid is 50% propylene glycol
- Assumption #2: The collector circuit has the minimum fluid capacitance rate (e.g., the value of ($8.01Dcf$) as defined for Equation 12.5)
- Assumption #3: Collector flow rate is 0.03125 gpm per ft.² of collector (This is equivalent to a flow rate of 1 gpm through a 4 foot × 8 foot collector.)
- Assumption #4: The $F_R U_L$ value for the collector is 0.865 Btu/hr./ft.²/°F

The author recommends a *minimum* effectiveness of 0.55 for the collector-to-storage heat exchanger, with a "preferred" effectiveness value of 0.7.

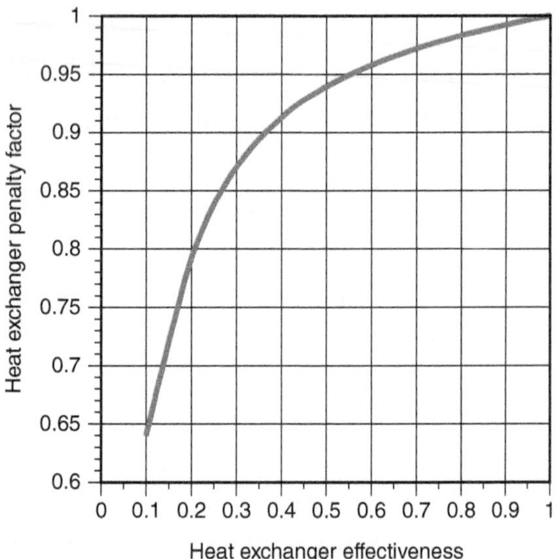

Figure 12-3 Heat exchanger penalty factor (F_R'/F_R) versus effectiveness of collector-to-storage heat exchanger and for stated assumptions.

The quantity (G) in Equation 12.2 is the ratio of the monthly average ($\tau\alpha$) value to the perpendicular ($\tau\alpha$) value of the collector's glazing transmissivity (τ), multiplied by its absorber plate absorptivity (α). It can be mathematically expressed as Equation 12.6.

(Equation 12.6)

$$G = \left(\frac{\overline{(\tau\alpha)}}{(\tau\alpha)_n}\right)$$

Calculating the value of G for each month involves several other equations and detailed data on the direct and diffuse solar radiation available at the site. It also requires a model of how the transmissivity of the collector's glazing and the absorptivity of its absorber plate coating vary as a function of the solar radiation's incident angle. Detailed methods for calculating this quantity are given in reference 1 at the end of the chapter. However, *the value of G can be estimated at 0.96 when the following assumptions apply:*

- Assumption #1: The collector array is made up of single glazed flat-plate collectors.
- Assumption #2: The collector array has a slope between (local latitude $-15°$) and (local latitude $+15°$).
- Assumption #3: The collector array faces within 15° of true south.

The quantity H_T used in Equation 12.2 is the total daily solar radiation incident on the plane of the collector. H_T must be computed as a monthly average value. It can be found for several different collector slope angles using the tables available at the website http://rredc.nrel.gov/solar/pubs/redbook/. These downloadable tables were discussed in Chapter 7. An example is given in Figure 7-32. Each table lists the total daily solar radiation, for an average day in each month, in units of (kWh/m²/day). To use this radiation data in Equation 12.2, these units must be converted to (Btu/ft.²/day). The conversion factor is as follows:

$$\left(\frac{\text{kWh}}{\text{m}^2 \cdot \text{day}}\right)(317) = \left(\frac{\text{Btu}}{\text{ft.}^2 \cdot \text{day}}\right)$$

RANGE OF INPUTS FOR THE F-CHART METHOD

The range of inputs used in developing the f-Chart method—which remain valid for the simplified version—are as follows:

- Normal ($\tau\alpha$) coefficient of collector: $0.6 \leq (\tau\alpha)_n \leq 0.9$
- F_R multiplied by gross collector area (A_c): $54 \leq (F_R A_c) \leq 1291$ (ft.²)
- Collector heat loss coefficient (U_L): $0.37 \leq (U_L) \leq 1.46$ (Btu/hr./ft.²/°F)
- Collector slope angle (B): local latitude $-15° \leq$ (B) $\leq$ local latitude $+15°$
- Building heat loss coefficient (U_H): $157 \leq (U_L) \leq 1265$ (Btu/hr./°F)
- Thermal storage volume per unit of collector area*: 1.84 gallons/ft.²
- Load heat exchanger size (LHES)**: 2.0 (unitless)

THERMAL STORAGE VOLUME CORRECTION FOR (X)

f-Chart was developed with an assumed thermal storage volume of 1.84 gallons of water storage per gross square foot of collector area. Thus, a solar thermal

*Other ratios of thermal storage volume to collector area can be used to modify the X value of Equation 12.1 and will be discussed later in this section.

**Other values of the load heat exchanger size (LHES) can be used to modify the Y value of Equation 12.2 and will be discussed later in this section.

system with 150 gross square feet of collector area is assumed to have a corresponding thermal storage volume of 150 × 1.84 = 276 gallons.

It is possible to correct the value of the dimensionless quantity X from Equation 12.1 based on different thermal storage volumes using Equation 12.7.

(Equation 12.7)

$$X_c = X\left(\frac{V_{actual}}{1.84}\right)^{-0.25}$$

where:

X_c = *corrected* value of dimensionless quantity X based on the actual storage volume used

X = dimensionless quantity X based on standard (assumed) size of storage

V_{actual} = size of storage used in system (gallons/ft.^{2} of gross collector area)

1.84 = storage volume assumed in f-Chart (gallon/ft.^{2} of gross collector area)

Example 12.1: Assume the standard value of X, determined using Equation 12.1, was 8.0. Determine the corrected value of X (e.g., X_c) if the system has 150 square feet of gross collector area and 450 gallons of thermal storage.

Solution: First, determine the ratio of storage volume to collector area:

$$V_{actual} = \frac{450 \text{ gallons}}{150 \text{ ft.}^2} = 3.0\frac{\text{gallon}}{\text{ft.}^2}$$

Next, substitute this number into Equation 12.7 and calculate the corrected value (X_c):

$$X_c = X\left(\frac{V_{actual}}{1.84}\right)^{-0.25} = 8.0\left(\frac{3.0}{1.84}\right)^{-0.25} = 7.08$$

Discussion: A quick look at Figure 12-2 shows that a lower value of dimensionless quantity X (e.g., the value X_c), when combined with a given value of dimensionless quantity Y, will yield a slightly higher solar fraction for the system, all other quantities being the same.

Parametric studies have shown that increasing the storage volume per unit of collector area beyond 3.0 gallons per gross square foot of collector yields very insignificant gains in solar fraction. Often these gains do not justify the added cost of using larger storage volumes.

LOAD HEAT EXCHANGER SIZE CORRECTION FOR (Y)

Within the f-Chart method, the term **load heat exchanger** refers to the device(s) that transfer heat from the water supplied by the solar collectors to heated space within the building. In this text, the term "heat emitter" has been used for such device(s). Examples of load heat exchangers (a.k.a. heat emitters) include a single large air handler or multiple smaller air handlers, all operating simultaneously under design load conditions. Other examples include a system of panel radiators, a radiant ceiling panel, or a radiant floor panel. Although such devices are often used within zones, f-Chart considers the load heat exchanger to be the *total of all zoned heat emitters operating simultaneously*, as is often assumed under design load conditions.

The importance of designing hydronic heating distribution systems supplied by solar collectors, and other renewable heat sources for low supply water temperatures, has been emphasized throughout this text. The greater the total surface area of the heat emitters, the lower the system's supply water temperature will be for a given rate of heat transfer. This relationship is implicit in how f-Chart defines the **load heat exchanger size**. Mathematically, the load heat exchanger size it is defined by Equation 12.8.

(Equation 12.8)

$$LHES = \frac{e_{load}(C_{min})}{UA_b}$$

where:

$LHES$ = load heat exchanger size (unitless)

e_{load} = effectiveness of the load heat exchanger (unitless)

C_{min} = minimum fluid capacitance rate of the load heat exchanger (Btu/hr./°F)

UA_b = heating load coefficient of the building (Btu/hr./°F)

The numerator of Equation 12.5 indicates the ability of the load heat exchanger to transfer heat. The larger its value, the lower the supply water temperature can be for a given rate of heat transfer. The denominator in Equation 12.5 indicates the building's ability to loss heat through its thermal envelope. The smaller its value, the lower the rate of heat loss from the building for a given difference between the inside and outside air temperature. The greater the ability of the load heat exchanger to transfer heat in comparison to the ability of the building to loss heat, the lower the supply water temperature to the distribution system can be. Lower distribution system temperatures imply lower collector operating temperatures and thus higher thermal efficiencies. This relationship was discussed in Chapter 8.

For the simplified version of f-Chart, the LHES can be calculated using Equation 12.9.

(Equation 12.9)

$$LHES_s = \frac{Q/\Delta T}{UA_b}$$

where:

$LHES_s$ = load heat exchanger size for the simplified method (unitless)

Q = rate of heat output from the hydronic distribution system at design load (Btu/hr.)

ΔT = difference between average water temperature of the distribution system and the room air temperature at design load conditions (°F)

UA_b = heat load coefficient of the building (Btu/hr./°F)

Example 12.2: A hydronic radiant ceiling panel in a building has an overall area of 1,200 square feet. When operating with an average water temperature of 120 °F, it can release 36 Btu/hr./ft.^{2} into the building when the interior air temperature is 70 °F. The heat loss of the building is 24,000 Btu/hr. when the inside temperature is 70 °F and the outside temperature is 10 °F. Determine the value of $LHES_s$ for this situation.

Solution: The output of the full radiant ceiling area is:

$$Q = \left(36\frac{\text{Btu}}{\text{hr.}\cdot\text{ft.}^2}\right)(1{,}200\text{ ft.}^2) = 43{,}200\frac{\text{Btu}}{\text{hr.}}$$

The ΔT between the average water temperature in the radiant ceiling panel and the room air is (120 °F − 70 °F) = 50 °F:

The UA_b of the building is:

$$UA_b = \frac{24{,}000\text{ Btu/hr.}}{(70°\text{F} - 10°\text{F})} = 400\frac{\text{Btu}}{\text{hr.}\cdot°\text{F}}$$

Substituting these values into Equation 12.5 and calculating yields:

$$LHES_s = \frac{(Q/\Delta T)}{UA_b} = \left[\frac{\left(\dfrac{43{,}200\frac{\text{Btu}}{\text{hr.}}}{50°\text{F}}\right)}{400\frac{\text{Btu}}{\text{hr.}\cdot°\text{F}}}\right] = 2.16$$

Discussion: The value of $LHES_s$ in this example is slightly higher than the default value of 2.0 assumed in the f-Chart method. Increasing the value of $LHES_s$ will increase the solar heating fraction (f), all other conditions being the same.

The dimensionless quantity Y of Equation 12.2 can be corrected for load heat exchanger sizes that are different from the 2.0 value assumed in f-Chart. The corrected value of Y can be determined using Equation 12.10.

(Equation 12.10)

$$Y_c = Y\left[0.39 + 0.65e^{\left(\frac{-0.139}{LHES_s}\right)}\right]$$

where:

Y_c = *corrected* value of dimensionless quantity Y based on actual load heat exchanger size

Y = dimensionless quantity Y based on standard (assumed) load heat exchanger size of 2.0

e = 2.718281828 (the base of the natural logarithm system)

Example 12.3: A well-insulated house has its entire interior ceiling area covered by a hydronic radiant panel. The load heat exchanger size calculated using Equation 12.9 is 5.0. The dimensionless quantity Y for the system was determined to be 1.0 using Equation 12.2. Determine the *corrected* value of Y (e.g., Y_c) based on the larger value of $LHES_s$.

Solution: Just substitute the value of LHESs into Equation 12.10 and calculate.

$$Y_c = Y\left[0.39 + 0.65e^{\left(\frac{-0.139}{LHES_s}\right)}\right]$$

$$= 1.0\left[0.39 + 0.65e^{\left(\frac{-0.139}{5.0}\right)}\right] = 1.022$$

Discussion: For a given value of the dimensionless quantity X, an increase in the dimensionless quantity Y will increase the solar heating fraction.

From a mathematical standpoint, the optimum value of the load heat exchanger size (LHES) is infinity. This would imply an infinitely large heat emitter system relative to the heat loss rate of the building, a scenario that's nice to contemplate but impossible to achieve. From a practical standpoint, when the value of (*LHES*) is 10 or more, the thermal performance of the system will be essentially the same as if the value of *LHES* were infinity, and thus no further improvement in performance would be seen by increasing the size of the load heat exchanger for that building. The f-Chart method was developed with a *LHES* value of 2.0, which represents a "reasonably generous" relationship between the size of the heat emitters in comparison to the heat loss of the building. Increasing the value of *LHES* above 2.0 produces a slight gain in solar heating fraction. Decreasing the *LHES* below 2.0 produces a relatively quick drop in the value of Y, and a corresponding drop in the solar heating fraction (f).

PUTTING IT ALL TOGETHER

The best way to fully understand the simplified f-Chart procedure is to use it to determine the monthly solar fraction for a given system. This will be demonstrated in Example 12.4.

Example 12.4: A house in Syracuse, NY, has a design heating load of 21,000 Btu/hr. when the inside air temperature is 70 °F and the outdoor air temperature is 0 °F. The occupants use 50 gallons of domestic hot water each day that must be heated from 50 °F to 120 °F. During an average February, Syracuse experiences 1,131 °F·days. The house is fitted with an array of five 4 foot × 8 foot single glazed flat-plate collectors, facing true south and sloped at local latitude (43°) plus 15° = 58°. The thermal storage tank has a volume of 400 gallons. The collector circuit operates at a flow rate of 1 gpm per collector (5 gpm total), using a 50 percent solution of propylene glycol antifreeze. The SRCC rating certificate for the collectors used indicate a Y intercept of 0.72, and a slope of 0.865 Btu/hr./ft.²/°F. The collector-to-storage heat exchanger has an effectiveness of 0.7. The home is heated by a radiant ceiling panel that covers 1,500 square feet of ceiling area and can release 28 Btu/ft.²/hr. when operating at an average water temperature of 110 °F. Use the simplified f-Chart method to estimate the monthly solar fraction of the total heating load supplied by this system in February.

Solution: There will be many calculations required before the solar fraction (f) can be calculated.

Start by establishing the space heating energy use for February. Use the degree day method with an assumed Cd value of 0.9 for February:

$$UA_b = \left(\frac{21{,}000\dfrac{\text{Btu}}{\text{hr.}}}{70°\text{F} - 0°\text{F}}\right) = 300\frac{\text{Btu}}{\text{hr.}\cdot°\text{F}}$$

$$E_{\text{heating}} = \left(300\frac{\text{Btu}}{\text{hr.}\cdot°\text{F}}\right)(1131°\text{F}\cdot\text{day})\left(\frac{24\text{ hr.}}{\text{day}}\right)0.9$$
$$= 7{,}329{,}00\text{ Btu}$$

Next, calculate the energy used for domestic water heating in February:

$$E_{\text{DHW}} = \left(50\frac{\text{gal}}{\text{day}}\right)\left(8.33\frac{\text{Btu}}{\text{gal}\cdot°\text{F}}\right)(120\text{ °F} - 50\text{ °F})(28\text{ day})$$
$$= 816{,}340\text{ Btu}$$

The total heating energy requirement for February is the sum of the energy required for space heating and that required for domestic water heating: 7,329,000 + 816,340 = 8,145,340 Btu. This is the value of (L) for Equations 12.1 and 12.2.

From the collector's SRCC certification report: $F_R(\tau\alpha)_n = 0.72$

From the collector's SRCC certification report: $F_RU_L = 0.865$

The gross area of the each collector is 4 ft. × 8 ft. = 32 ft.². The total collector array area is $A_c = 5 \times 32 = 160$ ft.².

The assumptions associated with estimating the value of G do apply, therefore $G = 0.96$.

The average outdoor air temperature for February can be calculated based on the accumulated degree days using Equation 12.4:

$$T_u = 65 - \frac{(DD_{65})_m}{N} = 65 - \frac{1131}{28} = 24.6°\text{F}$$

The hours in February are 24 × 28 = t_m = 672 hr.

Based on the stated system configuration, the assumptions required for using Figure 12-3 to estimate the heat exchanger penalty factor do apply. Therefore, the heat exchanger performance penalty factor is read from Figure 12-3 as 0.975.

The NREL table of solar radiation data for Syracuse, NY (Figure 7-32), shows that on an average February day, a surface facing true south and sloped at latitude + 15° receives a total daily solar radiation of 3.6 kWh/m²/day. This needs to be converted to Btu/ft.²/day.

$$\left(\frac{\text{kWh}}{\text{m}^2 \cdot \text{day}}\right)(317) = \left(\frac{\text{Btu}}{\text{ft.}^2 \cdot \text{day}}\right) = \left(\frac{3.6\text{kWh}}{\text{m}^2 \cdot \text{day}}\right)(317)$$

$$= 1{,}141\frac{\text{Btu}}{\text{ft.}^2 \cdot \text{day}}$$

There is now sufficient information to use Equations 12.1 and 12.2 to calculate the dimensionless quantities X and Y:

$$X = (F_R U_L)\left(\frac{F'_R}{F_R}\right)(212 - \overline{T}_a)(t_m)\left(\frac{A_c}{L}\right)$$

$$= (0.865)(0.975)(212 - 24.6)(672)\left(\frac{160}{8{,}145{,}340}\right)$$

$$= 2.09$$

$$Y = (F_R(\tau\alpha)_R)\left(\frac{F'_R}{F_R}\right)\left(\frac{\overline{(\tau\alpha)}}{(\tau\alpha)_n}\right)(\overline{H_T})N\left(\frac{A_c}{L}\right)$$

$$= (0.72)(0.975)(0.96)(1141)28\left(\frac{160}{8{,}145{,}340}\right) = 0.42$$

The values of X and Y are now established for the storage volume and load heat exchanger size assumed in f-Chart. However, *these X and Y values still need to be corrected for the actual storage size used in the system and the actual load heat exchanger size used in the building.*

The corrected value of X (e.g., X_c) is found using Equation 12.7. This requires the ratio of storage volume to collector area used in the system. That ratio is 400 gallons/160 ft.² = 2.5 gallons/ft.².

$$X_c = X\left(\frac{V_{\text{actual}}}{1.84}\right)^{-0.25} = 2.09\left(\frac{2.5}{1.84}\right)^{-0.25} = 1.936$$

The corrected value of Y (e.g., Y_c) is found using Equation 12.10; however, the value of LHESs must first be calculated using Equation 12.9.

Using Equation 12.9, with the total output of the radiant ceiling at design load conditions being (28 Btu/hr./ft.²) × 1,500 ft.² = 42,000 Btu/hr. and the ΔT being the average water temperature minus the room air temperature (110 °F − 70 °F) = 40 °F.

$$LHES_s = \frac{(Q/\Delta T)}{UA_b} = \left[\frac{\left(\dfrac{42{,}000\dfrac{\text{Btu}}{\text{hr.}}}{40°\text{F}}\right)}{300\dfrac{\text{Btu}}{\text{hr.} \cdot °\text{F}}}\right] = 3.5$$

Using Equation 12.10 to determine Y_c:

$$Y_c = Y\left[0.39 + 0.65e^{\left(\frac{-0.139}{LHES_l}\right)}\right]$$

$$= 0.42\left[0.39 + 0.65e^{\left(\frac{-0.139}{3.5}\right)}\right] = 0.426$$

The corrected values of (X_c and Y_c) are now used in Equation 12.3 to determine the monthly solar fraction (f).

$$f = 1.029Y - 0.065X - 0.245Y^2 + 0.0018X^2 + 0.0215Y^3$$
$$= 1.029(0.426) - 0.065(1.936) - 0.245(0.426)^2$$
$$+ 0.0018(1.936)^2 + 0.0215(0.426)^3 = 0.276$$

Another option is to use Figure 12-2 with the values of X_c and Y_c to find f, as shown in Figure 12-4.

Discussion: The simplified f-Chart method predicts that the specified solar combisystem will deliver a solar heating fraction of 0.276 (e.g., 27.6 percent) of the total

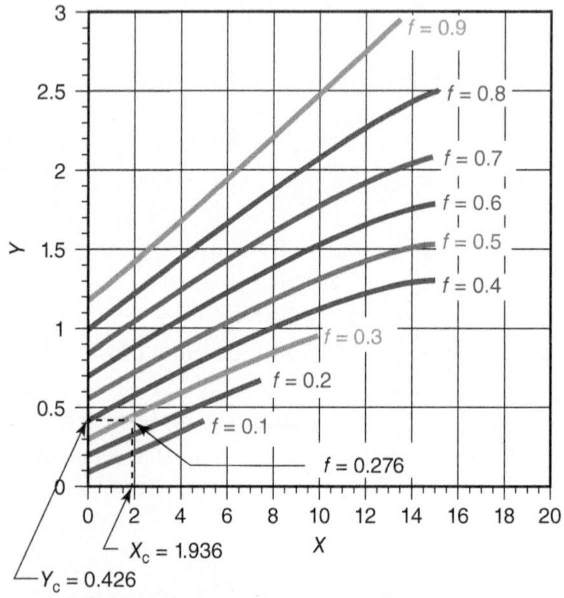

Figure 12-4 | Use of the f-Chart (Figure 12-2) with values of X_c and Y_c to find solar fraction (f) = 0.276.

space heating and domestic water-heating load during February. This is a significant contribution for this month in this cold, northern, and relatively cloudy location. It is partially attributable to a low space heating load and a low-temperature hydronic heat delivery system.

This example also demonstrates that manual calculations for executing the f-Chart method, even the simplified version, can be tedious. The next section demonstrates software that allows much faster execution of a wider ranging f-Chart method.

12.3 Using F-CHART Software

As previously mentioned, the f-Chart method was developed before personal computers were available. The previous section demonstrated that it is possible to execute the f-Chart method using a calculator. Still, few present-day system designers would be willing to spend the necessary time to manually calculate system performance based on one set of assumptions for all the variables involved. Even fewer would take the time to perform multiple sets of manual calculations to assess effects such as different collector areas, collector slopes, storage tank volumes, etc. Therefore, computer-based execution of the f-Chart method is now standard practice. Software appropriately named **F-CHART** is currently available as a Windows®-based tool from www.fchart.com.

F-CHART INPUTS

This section examines the inputs and results associated with using the Windows version of F-CHART software. More detailed information for all inputs and outputs can be found in the user manual, which is also downloadable for free at www.fchart.com.

Figure 12-5 shows the opening screen of F-CHART. The pull-down menus are visible at the top of the screen, as are the three floating input windows, which can be dragged around the main screen area.

There are several pull-down menus at the top of the F-CHART user interface. Some are used for inputs,

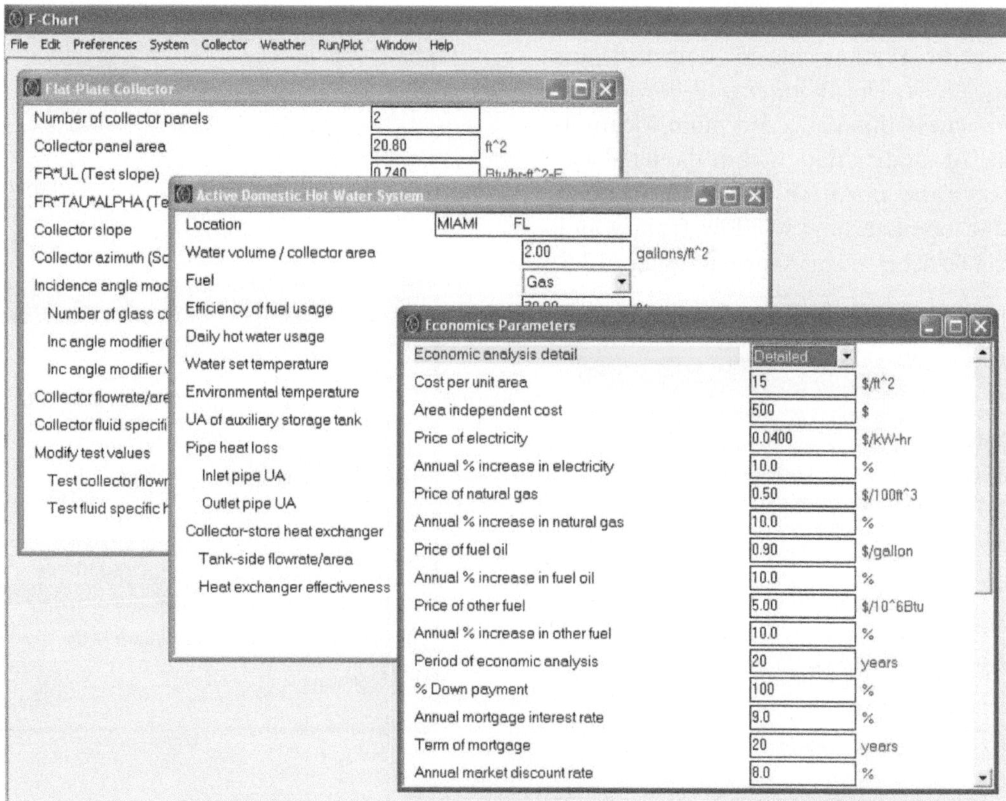

Figure 12-5 | Opening screen of F-CHART, showing the menu bar at top and three floating input windows.

others are used to display results, manage files, copy and paste information, or access the help system.

The ***preferences*** pull-down in F-CHART is used to configure the software for either metric (e.g., SI units), or standard North American (e.g., English units). *This configuration should be made before entering values for other inputs.*

The capabilities of F-CHART have been significantly expanded since the program's original offering. The types of systems it can analyze are shown in the ***system*** pull-down menu, seen in Figure 12-6. For solar combisystems, select *Water Storage and DHW* from this pull-down menu.

The type of collector used should also be selected before going on to other inputs. Figure 12-7 shows the open ***collector*** pull-down menu, with the types of collectors F-CHART can model.

For the systems discussed in this text, either flat-plate or evacuated tube collectors are the likely choices. F-CHART models both flat-plate and evacuated tube collectors based on the values input for the Y intercept and slope of the collector's efficiency line. The collector's **incident angle modifier** can also be specified.

The current version of F-CHART has a large internal database of weather data for both U.S. and international locations. The ***Weather*** pull-down menu can be used to access this data, add more locations to the database, or modify data within the database. Choosing *Select Data Location* under the ***Weather*** pull-down menu opens a new window from which weather data can be accessed, as shown in Figure 12-8.

Figure 12-6 | System-type selections pull-down menu from F-CHART. Select *Water Storage and DHW* option to analyze solar combisystems.

The data stored for each location includes **monthly average solar radiation on a horizontal surface**, average outdoor temperature, absolute humidity, cold water temperature in municipal water mains, **ground reflectivity**, and **°F-days**. Figure 12-9 shows an example of the default weather data for Albany, NY, in tabular form, which can be edited if necessary.

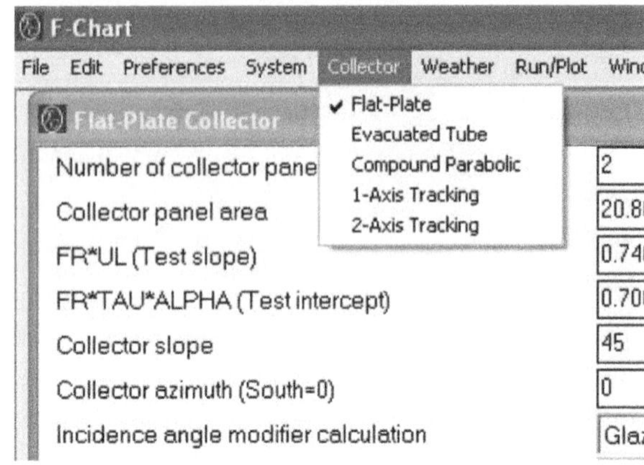

Figure 12-7 | Collector type selections pull-down menu from F-CHART.

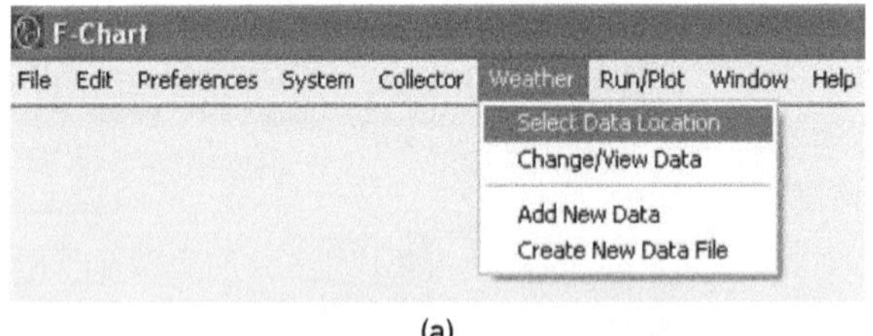

(a)

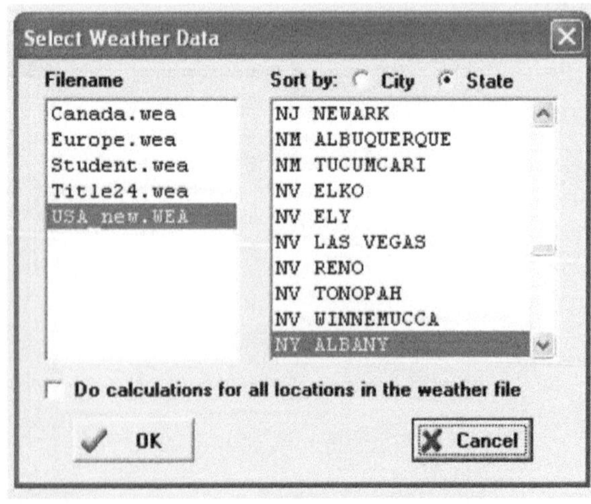

(b)

Figure 12-8 | Selecting *Select Weather Data* under the Weather menu opens a new window from which to select the location of the system.

After the pull-down menus have been used to select units, system type, collector type, and weather data, the user can move to more detailed inputs. Figure 12-10 shows the three data input windows that open when the F-CHART is launched.

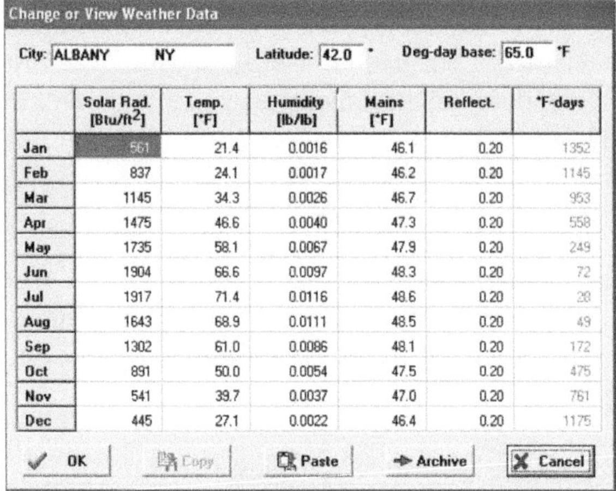

Change or View Weather Data

City: ALBANY NY Latitude: 42.0 ° Deg-day base: 65.0 °F

	Solar Rad. [Btu/ft²]	Temp. [°F]	Humidity [lb/lb]	Mains [°F]	Reflect.	°F-days
Jan	561	21.4	0.0016	46.1	0.20	1352
Feb	837	24.1	0.0017	46.2	0.20	1145
Mar	1145	34.3	0.0026	46.7	0.20	953
Apr	1475	46.6	0.0040	47.3	0.20	558
May	1735	58.1	0.0067	47.9	0.20	249
Jun	1904	66.6	0.0097	48.3	0.20	72
Jul	1917	71.4	0.0116	48.6	0.20	28
Aug	1643	68.9	0.0111	48.5	0.20	49
Sep	1302	61.0	0.0086	48.1	0.20	172
Oct	891	50.0	0.0054	47.5	0.20	475
Nov	541	39.7	0.0037	47.0	0.20	761
Dec	445	27.1	0.0022	46.4	0.20	1175

OK Copy Paste Archive Cancel

Figure 12-9 | Example of the weather data contained in F-CHART for Albany, NY.

One window is for collector information, another for system configuration information, and the third for information needed for economic analyses. The titles displayed at the top of these windows change based on the type of collector and system previously selected in the pull-down menus. Each of these windows opens with **default values** or selections, all of which can be changed as needed. If an economic analysis is not required, or the user just wants to investigate the thermal performance of systems, the *"Economics Parameters"* window can be temporarily closed.

The designer uses the collector input window to specify collector performance by entering values for the Y intercept ($F_R\tau\alpha$) and slope (F_RU_L) of the collector's instantaneous efficiency line, as discussed in Chapter 8. This window also allows the collector array to be specified in terms of gross collector area, number of collectors, slope angle, and azimuth angle. Inputs are also provided for the incident angle modifier. When flat-plate collectors are selected, the incident angle modifier can be specified in three possible ways: the number of glazings, the b_0 constant, or a table that

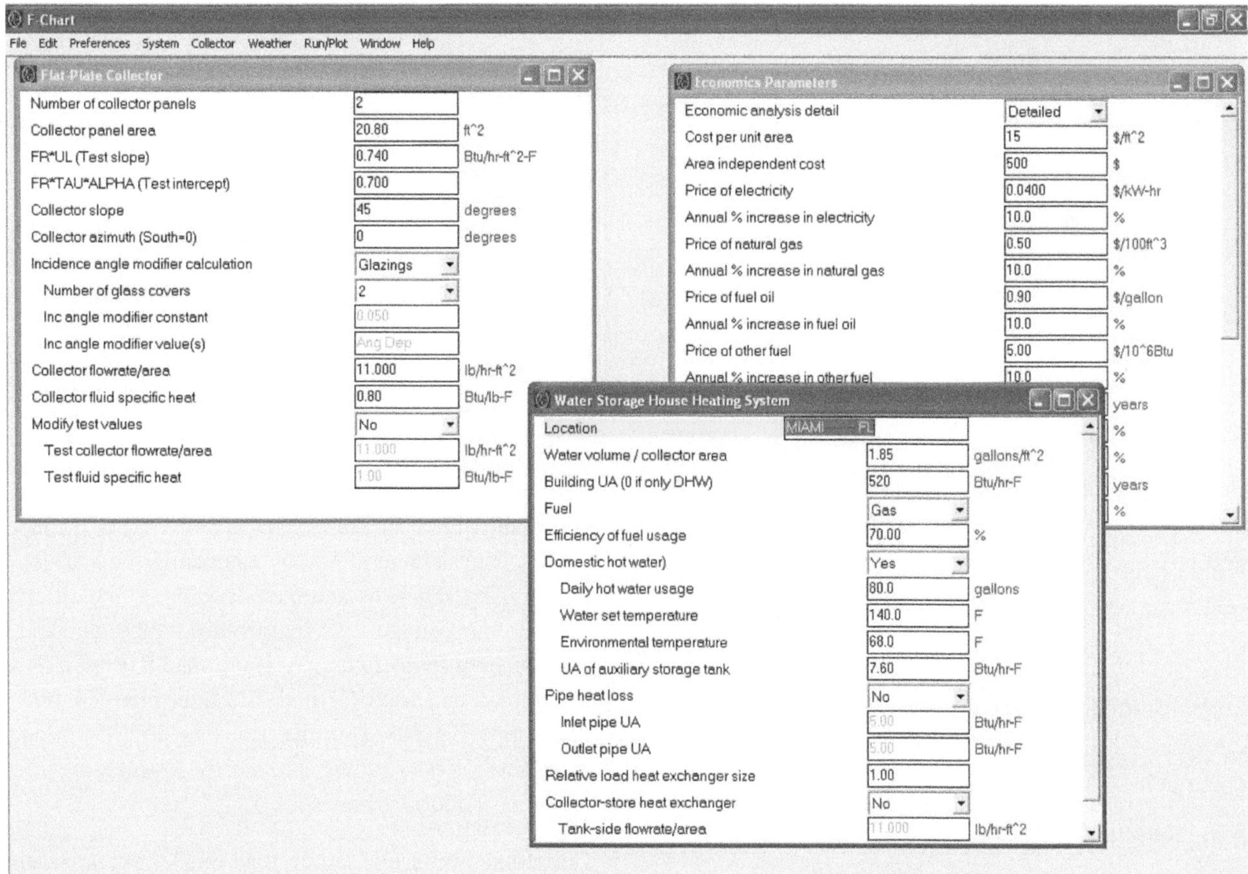

Figure 12-10 | F-CHART input windows configured for a liquid-type combisystem application.

associates a value of k for each 10° increment of the incident angle, as found on an SRCC rating certificate. When evacuated tube collectors are selected, the incident angle modifier can be specified in planes that are parallel and perpendicular to the long axis of tubes. Additional inputs are provided for specifying the collector's flow rate and the specific heat of the fluid used in the collector circuit.

Another window accepts information about the loads and balance of system. The size of the thermal storage tank is entered as gallons of storage per square foot of collector. The performance of the **collector-to-storage heat exchanger** is specified by its effectiveness, and the system's location is selected from the built-in database.

The Economics Parameters window accepts information for the cost of the system, the cost of conventional fuels, and how the system will be financed.

Example 12.5: Prepare the F-CHART inputs for a technical analysis of the following solar combisystem in Glens Falls, NY:

Hardware configuration:

- Drainback-protected system with five single-glazed collectors, each measuring 4 ft. × 8 ft.
- Y intercept of collector efficiency equation $(F_R\tau\alpha) = 0.70$
- Slope of collector efficiency equation $(F_R U_L) =$ 0.714 Btu/hr./ft.²/°F
- Collector array facing true south (Azimuth = 0°)
- Collector slope = 60°
- Steady state flow rate through each collector = 1.0 gpm
- Collector supply and return piping is 3/4-inch copper covered with 3/4-inch-thick foam rubber insulation
- Fluid circulated through collectors is water (specific heat = 1.0 Btu/lb./°F)
- Storage tank volume = 300 gallons
- Load information:
- Nearest weather data location (available within F-CHART) for Glens Falls, NY, is Albany, NY
- Building heat loss is 25,000 Btu/hr., based on 70 °F inside and −5 °F outdoor air temperatures
- Building has 1,500 square feet of the low-temperature radiant ceiling panel, as described in Chapter 5. The performance of this radiant panel heat emitter is given by Figure 5-81
- Domestic water load is 60 gallons per day heated from the local water main temperature to 115 °F

Solution: Some of the stated values, such as collector slope and azimuth, the Y intercept and slope of the collector efficiency equation, and the specific heat of the collector fluid, can be directly entered into the collector window of F-CHART. However, some of the remaining stated values must be converted into units compatible with F-CHART before they can be entered.

One of those values is collector flow rate. F-CHART requires collector flow rate to be entered in units of (lb./hr./ft.²) of collector. The following calculation can be used to convert to the required units:

$$\frac{\left(1\frac{\text{gallon}}{\text{min.}}\right)\left(\frac{60\text{ min.}}{\text{hr.}}\right)\left(\frac{8.33\text{ lb}}{\text{gallon}}\right)}{32\text{ ft.}^2} = 15.6\frac{\text{lb}}{\text{hr.}\cdot\text{ft.}^2}$$

F-CHART also requires the storage tank volume to be expressed in units of gallons per square feet of collector. This can be easily calculated from the stated number of collectors, their size, and the given storage tank volume:

$$\frac{300\text{ gallon}}{(5\text{ collectors})\left(32\frac{\text{ft.}^2}{\text{collector}}\right)} = 1.875\frac{\text{gallon}}{\text{ft.}^2}$$

The **building UA** is determined by dividing the design heat loss of the building by the temperature difference between the inside and outside air temperature at design load.

$$Building\ UA = \frac{25{,}000\text{ Btu/hr.}}{(70-(-5))°\text{F}} = 333.3\text{ Btu/hr./°F}$$

The **inlet pipe UA** and **outlet pipe UA** entries are for describing the heat loss characteristics of the collector piping. Each is determined by estimating the heat loss of those piping segments and then dividing by the difference between the average fluid temperature in the pipe and the outdoor area temperature. A value of 5.8 Btu/hr./°F was determined and used for both the Inlet pipe UA and the outlet pipe UA based on the heat loss of 50-foot lengths of 3/4-inch copper tubing covered by 3/4-inch-thick foam rubber insulation.

One final numerical input that must be prepared for F-CHART is called the ***relative load heat exchanger size***. For combisystems with hydronic heat emitters, this can be determined as follows:

1. Calculated the heat output from the hydronic distribution system, assuming all zones are operating, under design load conditions.
2. Divide the result of step 1 by the difference between the *average* water temperature in the distribution system and the room air temperature, both at design load conditions. The resulting number will have units of Btu/hr./°F.
3. Divide the result of step 2 by the *"building UA,"* which has already been determined for one of the inputs.
4. ***Divide the result of step 3 by 2.0 (because the f-Chart method was originally developed assuming a load heat exchanger size of 2.0).***

For the low-temperature radiant ceiling described in Chapter 5, steps 1 and 2 are completed in the following calculation, which uses information from the graph in Figure 5-81:

$$\frac{\left(36\frac{\text{Btu}}{\text{hr.}\cdot\text{ft.}^2}\right)}{(50°\text{F})}(1{,}500\ \text{ft.}^2) = 1{,}080\frac{\text{Btu}}{\text{hr.}\cdot°\text{F}}$$

Dividing this by the building UA yields:

$$\frac{1{,}080\frac{\text{Btu}}{\text{hr.}\cdot°\text{F}}}{333.3\frac{\text{Btu}}{\text{hr.}\cdot°\text{F}}} = 3.24$$

Finally, dividing this by 2.0 yields the *relative* load heat exchanger size needed by F-CHART:

Relative load heat exchanger size = 3.24/2 = 1.62

The values are now ready to be entered into their respective locations on the F-CHART input windows. Figure 12-11 shows all values entered into their proper input locations.

Discussion: As is true with any software, F-CHART requires gathering and preparation of data. Much of this is straightforward arithmetic that in some cases is combined with units conversion. The program is now ready to run. Users should refer to the F-CHART user manual for complete descriptions of inputs.

F-CHART OUTPUTS

After all inputs have been configured to describe the system, the software is run by pressing the F2 key. The results are quickly calculated and displayed in the ***Thermal Output window***, as shown in Figure 12-12.

The *Thermal Output* window lists several values calculated on a monthly basis. The first column, titled *"Solar,"* lists the total monthly solar radiation incident on the collector array (e.g., at the array's specified slope and azimuth), expressed in units of **MMBtu** (1 MMBtu = 1,000,000 Btu). The second column, titled *"Heat,"* is the monthly total space heating demand of

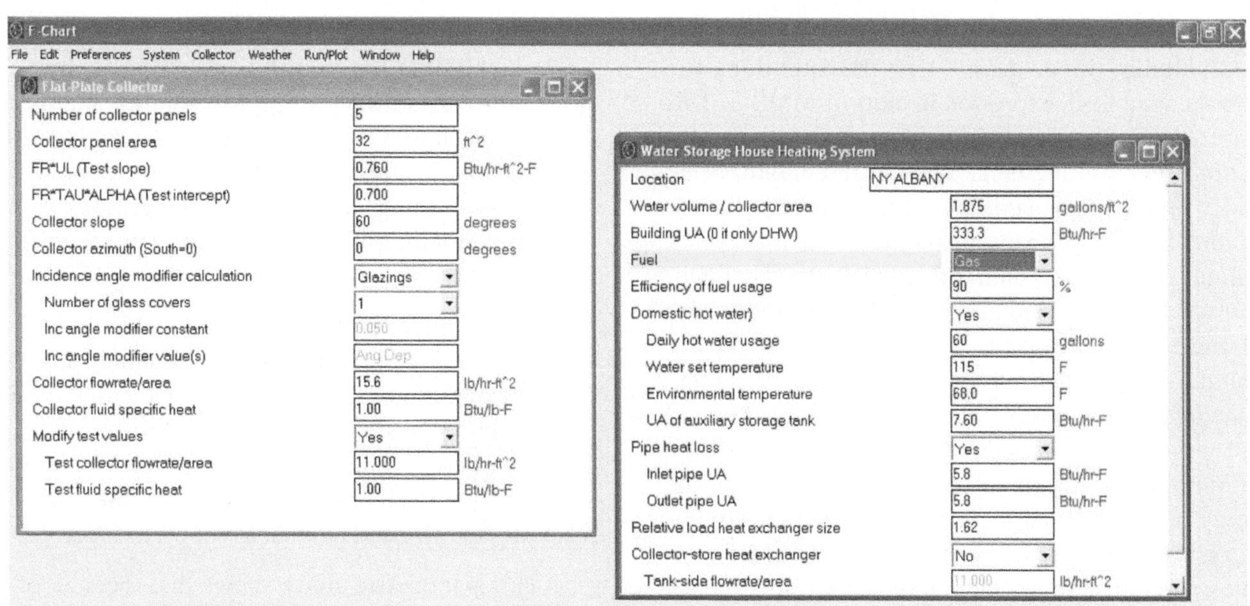

Figure 12-11 | All inputs for Example 12.5 entered into the appropriate F-CHART input windows.

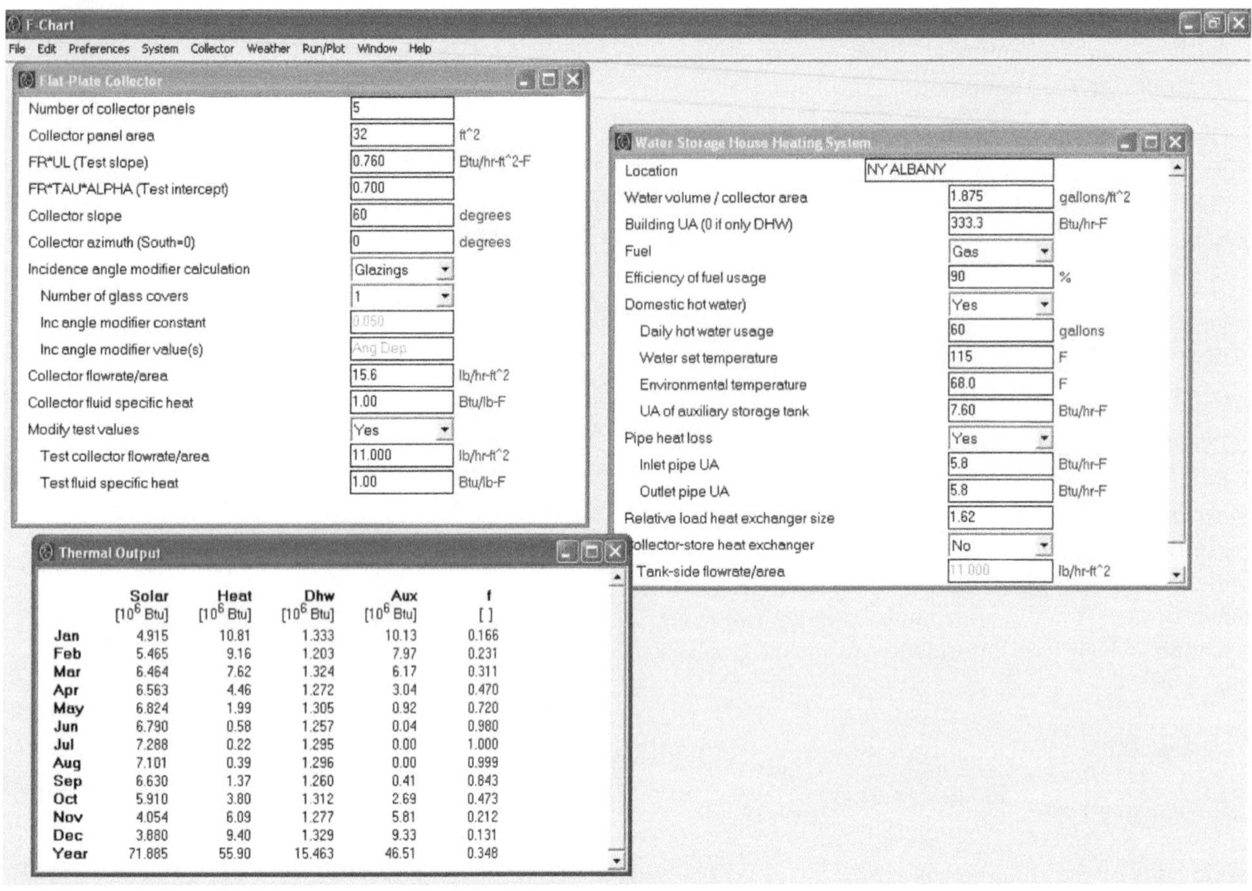

	Solar [10^6 Btu]	Heat [10^6 Btu]	Dhw [10^6 Btu]	Aux [10^6 Btu]	f [ ]
Jan	4.915	10.81	1.333	10.13	0.166
Feb	5.465	9.16	1.203	7.97	0.231
Mar	6.464	7.62	1.324	6.17	0.311
Apr	6.563	4.46	1.272	3.04	0.470
May	6.824	1.99	1.305	0.92	0.720
Jun	6.790	0.58	1.257	0.04	0.980
Jul	7.288	0.22	1.295	0.00	1.000
Aug	7.101	0.39	1.296	0.00	0.999
Sep	6.630	1.37	1.260	0.41	0.843
Oct	5.910	3.80	1.312	2.69	0.473
Nov	4.054	6.09	1.277	5.81	0.212
Dec	3.880	9.40	1.329	9.33	0.131
Year	71.885	55.90	15.463	46.51	0.348

Figure 12-12 F-CHART *Thermal Output* window (seen in lower left) for the system specified in Example 12.5. Input information is seen in the background windows.

the building, also expressed in units of MMBtu. The third column, titled *"Dhw,"* lists the total domestic water-heating load, expressed in units of MMBtu. This load includes the standby heat loss from the domestic hot water storage tank. The fourth column, titled *"Aux,"* lists the total auxiliary energy needed to satisfy the total load (e.g., space heating plus domestic water heating). The final column, titled *"f,"* is the fraction of the total load the system is estimated to supply with solar energy, expressed as a decimal percentage. For example, an f value of 0.170 indicates that 17 percent of the *total* (space heating plus domestic water heating) load was supplied by solar energy. The values at the bottom of each column are annual summations of the energy quantities, and the **annual solar fraction**. For the system modeled in Figure 12-12, F-CHART estimates an annual solar fraction of 0.348. This means that 34.8 percent of the total annual space heating plus domestic hot water load is estimated to be supplied by solar energy.

The user can now make changes to any of the inputs and quickly find the resulting changes in output. The user can also perform **parametric studies** on a specific system by varying just one input and generating results that shows the effect of this one parameter on overall system performance. This is very helpful for gaining an understanding of which changes have significant versus minor effects on overall system performance.

For example, multiple F-CHART runs were made for the combisystem described in Example 12.5. Only the size of the storage tank was varied from a low of 1.0 gallon per square foot of collector to a high of 3.0 gallons per square foot of collector. The resulting changes in annual solar fraction were determined for each storage size and plotted, as shown in Figure 12-13.

This parametric study shows that the size of the storage tank, within the range of 1.0 to 3.0 gallons of water per square foot of collector, has relatively little effect on annual solar fraction. The large thermal

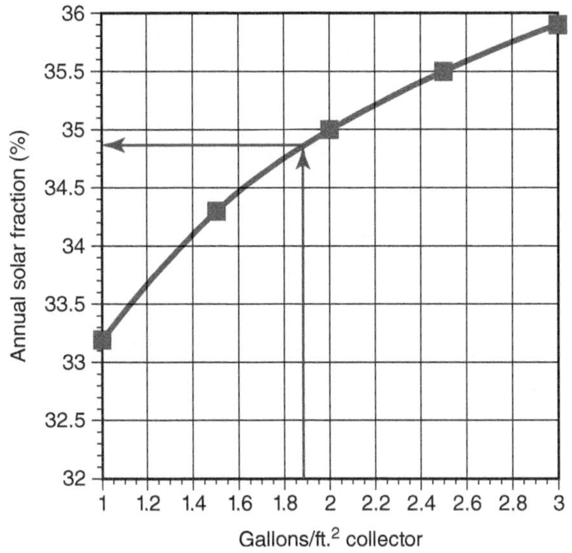

Figure 12-13 The effect of varying the ratio of storage volume per square foot of collector area for the combisystem described in Example 12.5. All other inputs remained unchanged.

storage tank (3.0 gallons per square foot of collector), which translates to a 480-gallon tank in the example system, does increase annual solar fraction from about 34.8 percent to 35.9 percent. Likewise, reducing the tank volume to 1.0 gallons per square foot of collector, which translates to a 160-gallon tank for the example system, lowers the annual solar fraction to about 33.2 percent. These changes are small compared to the potential logistic and cost differences involved with installing tanks of significantly different size.

F-CHART can also generate graphs showing one or more of the calculated results on a monthly basis. Figure 12-14 shows the monthly solar fraction for the system in Example 12.5. Figure 12-15 graphs monthly values for incident solar radiation, space heating load, domestic water-heating load, and auxiliary energy. Both of these graphs were quickly generated using F-CHART's ***plot*** pull-down menu.

Figure 12-14 shows that the system will meet (and likely exceeded) a 100 percent solar fraction during an average July. Unfortunately, F-CHART does not show how much excess energy was available in months where the solar fraction exceeds 100 percent.

Practical solar combisystem designs minimize **excess energy gain** in summer. *If F-CHART shows more than one summer month where the solar heating fraction is over 100 percent, the system is probably oversized.* This could lead to increased operation of over-temperature protection subsystems. It also decreases the economic return of the system relative to that obtained with systems that do not produce excess heat in summer.

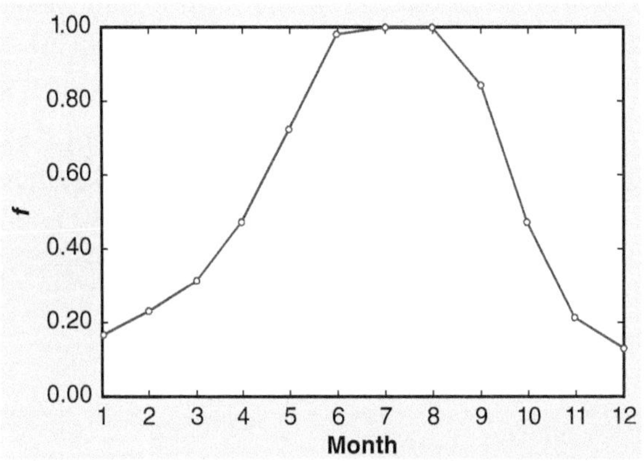

Figure 12-14 Monthly solar fraction for a solar combisystem in Albany, NY, as described in Example 12.5.

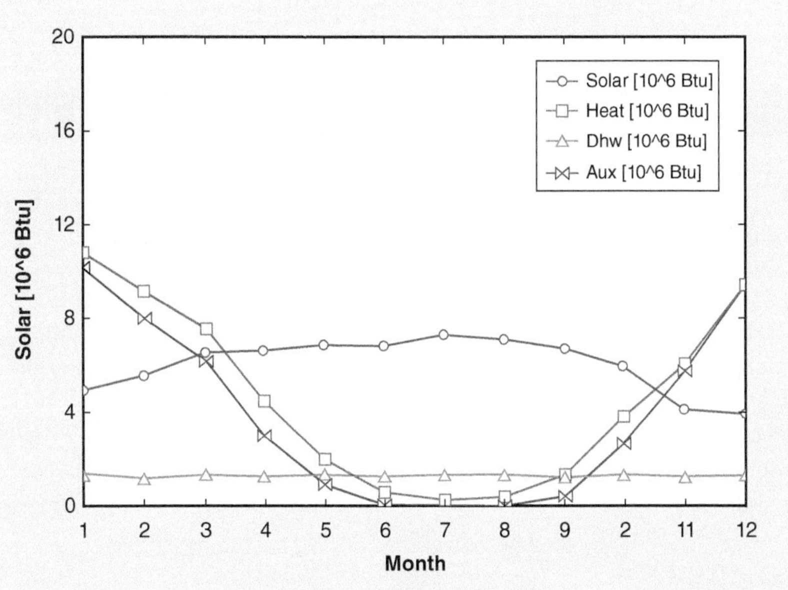

Figure 12-15 Monthly values for incident solar radiation, space heating load, domestic water-heating load, and auxiliary energy for system in Example 12.5.

SUMMARY

This chapter has provided an overview of the f-Chart method for assessing the monthly and annual thermal performance of solar thermal systems. The simplified version of the f-Chart method presented in Section 12.2 allows for manual calculations of monthly solar fractions. Example 12.4 demonstrated how this is done. Realistically, any heating professional involved in the design of solar thermal systems will likely choose to use a software-based version of the f-Chart method rather than manual calculations. Still, it is important to see the principles and mathematics underlying the f-Chart method.

Section 12.3 gave an overview of commercially available F-CHART software. Readers are encouraged to download and read the user manual for the current version of F-CHART for more detailed explanations of its capabilities. Those with access to F-CHART software are encouraged to perform parametric studies that help in understanding how sensitive a given design factor is to overall system performance. Those who would like to see a full derivation of the f-Chart method are encouraged to read the first reference listed below.

Finally, the calculations performed by the f-Chart method, as well as other sizing methods, assume that the collectors are clean and free of snow or other materials that could block sunlight. They also assume that the collectors are unshaded for the specified collector mounting position. Performance results will obviously be lower if shading is present. Accumulating dirt will also *slightly* degrade thermal performance as collectors age. Typical losses are in the range of 1 to 2 percent.

ADDITIONAL RESOURCES

The following references are recommended for additional reading on analysis of solar thermal systems using the f-Chart method:

1. Duffie, John A., and William A. Beckman. (2013). *Solar Engineering of Thermal Processes,* 4th edition. John Wiley & Sons. ISBN 978-0-470-87366-3.
2. Beckman, William A., Sanford A. Klein, and John A. Duffie. (1977). *Solar Heating Design by the f-Chart Method.* John Wiley & Sons. ISBN 0 471 03406-1.

KEY TERMS

°F-days
annual solar fraction
building UA
collector-to-storage heat exchanger
default value
dimensionless quantity
effectiveness
empirical correlation
excess energy gain
F-CHART (software)
f-Chart method
fluid capacitance rate
ground reflectivity
heat exchanger penalty factor (F_R'/F_R)
incident angle modifier
inlet pipe UA
instantaneous efficiency line
load heat exchanger
load heat exchanger size
mainframe computer
MMBtu
monthly average solar radiation on a horizontal surface
monthly solar fraction (f)
outlet pipe UA
parametric study
physical quantity
relative load heat exchanger size
slope [of efficiency line]
SRCC rating certificate
thermal output window
TRNSYS
Y-intercept [of efficiency line]

QUESTIONS AND EXERCISES

1. Explain why estimating the monthly or annual heat delivery from a solar combisystem is more complex than a hydraulic analysis of the piping circuits within the system.

2. What does the "f" in f-Chart stand for?

4. Heat is delivered to a building through a low-temperature panel radiator system that can deliver 36,000 Btu/hr. into 70 °F internal space when operating with a supply water temperature of 120 °F, and a design temperature drop of 20 °F. The building's design heat loss is 25,000 Btu/hr. based on an inside temperature of 70 °F and an outdoor temperature of −5 °F. Determine the load heat exchanger size ($LHES_s$).

5. The dimensionless quantity X for the f-Chart method has been determined as 3.0 for a given month. Find the corrected value of this quantity (X_c), assuming the $LHES_s$ value calculated in Exercise 4. Assuming the system's Y value for that month is 1.0, estimate the system's monthly solar fraction using the corrected X value (X_c) and Figure 12-2.

6. Describe what the quantity (F'_R/F_R) does in the equations for the dimensionless quantities X and Y.

7. A given location experiences 1,250 °F·days during an average January. What is the average outdoor temperature at this location during that month?

8. The collector-to-storage heat exchanger in a solar combisystem operates at the conditions shown in Figure 12-16. Determine:
   a. The rate of heat transfer across the heat exchanger
   b. The effectiveness of the heat exchanger
   c. The heat exchanger penalty factor (F'_R/F_R) associated with this heat exchanger

9. A house in Syracuse, NY, has a design heating load of 30,000 Btu/hr when the inside air temperature is 70 °F and the outdoor air temperature is 0 °F. The occupants use 60 gallons of domestic hot water each day that must be heated from 50 °F to 120 °F. During an average February, Syracuse experiences 1,131 °F·days. The house is fitted with an array of six 4-foot × 8-foot single-glazed flat-plate collectors, facing true south and sloped at local latitude (43°) plus 15° = 58°. The thermal storage tank has a volume of 450 gallons. The collector circuit operates at a flow rate of 1 gpm per collector (5 gpm total), using a 50 percent solution of propylene glycol antifreeze. The SRCC rating certificate for the collectors used indicates a Y-intercept of 0.72 and a slope of 0.865 Btu/hr./ft.²/°F. The collector-to-storage heat exchanger has an effectiveness of 0.6. The home is heated by a radiant ceiling panel that covers 1,500 square feet of ceiling area and can release 28 Btu/ft.²/hr. when operating at an average water temperature of 110 °F. Use the simplified f-Chart method to estimate the monthly solar fraction of the total heating load supplied by this system in February.

10. What is the difference between TRNSYS software and F-CHART software?

11. List at least six quantities that can be taken from an SRCC OG-100 collector performance report for use as input data to F-CHART software.

12. A flat-plate collector measures 47-inches wide by 119-inches tall. Water passes through this collector at a flow rate of 1.25 gpm. Convert this flow rate into units of lb./hr./ft.².

13. A building for which a solar combisystem is being designed has a heat loss of 22,000 Btu/hr. when the

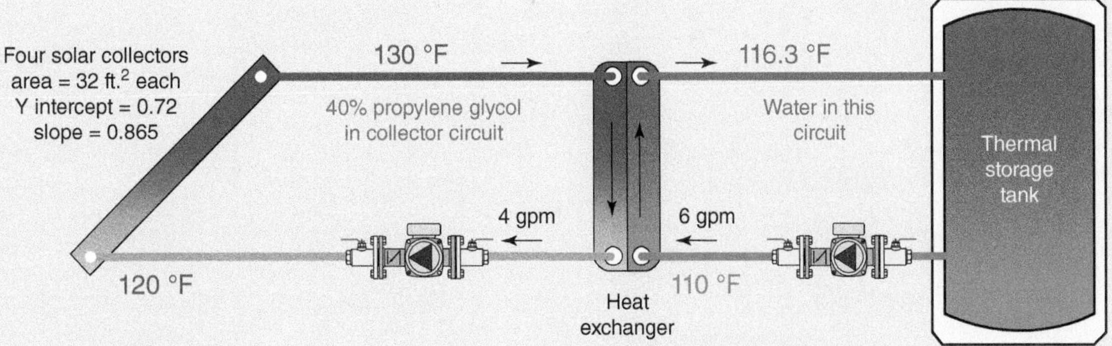

Figure 12-16 | Operating conditions for the collector-to-storage heat exchanger in Exercise 8.

inside temperature is 68 ºF and the outside design temperature is –5 ºF. Determine the "building UA" of this building.

14. Assume the building in Exercise 13 is equipped with a radiant ceiling heating system that covers 1,500 square feet of area. It can deliver 22,000 Btu/hr. when supplied with 115 ºF water and operating with a 20 ºF temperature drop. The inside air temperature is 68 ºF. Determine the *relative load heat exchanger size* for this situation.

The remaining exercises require the use of F-CHART software.

15. Use F-CHART software to determine the annual solar heating fraction for space heating plus domestic water heating for the following combisystem:
- system location: Denver, CO
- number of collectors: 5
- type of collector: single-glazed flat-plate collector
- collector size: 48 inches × 96 inches
- Y intercept of collector efficiency equation: 0.80
- slope of collector efficiency equation: 0.70
- collector flow rate: 0.85 gpm per collector
- collector fluid: water
- freeze protection: drainback with no heat exchanger between collectors and storage
- collector array tilt = local latitude +15º
- collector array azimuth: 170º
- storage tank volume: 350 gallons
- UA of collector supply piping = 5.0 Btu/hr./ºF
- UA of collector return piping = 4.0 Btu/hr./ºF
- building design load = 22,000 Btu/hr.
- building interior temperature at design load: 68 ºF
- outside air temperature at design load: –5 ºF
- building heat emitter: 1,500 square feet of radiant ceiling
- heat emitter output: 22,000 Btu/hr. at 105 ºF average water temperature
- Daily hot water load: 60 gallons heated from local cold water temperature to 115 ºF

16. After completing Exercise 15, use F-CHART software to perform a parametric study to determine the effect of varying the collector array azimuth angle between 45º east of true south, and 45º west of true south, using increments of 10º. Then make a plot showing the solar heating fraction as a function of the collector array azimuth angle.

17. After completing Exercise 15, set the collector azimuth angle to true south. Use F-CHART software to perform a parametric study for collector slopes from 30º to 90º, in increments of 10º. Then make a plot showing the solar heating fraction as a function of the collector slope angle.

18. Use F-CHART software to make a graph of the monthly solar energy on the collector array described in Exercise 15 as well as the heating load of the building described in Exercise 15.

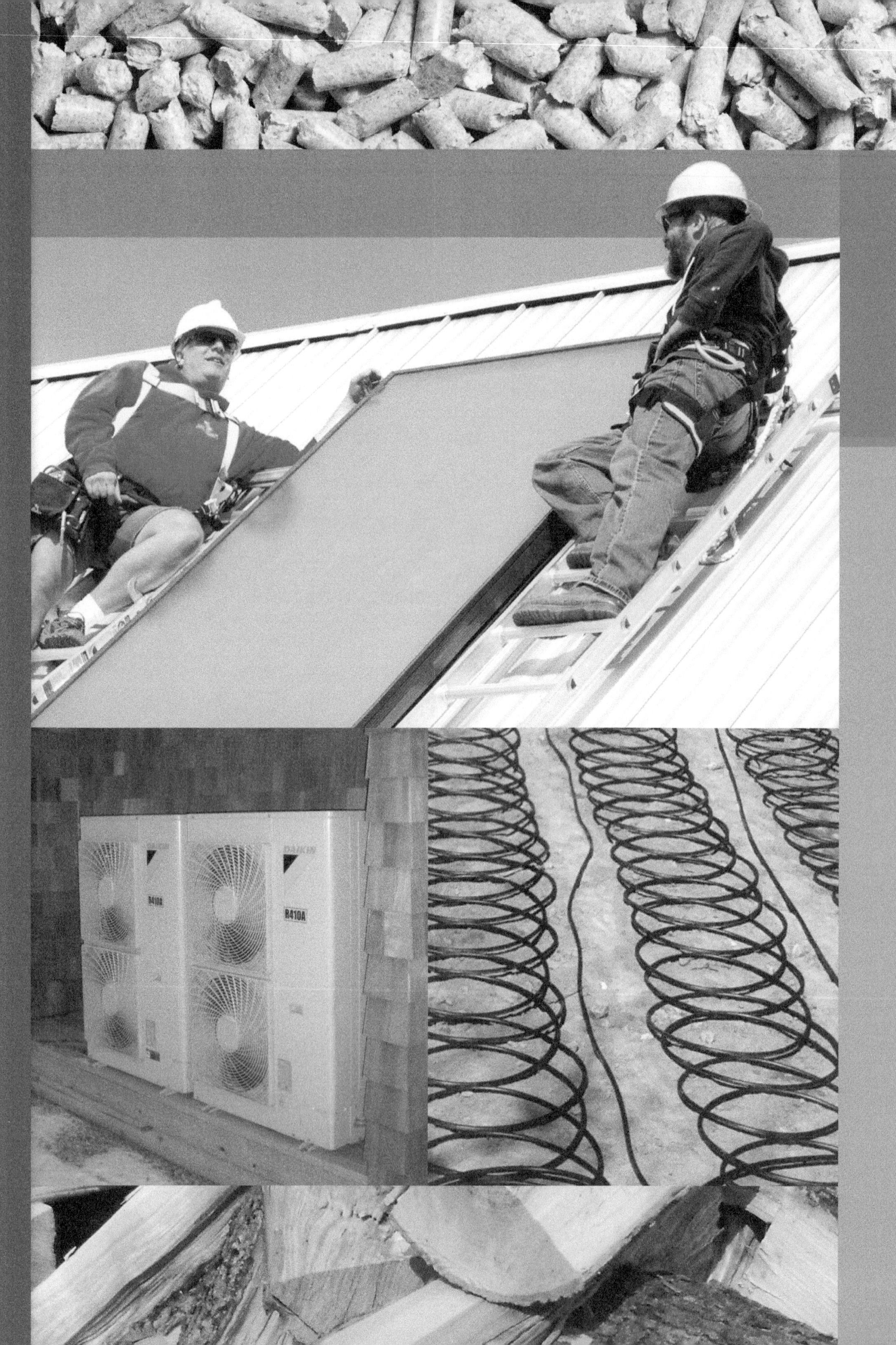
R410A
R410A

CHAPTER 13

Air-to-Water Heat Pump Systems

OBJECTIVES

After studying this chapter, you should be able to:

- Describe several advantages of air-to-water heat pumps.
- Explain the basic operation of a refrigerant cycle in an air-to-water heat pump.
- Describe how a reversing valve works in a heat pump.
- Describe the coefficient of performance (COP) of a heat pump.
- Explain how changes in operating conditions affect the COP of an air-to-water heat pump.
- Define energy efficiency ratio (EER).
- Explain how changes in operating conditions affect the EER of an air-to-water heat pump.
- Describe how changes in operating conditions affect the heating and cooling capacity of an air-to-water heat pump.
- Describe the operation of at least two heating-only systems using an air-to-water heat pump.
- Explain how air-to-water heat pumps can be used in combination with thermal storage.
- Describe the operation of several heating and cooling systems using air-to-water heat pumps.
- Explain options for heating domestic hot water with air-to-water heat pumps.
- Provide freeze protection for air-to-water heat pumps in cold climates.

13.1 Introduction

Heat pumps are devices that move heat from some material at a low temperature to another material at a higher temperature. The low-temperature material from which heat is absorbed is called the **source**. The higher temperature material to which the heat is delivered is called the **sink**.

When used to heat buildings, heat pumps can gather low-temperature heat from sources such as outdoor air, ground water, lakes or ponds, or tubing buried within the earth.

Heat pumps that extract low-temperature heat from outside air are relatively common in North America. Such units are called **air-source heat pumps**. Most heat pumps of this type are configured to deliver higher temperature heat through a forced-air distribution system within the building. This leads to the more specific classification of **air-to-air heat** pump. It is also possible to combine an air source heat pump with hydronic heat delivery. Such devices are called **air-to-water heat pumps** and are the main focus of this chapter.

Heat pumps that extract low-temperature heat from lakes, ponds, wells, or tubing buried in the earth use water, or a mixture of water and antifreeze, to convey heat from those sources to the heat pump. They are thus classified as "water source" heat pumps. Those that deliver this heat through a forced-air systems are more specifically called **water-to-air heat pumps**. Those that deliver their heat using a hydronic distribution system are known as **water-to-water heat pumps**. The latter are discussed in detail in Chapter 14.

HEAT PUMPS ARE RENEWABLE ENERGY HEAT SOURCES

Some people do not consider heat pumps, whether air-source or water source, to be renewable energy heat sources. They base their position on the fact that most heat pumps require electricity to operate, and that electricity may come from nonrenewable sources, such as coal-fired power plants. This is certainly a possibility. However, this view fails to recognize the ultimate source for most of the heat delivered by heat pumps. That source is the sun.

Solar energy is absorbed by the earth's atmosphere. Although the resulting outside air temperature can vary widely with location and season, the amount of heat in outside air is massive compared to all the heating energy used for buildings worldwide. That heat can be readily transferred to any material that is at a lower temperature than the surrounding air. Thus, even outdoor air at 0 °F can give up heat to a material that has a temperature of, say, −5 °F.

During warm weather, large amounts of solar energy are also absorbed into soil. The absorbed energy

Figure 13-1 The sun is the ultimate source for most of the heat delivered by both air source and ground-source heat pumps.

causes temperature increases in the soil that vary with depth. Shallower soils experience more temperature variation than deep soils. Below depths of 25 feet, the seasonal temperature variation in soil is very small. At this depth and lower, soil temperature remains closed to the average annual air temperature. The solar-derived heat that is stored in soil can be efficiently extracted by a geothermal heat pump and used to heat buildings as well as domestic water.

HEAT PUMPS COMPARED TO SOLAR THERMAL SYSTEMS

Consider the following comparison, which describes how certain types of heat pumps can deliver benefits that are equivalent to, and in some cases even superior to, other well-recognized renewable energy heat sources.

It is entirely possible that a properly designed air-to-water heat pump system, in a moderate winter climate, or a geothermal heat pump system in almost any location within the continental United States, can deliver at least three times more thermal energy to a building, compared to the electrical energy required to operate it. This means that at least 67 percent of the heat transferred to the building was extracted from outside air, or the earth, both of which act as storage for solar-derived heat. The remaining 33 percent of the energy was supplied from an electrical utility.

Contrast this situation with a typical solar water heating system that, on an average year, provides 67 percent of the heat required for domestic water heating, with the balance of the energy coming from an electrically operated auxiliary element in the upper portion of the storage tank.

Both systems provide essentially the same ratio of desired effect (e.g., total heat delivered to a load), divided by the necessary input (e.g., electrical energy required to operate the system). Few would argue that the solar water heating system is not a renewable energy source based on this performance. The same conclusion should be obvious for the heat pumps system.

As this text is being written, an increasing number of states within the United States are formally recognizing heat pumps as renewable energy sources. This allows them to qualify for various financial incentive programs. Some electric utilities are also offering incentive programs to encourage the use of heat pumps. The federal government already recognizes geothermal heat pumps that meet **Energy Star** requirements as eligible for a 30 percent **Residential Renewable Energy Tax Credit**. The latter credits are due to expire on December 31, 2016. Up-to-date information on various state and federal incentive programs applying to heat pumps and other renewable energy heat sources can be obtained at http://www.dsireusa.org.

Another benefit of electrically operated heat pumps is that they can be used in buildings that have a grid-connected **solar photovoltaic system**. In locations where **net metering** is available, surplus electrical energy generated by the solar photovoltaic system can be sent back to the electric utility grid. The utility meter records this reverse energy flow. In effect, the surplus electrical energy is "sold" back to the utility at full retail cost. This energy can then be "repurchased" at a later date when it is required. In heating-dominated climates, this allows surplus solar-derived electrical energy to be "stored" on the electrical grid, perhaps for several months, and later drawn back through the building's electrical meter without financial penalty. Thus, it is possible for surplus solar-derived electrical energy created during times when there is little if any operation of the heat pump to be used by the heat pump during the coldest portions of the year. This is a very significant and synergistic benefit—one that will become increasingly common as the demand for **net zero buildings** increases.

13.2 Heat Pump Fundamentals

The heat pumps discussed in this text all operate using the **vapor compression refrigeration cycle**. During this cycle, a chemical compound called the **refrigerant** circulates through a closed piping loop passing through all major components of the heat pump. These major components are named based on how they affect the refrigerant passing through them. They are as follows:

- Evaporator
- Compressor
- Condenser
- Thermal expansion valve (TXV)

The arrangement of these components that forms a complete refrigeration circuit is shown in Figure 13-2.

To describe how this cycle works, a quantity of refrigerant will be followed through the complete cycle.

The cycle begins at station (1) with cold liquid refrigerant in the **evaporator**. At this location, the refrigerant is at a lower temperature than the source media (e.g., air or water) passing across the evaporator. Because of this temperature difference, heat moves from the higher temperature source media into the lower temperature refrigerant. As the refrigerant absorbs heat, it changes from a liquid to a vapor (e.g., it evaporates). Large quantities of low-temperature heat are absorbed as the refrigerant changes from a liquid to a vapor. The vaporized refrigerant continues to absorb heat until it is slightly warmer than the temperature at which it evaporates. The additional heat required to raise the refrigerant's temperature above its **saturation temperature** (e.g., the temperature where it vaporizes) is called **superheat**, and it also comes from the source media.

The vaporized and slightly superheated refrigerant leaves the evaporator and flows into the **compressor** at station (2). Here, a reciprocating piston, or an orbiting scroll, both driven by an electric motor, compresses the vaporized refrigerant. This significantly increases both the pressure and the temperature of the refrigerant vapor.

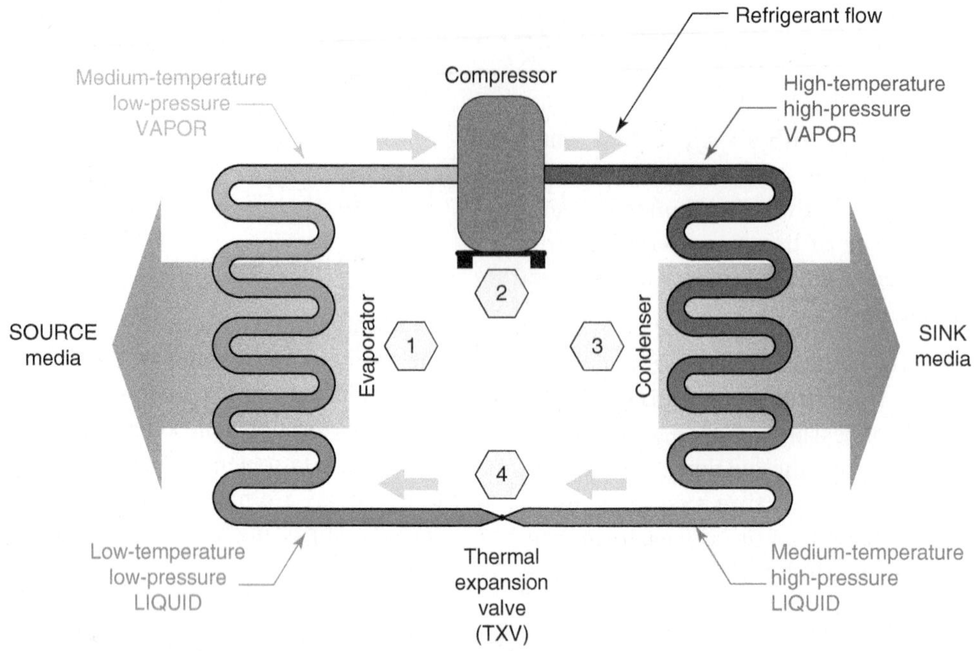

Figure 13-2 | Basic refrigeration cycle.

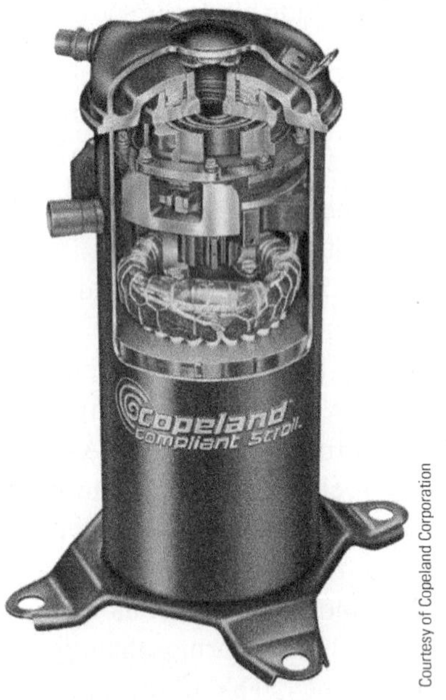

Figure 13-3 | Example of a scroll compressor.

The electrical energy used to operate the compressor is also converted to heat and added to the refrigerant.

Scroll compressors are now the more commonly used compressing device in the types of air-to-water and water-to-water heat pumps discussed in this text. Figure 13-3 shows an example of such a compressor.

Scroll compressors contain two precisely machined metal **scrolls**. One is fixed, and the other orbits, rather than rotates. This action creates a continuous flow of refrigerant vapor to be drawn into the compressor and progressively compressed as it passes through the tightening gaps between the two scrolls. The high-pressure refrigerant vapor leaves through a port that is centered with the inner portion of the fixed scroll, as shown in Figure 13-4.

The temperature of the refrigerant vapor leaving the compressor is usually in the range of 120 °F to 170 °F, depending on the other operating conditions of the heat pump.

The hot refrigerant vapor then flows into the **condenser** at station (3). In air-to-water heat pumps, and water-to-water heat pumps, the condenser is a refrigerant-to-water heat exchanger. The most commonly used configuration is a **coaxial tube-in-tube** design, as shown in Figure 13-5. Water flows through the inner copper tube, while hot refrigerant flows in the opposite direction through the space between the inner and outer tubes. The entire heat exchanger is typically wrapped with insulation within the heat pump enclosure.

The hot refrigerant vapor gives up heat to the cooler stream of water, which carries the heat away from the heat pump, either directly to a load, or in some systems, to a thermal storage tank.

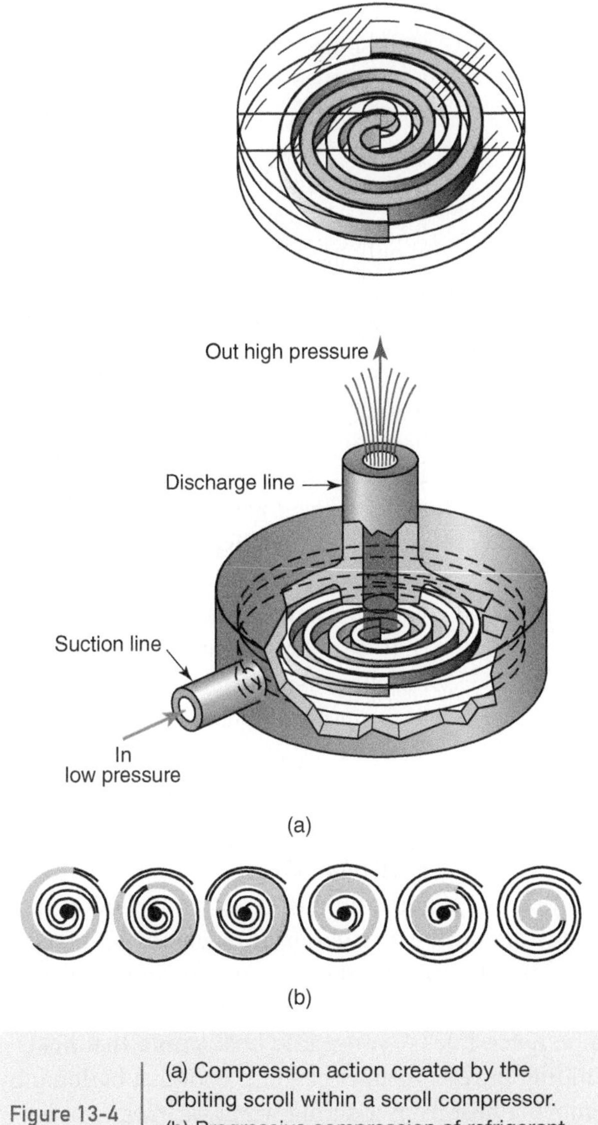

Figure 13-4 (a) Compression action created by the orbiting scroll within a scroll compressor. (b) Progressive compression of refrigerant between the fixed and orbiting scrolls.

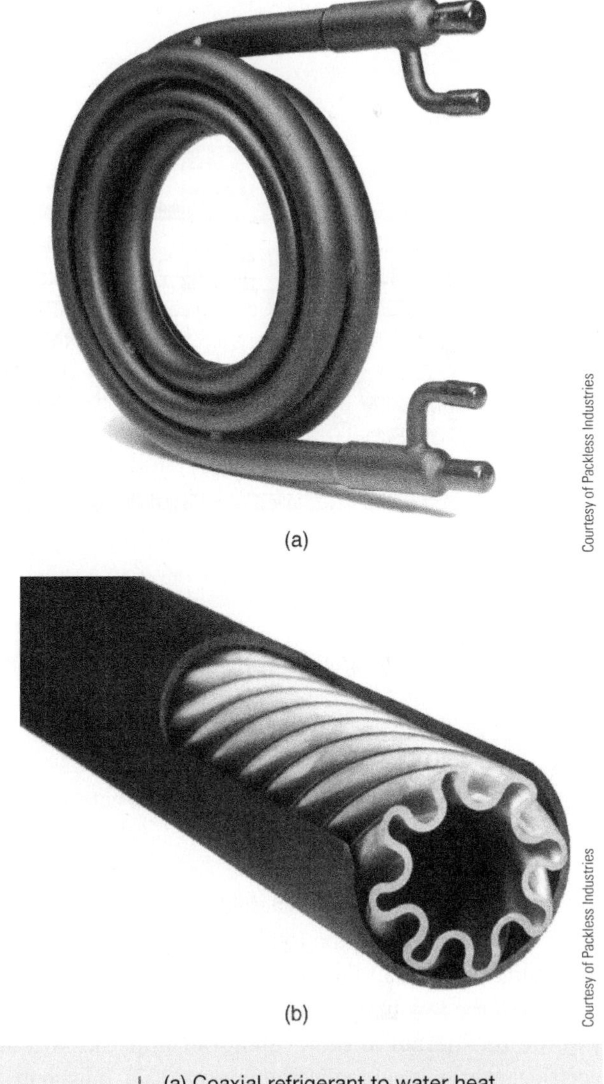

Figure 13-5 (a) Coaxial refrigerant to water heat exchanger that is often used as the condenser in an air-to-water or water-to-water heat pump. (b) Internal view of tube-in-tube design.

As it gives up heat within the condenser, the refrigerant condenses from a high-pressure, high-temperature vapor into a high-pressure and somewhat cooler liquid.

The high-pressure liquid refrigerant then flows through the **thermal expansion valve** at station (4), where its pressure is greatly reduced. The drop in pressure causes a corresponding drop in temperature, restoring the refrigerant to the same condition at which this description of the cycle began. The refrigerant is now ready to repeat the cycle.

The refrigeration cycle remains in continuous operation whenever the compressor is running. This cycle is not unique to heat pumps. It is used in refrigerators, freezers, room air conditioners, dehumidifiers, water coolers, soda vending machines, and other heat-moving machines. The average person is certainly familiar with these devices but often takes for granted how they operate.

Figure 13-6 shows the energy flows created by the refrigeration cycle.

The arrow labeled (Q1) represents the low-temperature heat absorbed by the evaporator. Arrow (Q2) represents the electrical energy used to operate the compressor. Arrow (Q3) represents the heat output from the heat pump's condenser.

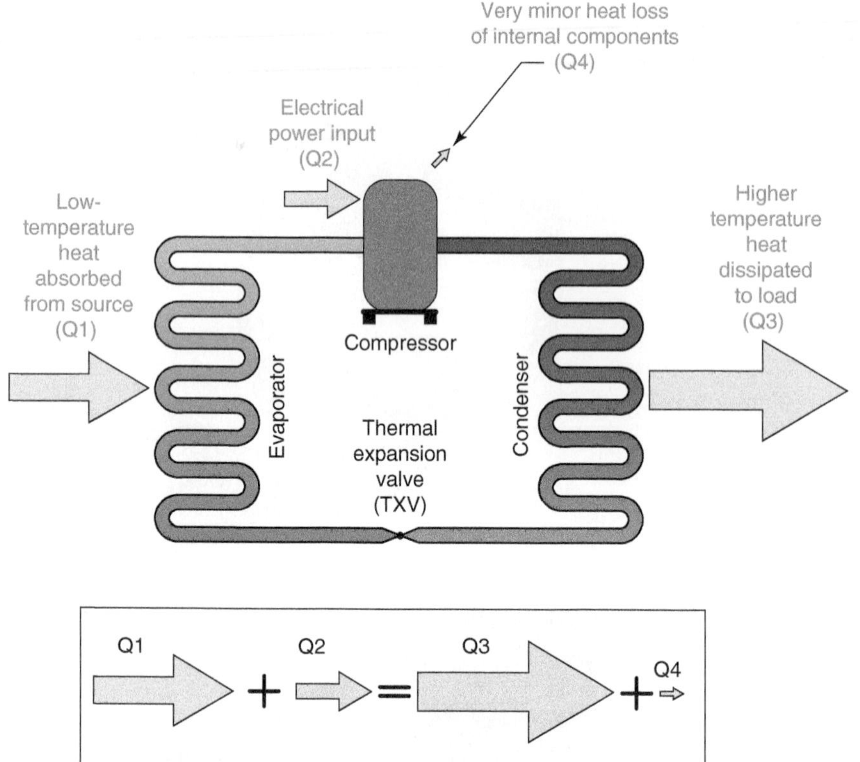

Figure 13-6 | Energy flow created by the refrigeration cycle within a heat pump.

The first law of thermodynamics dictates that, under steady state conditions, the total energy input rate to the heat pump must equal the total energy output rate. Thus, the sum of the low-temperature heat absorption rate (Q1) into the refrigerant at the evaporator, plus the rate of electrical energy input to the compressor (Q2), must equal the total rate of energy dissipation from the heat pump. The vast majority of this heat output will be at the condenser (Q3). A very small amount of heat output (Q4) occurs from the heat losses of components within the heat pump. The latter is typically ignored during routine performance testing. Thus, if any two of the three quantities (Q1, Q2, or Q3) are determined through measurements, the remaining quantity can be determined using the simplified relationship that Q1 + Q2 = Q3.

REVERSIBLE HEAT PUMPS

As previously discussed, all heat pumps move heat from a lower temperature material (e.g., the source) to another material at higher temperature (e.g., the sink). The **non-reversible heat pump** described thus far in the chapter can be used in a **heating-only application** or a **cooling-only application**.

In a heating-only application, the heat pump's evaporator always gathers low-temperature heat from some source where heat is freely available and abundant. The condenser always delivers higher temperature heat to a load. One example is a heat pump that is only used to heat a building. Another would be a heat pump that only delivers energy to heat domestic water.

In a cooling-only application, the heat pump's evaporator always absorbs heat from a media that is intended to be cooled. The condenser always dissipates the unwanted heat to some media that can absorb and dissipate it (i.e., outside air, ground water, or soil). Examples would include a common air conditioner that can only remove heat from a building. Another would be a water-to-water heat pump that is used to chill water as part of an ice-making process.

Although there are several applications where non-reversible heat pumps can be used, one of the most unique benefits of modern heat pumps is that the refrigerant flow can be reversed to immediately convert the heat pump from a heating device to a cooling device. Such heat pumps are said to be reversible. A **reversible heat pump** that heats a building in cold weather can also cool that building by removing heat from it during warm weather.

Reversible heat pumps contain an electrically operated device called a **reversing valve**. An example of such a valve is shown in Figure 13-7.

Figure 13-7 | Example of a pilot-operated reversing valve.

The type of reversing valves used in the heat pumps discussed in this text contain a slide mechanism that is moved by refrigerant pressure. The direction of movement is controlled by a small solenoid valve, called a **pilot solenoid valve**. When the heat pump is in heating mode, the magnetic coil of pilot valve is not energized. This allows the high-pressure refrigerant leaving the compressor to position the slide within the reversing valve so hot refrigerant gas from the compressor goes to the condenser, as shown in Figure 13-8a. In this mode, the condenser is the heat exchanger that transfers heat to a flowing stream of water, which carries the heat to the load.

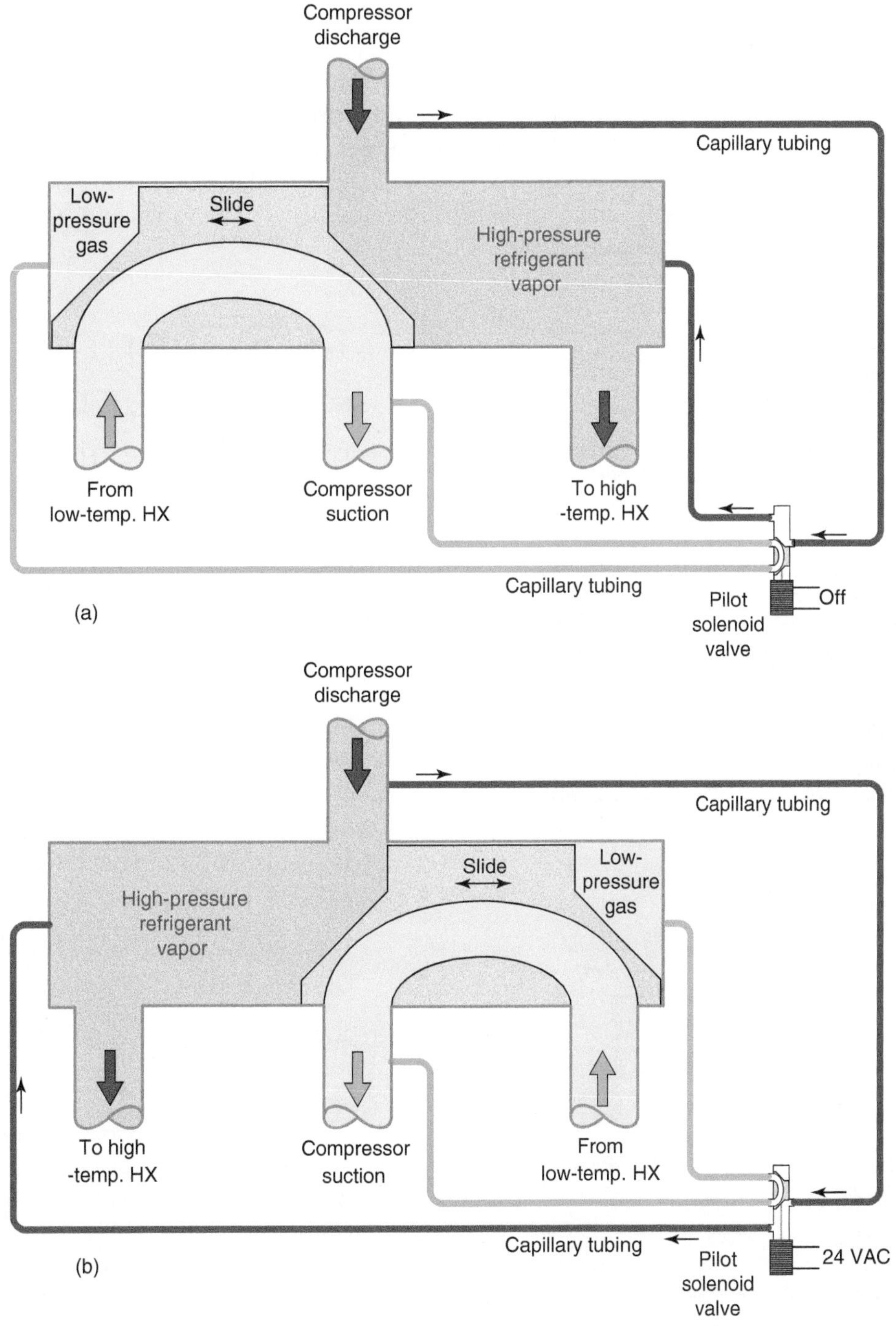

Figure 13-8 (a) In heating mode, the pilot valve is not energized and the slide is held to the left side of its chamber by refrigerant pressure. (b) In cooling mode, the pilot valve is powered on. Refrigerant pressure forces the slide to the right side of it chamber.

When the heat pump needs to operate in cooling mode, the pilot solenoid valve is energized by a 24 VAC signal. This allows the refrigerant pressure to immediately move the slide within the reversing valve to the opposite end of its chamber. Hot gas leaving the compressor is now routed to the heat pump's other heat exchanger (e.g., what was the evaporator now becomes the condenser). This is illustrated in Figure 13-8b.

Figure 13-9 shows a heat pump refrigeration system that includes a reversing valve. Notice that that the refrigerant heat exchangers on the left and right side of the diagram change their function between serving as the evaporator or condenser, depending on the operating mode. It is customary to reference the heating mode function when describing these heat exchangers as either the evaporator or the condenser.

Some reversible heat pumps use two thermal expansion valves in combination with two check valves. One thermal expansion valve functions in the heating mode while the other functions during the cooling mode. Other heat pumps use a single "bi-directional" thermal expansion valve. For simplicity, the heat pump refrigeration piping diagrams shown in this text assume a single, **bidirectional thermal expansion valve**.

Some of the most recently introduced heat pumps use electronically controlled expansion devices that allow precise metering of refrigerant flow rate in combination with a variable speed compressor. This combination allows the heating and cooling capacity of the heat pump to be varied over a wide range.

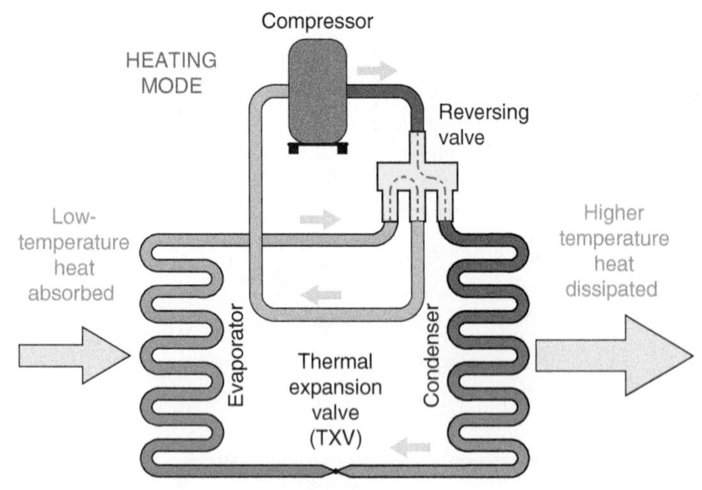

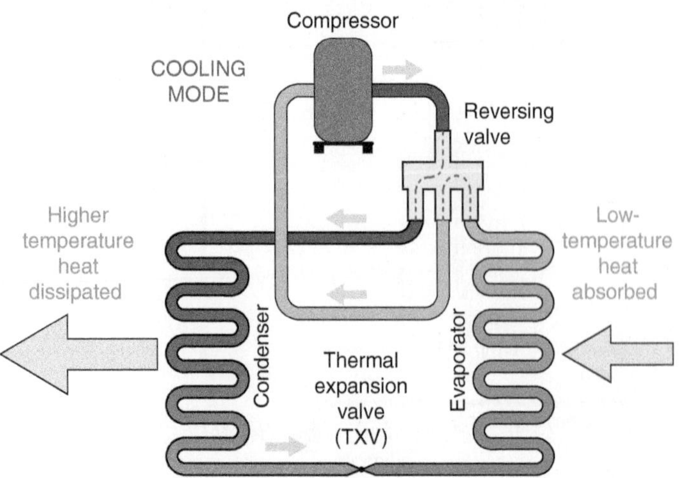

Figure 13-9 | Heat pump refrigeration diagram in both heating and cooling modes. Note refrigerant flow through the reversing valve.

13.3 Thermal Performance of Heat Pumps

There are several quantitative measurements used to describe heat pump performance. They are categorized based on operation in the heating or cooling mode. This section examines these performance indices in a general sense. Later sections and chapters will give specific information for each type of heat pump in each mode of operation.

HEATING MODE THERMAL PERFORMANCE

In the heating mode, the two indicators of thermal performance are:

a. Heating capacity

b. Coefficient of performance (COP)

The **heating capacity** of a heat pump is the *rate* at which it delivers heat to the load. As such, it is similar to the heating capacity of a boiler. However, *the heating capacity of any heat pump is highly dependent on its operating conditions, specifically the temperature of the source media and the temperature of the load media*. The greater the temperature difference between the source media and the sink media, the lower the heat pump's heating capacity.

Heating capacity also depends on the flow rate of both the source and sink media through the heat exchangers that serve as the heat pump's evaporator and condenser. The higher these flow rates are, the greater the heating capacity will be. However, the gains in heat capacity are not proportional to flow rate. Instead, capacity increases incrementally at high flow rates. Designers must be careful not to create situations in which slight gains in heating capacity come at the expense of significantly higher circulator or blower power requirements to create the higher flow rates. Later sections of this chapter and Chapter 14 show how heating capacity varies based on operating conditions.

The **coefficient of performance (COP)** of a heat pump is a number that indicates the *ratio of the beneficial heat output from the heat pump, divided by the electrical input required to operate the heat pump*. The higher the COP, the greater its rate of heat output per unit of electrical power input.

Equation 13.2 shows this relationship in mathematical form. The factor 3.413 in the denominator of this ratio converts watts into Btu/hr. Thus, the units in the numerator and denominator are the same and cancel out to give a unitless number for COP. All COPs must be expressed as unitless numbers.

(Equation 13.2)

$$COP = \frac{\text{heat output (Btu/hr.)}}{\text{electrical input (watt)} \times 3.413}$$

COP can be visualized as the ratio of the heat output arrow divided by the electrical power input arrow, as shown in Figure 13-10.

The ultimate thermal performance objective for any heat pump is to maintain its COP as high as possible.

Example 13.1: Assume that a heat pump is delivering heat to a load at a rate of 37,000 Btu/hr. The electrical power to operate the heat pump is measured as 3,000 watts. Determine the COP of the heat pump under its current operating conditions.

Solution: Using Equation 13.2:

$$COP = \frac{\text{heat output (Btu/hr.)}}{\text{electrical input (watt)} \times 3.413}$$

$$= \frac{37{,}000 \text{ Btu/hr.}}{3{,}000 \text{ watt} \times 3.413\left(\frac{\text{Btu/hr.}}{\text{watt}}\right)} = 3.61$$

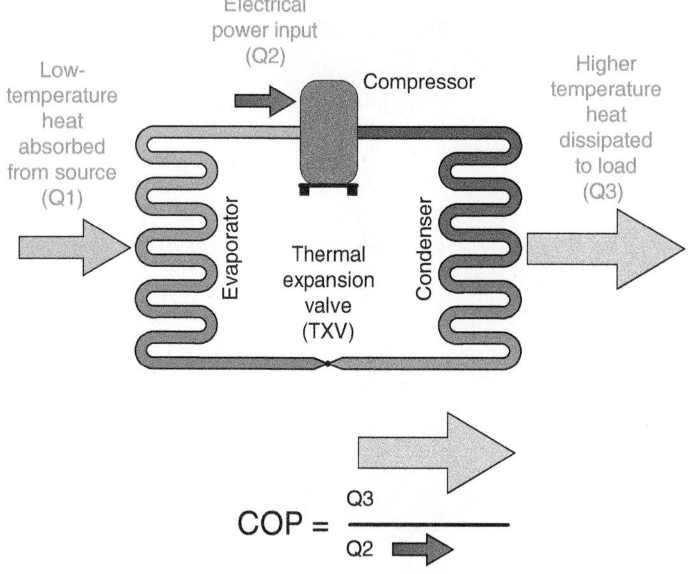

Figure 13-10 | Visualizing the COP of a heat pump.

Discussion: Notice that the units of *watt* and *Btu/hr.* both cancel out in this equation. The COP is always a unitless number.

Another way to think of COP is the number of units of heat output energy the heat pump delivers, per unit of electrical input energy. Thus, if a heat pump operates at a COP of 4.1, it provides 4.1 units of heat output energy *per unit* of electrical input energy.

COP can also be considered as a way to compare the thermal advantage of a heat pump to that of an electric resistance heating device that provides the same heat output. For example, if an electric resistance space heater is 100 percent efficient, then by comparison, a heat pump with a COP of 4.1 would be 410 percent efficient. Some would argue that no heat source can have an efficiency greater than 100 percent. This is true for any heat source that simply converts a fuel into heat. However, a heat pump is better described as a device that *moves* heat from lower temperature to higher temperature, rather than a device that creates heat. As such, its beneficial effect is equivalent to a hypothetical heat source that would have an efficiency significantly higher than 100 percent.

The COP of all heat pumps is highly dependent on operating conditions. This includes the temperature of the source media as well as the media to which the heat pump outputs heat. *The closer the temperature of the source media is to the temperature of the load media, the higher the heat pump's COP.*

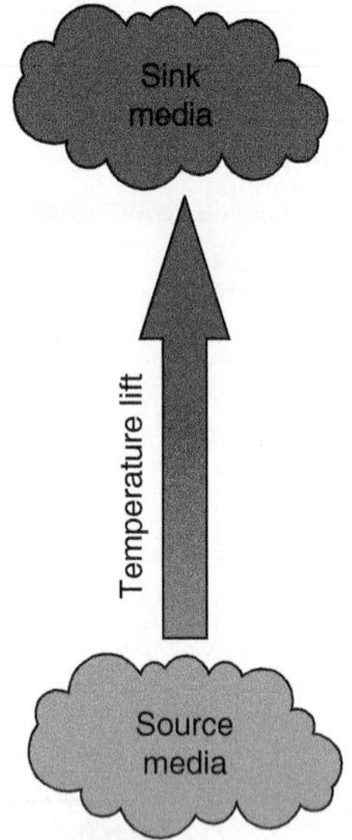

Figure 13-11 The smaller the temperature lift between the source media and the sink media, the higher the heat pump's COP.

One can visualize the difference between the source and sink temperatures as the "**temperature lift**" the heat pump must provide, as shown in Figure 13-11. The smaller the lift, the higher the heat pump's COP.

The theoretical maximum COP that any heat pump can attain was established by nineteenth-century scientist Sadi Carnot and is appropriately named the **Carnot COP**. It is based on the ***absolute temperatures*** of the source media and sink media and given by Equation 13.3.

(Equation 13.3)

$$COP_{Camot} = \frac{T_{sink}}{(T_{sink} - T_{source})}$$

where:

COP_{Carnot} = Carnot COP (the maximum possible COP of any heat pump)

T_{sink} = *absolute* temperature of the sink media to which heat is delivered (°R)

T_{source} = *absolute* temperature of the source media from which heat is extracted (°R)

°R = °F + 458°

Example 13.2: Determine the maximum possible COP of a heat pump extracting heat from lake water at 40 °F and delivering that heat to a stream of heated water that averages 100 °F within the heat pump's condenser.

Solution: The temperatures of 40 °F and 100 °F must be converted to degrees Rankine temperatures before using Equation 13.3.

$$100\ °F + 458\ °F = 558\ °F$$
$$40\ °F + 458\ °F = 498\ °F$$

These values can now be entered into Equation 13.3.

$$COP_{Camot} = \frac{T_{sink}}{T_{sink} - T_{source}} = \frac{558\,°F}{558\,°F - 498\,°R} = 9.3$$

Discussion: This theoretical maximum COP is based on a hypothetical heat pump that has no mechanical energy losses due to friction or electrical losses due to resistance. It is also based on "infinitely sized" source and sink that remain at the same temperature as they give up and absorb heat. No real heat pump operates under such idealized conditions, and thus *no real heat pump ever attains the Carnot COP*. The COP of currently available heat pumps, even when operated under very favorable conditions, is substantially lower than the Carnot COP. Still, the Carnot COP provides a means of comparing the performance of evolving heat pump technology to a theoretical limit. It also demonstrates the inverse relationship between the "temperature lift" of a heat pump and COP. Notice that the theoretical Carnot COP of a heat pump, calculated using Equation 13.3, approaches infinity as the temperature lift ($T_{sink} - T_{source}$) approaches zero.

COOLING MODE THERMAL PERFORMANCE

In the cooling mode, the two indicators of thermal performance are:

a. Cooling capacity

b. Energy efficiency ratio (EER)

Cooling capacity represents the *total* cooling effect (sensible cooling plus latent cooling) that a given heat pump can produce while operating at specific conditions.

Heat pumps that deliver cooling using a refrigerant-to-air heat exchanger, and forced air distribution systems, will have separate ratings for **sensible cooling capacity** and **latent cooling capacity**. However, because air-to-water and water-to-water heat pumps both deliver a stream of cool water as their output, there is only one rating for cooling capacity.

Cooling capacity is significantly affected by the temperature of the fluid streams passing through the evaporator and condenser. To a lesser extent, it's also affected by the flow rates of these fluid streams.

The cooling capacity of any heat pump will increase when the temperature of the source media (e.g., the material from which heat is being absorbed) increases. Cooling capacity will also increase when the temperature of the sink media (e.g., the material to which heat is rejected) decreases. So, as was true for both heating capacity and COP, the closer the source temperature is to the sink temperature, the higher the cooling capacity of the heat pump.

ENERGY EFFICIENCY RATIO

In North America, the common way of expressing the instantaneous cooling efficiency of a heat pump is an index called **energy efficiency ratio (EER)**, which is defined as by Equation 13.4:

(Equation 13.4)

$$EER = \frac{Q_c}{w_e}$$

where:

EER = energy efficiency ratio

Q_c = cooling capacity (Btu/hr.)

W_e = electrical power input to heat pump (watts)

The higher the EER of a heat pump, the lower the electrical power required to produce a given rate of cooling.

Like COP, the EER of a water-to-water heat pump is a function of the source and sink temperatures. The warmer the source media temperature is compared to the sink media temperature, the higher the heat pump's EER. To maximize EER, designers of chilled-water-cooling systems using either air-to-water or water-to-water heat pumps should use the highest possible chilled water temperature that still allows adequate dehumidification. EER is also slightly influenced by flow rates. Higher flow rates of either the source media or the sink media will incrementally increase EER.

TONNAGE

It is customary to describe the heating and cooling capacity of refrigeration-based equipment such as heat pumps using the units of **tons**. In this context, a ton describes a *rate* of heat flow. More specifically, 1 ton equals 12,000 Btu/hr. Thus, a "3-ton" heat pump implies a nominal heating or cooling capacity of 3 × 12,000 or 36,000 Btu/hr. The tonnage of a heat pump has nothing to do with heat pump's weight. The unit of "ton" as a description of cooling capacity originated during the transition from stored natural ice as a mean of cooling, to mechanical refrigeration. It represents the average heat transfer rate associated with melting one ton of ice over a 24-hour period.

A description of a heat pump based on tons is a *nominal* rating at some specific set of operating conditions. Thus, a "3-ton" rated heat pump could produce a heat output significantly higher than 3 tons when operated under more favorable conditions and significantly less than 3 tons when operated under less favorable conditions.

13.4 Air-to-Water Heat Pumps

Air-to-water heat pumps have an outside unit, similar to that of an air-to-air heat pump or central air conditioning system. However, the heat generated while operating in the heating mode is delivered to a hydronic distribution system within the building. When operating in the cooling mode, air-to-water heat pumps deliver a stream of chilled water to an interior hydronic distribution system. The use of a hydronic distribution system for both heating and cooling creates many possibilities that are not possible with forced air distribution systems.

Figure 13-12 shows the outside unit of a modern air-to-water heat pump. The insulated pipes, seen penetrating the building wall behind the unit, connect it to the interior portion of the system.

The outdoor unit can serve as either the evaporator (in the heating mode) or the condenser (in the cooling mode). When operating, variable speed fans pull outside air across the air-to-refrigerant heat exchanger on the rear of the unit. In heating mode, this heat exchanger serves as the evaporator. Low-temperature heat is absorbed from the air, and the cooled air is discharged through the fan grills seen at the front of the unit.

The unit shown in Figure 13-12 is an example of a **self-contained air-to-water heat pump**. This type of heat pump contains all the components necessary for the refrigeration cycle within the outdoor unit. It may

(a)

(b)

Figure 13-12 (a) Front of outdoor unit of an air-to-water heat pump. (b) Rear of same unit showing large air-to-refrigerant heat exchange coil.

Figure 13-13 All refrigeration cycle components in this self-contained air-to-water heat pump are located in the outdoor unit.

also contain devices such as a circulator, expansion tank, and system controller within the outdoor unit, as seen in Figure 13-13.

Systems designed using self-contained air-to-water heat pumps vary depending on the severity of the winter climate. In relatively mild climates, where outdoor temperatures only occasionally drop below freezing, it is generally acceptable to install the heat pump with water in the piping circuit between the outdoor unit and indoor distribution system. Most heat pumps of this type have a controller and an auxiliary electric heating element that automatically turn on, if necessary, to protect the unit from freezing. These components will add enough heat to maintain the water-filled portion of the heat pump above freezing, even if there is no load on the system.

However, in situations where a prolonged power outage is possible, and no backup generator capable of running the heat pump is available, it is advisable to install the unit as part of an antifreeze-protected hydronic circuit. Figure 13-14 shows two options for such an installation depending on the make and model of heat pump used. Some air-to-water heat pumps come equipped with an internal circulator and expansion tank, whereas other require these to be installed by others.

To minimize the thermal penalty associated with the interior heat exchanger, it should be sized for a maximum **approach temperature** of 5 °F while transferring the full rated heating output of the heat pump.

Some manufacturers also offer **split system** versions of air-to-water heat pumps. In a split system, the refrigerant-to-water heat exchanger, as well as the circulator, expansion tank, and any other components that contain water, are located within an indoor unit. The compressor, and other refrigeration components remain in the outdoor unit. Two copper refrigerant lines, as well as some electrical wiring, connect the

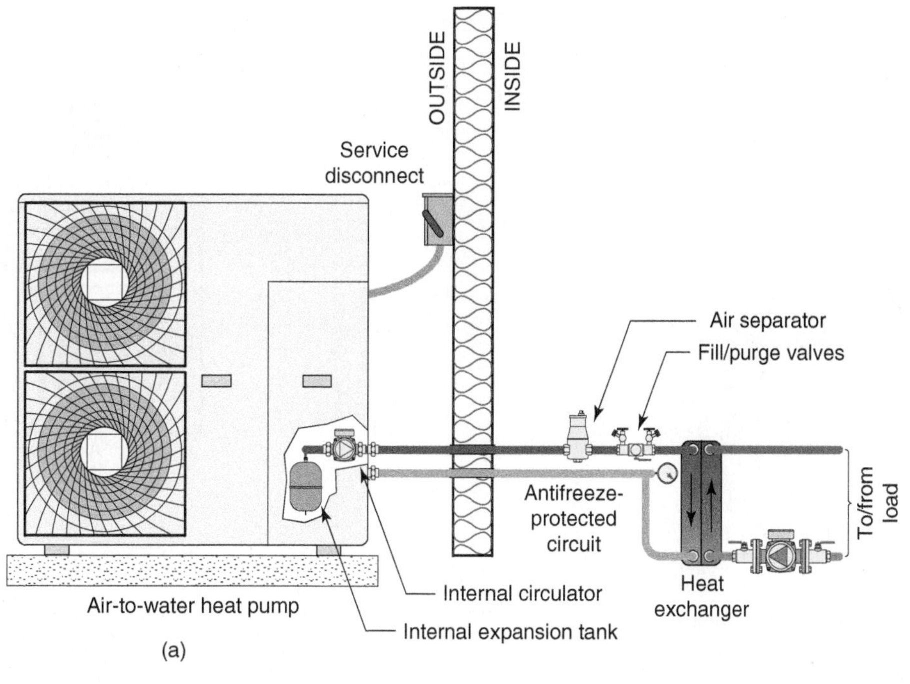

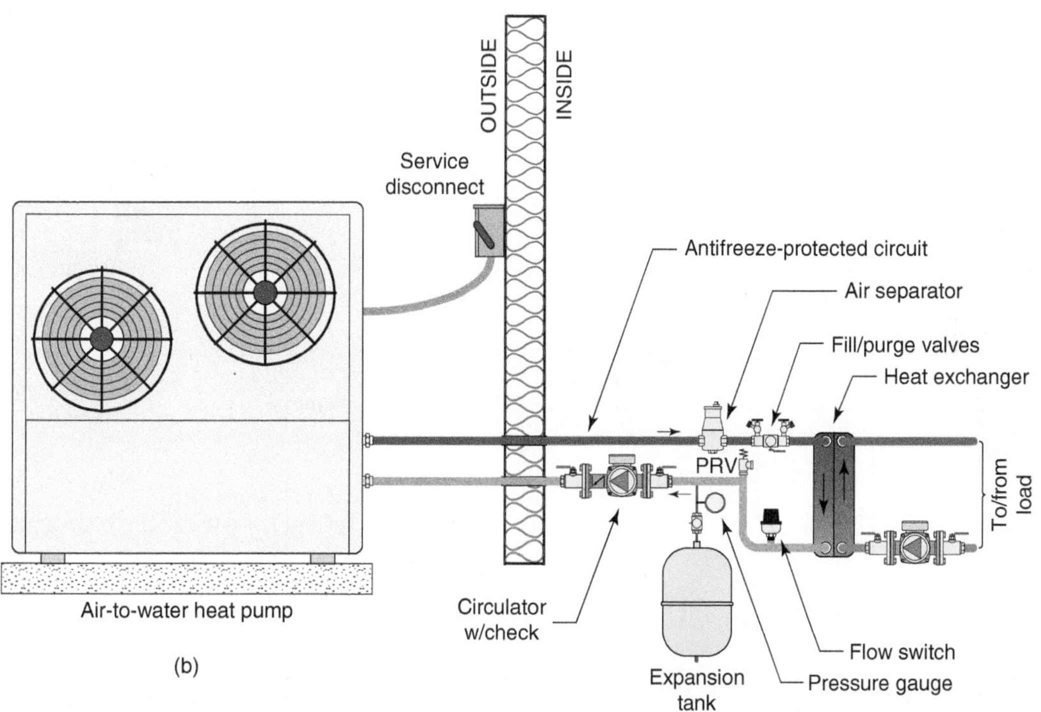

Figure 13-14 (a) Air-to-water heat pump with antifreeze-protected circuit. Circulator and expansion tank for this circuit are contained within the heat pump. (b) Antifreeze-protected circuit with circulator and expansion tank located in building.

indoor and outdoor units. Figure 13-15 shows the outdoor unit, indoor unit, and how they would be connected in an installation.

One advantage of a split system is the absence of water in the outdoor portions of the system. This eliminates any need to freeze protect the outdoor components. However, split systems require the installation of refrigerant piping. Such piping, even if pre-charged with refrigerant, requires a refrigeration technician during installation.

Figure 13-16 shows the major internal components in a self-contained air-to-water heat pump.

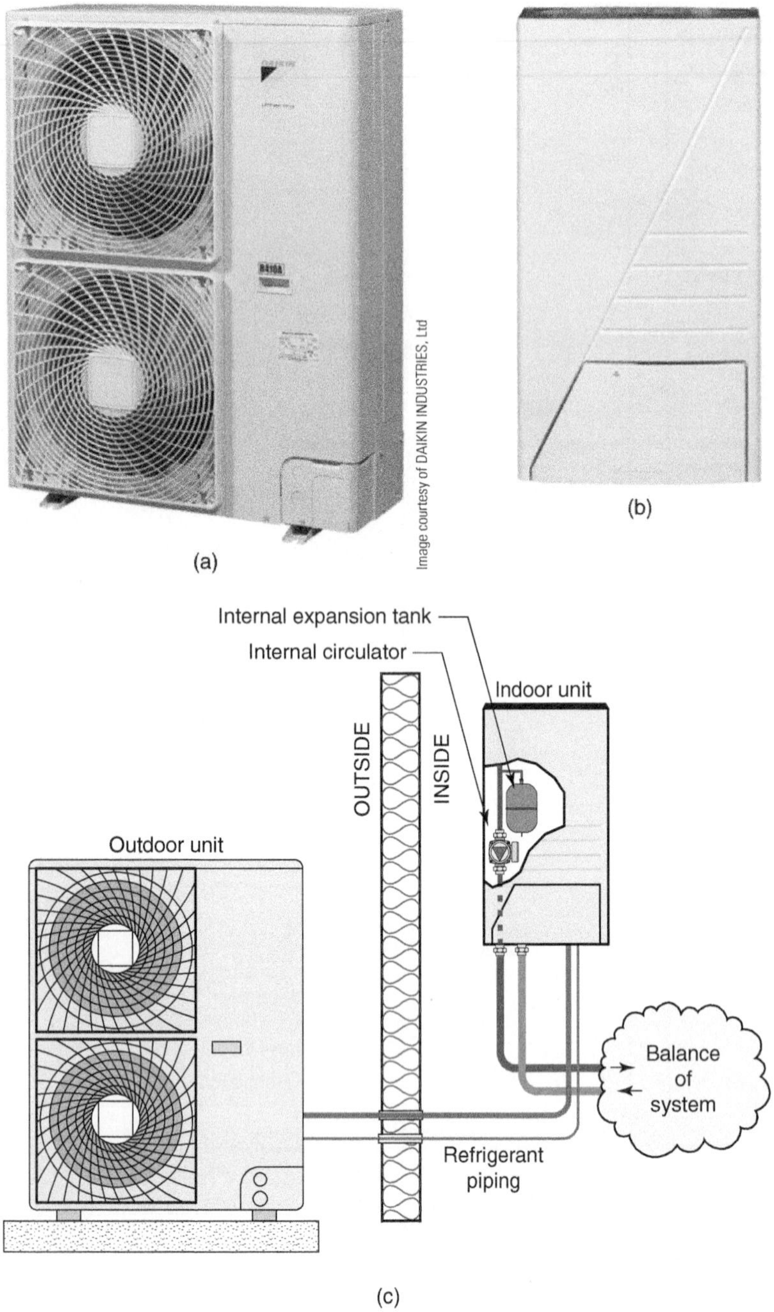

Figure 13-15 (a) Outdoor unit. (b) Indoor unit. (c) Outdoor and indoor unit connected with refrigeration piping.

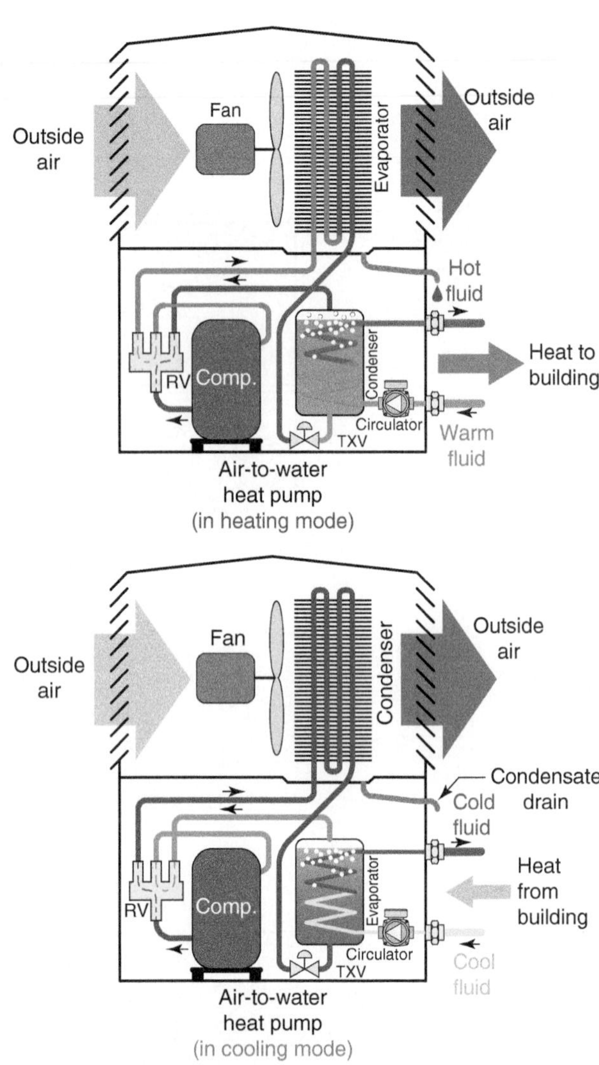

Figure 13-16 Major components in a self-contained air-to-water heat pump.

The depictions in Figure 13-16 show a circulator within the heat pump. Some manufacturers provide this circulator, while others do not. In cases where it is provided, designers should check the "net pump curve" available from the circulator. This would be the circulator's normal pump curve minus the head loss curve associated with flow through the refrigerant-to-water heat exchanger in the heat pump. In most cases, the included circulator is sufficient to move flow through the heat pump and a relatively small primary circuit, or buffer tank circuit, within the building. Never assume this circulator has a sufficient pump curve to create adequate flow through an entire hydronic distribution system.

INTERIOR AIR-TO-WATER HEAT PUMPS

There are also air-to-water heat pumps designed to be located inside buildings. They use ducting to bring outside air to their air-handling section as well as to discharge that air back outside. Having the heat pump inside has advantages as well as disadvantages:

Advantages:

- No outdoor equipment beyond air intake and discharge grills
- Less potential to freeze water contained within the heat pump
- Less environmental weathering effect on equipment
- Reduced potential for debris on heat transfer coil surfaces

Disadvantages:

- Requires more interior space
- Brings compressor sound inside the structure
- Requires careful coordination with building design to ensure that adequately sized ducting can be accommodated and terminated above snow level

Figure 13-17 shows an example of an interior air-to-water heat pump. The large-diameter insulated flexible ducting brings outdoor air to the unit and exhausts it back outside.

Interior versions of air-to-water heat pumps are currently used in Europe, but as of this writing, are not available in North America. [*note from author: I wish they were* ☺...].

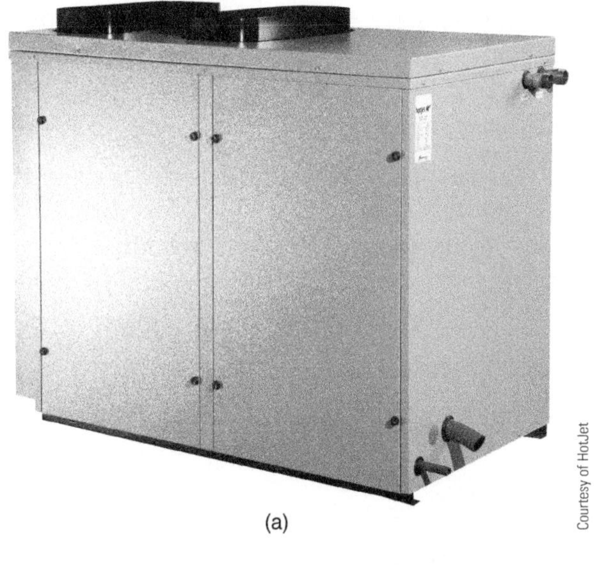

(a)

Courtesy of HotJet

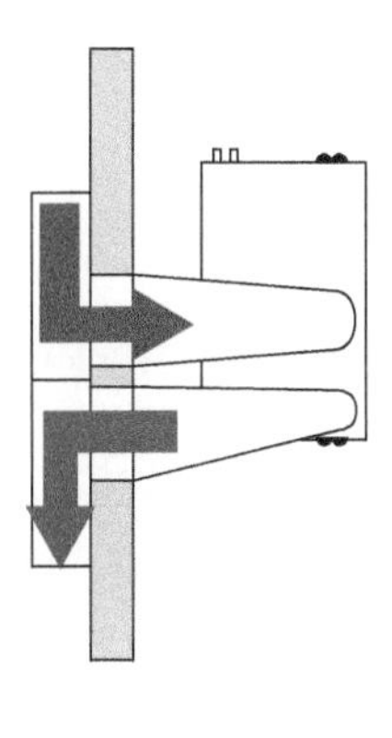

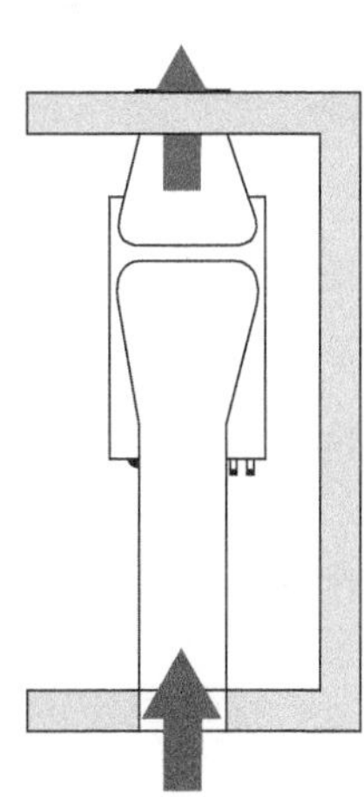

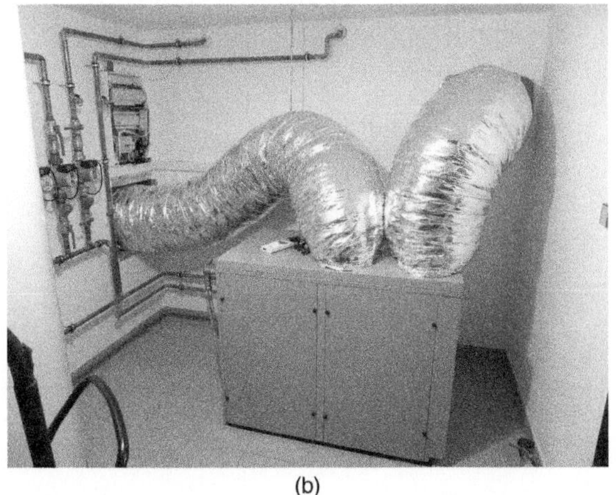

(b)

Courtesy of HotJet

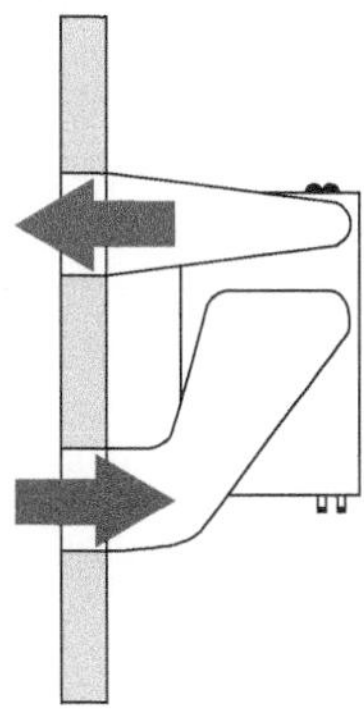

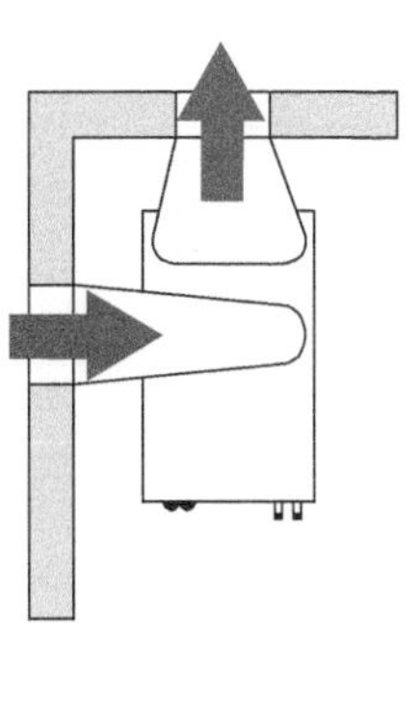

(c)

Courtesy of HotJet

Figure 13-17 | (a) Example of an interior air-to-water heat pump. (b) Installed interior air-to-water heat pump. (c) Possible arrangement of ducting for incoming and exhaust air.

13.5 Thermal Performance of Air-to-Water Heat Pumps

Air-to-water heat pumps are manufactured in a number of variations. These include:

- Single-speed on/off compressor
- Variable-speed compressor/modulating capacity
- Two-stage compressor/stepped capacity

The type of heat pump selected will, in part, determine how the balance of the system is designed. For example: A single-speed on/off heat pump will require the installation of a buffer tank if it is combined with a highly zoned hydronic distribution system. A modulating heat pump may not require such a tank. These factors will be discussed in later sections of this chapter.

HEAT CAPACITY

Most manufacturers list a **nominal heating capacity** and **nominal cooling capacity** for each heat pump model they offer. The word *nominal* means that the capacity is based on some set of reference conditions, which are typically listed with the nominal capacity ratings. Figure 13-18 shows an example of a rating table showing nominal heating and cooling capacities as well as the operating conditions upon which these capacities are based.

The listed heating capacity for each of the three heat pump models was established based on an entering outdoor dry bulb temperature of 44.6 °F and a corresponding outdoor wet bulb temperature of 42.8 °F. The water temperature *leaving* the condenser was assumed to be 95 °F, with a flow rate that corresponded to a temperature rise of 9 °F. Thus, the entering water temperature assumed was 86 °F.

Model Number	Heating Capacity (Btu/hr.)
EBLQ036	38,210
EBLQ048	47,770
EBLQ054	54,590

Operating conditions:
Outside air: DB/WB 44.6 °F/42.8 °F (7°/6 °C)
Temperature of water leaving condenser: 95 °F (35 °C) ($\Delta T = 9$ °F (5 °C))

Figure 13-18 Nominal heating capacities of three different models of air-to-water heat pumps, and the conditions under which these capacities were determined.

It's important to understand that the heating capacity, as well as the COP, can be significantly higher or lower than these nominal values based on actual operating conditions.

Figure 13-19 shows how the heat capacity of an air-to-water heat pump with a nominal rated output of about 54,590 Btu/hr. varies with both outdoor temperature and **leaving load water temperature**. The latter is the water temperature *leaving* the heat pump's condenser. These graphs assume a nominal 9 °F temperature increase as water flows through the heat pump's condenser. Thus, the entering water temperature can be estimated by subtracting 9 °F from the leaving load water temperature.

The heat pump's heating capacity is significantly influenced by outdoor air temperature, and to a lesser extent, by the temperature of the water flowing through its condenser. At an outdoor temperature of 10 °F, this unit

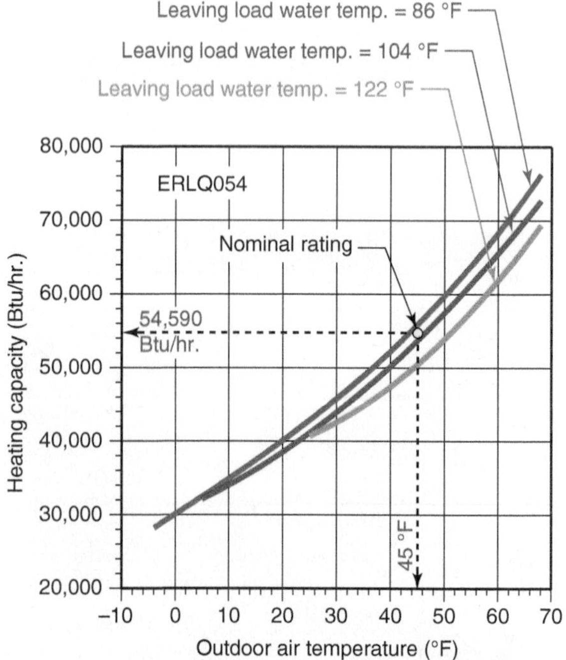

Figure 13-19 Variation in heating capacity based on outdoor air temperature and leaving condenser water temperature, for a specific make and model of air-to-water heat pump.

has a heating capacity of about 34,000 Btu/hr., assuming the same entering and leaving water temperatures as given in Figure 13-18, This is a 37 percent drop in heating capacity compared to the stated nominal heating capacity.

Figure 13-19 also shows that higher leaving water temperatures decrease heating capacity. The highest leaving water temperature shown, 122 °F, is only possible down to outside air temperatures of about 25 °F. At lower air outside air temperatures, the temperature of the water leaving the condenser drops steadily.

In cold winter climates, where outdoor design temperatures drop to less than 10 °F, the heat output from this type of heat pump will likely have to be supplemented by an auxiliary heat source. Some air-to-water heat pumps are equipped with supplemental electric heating elements that can provide up to 6 kW (20,478 Btu/hr.) of additional heating capacity. Figure 13-20 shows how the heat output of this auxiliary element, working in combination with the heat output of the heat pump, is used to meet the heating load.

The smooth, descending black curve represents the space heating load versus the number of hours in the heating season. It begins at design load in the upper left corner and eventually drops to no load at the right side of the graph. The duration of the heating season in this graph is assumed to be 5,000 hours. This would represent a relatively cold northern U.S. climate.

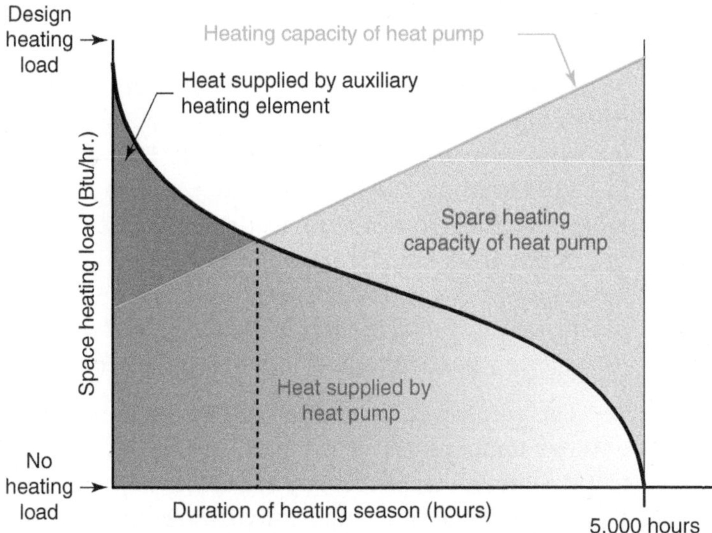

Figure 13-20 | How the heating load is supplied by a combination of an air-to-water heat pump, working simultaneously with an auxiliary heating element.

The portion of the graph to the left of the vertical dashed line represents times, during the heating season, when the output of the air-to-water heat pump is insufficient to meet the heating load. During these times, the electric resistance heating element within the heat pump provides the required additional heat. The refrigeration system with the heat pump is still operating during these times.

The shaded areas represented in **blue** and **orange** in Figure 13-20 represent the relative amounts of *energy* contributed by the air-to-water heat pump and electric resistance element, respectively. Ideally, the area represented by the heat pump would fill as much space as possible under the curve, while the area represented by the auxiliary heating element would be as small as possible. These areas will vary for each specific installation depending on the length of the heating season and the **temperature bin hours** associated with the climate.

In some systems, it is also possible use a gas-fired boiler rather than an electric element as a supplemental heat source. The heat pump would typically remain in operation down to some minimum outdoor air temperature below which its manufacture does not recommend operation. Some currently available air-to-water heat pumps can operate in outdoor air temperatures as low as −4 °F. At that point, the heat pump would be turned off and the boiler would handle all heat production. This scenario is represented in Figure 13-21.

Another way to view the heating ability of the air-to-water heat pump is shown in Figure 13-22.

In Figure 13-22, the heating load is assumed to increase linearly from where it begins, at an outdoor temperature of 65 °F, to a maximum of 40,000 Btu/hr., when the outdoor temperature is at its design value of 0 °F. The heat output available from the heat pump is assumed to be a maximum of 35,000 Btu/hr. when the outdoor air temperature is 65 °F, and decreases to 20,000 Btu/hr. at design load. The **balance point**, where heat pump output matches space heating load, occurs at an outdoor temperate of about 27 °F. At lower outdoor temperatures, some supplemental heat is needed. At outdoor temperatures above 27 °F, the heat pump has excess capacity. The design objective is to minimize supplemental heating where possible and cost-effective. This can be done by decreasing the design heating load in comparison to the nominal heating capacity of the heat pump. Warmer climates also move the balance point to the right and reduce supplemental heat requirements.

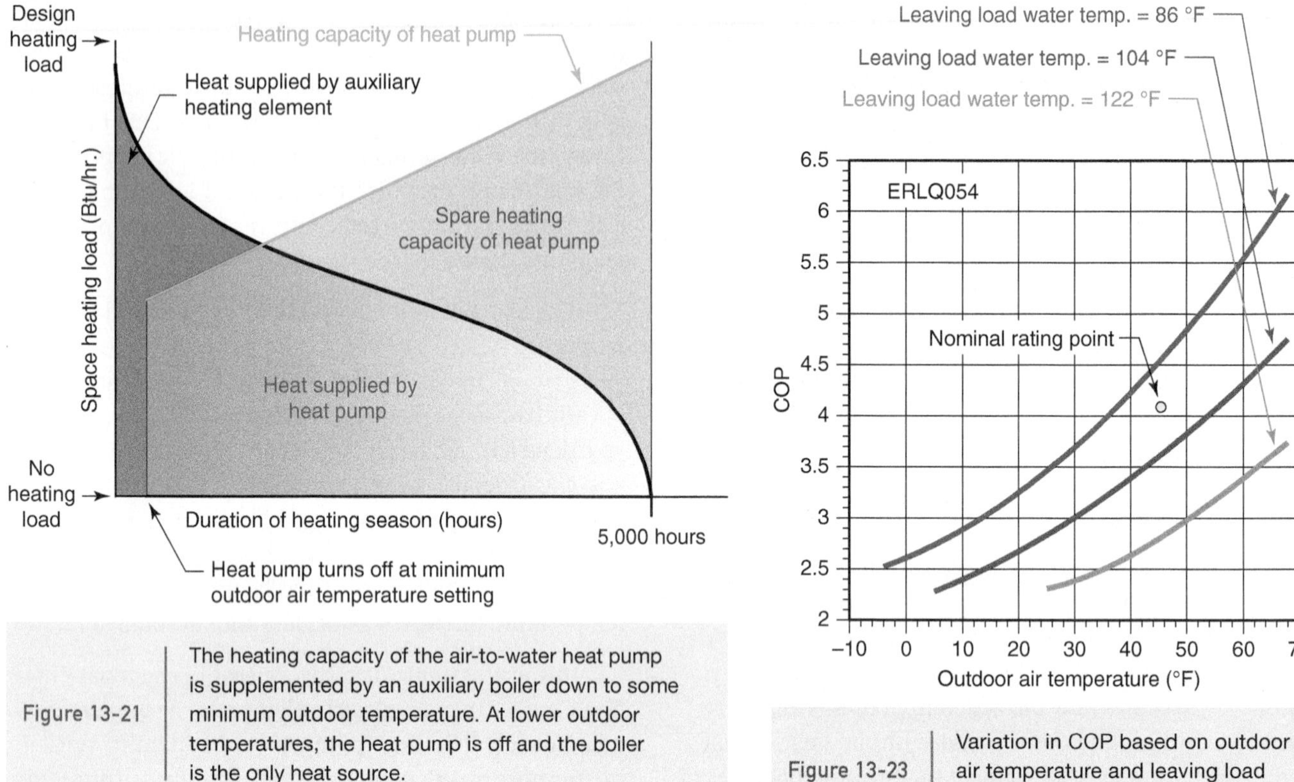

Figure 13-21 The heating capacity of the air-to-water heat pump is supplemented by an auxiliary boiler down to some minimum outdoor temperature. At lower outdoor temperatures, the heat pump is off and the boiler is the only heat source.

Figure 13-23 Variation in COP based on outdoor air temperature and leaving load water temperature.

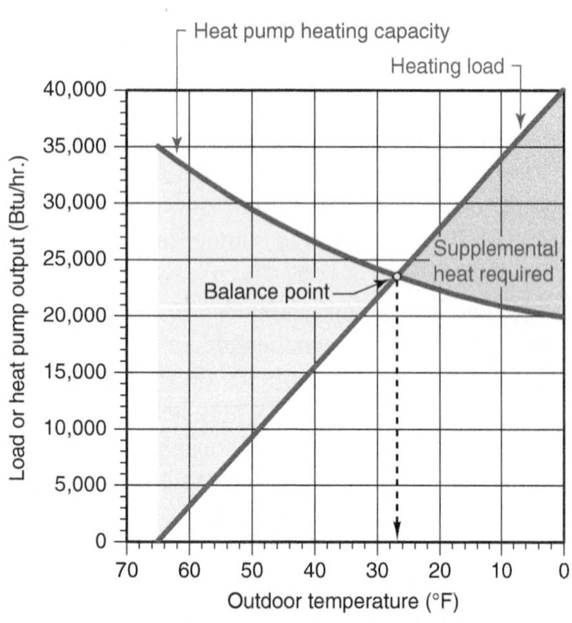

Figure 13-22 Heat pump output and supplemental heating required based on outdoor air temperature.

COEFFICIENT OF PERFORMANCE (COP)

Figure 13-23 shows how the COP of the same heat pump referenced in Figure 13-18 varies, based on outdoor air temperature and leaving load water temperature.

This graph shows that the heat pump's COP trends similar to its heating capacity. The lower the outdoor temperature, the lower the COP. The variation in COP based on leaving load water temperature is more pronounced than with heating capacity. Again, the implication for the designer is clear. *For good performance, it is imperative to match air-to-water heat pumps with hydronic distribution systems that operate at the lowest possible supply water temperature.*

The trend of improving COP with increasing outdoor temperature, and lower leaving load water temperature, is complementary to outdoor reset control. For example, assume that a hydronic distribution system requires a supply water temperature of 105 °F at an outdoor design temperature of 10 °F. The reset line for such a distribution system is shown in Figure 13-24.

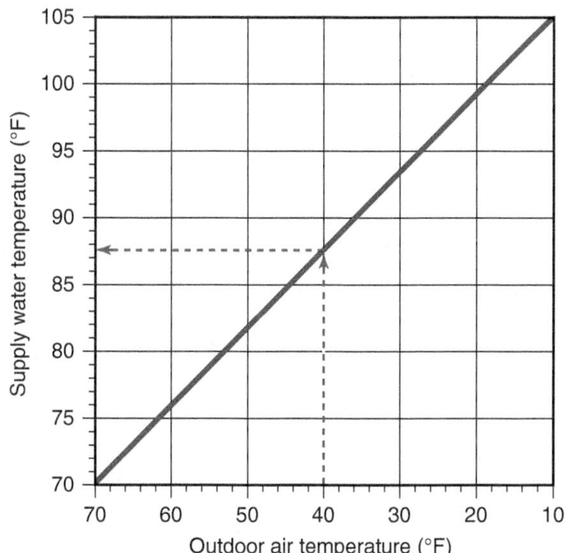

Figure 13-24 | Reset line for a low-temperature distribution system.

Assume this system is supplied by an air-to-water heat pump with a heating capacity described by Figure 3-19, and a COP described by Figure 13-23. At an outdoor design condition of 10 °F, the heating capacity is approximately 34,000 Btu/hr., and the COP is about 2.4. However, when the outdoor temperature rises to 40 °F, Figure 13-24 indicates that the distribution system only requires a supply temperature of 87.5 °F. Under this condition, the heat pump's COP is approximately 4.2.

At an outdoor temperature of 40 °F, the heat pump's heating capacity rises to about 52,000 Btu/hr. at the reduced supply water temperature of 87.5 °F. However, only a portion of this heating capacity is needed, because the heating load has decreased significantly.

Heat pumps with **variable-speed compressors** can respond to such situations by reducing their capacity to match the load. Heat pumps with one single-speed compressor would need to cycle on and off under these conditions. The duration of these cycles will be influenced by the size of the **buffer tank** used in the system. The larger the buffer tank, the longer the operating cycles of the heat pump. As with on/off boilers, longer and fewer on/off operating cycles are generally preferred over frequent and short on/off cycles. Longer cycles reduce stress on components and will likely increase the heat pump's life.

Model Number	Nominal Cooling Capacity (Btu/hr.)
EBLQ036	43,830
EBLQ048	54,570
EBLQ054	57,070

Outdoor air temperature: 95 °F (35 °C)
Water temperature leaving evaporator 64.4 °F (18 °C) (ΔT = 9 °F (5 °C))

Figure 13-25 | Nominal cooling capacity of three models of air-to-water heat pumps based on stated operating conditions.

COOLING CAPACITY

The cooling capacity of an air-to-water heat pump is also a strong function of both outdoor air temperature and chilled water temperature. Figure 13-25 shows the nominal cooling capacity of the same three air-to-water heat pump models represented (in heating mode) in Figure 13-18.

The nominal cooling capacities are based on an outdoor air temperature of 95 °F, and a water temperature of 64.4 °F *leaving* the evaporator. Figure 13-26 shows how the cooling capacity of the larger heat pump model (EBLQ054) heat pump is influenced by different outdoor air and chilled water temperatures.

To maximize cooling performance, designers should plan chilled-water-distribution systems to operate at the *highest* possible chilled water temperature that still allows for adequate latent cooling (e.g., moisture removal). Chilled water temperatures in the range of 50 °F to 60 °F will generally provide adequate performance, especially when combined with a deep cooling coil having multiple tube passes.

Figure 13-27 shows the energy efficiency ratio (EER) of the same heat pump.

The trending of EER mimics that of cooling capacity. Lower outdoor air temperatures and higher chilled water temperatures improve EER, and thus provide a given cooling effect using less electrical energy. These factors can be "exploited" by operating the heat pump under the most favorable conditions, which for cooling, is usually at night. Adding thermal storage can also help leverage these favorable operating conditions, as will be discussed later in this chapter.

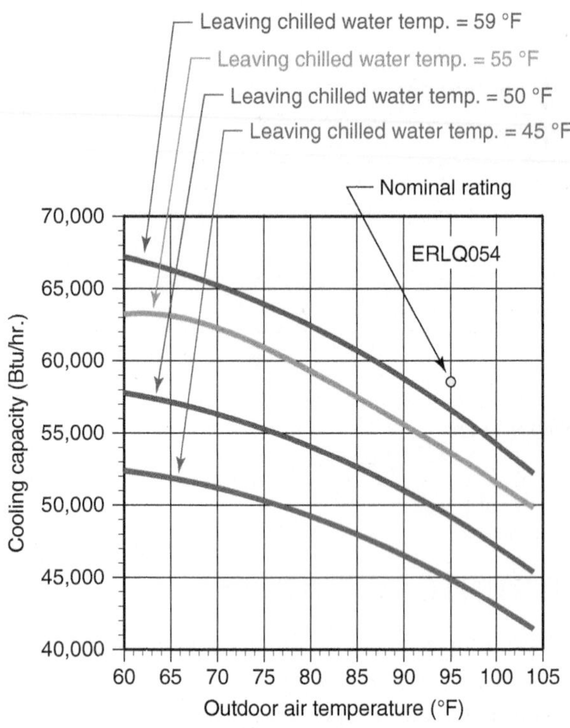

Figure 13-26 | Cooling capacity of a specific air-to-water heat pump. Yellow dot represents nominal cooling capacity rating of 57,070 Btu/hr.

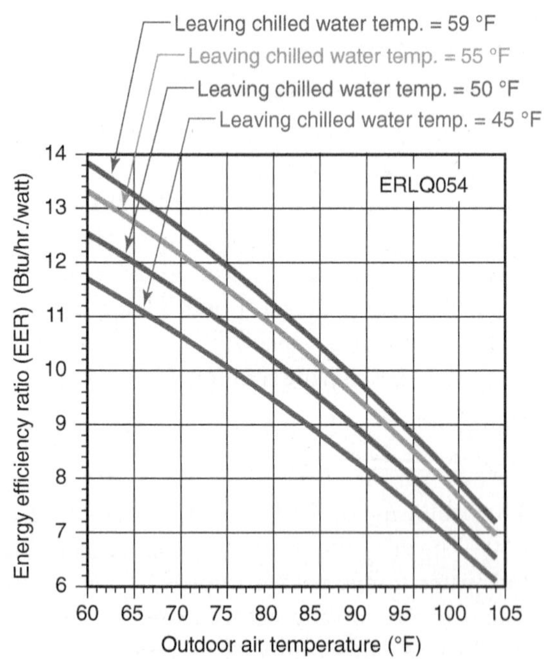

Figure 13-27 | Energy efficiency ratio (EER) of air-to-water heat pump as a function of outdoor temperature and temperature of chilled water leaving heat pump.

13.6 Design Details for Air-to-Water Heat Pumps

There are several details that system designers should address as they apply air-to-water heat pumps. This section provides an overview. Manufacturers installation and operation **(I/O) manuals** should always be referenced for specific installation requirements.

CAPACITY CONTROL

Some air-to-water heat pumps have an **inverter drive compressor**. Such compressors use powerful and permanent neodymium magnets in their rotor. The speed of the rotor, and hence the speed of the orbiting scroll, is controlled by high-speed switching of the magnetic poles in the motor's stator, similar to the ECM-based circulators discussed in earlier chapters.

The variable speed compressor allows the heat pump to regulate heat output between approximately 25 and 100 percent of its maximum possible capacity. This modulation allows the heat pump's capacity to better match the heating or cooling load as it varies throughout a day.

Depending on how extensively the distribution system is zoned, systems with modulating capacity heat pumps *may not* need buffer tanks between the heat pump and the distribution system. Even in heavily zoned systems, a modulating heat pump often reduces the size of the buffer tank compared to that required in systems using on/off heat pumps. Thus, modulating output can simplify design and lower installation cost compared to systems using on/off heat pumps.

A two-stage heat pump, or multiple two-stage heat pumps controlled as a group, also provides a good degree of capacity control, and thus provides design flexibility similar to that of a modulating heat pump. Multiple/staged on/off heat pumps are particularly useful in larger homes or light commercial buildings where design loads typically exceed a nominal 5 tons.

CIRCULATORS

Figures 13-14a and 13-15c show a circulator within the heat pump. Some manufacturers provide this circulator, while others do not. When it is provided, designers should always check the "net pump curve"

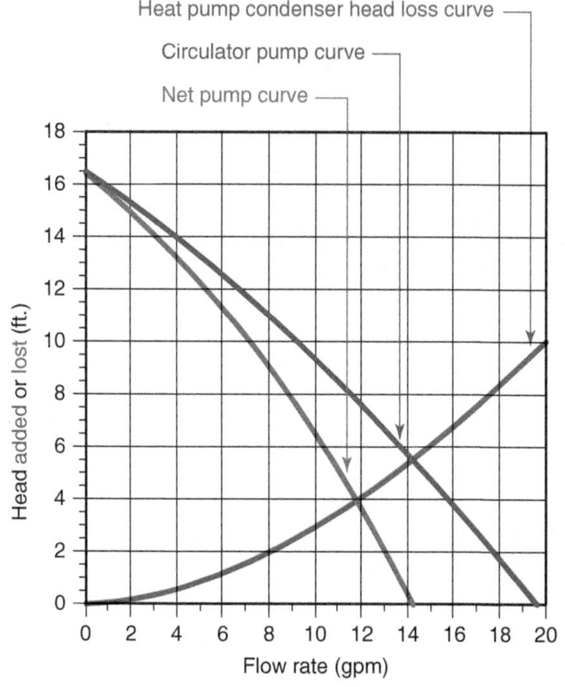

Figure 13-28 The net pump curve available from an internal circulator is that circulator's pump curve, minus the head loss curve of the heat pump's condenser.

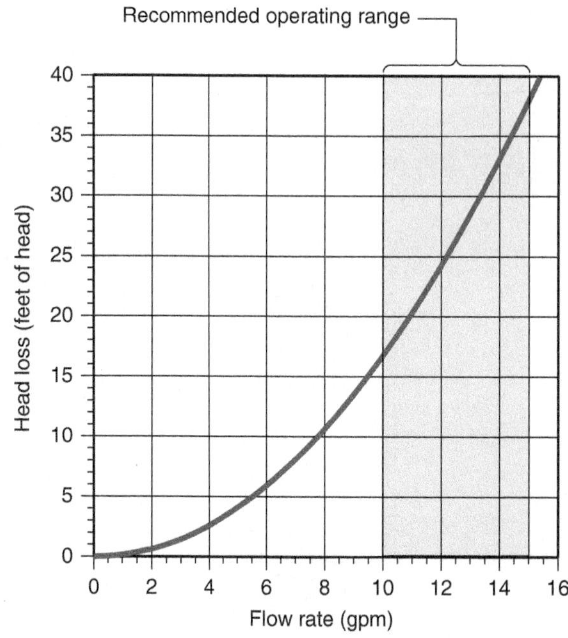

Figure 13-29 Head loss versus flow rate for a specific make and model of air-to-water heat pump.

available from the circulator. The net pump curve is the circulator's normal pump curve minus the head loss curve associated with moving flow through the heat pump's refrigerant to water heat exchanger, as illustrated in Figure 13-28.

In most cases, an included circulator is sufficient to move flow through the heat pump and a relatively small primary circuit, or buffer tank circuit, within the building. Never assume this circulator has a sufficient pump curve to create flow through an entire hydronic distribution system.

HEAD LOSS

As is true with all hydronic components, there is head loss associated with flow through the refrigerant-to-water heat exchanger in an air-to-water heat pump. This head loss will be specific to each make and model. Designers must factor this head loss in with that of the other piping and components in the circuit between the heat pump and balance of the system. Figure 13-29 shows an example of the head loss through one specific air-to-water heat pump with a nominal heating capacity of 60,000 Btu/hr.

Designers should also verify the minimum and maximum flow rates recommended for a specific make and model of heat pump. Absent any specified requirements, general guidelines are for flow rates in the range of 1.5 to 2.5 gpm *per nominal ton* of heating capacity.

EXPANSION TANKS

Some air-to-water heat pumps contain a "pancake"-style expansion tank within their outdoor unit, or in the case of a split system, within the indoor unit. This expansion tank is sufficient to accommodate the expansion of the fluid in the heat pump and some assumed indoor piping. Do not assume this tank has sufficient capacity to handle the expansion needs of a large residential hydronic distribution system. In such cases, it may be necessary to size a **supplemental expansion tank** for the system. The connection point for this supplemental tank should be close to the connection point for the internal expansion tank and upstream of the circulator, as shown in Figure 13-30. Also be sure to adjust the air pressure in the diaphragms of both tanks to the same value.

FLOW SWITCHES

Some air-to-water heat pumps operate with water, rather than an antifreeze solution in their refrigerant-to-water heat exchanger. In such applications, a **flow switch** should always be used to verify sufficient water

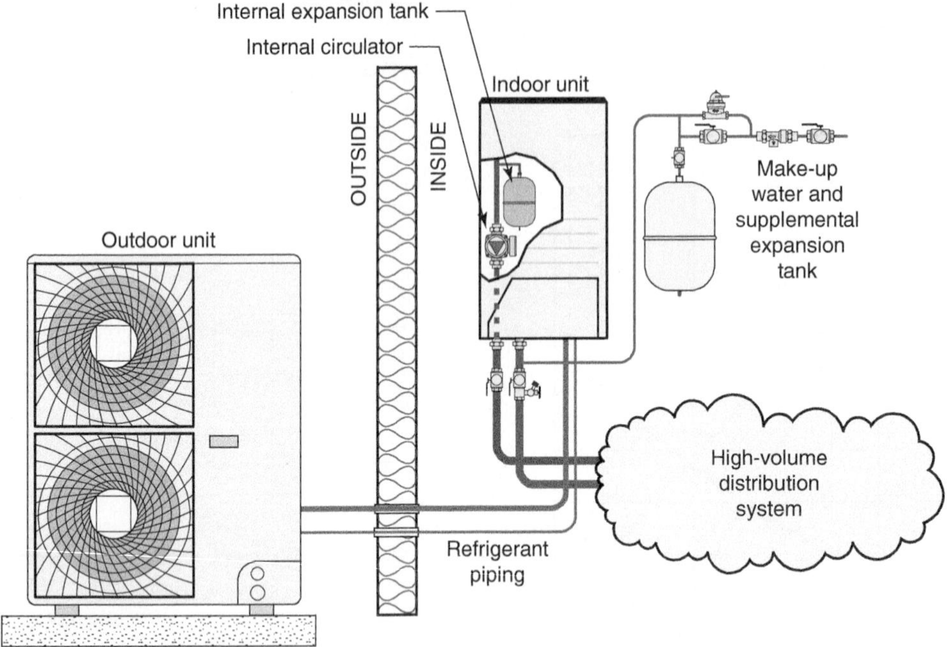

Figure 13-30 | Placement of a supplemental expansion tank in system when the heat pump's internal expansion tank is not large enough to handle the system's total expansion volume.

flow through the refrigerant-to-water heat exchanger in the heat pump when it is operating as an evaporator. Without proper flow, the water within the heat pump's evaporator could freeze due to refrigerant flowing through the evaporator at subfreezing temperatures. Such a freeze could eventually rupture the heat exchanger. The flow switch can prevent this by turning of the refrigeration system whenever insufficient flow is detected.

Some manufactures provide an internal flow switch within their heat pump, as seen in Figure 13-31. Others require the installation of an external flow switch, as seen in Figure 13-32.

Most flow switches are mounted on a tee, as shown in Figure 13-33.

The flow switch shown in Figure 13-33 is composed of a blade (1) fastened to a control rod (2) connected at the top to an adjustable counter spring (3). This assembly pivots when the flow rate past the blade reaches or exceeds the set value. The pivoting action operates a microswitch contained in a protective casing (4). At rest, the counter spring keeps the microswitch contact open. When flow decreases to the adjustable opening value, the internal spring force overcomes the fluid thrust against the blade, and the microswitch contacts open. The flow rates for closing (increasing flow) and opening (decreasing flow) can be modified by means of the adjusting screw (6). A stainless steel bellows (7) separates the electric and the hydraulic parts, preventing any contact between the fluid and the electric components. The length of the blade can be changed to accommodate a wide range of pipe sizes.

Figure 13-31 | Internal flow switch within an air-to-water heat pump.

Figure 13-34 shows how a flow switch should be located between the heat pump and the balance of system.

The flow switch should be adjusted to detect a minimum acceptable water flow rate through the heat exchanger. A typical control sequence for this system, operating in cooling mode, would be as follows: (1) A temperature setpoint controller determines when the buffer tank requires cooling and closes its electrical contacts in response. (2) A low voltage control signal passes through these contacts to a relay that turns on the circulator to create water flow through the evaporator. (3) The electrical contacts within the flow switch close when a sufficient flow rate is established through the flat-plate heat exchanger. (4) The closed contacts in the flow switch complete a low voltage control circuit that allows the refrigeration system to operate. If the flow rate falls below a minimum value, the refrigeration system is immediately turned off.

External flow switches

Figure 13-32 | Flow switches mounted external to heat pump. Note that all piping is insulated to prevent condensation.

FREEZE PROTECTION

Some air-to-water heat pumps contain control logic that monitors the water temperature within the unit. If that temperature approaches freezing (typically 34 °F or 35 °F),

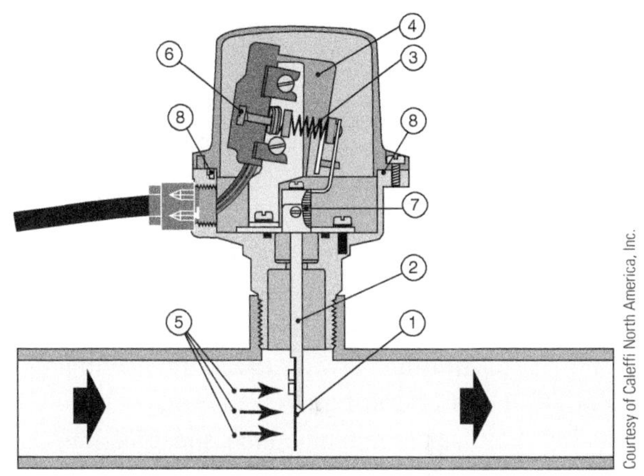

Courtesy of Caleffi North America, Inc.

Figure 13-33 | Internal components of a flow switch.

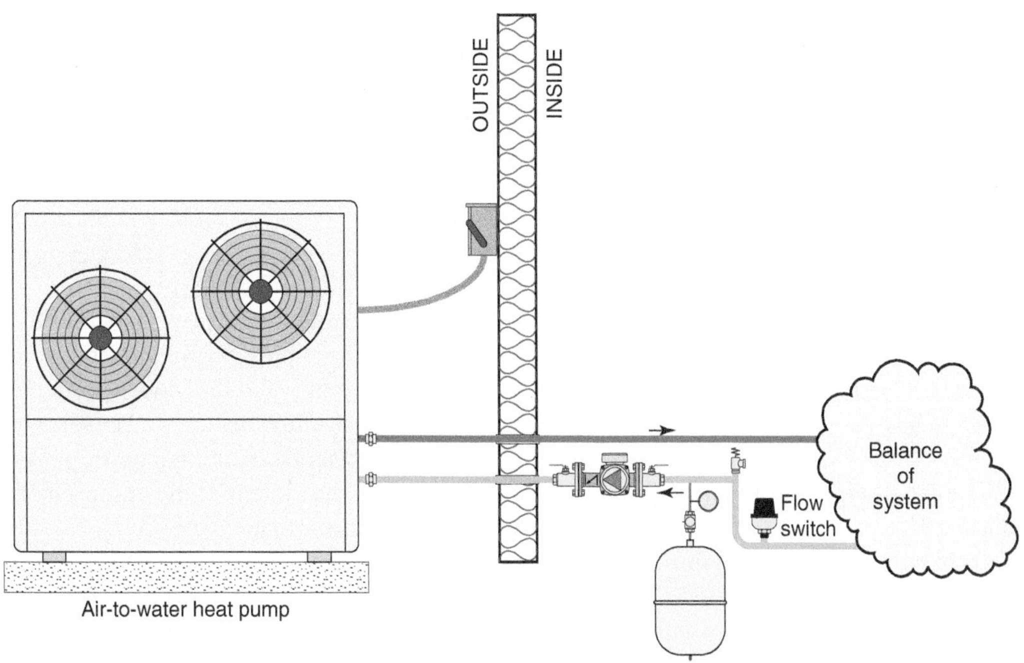

Figure 13-34 | Placement of flow switch in water-filled circuit supplied by an air-to-water heat pump.

the heat pump's internal controller turns on an internal electric heating element that maintains the water at a temperature of about 38 °F, even when the refrigeration system is not operating. The circulator that creates flow through the heat pump is also turned on in this mode. If heat from the electric resistance element is unable to maintain the water temperature safely above freezing, the heat pump may turn itself on in the heating mode to provide additional heat input.

In mild climates, where subfreezing temperatures are possible but limited to a few days per year, it is generally acceptable to operate a self-contained air-to-water heat pump with water flowing through the outdoor. This is beneficial because it eliminates the need to either fill the entire system with an antifreeze solution or install a heat exchanger between the heat pump circuit and the remainder of the system. Such a heat exchanger adds to installation cost and imposes a thermal penalty. The latter will reduce the heat pump's heating and cooling capacity, as well as its COP and EER. Some manufacturers *require* their heat pump be operated with an antifreeze solution. Always check the manufacture's installation and operating (I/O) manual to ensure the installation is "as per manufacture's specifications."

In colder climates, where subfreezing temperatures are persistent throughout the heating season, any self-contained air-to-water heat pump mounted outside the building should be protected by an antifreeze solution. In systems with limited interior piping volume, the entire system is typically filled with a non-toxic solution of inhibited propylene glycol antifreeze. In systems with a substantial interior water volume, a brazed-plate heat exchanger is typically installed between the heat pump circuit and remaining portions of the system, as shown in Figure 13-14. *This heat exchanger should sized for an approach temperature difference no higher than 5 °F, while transferring the heat pump's full rated heating capacity. This minimizes the thermal penalty associated with its use.*

DIRT SEPARATION

All air-to-water heat pumps are adversely affected by any dirt that may be carried into their refrigerant-to-water heat exchangers. This can be minimized by ensuring that all water piping connected to the unit is thoroughly flushed and preferably washed with a suitable **hydronic detergent** that can remove any solder flux, cutting oils, or other impurities that may be in the piping as it is assembled.

(a)

Courtesy of Caleffi North America, Inc.

(b)

Courtesy of Caleffi North America, Inc.

Figure 13-35 (a) Dirt separator to remove particles within heat pump piping circuit. (b) Cross section of a combined air and dirt separator.

An additional recommendation is to install a **dirt separator**, such as shown in Figure 13-35a, which can capture very small particles within the flow stream and provide a means of flushing them from the system. This separator should be installed on the piping leading into the heat pump. It should also be installed so that its drainage valve is fully accessible. Another option is to install a combined air and dirt separator as shown, in cross section, in Figure 13-35b. This device combines high-efficiency air separation with high-efficiency dirt separation. Both devices shown in Figure 13-35 also have a removable magnet collar that enhances capture of any ferrous metal particles in the system.

OTHER HYDRONIC TRIM

It is good practice to provide isolation and purging valves at the piping between a self-contained air-to-water heat pump and the indoor portion of the system. These valves allow the unit to be isolated, drained, and removed, if necessary, without disrupting the remaining portions of the system. If the unit is piped to an interior heat exchanger, that circuit should be equipped with an expansion tank, pressure-relief valve, pressure gauge, and means of filling and purging the circuit. These details are shown in Figure 13-36. This circuit should *not* be equipped with a make-up water system. The latter, if present, could potentially dilute the antifreeze if a small leak were to develop within the circuit.

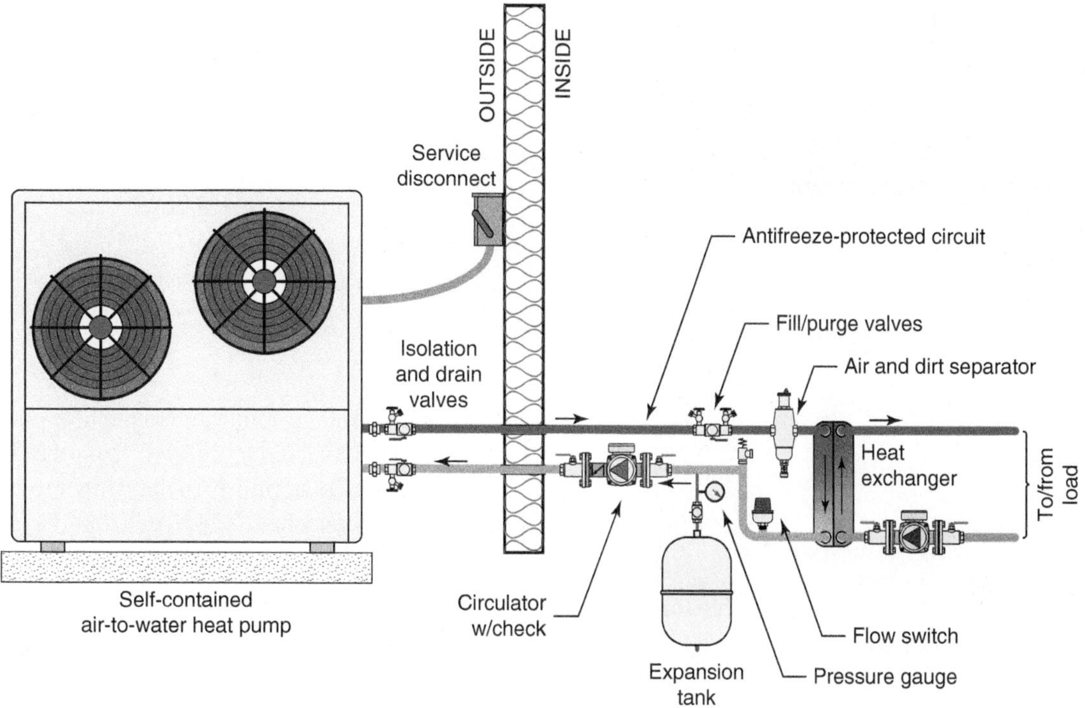

Figure 13-36 | Self contained air-to-water heat pump with antifreeze protection and other hydronic trim.

OUTDOOR UNIT PLACEMENT

All air-to-water heat pumps change the temperature of the air as it passes through them. In the heating mode, the air is substantially cooled as it passes through the unit. In the cooling mode, the existing air is substantially warmed as it leaves the unit. Designers should always consider this when planning placement of the outdoor unit.

The following general guidelines are appropriate. Consult the installation and operating (I/O) manual for each specific heat pump to verify any other specific installation requirements.

- Mount outdoor unit so that there is full access to all front and side service panels.
- Do not place the unit so that its air intake side faces directly into the prevailing wind. The preferred orientation is with the air intake side facing perpendicular to prevailing winds.
- Do not place the unit close to windows, especially those in sleeping areas or other locations where the sound of the unit's compressor and fans would be objectionable.
- Do not place the unit where the air intake side will be subject to accumulating grass clipping, leaves, or other debris.
- Do not place the unit directly under the edge of a roof because rain runoff will impinge on it.
- Allow at least 12 inches of air space between the air intake side of the unit and any adjacent wall or other structure.
- Always allow for drainage of condensate from the unit. Provide a suitable runoff so that condensate will not form puddles under or near the unit.
- Do not install multiple heat pumps in such a way that the air discharging from one unit is directed toward the intake of another unit that is within 8 feet of it.
- If two outdoor units are installed with their air *discharge* sides facing each other, allow at least 7 feet of space between the units.
- When two outdoor units are installed with their air *intake* sides facing each other, allow at least 2 feet of space between units.
- In cold climates, the unit must be elevated above accumulating snow.
- The preferred installation in cold climates is in a semi-protected alcove, as shown in Figure 13-37. Such an installation partially shelters the unit from precipitation and wind while still allowing good air flow space on the air intake side. It also elevates the units above snow accumulation. Installing the units with parallel air flow, as shown in Figure 13-37, also prevents any significant interference between their air flow patterns.

Courtesy of ReVision Energy

Figure 13-37 | Installation of two air-to-water heat pumps in alcove that provides partial protection from the elements.

In cold climates, where winter heating dominates over summer cooling, placing the heat pump in proximity to southerly facing walls provides several benefits. First, it creates a microclimate that tends to slightly warm the air around the units in winter. This will improve thermal performance. Southern exposure will also help reduce snow accumulation. The overhang of the alcove, such as shown in Figure 13-37, can also be sized to provide shading of the heat pumps in summer while still allowing them to be exposed to low sun angles in winter.

LOW-VOLTAGE ELECTRICAL INTERFACING

There are several variations of air-to-water heat pumps on the market. Some are single-speed on/off units, others have 2-stage refrigeration systems, and still others have variable-speed compressors. Depending on the configuration of their internal control circuits, different heat pumps will require different types of external control inputs to initiate operation in heating or cooling mode. *The electrical schematics shown in the example systems presented later in this chapter assume the generic low-voltage terminal strip interface shown in Figure 13-38.*

The (R) terminal is a source of 24 VAC from a transformer inside the heat pump. The (C) terminal is the common side of that transformer. There should always be 24 VAC between the (R) and (C) terminals whenever

R 24 VAC
Y Compressor
O Reversing valve
C Common

Figure 13-38 | Typical low-voltage terminal strip in a single speed on/off air-to-water heat pump.

there is line voltage connected to the heat pump. The (Y) terminal, if connected to 24 VAC from the (R) terminal, turns on the heat pump in heating mode. The (O) terminal, if connected to 24 VAC from the (R) terminal, energizes the heat pump's refrigerant reversing valve to enable the unit for cooling operation.

To turn the heat pump on in the heating mode, a "dry" (e.g., unpowered) contact closure must be made between the (R) and (Y) terminals. This must also be the case to operate the unit in cooling. However, in the cooling mode, the reversing valve must also be energized with 24 VAC. This requires the wiring shown in Figure 13-39. Here, an extra 24 VAC relay is used to energize both (Y) and (O) when there is a cooling demand.

Designers should verify the exact electrical control interfacing required for the heat pump they select and ensure the system wiring coordinates with these requirements. The manufacturer's installation and operating (I/O) manual for the heat pump typically shows a detailed electrical schematic for the internal wiring of the heat pump and describes what is necessary to turn the unit on in heating or cooling mode.

LINE VOLTAGE ELECTRICAL INTERFACING

Most of the air-to-water heat pumps designed for residential and light commercial buildings are operated by a 240 VAC, single-phase power circuit. The ampacity of this circuit is dictated by the capacity of the heat pump and typically specified in the manufacturer's installation and operating (I/O) manual. The **National Electrical Code (NEC)** requires a weatherproof outdoor disconnect adjacent to the outdoor portion of the heat pump. Pulling the handle of this disconnect to the off position turns off all electrical power to the outdoor unit. The NEC and/or local electrical codes may require addition disconnects or service switches for the indoor portion of the system.

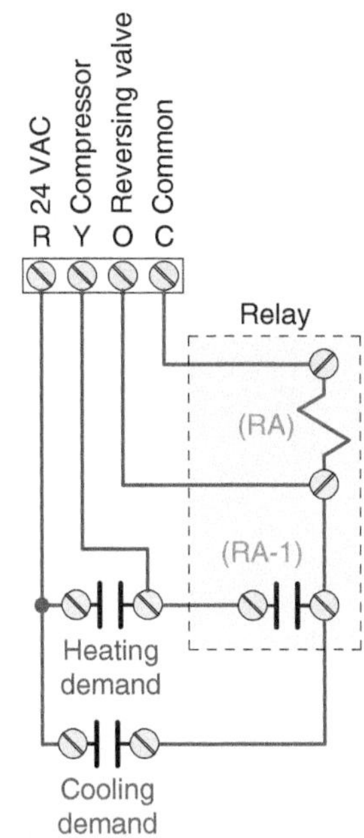

Figure 13-39 Wiring between dry contacts that call for heating or cooling, and the typical low-voltage terminal strip in a single-speed on/off heat pump. The relay enables 24 VAC from the R terminal to pass to both the Y terminal and O terminal for cooling operation.

Some air-to-water heat pumps contain electrical resistance heating elements that add heating capacity when needed. Check with the manufacturer's installation and operating (I/O) manual for specific wiring requirements for these elements. Some heat pumps may require a separate, dedicated 240 VAC circuit for the heating elements. Others may draw element power from a single 240 VAC circuit serving the entire heat pump.

13.7 Thermal Storage and Air-to-Water Heat Pumps

There are several applications in which air-to-water heat pumps can benefit from being combined with thermal storage. The following are the most common applications:

- Thermal storage is used as a buffer to prevent short cycling a heat pump that supplies a highly zoned distribution system.
- High-volume thermal storage is to provide **diurnal operation** of the heat pump under favored conditions.

The use of thermal storage for buffering is especially important in systems that have an extensively zone distribution system supplied by a single speed or two-speed heat pump. *Buffer tanks allow the rate of heat production by the heat pump to be significantly different from the rate of heat delivery by the distribution system.* Their use prevents the heat pump from short cycling. This undesirable condition causes the heat pump's compressor to turn on and off in rapid succession. Some of the on-cycles might be less than a minute in duration. This leads to premature wear on components such as compressor contactors, and the compressor motor itself.

Most air-to-water heat pumps are designed for use with closed pressurized thermal storage tanks. A typical installation, including an isolating heat exchanger for the heat pump, is shown in Figure 13-40.

In this system, the air-to-water heat pump is tuned on and off by an outdoor reset controller. This controller monitors the temperature of the water inside the buffer tank as well as outdoor temperature. When active, the outdoor reset controller uses the current outdoor temperature, along with its settings, to calculate the target supply water temperature required by the distribution system. It then operates the heat pump as necessary to maintain the tank water within a few degrees of this target temperature. This operating logic, discussed in several previous chapters, is illustrated in Figure 13-41. This logic is in effect whenever one or more zones of the space heating system are calling for heat.

SIZING A BUFFER TANK

For systems with an on/off heat pump, the size of the buffer tank is determined by the desired "**on-time**" of the heat pump, and the acceptable temperature rise of the water in the buffer tank during this on-time. The on-time is the time between when the heat pump turns on to begin warming the tank and when it turns off after increasing the tank temperature through the selected temperature rise. Equation 13.5 can be used to determine tank size.

(Equation 13.5)

$$V_{\text{btank}} = \frac{t(Q_{\text{HP}} - Q_{\text{L}})}{500(\Delta T)}$$

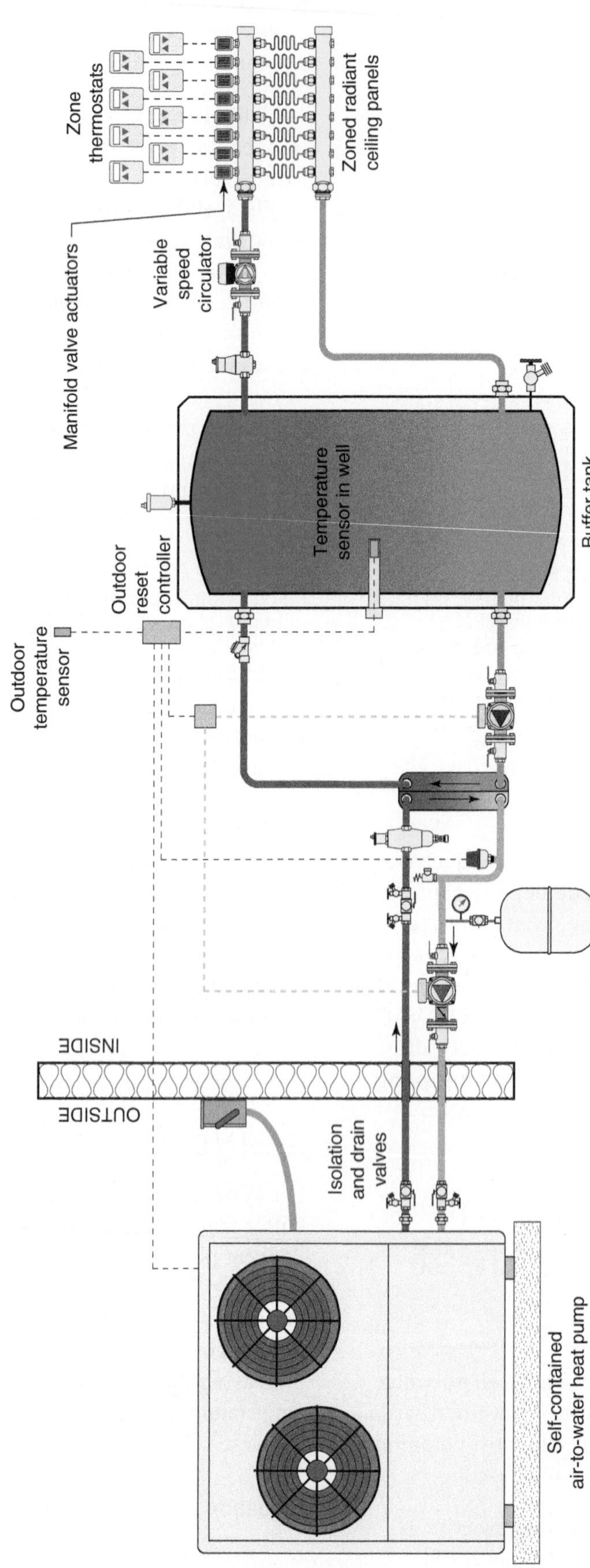

Figure 13-40 | Connecting an air-to-water heat pump with a pressurized buffer tank.

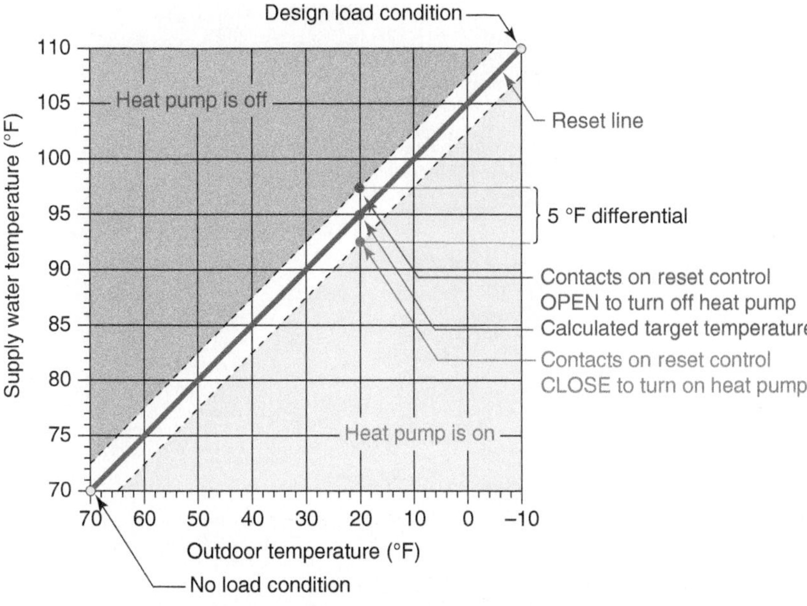

Figure 13-41 Temperature in buffer tank is adjusted by an outdoor reset controller.

where:

V_{btank} = required volume of buffer tank (gallons)

t = desired on-time for heat source (minutes)

Q_{HP} = heating capacity of heat pump (Btu/hr.)

Q_L = any heating load served by buffer tank while charging (Btu/hr.)

ΔT = selected temperature rise of tank during heat pump on-time (°F)

Example 13.2: Determine the size of a buffer tank that will absorb 48,000 Btu/hr. from the heat pump while increasing in temperature from 90 °F to 110 °F, during a heat pump on-cycle of 10 minutes. Assume there is no concurrent heating load removing heat from the tank during this on-cycle.

Solution: The selected temperature rise (ΔT) is 110 − 90 = 20 °F. Putting this and the remaining data into Equation 13.5 yields:

$$V_{btank} = \frac{t(Q_{HP} - Q_L)}{500(\Delta T)} = \frac{10(48{,}000 - 0)}{500(20)} = 48 \text{ gallons}$$

Discussion: If the allowed temperature rise was 10 °F, rather than 20 °F, the required tank volume would double to 96 gallons. If the desired on-time was only 5 minutes rather than 10 minutes, the volume would be cut in half. Anything that increases the desired on-time, or decreases the selected temperature rise during this on-time, will increase the required tank volume, and vice versa.

The most conservative estimate for buffer tank size is made when the value of Q_L is zero (e.g., there is no load removing heat from the buffer tank when the heat pump is operating). However, in most systems, when there is no load "calling" for heat, the outdoor reset controller will not be energized, and the heat pump will not be running. This makes the tank volume calculated using Equation 13.5 conservatively large.

Equation 13.5 can also be used to determine the size of a buffer tank used for cooling. The value of Q_{HP} would be the *cooling capacity* of the heat pump. The value of ΔT would be the allowed temperature drop in the tank during this on-time.

In systems with a *modulating* heat pump, Equation 13.5 can be slightly modified into Equation 13.6:

(Equation 13.6)

$$V_{btank} = \frac{t(Q_{hpmin} - Q_{Lmin})}{500(\Delta T)}$$

where:

V_{btank} = required volume of buffer tank (gallons)

t = desired on-time for heat source (minutes)

Q_{hpmin} = minimum stable output capacity of heat pump (Btu/hr.)

Q_{Lmin} = smallest zone load that may be active (Btu/hr.)

ΔT = allowed temperature rise of tank during heat pump on-time (°F)

Example 13.3: A heat pump has a nominal output of 55,000 Btu/hr. and can modulate down to 25 percent of this output. Assume the system it supplies has several zones, and the smallest of these zones requires 4,000 Btu/hr. under the same outdoor conditions at which the nominal heating capacity of the heat pump was established. The designer decides that a 12 °F temperature swing in the buffer tank is acceptable and that the minimum run time for the heat pump is 10 minutes. Size a buffer tank for this application.

Solution: The minimum heating capacity of the modulating heat pump is 0.25 (55,000) = 13,750 Btu/hr. Substituting this and the other given information into Equation 13.6 yields:

$$V_{\text{btank}} = \frac{t(Q_{\text{hpmin}} - Q_{\text{Lmin}})}{500(\Delta T)} = \frac{10(13{,}750 - 4{,}000)}{500(12)}$$
$$= 16.3 \text{ gallons}$$

Discussion: This is a relatively small tank compared to that in the previous example. The primary reason is the ability of the heat pump to modulate to a much lower output compared to the minimum load requirement. In some systems, this relatively small volume might be achieved through upsized piping and other devices such as a hydraulic separator. In such cases, there would be no need to install a separate buffer tank.

Figure 13-42 | A 1,200-gallon unpressurized thermal storage tank.

DIURNAL HEAT STORAGE

Another application for air-to-water heat pumps paired with thermal storage allows the heat pump to operate under the most favorable ambient conditions, independent of the current demand for space heating. This is called **diurnal heat storage**. One example is operating the heat pump during the daytime hours of a winter day, when the ambient air temperatures are usually higher than at night. This improves the heat pump's heating capacity and COP. The heat delivered by the heat pump would be temporarily stored in a large, well-insulated thermal storage tank.

In cooling mode, the heat pump would be operated during the night, when outdoor temperatures are usually lower. This improves the heat pump's cooling capacity as well as its EER. The chilled water is stored in the tank for use the following day.

Diurnal heat storage can require large-volume tanks, in some cases over 1,000 gallons. It also requires very well-insulated tanks, because heat will often be stored for several hours, perhaps even two or three days, until needed by the load.

Figure 13-42 shows an example of a large, unpressurized thermal storage tank, with a volume of 1,200 gallons and insulated with 8 inches of polystyrene foam.

These large thermal storage tanks are typically located in basements. Their size mandates well-planned placements.

Designers should also remember that these tanks are vented to the atmosphere and are therefore "open" devices. Thus, all the design details related to unpressurized tanks, and discussed in Chapter 9, apply to their use.

Figure 13-43 shows how a large, well-insulated thermal storage tank could be combined with an air-to-water heat pump in a heating-only application.

Note the piping details for moving heat between the heat exchanger and the unpressurized tank. First, the piping connections on the tank penetrate through its sidewall and above the high water level. This eliminates potential leakage at the piping penetrations. The **gooseneck piping** above the water level in the tank will be under slight sub-atmospheric pressure. This portion of the piping should not contain air vents, valves, or other devices that could allow air to be sucked in.

Second, a stainless steel circulator is used because the tank is vented to the atmosphere. This circulator is mounted as low as practical to maximize static pressure on the circulator, which reduces the potential for cavitation.

Third, the heat exchanger is piped for counterflow to maximize the heat transfer rate for a given set of operating conditions.

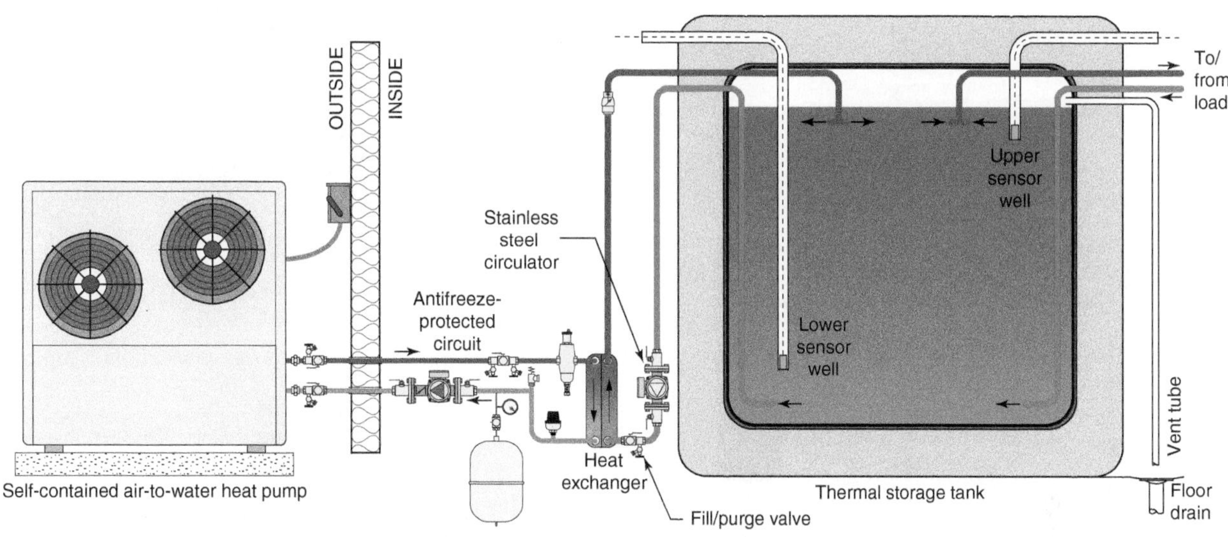

Figure 13-43 | Coupling an air-to-water heat pump to a large thermal storage tank for diurnal heat storage.

Finally, the external piping is connected with the piping inside the storage tank so that incoming water is released horizontally rather than vertically. This minimizes vertical mixing within the tank and thus helps preserve temperature stratification.

DIURNAL STORAGE FOR HEATING AND COOLING

Because most air-to-water heat pumps are reversible, it is possible to use diurnal heat storage for heating as well as cooling. However, this requires piping and control details that allow the thermal storage tank to remain properly stratified in each mode. When the system is in the heating mode, these details must allow heated water from the heat exchanger to be introduced into the upper portion of the tank, while cooler water in the lower portion of the tank is routed back to the heat exchanger. On the load side of the system, the hottest water in the upper portion of the tank must be sent to the load, while cooler water returning from the load is reintroduced near the bottom of the tank. When the system operates in cooling mode, these flow directions must all be reversed.

One way to create this **flow reversal** is though use of **4-way valves** coupled to **2-position spring-return actuators**. The type of 4-way valve needed is the same valve that is often used for mixing hot water from a conventional boiler with cooler water returning from a low-temperature distribution system. An example of such a valve is shown in Figure 13-44.

Figure 13-44 | Example of 4-way valve.

Figure 13-45 shows how the internal vane within this valve is used to direct flow when the valve is used as a flow-reversing device.

In heating mode, the vane is at the 0° position. Flow is directed upward within the thermal storage tank. This routes the hottest water to the load. In the cooling mode, the vane rotates 90°. This directs flow downward within the tank and sends the coolest available water to the load.

Figure 13-46 shows how a 2-position spring return actuator can be mounted to the 4-way valve.

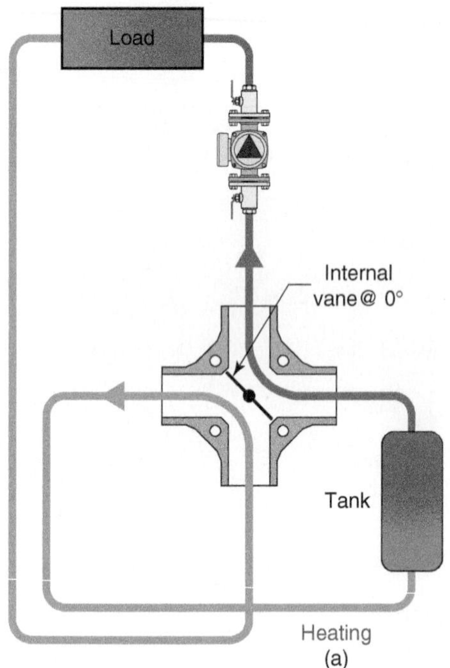

(a)

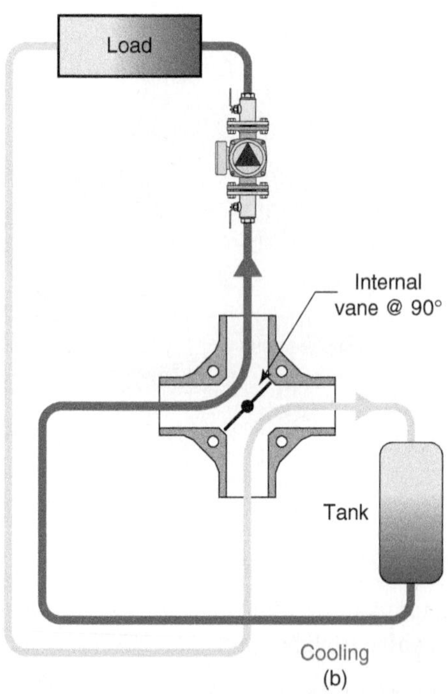

(b)

Figure 13-45 Internal vane of 4-way valve rotates between 0° and 90° position to reverse flow between the thermal storage tank and load. (a) Heating mode requires upward flow through tank. (b) Cooling mode requires downward flow through tank.

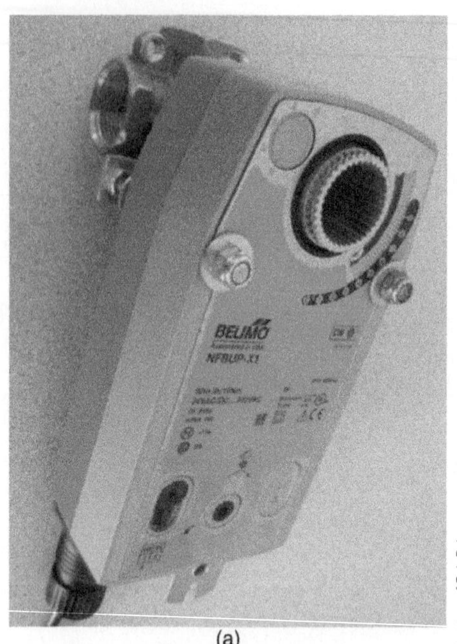

(a)

Courtesy of Bob Rohr

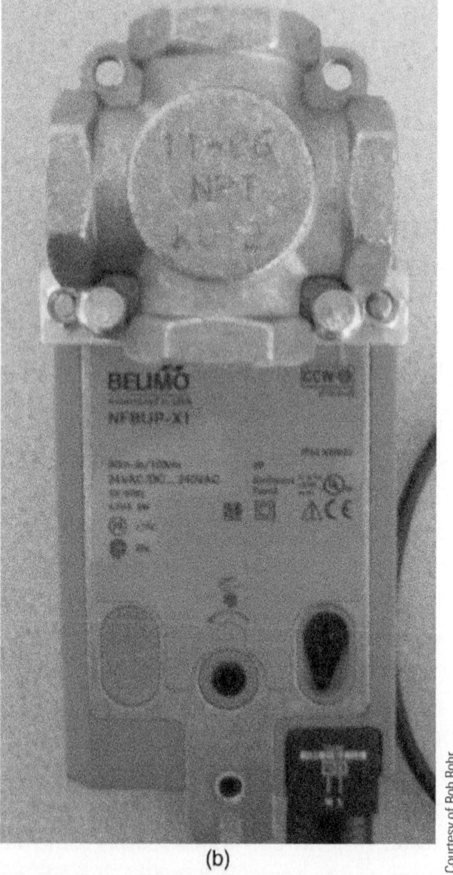

(b)

Courtesy of Bob Rohr

Figure 13-46 Mounting a 2-position spring-return actuator to a 4-way valve. (a) Top view. (b) Bottom view. Note that an adapter plate is used to bolt the valve's mounting holes to the actuator.

The 2-position actuator provides the shaft rotation necessary to move the vane between the 0° and 90° positions. A typical configuration has the vane at the 0° position when the actuator is not powered. When 24 VAC is supplied to the actuator, it rotates the valve's shaft so that the vane is at the 90° position. The time required for this rotation can sometimes be adjusted on the actuator. When power is removed from the actuator, a spring within the actuator rotates the valve's shaft back to the 0° position. Thus, a simple relay contact closure can be used, in combination with this actuator and valve, to control flow reversal.

Figure 13-47 shows how *two* 4-way motorized valves, configured for flow reversal, can be used in combination with diurnal thermal storage and a split system air-to-water heat pump.

The heat exchanger is required between the split system heat pump and thermal storage because the refrigerant-to-water heat exchanger within the indoor portion of the heat pump is constructed of carbon steel. As such, it is not suitable for a direct connection to an open thermal storage tank. Note that all circulators used in circuits connected to the unpressurized thermal

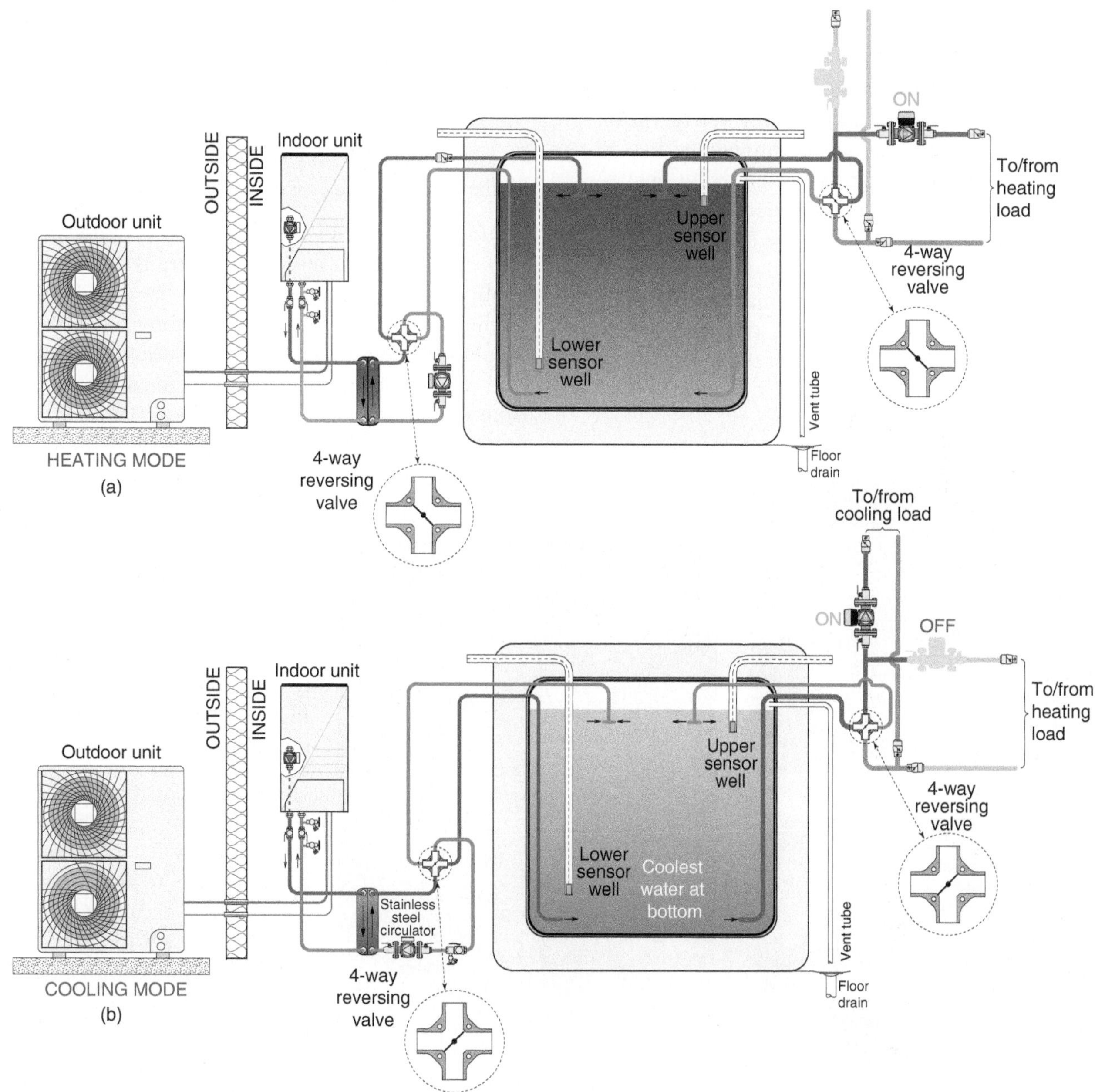

Figure 13-47 (a) Two 4-way valves used for flow reversal on heat pump side and load side of thermal storage. System shown in heating mode. (b) Same system operating in cooling mode.

storage tank must be of stainless steel or bronze construction. No ferrous metal components can be used in these circuits due to corrosion potential.

SPLIT SYSTEM AIR-TO-WATER HEAT PUMP MODULE

Another approach that combines the features of an air source heat pump with hydronic heating and cooling delivery is shown in Figure 13-48.

The DX2W module allows several readily available heat pump condenser units to provide heating and cooling through a hydronic distribution system.

In preferred heating mode operation, the outdoor unit serves as the evaporator for the refrigeration system. Hot refrigerant gas passes through the coaxial condenser within the module, transferring heat to water. This water is circulated to a nearby buffer tank using the module's internal circulator. The temperature of the buffer tank is regulated based on outdoor reset control. This maximizes the thermal performance of the heat pump by not raising the buffer tank temperature above what is needed for the current space heating load.

The buffer tank prevents the heat pump refrigeration system from short cycling when supplying a highly zoned hydronic distribution system operating under partial load conditions.

Upon a demand for space heating, heated water from the buffer tank passes through the motorized 3-way mixing valve and on to a distribution system serving low-temperature hydronic heat emitters such as radiant panels. On its way to these heat emitters, the heated water passes through an electric boiler, which is also controlled based on outdoor reset. This boiler provides supplemental heating, if necessary, on very cold days. Water returning from the distribution system passes back into the buffer tank. This mode of operation is shown in Figure 13-49.

If the heat pump is not operational, or is operating in defrost mode, the full space heating load is provided by the electric boiler, as shown in Figure 13-50.

In cooling mode, the heat pump operates to chill water in the buffer tank. When there is a demand for cooling, some chilled water from the tank is routed through the coil of an air handler to provide a portion of the sensible cooling load and the entire latent cooling load (e.g., moisture removal). Additional chilled water from the buffer tank enters the 3-way mixing valve, where its temperature is adjusted based on the current dewpoint of the interior air. The temperature of the water leaving the mixing valve remains slightly above the current interior dewpoint temperature. This prevents surface condensation when this water is routed through the radiant panel distribution system to provide radiant cooling. Figure 13-51 shows the system in cooling mode operation.

In space heating mode, the COP of this system varies from approximately 2.1 at 0 °F outdoor temperature to 4.6 at 60 °F outdoor temperature. Computer modeling for a 5-ton heat pump supplying a house with a 50,000 Btu/hr. design load in a Canadian Maritime climate (Nova Scotia) shows an average seasonal COP of about 3.4. This is based on a low-temperature distribution system operating with a supply water temperature of 110 °F at

Courtesy of Benoit Maneckjee, ThermAtlantic Energy Products Inc.

(a)

Courtesy of Benoit Maneckjee, ThermAtlantic Energy Products Inc.

(b)

Courtesy of Benoit Maneckjee, ThermAtlantic Energy Products Inc.

(c)

Figure 13-48 (a) DX2W module (white cabinet) in system with buffer tank and electric boiler. (b) Internal components in DX2W module. (c) Outdoor heat pump unit.

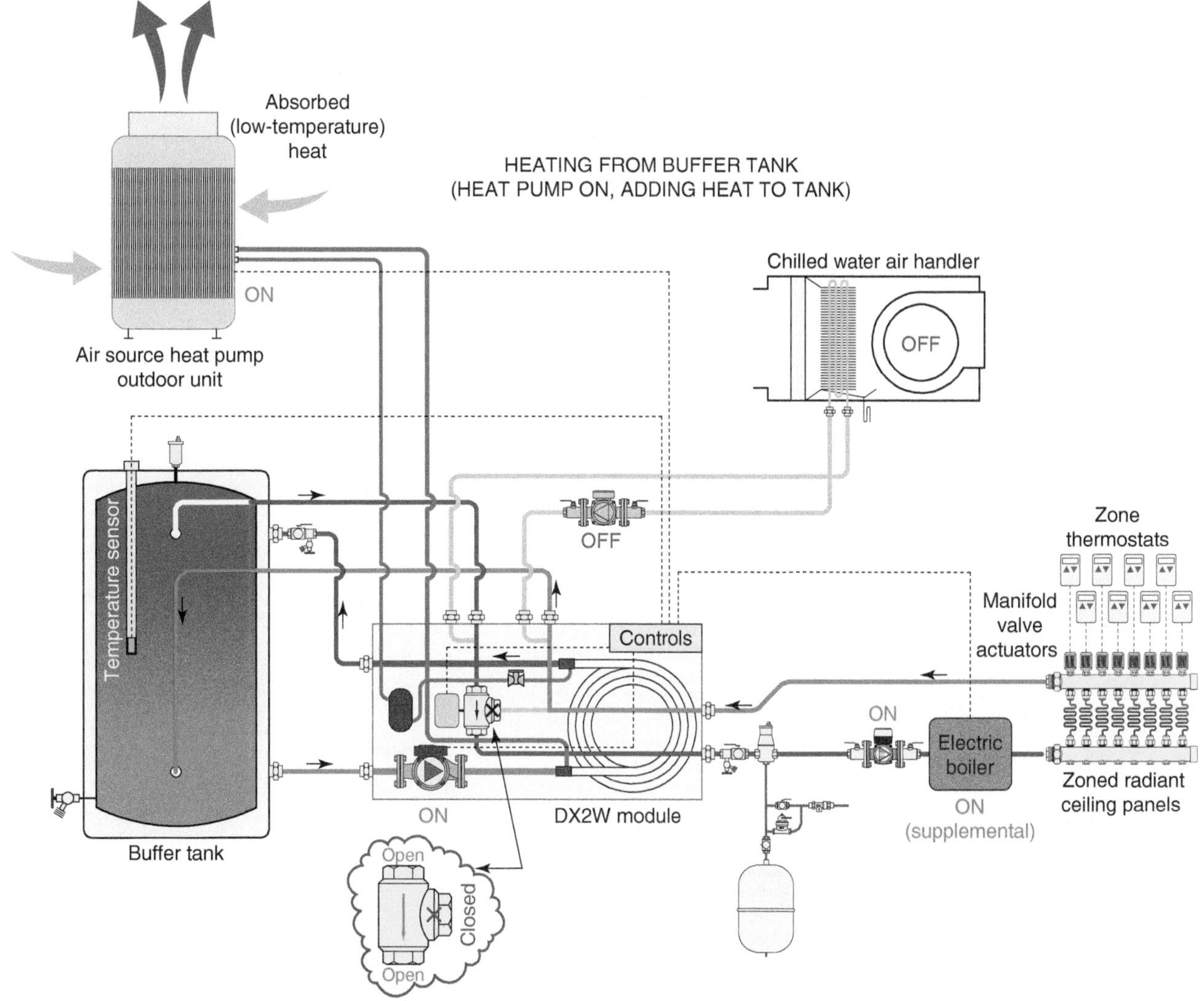

Figure 13-49 | Heat pump delivers heat to buffer tank. Buffer tank delivers heat to space heating load. Electric boiler provides supplemental heat if necessary.

design load. As with all heat pump systems discussed in this text, a distribution system that can operate at low water temperatures (e.g., maximum 120 °F supply water temperature at design load) is crucial for good performance.

Figure 13-52 shows two additional details that could be used with this system.

The detail in Figure 13-52a is the on-demand domestic water preheating subsystem discussed in Chapter 9. This subsystem uses water within the buffer tank, which has been heated by the heat pump, to preheat domestic water. The thermostatically controlled tankless water heater provides any necessary temperature boost. During cooling mode operation, the buffer tank contains chilled water. The heat exchanger for domestic water preheating should be isolated from the tank during this mode. The tankless electric water heater must be sized to provide the full domestic water heating load when no preheating is available.

The detail in Figure 13-52b addresses possible requirements for the electric boiler used to supplement the output of the heat pump. Some electric boilers have minimum flow rates below which they should not be operated. This requirement can be met by installing the electric boiler with its own circulator and coupling it to the distribution system using a pair of closely spaced tees, as shown. This arrangement provides hydraulic separation between the boiler circulator and the variable-speed pressure-regulated distribution circulator.

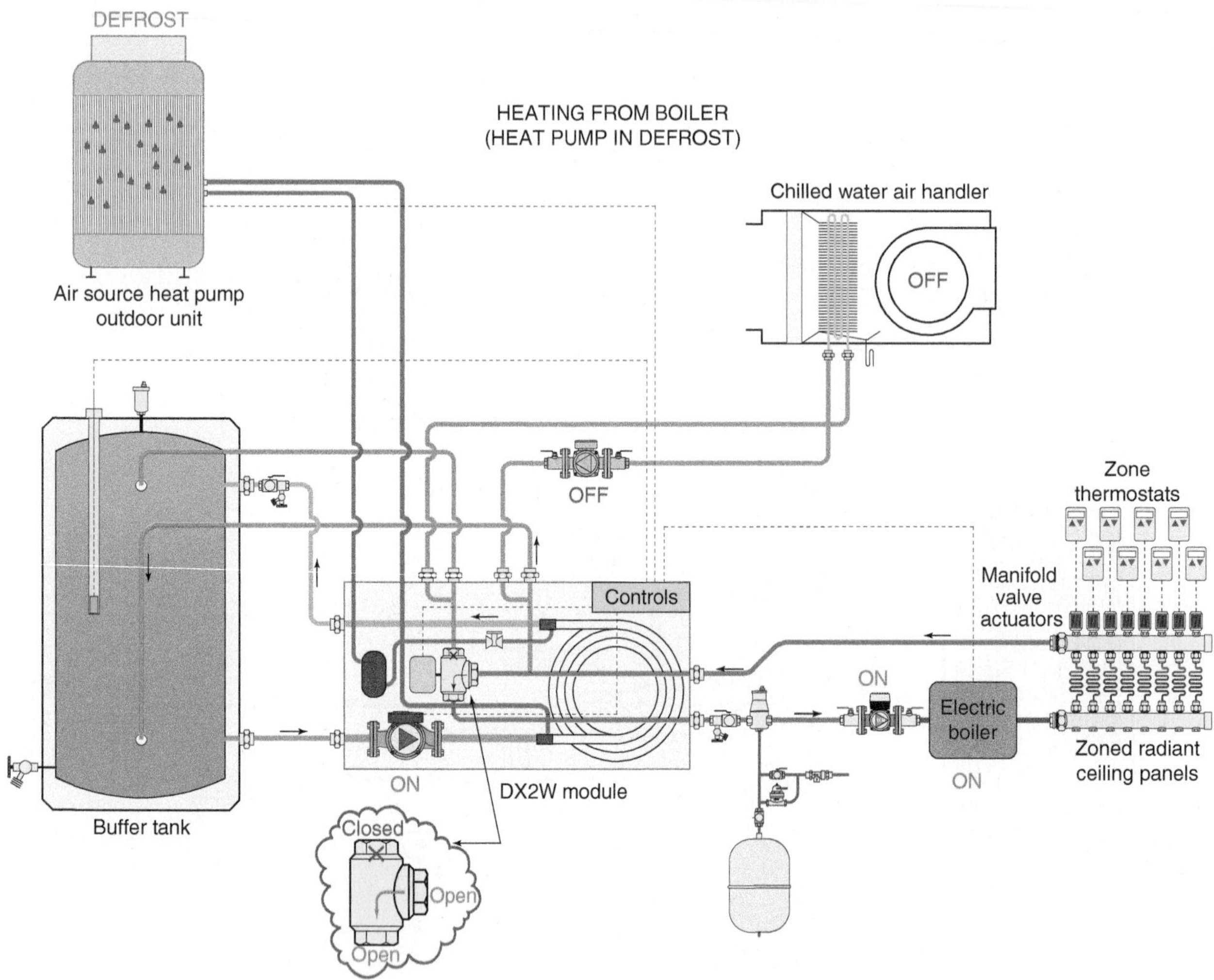

Figure 13-50 | Full space heating load handled by electric boiler if heat pump is not operational, or is in defrost mode, with buffer tank below usable temperature for space heating.

13.8 Examples of Combisystems Using Air-to-Water Heat Pump Systems

This section gives several examples of complete air-to-water heat pump systems. Two of the systems will be for "heating-only" applications. These are appropriate for climates where heat loads dominate and cooling is not needed. The remaining systems provide heating and cooling. These systems represent a mix of split-system heat pumps and self contained heat pumps. Many of the details used in these systems will be similar to those used in some of the solar thermal combisystems discussed in previous chapters. Readers should pay particular attention to such details because of their potential use in a wide variety of application.

AIR-TO-WATER HEAT PUMP COMBISYSTEM #1

The combisystem shown in Figure 13-53 is a heating-only application. Because a split system heat pump is used, water is only present within the indoor unit. Thus, no antifreeze is required to protect the outdoor unit. The indoor unit is also assumed to contain both a circulator and expansion tank. It is also assumed that the designer has verified that the *net* pump curve for this

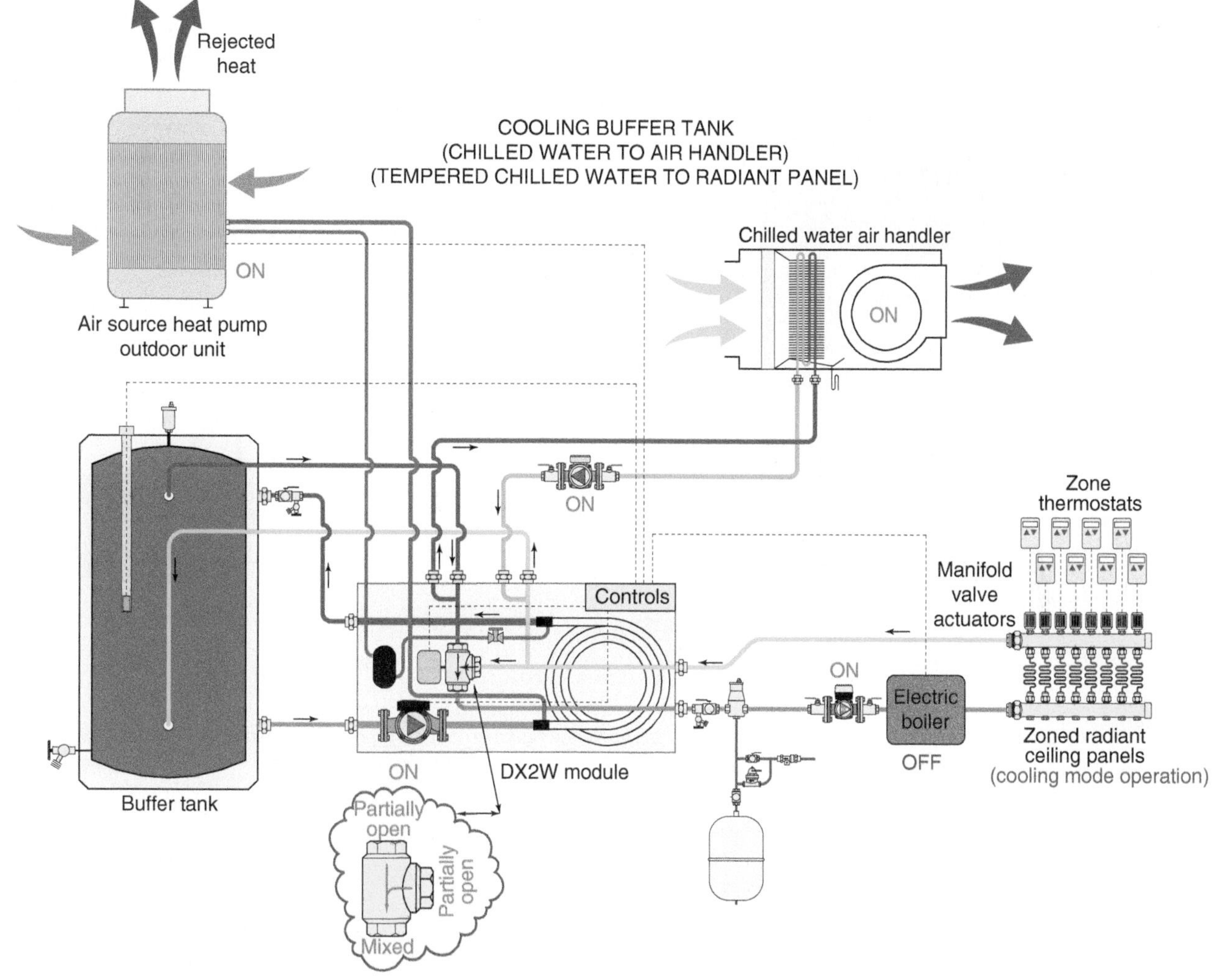

Figure 13-51 | System in cooling mode operation. Cooling is provided by both cool radiant panel and chilled water air handler.

circulator will provide adequate flow through either the space heating distribution system or the heat exchanger coil of the indirect water heater. The designer has also verified that the size of the expansion tank within the indoor unit of the heat pump has sufficient volume to handle the expansion of all water in the system.

This system uses a 3-way diverter valve to route heated water leaving the heat pump's indoor unit to either the *single-zone*/low-temperature heating distribution system or to the coil in the indirect water heater. The actuator of this valve is energized upon a call for domestic water heating. If there is a simultaneous call for space heating and domestic water heating, the latter gets priority. When the domestic water heating load is satisfied, the heat pump's output is directed to space heating.

Piping details include spring-loaded check valves to minimize thermal migration. A make-up water subassembly, air separator, and purging valve are also included. A dirt separator is also shown.

The control logic and hardware required to operate this system is completely contained in the heat pump.

AIR-TO WATER HEAT PUMP COMBISYSTEM #2

The heating-only system shown in Figure 13-54a uses a self-contained heat pump that supplies a heavily zoned space heating distribution system. Although the heat pump is capable of producing chilled water for cooling, that function is not being used in this system.

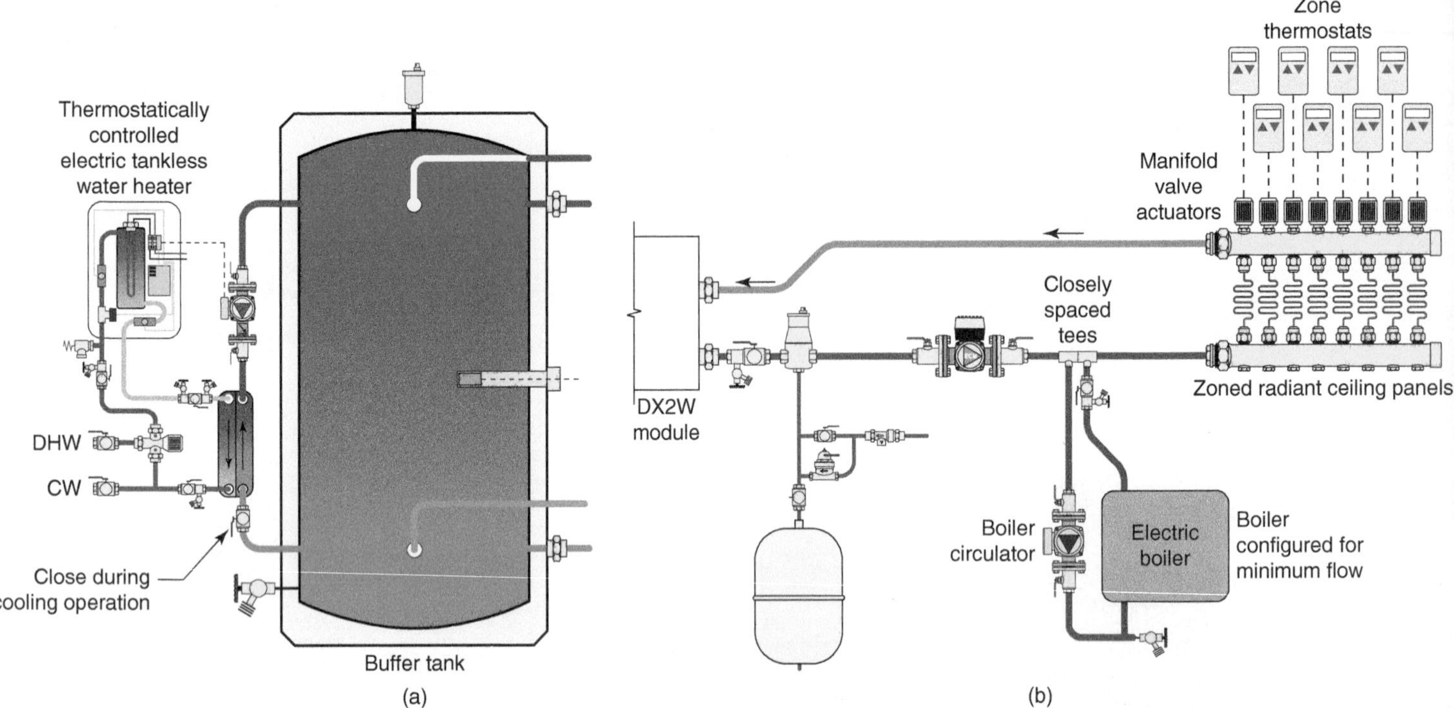

Figure 13-52 | (a) Possible details for providing domestic hot water preheating when operating heat pump in heating mode. (b) Detail providing minimum boiler flow rate in a system using ThermAtlantic DX2W air-to-water heat pump.

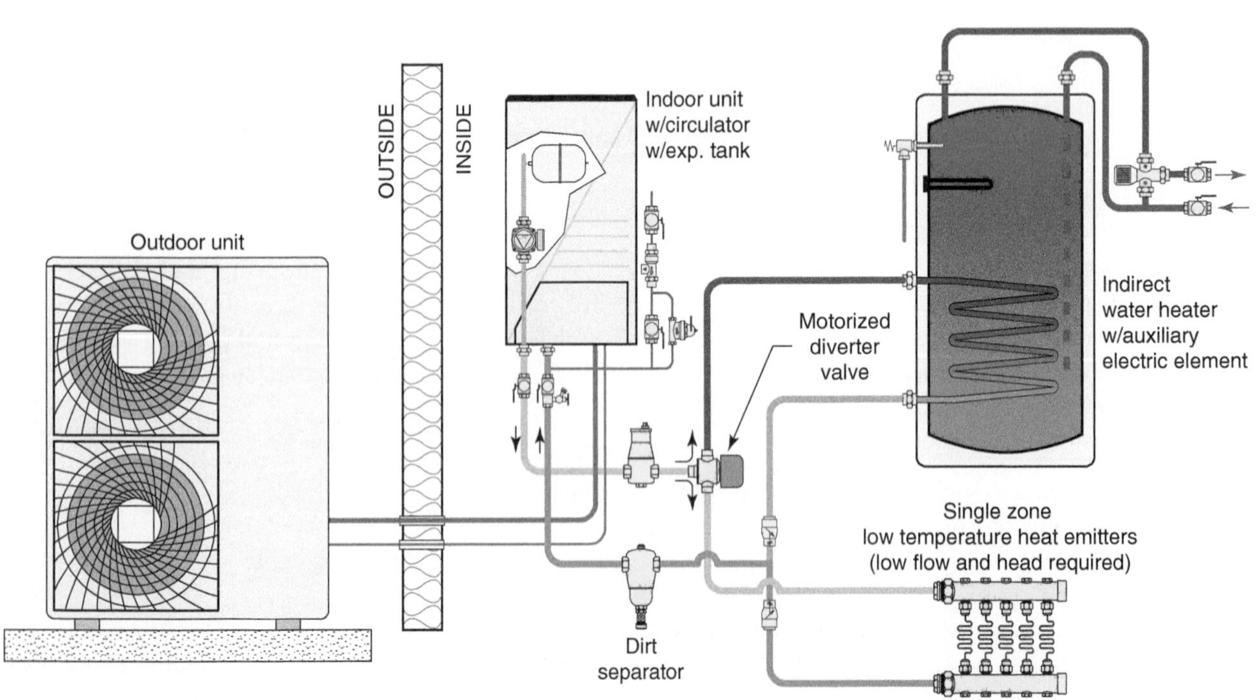

Figure 13-53 | Heating only combisystem using split system air-to-water heat pump.

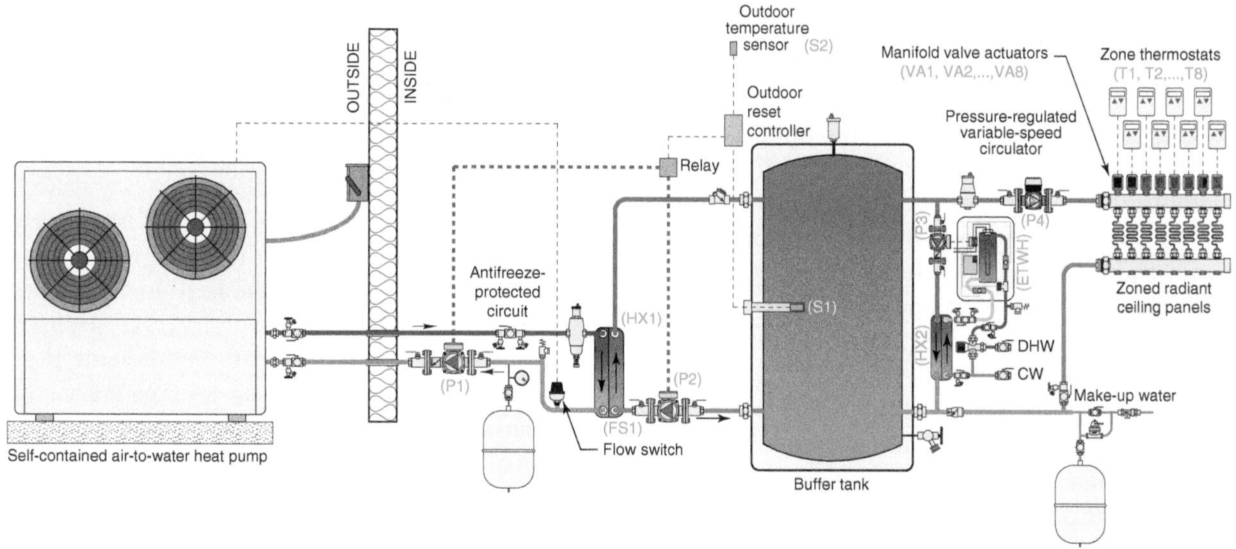

Figure 13-54a | Self-contained air-to-water heat pump in heating only application. The distribution system is extensively zoned. External heat exchanger used for domestic water preheating.

This system uses an on/off heat pump. It cannot reduce its output to match the potential low heating needs of the smaller space heating zones. Hence, a properly sized buffer tank is installed to prevent the heat pump from short cycling.

The water temperature within the buffer tank is regulated by an outdoor reset controller, with settings that produce the reset line shown in Figure 13-54b.

The outdoor reset controller is energized whenever there is a call for heat from any of the space heating thermostats. Its settings are configured for a temperature differential of 5 °F and a design load target water temperature of 110 °F, corresponding to an outdoor design temperature of 10 °F.

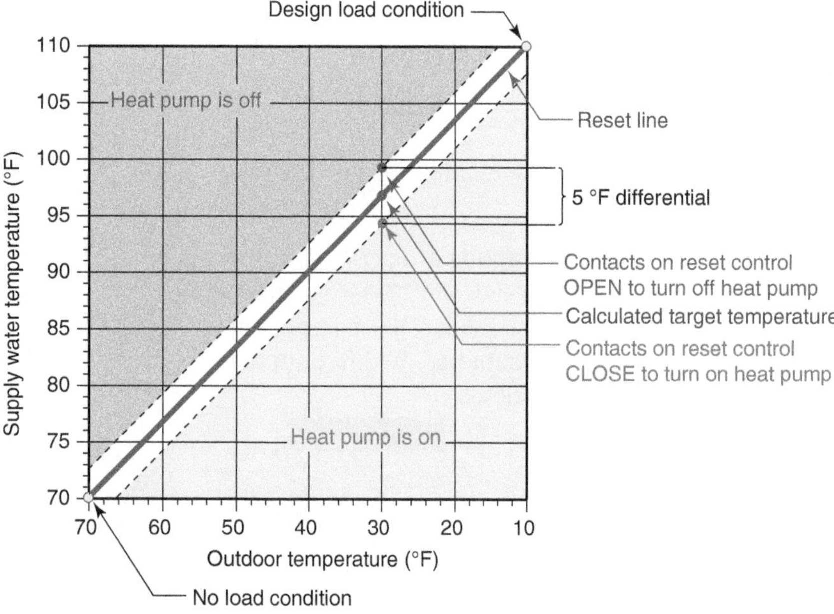

Figure 13-54b | Outdoor reset control of the water temperature in the buffer tank of system shown in Figure 13-54a.

Domestic water is heated "on demand" using an external stainless steel brazed-plate heat exchanger. The details for this approach were discussed in Chapter 9. Whenever there is a demand for domestic hot water of 0.6 gpm or higher, the flow switch inside the tankless electric water heater closes. This closure is used, in combination with a relay, to turn on the circulator that routes water from the upper portion of the buffer tank through the primary side of this heat exchanger. Closure of the flow switch also energizes an electrical contactor within the tankless heater to supply 240 VAC to triacs, which regulate the electrical power supplied to the heating elements. The power supplied to the elements is limited to that required to provide the desired outgoing water temperature. All heated water leaving the tankless heater flows into an ASSE 1017–rated mixing valve to ensure a safe delivery temperature to the fixtures.

Because the thermostatically controlled tankless heater can provide a boost in water temperature, the buffer tank does not have to be maintained at a temperature suitable to provide the *full* temperature rise of the domestic hot water. Instead, its temperature is regulated by an outdoor reset controller. If space heating is provided by low-temperature radiant panels, the maximum (e.g., design load) water temperature at the sensor in the buffer tank will only be a few degrees above the target temperature of 110 °F. On partial load days, the outdoor reset controller will limit the tank temperature to that required to meet the space heating load. This improves the heat pump's heating capacity and COP.

Because domestic water preheating is done by the heat pump, heat is imparted to the domestic water at whatever the COP of the heat pump happens to be. In most applications, that COP will be in the range of 2.5 to perhaps 4.0, depending on the operating conditions of the heat pump. This is beneficial because it displaces heat that will otherwise be added by the tankless electric water heater, which, being an electric resistance heating device, always operates at a COP of 1.0.

The electrical wiring diagram for combisystem #2 is shown in Figure 13-54c.

The following is a description of operation of the system shown in Figures 13-54a and 13-54c. Readers are encouraged to cross-reference the component designations, shown in parentheses, within both the piping and electrical schematics.

Description of Operation:

1. *Space heating mode:* The distribution system has several zones, each equipped with a 24 VAC thermostat (T1, T2, T3, etc.). Upon a demand for heat from one or more of these thermostats, the associated 24 VAC manifold valve actuators (VA1, VA2, VA3, etc.) are powered on. When a valve actuator is fully open, its end switch closes. This supplies 24 VAC to the coil of relay (R2). This closes the normally open contacts (R2-1), which supply 120 VAC to circulator (P4). Circulator (P4) operates in constant differential pressure mode to supply the flow required by the distribution system, based on how many zones are calling for heat. A call for heat also supplies 24 VAC to the outdoor reset controller (ORC). This controller measures outdoor temperature from sensor (S2). It uses this temperature, in combination with its settings, to calculate the target supply water temperature for the buffer tank. If the tank requires heating, the normally open relay contacts within the (ORC) close. This energizes the coil of relay (R3). One set of normally open contacts (R3-1) close to turn on circulators (P1) and (P2). This establishes flow through the heat pump and between the heat exchanger (HX1) and the buffer tank. When the flow rate between heat exchanger (HX1) and the heat pump reaches the activation setting of flow switch (FS1), it closes its contacts to complete the circuit between the R and Y terminals in the heat pump. This turns on the heat pump in heating mode and delivers heat to the buffer tank. When all thermostats are off, circulator (P4) is turned off along with the outdoor reset controller (ORC). The heat pump is also turned off if it is operating.

2. *Domestic water heating mode:* Whenever there is a demand for domestic hot water of 0.6 gpm or higher, the flow switch inside the tankless electric water heater closes. This closure applies 240 VAC to the coil of relay (R1). The normally open contacts (R1-1) close to turn on circulator (P3), which circulates heated water from the upper portion of the buffer tank through the primary side of the domestic water heat exchanger (HX2). The domestic water leaving (HX2) is preheated to a temperature that is a few degrees less than the buffer tank temperature. This water passes into the thermostatically controlled tankless water heater, which measures its inlet temperature. The electronics within this heater control electrical power supplied to the heat elements based on the necessary temperature rise to achieve the set domestic hot water supply temperature. All heated water leaving the tankless heater flows into an ASSE 1017–rated mixing valve to ensure a safe delivery temperature to the fixtures. Whenever the demand for domestic hot water drops below 0.4 gpm, circulator (P3) and the tankless electric water heater are turned off.

The control logic just described is "generic." Some heat pumps may require different wiring. For example, some heat pumps may provide an internal relay to operate circulators (P1) and (P2). The operation of these circulators may be "coordinated" with other actions within the heat pump, such as the defrost cycle. Some heat pumps may also have an internal flow switch that eliminates the need for the external flow switch (FS1). Designers must be familiar with the exact control sequences used in the heat pumps they select. They must then be sure these control sequences are coordinated with the balance of the system.

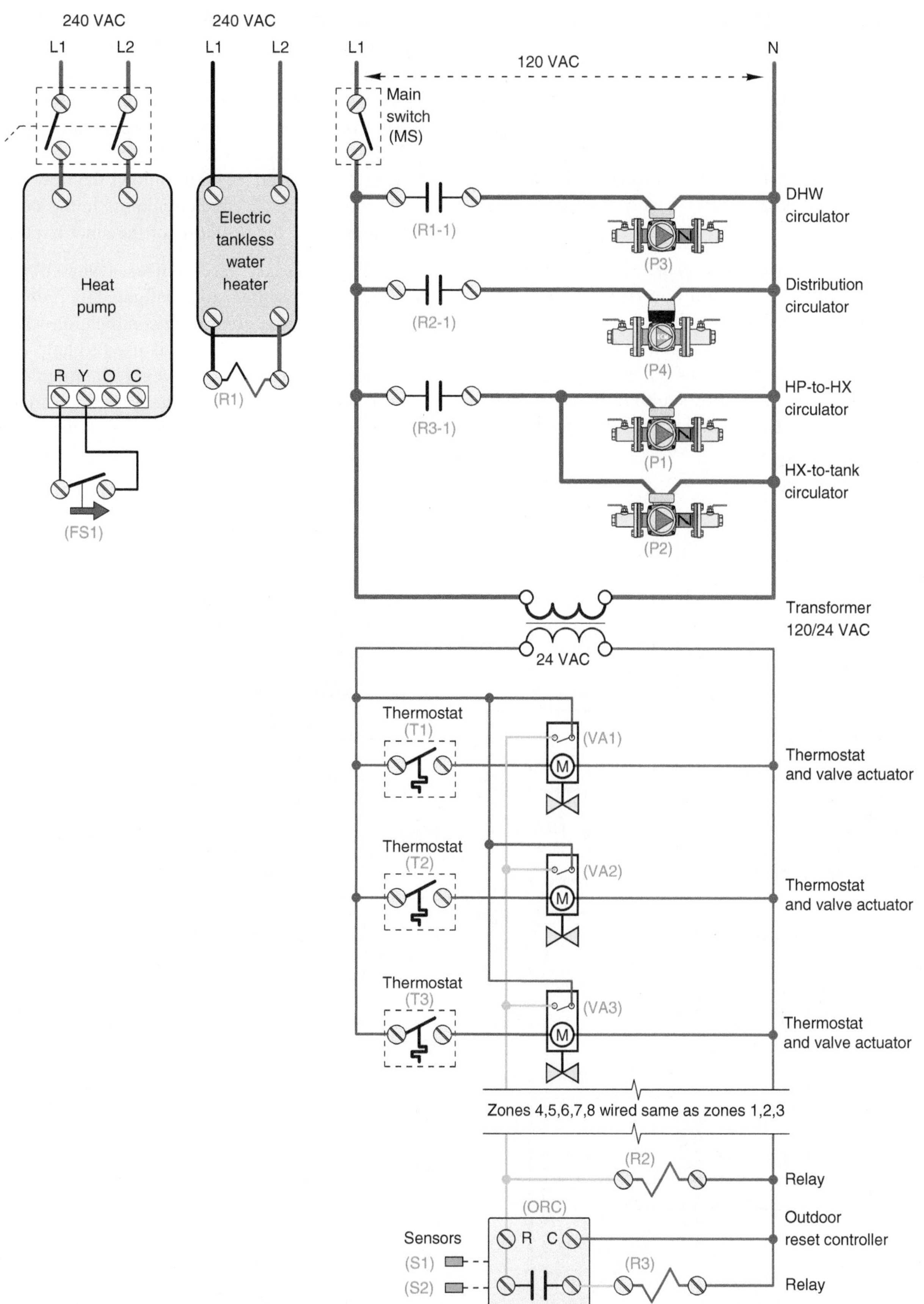

Figure 13-54c | Electrical wiring diagram for combisystem #2 shown in Figure 13-54a.

AIR-TO-WATER HEAT PUMP COMBISYSTEM #3

The combisystem shown in Figure 13-55a provides heating, cooling and domestic hot water. It uses a modulating air-to-water heat pump capable of modulating its heating and cooling capacity to approximately 25 percent of its maximum capacity.

A comparison of this system, and combisystem #1, shows similarities. The principle difference is that combisystem #3 has multiple zones of heating, and includes two zones of chilled water cooling. The distribution subsystems required for this zoning have flow and head requirements beyond what can be supplied by the circulator supplied with the heat pump's indoor unit. The situation is handled by creating a relatively small and short **primary loop**, which has a flow and head requirement within the capacity of the heat pump's circulator. This primary loop supplies two separate **secondary circuits**, one for heating and the other for cooling. Each secondary circuit is hydraulically separated from the primary circuit by a set of closely spaced tees. This allows the flow rate and head produced by all three circulators in the system to be unaffected by the operation of the other circulators.

Each secondary circuit uses a variable-speed pressure-regulated circulator, configured for proportional differential pressure control, in combination with zone valves. Reverse return piping is used to help balance flow through each zone. However, balancing valves are still included to allow accurate setting of the flow through each zone. These balancing valves can also

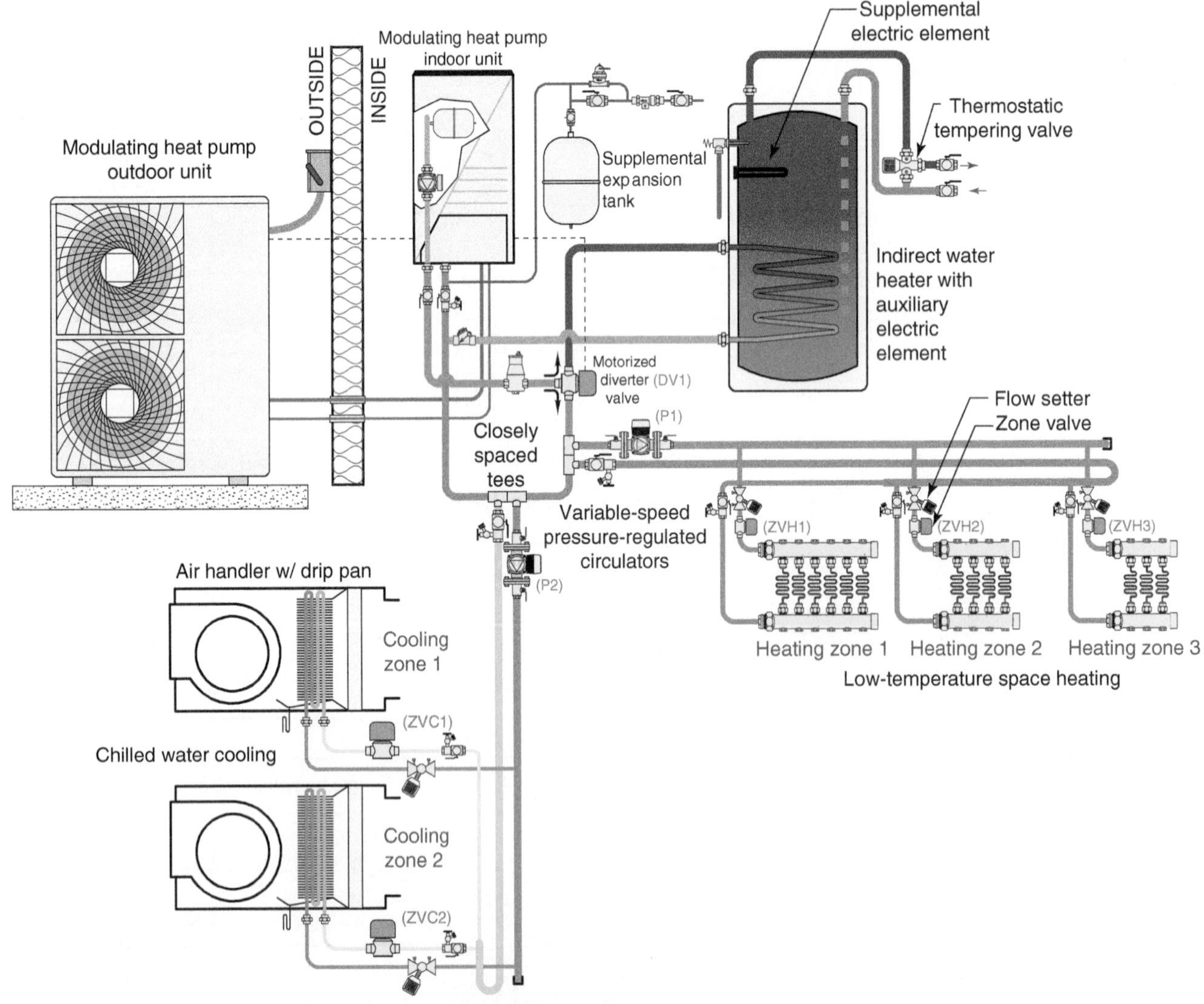

Figure 13-55a | Combisystem #3 provides heating, cooling, and domestic hot water using a modulating split system air-to-water heat pump.

serve as isolation valves, if necessary. The zone valves for the chilled water air handlers are located on the outlet side of the air handler coil. The slightly warmer water temperature at this location reduces the possibility of condensation forming on the zone valves. All piping components that could convey chilled water are insulated and vapor sealed.

Domestic water is heated by the heat pump and controlled as a priority load. Whenever the thermostat in the domestic hot water tank calls for heating, the diverter valve is energized. Heated water is directed from the heat pump's indoor unit through the coil in the indirect water heater. An electric heating element in the upper portion of the indirect tank provides any necessary temperature boost.

The ability of the heat pump to modulate its heating and cooling capacity, combined with the limited zoning used in both the heating and cooling secondary circuits, eliminates the need of a buffer tank. This is a significant advantage.

The internal control logic within the heat pump handles the priority domestic water heating. This logic may be the same or different, depending on the current operating mode of the system (e.g., is it off, operating in heating mode, or operating in cooling mode?).

A supplemental expansion tank is used in this system. This tank is needed because the expansion volume of the water in the system is greater than what the expansion tank within the heat pump's indoor unit can accommodate. A standard expansion tank sizing calculation should be made to determine the total expansion tank volume required. The minimum volume of the supplemental expansion tank is then found by subtracting the volume of the internal tank from the total expansion tank volume requirement. The air pressure within the two expansion tanks should be adjusted to be equal. The two tanks should also be mounted with their inlet connections at approximately the same height. This provides equal static pressure at both tanks. Another option is to disconnect the internal expansion tank and size the supplemental expansion tank for the full expansion requirement of the system.

The electrical wiring diagram for combisystem #3 is shown in Figure 13-55b. It assumes that the flow switch is supplied as an internal component of the heat pump.

The control strategy uses an SPDT c/o **mode selection switch** to determine if the system is enabled for heating or cooling, or remains off. It cannot be enabled for simultaneous heating and cooling. This switch ensures that 24 VAC power will only reach electrical components associated with the selected operating mode.

When enabled for heating, 24 VAC is applied to the (RH) terminal of each thermostat. When a thermostat calls for heat, it switches 24 VAC to its (W) terminal. This energizes the associated heating zone valve. When the end switch in an energized zone valve closes, 24 VAC is passed on to the coil of the heating relay (RH1). One normal open contact in this relay (RH1-1) operates the heating secondary circulator (P1). Another normally open contact in (RH1-2) closes between the (R) and (Y) terminals on the heat pump, enabling it to operate in heating mode. The control circuitry within the heat pump turns on the internal circulator (PHP). The control circuitry within the heat pump also monitors the temperature of the domestic water heater using aquastat (TW1) and operates the heat pump circulator and diverter valve as needed to maintain this temperature.

When enabled for cooling, 24 VAC is applied to the (RC) terminal of thermostats (T1) and (T2). Thermostat (T3) cannot call for cooling. When either (T1) or (T2) call for cooling, an associated relay coil (RC1) or (RC2) is energized, as are cooling zone valves (ZVC1) or (ZVC2). One set of relay contacts is used to turn on the cooling secondary circulator (P2). A second set of relay contacts is used to supply line voltage to operate the blowers in the air handlers. The end switches of the cooling zone valves are wired in parallel and used to enable the heat pump to operate in cooling mode. When the end switch of either cooling zone valve closes, 24 VAC from the heat pump's internal transformer passes through the end switch and energizes the coil of relay (RA) and reversing valve terminal (O) to enable cooling mode operation. A normally open contact (RA-1) closes to enable supply 24 VAC from the heat pump's internal transformer to terminal (Y), which turns on the compressor.

The following is a description of the operation of the system shown in Figures 13-55a and 13-55b. Readers are encouraged to cross-reference the component designations, shown in parentheses, within both the piping and electrical schematics.

Description of Operation:

1. *Space heating mode:* The mode selection switch (MSS) must be set for heating. This supplies 24 VAC to the (RH) terminals of each thermostat. If a thermostat

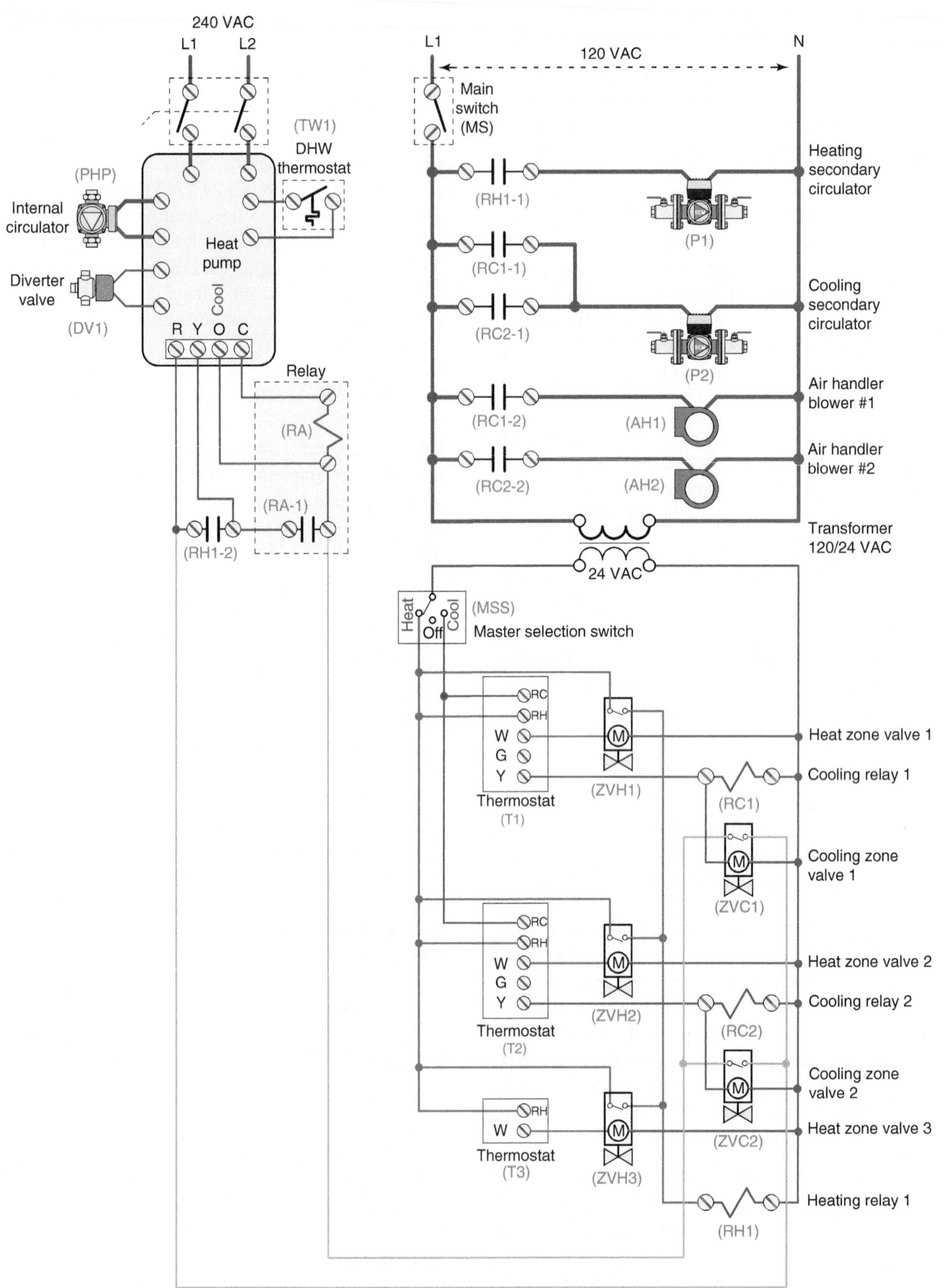

Figure 13-55b | Electrical wiring for combisystem #3 shown in Figure 13-55a.

is set for heating mode, and it calls for heat, 24 VAC is switched to the thermostat's (W) terminal. This supplies 24 VAC to the associated heating zone valve (ZVH1, ZVH2, or ZVH3). When the end switch in that zone valve closes, 24 VAC passes to the coil of relay (RH1). One set of normally open contacts (RH1-1) closes to energize circulator (P1), which operates in proportional differential pressure control mode. Another set of normally open contacts (RH1-2) close to provide a heating demand to the heat pump (connecting the (R) terminal to the (Y) terminal). The heat pump then operates based on its own internal control system. It monitors the status of the domestic hot water tank thermostat (TW1) and, when necessary, energizes the diverter valve (DV1) to direct hot water to the coil of the indirect water heater. It also operates its internal circulator (PHP) to create flow through the primary loop.

2. *Space cooling mode:* The mode selection switch (MSS) must be set for cooling. This supplies 24 VAC to the (RC) terminals of thermostats (T1) or (T2). If either of these thermostats is set for cooling mode, and calls for cooling, 24 VAC is switched to the thermostat's (Y) terminal. This supplies 24 VAC to the associated cooling relay (RC1) or (RC2) and cooling zone valves (ZVC1) or (ZVC2). One set of contacts (RC1-1) or (RC2-1) closes to provide line voltage to the secondary cooling circulator (P2), which operates in proportional differential pressure mode. A second set of contacts (RC1-2) or (RC2-2) close to provide line voltage to the associated air handler blowers (AH1) or (AH2). The end switches in the cooling zone valves close when those valves reach their fully open position. This closure passes 24 VAC from the heat pump's internal transformer to the coil of relay (RA) and to terminal (O), which energizes the heat pump's reversing valve for cooling mode operation. Normally open contact (RA-1) closes to provide 24 VAC to terminal (Y) to enable compressor operation. The heat pump controls its compressor speed based on its own internal control system. It also operates its internal circulator (PHP) to create flow through the primary loop.

3. *Domestic water heating mode:* The heat pump monitors the status of the domestic hot water tank thermostat (TW1) and, when necessary, turns itself on in heating mode. It also energizes the diverter valve (DV1) to direct hot water to the coil of the indirect water heater. It also operates its internal circulator (PHP) to create flow through the primary loop. When necessary, the heat pump's internal controller also operates the auxiliary heating element in the hot water storage tank.

Designers should note that there are several possible variations of this wiring diagram. For example, the air handlers selected may have their own internal electrical system, and only require a dry contact closure to enable their operation. This is easily handled by using the relay contacts (RC1-3) and (RC2-3) as dry contacts. It would also be possible to modify this electrical wiring so that thermostats (T1) and (T2) could turn on the blowers in their respective air handlers at times when there is no call for cooling. This would be done by adding two more relays that are energized by the (G) terminals in thermostats (T1) and (T2). The contacts of these relays would then connect line voltage to the blowers in the respective air handlers.

More heating and cooling zones are easily added in either of the secondary circuits. Designers should verify the ability of the modulating heat pump to match the minimum heating or cooling load of each zone. With a reasonable match in capacity, there is no need for a buffer tank. However, extensive use of small capacity zones may require the use of a buffer tank within the primary circuit. That circuit may also have to be reconfigured so that only the space heating and cooling loads are buffered. The domestic water heating load is "self-buffering" and therefore should not be supplied through a buffer tank.

AIR-TO-WATER HEAT PUMP COMBISYSTEM #4

In very cold climates, air-to-water heat pumps may not have sufficient capacity to supply full design heating load. One option is to disable the heat pump's operation when the outdoor temperature is below a specific value and switch to a boiler as the heat source. The system in Figure 13-56a was designed for this option.

This system uses a 2-stage self-contained air-to-water heat pump. In cooling mode, the heat pump's cooling capacity is assumed to be well matched to the two chilled water air handlers. The heat pump will operate on stage 1 cooling when only one of these air handlers is active. If the second air handler turns on, the heat pump will switch to two-stage operation to maintain adequately low chilled water temperature. Because the capacity of the air handlers is reasonably matched to the cooling capacity of the heat pump, there is no need to include a chilled water buffer tank in the system.

The space heating distribution system has eight zones. The 2-stage heat pump cannot modulate its heat

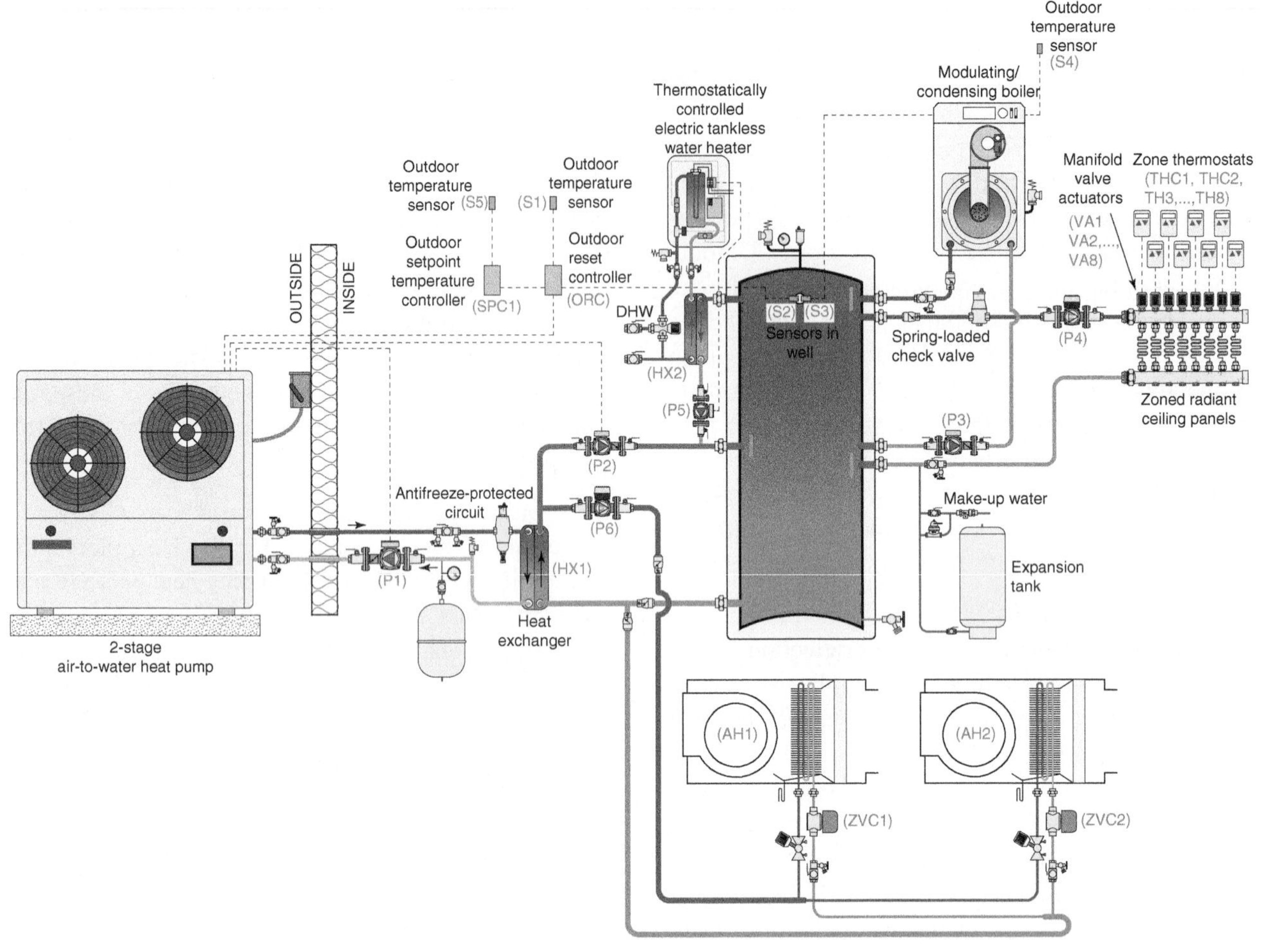

Figure 13-56a | 2-stage air-to-water heat pump with zoned heating and cooling, and auxiliary boiler.

output low enough to match the heating load when only one or two small zones are active. Thus, a buffer tank is needed to prevent heat pump short cycling during heating mode operation.

When outdoor temperatures are high enough to allow the heat pump to meet the heating load, it serves as the sole heat source for the buffer tank. The water temperature in the buffer tank is maintained based on outdoor reset control. This allows the heat pump to operate at lower condenser temperatures during mild weather, increasing its heating capacity and COP.

When the outdoor temperature drops below a preset value, the heat pump is disabled, and the boiler serves as the sole heat source for the buffer tank.

Domestic water is preheated through the external brazed-plate stainless steel heat exchanger. Final heating to setpoint is accomplished by a thermostatically controlled tankless electric water heater.

The electrical wiring diagram for combisystem #4 is shown in Figure 13-56b. It assumes that the flow switch is supplied as an internal component of the heat pump.

Portions of this electrical schematic are similar to that shown for combisystem #3. Of necessity, the electrical wiring for combisystem #4 is more extensive because there are more devices to be controlled. To reduce the size of the drawing, only two of the heating-only thermostats are shown. The wiring for the other heating-only thermostats not shown would be identical to that for thermostat (TH3).

In this system, thermostat (THC1) controls heating in zone 1 and one zone of cooling. Thermostat (THC2) controls heating in zone 2, and a second zone of cooling. Thermostats (TH3, TH4, TH5, TH6, TH7, TH8) only control heating in zones 3, 4, 5, 6, 7, and 8.

The following is a description of operation of the system shown in Figures 13-56a and 13-56b. Readers

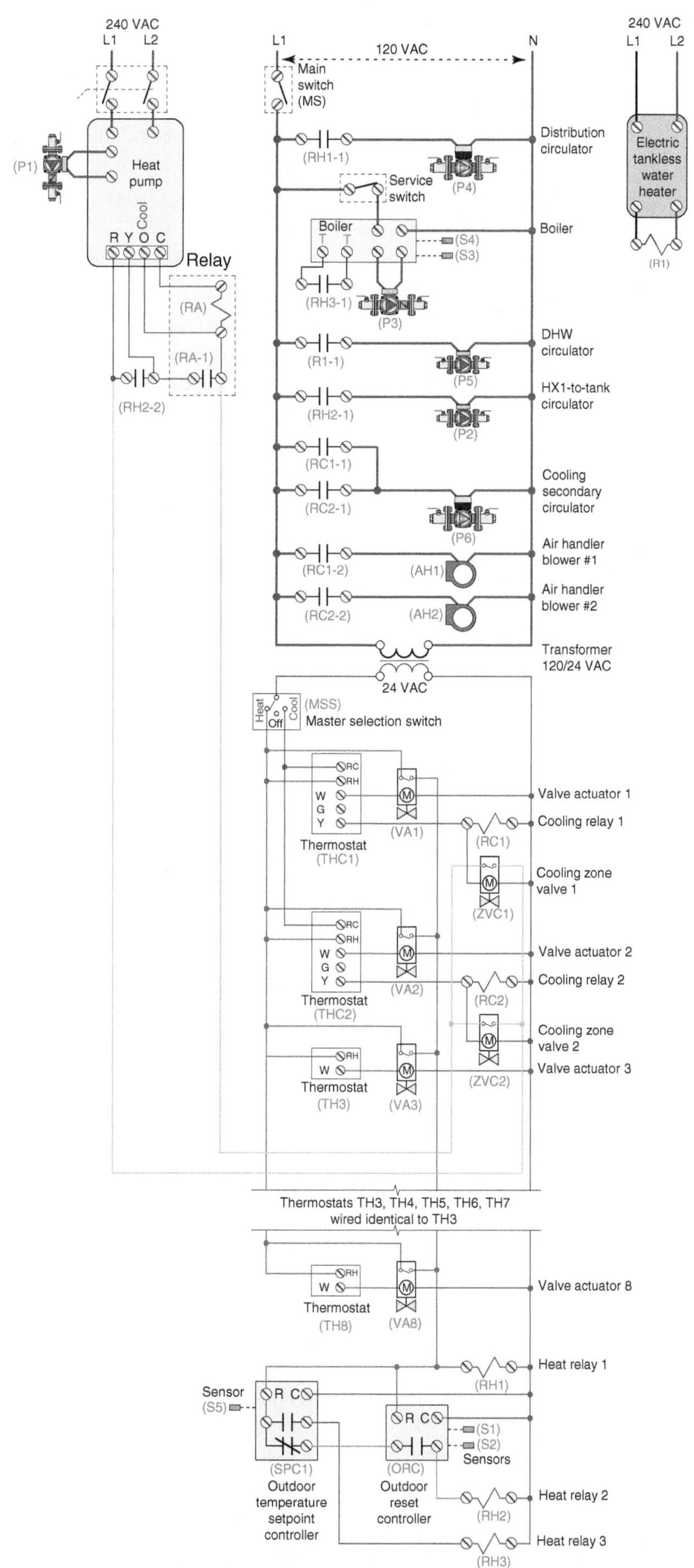

Figure 13-56b | Electrical wiring for combisystem #4 shown in Figure 13-56a.

are encouraged to cross-reference the component designations, shown in parentheses, within both the piping and electrical schematics.

Description of Operation:

1. *Space heating mode:* The mode selection switch (MSS) must be set for heating. This supplies 24 VAC to the (RH) terminals of thermostats (THC1) and (THC2). It also supplies 24 VAC to the (R) terminals of the heating-only thermostats shown (TH3, TH4, TH5, TH6, TH7, TH8). If a thermostat is set for heating mode, and it calls for heat, 24 VAC is switched to the thermostat's (W) terminal. This supplies 24 VAC to the associated heating valve actuator (VA1, ..., VA8). When the end switch in a valve actuator closes, 24 VAC is also sent to the coil of relay (RH1). One set of normally open contacts (RH1-1) closes to energize circulator (P4), which then operates in proportional differential temperature control mode.

Upon a call for heating from any thermostat, 24 VAC is also supplied to the outdoor temperature setpoint controller (SPC1) and the outdoor reset controller (ORC). If the outdoor temperature is above the minimum value set on (SPC1), which in this system is 0 ºF plus a 4 ºF differential, the heat pump will be the heat source. In this case, 24 VAC passes through the normally closed contact within (SPC1), and on to the normally open contact in the outdoor reset controller (ORC). The (ORC) measures the outdoor temperature using sensor (S1) and calculates the target temperature for the buffer tank. It measures the temperature in the upper portion of the buffer using sensor (S2). If the buffer tank temperature is too low to supply the load, the normally open contact in the (ORC) closes. This energizes relay coil (RH2). One normally open contact (RH2-1) closes to energize circulator (P2). Another normally open contact (RH2-2) closes to enable the heat pump to operate in heating mode, (e.g., contact closure between the R and Y terminals in the heat pump). The heat pump then turns on circulator (P1) and operates under its own internal control logic.

If the outdoor temperature detected by (SPC1) is below the minimum value (0 ºF), the normally open contact in (SPC1) closes and its normally closed contact opens. This turns off relay (RH2) and thus disables operation of the heat pump as well as circulator (P2). It also applies 24 VAC to the coil of relay (RH3). Normally open contact (RH3-1) closes as a dry contact across the (T T) terminals in the boiler, enabling it to operate. The boiler turns on circulator (P3) and begins operating under its own internal outdoor reset controller settings. It uses these settings, in combination with the outdoor temperature measured by sensor (S4) to calculate the target temperature in the buffer tank. It measures the temperature in the upper portion of the buffer tank using sensor (S3). When necessary, the boiler fires to raise the buffer tank to a temperature that can supply the heating load.

2. *Space cooling mode:* The mode selection switch (MSS) must be set for cooling. This supplies 24 VAC to the (RC) terminals of thermostats (THC1) and (THC2). If either of these thermostats is set for cooling mode, and calls for cooling, 24 VAC is switched to the thermostat's (Y) terminal. This supplies 24 VAC to the associated cooling relay (RC1) or (RC2). One set of contacts (RC1-1) or (RC2-1) closes to provide line voltage to cooling circulator (P6), which then operates in proportional differential temperature control mode. Another set of contacts (RC1-2) or (RC2-2) closes to provide line voltage to the associated air handler blowers (AH1) or (AH2). The end switches in the cooling zone valves close when those valves reach their fully open position. This closure passes 24 VAC from the heat pump's internal transformer to the coil of relay (RA), and to terminal (O), which energizes the heat pump's reversing valve for cooling mode operation. Normally open contact (RA-1) closes to provide 24 VAC to terminal (Y) to enable compressor operation. The heat pump controls its compressor speed based on its own internal control system.

3. *Domestic water heating mode:* Whenever there is a demand for domestic hot water of 0.6 gpm or higher, the flow switch inside the tankless electric water heater closes. This closure applies 240 VAC to the coil of relay (R1). The normally open contacts (R1-1) close to turn on circulator (P5), which circulates heated water from the upper portion of the buffer tank through the primary side of the domestic water heat exchanger (HX2). The domestic water leaving (HX2) is preheated to a temperature a few degrees less than the current buffer tank temperature. The domestic water leaving (HX2) passes into the thermostatically controlled tankless water heater, which measures its inlet temperature. The electronics within this heater control electrical current flow to the heat elements based on the necessary temperature rise to achieve the set domestic hot water supply temperature. All heated water leaving the tankless heater flows into an ASSE 1017–rated mixing valve to ensure a safe delivery temperature to the fixtures. Whenever the demand for domestic hot water drops below 0.4 gpm, circulator (P5) and the tankless electric water heater are turned off.

When the system remains in cooling mode, the buffer tank will eventually cool to a temperature close to room air temperature. Minimal domestic water preheating will occur. The electric tankless water heater must therefore be sized to provide the full temperature rise required for domestic water heating under this condition.

The VA Rating of the 120/24 VAC transformer that supplies power the lower portion of the ladder diagram should be selected based on the total VA rating of all devices that could be simultaneously active. The minimum VA rating of the transformer should be this total VA rating *plus 10 VA*.

AIR-TO-WATER HEAT PUMP COMBISYSTEM #5

The lower the supply water temperature at which a heating distribution system operates, the higher the heating capacity and COP of the heat pump supplying that system. One way to lower the supply water temperature is to increase the total heat transfer surface area of the heat emitters.

Consider a system in which low-temperature radiant panels are selected for space heating, and chilled water air handlers are selected for cooling. If the chilled water air handlers were also operated during the heating mode, they would increase the heat emitter surface area of the overall distribution system and thus lower the water temperature at which the distribution system can supply a given heat output. The system shown in Figure 13-57a shows how this can be done.

For simplicity, only one of three radiant panel manifold stations, and one of three air handlers, are shown in the schematic. The three manifold stations could be of different size. Likewise, the three air handlers could have different heating and cooling capacities. Whatever the case, all of these devices will operate at the same supply temperature in the heating mode. The three air handlers will operate at the same chilled water supply temperature in cooling mode.

When the system is operating in heating mode, warm water from the buffer tank is available to all six zone valves. The system is wired so that the zone valves operate in pairs. The zone valve that controls flow to a given radiant panel manifold station, is paired with the zone valve that controls flow to the air handler in the same zone. During the heating mode, the objective is to allow flow to both the radiant panel circuits and the air handler in a given zone, whenever the thermostat in that zone calls for heat.

When the system is operating in the cooling mode, chilled water can only be delivered to the air handlers, which are equipped with condensate drip pans. Flow to

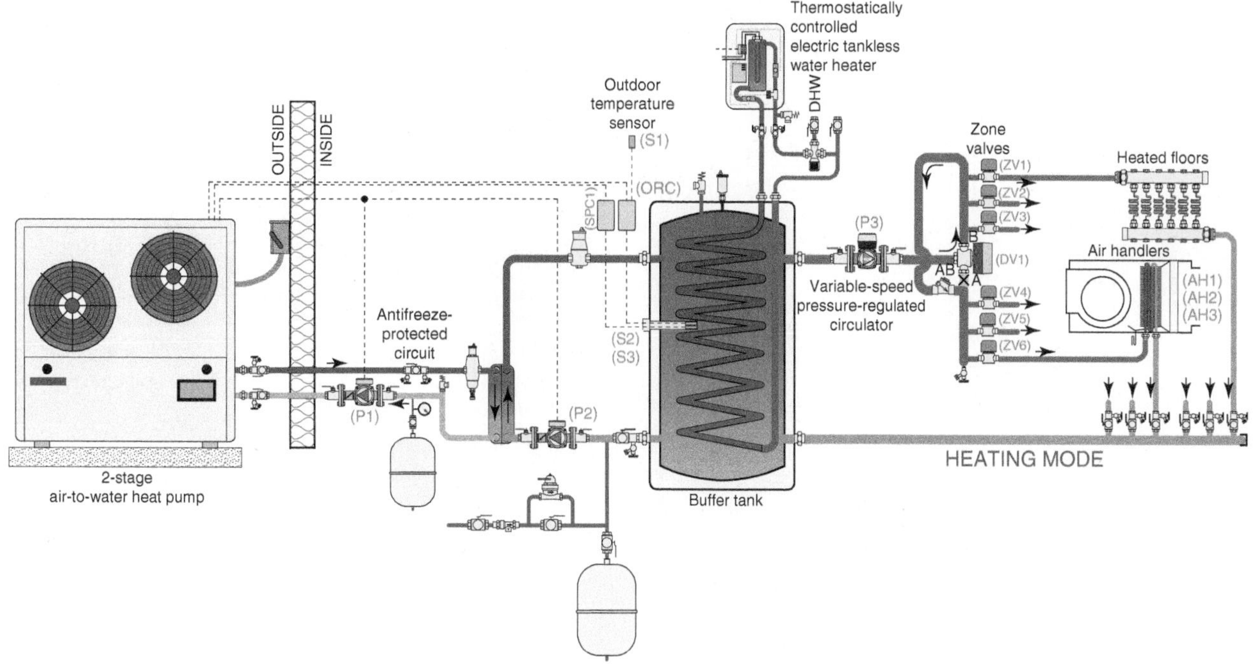

Figure 13-57a | Combisystem #5 shown in heating mode.

the radiant panel manifold stations is prevented by the 3-way diverter valve and a check valve. Figure 13-57b shows the system in cooling mode.

Whenever there is a demand for domestic hot water, cold water is drawn through the stainless steel coil inside the buffer tank. During the heating mode, domestic water is preheated as it is drawn through the coil. The temperature to which it is heated depends on the temperature of the buffer tank, which is regulated by an outdoor reset controller. Any necessary temperature boost is provided by the thermostatically controlled tankless electric water heater.

In the cooling mode, domestic cold water is also drawn through the internal stainless steel coil. If the entering water is cooler than the water in the tank, it absorbs heat from the tank water and thus displaces some of the cooling capacity required of the heat pump. The tankless water heater then provides nearly all the required temperature rise. In warm climates, where the entering cold water temperature is higher than the chilled water temperature being maintained by the heat pump, manually operated bypass valves (not shown in Figure 13-57b) could be installed to prevent domestic water flow through the coil. Instead, the cold water would flow directly to the tankless water heater.

The electrical wiring diagram for this system is shown in Figure 13-57c. It assumes that the flow switch is supplied as an internal component of the heat pump.

The following is a description of operation for combisystem #5.

Description of Operation:

1. *Space heating mode:* The mode selection switch (MSS) must be set for heating. This supplies 24 VAC to the (RH) terminals of all three thermostats (T1, T2, and T3). When any thermostat calls for heat, it passes 24 VAC from its first stage (W) terminal to an associated zone valve (ZV1, ZV2, or ZV3), which opens to allow flow to the associated radiant panel manifold station. When that zone valve is fully open, the zone valve's end switch closes. This passes 24 VAC to the paired zone valve (ZV4, ZV5, or ZV6), which then allows flow through the air handler associated with that zone. The end switch in the zone valves associated with the air handlers completes a 24 VAC circuit that energizes the coil of relay (R1). A normally open contact (R1-1) closes to connect line voltage to circulator (P3), which then operates in proportional differential pressure mode.

Whenever the mode selection switch is set to heating 24 VAC is passed to the outdoor reset controller (ORC),

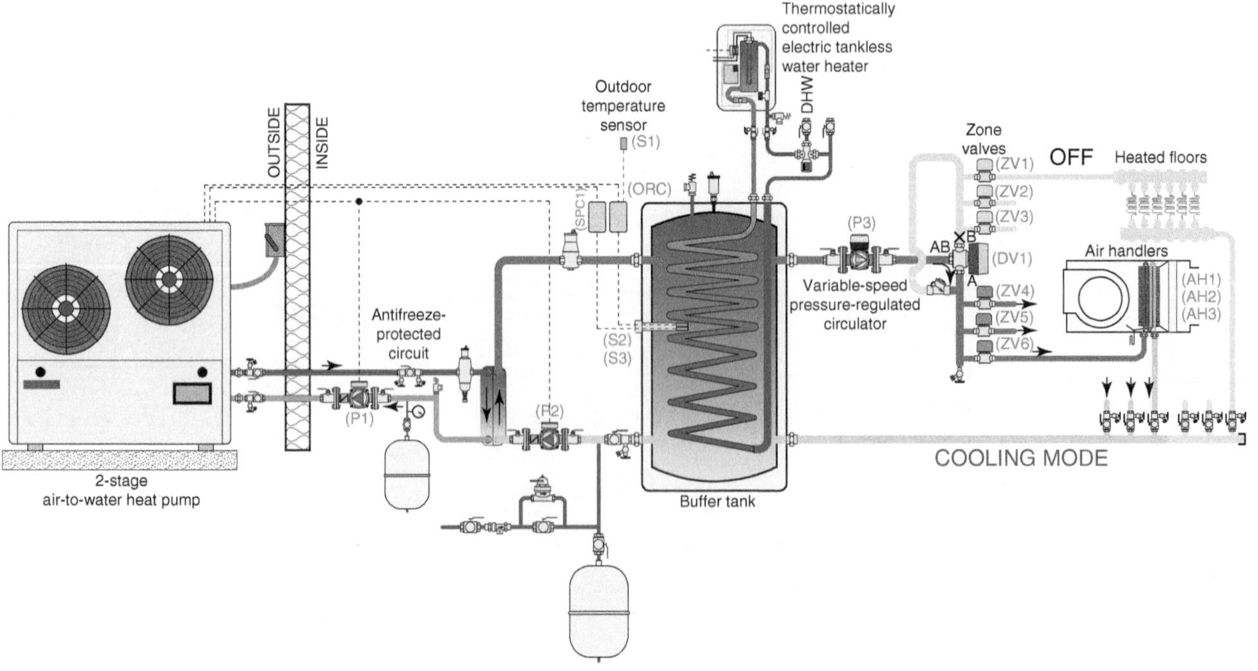

Figure 13-57b | Combisystem #5 operating in cooling mode. Note that chilled water is blocked from the heating zones by the diverter valve and swing check valve.

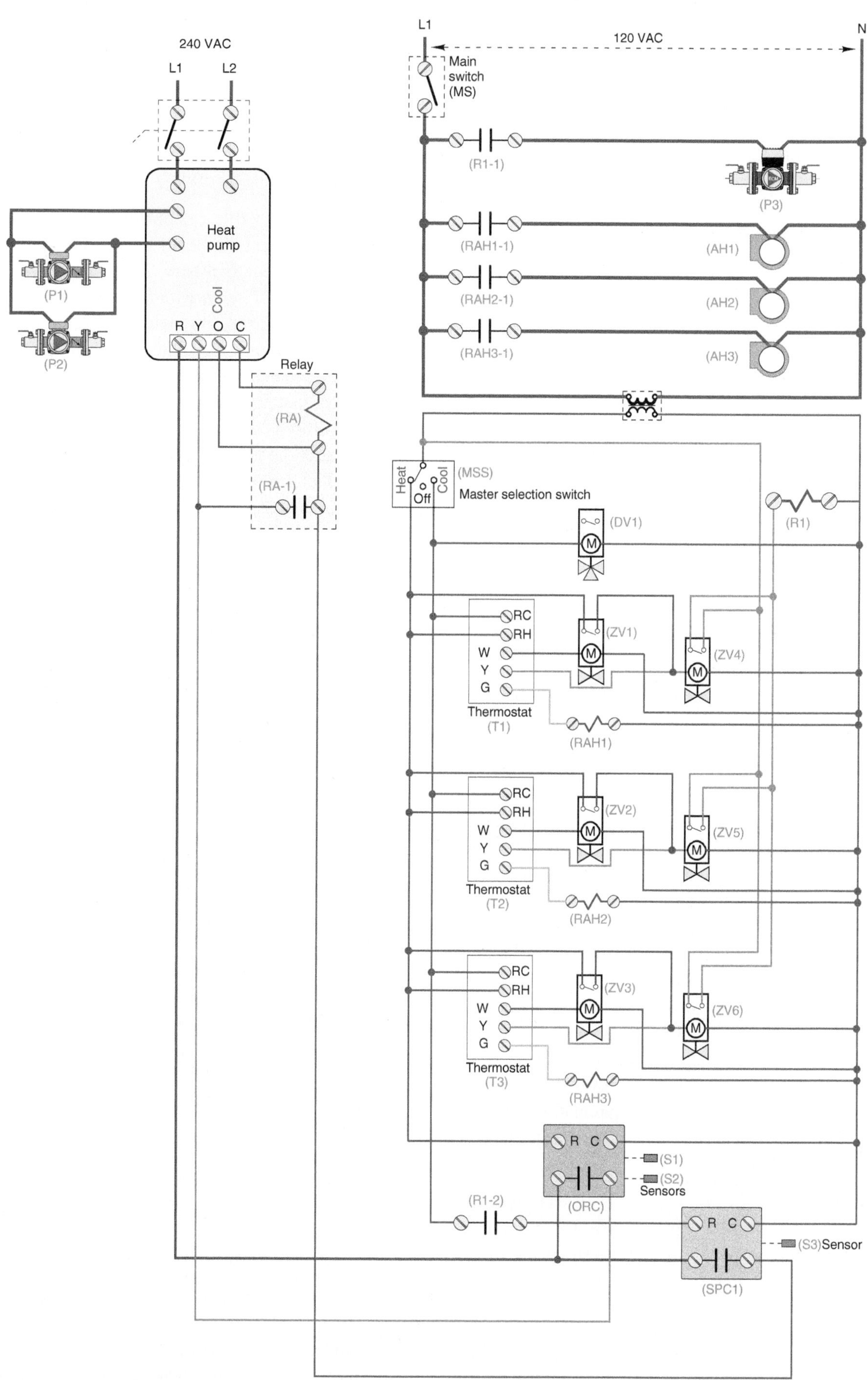

Figure 13-57c | Electrical wiring schematic for combisystem #5 shown in Figure 13-57a.

which monitors the temperature in the buffer tank at sensor (S2). This controller measures outdoor temperature at sensor (S1), and uses this temperature along with its settings to calculate the target temperature for the heating distribution system. If the temperature at sensor (S2) is too low to supply the heating load, the normally open contact in the (ORC) closes. This turns on the heat pump in heating mode (completing a circuit between the heat pump's R and Y terminals). The heat pump supplies line voltage to operate circulators (P1) and (P2) and thus transfers heat to the buffer tank. NOTE: Whenever the mode selection switch (MSS) is set to heating, the buffer tank temperature is maintained close to the target water temperature, as determined by the (ORC). This allows for domestic water preheating regardless of the space heating load status.

2. *Space cooling mode:* The mode selection switch (MSS) must be set for cooling. This supplies 24 VAC to the (RC) terminals of all three thermostats (T1, T2, and T3). If any thermostat is set for cooling mode, and calls for cooling, 24 VAC is switched to the thermostat's (Y) terminal. This supplies 24 VAC to the associated cooling zone valve (ZV4, ZV5, or ZV6). The zone valve opens to allow flow to the associated zone air handler. The end switch in the zone valve completes a 24 VAC circuit that energizes the coil of relay (R1). A normally open contact (R1-1) closes to connect line voltage to circulator (P3), which then operates in proportional differential pressure mode. Another normally open contact (R1-2) closes to pass 24 VAC to the cooling setpoint controller (SPC1). This controller measures the temperature within the buffer tank at sensor (S3). If the water temperature is above the upper limit where effective cooling is possible, the normally closed contact in (SPC1) closes. This connects 24 VAC from the heat pump's internal transformer to the coil of relay (RA), and to the heat pump's (O) terminal to energize the reversing valve for cooling operation. Normally open contact (RA-1) closes to connect 24 VAC from the heat pump's internal transformer to the (Y) terminal to turn on the compressor. The heat pump applies line voltage to operate circulators (P1) and (P2) and thus transfers chilled water to the buffer tank.

3. *Domestic water heating mode:* When domestic hot water is drawn from a fixture, cold water passes through the internal stainless steel coil in the buffer tank and is preheated. It then passes into the thermostatically controlled tankless electric water heater. Whenever the flow rate reaches 0.6 gpm or higher, the flow switch in the heater enables the elements to operate. The electronics within the heater control electrical power supplied to the elements based on the necessary temperature rise to achieve the set domestic hot water supply temperature. All heated water leaving the tankless heater flows into an ASSE 1017–rated mixing valve to ensure a safe delivery temperature to the fixtures. Whenever the demand for domestic hot water drops below 0.4 gpm, the electric water heater is turned off.

It is also possible to configure the control wiring for this system so that the air handlers would operate as **second-stage heat emitters**. *This requires thermostats with two stages of heating and a single stage of cooling.* In such a mode, the **first-stage contacts** of a 2-stage thermostat turn on the radiant panel circuits. If the air temperature then stabilizes to the desired setpoint, the air handler for that zone remains off. However, if heat output from the floor is unable to maintain the desired setpoint temperature, the **second stage contacts** in the thermostat close, which turns on that zone's air handler to provide additional heat input. The cooling operation remains the same. A wiring diagram for this system configuration is shown in Figure 13-58. No changes in the piping schematic are required for the system to operate in this configuration.

13.9 Performance Simulation Software and Comparison with Geothermal Heat Pumps

As is true for solar thermal systems, the seasonal performance of an air-to-water heat pump system is influenced by many variables, not the least of which is the weather at the installation site. Making an accurate estimate of seasonal performance involves calculations that can track the change in performance indices, such as heating capacity, cooling capacity, COP, and EER, over an entire heating and cooling season. Operating cost can be influenced by project-specific conditions such as **off-peak electrical rates**, or owner setback schedules. It would be very tedious to attempt to factor all these variables into manual calculations. Thus, software-based analysis is the only practical way to assess seasonal performance of air-to-water heat pumps systems.

Some air-to-water heat pumps manufacturers provide software that enables designers to assess the expected performance of a their heat pumps, in specific applications. Figure 13-59 shows examples of the screens from one of these software offerings.

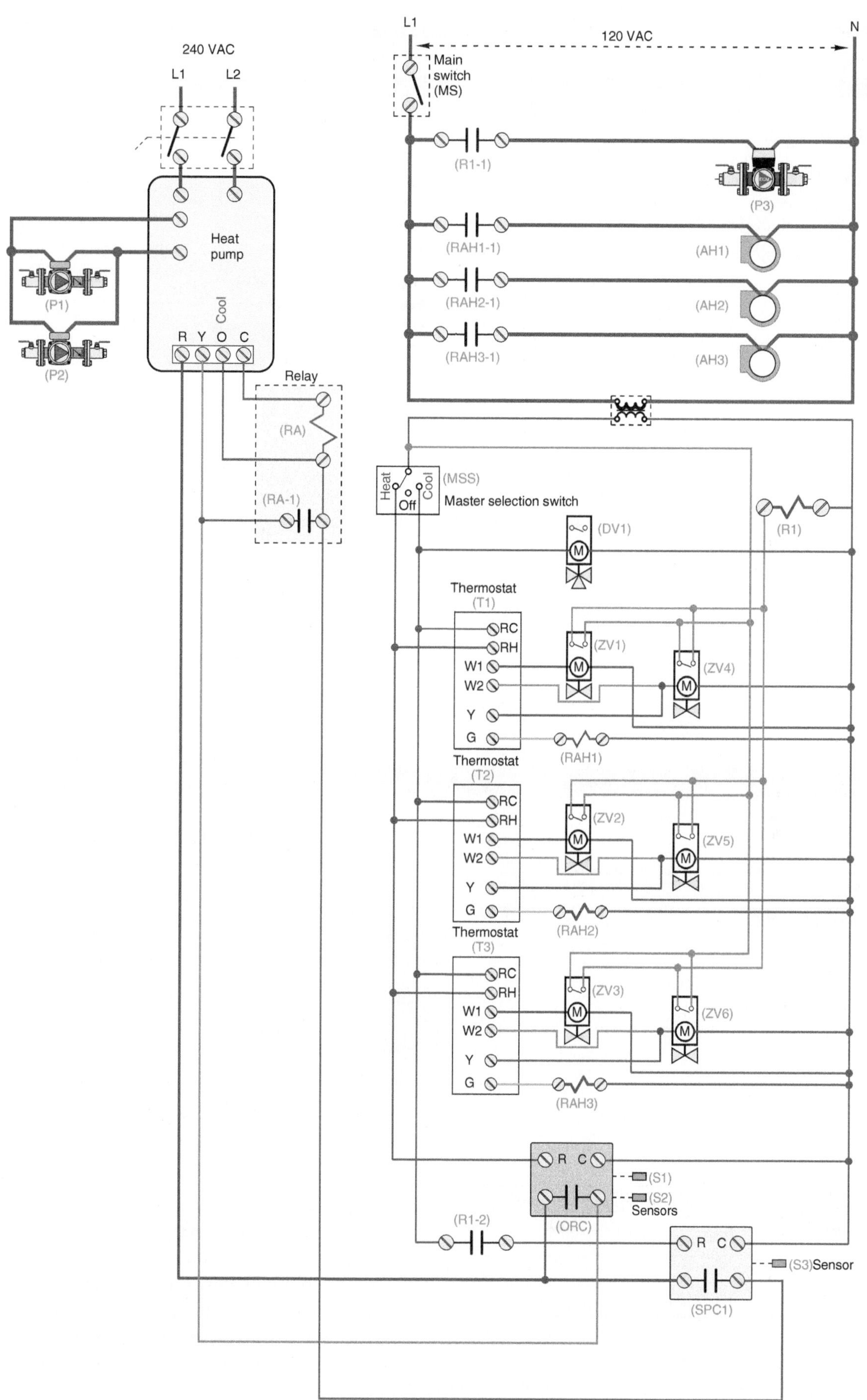

Figure 13-58 | Electrical wiring to operate the air handlers in combisystem #5 shown in Figures 13-57a and 13-57b on second-stage heating using a 2-stage thermostat.

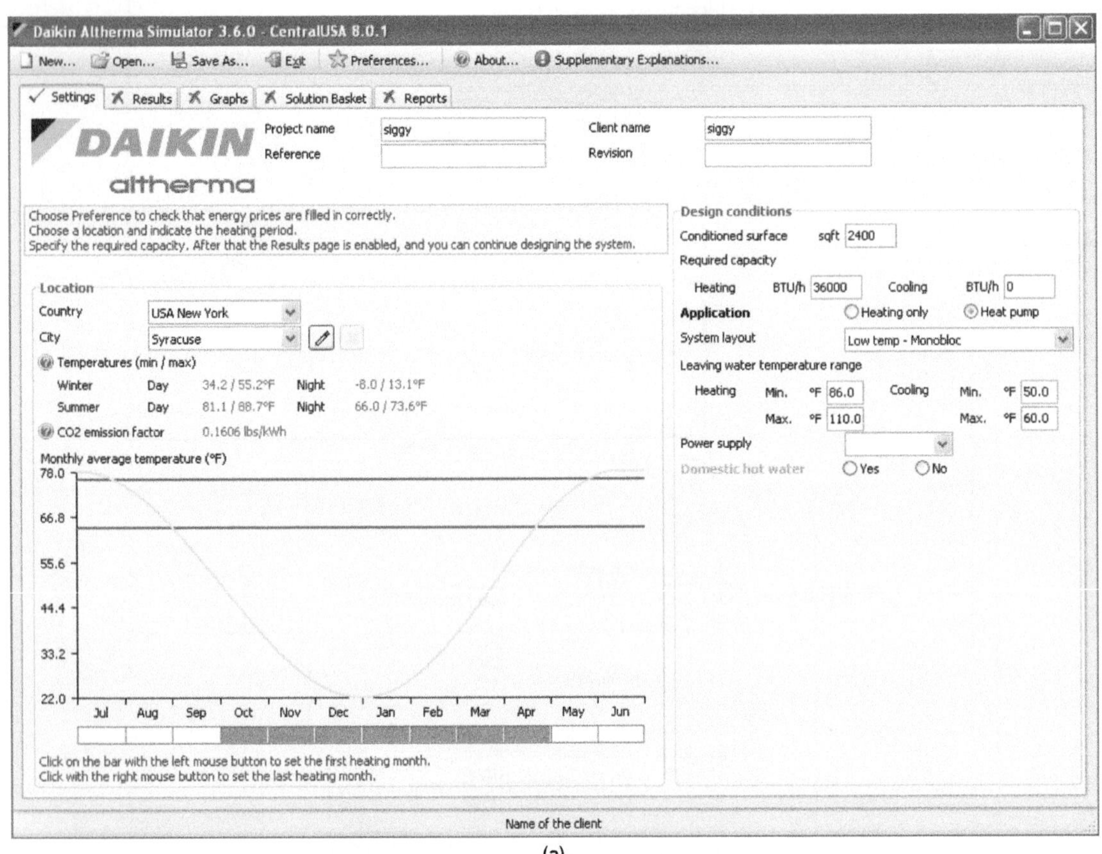

(a)

(b)

Figure 13-59 Example of design assistance software available for air-to-water heat pump systems from Daikin. (a) Main screen. (b) Matrix to specify day and night utility rates.

The software shown allows the designer to specify costs for electrical energy during both on-peak and off-peak periods. It also allows specification of a temperature setback schedules.

COMPARISON WITH GEOTHERMAL HEAT PUMPS

The installation and operating cost of any heat pump can vary considerably from one project to another. It is important to determine accurate estimates for both of these costs when making comparisons.

When discussing heat pump performance, it's common to hear the comment that geothermal heat pumps offer better thermal performance than air-to-water heat pumps. In cold climates, a well-designed and properly installed geothermal heat pump usually does offer a higher **seasonal average COP** compared to an air-to-water heat pump. However, one should not judge the economic merit of a heat pump system based solely on the seasonal average COP it delivers. A proper analysis must also factor in the installation cost as well as other factors that could add or remove cost over the life of the system.

The following is a relatively simple comparison between two heating systems: One is supplied by an air-to-water heat pump, and the other by a geothermal water-to-water heat pump. Both heat pumps supply a low temperature hydronic radiant panel distribution system in a modest house. The comparison is only made for heating operation, which dominates in the Syracuse, NY, location where the systems are assumed to be located. The comparison was made using cost estimates and utility rates in effect in early 2013. The installed costs do not include the low-temperature distribution systems, which are assumed to be identical. The comparison includes estimates for the thermal performance and associated operating cost of both systems. It assumes equal design life for each heat pump and doesn't include maintenance costs. As such, it is not a complete life cycle cost analysis, but rather an initial "sketch" that can be used to determine if a more detailed analysis is justified. Chapter 16 presents methods for making more detailed economic comparisons.

The project specifics were as follows:

- *Example house: 36,000 BTU/hr. design load at 70 °F inside & 0 °F outside*
- *Location: Syracuse, NY (6,720 heating degree days)*
- *Total estimated heating energy required: 49.7 MMBTU/season*
- *Average cost of electricity: $0.13/kwhr*
- *Distribution system: radiant panels with a supply temperature = 110 °F at design conditions*

Thermal performance estimates where made using two different software packages, one for the air-to-water heat pump and the other for the geothermal heat pump. The analysis software used was provided by the heat pump manufacturers.

The following results were generated:

Air-to-water heat pump option:

- *Heat pump: Nominal 4.5-ton split system air-to-water heat pump supplying this load*
- *Seasonal average COP = 2.8.*
- *Estimated installed cost = $10,600 (not including distribution system)*

Geothermal heat pump option:

- *Heat pump: Nominal 3-ton water-to-water heat pump with vertical earth loop*
- *Seasonal average COP = 3.28.*
- *Estimated installed cost = $20,550 (not including distribution system)*

($11,800 for earth loop, $8,750 for balance of system, not including distribution system)

- *Deduct for 30% federal tax credit: (–$6,165)*
- *Net installed cost: $14,385 (not including distribution system)*

Annual space heating cost comparison:

AIR-TO-WATER HEAT PUMP (COP_{ave} *= 2.8) = $676/year*

GEOTHERMAL HEAT PUMP (COP_{ave} *= 3.28) = $578/year*

Difference in annual heating cost: $98/year

Difference in net installed cost: $3,785

Simple payback on higher cost of geothermal heat pump: 3785/98 $\approx$ *38 years*

This comparison was based on the relatively cold winter climate of Syracuse, NY. In such climates, it is common for the seasonal average COP of an air-to-water heat pump to be lower that of a well-designed and properly installed geothermal heat pump system. However, the installed cost of the air-to-water heat pump will often be several thousands of dollars less because there is no earth loop. Thus, the incremental savings associated with the slightly higher seasonal average COP of the geothermal heat pump will produce very marginal returns on the extra investment. Furthermore, this comparison includes the 30 percent federal income tax credit available on residential geothermal heat pump installations in the United States, at the time this text was written. This reduces the cost of the geothermal heat pump system by $6,165. As of 2013, this credit is not available for air-to-water heat pump systems. This residential tax credit is scheduled to expire on December 31, 2016. If this credit did not apply, the incrementally higher installation cost of the geothermal heat pump system would be $9,950. The first-year operating cost savings of the geothermal heat pump system, compared that of the air-to-water heat pump, would remain $98. This makes the simple payback for the extra investment in the geothermal heat pump system at just over 100 years! Since most currently available heat pump systems would have useful lives of perhaps 15 to 25 years, such a long payback is unattainable and of no practical interest.

Keep in mind that this is a specific comparison, and the results should not be generalized to other systems. Still, this comparison demonstrates that the economics of an air-to-water heat pump can be very competitive to those of geothermal heat pumps, even when the latter is heavily subsidized.

SUMMARY

This chapter has introduced air-to-water heat pumps as a unique renewable energy heat source that can also provide cooling. As with other renewable energy heat sources, the key to good performance is matching the heat pump with a low-temperature hydronic distribution system.

Many of the system details shown in this chapter can also be used in combination with geothermal heat pumps, which are covered in the next chapter.

Designers are encouraged to familiarize themselves with the details of the heat pumps they select and ensure that the balance of the system design, and controls configuration, is properly coordinated with those details.

KEY TERMS

2-position spring-return actuator
4-way valve
absolute temperature
air-source heat pump
air-to-air heat pump
air-to-water heat pump
approach temperature
balance point
bidirectional thermal expansion valve
buffer tank
Carnot COP
coaxial tube-in-tube heat exchanger
coefficient of performance (COP)
compressor
condenser
cooling capacity
cooling-only application
dirt separator
diurnal heat storage
diurnal operation
energy efficiency ratio (EER)
Energy Star
evaporator
first-stage contact
flow reversal
flow switch
gooseneck piping
heat pump
heating capacity
heating-only application
hydronic detergent
inlet diffuser
installation and operation (I/O) manual
inverter drive compressor
latent cooling capacity
leaving load water temperature
mode selection switch
National Electrical Code (NEC)
net metering
net zero buildings
nominal cooling capacity
nominal heating capacity
non-reversible heat pump
off-peak electrical rate
on-time [of heat pump]
pilot solenoid valve
primary loop
refrigerant
Residential Renewable Energy Tax Credit
reversible heat pump
reversing valve
saturation temperature
scroll compressor
scroll
seasonal average COP
second-stage contact
second-stage heat emitter
secondary circuit
self-contained air-to-water heat pump
sensible cooling capacity
sink (for a heat pump)
solar photovoltaic system
source (for a heat pump)
split system
superheat
supplemental expansion tank
temperature bin hour
temperature lift
thermal expansion valve
tons [refrigeration capacity]
vapor compression refrigeration cycle
variable speed compressor
water-to-air heat pump
water-to-water heat pump

QUESTIONS AND EXERCISES

1. Explain the difference between the source and sink for a heat pump.
2. Explain why air-to water heat pumps are renewable energy heat sources.
3. Describe the function of a pilot solenoid valve within a refrigerant reversing valve.
4. What does it mean when a refrigerant is superheated? Be specific.
5. A heat pump is being monitored with instruments as it operates. These instruments show that the electrical power supplied to the compressor is 4,000 watts. Water is flowing into the condenser of the heat pump at 10 gpm and 95 ºF. It leaves the condenser at 106 ºF. Estimate:
   a. The rate at which heat is being absorbed by the evaporator
   b. The COP of the heat pump under these conditions
6. The heat output of an air-to-water heat pump is measured at 55,000 Btu/hr. Assume the heat pump is operating with a COP of 3.5. Estimate the electrical power supplied to the compressor.

7. A heat pump moves heat from a source material having a temperature of 25 °F. The heat is delivered to a sink material at a temperature of 115 °F. The heat pump's manufacturer states it operates at a COP of 7.0 under these conditions. Explain and justify why this is impossible.

8. Explain why real heat pumps cannot attain performance as high as the Carnot COP.

9. An air-to-water heat pump is operating in cooling mode. Water flows into its evaporator at 65 °F and leaves at 50 °F. The water is flowing at 9 gpm. A watt meter indicates that the compressor is demanding 5KW. Determine the EER of the heat pump under these conditions.

10. Discuss an advantage of a split system air-to-water heat pump for a system installed in a cold northern climate.

11. Describe what happens to the heating capacity of an air-to-water heat pump under each of the following conditions.
    a. The outdoor temperature decreases
    b. The entering water temperature increases
    c. The flow rate of water through the heat pump increases

12. Describe an application in which an on/off air-to-water heat pump does not require a buffer tank.

13. Describe a situation in which a modulating heat pump capable of reducing its output to 25 percent of maximum output still requires a buffer tank.

14. The pump curve for a circulator that is built into a self-contained air to water heat pump is shown in Figure 13-60. Also shown is the head loss curve for water flowing through that heat pump. Sketch out the "net pump curve" for this situation.

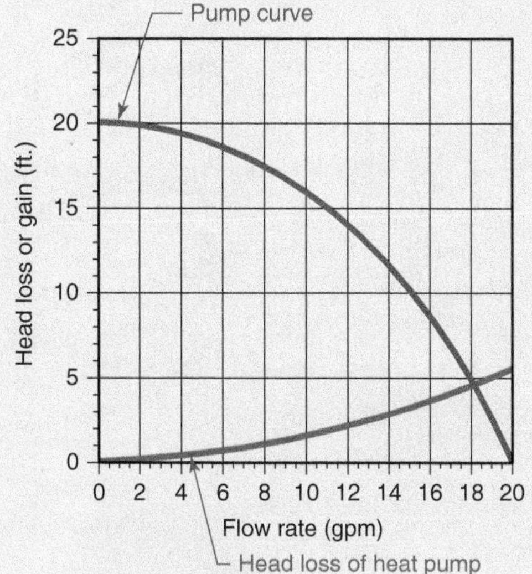

Figure 13-60 | Pump curve and head loss curve for heat pump for Exercise 14.

15. Why is it necessary to have a flow switch that monitors water flow through an air-to-water heat pump? Describe a possible failure mode if the flow switch were not present.

16. Describe what could happen if two air-to-water heat pumps are mounted two feet apart, and the air discharge from one heat pump is directed toward the air intake of the other heat pump. Assume that both units are operating in heating mode.

17. An on/off heat pump has a maximum output of 52,000 Btu/hr. It will be connected to a system where the smallest zone load will be 2,000 Btu/hr. Determine the volume of a buffer tank such that this heat pump will run for a minimum on-time of 12 minutes while raising the water temperature in the tank from 90 °F to 115 °F.

18. What is the reason to use flow reversal when adding heat versus removing heat from a large thermal storage tank?

19. What is the advantage of using an outdoor reset controller to regulate the temperature in a buffer tank rather than having the heat pump maintain the water temperature close to its design load value?

20. What is the function of the closely spaced tees in Figure 13-55a?

21. Sketch out a modification of the system shown in Figure 13-55a that would use an on/off heat pump in combination with a buffer tank to replace the modulating heat pump and closely spaced tees. Include detailing for domestic water heating.

22. Describe a situation in which it would be necessary to add a supplemental expansion tank to a system that has a heat pump with an internal expansion tank. Why is it important to have both expansion tanks at the same air-side pressurization, and with their inlet connections at approximately the same elevation?

23. Why shouldn't the seasonal average COP of two heat pump systems be the sole criteria used to determine which heat pump provides the best overall value?

R410A

CHAPTER 14

Water-to-Water Geothermal Heat Pump Systems

OBJECTIVES

After studying this chapter, you should be able to:

- Identify the major components of a water-to-water heat pump.
- Predict how changes in operating conditions affect the heating capacity and COP of water-to-water heat pumps.
- Evaluate how changes in operating conditions affect the cooling capacity and EER of water-to-water heat pumps.
- Understand the rationale behind geothermal heat pump systems.
- Describe the strengths and limitations of both open-loop and closed-loop geothermal heat pump systems.
- Describe configurations for horizontal and vertical earth loops.
- Explain the procedures used to join HDPE piping using butt fusion and socket fusion.
- Use reference information to size both horizontal and vertical earth loops.
- Describe the operation of heating-only systems using water-to-water heat pump.
- Understand how water-to-water heat pumps can be used in combination with thermal storage.
- Discuss how multiple water-to-water heat pumps can be controlled in stages.
- Describe the operation of several heating and cooling systems using water-to-water heat pumps.

14.1 Introduction

Chapter 13 provided a general discussion of heat pumps as well as the specifics associated with air-to-water heat pumps. This chapter continues the discussion of heat pumps, but for types that absorb low-temperature heat from water rather than air. These heat pumps also deliver higher temperature heat to a stream of water. As such, they are called **water-to-water heat pumps**. This type of heat pump can be used in a wide variety of applications for building heating and cooling as well as applications such as domestic water heating, pool heating, or other processes where heated water, chilled water, or both are required.

One of the premier applications for water-to-water heat pumps is extracting low-temperature heat from the ground and delivering it to a building through a hydronic distribution system. The heat pumps in these systems are referred to by various names, including **geothermal heat pump**, **ground-source heat pump**, and **earth-coupled heat pump**. Although the author prefers the label *earth-coupled heat pump* as best describing how these devices interact with the ground, few people are familiar with this name. Instead, the designation "geothermal" heat pump has been widely popularized by manufacturers, utilities, and regulators. With that in mind, this text will describe systems in which the heat pump adds or removes heat from the earth as geothermal heat pump systems.

Courtesy of ClimateMaster, Inc.

Figure 14-1 | A modern water-to-water heat pump.

14.2 Water-to-Water Heat Pumps

Water-to-water heat pumps use many of the same components as air-to-water heat pumps. They include a compressor, refrigerant-to-water heat exchanger, reversing valve, and thermal expansion valve. The primary *difference* is that the air-to-refrigerant heat exchanger in an air-to-water heat pump, which serves as the evaporator during heating mode operation, and its associated fans, are replaced by a second water-to-refrigerant heat exchanger. This allows the heat pump components to fit into a smaller cabinet that is usually installed inside a building. Figure 14-1 shows an example of a modern water-to-water heat pump.

Figure 14-2 shows how the refrigeration components are configured in a reversible water-to-water heat pump.

The refrigeration cycles operates in the same manner as previously described for air-to-water heat pumps. In the heating mode, low-temperature heat is transported to the evaporator by a stream of fluid. That fluid may be water or a water-based antifreeze solution. It is pushed through the evaporator by a pump or circulator. Cold liquid refrigerant enters the evaporator at a temperature significantly lower than that of the water or antifreeze solution. Heat is absorbed into the refrigerant, causing it to vaporize and attain some degree of superheat. The low-temperature superheated gas then passes through the refrigerant reversing valve and on to the compressor, where its temperature and pressure are greatly increased. The hot gas leaves the compressor, passes through the reversing valve, and flows on to the condenser. Here, it transfers heat to another fluid stream being pushed through the condenser by a circulator. The hot refrigerant gas condenses back to a cooler liquid but remains at a relatively high pressure. The liquid refrigerant then passes into a thermal expansion valve, which lowers its pressure to the point where it returns to the condition from which this description began. The refrigeration cycle remains in continuous operation whenever the compressor is running.

As this text is being written, most water-to-water heat pumps used for residential and light commercial building applications are on/off devices. The combisystems shown later in this chapter are configured based on this operating characteristic. The author anticipates that **modulating water-to-water heat pumps** with variable-speed compressors will soon become available. Like its air-to-water heat pump equivalent, a modulating water-to-water heat pump will be able to better match its thermal output to a varying heating or cooling load.

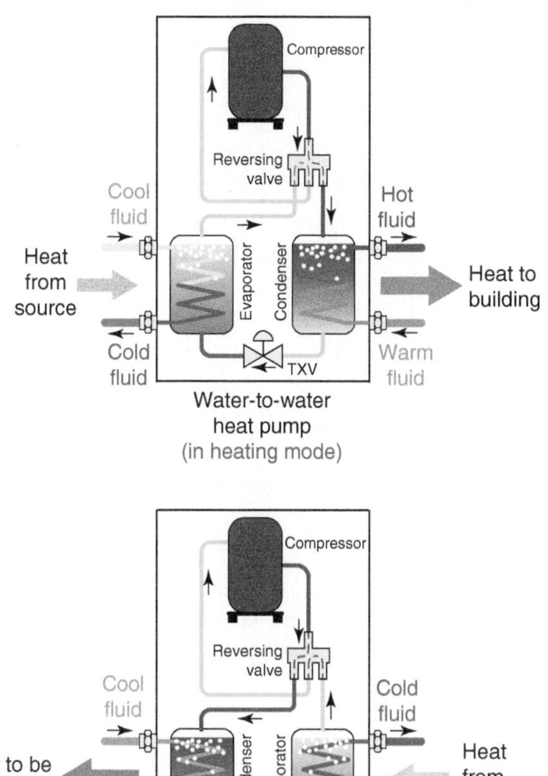

Figure 14-2 | Typical arrangement of refrigeration system components in a reversible water-to-water heat pump.

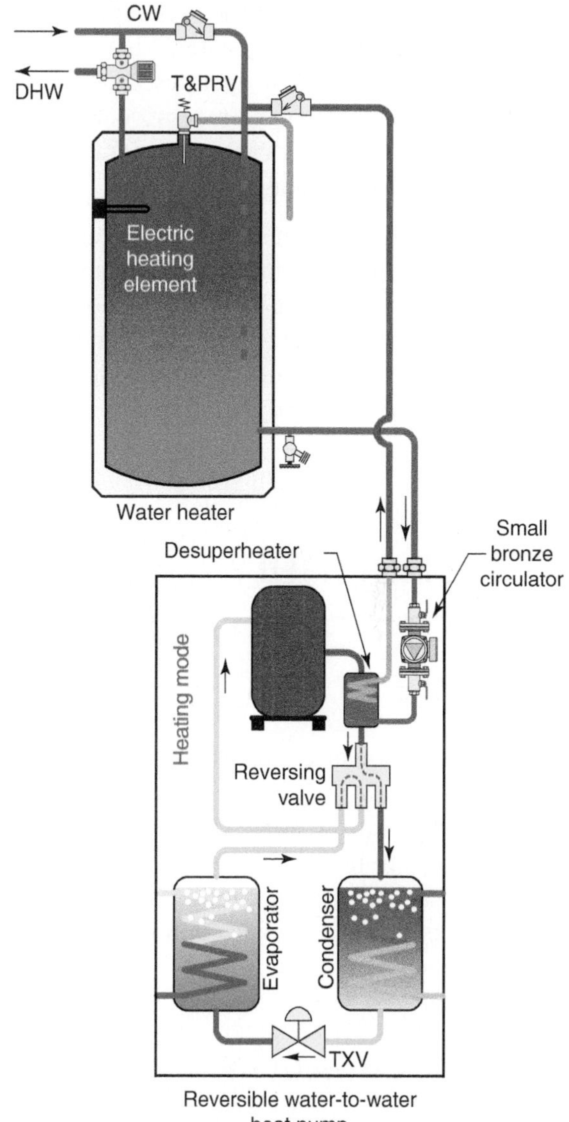

Figure 14-3 | Desuperheater heat exchanger within a reversible water-to-water heat pump, along with piping to a domestic hot water tank.

DESUPERHEATERS

Many water-source heating pumps can be purchased with an optional **desuperheater**. This is a small refrigerant-to-water heat exchanger that is factory-installed between the discharge port of the compressor and the inlet of the reversing valve, as shown in Figure 14-3.

A desuperheater absorbs heat from the hot refrigerant gas leaving the compressor and transfers that heat to a stream of domestic water. This allows the heat pump to contribute to domestic water heating as well as space heating. Desuperheaters are designed to extract the superheat of the refrigerant. This cools the refrigerant vapor, but not to the point where it begins to condense into a liquid. This process is called **desuperheating**. When the heat pump is operating in heating mode, heat transferred to domestic water at the desuperheater is not available for space heating. However, desuperheaters are designed by the heat pump manufacturer so that only a fraction (usually 5 to 10 percent) of the total thermal energy in the hot refrigerant gas leaving the compressor is transferred to domestic hot water. When the heat pump operates in cooling mode, all heat transferred to domestic water is "free" heat. That's because this heat would otherwise be dissipated to an earth loop, or another fluid stream that is cooling the heat pump's condenser.

Heat pumps equipped with desuperheaters are also usually supplied with a small bronze or stainless steel circulator that creates flow between the hot water storage tank and the desuperheater. This circulator is typically wired so that it operates whenever the compressor is on. However, in some heat pumps, the desuperheater circulator can also be turned on and off depending on the operating mode.

Figure 14-3 shows water from the lower portion of tank-type water heater flowing into the desuperheater. Water leaving the desuperheater flows back to the dip tube in the cold water connection to the water heater. This allows the desuperheater to transfer heat to the cooler water in the tank and thus maintain maximum heat transfer rates.

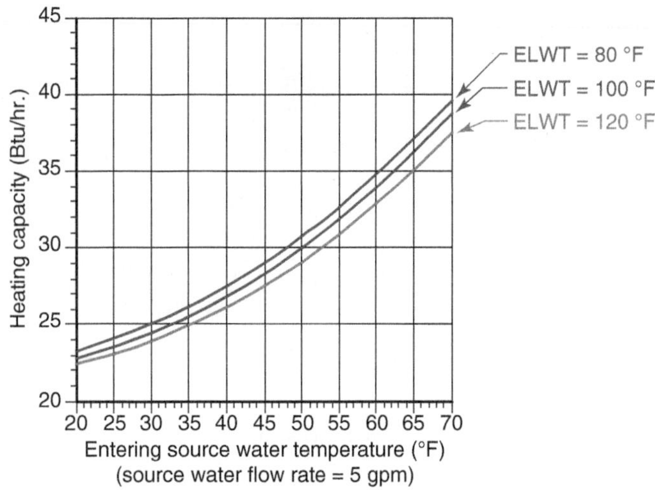

Figure 14-4 Heating capacity of a nominal 3-ton water-to-water heat pump as a function of entering source water temperature and entering load water temperature.

14.3 Thermal Performance of Water-to-Water Heat Pumps

The thermal performance indices used with water-to-water heat pumps are the same as those used for ATW heat pumps. In heating mode, thermal performance is expressed using heating capacity and coefficient of performance (COP). In cooling mode, thermal performance is expressed using cooling capacity and energy efficiency ratio (EER).

HEATING CAPACITY

The heating capacity of a water-to-water heat pump is the rate at which it delivers heat to the stream of water flowing through its condenser. As is true with air-to-water heat pumps, the heating capacity of a water-to-water heat pump is highly dependent on its operating conditions, specifically the temperature of the **source fluid** as well as the temperature of the **sink fluid**. The latter is the fluid stream, which carries heat away from the heat pump's condenser. Heating capacity also depends on the flow rate of both the source and sink fluid streams.

Figure 14-4 shows the heating capacity of a modern water-to-water heat pump with a nominal rating of 3 tons.

The heating capacity is plotted as a function of **entering source water temperature** and three different **entering load water temperatures (ELWT)**. The entering source water temperature is that of the water or antifreeze solution entering the heat pump's evaporator. Likewise, the entering load water temperature is that of the water entering the heat pump's condenser.

This graph shows that heating capacity drops off significantly with decreasing source water temperature. This relationship is crucially important in applications where source water temperature varies over a wide range. For example, in October, the fluid temperature entering the evaporator of a water-to-water heat pump connected to a horizontal earth loop might be 60 °F. If the water temperature entering the condenser at the time was 80 °F, Figure 14-4 indicates that the heat pump's heating capacity would be approximately 35,000 Btu/hr. However, by early March, in a northern location, the fluid temperature supplied by the same horizontal earth loop might only be 35 °F. This would reduce the heating capacity of the heat pump to about 26,500 Btu/hr. (assuming the same entering load water temperature of 80 °F). This 24 percent drop in heating capacity is significant and must be accommodated during system design. In some systems, supplemental heat from a different heat source may be necessary.

Heating capacity is also affected by the entering *load* water temperature. The higher this temperature, the lower the heating capacity of the heat pump. Figure 14-4 shows three representative entering load water temperatures: 80, 100, and 120 °F. Although the drop in heating capacity, per degree Fahrenheit change in load water temperature, is not as pronounced as with decreasing source water temperature, it still must be recognized during system design. This characteristic reinforces the need to design hydronic heating distribution systems around the lowest possible supply water temperatures. As has been the recommendation in earlier chapters, no hydronic distribution system supplied by a renewable energy heat source should require a supply water temperature higher than 120 °F under design load conditions.

The temperature of the water *leaving* the heat pump depends on water flow rate and the heating capacity. It can be determined using Equation 14-1.

(Equation 14-1)

$$T_{LLWT} = T_{ELWT} + \frac{Q_{HP}}{(8.01Dc)f_c}$$

where:

T_{LLWT} = temperature of the load water leaving heat pump (°F)

T_{ELWT} = temperature of load water entering heat pump (°F)

Q_{HP} = heating capacity of heat pump (Btu/hr.)

D = density of fluid flowing into condenser (lb./ft.^{3})

c = specific heat of fluid flowing into condenser (Btu/lb./°F)

f_c = water flow rate through heat pump condenser (gpm)

Example 14.1: Assume a heat pump has the heating capacity shown in Figure 14-4. It is operating with 80 °F entering load water temperature and 35 °F entering source water temperature. The flow rate through the condenser is 8 gpm. Determine the outlet water temperature from the condenser.

Solution: At the indicated entering source water temperature and entering load water temperature (ELWT), the heating capacity of the heat pump is 26,500 Btu/hr. Using this value along with the stated operating conditions yields the following condenser outlet temperature.

$$T_{LLWT} = T_{ELWT} + \frac{Q_{HP}}{(8.01Dc)f_c}$$

$$= 80 + \frac{26500}{(8.01 \times 62.1 \times 1.00)(8)} = 86.7 \text{ °F}$$

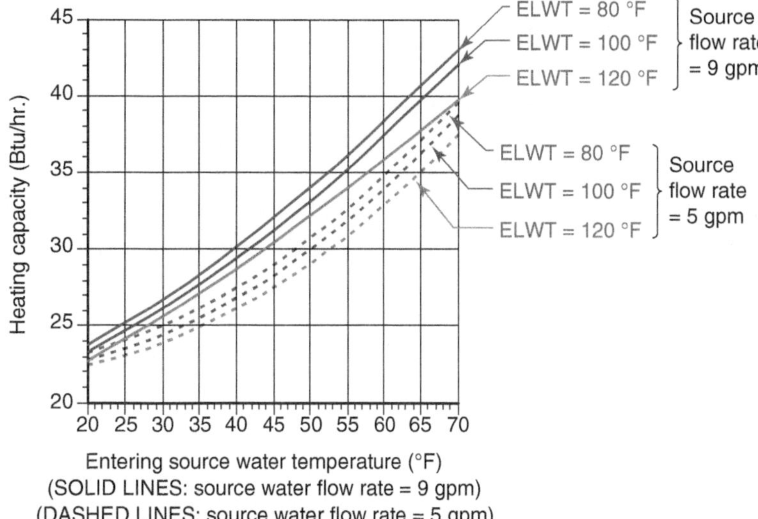

Figure 14-5 | Effect of source water flow rate on heating capacity for a nominal 3-ton water-to-water heat pump.

Discussion: The increase in water temperature is only 6.7 °F. Equation 14.1 is based on the sensible heat rate equation introduced in Chapter 4. Although the temperature rise of the water stream is only 6.7 °F, this stream is carrying a substantial rate of heat away from the heat pump.

The flow rate through both the evaporator and condenser also affects the heat pump's heating capacity. Figure 14-5 shows this effect for the same water-to-water heat pump represented in Figure 14-4, but for specific source water flow rates of 5 gpm and 9 gpm.

Notice that heating capacity *decreases* slightly as the flow rate through the evaporator decreases. This is also true for flow rate through the condenser. Lower flow rates lower the average fluid temperature within the evaporator. They also reduce turbulence within heat exchangers, which slightly decreases convective heat transfer between the moving fluid and metal walls of the heat exchanger.

Given the increase in heating capacity at higher flow rates, it may seem intuitive to design heat pumps to operate at the highest possible flow rates. However, designers should always consider the increase in circulator power required to establish these higher flow rates and balance the cost of the increased electrical energy consumption of the circulators against the gain in capacity. Flow rates over 3 gpm per ton (e.g., 12,000 Btu/hr.) of heat transfer rate are seldom necessary or justified.

COEFFICIENT OF PERFORMANCE (COP)

The **coefficient of performance (COP)** of a water-to-water heat pump is defined the same as for an air-to-water heat pump. It is the ratio of the instantaneous heat output divided by the electrical power input needed to operate the heat pump. This can be expressed as Equation 14.2.

(Equation 14.2)

$$\text{COP} = \frac{\text{heat output (Btu/hr.)}}{\text{electrical input (watt)} \times 3.413}$$

The factor 3.413 in the denominator of Equation 14.2 converts watts to Btu/hr. This is necessary so that COP remains a unitless number.

The COP of a water-to-water heat pump is strongly dependent on the temperatures of the fluids entering its evaporator and condenser. Figure 14-6 shows the relationship between these temperatures and the heat pump's COP.

It is always desirable to operate a heat pump at the highest possible COP. Figure 14-6 shows that the higher the source water temperature, and the lower the entering load water temperature (ELWT), the higher the COP. Depending on the application, designers have limited ability to affect the entering source water temperature. For example, if water is drawn from a well, there is nothing the designer can do to change its temperature. However, if the fluid entering the heat pump flows through a closed tubing circuit buried in the earth, the depth of that circuit, and the amount of tubing it contains, will significantly affect the fluid temperature entering the heat pump's evaporator.

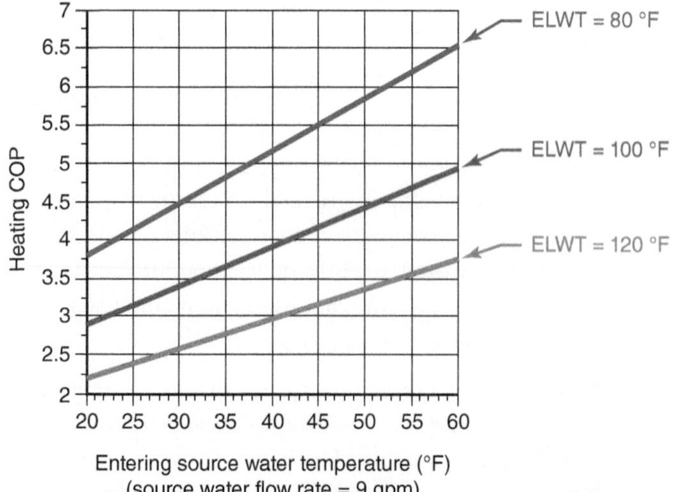

Figure 14-6 COP of a specific water-to-water heat pump as a function of the entering source water temperature and entering load water temperature.

The designer has considerable control over the entering *load* water temperature. In space heating applications, this is the temperature of water returning to the heat pump from the distribution system. Anything that lowers this temperature will improve the COP of the heat pump. Thus, to maximize COP, heating distribution systems should be designed around the lowest operating temperature possible.

HEATING PERFORMANCE STANDARDS

In North America, the heating capacity and COP of water source heat pumps is currently represented by a standard entitled: ***ANSI/AHRI/ASHRAE/ISO Standard 13256-2*** *Water-to-Water and Brine-to-Water Heat Pumps—Testing and Rating for Performance.* This standard makes assumptions about the power required by the circulator(s) used to create flow through the heat pump's evaporator (in heating mode). Specifically, it adds an assumed power demand for the circulator to the measured power consumption of the heat pump when defining how the heat pump's COP is calculated. Thus, the COP under this standard would be calculated as follows:

(Equation 14.3)

$$\text{COP}_{13256} = \frac{Q}{(P_i + P_c)3.413}$$

where:

COP_{13256} = COP of heat pump as defined by ANSI 13256-2 standard (unitless)

Q = measured heat output from heat pump condenser (Btu/hr.)

P_i = measured electrical power to operate heat pump (watts)

P_c = assumed power demand of evaporator circulator (watts)

The assumed power draw of the evaporator circulator, according to the ANSI 13256-2 standard, can be calculated using Equation 14.4.

(Equation 14.4)

$$w_c = 0.6289(f)H_L$$

where:

w_c = assumed power demand of evaporator circulator (watts)

f = flow rate through heat pump evaporator (gpm)

H_L = head loss created by flow through evaporator at flow rate (f) (ft. of head)

Example: 14.2: The technical data sheet for a water-to-water heat pump indicates that a water flow rate of 9 gpm through the heat pump's evaporator will produce a corresponding head loss of 13.9 feet. Estimate the assumed circulator power included in the heat pump's COP-based Equation 14.4.

Solution: Inserting these values into Equation 14.4 yields:

$$w_c = 0.6289(f)H_L = 0.6289(9.0)13.9 = 78.7 \text{ watts}$$

Discussion: The intent of adding an assumed circulator power demand into the COP calculation is to create a more realistic "**net COP**" of the heat pump system, which requires a circulator to create flow through the evaporator. However, Equation 14.4 still yields an estimated power requirement. It does not account for differences in the type of circulator used or the wire-to-water efficiency of that circulator as it is applied. In the case of a water-to-water heat pump, it only accounts for the circulator power on the evaporator side of the unit, and not for the circulator power used to create flow through the heat pump's condenser.

Example 14.3: The measured power demand of a water-to-water heat pump is 3,000 watts. Its heat output is 40,000 Btu/hr. Water flows through its evaporator at 11 gpm and creates a corresponding head loss of 20 feet. Determine the COP of the heat pump itself as well as the *net COP* of the heat pump based on including the assumed power demand of the evaporator circulator as per the ANSI 13256-2 standard.

Solution: The COP of the heat pump itself is:

$$\text{COP} = \frac{Q}{(P_i)3.413} = \frac{40{,}000}{(3000)3.413} = 3.91$$

The net COP requires the assumed power demand to be calculated using Equation 14.4.

$$w_c = 0.6289(f)H_L = 0.6289(11)20 = 138 \text{ watts}$$

The net COP can now be calculated using Equation 14.3.

$$\text{COP}_{13256} = \frac{Q}{(P_i + P_c)3.413} = \frac{40{,}000}{(3000 + 138)3.413} = 3.73$$

Discussion: The net COP, based on the ANSI 13256-2 standard is about 4.6 percent lower than the COP of the heat pump itself.

Two circulators are required to operate whenever a water-to-water heat pump is operating. A more accurate net COP of this situation could be calculated by adding the total wattage of both circulators to the power demand of the heat pump itself, as represented by Equation 14.5.

(Equation 14.5)

$$\text{COP}_{\text{WTWnet}} = \frac{Q}{(P_i + P_{evap} + P_{cond})3.413}$$

where:

$\text{COP}_{\text{WTWnet}}$ = net COP of a water-to-water heat pump (unitless)

Q = measured heat output from heat pump condenser (Btu/hr.)

P_i = measured power demand of heat pump (watts)

P_{cond} = measured power demand of condenser circulator (watts)

P_{evap} = measured power demand of evaporator circulator (watts)

The ANSI 13256-2 standard also establishes values for the water temperatures associated with the heating capacity ratings of water-to-water heat pumps. These temperatures differ depending on the source of water supplied to the heat pump's evaporator. The two categories pertaining to water-to-water heat pumps include ground water supplied directly to the heat pump (**GWHP**) and a heat pump supplied by a closed ground loop (**GLHP**). Those water temperatures are as follows:

Water source and mode	Water temp. entering evaporator	Water temp. entering condenser
(GWHP) heating	50 °F	104 °F
(GLHP) heating	32 °F	104 °F

COOLING PERFORMANCE

The cooling performance of a water-to-water heat pump is given by the same two indices used for air-to-water heat pumps:

- Cooling capacity
- Energy efficiency ratio (EER)

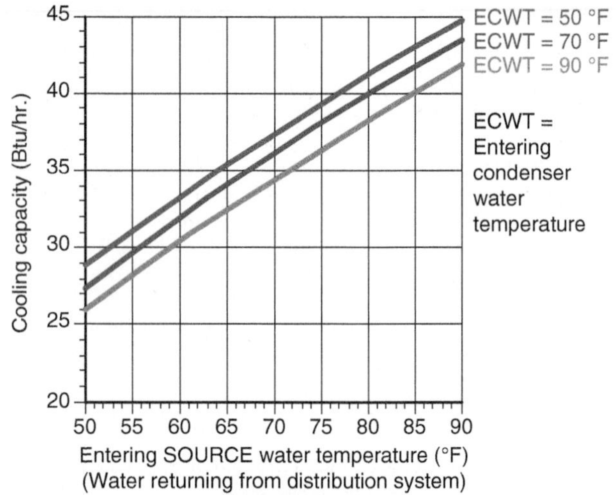

Figure 14-7 Cooling capacity of a water-to-water heat pump as a function of the entering source water temperature and entering condenser water temperature.

COOLING CAPACITY

Cooling capacity is the rate at which the water-to-water heat pump absorbs heat from the fluid stream passing through its evaporator. This rate depends the temperature of the fluid streams passing through the evaporator as well as through the condenser. To a lesser extent, it's also depends on the flow rates of these two streams.

The cooling capacity of the water-to-water heat pump discussed earlier in this section is represented graphically in Figure 14-7.

The horizontal axis shows entering source water temperature. This is the temperature of the water *returning from the cooling distribution system and flowing into the heat pump's evaporator.* The three sloping curves represent three entering condenser water temperatures (ECWT). This is the entering temperature of the fluid stream to which heat is being added. For example, the blue line showing an ECWT of 50 °F may represent fluid returning from an earth loop and entering the heat pump at 50 °F.

As the temperature of the entering *source* water goes up, so does the heat pump's cooling capacity. Thus, a water-to-water heat pump has a higher cooling capacity when supplying a cooling distribution system that operates at 55 °F chilled water supply temperature, compared to one that requires a 45 °F chilled water supply temperature. This graph also shows that lowering the temperature of the fluid entering the heat pump's condenser increases the heat pump's cooling capacity. This temperature may or may not be under the control of the designer depending on the source of the fluid. For example, if the condenser is being cooled by a stream of ground water, there is very little, if anything, the designer can do to affect the temperature of that water. However, if the fluid entering the condenser comes from a buried earth loop, the designer has somewhat more control over its temperature depending on how that earth loop was designed. Larger earth loops generally lower the temperature of the fluid entering the condenser when the heat pump is operating in cooling mode.

ENERGY EFFICIENCY RATIO (EER)

The instantaneous cooling efficiency of a water-to-water heat pump is expressed as an index called the **energy efficiency ratio (EER)**, which is defined by Equation 14.6:

(Equation 14.6)

$$\text{EER} = \frac{Q_c}{w_e} = \frac{\text{cooling capacity (Btu/hr.)}}{\text{electrical input wattage}}$$

where:

EER = energy efficiency ratio

Q_c = cooling capacity (Btu/hr.)

W_e = electrical power demand of heat pump (watts)

The higher the EER of a heat pump, the lower the electrical power required to produce a given rate of cooling.

Notice that the EER of a heat pump is not a unitless number, like COP. It has units of Btu/hr./watt. This unit can be interpreted as the number of Btu/hr. the heat pump supplies to the cooling load, per watt of electrical power demand. It can also be thought of as the number of thousands of Btu/hr. the heat pump supplies to the cooling load, per kilowatt of electrical power demand. Thus, a heat pump with an EER of 18.0 would supply 18,000 Btu/hr. of cooling capacity, per kilowatt (kW) of electrical power demand.

In *Europe*, the cooling efficiency of heat pumps is indicated by **cooling COP**. This would be the EER of the heat pump divided by 3.413. The latter value converts the denominator of the EER ratio to Btu/hr., and thus produces a unitless cooling COP number.

Like COP, the EER of a water-to-water heat pump is a function of the source and condenser water temperature. This variation is shown in Figure 14-8.

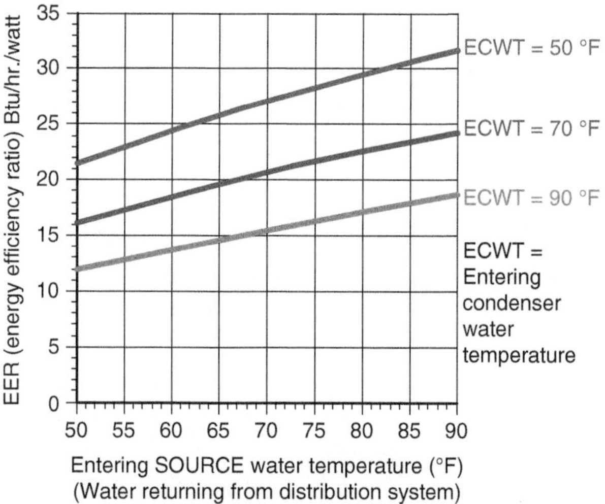

Figure 14-8 The energy efficiency ratio (EER) of a water-to-water heat pump as a function of the entering source water temperature and entering condenser water temperature.

This graph shows that EER increases as the temperature of the entering source water increases. It can also be seen that the lower the entering condenser water temperature (ECWT), the higher the EER.

As is true with heating, design decisions that reduce the temperature difference between the entering source water and entering condenser water improve both the heat pump's cooling capacity and EER. Higher fluid flow rates through the evaporator, the condenser, or both, also increase cooling capacity and EER. However, increased flow typically requires higher electrical power to the circulator(s) creating this flow. Flow rates over 3 gpm per ton (e.g., 12,000 Btu/hr.) of cooling capacity should be avoided.

COOLING PERFORMANCE STANDARDS

In North America, the cooling capacity and EER of water source heat pumps, which includes both water-to-air and water-to-water heat pumps, is currently represented by the standard *ANSI/AHRI/ASHRAE/ISO Standard 13256-2 Water-to-Water and Brine-to-Water Heat Pumps—Testing and Rating for Performance.* This standard makes assumptions about the power required by the circulator(s) used to create flow through the heat pump's condenser (in cooling mode). Specifically, it adds an assumed power demand for the circulator creating flow through the heat pump's condenser, to the measured power consumption of the heat pump, for purposes of determining the heat pump's EER. Thus, the EER under this standard can be calculated using Equation 14.7.

(Equation 14.7)

$$EER_{13256} = \frac{Q}{(P_i + P_c)}$$

where:

EER_{13256} = energy efficiency ratio based on ANSI 13256-2 (Btu/hr./watt)

Q_c = cooling capacity (Btu/hr.)

P_i = measured electrical power demand of heat pump (watts)

P_c = assumed electrical power demand of condenser circulator (watts)

The assumed power demand of the condenser circulator can be calculated based on Equation 14.4.

The flow rate and associated head loss of the condenser circulator can be found in technical specifications provided by the heat pump's manufacturer. As is true with COP, adding an assumed power demand for the condenser circulator will lower the net EER, compared to an EER calculated solely on the power demand of the heat pump. In the case of a water-to-water heat pump, the net EER calculated based on Equation 14.7 only includes the power demand of the condenser circulator. All water-to-water heat pumps also require a circulator to create flow through the evaporator whenever they are operating. A true "net EER" calculation would add the measured power demand of both the condenser circulator and the evaporator circulator to the power demand of the heat pump.

The ANSI 13256-2 standard also establishes values for the water temperatures associated with the cooling capacity and EER ratings of water-to-water heat pumps. Those temperatures differ depending on the source of water supplied to the heat pump's evaporator. The two categories pertaining to water-to-water heat pumps include ground water supplied directly to the heat pump (GWHP) and a heat pump supplied by a closed ground loop (GLHP). The water temperatures are as follows:

Water source and mode	Water temp. entering evaporator	Water temp. entering condenser
(GWHP) cooling	53.6 °F	59 °F
(GLHP) cooling	53.6 °F	77 °F

TONNAGE

It is customary to describe the heating and cooling capacity of refrigeration-based equipment such as heat pumps using the units of **tons**. In this context, a ton describes a *rate* of heat flow—specifically, 1 ton equals 12,000 Btu/hr. Thus, a "3-ton" heat pump implies a nominal heating or cooling capacity of 3 × 12,000, or 36,000 Btu/hr. The tonnage of a heat pump has nothing to do with the heat pump's weight. A description of a heat pump based on tons is also a *nominal* rating at some specific set of operating conditions, such as those prescribed by the ANSI 13256-2 standard. Thus, a "3-ton" rated heat pump could produce a heat output that is significantly higher than 3 tons when operated under more favorable conditions and significantly less than 3 tons when operated under less favorable conditions.

14.4 Design Details for Water-to-Water Heat Pump Systems

There are several details that system designers should address when applying water-to-water heat pumps. This section provides an overview of those details. Manufacturers' installation and operation (I/O) manuals should always be consulted for other specifics required by their products.

BUFFER TANKS

Most water-to-water heat pumps currently available in North America are on/off devices. They cannot modulate their heating or cooling capacity.

When the heating capacity of the heat pump closely matches the heat dissipation ability of the distribution system, such as under design load conditions, the on/off characteristic of the heat pump is not an issue. However, if an on/off heat pump is connected to a *zoned* distribution system, the heat dissipation ability of that distribution system may be far less than the heat output of the heat pump. This situation, if uncorrected, can lead to short cycling of the heat pump. Such operation can significantly shorten the life of components such as electrical contactors, start capacitors, and compressors. A similar mismatch between the cooling capacity of an on/off heat pump and a zoned cooling distribution system can also create undesirable short cycling and associated premature component failure.

The solution is to add a **buffer tank** between the output side of the water-to-water heat pump and the zoned distribution system, as shown in Figure 14-9.

For heating mode operation, the size of the buffer tank is determined by the desired "on-time" of the heat pump and the acceptable temperature rise of the water in the buffer tank during this on-time. The on-time is that time between when the heat pump turns on to begin warming the tank and when it turns off after raising the tank's temperature through the allowed ΔT.

Equation 14.8 can be used to determine tank size.

(Equation 14.8)

$$V_{\text{btank}} = \frac{t(Q_{\text{HP}} - Q_{\text{L}})}{500(\Delta T)}$$

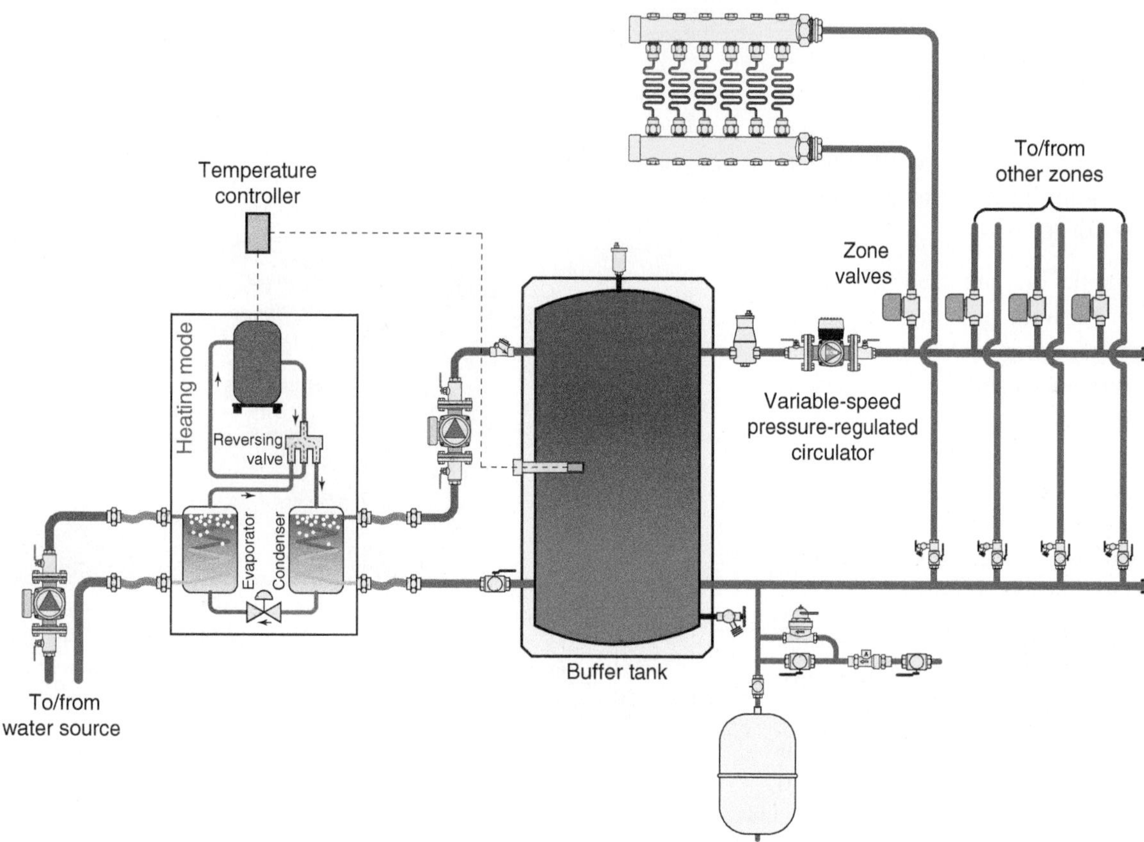

Figure 14-9 | Adding a buffer tank between a water-to-water heat pump and a zoned distribution system.

where:

V_{btank} = required volume of buffer tank (gallons)

t = desired on-time for heat source (minutes)

Q_{HP} = heating capacity of heat pump (Btu/hr.)

Q_L = any heating load served by buffer tank while charging (Btu/hr.)

ΔT = allowed temperature rise of tank during heat pump on-time (°F)

Example 14.4: Determine the size of a buffer tank that will absorb 48,000 Btu/hr. from the heat pump while increasing in temperature from 90 °F to 110 °F, during a heat pump on-cycle of 10 minutes. Assume there is no heating load on the tank during this charging.

Solution: The temperature rise (ΔT) is 110 – 90 = 20 °F. Putting this and the remaining data into Equation 14.8 yields:

$$V_{btank} = \frac{t(Q_{HP} - Q_L)}{500(\Delta T)} = \frac{10(48{,}000 - 0)}{500(20)} = 48 \text{ gallons}$$

Discussion: If the allowed temperature rise were 10 °F, rather than 20 °F, and the other conditions remained the same, the required tank volume would double to 96 gallons. If the desired on-time were only 5 minutes rather than 10 minutes, and the other conditions remained the same, the volume would be cut in half. Anything that increases the desired on-time or decreases the allowed temperature rise during this on-time, increases the required tank volume, and vice versa. The most conservative estimate for buffer tank size is made when the value of Q_L is zero (e.g., there is no load-removing heat from the buffer tank as heat is being added to it). However, in most systems, when there is no load "calling" for heat, the heat pump will not be running.

Equation 14.8 can also be used to determine the size of a buffer tank used for cooling. The value of Q_{HP} would be the *cooling capacity* of the heat pump. The value of ΔT would be the allowed temperature drop in the tank during this on-time.

In systems that provide both zoned heating and zoned cooling, the size of the buffer tank should be calculated for each mode of operation and the larger of the two volumes selected. In systems that supply simultaneous, or near-simultaneous heating and cooling, two separate buffer tanks are often used. Each tank is maintained at a suitable temperature for its load.

Buffer tanks that are used for chilled water storage must have a continuous layer of spray foam insulation between the inner pressure vessel and the tank's jacket. This is necessary to prevent moisture-laden air from contacting the pressure vessel, which allows condensation to form. Figure 14-10 shows an example of a foam-insulated buffer tank that can be used for chilled water storage.

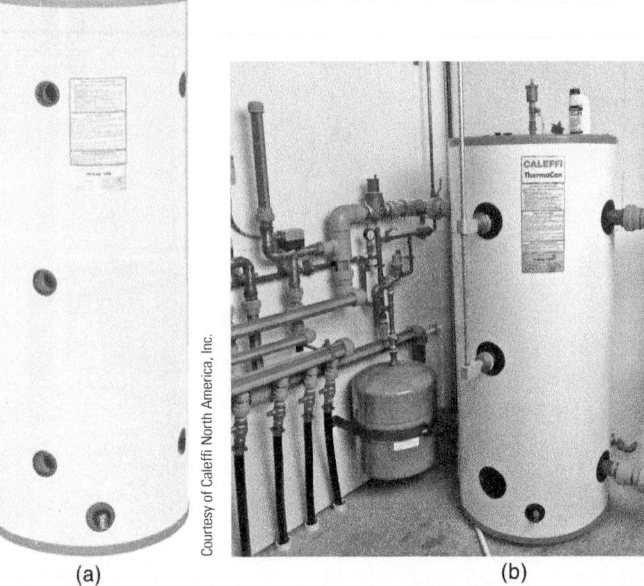

(a) Courtesy of Caleffi North America, Inc.

(b) Courtesy of Chris Cook

Figure 14-10 (a) Example of a buffer tank suitable for chilled water storage. (b) Installed buffer tank. Note electrical conduit leading to sensor wells at front of tank.

This tank provided multiple large-diameter openings that allow the tank to provide both buffering and hydraulic separation between the circulator that provides flow from the heat pump to the tank, and a variable-speed circulator that regulates flow to the zones. The positioning of these openings allows the tank to maintain reasonable temperature stratification. The large diameter piping connections provide minimal head loss.

FLOW RATE AND HEAD LOSS CONSIDERATIONS

Whenever a heat pump is operating, refrigerant is removing heat from the water passing through the evaporator and releasing heat into the water passing through the condenser. If the flow of water through either the evaporator or condenser is stopped, or significantly slowed, the heat pump must quickly and automatically shut down to prevent physical damage.

A flow restriction or stoppage in the evaporator can lead to rapid formation of ice crystals on the cold heat exchanger surfaces separating the water and refrigerant. Since the evaporator contains only a small amount of water, the ice build-up can quickly choke off flow. A "hard freeze" can occur in less than 1 minute of operation without water flow. A number of such hard freezes will eventually rupture the copper tubing in the evaporator and result in a costly repair. Because of this possibility, many manufacturers install a temperature-sensing safety switch called a **freezestat** near the water outlet of the evaporator. This switch monitors the water temperature leaving the evaporator, and stops the compressor if that temperature drops below a preset limit. Some modern heat pumps may use a solid-state temperature sensor, such as a **thermistor**, rather than a freezestat. This sensor is wired to the heat pump's internal controller, which can also turn off the compressor.

Maintaining water flow through the condenser whenever the heat pump is operating is equally important. A flow restriction in this stream can cause the refrigerant head pressure to increase rapidly. Eventually this high pressure will trip a **high-pressure safety switch**, or electronic pressure sensor, that shuts off the compressor.

Heat pump manufacturers usually recommend specific minimum flow rates for both the evaporator and condenser. If no flow rate is specified, the system should provide 2 to 3 gallons per minute of water flow per ton (12,000 Btu/hr.) of heat transfer in both the evaporator and condenser.

Designers must be careful not to install devices such as zone valves or three-way mixing valves in series with the heat pump condenser. If such valves are present, they can restrict or totally stop water flow through the condenser causing an automatic shutdown of the heat pump's compressor.

FREEZE PROTECTION

Modern water-to-water heat pumps can extract heat from the fluid passing through their evaporator, down to temperatures well below the freezing point of water. Such operation can occur in heat pumps connected to horizontal earth loops in northern climates. By late winter, the operating temperature of the earth loop can be lower than 32 °F. This mandates the use of an antifreeze solution in the earth loop. The most common antifreeze is a solution of **inhibited propylene glycol**. Some systems also use antifreeze solutions based on **ethanol** and **methanol**. Designers should keep in mind that the refrigerant temperature within an evaporator can be several degrees lower than the fluid temperature entering the evaporator. If water, without antifreeze, is used in the earth loop, ice crystals can form even when the entering water temperature is in the range of 35 °F.

VIBRATION ISOLATION

The compressor in a heat pump creates some operating noise and vibration as it operates. Modern **scroll compressors** emit less noise and vibration compared to older heat pumps using **reciprocating compressors**. Manufacturers use special vibration isolation mounts to minimize transfer of vibration from the compressor to other components in the heat pump. Designers should take further steps to reduce transfer of any remaining noise and vibration from the heat pump to the piping it connects to.

One common detail that reduces vibration transfer is to install **reinforced hose assemblies** between the connections on the heat pump and any rigid piping that carries flow to and from the heat pump. Example of such a hose assemblies is shown in Figure 14-11.

Flexible hoses are available in a range of pipe sizes, lengths, and pressure ratings. A typical hose has a fixed MPT fitting at one end and a swivel MPT fitting at the other end. The latter allows the hose to be installed between two fixed FPT connections, without stressing the hose due to twisting. The internal portion of the hose is usually made of EPDM rubber. It is encased within a braided stainless steel jacket, which prevents bulging of the rubber hose due to pressure. Common hose assembly lengths range from 12 inches to 36 inches. The hose assembly can be curved or otherwise offset between its connections. This is the preferred mounting because it helps absorb vibration as well as reducing stress on piping due to thermal expansion and contraction.

AIR AND DIRT SEPARATION

All water-to-water heat pumps are adversely affected by any dirt that might be carried into their heat exchangers. This can be minimized by ensuring that all water piping connected to the unit is thoroughly flushed and washed with a suitable **hydronic detergent** that can remove any solder flux, cutting oils, or other impurities that may be deposited on the piping as it is assembled.

Dirt is especially likely if the water-to-water heat pump is connected to an earth loop. Such loops are often constructed by joining several segments of HDPE tubing together within trenches, or other excavations in the earth. It is very likely that some dirt or mud will inadvertently enter the tubing as the earth loop is assembled. Insects or other debris may also get into the tubing during handling or storage. Although the majority of this debris should be purged out of the earth loop when the system is commissioned, fine dirt particles can still remain in the loop. A high-performance dirt separator is recommended by the author as a means of capturing residual dirt particles down to 5-micron particle size.

It is also important to remove as much air as possible from closed piping circuits connected to water-to-water heat pumps. Initial purging will remove most of the **bulk air** from the earth loop. However, fresh water contains dissolved molecules of oxygen, nitrogen, and other gases, which will not be removed by purging. A high-performance air-separating device with an internal coalescing media can capture these dissolved gases as the system operates and expel them from the piping circuit.

High-performance air and dirt separation will enhance the performance of any earth loop. These functions can be accomplished using two separate devices, or both functions can also be handle by a single device, appropriately called an **air and dirt separator**. Figure 14-12 shows an example of such a device.

Low-voltage electrical interfacing with heat pump

There are two types of water-to-water heat pumps currently on the North American market. The most common is a single-speed on/off unit. Another type uses a two-stage refrigeration system to provide two different heating capacities, as well as two different cooling capacities.

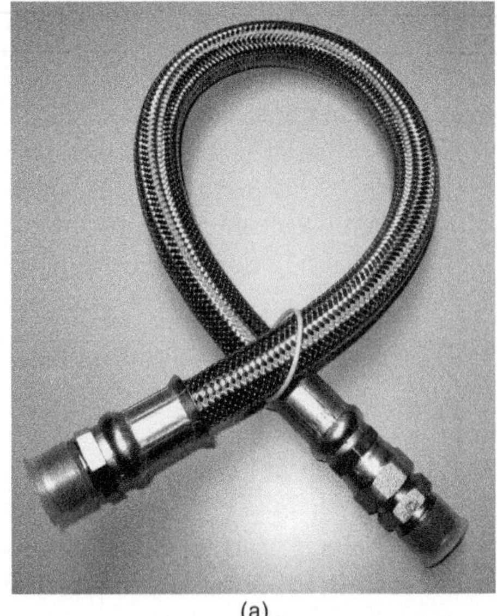

(a)

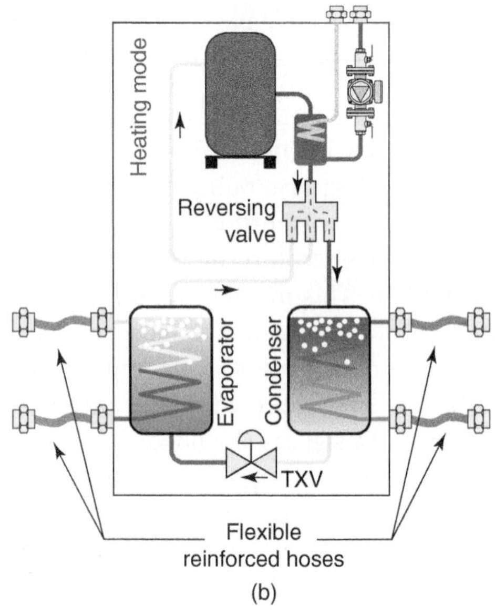

(b)

(c)

Figure 14-11 (a) Example of a flexible reinforced hose which reduces noise and vibration transfer from a water-to-water heat pump to rigid piping. Product by Chamberlin Rubber Company. (b) Typical installation locations for flexible reinforced hoses connected to a water-to-water heat pump. (c) Flex hoses connecting from PP-R polypropylene pipe to heat pumps. Two hoses serve the evaporator, two serve the condenser, and two serve the DHW desuperheater.

Depending on the configuration of their internal control circuits, different heat pumps can require different types of external control inputs to initiate operation in heating or cooling mode. A single-speed water-to-water heat pump will have a low-voltage terminal strip, as shown in Figure 14-13.

The (R) terminal is a source of 24 VAC from a transformer contained within the heat pump. The (C) terminal is the common side of that transformer. There should always be 24 VAC between the (R) and (C) terminals whenever there is line voltage connected to the heat pump. The (Y) terminal, if connected to

(a)

(b)

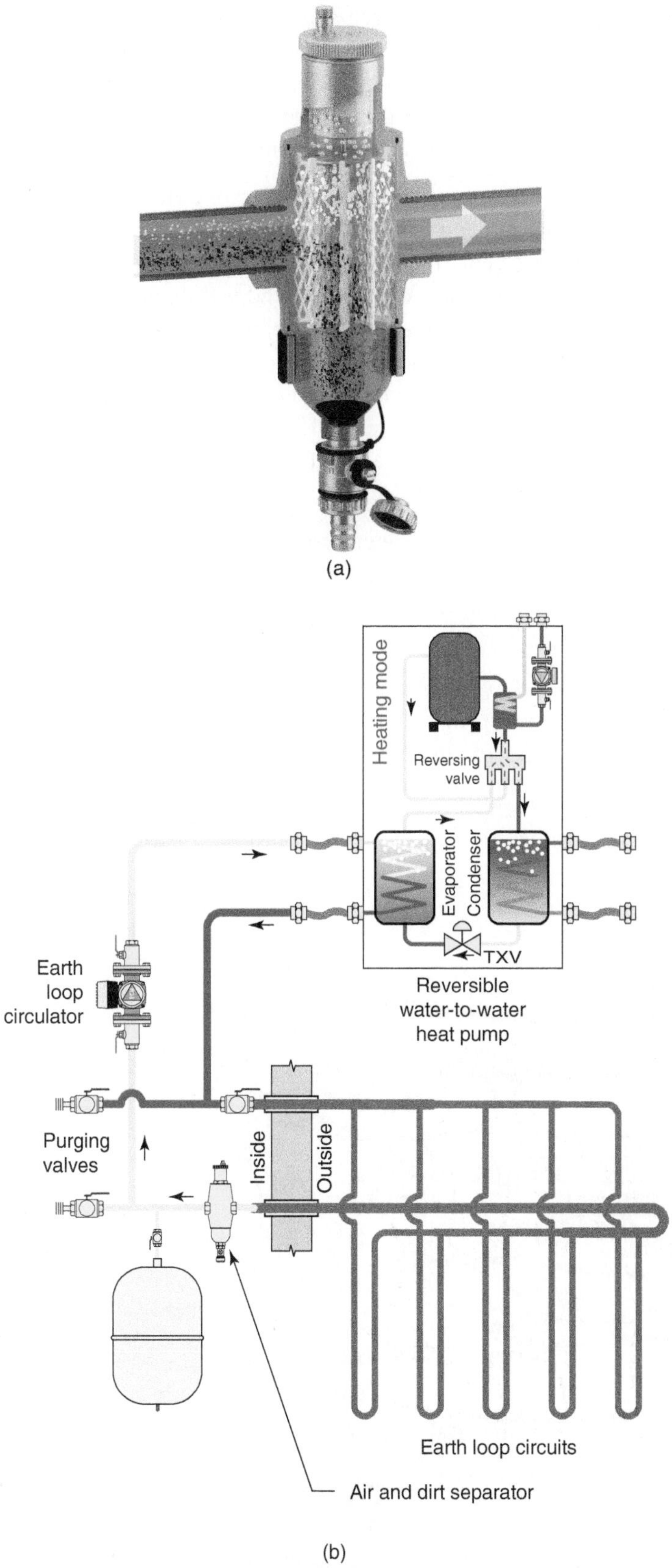

Figure 14-12 (a) Example of a combined air and dirt separator. (b) Installation of a combined air and dirt separator within an earth loop.

24 VAC from the (R) terminal, will turn on the heat pump's compressor, enabling it to operate in heating mode. The (O) terminal, if connected to 24 VAC from the (R) terminal, will energize the refrigerant reversing valve, which configures the refrigerant circuit for cooling. However, the compressor must also be turned on for the heat pump to operate in cooling mode.

Figure 14-14 shows the necessary **dry contacts** (e.g., unpowered contacts) necessary to turn the heat pump on in heating or cooling mode. These dry contacts could be contained within a thermostat or another control device, such as an outdoor reset controller. In animated terms, the heat pump "doesn't care" which device creates the contact closure. Thus,

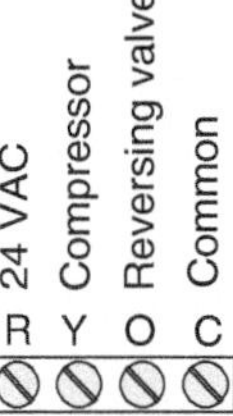

Figure 14-13 Typical low-voltage terminal strip in a single-speed on/off water-to-water heat pump.

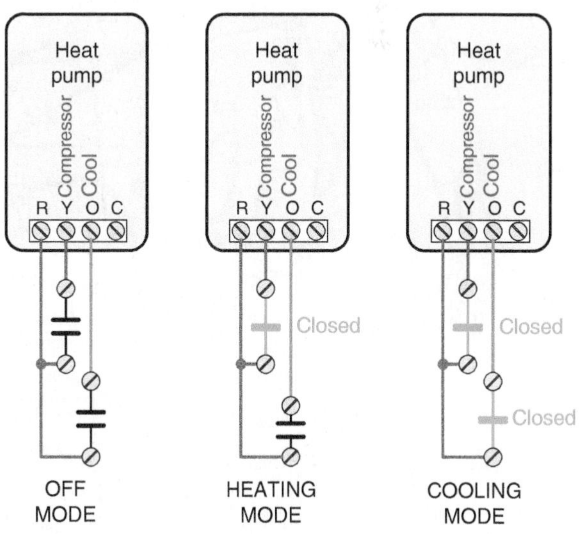

Figure 14-14 Typical electrical interface for a single speed on/off water-to-water heat pump. A dry contact closure between (R) and (Y) turns the heat pump's compressor, enabling it to operate in heating mode. A dry contact closure between (R) and (O) as well as (R) and (Y) is needed to operate the heat pump in cooling mode.

there are many possible ways to turn the heat pump on in either heating or cooling mode. These will be shown in the example systems presented later in the chapter.

Designers should always verify the exact electrical control interface required by the heat pump they select and ensure that the system wiring coordinates with these requirements. The manufacturer's installation and operating (I/O) manual for the heat pump will typically show a detailed electrical schematic for the internal wiring of the heat pump and describe what is necessary to turn the unit on in heating or cooling mode.

14.5 Water-to-Water Heat Pumps Supplied from Ground Water

In most areas of North America, the temperature of ground water is such that low-temperature heat can be readily extracted by a water-to-water heat pump. This is also true for water in large ponds or lakes, even those that completely freeze oven in winter. The water near the bottom of frozen lakes and large ponds has a temperature of about 4 °C (39.2 °F). Water attains its maximum density at this temperature and thus settles to the bottom of the lakes and ponds.

The temperature of ground water in wells that are at least 25 feet deep only varies a few degrees Fahrenheit during the year. The solar heat gains of summer have very little influence on deep ground water temperature. Neither do the extreme low air temperatures of winter. Thus, *the ground water temperature in most drilled wells remains close to the average annual outside air temperature.* Figure 14-15 gives the approximate ground water temperature for locations within the continental United States.

Ground water temperatures in most areas of North America are also low enough to serve as an excellent heat sink for a water-to-water heat pump operating in the cooling mode.

Two key issues must be examined whenever ground water, or water from a lake or pond, is being considered for use with a water-source heat pump. They are (1) **water quantity** and (2) **water quality**.

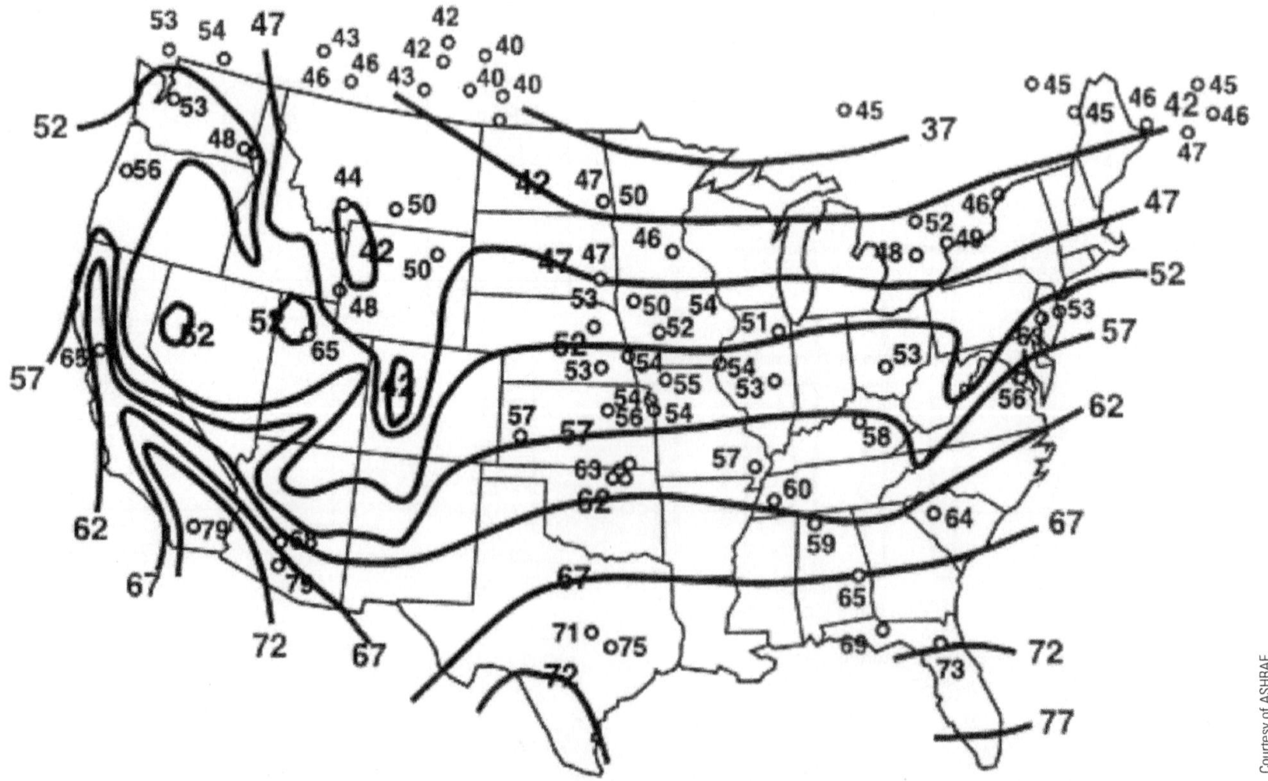

Figure 14-15 | Ground water temperatures in the continental United States.

WATER QUANTITY

The term *water quantity* describes the need for sufficient ground water to be available from the intended source whenever the water-to-water heat pump is operating. In some cases, this could be 24 hours per day and 7 days a week. If the source is unable to maintain the necessary water flow, the heat pump cannot operate and heating or cooling will not be provided.

An estimate for water usage is 2 to 3 gallons per minute per ton (12,000 Btu/hr.) of heat transfer at the evaporator or condenser. When the heat pump is operating in *heating* mode, the water flow rate must be sufficient to supply the evaporator load of the heat pump. The evaporator load can be calculated using Equation 14-9.

(Equation 14.9)

$$Q_{evap} = Q_{load}\left(\frac{COP - 1}{COP}\right)$$

where:

Q_{evap} = rate of heat absorption from ground water at evaporator (Btu/hr.)

Q_{load} = rate of heat output from condenser (Btu/hr.)

COP = coefficient of performance of the heat pump

Example 14.5: A water-source heat pump is operating at a COP of 3.61 while delivering heat to its load at 38,000 Btu/hr. Determine the required water flow rate through the evaporator, assuming 3 gpm/ton.

Solution: The rate of heat absorption at the evaporator is:

$$Q_{evap} = Q_{load}\left(\frac{COP - 1}{COP}\right) = 38{,}000\left(\frac{3.61 - 1}{3.61}\right)$$
$$= 27{,}470 \text{ Btu/hr.}$$

The required water flow rate is:

$$\left(\frac{27{,}470 \text{ Btu/hr.}}{12{,}000\frac{\text{Btu/hr.}}{\text{ton}}}\right)\left(\frac{3 \text{ gpm}}{\text{ton}}\right) = 6.87 \text{ gpm}$$

Discussion: Although this may not seem large a large flow rate, it equates to 6.87 × 60 = 412 gallons per hour, or 412 × 24 = 9,888 gallons per day, if the heat pump remains in continuous operation. Most large ponds or lakes could easily supply a nominal 10,000 gallons per day—assuming the water returns to the pond or lake after passing through the heat pump. However, many residential wells could *not* supply this quantity of water, especially on a sustained basis. In some cases, heat pumps sourced from residential wells have exceeded the well's ability to supply water. This not only causes the heat pump to automatically shut down, it may also leave the building without a source of fresh water, if that water were supplied from the same well.

When operating in the *cooling* mode, the rate of heat rejection to the ground water stream is based on the total heat output of the heat pump. Again, an estimate of 2 to 3 gpm per ton of heat rejection can be used to determine total water requirements.

To conserve water, especially when it is supplied from a well, the water-to-water heat pump should operate at or near the minimum required water flow rate. This requirement should always be verified with the heat pump manufacturer.

After water has passed through a heat pump, it must be properly returned to the environment. If extracted from a lake or pond, it is possible that it can be returned to the same body of water. However, environmental regulations may constrain such situations. Always check on local and state environmental codes and regulations to see if ground water from any source can be used to supply a heat pump system.

Water extracted from a well may require a more elaborate means of disposal. Although the water is not contaminated as it passes through the polymer tubing and copper or stainless steel piping components of the system, it cannot can be disposed of in a ditch, storm sewer, or dry well. Most states now have regulations regarding if, when, and how water can be returned to an underground aquifer. Again, always investigate the options based on the codes and regulations in effect at the installation site.

In areas where water can be reinjected into the ground, it may be necessary to drill a separate well within a minimum prescribed distance from the source well. In other cases, it may be possible to reinject water into the same well that supplied it. The latter is more common with deep wells. A detailed evaluation of such options is best done by design professionals familiar with ground water regulations and subsurface conditions. Local well drillers are a good place to begin investigating options and related costs.

WATER QUALITY

Another important issue regarding use of ground water with heat pumps is water *quality*. It is critically important to determine what impurities may be present in the anticipated ground water source, before committing to its use. Water that contains, silt, calcium carbonate, sulfur compounds, salt, iron, bacteria, or other contaminants can quickly create deposits within the water-to-refrigerant heat exchanger of a water-source heat pump. This can cause a rapid deterioration in both flow rate and heat transfer. In some cases, it can completely plug the heat exchanger. Such conditions must be avoided.

The table in Figure 14-16 lists the water quality standards recommended by one manufacturer of water source heat pumps. Other manufacturers may differ in their minimum recommended standards for ground water. *Always reference the water quality standards required by the heat pump manufacturer, and always have the anticipated water source professionally tested*

Water Quality Parameter	HX Material	Closed Recirculating	Open Loop and Recirculating Well		
Scaling Potential - Primary Measurement					
Above the given limits, scaling is likely to occur. Scaling indexes should be calculated using the limits below					
pH/Calcium Hardness Method	All	-	**pH < 7.5 and Ca Hardness <100ppm**		
Index Limits for Probable Scaling Situations - (Operation outside these limits is not recommended)					
Scaling indexes should be calculated at 150°F [66°C] for direct use and HWG applications, and at 90°F [32°C] for indirect HX use. A monitoring plan should be implemented.					
Ryznar Stability Index	All	-	**6.0 - 7.5** If >7.5 minimize steel pipe use.		
Langelier Saturation Index	All	-	**-0.5 to +0.5** If <-0.5 minimize steel pipe use. Based upon 150°F [66°C] HWG and Direct well, 85°F [29°C] Indirect Well HX		
Iron Fouling					
Iron Fe^{2+} (Ferrous) (Bacterial Iron potential)	All	-	**<0.2 ppm (Ferrous)** If Fe^{2+}(ferrous)>0.2 ppm with pH 6 - 8, O2<5 ppm check for iron bacteria		
Iron Fouling	All	-	**<0.5 ppm of Oxygen** Above this level deposition will occur.		
Corrosion Prevention					
pH	All	**6 - 8.5** Monitor/treat as needed	**6 - 8.5** Minimize steel pipe below 7 and no open tanks with pH <8		
Hydrogen Sulfide (H_2S)	All	-	**<0.5 ppm** At H_2S>0.2 ppm, avoid use of copper and copper nickel piping or HX's. Rotten egg smell appears at 0.5 ppm level. Copper alloy (bronze or brass) cast components are OK to <0.5 ppm.		
Ammonia ion as hydroxide, chloride, nitrate and sulfate compounds	All	-	**<0.5 ppm**		
Maximum Chloride Levels			Maximum Allowable at maximum water temperature.		
			50°F (10°C)	75°F (24°C)	100YF (38YC)
	Copper	-	<20ppm	NR	NR
	CuproNickel	-	<150 ppm	NR	NR
	304 SS	-	<400 ppm	<250 ppm	<150 ppm
	316 SS	-	<1000 ppm	<550 ppm	< 375 ppm
	Titanium	-	>1000 ppm	>550 ppm	>375 ppm
Erosion and Clogging					
Particulate Size and Erosion	All	<10 ppm of particles and a maximum velocity of 6 fps [1.8 m/s] Filtered for maximum 800 micron [800mm, 20 mesh] size.	<10 ppm (<1 ppm "sandfree" for reinjection) of particlesand a maximum velocity of 6 fps [1.8 m/s]. Filtered for maximum 800 micron [800mm, 20 mesh] size. Any particulate that is not removed can potentially clog components.		

Notes:

- Closed Recirculating system is identified by a closed pressurized piping system.
- Recirculating open wells should observe the open recirculating design considerations.
- NR - Application not recommended.
- "-" No design Maximum.

Rev.: 01/21/09B

Courtesy of ClimateMaster, Inc.

Figure 14-16 | Minimum water quality standards required for use of ground water circulated directly through heat pumps.

to determine if it meets or exceeds these standards. Failure to do so could render an otherwise well-planned system virtually useless or void the warranty on the heat pump. Also remember that water quality may fluctuate with time of year due to surface runoff or other environmental factors.

Any accumulation of sediment within the heat pump's water-to-refrigerant heat exchanger will create a **fouling factor** that lowers the rate of heat transfer and thus the performance of the heat pump. Depending on the source of the ground water, it may be necessary to provide filtration to remove particulates before the water passes through the heat pump. The specifications listed in Figure 14-16 call for a particulate concentration of less than 10 **ppm** (parts per million) and particulate filtration that can remove all particles greater than 800-**micron** size. Other heat pump manufacturers may require different filtration requirements.

One device that removes sediment from water is a **spin-down filter**, as shown in Figure 14-17.

Although the element in a spin-down sediment filter captures dirt particles, it is not designed as a "throwaway" cartridge, as found in some potable water filtering devices. Instead, sediment filters are design to be periodically "blown down" using pressurized water. The ball valve downstream of the filter (e.g., on the lower pressure side) is closed, and the ball valve at the base of the filter is opened. This allows the pressurized water source to blow the captured dirt out the bottom of the filter bowl. Installing pressure gauges on each side of the sediment filter provides a reference pressure drop at which the filter can be blown down. This pressure drop must be observed when the heat pump is operating. Sediment filter manufactures typically offer a range of element mesh size options. For example, a 60-mesh filter element will capture particles down to 250-micron size, and a 115-mesh element will capture particle down to 125 microns. The mesh size should be determined based largest particle size "allowed" through the heat pump, as well as any other hardware piped inline with it, based on the manufacturer's requirements. Filter elements that capture smaller particles are preferable from the standpoint of reducing potential fouling of the heat pump and other hardware. However, filter elements that capture very small particles will "load up" more quickly and thus require more frequent blow downs.

Another filter option is an **automatic backwash filter** with permanent stainless steel filter media. An example of such a filter is shown in Figure 14-18.

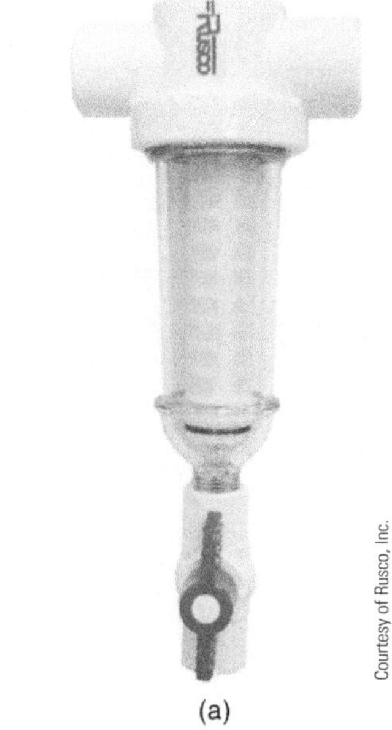

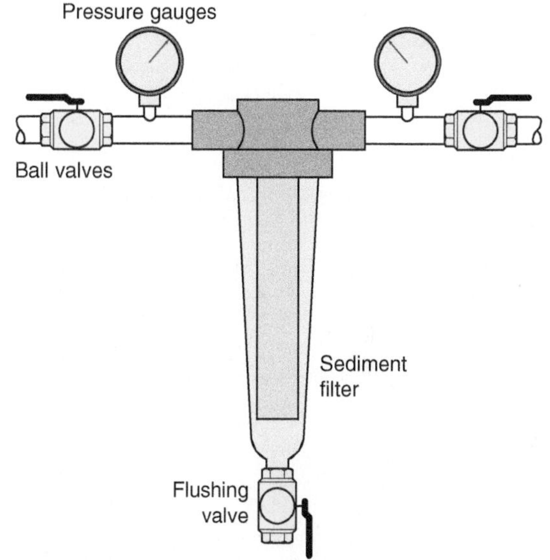

Figure 14-17 (a) Example of a cartridge filter assembly. (b) Suggested installation of the cartridge filter assembly with isolating ball valves and pressure gauges.

Automatic backwash filters are considerably more expensive than cartridge filters. However, they provide fully automatic operation and don't require replacement filter cartridges. Their stainless steel filter media is meant to last for years. They are available with various particle size ratings down to approximately 5 microns.

Automatic backwash filters have a controller that monitors differential pressure across the filter media. When it reaches a specific setting (such as 7 psi), the

(a)

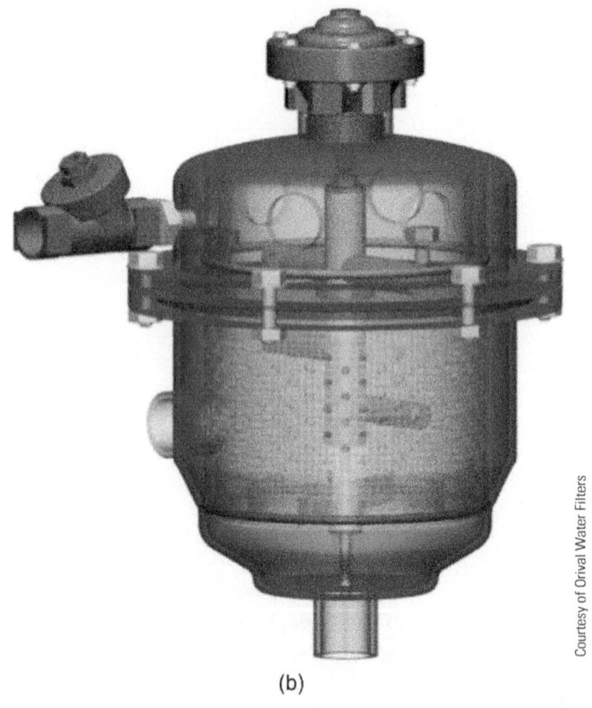

(b)

Figure 14-18 (a) Example of an automatic filter that backwashes itself when the differential pressure across it reach a specific setting. (b) Internal stainless steel filter basket.

controller energizes a rinse valve through which the accumulated particulates will be flushed. Rinse cycles are short, typically lasting 10 to 15 seconds. *During this time the water supplied to the filter must reach a specified flow rate, which may be significantly higher than the normal flow rate through the filter.* A 1.5-inch pipe size filter can require a minimum rinse flow rate of 35 gpm. This higher flow can be provided from another pressurized water supply. It might also be provided by operating a variable-speed pump at high speed or by energizing a secondary pump. The filter can continue to provide flow to the downstream components during the rinse cycle, provided that the water source can supply the necessary flow rate plus the rinse flow rate. Another alternative would be to provide sufficient thermal mass in the water source circuit supplying the heat pump(s) so that the heat pump(s) can continue to operate during the rinse cycle.

In some cases, the chemical nature of the water, rather than its particulate content, will be the limiting factor in its use to supply water-to-water heat pumps. Ground water with high concentrations of hydrogen sulfide, other sulfur compounds, iron compounds, or salt, can be corrosive to the standard copper heat exchangers used in many heat pumps. Some manufacturers offer **cupronickel** or titanium heat exchangers that can tolerate more aggressive water chemistry. Again, refer to the manufacturer's specifications and I/O manuals for specific recommendations on heat exchanger materials based on analysis of the water available at a site.

GROUND WATER SUPPLED FROM WELLS

When a suitable source of ground water is available, the system must be designed to provide adequate water flow through the water-to-water heat pump and then properly return the water to its source.

In situations where the ground water is supplied from a well, it is common to use a **submersible well pump** to provide flow and lift. The latter must be evaluated in situations where the water table may be well below the surface. Under such conditions, the pump must be sized to lift the water as well as overcome the frictional head losses of the supply piping, heat pump heat exchanger, and return piping. When large lifts are required, a more powerful submersible pump must be used relative to applications where the lift is only a few feet. The operating cost of a large submersible well pump may significantly lower from the "net" COP and EER of the heat pump *system*, even in situations where the heat pump itself operates at high COP and EER values.

One method of using a drilled well to supply a water-to-water heat pump is shown in Figure 14-19. The well serves as the ground water supply as well as return. The well may also provide domestic water to the building served by the heat pump.

In this system, the submersible pump is tuned on and off by a **pressure switch** that monitors pressure within the compression tank. The pressure switch turns on the pump when the pressure in the tank drops to or below a lower threshold value and off when pressure is restored to an upper limit. The compression tank should be sized, and its air-side pressure adjusted, so that it allows the well pump to remain *on* for a minimum of 3 minutes, or as otherwise recommended by the pump manufacturer. This minimum on-time allows the heat generated by the pump motor's start windings to be adequately dissipated before the motor is turned off. If the heat pump is not operating during this time, and there is no other demand for water, all flow from the well pump must be stored in the compression tank.

Equation 14.10 can be used to size the compression tank for such a requirement.

(Equation 14.10)

$$V_T = \frac{f(t)(P_H + 14.7)}{0.9(P_H - P_L)}$$

where:

V_t = *minimum* tank shell volume of compression tank (gallons)

t = minimum ON-time of well pump (minutes)

f = flow rate produced by well pump at *lower* pressure (P_L) (gpm)

P_L = lower limit pressure of pressure switch (psi)

P_H = higher limit pressure of pressure switch (psi)

0.90 = a safety factor that allows 10 percent of the compression tank volume to remain filled with water when the internal air pressure reaches its lower limit

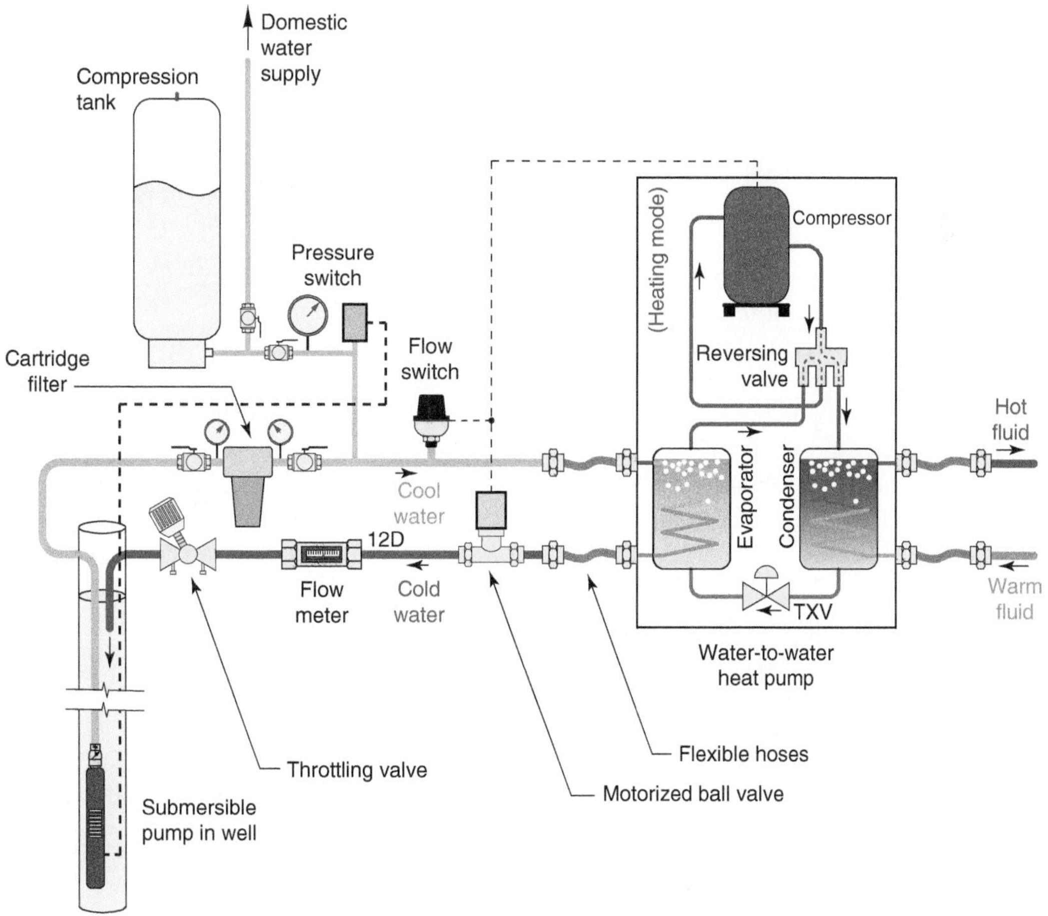

Figure 14-19 | Use of a single drilled well for ground water supply to a water-to-water heat pump.

Example 14.6: A submersible well pump requires a minimum on-time of 3 minutes so it can dissipate heat from its start windings. Assuming the pressure switch is set for 40 psi = ON, 60 psi = OFF. The submersible pump can produce a flow of 15 gpm when operating at the lower pressure setting. Determine the minimum volume of the compression tank so that it can accept water from the well pump without any simultaneous flow to the heat pump or domestic water load.

Solution: Substituting the given information into Equation 14.10 yields:

$$V_T = \frac{f(t)(P_H + 14.7)}{0.9(P_H - P_L)} = \frac{15(3)(60 + 14.7)}{0.9(60 - 40)}$$
$$= 187 \text{ gallons}$$

Discussion: Equation 14.10 sizes the compression tank so that it can accept the flow from the submersible pump for 3 minutes, without any water demand by either the heat pump or domestic plumbing. In this example, the volume of the required tank is substantial. Keep in mind that the factor 0.9 in Equation 14.10 allows 10 percent of the tank's volume to remain in the tank when the pressure reaches it lower value (e.g., where the well pump is turned on). This sizing is also conservative in that the flow rate from the well pump will slightly decrease as pressure builds in the compression tank. The size of the compression tank could be reduced by lowering the lower pressure limit and/or increasing the upper pressure limit. It could also be reduced by allowing only 5 percent of the tank volume to be filled with water when the lower pressure is reached (versus the 10 percent assumed). Finally, when a large volume is required, it can be met by using two or more tanks that total up to the same volume. In such cases, be sure the air-side pressure in all the tanks is equally adjusted to the lower pressure setting of the pressure switch and that all tanks are mounted at the same elevation.

For the system shown in Figure 14-19, water flow through the heat pump is controlled by a motorized ball valve. This valve is located downstream of the heat pump. It opens whenever the heat pump is operating and closes when the heat pump shuts off. The valve body must be rated for contact with potable water. It could be made of lead-free brass, stainless steel, or PVC. The valve must also be capable of closing off against the maximum water pressure in the system and should do so immediately after the heat pump turns off.

Courtesy of Hayward Flow Control, Inc.

Figure 14-20 | Motorized ball valve with CPVC, or PVC body.

Figure 14-20 shows an example of a CPVC ball valve with motorized actuator. This valve operates on 120 VAC. It needs to be powered open and powered close. The actuator requires about 2.5 seconds to rotate the internal ball by 90° and thus move it from fully closed to fully open, or vice versa. This time is beneficial in that it helps eliminate water hammer effects that may be created by valves with very fast closing times.

When selecting this valve, its **flow coefficient**, which is also known as the valve's **Cv value**, should be at least equal to the flow rate of water that needs to pass through the heat pump while it operates. The Cv rating of a valve is the flow rate of 60 °F water, measured in gallons per minute, that creates a pressure drop of 1 psi across the valve. Full port ball valves induce less pressure drop than those with standard port balls and are thus preferred from the standpoint of reducing pumping energy. In heat pump applications, it is likely that hundreds of thousands of gallons of ground water will move through such a valve over several years of operation. Thus, it is desirable to use a valve that is fully serviceable and for which replacement parts are readily available.

Other components in the piping include a flow meter, throttling valve, and flow switch. The throttling valve, in combination with a flow meter, allows the water flow rate to be adjusted to the target value required by the heat pump.

Some flowmeters, such as the **rotameter** style product shown in Figure 14-21, must be mounted in a vertical pipe, and should have at least 12 diameters of straight pipe upstream of their inlet to help suppress turbulence. Flowmeters of other designs, such as those using turbines or spring-balanced indicators, can be mounted in a horizontal pipe. Always verify the proper mounting of the specific flow meter used.

It is also possible to *estimate* the flow rate through a heat pump using **Pete's Plugs** in combination with pressure gauges, and a graph of pressure drop versus flow rate for the heat pump, as illustrated in Figure 14-22.

Pete's Plugs are small piping components that allow a needle-like probe on either a thermometer, or pressure gauge, to be inserted through a "self-sealing" synthetic packing. They are often installed in tees located close to the inlet and output piping of a water

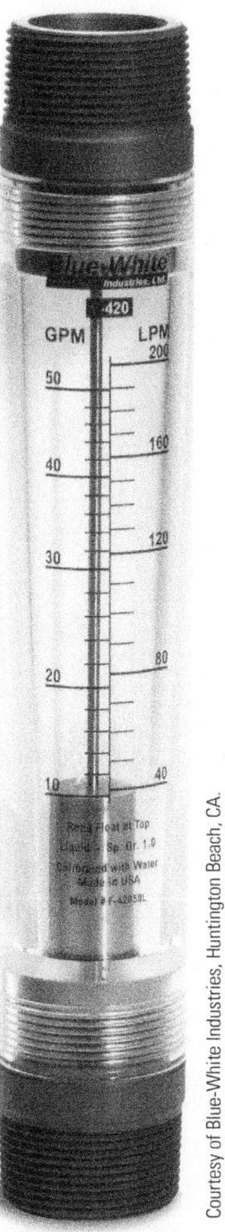

Courtesy of Blue-White Industries, Huntington Beach, CA.

Figure 14-21 | Example of a rotameter type flow meter that must be mounted in a vertical pipe.

source heat pump. A pressure gauge is inserted into both Pete's Plugs to measure the pressure drop between the inlet and outlet of the heat pump's heat exchanger. The flow rate through the heat pump is then estimated using a graph of pressure drop versus flow rate supplied by the heat pump manufacturer. For example, if the pressure drop between the two Pete's Plugs shown on the left side of the heat pump in Figure 14-22b happens to be 5 psi, and the graph in Figure 14-22c represents the flow characteristic of that heat pump, the pressure drop corresponds to a water flow rate of 5.7 gpm.

If the heat pump manufacturer provides a graph of *head loss* versus flow rate, rather than pressure drop versus flow rate, the pressure drop can be calculated based on head loss using Equation 14.11.

(Equation 14.11)

$$\Delta P = H_L\left(\frac{D}{144}\right)$$

where:

H_L = head added or lost from the fluid (feet of head)

ΔP = pressure change corresponding to the head added or lost (psi)

D = density of the fluid at its corresponding temperature (lb./ft.^{3})

The pressure drop versus flow rate graph for a water source heat pump is usually established based on test data obtained with cold water flowing through the heat pump's heat water-to-refrigerant heat exchanger. If the pressure drop across the heat pump is measured for a system using a fluid other than water, within the temperature range of 50 °F to 70 °F, the pressure drop curve should be corrected based on Equation 14.12.

(Equation 14.12)

$$\Delta P_c = \Delta P\left(\frac{\alpha D}{3.6757}\right)$$

where:

ΔP_c = corrected pressure drop at any given flow rate (psi)

ΔP = pressure drop at any given flow rate based on 60 °F water in circuit (psi)

α = alpha value of fluid in circuit (see Figure 3-24, or Equation 3.10)

D = density of fluid in circuit (lb./ft.^{3})

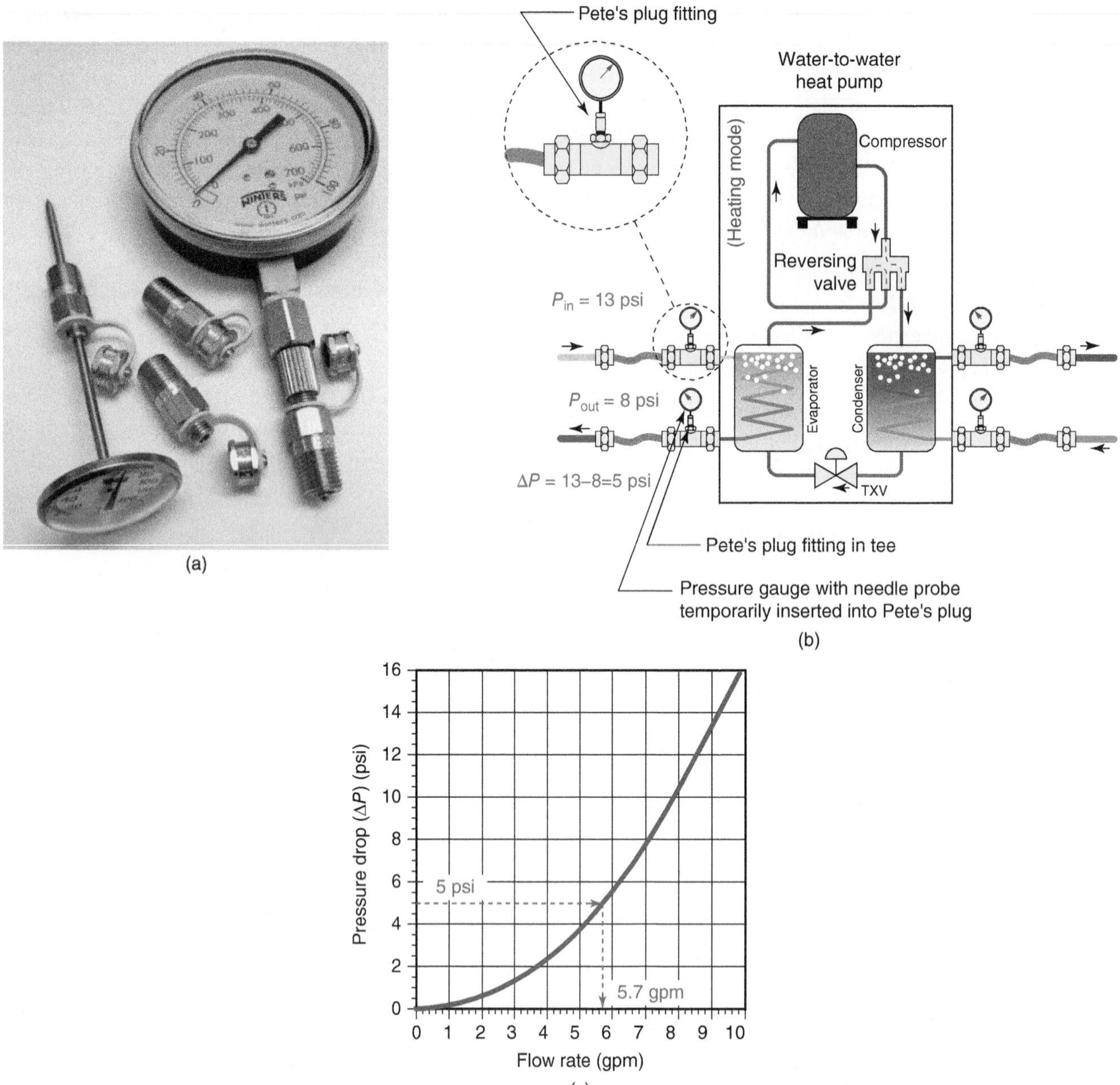

(a) (b) (c)

Figure 14-22 (a) Pete's Plugs with thermometer and pressure gauge in position. (b) Mounting locations for Pete's Plugs near the inlet and outlet of the heat exchangers on the water-to-water heat pump. (c) Graph of pressure drop versus flow rate provided by the heat pump manufacturer.

Example 14.7: The pressure drop versus flow rate for the water-to-refrigerant heat exchanger in a water-to-water heat pump is shown in Figure 14-23. This graph was based on testing done with 60 °F water. The pressure drop across the heat exchanger is measured during late summer, when the average fluid temperature within the heat exchanger is 85 °F. The fluid passing through the heat exchanger is a 20 percent solution of propylene glycol. Correct the pressure drop versus flow rate graph based on this fluid and estimate the flow rate through the heat exchanger when the pressure drop measured between the Pete's Plugs on the inlet and outlet of the heat exchanger is 6.5 psi.

Solution: The alpha value for a 20 percent solution of propylene glycol at a temperature of 85 °F is estimated from Figure 3-23: The density of the same solution is estimated from manufacturer's data:

$$\alpha = 0.0630$$

$$D = 63.4 \text{ lb./ft.}^3$$

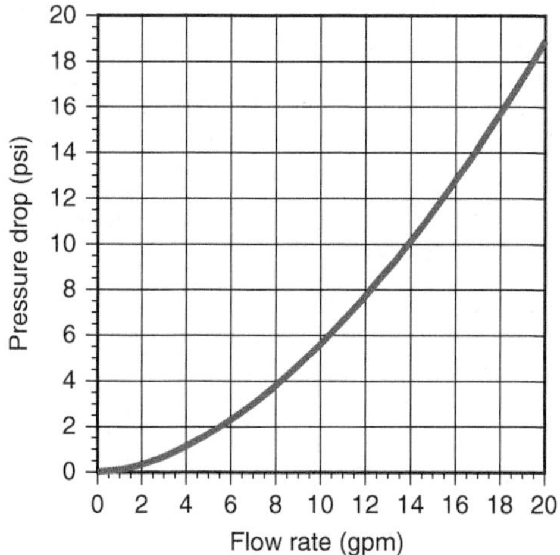

Figure 14-23 | Pressure drop versus flow rate for heat pump in Example 14.7, based on 60 °F water.

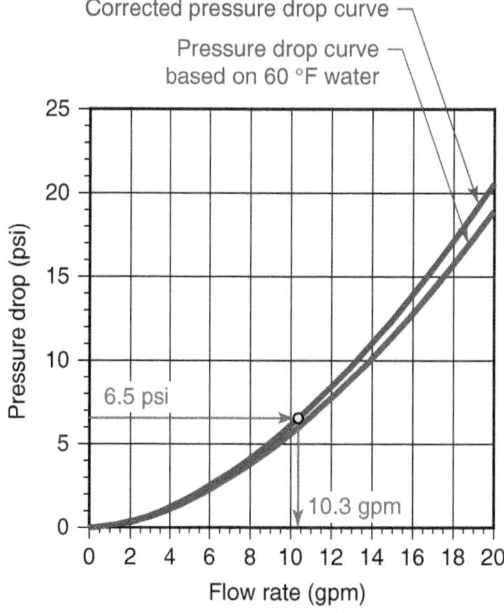

Figure 14-24 | Red curve is pressure drop versus flow rate for 20 percent solution of propylene glycol at 85 °F. At 6.5 psi measure pressure drop, the corresponding flow rate through the heat pump is 10.3 gpm.

The correction factor for pressure drop can be determined using Equation 14.12.

$$\Delta P_c = \Delta P\left(\frac{\alpha D}{3.6757}\right) = \Delta P\left(\frac{0.0630 \times 63.4}{3.6757}\right)$$
$$= \Delta P(1.087)$$

The curve shown in figure 14–23 can be modified by increasing the vertical value of several points along the curve by multiplying by 1.087.

Discussion: Had the measured pressure drop of 6.5 psi been applied to the blue curve in Figure 14-23, which represents 60 °F water in the circuit, the implied flow rate would have been about 11.0 gpm. The corrected (red) curve shows the flow rate of the 20 percent solution of propylene glycol at 85 °F to be 10.3 gpm. Designers should always remember to correct flow rates and associated head loss or pressure drop for hydronic circuits operating with fluids other than water, or fluid temperatures significantly different from those at which the data was taken. If the heat pump manufacturer supplied a graph of head loss versus flow rate, rather than pressure drop versus flow rate, the head loss values on the vertical axis could be converted to pressure drop values using Equation 14.11.

Although not as accurate as most inline flow meters, using pressure drop measured between two Pete's Plugs, along with a pressure drop versus flow rate graph, is sufficiently accurate to determine if the flow rate through the heat pump is reasonably close to the desired value. Pete's Plugs can also be used to verify the fluid temperature between the inlet and outlet of the heat pump. The temperature change across the heat pump's water flow stream, in combination with the flow rate, can then be used to estimate the heat transfer using either Equation 3.2 or 3.3.

Equation 3.2 (repeated)

$$Q = (8.01Dc)f(\Delta T)$$

where:

Q = rate of heat transfer into or out of the water stream (Btu/hr.)

8.01 = a constant based on the units used

D = density of the fluid (lb./ft.^{3})

c = specific heat of the fluid (Btu/lb./°F)

f = flow rate of water through the device (gpm)

ΔT = temperature change of the water through the device (°F)

For *cold water only,* this equation simplifies to:

Equation 3.3 (repeated)

$$Q = 500f(\Delta T)$$

where:

Q = rate of heat transfer into or out of the water stream (Btu/hr.)

f = flow rate of water through the device (gpm)

500 = constant rounded off from 8.33 × 60

ΔT = temperature change of the water through the device (°F)

Equation 3.3 is technically only valid for water in the temperature range of 60 °F. Equation 3.2 is valid for both water and antifreeze solutions. It requires data for the specific heat and density of whatever fluid is used, at the average temperature of that fluid as is passes through the heat pump.

Keep in mind that the flow rate through a heat pump supplied with ground water from a compression tank will vary as the air pressure within the compression tank varies between the upper and lower settings of the pressure switch.

The flow switch shown in Figure 14-19 verifies that suitable water flow exists before the heat pump is allowed to operate. The throttling valve shown in Figure 14-19 also suppresses cavitation within the motorized ball valve by keeping the pressure in the return piping above the vapor pressure of the water. This is especially important if there is a large vertical distance between the valve and the water level in the well.

Figure 14-25 shows the typical wiring for the electrical components shown in Figure 14-19.

The following description of operation is based on Figure 14-25a, in which the motorized valve opens when 120 VAC is applied and closes using an internal spring when there is no voltage applied. Upon a demand for heating, 24 VAC from the transformer in the heat pump passes to the coil of relay (R1). A contact in that relay (R1-1) closes to provide 120 VAC to the motorized valve, causing it to open. 24 VAC also passes to one terminal of the flow switch (FS1). The

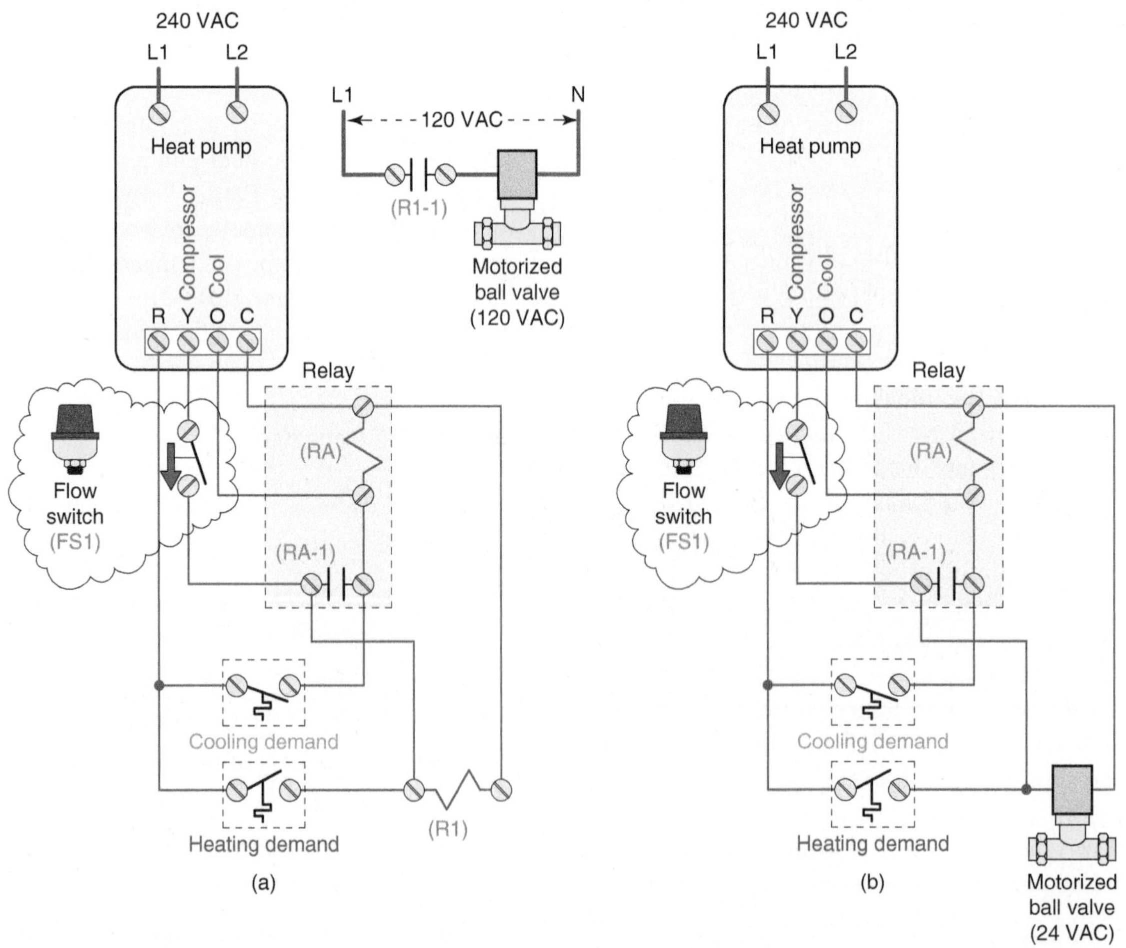

Figure 14-25 | Electrical wiring for the components shown in Figure 14-19. (a) Wiring if motorized valve operates on 120 VAC. (b) Wiring if motorized valve operates on 24 VAC.

flow switch closes when the flow rate through it, and the heat pump, reaches a minimum value to which the switch is set. This passes 24 VAC to the (Y) terminal of the heat pump, turning on the compressor and allowing the heat pump to operate in heating mode. Upon a demand for cooling, 24 VAC passes to the coil of relay (RA). Contact (RA-1) closes supplying 24 VAC to the coil of relay (R1), the flow switch, and terminal (O) on the heat pump. When the motorized valve opens, and flow through the flow switch and heat pump have reached a minimum set value, the flow switch contacts close, supplying 24 VAC to the (Y) terminal of the heat pump, allowing the compressor to operate. With 24 VAC also present at the (O) terminal of the heat pump, the refrigerant reversing valve will be on and the heat pump will operate in cooling mode.

If the motorized valve operates on 24 VAC, it is possible to eliminate relay (R1), as shown in Figure 14-25b. However, the designer should always verify that the heat pump's transformer is adequate to supply the power draw of the motorized valve. If it is not, the motorized valve will need to be supplied from a separate 24 VAC power source and relay (R1) would again be required.

The wiring shown in Figure 14-25a would have to be modified if the motorized ball valve required power for both opening and closing. Line voltage would have to be supplied to the close terminal of the valve's actuator whenever the heat pump was not operating. This is easily done using a normally closed contact on relay (R1).

Operation of the submersible well pump is controlled by the pressure switch mounted near the compression tank. It is not directly linked to the electrical wiring that controls the heat pump. Thus, the submersible pump runs only as necessary to keep the pressure within the compression tank between its upper and lower limits. The pressure within the compression tank will vary depending on water demand of the heat pump as well as domestic water demand.

It is also possible to use a **variable-speed submersible pump** to supply water from a well to a heat pump. The motor of such a pump is driven by a **variable frequency drive (VFD)**, which monitors a pressure sensor. The controller varies the speed of the pump to maintain a nearly constant pressure at the pressure sensor. The compression tank required for this type of pump is much smaller than that required for an on/off submersible pump. Figure 14-26 shows an example of a variable-speed submersible pump along with its associated controller, pressure sensor, and compression tank.

The graph shows the range over which the pump curve can be varied from some maximum speed to a minimum speed. This graph is for a specific pump model. Other models are available with higher and lower head capabilities.

WATER SUPPLIED FROM A LAKE

If a building is located near the shore of a lake, it may be possible to use water from the lake to supply a water-to-water heat pump. The first step is to determine what regulations are in effect regarding the use of lake water for this purpose. Although water passing through a heat pump should not be chemically altered, its temperature will be slightly reduced or increased, depending on the operating mode of the heat pump. If the water is supplied from a lake or a very large pond,

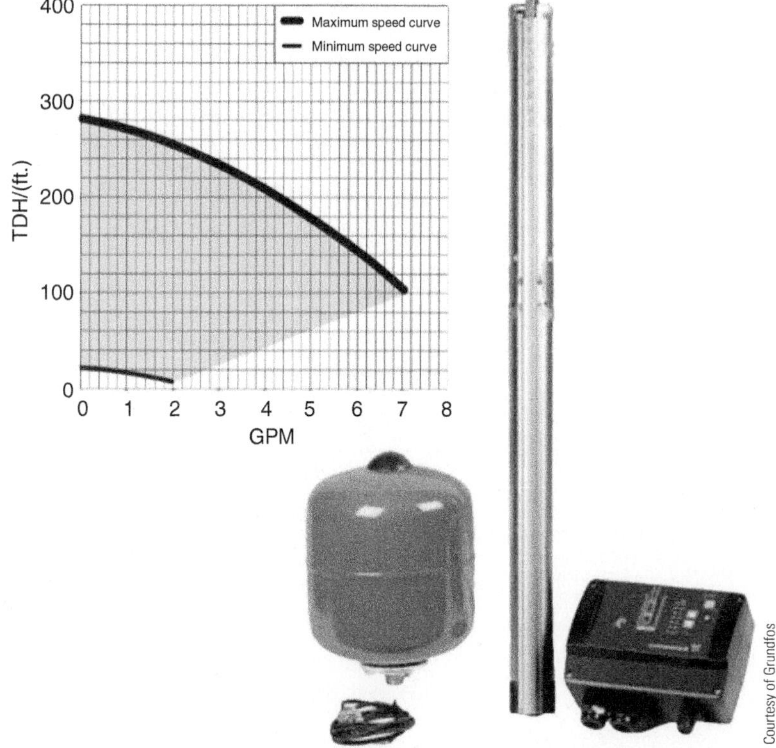

Figure 14-26 Example of a variable-speed submersible pump along with controller, pressure sensor, and compression tank.

this temperature change will have very little effect on the aquaculture within that body of water. However, smaller bodies of water may be more affected. In some cases, local, state, or even federal level regulations may not allow water from a lake to be used for supplying a heat pump.

Assuming the regulations in effect *do* allow for such use, the next step is to investigate a means of getting water from several feet below the surface of the lake, to the building, and returning it to the lake. Depending on location and shore conditions, this may also involve permitting for shoreline excavation, erosion control, or other detailing.

One approach is to install a suitably sized HDPE pipe equipped with a **foot valve**, on a concrete pedestal, that sits on the lake bed, several feet below the surface, as shown in Figure 14-27. The pedestal must elevate the foot valve above any silt on the lake bed. The foot valve is a special type of check valve that prevents water that has passed into the pipe from draining out when the pump turns off.

Another pipe is needed to return water to the lake in a suitable manner. The end of the return pipe should also be elevated above the silt layer of the lake bed, and located so that it discharges several feet away, and in a different direction, from the intake pipe.

The density of HDPE pipe is slightly lower than that of water. This will create a slight buoyancy effect, even when the pipe is completely filled with water. Thus, it will be necessary to add some **ballast** along the piping to assure that it does not float up from the lake bed. Concrete blocks are often used for such purpose. The piping should be firmly attached to such ballast using non-corrodible straps.

The supply and return pipes must run from the pedestal location within the lake, across the shore, and eventually to the location of the heat pump. Both pipes must be buried deep enough to prevent any possibility of freezing. Depending on the site, the logistics of getting these pipes across the shore area can be formidable and must be considered early in the process, as a prerequisite to continuing with this approach.

If a shallow well pump will be used to create flow in this system, it must be evaluated for the lift required between the water level in the lake, and the pump inlet. In theory, the maximum height over which cold water can be drawn upward by a shallow well pump is approximately 33.9 feet (at sea level). In practice, this distance decreases due to head loss in the piping supplying the pump inlet, as well as increasing vapor pressure of water as its temperature increases. It also decreases for sites at higher elevations.

Another method that has been successfully used in systems requiring multiple water source heat pumps is the installation of a submersible pump within a sleeve that runs from a shoreline location, to a point in the lake where the pump will be located. Figure 14-28 illustrates the concept.

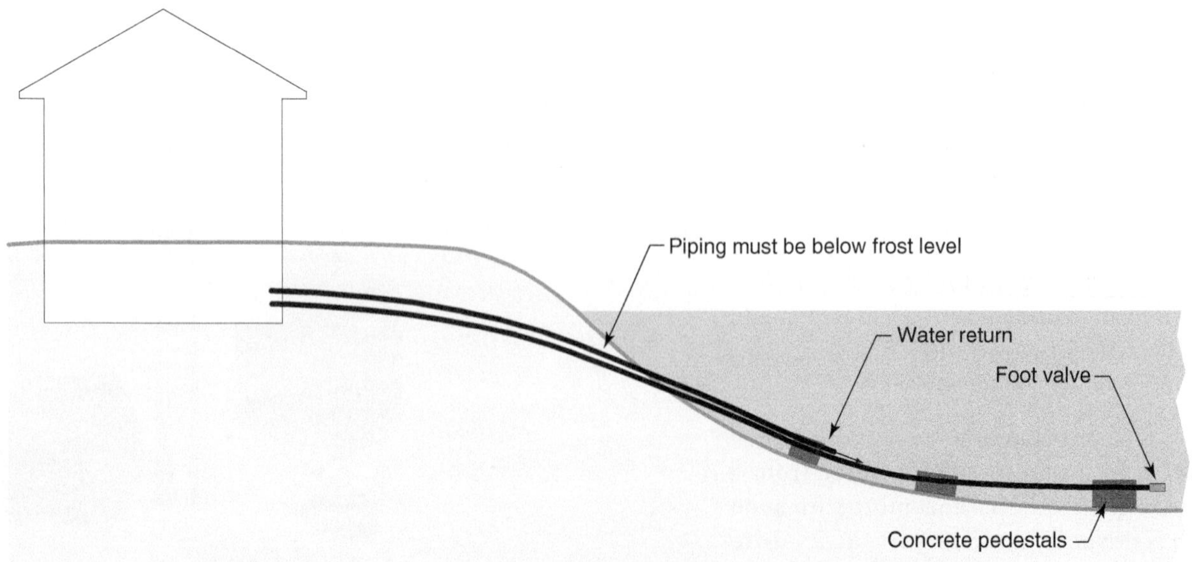

Figure 14-27 | Piping from building to foot valve in lake. A foot valve is installed at the end of the inlet pipe.

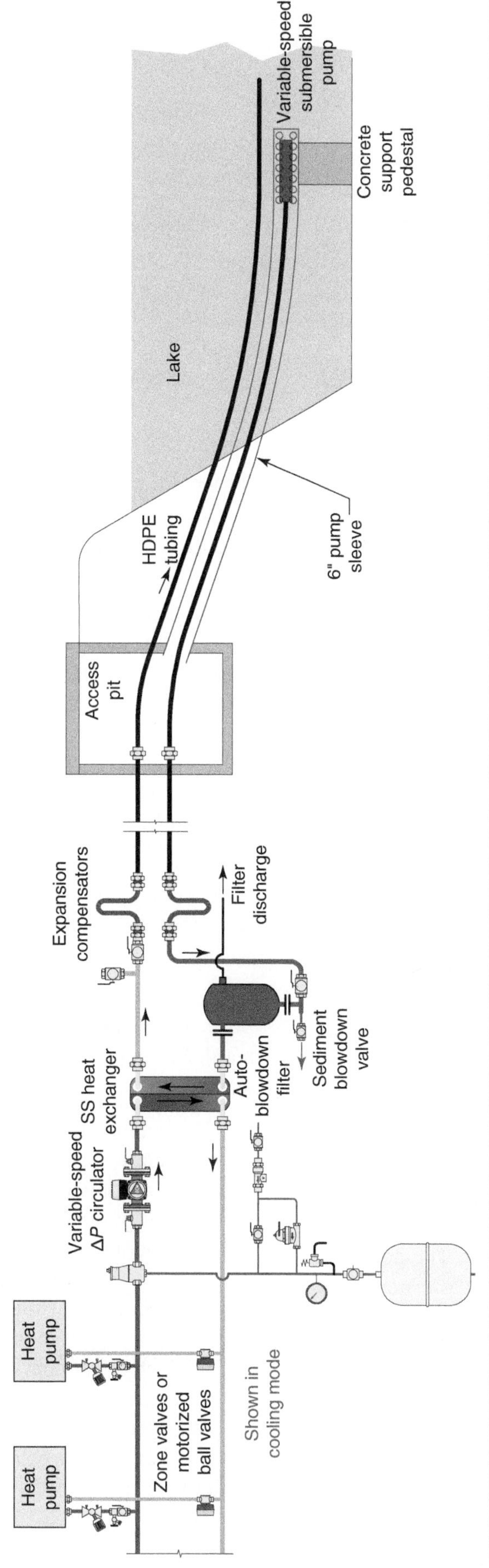

Figure 14-28 | Use of a submersible pump within a sleeve to provide lake water to a heat pump system.

The sleeve is constructed of 6-inch diameter HDPE or heat fused polypropylene tubing. It runs from a service vault on the shore, to a pedestal located on the lake bed several feet below the surface. The end of the sleeve is capped. Several large holes are drilled through the sleeve near its end to allow lake water to freely flow into the sleeve and then into the submersible pump.

The shore end of the sleeve opens into a service vault. The design and placement of the vault and sleeve must be such that the submersible pump and its attached piping and wiring can be pulled back through the sleeve, and up from the vault, if maintenance is required. The upper portions of the vault can be covered with extruded polystyrene insulation to minimize heat loss. The vault should be located in well-drained soil and be as watertight as possible. A floor drain or drainage sump with pump is also recommended to prevent water accumulation within the vault.

The system in Figure 14-28 includes an automatic backwash filter, which filters particulates down to 100-micron size from the incoming water. The intent is to minimize sediment accumulation on the internal surfaces of the downstream heat exchanger. The filter is equipped with pressure sensors and controls that monitor the resistance to flow through the stainless steel filter media. When the differential pressure reaches a preset value, the filter automatically initiates a **backwash cycle**, which ejects the collected sediment through a drainage port on the filter.

A flat-plate stainless steel heat exchanger is used to separate the lake water from the water passing through the heat pumps. This prevents any particulates or microorganisms in the lake water from passing into the remainder of the hydronic system. The heat exchanger should be sized for a maximum approach temperature of 5 °F when all heat pumps are operating at design load conditions. This minimizes the thermal penalty associated with the heat pump.

The submersible pump is operated by a variable-speed controller, which monitors temperature differential across the heat exchanger. The goal is to maintain a preset temperature drop. An increasing temperature differential across the heat exchanger implies an increasing thermal load. The controller responds by increasing the speed of the submersible pump, which increases flow rate as necessary to restore the desired temperature differential. If the temperature differential decreases, the speed of the submersible pump also decreases to reduce power consumption under reduced load. The submersible pump is driven at full speed whenever the automatic filter is backwashing.

Flow through each heat pump is controlled by a zone valve, or motorized ball valve, that opens only when its associated heat pump is operating. A variable-speed pressure-regulated circulator regulates the flow rate to the heat pumps based on the status of these valves. This "demand-based" flow control reduces pumping power under partial load conditions.

14.6 Water-to-Water Heat Pumps Supplied by Earth Loops

There are many locations where ground water is either not available or not suitable for use with water-source heat pumps. In such situations, an alternative approach is to extract low-temperature heat from soil using a closed piping circuit buried several feet below the earth's surface. Water, or a mixture of water and antifreeze, is circulated through this closed-loop earth heat exchanger whenever the heat pump is operating. Such systems are best described as earth-coupled heat pumps.

There are several ways in which tubing can be embedded in soil. Most current installation methods are classified as either **horizontal earth loops** or **vertical earth loops**.

Figure 14-29 shows the concept of a water-to-water heat pump extracting low-temperature heat from the earth by circulating a fluid through a closed horizontal earth loop.

As long as the fluid circulating through the earth loop is cooler than the surrounding soil, heat will move from the soil into the fluid. Modern water-to-water heat pump are capable of operating with fluid inlet temperatures down to about 25 °F (provided the fluid in the earth loop contains a suitable antifreeze). The temperature of the fluid leaving the heat pump and going back out to the earth loop could be as low as 20 °F. Such lower operating temperatures are typically only encountered in horizontal earth loops, near the end of winter, and in cold locations. Thus, even soil that is at or slightly below freezing can still contribute heat to the earth loop.

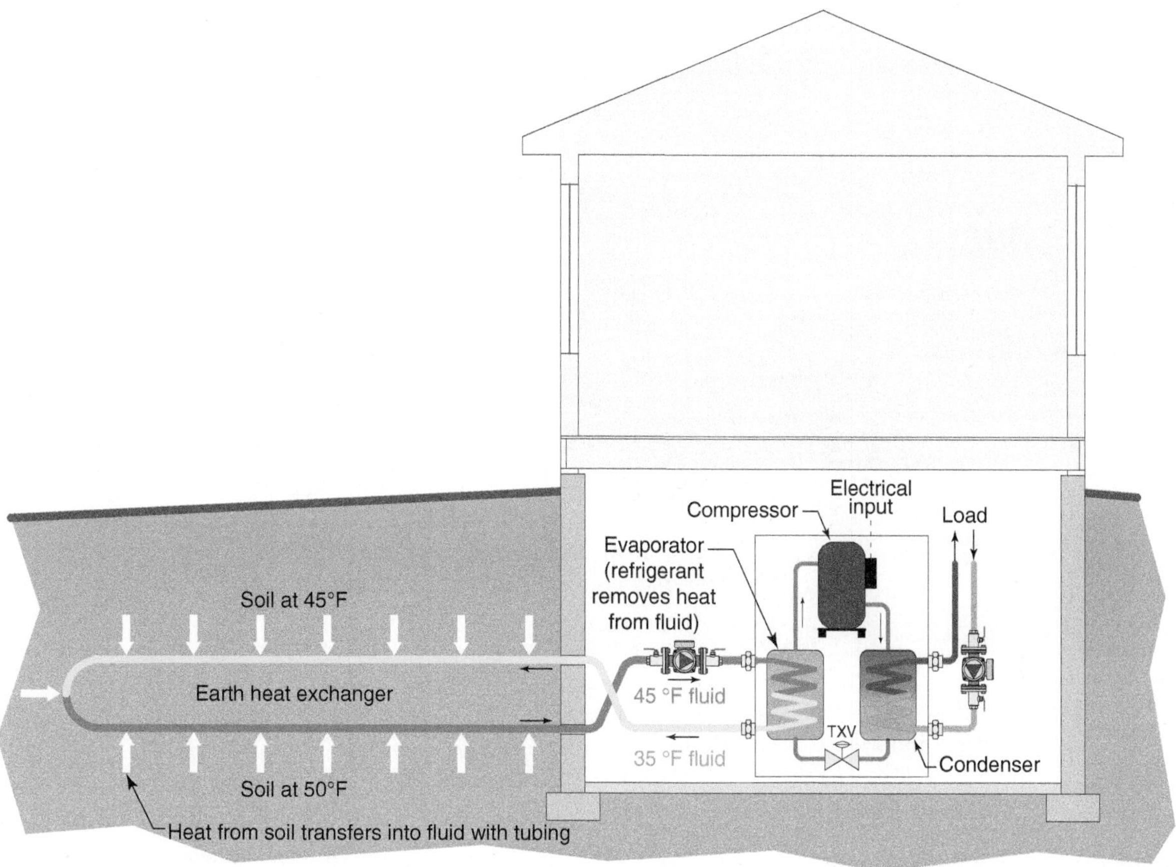

Figure 14-29 | Heat from soil transfers into cooler fluid circulating through the closed earth loop.

HORIZONTAL EARTH LOOP OPTIONS

Horizontal earth loops place tubing in trenches or other open excavations, which typically range from 4 to 8 feet deep. Placement is often determined by soil conditions, available land area, and available excavating equipment.

Figure 14-30 shows the use of a **chain trencher** to create a narrow (6-inch wide) trench into which earth loop piping is placed. Such trenchers work best in soils having rocks no larger than fist size. A single pipe can be laid in a continuous circuit at the bottom of such a trench.

A coil of either HDPE or PEX tubing is rolled along-side the trench. Pipe coming off the coil is lowered to the bottom of the trench. The sand fill shown at the bottom of Figure 14-30b is only necessary if the excavated **tailings** contain chips or shards with sharp edges. Such material may create undesirable voids against the tubing if used for backfill directly over the tubing. Granular soil particles provide the best backfill. Once the tubing is covered with several inches of fine soil, the balance of the trench can be backfilled with the tailings from the excavation.

The trench shape can vary depending on available space. Serpentine shaped trenches with multiple return bends are possible where placement is restrictive, as illustrated in Figure 14-31. Longer trenches that reduce return bends are generally preferred where space is available. A general recommendation is to allow at least 15 feet between parallel trenches to minimize thermal interaction.

It is also possible to include *two* runs of piping in a single narrow trench, as shown in Figure 14-32. In such cases, the trench is partially backfilled after the first pipe is laid in place. The piping makes a U-bend at the end of the trench and returns along the top of the backfill approximately 2 feet above the other pipe. Methods for determining the thermal performance associated with multiple piping runs in a single trench have been developed and are presented later in the chapter.

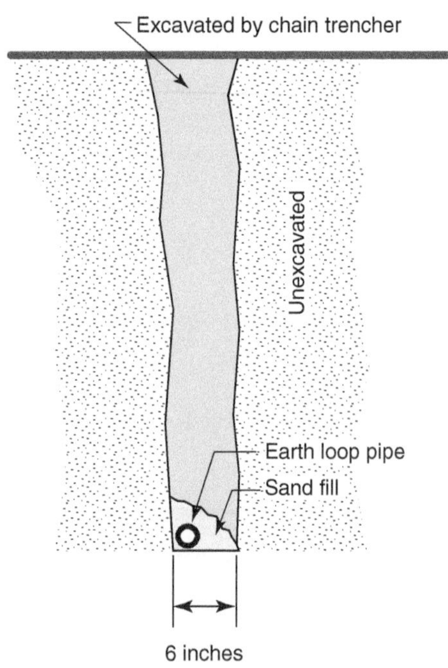

Figure 14-30 | (a) A chain trencher being used to create a narrow trench. (b) Cross section of a nominal 6-inch wide trench with 6-foot depth and single pipe at bottom.

Figure 14-31 | Installation of horizontal tubing in a single tube per trench arrangement.

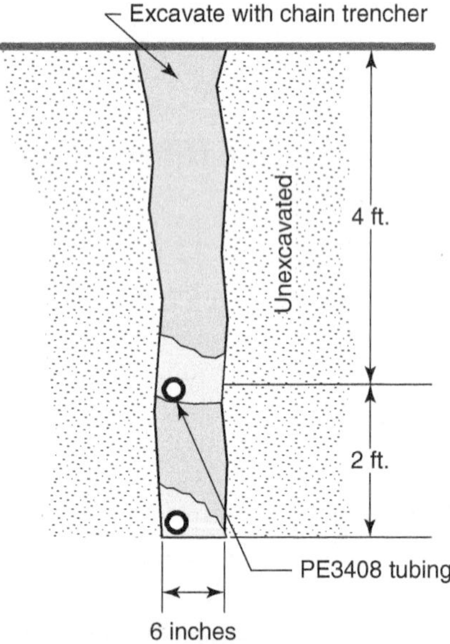

Figure 14-32 | Use of two tubing runs within a single trench.

When chain trenchers are not available, or when soil conditions are not conducive to their use, an excavator with a 2-foot (or wider) bucket is another option. The wider trenches it creates can be used to place two pipes side by side, or to duplicate the placement shown in Figure 14-32 to create what is known as a "**4-pipe square**" earth loop installation. These options are shown in Figure 14-33.

Other horizontal trench options includes 4 pipes in a side-by-side arrangement, as seen in Figure 14-34a, and even 10 side-by-side pipes, as seen in Figure 14-34b.

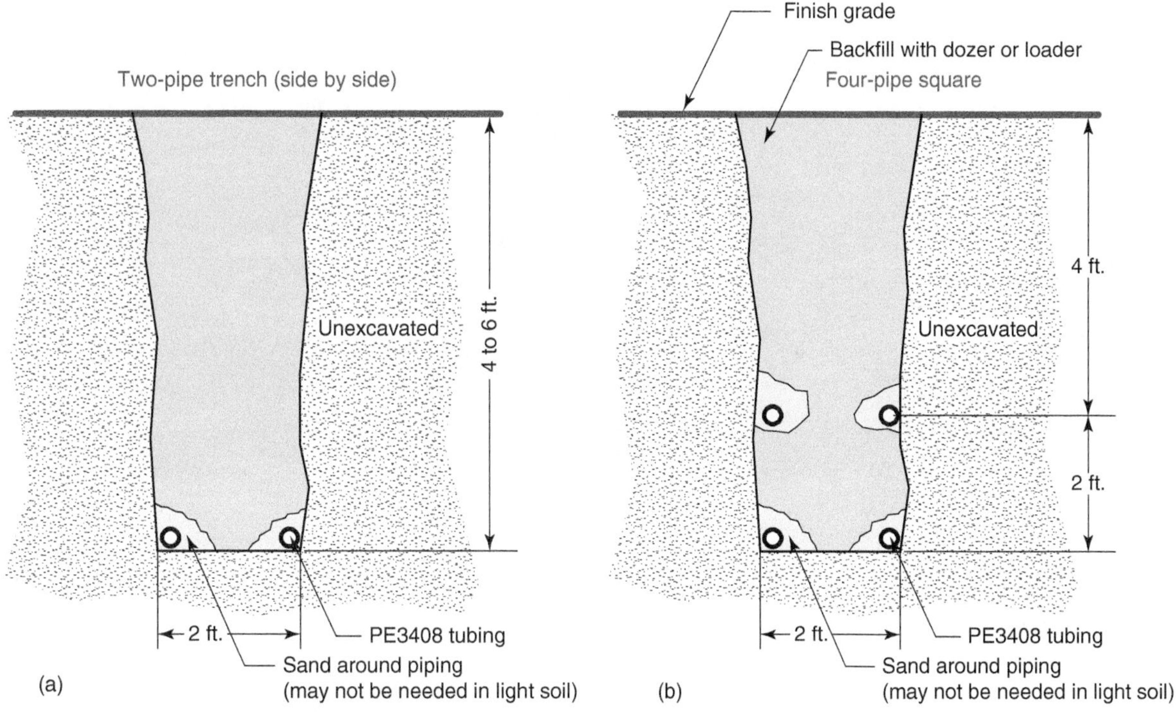

Figure 14-33 | (a) Two pipes, side-by-side, in excavated ditch. (b) 4-pipe square in excavated ditch.

Figure 14-34 | (a) 4-tube side-by-side tubing arrangement. (b) 10-tube side-by-side tubing arrangement.

It is also possible to install tubing in shapes known as "**slinkies**." The tubing coil is manipulated and refastened to yield the slinky shapes shown in Figure 14-35.

The number of slinkies required is based on location, tubing depth, and heat pump capacity. It is common to size each slinky coil to handle approximately 1 ton of evaporator load. Several individual coils are then placed in the excavation and manifolded together in a parallel piping configuration, as seen in Figure 14-35a. The thermal performance of slinkies can be varied by adjusting the amount each coil overlaps the previous coil. This dimension is known as the **pitch [of the slinky]**.

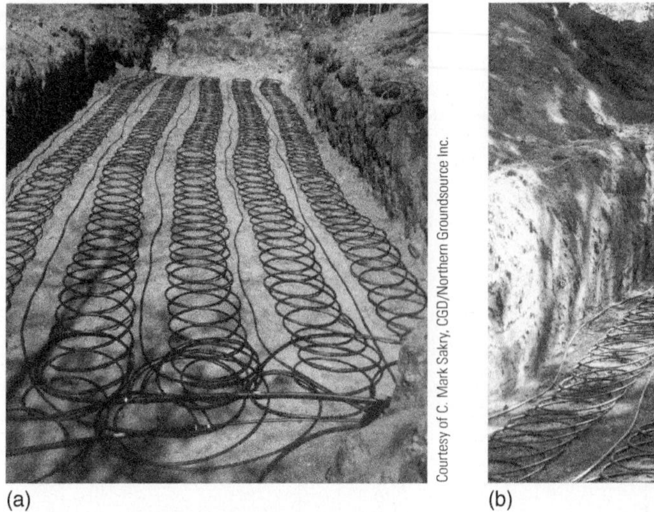

(a)

(b)

Figure 14-35 | (a) Five slinky coils manifolded together in parallel piping arrangement. (b) Slinky coils being backfilled.

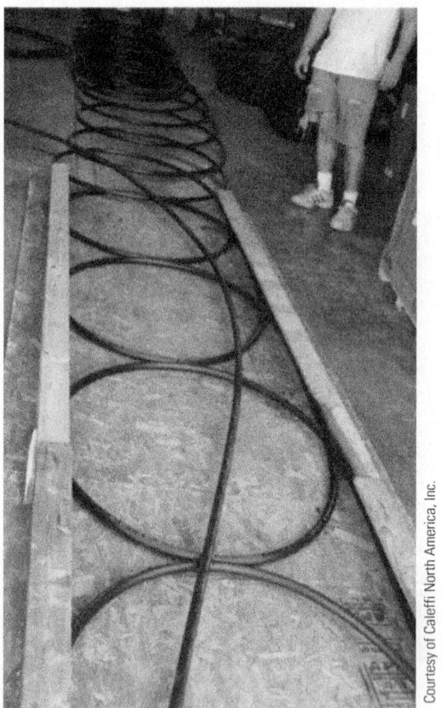

Figure 14-36 | Temporary jig used to assemble slinky.

Figure 14-37 | Slinky being rolled for transport. Note the return pipe from the end of the slinky fastened to the right side of the coils.

Slinky coils are usually assembled in a temporary jig, as shown in Figure 14-36. Individual coils are "slid" from the top of the tubing coil. The coil is not unwound as it would be in a normal linear piping installation. Once the pitch of the slinky is determined, the overlapping coils are held in position using nylon pull ties. The straight return run of the slinky is also fastened to the coils using these ties.

Once the slinky is assembled, it can be rolled up for easier transport to the excavation, as seen in Figure 14-37.

The amount of tubing required in a horizontal earth loop depends on several factors. For heating mode operation, the following factors must be considered:

- heating capacity and COP of the heat pump(s) served by the earth loop
- geographic location (as it affects soil temperature and running time of heat pump)
- thermal properties of the soil
- average depth of the tubing
- size of the tubing
- minimum desired fluid temperature to be supplied to heat pump

Evaluating these factors requires considerable mathematics and is often handled through the use of software. Procedures for sizing earth loops will be discussed later in this chapter.

VERTICAL EARTH LOOP OPTIONS

There are many properties, especially in urban or suburban locations, that lack sufficient land area for the installation of horizontal earth loops. The alternative is the use of a vertical earth loop. This type of earth loop requires drilling equipment similar or identical to that used for creating a water well. The choice of drilling equipment will often be dictated by local availability as well as soil conditions. Most boreholes are 6 inches in diameter and can be as deep as 500 feet. As a guideline, 125 to 175 feet of borehole is required *per ton* of heat pump evaporator capacity.

Once bored, a **U-tube assembly** of HDPE or PEX tubing, which is also sometimes called a **probe**, is inserted down the full depth of each borehole. The end of the U-tube assembly is a tight 180° return bend. It is made by heat fusing a special fitting to HDPE tubing, or supplied as a pre-formed shape using PEX tubing, as shown in Figure 14-38.

In some cases, two U-tube assemblies are inserted into the same borehole. Although this requires twice as much pipe per foot of borehole, it can also reduce the required length of borehole by up to 30 percent. Figure 14-39 shows examples of a double U-tube assembly being prepared for, and then inserted into a borehole.

All U-tube assemblies must be pressure tested before they are inserted into a borehole. A common recommendation is to pressurize the piping assembly to 75 psi and verify that it maintains this pressure for several hours to confirm there are no leaks.

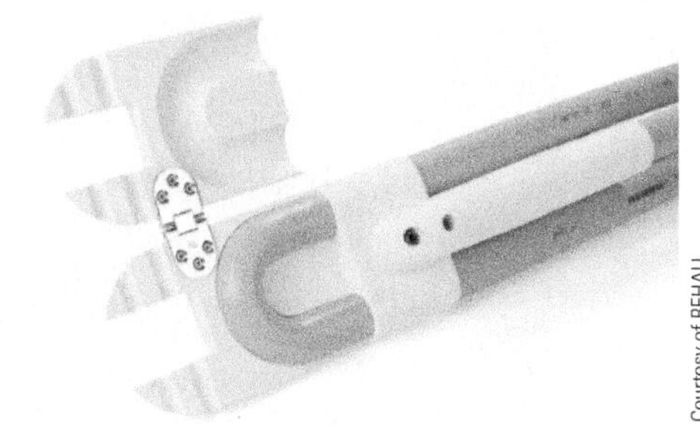
Courtesy of REHAU

Figure 14-38 | Preformed U-bend in PEX tubing mounted within a polymer probe end piece.

If the U-tube assembly is empty, and the borehole contains water, the buoyancy of the air-filled pipe will resist efforts to push it downward into the standing water column. This can be compensated for by attaching a heavy weight, shaped to fit into the borehole, to the bottom of the U-bend. The total weight should slightly exceed the weight of water that would fill the volume of all piping that needs to be pushed below the water level. Another option is to fill the U-tube assembly with water before inserting it in the borehole. Keep in mind that HDPE and PEX tubing are both slightly less dense than water and will thus still produce some buoyant lift when submerged.

After insertion, the spaces between the piping and the walls of the borehole are filled with **grout**. Although several grout formulations have been developed, most are based on an expansive clay called **bentonite** mixed with a fine aggregate such as sand. The objective of grouting is to fill any voids between the tubing and soil and thus maximize heat conduction. Grouting also seals the borehole so that surface contamination cannot flow down into the water aquifer. Some state and local governments have specific requirements on grouting boreholes to protect the integrity of ground water. Be sure to verify and comply with any such local requirements.

The grouting material is mixed to form a slurry, which is then pumped down into the lower portion of the borehole through a pipe called a **tremie**. The tremie is pulled upward as the grout is added. Eventually, the tremie is pulled completely out of the borehole as the grouting procedure is completed.

After all U-bend assemblies have been inserted and grouted, they need to be connected to header

Figure 14-39 (a) End of double U-tube assembly. (b) Tube spacers used with double U-bend assemblies. (c) Iron weight used to help pull U-tube assembly into borehole. (d) Double U-tube assembly being lowered into borehole.

piping. In cold climates, the headers should be installed several feet below final grade to protect them against excessive heat loss to cold soil. Typically, the headers are installed in a trench that is excavated along side the boreholes. Another possibility is to include the headers in a trench that straddles the boreholes, as seen in Figure 14-40. The ends of the U-tubes are carefully bent over as necessary and fusion joined to the header piping.

When multiple boreholes are used, it is very important that the header assembly is constructed in a reverse return arrangement, with stepped pipe sizes, as illustrated in Figure 14-41.

Assuming each U-tube assembly is approximately the same length, this header arrangement allows the overall flow to divide equally between U-tube assemblies. The stepped pipe sizes should be selected to provide approximately *equal head loss per foot of pipe*, based on the flow rate in a given section of header. This detailing is important, because there are no valves or other means of balancing flow through the individual U-tube assemblies.

After all header joints are complete, the entire earth loop assembly should be pressure tested before it is backfilled.

SUBMERGED CLOSED LOOP HEAT EXCHANGERS

It is also possible to extract low-temperature heat from a large pond or lake using a **submerged heat exchanger**. This heat exchanger is part of a closed earth loop circuit. It does not take on any water directly from the pond or lake nor discharge water to the pond or lake. Instead, it operates like the previous described closed-loop systems.

Various approaches have been used in creating submerged heat exchangers. Some involve creating an assembly of several coils of HDPE tubing, piped in parallel to a header system, and then sinking this assembly in the pond or lake using concrete ballast or galvanized steel chain-link fencing. Figure 14-42 shows an example of such an assembly on its way into a pond.

Courtesy of C. Mark Sakry, CGD/Northern Groundsource Inc.

Figure 14-42 Pond heat exchanger fabricated from several HDPE slinky coils and galvanized steel fencing being floated into a pond. When filled with fluid, this assembly will sink to the bottom of the pond.

Courtesy of The Charles Machine Works, Inc.

Figure 14-40 Header piping connecting several U-tube assemblies that have been inserted and grouted into vertical boreholes.

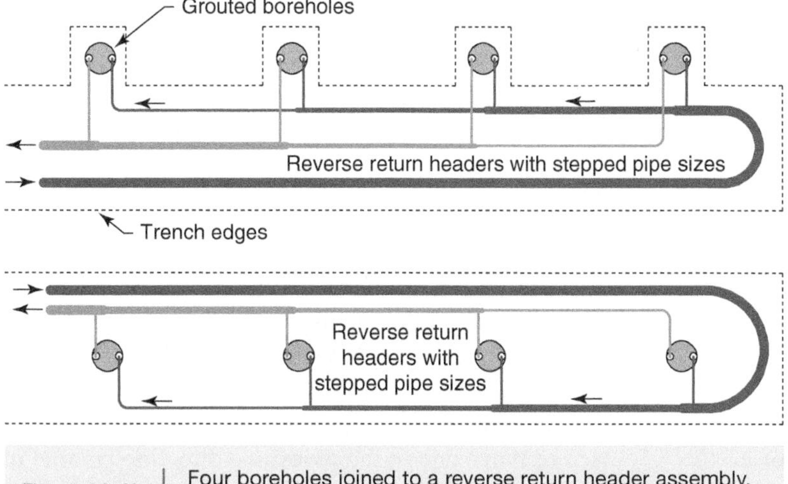

Figure 14-41 Four boreholes joined to a reverse return header assembly. Note the stepped pipe sizes used in the headers.

This approach requires considerable logistics in both fabricating the heat exchanger assembly, moving it into the pond or lake, and submerging it in a way that ensures it stays at the bottom for many years. As is true with other geothermal heat pump systems involving ponds or lakes, permits may be necessary. Restrictions in some areas may also prevent use of this approach.

Another approach uses an assembly of prefabricated stainless steel heat exchange plates, such as shown in Figure 14-43. This heat exchanger assembly is joined to suitably sized HDPE tubing and typically placed on the bottom of a pond or lake. The skids at the bottom of the unit keep the heat-exchanger plates above the silt level and thus fully enabled for convective heat transfer.

This type of heat exchanger has a very high ratio of surface area to volume and thus provides good performance. It is especially effective in applications where there is a significant cooling load. In cold climate heating applications, the water temperature at the bottom of a lake or large/deep pond will often be in the range of 39.2 °F to perhaps 45 °F. The fluid flowing through the heat exchanger must be several degrees cooler than the surrounding water to create the necessary rate of heat transfer. Thus, the fluid temperature coming into the heat pump will usually be relatively cool, often in the range of 25 to perhaps 40 °F. The vast thermal mass of a lake or large pond

(a)

(b)

Courtesy of AWEB Supply, Inc.

Figure 14-43 (a) Example of a stainless steel plate heat exchanger assembly, with support skids and attached piping. (b) A plate heat exchanger, with attached piping, being lowered into a small lake by a crane.

will also keep these temperatures relatively consistent throughout the heating season, especially when the heat exchanger is placed deeper than 25 feet below the surface. The heating capacity and COP of the heat pump need to be evaluated at these temperatures to determine the overall system performance.

14.7 Earth Loop Piping and Pipe Joining Procedures

There are two types of pipe currently used for earth loop heat exchangers: **high-density polyethylene (HDPE)** and cross-linked polyethylene (PEX). Both have a well-proven record of durability in these applications.

HIGH-DENSITY POLYETHYLENE (HDPE) PIPE

Currently, HDPE is the most common earth loop piping material. The specific type of HDPE is designated as PE3608 based on the ASTM F-412 standard. The pressure rating of this pipe is determined by its **diameter ratio (DR)**, which is the ratio of the *outside* diameter divided by the wall thickness. Common DRs for HDPE tubing are 7, 9, 11, 13.5, 15.5, and 17.5. A diameter ratio (DR) of 11 or *lower* is often recommended for earth loop applications. DR-11 PE3608 piping has a pressure rating of 160 psi. The pressure rating of DR-9 PE3608 tubing is 200 psi. The lower the DR value, the thicker the pipe wall. Although thicker wall pipes are more durable, they also add more thermal resistance in the path of heat conduction from the soil into the fluid within the pipe, or vice versa. The latter is undesirable from the standpoint of heat pump performance.

Figure 14-44 lists the dimensions of DR-11 HDPE tubing in sizes used for systems in this text.

HDPE is a **thermoplastic**. As such, it can be repeatedly melted and reformed. This allows HDPE pipe to be joined using **heat fusion welding**. When done correctly, this technique produces joints that are as strong as or stronger than the pipe itself. Fusion welding has been used to create permanent, leak-proof joints in several applications of HDPE piping, including buried natural gas piping.

There are three specific types of fusion welding:

a. Butt fusion welding
b. Socket fusion welding
c. Electrofusion fittings

BUTT FUSION WELDING

Butt fusion welding describes a process where the *ends* of two pieces of HDPE piping are permanently bonded to each other using heat. It requires the use of several specialized tools. One is a **butt fusion machine**, shown in Figure 14-45. This machine holds the two pipe ends in precise alignment. The sloping handle is used to move one of the pipes in a direction parallel to its centerline. The other pipe end does not move, but is held rigidly in place and parallel to the other tube.

Nominal size (inch)	Inside diameter (inch)	Outside diameter (inch)	Wall thickness (inch)	Internal volume (gallons/100 feet)
3/4	0.859	1.05	0.095	3.01
1	1.076	1.315	0.12	4.72
1.25	1.358	1.66	0.151	7.53
1.5	1.555	1.9	0.173	9.86
2	1.943	2.375	0.216	15.41
3	2.864	3.5	0.318	33.5

Figure 14-44 | Dimensional and volumetric data for DR-11 HDPE tubing.

Figure 14-45 | Butt fusion machine.

The following procedure describes how a butt fusion joint is made.

STEP 1: In preparation for welding, the ends of the two pipes should be wiped clean and squarely cut using a sharp tube cutter. Each end is then clamped into the opposing jaws of the butt fusion machine. Once clamped, the handle of the machine is moved so that the two ends come together. At this point they should be checked for alignment. If necessary, one of the pipes can be loosened and rotated so that it provides the best possible alignment with the other end.

STEP 2: With the two ends of the piping now aligned and clamped into the butt fusion machine, the handle of the machine is moved to bring the two pipe ends apart far enough to insert a **facing tool** between them. An example of this tool is shown in Figure 14-46.

The handle of the butt fusion machine is now used to press both ends of the pipe against the cutter head of the facing tool. The handle of the facing tool is manually rotated, while slight pressure is applied to the handle of the butt fusion machine to force the pipe ends against the cutter heads of the facing tool. The blades on the facing tool will precisely trim the ends of the pipes so that they are perfectly square with each other and clean. At this point, the facing tool is lifted out of the butt fusion machine and the two pipe ends brought together for a final alignment check. It is important not to touch the ends of the pipe at this point. Any dirt of oil from fingers could interfere with proper joining of the pipe ends.

STEP 3: A thermostatically controlled **electric heating tool**, which has already heated up so that its surfaces are approximately 500 °F, is now placed between the two pipe ends. Figure 14-47 shows an example of this heating tool.

The heating tool is supported in the proper orientation by the rails of the butt fusion machine. The sloping handle of the fusion machine is now moved to bring the pipe ends into gentle contact with the faces of the heater tool. The time during which the pipe ends are held against the heater tool varies with pipe size and ambient temperature conditions. It is typically specified by the manufacturer of the fusion welding tools and generally ranges from 30 seconds to perhaps 2 minutes for the pipe sizes described in this text. The machine operator watches for a **bead** to form on the ends of the pipes where they contact the face of the heating tool.

STEP 4: Once the specified heating time has elapsed, the butt fusion machine is operated to pull the pipe ends away from the heating tool. The heating tool is lifted out of position and placed into an insulated bag or cradle to prevent it from damaging other materials or inadvertently burning someone. The two molten pipe ends are then pressed together by the butt fusion machine, until a **double roll-back bead** is, as seen in Figure 14-48 is formed. The butt fusion machine should then be locked in position to prevent further movement. The pipes should remain in the machine until the fusion joint has cooled to the point where it can be safely touched. The manufacturer of the pipe fusion equipment typically provides a table of recommended joint cooling time based on pipe size and ambient conditions. These times typically range from 2 to 5 minutes for the pipe sizes discussed in this text.

(a) Courtesy of McElroy Manufacturing, Inc.

(b) Courtesy of Delta Geothermal

Figure 14-46 | (a) A rotary facing tool. (b) Rotary facing tool locked into position between tube ends in the butt fusion machine.

(a) Courtesy of McElroy Manufacturing, Inc.

(b) Courtesy of Delta Geothermal

Figure 14-47 | (a) Example of a heater tool. (b) Heater tool inserted between pipe ends in butt fusion machine.

Following the cooling period, the jaws of the machine are opened and the completed joint is lifted out and ready for service.

Figure 14-48 | A double rollback bead forms as the molten ends of the pipe are pressed together during butt fusion.

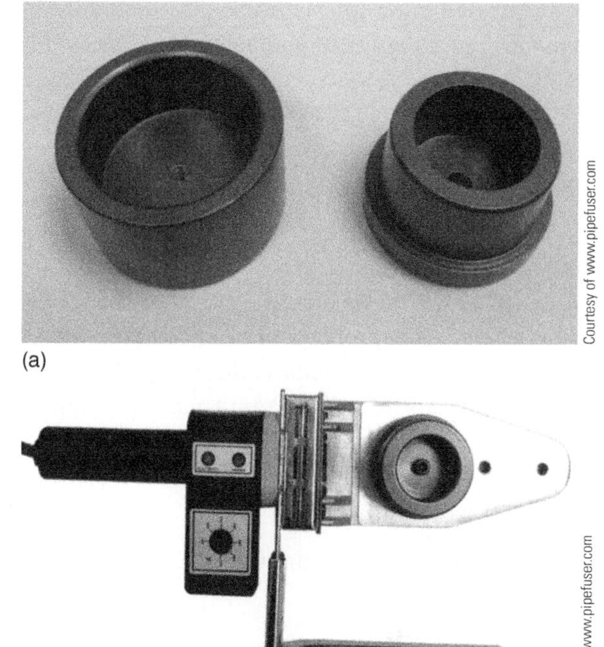

(a)

Courtesy of www.pipefuser.com

(b)

Courtesy of www.pipefuser.com

Figure 14-50 | (a) A matched set of socket fusion adapters. (b) A heating tool with fusion adapters bolted to it.

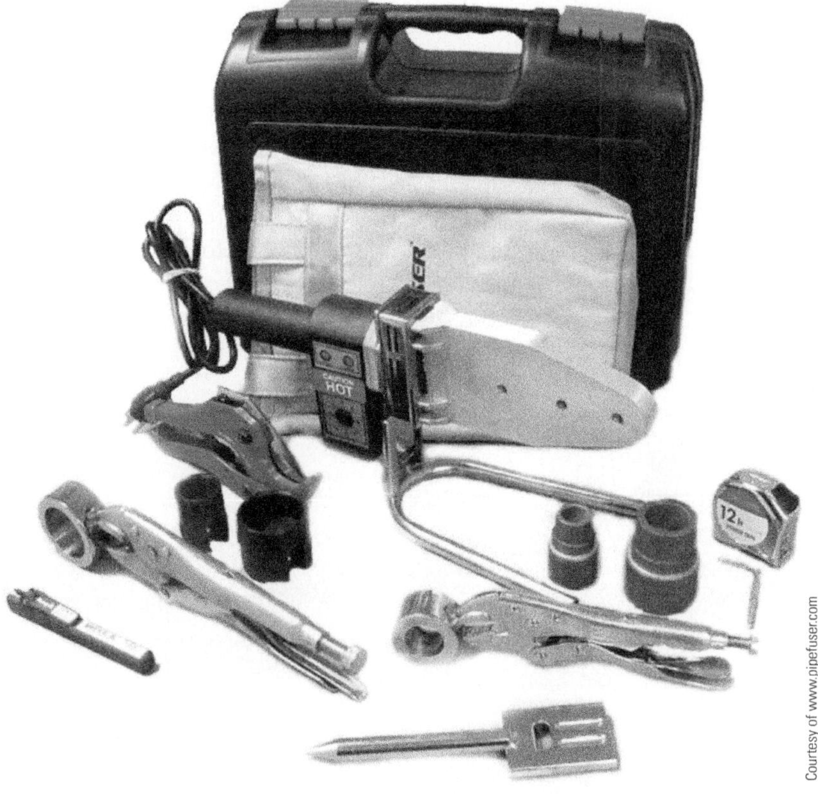

Courtesy of www.pipefuser.com

Figure 14-49 | Example of a toolkit for socket fusion welding.

SOCKET FUSION

Another method of fusion welding HDPE pipe is called **socket fusion**. It involves joining pipe with fittings. As with butt fusion welding, special tools are required. These tools are usually supplied as a kit, an example of which is shown in Figure 14-49.

The components in this kit include an electric heating tool and several **socket fusion adapters** that bolt to the heating tool to accommodate different size pipe and fittings. These adapters are supplied in pairs, as shown in Figure 14-50. One fits around the outside diameter of the pipe, and the other fits the inside diameter of the fitting socket. They are mounted on opposite sides of the heating tool.

Another part of the kit is a **chamfer tool**, as seen in Figure 14-51. This tool is placed over the end of the pipe and rotated by hand to cut a small chamfer (e.g., beveled edge) on the tube. This helps the tube easily slip into the heating adapter as well as into the fitting it will be fused to. Some chamfer tools are also used as a depth gauge to determine the proper insertion depth of the tube into the fitting.

Proper socket fusion requires careful control of the tube insertion depth. This is provided by attaching a **cold ring clamp** to the outside of the pipe. This clamp, shown in Figure 14-52, is specifically built for the tube size being welded. It is properly positioned by first placing the depth gauge over the end of the pipe until it bottoms out and then sliding the clamp rings up against the depth gauge. Once the cold ring clamp is secured to the pipe, it can be used to safely handle the pipe near the heating tool.

The following procedure is used of socket fusion.

Courtesy of www.pipefuser.com

Figure 14-51 | Chamfer tool.

STEP 1: Prepare the heating tool by first bolting on the appropriate size heating adapters for the pipe size being fused. Place the heating tool on its stand and plug it in. Allow sufficient time for the surfaces of the socket fusion adapters to reach a temperature of 500 °F. Verify this temperature with a surface thermometer.

STEP 2: Be sure the end of the tube is cut square. Wipe the end of the tube and the socket of the fitting with a clean cloth to remove any dirt of grease. **MEK (methyl ethyl ketone)** can be used as a solvent, if needed, to remove residue from the pipe or fitting. If used, allow the MEK to fully evaporate from the surfaces to be joined. This should only take a few seconds.

STEP 3: Place the chamfer tool over the end of the pipe and manually rotate it until it spins easily. This cuts the necessary bevel on the tube. If the chamfer tool also serves as the depth gauge, leave it on the end of the pipe and attached the cold clamp ring next to it. Remove the depth gauge and recheck that the end of the tube is clean.

STEP 4: With gloves on both hands, simultaneously slide the end of the pipe, and the fitting, into and onto their respective socket fusion adapters on opposite sides of the heating tool. *Do not twist* either of them as they are pushed into place. Hold them in position for the required heating time based on the pipe size.

STEP 5: When the required heating time has elapsed, simultaneously pull the end of the pipe and the fitting away from the heating tool. Immediately slide them together, *without twisting*, until the cold ring clamp contacts the fitting socket. This ensures proper insertion depth of the pipe into the fitting. Hold the assembly in this position for the recommended cooling time, which is usually 2 to 4 minutes, depending on pipe size. Allow

Courtesy of www.pipefuser.com

(a)

Courtesy of Delta Geothermal

(b)

Courtesy of Delta Geothermal

(c)

Figure 14-52 | A cold ring clamp for a specific pipe size. (b) A cold ring clamp mounted to a HDPE pipe. The portion of the pipe on the top side of the clamp will be inserted into the fitting during welding. (c) Pipe with cold ring clamp inserted on left side of heater tool, while fitting is heated on right side of heater tool.

Courtesy of ClimateMaster, Inc.

Figure 14-53 | Example of a header assembled with socket fusion fittings and HDPE pipe.

the joint to cool to the touch before handling it or otherwise stressing the warm joint.

Figure 14-53 shows an example of a buried earth loop header that has been assembled using socket-fused fittings and HDPE pipe.

ELECTROFUSION

The final method of pipe fusion that is suitable for HDPE earth loops is called **electrofusion**. This technique combines HDPE tubing with special fittings that are manufactured with internal heating wires surrounding the inner face of their sockets. These heating wires connect to the electrodes projecting from the sides of fitting, as shown in Figure 14-54.

Before electrofusion takes place, the sockets of the fitting is wiped down with a MEK (methyl ethyl ketone) solvent to remove any dirt or grease. The exterior surface of the pipe is then shaved with a special tool to remove any surface oxidation or other contaminants. The tube ends are then inserted into the fitting. This assembly is clamped into a fixture that ensures proper insertion depth, alignment, and clamping pressure. A specialized electrical power supply is connected to the electrodes on the fitting to provide low-voltage/high-amperage power to the electrodes. The heating wires within the fitting quickly heat the HDPE to its melting point, and the outer surface of pipe is fused to the socket of the fitting.

Courtesy of Craig Johnson/VARI-TECH LLC

(a)

Courtesy of Mr. Renzo Bortoli

(b)

Figure 14-54 | (a) Example of an electrofusion fitting. (b) Electrofusion welder.

HEAD LOSS OF HDPE TUBING

The head loss of DR-11 HDPE tubing, which is commonly used for earth loops, can be determined using Equation 14.13. The data needed for this equation is given in Figure 14-55.

(Equation 14.13)

$$H_{L/100'} = cR(f)^{1.75}$$

where:

$H_{L/100'}$ = head loss of piping operating with 40 °F water (feet of head *per 100 ft of tubing*)

c = correction factor for various antifreeze solutions and fluid temperatures (from Figure 14-55b,c,d)

Tubing	R (water @ 40 °F)
3/4″ DR-11 HDPE tubing	$R = 0.269$
1″ DR-11 HDPE tubing	$R = 0.0939$
1.25″ DR-11 HDPE tubing	$R = 0.0316$
1.5″ DR-11 HDPE tubing	$R = 0.0168$
2″ DR-11 HDPE tubing	$R = 0.0059$

(a)

PROPYLENE GLYCOL	40 °F	30 °F	20 °F
15% propylene glycol	c = 1.19	c = 1.25	*
20% propylene glycol	c = 1.24	c = 1.31	c = 1.39
25% propylene glycol	c = 1.34	c = 1.42	c = 1.52

*indicates insufficient freeze protection

(b)

METHANOL	40 °F	30 °F	20 °F
10% methanol	c = 1.13	c = 1.18	*
15% methanol	c = 1.17	c = 1.23	c = 1.31
20% methanol	c = 1.20	c = 1.26	c = 1.35

*indicates insufficient freeze protection

(c)

ETHANOL	40 °F	30 °F	20 °F
15% ethanol	c = 1.26	c = 1.36	*
20% ethanol	c = 1.31	c = 1.42	c = 1.56
25% ethanol	c = 1.35	c = 1.47	c = 1.64

*indicates insufficient freeze protection

(d)

Figure 14-55 (a) Values of R used in Equation 14-13. (b) Correction factors for propylene glycol. (c) Correction factors for methanol. (d) Correction factors for ethanol.

R = number for given size of DR-11 HDPE tubing (from Figure-14-55a)

f = flow rate (gpm)

Example 14.8: Determine the head loss of 900 feet of 3/4-inch DR-11 HDPE tubing operating with a 20 percent solution of propylene glycol at 30 °F, at a flow rate of 4 gpm.

Solution: The head loss of the tubing per 100 feet of length is found using Equation 14.13 along with the value of $R = 0.269$ and the antifreeze correction factor $c = 1.31$. The latter values were obtained from Figure 14-55.

$$H_{L/100'} = cR(f)^{1.75} = (1.31)(0.269)(4)^{1.75} = 3.99 \text{ ft}/100'$$

Thus, the total head loss of 900 feet of this tubing would be $9 \times 3.99 = 36$ feet of head.

Discussion: Cooler fluids and higher concentrations of antifreeze will increase the head loss of a given earth loop circuit.

CROSS-LINKED POLYETHYLENE (PEX) TUBING EARTH LOOPS

Cross-linked polyethylene tubing manufactured using the high-pressure peroxide method of **cross-linking** (e.g., **PEX-a**) can also be used for earth loops. In North America, the use of PEX-a tubing for earth loops is relatively new in comparison to non-crosslinked PE3608 tubing. PEX-a tubing does offer better resistance to abrasion and stress cracking compared to HDPE. Its pressure/temperature ratings also exceed those required for HDPE piping in earth loop applications. Another advantage of PEX-a tubing is that it can be bent without heating to a radius of five times its outer diameter. HDPE pipe should only be bent to a minimum radius of 25 times its outer diameter.

Some systems that use PEX-a tubing for earth loops are planned so that continuous lengths of tubing, free of any joints, can be used for the buried portion of the earth loop. The ends of the tubing are joined to a **manifold station** mounted in an accessible location, either inside or outside the building. This type of piping will be discussed in the next section.

It is also possible to join lengths of PEX-a tubing in ways that are fully compatible with earth loop applications. Being a thermoset polymer, PEX-a tubing *cannot* be joined by fusion welding, as is commonly used for HDPE. Any buried joints must be made using mechanical couplings approved for such purpose by the tubing manufacturer. Figure 14-56 shows a cutaway example of a **cold expansion compression sleeve** used to join PEX-a piping for earth loop applications.

Figure 14-57 shows the sequence and tools needed to make a cold expansion joint in PEX-a tubing.

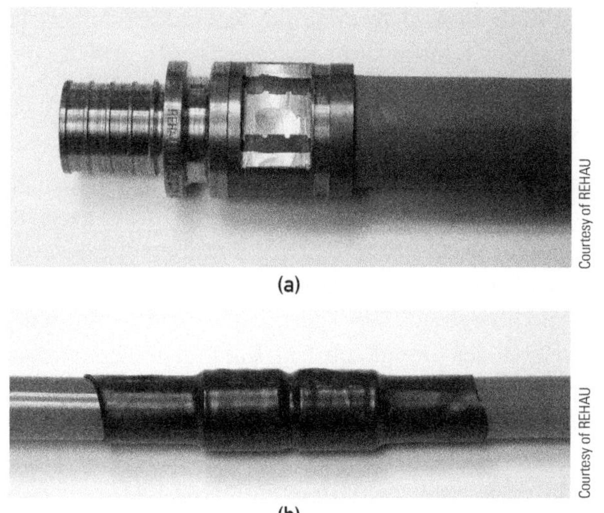

Figure 14-56 (a) Cut-away of a cold expansion compression sleeve used to join PEX-a tubing. (b) Heat shrink wrap over completed fitting protects it from soil.

The head loss of PEX-a tubing can be calculated using the methods and data presented in chapter 3.

14.8 Earth Loop Design

There are many ways to design earth loop circuits. All earth loop these designs must be based on the specific method used to place piping in the earth. They also need to provide reasonably low head loss. The latter, in combination with the flow rate through the earth loop, will determine the circulator power required to provide adequate flow through the earth loop and heat pump.

Most earth loops used in current generation geothermal heat pump systems use a fixed speed circulator. The flow rate within such earth loops will vary slightly depending on the temperature of the fluid within the loop. During the heating season, when earth loop fluid temperature decreases to perhaps as low as 25 °F, the flow rate produced by a fixed speed

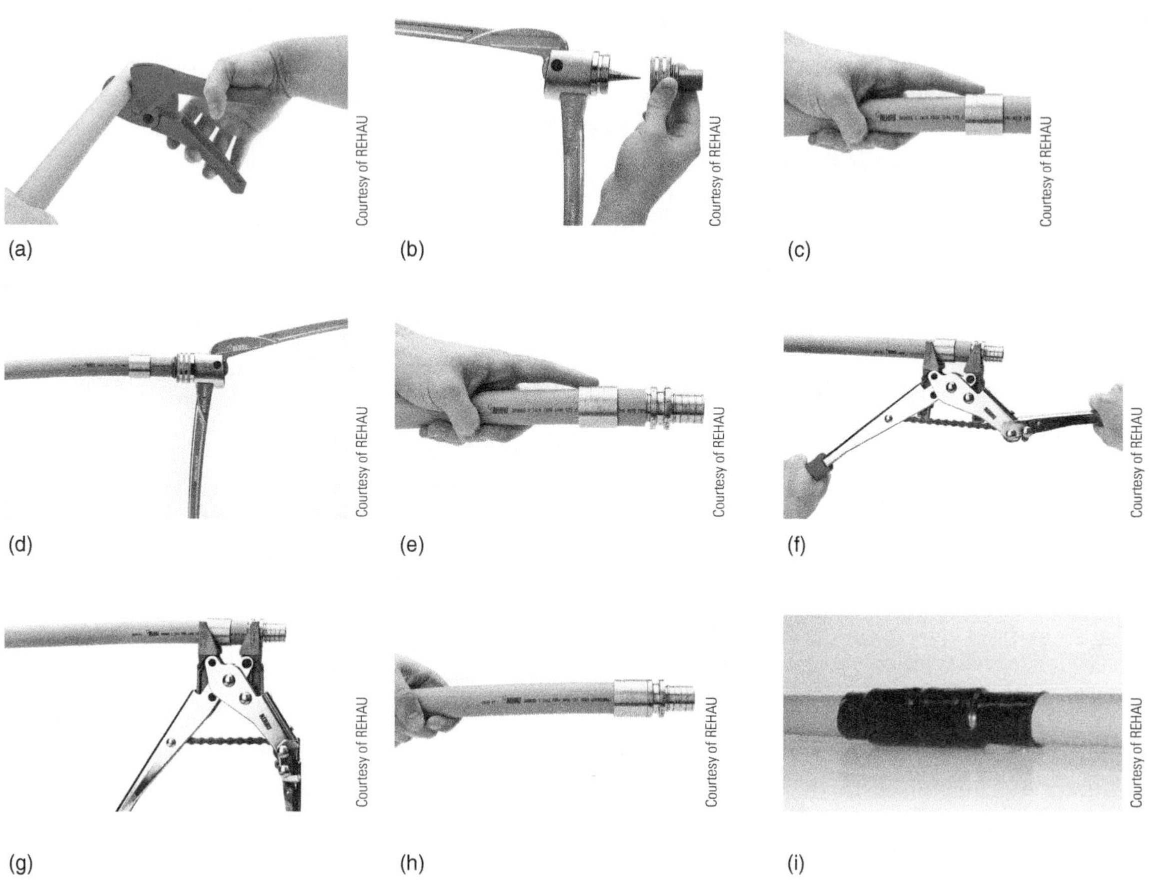

Figure 14-57 (a) Cut tube square. (b) Install expansion head on tool. (c) Slide brass sleeve over tube. (d) Expand tube. (e) Insert brass coupling fitting. (f) Attach compression tool. (g) Keep ratcheting compression tool to pull sleeve all the way to stop on brass coupler. (h) One side of joint completed. (i) Complete other side of joint using same technique and then install heat shrink sleeve to protect brass against corrosion in soil.

circulator will decrease. This is the result of increased head loss due to higher density and increased viscosity of the earth loop fluid. In summer, when the heat pump is operating in cooling mode, the temperature of the earth loop fluid will increase, perhaps as high at 95 °F. These elevated temperatures will lower both the density and viscosity of the earth loop fluid and thus allow the fixed-speed circulator to create higher flow rates.

One of the often-cited criteria for earth loop design is to maintain **turbulent flow** within the earth loop. Turbulent flow increases convective heat transfer between the inside of the pipe wall and the fluid, compared to **laminar flow**. In the latter, fluid moves through piping in mostly parallel streamlines and with very little internal mixing. The limiting condition under which turbulent flow must be maintained occurs when the earth loop fluid is at its lowest temperature.

REYNOLDS NUMBER

The **Reynolds number** of a flow stream determines if that flow stream is laminar or turbulent. *For turbulent flow, the Reynolds number must be 2,500 or higher.* The Reynolds number is calculated by combining several other physical quantities. For a flow stream through a pipe, the Reynolds number is based on the fluid's density and viscosity, its flow velocity, and the inside diameter of the pipe. Because the density and viscosity of fluids depend on their temperature, the Reynolds number is also a temperature dependent number.

The Reynolds number for a fluid passing through a round pipe can be calculated using Equation 14.14.

(Equation 14.14)

$$Re\# = \frac{vdD}{\mu}$$

where:

v = average flow velocity of the fluid (ft./sec.)

d = internal diameter of pipe (ft.)

D = fluid's density (lb./ft.^{3})

μ = fluid's dynamic viscosity (lb./ft./sec.)

The Reynolds number is always a ***dimensionless number***. All units of the physical quantities used to calculate the Reynolds number must mathematically cancel out when combined in Equation 14.14. If they don't, the calculation is missing one or more unit conversion factors. *Any calculated value for a Reynolds number, that is not unitless, is invalid.*

Example 14.9: Determine the Reynolds number for a 20 percent solution of propylene glycol and water passing through a 1" DR-11 HDPE pipe at a flow rate of 5 gpm. Assume the temperature of the solution is 30 °F.

Solution: There are several values that must be calculated or referenced before using Equation 14.14. First, it is necessary to find the density and dynamic viscosity of the 20 percent solution of propylene glycol at its temperature of 30 °F. These values can be referenced in data tables provided by the fluid manufacturer. *When referencing viscosity, be sure the units are the same as those listed for Equation 14.14 (e.g., lb./ft./sec.).* If they are not, first convert the stated viscosity into units to lb/ft/sec before using it in Equation 14.14.

One manufacture's reference tables provide the following data for a 20 percent solution of inhibited propylene glycol (e.g., Dowfrost).

D = 64.14 lb./ft.^{3}

Viscosity = μ = 4.23 cps 5 4.23 **centipoises**

The stated viscosity can be converted to the necessary units as follows:

$$4.23\ \text{cps}\left(\frac{0.000672\frac{\text{lb.}}{\text{ft.}\cdot\text{sec.}}}{1\ \text{cps}}\right) = 0.002843\frac{\text{lb.}}{\text{ft.}\cdot\text{sec.}}$$

Thus, μ = 0.002843 lb./ft./sec.

The average flow velocity can be found using Equation 3.5 and the exact inside diameter of the 1-inch DR-11 HDPE pipe from Figure 14-44.

d = 1.076 inch

Using this diameter in Equation 3.5 yields the average flow velocity:

$$v = \left(\frac{0.408}{d^2}\right)f = \left(\frac{0.408}{1.076^2}\right)5 = 1.76\ \text{ft./sec.}$$

The data is now ready to be entered into Equation 14.14 for the Reynolds number calculation.

$$Re\# = \frac{vdD}{\mu}$$

$$= \frac{\left(1.76\frac{\text{ft.}}{\text{sec}}\right)(1.076\ \text{inch})\left[\frac{1\ \text{ft.}}{12\ \text{inch}}\right]\left(64.14\frac{\text{lb.}}{\text{ft.}^3}\right)}{\left(0.002843\frac{\text{lb.}}{\text{ft.}\cdot\text{sec}}\right)}$$

$$= 3{,}560$$

Discussion: The Reynolds number of 3,560 is well above the minimum value of 2,500, and thus the flow will be turbulent under these conditions. Calculating the Reynolds number requires several steps in gathering data and making sure that data is in the correct units. Remember, the Reynolds number must always be a dimensionless number. The best way to ensure that a valid Reynolds number is being calculated is to include the units along with the numerical values for all the input data and make sure the units mathematically cancel prior to running the numbers through a calculator. This type of **units check** revealed that the conversion factor of (1 ft./12 inches), shown in red in the Reynolds number calculation for this example, was necessary. Without including this conversion factor, the overall units would not cancel out and the calculated value for Reynolds number would have been high by a factor of 12.

The tables in Figure 14-58 show the minimum flow rate (in gpm) required in a given size of DR-11 HDPE tubing to maintain turbulent flow. A range of fluid temperatures and concentrations of propylene glycol antifreeze are given. Reference 1, cited at the end of the chapter, contains similar tables for other antifreezes.

These tables show that the lower the fluid temperature, and the higher the concentration of propylene glycol, the higher the required flow rate to maintain turbulent flow. Any design factors that keep the earth loop operating at higher minimum temperatures, and thus compatible with lower percentages of propylene glycol antifreeze, will reduce the minimum flow rates required for turbulence. These factors will also likely reduce the required circulator size and operating cost.

EARTH LOOP PIPING LAYOUT

There are many possible earth loop topologies. Piping layout depends on whether the earth loop is installed horizontally or vertically. This orientation is often dictated by available land area and/or the type of excavation equipment available. Once a layout has been determined, the hydraulics of the piping arrangement can be analyzed to determine the necessary flow rate through the earth loop, and the associated head loss. A suitable circulator can then be selected.

One of the simplest earth loop arrangements is a single pipe circuit that begins at the heat pump, passes out through the building wall, continues through several hundred feet of buried piping, reenters the building, and finally reconnects to the heat pump. This type of **series circuit** is typically used in combination with a chain trencher for the excavation. The tubing used must be carefully selected to maintain turbulent flow under the lowest operating fluid temperature. It should also provide relatively low head loss and thus minimize circulator power requirements.

Most earth loops used for residential and light commercial geothermal heat pump systems contain two or more parallel branch circuits. When HDPE pipe is used for the earth loop, it is common practice to fusion weld the branch circuits to an HDPE header that will be buried in the soil outside the building, such as shown in Figure 14-59.

When the parallel branches join a header at closely spaced connections, such as seen in Figure 14-59, the head loss in the branches is far greater than the head loss along the header. If the branches are close to the same length, they will all operate at essentially equal flow rates. In such cases, the header can be made of same piping material that connects the earth loop to the building. The branches should connect to the headers in a reverse return arrangement, as shown in Figure 14-60.

When the headers will be more than a few feet long, the best approach is to use stepped pipe sizes along both the supply and return headers, and reverse return piping, as shown in Figure 14-61.

The objective of stepped piping sizes is to make the *head loss per foot of length* approximately the same along the length of the header. The pipe sizes need to be selected based on the change in flow rate through the header each time it passes a branch connection. The degree to which this sizing objective can be achieved depends on the number of branches and the availability of different pipe sizes. In many cases, there are not enough available pipe sizes to provide exactly the same head loss per unit of length. Still, any changes in pipe diameter that approximates this condition are beneficial. The head loss per unit of length can be calculated using Equation 14.13 and its associated data.

The objective of the reverse return header configuration, in combination with the stepped pipe sizes, is the equally divide the total flow entering the supply header among all the branch circuits connected to it. In theory, perfectly balanced flow will only occur when all three of the following conditions are met:

1. All branch circuits are the same pipe type, size, and length.

3/4″ DR-11 HDPE	20 °F	25 °F	30 °F	35 °F	40 °F
Water	-	-	-	-	1.05 gpm
20% propylene glycol	3.54 gpm	3.17 gpm	2.79 gpm	2.53 gpm	2.26 gpm
25% propylene glycol	5.00 gpm	4.43 gpm	3.85 gpm	3.43 gpm	3.02 gpm
30% propylene glycol	6.47 gpm	5.67 gpm	4.88 gpm	4.33 gpm	3.77 gpm

(a)

1″ DR-11 HDPE	20 °F	25 °F	30 °F	35 °F	40 °F
Water	-	-	-	-	1.31 gpm
20% propylene glycol	4.43 gpm	3.96 gpm	3.50 gpm	3.17 gpm	2.82 gpm
25% propylene glycol	6.26 gpm	5.54 gpm	4.81 gpm	4.29 gpm	3.78 gpm
30% propylene glycol	8.09 gpm	7.10 gpm	6.11 gpm	5.42 gpm	4.71 gpm

(b)

1.25″ DR-11 HDPE	20 °F	25 °F	30 °F	35 °F	40 °F
Water	-	-	-	-	1.66 gpm
20% propylene glycol	5.59 gpm	5.00 gpm	4.41 gpm	4.00 gpm	3.56 gpm
25% propylene glycol	7.90 gpm	7.00 gpm	6.07 gpm	5.42 gpm	4.77 gpm
30% propylene glycol	10.21 gpm	8.96 gpm	7.71 gpm	6.84 gpm	5.95 gpm

(c)

1.5″ DR-11 HDPE	20 °F	25 °F	30 °F	35 °F	40 °F
Water	-	-	-	-	1.90 gpm
20% propylene glycol	6.40 gpm	5.72 gpm	5.05 gpm	4.58 gpm	4.08 gpm
25% propylene glycol	9.05 gpm	8.01 gpm	6.96 gpm	6.20 gpm	5.46 gpm
30% propylene glycol	11.69 gpm	10.26 gpm	8.82 gpm	7.83 gpm	6.81 gpm

(d)

2″ DR-11 HDPE	20 °F	25 °F	30 °F	35 °F	40 °F
Water	-	-	-	-	2.38 gpm
20% propylene glycol	7.99 gpm	7.15 gpm	6.31 gpm	5.72 gpm	5.10 gpm
25% propylene glycol	11.31 gpm	10.01 gpm	8.69 gpm	7.75 gpm	6.82 gpm
30% propylene glycol	14.61 gpm	12.82 gpm	11.03 gpm	9.78 gpm	8.51 gpm

(e)

Figure 14-58 | Minimum flow rates required for turbulent flow based on volume percent propylene glycol, pipe size, and fluid temperature.

Courtesy of Stage 3 Renewables Inc./www.stage3renewables.com/

(a)

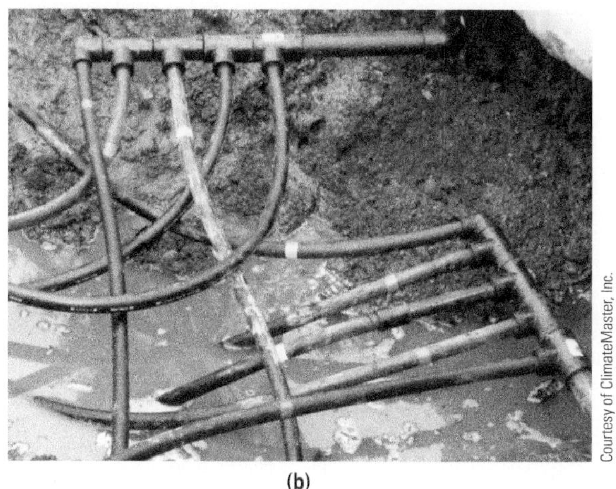
Courtesy of ClimateMaster, Inc.

(b)

Figure 14-59 | (a) Parallel earth loop circuits of HDPE tubing being joined by socket fusion. (b) A completed header with five parallel branch circuits. Note stepped pipe sizes on header.

Short supply header
Short return header
Cap
Cap
Parallel earth loop circuits

Figure 14-60 | Short headers, arranged for reverse return flow, do not require stepped pipe sizes.

2. The header piping is sized so that it creates equal head loss per unit of length over its entire length.
3. The headers are symmetrical arrange using reverse return piping.

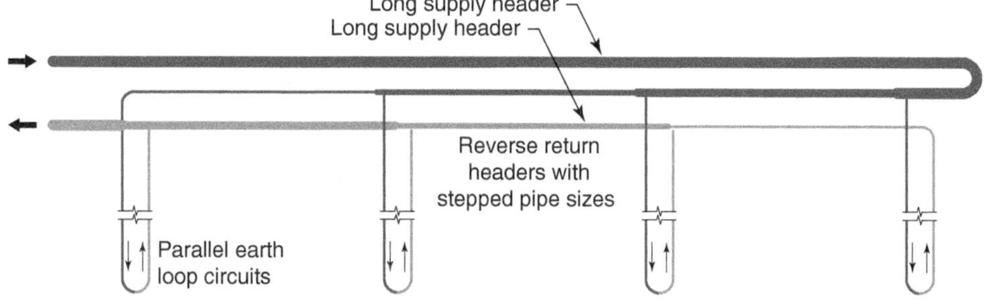

Figure 14-61 | Use of stepped piping sizes along a long header.

In practice, it is virtually impossible to achieve all of these conditions. For example, there may not be pipe sizes or fittings available that allow a long header to have exactly the same head loss per foot of length. In practice, the length of the parallel branches may also vary slightly. A variation of up to 5 percent in length (between the longest and shortest branch) is generally accepted common practice.

Example 14.10: Assume an earth loop has been sized to include three parallel branches of 1-inch DR-11 HDPE, each 1,000 feet long. The connection points for each branch are located 25 feet apart. The lowest anticipated operating temperature of the 20 percent solution of propylene glycol in the earth loop is 30 °F. 100 feet of piping is required from the start of the earth loop to the inlet of the supply header, and 50 feet of piping is required from the outlet of the return header to the end of the earth loop, as shown in Figure 14-60. The minimum flow required by the heat pump supplied by the earth loop is 7.5 gpm. Determine the following:

(a) The minimum flow rate for turbulent flow in each branch

(b) The total flow rate of all three parallel branches

(c) The sizes of the stepped headers

(d) The estimated head loss of the entire earth loop

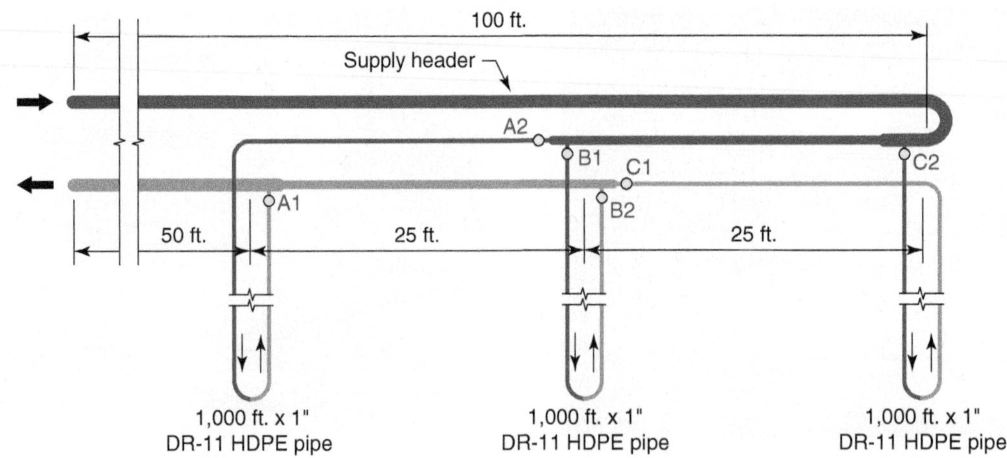

Figure 14-62 | Piping layout for Example 14.10.

Solution:

(a) The minimum flow rate for turbulent flow in the 1-inch DR-11 HDPE tube, with 20 percent propylene glycol at 30 °F is found in Figure 14-58. f_{min} = 2.79 gpm.

(b) The minimum flow required for turbulence is greater than the 2.5 gpm per branch (e.g., 7.5 gpm total flow rate) required by the heat pump. Thus, the total flow rate of the earth loop is constrained by the minimum flow rate required for turbulence and will be 3 × 2.79 = 8.37 gpm.

(c) The section of pipe between points A1 and B2 will carry 2 × 2.79 = 5.58 gpm. A 1.25-inch size DR-11 HDPE tube requires 4.41 gpm for turbulence while carrying 20 percent propylene glycol at 30 °F. Thus, if operating at 5.58 gpm, the flow would be turbulent. The head loss per 100 feet can be estimated using Equation 14.13, along with the glycol correction factor (c), and the (R) value for 1.25-inch piping:

$$H_{L/100'} = cR(f)^{1.25} = 1.31 \times 0.0316(5.58)^{1.25}$$
$$= 0.84 \text{ ft.}/100'$$

The piping from the start of the earth loop to point C2, and from point A1 to the end of the earth loop, must carry the total flow rate of 3 × 2.79 = 8.37 gpm. A 1.5-inch DR-11 HDPE pipe requires a minimum flow rate of 5.05 gpm of 20 percent propylene glycol at 30 °F for turbulence. Thus, at a flow rate of 8.37 gpm, the flow would be turbulent. The head loss per 100 feet can be estimated using Equation 14.13, along with the glycol correction factor (c) and (R) value for 1.5-inch piping:

$$H_{L/100'} = cR(f)^{1.25} = 1.31 \times 0.0168(8.37)^{1.25}$$
$$= 0.906 \text{ ft.}/100'$$

The head loss per 100 feet of pipe is quite close to 0.84 ft./100 feet for the section of header between points A1 and B2 as well as between B1 and C2. Thus, the objective of maintaining approximately equal head loss per unit of length along the reverse return header is reasonably attained with this choice of pipe sizes.

(d) Since the flow resistance is approximately the same through all three branches, the head loss of the overall earth loop can be determined by totaling the head loss along one path from the beginning of the earth loop to the end of the loop. The path the will be analyzed is shown in gray in Figure 14-63, along with the selected pipe sizes.

Head loss needs to be determined for the following segments along the gray path.

- 100 feet of 1.5-inch pipe at 8.37 gpm
- 25 feet of 1.25-inch pipe at 5.58 gpm
- 1,000 feet of 1-inch pipe at 2.79 gpm
- 50 feet of 1.5-inch pipe at 8.37 gpm

Head loss can be determined using Equation 14.13 along with the appropriate correction factor for 20 percent propylene glycol at 30 °F (c = 1.31), and the associated value of (R) based on pipe size. Also, because they operate at the same flow rate, the 50- and

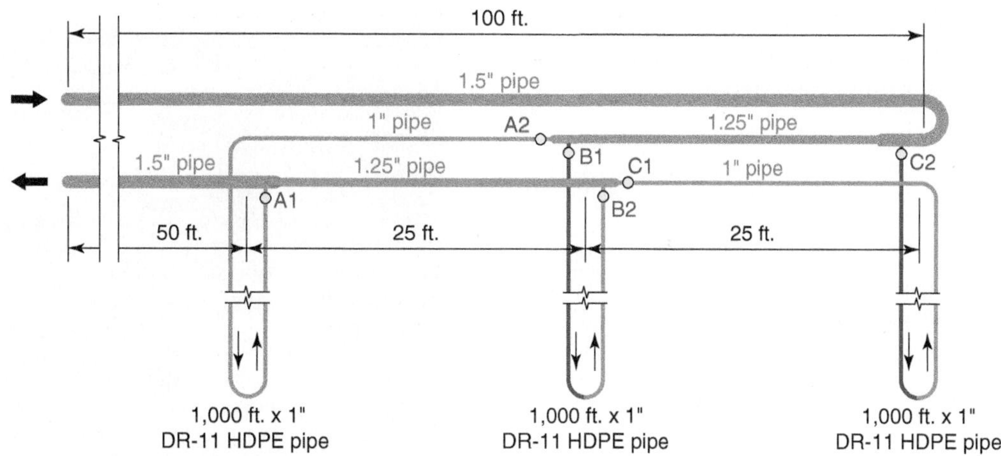

Figure 14-63 | The piping path to be analyzed for head loss is shown in gray.

100-foot lengths of 1.5-inch pipe can be combined into a single length of 150 feet.

The head loss of 150 feet of 1.5-inch pipe at 8.37 gpm is:

$$H_{L/100'} = cR(f)^{1.25} = 1.31 \times 0.0168(8.37)^{1.25}\left(\frac{150}{100}\right)$$

$$= 1.36 \text{ ft.}$$

The head loss of 1,000 feet of 1-inch pipe at 2.79 gpm is:

$$H_{L/100'} = cR(f)^{1.75} = 1.31 \times 0.0939(2.79)^{1.75}\left(\frac{1000}{100}\right)$$

$$= 7.41 \text{ ft.}$$

The head loss of the 25 feet of 1.25-inch pipe at 5.58 gpm is:

$$H_{L/100'} = cR(f)^{1.75} = 1.31 \times 0.0316(5.58)^{1.75}\left(\frac{25}{100}\right)$$

$$= 0.21 \text{ ft.}$$

The total head loss is thus 1.36 + 7.41 + 0.21 = 8.98 ft.

Discussion: The ratios shown in parentheses convert the head loss, as calculated in feet of head *per 100 feet of pipe*, to the head loss of the actual length of each piping segment. This analysis has determined that the earth loop configuration, using the selected pipe size, and operating at 8.37 gpm, has a corresponding head loss of 8.98 ft. This does not include the head loss of the water-to-refrigerant heat exchanger within the heat pump, or the head loss of any other piping components used to connect the heat pump to the earth loop.

Example 14.11: Assume that the heat pump used with the earth loop analyzed in Example 14.10 has the pressure drop versus flow rate curve shown in Figure 14-23. The interior piping between the earth loop and the heat pump has been determined to be equivalent to 50 feet of 1.25" DR-11 HDPE pipe. Determine the overall head loss of the complete heat pump/earth loop circuit.

Solution: The information given in Figure 14-23 gives pressure drop, rather than head loss, as a function of flow rate. At the previously determined earth loop flow rate of 8.37 gpm, the pressure drop across the heat pump is read from Figure 14-23 as 4.2 psi. This can be converted to head loss using Equation 3.7.

Equation 3.7 (repeated)

$$H = \frac{144(\Delta P)}{D}$$

where:

H = head added or lost from the liquid (feet of head)

ΔP = pressure change corresponding to the head added or lost (psi)

D = density of the liquid at its corresponding temperature (lb./ft.^{3})

The density of a 20 percent solution of propylene glycol at 30 °F is read from manufacturer's literature as:

$$D = 64.14 \text{ lb./ft.}^3$$

Thus, the head loss across the heat pump at a flow rate of 8.37 gpm is:

$$H = \frac{144(\Delta P)}{D} = \frac{144(4.2)}{64.14} = 9.43 \text{ ft.}$$

The head loss of the addition piping (e.g., 50 feet of 1.25-inch DR-11 HDPE tubing is found using Equation 14.13 along with the appropriate data.

$$H_{L/100'} = cR(f)^{1.75} = 1.31 \times 0.0316(8.37)^{1.75}\left(\frac{50}{100}\right)$$
$$= 0.853 \text{ ft.}$$

The total head loss of the circuit is the sum of the head loss across the earth loop, heat pump, and miscellaneous connecting piping:

$$H = 8.98 \text{ ft.} + 9.43 \text{ ft.} + 0.853 \text{ ft.} = 19.3 \text{ feet of head.}$$

Discussion: A circulator can now be selected that would provide at least 19.3 feet of head at a flow rate of 8.37 gpm. Note that the head loss across the heat exchanger within the heat pump is greater than the head loss of the earth loop. Although this is not always the case, the heat pump's heat exchanger will usually be a significant portion of the circuit's total head loss.

Courtesy of REHAU

Figure 14-64 | Less-than-ideal trench conditions.

MANIFOLD-BASED EARTH LOOPS

Many earth loops have multiple parallel branches that are fusion welded to a common header system, all of which are eventually buried. The required fusion joints are done in the excavated trenches. Although this approach has been successfully used on thousands of installations, it does, at times, have to be performed under less than ideal conditions, such as shown in Figure 14-62. Working with fusion welding equipment under cold, wet, and muddy conditions requires special care to ensure the joints being made are clean, dry, and properly heated.

An alternative method of constructing parallel earth loops involves bringing all circuits to a specially designed **geothermal manifold station**. This manifold station can be located within an enclosed space, such as a basement, or in an exterior vault.

In some respects, this approach is similar to routing multiple radiant panel heating circuits to a common manifold station. However, the manifold stations used for earth loops are usually larger in diameter to accommodate higher flow rates. Most of them are constructed of polymer materials since the operating temperature and pressure range of earth loops is well within what these materials can withstand. Figure 14-65 shows two examples of geothermal manifold stations.

The use of manifold-based earth loops provides several advantages relative to earth loops created using buried fusion joints:

- There is no need to fabricate a header within excavated trenches.
- It is possible to install several different earth loop configurations without need of fusion joints, and thus no need of fusion tools.
- This approach eliminates the need to use reverse return piping to help balance flows through all parallel branches.
- This approach eliminates the need to step pipe sizes along site-built extended headers. This precludes having to work with multiple sizes of earth loop piping and the need for associated tooling.
- The outgoing and return temperatures of the earth loop are easily monitored by thermometers on the manifolds.

When the manifold station includes *isolation and flow indicating/balancing valves,* additional benefits include:

- Each circuit of the earth loop can be independently flushed during filling and purging. This significantly

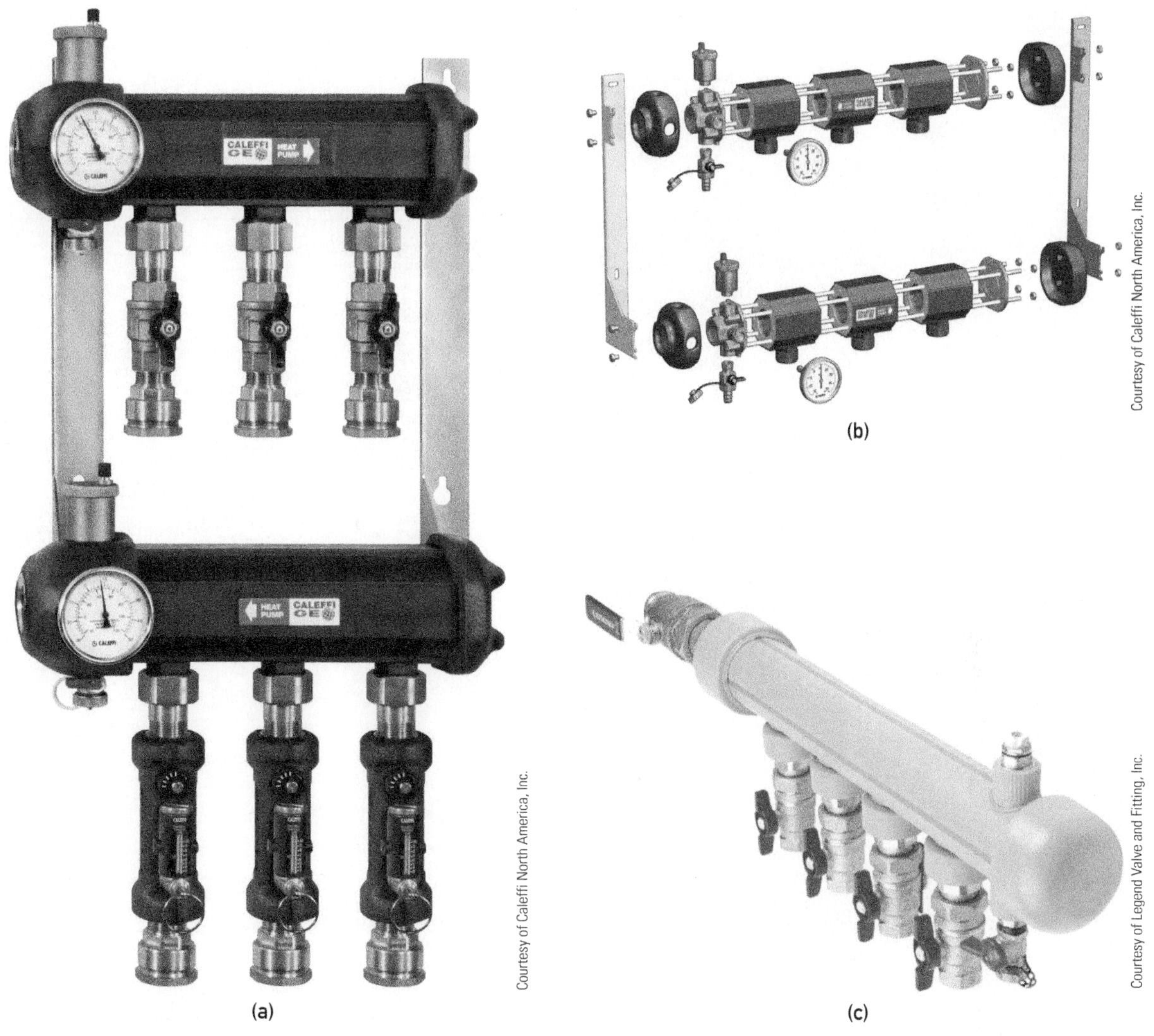

Figure 14-65 | (a) Example of a sectional geothermal manifold station. (b) Exploded view of a sectional geothermal manifold station. (c) Example of a polypropylene geothermal manifold station.

reduces the size of the flush pump required to purge the system of air.

- The flow rate through each earth loop circuit can be verified and adjusted if necessary.
- When necessary, circuits of different length and/or pipe size can be used and properly balanced at the manifold station.
- If a buried branch circuit ever fails due to future excavation, drilling, etc., it can be completely isolated from the remainder of the system at the manifold station.

Geothermal manifold stations can be located inside the building served by the heat pump system. Figure 14-66 shows examples of two such installations.

Manifold stations located inside buildings require two penetrations of the foundation wall for each earth loop circuit. These penetrations must be properly detailed to prevent entry of water or insects. One method for creating a watertight seal uses a **mechanically compressed collar** between the hole through the foundation wall and the outer surface of the pipe passing through it. An example of such a collar is shown in Figure 14-67.

Properly installed compression collars can resist water pressures up to 20 psi, which is far greater than what any foundation wall should be exposed to. Attempting to create a reliable seal by filling the space between the pipe and the hole in the foundation with mortar or caulk are usually futile, often resulting in water seepage.

Courtesy of REHAU

(a)

Courtesy of Harvey Youker

(b)

Figure 14-66 (a) Installed polypropylene geothermal manifold with isolation valves and flow meters on each circuit. (b) Sectional geothermal manifold during installation.

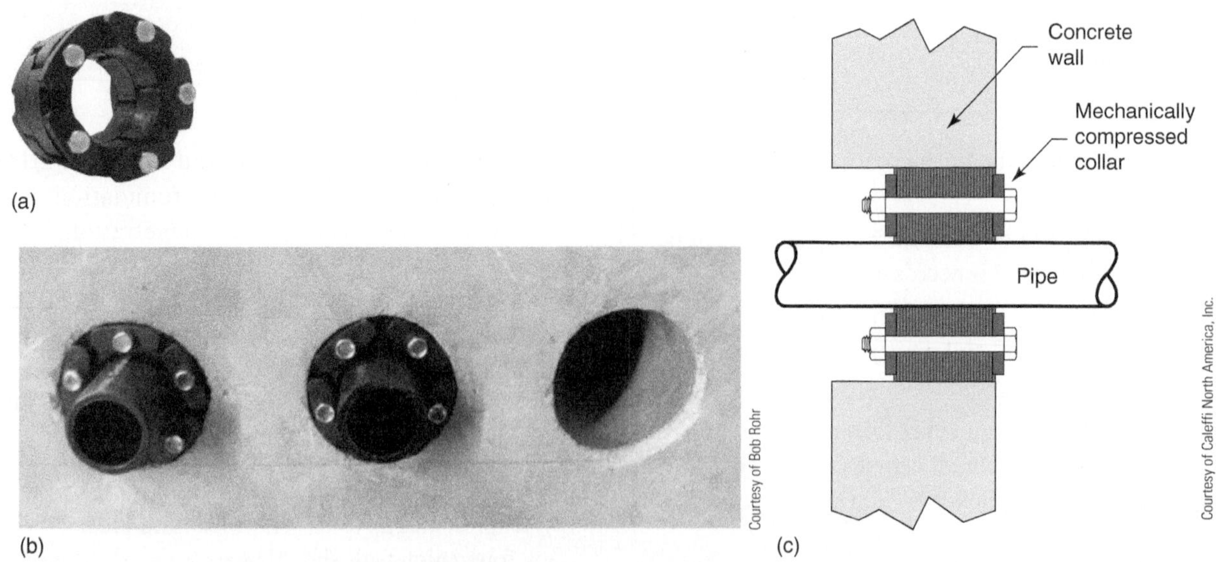

Figure 14-67 (a) Example of a mechanically compressed sealing collar. (b) Pipe installed in collar through foundation wall. (c) Cross section of compression collar in foundation wall.

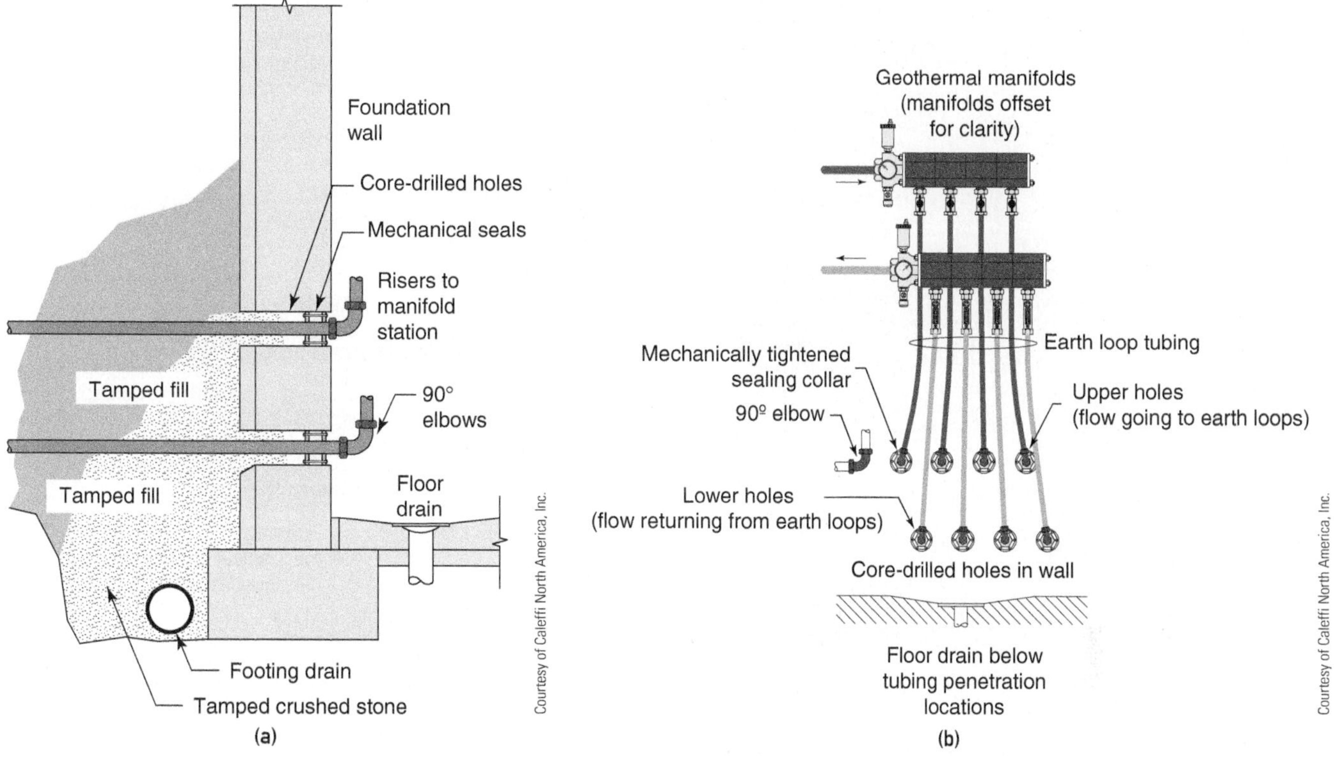

Figure 14-68 | (a) Cross section of foundation wall penetration for vertically spaced earth loop piping. Note use of 90° elbows to turn piping up to manifold. (b) Installation of manifold above piping penetrations.

In situations where all earth loop circuits come to the foundation wall at the same elevation, the holes through the foundation wall can also be at that level. The holes can be core-drilled through poured concrete or concrete blocks walls. They should be spaced far enough apart to prevent structural damage to the wall and its steel reinforcement. They must be large enough to allow compression collars to fit between the outside of the tube and the penetration hole. Manufacturers of compression collars provide tables for selecting the proper seal and penetration hole size for a given size of pipe.

In situations where the earth loop tubes are vertically stacked within the trench and come to the wall at different elevations, it may be more convenient to vertically and horizontally stagger the penetration holes, as shown in Figure 14-68.

The installation shown in Figure 14-68 requires 90° elbows to turn the pipe from its horizontal passage through the foundation wall to a vertical direction to meet with the manifold connection. If HDPE tubing is used for the earth loop, a suitable polymer elbow with compression type connections, as seen in Figure 14-69, can be used. If the earth loops are PEX-a tubing, a cold expansion brass elbow, as seen in Figure 14-66a, can be used. Some manufacturers also provide stab fittings rated for geothermal applications.

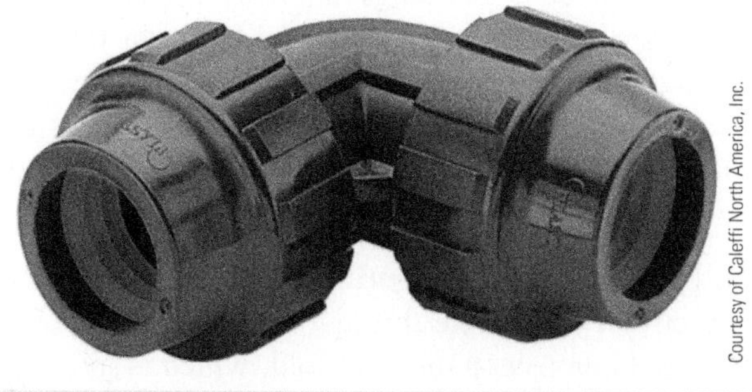

Figure 14-69 | Example of a compression elbow for HDPE pipe.

The exterior area around the piping penetrations should be backfilled with clean, fine, crushed stone to allow rapid drainage of any water that may migrate into that area. The crushed stone and soil in this area should be thoroughly tamped when placed to prevent any future settlement that could strain the piping where it penetrates the wall. A floor drain located under the manifold is convenient during commissioning and possible future maintenance.

Figure 14-70 | Use of a utility vault houses a geothermal manifold station.

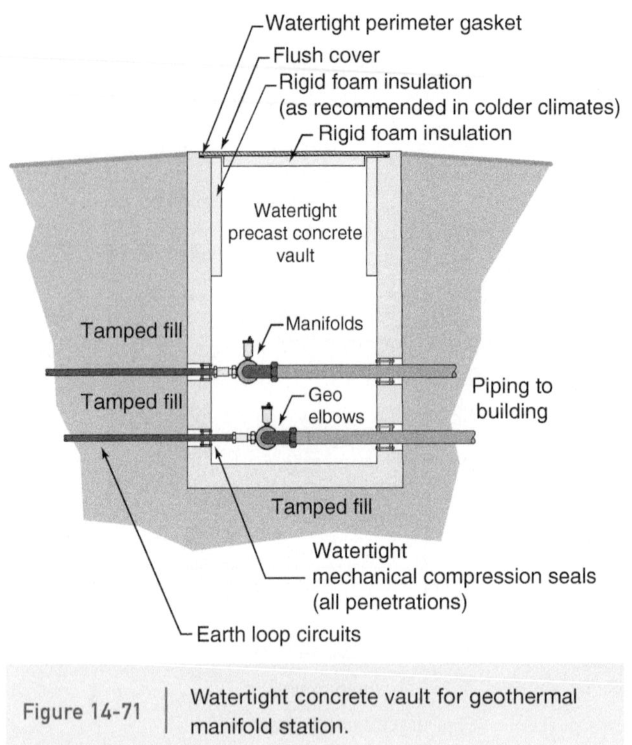

Figure 14-71 | Watertight concrete vault for geothermal manifold station.

It is also possible to mount geothermal manifold stations outside of buildings and route a single supply and return pipe from the manifold station into the building. In all cases, the manifold should be accessible for adjustment and maintenance. This requires the manifold to be mounted in some type of **vault enclosure**.

On option for such a vault is shown in Figure 14-70. It is based on the use of a **utility vault**. Individual earth loop circuits penetrate through the side of the utility vault and connect to the manifold, which is mounted to the inside wall of the vault. A set of larger diameter pipes connect the manifold station to the remained of the system inside the building.

The utility vault must be located at the proper elevation so that its top is flush with finish grade. This type of vault is generally not watertight. Water may enter around the perimeter of the cover as well as through piping penetrations. Because of this, it's important to provide drainage from the bottom of the vault to prevent water accumulation.

The use of this type of vault is also more suited to milder climates. It is not recommended in colder climates due to the proximity of the tubing and manifold to finish grade. Subfreezing temperatures could cause excessive heat loss from these components. In extreme winter climates, the fluid in the earth loop, even when protected with a 20 percent solution of propylene glycol, could freeze, especially when there is no flow through the earth loop.

Figure 14-71 shows an alternative vault structure. It is based on a precast concrete structure. All piping penetrations through the walls of this structure are sealed using the compression collar shown in Figure 14-67.

The vault structure is placed so that adjacent grade slopes away from it. It also has a cover plate with a full gasket to prevent rain from leaking in. The underside of the cover, as well as the upper portion of the vault walls, are clad with extruded polystyrene to minimize heat loss to cold ambient air. The vault is also deep enough to allow earth loop circuits to enter at their respective depths. The weight of the concrete structure prevents the possibility of uplift due to buoyancy created by water-saturated soils surrounding it.

Other products, such as the molded polyethylene manifold chamber shown in Figure 14-72, are available in some markets.

These watertight chambers are provided with stubs of HDPE tubing for both the parallel branch circuits and piping mains to the building. These stubs are fusion welded to the sides of the vault and are therefore watertight. After the vault has been placed, the stubs are

(a)

(b)

Figure 14-72 | (a) Example of a polyethylene vault in which a manifold station is located. Note piping stubs for branch and main circuit connections. (b) Valved manifold station supplied with vault enclosure.

fusion welded to the branch piping and mains using either socket fusion or electrofusion tools. The vault is placed so that its upper lid is flush with the finish grade. The cover rests on a gasket that provides a watertight seal.

CLOSED EARTH LOOPS

Traditionally, all earth loops were designed as closed, slightly pressurized hydronic circuits. When properly detailed, such loops can provide decades of operation with very little maintenance.

A **closed earth loop** is one in which the entire piping assembly is sealed from the atmosphere and designed to operate under slight pressure. Such an approach leverages the same principles used in all closed-loop hydronic systems. It also provides the following benefits:

- Virtually no loss of fluid over time due to evaporation
- Ability to provide higher static pressure at circulator inlet to protect against cavitation
- Ability to maintain fluid relatively free of dissolved air
- Reduced potential for corrosion of ferrous metal components

Figure 14-73 shows a *traditional* earth loop configuration for a single water-to-water heat pump.

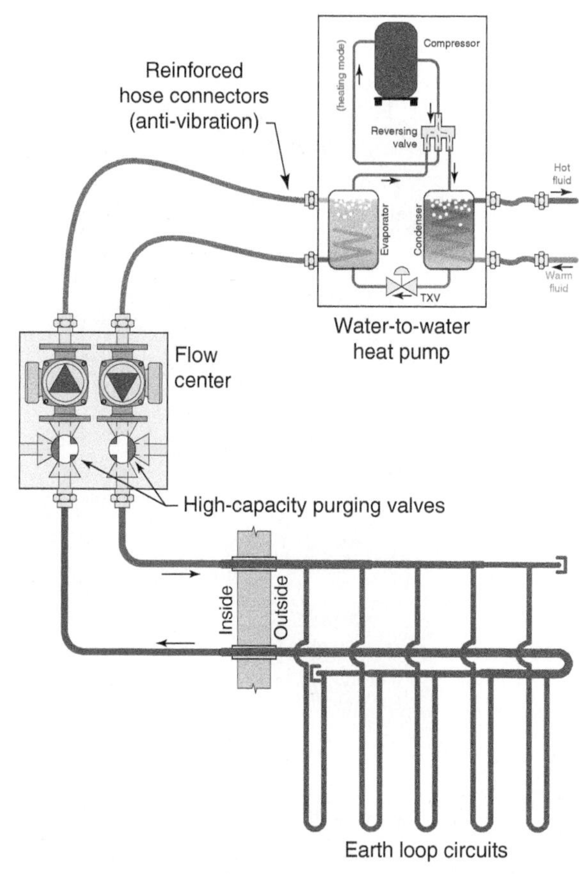

Figure 14-73 | A typical closed earth loop serving a single water-to-water heat pump.

The earth loop is fabricated using fusion welding of branch circuits into a reverse return header that is buried outside the building. Piping is run from the interior connections of the earth loop to a **flow center**. Figure 14-74 shows examples of two flow centers.

(a)

(b)

Figure 14-74 (a) Example of a flow center for a closed loop geothermal heat pump system that uses two standard circulators in series. (b) Example of a flow center that uses a single high-efficiency ECM-based circulator.

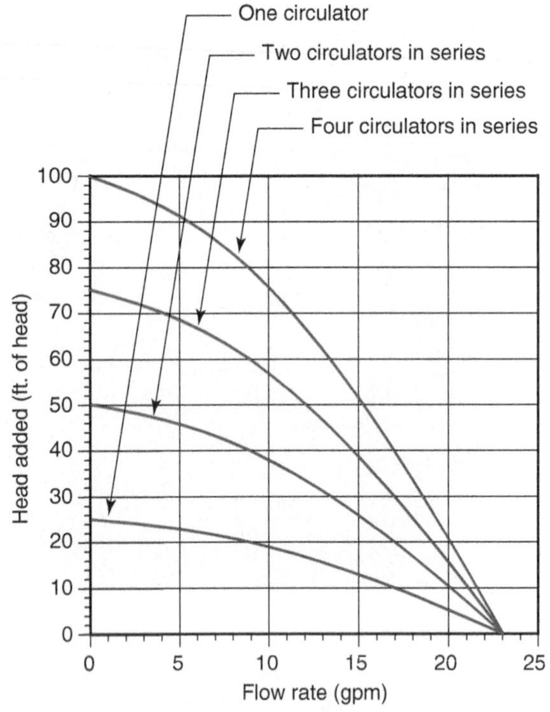

Figure 14-75 Effective pump curves when two or more circulators are combined in series.

A standard flow center contains from one to four circulators depending on its rated flow and head capacity. The circulators are arranged so they pump in series with each other. Two identical circulators in series produce twice the head at any given flow rate compared to that of a single circulator. Four identical circulators in series produce four times the head at any given flow rate compared to that of a single circulator. This effect is illustrated in Figure 14-75.

Series connected circulators provide an **effective pump curve** that is quite steep. Such a curve often provides a good match to the relatively high head requirements of residential earth loops.

Recently introduced flow centers equipped with ECM-based circulators can be operated by controllers that measure temperature differential across the earth loop connections at the heat pump. The speed of the circulator is regulated to maintain a user-set temperature differential. These flow centers are especially well suited to 2-stage heat pumps. They can increase flow rate when both stages are operating, while reducing circulator power demand when only one stage is operating. Variable speed ECM earth loop circulators will also become more common in combination with modulating heat pumps having variable speed compressors. The higher wire-to-water efficiency of the ECM-based circulator, compared to a standard circulator, significantly reduces the power required to maintain adequate earth loop flow. This is important for achieving high *system* COP.

Many flow centers also contain special high flow capacity 3-way valves that are used to fill and purge both the earth loop, and the heat pump circuit. These valves are operated using an Allen key. Figure 14-76 shows typical positioning for various fill and purge tasks.

One setting position of the 3-way valves allows the earth loop circuit to be isolated, filled, and purged. Another allows the heat pump portion of the circuit to be filled and purged. Another allows simultaneous flow through both the earth loop and heat pump portions of the circuit. This is typically the last setting of the valves during the purging process. This final purging configuration is often assisted by turning on the circulators in the flow center to provide additional flow, which helps dislodge and purge air bubbles in the circuit. When all purging is completed, the valves are returned to the setting labeled "normal operation" in Figure 14-76. The flow necessary to fill and purge the earth loop circuit is provided by a flush cart. This procedure and the necessary hardware will be discussed later in this chapter.

The flow center is typically connected to the heat pump using flexible reinforced hoses. This allows for fast installation. It also provides vibration and noise dampening between the heat pump and the flow center.

PREFERRED CONFIGURATION OF CLOSED EARTH LOOPS

Figure 14-77 shows the author's preferred configuration for a closed earth loop circuit serving a single heat pump, based on currently available technology.

This system uses a manifold-type earth loop. The manifold station could be located either inside the building or in an exterior vault. The manifold is equipped with isolation valves at the beginning and end of each branch circuit. This allows the branch circuits to be filled and flushed one at a time, which is a significant advantage that will be discussed in the next section. Isolation valves also allow each circuit to be completely isolated in the unlikely event of a leak or a crushed tube. This is not an option when the branches are directly fused or otherwise connected to valveless headers.

Flow from the earth loop passes through a high efficiency air and dirt separator. This device continuously scavenges and ejects dissolved air in the circuit. After multiple passes, it can also capture particulates in the flow stream down to 5-micron particle size. The combined air and dirt separator does not replace the need to flush the system at startup. Instead, it provides

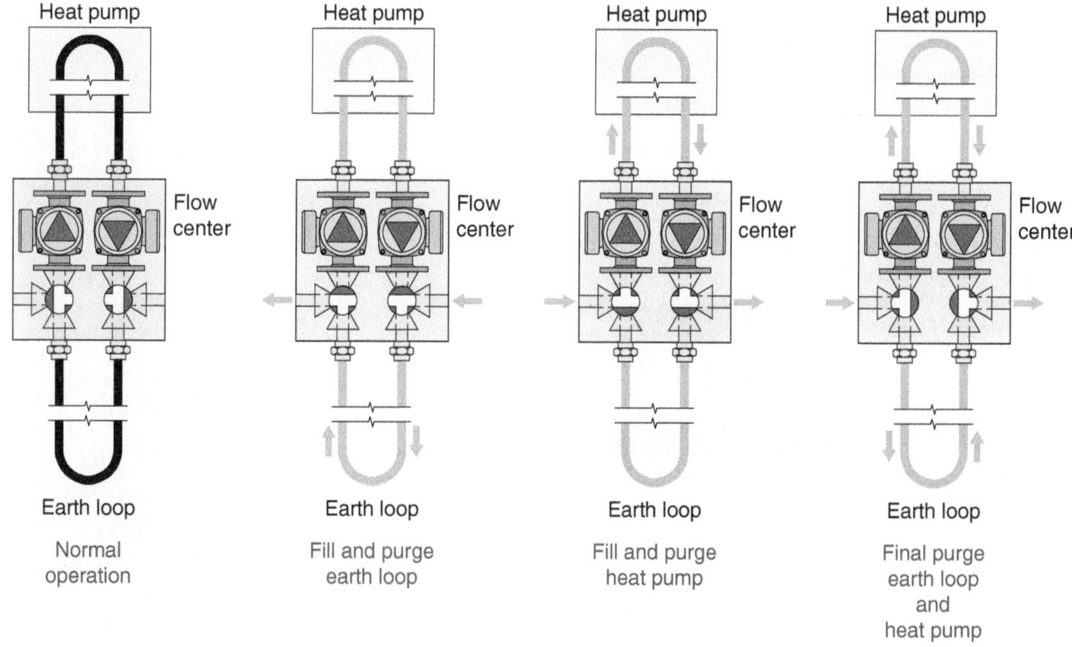

Figure 14-76 | Operation of 3-way valves for filling and purging earth loop and piping through heat pump.

a permanent means of keeping the system's fluid free of air and dirt. This reduces the potential for fouling of heat exchanger surfaces within the heat pump. It also enhances the life of other components in the circuit.

The earth loop circuit also contains a diaphragm-type expansion tank. This component significantly reduces pressure variations in the circuit as the fluid and pipe change temperature. Wide pressure fluctuations increase the potential for leaks. Under the right conditions, they can also cause circulator cavitation. As is true with air separators, expansion tanks are recognized as a fundamental component in nearly all modern closed-loop hydronic systems. An earth loop should not be an exception. A pressure gauge is also shown, along with an isolation valve for the expansion tank. The latter allows easy replacement of the tank if ever necessary.

A single ECM-based circulator is used in the circuit. There are now several manufacturers offering various models of such circulators, with pump curves suitable to geothermal heat pump systems. At full speed, an ECM circulator has about half the electrical power demand of a PSC (permanent split capacitor) wet-rotor circulator. Using an ECM-based circulator significantly reduces the parasitic power use of the system. Some models of ECM circulators are also equipped with control interfaces that allow the possibility of intelligent variable flow-rate control within the earth loop depending on operating conditions, which would lead to further savings. Variable flow rate control will be increasingly used in future geothermal heat pump applications.

The earth loop circuit also contains a flow meter, in this case integrated along with an isolation valve. The flow meter, in combination with the two thermometers seen on the manifolds, allows technicians to estimate the rate of heat delivery from the earth loop using the sensible heat rate equation (e.g., Equation 3.2).

OPEN EARTH LOOPS

An open earth loop is any earth loop that is not sealed from the atmosphere. In recent years, several manufacturers have developed products called standing column flow centers that replace the type of flow center previously described for a pressurized closed earth loop. An example of a standing column flow center is shown in Figure 14-78.

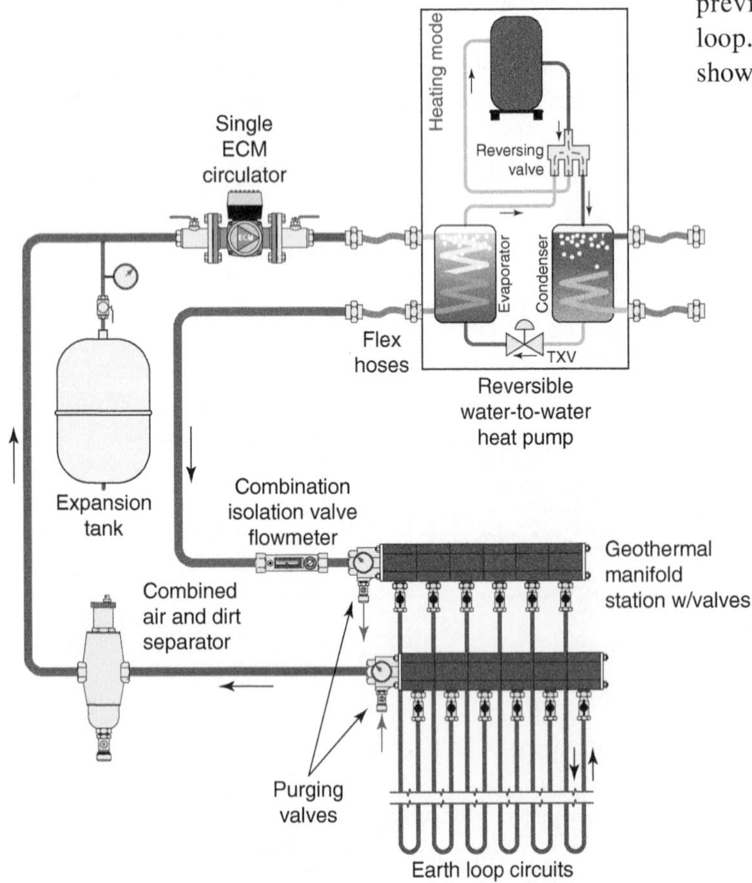

Figure 14-77 | Author's preferred configuration for a closed earth loop circuit configuration.

Courtesy of B&D Mfg, Inc.

Figure 14-78 | Example of an unpressurized flow center.

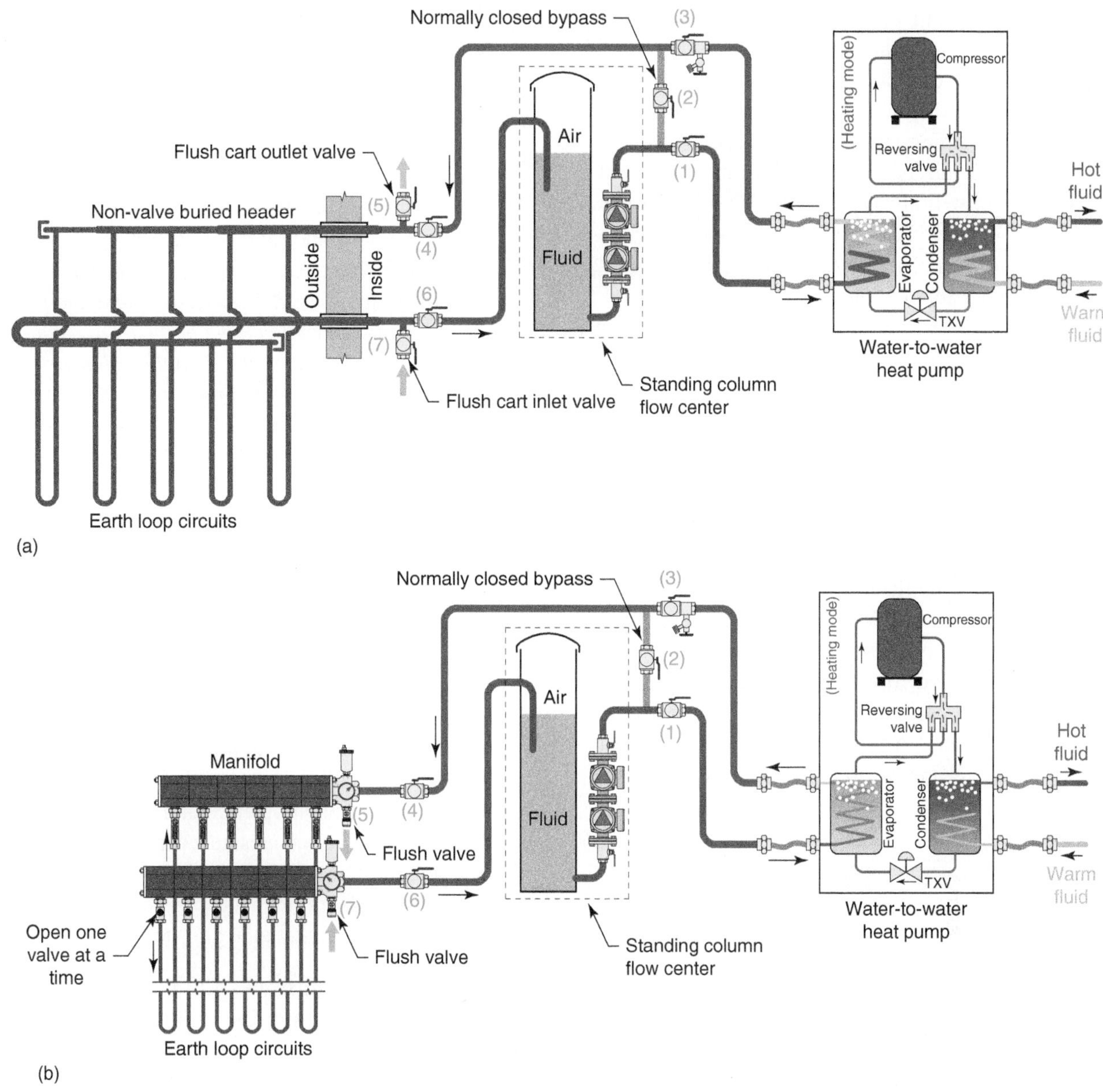

Figure 14-79 (a) Typical installation of an unpressurized flow center when used with non-valved earth loop headers. (b) Typical installation when using a valved earth loop manifold.

Figure 14-79 shows how a standing column flow center would be installed in a residential or light commercial geothermal heat pump system.

A standing column flow center has a vertical chamber called a **canister**. It is usually made of a polymer or of stainless steel. When the system is commissioned, the canister is partially filled with earth loop fluid. The space within the canister that is above the fluid level serves as an expansion volume. A vented lid prevents any positive air pressure or vacuum from developing inside the canister. The earth loop circulators draw fluid from the bottom of the canister; thus, the circulator's inlet remains flooded and under a *slight* positive pressure. As with closed-loop flow centers, standing column flow centers are available with one to four circulators depending on the flow rate and head requirement to be met.

The flow center's canister performs several functions. It allows fluid to be quickly poured into the system as the earth loop is filled. It also allows air bubbles to rise to the surface of the standing fluid column and thus helps deaerate the system's fluid. It also provides a space to accommodate the expansion and contraction volume of the earth loop fluid as its temperature changes.

Designers should keep in mind that *any* open hydronic circuit allows the fluid to remain in contact with the atmosphere on a continuous basis. As such, the water in the earth loop circuit will be constantly releasing and absorbing molecules of oxygen, nitrogen, and other gases in air as its temperature increases and decreases. There will always be some dissolved air in the fluid. The dissolved oxygen molecules in the water are the main concern. They will cause oxidation (e.g., corrosion) of any ferrous metals in the system, such as the volutes of cast-iron circulators. *Most circulator manufacturers specifically disclaim warranty coverage on cast-iron circulators used in any type of open-loop system.* Only circulators with bronze or stainless steel volutes are typically rated, by their manufacturer, for use in open-loop systems.

The presence of oxygen in the earth loop fluid can also cause degradation of glycol-based antifreeze fluids as well as allow growth of certain aerobic bacteria. There are additives called **oxygen scavengers** that can be used to limit oxidation. Other additives called **biocides** can be used to control biological growth. Such additives do compensate for these undesirable conditions; however, to remain effective, they must be maintained at proper levels. This requires periodic testing of the fluid and subsequent adjustment of additive levels.

There will also be some evaporation of system fluid over time. Although additional fluid can be easily added to the canister, the evaporation of alcohol-based antifreeze chemicals (such as methanol or ethanol) may be a concern within occupied spaces. Designers should carefully consider these issues.

EXPANSION COMPENSATION IN CLOSED EARTH LOOPS

There are varied opinions regarding the use of expansion tanks within closed geothermal earth loops. Some say the tank is not needed because the HDPE tubing will itself expand and contract to absorb the expansion and contraction of the earth loop fluid. Thousands of earth loops have been installed without expansion tanks, primarily in North America. Some have operated acceptably, while others have experienced pressure variations that have caused problems.

Most European geothermal heat pump installations include an expansion tank as a standard part of the earth loop. The reasoning is that such a tank reduces pressure fluctuations and averts some of the problems associated with "tankless loops."

It is the author's opinion that an expansion tank *should* be used in every closed earth loop circuit. The following observations support this recommendation.

1. One characteristic of earth loops is that the fluid within them expands and contracts at a different rate than the volumetric expansion and contraction of HDPE piping. *As the earth loop fluid and piping cool, the internal volume of the HDPE piping decreases at a greater rate than the volumetric contraction of the fluid within the piping.* This causes a pronounced increase in loop pressure in systems not equipped with expansion tanks. This pressure rise increases the potential for fluid leaks at piping connections.
2. When the heat pump is operating in cooling mode, the temperature of the earth loop fluid and piping increase. *Under such conditions, the internal volume of the piping increases at a greater rate than the volumetric expansion of the fluid.* This causes a drop in earth loop pressure. In systems without an expansion tank, the fluid pressure at the inlet of the earth loop circulator can drop low enough to cause cavitation. A cavitating circulator will not produce proper flow rates, creates unacceptable noise levels, and will eventually fail due to erosion of the impeller or volute. Installing an expansion tank in the earth loop is one way to avoid such problems.

Figure 14-80a illustrates fluid being pushed into the expansion tank from the contracting earth loop when the heat pump is operating in heating mode (e.g., earth loop temperature is decreasing).

Figure 14-80b illustrates fluid flowing out of the expansion tank into an expanding earth loop when the heat pump is operating in the cooling mode (e.g., earth loop temperature is increasing). The minimum and maximum earth loop temperatures indicated are typical.

Equation 14.15 can be used to determine the volume of fluid *drawn out* of the expansion tank when the earth loop temperature is increasing during cooling mode operation:

(Equation 14.15)

$$V_{\text{out}} = V_{\text{loop}}\left[(\alpha(\Delta T) + 1)^3 - \left(\frac{D_{\text{fill}}}{D_{\text{high}}}\right)\right]$$

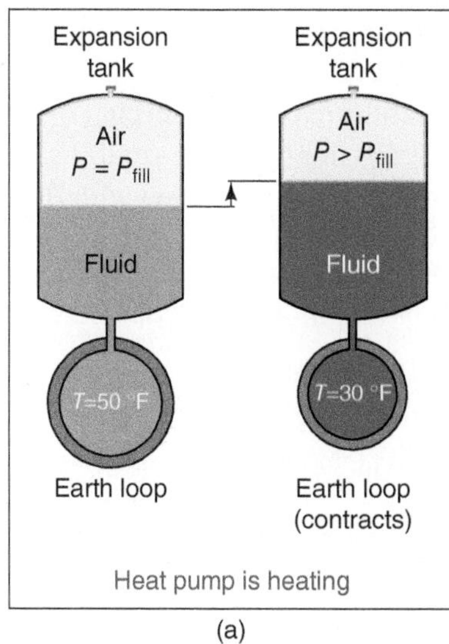

(a)

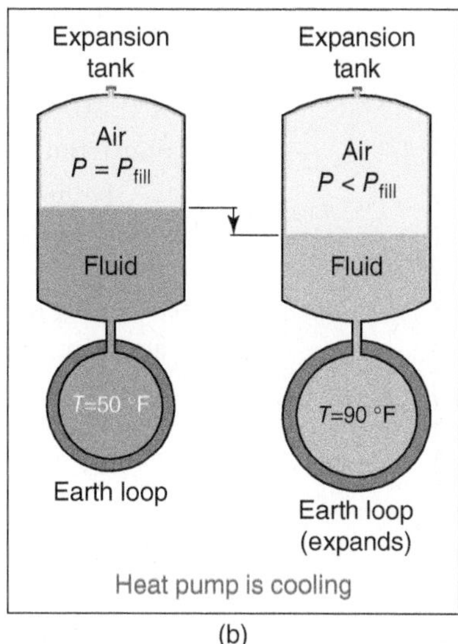

(b)

Figure 14-80 (a) Fluid being pushed into the expansion tank during heat mode operation, causing an increase in circuit pressure. (b) Fluid leaving the expansion tank during cooling mode operation, causing a drop in circuit pressure.

where:

V_{out} = volume of fluid flowing out of expansion tank (gallons)

V_{loop} = total volume of earth loop when filled and purged (gallons)

α = coefficient of linear expansion of earth loop piping (in/in/°F)

ΔT = absolute value (e.g., always a positive number) of temperature change of loop (°F)

D_{fill} = density of fluid when loop is filled and purged (lb./ft.^{3})

D_{high} = density of fluid when loop is at *maximum* temperature (lb./ft.^{3})

Equation 14.16 can be used to determine the volume of fluid *added* to the expansion tank when the earth loop temperature is decreasing during heating mode operation:

(Equation 14.16)

$$V_{in} = V_{loop}\left[\left(\frac{D_{fill}}{D_{low}}\right) - (1 - \alpha(\Delta T))^3\right]$$

where:

V_{in} = volume of fluid flowing *into* of expansion tank (gallons)

V_{loop} = total volume of earth loop when filled and purged (gallons)

α = coefficient of linear expansion of earth loop piping (in./in./°F)

ΔT = absolute value (e.g., always a positive number) of temperature change of loop (°F)

D_{fill} = density of fluid when loop is filled and purged (lb./ft.^{3})

D_{low} = density of fluid when loop is at *minimum* temperature (lb./ft.^{3})

Example 14.12: An earth loop contains the following tubing:

5,000 feet of 1-inch DR-11 HDPE tubing

200 feet of 1.5-inch DR-11 HDPE tubing

The earth loop operates with a 20 percent solution of propylene glycol. The system is filled with this solution when the earth loop piping and fluid are both at 50 °F. Determine the amount of fluid:

a. Extracted from the expansion tank when the loop temperature increases to 90 °F.
b. Added to the expansion tank when the loop temperature drops to 30 °F.

Solution: The total volume of the earth loop is found using data from Figure 14-44:

$$5000\text{ ft.}\left(\frac{4.72\text{ gal}}{100\text{ ft.}}\right) + 200\text{ ft.}\left(\frac{9.86\text{ gal}}{100\text{ ft.}}\right) = 256\text{ gallons}$$

The density of the 20 percent propylene glycol fluid is found at the required temperatures based on the manufacturer's specifications. It can also be referenced in Figure 14-81.

For 20 percent propylene glycol at 90 °F: $D = 63.37$ lb./ft.^{3}

For 20 percent propylene glycol at 50 °F: $D = 63.92$ lb./ft.^{3}

For 20 percent propylene glycol at 30 °F: $D = 64.14$ lb./ft.^{3}

The **coefficient of linear expansion** for HDPE tubing is 0.0001 in./in./°F.

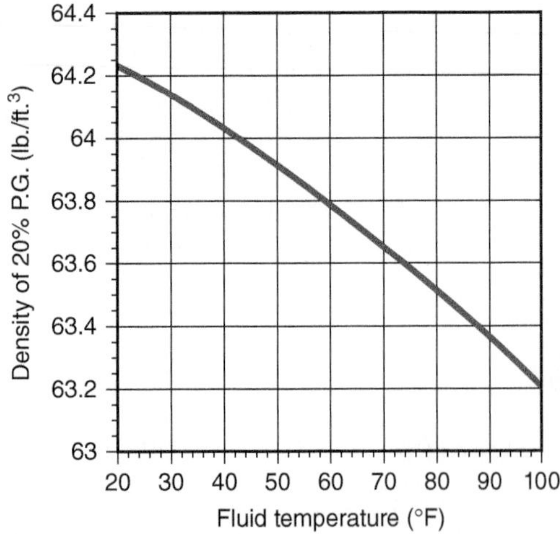

Figure 14-81 | Density of 20 percent solution of propylene glycol versus temperature.

The amount of fluid removed from the expansion tank when the loop is raised from 50 °F to 90 °F would be:

$$V_{out} = V_{loop}\left[(\alpha(\Delta T) + 1)^3 - \left(\frac{D_{fill}}{D_{high}}\right)\right]$$

$$= 256\left[(0.0001(40) + 1)^3 - \left(\frac{63.92}{63.37}\right)\right]$$

$$= 0.86 \text{ gallons}$$

The amount of fluid added to the expansion tank when the loop is lowered from 50 °F to 30 °F would be:

$$V_{in} = V_{loop}\left[\left(\frac{D_{fill}}{D_{low}}\right) - (1 - \alpha(\Delta T))^3\right]$$

$$= 256\left[\left(\frac{63.92}{64.14}\right) - (1 - 0.0001(20))^3\right]$$

$$= 0.66 \text{ gallons}$$

Discussion: These values are the *net* volume changes resulting from the combination of expansion and contraction of the fluid as well as that of the earth loop tubing.

SIZING AN EXPANSION TANK FOR A CLOSED EARTH LOOP

Once the net change in fluid volume is determined, an expansion tank can be selected that accommodates this volume change while undergoing an acceptable change in air-side pressure.

Example 14.13: Determine the minimum expansion tank volume required to accommodate the fluid extraction of 0.86 gallons in the previous example, while dropping from an initial air-side pressure of 20 psi to 10 psi. Assume the tank is initially half-filled with fluid as the loop begins heating up.

Solution: **Boyle's law** is set up and solved based on these assumptions:

$$V_{tank} = \frac{2(V_{out})}{\left(\frac{P_i + 14.7}{P_f + 14.7}\right) - 1}$$

where:

V_{tank} = required volume of expansion tank (gallons)

V_{out} = volume of fluid leaving tank as loop heats to max operating temperature (from Equation 14.15) (gallons)

P_i = initial air-side pressure in tank (when loop is charged) (psig)

P_f = final air-side pressure in tank (when loop is at max. temperature) (psig)

$$V_{tank} = \frac{2(V_{out})}{\left(\frac{P_i + 14.7}{P_f + 14.7}\right) - 1} = \frac{2(0.86)}{\left(\frac{20 + 14.7}{10 + 14.7}\right) - 1}$$

$$= 4.25 \text{ gallons}$$

Discussion: An expansion tank of 4.25 gallons total shell volume initially at 20 psi air-side pressure, and half-filled with fluid, will allow the system loop pressure to drop from 20 to 10 psi as the earth loop warms to its maximum operating temperature. This is a reasonable change in pressure that should not adversely affect the operation of a properly installed circulator (e.g., circulator positioned so that it pumps away from the expansion tank connection point). Selecting an expansion tank slightly larger than the calculated 4.25 gallons will further reduce the pressure change in the earth loop.

Note: When setting up this tank, the air-side pressure should be adjusted to 10 psig. When the loop is charged to 20 psi, the air volume in the tank will be compressed to half its original volume and thus be at 20 psi. The tank will be half-filled with fluid under this condition. The extra fluid volume in the expansion tank is desirable in systems with air separators. It provides some "reserve" fluid that eventually fills the void left by air being separated and ejected from the earth loop circuit.

Example 14.14: Assume that a designer selected a 6-gallon expansion tank for the system described in Example 14.13. The tank has an initial air charge of 10 psi, but is half-filled with fluid when the loop is initially charged to 20 psi. Determine the maximum pressure in the air side of the tank when the earth loop drops to its lowest temperature of 30 °F.

Solution: Again, Boyle's law yields the following:

$$P_f = \frac{(P_i + 14.7)}{\left(1 - \frac{2V_{in}}{V_{tank}}\right)} - 14.7$$

where:

P_f = final air-side pressure in tank (when loop is at max. temperature) (psig)

P_i = initial air-side pressure in tank (when loop is charged) (psig)

V_{tank} = selected volume of expansion tank (gallons)

V_{in} = volume of fluid entering tank as loop drops to min. operating temperature (gallons)

$$P_f = \frac{(P_i + 14.7)}{\left(1 - \frac{2V_{in}}{V_{tank}}\right)} - 14.7 = \frac{(20 + 14.7)}{\left(1 - \frac{2(0.66)}{6}\right)} - 14.7$$

$$= 29.8 \text{ psi}$$

Discussion: The pressure rise from 20 to almost 30 psi should not cause any operational problems.

In systems where the earth loop is warmed during the cooling season and cooled during the heating season, the net volume change of the expansion tank should be evaluated in both modes. The expansion tank can then be sized based on the more constraining condition. Typically, this will be the acceptable drop in loop pressure when the heat pump is operating in cooling mode.

Designers can use the equations presented to experiment with different pressure variations, fluid properties, and loop volumes to see the effect on expansion tank sizing.

EARTH LOOP FLUIDS

Because the evaporators in most geothermal heat pumps can operate at refrigerant temperatures well below freezing, it is common to use an antifreeze solution rather than just water in earth loops. Several types of antifreeze solutions based on salts, glycols, and alcohol additives have been used in geothermal systems. Each of these solutions has strengths and limitations.

Salt-based solutions of **calcium chloride** and **potassium acetate** have been used in some earlier generation geothermal heat pump systems. They were referred to as **brine** solutions. While offering acceptable environmental characteristics, salt-based solutions often prove corrosive to metal components, including cast iron and copper. These solutions have also shown a propensity to leak through certain pipe joints due to their low surface tensions. Currently, salt-based solutions are not widely used nor recommended in geothermal heat pump applications.

Alcohol-based fluids include diluted solutions of **methanol** and **ethanol**. Methanol, although good from the standpoint of relatively high specific heat, good freezing point depression, and low viscosity has the negative of high oral toxicity. It is also considered a flammable substance, even in 20 percent methanol/80 percent water concentrations. Because of this, some municipalities have specifically banned its use in geothermal earth loops. Methanol has also proven to cause corrosion in open-loop circuits.

Ethanol solutions as low as 20 percent concentration are also considered flammable liquids according to NFPA standard 325. Any ethanol used for antifreeze purposes is **denatured** (e.g., rendered undrinkable through additives, some of which may be toxic). Premixed solutions of ethanol and deionized water are commercially available for geothermal applications in North America. Installers should follow all information provided by suppliers regarding handling, storage, and disposal of such fluids.

Although both types of alcohol-based solutions have been successfully used in geothermal heat pump systems, it is imperative to check local ordinances or OSHA regulations that may constrain or restrict their use. Safety regulations may require a separation of at least 10 feet between the alcohol-based fluid and any potential ignition source. They may also require that any open container containing an alcohol-based fluid, such as a flushing cart, remain outside the building. Other precautions include electrical bonding and grounding of containers during fluid transfer to prevent the possibility of arcs due to static electricity.

It is also imperative that any air-venting equipment in piping containing alcohol-based solutions be equipped with vent discharge piping that can carry

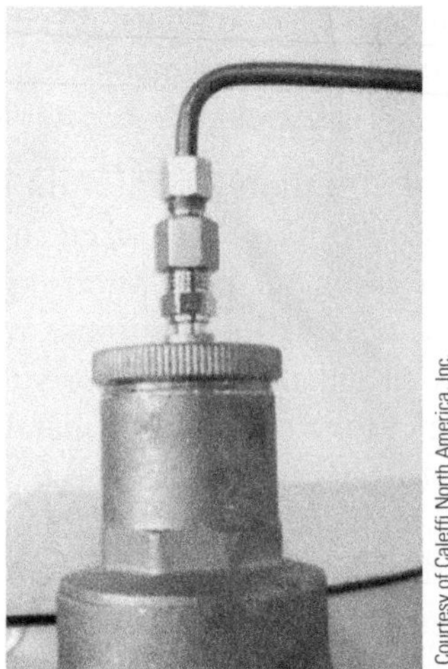

Figure 14-82 | Example of an adapter fitting that allows a copper tube to be connected to the top of an air separator.

any vapors outside the building and discharge them to open air away from any electrical equipment or other potential sources of ignition. Special fittings are available from manufacturers of air vents and air separating devices for this purpose. An example of such a fitting is shown at the top of the air separator in Figure 14-82.

The most widely used antifreeze in geothermal heat pump applications is a water-based solution of inhibited propylene glycol. Concentrations of 20 percent are common. Propylene glycol is not toxic. Commercially available propylene glycol sold for use in HVAC systems contains small amounts of other chemicals called **inhibitors**. These chemicals make the solution less acidic, discourage biological growth, and minimize corrosion potential. Because it is nonflammable and nontoxic, it is acceptable to allow air-venting devices in systems containing propylene glycol to discharge directly into mechanical rooms.

14.9 Earth Loop Sizing

The amount of tubing required for a given earth loop depends on several variables, including:

- The heat pump's capacity compared to the building's design heating and cooling load
- The minimum allowed temperature of the earth loop fluid during the heating season
- The maximum allowed temperature of the earth loop fluid during the cooling season
- The arrangement of the tubing within trenches or boreholes
- The diameter and wall thickness of the tubing
- The average depth of the tubing
- The thermal conductivity, thermal diffusivity, density, and moisture content of the soil

Some general observations about earth loop length are as follows:

- Horizontal earth loops often require a much larger land area than vertical earth loops.
- Wet and dense soils are preferable to dry and light soils. Water-saturated soils allow for good thermal diffusion and thus tend to reduce the amount of buried tubing.
- Horizontal earth configurations with greater average piping depth tend to require less tubing.
- Multiple pipes placed close to each other are not as effective at gathering surrounding heat as single tubes placed several feet apart. However, multiple pipes within a single trench can significantly reduce the amount of excavation required and often provide a more economical alternative relative to single-tube trenches.
- Many horizontal earth loops in cold climates operate with an antifreeze solution that remains free of ice crystals at temperatures of 15–20 °F. A common minimum "design" earth loop temperature is 30 °F. Increasing this temperature 3 °F to 5 °F will significantly increase the amount of tubing required in the earth loop. Decreasing this temperature reduces the amount of tubing required, but at the expense of reduced heat pump performance.
- Horizontal earth loops will experience greater temperature variation between fall and spring, compared to vertical earth loops. This allows the heat pump to achieve relatively high heating capacity and COPs in fall. However, both of these performance indices decrease as winter progresses. Heat pumps supplied from horizontal earth loops tend to be at their minimum heating capacity and COP in late winter or early spring.
- Because of relatively minor variations in deep soil temperature, heat pumps supplied from vertical earth loops will have relatively consistent heating capacity and COP over the entire winter.

- Most earth loops should be designed to ensure that flow through them remains turbulent at their minimum fluid temperature. Turbulent flow provides better heat transfer between the tube wall and loop fluid.

The analytical methods used to accurately size an earth loop are quite involved. This section presents the relevant equations, data, and procedures required for selected types of horizontal earth loops. All designers planning to size an earth loop should be familiar with this process. They should also be capable of performing the manual sizing procedure given in this section. That said, most geothermal heat pump system designers now rely on software to process the necessary mathematics and provide quick answers to the "what if" questions such as, What if the earth loop was installed 1 foot lower than originally planned?

After the manual sizing procedure is presented, examples of software for performing equivalent sizing will be discussed.

Designers should remember that both the manual calculation procedure and sizing software contain inherent assumptions. As such, they should be used as guidelines, rather than a method that produces a precise earth loop length that must be adhered to if the system is to operate correctly. Designers should also consider that a given geothermal heat pump system only gets "one shot" at an earth loop installation. It's unlikely, due to cost and logistics, that an undersized earth loop will ever be modified by adding more piping if it fails to perform as expected.

SIZING PROCEDURE FOR HORIZONTAL EARTH LOOPS (HEATING MODE OPERATION)

STEP 1: Determine the design heating load of the building to be served by the water-to-water heat pump. The design heating load should be based on methods presented in Chapter 2. For residential applications, the load will be based on the desired indoor air temperature, (which is typically in the range of 68 °F to 72 °F based on comfort preferences), and the 97.5 percent outdoor design dry bulb temperature.

STEP 2: Decide what portion of the design heating load the water-to-water heat pump will be providing. In the past, it was common to select *air-to-air* heat pumps with a heating capacity of about 75 percent of design heating load. The intent was to prevent excessive cooling capacity, which would cause the heat pump to short cycle during cooling mode operation. This old "rule of thumb" does not apply to properly designed hydronic systems using water-to-water (water-to-water) heat pumps.

It is possible to select a water-to-water heat pump to supply the full design heating load of a building. However, there may also be circumstances in which it is not necessary or desirable to do so. For examples, some designers may choose a **dual-fuel approach** in which the water-to-water heat pump supplies heat to the building under certain conditions, but is then supplemented by a different heat source when those conditions are not present. An example would be selecting a heat pump that can handle the full heating load of a building down to an outdoor temperature of 0 °F and then enabling a supplemental boiler to supply the additional heat output when the outdoor temperature drops below 0 °F. The latter condition may only occur a few hours in a typical year. Another possibility is where the building owner wants a system with a back-up heat source such as a boiler in case the heat pump system needs to be turned off for repair. Both of these scenarios are viable and often provide favorable economics over rigid insistence that the heat pump *must* provide the full heating load under all conditions.

STEP 3: Select a **minimum entering loop fluid temperature** at which the heat pump must provide the capacity requirements established in step 2. A good "starting point" for this minimum entering fluid temperature, based on field experience in moderately cold climates, is 30 °F +/− 2 °F. Choosing a minimum entering loop temperature lower than 30 °F will decrease the amount of pipe and excavation required for the earth loop. It will also decrease the heating capacity and **seasonal average COP** of the heat pump. Choosing a value higher than 30 °F will increase the amount of tubing in the earth loop and increase the heat pump's heating capacity and seasonal average COP.

STEP 4: Select a heat pump that can match the capacity requirements from step 2 while operating at the minimum entering fluid temperature selected in step 3. This requires a review of manufacturers' data sheets. Perhaps, a heat pump is available that can provide the exact heating capacity determined in step 2, while operating at the minimum entering loop fluid temperature selected in step 3. Often-times, this ideal match will not occur. In such cases, the designer can choose to go with a higher capacity heat pump, or a lower capacity heat pump, depending on available models, costs, etc. Choosing a higher capacity heat pump will typically mandate use of a larger and more expensive earth loop. Choosing a small capacity heat pump will typically mandate some form of supplemental heating for the system. Both are plausible options that should be considered, keeping in mind that the model selected might only be a tentative choice pending further determination of earth loop requirements and installation cost. Once the heat pump model is selected, an earth loop can be sized for it. Be sure to note the COP of the selected heat pump at the minimum entering loop fluid temperature. This data is typically listed on the manufacturer's rating sheet.

STEP 5: The designer must now choose the type of earth loop to be used. There are many possibilities, including both horizontal and vertical loops. Often the choice comes down to simultaneous consideration of available land area, available excavation or drilling equipment, estimated cost of excavation, experience, and simple preferences.

The remainder of this section will focus on *horizontal* earth loops, and the procedures and data relevant to sizing them. Procedures and data for sizing vertical earth loops are given in reference #1, cited at the end of the chapter. Because this text is focused on heating, the remainder of this section will also base earth loop sizing on heating demand rather than cooling demand. Again, full information for evaluating the earth loop requirements for cooling are presented in reference #1.

STEP 6: The length of pipe required for a specified heat pump and selected horizontal earth loop installation method is determined by evaluating Equation 14.17:

(Equation 14.17)

$$L_H = \frac{(HC)_D\left(\frac{COP_D - 1}{COP_D}\right)(R_P + [R_s f_H]P_m)}{(T_{SL} - EMFT_{min})}$$

where:

L_H = total length of pipe required for heating (ft)

$(HC)_D$ = heat pump heating capacity at design conditions (Btu/hr.)

COP_D = heat pump COP at design conditions (unitless)

R_p = pipe thermal resistance (hr.·ft.·°F/Btu)

R_s = soil thermal resistance (hr.·ft.·°F/Btu)

f_H = heat pump run fraction in design month (January)

P_m = multiplier to account for pipe diameter (Pm for 3/4-inch pipe = 1.0) (unitless)

T_{SL} = soil temperature at average loop depth on design day (°F)

$EMFT_{min}$ = *average* fluid temperature in heat pump evaporator at design conditions (°F)

Evaluating Equation 14.17 requires several tasks in which values for the inputs are either referenced or calculated. Once all these values are determined, they can be processed through Equation 14.17 to determine the necessary total length of pipe in the earth loop. In the case of earth loops with parallel branches, the total length of pipe determined by Equation 14.17 typically only includes the parallel branch piping and not any buried header piping required to connect these branches into a complete closed piping assembly. While the header piping does help gather heat from the soil, its overall contribution to the total heat gain is typically small, especially if a compact header is located close to the building. In cases where the branch circuits connect to an interior manifold, there are no headers and thus no heat gain from them.

Examination of Equation 14.17 shows that the earth loop length will be inversely proportional to the "driving ΔT" between (T_L) (e.g., the lowest **undisturbed soil temperature** at the average depth of the earth loop), and the lowest fluid temperature within the heat pump's evaporator (EFT_{min}). The latter temperature is based on the lowest entering loop fluid temperature selected in step 3. The "logic" used in Equation 14.17 assumes that EFT_{min} occurs simultaneous with T_L.

As a reminder, we are still discussing step 6. This step encompasses many "sub-steps," which themselves require new equations and data. This new information will be introduced, and the step-by-step process will resume as necessary.

Annual variation in soil temperature

The temperature of soil, at a given depth below the surface, depends on the **average soil surface temperature** at a given location, the thermal properties of the soil, and the time of year. Figure 14-83 shows the theoretical variation in undisturbed soil temperature based on a soil with average thermal properties and in a location where the average soil surface temperature is 50 °F.

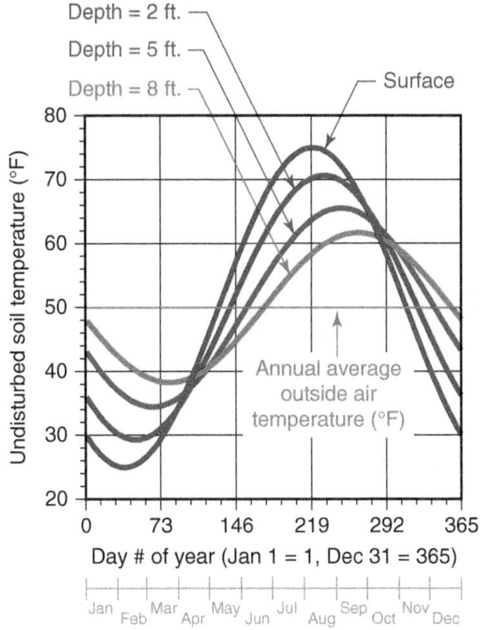

Figure 14-83 Undisturbed soil temperature at different depths, as a function of time of year. Prepared for a location where the annual average soil surface temperature is 50 °F.

The curves shown in Figure 14-83 are based on Equation 14.18.

(Equation 14.18)

$$T(y,t) = T_m - A_s\left(e^{-y\left(\frac{\pi}{365\alpha}\right)^{0.5}}\right)\cos\left(\frac{2\pi}{365}\left[t - t_0 - \frac{y}{2}\left(\frac{365}{\pi\alpha}\right)^{0.5}\right]\right)$$

where:

$T(y,t)$ = soil temperature at depth y and day# t

A_s = amplitude of temperature variation at soil surface (°F)

α = soil thermal diffusivity (ft.²/day)

t_o = phase constant (day# of minimum soil surface temperature)

t = day number (Julian date)

y = depth below surface (ft.)

T_m = mean soil surface temperature (°F)

Below a depth of 25 feet, there is very little change in soil temperature on an annual basis. This is why water from drilled wells has a relatively constant temperature throughout the year. It's also the reason that vertical earth loops tend to provide more consistent temperatures compared to those provided from horizontal earth loops.

The *lowest* soil temperature at a given depth can be determined using Equation 14.19.

(Equation 14.19)

$$T_{SL} = T_m - A_s\left(e^{-y\left(\frac{\pi}{365\alpha}\right)^{0.5}}\right)$$

where:

T_{SL} = lowest soil temperature at average loop depth on design day (°F)

T_m = mean soil temperature (°F) (see Figure 14-84)

A_s = amplitude of the earth surface temperature swing, above and below its mean value (°F) (see Figure 14-84)

y = depth below surface, or average depth of earth loop piping (ft.)

α = thermal diffusivity of soil ft.²/day (see Figure 14-85)

The **amplitude** of the temperature swing (As) at the soil surface, as well as the **annual average soil temperature** (T_m) and number of days into year when the minimum surface temperature occurs (T_0), are given in Figure 14-84 for selected locations. Reference 1, cited at the end of this chapter, contains similar data for more locations.

Thermal diffusivity of soil

The **thermal diffusivity** of soil is the ratio of the soil's thermal conductivity, to its heat storage ability. Both properties are important in geothermal heat pump applications. Soils with low thermal diffusivity tend to have less variation between their annual minimum and maximum temperatures over the first 12 feet below the surface, compared to soils with higher values of thermal diffusivity.

The thermal diffusivity of soil (α) can be calculated based on the soil's thermal conductivity, density, and specific heat. All three of these values depend on the **mineralogy** of the soil, and it moisture content. If a soil sample is taken, preserved, and tested in a laboratory to determine its thermal properties (e.g., thermal conductivity, density, and specific heat), its thermal diffusivity can be calculated using Equation 14.20.

State	City	T_m (°F)	As (°F)	(T_0) (days)
Colorado	Grand Junction	55	25	32
DC	Washington	57	22	36
Idaho	Boise	53	21	34
Illinois	Chicago	51	25	37
Illinois	Urbana	53	26	35
Indiana	Fort Wayne	53	24	35
Indiana	Indianapolis	55	24	34
Indiana	South Bend	52	25	37
Iowa	Des Moines	52	28	35
Iowa	Sioux City	51	29	34
Kansas	Topeka	56	26	35
Kentucky	Louisville	60	22	33
Maine	Portland	48	22	39
Massachusetts	Plymouth	51	21	43
Michigan	Detroit	50	25	39
Minnesota	Duluth	41	28	37
Minnesota	International Falls	39	31	34
Minnesota	Minneapolis	47	29	35
Missouri	Kansas City	58	26	35
Missouri	Springfield	58	23	34
Montana	Billings	49	23	37
Montana	Great Falls	48	23	36
Montana	Missoula	46	21	32
Nebraska	Lincoln	53	28	34
Nevada	Ely	47	22	35
New Jersey	Trenton	55	22	38
New Mexico	Albuquerque	59	22	31
New York	Albany	50	25	38
New York	Binghamton	48	24	38
New York	Niagara Falls	50	24	24
New York	Syracuse	50	24	38
North Carolina	Greensboro	60	20	31

Figure 14-84 | Annual mean soil temperature (T_m), amplitude of temperature variation (As), and day # of lowest surface temperature (T_o), for selected U.S. locations.

State	City	T_m (°F)	As (°F)	(T_0) (days)
North Dakota	Grand Forks	42	33	35
Ohio	Akron	52	23	37
Ohio	Columbus	55	22	34
Ohio	Dayton	56	24	35
Ohio	Toledo	51	25	36
Oklahoma	Oklahoma City	62	23	34
Oklahoma	Tulsa	62	23	34
Oregon	Medford	55	17	34
Oregon	Portland	54	13	37
Pennsylvania	Philadelphia	55	22	34
Pennsylvania	Pittsburg	52	23	36
Pennsylvania	Wilkes-Barre	52	23	36
South Dakota	Rapid City	50	25	38
Tennessee	Knoxville	61	21	31
Tennessee	Memphis	63	21	32
Tennessee	Nashville	60	21	32
Utah	Salt Lake	53	24	35
Vermont	Burlington	46	26	37
Virginia	Norfolk	61	20	37
Virginia	Richmond	60	19	33
Virginia	Roanoke	59	20	33
Washington	Seattle	53	12	36
Washington	Spokane	49	21	32
West Virginia	Charleston	58	20	33
Wisconsin	Green Bay	46	26	37
Wisconsin	Madison	49	27	36
Wyoming	Cheyenne	48	21	39

Figure 14-84 | *(continued)*

(Equation 14.20)

$$\alpha = \frac{24k}{Dc}$$

α = thermal diffusely (ft.^{2}/day)

k = thermal conductivity (Btu/hr./ft./°F)

D = density (lb./ft.^{3})

c = specific heat (Btu/lb./°F)

On larger geothermal heat pump projects, designers often specify on-site thermal testing of soils to determine these properties. However, such tests

Soil description	Density (lb./ft.³)	Thermal diffusivity (ft.²/day)
Heavy wet soil	131	0.60
Heavy dry soil	125	0.48
Light wet soil	100	0.48
Light dry soil	90	0.26

Figure 14-85a | Thermal properties of generic soil types.

often costs several thousand dollars and are therefore seldom used for smaller systems. The next best option is a qualitative assessment of the soil (e.g., its clay, sand, loam). Figure 14-85 provides some qualitative correlation between soil type and typical thermal diffusivity valves. Reference 1, cited at the end of the chapter, contains additional soil classification and associated thermal diffusivity data.

Example 14.15: Determine the minimum undisturbed soil temperature at a depth of 5 feet for Syracuse, NY. Assume a heavy wet soil with a thermal diffusivity of 0.60 ft²/day.

Solution: First, determine the values of Tm, and As for Syracuse using Figure 14-84:

$T_m = 50$ °F

$A_s = 24$ °F

From Figure 14-85a, the thermal diffusivity (α) value for heavy wet soil is 0.60 ft.²/day. This data can now be processed through Equation 14.19:

$$T_{SL} = T_m - A_s\left(e^{-y\left(\frac{\pi}{365\alpha}\right)^{0.5}}\right) = 50 - 24\left(e^{-5\left(\frac{\pi}{365(0.60)}\right)^{0.5}}\right)$$
$$= 36.8°F$$

Discussion: The lowest undisturbed soil temperature at a depth of 5 feet is calculated as 36.8°F. The word *undisturbed* implies that the temperature is solely the result of natural processes and not influenced by the presence of buried tubing carrying fluid that is absorbing heat. Such a situation would further reduce the soil temperature in the proximity of the tubing. This effect is accounted for through a factor called **soil resistance (R_s)**, which will be presented shortly. Mathematically minded observers will also notice that Equation 14.19 is identical to the first group of terms in Equation 14.18. This occurs because the maximum absolute value of the cosine term in the latter portion of Equation 14.18 would be 1.0.

Pipe resistance (R_p)

Equation 14.17 requires a value for **pipe resistance (R_p)**. This is the thermal resistance of the piping used for the earth loop. It can be calculated using Equation 14.21.

(Equation 14.21)

$$R_p = \left(\frac{1}{2\pi k_p}\right)\ln\left[\frac{D_o}{D_i}\right]$$

where:

k_p = thermal conductivity of pipe material (Btu/hr./ft./°F)

D_o = outside diameter of pipe (in.)

D_i = inside diameter of pipe (in.)

HDPE pipe, which is commonly used for earth loops, has a k_p value of 0.225 Btu/hr./ft./°F.

The ratio of outside to inside diameter for pipes in a given DR category remains the same. Thus, for a given DR category, such as DR-11, and a given material (such as HDPE), the value of pipe resistance (R_p) remains constant. For DR-11 HDPE pipe, $\boldsymbol{R_p}$ = ***0.141 (hr.·ft.·°F/ Btu)***.

Soil Resistance (R_s)

Equation 14.17 also requires a value for the soil resistance (R_s). This value is based on complex numerical analysis of how heat is transferred from undisturbed soil

Soil type	Range of thermal conductivity (k) (Btu/hr.·ft.·°F)	Suggested k for heating design	Suggested k for cooling design
Sands	0.35–1.45	1.30	0.45
Silts	0.50–1.45	1.30	0.95
Clays	0.50–0.95	0.90	0.65
Loams	0.50–1.45	1.30	0.55

Figure 14-85b | Selected values for thermal conductivity of various soils.

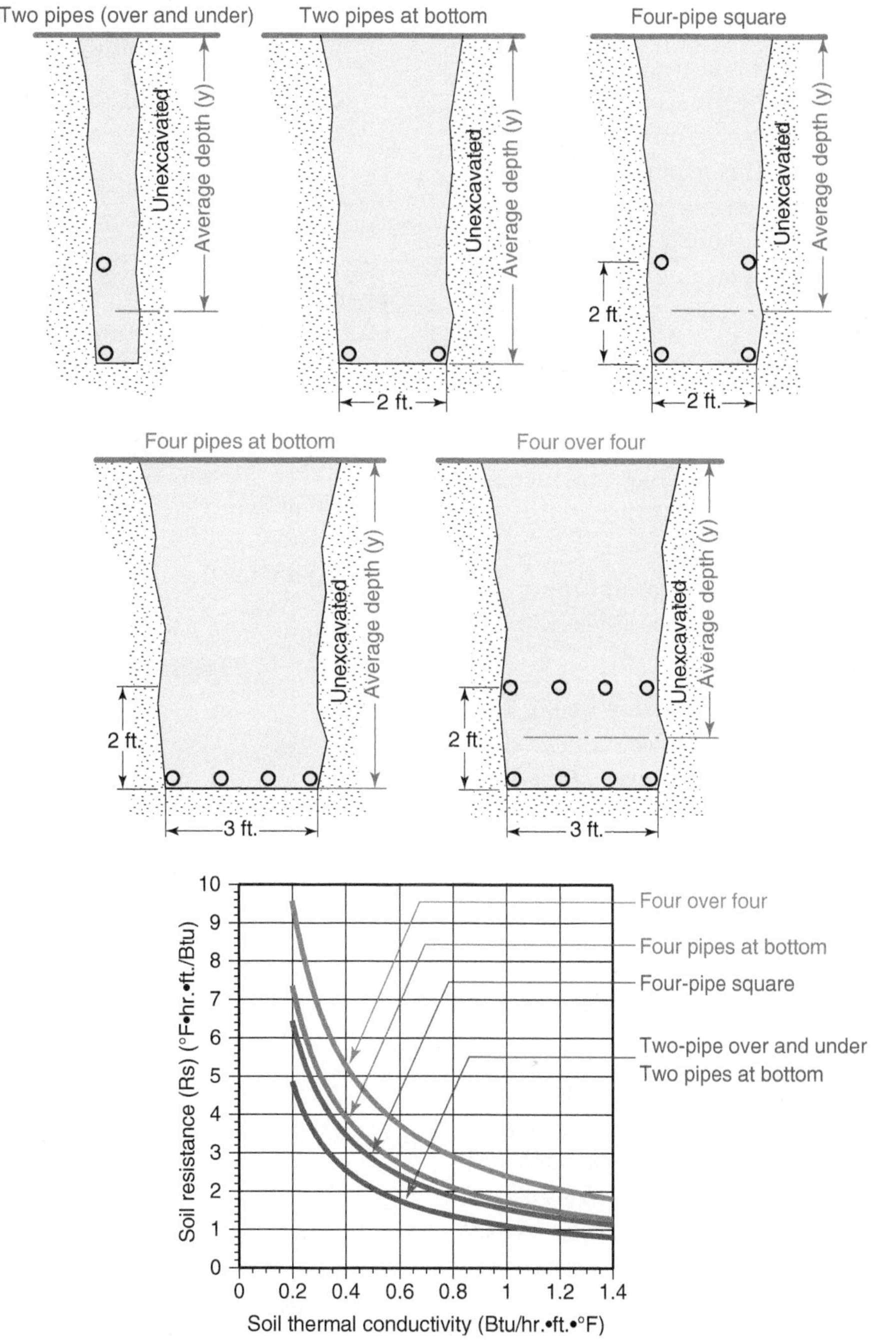

Figure 14-86 | Soil resistance (Rs) as a function of thermal conductivity of soil and a specific piping configuration.

near the earth loop to the outer surface of the earth loop piping. Its value depends on how the tubing is arranged in the earth as well as the thermal conductivity of the soil. Figure 14-86 provides a guideline for estimating the thermal conductivity of various soils as well as a suggested value of thermal conductivity to be used in calculations for sizing the earth loop for heating mode operation.

Figure 14-86 gives the soil resistance (R_s) as a function of the soil's thermal conductivity and for several common methods of installing piping within horizontal earth loops. Reference 1 provides more extensive tables to determine soil resistance as well as ways to modify the value of soil resistance in cases where multiple side-by-side trenches are installed close to each other.

Pipe Size Multiplier (P_m)

The **pipe size multiplier (P_m)** is also needed to evaluate Equation 14.17. It adjusts the results for situations in which piping larger than 3/4-inch is used for the parallel branches of the earth loop. This multiplier is based on the number of pipes in the trench (N) and the nominal size (e.g,. Inside diameter) of those pipes. For trenches containing up to 8 pipes, P_m is determined as follows:

- for 3/4-inch pipe: $P_m = 1.0$
- for 1-inch pipe: $P_m = 0.9492 + (0.0093)N - (0.0005)N^2$
- for 1.25-inch pipe: $P_m = 0.8877 + (0.0254)N - (0.0026)N^2 + (0.0001)N^3$

N = number of pipes in trench.

Reference 1 contains pipe size multiplier for situations where more than eight pipes are installed in a single trench.

As a reminder, we are still discussing step 6. This step encompasses many "sub-steps," which themselves require new equations and data. This new information will be introduced, and the step-by-step process will resume as necessary.

Heating Run Fraction (f_H)

The **heating run fraction (f_H)**, is the percentage of time, during the design month of January, that the heat pump is operating.

There are various methods for estimating the run fraction of a heat pump in the chosen heating design month of January. One of the simplest is based on the number of **degree days (°F·days)** that occur in January. This data is usually available for many locations, in printed as well as online references such as http://www.climate-zone.com.

Equation 14.22 can be used to estimate the run fraction for the heat pump in January.

(Equation 14.22)

$$f_H = \frac{(L_D)\left[\dfrac{Ti_D - \left(65 - \dfrac{DD_{jan}}{31}\right)}{T_{iD} - T_{vD}}\right]}{(HC)_D}$$

where:

f_h = heating run fraction for January (unitless)

L_D = design heating load of building (Btu/hr.)

Ti_D = indoor air temperature upon which design heating load is based (°F)

To_D = outdoor air temperature upon which design heating load is based (°F)

DD_{jan} = number of (base 65 °F) heating degree days in January (unitless)

HC_D = heating capacity of heat pump at minimum entering loop fluid temperature (Btu/hr.)

Example 14.16: A heat pump with a heating capacity of 32,000 Btu/hr. at a corresponding minimum entering loop fluid temperature of 30 °F has been selected for a house with a design heating load of 30,000 Btu/hr. The load is based on an indoor temperature of 68 °F, and an outdoor design temperature of –5 °F. The house is located in Syracuse, NY, where 1,321 (base 65) heating degree days occur during a typical January. Estimate the heating run fraction of the heat pump in this application:

Solution: Just enter the appropriate values into Equation 14.22 and calculate.

$$f_H = \frac{(L_D)\left[\dfrac{T_{iD} - \left(65 - \dfrac{DD_{jan}}{31}\right)}{T_{iD} - T_{vD}}\right]}{(HC)_D}$$

$$= \frac{(30{,}000)\left[\dfrac{68 - \left(65 - \dfrac{1321}{31}\right)}{68 - (-5)}\right]}{32{,}000} = 0.586$$

Discussion: This analysis estimates that the heat pump will be operating (58.6 percent) of the time during January. It is only an estimate because it doesn't account for internal heat gains from sunlight, occupants, equipment, or other sources. It also doesn't account for reduction in heating load due to thermostat setbacks. Internal heat gains and temperature setbacks would reduce the heating run fraction, which, in theory, would allow for a smaller earth loop. However, a conservative approach ignores these factors and thus sizes the earth loop to supply the system without the benefit of heat gains or thermostat setbacks.

Evaporator mean fluid temperature at design (EMFT)$_d$

The **evaporator mean fluid temperature at design conditions** *is the average earth loop fluid temperature within the heat pump's evaporator when the loop is operating at the lowest entering fluid temperature* *selected in step 3.* It is the average of the fluid inlet temperature and fluid outlet temperature under this condition. Its value depends on the heat pump's heating capacity as well as its COP. It also depends on the properties of fluid used in the earth loop and the flow rate of this fluid through the heat pump. Equation 14.23 can be used to determine this value:

(Equation 14.23)

$$(\text{EMFT})_{\min} = T_{\text{loopmin}} - \frac{(\text{HC})_{\text{D}}\left(\frac{\text{COP}_{\text{D}} - 1}{\text{COP}_{\text{D}}}\right)}{(16.02Dc)f}$$

where:

$(\text{EMFT})_{\min}$ = *average* fluid temperature in heat pump evaporator at minimum entering loop fluid temperature (°F)

T_{loopmin} = minimum entering loop fluid temperature to heat pump (°F)

$(\text{HC})_{\text{D}}$ = Heat pump's heating capacity at minimum entering loop fluid temperature (Btu/hr.)

COP_{D} = COP of heat pump at minimum entering loop fluid temperature (unitless)

f = flow rate of earth loop fluid through heat pump evaporator (gpm)

D = density of earth loop at minimum entering loop fluid temperature (lb./ft.³)

c = specific heat of earth loop at minimum entering loop fluid temperature (Btu/lb./°F)

Example 14.17: A heat pump with a heating capacity of 32,000 Btu/hr. at a corresponding minimum entering loop temperature of 30 °F has a corresponding COP of 2.7. The earth loop fluid is a 20 percent solution of propylene glycol, which flows through the heat pump at a rate of 9 gpm. Determine the average fluid temperature within the evaporator under these conditions.

Solution: The fluid properties can be referenced in manufacturer's literature.

The density of the 20 percent propylene glycol solution at 30 °F is 64.14 lb./ft.³.

The specific heat of this solution, at the same conditions, is 0.938 Btu/lb./°F.

All values are now available to process through Equation 14.23:

$$(\text{EMFT})_{\min} = T_{\text{loopmin}} - \frac{(\text{HC})_{\text{D}}\left(\frac{\text{COP}_{\text{D}} - 1}{\text{COP}_{\text{D}}}\right)}{(16.02Dc)f}$$

$$= 30 - \frac{(32{,}000)\left(\frac{2.7 - 1}{2.7}\right)}{(16.02 \times 64.14 \times 0.938)9.0}$$

$$= 27.7\ °\text{F}$$

Discussion: The temperature of 27.7 °F is the average fluid temperature at which the earth loop operates under the "worst case" condition. The lower this temperature is, compared to the lowest soil temperature at the average depth of the earth loop, the greater the "driving ΔT" that forces heat from the soil into the earth loop fluid. The portion of Equation 14.23 that follow the T_{loopmin} term is simply half the temperature drop of the earth loop fluid as it passes through the heat pump under the stated conditions.

This concludes the intermediate steps needed for step 6 of the procedure.

STEP 7: Once all the inputs for Equation 14.17 have been determined, the total length of piping for the earth loop is calculated using that equation.

Example 14.18: A horizontal earth loop is to be sized for a water-to-water heat pump system located in Syracuse, NY. The heat pump will supply a low temperature radiant ceiling panel that operates with a supply water temperature of 115 °F, and a return temperature of 100 °F, under design load conditions. The design heating load of the building is 30,000 Btu/hr., and is based on an indoor temperature of 68 °F and an outdoor design temperature of –5 °F. The designer plans to size the heat pump for the full design heating load of the building. The designer reviews literature from several heat pump manufacturers and finds a heat pump that provides 36,000 Btu/hr. when operating with an entering loop fluid temperature of 30 °F and an entering load water temperature of 100 °F. The COP of the heat pump under these conditions is 2.8. The earth loop fluid is a 20 percent solution of propylene glycol, which flows through the heat pump at a rate of 9 gpm. The soil is judged to be wet clay. A 4-pipe square earth loop using 1" DR-11 HDPE tubing for the branch circuits is selected. The piping will have an average burial depth of 6 feet. Determine the total length of pipe required.

Solution: The answer requires the evaluation of Equation 14.17. This requires the gathering or calculation of several other values, some of which are already stated or easily found within tables or graphs in this section:

Step 1: Design heating load 5 30,000 Btu/hr., at 68 °F inside, and –5 °F outside

Step 2: Design for full heating load

Step 3: Select minimum entering fluid temperature as 30 °F

Step 4: Selected heat pump capacity = 36,000 Btu/hr., and corresponding COP = 2.8

Step 5: Horizontal 4-pipe square earth loop with average pipe depth of 6 feet

Step 6: Minimum soil temperature at average pipe depth (using Equation 14.19 and Figure 14-84):

$T_m = 50$ °F

$A_s = 24$ °F

$$T_{SL} = T_m - A_s\left(e^{-y\left(\frac{\pi}{365\alpha}\right)^{0.5}}\right) = 50 - 24\left(e^{-6\left(\frac{\pi}{365(0.6)}\right)^{0.5}}\right)$$
$$= 38.3\ °F$$

Look up thermal diffusivity of soil (Figure 14-85): $\alpha =$ 0.60 ft.²/day

Determine pipe resistance ($R_p = 0.141$ °F·hr.·ft./Btu (for DR-11 HPDE pipe)

Determine soil resistance (based on $k = 0.9$ (Btu/hr./ft./°F))

Using Figure 14-86, $R_s = 1.70$ °F·hr.·ft./Btu

Determine pipe size multiplier for 1-inch pipe

$P_m = 0.9492 + (0.0093)N - (0.0005)N2 = 0.9492 + (0.0093)4 - (0.0005)42 = 0.978$

Calculate heating run fraction for January (using Equation 14.22):

$$f_H = \frac{(L_D)\left[\dfrac{T_{iD} - \left(65 - \dfrac{DD_{jan}}{31}\right)}{T_{iD} - T_{oD}}\right]}{(HC)_D}$$

$$= \frac{(30{,}000)\left[\dfrac{68 - \left(65 - \dfrac{1321}{31}\right)}{68 - (-5)}\right]}{36{,}000} = 0.521$$

Calculate evaporator mean fluid temperature at design conditions (using Equation 14.23)

$$(EMFT)_D = T_{loopmin} - \frac{(HC)_D\left(\dfrac{COP_D - 1}{COP_D}\right)}{(16.02Dc)f}$$

$$= 30 - \frac{(36{,}000)\left(\dfrac{2.8 - 1}{2.8}\right)}{(16.02 \times 64.14 \times 0.938)9.0} = 27.3°F$$

Step 7: Entering all values into Equation 14.17 and calculate:

$$L_H = \frac{(HC)_D\left(\dfrac{COP_D - 1}{COP_D}\right)(R_P + [R_s f_H]P_m)}{(T_{SL} - EMFT_{min})}$$

$$= \frac{36{,}000\left(\dfrac{28.1 - 1}{2.8}\right)(0.141 + [1.7 \times 0.521] \times 0.978)}{(38.3 - 27.3)}$$

$$= 2119 \text{ ft}$$

Discussion: A total of 2,119 feet of 1″ DR-11 HDPE pipe, buried in a 4-pipe square (e.g., two pipes at 7 foot depth, and two at 5 foot depth) are required. This would require about 2119 / 4 = 530 feet of trench.

COMPUTER-AIDED EARTH LOOP DESIGN

There are several software packages now available for analyzing the performance of closed earth loops. Some are generic, while others are supplied by heat pump manufacturers. Figure 14-87 shows an example of one such software that has been configured to design a 4-pipe square earth loop.

Figure 14-88 shows the results of running the software for the 4-pipe earth loop system. In this case, the software is stating that 440 feet of trench will be required based on a minimum entering earth loop temperature of 30 °F and an average pipe depth of 6 feet.

As is true with design of solar thermal systems, and air-to-water heat pump systems, software allows rapid evaluation of design options. The ability to get immediate results on the consequences of changing one or more design parameters allows the designer to quickly bracket a reasonable solution. This process can often be described as design by repeated analysis.

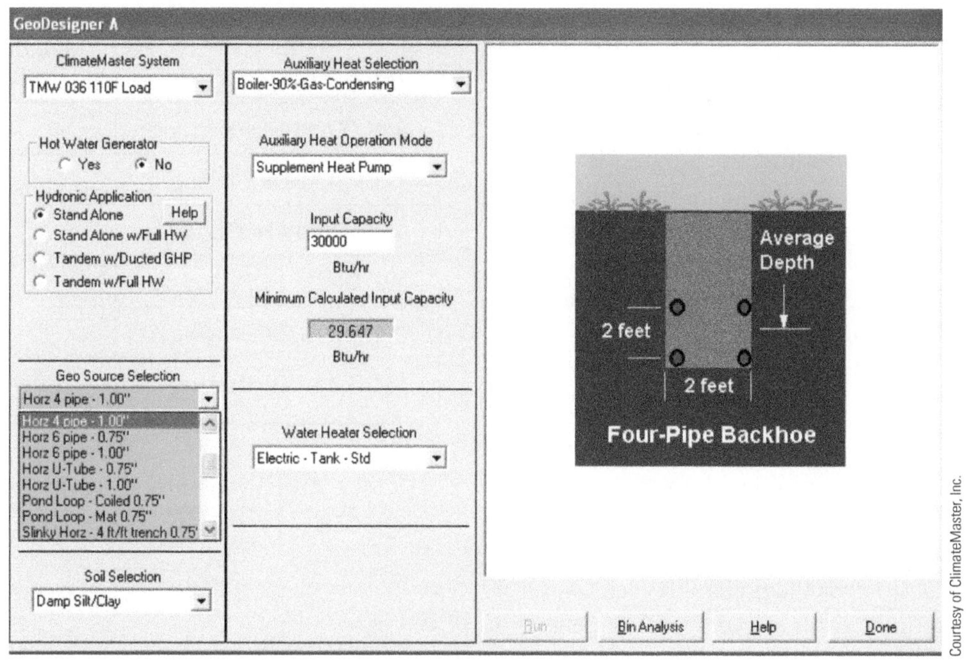

Courtesy of ClimateMaster, Inc.

Figure 14-87 | Example of software configured to analyze a 4-pipe square earth loop.

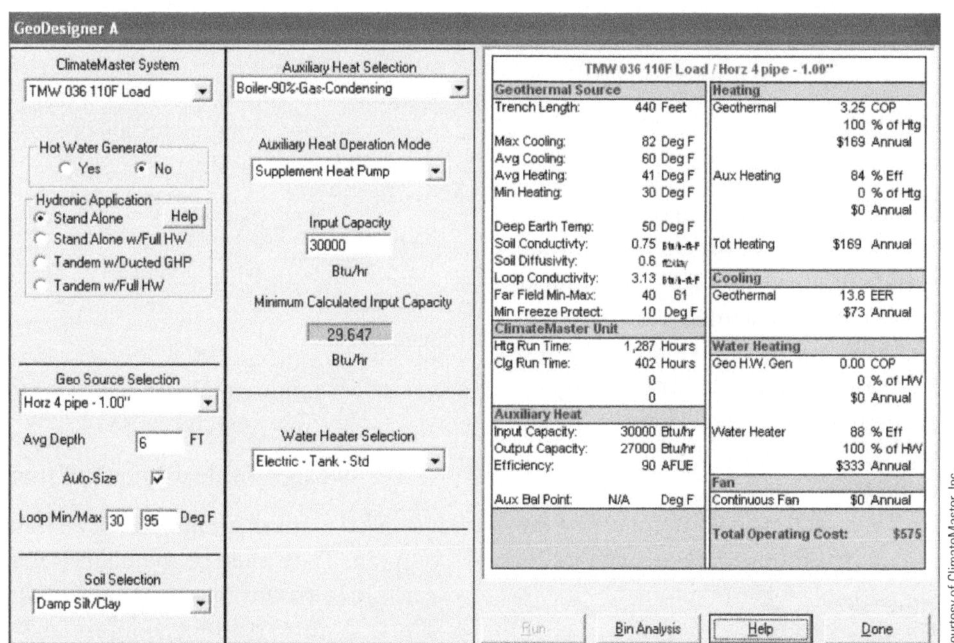

Courtesy of ClimateMaster, Inc.

Figure 14-88 | Results for a specific analysis of a 4-pipe square earth loop with an average depth of 6 feet and a minimum entering fluid temperature of 30 °F.

14.10 Filling and Purging Earth Loops

Filling the earth loop with fluid and purging it of air are critically important steps when commissioning any closed-loop geothermal heat pump system. Proper purging helps ensure good system performance, quiet operation, and long system life. This section discusses how to do this for several types of earth loops and flow center configurations.

During construction, there are many opportunities for dirt and other contaminants, such as sawdust and insects, to find their way into the earth loop piping. This debris needs to be removed in a way that does not send it into the heat pump or the earth loop circulators. A forced stream of water is the preferable material for flushing debris from the earth loop piping. The general procedure for flushing is as follows:

1. Fill and flush the buried portions of the earth loop.
2. Fill and flush the interior piping components of the earth loop (interior piping and heat pump heat exchanger).
3. Simultaneously purge both the exterior and interior portions of the earth loop.
4. Once this flushing is completed, the necessary quantity of antifreeze (if used) can be added to the earth loop.

Most earth loops contain multiple parallel branches. This reduces the head loss of the overall earth loop while providing sufficient pipe surface area for proper heat transfer. If these branches are joined to a **valveless header**, as is the common practice when the header is buried with the branch piping, it is necessary to purge all the branches simultaneously. However, if the branches are connected to a manifold with individual valves for each branch, it is possible to individually purge each branch. There are major differences in the purging equipment required for these two situations.

Experience has shown that a minimum flow velocity of 2 feet per second is necessary to consistently carry air and low-density dirt particles along piping that may be oriented vertically, horizontally, or at any arbitrary angle. Thus, it is necessary to establish and maintain a flow rate within each earth loop circuit being purged that ensures this minimum flow velocity is maintained. Larger or higher-density particles, such as pebbles, will not necessarily be moved along piping operating at this flow velocity. Special care should be taken during installation to keep larger solid contaminants out of piping.

Tubing size/type	Min. purge flow rate (gpm)
3/4″ DR-11 HDPE tubing	3.6
1″ DR-11 HDPE tubing	5.7
1.25″ DR-11 HDPE tubing	9.0
1.5″ DR-11 HDPE tubing	11.8
2″ DR-11 HDPE tubing	18.5
3″ DR-11 HDPE tubing	40.2

Figure 14-89 Minimum flow rates required to achieve flow velocity of 2 feet per second in DR-11 HDPE pipe.

The table in Figure 14-89 shows the minimum required flow rates for achieving a flow velocity of 2 feet per second in the DR-11 HDPE 3608 tubing commonly used for earth loop circuits.

Use Equation 14.24 to determine the minimum purging flow rates needed for a flow velocity of 2 feet per second in tubing materials or sizes other than those listed in Figure 14-89.

(Equation 14.24)

$$f_{\text{min}} = 4.896(d_i)^2$$

where:

f_{min} = minimum flow rate needed to achieve 2 ft./sec. flow velocity (gpm)

d_i = exact inside diameter of tubing (in.)

These are the *minimum* acceptable flow rates for purging. They must be maintained within *each* parallel earth loop circuit that is being simultaneously purged. Thus, if an earth loop contained five (non-valved) parallel circuits of 1-inch DR-11 HDPE piping, the minimum purging flow rate for the earth loop would be 5 × 5.7 gpm = 28.5 gpm. This is a substantial flow rate, which will require a pump significantly more powerful than the earth loop circulator.

Once the minimum purging flow rate is determined, the corresponding head loss of the earth loop should be calculated. This is found by combining the head loss of the longest earth loop circuit along with an estimated head loss for the header piping.

Equation 14.25 can be used to estimate the head loss of DR-11 HDPE tubing operating with cold water, which is typical during filling and purging. The corresponding values of R are given in Figure 14-90.

Tubing	R
3/4″ DR-11 HDPE tubing	$R = 0.269$
1″ DR-11 HDPE tubing	$R = 0.0939$
1.25″ DR-11 HDPE tubing	$R = 0.0316$
1.5″ DR-11 HDPE tubing	$R = 0.0168$
2″ DR-11 HDPE tubing	$R = 0.0059$
2″ DR-11 HDPE tubing	$R = 0.0009623$

Figure 14-90 | Values for R for use in equation 14.25, based on cold water flowing through DR-11 HDPE piping.

Equation 14.25

$$H_{L/100'} = R(f)^{1.75}$$

where:

$H_{L/100'}$ = head loss of piping (feet of head per 100 ft. of tubing)

R = number from table in Figure 8-2

f = flow rate (gpm)

Example 14.19: Determine the head loss of 900 feet of 3/4-inch DR-11 HDPE tubing operating with cold water at a flow rate of 4 gpm.

Solution: The head loss of the tubing per 100 feet of length is found using Equation 14.25 with the value of $R = 0.269$:

$$H_{L/100'} = R(f)^{1.75} = 0.269(4)^{1.75} = 3.04 \text{ ft./100'}$$

Thus, the total head loss of 900 feet of this tubing would be 9 × 3.04 = 27.4 feet.

Discussion: Remember that Equation 14.25 gives the head load per 100 feet of pipe. Be sure to adjust the calculated heat loss per 100 feet of pipe by the actual length of the pipe. This is why the value 3.04 was multiplied by 9 as the final step in this example. Also, note that head loss increases with the 1.75 power of flow rate. Doubling the flow rate increases head loss by a factor of 3.36.

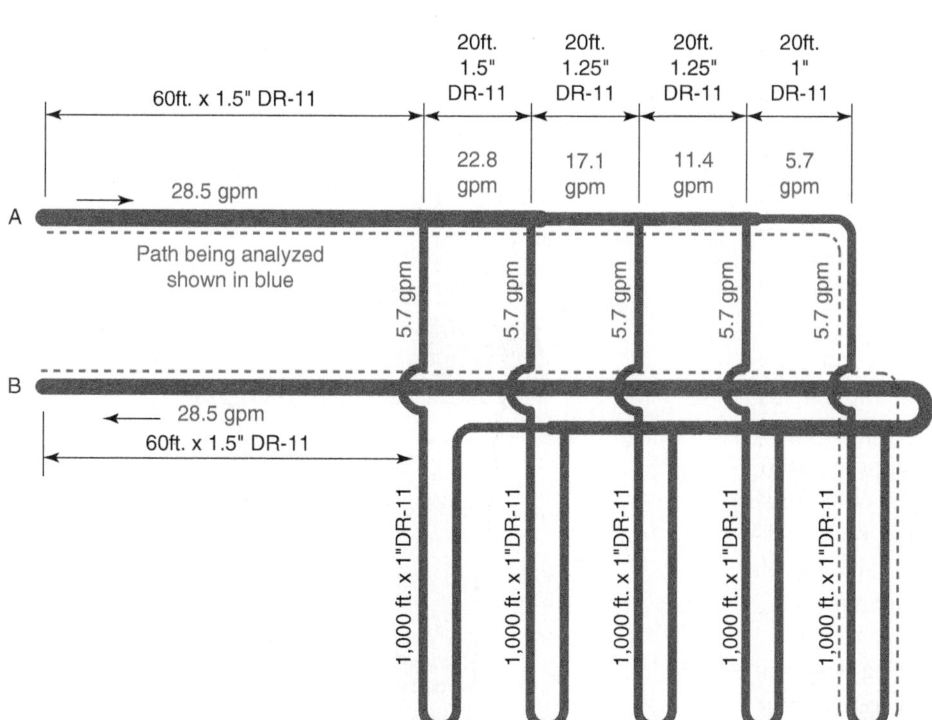

Figure 14-91 | Non-valved earth loop, with stepped pipes in header, for Example 14.20.

Example 14.20: Estimate the total head loss of the non-valved earth loop shown in Figure 14-91. Assume each parallel circuit is 1-inch DR-11 HDPE tubing operating at its minimum purging flow rate of 5.7 gpm. Note the stepped header sizes and associated flow rates.

Solution: Since each earth loop circuit is the same length, and the headers use stepped pipe sizes and reverse return, the flow rate through each branch will be approximately the same. Thus, the head loss needs to be analyzed along the path shown by the blue dashed line in Figure 14-91. This requires the head loss of each piping segment to be separately calculated based on pipe size and flow rate. The resulting head losses can then be added to determine the total head loss.

There is approximately (60 + 20 + 80 + 80) = 220 feet of 1.5″ DR-11 pipe in the flow path indicated by the blue dashed line. The head loss of this pipe is calculated using Equation 14.25:

$$H_{1.5} = 0.0168(28.5)^{1.75}\left(\frac{220}{100}\right) = 12.99 \text{ ft.}$$

The head loss of the two segments of 1.25" pipe is as follows:

$$H_{1.25*} = 0.0316(17.1)^{1.75}\left(\frac{20}{100}\right) = 0.909 \text{ ft.}$$

$$H_{1.25*} = 0.0316(11.4)^{1.75}\left(\frac{20}{100}\right) = 0.447 \text{ ft.}$$

The head loss of the 1-inch DR-11 pipe is as follows:

$$H_{1*} = 0.0939(5.7)^{1.75}\left(\frac{1020}{100}\right) = 20.14 \text{ ft.}$$

Thus, the total head loss is 12.99 + 0.909 + 0.447 + 20.14 = 34.5 ft.

Discussion: The text shown in parentheses adjusts the head loss calculated in feet of head loss per 100 feet of pipe to the head loss for the amount of piping in each segment. The resulting total head loss is substantial. It implies that proper purging of this non-valved earth loop will require a purging pump capable of producing a minimum flow rate of 28.5 gpm at a corresponding head of 34.5 feet. This is well beyond the capability of small wet-rotor hydronic circulators or small submersible transfer pumps. This flow/head requirement will require a submersible well pump or swimming pool pump. A pump with a minimum 1.5-horsepower motor is often recommended for flushing earth loops serving up to 6-ton systems. Larger non-valved earth loops will require even larger purging pumps.

Example 14.20 demonstrates that a powerful purging pump is necessary to simultaneously purge several parallel (non-valved) earth loop circuits. It is also convenient to have a fluid reservoir from which fluid can be pumped as well as purging hardware that is easy to set up and use. These requirements can be met using a commercially available **purge cart**, an example of which is shown in Figure 14-92.

It is also possible to *build* a purge cart, albeit without all the features of a commercial product, as shown in Figure 14-93. This flush cart is based on a

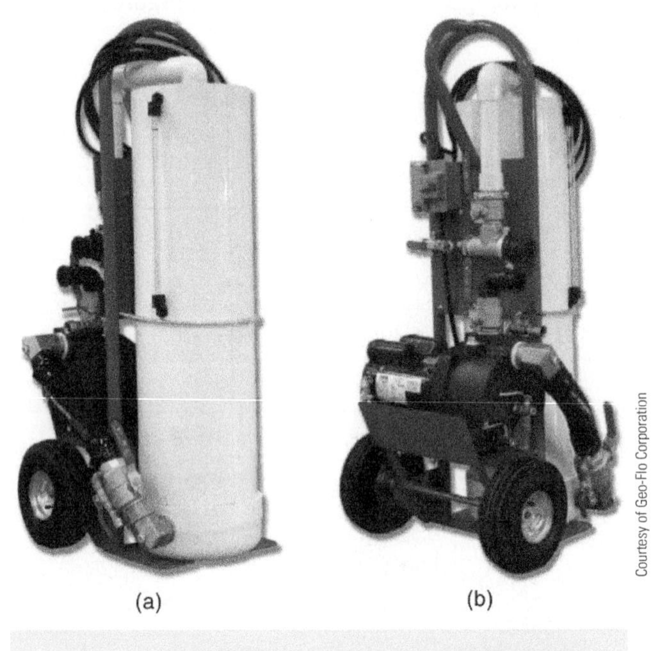

Figure 14-92 | Example of a commercially available purge cart.

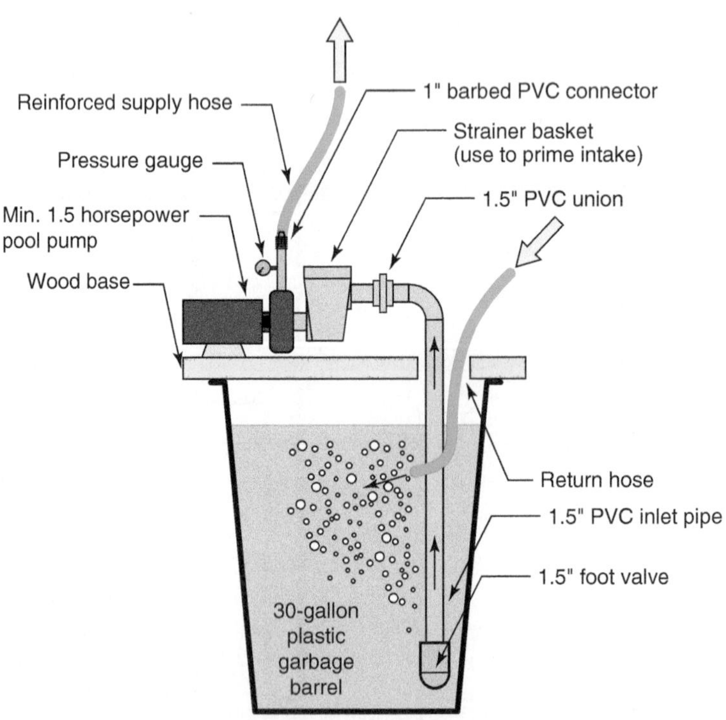

Figure 14-93 | Example of a "home-made" flush cart.

1.5-horsepower or larger swimming pool pump. The pump is mounted to a wooden base that suspends it over a 30-gallon plastic garbage barrel. Water is drawn into the circulator through a 1.5-inch PVC inlet pipe equipped with a foot valve. The pump volute and inlet pipe are primed by adding water to the strainer basket attached to the pool pump.

Figure 14-94 shows how this purge cart would be connected to fill and purge a non-valved earth loop.

The valves shown in Figure 14-94 allow the earth loop to be flushed *without passing flow through the heat pump*. This is preferable because the earth loop is the most likely location for dirt. It is better to get this dirt out of the system rather than force it into and (hopefully) through the heat pump and loop circulator.

It is very important to use *reinforced* rubber tubing between the purging pump and earth loop valving. This hose will operate under significant pressure during purging. An unreinforced hose could easily expand and burst under this pressure. A hose with a minimum pressure rating of 60 psi is suggested. It is also important to clamp or hold the end of the hose returning to the barrel so that the thrust of the water flowing through it does not lift it out of the barrel. The end of the return hose should be directed so that air bubbles returning to the barrel are not jetted directly toward the foot valve.

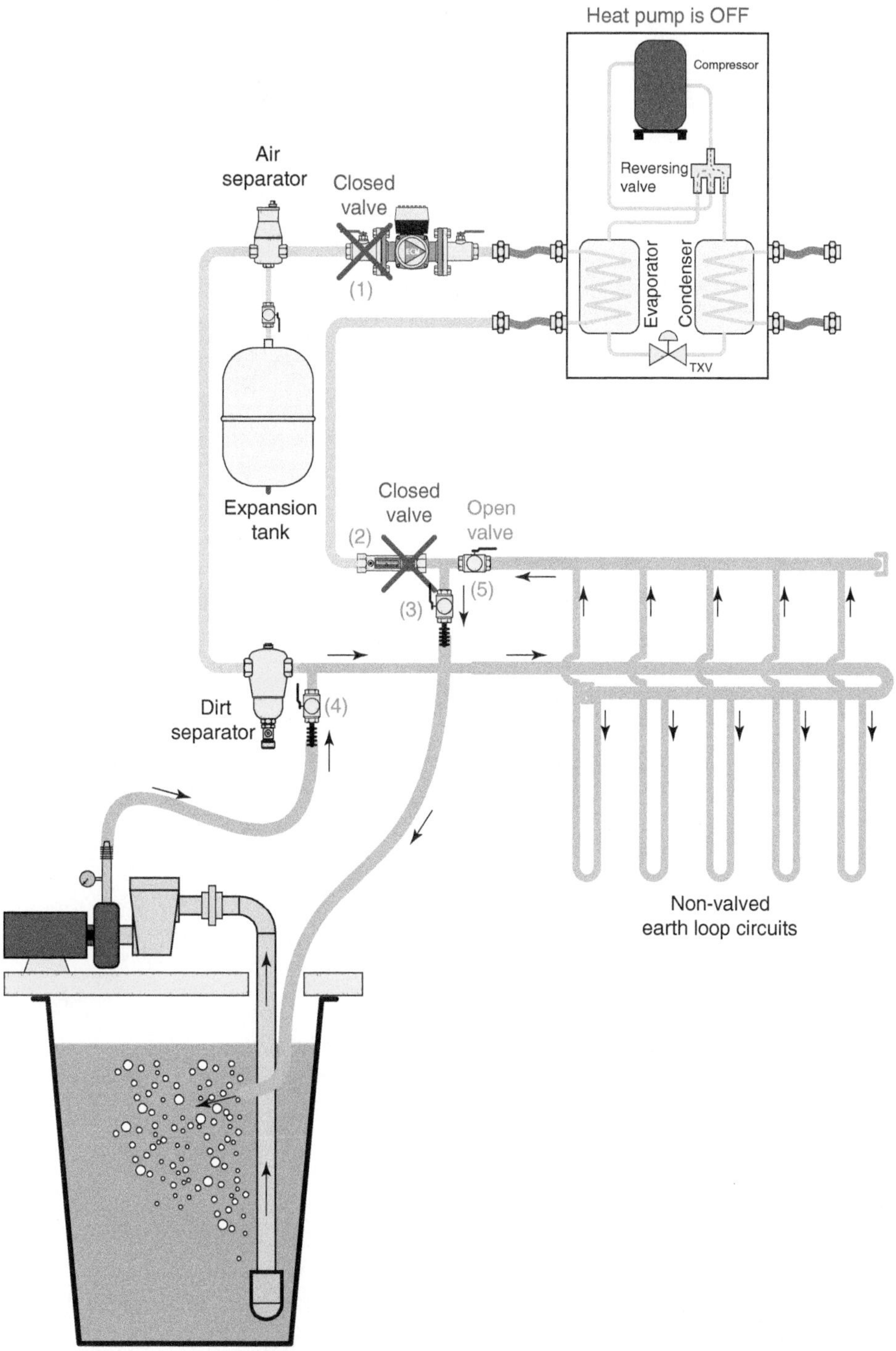

FILLING AND PURGING CLOSED-LOOP, NON-VALVED EARTH LOOPS

Most earth loops can be operated for a few days with just water. This provides time for small dirt particles to be removed by the dirt separator. It also allows time to locate and repair any minor leaks before adding antifreeze to the system.

The following procedure can be used to fill and purge the non-valved earth loop shown in Figure 14-94.

Figure 14-94 | Connecting flush cart to fill and purge to non-valved earth loop. Note valve placement and settings.

STEP 1: Close the isolation flange valve (1) of the earth loop circulator as well as the inline valve in the flowmeter/isolation valve (2). This isolates the circulator and heat pump from the initial purging flow. Connect the purging hoses, as shown in Figure 14-94. Open ball valves (3) and (4). Also be sure that ball valve (5) is fully open.

STEP 2: Prime the purging pump and inlet pipe by filling the strainer basket on the pump and then reinstalling the basket's cover. Fill the barrel with clean water.

STEP 3: Turn on the purging pump. The water level in the barrel will drop rapidly as water is pumped into the system. It is handy to have several 5-gallon pails of water that can be quickly poured into the purge cart's barrel as water is pumped into the system. If the pumping rate exceeds the ability to add water to the barrel, the pump can be temporarily turned off while the barrel is refilled. As the system fills with water, air will be rapidly forced through the return hose. Soon, this hose will return a mixture of air and water to the barrel. Hold or clamp the return hose so that the returning air bubbles are not aimed directly at the foot valve location, where they could be entrained and reinjected into the system. Eventually the return flow will be free of air bubbles. Allow the purging to continue a few more minutes to help capture as much bulk air as possible.

STEP 4: When no more bubbles are returning to the barrel, it's time to purge the remainder of the circuit. Open the ball valve in the flow meter/isolation valve (2). Also open the isolation flange on the earth loop circulator (1). Close the inline ball valve (5) to block purging flow through the earth loop. Flow will now pass through the circulator and heat pump, as shown in Figure 14-95.

More air will be displaced from the piping, heat pump, and other components and quickly be returned to the purge cart's barrel. Allow the purging flow to continue

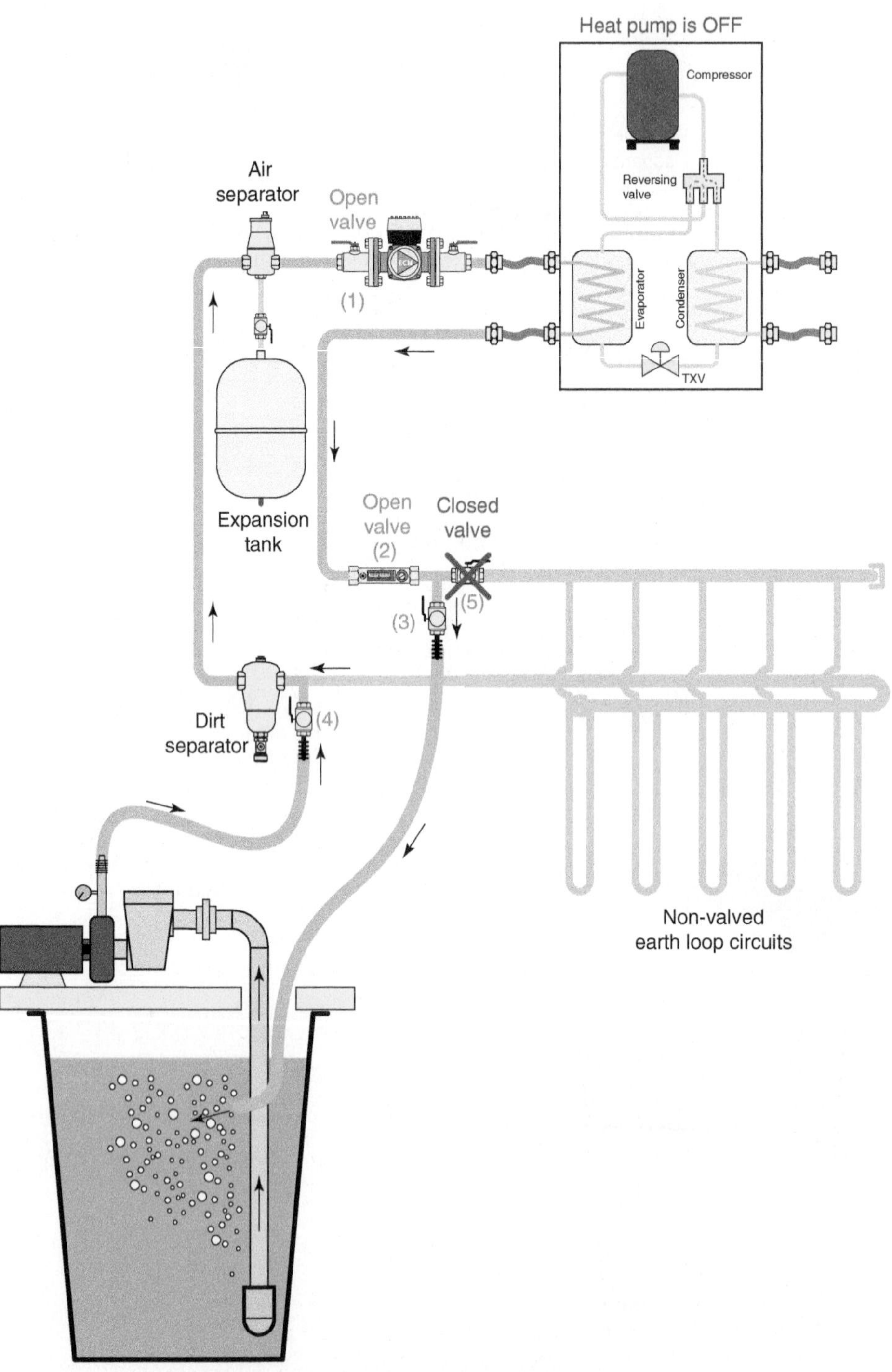

Figure 14-95 | Filling and purging the remaining portion of the circuit.

in this mode until the returning stream is free of all visible bubbles.

STEP 5: Finally, starting with the configuration shown in Figure 14-95, open the ball valve (5) leading to the earth loop while the purging pump continues to operate. This allows simultaneous flow through the earth loop and interior piping passing through the heat pump. The purging flow rate will rise slightly due to lower overall flow resistance. If the earth loop circulator is wired, turn it on to increase flow and displace air bubbles.

STEP 6: When no further air is returning to the barrel, partially close the outlet purging valve while the purging pump remains on. This will add fluid to the loop, partially compressing the air sealed in the expansion tank, and increase the pressure within the loop. A static pressure of 10 to 20 psi is adequate in nearly all earth loops equipped with an expansion tank. Ideally, charge the system to twice the initial air pressure in the expansion tank. This will leave the expansion tank half-filled with fluid and able to perform as predicted based on the sizing calculations covered earlier in this chapter. When this pressure is achieved, close the *inlet* ball valve between the purging pump and earth loop then turn off the purging pump.

ADDING ANTIFREEZE TO THE EARTH LOOP

Assuming the earth loop is equipped with a dirt separating device, as shown in Figures 14-94 and 14-95, and that the initial charge of water has been circulating for a few days, the water should be relatively clean and deareated. It is now ready for antifreeze.

Start by opening the blow down valve at the bottom of the dirt separator to flush out any collected dirt particles. This is best done with the earth loop circulator off.

Next, calculate the necessary volume of antifreeze based on a reasonably accurate estimate of the total earth loop volume. Figure 14-4 lists the volume for common sizes of DR-11 HDPE pipe. The volume of the heat pump's heat exchanger and other small components is usually small compared to the earth loop volume. Still, estimates for the volume of these components can be made or referenced in manufacturer's literature, and added to the volume of the piping.

Once the total volume of the earth loop has been estimated, the required volume of antifreeze is a simple percentage—typically between 15 percent and 25 percent. A 15 percent (by volume) solution of propylene glycol provides protection against ice crystal formation (e.g., slush) down to a temperature of about 23 °F. A 25 percent solution will prevent ice crystal formation down to approximately 15 °F. Some manufacturers of inhibited propylene glycol recommend a minimum 20 percent concentration for the inhibitors and biocides to be effective.

The barrel of the purge cart can be used to mix and inject antifreeze into a system. The following procedure is suggested:

STEP 1: Calculate the amount of antifreeze required using Equation 14-26:

Equation 14.26

$$V_{antifreeze} = (\%)(V_s + V_{b\,min})$$

where:

$V_{antifreeze}$ = volume of (100 percent) antifreeze required (gallons)

$\%$ = desired volume percentage of antifreeze required for a given freeze point (decimal percentage)

V_s = calculated volume of the earth loop (gallons)

$V_{b\,min}$ = minimum volume of fluid required in the barrel to allow proper pumping (gallons)*

*The minimum volume of fluid required in the barrel of the purge cart will depend on how the foot valve is located. It can be determined by experimenting with the purge cart (using water). Reduce the water level in the barrel until the pump begins to draw air into the foot valve. Make sure the return hose is positioned to minimize turbulence around the foot valve. Using a permanent marker, draw a prominent line on the side of the barrel at least 1 inch above this level. This is the minimum operating level of the barrel. Write "Minimum Operating Level" on the outside of the barrel just above this line. Fill the barrel with water up to this line and then carefully pour it out into containers of known volume. Once the total volume is measured, write it on the side of the barrel, as shown in Figure 14-96.

STEP 2: Drain an amount of water from the earth loop equal to the required volume of antifreeze calculated in step 1. This water can either be drained from a low

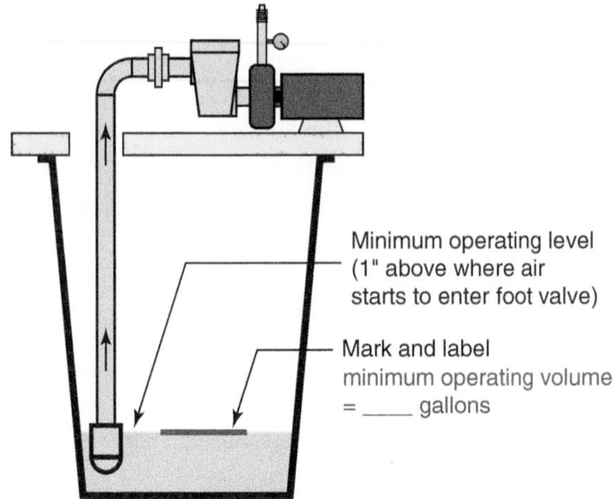

Figure 14-96 | Marking the minimum operating level on the flush cart barrel.

Figure 14-97 | Example of a small hand pump that can be used to add fluid to an earth loop.

point valve or, if necessary, forced from the earth loop by adding compressed air through another valve.

STEP 3: Attach the purging hoses to the system, as shown in Figure 14-94. Keep both inlet and outlet valves (4) and (5) closed.

STEP 4: Fill the barrel to its minimum operating level with water and then pour in the required volume of antifreeze calculated in step 1. Turn on the purging pump. Open the inlet and outlet valves (3) and (4). If the barrel will not hold this volume, pour more antifreeze in as the fluid level drops.

STEP 5: Follow the same purging procedures described earlier. Allow the mixture to circulate through the system for 60 minutes to thoroughly mix the water and antifreeze.

STEP 6: When no further air is returning to the barrel, partially close the outlet purging valve while the purging pump remains on. This will add fluid to the loop, partially compress the air sealed in the expansion tank, and increase the pressure within the loop. A static pressure of 10 to 20 psi is adequate in nearly all earth loops equipped with an expansion tank. Ideally, charge the system to twice the initial air pressure in the expansion tank. This will leave the expansion tank half-filled with fluid and able to perform as predicted based on the sizing calculations covered earlier in this chapter. When this pressure is achieved, close the *inlet* ball valve (4) and the outlet ball valve (5) between the purging cart and earth loop, then turn off the purging pump. If the desired static pressure cannot be achieved by the purge cart pump, use a hand-operated pump, such as the one shown in Figure 14-97, to boost the pressure.

STEP 7: The system should now contain the desired concentration of antifreeze and be purged of all but dissolved air. Turn off the purge cart and disconnect the hoses. The fluid remaining in the purge barrel will have the same concentration of antifreeze as the fluid in the earth loop. It can be poured into clean containers and saved for use in future systems. Be sure to label the storage containers with a description of the fluid, as well as the date it was blended.

FILLING AND PURGING CLOSED-LOOP, *VALVED*, EARTH LOOPS

One distinct advantage of a valved earth loop is the ability to purge each earth loop circuit, one at a time. This significantly reduces the capacity of the purging pump relative to non-valved systems.

The suggested piping and purging connections for a valved manifold earth loop system are shown in Figure 14-98.

As with non-valved earth loops, it is preferable to purge the buried portion of the earth loop before purging the interior piping passing through the heat pump and circulator. It is also necessary to select a purging pump with sufficient flow and head capacity to purge each earth loop circuit individually. For example, if each earth loop circuit was a 1,000-foot-long loop of 1-inch DR-11 HDPE pipe, the minimum purging flow

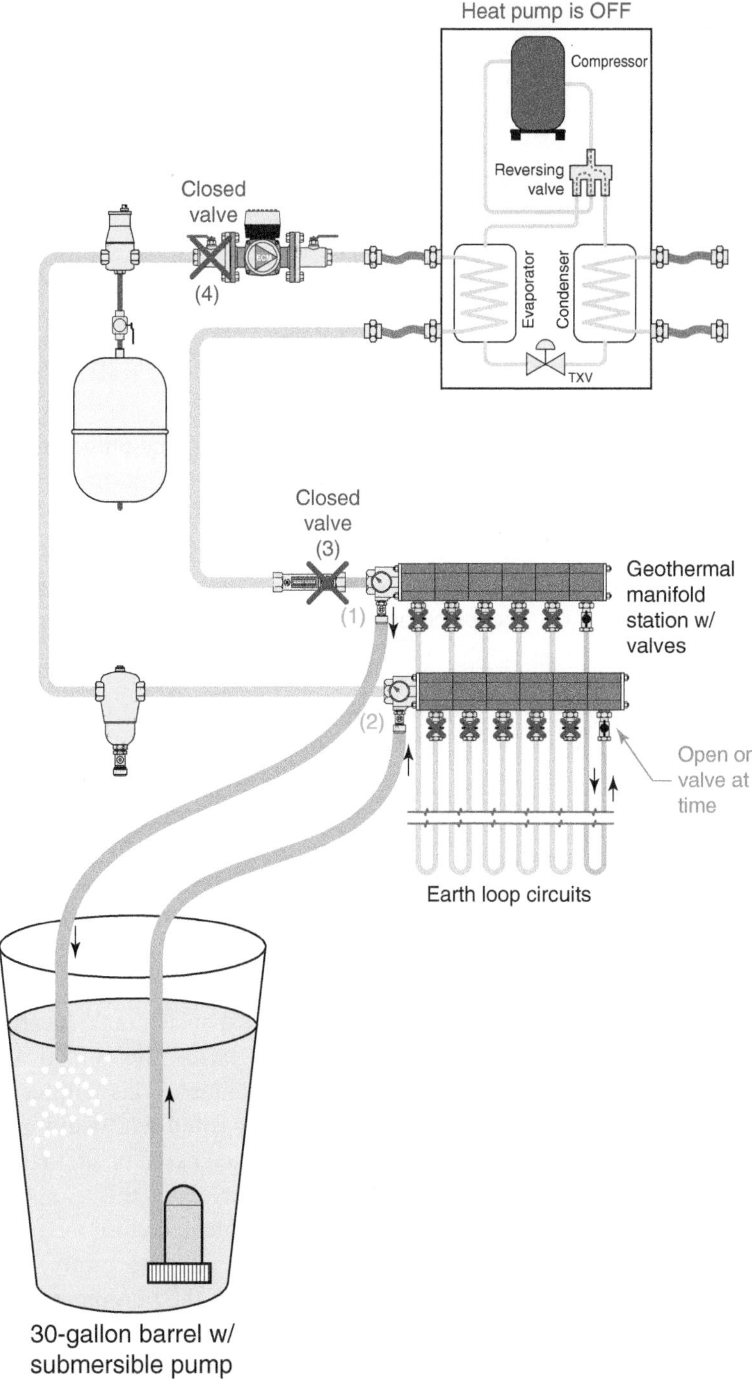

Figure 14-98 | Purging the piping branches in an earth loop with a valved manifold.

rate would be 5.7 gpm and the required head would be 10 times the head loss per 100 feet.

The head loss per 100 feet of 1-inch DR-11 HDPE pipe is found using Equation 14.25:

$$H_{L/100} = R(f)^{1.75} = 0.0939(5.7)^{1.75} = 1.97 \text{ ft./100}'$$

Thus, the total head loss of 1000 feet of this pipe is $10 \times 1.97 = 19.7$ feet.

A nominal 5 percent should be added to the head loss to account for the flow resistance created by the manifold components and purging hoses. Thus, the minimum purging pump requirement for the earth loop would be a flow rate of 5.7 gpm with corresponding head loss of $(1.05)19.7 = 20.7$ feet.

The capacity of the purging pump should also be checked against the minimum flow rate required to purge the interior portion of the earth loop. The purging pump must be able to create a flow velocity of at least 2 feet per second within the largest pipe size used in the earth loop circuit.

The purging procedure is the same as would be used for a manifold-type hydronic distribution system, such as used for radiant panel heating.

STEP 1: Connect the purging hoses, as shown in Figure 14-98.

STEP 2: Fill the purging barrel with water.

STEP 3: Open the valves for the circuit at one end of the earth loop manifold and close all other circuit isolation valves.

STEP 4: Open the purging inlet and outlet valves (1) and (2) on the manifold.

STEP 5: Close the inline ball valve in the flowmeter/isolation valve (3) as well as the valve on the loop circulator inlet flange (4).

STEP 6: Hold or fasten the return hose near the top of the barrel.

STEP 7: Turn on the purging pump. Water will flow into the supply manifold and push air through the one open earth loop circuit. This air will return to the barrel. Eventually, the return stream will transition from air to water. When the return stream is running free of visible air, open the isolation valve for the adjacent circuit on the manifold and then close the isolation valve for the circuit just purged. Repeat this procedure for each circuit until every circuit on the manifold has been purged.

STEP 8: Once the buried portion of the earth loop has been purged, open the flow meter/isolation valve

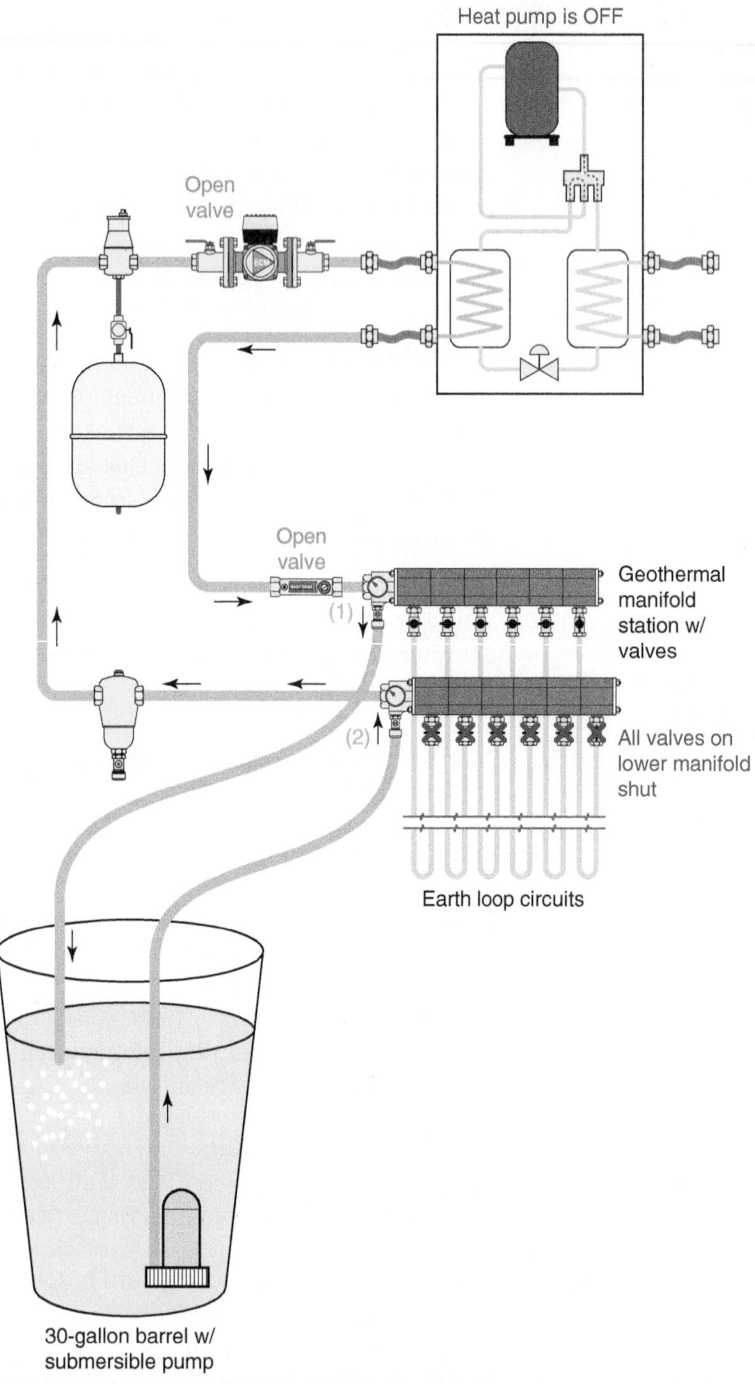

Figure 14-99 | Purging the interior portion of the earth loop circuit.

(3) as well as the isolation valve on the circulator inlet flange (4). Close all branch circuit valves on the return manifold (e.g., the lower manifold in Figure 14-98). The purging flow is now directed through the heat pump's heat exchanger, circulator, and other interior portions of the earth loop. Continue purging this portion of the system until the return stream is free of air bubbles. If the circulator is wired, it can be turned on to further increase purging flow and dislodge air bubbles. This purging mode is shown in Figure 14-99.

STEP 9: Finally, open all the branch circuit manifold valves while the purging pump continues to operate. This allows simultaneous flow through the earth loop and interior piping through the heat pump. The purging flow rate will rise slightly due to lower overall flow resistance. If the earth loop circulator is wired, turn it on to increase flow and displace air bubbles. At this point the earth loop circuit should be completely filled with water. Close the manifold purging valves (1) and (2) and turn off the purging pump.

STEP 10: Most newly filled and purged earth loops can be operated for a few days with just water. This provides time for small dirt particles to be removed by the dirt separator. It also allows time to locate and repair any minor leaks before adding antifreeze to the system.

STEP 11: Follow the previously described procedure for adding antifreeze to the earth loop. Once the antifreeze has been added, the static pressure of the loop can be adjusted. Partially close the outlet purging valve while the purging pump remains on. This will add fluid to the loop, partially compress the air sealed in the expansion tank, and increase the pressure within the loop. A static pressure of 10 to 20 psi is adequate in nearly all earth loops equipped with an expansion tank. Ideally, charge the system to twice the initial air pressure in the expansion tank. This will leave the expansion tank half-filled with fluid and able to perform as predicted based on the sizing calculations covered earlier in this chapter. When this pressure is achieved, close the *inlet* ball valve between the purging pump and manifold and then turn off the purging pump. If the desired static pressure cannot be achieved by the purge cart pump, use a hand-operated pump to boost the pressure.

14.11 Examples of Combisystems Using Geothermal Water-to-Water Heat Pumps

This section shows several examples of complete water-to-water heat pump systems supplied by closed earth loops. One of the systems is for a "heating-only" application. It is appropriate for climates where heat loads dominate and cooling is not needed. The remaining systems provide heating and cooling.

Keep in mind that several of the combisystems shown in section 8 of Chapter 13, dealing with air-to-water (ATW) heat pumps, can be easily modified for use with water-to-water heat pumps. The one difference is that water-to-water heat pumps with variable-speed compressors are currently not available in North America. Thus, any time a zoned heating or cooling distribution system is used with a single-speed water-to-water heat pump, a suitably sized buffer tank must be included between the heat pump and distribution system. Sizing methods for buffer tanks were given earlier in this chapter.

Given that most of the example systems shown in Chapter 13 can be adapted for use with water-to-water heat pumps, this section shows new system configurations. Some of the details used will be similar or identical to those used with the solar thermal combisystems discussed in previous chapters. Readers should pay particular attention to such details because of their potential use in a wide variety of systems.

The closed earth loop configurations will vary from one system to another. These variations include valved manifold type earth loops as well as non-valved header earth loops, as discussed earlier in this chapter. There is no intent to show a specific earth loop configuration as a prerequisite to a specific balance-of-system design. The variations are simply to show different possibilities. The only prerequisite is that each earth loop is sized to the requirements of the heat pump in the system and equipped with the proper hardware for purging, air and dirt elimination, and expansion compensation.

GEOTHERMAL WATER-TO-WATER HEAT PUMP COMBISYSTEM #1

The combisystem shown in Figure 14-100a is a simple, heating-only system in which a single heat pump supplies a *single zone*, slab-type radiant panel for heating. Domestic hot water is heated through use of a stainless steel brazed-plate heat exchanger. A setpoint controller monitors the temperature in the domestic hot water tank and turns on the heat pump when the tank requires heat. Both the space heating load and domestic water heating load can independently call for heat pump operation.

This system does not have a buffer tank. Therefore, *it is critically important that the heating capacity of the heat pump be matched with the heat dissipation capability of the single-zone space heating system as well as that of the domestic water heating subsystem.* When the system is turned on in either of these modes, it will immediately seek to establish thermal equilibrium between the rate of heat production by the heat pump and the rate at which the load absorbs heat. If the ability of the load to absorb heat is well matched to heat production rate of the heat pump, thermal equilibrium should occur at water temperatures that provide good heat pump performance. The *maximum* supply water temperature at which this should occur is 120 °F. Lower supply water temperatures are preferred when possible. *If the heat transfer ability of the load is not well matched to the rate of heat production, short cycling and/or automatic shut down of the heat pump is likely. This must be avoided.*

The space heating distribution system must be designed so that it can dissipate the maximum output from the heat pump, using a supply water temperature no higher than 120 °F. Even lower supply water temperatures are preferable when possible. The lower the supply water temperature, the higher the heating capacity and COP of the heat pump. Standard hydronic trim including an air separator, pressure-relief valve, purging valve, and make-up water assembly are used in the distribution system. These components, and the way they are arranged, allow them to perform the same function they would in a system supplied by a conventional heat source such as a boiler.

The heat pump shown in Figure 14-100a is a "heating-only" unit. It does have a reversing valve and thus cannot operate as a chiller. Although some heat pump manufacturers do offer heating-only units, the decision to specify such a unit should carefully consider any possibility of a future system modification to include chilled water cooling. If there is any possibility of such a modification, the heat pump should be specified with a reversing valve. The difference in cost between a heating only heat pump and a standard heat pump with a reversing valve is usually small.

The electrical wiring diagram for combisystem #1 is shown in Figure 14-100b.

The electrical control system uses a **2-stage temperature setpoint controller** to continuously monitor the temperature of the domestic hot water.

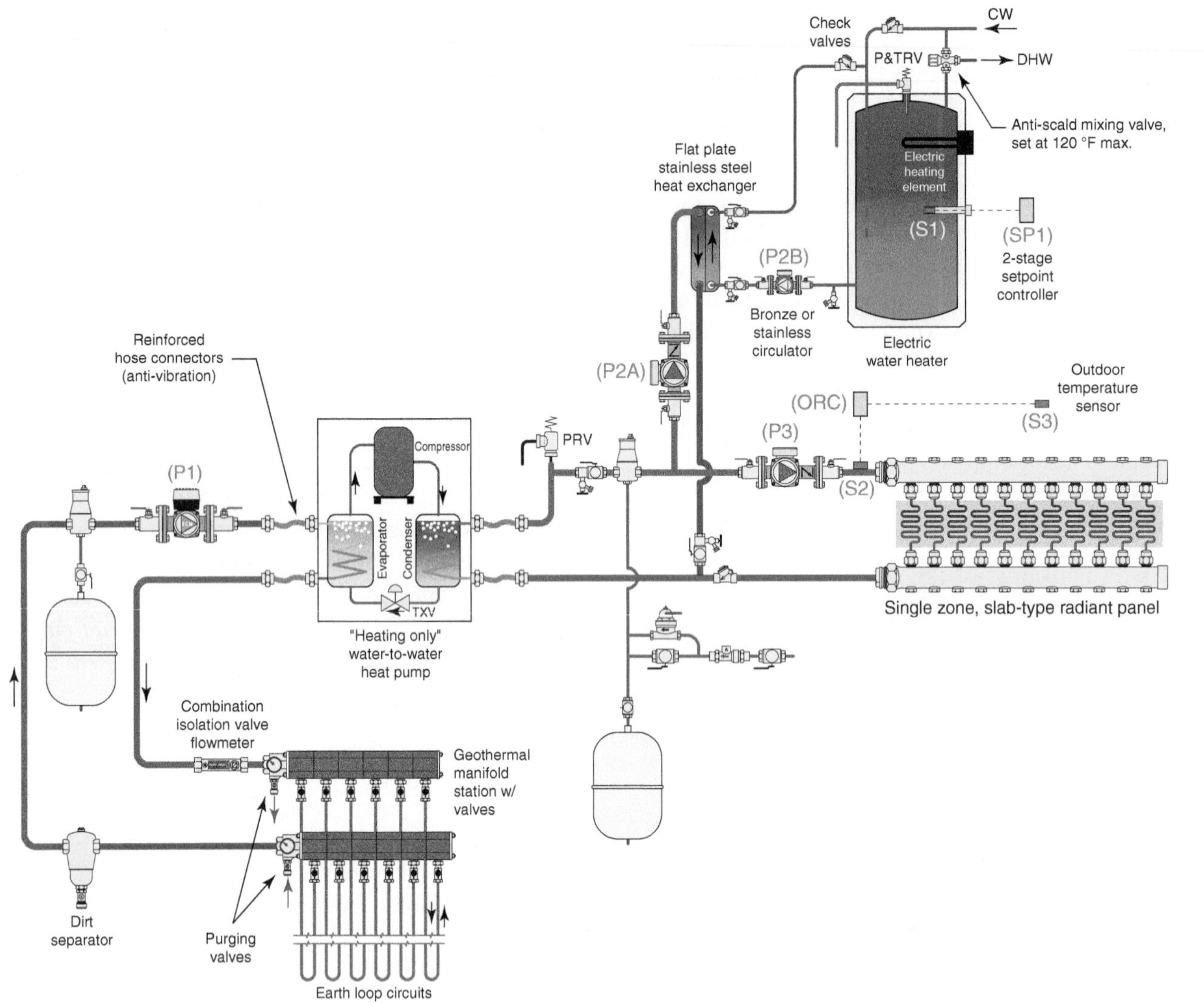

Figure 14-100a | Combisystem #1. For space heating and domestic water heating.

This controller's objective is to allow the water-to-water heat pump to heat domestic water whenever possible, but also allow the electric heating element at the top of the tank to add heat if the heat pump cannot maintain the desired domestic hot water delivery temperature. The first stage contacts of the setpoint controller are set to close when the temperature at the tank sensor (S1) drops to or below 110 °F, and open when the temperature reaches 120 °F. The second stage contacts are set to close when the temperature at sensor (S1) is 100 °F or less, and open when the sensor temperature reaches 120 °F. Thus, the electric heating element is only turned on if the water temperature in the tank drops to or below 100 °F, as might occur during a period of high demand. *The heating thermostat supplied with the electric water heater is set for a minimum of 140 °F and acts as a safety device (rather than an operating controller).* If the 2-stage setpoint controller failed to interrupt power to the heating element when sensor (S1) reaches 120 °F, the tank's own internal thermostat would eventually open to interrupt power to the element when and if the water temperature reached 140 °F. Keep in mind that many of the internal thermostats supplied with electric water heaters are not highly accurate; thus, the 140 °F setting was selected to keep the thermostat closed unless there is truly a failure of the 2-stage setpoint controller (e.g., the tank temperature was able to climb substantially above 120 °F). The ASSE 1017 thermostatic mixing valve on the outlet of the water heater provides a final safety device that prevents water at temperatures above 120 °F from being delivered to the hot water taps.

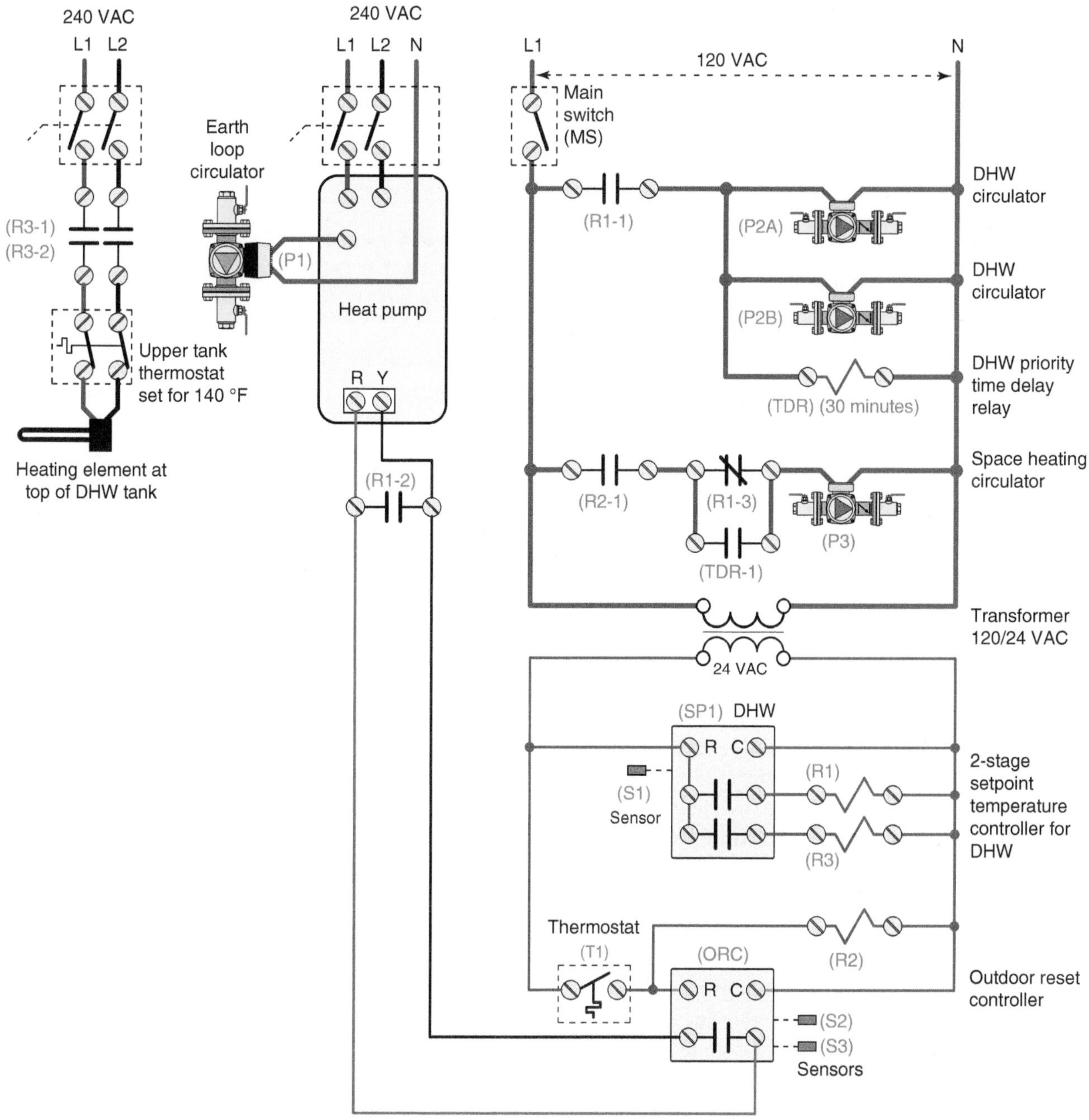

Figure 14-100b | Electrical wiring diagram for water-to-water heat pump combisystem #1.

If the tank has two electric heating elements, the low element should be disconnected so that it does not heat the water in the lower two-thirds of the tank.

The following is a description of operation of combisystem #1. Readers are encouraged to cross-reference the component designations, shown in parentheses, within both the piping and electrical schematics.

Description of Operation:

1. *Space heating mode:* Upon a demand for heat from room thermostat (T1), 24 VAC is passed to the coil of relay (R2). A normally open contact (R2-1) closes to supply line voltage through a normally closed relay contact (R1-3), to turn on space heating circulator (P3). 24 VAC is also applied to power up the outdoor reset controller (ORC). This controller measures outdoor

temperature at sensor (S3). It uses this temperature, in combination with its settings, to calculate the target supply water temperature for the distribution system. It also measures the supply water temperature at sensor (S2). If the temperature at (S2) is lower than half the set differential below the target temperature, the normally open contacts in the (ORC) closes to complete a circuit between terminals (R) and (Y) on the heat pump. This turns on the heat pump in heating mode. An internal relay within the heat pump applies line voltage to operate earth loop circulator (P1). If the temperature at sensor (S2) is more than half the differential above the target temperature, the (ORC) opens its contacts to turn off the heat pump and circulator (P1). Circulator (P3) continues to operate until thermostat (T1) is satisfied.

2. *Domestic water heating mode:* Domestic water heating is treated as the ***priority load***. If there is a call for domestic water heating when the space heating mode is active, the latter mode is temporarily interrupted for up to 30 minutes to allow the full output of the heat pump to be delivered to the domestic water heating load. A call for domestic water heating is initiated by the 2-stage temperature setpoint controller (SP1), which monitors the tank temperature at sensor (S1). This controller is powered on whenever the main switch (MS) is closed. When the temperature of sensor (S1) drops to or below 110 °F, the first stage contacts in (SP1) close. This energizes the coil of relay (R1). A set of contacts (R1-1) close to supply line voltage to circulators (P2A) and (P2B). Line voltage is also supplied to the coil of a time delay relay (TDR). A second set of relay contacts (R1-2) close across the (R) and (Y) terminals of the heat pump to turn it on in heating mode. An internal relay within the heat pump applies line voltage to operate earth loop circulator (P1). A third set of normally closed contacts (R1-3) opens to interrupt line voltage to circulator (P3), which suspends space heating. The time delay relay (TDR) is configured to close its normally open contacts (TDR-1) after an elapsed time of 30 minutes from when its coil was energized. If this occurs, contacts (TDR-1) close to reestablish line voltage to circulator (P3), allowing space heating to resume. If the temperature at sensor (S1) drops to 100 °F or less, the second stage contacts in (SP1) close. This energizes the coil of contactor (R3), which applies 240 VAC to the thermostat within the electric water heater through contacts (R3-1) and (R3-2). This thermostat is set to 140 °F and is thus closed. Its sole purpose is to provide a safety device that would interrupt power to the heating element if the water temperature in the tank reached 140 °F. Thus, power is applied to the electric element to provide supplemental heating of the tank. The heat pump remains on during this time. This element remains on until the temperature at sensor (S1) reaches 120 °F, at which point the element is turned off.

GEOTHERMAL WATER-TO-WATER HEAT PUMP COMBISYSTEM #2

The system in Figure 14-101a provides multiple zones of heating, in combination with a *single zone of cooling*. An on/off water-to-water heat pump, equipped with a desuperheater, is the sole heat source of space heating and the sole source of chilled water for space cooling. Depending upon how it is controlled, this heat pump can also provide a significant portion of the domestic water heating load.

Because an on/off heat pump is used with multiple zones of heating, a buffer tank is required. This prevents the heat pump from short cycling when only one or two zones are operating as well as under other partial load conditions.

The earth loop is a slightly pressurized closed loop, with buried non-valved headers. Flow is provided by two series-arranged circulators within the flow center. The earth loop is filled and purged using the two 3-way valves in the flow center. The earth loop is equipped with an expansion tank, as well as a combined air and dirt separator. Flexible, reinforced hoses are used to connect the flow center to the heat pump.

During heating mode operation, the heat pump is turned on and off by an outdoor reset controller that measures the outdoor temperature and uses this temperature, along with its settings, to determine the appropriate target temperature for the buffer tank. It operates the heat pump as necessary to maintain the buffer tank within a narrow range of this target temperature. Figure 14.101b shows how the outdoor reset controller is configured so that the target supply water temperature to the distribution system is 110 °F at design load conditions, which in this case correspond to an outdoor temperature of 10 °F.

Individual space heating zones are controlled by manifold valve actuators that respond to their associated thermostats. When any one or more zones of heat are active, the variable-speed pressure-regulated circulator (P4) operates in constant differential pressure mode.

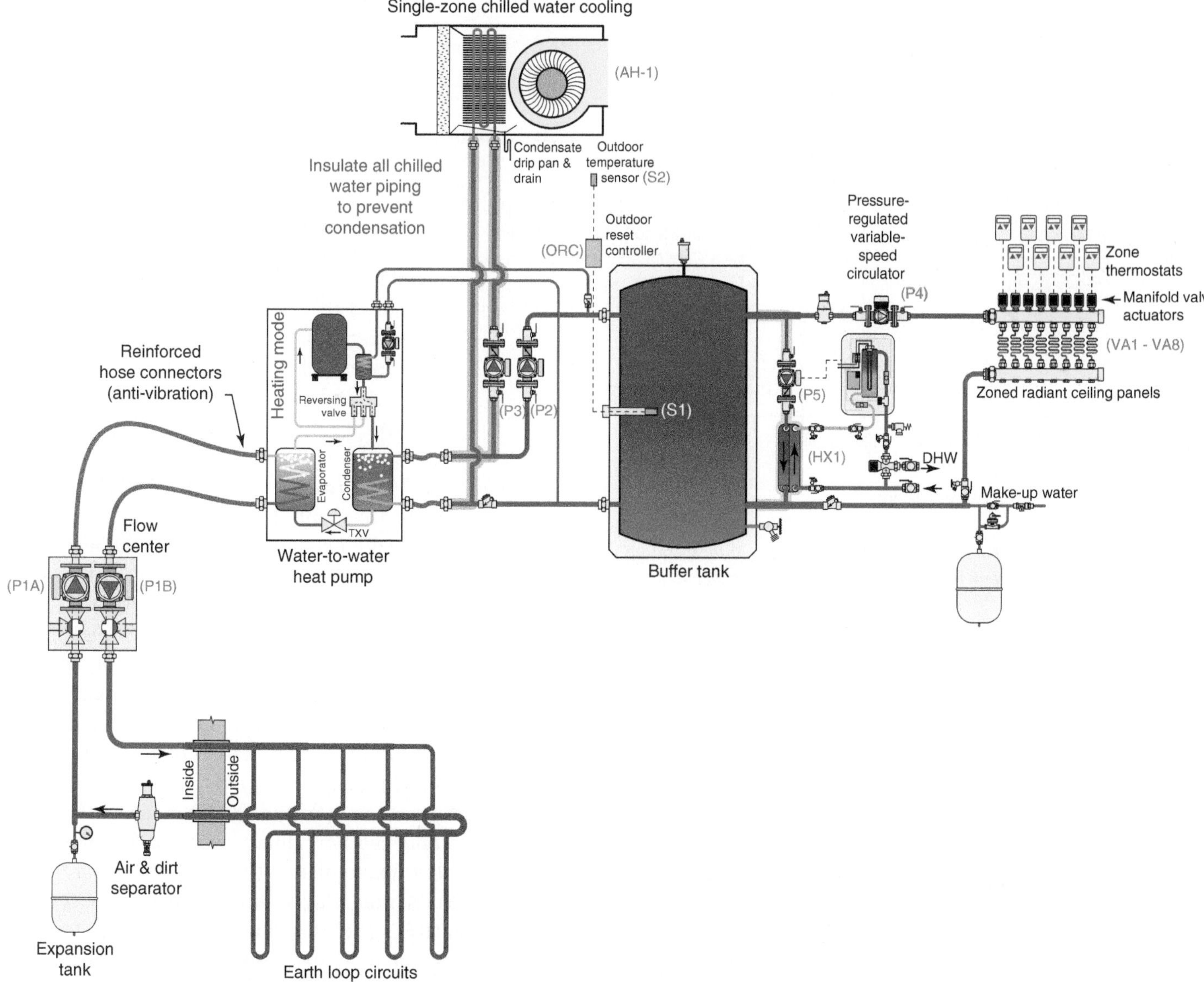

Figure 14-101a | Combisystem #2. Single water-to-water heat pump provides multiple zones of heating, single zone cooling, and a portion of the domestic water heating load.

Domestic water is heated "on-demand" using an external stainless steel brazed-plate heat exchanger. The details for this approach were discussed in Chapter 9. Whenever there is a demand for domestic hot water of 0.6 gpm or higher, the flow switch inside the tankless electric water heater closes. This closure is used, in combination with a relay, to turn on the circulator that moves water from the upper portion of the buffer tank through the primary side of this heat exchanger. Closure of the flow switch also energizes an electrical contactor within the tankless heater closes to supply 240 VAC to triacs, which regulate electrical power to the heating elements. The power delivered to the elements is limited to that required to provide the desired domestic hot water delivery temperature. All heated water leaving the tankless heater flows into an ASSE 1017–rated mixing valve to ensure a safe delivery temperature to the fixtures.

Cooling is provided by a single chilled water air handler that has been selected to match the cooling capacity of the heat pump at a chilled water supply temperature of 50 °F. *This eliminates the need for a buffer tank in cooling mode operation.* Circulator (P3) operates in combination with the blower in the air handler, and the heat pump, whenever cooling is

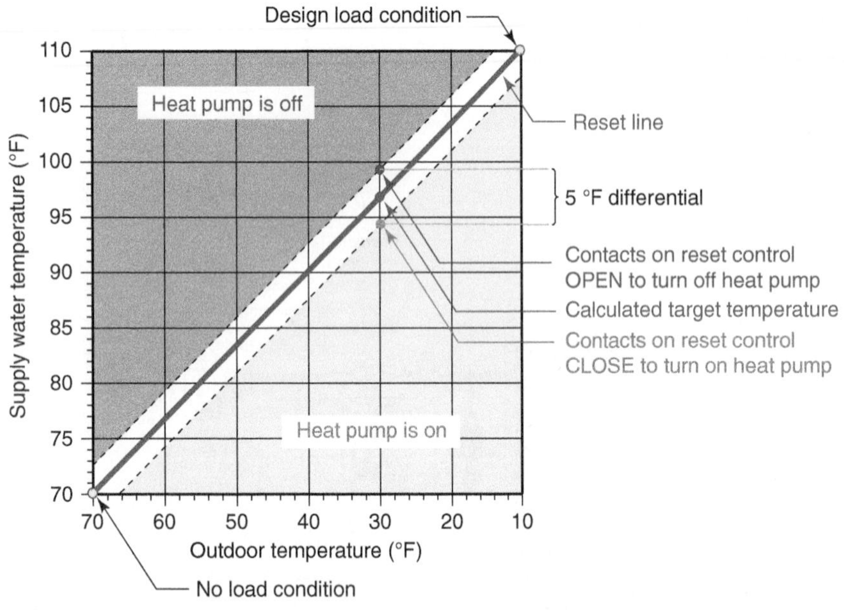

Figure 14-101b | Outdoor reset control of the water temperature in the buffer tank for combisystem #2 in Figure 14-101a.

required. All piping and components in the chilled water portion of the system are insulated and vapor sealed to prevent condensation formation. The air-handler is equipped with a drip pan to capture and dispose of condensate that forms on its coil.

The desuperheater within the heat pump absorbs heat from the hot refrigerant gas leaving the compressor and transfers it to the buffer tank, where it can be used by the on-demand domestic water heating sub-system. When the heat pump is operating in cooling mode, the heat transferred to the buffer tank by the desuperheater is truly "free heat" that would otherwise be dissipated to the earth loop. When the heat pump operates in heating mode, the combined heat transferred from the refrigerant to system water through the desuperheater and the condenser slightly increases the COP of the heat pump, compared to a situation in which only the condenser is removing heat from the refrigerant.

The electrical wiring diagram for combisystem #2 is shown in Figure 14-101c.

The following is a description of operation for combisystem #2. Readers are encouraged to cross-reference the component designations, shown in parentheses, within both the piping and electrical schematics.

Description of Operation:

1. *Space heating mode:* The mode selection switch (MSS) must be set for heat. This supplies 24 VAC to the 24VAC input on all thermostats (T1 through T8). Thermostat (T1) controls heating in zone 1 and cooling for the entire house. Thermostats (T2 through T8) are heating-only thermostats. If any thermostat calls for heating, 24 VAC is passed to the associated manifold valve actuator for that zone, causing it to open. When the actuator is fully open, its end switch closes. This passes 24 VAC to heating relay (RH1) and powers on the outdoor reset controller (ORC). One relay contact (RH1-1) closes to supply line voltage to heating distribution circulator (P4). The outdoor reset controller (ORC) measures the outdoor temperature at sensor (S2). It uses this temperature, in combination with its settings, to calculate the target temperature for the buffer tank. It also measures the current temperature of the buffer tank at sensor (S1). If the measured temperature is more than half the set differential below the target temperature, 24 VAC is passed to the coil of relay (RH2). One contact (RH2-1) closes across the (R) and (Y) terminals of the heat pump, turning the heat pump on in the heating mode. The earth loop circulators (P1A) and (P1B) are turned on by an internal relay within the heat pump. A second contact (RH2-2) closes to apply line voltage to circulator (P2), which creates flow between the heat pump and the buffer tank. The heat pump continues to run, assuming a heat demand is still present, until the temperature at sensor (S1) reach one half the differential of the (ORC) above the target temperature. When all thermostats are satisfied, relay coil (RH1) and the outdoor reset controller (ORC) are turned off.

2. *Space cooling mode:* The mode selection switch (MSS) must be set for cool. This supplies 24 VAC to the (RC) terminals of thermostat (T1). When the mode selection switch (MSS) is set for cooling, 24 VAC is interrupted to the remaining thermostats (T2) through (T8). If thermostat (T1) calls for cooling, 24 VAC is switched to the thermostat's (Y) terminal as well as its (G) terminal. This energizes the coil of relay (RC1). One contact (RC1-1) closes to connect the heat pump's (R) terminal to its (Y) terminal, which turns on the compressor. Another contact (RC1-3) closes to connect

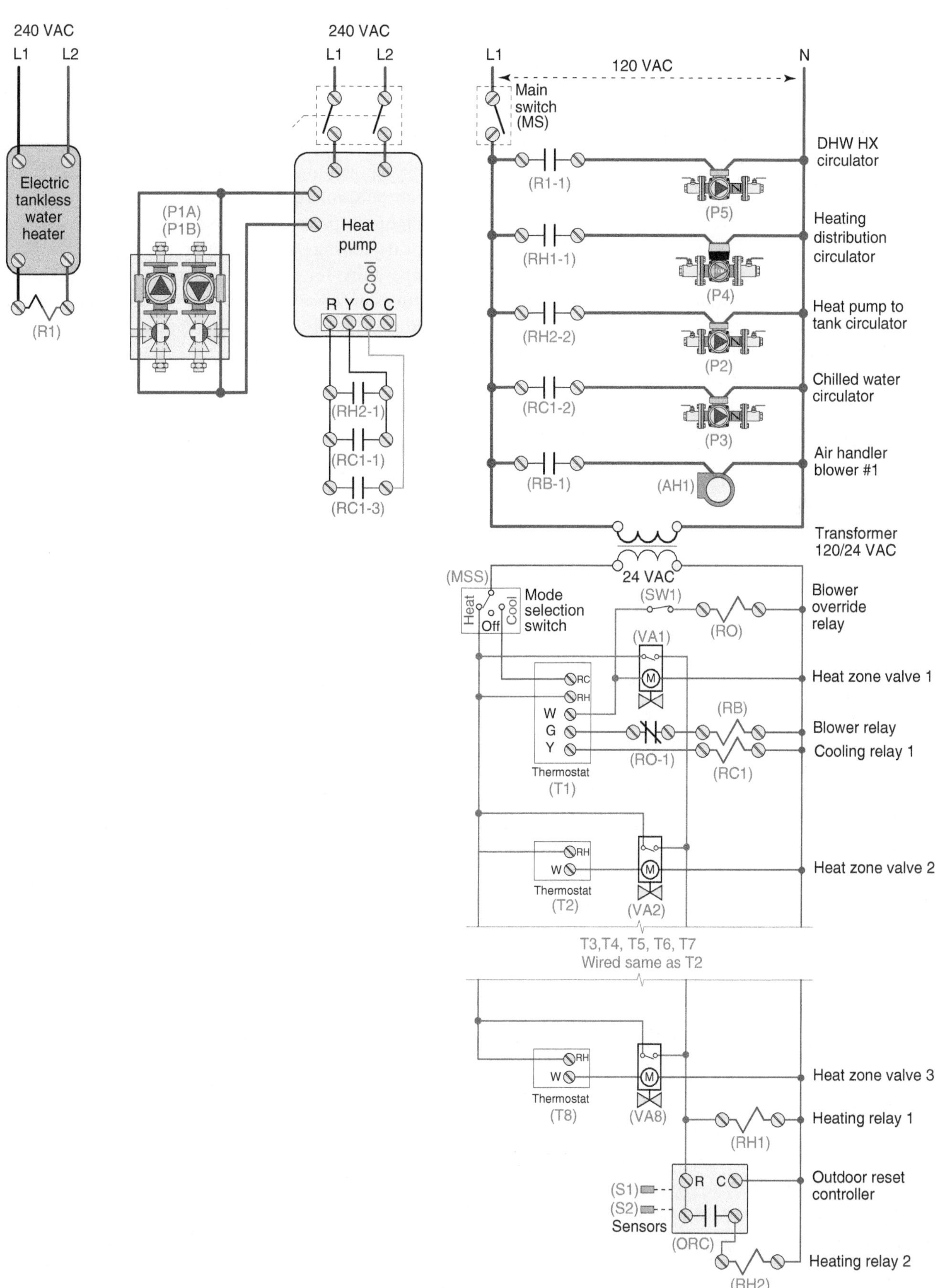

Figure 14-101c | Electrical wiring diagram for water-to-water heat pump combisystem #2.

the heat pump's (R) terminal to its (O) terminal. This turns on the reversing valve so that the heat pump is operating in cooling mode. Another contact (RC1-2) closes to supply line voltage to cooling distribution circulator (P3). The earth loop circulators (P1A) and (P1B) are turned on by an internal relay in the heat pump. The 24 VAC applied to the thermostat's (G) terminal energizes the coil of relay (RB). Contact (RB-1) closes to supply line voltage to blower in the air handler.

3. *Domestic water heating mode:* Whenever there is a demand for domestic hot water of 0.6 gpm or higher, the flow switch inside the tankless electric water heater closes. This closure applies 240 VAC to the coil of relay (R1). The normally open contacts (R1-1) close to turn on circulator (P5), which circulates heated water from the upper portion of the buffer tank through the primary side of the domestic water heat exchanger (HX1). The domestic water leaving (HX1) is preheated to a temperature a few degrees less than the buffer tank temperature. This water passes into the thermostatically controlled tankless water heater, which measures its inlet temperature. The electronics within this heater control electrical power supplied to the heat elements based on the necessary temperature rise to achieve the set domestic hot water supply temperature. All heated water leaving the tankless heater flows into an ASSE 1017–rated mixing valve to ensure a safe delivery temperature to the fixtures. Whenever the demand for domestic hot water drops below 0.4 gpm, circulator (P5) and the tankless electric water heater are turned off.

4. *Blower operation:* The blower in the air handler will always operate in cooling mode. It may also be set for operation in either heating or cooling mode by setting the fan switch on thermostat (T1) to "on." If the blower is not to run in the heating mode, switch (SW1) should be closed. This will allow 24 VAC to energize the coil of relay (RO) when thermostat (T1) calls for heat. A normally closed contact (RO-1) opens, interrupting 24 VAC to the blower relay (RB) and thus preventing the blower from operating. If switch (SW1) is left open, the blower will operate whenever thermostat (T1) calls for heating.

GEOTHERMAL WATER-TO-WATER HEAT PUMP COMBISYSTEM #3

When multiple zones of heating and cooling are supplied from an on/off water-to-water heat pump, the system must have at least one buffer tank. The system shown in Figure 14-102a uses a single buffer tank for both heating and cooling operation.

Figure 14-102a shows the system in heating mode operation. The cooling subsystem, which is inactive, is shown in gray. When there is a demand for space heating, the heat pump is turned on and off by an outdoor reset controller, which measures the outdoor temperature and uses this temperature, along with its settings, to determine the appropriate target temperature for the buffer tank. It operates the heat pump as necessary to maintain the buffer tank within a narrow range of this target temperature.

Zone thermostats call for heat by energizing their associated zone valves. The variable-speed pressure-regulated circulator (P3) operates in constant differential pressure mode whenever any of the zones are calling for heat. The thermal mass of the buffer tank prevents short cycling of the heat pump when only one zone is active as well as under other partial load conditions.

This system uses a closed, slightly pressurized earth loop, with buried non-valved headers. It is equipped with an expansion tank as well as a combined air and dirt separator. Appropriate valves are provided for high-capacity filling and purging. Earth loop flow is provided by a single ECM-based circulator. This circulator can operate in fixed-speed mode or in a variable-speed mode if a suitable controller is provided that can optimize earth loop flow rate for a given operating condition.

Domestic water is partially heated by the desuperheater within the heat pump. The desuperheater operates whenever the compressor is on. Cool domestic water from near the bottom of the storage tank is drawn through the desuperheater and routed back through the dip tube of the electric storage water heater. This allows the desuperheater to operate on the lower temperature water in lower portion of the tank, thus preserving temperature stratification. The electric heating element provides any necessary supplemental heating to achieve the desired domestic hot water delivery temperature.

Figure 14-102b shows the same system in cooling mode. The inactive heating portions of the system are shown in gray.

In the cooling mode, the heat pump is operated by a temperature setpoint controller that monitors the temperature of the buffer tank. The objective is to keep the tank temperature between 45 °F and 60 °F, *whenever the mode selection switch is set for cooling.* Each air handler is operated by an associated zone

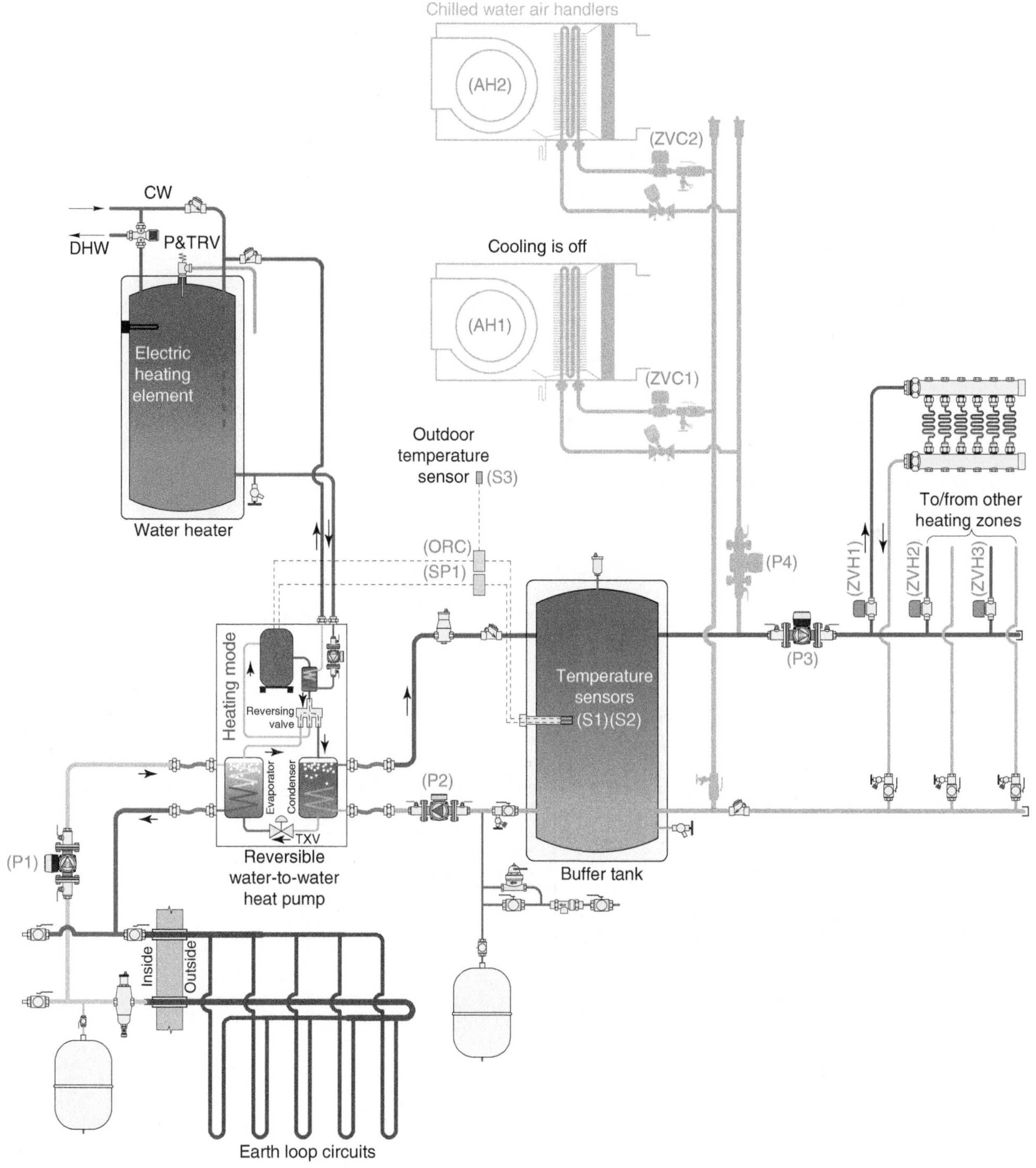

Figure 14-102a | Combisystem #3. Heating and cooling supplied through a single buffer tank. Grayed components in distribution system are inactive in heating mode.

thermostat (e.g., the same thermostat that controls that zone's radiant panel during heating mode). Upon a call for cooling from any zone thermostat, the associated zone valve opens and the variable-speed pressure-regulated circulator (P4) is turned on in proportional differential pressure mode.

This system is not designed to provide simultaneous heating and cooling. This system is also not appropriate for situations where heating may be required during morning hours, followed by cooling during the afternoon. It makes little sense to warm up the buffer tank for heating, only to have to cool it down

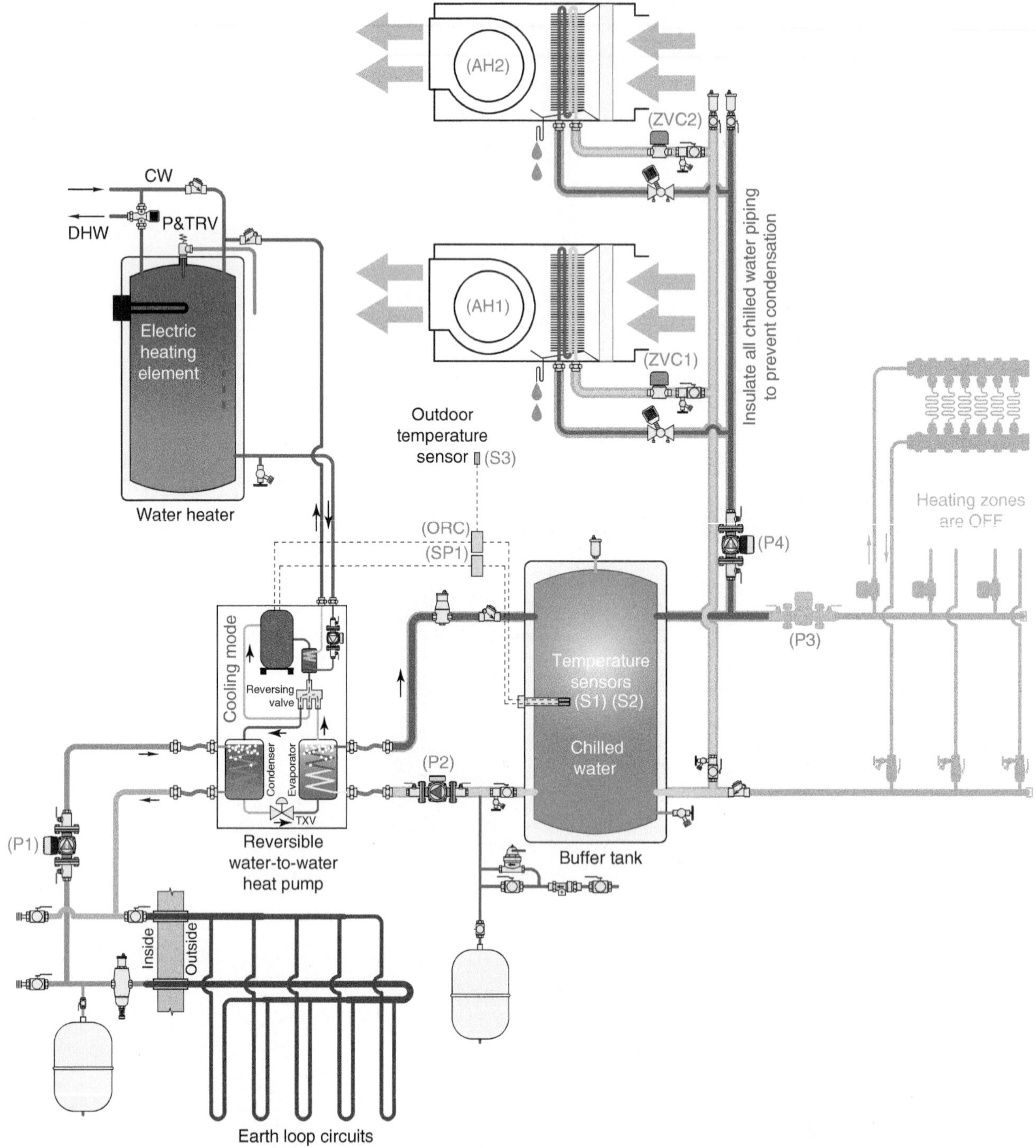

Figure 14-102b | Water-to-water heat pump combisystem #3 in cooling mode operation. Inactive portion of distribution system shown in gray.

a few hours later for cooling. Therefore, this system should only be used in climates where there is at least several days between the end of the heating season and the beginning of the cooling season. During this time, the buffer tank will eventually stabilize to room temperature and await the onset of the next operating mode.

There are buildings where morning heating is often followed by afternoon cooling. If zoned heating and cooling are required, these situations should be handled by using two buffer tanks in the system: one for heating and the other for cooling.

The electrical wiring diagram for combisystem #3 is shown in Figure 14-102c.

The following is a description of operation of combisystem #3. Readers are encouraged to cross-reference the component designations, shown in parentheses, within both the piping and electrical schematics.

Description of Operation:

1. *Space heating mode:* The mode selection switch (MSS) must be set for heat. This supplies 24 VAC to all thermostats. Thermostat (T1) controls heating in zone 1 and cooling for zone 1. Thermostat (T2) controls heating for zone 2 and cooling for zones 2 and 3. If any thermostat calls for heating, 24 VAC is passed from the (W) terminal of the thermostat to the associated zone valve for that zone, causing it to open. When the zone valve is fully open, its end switch closes. This passes 24 VAC to heating relay (RH1) and powers up the outdoor reset controller (ORC). One relay contact (RH1-1) closes to supply line voltage to operate the heating distribution circulator (P3). The outdoor reset controller (ORC) measures the outdoor temperature at sensor (S3). It uses this temperature, in combination with its settings, to calculate the target temperature for the buffer tank. It also measures the current temperature of the buffer tank at sensor (S1). If the measured temperature is more than half the differential below the target temperature, 24 VAC is passed to the coil of relay (RH2). One contact (RH2-1) closes across the (R) and (Y) terminals of the heat pump, turning the heat pump on in the heating mode. The earth loop circulator (P1) is turned on by an internal relay within the heat pump. A second contact (RH2-2) closes to apply line voltage to circulator (P2), which creates flow between the heat

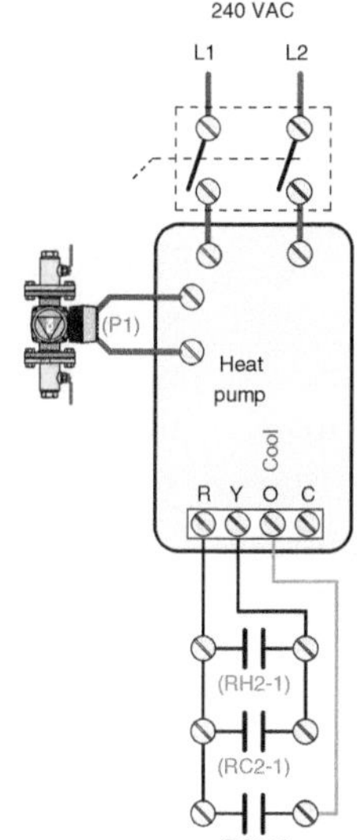

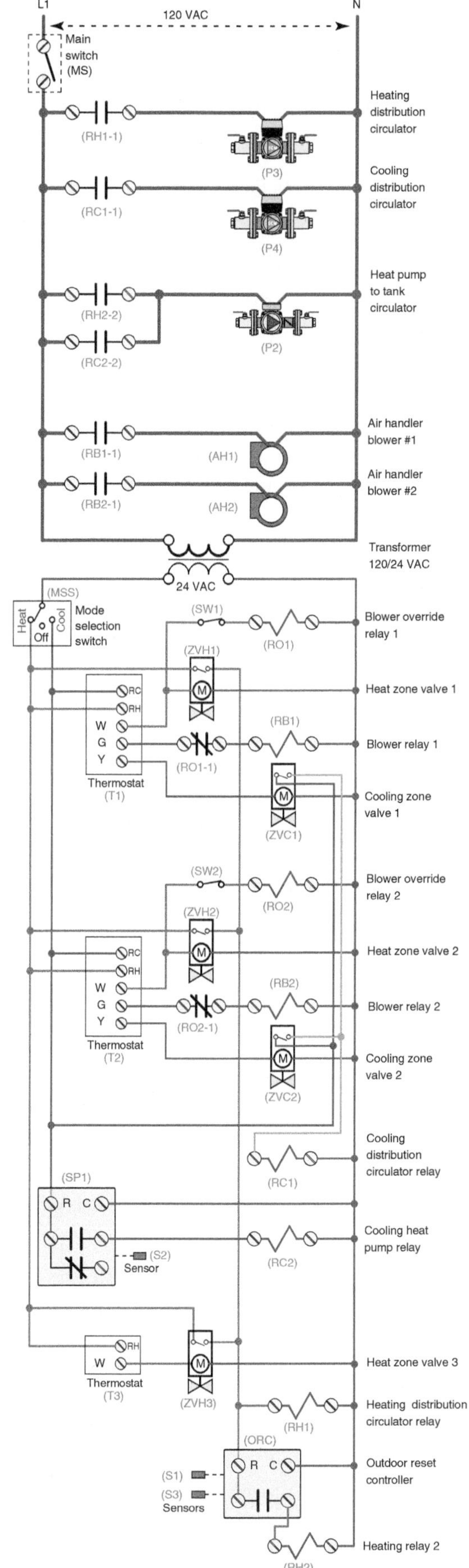

Figure 14-102c | Electrical wiring diagram for water-to-water heat pump combisystem #3.

pump and the buffer tank. The heat pump continues to run, assuming a heat demand from a thermostat is still present, until the temperature at sensor (S1) reach one half the differential above the target temperature. When all thermostats are satisfied, relay coil (RH1) and the outdoor reset controller (ORC) are turned off.

2. *Space cooling mode:* The mode selection switch (MSS) must be set for cool. This supplies 24 VAC to the (RC) terminals of thermostats (T1) and (T2). It also supplied 24 VAC to power on the cooling setpoint controller (SP1). (SP1) measure the temperature of the water in the buffer tank at sensor (S2). If the water temperature is above 60 °F, the relay contacts within (SP1) close. The passes 24 VAC to cooling relay (RC2). One set of contacts (RC2-1) close to connect the heat pump's (R) terminal to its (Y) terminal, which turns on the heat pump's compressor. Another set of contacts (RC2-3) close to connect the heat pump's (R) terminal to its (O) terminal. This turns on the heat pump's reversing valve to enable cooling mode operation. Another set of contacts (RC2-2) close to supply line voltage to circulator (P2) to provide flow between the heat pump and buffer tank. The heat pump continues to operate until the temperature of sensor (S2) drops to 45 °F, at which point (SP1) turns off the heat pump and circulator (P2).

If thermostat (T1) or (T2) call for cooling, 24 VAC is switched to the thermostat's (Y) terminal and to its (G) terminal. The 24 VAC at the (Y) terminal energizes the associated chilled water zone valves (ZVC1) or (ZVC2). When either of these zone valves reach their fully open position, their internal end switch closes, which applies 24 VAC to relay (RC1). A set of contacts (RC1-1) close to supply line voltage to cooling distribution circulator (P4). The 24 VAC supplied to the (G) terminal on either thermostat energizes the associated blower relay (RB1) or (RB2). A contact (RB1-1) closes to supply line voltage to the blower in air handler (AH1). A contact (RB2-1) closes to supply line voltage to the blower in air handler (AH1).

3. *Blower operation:* The blower in either air handler will always operate in cooling mode. It may also be set for operation in either heating or cooling mode by setting the fan switch on thermostat (T1) to "on." If the blower is not to run in the heating mode, switch (SW1) or (SW2) should be closed. This allows 24 VAC to energize the coil of relay (RO1) or (RO2) when the associated thermostat (T1) or (T2) calls for heat. A normally closed contact (RO1-1) or (RO2-1) opens, interrupting 24 VAC to the blower relay (RB1) or (RB2) and thus preventing the blower from operating. If switch (SW1) or (SW2) is left open, the blower will operate whenever the associated thermostat calls for heating.

4. *Domestic Water Heating:* Domestic water is circulated through the desuperheater in the heat pump whenever the compressor is operating. This provides preheating. The electric heating element in the upper part of the water heater provides any needed supplemental heating.

GEOTHERMAL WATER-TO-WATER HEAT PUMP COMBISYSTEM #4

Although it's possible to size a geothermal heat pump to provide the full design heating load of a house, there are other options. One that lends itself to modern hydronics technology is to combine the heat pump with a boiler. This is often called a "dual fuel" approach, and there are several benefits associated with it.

For example, in some locations, utilities offer steeply discounted **time-of-use electrical energy rates** during periods of low demand. These rates can significantly reduce the operating cost of a heat pump during off-peak hours, while allowing a gas-fired boiler to meet demand during "on-peak" periods when electrical rates are significantly higher.

A dual-fuel system also provides the security that one heat source can cover some or all of the heating load when the other heat source is down for maintenance.

Finally, a dual-fuel approach allows for the possibility of sizing the water-to-water geothermal heat pump to less than the design heating load of the building. This may be necessary due to limited land area for installation of the earth loop. If may also be necessary in situations where earth loop installation costs are high.

The schematic in Figure 14-103a shows how a water-to-water heat pump can be combined with a gas-fired or propane-fired mod/con boiler to provide a dual fuel combisystem for space heating, cooling, and domestic water heating.

This system uses a closed, slightly pressurized earth loop, with a valved manifold station as the beginning and ending point for all parallel earth loop circuits. The earth loop has an expansion tank as well as a combined air and dirt separator. Flow is provided by a single ECM-based circulator, which operates

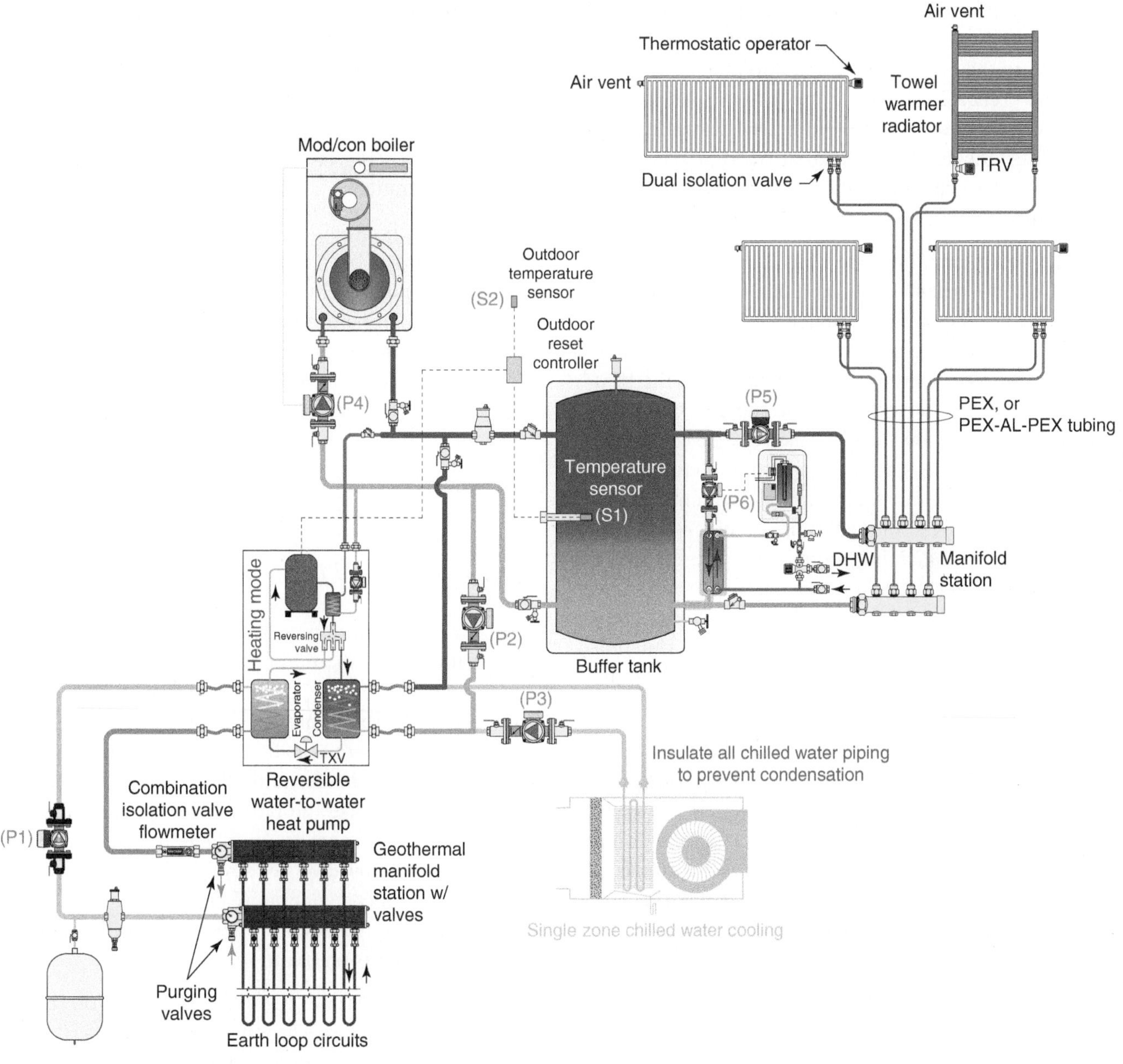

Figure 14-103a | Combisystem #4. Example of a dual fuel system (shown in heating mode).

whenever the heat pump's compressor is on. This circulator can operate in fixed-speed mode, or in a variable-speed mode if a suitable controller is provided that can optimize earth loop flow rate for a given operating condition.

The buffer tank receives heat from either the heat pump or the gas-fired boiler. This tank prevents short cycling of either heat source. It also allows both heat sources to operate simultaneously if needed under high-demand situations. Finally, this tank provides hydraulic separation between the various circulators used in the system.

Domestic water is heated "on-demand" using an external stainless steel brazed-plate heat exchanger. The details for this approach were discussed in Chapter 9. Whenever there is a demand for domestic hot water of 0.6 gpm or higher, the flow switch inside the tankless electric water heater closes. This closure, in combination with a relay, turns on circulator (P6) that moves water from the upper portion of the buffer tank through the primary side of this heat exchanger. Closure of the flow switch also energizes a contactor within the tankless heater. This supplies 240 VAC to triacs, which regulate the electrical current supplied to the heating elements. The triacs are under the control of electronics within the tankless heater that measure incoming and outgoing water temperature. The power delivered to the elements is limited to that required to boost the preheated domestic water to the desired hot water delivery temperature. All heated water leaving the tankless heater flows into an ASSE 1017–rated mixing valve to ensure a safe delivery temperature to the fixtures.

The heat pump's desuperheater also adds heat to the buffer tank whenever the compressor is operating. This heat can be used for space heating, or by the on-demand domestic water heating subsystem. When the heat pump is operating in cooling mode, the heat transferred to the buffer tank from the desuperheater is truly "free heat" that would otherwise be dissipated to the earth loop. This allows the earth loop to operate at slightly lower fluid temperatures, which slightly improves the EER of the heat pump. When the heat pump operates in heating mode, heat is transferred from the refrigerant to system water through the condenser as well as the desuperheater. This slightly increases the heat pump's COP, compared to a situation in which only the condenser is removing heat from the refrigerant.

The space heating distribution system uses panel radiators, sized for a supply water temperature of 120 °F at design load. Each panel radiator has a wireless thermostat radiator valve that allows room-by-room temperature control. All panel radiators are supplied through a homerun distribution system using 1/2-inch PEX or PEX-AL-PEX tubing from a common manifold station. Flow is created by a variable-speed pressure-regulated circulator set for constant differential pressure operation. For simplicity, only four radiators are shown. However, a larger manifold could be used to expand the distribution system to supply several more radiators.

The operation of space heating distribution circulator (P5) is controlled by the **master thermostat** (T1). If this thermostat is satisfied, circulator (P5) is turned off, and no heat flows to the distribution system. This allows the entire building to be put into a reduced temperature **setback mode** from a single thermostat. During the heating season, the setting of the master thermostat should be 2 °F or 3 °F above the normal desired air temperature. This maintains circulator (P5) in operation and allows the individual thermostatic radiator valves to "fine-tune" the comfort level in their respective spaces.

Notice that the piping turns downward from the outlet of circulator (P5) into the manifold station. This drop creates a **thermal trap** that discourages warm water migration into the manifold station when circulator (P5) is off. It eliminates the need for a spring-loaded check valve to prevent thermal migration. This is especially important when the buffer tank is warm and space heating is not needed, such as during the summer operation.

Cooling is provided by a single chilled water air handler that has been matched to the cooling capacity of the heat pump. Thus, no buffer tank is required in cooling mode operation. The air handler is equipped with a drip pan and drain to dispose of condensate that form on its coil. All portions of the piping system that handle chilled water are insulated and vapor sealed to prevent condensation.

One electrical schematic for combisystem #4 is shown in Figure 14-103b. *This schematic uses a manually operated switch that selects either the heat pump, or gas-fired boiler, as the system's sole heat source.* The selected heat source is always under the control of the outdoor reset controller. The boiler's internal controls should be configured to act solely as high temperature limit. On many boilers, this operating mode is used for domestic water heating. It allows the boiler to operate at full capacity whenever heat is needed in the buffer tank. This is acceptable with this system because the buffer tank provides thermal mass to prevent short cycling. The temperature limit of the boiler should be set several degrees above the maximum supply temperature from the buffer tank so that the outdoor reset controller retains full control of when the boiler is on or off.

An alternative electrical schematic for combisystem #4 is shown in Figure 14-103c. It replaces the heat source selection switch and the 1-stage outdoor reset controller with a **2-stage outdoor reset controller**. This controller operates the heat pump as the preferred (e.g., first stage) heat source for the buffer tank. If the heat output from the heat pump is sufficient to keep the buffer tank temperature at or close to the target temperature, the boiler will not operate. However, if heat output from the heat pump cannot keep pace with heat removal from the buffer tank, the boiler will be turned on as the second stage heat source. In this scenario, both the heat pump and boiler can simultaneously add heat to the buffer tank.

The following is a description of operation of the 2-stage operating mode of combisystem #4. Refer to Figure 14-103a for the piping arrangement and

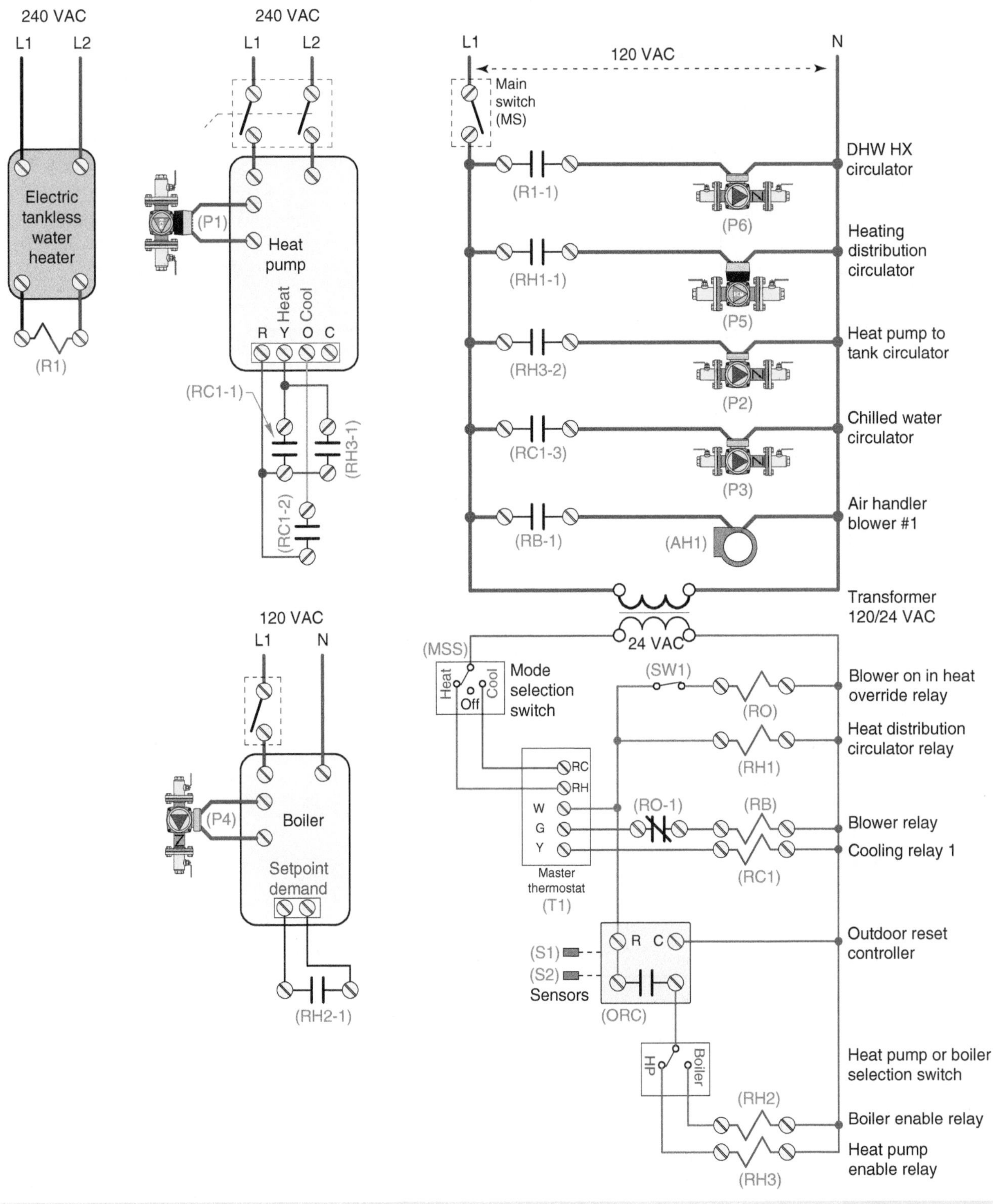

Figure 14-103b | Electrical wiring diagram for combisystem #4, using manually operated switch to select either the heat pump or the boiler as the sole heat source.

Figure 14-103c for the electrical schematic. Readers are encouraged to cross-reference the component designations, shown in parentheses, within both the piping and electrical schematics.

Description of Operation:

1. *Space heating mode:* The mode selection switch (MSS) must be set for heat. This supplies 24 VAC to the (RH) terminal on master thermostat (T1). Upon

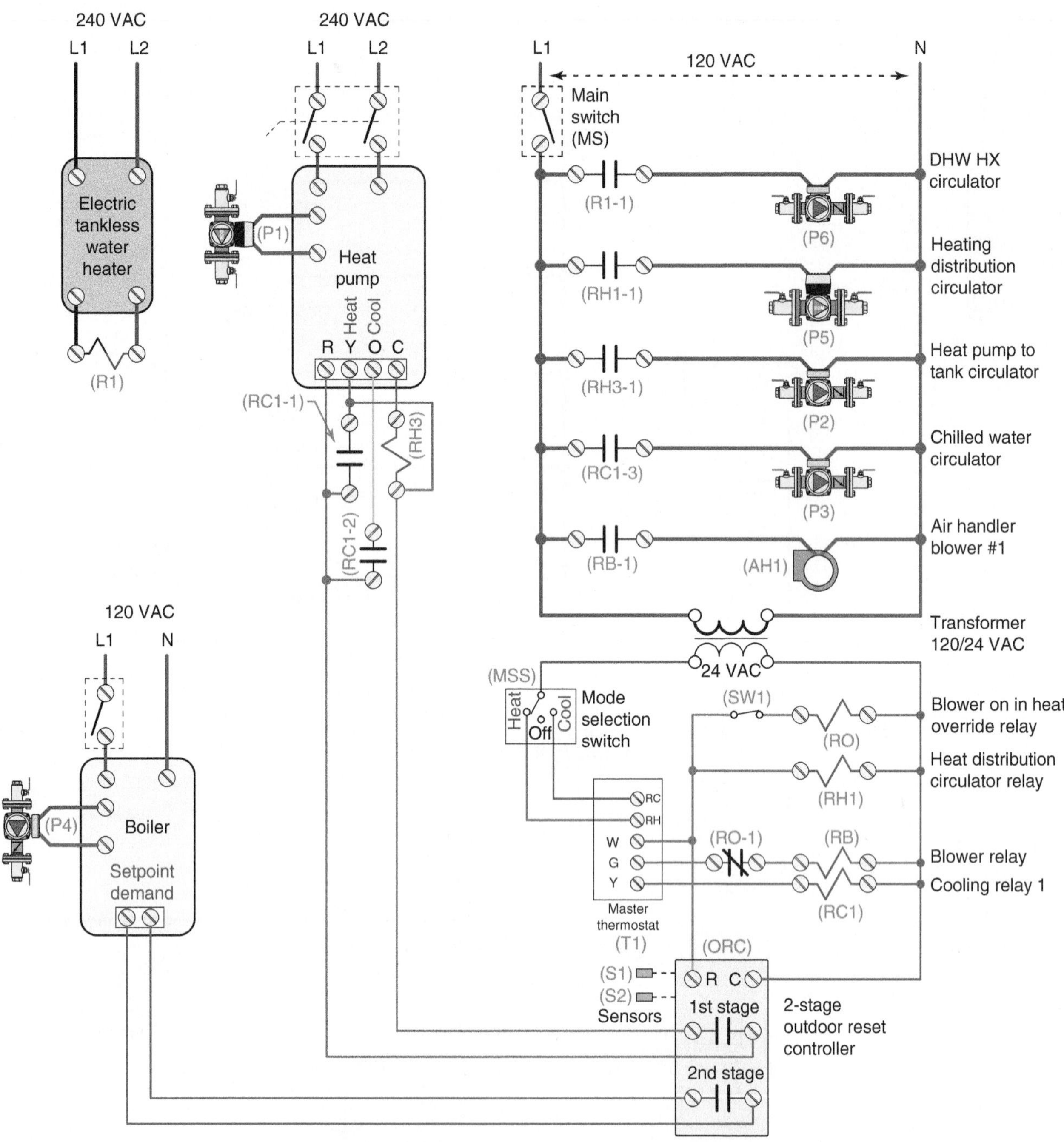

Figure 14-103c | Electrical wiring diagram for water-to-water heat pump combisystem #4, using 2-stage outdoor reset controller.

a call for heat from (T1), 24 VAC is passed from the thermostat's (W) terminal to energize the coil of relay (RH1). Relay contact (RH1-1) closes to supply line voltage to space heating distribution circulator (P5), which then operates in constant differential pressure mode. 24 VAC is also supplied to power up the 2-stage outdoor reset controller (ORC). This controller measures outdoor temperature at sensor (S2) and uses this temperature, along with its settings, to calculate the target water temperature in the buffer tank. If the actual temperature in the buffer tank, measured by sensor (S1), is slightly below the calculated target temperature, the first stage contacts in the (ORC) close. This completes a circuit between the (R) and (Y) terminals of the heat pump, causing it to operate in heating mode. It also supplies 24 VAC from the heat pump's transformer to the coil of relay (RH3). A contact (RH3-1) closes to supply line voltage to circulator (P2), which creates flow between the heat pump and buffer tank. Earth loop circulator (P1) is turned on by the heat pump's internal circulator relay whenever the heat pump operates in either heating or cooling mode.

If the heat pump is unable to maintain the buffer tank temperature near the target value, the 2nd stage contacts in the (ORC) close. The completes a 24 VAC circuit within the boiler that allows it to operate in setpoint mode. The boiler supplies 120 VAC to operate circulator (P4) whenever boiler operation is enabled. Under this condition, both the boiler and heat pump are adding heat to the buffer tank.

The heat output from each panel radiator is regulated by a wireless thermostatic radiator valve. Each valve varies the flow rate through its associated panel radiator based on its settings and the current room temperature.

2. *Space cooling mode:* The mode selection switch (MSS) must be set for cool. This supplies 24 VAC to the (RC) terminals of thermostat (T1). When this thermostat demands cooling, 24 VAC is passed from the thermostat's (Y) terminal to the coil of relay (RC1). One set of contacts (RC1-1) closes to complete a 24 VAC circuit from the heat pump's (R) terminal to its (Y) terminal. This enables the heat pump's compressor to operate. Another set of contacts (RC1-2) closes to complete a 24 VAC circuit from the heat pump's (R) terminal to its (O) terminal. This powers on the heat pump's reversing valve, allowing the heat pump to operate in cooling mode. Another set of contacts (RC1-3) closes to apply 120 VAC to circulator (P3), which creates flow between the heat pump and air handler (AH1).

Whenever thermostat (T1) calls for cooling, 24 VAC from terminal (G) is also passed to blower relay (RB). A set of contacts (RB-1) closes to supply 120 VAC to the blower in air handler (AH1).

During cooling mode, heat from the heat pump's desuperheater is also transferred to the buffer tank. A small circulator within the heat pump creates the necessary flow.

3. *Blower operation:* The blower in the air handler will always operate in cooling mode. It can also operate in either heating or cooling mode by setting the fan switch on thermostat (T1) to "on." If the blower is not to run during the heating mode, switch (SW1) should be closed. This allows 24 VAC to energize the coil of relay (RO), when thermostat (T1) calls for heat. A normally closed contact (RO-1) opens, interrupting 24 VAC to the blower relay (RB) and thus preventing the blower from operating. If switch (SW1) is left open, the blower will operate whenever thermostat (T1) calls for heating.

4. *Domestic water heating mode:* Whenever there is a demand for domestic hot water of 0.6 gpm or higher, the flow switch inside the tankless electric water heater closes. This closure applies 240 VAC to the coil of relay (R1). The normally open contact (R1-1) closes to turn on circulator (P6), which circulates heated water from the upper portion of the buffer tank through the primary side of the domestic water heat exchanger (HX1). The domestic water leaving (HX1) is preheated to a temperature a few degrees less than the buffer tank temperature. This water passes into the thermostatically controlled tankless water heater, which measures its inlet temperature. The electronics within this heater control the electrical power supplied to the heat elements based on the temperature rise needed to achieve the set domestic hot water delivery temperature. All heated water leaving the tankless heater flows into an ASSE 1017–rated mixing valve to ensure a safe delivery temperature to the fixtures. Whenever the demand for domestic hot water drops below 0.4 gpm, circulator (P5) and the tankless electric water heater turn off.

GEOTHERMAL WATER-TO-WATER HEAT PUMP COMBISYSTEM #5

It is possible to combine the heat output from a water-to-water heat pump with the heat supplied from a solar collector array. There are several ways to do this. For example, the solar collector array and the earth loop can both be connected to a low-temperature thermal storage tank. The heat pump would extract low-temperature heat from this tank and deliver higher temperature heat to the load or another buffer tank. This approach, while viable for space heating, fails to use the solar collectors to their full potential for domestic water heating, especially in warmer months.

Figure 14-104a shows an arrangement that avoids this limitation. Here, the solar collector array can supply heat to a domestic hot water storage tank, or to the earth loop, depending on the currently available solar energy and the needs of the system.

This design allows for several operating modes. For example, when there is sufficient solar radiation

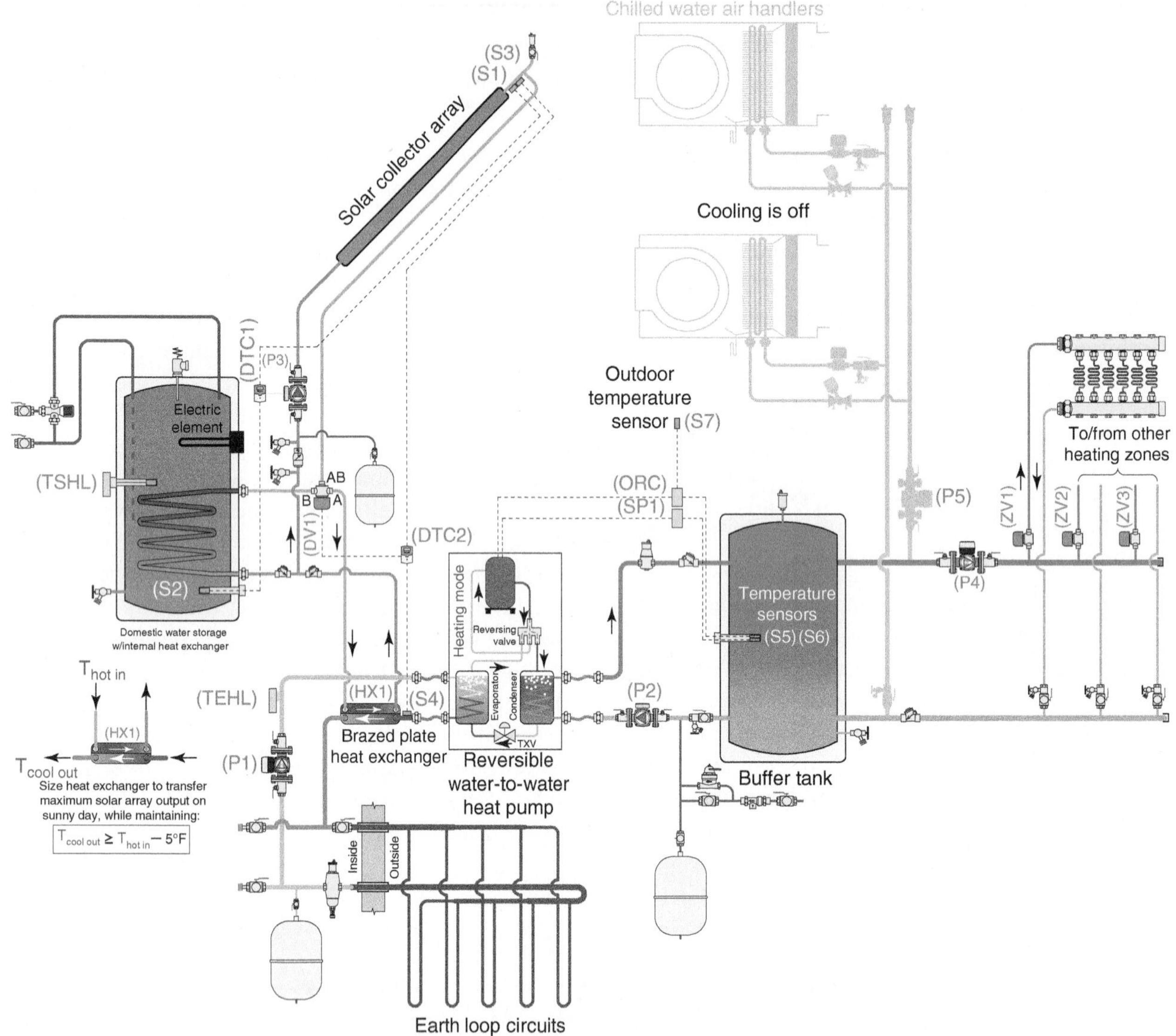

Figure 14-104a | Combisystem #5. Combining a water-to-water heat pump with solar collectors.

to operate the collector array for domestic water heating, that mode should have priority because it uses solar energy to offset heat that would other wise be provided by an electric heating element. There will be times, however, when the intensity of the solar radiation is not high enough to maintain energy flow into the domestic storage tank. Under this condition, it may still be possible to operate the collectors at much lower temperatures, and route the captured heat into the earth loop. In such a case, the diverter valve would be powered on to direct the flow of antifreeze solution from the solar collector array to a brazed-plate heat exchanger located in the earth loop circuit, on the *outlet* side of the heat pump. This arrangement provides the coolest fluid through the low temperature side of the heat exchanger and thus maximizes the approach temperature difference across the heat exchanger for optimal heat transfer.

Using this heat exchanger also allows the fluid in the earth loop to be different from that in the collector array. For example, in cold climate applications, the collector array may use a 50 percent solution of propylene glycol and water, while the earth loop only requires a 20 percent solution of propylene glycol. One could argue that using a 50 percent solution of propylene glycol in both the collector array and earth loop would eliminate the need for the heat exchanger. However, this would significantly increase installation cost based on the additional antifreeze required. It also would require substantially higher earth loop flow rates to maintain the turbulent flow necessary for good heat transfer when the earth loop is operating at or near its minimum temperature. The latter is due to the significantly increased viscosity as the percentage of propylene glycol in solution is increased. A high concentration of propylene glycol in the earth loop would also increase the power required for the earth loop circulator.

The brazed-plate heat exchanger is sized to transfer the anticipated heat output from the collector array, to the earth loop, using an approach temperature difference no higher than 5 °F. This minimizes the thermal penalty of having the heat exchanger in the system. The output from the collector array should be based on a sunny late winter or early spring day when the earth loop temperature is likely near it minimum value. This condition would tend to keep the collectors operating at relatively low inlet temperatures and thus high thermal efficiency and high heat output. It may even be possible for the solar collector to operate at inlet fluid temperatures *lower* than outdoor air temperature. Under such a condition, the collector would be absorbing solar energy as well as heat from the air surrounding the collector.

The piping arrangement shown in Figure 14-104a also allows the possibility of using the earth loop as a heat dump for the solar collector array. This is practical in systems where there is a relatively small cooling load, and thus the earth loop temperatures are not highly elevated due to heat dissipation from the heat pump. Some criteria would need to be established that prevents solar heat dissipation to the earth loop from significantly degrading the cooling performance of the heat pump. In northern climates, with short cooling seasons, this is not likely to be a problem. However, it is more likely to cause problems in southern climates with long cooling seasons and thus elevated earth loop temperatures.

A question that often arises is, Can this approach be used to store excess solar heat from summer, in the soil surrounding the earth loop? The intent of doing this is to extract the stored heat when it is needed in fall or winter. This is a hard question to answer with any degree of assurance. It is also one that is highly specific to the installation site. For example, if a site happened to have moving groundwater water in proximity to the earth loop, it is very likely that the groundwater would carry away any solar-derived heat dissipated into the soil. Thus, heat storage would not be possible. However, if the geology in proximity to the earth loop had minimum interaction with groundwater, the prospects of heat storage are better. At this point, there is no cost-effective way to evaluate the potential for such heat storage, especially in residential scale systems. Thus, the only conclusion is that it is a possibility, and its cost effectiveness could only be determined by experimenting with the system.

The balance of the system is similar to combisystem #3. During the heating season, the buffer tank is maintained at a suitable target water temperature by an outdoor reset controller. Flow is delivered to any active heating zone by a variable-speed pressure-regulated circulator set to operate in constant differential pressure mode. During the cooling season, the buffer tank temperature is maintained between 45 °F and 60 °F by a temperature setpoint controller. Chilled water is delivered to the air handler in any zone calling for cooling. Because a single buffer tank is used for heating and cooling, this system is *not* appropriate for climates where morning heating could be followed by a need for afternoon cooling.

The electrical wiring diagram for combisystem #5 is shown in Figure 14-104b.

A challenge with this system is determining the "optimal" control strategy for the solar subsystem. The goal is to use whatever solar energy can be captured to displace the greatest amount of electrical energy that would otherwise be used for space heating and domestic water heating. Solar energy used for domestic water heating offsets electrical resistance heating, which operates at a COP of 1.0. Solar energy directed into the earth loop will slightly raise the COP of the heat pump. For example, on an annual basis it *might* raise the average COP of the heat pump from 3.0 to 3.1. While any increase in COP is desirable, it's very hard to estimate what that increases might be based on solar energy input. Such a situation could be evaluated through computer simulation using specific performance models for all the components and loads involved. This would include the building heating load, the domestic hot water load, the thermal characteristics of the earth loop, the solar collector array, the two heat exchangers, air temperatures, and solar energy availability. While such simulation is possible using software tools such as TRNSYS, it is currently too expensive and time-consuming to be used for routine design purposes.

The control system shown in Figure 14-104b attempts to use available solar energy for domestic

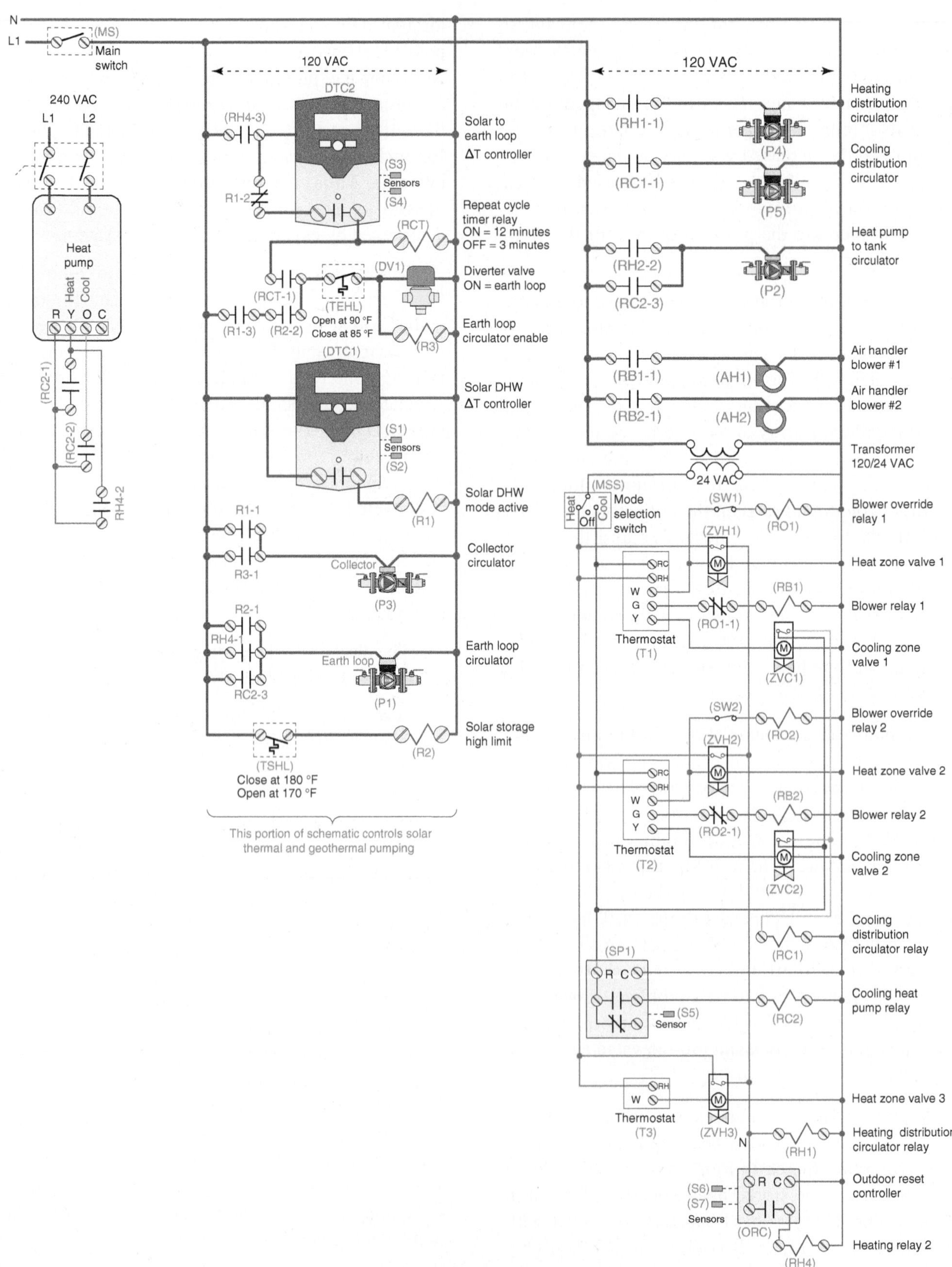

Figure 14-104b | Electrical wiring diagram for water-to-water heat pump combisystem #5.

water heating whenever possible. It otherwise routes available solar energy to the earth loop to increase the COP of the heat pump.

When the system is operating in the solar-to-earth loop mode, the collector operating temperature will be relatively low, even under bright sun conditions. Thus, it is possible that conditions that would otherwise allow solar energy to be added to the domestic hot water storage tank may occur, but would go "undetected" by a standard differential temperature controller that compares the collector temperature to the storage tank temperature.

Although it might be possible to detect such a condition using a sophisticated controller that accepts input from a solar radiation sensor, along with sensor inputs for ambient air temperature and tank temperature, and then combines these measurements with performance indices for the collectors and tank heat exchanger, such controllers are currently not available. In lieu of this, the control system in Figure 14-104b uses a **repeat cycle timer relay**. An example of such a device is shown in Figure 14-104c.

The repeat cycle timer relay is configured so that when its coil is powered on, a set of normally open contacts close for a user-determined time, and then open for another, independently set, user-determined time. For example, the contacts can close for 12 minutes and then open for 3 minutes. This cycling continues as long as the relay's coil remains energized.

(c)

Figure 14-104c | Repeat cycle timer relay and mounting socket.

The control system in Figure 14-104b is configured so that the repeat cycle timer relay (RCT) is energized each time a collector-to-earth loop heat transfer process is initiated. Its contacts close for 12 minutes, allowing the diverting valve (DV1) to route flow of antifreeze fluid from the collector array to heat exchanger (HX1), where the solar-derived heat is transferred to the earth loop. After 12 minutes, the contacts of the repeat cycle timer open. This turns off the diverter valve, as well as the collector circulator, for 3 minutes. During this time, the collector is stagnating. If the solar radiation intensity is moderate to high, the collector's absorber plate temperature will rise rapidly. This would allow the differential temperature controller that controls the collector-to-storage tank mode to initiate the process of moving heat from the collectors to the storage tank. Once initiated, this mode would "lock out" the solar-to-earth loop mode until the collector could no longer supply heat to the domestic hot water tank. At that point, control would be returned to the differential temperature controller that handles the solar-to-earth loop mode. Thus, whenever the heat pump is operating in heating mode, the system creates a 3-minute "sampling period" every 15 minutes, in which the collectors are "unloaded" from the earth loop to see if they could operate in the domestic water heating mode. The length of the on-cycle and off-cycle of the repeat cycle timer can be adjusted, as necessary, to improve system response.

The following is a description of operation of the combisystem #5. Readers are encouraged to cross-reference the component designations, shown in parentheses, within both the piping and electrical schematics. Readers are also encouraged to "hang in there." This is a relatively complicated control system to explain, but also one that shows the possibilities of relatively simple control devices.

Description of Operation:

1a. *Space heating mode:* The mode selection switch (MSS) must be set for heat. This supplies 24 VAC to all thermostats . Thermostat (T1) controls heating in zone 1 and cooling for zone 1. Thermostats (T2) controls heating for zone 2 and cooling for zones 2 and 3. If any thermostat calls for heating, 24 VAC is passed to the associated zone valve for that zone, causing it to open. When the zone valve is fully open, its end switch closes. This passes 24 VAC to heating relay (RH1) and powers on the outdoor reset controller (ORC). One relay contact (RH1-1) closes to supply line voltage to operate the heating distribution circulator

(P4). The outdoor reset controller (ORC) measures the outdoor temperature at sensor (S7). It uses this temperature, along with its settings, to calculate the target temperature for the buffer tank. It also measures the temperature of the water in the buffer tank at sensor (S6). If that temperature is more than half the set differential below the target temperature, 24 VAC is passed to the coil of relay (RH4). One contact (RH4-1) closes to supply line voltage to the earth loop circulator (P1). Another contact (RH4-2) closes across the (R) and (Y) terminals of the heat pump, turning the heat pump on in the heating mode.

1b. *Solar-to-earth loop mode:* During a demand for heat pump operation, in heating mode, a third contact (RH4-3) closes to supply line voltage to the differential temperature controller (DTC2). This controller measures the temperature at the top of the solar collector array using sensor (S3). It also measures the temperature of the water coming out of the heat pump evaporator at sensor (S4). If the collector temperature at (S3) is at least 5 degrees above the earth loop temperature at (S4), a relay contact within (DTC2) closes. This supplies line voltage to a repeat cycle timer (RCT). When powered on, the (RCT) closes its normally open contacts (RCT-1) for 12 minutes. This allows line voltage to pass through the normally closed contacts of thermostat (TEHL) and power on diverter valve (DV1) and the coil of relay (R3). The diverter valve directs heated antifreeze fluid leaving the collector array through the earth loop heat exchanger (HX1). Relay contact (R3-1) closes to supply line voltage to the collector circulator (P3). Captured solar energy is now being delivered to the earth loop. However, if the earth loop temperature exceeds 90 °F, as measured by thermostat (TEHL), the normally closed contacts within (TEHL) open, which turns of the diverter valve and relay (R3), and thus stops further solar heat input to the earth loop.

1c. *Solar-to-storage tank mode:* After 12 minutes have elapsed, the repeat cycle time (RCT) opens contacts (RCT-1) to turn of diverter valve (DV1) and relay (R3). This stops flow through the collector array for 3 minutes. If, during this 3-minute interval, the collector temperature increases to a value that is at least 10 °F above the storage tank temperature measured at sensor (S2), differential temperature controller (DTC1) passes 120 VAC to relay (R1). A relay contact (R1-1) closes to turn on the collector circulator (P3). Another normally closed contact (R1-2) opens to interrupt power through the output relay in (DTC2). The diverter valve (DV1) is now unpowered and, as such, directs flow leaving the collector array to the internal heat exchanger in DHW storage tank. The system is now in the solar-to-storage tank operating mode and will remain in that mode until the temperature difference between (S1) and (S2) drops to 3 °F or less, at which point power to relay (R1) is interrupted and the collector circulator is turned off.

1d. *Solar over-temperature mode:* This mode allows the earth loop to serve as a heat dump for the solar collector array if necessary. If the solar-to-storage tank mode is active (e.g., Relay R1 is energized), and the temperature of the solar storage tank reaches 180 °F, as detected by thermostat (TSHL), relay (R2) is turned on. One contact, (R2-1) closes to supply 120 VAC to earth loop circulator (P1). Another contact (R2-2) closes to turn on the diverter valve (DV1). Heat is now directed from the collector array to the earth loop heat exchanger (HX1). This mode continues until the storage tank temperature drops to 170 °F or until (DTC1) turns off relay (R1).

1e. *Earth loop over-temperature mode:* If the earth loop temperature entering the heat pump reaches 90 °F, thermostat (TEHL) opens. This prevents any further heat transfer from the solar collectors to the earth loop. This operating mode should be rare for systems with properly sized earth loops, especially in northern climates. The contacts in thermostat (TEHL) close when the earth loop cools to 85 °F. These settings are adjustable.

2. *Space cooling mode:* The mode selection switch (MSS) must be set for cool. This supplies 24 VAC to the (RC) terminals of thermostats (T1) and (T2). It also supplies 24 VAC to power on the cooling setpoint controller (SP1). (SP1) measure the temperature of the water in the buffer tank at sensor (S5). If the water temperature is above 60 °F, the relay contacts within (SP1) close. This supplies 24 VAC to cooling relay (RC2). One set of contacts (RC2-1) closes to connect the heat pump's (R) terminal to its (Y) terminal, which turns on the heat pump's compressor. Another set of contacts (RC2-2) closes to complete a circuit from the heat pump's (R) terminal to its (O) terminal, which turns on the reversing valve so the heat pump is operating in cooling mode. Another set of contacts (RC2-3) closes to supply line voltage to circulator (P2) to provide flow between the heat pump and buffer tank. The heat pump continues to operate until the buffer tank sensor (S5) drops to 45 °F.

If thermostat (T1) or (T2) call for cooling, 24 VAC is switched to the thermostat's (Y) terminal, and to its

(G) terminal. The 24 VAC at the (Y) terminal energizes the associated zone valves (ZVC1) or (ZVC2). When either zone valve reaches its fully open position, its internal end switch closes, which supplies 24 VAC to relay (RC1). A set of contacts (RC1-1) closes to supply line voltage to cooling distribution circulator (P5). The 24 VAC supplied to the (G) terminal on either thermostat energizes the associated blower relay (RB1) or (RB2). A contact (RB1-1) closes to supply line voltage to the blower in air handler (AH1). A contact (RB2-1) closes to supply line voltage to the blower in air handler (AH2).

3. *Blower operation:* The blower in either air handler will always operate in cooling mode. It can also be set to operate in heating mode by setting the fan switch on thermostat (T1) to "on." If the blower is not to run in the heating mode, switch (SW1) or (SW2) should be closed. This allows 24 VAC to energize the coil of relay (RO1) or (RO2), when the associated thermostat (T1) or (T2) calls for heat. A normally closed contact (RO1-1) or (RO2-1) opens, interrupting 24VAC to the blower relay (RB1) or (RB2) and thus preventing the blower from operating. If switch (SW1) or (SW2) is left open, the associated blower will operate whenever the associated thermostat calls for heating.

4. *Domestic Water Heating:* Domestic water is preheated by the solar collector array. Any required supplemental heating is provided by the electric heating element in the upper port of the domestic hot water storage tank.

SUMMARY

This chapter has discussed the operating characteristics of modern water-to-water heat pumps supplied from geothermal heat sources. It has also shown the versatility they bring to combisystems designed for residential and light commercial buildings. Several of the details discussed are similar or identical to those used with air-to-water heat pumps.

Designers should always keep the following points in mind:

- To maximize the heating capacity and COP of water-to-water heat pumps, design space heating distribution systems for a maximum supply water temperature of 120 °F under design load conditions. Even lower supply water temperatures are preferred when possible.
- Always ensure there is adequate flow through the heat pump's evaporator and condenser whenever the compressor is operating.
- Always use a buffer tank between an on/off heat pump and a zoned heating or cooling distribution system.
- Make use of outdoor reset control to keep the supply water temperatures to space heating distribution systems as low as possible during partial load conditions.
- Whenever possible, specify high-efficiency ECM-based circulators. They can be used for fixed or variable flow in earth loops as well as for variable flow in distribution systems using valved-based zoning.
- Use an appropriately sized expansion tank in every closed earth loop.
- Equip earth loops with high-efficiency air and dirt separators.

ADDITIONAL RESOURCES

1. (2009). *Ground Source Heat Pump Residential and Light Commercial Design and Installation Guide*. International Ground Source Heat Pump Association/Oklahoma State University. ISBN 978-0-929974-07-1.
2. Dickie, Eric. *Energy Exchange, Geothermal Exchange and Beyond*, 3rd edition. ISBN 978-0-9782997-9-8, Simonne Dickie.

KEY TERMS

2-stage outdoor reset controller
2-stage temperature set-point controller
4-pipe square
air and dirt separator
amplitude
annual average soil temperature
ANSI/AHRI/ASHRAE/ISO Standard 13256-2
automatic backwash filter
average soil surface temperature
backwash cycle
ballast
bead
bentonite
biocide
Boyle's law
brine
buffer tank
bulk air
butt fusion machine
butt fusion welding
calcium chloride
canister
cartridge filter
centipoises
chain trencher
chamfer tool
closed earth loop
coefficient of linear expansion
coefficient of performance (COP)
cold expansion compression sleeve
cold ring clamp
cooling COP
cross-linking
cupronickel
Cv value
degree days (°F·days)
denatured [alcohol]
desuperheater
desuperheating
diameter ratio (DR)
dimensionless number
double rollback bead
dry contact
dual-fuel approach
earth-coupled heat pump
effective pump curve
electric heating tool
electrofusion
energy efficiency ratio (EER)
entering load water temperatures (ELWT)
entering source water temperature
ethanol
evaporator mean fluid temperature at design conditions
facing tool
flow center
flow coefficient
foot valve
fouling factor
freezestat
geothermal heat pump
geothermal manifold station
GLHP
ground-source heat pump
grout
GWHP
heat fusion welding
heating run fraction (f_H)
high-density polyethylene (HDPE)
high-pressure safety switch
horizontal earth loop
hydronic detergent
inhibited propylene glycol
inhibitors
laminar flow
manifold station
master thermostat
mechanically compressed collar
MEK (methyl ethyl ketone)
methanol
micron
mineralogy
minimum entering loop fluid temperature
modulating water-to-water heat pumps
net COP
oxygen scavenger
Pete's Plugs
PEX-a
pipe resistance (R_p)
pipe size multiplier (P_m)
pitch [of the slinky]
potassium acetate
ppm
pressure switch
priority load
probe
purge cart
reciprocating compressor
reinforced hose assemblies
repeat cycle timer delay
Reynolds number
rotameter
scroll compressor
seasonal average COP
series circuit
setback mode
sink fluid
slinkies
socket fusion
socket fusion adapters
soil resistance (Rs)
source fluid
spin-down filter
standing column flow center
submerged heat exchanger
submersible well pump
tailings
thermal diffusivity
thermal trap
thermistor
thermoplastic
time-of-day electrical rates
tons [of capacity]
tremie
turbulent flow
U-tube assembly
undisturbed soil temperature
units check
utility vault
valveless header
variable frequency drive (VFD)
variable-speed submersible pump
vault enclosure
vertical earth loop
water quality
water quantity
water-to-water heat pump

QUESTIONS AND EXERCISES

1. Describe the differences between the principal components in a water-to-water heat pump versus an air-to-water heat pump.

2. Describe what could happen if the fluid flow rate from the earth loop to the *evaporator* of a water-to-water heat pump is too low.

3. Describe what could happen if the water flow rate through the *condenser* of a water-to-water heat pump was too low.

4. What load is supplied by the desuperheater within a water-to-water heat pump?

5. Describe the difference in the condition of the refrigerant from when it exits the compressor, to when it exits the desuperheater.

6. Describe what happens to the COP of a water-to-water heat pump under the following conditions, when all other operating conditions remain the same:
   a. The temperature of the fluid entering the heat pump's evaporator drops
   b. The temperature of water entering the heat pump's condenser increases
   c. The flow rate through the heat pump's evaporator decreases

7. Water flows into the condenser of a water-to-water at 100 °F and 8 gpm. The heat pump is operating at a heating capacity of 36,000 Btu/hr. Determine:
   a. The water temperature leaving the condenser
   b. The temperature rise of the water through the condenser

8. A water-to-water heat pump has the following conditions:

   Water flow rate through condenser = 9 gpm

   Water temperature rise across condenser = 10 °F

   Wattage to operate compressor = 3 KW

   Head loss across the condenser = 15 feet at flow of 9 gpm

   Determine:
   a. The current heating capacity of the heat pump
   b. The current COP of the heat pump
   c. An estimate of the rate at which low temperature heat is being absorbed by the heat pump's evaporator.

9. Calculate the "net COP" for the heat pump described in Exercise 8 based on the ANSI 13256-2 standard. Describe the difference between the COP computed in Exercise 8 and the "net COP" of the same heat pump as it would be calculated based on the ANSI 13256-2 standard.

10. Which of the following will *improve* the EER of a water-to-water heat pump?
    a. Increasing the condenser water temperature
    b. Decreasing the chilled water temperature
    c. Increasing the earth loop temperature
    d. Increase the chilled water supply temperature.

11. Why is a buffer tank necessary when connecting an on/off water-to-water heat pump to a zoned distribution system?

12. A water-to-water heat pump system includes a 50-gallon buffer tank. The controller operating the heat pump is set to turn it on when the buffer tank temperature drops to 85 °F, and off when the buffer tank temperature reaches 110 °F. Determine the length of the heat pump on-time if that heat pump delivers 38,000 Btu/hr. to the tank, when there is no concurrent space heating load on the system.

13. What happens to the on-time of a heat pump when each of the following modifications is made, while all other operating conditions remain the same:
    a. The volume of the buffer tank is increased from 30 to 60 gallons?
    b. The temperature increase of the buffer tank is changed from 15 to 30 °F?
    c. Both A and B modifications are made?

14. What are the functions of the inhibitor in inhibited propylene glycol?

15. What are the functions of reinforced hoses connecting a water-to-water heat pump to rigid piping?

16. What type of air cannot be eliminated from the earth loop circuit by forced-water purging?

17. A water-to-water heat pump operates with a heat output of 43,000 Btu/hr. from its condenser, and a COP of 3.6. Determine the required water flow rate through the *evaporator* based on the goal of 2.5 gpm per ton.

18. Describe the function of the flow switch in a system that supplies well water to a water-to-water heat pump.

19. Pete's Plugs are installed on the earth loop side of a water-to-water heat pump. At the time, the earth loop is operating with water at a temperature of about 60 °F. When gauges are inserted into the Pete's Plugs, one reads 15 psi, and the other reads 9.0 psi. Assuming the heat pump has the pressure drop curve shown in Figure 14-23, determine the water flow rate through the earth loop.

20. Describe a soil condition in which a standard chain trencher would not be a good choice for installing earth loop piping.

21. Why can't PEX tubing be joined using butt fusion? Can PEX tubing be joined using socket fusion?

22. What is the purpose of grouting an earth loop piping circuit after it has been inserted into a borehole. What might happen if the grouting was omitted?

23. When connecting several earth loop circuits to a common header assembly, describe:
    a. A situation in which the header piping size can remain constant.
    b. A situation in which the header piping sizes should be stepped.

24. An earth loop contains 4,000 feet of 3/4-inch piping, 125 feet of 1.25-inch piping, and 50 feet of 2-inch piping. All piping is DR-11 HDPE. Determine:
  a. The total volume of the earth loop piping
  b. The total surface area of the earth loop piping
  c. The amount of inhibited propylene glycol required if a 20 percent solution is desired in this earth loop.

25. Why should a cold ring clamp always be used when making a socket fusion weld in HDPE tubing?

26. An earth loop consists of three parallel branches of 3/4-inch by 1000 foot long HDPE DR-11 tubing connected to a non-valved header. The header is very short. The mains piping from that header to the heat pump is 60 feet of 1.25-inch HDPE DR-11 tubing. The earth loop will operate with a 20 percent solution of inhibited propylene glycol that will have a minimum temperature of 30 °F. Determine:
  a. The minimum flow rate through the mains piping, and the branch piping, based on maintaining turbulent flow (e.g., a Reynolds number of at least 2,500).
  b. Once these flow rate are determined, estimate the head loss of the overall earth loop when operating at these flow rates while carrying the 20 percent propylene glycol solution at an average temperature of 30 °F.

27. Describe at least three benefits of including an expansion tank in a closed earth loop circuit.

28. Describe why the pressure in a closed earth loop circuit increases when the heat pump is operating in a heating mode. Why does this pressure decrease when the heat pump is operating in a cooling mode?

29. An earth loop contains 4,000 feet of 1-inch DR-11 HDPE tubing and 100 feet of 1.5-inch DR-11 HDPE tubing. The earth loop operates with a 20 percent solution of propylene glycol. The system is filled with this solution when the earth loop piping and fluid are both at 50 °F. Determine the amount of fluid:
  a. Extracted from the expansion tank when the loop temperature increases to 90 °F.
  b. Added to the expansion tank when the loop temperature drops to 30 °F.

30. Determine the minimum undisturbed soil temperature at a depth of 5 feet for Seattle, WA. Assume a light wet soil.

31. A heat pump with a heating capacity of 36,000 Btu/hr. at a corresponding minimum entering loop fluid temperature of 30 °F, has been selected for a house with a design heating load of 40,000 Btu/hr. The load is based on an indoor temperature of 68 °F, and an outdoor design temperature of 0 °F. The house is located in Buffalo, NY, where 1283 (base 65) heating degree days occur during a typical January. Estimate the heating run fraction of the heat pump in this application:

32. Why is it preferable to fill and flush the earth loop before filling and flushing the heat pump?

33. Why should the expansion tank on a closed earth loop system be partially filled with fluid at the conclusion of the purging process?

34. Describe a system in which a buffer tank is not needed between the water-to-water heat pump and heating or cooling distribution system.

35. Describe at least three benefits of a "dual-fuel" system that combines a water-to-water heat pump with a gas-fired mod/con boiler.

36. For the system shown in Figure 14-101a, why is a buffer tank not required in cooling mode operation?

37. What are the advantages of using an outdoor reset controller to determine the water temperature in a buffer tank rather than a simple temperature setpoint controller?

38. Describe how the system shown in Figure 14-103a uses heat from the desuperheater to help heat domestic water when the system in operating in:
  a. Cooling mode
  b. Heating mode

39. Describe the purpose of the repeat cycle timer relay in the system shown in Figure 14-104b.

40. Describe two situations in which it would be detrimental to dissipate heat from solar collectors into an earth loop.

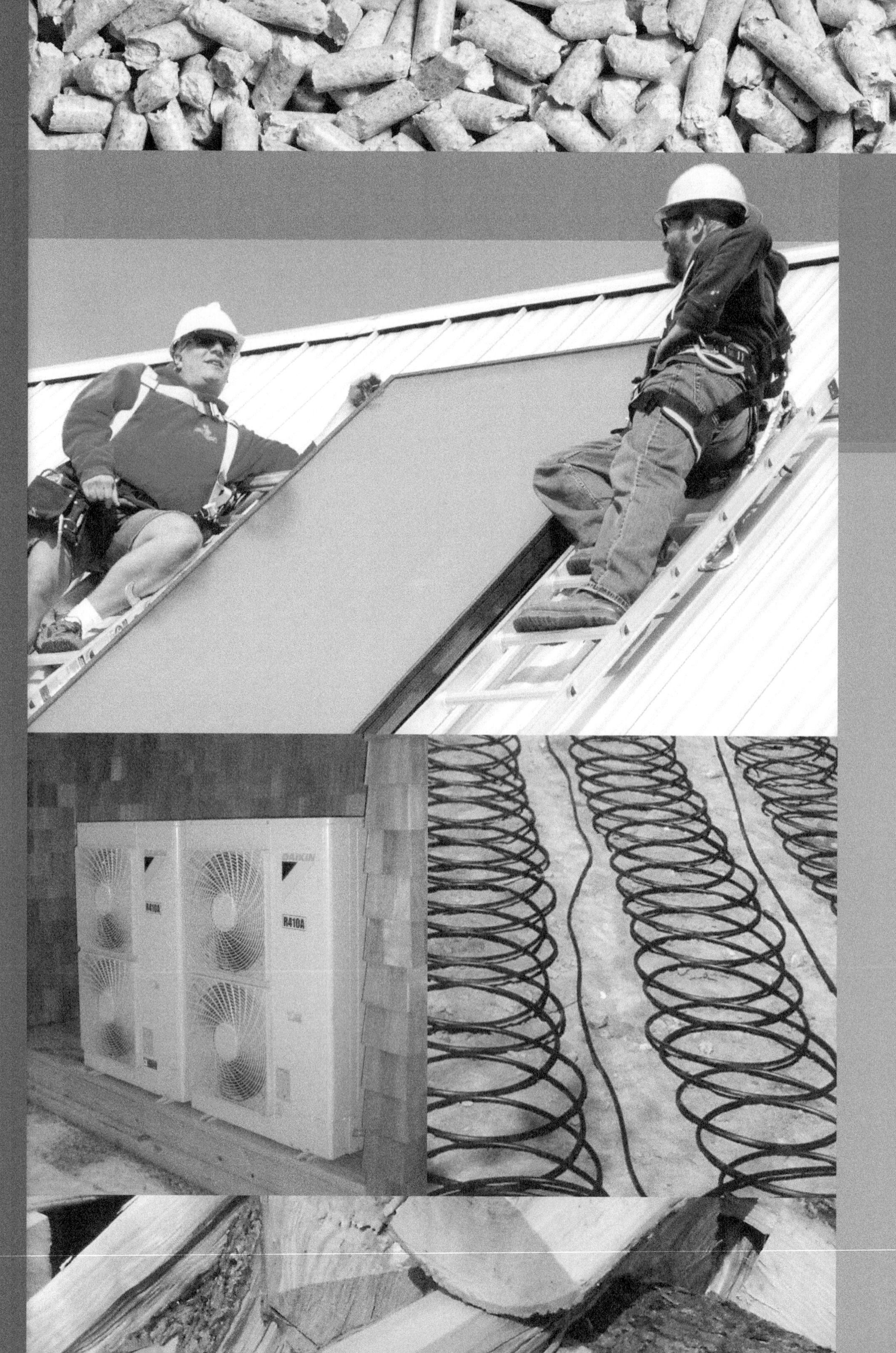
R410A
DAIKIN
R410A

CHAPTER 15

Wood-Fired and Pellet-Fired Hydronic Systems

OBJECTIVES

After studying this chapter, you should be able to:

- Describe some early attempts at wood-fired hydronic heating.
- Determine the heating value of various types of wood.
- Compare the heating energy available from wood-based fuels to those of other fuels.
- Understand the need to protect wood-fired boilers from low water temperatures.
- Describe several methods of providing boiler protection.
- Explain the operation of wood-gasification boilers.
- Discuss the preferred method of installing outdoor wood-fired heaters.
- Describe several design details for systems using wood-gasification boilers.
- Size a thermal storage tank for wood-fired and pellet-fired boiler systems.
- Design combisystems using wood-fired and pellet-fired boilers.
- Combine wood-fired hydronic heat sources with conventional heat sources.
- Explain the concept of temperature stacking in a thermal storage tank.
- Size a wood-gasification boiler for a specific load.
- Size a pellet-fired boiler for a specific load.
- Understand the operation of several example systems.
- Avoid common design and installation errors associated with wood-fired hydronic systems.

15.1 Introduction

Wood has been used as a heating fuel for centuries. One estimate is that humans have used wood for warming shelters and cooking for over 10,000 years. Its availability, or lack of availability, has even shaped the development of civilization. Recorded history shows a recurring cycle based on the availability of wood for fuel. The pattern begins with relatively abundant sources of wood growing close to where the wood was needed. Then, as populations expanded, the demand for wood outpaced local supply. The logistics of supply and transportation became more tenuous. This forced civilizations to adapt, either through better use of other energy sources such as the sun or by moving on to another location where wood was more accessible.

Until methods for extracting and transporting fossils fuels and electricity became available in the 1900s, wood was the primary heating fuel in American homes. During this time, nearly all North American homes were constructed with one or more wood-burning fireplaces or stoves. Many homes had "woodsheds" attached to them, or located close by, which were filled and emptied on an annual basis. Harvesting wood for fuel was a common as harvesting crops for food.

TRANSITION TO FOSSIL-BASED ENERGY

As interstate transportation systems were developed for petroleum and natural gas, wood was increasingly viewed as an antiquated heating fuel. This perception was reinforced by the fact that oil, gas, and electrical heat sources provided *automatic operation*, freeing occupants to pursue other activities rather than gathering wood for fuel, or tending fires. For many people, home heating became as simple as turning a thermostat dial to the desired temperature and paying a monthly fuel bill. This convenience appealed to those who had spent years cutting, splitting, and carrying firewood. Petroleum companies and utilities focused their marketing on persuading homeowners to "modernize" to fully automatic heating using oil, gas, and electricity. At the time, there was virtually no concern about long-term sustainability or the ecological consequences associated with heating buildings using fossil fuels.

Today, the worldwide effects of energy options are better understood, rapidly communicated, and more accurately extrapolated into the future. Looking beyond political or organizational agendas, it seems obvious that increasing population imposes constraints that must be respected if quality of life is to be maintained or improved. One of the most pressing of those constraints is establishing sustainable energy supplies. This has led to renewed interest in wood as a heating fuel.

REEMERGENCE OF WOOD AS A HEATING FUEL

Wood is considered a **carbon-neutral** fuel. Burning wood only releases the carbon that was already embodied in it through photosynthesis. Unlike burning fossil fuels, and on a long-term basis, no additional carbon is brought to the earth's surface or into its atmosphere by burning wood. This characteristic is now seen as highly desirable. Carbon neutrality, combined with modern technologies that can convert the fuel energy of wood into heat with efficiencies upward of 90 percent, has sparked global interest in wood as a "rediscovered" fuel for central heating.

Estimates based on the 2012 U.S. Census indicate that approximately 2.5 million American households now use wood as a primary heating fuel. This makes wood the nation's most used renewable heating energy source. The use of firewood and wood pellets for heating has increased by approximately 34 percent between 2000 and 2010. As this text is being written, there is pending federal legislation that would further accelerate this growth through tax incentives to homeowners who buy qualifying high-efficiency wood-fired or pellet-fired heat sources. This legislation provides incentives that parallel those available for solar thermal and geothermal heat pump systems.

This chapter describes how to use modern hydronics technology to make the most of what wood, as a heating fuel, can offer. It introduces modern wood-fired and pellet-fired hydronic heat source that extract as much heat as possible from wood. It goes on to show how to use other hydronic hardware, much of which was introduced in previous chapters, to deliver this heat in a very controlled manner that produces excellent comfort and reliable operation.

15.2 The Evolution of Wood-Fueled Hydronic Heating

Early attempts at combining wood-derived heat with hydronic distribution systems produced mixed results. Following the oil embargo of the early 1970s, many individuals and companies quickly began exploring ways in which wood-derived heat could be combined with the hydronic distribution systems of that era. Although some successful products were developed, there were also many devices put on the market that had little, if any, engineering or testing to back them up.

RUDIMENTARY DEVICES

Many of these devices could be described as rudimentary heat exchangers attached to or placed within fireplaces or wood stoves. The fireplace grate depicted in Figure 15-1 is one example. The water jacket installed on a wood stove flue in Figure 15-2 is another.

The thermal performance of these devices was often limited by existing materials, manufacturing methods, shape constraints, or fuel choices. In many cases, the formation of **creosote**, or accumulation of ash, rapidly diminished the rate at which these devices could transfer heat to water passing through them. They appealed to those who were looking for quick solutions but unaware of the principals necessary for efficient combustion of firewood, good heat transfer, and product longevity. The marginal performance of some devices, combined with safety issues, discouraged further expansion of the market, especially when fossil fuel prices decreased and supplies were abundant.

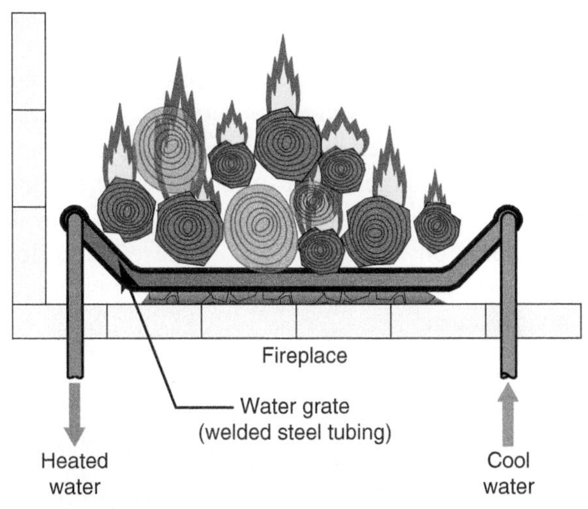

Figure 15-1 | Concept of a water-cooled fireplace grate.

Figure 15-2 | A water heating coil wrapped around a section of flue pipe.

OUTDOOR WOOD-FIRED HYDRONIC HEATERS

Another wood-burning product that developed in North America over the last three decades is known as an **outdoor wood-fired hydronic heater**. These unpressurized devices resemble a small, steel storage building, as shown in Figure 15-3. They are usually located several yards away from the building(s) to which they supply heat. During cold weather, they operate with a continuously maintained fire and transfer heat into a water-filled compartment that surrounds the combustion chamber. The heated water flows through underground piping to a heat exchanger inside the building. The large combustion chamber can hold enough wood to sustain the fire for several hours.

The combustion chamber in outdoor wood-fired heaters is largely surrounded by a steel **water jacket**. The water that circulates through the heater passes through this water jacket and absorbs heat from the combustion chamber. This arrangement, while straightforward to manufacture, does not present optimal conditions for efficient combustion. The surfaces of the water facing into the combustion chamber are relatively cool in comparison to the 1000+ °F temperatures needed for efficient combustion of the wood.

Figure 15-3 | Example of an outdoor wood-fired heater.

This leads to incomplete combustion along with creosote formation on those surfaces. These conditions are worsened when the boiler is in **stand-by mode**, which occurs after the heater's **aquastat** detects a high-temperature limit and responds by closing a **damper** or turning off a blower to reduce air flow to the combustion chamber. Low combustion efficiency leads to significant smoke generation, as shown in Figure 15-4. It also results in thermal efficiencies in the range of 40 percent or lower, depending on fuel quality and the size of the heater relative to the design heating load of the building it serves. The more oversized the heater, the lower its average thermal efficiency.

Outdoor wood-fired heaters have recently come under considerable scrutiny from regulators such as the **Environmental Protection Agency (EPA)**. The primary concern is relatively high **$PM_{2.5}$ particulate emissions** (e.g., emission of particulates in sizes equal to or smaller than 2.5 microns). Such particles can penetrate the innermost tissue structure of human lungs and have been linked to several adverse health effects.

Although not sanctioned by manufacturers, these units have combustion chambers large enough to burn materials other than properly dried firewood, which can also lead to excessive emissions. In some jurisdictions, the installation of these devices is prohibited or severely restricted due to their potential to negatively affect both local and regional air quality.

Still, there are tens of thousands of these units installed in North America, and more continue to be installed where permitted. This chapter addresses the market reality of such units and shows preferred methods of installation that make the most of what they offer.

SINGLE-STAGE WOOD-FIRED BOILERS

There are also **pressure-rated wood-fired boilers** built of welded steel or cast iron, as seen in Figure 15-5.

Pressure-rated boilers allow the entire hydronic system to be designed as a closed loop. These boilers are sometimes described as **natural draft wood-fired boilers** or **single-stage wood-fired boilers**. They rely on the draft created by the chimney to draw surrounding air into their combustion chamber through a **thermostatically controlled damper**. As the temperature of the water inside the boiler decreases due to less heat from the combustion process, the damper modulates open to allow greater air (oxygen) flow into the combustion chamber. This increases the rate of combustion, assuming there is adequate firewood in the combustion chamber.

Figure 15-4 | Emissions from outdoor wood-fired heaters, especially during stand-by operation, are significantly higher than those from other methods of wood fueled combustion.

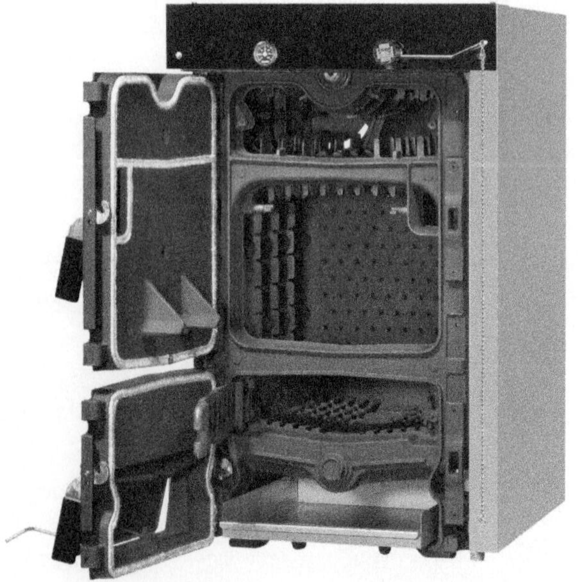

Courtesy of OHT, Inc.

Figure 15-5 | Example of a cast-iron pressure-rated wood fired boiler.

WOOD-GASIFICATION BOILERS

The current state-of-the art in cordwood fired hydronic heat sources is a **wood-gasification boiler**. Wood gasification involves heating wood in the absence of sufficient oxygen for complete combustion. The pyrolytic gases emitted by the wood under these conditions contain chemical energy that can be efficiently extracted through controlled combustion in a **secondary combustion chamber** supplied with superheated air.

An example of a modern wood-gasification boiler is shown in Figure 15-6a. Figure 15-6b shows the secondary combustion chamber of this boiler when it is operating. The flame in this chamber can be likened to that of a blowtorch. Combustion temperatures over 2,000 °F can be achieved. The result is a highly efficient conversion from chemical to thermal energy, with minimal ash accumulation. The wood-gasification process, and wood-gasification boilers, are discussed in more detail later in this chapter.

PELLET-FIRED BOILERS

Another example of a modern wood-fueled heating device is the **pellet-fired boiler**, shown in Figure 15-7.

One advantage of most pellet-fired boilers is that the combustion process can be *automatically* turned on and off. Some pellet-fired boilers can also modulate their heat output when necessary. This allows the output of a pellet-fired boiler to better match the heating needs of a building without need of a large thermal storage tank, although some thermal storage is still preferable. High-quality "premium" wood pellets also reduce particulate emissions and ash compared to firewood. Pellet boilers can be configured for automatic fuel loading from a bulk storage device. The details of pellet boilers and fuel supply systems are discussed later in this chapter.

Another category of wood-burning hydronic heaters that has recently become available in North America is known as a **hydronic wood stove**. Examples of such products are shown in Figure 15-8.

At present, most of these devices are manufactured in Europe, and some are imported to North America. They are natural draft devices and do not require fans or blowers to maintain operation. The combustion process, including secondary combustion in some units, is driven by chimney draft. These devices are designed to deliver 50 to 75 percent of their heat production to a stream of water circulated through their internal heat exchanger. The remaining heat output is delivered directly to the room. Hydronic wood stoves are often used in combination with a thermal storage tank. This allows for direct heating of the space where the stove

Courtesy of Mark Odell

(a)

Courtesy of Mark Odell

(b)

Figure 15-6 (a) Example of a modern wood-gasification boiler. (b) Flame within the secondary combustion chamber of a wood-gasification boiler.

Courtesy of Maine Energy Systems (MESYS)

Figure 15-7 | A modern pellet-fired boiler installation. Pellets are automatically supplied from the bag silo at left to the boiler.

Courtesy of CALUWE Inc. - Biomass Heating Solutions

(a)

Courtesy of CALUWE Inc. - Biomass Heating Solutions

(b)

Figure 15-8 | (a) Example of hydronic wood stove. (b) Example of a hydronic wood stove operating with wood gasification and secondary combustion in lower chamber.

is located, while also storing heat for later use or for delivery to other areas of the building. This **heat output ratio** between hydronic and direct-to-room heating also allows these devices to operate without overheating houses that have very low heating loads, such as those built to the German PassivHaus standard.

WOOD CHIP-FIRED BOILERS

Another category of wood-fired boiler is specifically designed to burn **wood chips**. These chips are often produced at the site of logging operations by grinding logs that are not suitable for lumber, or in some cases entire trees. Figure 15-9 shows an example of wood chips being produced at a harvesting site. A large chipper can fill a 53-foot-long truck box in about 10 minutes.

Wood chips typically have higher moisture content relative to dried firewood or pellets. Chips in the range of 25 to 35 percent moisture content are common. Some industrial-scale boilers can operate with chips having moisture content as high as 45 percent. The high moisture content requires specially designed combustion systems. Higher moisture levels also mean less available heat from each pound of chips, because all the liquid moisture in the chips has to be vaporized during the combustion process. Wood chips are also available with lower moisture content in the range of 15 to 30 percent. These chips are more suitable to smaller scale heating systems with design heating loads in the range of 85,000 Btu/hr. and larger. Boiler manufacturers provide specifications on allowable dimensions, moisture content, and bark or leafy matter content for the chips to be burnt in their boilers.

(a)

Courtesy of Messersmith Mfg., Inc.

(b)

Courtesy of NREL

Figure 15-9 (a) Production of wood chips using a large chipper at a wood harvesting site. (b) Close-up of wood chips. Note that some bark and leafy matter are present.

Wood chips are typically delivered by truck and dumped into a storage bin. From there, they are transported to the boiler by a **motorized auger** system, as shown in Figure 15-10.

The chips are typically "handed off" from the storage transport auger to another "stoker" auger that carries them into the boiler's combustion chamber. These auger systems can be seen on the left side of Figure 15-11. Additional mechanisms within the boiler are automatically operated to separate ash from burning chips and scrape ash from heat-transfer surfaces. The ash is moved by a separate auger to an **ash barrel** located next to the boiler.

Combustion is managed by a microprocessor-based controller that measures boiler temperature and the oxygen content in the exhaust gases. The controller operates the electric ignition system, combustion air blowers, motorized augers, and ash scrapers as needed to regulate heat production.

Traditionally, the logistics involved in purchasing and handling wood chips have favored their use in

Courtesy of CALUWE Inc. - Biomass Heating Solutions

Figure 15-10 A motorized auger moving wood chips from a storage bin to the boiler.

larger commercial or institutional installations. However, smaller wood chip boilers suitable for residential and light commercial buildings are now available in North America. One such boiler can modulate its heat production rate between 7 and 25 kilowatts, (23,900–85,300 Btu/hr.).

EXPANDING OPPORTUNITIES

As this text is written, research and development continues on high-efficiency boilers that burn firewood, pellets, and wood chips. Advanced combustion controls using **lambda sensors** to monitor the oxygen content of exhaust gases are being integrated in these boilers. They allow for continuously adjusted air supply that helps optimize combustion. Steady state thermal efficiencies of 85 to 88 percent are possible in boilers designed to operate without flue gas condensation. There are also condensing capable pellet-fired boilers available in Europe that can achieve steady-state thermal efficiencies above 90 percent. Progress is also being made toward reduced emissions. These developments will further encourage the use of wood and pellet-fired heating systems in residential and light commercial buildings. Modern hydronics technology further leverages the environmental and operating cost advantages of these boilers by allowing them to deliver superior comfort in a wide variety of applications.

Figure 15-11 | Example of a 35 kWh (120,000 Btu/hr.) wood chip boiler. Chip feeding augers seen lower left.

15.3 Wood as a Heating Fuel

Wood is a form of **biomass**. In essence, it is stored solar energy. Plants convert solar energy into stored chemical energy through a process known as **photosynthesis**. The green leaf pigment chlorophyll uses sunlight to convert carbon dioxide from the atmosphere, as well as water and minerals absorbed from soil, into carbohydrates, and oxygen. The oxygen is released to the atmosphere and the carbohydrates (sugars and starches) are used to build new plant cells.

With proper forest management, wood is a fully renewable energy source. In some regions, there is even an excess of non-lumber grade wood accumulating in forests. Harvesting this wood for fuel reduces the risk of forest fires and improves forest health. Wood for fuel can also be produced by growing specific species of trees, such as shrub willow, as short-term rotational crops.

In addition to being carbon neutral, wood has very low sulfur content. Its combustion does not contribute to atmospheric accumulation of sulfur oxides, a cause of acid rain.

In many locations, wood is an indigenous energy source. It seldom requires interstate transportation, as do other energy sources. Its use as a fuel supports local economies and reduces dependency on foreign energy suppliers.

ENERGY CONTAINED IN WOOD

One of the most important measures of any fuel is the chemical energy contained within a given mass or volume. The fuel energy contained in *oven-dried*

Figure 15-12 | Wood is a carbon-neutral fuel.

mature wood is approximately 7,950 Btu per pound. This value is based on *weight* rather than volume and is *valid for both hardwoods and softwoods*. The latter tend to be less dense, and a given *volume* of softwood thus contains less fuel energy than an equal *volume* of hardwood.

The fuel energy of 7,950 Btu/lb. is only attainable in wood that has been oven dried to zero moisture content. The **lower heating value** of wood with other moisture contents can be approximated by Equation 15.1.

(Equation 15.1)

$$LHV = 7950 - 90.34(w)$$

Where:

LHV = lower heating value (Btu/lb.)*

w = moisture content (percent)

The lower heating value of wood, or other fuels, excludes the latent heat of water vapor produced during combustion. In the case of wood, it is an appropriate indicator of potential heat content because nearly all current-generation wood-burning boilers used in residential and small commercial heating systems are *not* intended to operate with sustained flue gas condensation.

Example 15.1: Estimate the lower heating value of dried firewood with a moisture content of 20 percent.

Solution: Using Equation 15.1:

$$LHV = 7950 - 90.34(w) = 7950 - 90.34(20) = 6143\frac{\text{Btu}}{\text{lb.}}$$

Discussion: The value of 6,143 Btu/lb. is the theoretically available heating energy in the wood. Not all of this energy can be recovered in any practical combustion process. Modern wood-gasification boilers are now capable of achieving steady state combustion efficiencies in the range of 80 to 85 percent. However, to attain this efficiency it is critical for them to be fired using wood at a maximum moisture content of 20 percent. Even lower moisture content is preferred if available. Recent testing of wood-gasification boilers indicates that their *average* thermal efficiency, over a *complete burn cycle*, is often in the range of 70 to 75 percent.

The higher the wood's moisture content, the lower its available heating energy. This is a result of having

Figure 15-13 | Examples of covered open-air wood storage that encourage drying. Wood-gasification boiler is also housed in this building.

to evaporate moisture during the combustion process. Evaporating water requires a substantial amount of energy (970 Btu/lb.). That energy must come from the heat created as the wood is burned and, as such, is not available for space heating or other heating loads. To attain the greatest fuel value, all firewood should be thoroughly dried. Freshly cut "green" wood can have a moisture content over 60 percent. A minimum suggested guideline for achieving 20 percent moisture content is to dry split wood outside, *under cover*, for at least 12 months. The time required to properly dry firewood varies with local climate and species. Drying is enhanced by exposing split firewood to breezes as well as sunlight. Protecting the stacked wood from precipitation is also essential for expedient drying. Figure 15-13 shows examples of stacked firewood that encourage drying.

The moisture content of firewood can be checked using a **moisture meter**, such as the one shown in Figure 15-14. Moisture meters work by measuring the electrical resistance between two needle probes that are a fixed distance apart. This resistance is then correlated to a moisture content by the meters internal processor.

After turning the meter on, be sure it is set for "firewood mode." The two needle probes at the top of the meter should then be firmly embedded in the wood sample being tested.

When using a moisture meter to test the condition of a stack of wood, several pieces of wood should be selected from different areas of the pile. If the pile contains more than one species of wood, a good practice is to check the moisture content of pieces from each species. Each piece should then be split open to expose the interior fibers, which tend to dry slower than those

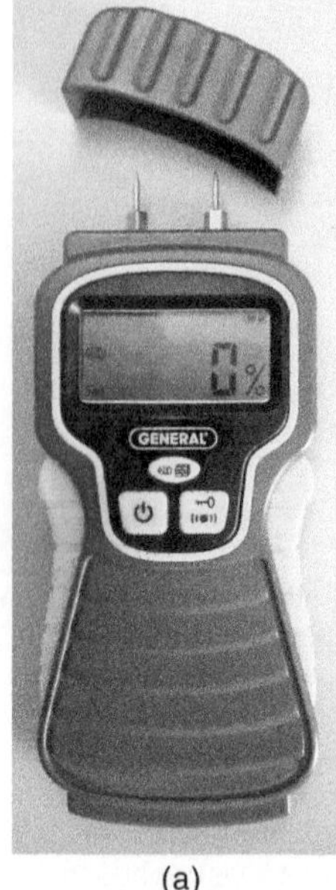

(a)

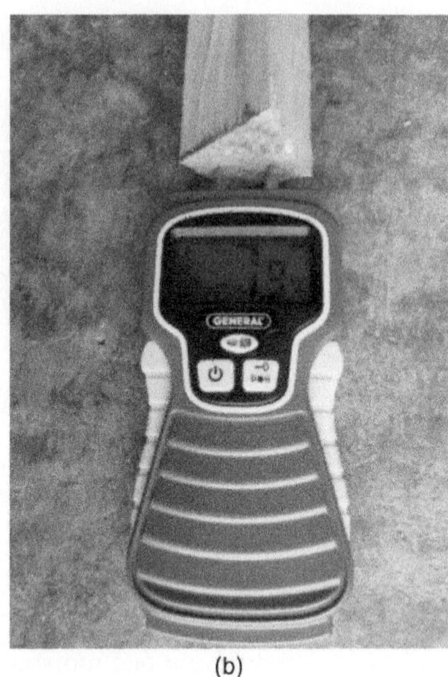

(b)

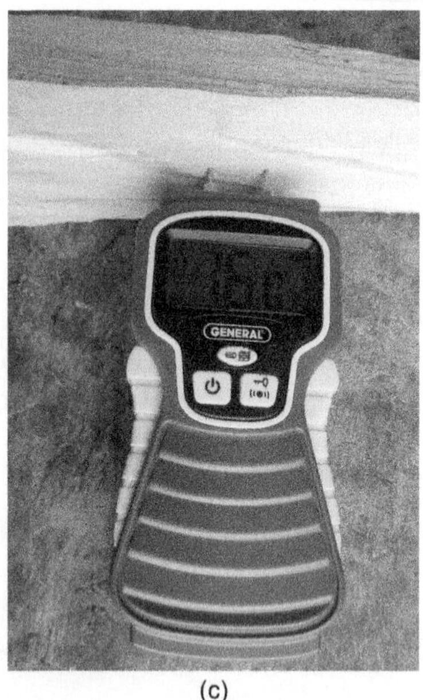

(c)

Figure 15-14 (a) Moisture meter with two needle probes at top. (b) Use of a moisture meter to check moisture content of firewood at end grain (reading = 7.9 percent). (c) Same meter, same piece of wood, but split open to check interior moisture content (reading = 16 percent).

near the surface of the wood. As Figures 15-14b and 15-14c show, there can be large differences between surface and interior moisture content of the same piece of wood.

A somewhat less precise moisture test can be performed using a standard multimeter. This test does not give the exact percentage moisture content of the wood, but it can be used to determine if the moisture content is lower than 20 percent. The process involves the use of a finish nail with the same diameter as the multimeter probes. Drive the nail into the wood to create two parallel holes in the wood that are 1.25 inches apart, and at least 3/8 inch deep. Remove the nail and insert the multimeter probes snugly into these holes. If the meter indicates a resistance of 3 megohms (3,000,000 ohms) or higher, the wood's moisture content is at or below 20 percent. If the resistance is less than 3 megohms, the wood requires additional drying.

FIREWOOD MEASUREMENT UNITS

In North America, firewood is typically sold by volume. Two common units are used:

- **Full cord** refers to a volume of neatly stacked, split wood that measures 4 feet high by 8 feet long by 4 feet deep.
- **Face cord** refers to a volume of neatly stacked, split wood that measures 4 feet high by 8 feet long by 16 inches deep. Three face cords stacked tightly together would constitute a full cord.

Neither of these definitions is precise, because repeatedly stacking the same firewood will always produce slightly different overall stack dimensions. Still, these units have been used in the North American firewood industry so long that some people refer to firewood as **cordwood**.

Although the fuel value of a *pound* of hardwood versus a pound of softwood having the same moisture content is essentially equal, the *density* of various woods will significantly affect the energy contained in the same volume (e.g., full cord or face cord). Figure 15-15 lists the approximate lower heating value contained in a *full cord* of several different species of wood, all with an assumed 20 percent moisture content.

WOOD PELLETS

Wood pellet technology was first developed in Sweden during the early 1980s.

Pelletization is a common process for turning powdered or fibrous materials into a mechanically stable form. Pelletized animal feed is one example. This process can be used to convert clean wood sawdust,

Species	Approximate weight (lb./full cord)	Lower heating value (MMBtu/full cord)*
White oak	3689	23.6
American beech	3757	24.0
Sugar maple	3757	24.0
Black cherry	3120	20.0
Eastern white pine	2236	14.3
Red or white spruce	2100	14.5

*1 MMBtu = 1,000,000 Btu

Figure 15-15 | Approximate weight and estimated energy content of different wood species, all with an assumed 20 % moisture content.

shavings, and chips into pellets. The feed stocks used to make high quality (e.g., "premium") wood pellets *cannot* contain tree bark, which if present, would lower the heating value of the pellet and increases ash formation. Wood pellets are much denser and drier than other forms of wood fuel such as chips and shavings. They are also easier to store, transport, and ignite relative to other forms of wood fuel.

The production process starts by converting the wood feed stock into small fibers using a **hammermill**. These fibers are then dried to approximately 8 percent moisture content. In many production facilities, a portion of the wood fibers is used to fuel the drying process. Next, the dried fibers are mechanically compressed in a **pellet mill**, which can exert pressures of approximately 60,000 psi on the fibers, forcing then through 1/4-inch diameter holes in a thick steel die. This compression generates heat that raises the temperature of the wood fiber to about 190 °F. This causes natural resins, called **lignins**, within the wood cells to melt and bond the fibers together. No other bonding agents are required. The extruded pellets emerge from the extrusion die and randomly break into small pieces (e.g., pellets) that typically are 0.5 to 1.25 inches in length, as seen in Figure 15-16.

Figure 15-16 | Wood pellets.

After manufacturing, pellets are easily handled by **pneumatic** and **auger** transport systems. They can be stored and transported by equipment similar to that used for pelletized animal feed. As such, they lend themselves to existing infrastructures for mass production and transportation. Figure 15-17 shows how pellets are bulk delivered and pneumatically conveyed from the delivery truck into an interior storage bin.

Wood pellets have a density of about 40.6 lb./ft.^{3}, which is comparable to the dry density of maple, at about 42 lb./ft.^{3}. However, their smaller size allows for a significantly greater amount of fuel energy to be stored in a given volume compared to stacked firewood. Their size and shape also allow for excellent mixing with air during the combustion process. Pellets also produce less ash and lower emissions compared to those produced by other wood-based fuel stocks.

The lower heating value for pellets ranges from approximately 7,600 to 8,400 Btu/lb. This is significantly greater than the lower heating value of firewood (at 20 percent moisture content), calculated in Example 15.1, at about 6,143 Btu/lb.

The total energy required to manufacture pellets is usually below 2 percent of their final energy content. This compares favorably to the 10 to 12 percent embodied energy required to refine common fossil fuels.

Wood pellets are typically purchased in either 40-pound bags, or by the ton. The bags are intended for use with manually feed appliances. It is common for 50 40-pound bags to be shrink-wrapped together on a standard shipping pallet. When purchased "in bulk," wood pellets are delivered to the site by truck and pneumatically conveyed into an interior bin or exterior storage silo, as seen in Figure 15-18.

There are many pellet production facilities now operating in North America. The chips and sawdust used to produce pellets comes from a variety of sources

Courtesy of Maine Energy Systems (MESYS)

Figure 15-17 | Truck making a bulk delivery of pellets.

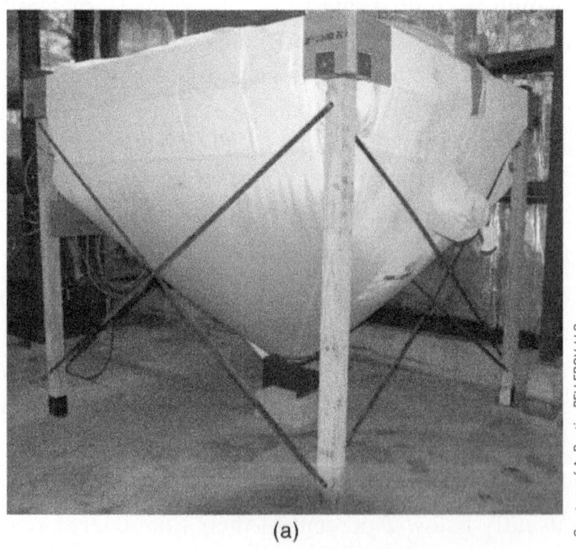

(a)

Courtesy of A. Boutin, PELLERGY, LLC.

(b)

Courtesy of A. Boutin, PELLERGY, LLC.

Figure 15-18 | Examples of bulk pellet storage options (s) interior fabric bin (a.k.a. Bag silo). (b) Exterior storage silo.

and includes both hardwood and softwood feedstocks. Given these conditions, and the consumer's desire for consistency, the **Pellet Fuels Institute (PFI)** (www.pelletheat.org) has developed a standard titled *Pellet Fuels Institute Standard Specification for Residential/Commercial Densified Fuel, June 1, 2011.* This standard defines three grades of pellets: PFI Premium, PFI Standard, and PFI Utility. Several properties for each of these grades are given in Figure 15-19, which is part of the PFI standard.

Examining the requirements listed in Figure 15-19 shows that PFI premium pellets have preferred qualities such as lower ash content, lower percentage of fines, and slightly better durability. Designers should verify which pellet grade is required by the boiler they are specifying.

The PFI standard also specifies how pellet samples are to be acquired and tested by independent laboratories. Many of these test procedures reference ASTM standards. The standard *New Source Performance*

Fuel Property	Residential/Commerical Densified Fuel Standards See Notes 1–3		
	PFI Premium	PFI Standard	PFI Utility
Normative Information - Mandatory			
Bulk Density, lb./cubic foot	40.0–46.0	38.0–46.0	38.0–46.0
Diameter, inches Diameter, mm	0.230–0.285 5.84–7.25	0.230–0.285 5.84–7.25	0.230–0.285 5.84–7.25
Pellet Durability Inbox	≥96.5	≥95.0	≥95.0
Fines, % (at the mill gate)	≤0.50	≤1.0	≤1.0
Inorganic Ash, %	≤1.0	≤2.0	≤6.0
Length, % greater than 1.50 inches	≤1.0	≤1.0	≤1.0
Moisture, %	≤8.0	≤10.0	≤10.0
Chloride, ppm	≤300	≤300	≤300
Heating Value	NA	NA	NA
Informative Only - Not Mandatory			
Ash Fusion	NA	NA	NA

Courtesy of Pellet Fuels Institute

Figure 15-19 | Requirements for PFI pellet grades.

Standard for Residential Wood Heaters (NSPS) recently enacted by the U.S. Environmental Protection Agency (EPA) requires residential pellet-burning appliances to use pellets that the meet the PFI standard.

Pellet producers who meet the production and quality assurance standards established by the Pellet Fuel Institute are allowed to label their product with a **quality mark**, an example of which is shown in Figure 15-20. This quality mark is printed on the bags or on paperwork accompanying a bulk delivery of pellets.

WOOD BRIQUETTES

Another recently developed form of densified wood fuel is known as **briquettes**. An example of wood briquettes is shown in Figure 15-21.

Wood briquettes are made of shredded and dried wood fiber similar to that used for pellets. The fiber is compressed under very high pressure. This causes the lignins in the wood to bond the fibers together, without need of other bonding agents. The resulting briquettes are very hard and dense. Depending on size, a single briquette weighs between 3 and 4 pounds. The lower heating value of the compressed wood fiber in a briquette is approximately 8,000 Btu/lb. When burned, they produce significantly less particulate emissions and ash compared to dried firewood.

Wood briquettes are typically sold by the pallet. This allows for efficient handling and transportation. A typical pallet measures 4 feet by 4 feet by 3 feet high and contains 300 briquettes. Wood briquettes must be stored away from sources of moisture.

The size of the briquettes allows them to be easily ignited while also burning at a rate that allows fire to last for several hours. They have less dust, dirt, and insects associated with their storage and use, compared to firewood.

COMBUSTION OF WOOD

Unlike common fossil fuels such as natural gas and oil, the combustion of firewood involves several phases. These include:

1. Warming the wood
2. Drying the wood

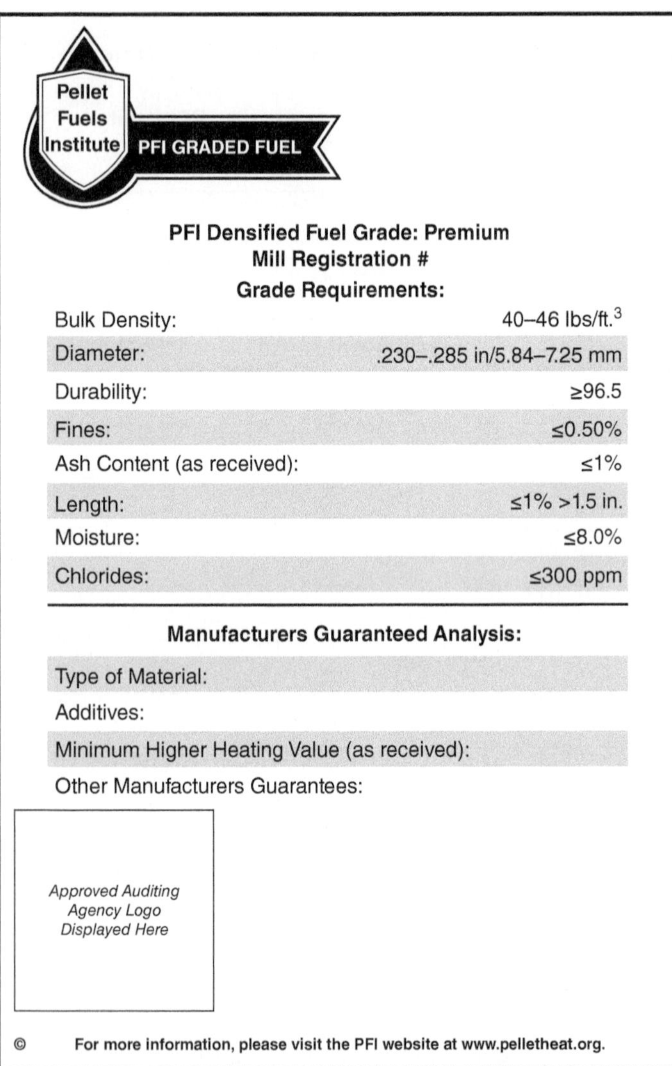

Pellet Fuels Institute — PFI GRADED FUEL

PFI Densified Fuel Grade: Premium
Mill Registration #
Grade Requirements:

Bulk Density:	40–46 lbs/ft.^{3}
Diameter:	.230–.285 in/5.84–7.25 mm
Durability:	≥96.5
Fines:	≤0.50%
Ash Content (as received):	≤1%
Length:	≤1% >1.5 in.
Moisture:	≤8.0%
Chlorides:	≤300 ppm

Manufacturers Guaranteed Analysis:

Type of Material:
Additives:
Minimum Higher Heating Value (as received):
Other Manufacturers Guarantees:

Approved Auditing Agency Logo Displayed Here

© **For more information, please visit the PFI website at www.pelletheat.org.**

Figure 15-20 | Example of a quality mark from the Pellet Fuels Institute.

Courtesy of Pellet Fuels Institute

3. Pyrolytic decomposition
4. Gasification of fuel
5. Gasification of carbon
6. Oxidation of combustible gases

When a fire begins, the wood is usually at the temperature of it surroundings. It is necessary to warm the wood to the point where any water it contains will be boiled off. Finely split pieces of very dry wood, in combination with easily combustible materials such as newsprint, or dry wood shavings, are used as **kindling**, to initiate a small fire within the combustion chamber of a manually fed wood-fired boiler. This small fire supplies the heat needed to raise the temperature of larger pieces of wood. As it is heated, water within the

Figure 15-21 | Wood briquettes made by compacting wood fiber.

cellular structure of the wood eventually vaporizes and is forced out of the wood by the internal pressure it generates. This process can often be seen (and even heard) as water vapor hisses out the cut ends of firewood that has reached a temperature over 212 °F. It is especially noticeable in wood that has a moisture content over 20 percent. This phase of combustion also produces dense smoke, which is a mixture of uncombusted **particulates** and the escaping water vapor.

The initial warming and drying of firewood is a prerequisite to further combustion. These processes are called **endothermic**, because they *absorb* heat. The colder the wood, and the higher its moisture content, the greater the amount of heat required to transition the wood to the higher temperature phases of combustion.

When wood reaches a temperature of about 300 °F, the pyrolytic decomposition of wood begins. Long-chain molecules within the wood are broken down into smaller gaseous hydrocarbon molecules. The gases given off by the thermal decomposition of wood, with minimal amounts of oxygen present, are called **pyrolytic gases**.

When the pyrolytic gases reach a temperature of about 450 °F, they react with available oxygen, creating **exothermic** (heat-generating) chemical reactions. The efficacy of this process depends on the amount of oxygen available, the degree to which the pyrolytic gases are mixed with that oxygen, and the time this mixture remains in the combustion chamber. In a gasification boiler, the reaction between oxygen and pyrolytic gases can generate temperatures upward of 2,000 °F. The heat from this reaction is sufficient to cause gasification of solid carbon. This gas, when reacting with oxygen, creates the orange and yellow flames seen in a wood-fueled fire. The combustion zone is now at a temperature where most of the combustible gases are consumed. When this process is properly controlled, minimal amounts of ash or **clinkers** are formed. The latter are partially burned, carbon-rich solids that result from a less than perfect combination of heat, fuel, and oxygen within the combustion chamber.

As the wood continues to be gasified, the average combustion zone temperature declines to a relatively stable plateau phase and then drops rapidly as the last portions of the fuel charge are consumed.

At any given time, various amounts of wood within a combustion chamber are undergoing different phases of combustion. While a bed of glowing coals operates at the exothermic higher temperature carbon pyrolysis phase, a piece of firewood just thrown into the combustion chamber is still warming and drying and thus operating in the endothermic phases. The stability and combustion efficiency of the overall fire depend on the balance between these phases.

Figure 15-22 illustrates the temperatures and heat transfer involved in the combustion of wood as a function of time.

CREOSOTE

Depending on the design of the wood-burning device, and how it is operated, some of the pyrolytic gases formed as wood is heated may be carried away from the combustion zone before they can react with oxygen and release their embodied chemical energy. As these unburned gases and vapors come in contact will cooler surfaces, they condense into a sticky, tar-like substance called creosote. This can occur within the wood-burning device as well as in the vent piping or chimney. Figure 15-23 shows an extreme case of creosote accumulation against the walls of a vent pipe.

Creosote accumulation is extremely dangerous. Because it's formed from unburned hydrocarbons, creosote has considerable fuel value. If sufficiently reheated by other combustion products or flames, that fuel value can quickly reappear as a chimney fire. Such fires can create extremely powerful convective air flow within the chimney that further increase the fire's intensity. Some chimney fires can destroy the chimney's integrity within minutes and then quickly spread to the building structure. They obviously must be avoided, and such avoidance starts with not creating conditions that form creosote.

One of the best ways to minimize creosote formation and boost efficiency is to burn well-seasoned firewood, as hot as possible. Creosote formation within a wood-fired boiler can also be reduced by maintaining a minimum inlet water temperature to the boiler. This allows the surfaces of the water-cooled heat exchanger to remain above the temperature at which pyrolitic gases condense into creosote. Details for this are shown and discussed later in this chapter.

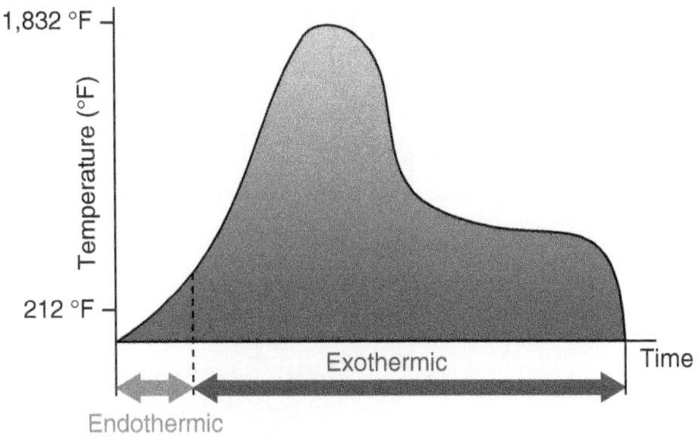

Figure 15-22 | Temperature versus time for well-controlled wood combustion.

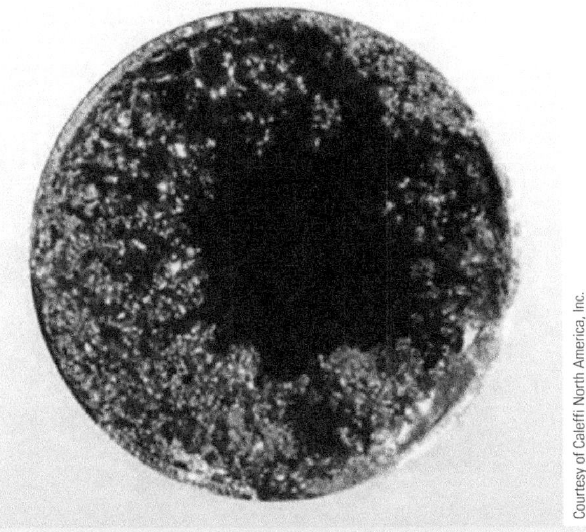

Figure 15-23 | Extreme creosote formation inside a vent connector pipe.

15.4 Outdoor Wood-Fired Hydronic Heaters

One approach to heating with wood advocates keeping both the firewood and the fire outside the building being heated. This approach has the following benefits:

- It is inherently safer since the fire, fuel, and emissions are not inside the building.
- The dirt, smell, and insects associated with bringing firewood inside are eliminated.
- Placement of relatively large and heavy heat sources is simpler outside than inside.
- Minimal interior space is required for other components used in combination with the outdoor heater.

One device that is designed to burn firewood outside and transfer some of the heat produced into a building through buried hydronic piping is called an **outdoor wood-fired hydronic heater**. Most of these heaters are *unpressurized* devices in which a water-filled compartment surrounding the combustion chamber is vented to the atmosphere. An example of an outdoor wood-fired hydronic heater is shown in Figure 15-24.

Most outdoor wood-fired hydronic heaters have thermostatically controlled blowers and/or air dampers that regulate air flow into the combustion chamber. These devices operate based on the water temperature in the heater. When that temperature reaches an upper limit, the heater goes into stand-by mode. The damper moves to a position that allows minimal air into the combustion chamber. If the heater has a combustion air blower, it either stops or decreases speed to some minimum setting. Although this action decreases heat output, it does so by significantly changing the chemistry of the combustion reaction. The fire becomes "starved" for oxygen. This leads to **incomplete combustion**, significant smoke generation, increases in carbon monoxide generation, and increased creosote formation. This condition is worsened by oversizing the heater relative to the heating load. The more oversized the heater, the more time it spends in stand-by mode, with low combustion efficiency and high emissions.

(a)

(b)

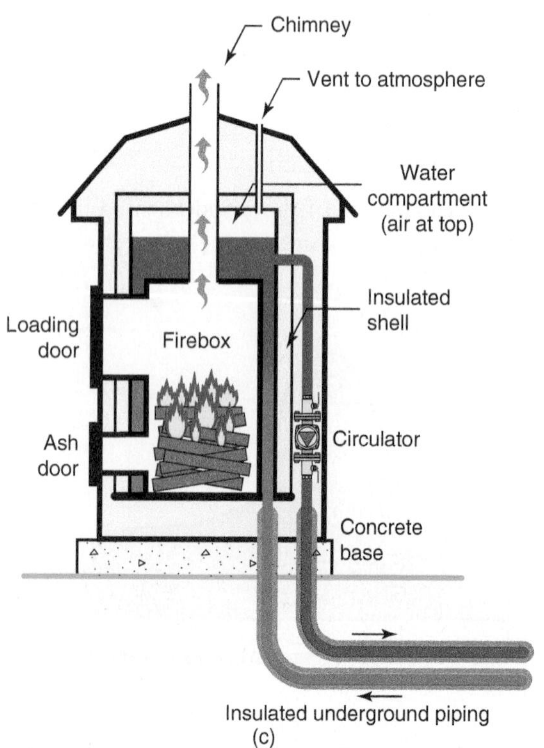

(c)

Figure 15-24 (a) Example of an outdoor wood-fired heater being fueled. (b) Outdoor wood-fired heater installed with wood storage canopy. (c) Generic internal design of an outdoor wood-fired heater.

Figure 15-25 | Creosote formation within the combustion chamber of an outdoor wood-fired heater.

Figure 15-26 | Under certain conditions, the low vents on some outdoor wood-fired heaters allow smoke to drift sideways toward buildings and neighboring properties.

The fireboxes in outdoor wood-fired hydronic heaters can hold a large amount of wood, which allows for several hours of burn time. On larger units, the water compartment may hold *several hundred* gallons. This gives the unit considerable thermal mass that can supply heat to a building for several hours after the fire has subsided. However, the proximity of the large quantity of water to the combustion chamber also keeps the steel walls that separate the water from the combustion chamber relatively cool. This leads to creosote formation, as shown in Figure 15-25.

The thermal efficiency of outdoor wood-fired hydronic heaters is significantly lower than other wood-burning heat sources discussed in this chapter. The thermal efficiency of a given unit depends on how it is operated, but average efficiencies in the range of 30 to 45 percent are typical when the unit's combustion blower (if present) is operating. Thermal efficiency will decrease when the unit is in stand-by mode and the combustion blower (if present) is off and the air damper is closed. The use of cold firewood with moisture content greater than 20 percent also decreases thermal efficiency and increases creosote formation.

Some municipalities have enacted laws that significantly restrict or even ban the use of outdoor wood-fired hydronic heaters. The primary concern is avoiding smoke plumes that, under certain conditions, do not rise sufficiently above the low chimneys often used with these devices, as seen in Figure 15-26.

Low-hanging smoke can drift toward buildings, move over neighboring properties, or drift across roads, causing pollution as well as safety issues. In some cases, it is necessary to install a tall chimney near the outdoor heater to lift smoke to a height where it will not affect adjacent buildings. Minimum chimney heights and property line setback distances are now mandated in some state regulations. Due to such concerns and requirements, outdoor wood-fired hydronic heaters are only practical in rural locations, where sufficient space is available.

Although outdoor wood-fired hydronic heaters do *not* represent state-of-the-art wood combustion technology, there are tens of thousands of them installed and operating in North America, and they remain available in many locations. These heaters are addressed in this chapter so that their installation can take advantage of the best available hydronics technology, enabling them to deliver the heat they generate when and where it is needed.

OPEN-LOOP SYSTEMS USING OUTDOOR WOOD-FIRED HEATERS

Nearly all outdoor wood-fired hydronic heaters have a nonpressurized water compartment that is vented to the atmosphere. As such, the piping circuit they connect to is considered an **open loop** and must be designed accordingly.

Courtesy of Bob Rohr

Figure 15-27 | Sludge within a cast-iron circulator due to oxygen-based corrosion.

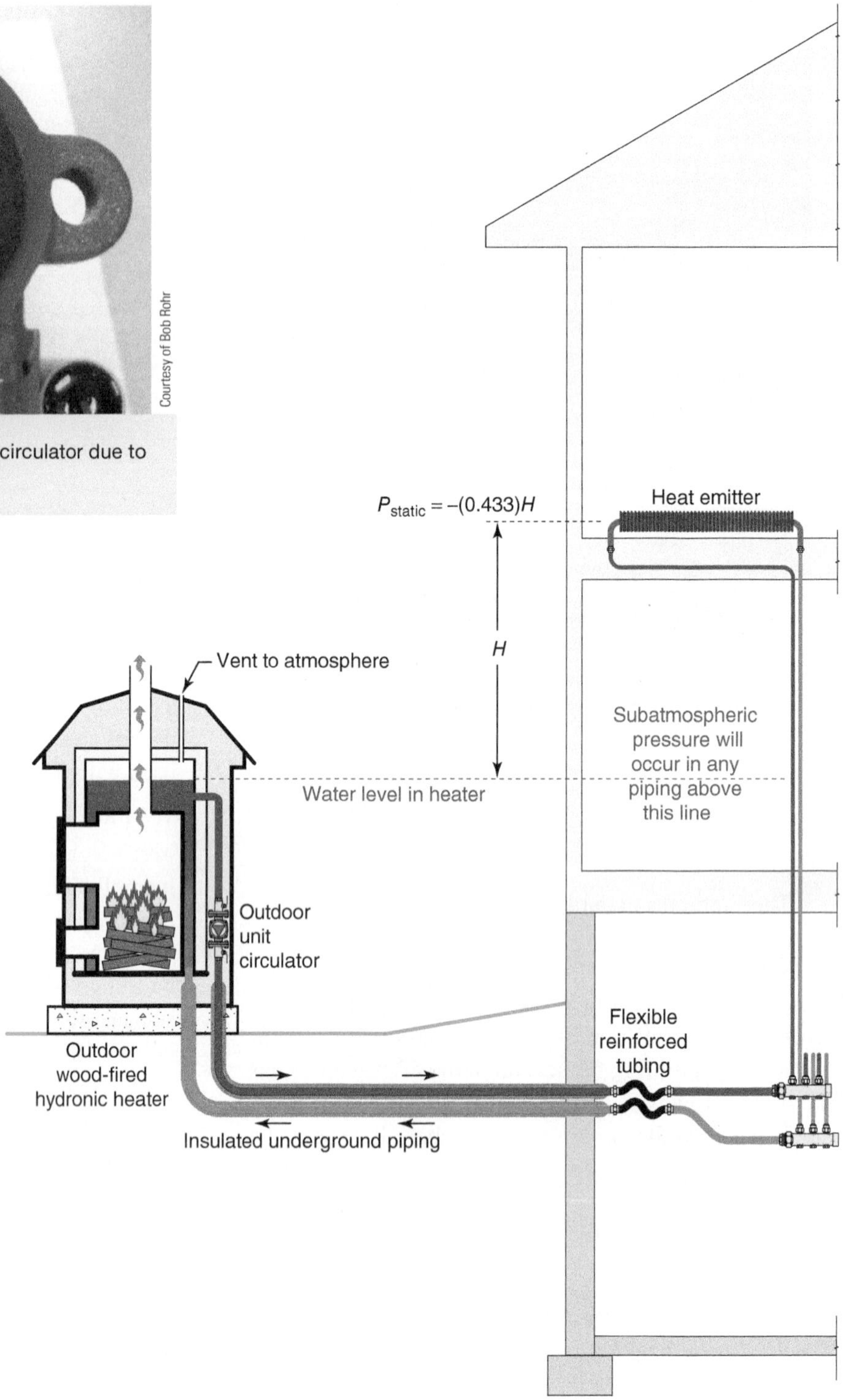

One concern of any open-loop system is the ability of water to reabsorb oxygen into solution as it cools. This phenomena was discussed in Chapter 3. Oxygen is continually available from the atmosphere through the vent tube, even if there is a slight amount of water trapped in that tube. Over time, the availability of oxygen can "feed" corrosion reactions within the system, especially if that system contains carbon steel or cast-iron components. Such reactions can form sludge, as seen within the circulator volute in Figure 15-27.

Most circulator manufacturers do *not* warrant cast-iron circulators for use in open-loop systems. Instead, they require circulators with stainless steel or bronze volutes. Although readily available, such circulators are significantly more expensive than their cast-iron equivalents. Other ferrous metal components such as steel panel radiators, cast-iron radiators, black iron piping, cast-iron valves, air scoops, and steel expansion tanks are also *not* intended for use in open-loop systems.

$P_{static} = -(0.433)H$

Heat emitter

H

Vent to atmosphere

Subatmospheric pressure will occur in any piping above this line

Water level in heater

Outdoor unit circulator

Flexible reinforced tubing

Outdoor wood-fired hydronic heater

Insulated underground piping

Figure 15-28 | Subatmospheric pressure develops above the water level in the heater whenever the circulator is off.

Another issue in open-loop systems is that water in piping located above the water level in the outdoor wood-fired hydronic heater will be under subatmospheric pressure when the system circulator is off. The higher the system piping rises above the water level in the outdoor wood-fired hydronic heater, the greater the negative pressure in the piping. This situation is shown in Figure 15-28.

Although it is possible to operate a hydronic system under certain negative pressure conditions, there are two details that must be observed to avoid problems. First, *there cannot be any automatic venting devices, air separators, or valves with packings located above the water level in the system.* Such devices, if present, can admit air into the system under negative pressure. That air will then make gurgling sounds as it travels through piping when the circulator is on. Second, *the height of the system piping must be limited based on maximum water temperature.* The goal is to avoid conditions that can cause the water to boil within the piping. Doing so requires that the water always remain below its **vapor pressure** in all locations. The most common problem is **steam flash** created by hot water high in system piping. When the circulator turns off, the static pressure on this water can instantly drop below the vapor pressure corresponding to the water's temperature. If this occurs, the water immediately flashes to steam and creates very noticeable "banging" sounds within the piping.

The curve in Figure 15-29 shows the ***absolute* pressure** at which water will boil based on its temperature. For example: Water at 212 °F boils at an *absolute* pressure of 14.7 psi. This corresponds to typical atmospheric pressure at sea level (0 psi gauge pressure).

The static **gauge pressure** of water located above the water level in the outdoor wood-fired hydronic heater can be found using Equation 15.2:

(Equation 15.2)

$$P_{\text{static}} = -\left(\frac{D}{144}\right)H$$

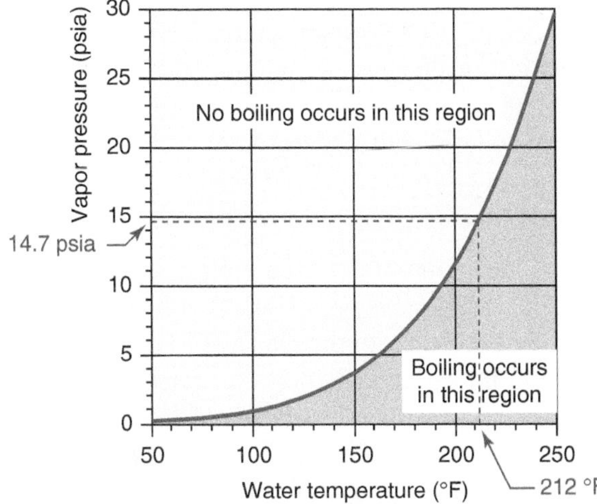

Figure 15-29 | Vapor pressure of water.

where:

P_{static} = static gauge pressure of the water at a given location (with circulator off) (psig)

D = density of water based on its current temperature (lb./ft.^{3})

H = height from top of system piping down to water level in outdoor wood-fired hydronic heater (ft.)

Example 15.2: Determine the static gauge pressure of 190 °F water located 16 feet above the water level in the outdoor heater. Will it boil at the top of the system when the circulator turns off?

Solution: First, determine the density of water at 190 °F using Figure 3-11.

$$D = 60.2 \text{ lb./ft.}^3$$

The negative pressure at the top of the system is then found using Equation 15.2:

$$P_{\text{static}} = -\left(\frac{D}{144}\right)H = \left(\frac{60.2}{144}\right)16 = -6.69 \text{ psi}$$

Now, convert the negative gauge pressure to absolute pressure. Just subtract 6.69 psi from atmospheric pressure (14.7 psi).

$$P_{\text{absolute}} = 14.7 - 6.69 = 8.01 \text{ psia}$$

Water will boil whenever the absolute pressure on it is lower than its vapor pressure. Figure 15-29 shows the vapor pressure of 190 °F water to be 9.5 psia. Thus, boiling will occur.

Discussion: One way to prevent boiling is to lower the water temperature. Another option might be to design the distribution system with a lower overall height. Still another possibility is to connect the outdoor wood-fired hydronic heater to a closed/pressurized distribution system. This is discussed next.

HYBRID OPEN/CLOSED SYSTEMS

One way to avoid the corrosion and negative pressure issues associated with an open-loop hydronic system is to connect the outdoor wood-fired hydronic heater to a closed hydronic distribution system using a stainless steel brazed-plate heat exchanger. An example of this approach is shown in Figure 15-30.

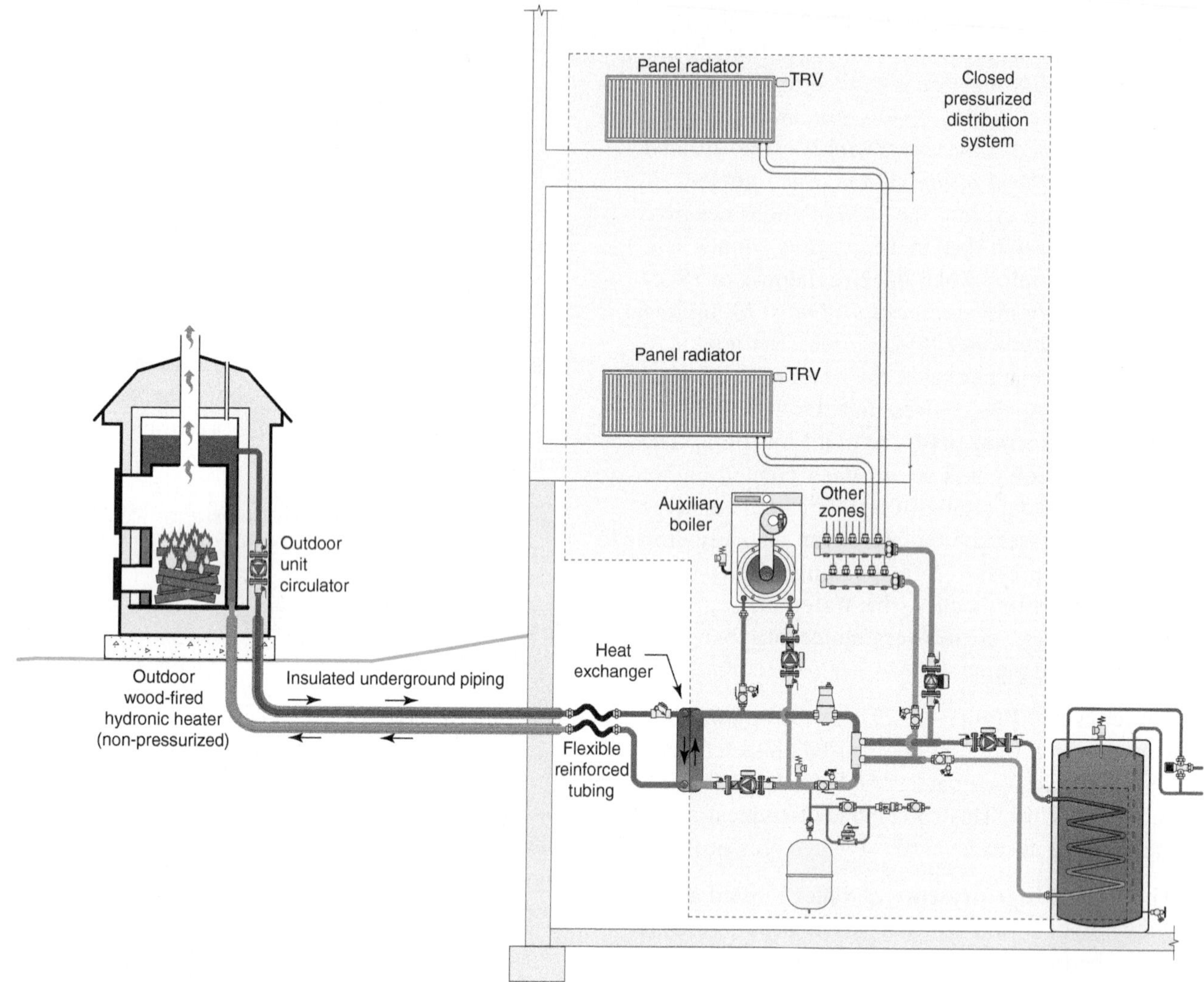

Figure 15-30 | Using a stainless steel brazed plate heat exchanger to separate the open-loop circuit of the outdoor wood-fired heater from the closed-loop distribution system.

All piping components on the secondary (e.g., lower temperature) side of the brazed-plate heat exchanger are part of a closed-loop distribution system that can operate under a positive pressure. This portion of the system can use cast-iron circulators and other ferrous metal components. It can also be equipped with an auxiliary heat source, if needed.

The heat exchanger should be sized to transfer the design heating load of the building, plus the maximum domestic hot water load, using an approach temperature difference no higher than 10 ºF.

UNDERGROUND PIPING

It is very important that underground piping between an outdoor wood-fired hydronic heater and the building(s) it serves is properly selected and detailed. Use of improper materials or installation methods can allow excessive heat loss from the tubing, which further lowers the net thermal efficiency of the heat source. Unfortunately, there have been many installations of underground piping using inadequate insulation methods. One example is shown in Figure 15-31.

The most commonly used tubing for underground piping installations is cross-linked polyethylene manufactured according to the ASTM F876 standard. The minimum tube size is 1 inch. *Larger tube sizes may be necessary and justified base on flow rate and the distance between the outdoor wood-fired hydronic heater and the building. This tubing must be insulated using materials and methods specifically intended for underground installation.* The insulation system must prevent ground moisture from entering through

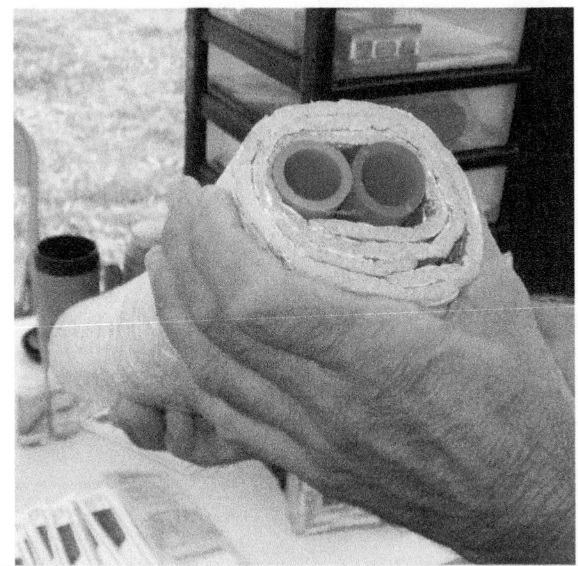

Figure 15-31 Example of underground insulated tubing that is highly vulnerable to moisture gain or physical damage from surrounding soil.

its outer jacket. It must also be strong enough to absorb stresses created by soil expansion and contraction.

Pre-insulated direct burial piping products, such as that shown in Figure 15-32, are now widely available and well suited for use with outdoor wood-fired heaters. These products are available in both single-tube and dual-tube configurations. Special fittings along with specially designed compression sealing collars are available to create watertight wall penetrations.

Pre-insulated tubing should be buried in trenches *at least* 3 feet deep, and not subject to water accumulation or settling. The pre-insulated tubing should be placed on smooth undisturbed soil and bedded in sand or other fine soil. The backfill material should not contain large or sharp rocks that could potentially damage the insulation shell. *Special care should be taken to ensure that settling cannot occur near wall penetrations, which can impose large shear forces on the tubing.*

THERMAL EXPANSION

The fluid transferring heat from an outdoor wood-fired hydronic heater to a building is subject to wide temperature variations. This, combined with the thermal expansion characteristics of PEX tubing, demands proper detailing where the piping enters the building and where it connects to the outdoor wood-fired hydronic heater.

Equation 15.3 can be used to estimate the linear expansion of PEX tubing based on its length and temperature change.

(Equation 15.3)

$$\Delta L = 0.000094(L)(\Delta T)$$

where:

ΔL = change in length (inches)

L = length of tubing before temperature change (inches)

ΔT = change in temperature of tubing (°F)

Example 15.3: PEX tubing runs 150 feet from an outdoor wood-fired heater to its termination just inside a basement wall. The tubing was placed when its temperature was 60 °F. Determine how much expansion will occur if the tubing reaches a temperature of 190 °F.

Solution: Putting the stated values into Formula 15.3 yields:

$$\Delta L = 0.000094(L)(\Delta T)$$

$$= 0.000094(150)\left(\frac{12\text{ inch}}{1\text{ foot}}\right)(190 - 60) = 22\text{ inches}$$

Discussion: This is a large amount of movement. If provisions are not made to accommodate it, the tubing is subject to deformation, either of itself or in combination with the interior tubing it connects to.

The expansion and contraction movement of PEX tubing can be accommodated by either reinforced flexible hoses that connect it to rigid tubing inside the building, or by other expansion compensators that are sized for the required movement. These should be located so that piping expansion movement is absorbed by the compensator while imposing minimal stress on interior rigid piping. The latter should be rigidly fixed near the compensator, as shown in Figure 15-33. This isolates the movement and stress from rigid interior piping.

Manufacturers of outdoor wood-fired hydronic heaters may have specific limitations on how PEX tubing should be connected to their unit. A minimum length of metal piping may be required between the heater's water compartment and the PEX tubing.

Figure 15-32 (a) Single PEX tube in insulated HDPE sleeve. (b) Twin PEX tubes in insulated sleeve. (c) Compression collar designed to provide water-tight seal where preinsulated tube passes through concrete foundation wall. (d) Insulated double tube assembly in ditch.

USING AN AUXILIARY BOILER WITH AN OUTDOOR WOOD-FIRED HYDRONIC HEATER

Many building owners desire systems in which a conventional heat source automatically turns on whenever the outdoor wood-fired hydronic heater cannot supply either the space-heating load or the domestic water heating load. The versatility of hydronics allows for such systems. One example is shown in Figure 15-34.

All components on the lower temperature side of the heat exchanger form a closed-loop pressurized distribution system. Heat can be supplied from the outdoor wood-fired hydronic heater through the heat exchanger, or from the auxiliary boiler. This piping also allows the possibility of *simultaneous* heat input from both sources. The latter scenario must be carefully managed to avoid inadvertently sending heat generated by the auxiliary boil to the outdoor heater (other than if required for freeze protection). This can be done by using a **differential temperature controller** to monitor the temperature rise between the incoming water from the outdoor wood-fired heater and the water returning from the load side of the system, as shown in Figure 15-35.

If the auxiliary boiler is operating, but the temperature difference (T_1-T_2) approaches a minimum detectable value of 2 °F or 3 °F, the outdoor wood-fired hydronic heater is contributing very little heat. *Under such conditions, the circulator providing flow through load side of the heat exchanger should be turned off.* This prevents heat generated by the auxiliary boiler from being transferred outside in systems where the circulator in the outdoor wood-fired hydronic heater operates continuously. If the temperature differential eventually rises to perhaps 5 °F, as it might if the wood-fired heater were refueled, the load side circulator can be turned back on.

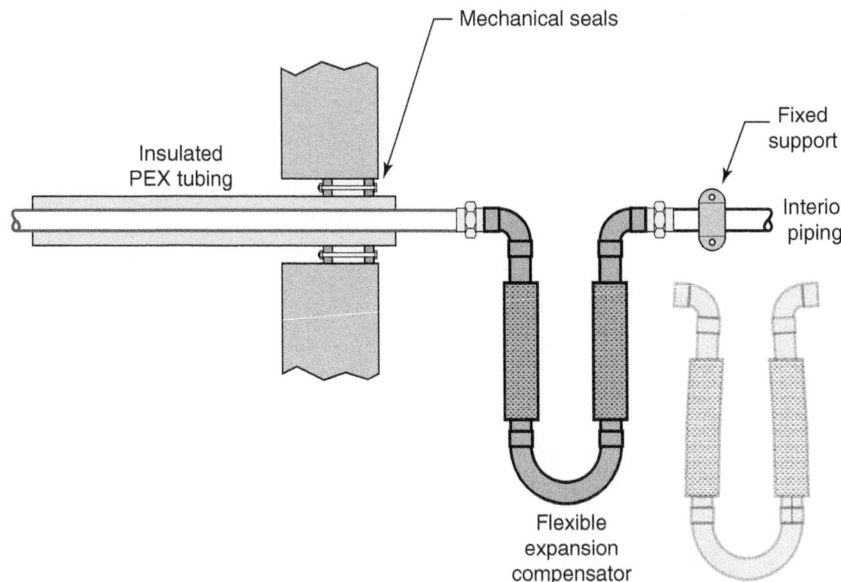

Figure 15-33 | Use of an expansion compensator to absorb movement of buried PEX tubing where it enters a building.

FREEZE PROTECTION OPTIONS

One common method of protecting an outdoor wood-fired hydronic heater from freezing is maintaining continuous water circulation between the outdoor unit and the building it serves. This approach assumes that electricity is available to operate the circulator and that sufficient heat from within the building(s) can be transferred to the circulating water as it passes though the interior portion of the system. In systems that use a water-to-air heat transfer coil mounted in the plenum of an interior furnace, heat is removed from the air stream being delivered to the building. This could significantly cool that air, leading to uncomfortable conditions. Antifreeze could be used to prevent freezing of the outdoor unit, but the large water compartment in many such units makes this an expensive option.

Another possibility, assuming the system includes an auxiliary hydronic heat source and that electricity remains available, is to circulate slightly warmed water through the outdoor wood-fired hydronic heater in such a way that its water compartment remains above freezing. One possible configuration is shown in Figure 15-36.

When the temperature sensor in the outdoor heater reaches a temperature a few degrees above freezing, a small **shunt circulator** routes warm water returning from the building's heating distribution system through the secondary side of the flat-plate heat exchanger. The setpoint temperature controller must also ensure that the circulator in the outdoor heater is operating. Flow through the secondary side of the heat exchanger should be adjusted using the flow setter valve so that excessive heat is not transported to the outdoor heater and such that indoor comfort is not compromised. With the proper controls, the piping in Figure 15-36 allows the auxiliary boiler to fire to provide heat to the outdoor unit, even if there is no other call for heat in the system.

15.5 Natural Draft Wood-Fired Boilers

There are several types of pressure-rated wood-fired boilers available in North America. These can be broadly categorized as:

1. Natural draft wood-fired boilers (using single-stage combustion)
2. Wood-gasification boilers (using two-stage combustion)

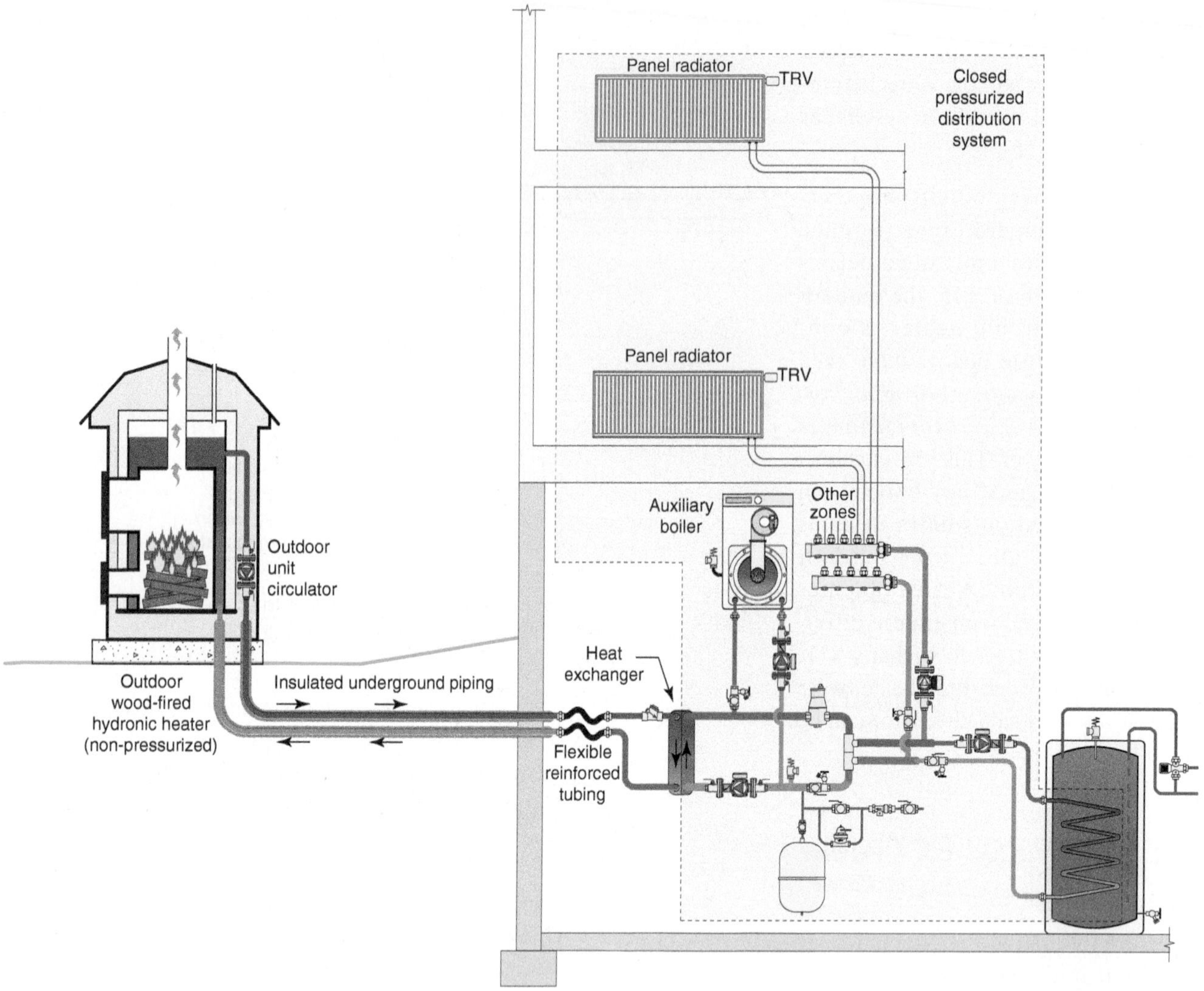

Figure 15-34 | Auxiliary boiler automatically supplies heat to building and indirect domestic water heater whenever the output of the outdoor wood-fired heater cannot.

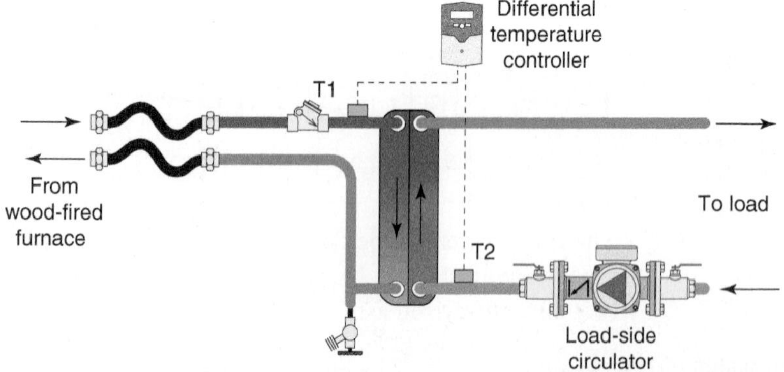

Figure 15-35 | Using a differential temperature controller to monitor the temperature difference between fluid arriving at the heat exchanger from the outdoor wood-fired hydronic heater and water returning from the distribution system.

This section deals with the first category. Wood-gasification boilers are covered in the next section.

Natural draft wood-fired boilers draw air from the space around them using only the draft created by their chimney. Some are constructed of cast iron, as seen in Figure 15-37a, while others are made of welded steel, as seen in Figure 15-37b.

CAST-IRON WOOD-FIRED BOILERS

Natural draft boilers with cast-iron sectional heat exchangers typically introduce **primary**

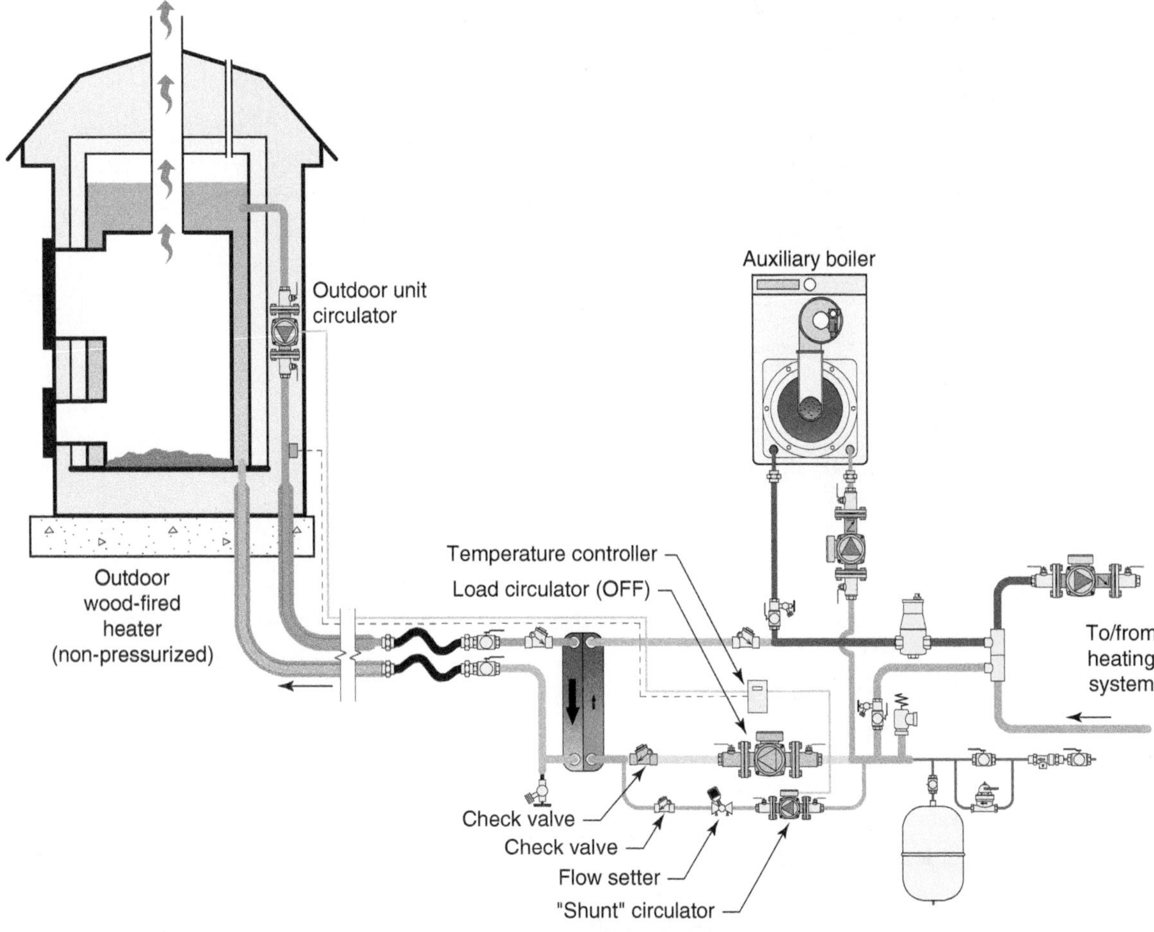

Figure 15-36 | Using a small shunt pump to meter warm water through heat exchanger to prevent freezing in outdoor circuit.

combustion air through a shutter in the lower front portion of the boiler. This air flows upward through a cast-iron grate that supports the coals and firewood, as shown in Figure 15-38.

Because air is only introduced to the combustion process in one location, this type of boiler is said to use **single-stage combustion**. Heat from the burning wood and coals is transferred directly to the cast-iron sections that form the combustion chamber. Exhaust products are carried upward through a vent connector to the chimney. A barometric damper should NOT be installed in the venting system of a natural draft wood-fired boiler. Doing so would allow a chimney to readily draw air if a chimney fire should occur. This could greatly intensify an already dangerous situation. Thus, all air regulation is controlled by the air intake shutter on the boiler.

As the charge of firewood is consumed, the coals accumulate on the lower grate, eventually burning down to a size where some of them fall through the grate and into the ash pan. When the fire is out, and the stove has fully cooled, the remaining **clinkers** can be cleaned from the grate, and the ash pan pulled out through the lower door.

Many boilers of this design can also be configured to burn coal. This usually requires the lower grating to be changed so that smaller coal embers are supported.

WELDED-STEEL WOOD-FIRED BOILERS

There are several variations in natural draft wood-fired boilers built of welded steel. There are also differences in the type of electrical controls used on these boilers. A common design uses a welded-steel shell that is lined on the bottom and sides with heat-resistant **refractory** bricks. These bricks protect the steel from very hot temperatures generated by the embers and coals at the base of a well-established fire.

Figure 15-37 | (a) Example of a natural draft wood-fired boiler constructed of cast-iron sections. (b) Natural draft wood-fired boiler constructed of welded steel.

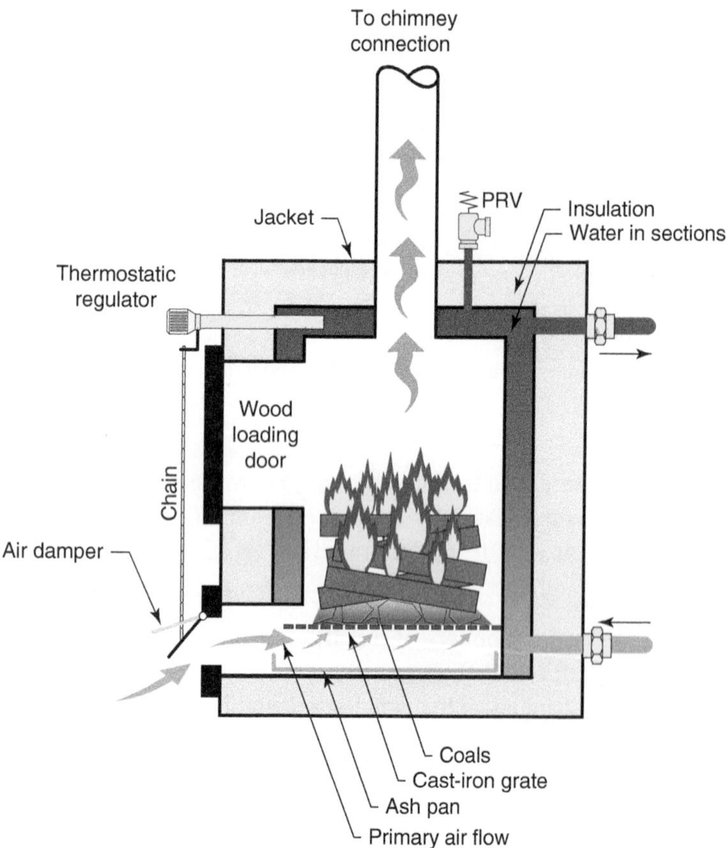

Figure 15-38 | Construction of a typical natural draft wood-fired boiler.

Heat is extracted from the fire through serpentine-shaped stainless steel tubes positioned at the upper sides and top of the combustion chamber. The stainless steel is resistant to corrosion from creosote and other chemical compounds that are generated in combustion. These tube assemblies penetrate the rear wall of the combustion chamber and are piped together outside the boiler. Primary combustion air enters through a shutter in the lower front of the combustion chamber. Figure 15-39 shows this construction.

Some boilers of this design use an internal damper that is positioned and shaped to encourage the pyrolytic gases produced as the wood is heated to recirculate over the fire, where they can mix with oxygen and combust, rather than being quickly pulled up the stack. Secondary air may also be introduced to enhance this process. If so, this would be called **two-stage combustion,** since air is supplied for both primary and secondary combustion. The internal damper is configured so that it allows for updraft operation as the fire is beginning and is then repositioned to encourage pyrolytic gas recirculation and secondary combustion once the fire has grown in intensity.

Boilers of this type are sometimes equipped with safety devices such as **stack temperature switches**, aquastats, and motorized dampers. These devices are usually wired in series so that abnormally high temperatures in either the stack, or the water in the boiler, cause the primary air shutter to quickly move to an almost closed position and thus quickly reduce the intensity of the fire. The steady-state thermal efficiency of natural draft-type boilers, burning well-seasoned wood, is typically in the low-to-mid 70 percent range.

15.6 Wood-Gasification Boilers

The current state-of-the-art combustion method for boilers that burn cordwood is called wood gasification. An example of a wood-gasification boiler is shown in Figure 15-40.

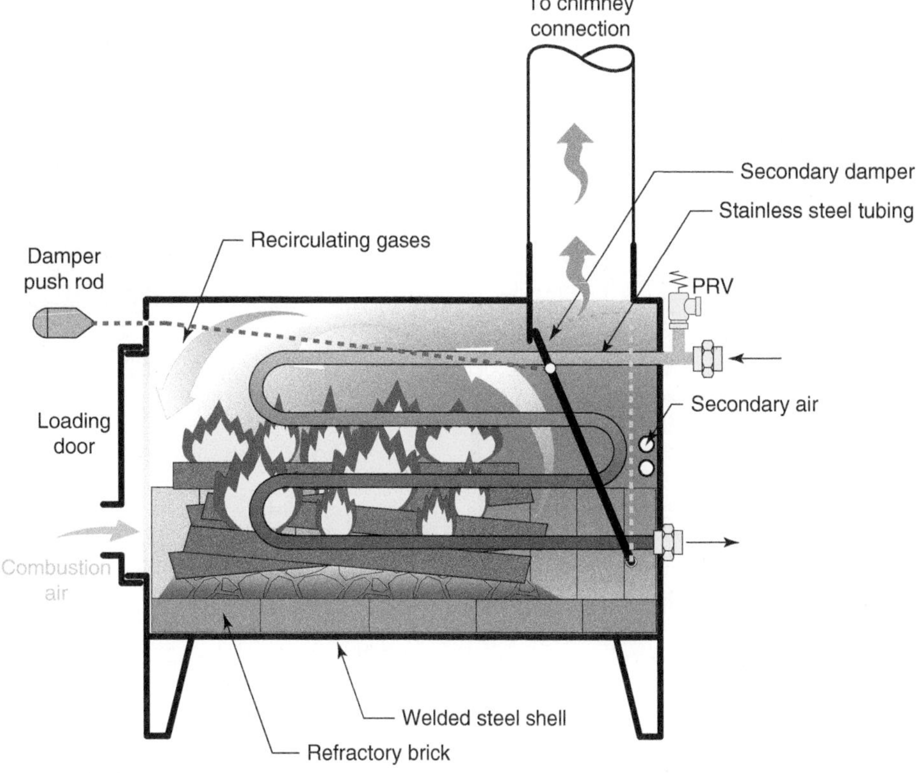

Figure 15-39 | Construction of a welded steel natural draft wood-fired boiler.

The fire in a wood-gasification boilers is kindled much like that in a natural draft wood boiler, using kindling and smaller pieces of firewood in the boiler's upper chamber. When the fire is started, the flue damper near the top of the upper combustion chamber is open and the boiler operates in a natural updraft mode. Once the fire in the upper chamber is stable, the firebox is fully loaded with split firewood, the upper damper is closed, and a blower turns on to redirect the pyrolitic gases emitted by the heated wood in a downward direction, as shown in Figure 15-41.

The pyrolytic gases pass through a slot in a ceramic grate at the base of the combustion chamber. Air, pressurized by the blower and heated by the fire, is forced through holes in the side of this slot and mixes with the hot pyrolytic gases. This creates a region of intense secondary combustion in which temperatures can exceed 2,000 °F. This secondary combustion zone has the appearance of a blowtorch, as seen in Figure 15-42a. This highly efficient two-stage combustion leaves very little residue.

The variable-speed combustion air blower in a wood-gasification boiler regulates the rate of combustion based on the water temperature in the boiler. As the water temperature approaches a set upper limit, the blower's speed is reduced.

Wood-gasification boilers must be vented to an approved chimney. Factory-built "all-fuel" chimneys that meet UL standard 103HT are commonly used. An example of such a chimney system is shown in Figure 15-42b.

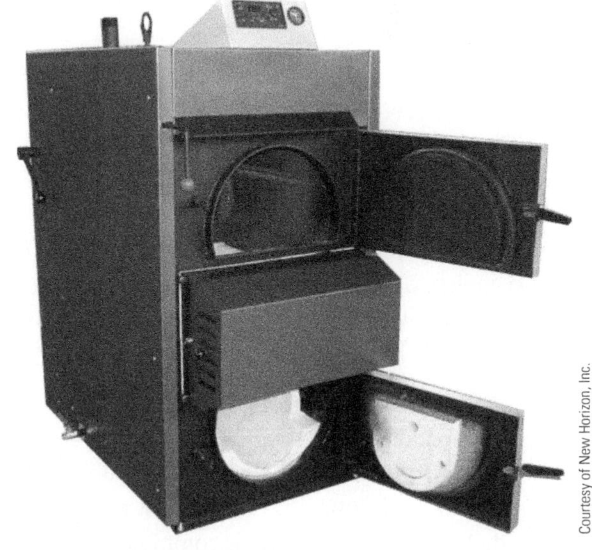

Figure 15-40 | Example of a wood-gasification boiler.

These chimneys are assembled from twist-lock sections of double-wall insulated piping. The inner wall is stainless steel, which is compatible with the water vapor formed during combustion. They outer wall may be either stainless steel or plated steel. The space between the inner and outer wall is filled with

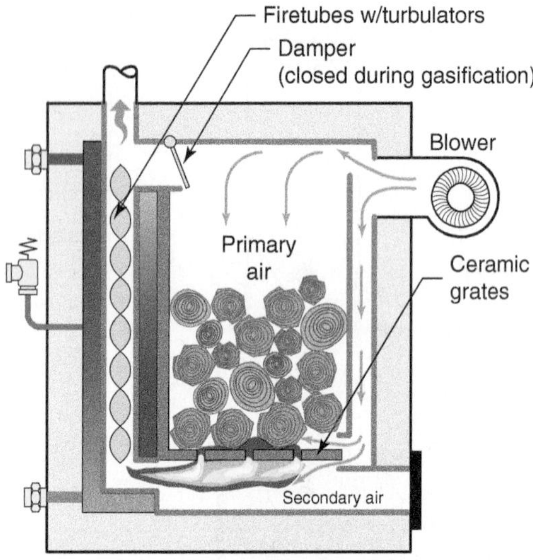

Figure 15-41 | Wood-gasification boiler operating in downdraft (e.g., gasification mode).

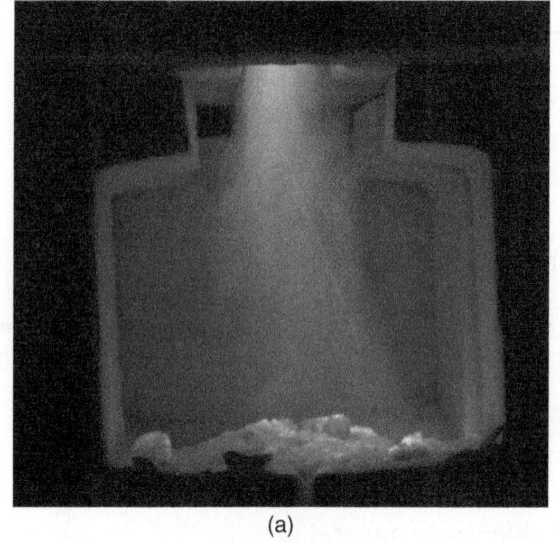
(a)

Courtesy of New Horizon, Inc.

(b)

Courtesy of Selkirk

Figure 15-42 | (a) Secondary combustion in the lower combustion chamber of a wood-gasification boiler. (b) Example of a factory-build "all fuel" chimney.

high temperature compatible insulation. This type of chimney is capable of continuous operation with flue gas temperatures as high as 1,000 °F. It can also handle intermittent temperature rises to 1,400 °F for 60 minutes, and 1,700 °F for 10 minutes. The latter temperatures do not occur in normal operation of wood-gasification boilers.

All wood-gasification boilers should be installed with a **barometric damper** in the vent connector pipe between the boiler and chimney, as shown in Figure 15-43. The weight on this damper should be adjusted to maintain the manufacturer's recommended **breach draft**. Typically, this is in the range of –0.02 to –0.05 inches of water. This slight negative pressure helps protect the venting system from releasing exhaust products into the building.

Wood-gasification boilers, when operating with properly seasoned firewood (20 percent or lower moisture content), routinely achieve steady-state thermal efficiencies in the low-to-mid 80 percent range. Their average thermal efficiency, over a complete burn cycle, is typically in the range of 70 to 75 percent. Ongoing developments using **lambda sensors** to monitor the oxygen level in the exhaust gases, and adjust secondary air flow based on these levels, show the potential to increase both steady-state and average (e.g., complete cycle) efficiencies.

The highest combustion efficiency is achieved by burning a full load of wood at the maximum possible rate. Wood-gasification boilers are intended to be operated as "batch burners," rather than devices that are frequently stoked with wood or operated at reduced capacity. This is quite different from what many people are accustomed to when operating woodstoves, or other non-gasification wood burning appliances. Operating a wood-gasification boiler in the same manner as a quasi-continuously burning woodstove will reduce its efficiency and increase its emissions.

Proper operation of a wood-gasification boiler usually results in heat being produced at a rate greater than that required by the load. This difference between the rate of heat production and rate of heat transfer to the load is further exacerbated by partial load conditions and zoned hydronic distribution systems.

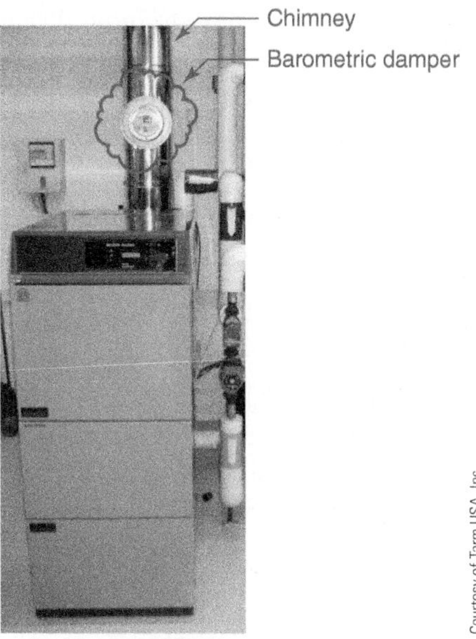

Figure 15-43 Installation of a barometric damper in the vent connector pipe of a wood-gasification boiler.

To maintain optimal efficiency and comfort, thermal storage must be used between any wood-gasification boiler and the load(s) it serves. This storage is provided as a well-insulated water-filled tank, which absorbs boiler heat output that exceeds the current load. Under some conditions, this thermal storage tank can also release heat to the load at rates significantly higher than the current output from the boiler. Some of the thermal storage tank options discussed in Chapter 9 can be used in systems supplied by wood-gasification boilers. Designers should always verify that the thermal storage tank can withstand the maximum possible water temperature that might be produced by the boiler. A *minimum* upper temperature rating of 200 °F is suggested.

Another benefit of wood-gasification boilers is that they operate with lower particulate emissions compared to non-gasification wood-burning boilers or outdoor wood-fired hydronic heaters. This is based on their ability to combust a higher percentage of the pyrolytic gases relative to devices using single stage combustion. Figure 15-44 shows the difference in visible smoke production from a wood-gasification boiler as it operates in both non-gasification mode (at startup), and gasification mode (a few minutes after startup).

Because of their higher thermal efficiencies, and lower emissions, it is likely that wood-gasification boilers will continue to gain market share against natural draft wood-fired boilers or outdoor wood-fired heaters.

Figure 15-44 Difference in visible smoke between (a) non-gasification operation and (b) gasification operation.

15.7 Pellet-Fired Boilers

Because of the size of wood pellets, and the way they are handled, pellet-fueled boilers are very different from boilers that burn cordwood. Most have two major subassemblies: one consisting of a **day hopper** and auger system that automatically feed pellets into the **burn pot** and another consisting of a combustion chamber, fire-tube heat exchanger, and ash management hardware. A cutaway of a modern residential scale pellet-fired boiler is shown in Figure 15-45.

The day hopper seen on the right side of Figure 15-45 can hold enough pellets for several hours of unattended operation. The hopper can be manually loaded by pouring in two or three 40-pound bags of pellets or automatically loaded with pellets from a bulk storage container such as the reinforced fabric **bag silo** seen in Figure 15-46.

Bulk storage systems use motor-operated augers or pneumatic tubing to move pellets from the storage

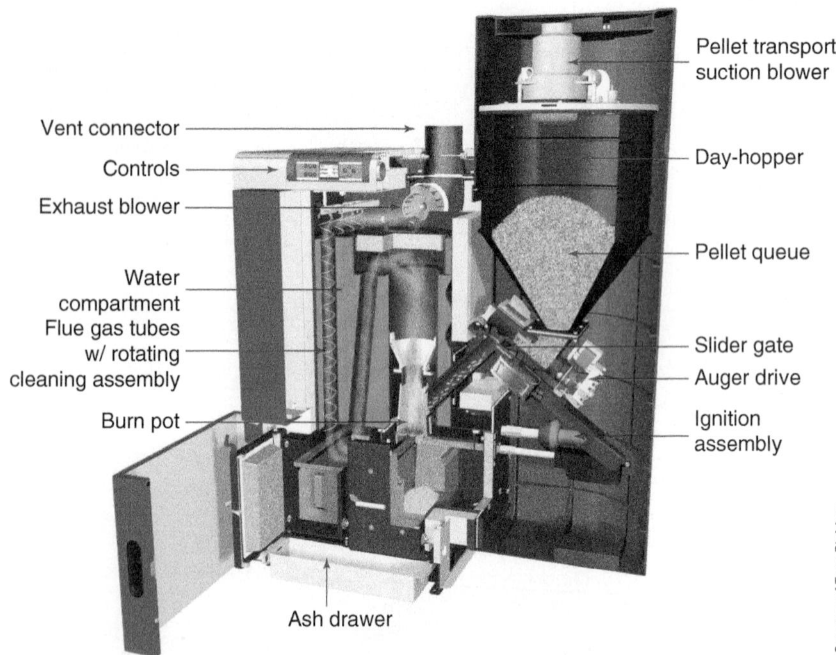

Figure 15-45 | Modern pellet-fired boiler.

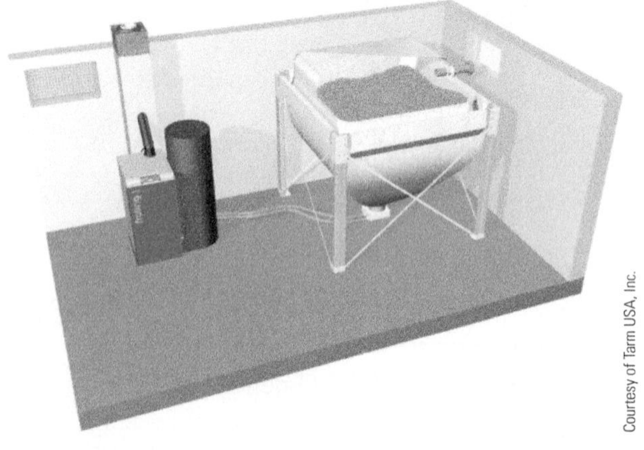

Figure 15-46 | A bag silo pellet storage system with pneumatic pellet transport to day hopper.

container to the day hopper attached to the boiler. The air flow needed to move pellets through pneumatic tubing is created by a blower in the day hopper.

Studies conducted in Europe have shown that freshly produced pellets, especially those made of softwood, can **outgas** minor amounts of carbon monoxide. The exact chemical process by which carbon monoxide is formed is not fully understood at present. It is affected by the time over which the pellets have been stored, the species of wood used, and the storage temperature of the pellets. The outgassing is strongest in the first week or two after the pellets have been produced and then typically drops to 10 percent or less of the initial rate. Mitigation of this carbon monoxide can usually be accomplished by passive ventilation of the storage bin. Some sources indicate than an air flow rate of 4 ft.^{3}/minute is usually adequate to protect a storage bin holding up to 5 tons of pellets. This ventilation can be accomplished using special caps on the fill and air relief tubes supplying the pellet storage bin. It can also be accomplished using a very small ventilation fan to maintain a slight negative air pressure within the pellet storage bin.

For large buildings, or in situations where interior storage is not possible, pellets can be stored in an outdoor silo such as shown in Figure 15-47.

Pellet storage silos are similar to the exterior grain bins used to store pelletized animal feed. In some cases, they use a motorized auger to move pellets from the bottom outlet of the silo to a day hopper adjacent to the boiler. In other cases, a pneumatic tube system is used to convey pellets into the building as needed. The top of a pellet storage silo is closed to prevent precipitation from entering.

Some companies also offer **energy boxes,** which combine bulk pellet storage, pellet transport system, boiler, and related hardware in a preassembled system, as seen in Figure 15-48.

The energy box is transported to the site and placed on a concrete pad. Electrical and insulated piping connections are routed underground to the building. These systems allow pellet heating to be used in situations where there is inadequate interior space for the boiler or pellet storage. It also keeps all combustion and fuel outside the building, which may be necessary in certain building environments.

Bulk pellet storage systems are periodically filled from a delivery truck. They allow a pellet-fueled boiler to operate unattended for weeks.

Unlike boilers that burn cordwood, most pellet boilers can automatically initiate combustion when there is a demand for heat. The startup process begins when a **motorized auger** loads a quantity of pellets into the burn pot. They are then ignited using superheated air or a ceramic heating element. Once

(a)

Courtesy of AHONA.com

(b)

Courtesy of Tarm USA, Inc.

Figure 15-47 (a) Example of an exterior pellet storage silo using an auger to transport pellets into a building where the boiler is located. (b) A pellet storage silo using buried pneumatic tubes to carry pellets into the adjacent building.

Courtesy of Tarm USA, Inc.

Figure 15-48 Example of an energy box that combines pellet storage, boiler, chimney, and related piping in a preassembled unit.

started, the combustion process is regulated using oxygen-sensing devices called lambda sensors in the combustion stream to control a variable-speed combustion air blower speed. This technology optimizes the **fuel/air ratio** at which combustion occurs, and can yield combustion efficiencies of 85 percent or higher.

Modern pellet-fueled boilers can also modulate their heat output over a limited range, typically from 30 to 100 percent of rated output. However, even with modulation, a properly sized thermal storage tank is still recommended in all pellet-fueled boiler systems.

A typical pellet-fueled boiler, starting at room temperature, requires 10 to 20 minutes between the when it is "called" to operate, and the time it reaches steady-state heat output. A thermal storage tank, installed between the boiler and the heating distribution system, can supply the required heat output to the load during this startup time.

The only manual intervention required with most modern pellet boilers is occasional removal of **fly ash** from the lower compartment on the boiler. The ash produced by high-quality **premium wood pellets** is approximately 1 percent of the pellet's original weight. Ash is typically removed from the boiler every 2 to 5 weeks depending on the quality of the pellets burned, the operating time of the boiler, and the ash collection/storage components in the boiler.

Most pellet-fueled boilers are vented to a chimney approved for solid fuel appliances. Factory-built "all-fuel" chimneys that meet UL standard 103HT are commonly used. These chimneys are assembled from twist-lock sections of double wall insulated piping. The inner wall is stainless steel, which is compatible with the water vapor formed during combustion. They outer wall may be either stainless steel or galvanized steel. The space between the inner and outer wall is filled with high temperature compatible insulation. This type of chimney is capable of continuous operation with flue gas temperatures as high as 1,000 °F.

The vent connector piping installed between the pellet-fueled boiler and chimney should be longitudinally seamless, and specifically made for venting solid fuel appliances. This type of vent connector piping is available in either black painted steel or stainless steel.

Courtesy of A. Boutin, PELLERGY, LLC.

(a)

Courtesy of Vermont Renewable Fuels

(b)

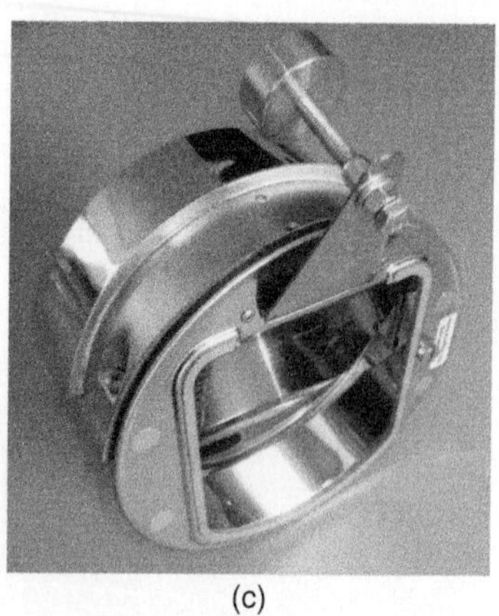

(c)

Figure 15-49 | (a) Pneumatic pellet delivery to an interior bag silo at a house. (b) Pneumatic pellet delivery to an exterior pellet silo. (c) A positive pressure sealing draft regulator with gasketed air passageway.

All joints between the vent connector piping and fittings should be sealed with high temperature (1,000 °F rated) flexible sealant. This prevents leakage of flue gases or fly ash through gaps when the boiler is starting and the vent connector piping is potentially under slight positive pressure. Any male crimped ends of the vent connector piping or fittings should face the direction of flue gas flow.

The author suggests use of a positive pressure sealing draft regulator installed in the vent connector piping at least 1 foot below where the vent connector piping from the boiler tees into a vertical chimney, or vertical vent connector pipe leading to the chimney. An example of a positive pressure sealing draft regulator is shown in Figure 15-49c.

This type of draft regulator is different from a standard barometric damper. It has a gasket where the hinged damper blade closes against the air passage of the regulator. This gasket provides a seal that inhibits leakage of flue gases or fly ash if the vent connector piping experiences temporary positive pressure. The blade of the draft regulator can also be adjusted to swing partially or fully open to limit negative draft pressure in the venting system under normal operating conditions.

Some pellet-fueled boilers can also be sidewall vented. This type of vent relies on a small induced-draft blower to create the proper negative pressure in the vent system. It also uses a small vertical section of vent piping between the boiler and sidewall termination to maintain a minimum draft during a power outage.

Always follow the boiler manufacturer's recommendations on boiler venting as well as any necessary combustion air requirements for the boiler.

15.8 Design Details for Wood- and Pellet-Fired Boiler Systems

PROTECTING AGAINST CORROSION

Like other boilers constructed of cast iron or carbon steel, wood-fired and pellet-fired boilers are designed to be used in *closed-loop* pressurized hydronic systems. The dissolved oxygen that initially enters these systems when they are filled with water quickly reacts with the wetted surfaces of ferrous metal components to form a superficial oxide layer that is not detrimental to the boiler or other metal components. At that point, the potential for oxygen-based corrosion is greatly reduced. However, to sustain this condition, it is imperative to minimize any source of air entry into the system. This includes proper detailing for air separation and venting, forced-water purging, proper expansion tank sizing and placement, and leak-free piping.

PROTECTION AGAINST SUSTAINED FLUE GAS CONDENSATION

Another similarity with conventional cast-iron or steel boilers is the need to protect wood-fired boilers and pellet-fired boilers against the corrosive effects of **sustained flue gas condensation**. Without proper detailing, this condensation will form on the fire-side surfaces of the boiler's heat exchanger, or inside the flue. This condensate contains water and other chemical compounds. It is acidic and will aggressively corrode steel and cast-iron surfaces. Creosote is also likely to form on surfaces that allow flue gases to condense.

The key to avoiding flue gas condensation is to maintain the interior surface of the combustion chamber and flue passes above the **dewpoint temperature** of water vapor or other chemical compounds in the flue gas stream. The dewpoint temperature depends on the moisture content of the firewood and the amount of air passing through the combustion chamber. Figure 15-50 shows how the dewpoint temperature varies within these conditions.

The **lambda ratio** is the actual amount of air supplied to the combustion process divided by the **stoichiometric air requirement**. The latter is the *minimum* amount of air needed to completely combust the fuel under ideal conditions. Any lambda ratio greater than 1 indicates that **excess air** is being supplied to the combustion process. This is generally desirable because an appropriate amount of excess air helps ensure proper mixing of the pyrolytic gases with oxygen and thus improves combustion efficiency. Excess air also creates a drying effect that lowers the dewpoint of the flue gases. However, as more air is supplied, more heat is carried away in the exhaust stream. Thus, there are optimal values of the lambda ratio for a given combination of other combustion conditions, such as fuel type, moisture content, and combustion zone temperature. Some modern wood- and pellet-fueled boilers have controls that continually measure the oxygen content of the exhaust gases and use this information to adjust the lambda ratio of the combustion process.

The higher the moisture content of the firewood, the higher the dewpoint of the flue gases. Properly seasoned firewood should have a moisture content no higher than 20 percent. Burning this wood with a typical lambda ratio of 1.5 results in theoretical dewpoint of 123 °F. Allowing for a slight safety factor, the minimum inlet water temperature to the boiler under these conditions should be 130 °F.

ANTI-CONDENSATION VALVES

One way to protect a wood-fired boiler against low inlet water temperature is to install a thermostatic mixing valve between the wood-fired or pellet-fired boiler, and its load. One example of such a valve is shown in Figure 15-51.

When used to protect wood-fired or pellet-fired boilers, these valves are often referred to as **anti-condensation valves**. They use a nonelectric thermostatic element to regulate the mixing of hot water from the boiler outlet and cooler water returning from the load or the lower portion of the thermal storage tank. These flows are automatically proportioned by the valve so that the inlet temperature to the boiler remains high enough to prevent sustained flue gas condensation.

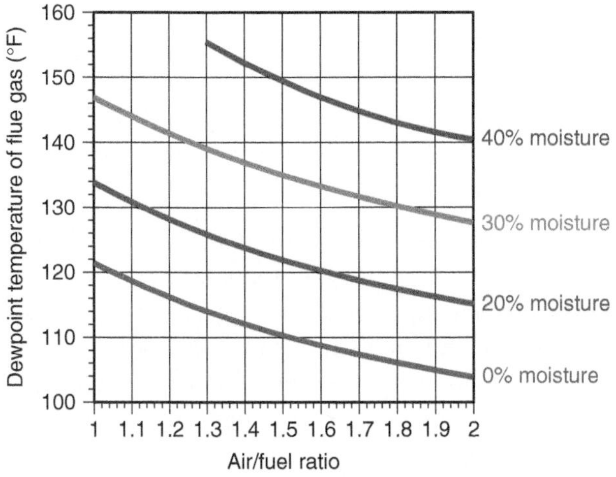

Dewpoint temperature of flue gas (°F)
160
150
140
130
120
110
100
40% moisture
30% moisture
20% moisture
0% moisture
1 1.1 1.2 1.3 1.4 1.5 1.6 1.7 1.8 1.9 2
Air/fuel ratio

Figure 15-50 Dewpoint of flue gases emitted by firewood as a function of lambda ratio and moisture content.

Courtesy of Caleffi North America, Inc.

Figure 15-51 Example of a 3-way thermostatic mixing valve designed to protect a wood-fired or pellet-fired boiler from low entering water temperatures.

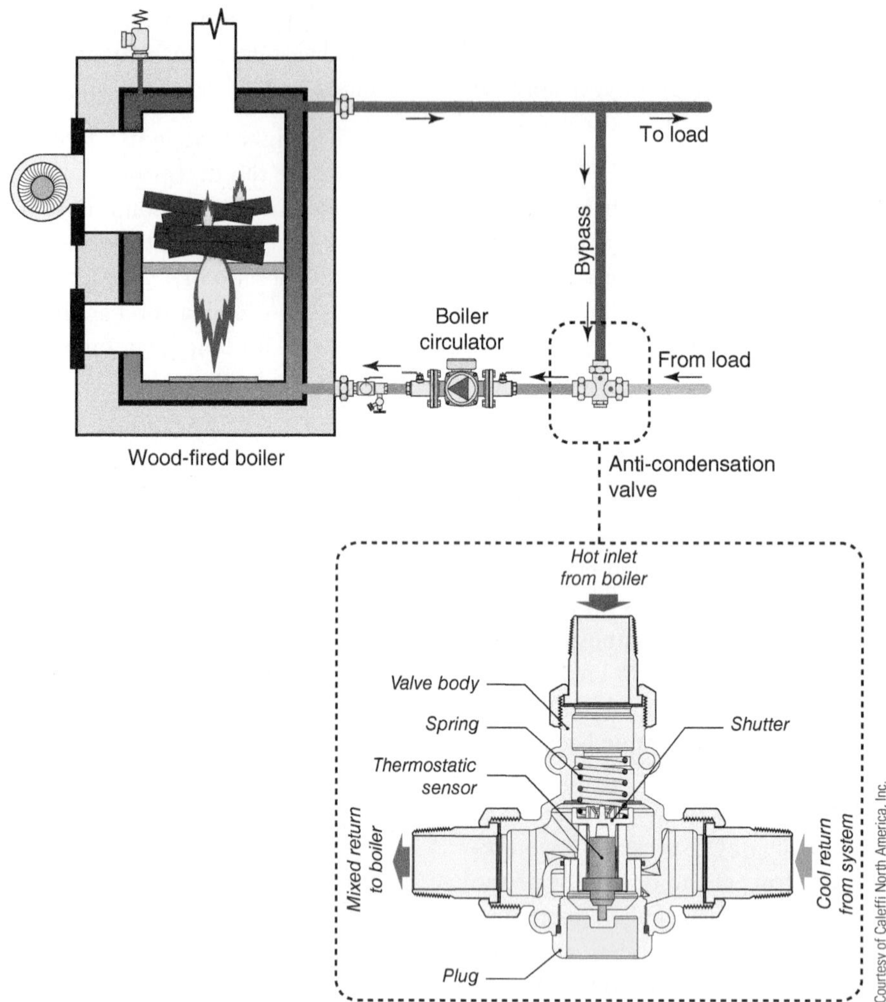

Figure 15-52 | Internal design, and placement of a 3-way thermostatic anti-condensation valves.

Unlike the 3-way thermostatic valves used to control supply water temperature in other hydronic heating applications, these valves do not have an adjustable dial to set the desired outlet temperature. Instead, they use an internal cartridge that has a fixed temperature setting and proportional operating range. These valves are ordered with a specific temperature setting. If that setting needs to be changed in the future, the internal cartridge can be removed and replaced with one that has a different setting.

The cartridges are available over a range of temperatures from about 110 °F to about 160 °F. The lower temperature is appropriate for very dry firewood (10 percent to 15 percent moisture content) being burned with relatively high excess air. The upper setting would be appropriate for wood with higher moisture content (more than 20 percent) being burned with lower excess air, which would typically only be acceptable in a properly configured wood chip boiler.

Figure 15-52 shows the internal design of an anti-condensation valve and its placement in relation to the boiler and the boiler circulator.

When the boiler is first fired, the water within the mixing valve will be close to room temperature. Under this condition, the valve's cool water inlet port will be fully closed, and the bypass port fully open. All water leaving the boiler will be routed back to the boiler's inlet. No water is routed to the load or thermal storage tank, as seen in Figure 15-53a. This allows the boiler temperature to rise quickly, and thus minimizes condensing mode operation.

As the water temperature leaving the boiler rises, the thermostatic element within the valve begins closing the bypass (hot) water inlet port and simultaneously opening the cool water inlet port. This allows some heated water to flow to the load, as shown in Figure 15-53b.

The bypass port on most anti-condensation valves allows some flow of hot bypass water when the temperature leaving the valve reaches the temperature setting of the thermostatic cartridge. The bypass port will continue closing as the temperature of the water leaving the valve climbs above the setting of the thermostatic cartridge. Assuming the water temperature leaving the boiler continues to rise, the bypass port will eventually close, and all water returning from the load, or lower portion of the thermal storage tank, will be routed directly back to the boiler. As the wood-fired boiler outlet temperature drops at the end of a firing cycle, the bypass port will begin opening in an attempt to keep the boiler inlet temperature above the dewpoint of the flue gases. Designers are encouraged to verify the temperature range between the setting of the thermostatic element, and the temperature at which the valve's bypass port is fully closed. On some valves this can be as high as 18 °F.

Some manufacturers also offer boiler **loading units** that consist of a 3-way thermostatic valve, circulator, and flapper check valve combined into a single unit, as seen in Figure 15-54.

During normal operation, when electrical power is available, the loading unit performs the same functions as the previously discussed anti-condensation valve, in combination with the boiler circulator. However, a unique feature of most loading units is their ability to allow **thermosyphon flow** between the boiler and load during power outages. This operating mode helps prevent excessive heat buildup within the wood-fired boiler, which could eventually cause the pressure-relief valve to open. During a power outage, a lightly loaded flapper valve within the loading unit is pushed open by the slight pressure differential created by buoyancy effects of heated water in the boiler and cooler water in other portions of the system, as shown in Figure 15-55.

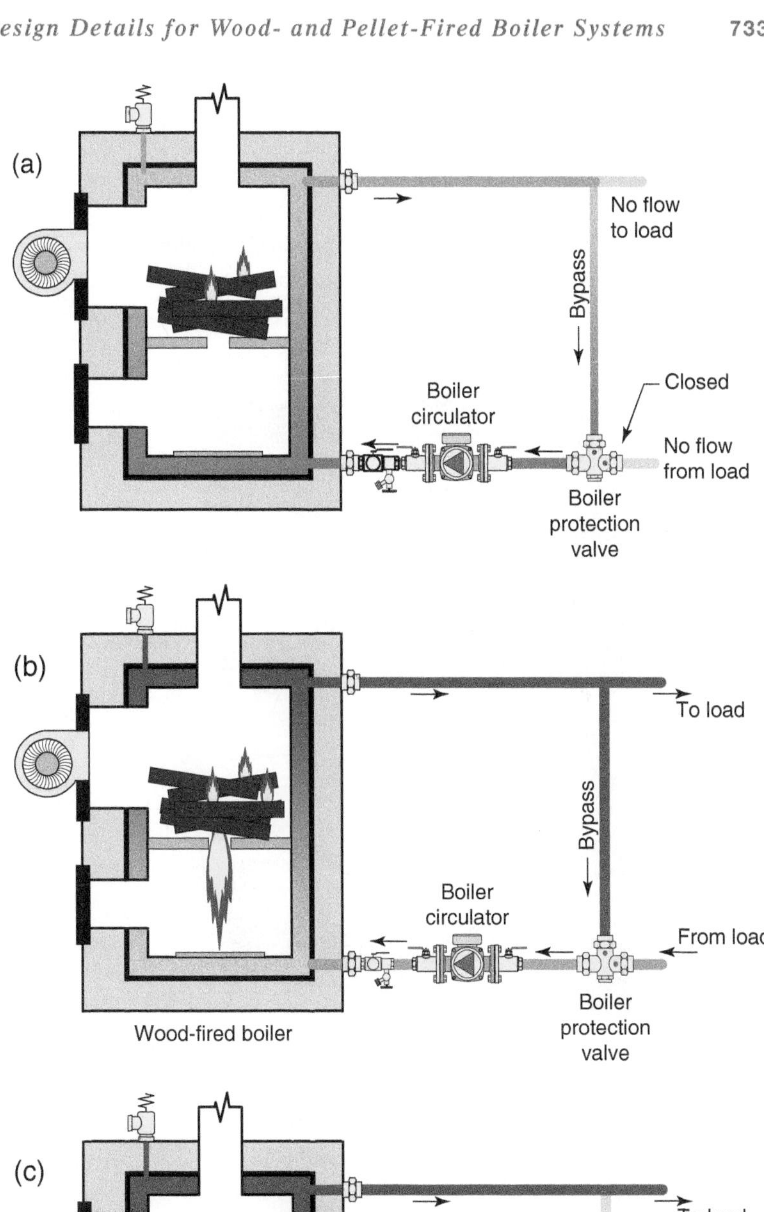

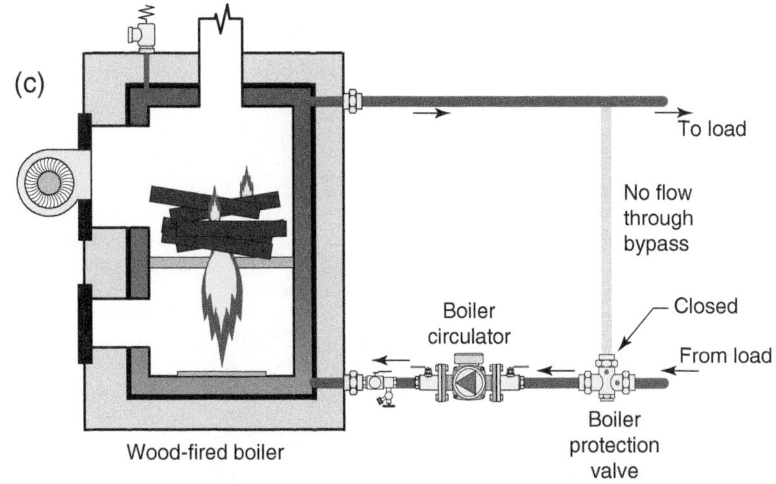

Figure 15-53 (a) When boiler water is cool, the bypass port is fully open, and the return from the load is fully closed. No water flows from the boiler to the load (b). As the water temperature entering the bypass port increases, the bypass port begins to close, and return port from the load begins to open. (c) When the water temperature exiting the valve is a few degrees above the valve's setpoint temperature, the bypass port is fully closed, and the return port from the load is fully open.

During normal operation, this flapper valve is held shut by the pressure differential created by the circulator.

When a power outage occurs, most wood-gasification boilers and pellet-fired boilers immediately reduce the air supplied to their combustion chamber. In some boilers this happens because the combustion air blower stops. In other boilers, the air intake shutter is closed by a spring-load actuator. Although this does not instantly stop further heat production, it significantly reduces the *rate* of further heat production to a level where thermosyphon flow between the boiler and its load can safely dissipate any further heat production. In systems without such protection, it is usually necessary to install a heat dump subsystem. This subsystem will be discussed shortly.

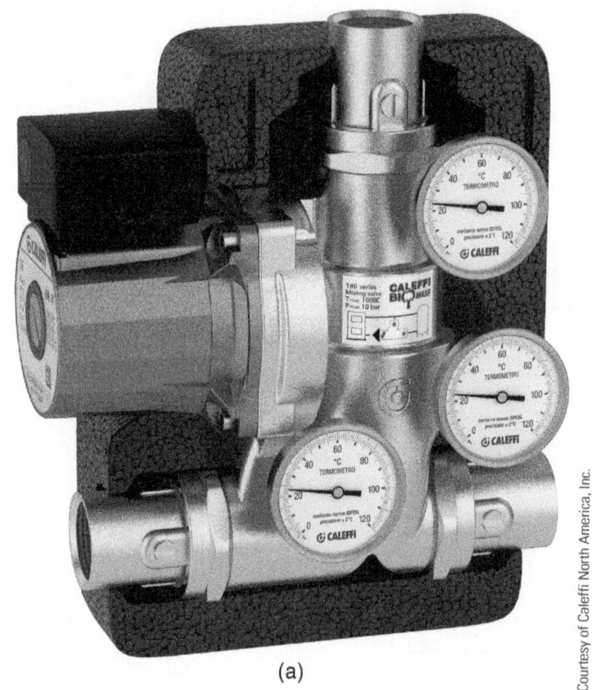

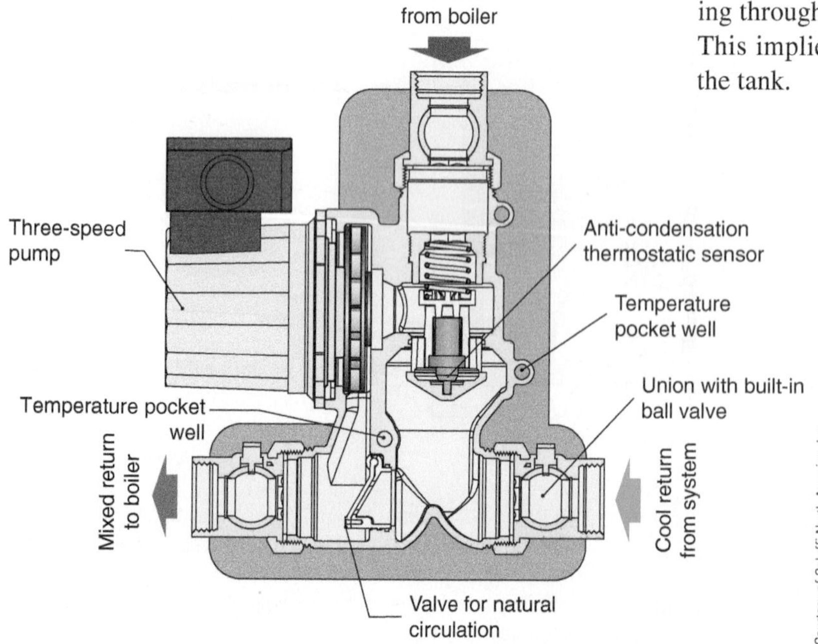

Figure 15-54 | (a) Example of a loading unit for a wood-fired or pellet-fired boiler. (b) Internal design of loading unit.

BOILER PROTECTION USING VARIABLE-SPEED PUMPING

Another method of protecting a wood-fired or pellet-fired boiler from low inlet water temperature is through use of a **variable-speed setpoint circulator**. The speed of this circulator is automatically adjusted to regulate the rate of heat transfer between the boiler and its thermal storage tank. One piping arrangement for this approach is shown in Figure 15-56.

In this system, circulator (P1) operates whenever the boiler is being fired. The closely spaced tees in the boiler circuit **hydraulically separate** the pressure differential created by circulator (P1) from the piping path leading through circulator (P2) and the thermal storage tank. Thus, even though circulator (P1) is operating, there is very little induced flow in the piping path leading through circulator (P2) and the thermal storage tank. This implies very little heat transfer from the boiler to the tank.

Circulator (P2) has onboard electronics that allow it to monitor a temperature sensor located near the inlet of the boiler. When the temperature at this sensor is below a user-set minimum value, (P2) remains off. As the temperature of sensor increases above the minimum value, the speed of the circulator (P2) increases. When the sensor temperature reaches 5 °F or more above the set minimum value, circulator (P2) is operating at full speed. The higher the speed of circulator (P2), the greater the flow rate of heated water from the boiler loop into the storage tank. Thus, by controlling the speed of circulator (P2), this system ensures that the rate of heat absorption by the tank never exceeds the rate of heat production within the boiler. The rate of heat

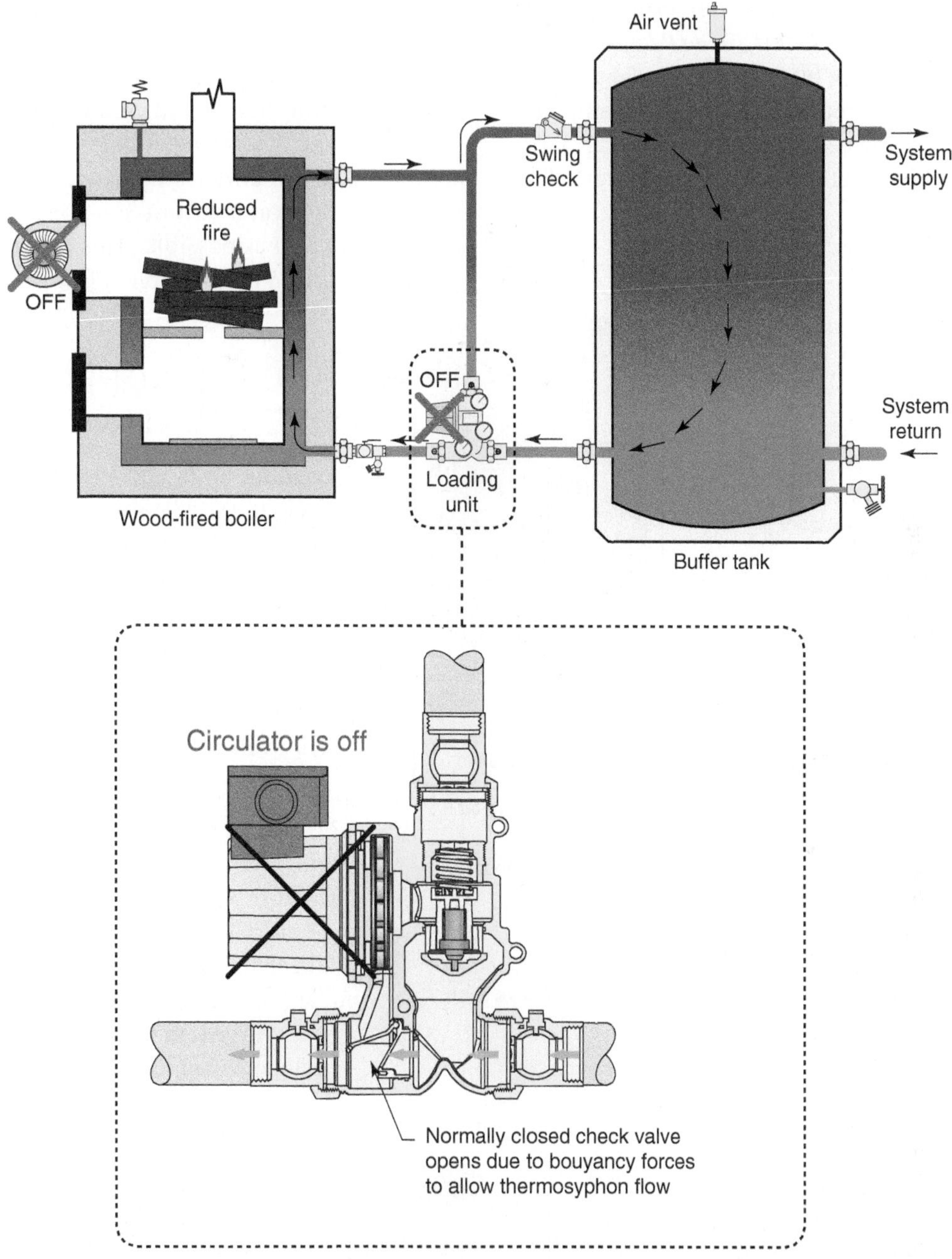

Figure 15-55 | During a power failure, the flapper valve within the loading unit opens to allow natural convection through the boiler to dissipate heat.

transfer is adjusted based on maintaining a minimum boiler inlet temperature, thus preventing sustained flue gas condensation within the boiler.

Figure 15-57 shows one example of a currently available variable-speed circulator with onboard electronics that can be configured for this type of boiler inlet temperature protection.

On larger systems, it is possible to configure a circulator that has an AC induction motor so that it is supplied through a **variable frequency drive (VFD)**. The VFD controls the speed of the circulator's motor by responding to a signal from a controller that monitors boiler inlet temperature.

The piping design shown in Figure 15-56 allows for natural thermosyphoning between the boiler and storage tank if a power outage occurs when the boiler is operating. Because the buoyancy-induced pressure differentials that drive thermosyphon flow are relatively week, *it is very important to remove any spring-loaded check valves from circulators (P1) and (P2)*. It is also important to use only a *swing* check valve where the piping downstream of circulator (P2) enters the upper left connection on the storage tank. This valve is intended to stop **reverse thermosyphoning** from the upper portion of the thermal storage tank, back through the piping assembly, and into to the lower portion of the tank. If the check valve was not present, this reverse thermosyphon flow would create a constant heat loss effect through this piping. The insulation shown on the piping from the boiler output to the tank inlet keeps the water in this portion of the circuit as warm as possible. This enhances buoyancy-driven thermosyphon flow.

Insulated piping
Air vent
Variable speed setpoint-controlled circulator
Swing check
Closely spaced tees
(P2)
(P1)
Wood-gasification boiler
Thermal storage tank
Boiler inlet temperature sensor
NOTE: Remove any internal check valves in circulators

Figure 15-56 Using of a variable-speed injection circulator to regulate the rate of heat transfer between a wood-fired boiler and storage tank.

PROTECTION AGAINST OVERHEATING

Loading units and variable-speed setpoint circulators, when properly installed, allow thermosyphon flow to develop between the boiler and its associated thermal storage tank during a power failure. This flow carries heat generated within the boiler to the thermal storage tank without raising system pressure to the point where the pressure relief valve opens.

However, not all wood-fired boilers can be adequately protected using these subsystems. An example would be a natural draft wood-fired boiler that is not equipped with a means of preventing air flow into the combustion chamber during a power outage. In such cases, it is vital to install an **overheat protection subsystem** that can safely dissipate the heat generated in the boiler during the power

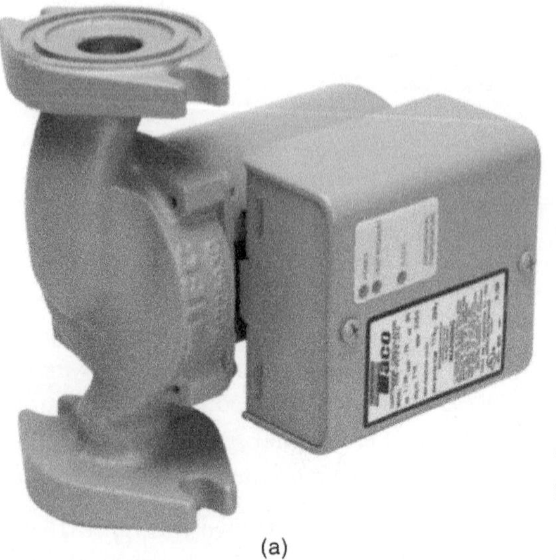

(a)

Courtesy of Taco, Inc.

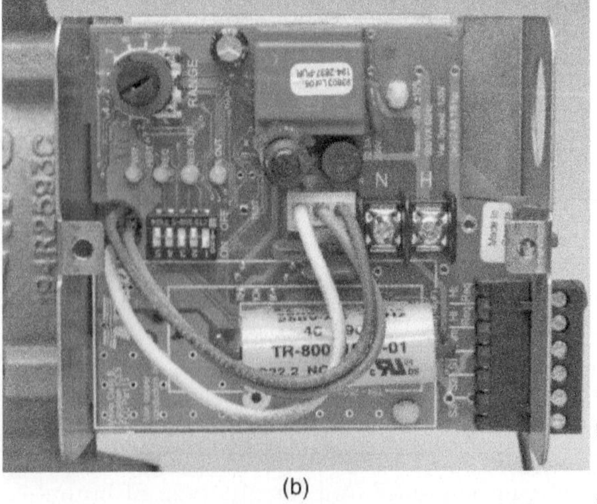

(b)

Courtesy of Taco, Inc.

Figure 15-57 (a) Example of a variable-speed circulator that can be configured for boiler inlet temperature protection. (b) Internal variable speed drive electronics.

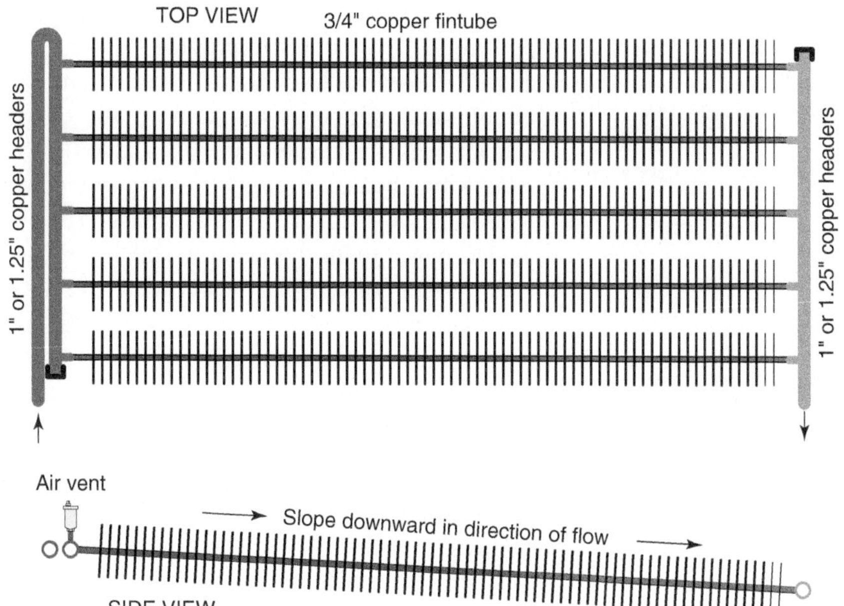

(a)

(b)

Courtesy of Mark Odell

Figure 15-58 (a) Schematic for a heat dump constructed of parallel fin-tube elements connected in reverse return piping. (b) An installed heat dump.

outage. There are several ways to do this depending on the type of wood-fired boiler (e.g., natural draft or wood gasification) used in the system. These methods require a **heat dump**—which is an arrangement of hydronic heat emitters, piping, and, in some cases, a low power circulator—that can safely dissipate the residual heat produced by the boiler.

One commonly used **passive heat dump** uses an array of fin-tube piping as the heat emitter. Multiple fin tubes are assembled in a parallel reverse-return piping array, as seen in Figure 15-58.

The fin-tube array in a passive heat dump *must* be installed *above* the boiler to allow for buoyancy-driven-flow between the boiler and the array. The fin-tube array should also be installed with a slight downward slope in the direction of flow. It is connected to the supply and return piping close to the boiler, as shown in Figure 15-59.

During normal operation, the **normally open zone valve** will be powered and thus held in its *closed* position. When a power outage occurs, this valve immediately opens, allowing thermosyphon flow to develop between the boiler and the fin-tube array.

When power is restored, the normally open zone valve immediately closes to block flow into the heat dump. The return side of the heat dump should remain unblocked so that the expansion and contraction of fluid within the heat dump can be absorbed by the system's expansion tank. It is also important to install a means of purging the air from the high point of the heat dump. A float vent is a good choice for this task. Figure 15-59 also shows a simple thermal trap where the return side piping from the heat dump tees into the boiler inlet pipe. This trap should be at least 12 inches deep. Its purpose is to discourage thermosyphoning in the vertical return piping of the heat trap during normal boiler operation.

When this type of heat dump is used, the operation of the normally open zone valve should be tested at least twice yearly to verify that is opens immediately after power is removed. A spring-return pushbutton switch can be installed to temporarily interrupt power to the zone valve to verify that it works correctly.

In systems using wood-gasification boilers, the combustion air blower will stop when power is lost. In some boilers, a weighted air intake damper, which is normally held open by the pressure differential created by the blower, will also close. This starves the fire for oxygen and quickly reduces further heat generation. One estimate is that heat output drops to about 10 percent of the pre-power outage rate within 3 minutes after the outage occurs. In these situations, the heat dump is sized to prevent water in the boiler from boiling. Given that most boilers of this type are equipped with 30 psi rated pressure-relief valves, it is possible for the water temperature in the boiler to climb several

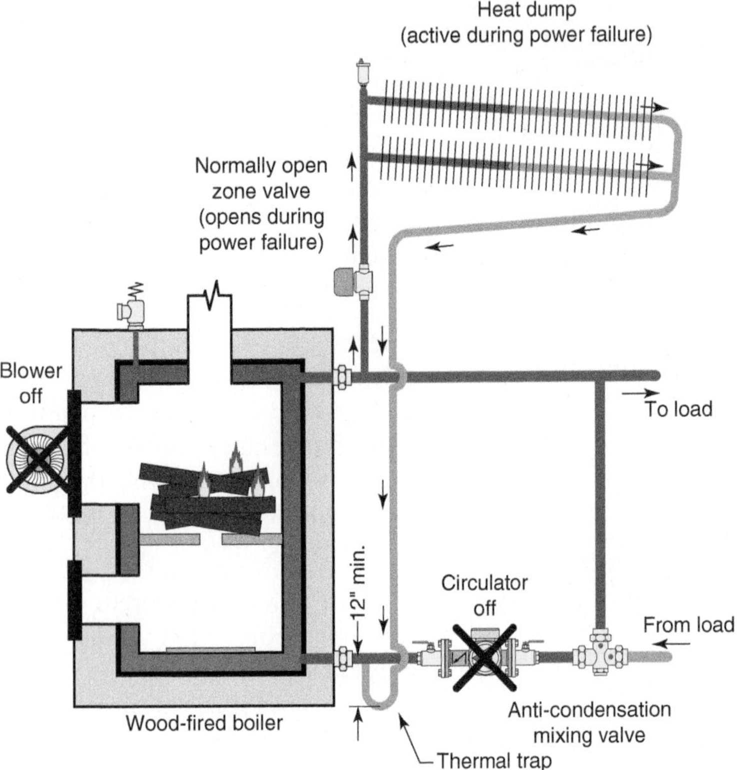

Figure 15-59 | Example of a passive heat dump using a normally open zone valve and fin-tube elements.

degrees above its atmospheric boiling temperature (e.g., above 212 °F) and not boil due to the system's positive pressure. *A conservative guideline is to size the fin-tube assembly to dissipate 10 percent of the boiler's full-rated heat output at an assumed dissipation rate of 300 Btu/hr. per foot.* This assumes common residential fin-tube, which typically has ¾-inch copper piping and 2.25-inch square aluminum fins. Thus, a boiler with a rated output of 150,000 Btu/hr. would require about 50 feet of this fin tube, which could be arranged as five parallel branches, each 10 feet long.

If the boiler uses atmospheric combustion, the amount of fin tube required for the heat dump depends on the ability of the boiler to suppress the fire at the onset of a power outage. If the boiler is equipped with an electrically operated **failsafe air intake damper**, the heat dump can be sized using the same guidelines as for a wood-gasification boiler. However, if the boiler is only equipped with a thermostatically operated air intake damper, it cannot suppress the fire. In this case, the fin-tube array should be sized for the maximum-rated heat output of the boiler. This will usually require substantially more fin tube relative to boilers that can suppress combustion at the onset of a power outage.

It is also possible to install a small battery-powered DC circulator to provide flow through an **active heat dump** during a power failure. The circulator is turned on by the normally closed contacts within a line voltage relay. The coil of this relay would be energized whenever utility power is on, and the normally closed contacts carrying the DC power to the circulator would thus remain open. A strap-on aquastat turns on the circulator if the boiler outlet temperature reaches 180 °F or higher and off when this temperature is below 160 °F. Figure 15-60 illustrates this approach.

The fin-tube-element array should be sized to dissipate the full output of the boiler using a temperature drop of 20 °F to 30 °F. The DC circulator should be sized to provide the flow necessary for this rate of heat dissipation at the selected temperature drop. The battery should be maintained at full charge using a trickle charger and sized to operate the circulator for at least 2 hours.

Still another possibility is use of a small AC-powered circulator driven by an **uninterruptible power supply (UPS)**. The latter are ready available to provide temporary power to computers during a power outage. A small circulator, such as the one shown in Figure 15-61, when adjusted for 30 watts power input, could run for several hours using a UPS with an 168 watt hour battery set (e.g., 2, 7AH/12V batteries).

Figure 15-62a shows how the UPS, circulator, AC relay, and heat dump piping should be arranged, and the flow direction when the heat dump is *active* during a power outage. The heat dump circulator is installed on the return side of its circuit. This reduces the temperature to which the circulator is exposed during normal boiler operation. Figure 15-62b shows the heat dump assembly when it is *inactive* during normal system operation.

Figure 15-63 shows the electrical circuit for the active heat dump subsystem of Figure 15-62.

Upon loss of utility power, the UPS immediately engages its internal battery and inverter to provide 120VAC power to any devices connected to it. The 120 VAC is routed through the normally open contacts in a **strap-on aquastat** that is fastened to the boiler's outlet pipe. If the temperature at the aquastat location reaches 180 °F, the normally open contacts in the aquastat close. This passes 120 VAC to the normally closed contacts of an AC relay. When utility power is lost, these contacts will be closed, allowing 120 VAC to pass to the circulator. If the temperature at the strap-on aquastat drops to

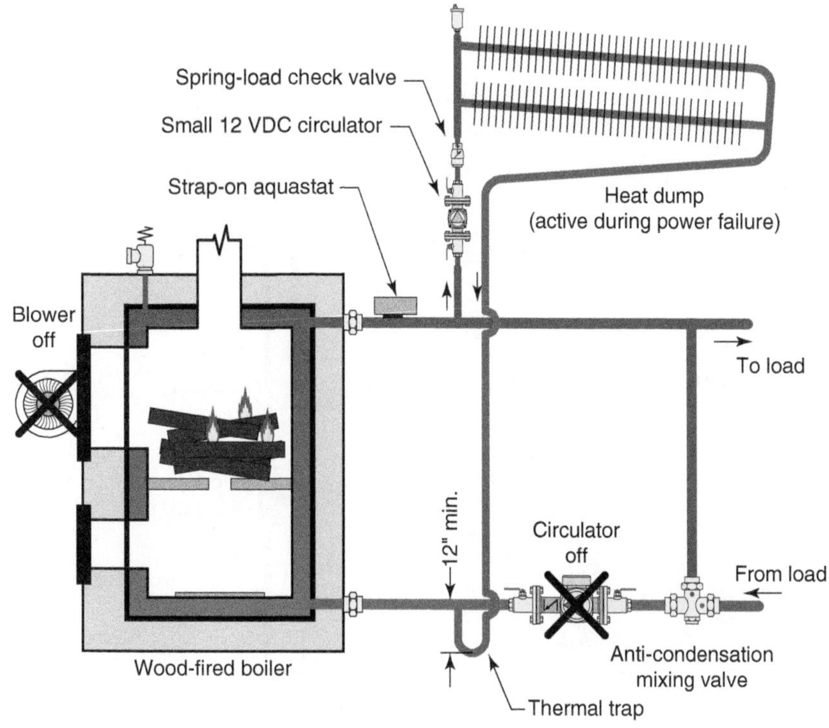

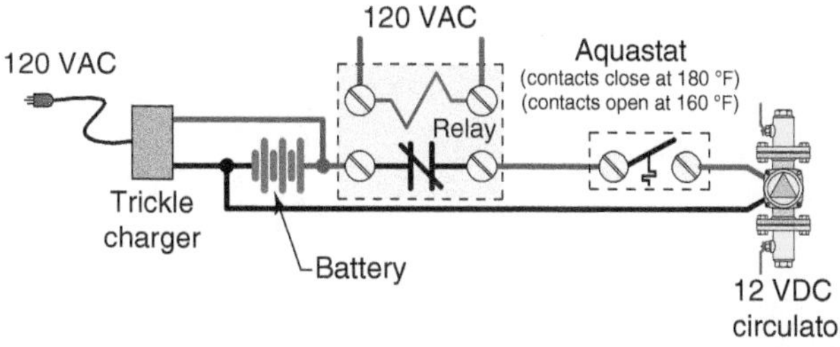

Figure 15-60 | Using a small DC pump to operate a heat dump circuit upon a power failure. Electrical schematic shown at bottom.

Figure 15-61 | Example of a small AC-powered circulator that can be adjusted to low operating wattage for use in a heat dump application.

160 °F or less, which does not represent a dangerously high temperature, the aquastat contacts open, turning off the heat dump circulator. This conserves battery energy within the UPS. The aquastat setting and differential can be adjusted to other temperatures if needed.

An active heat dump does not require the fin-tube elements, or other heat emitters, to be located above the boiler, which is necessary with passive heat dumps. Because the circulator creates forced convection within the fin tubes, their heat output is significantly higher. Assuming a flow rate of 1 gpm per tube, the fin tube could be conservatively sized for a nominal 500 Btu/hr. per foot of length. Be sure to include the spring-loaded check valve or weighted plug flow check valve in the supply side of the heat dump circuit to prevent forward convective flow during normal boiler operation as well as reverse flow when the boiler circulator is active.

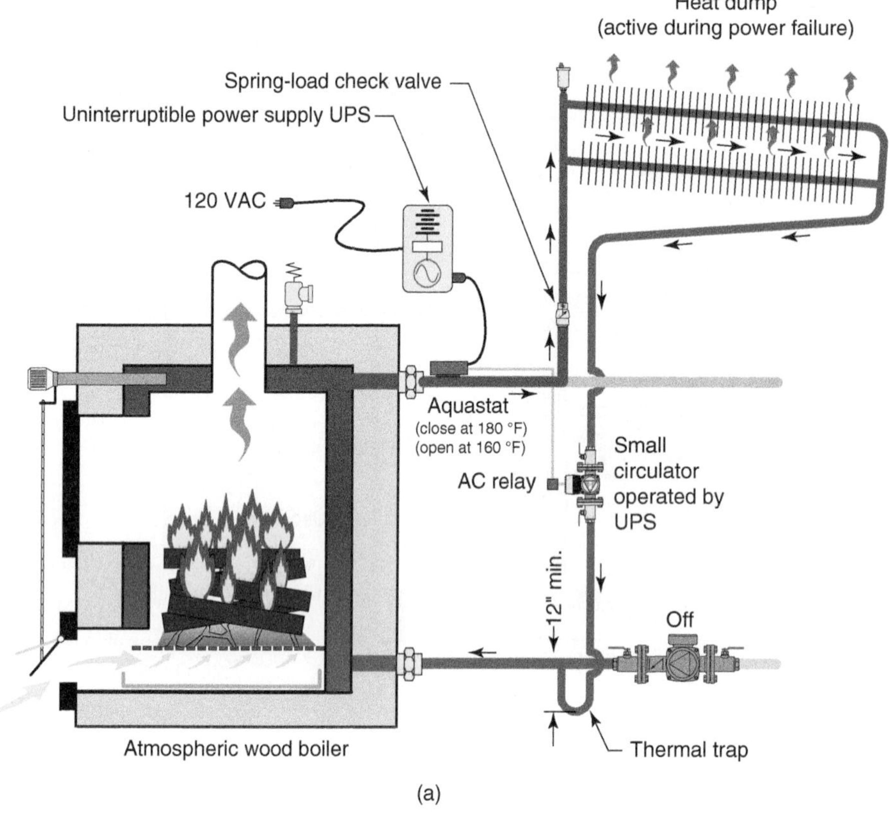

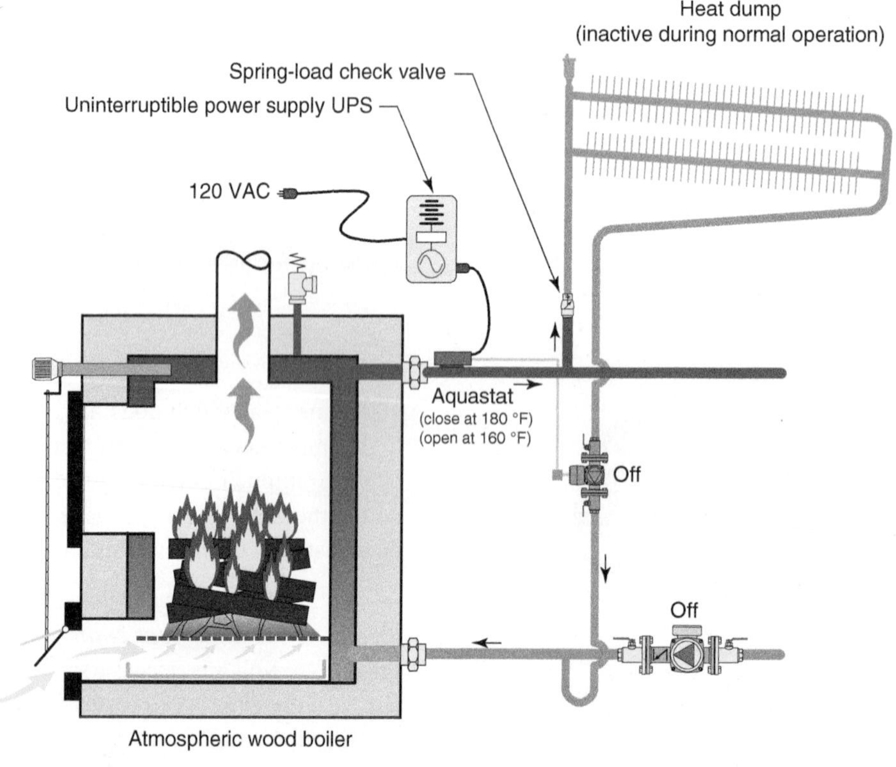

Figure 15-62 | Using a small AC circulator powered by an uninterruptible power supply (UPS) to provide circulation through a heat dump during a power failure. (a) Heat dump in operation. (b) Inactive heat dump during normal operation.

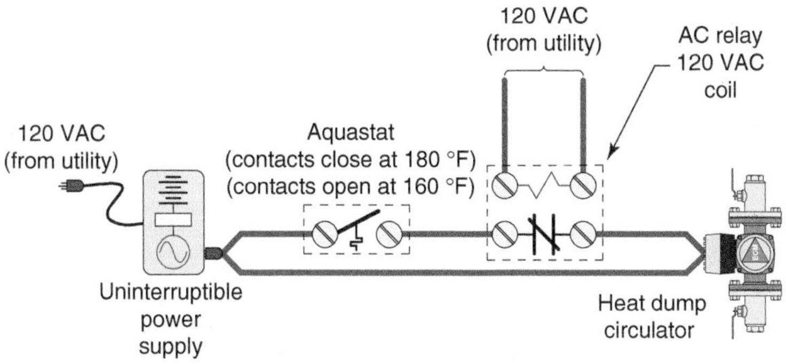

Figure 15-63 | Electrical diagram of circulator power system for heat dump.

THERMAL STORAGE TANKS

The rate of heat production by a wood-fired boiler can vary over a wide range. Sometimes, it will be much higher than the rate of heat transfer required by the load. At other times, it may be much lower, or even 0. Rarely will the rate of heat production match the rate of heat transfer required for space heating or domestic water heating.

To achieve stable operation, thermal mass must be present in the system. Most outdoor wood-fired heaters have large water volumes and thus inherently provide sufficient thermal mass to stabilize system operation. However, most wood-gasification and pellet-fired boilers do *not* have sufficient water volume to absorb all the heat they could generate duration several hours of operation. The solution is to add water to the system using a well-insulated thermal storage tank.

Thermal storage tanks can be either pressurized or unpressurized. Both approaches have their strengths and limitations. Both types of tank were discussed in detail in Chapter 9. What follows is a summary on how to use both types of tanks in combination with wood-fired and pellet-fired boilers.

UNPRESSURIZED THERMAL STORAGE TANKS

Heat is often added to or removed from unpressurized thermal storage tanks using a helical coil of copper or stainless steel tubing suspended within the tank. A typical configuration is shown in Figure 15-64.

These internal heat exchanger coils allow the boiler circuit to operate as a closed circuit. As with any other closed circuit, this requires the installation of an expansion tank, pressure-relief valve, purging valves, and air separator. These components will also be required on the closed loop formed by the heat extraction coil and the distribution system.

The boiler circuit shown in Figure 15-64 also includes an anti-condensation valve to protect the boiler from low inlet temperatures. It also has an overheat protection subsystem consisting of a small AC circulator power by an uninterruptible power supply.

A swing check valve is installed in the pipe carrying heated water to the heat input coil. Its purpose is to prevent reverse thermosyphoning through the piping outside the tank when the boiler is not operating.

One limitation of an unpressurized tank with heat extraction coil is that the water supplied to the distribution system will typically be several degrees Fahrenheit cooler than the water at the top of the tank. This temperature differential is necessary to drive heat from the tank water into water passing through the coil. The larger the coil surface area, the lower this temperature difference will be. The author suggests limiting this difference to 5 °F under design load conditions. Designing a distribution system that can operate at the lowest possible supply water temperature will also improve the performance of the heat exchanger coil. Consult coil manufacturers for specific heat transfer data on their coils.

Another possible configuration for systems with unpressurized tanks uses external brazed-plate stainless steel heat exchangers, as shown in Figure 15-65.

The piping circuit between the boiler and heat exchanger is a closed loop and, as such, is detailed the same as the system in Figure 15-64. The circuit between the boiler heat exchanger (HX1) and tank is an open circuit. Therefore, it requires a stainless steel or bronze circulator. That circulator should be mounted as low as possible to maximize the static pressure at its inlet. The "gooseneck" piping from this heat exchanger passes through the tank wall above the water level. The water pressure in the piping above the water level will, at times, be slightly less than atmospheric pressure. This piping must not contain any air vents, valves with packings, or other components that might allow air to seep into the piping and thus break the siphon that is established when the system is commissioned. Similar detailing applies to the load-side heat exchanger and its piping, seen on the right side of the thermal storage tank.

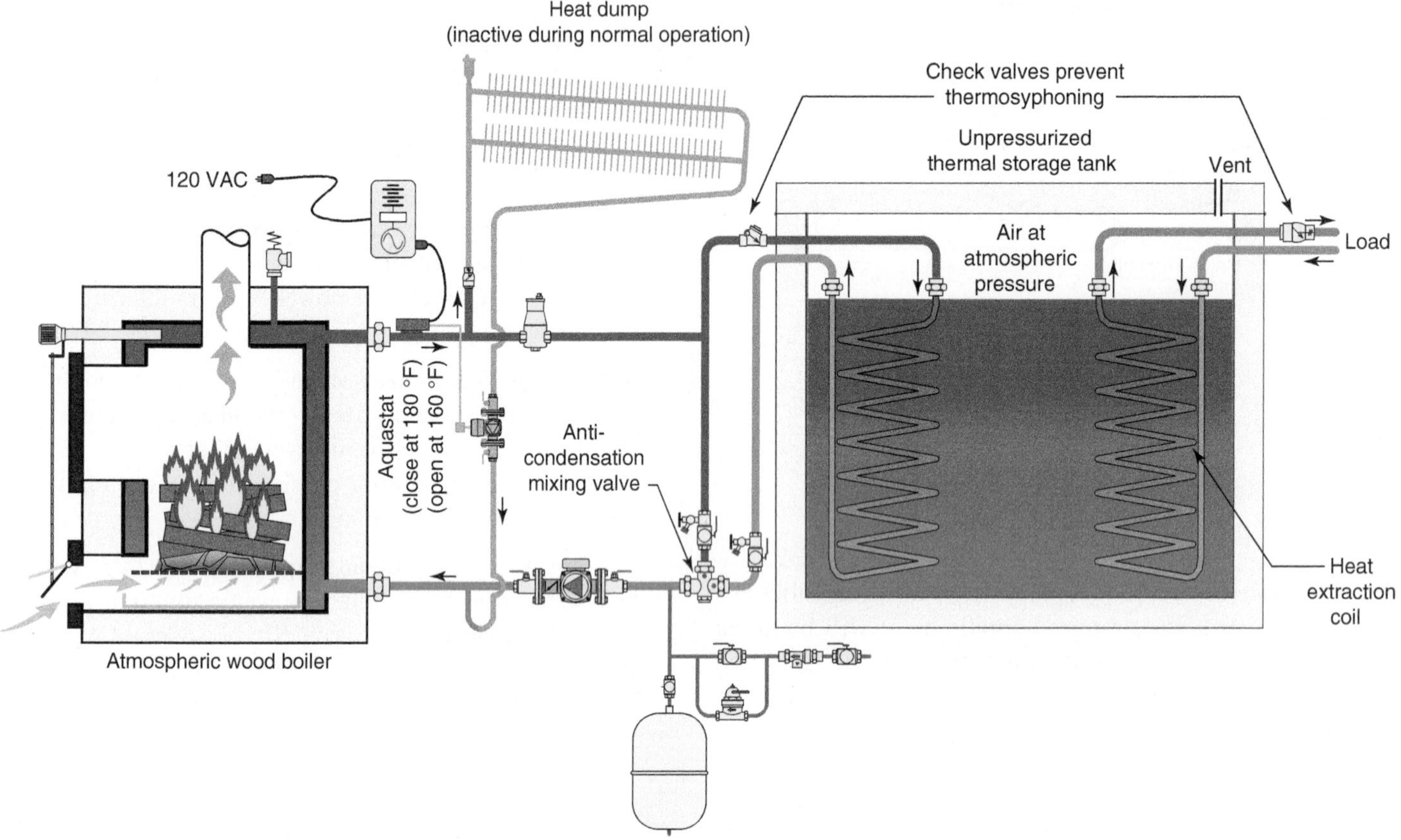

Figure 15-64 | Connecting a wood-gasification boiler to an unpressurized thermal storage tank using a helical coil heat exchanger made of copper or stainless steel tubing.

Because there is *forced* convection heat transfer on both sides of a brazed-plate heat exchanger, it requires significantly less surface area compared to the coiled heat exchanger used in the previous system. The external heat exchangers are also easily maintained or replaced if necessary. The author recommends that the boiler heat exchanger be sized to transfer the full-rated output of the wood-fired boiler using a maximum approach temperature difference of 5 °F. The load-side heat exchanger should also be sized for an approach temperature difference not higher than 5 °F.

Notice that the piping within the thermal storage tank ends with tees. For piping carrying flow into the tank, these tees allow the flow to be directed horizontally and thus help preserve temperature stratification. The tees on piping carrying water out of the tank draw in flow horizontally, which also helps preserve stratification. The author suggests that these tees be sized so that the flow velocity of water leaving the ends of the tees is not higher than 1 foot per second.

PRESSURIZED THERMAL STORAGE TANKS

Because they eliminate a need for heat exchangers, pressurized thermal storage tanks allow for a somewhat simpler design and improved thermal performance. Because they become part of a closed hydronic system, tanks constructed of carbon steel are acceptable.

Figure 15-66 shows a wood-gasification boiler connected to a pressurized thermal storage tank.

A loading unit regulates flow between the boiler and the thermal storage tank so that the boiler inlet water temperature quickly rises and remains above a specified minimum value to avoid sustained flue gas condensation within the boiler. The internal flapper valve within the loading unit opens during a power failure to allow thermosyphoning between the boiler and tank.

Non-ASME-rated pressurized thermal storage tanks are available in volumes up to 119 gallons. For volumes

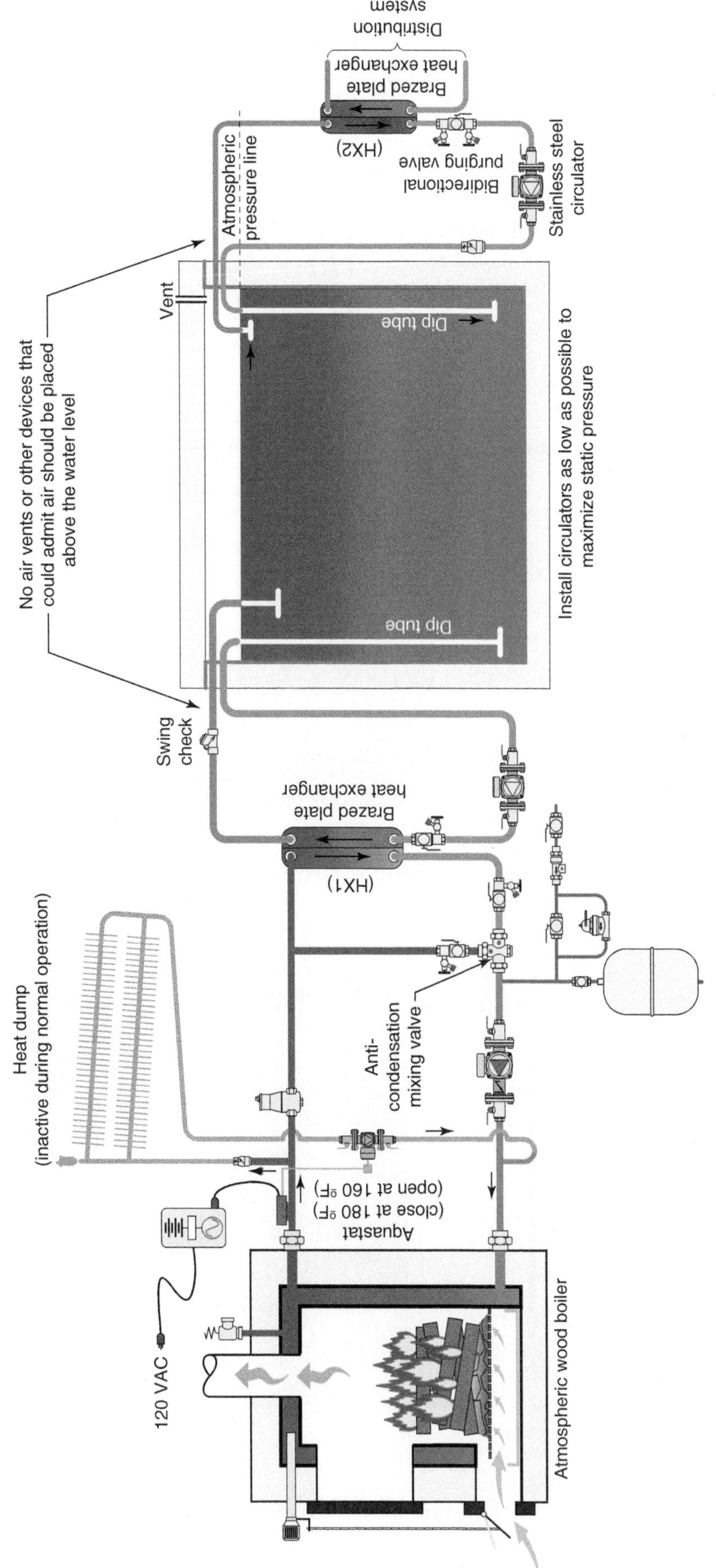

Figure 15-65 | Use of stainless steel brazed-plate heat exchangers in combination with an unpressurized storage tank.

of 120 gallons or more, many mechanical or building codes, especially those applying to publically accessible buildings, require pressurized thermal storage tanks to be **ASME rated**. Such tanks are available, as discussed in Chapter 9, but often cost significantly more than non-ASME-rated tanks on a dollar-per-gallon basis.

When larger storage volumes are needed, it is possible to use multiple parallel-connected thermal storage tanks, as shown in Figure 15-67. Care should be taken to design the interconnecting piping so that flow divides equally between all the tanks. See Chapter 9 for more information on how this can be done.

Another possibility for large thermal storage tanks is the use of a repurposed propane storage tank, as shown in Figure 15-68 and further discussed in Chapter 9.

THERMAL STORAGE TANK SIZING FOR WOOD-GASIFICATION BOILERS

The size of the thermal storage tank used with a wood-gasification boiler depends on several factors. They include:

- How the boiler will be operated
- How much wood can be added to the boiler's firebox
- The energy content of the firewood being used
- The minimum operating temperature of the heating distribution system
- The temperature and pressure ratings of the thermal storage tank

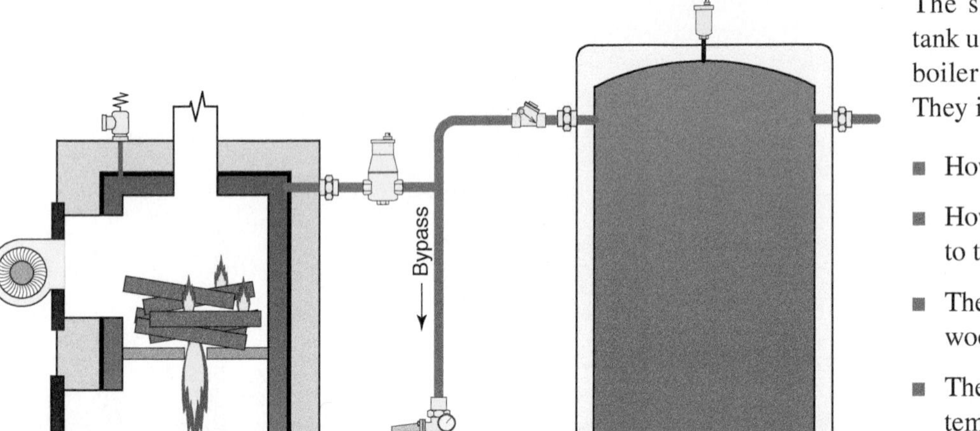

Figure 15-66 | Wood-fired boiler connected to pressurized thermal storage tank.

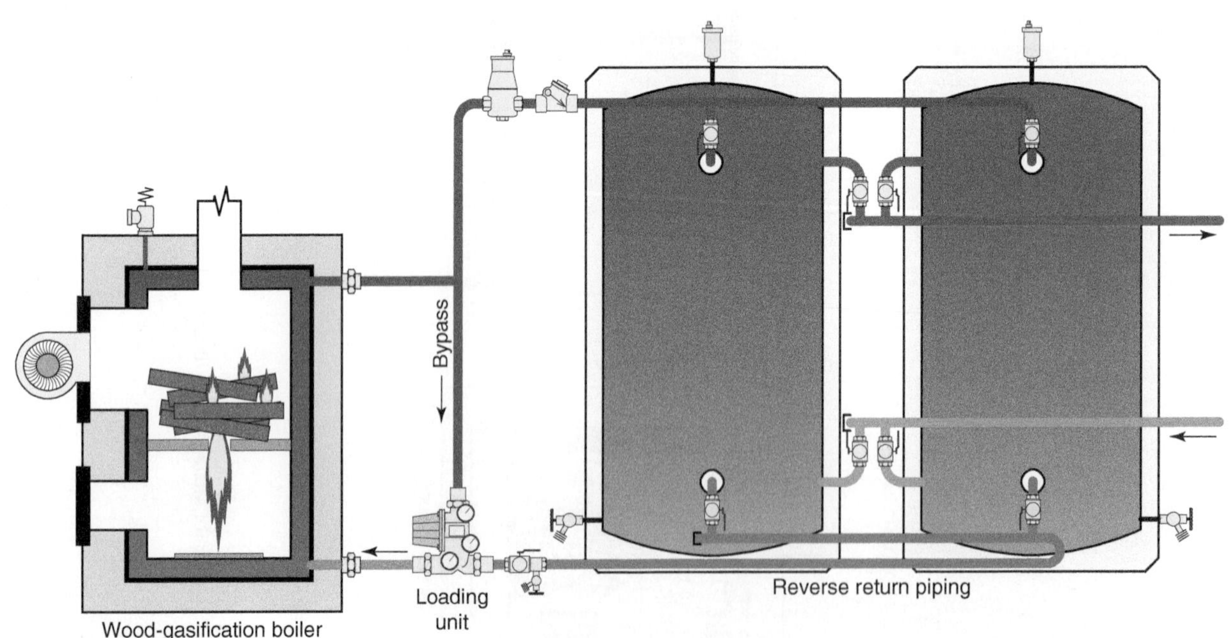

Figure 15-67 | Wood-fired boiler connected to multiple pressurized thermal storage tanks.

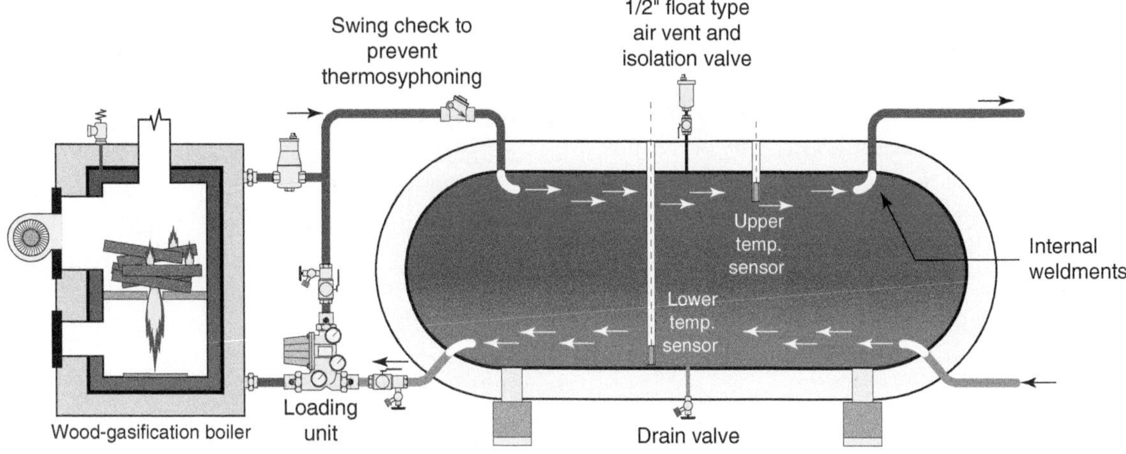

Figure 15-68 | Use of a repurposed propane storage tank as a buffer tank. Note internal detailing to help preserve stratification.

Wood-gasification boilers attain their best efficiency when a firebox full of cordwood is burned as a batch. This usually generates much more heat than what is needed by the load(s) while the boiler is firing. The solution to this mismatch is a substantial amount of thermal storage.

The operating temperature of the heating distribution system also affects the size of thermal storage tank. The lower the water temperature at which the distribution system can operate, the longer a given tank volume can supply the load before the wood-gasification boiler must be refired, or the system must shift to an auxiliary boiler. Low-temperature distribution systems with heat emitters, such as radiant panels, panel radiators, extended surface fin-tube convectors, or air handlers with large surface area coils, were discussed in Section 5 and are recommended for systems with wood-fired heat sources.

The upper temperature limit of the thermal storage tank also plays a role in determining its size. Unpressurized storage tanks typically have temperature limits of 160 to 180 °F. Pressurized metal storage tanks could theoretically contain water at temperatures above the atmospheric boiling point (212 °F); however, this is seldom done due to the safety considerations associated with storing **superheated water**. The author recommends an upper temperature limit no higher than 200 °F.

The sizing procedure that follows is appropriate for wood-gasification boilers. It assumes that the thermal storage tank absorbs a 95 percent of the heat released from burning a full charge of firewood, (e.g., the boiler's primary combustion chamber is fully loaded) *without any concurrent heating load.* The volume of the required thermal storage tank can be determined using Equation 15.4.

(Equation 15.4)

$$v = \frac{701(w)(n)}{\Delta T}$$

where:

v = required thermal storage tank volume (gallons)

w = weight of firewood that can be loaded in the combustion chamber (lb.)

n = average efficiency of the combustion process (decimal percent)

ΔT = temperature rise of the tank based on absorbing all heat from the combustion (°F)

701 = a constant based on the heating fuel value associated with 20 percent moisture content firewood

Example 15.4: Assume that the firebox of a wood-gasification boiler, when fully loaded, can hold 65 pounds of seasoned cordwood. The boiler's average combustion efficiency is 75 percent. Determine the thermal storage tank volume needed assuming the water in the tank will rise 60 °F as it absorbs heat from burning the full charge of wood.

Solution: Putting the data into Equation 15.4 yields:

$$v = \frac{701(w)(n)}{\Delta T} = \frac{701(65)(0.75)}{60} = 570 \text{ gallons}$$

Discussion: This result shows that a substantial thermal storage tank volume may be required in systems using wood-gasification boilers. This volume can be achieved with a single tank or by combining multiple tanks piped in parallel. One way to reduce the required volume is to implement a control strategy that widens the temperature range over which the tank is used. With the upper temperature typically limited to 200 °F, the temperature range of the thermal storage tank can be widened by reducing the temperature at which the space heating distribution system operates.

THERMAL STORAGE TANK SIZING FOR PELLET BOILERS

Unlike wood-gasification boilers, which usually consume a batch of firewood at a high rate, most modern pellet-fired boilers can start, stop, and in some cases modulate their heat output when required to better match the heating load. Still, given their limited range of modulation, as well as the possibility of being connected to a highly zoned distribution system, some amount of thermal mass is needed to avoid short cycling. Equation 15.5 can be used to size this tank.

(Equation 15.5)

$$V_{min} = \frac{0.9(q_{max} - q_{min})}{500(\Delta T)}$$

where:

V_{min} = minimum required thermal storage tank volume (gallons)

q_{max} = maximum heat output rate of boiler (Btu/hr.)

q_{min} = minimum concurrent heating load when boiler is operating (Btu/hr.)

t = minimum desired on-cycle time for boiler (minutes)

ΔT = temperature rise of the tank based on absorbing all heat from the combustion (°F)

0.9 = factor adjusting heat output for ignition, warm-up, and residual burn at end of cycle

The minimum on-cycle for a pellet-fired boiler should consider the time required to initiate and stabilized combustion, the steady burn time, and the time to burn the pellets remaining in the burn pot after the auger stops feeding. The recommended *minimum* burn time varies among manufacturers and models. A suggested guideline is 60 to 120 minutes.

Example 15.5: Determine the minimum volume of a thermal storage tank for a pellet-fueled boiler that has a maximum firing rate of 85,000 Btu/hr., when the minimum desired operating cycle length is 60 minutes and the allowed temperature change in the thermal storage tank is 50 °F. The minimum concurrent heating load is 5,000 Btu/hr.

Solution: Putting the given values into Equation 15.5 yields:

$$V_{min} = \frac{0.9(q_{max} - q_{min})t}{500(\Delta T)} = \frac{0.9(85000 - 5000)60}{500(50)}$$
$$= 173 \text{ gallon}$$

Discussion: The factor of 0.9 in Equation 15.5 slightly reduces the rated capacity of the boiler based on its output during the combustion start-up process and the residual pellet burn-off after the auger stops feeding pellets at the end of a firing cycle. These times intervals are included in the minimum on-cycle time. If the pellet boiler is operated to maintain a given thermal storage tank temperature, independent of the loads, the minimum concurrent load can be set to 0. Because of how it is operated, the size of the thermal storage tank required for a pellet fueled boiler is significantly smaller than that required for a wood-gasification boiler.

EXPANSION TANKS

All closed-loop hydronic systems or subsystems require a properly sized expansion tank. The procedure for sizing a diaphragm-type expansion tank was covered in Chapter 4. It accounts for the temperature change of the water in the system, from the coolest possible condition to the hottest possible condition. It also accounts for the total volume in the system and the system's pressure-relief valve rating.

In systems with a wood-gasification boiler, the volume of a pressurized thermal storage tank often represents a high percentage of the overall system volume. As shown in a previous example, the volume of the thermal storage tank can be several hundred gallons. Furthermore, the water in this tank is often subject to a wide temperature variation. This situation requires a relatively large expansion tank in comparison to a heating system of comparable heating output, but without large thermal storage tanks.

Figure 15-69 | Example of a floor-mounted diaphragm-type expansion tank.

In systems using a wood-gasification boiler and pressurized thermal storage tank, the size of a diaphragm-type expansion tank will often be in the range of 10 percent of the volume of the thermal storage tank. Thus, a system with a 500-gallon thermal storage tank will likely require an expansion tank with a shell volume of about of 50 gallons. Tanks of this size need to be floor mounted due to the weight of the tank and the water it may contain. Figure 15-69 shows an example of such a tank.

It is also possible to use multiple tanks with smaller volumes that add up to the total required expansion tank volume. However, all such tanks must have their air-side pressure adjusted equally. The tanks should also be mounted at the same elevation.

It is also good practice to install a ball valve near the inlet of any expansion tank that can isolate the tank from the system if it has be replaced. The handle of this ball valve should be removed and stored so that the valve is not inadvertently closed.

Given that wood-fired boilers are often combined with low-temperature hydronic distribution systems, it is important to recognize that not all the water in the system will reach the same maximum temperature. For example, the water contained within the boiler and storage tank may reach a temperature of 200 °F, while the maximum temperature supplied to the distribution system is 120 °F. In such cases, it is possible to size two *hypothetical* expansion tanks: one for the higher temperature portion of the system and the other for the lower temperature portion. The shell volume of these two hypothetical tanks can then be added together to size a single tank for the entire system.

Example 15.6: A system uses a wood-gasification boiler with a 500-gallon pressurized thermal storage tank. It supplies a low-temperature radiant panel distribution system, as shown in Figure 15-70. The water within the thermal storage tank and boiler could reach a maximum temperature of 200 °F. The boiler and piping to the thermal storage tank have an estimated volume of 15 gallons. The piping from the tank to the mixing valve has an estimated volume of 5 gallons. The low-temperature portion of the distribution system is estimated to contain 90 gallons of water that will reach a maximum temperature of 110 °F. The uppermost portion of the piping system is 20 feet above the connection to the expansion tank. Determine the size of the required expansion tank assuming the entire system was initially filled with water at 50 °F.

Solution: Two hypothetical expansion tanks will be sized, one for the high temperature portion of the system and the other for the lower temperature portion. The final expansion tank volume will be determined by adding these two volumes.

Start by determining the total volume in each portion of the system:

High-temperature volume = 15 + 500 + 5 = 520 gallons (heated from 50 °F to 200 °F)

Low-temperature volume = 90 gallons (heated from 50 °F to 110 °F)

Next, use Figure 3-11 to determine this density as well as the density of water at 50 °F, 110 °F, and 200 °F:

$$D_{50°F} = 62.4 \text{ lb./ft.}^3$$

$$D_{110°F} = 61.8 \text{ lb./ft.}^3$$

$$D_{200°F} = 60.1 \text{ lb./ft.}^3$$

Next, use Equation 4.5 to determine the air-side pressurization for the expansion tank based on maintaining a 5 psi positive pressure at the top of the system. This equation requires the density of water at the cold-fill temperature.

Equation 4.5 (repeated):

$$P_a = H\left(\frac{D_c}{144}\right) + 5$$

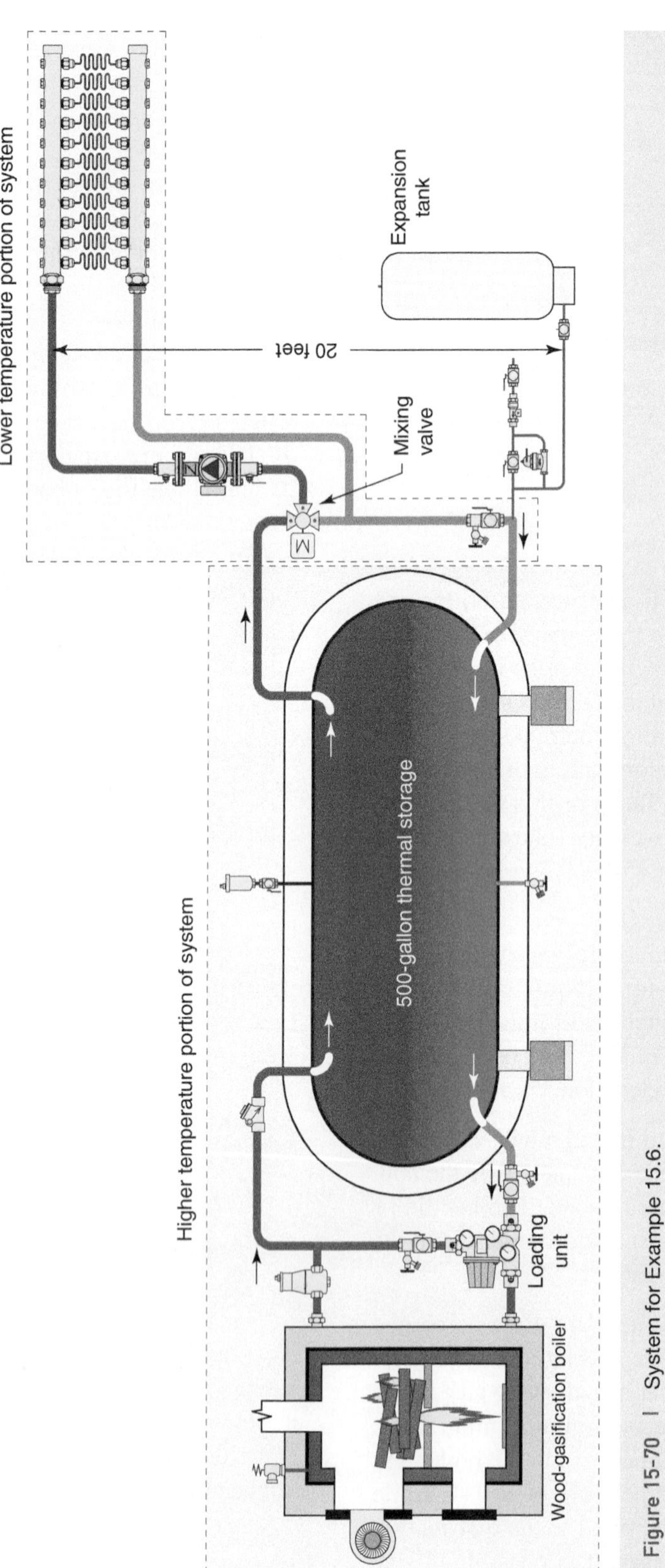

Figure 15-70 | System for Example 15.6.

where:

P_a = air-side pressurization of the tank (psi)

H = distance from inlet of expansion tank to top of system (ft.)

D_c = density of the fluid at its initial (cold) temperature (lb./ft.^{3})

$$P_i = 20\left(\frac{62.4}{144}\right) + 5 = 13.7 \text{ psi}$$

Now use Equation 4.6 to determine the size of the expansion tank for the hot portion of the system:

Equation 4.6 (repeated):

$$V_t = V_s\left(\frac{D_c}{D_h} - 1\right)\left(\frac{P_{RV} + 9.7}{P_{RV} - P_a - 5}\right)$$

where:

V_t = minimum required tank volume (gallons) (not acceptance volume)

V_s = fluid volume in the system (gallons)

D_c = density of the fluid at its initial (cold) temperature (lb./ft.^{3})

D_h = density of the fluid at the maximum operating temperature of the system (lb./ft.^{3})

P_a = air-side pressurization of the tank found using Equation 4.5 (psi)

P_{RV} = rated pressure of the system's pressure-relief valve (psi)

The expansion tank required for the "hot" portion of the system is:

$$V_t = V_s\left(\frac{D_c}{D_h} - 1\right)\left(\frac{P_{RV} + 9.7}{P_{RV} - P_a - 5}\right)$$
$$= 520\left(\frac{62.4}{60.1} - 1\right)\left(\frac{30 + 9.7}{30 - 13.7 - 5}\right) = 69.9 \text{ gallons}$$

The expansion tank required for the "warm" portion of the system is:

$$V_t = V_s\left(\frac{D_c}{D_h} - 1\right)\left(\frac{P_{RV} + 9.7}{P_{RV} - P_a - 5}\right)$$
$$= 90\left(\frac{62.4}{61.8} - 1\right)\left(\frac{30 + 9.7}{30 - 13.7 - 5}\right) = 3.07 \text{ gallons}$$

The total volume for a single expansion tank is thus 69.9 + 3.07 = 72.97 gallons = 73 gallons.

Discussion: The total water volume of this system is 610 gallons. The required expansion tank volume of 73 gallons is about 12 percent of this system volume. This shows that expansion tank sizing is typically dominated by the volume of the thermal storage tank, in combination with the large temperature swing of the water in the "hot" portion of the system. If the expansion tank had been sized assuming all the water in the system reached an upper temperature of 200 °F, the required expansion tank volume would be 82 gallons. However, even the 73-gallon size is conservative in that not all the water in the boiler and thermal storage tank is likely to reach 200 °F simultaneously. Likewise, not all the water in the radiant panel distribution will be at 110 °F simultaneously (because the water temperature has to decrease to release heat from the system). This example demonstrates the need to revert to fundamental design procedures, covered in previous chapters, to properly size components. It is irrelevant that a wood-fired heat source heated the water to 200 °F. Only the volumes, temperatures, heights, and pressures involved affected the outcome of the calculations.

COMBINING A WOOD-FIRED BOILER WITH AN AUXILIARY BOILER

Many who chose a wood-fired boiler as their primary heat source want a system that can automatically revert to an auxiliary heat source if the output from the wood-fired boiler cannot supply the heating load. A boiler fired by propane is a common choice for the auxiliary heat source, especially in rural areas that may not have access to natural gas.

There are several ways to combine both boilers into a system. The "best" approach will depend on how extensively the distribution system is zoned as well as the size of the thermal storage tank.

One seemingly simple method is to connect the boilers in series, as shown in Figure 15-71.

Series piping requires system water to flow through both boilers, even when one of them is not operating. Air currents moving through and around the unfired boiler can absorb heat from this water and carry it out the venting system. Piping the boilers in series also increases the head loss and pressure drop against which the circulator must operate. Although the head loss of most wood-fired boilers is small, this is not the case for some mod/con boilers. Series piping also precludes

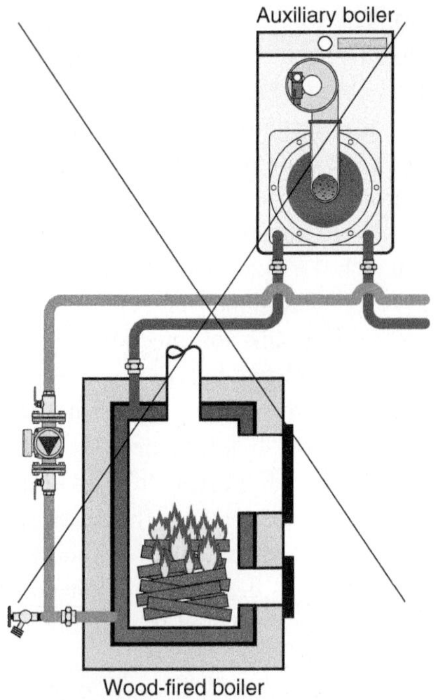

Figure 15-71 | Do not connect a wood-fired boiler and auxiliary boiler in series.

use of proper anti-condensation protection. Because of these limitations, *series piping is not recommended.*

Multiple boilers, regardless of their fuel type, should be piped in parallel. Figure 15-72 shows a wood-gasification boiler piped in parallel with a mod/con auxiliary boiler.

This arrangement places the auxiliary boiler *upstream* the thermal storage tank. It also allows the tank to accept heat from *both* boilers, and thus buffer each boiler against short cycling.

This arrangement is appropriate when the auxiliary boiler has low thermal mass, or where the space-heating distribution system is extensively zoned. It is also useful when the thermal storage tank is maintained at a temperature suitable for domestic water heating, such as when using the instantaneous DHW assembly discussed in Chapter 9 and shown in Figure 15-72.

Note that the piping connections between the auxiliary boiler and thermal storage tank are such that the auxiliary boiler only interacts with water in the upper portion of the tank. Temperature stratification within

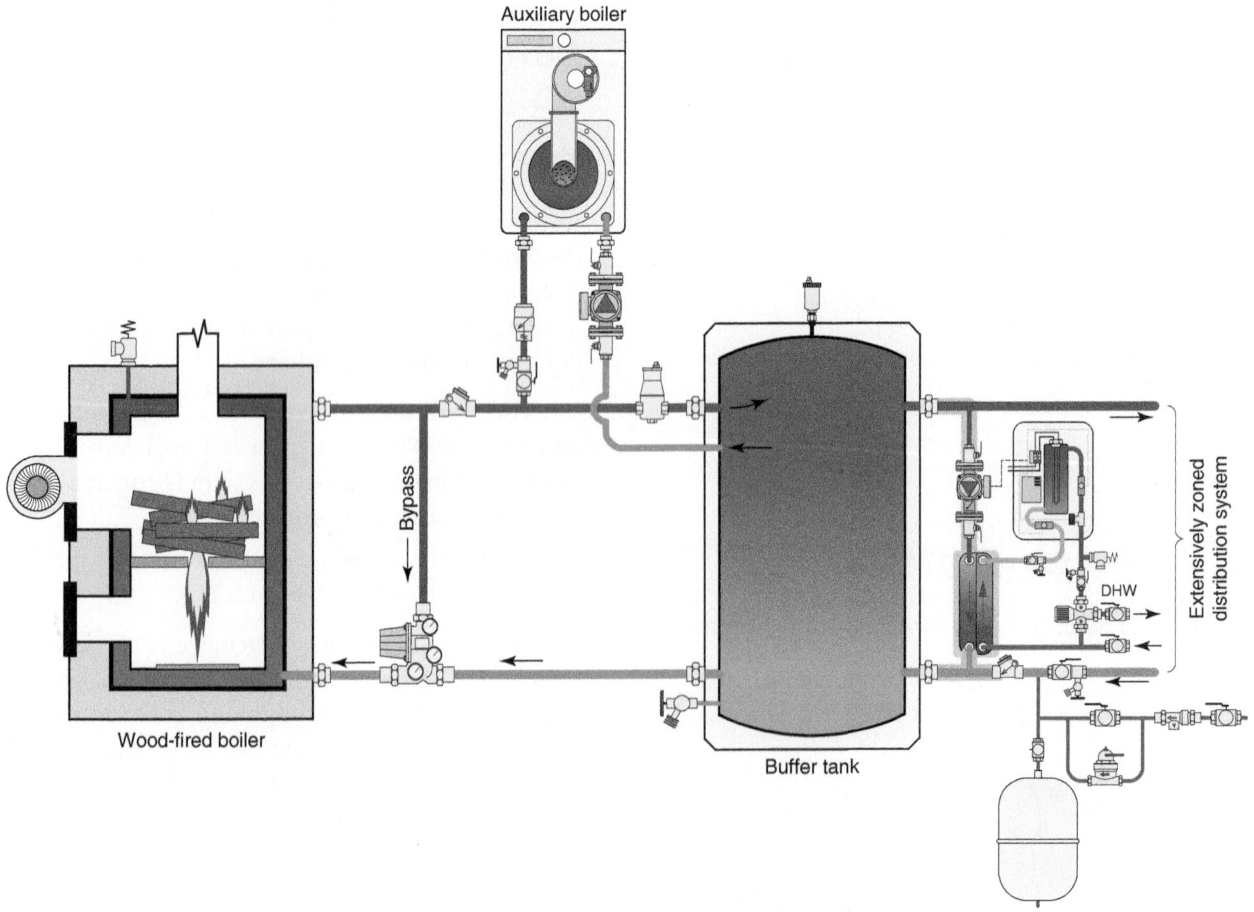

Figure 15-72 | Auxiliary mod/con boiler piped in parallel with wood-fired boiler, and upstream of the buffer tank.

the tank keeps the cooler water in the lower portion of the tank from being directly heated by the auxiliary boiler. This allows the upper portion of the tank to buffer the auxiliary boiler to protect it against short cycling, but at the same time limits the amount of thermal mass between the auxiliary boiler and the distribution system. It is not necessary or desirable to use the auxiliary boiler to heat the entire volume of a thermal storage sized based on previously discussed methods.

Figure 15-73 shows the auxiliary boiler piped *downstream* of the thermal storage tank. This allows the tank to cool toward room temperature while the auxiliary boiler supplies the load. It also prevents heat generated by the auxiliary boiler from warming the thermal storage tank, which, based on the second law of thermodynamics, is undesirable. The chemical energy stored in fuel can be contained for months or years, whereas that energy converted to heat can only be stored for a matter of hours. When the auxiliary boiler does not need buffering to prevent short cycling there is no point in converting fuel into heat until the latter is required by the load.

The configuration shown in Figure 15-73 is appropriate in systems where the auxiliary boiler's heat output is well matched to the zoning of the distribution system so that short cycling will not occur. It's also appropriate in systems where other low-temperature heat sources such as solar collectors or geothermal heat pumps may add heat to the thermal storage tank during times when the wood-fired boiler is not operating. Keeping heat produced by the auxiliary boiler out of the thermal storage tank allows that tank to cool when the wood-fired boiler is not being used. When the tank is located within the building's thermal envelope, the heat it gives up while cooling down to room temperature contributes to space heating. Lower tank temperatures also improve the heat gathering efficiency of both solar collectors and hydronic heat pumps.

Yet another arrangement for connecting an auxiliary boiler into a system with a wood-fired boiler is shown in Figure 15-74a.

This system has a **2-pipe thermal storage tank** (e.g., only two pipes connect the tank to the balance of the system). It also uses a variable speed injection circulator to transfer heated water from either the wood-gasification boiler or the thermal storage tank into the

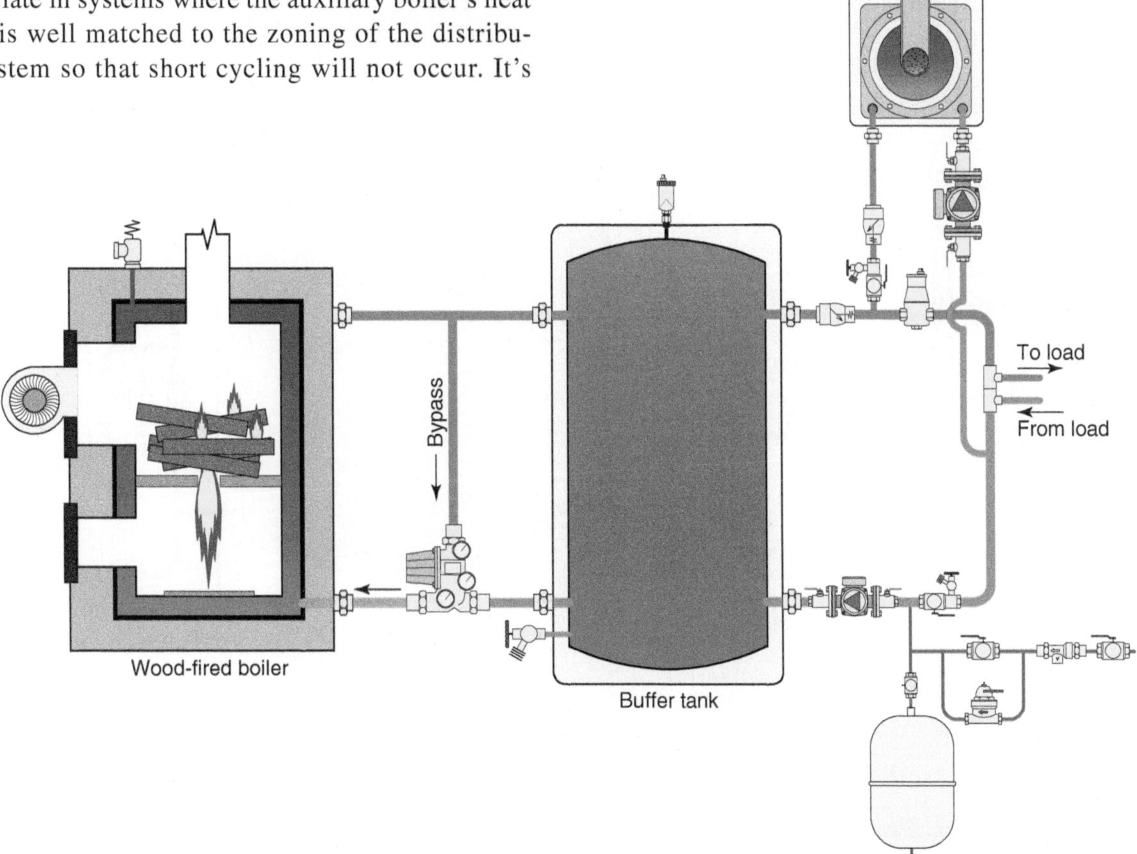

Figure 15-73 | Locating the auxiliary boiler after the buffer tank.

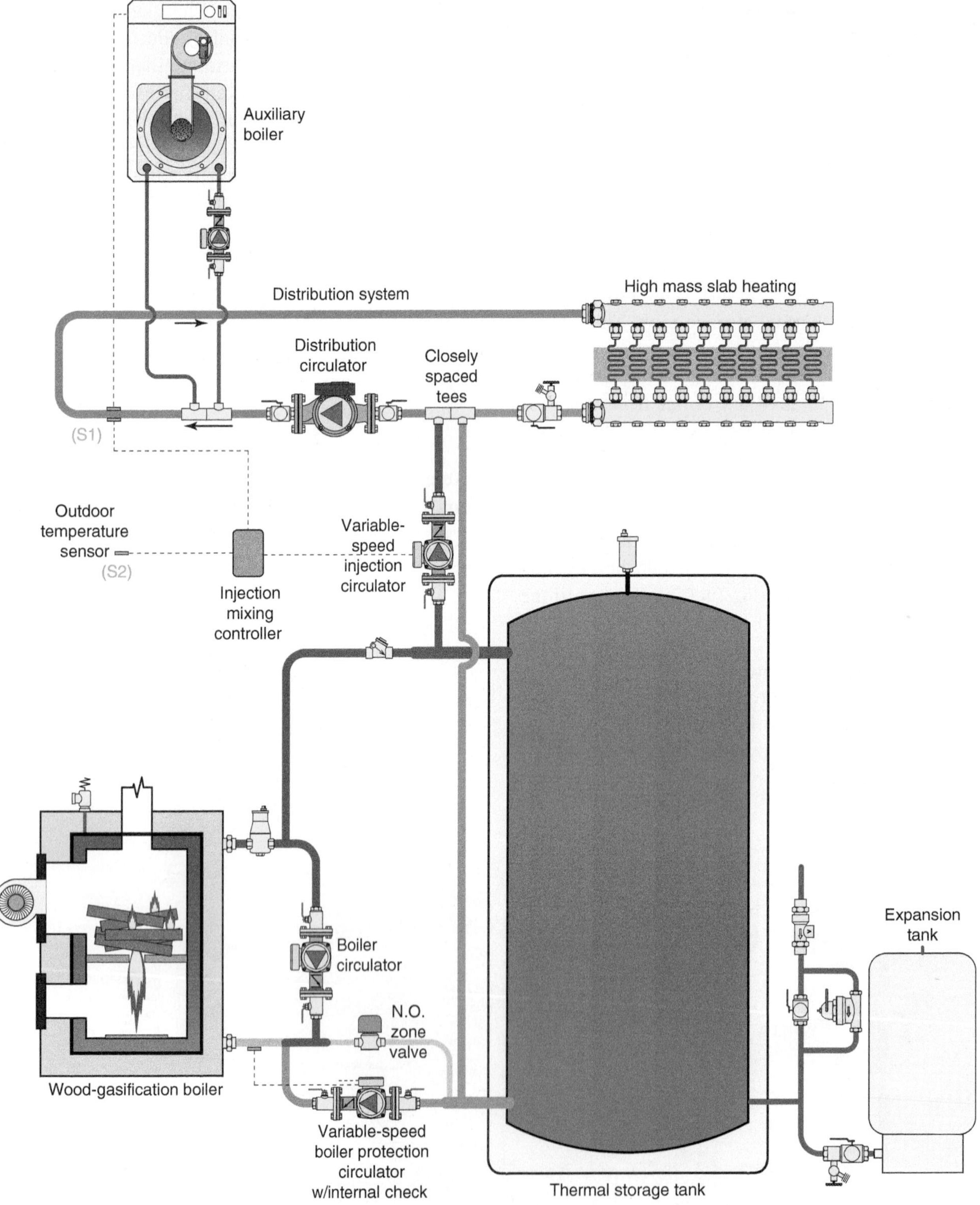

(a)

Figure 15-74a | Connecting a wood-fired boiler to a thermal storage tank using a 2-pipe tank configuration.

distribution system. The speed of the injection circulator is regulated by a controller that measures outdoor temperature and calculates a target supply water temperature based on outdoor reset. It then measures the water temperature supplied to the distribution system and compares it to this calculated target temperature. If the measured temperature needs to be increased, the speed of the injection circulator is increased, and vice versa.

The piping from the injection circulator to the distribution system, and the corresponding return pipe, connect to the distribution system using a pair of closely spaced tees. This provides hydraulic separation between the variable speed injection circulator and the circulator in the distribution system.

Similarly, the auxiliary boiler is connected into the distribution system using a pair of closely spaced tees. This boiler operates when necessary to supplement heat input from the wood-gasification boiler or storage tank.

In this system, the heat emitter is a single-zone, high-thermal-mass heated floor slab. It provides sufficient thermal mass to fully buffer operation of the auxiliary boiler. Thus, there is no need of using the upper portion of the thermal storage tank for buffering the auxiliary boiler.

If the wood-gasfication boiler is not fired for several days, both it and the thermal storage tank will eventually cool to the point where no further heat input to the distribution system is possible. If this occurs, the injection circulator remains off and the auxiliary boiler provides all heat required by the system.

The wood-gasification boiler is protected against sustained flue gas condensation by the variable speed circulator, which only begins to circulate water when the boiler inlet temperature rises to 130 °F.

The normally open zone valve (ZV1) opens upon a power failure to allow thermosyphon flow between the thermal storage tank and the wood-gasification boiler. It is closed at all other times. This valve is necessary to provide a "detour" around the spring-loaded check valve in the boiler protection circulator. The spring-loaded check valve is needed to prevent flow that is returning from the distribution system from passing through the wood-gasification boiler when the boiler protection circulator is not operating.

The system shown in Figure 15-74a can be modified for a situation where the distribution system consists of several zones of low-thermal-mass heat emitters. Figure 15-74b shows these modifications.

Only the upper portion of the schematic has changed. A 25-gallon buffer tank has been added to prevent short cycling of the auxiliary boiler during partial load conditions. The variable-speed injection circulator injects heat from the wood-gasification boiler or thermal storage tank into this small buffer tank. The small tank also provides hydraulic separation between the variable-speed injection circulator, the auxiliary boiler circulator, and the variable-speed pressure-regulated distribution circulator.

An example system, presented later in this chapter, provides full details for controlling the operation of this system.

USE OF CONVENTIONAL BOILERS FOR AUXILIARY HEAT

In some systems, a conventional boiler, rather than a mod/con boiler, is used as the auxiliary heat source for a wood-fired boiler. Examples of conventional boilers include oil-fired boilers and atmospheric gas-fired boilers using cast-iron, steel, or copper heat exchangers. In such cases, it is important to protect the conventional boiler from low entering water temperature. This can be done using a high-flow-capacity 3-way thermostatic valve, such as those shown in Figure 15-51. The near-boiler piping for such a valve is shown in Figure 15-75.

OTHER PIPE DETAILING

Always install a mixing device, such as a 3-way motorized mixing valve, between a thermal storage tank supplied by a wood-fired boiler or pellet-fired boiler and a low-temperature distribution system. As with solar thermal systems, this mixing device protects the low-temperature heat emitters from what could be very hot water delivered from the thermal storage tank. When an electronically controlled mixing device is used, it can also provide outdoor reset control of the water temperature supplied to the distribution system.

Only copper water tube or black iron piping should be used to connect a wood-fired or pellet-fired boiler to a thermal storage tank. The potential for temporary water temperatures well above 200 °F exists if a power outage occurs when there is a high-output fire in the boiler. Such temperatures are above the operating limits of PEX and most other types of polymer tubing.

It is not good practice to thread copper tubing adapters directly into steel connections on boilers or

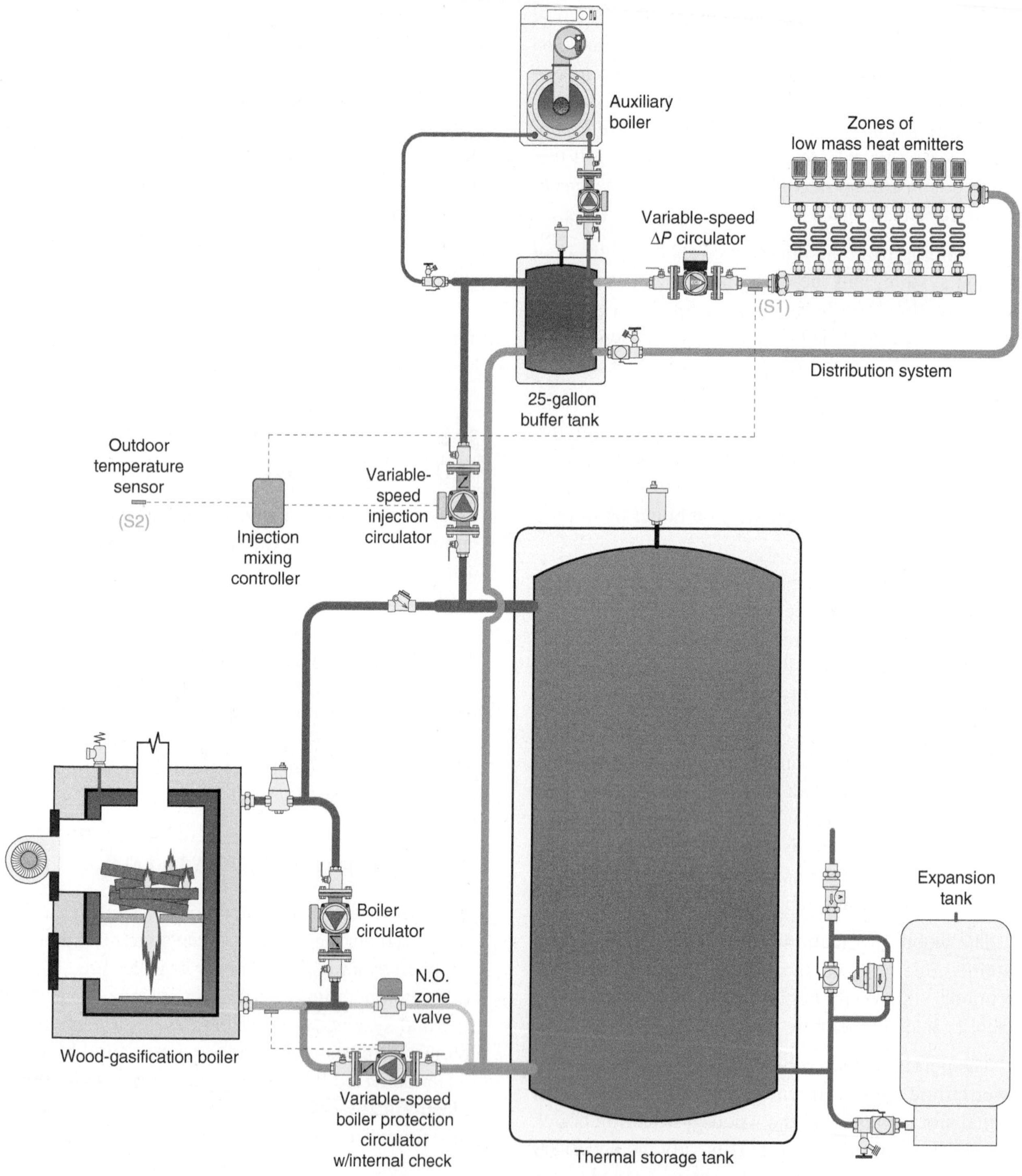

(b)

Figure 15-74b | Use of a small 25-gallon buffer tank to protect the auxiliary boiler against a short cycling and provide hydraulic separation between two variable speed circulators.

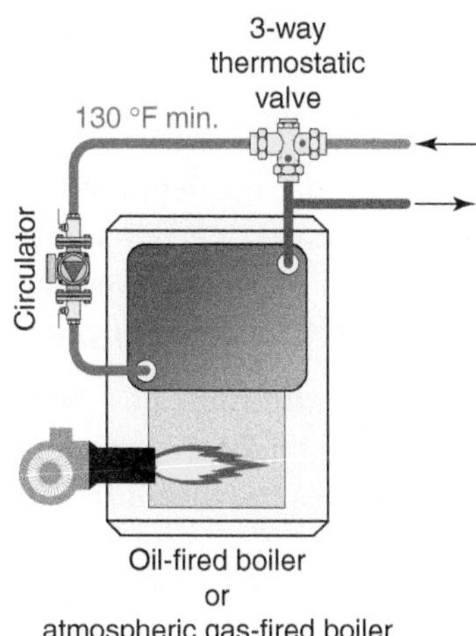

Figure 15-75 Using a high flow capacity 3-way thermostatic valve to protect a conventional boiler from low inlet water temperatures.

tanks. Doing so can lead to galvanic corrosion on the steel surfaces. Although dielectric unions are available for such connections, the author has found that the polymer portions of such unions tend to crack after several years of exposure to elevated temperatures. A standard brass union has proven to be a good choice for connecting copper tubing to steel or iron components in closed-loop systems. Such unions should be used if copper tubing needs to connect to a steel or iron boiler, or to a carbon steel tank. Copper adapters can be directly connected to any stainless steel components; however, the use of brass unions at such connections allow for easy disassembly of the piping if required.

Always install a pressure-relief valve in any portion of the system that forms a closed loop and contains a heat source. Thus, the system shown in Figure 15-64 would require two pressure-relief valves: one for the boiler circuit and one for the closed distribution system that connects to the load-side heat exchanger. Always be sure that all water within a piping assembly containing a pressure-relief valve can "communicate" with that valve. Do not create piping assemblies where combinations of check valves, zone valves, motorized valves, or other automatically operated devices can inadvertently isolate a portion of the piping system from the relief valve that is supposed to be protecting it. Be sure that all pressure-relief valves are fitted with piping that routes any discharge close to the floor level within a heated space, preferably over a floor drain. Never route the piping from a pressure-relief valve to or through a space that is subject to freezing temperatures.

Locate expansion tanks so that their inlet connection is relatively close to the intake side of circulators. This places the point of no pressure change near the circulator intake and allows the circulator to create positive differential pressures within the circuit.

TEMPERATURE STACKING IN THERMAL STORAGE

It is important not to short cycle pellet-fired boilers or wood chip–fired boilers. Thermal storage is usually necessary to obtain acceptably long operating cycles, especially if the pellet-fired or chip-fired boiler cannot modulate its heat output, or if the distribution system is highly zoned. It's also important to maximize the temperature cycling range of the thermal storage tank, from some reasonably high temperature that the boiler can achieve while still maintaining good efficiency, down to the lowest possible temperature useable by the distribution system.

Since the pellet-fired and chip-fired boilers currently available in North America are not designed to operate with sustained flue gas condensation, the minimum sustained inlet temperature to the boiler needs to be maintained in the range of 130 °F to 150 °F, depending on the moisture content of the fuel. The lower end of this range is suitable for use with very dry wood, such as wood pellets. The upper end is appropriate for higher moisture wood, such as typical wood chips. Operating these boilers with outlet water temperatures as high as 180 °F is usually possible, while still allowing them to operate at good thermal efficiency.

In consideration of these objectives and limitations, *it is prudent to "stack" the thermal storage tank with hot water before turning off a pellet-fired or wood chip-fired boiler.* This is true even in systems with low-temperature heat emitters, such as those discussed throughout this text. These systems can use a 3-way motorized mixing valve to mediate between the sometimes high water temperature in the thermal storage tank and the lower supply water temperature required by the distribution system.

Consider a thermal storage tank that has been previous charged with heat from a pellet-fired boiler but has had no heating load placed upon it for perhaps 2 hours or more. Under this condition, temperature stratification will be well established within the tank, as shown in Figure 15-76.

The transition zone between the cool water at the bottom and higher temperature water at the top is called a **thermocline**. The height of the thermocline will be determined by many factors, such as tank insulation, thermal conductivity of the tank shell, inlet flow velocity and direction, and overall temperature difference between the top and bottom of the tank.

When the next call for space heating occurs, hot water is drawn from the upper portion of the tank. Cool water returning from the distribution system enters the lower portion of the tank. This causes the thermocline to move upward, as shown in Figure 15-77.

Under this condition, the depth of the thermocline will vary based on mixing currents within the tank. Careful inlet detailing minimizes vertical mixing and helps preserve a well-defined thermocline.

Assuming the load continues, the pellet-fired boiler should be turned on *before* the temperature at the upper piping outlet decreases significantly. This condition can be detected using a sensor mounted within a well located *below* the outlet piping connection, as shown in Figure 15-78.

The configuration shown in Figure 15-78 assumes that the minimum usable water temperature of the distribution system, under design load conditions, is 120 °F. The pellet-fired boiler is turned on when the temperature several inches below the outlet connection drops to or below 120 °F. This allows the tank to continue supplying heat to the load at a reasonable stable temperature while the pellet-fired boiler is warming up.

Low-mass pellet-fired boilers may only require 10 minutes to reach suitable conditions where heat can be supplied to the thermal storage tank. Higher mass boilers may require 20 to 30 minutes to initiate combustion, raise their inlet water temperature to approximately 130 °F, and only then begin delivering heat to the tank. The height between the upper sensor well and outlet piping connection should be determined based on this warm-up time. The longer the warm-up, the greater the

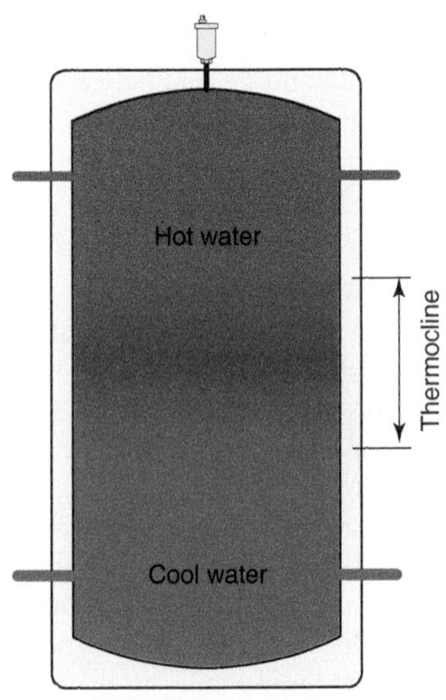

Figure 15-76 | Temperature stratification and thermocline within a thermal storage tank.

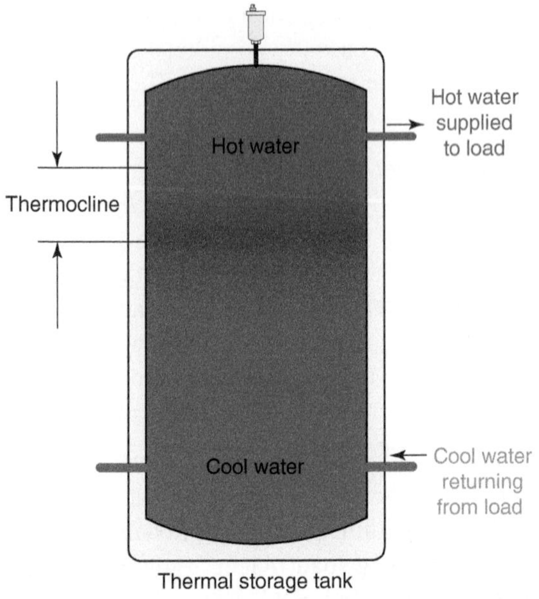

Figure 15-77 | As hot water is extracted from the upper right connection of the tank, and cooling water enters at the lower right connection, the thermocline moves upward within the tank.

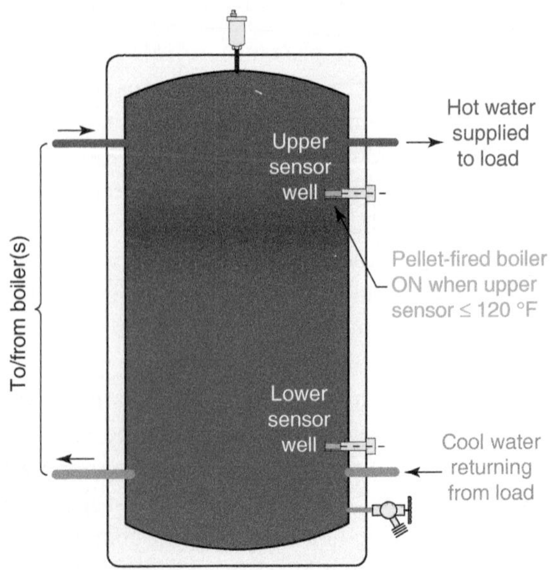

Figure 15-78 | Pellet boiler fired when temperature at upper sensor ≤ 120 °F.

height difference should be. Some tanks may allow for vertical sensor wells, as shown in Figure 15-79. The position of the sensor that initiates pellet-fired boiler operation can be adjusted base on the length of the well.

If the pellet-fired boiler's output is less than the load, the temperature of the water in the upper portion of the tank will decrease. This indicates that additional heat input from the auxiliary boiler is required. Depending on the controller(s) being used, this condition can be detected by the same temperature sensor used to initiate firing of the pellet-fired boiler, a second temperature sensor in the same sensor well, or a second sensor in another well located higher in the tank, as shown in Figure 15-79. A suggested criteria for turning on the auxiliary boiler is when the temperature in the upper portion of the tank drops to 5 ºF below the temperature at which the pellet-fired boiler was turned on. Figure 15-80 shows a situation where the auxiliary boiler is turned on when the temperature at the upper sensor location is 115 ºF.

With both the pellet-fired boiler and auxiliary boiler operating, heat input to the tank will typically be greater than the load. Hot water will begin stacking within the tank, and the thermocline will move downward.

To minimize fuel consumption by the auxiliary boiler, it will be turned off when the temperature at the upper sensor reaches 130 ºF. The allows the water in the upper portion of the tank to absorb some heat, preventing the auxiliary boiler from short cycling while meeting the output requirement to meet the load. Provided that a low temperature distribution system is used, this also allows the auxiliary boiler to operate at temperatures that allow sustained flue gas condensation and thus provide high thermal efficiency. However, it doesn't allow the auxiliary boiler to heat the full volume of the thermal storage tank. The pellet-fired boiler continues to operate during this time. If, after the auxiliary boiler has turned off, the load remains higher than the heat output of the pellet-fired boiler, the auxiliary boiler will again turn on to repeat this cycle.

As the load decreases, or ends, it is desirable to keep the pellet-fired boiler operating until the majority of the thermal storage tank has reached a high temperature. Doing so increases the cycle duration of the pellet-fired boiler, allowing it to operate at high thermal efficiency and with minimal emissions. This strategy is known as **temperature stacking**. The tank is considered fully stacked when a relatively high temperature is detected near the bottom of the tank, as shown in Figure 15-81.

The control logic required to regulate the temperature stacking process may be provided by the controller supplied with the pellet-fired boiler. If not, it can be created using a 2-stage temperature setpoint controller in combination with a 1-stage temperature setpoint controller. Such controllers are readily available from several suppliers. Figure 15-82 shows a ladder diagram that combines these controllers with relays to provide the required control logic.

With this configuration, a signal indicating that space heating should be enabled when contact (T1) closes. This supplies 24 VAC to the coil of relay (R1). Contact (R1-1)

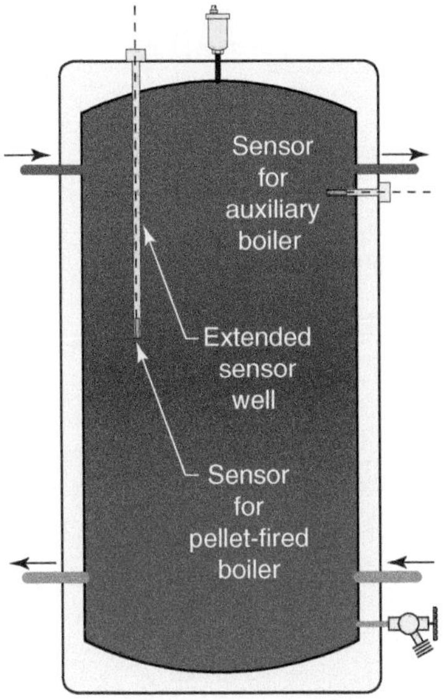

Figure 15-79 Using a separate well to lower the position of the sensor that controls when the pellet-fired boiler is turned on.

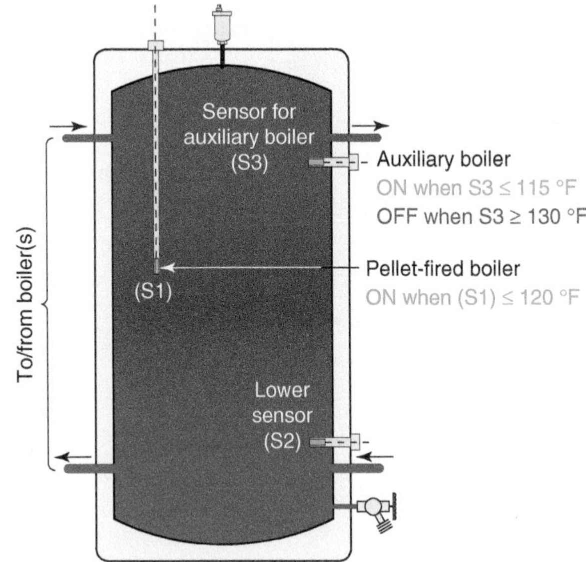

Figure 15-80 Auxiliary boiler turned on when temperature at upper sensor ≤ 115 °F. Pellet-fired boiler remains on under this condition.

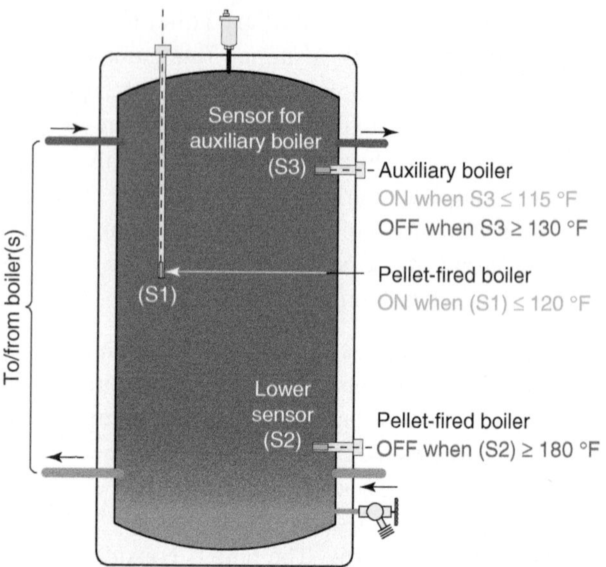

Figure 15-81 The thermal storage tank is considered fully "stacked" when a relatively high temperature (in this case, 180 °F) is detected near the bottom of the tank.

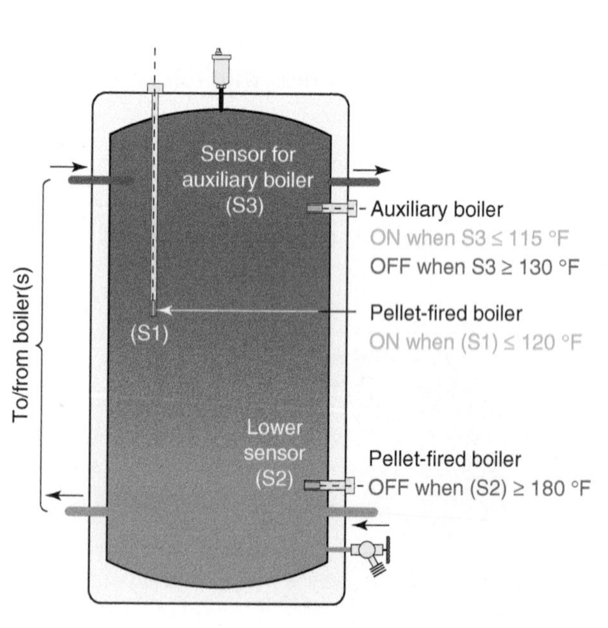

closes to power on the 2-stage setpoint controller as well as the 1-stage setpoint controller.

The 2-stage controller examines the temperatures at sensors (S1) and (S2). If the temperature at sensor (S1) in the upper portion of the tank is less than or equal to 120 °F, the stage 1 contact in the 2-stage controller closes. This passes 24 VAC to the coil of relay (R2). Contact (R2-2) closes to complete a circuit between the (T T) terminals (e.g., the heat demand terminals) on the pellet boiler, enabling it to fire. Another contact, (R2-1) also closes to pass 24 VAC to one side of the stage 2 contact. *This stage 2 contact will be closed because the temperature at the bottom of the tank is much lower than 170 °F (because the top of the tank is only 120 °F or less).* If the temperature at sensor (S1) increases to 130 °F or higher, the stage 1 contacts will open. However, there is still a path for 24 VAC through (R2-1) and the stage 2 contact to keep the coil of relay (R2) energized. This keeps the pellet-fired boiler operating. When

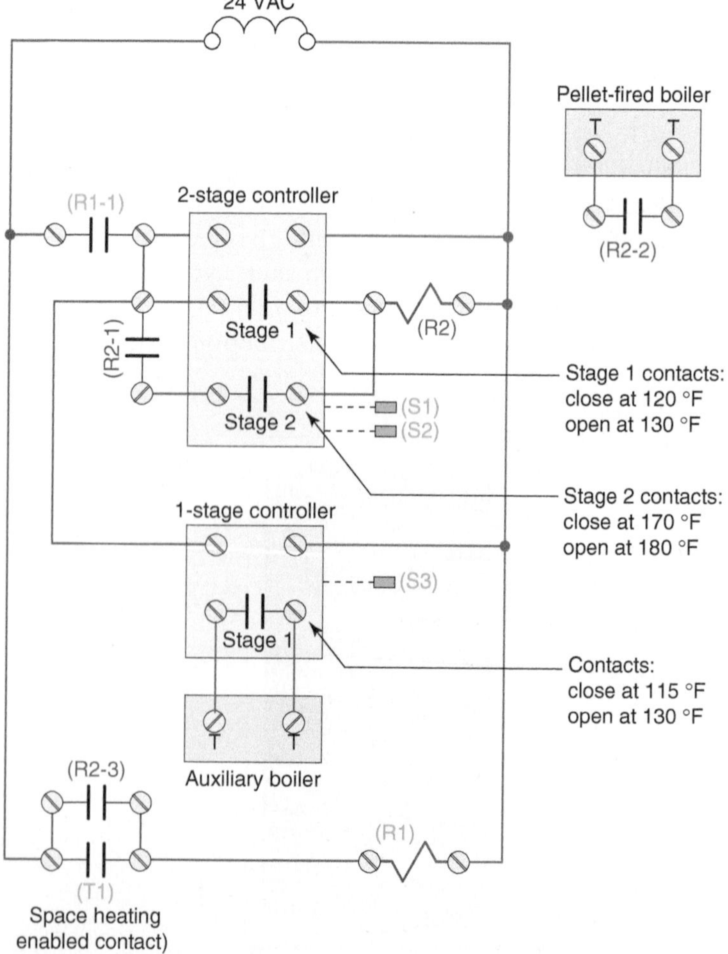

Figure 15-82 Control logic for temperature stacking using a 2-stage temperature setpoint controller along with a 1-stage temperature controller.

the temperature detected by sensor (S2) in the lower portion of the tank reaches 180 °F, the stage 2 contact in the 2-stage controller opens, and the pellet-fired boiler turns off. A third pole of relay (R2), contact (R2-3), closes in parallel with the dry contact (T1) that enabled space heating oepration. This allows the pellet-fired boiler to remain on until the temperature stacking process is complete, even if the demand for space heating is no longer present.

If the pellet-fired boiler is unable to maintain the temperature in the upper portion of the tank above 115 °F, the 1-stage setpoint controller turns on the auxiliary boiler to supplement the heat output of the pellet-fired boiler. The auxiliary boiler remains on until the temperature detected at sensor (S3) in the upper portion of the tank reaches 130 °F.

When using this scenario, it is important that both boilers are solely under the control of the controllers shown in Figure 15-82. They should not be operating based on outdoor reset control, which could interfere with the objective of the setpoint controllers just described.

15.9 Sizing Wood-Gasification and Pellet-Fired Boilers

The wood-gasification boilers and pellet-fired boilers discussed in this chapter have very different operating characteristics compared to boilers of similar capacity but fired by natural gas, propane, or fuel oil. The latter are capable of igniting their fuel and producing heat very quickly upon receiving a signal to operate. This is not true with wood-fired and pellet-fired boilers.

With a wood-gasification boiler, the fire must be manually started and the boiler's firebox subsequently loaded full of firewood. The highest thermal efficiency is achieved by burning a firebox full of wood as hot and fast as possible. This usually creates a heat output rate that is significantly higher than the concurrent heating load. Thermal storage must be used to absorb the difference between the rate of heat production and the rate at which heat is supplied to the load.

In the case of boilers fired by pellets and wood chips, several minutes are required to initiate combustion and bring the boiler to a point where it is operating without sustained flue gas condensation and ready to transfer heat to a thermal storage tank. These characteristics make sizing a wood-gasification boiler, pellet-fired boiler, or wood chip–fired boiler quite different compared to sizing conventional boilers.

SIZING A WOOD-GASIFICATION BOILER

The sizing of a wood-gasification boiler is based on the number of firing cycles the owner is comfortable with during a design day. In most cases, this will vary from one to three cycles per day, with two being a common compromise. Boiler sizing is determined based on the total space heating and domestic water heating energy needed during a design day and the number of firing cycles that will be used to produce this energy. Equation 15.6 can be used to estimate the size of a wood-gasification boiler based on the weight of firewood required to fill or nearly fill its firebox.

(Equation 15.6)

$$W = \frac{[T_{\text{inside}} - (T_{\text{d}} + 5)](UA_{\text{b}})24 + E_{\text{daily}}}{eCN}$$

Where:

W = weight of firewood required to fill firebox of wood-gasification boiler (lb.)

T_{inside} = indoor air temperature for design load conditions (°F)

T_d = outdoor design air temperature (°F)

UA_b = heat loss coefficient of building (Btu/hr./°F)

24 = hours in day

E_{daily} = daily heat required for domestic hot water (Btu) (Equation 2.11)

e = *average* (full burn cycle) efficiency of wood-gasification boiler while operating (decimal percent)

C = lower heating value of firewood being used (Btu/lb.) (Equation 15.1)

N = number of complete firing cycles per day under design load conditions

5 = the 24-hour average outdoor temperature is assumed to be 5 °F above the outdoor design temperature

Example 15.7: Estimate the firebox size, based on the weight of wood it can contain, for a wood-gasification boiler that supplies a building with a design heating load of 50,000 Btu/hr. in a climate where the outdoor design temperature is –5 °F and the desired indoor temperature is 70 °F. The building also requires 60 gallons per day of domestic hot water heated from 50 to 120 °F. The owner wishes to have no more than two complete firing cycles during a design day. Assume the wood-gasification boiler will be burning sugar maple at an average moisture content of 20 percent and has an average combustion efficiency of 75 percent.

Solution: The value of UA_b is found by dividing the design heat load by the design temperature difference:

$$UA_b = \frac{50000\dfrac{\text{Btu}}{\text{hr}}}{70°\text{F} - (-5°\text{F})} = 667\frac{\text{Btu}}{\text{hr}\cdot°\text{F}}$$

The daily energy required for domestic water heating is estimated using Equation 2.11:

$$E_{daily} = (G)(8.33)(T_{hot} - T_{cold})$$
$$= (60)(8.33)(120 - 50) = 34986\text{ Btu}$$

The lower heating value of the specified wood is estimated using Equation 15.1:

$$LHV = 7950 - 90.34(w) = 7950 - 90.34(20)$$
$$= 6143\frac{\text{Btu}}{\text{lb.}}$$

These calculated values, along with the assumed values, can now be substituted into Equation 15.6:

$$W = \frac{[T_{inside} - (T_d + 5)](UA_b)24 + E_{daily}}{eCN}$$
$$= \frac{[70 - (-5 + 5)](667)24 + 34986}{(0.75)(6143)(2)} = 125\text{ lb.}$$

The result indicates that the boiler's firebox, when fully loaded, should be capable of holding about 125 pounds of firewood.

Although this number is valid, it is not very useful. An alternative is to convert this weight into an estimated firebox volume. The information in Figure 15-15 can be used for this. In this case, the a full cord of sugar maple weighs 3,757 lb. Therefore, the density of this wood, stacked as it would be in a firewood cord, is:

$$D = \frac{3757\text{ lb.}}{4\text{ ft.} \times 4\text{ ft.} \times 8\text{ ft.}} = 29.4\frac{\text{lb.}}{\text{ft.}^3}$$

However, it is unlikely that the firewood will be stacked into the active primary combustion chamber of the wood-gasification boiler as neatly and compactly as it might be stacked in a cordwood pile. The author suggests using a **firebox stacking density** of two-thirds that of the density of the wood when stacked in a cordwood pile. Thus, the density of the wood when stacked into the boiler's primary combustion will be about 20 lb./ft.^{3}.

The firebox volume required to contain 125 lb. of wood at this assumed firebox stacking density is thus:

$$V_{chamber} = \frac{125\text{ lb.}}{20\dfrac{\text{lb.}}{\text{ft.}^3}} = 6.25\text{ ft.}^3$$

So, a boiler with a firebox having an internal volume of at least 6.25 cubic feet would be needed to meet the stated requirements and assumptions of this example.

Discussion: Notice that the Btu/hr. rating of the boiler was not used in this sizing method. *Matching a Btu/hr. rating of a boiler to the design load of the building is not appropriate in the case of wood-gasification boilers* because they are not intended to operate at a quasi-steady heat output rate.

If the owner had wanted to meet the design day heating requirement with only a single firing cycle, a boiler with twice the firebox volume would be required. This would be a substantially larger and more expensive boiler, but perhaps justified based on the owner's intended operating mode. Likewise, if the owner were willing to fuel the boiler three times on a design day, the required firebox volume would be two-thirds of that calculated in this example.

SIZING A PELLET-FIRED BOILER

Of all the wood-fired boilers discussed in this chapter, pellet-boilers come closest to mimicking the operating characteristics of boiler fired by conventional fuels. Modern pellet boilers can be configured to automatically load their fuel, ignite it, regulate the combustion process, maintain their internal surfaces free of sustained flue gas condensation, and even compress the small amount of ash they generate to lengthen time between ash removals. Several currently available pellet-fired boilers are capable of modulating their heat output. Turn-down ratios range from 3:1 to as high as 10:1. Modulation allows the boiler to better track the heating load as it continually varies.

Given these characteristics, it may seem intuitive that pellet-fired boilers should be sized the same as conventional boilers. The process would be as simple as determining the design heating load of the building and selecting a boiler with a rated heat output that at least equals, or possibly exceeds, that load by some safety factor. However, experience has proven this is not optimal. To understand why, the nature of the seasonal heating load needs to be examined.

As discussed in Chapter 2, most heating systems are sized based on a "snapshot" condition commonly referred to as **design load**. This is the theoretical rate of heat loss from the building when the indoor temperature is at an assumed comfort condition such as 70 °F, and the outdoor temperature is at a value called the 97.5 percent or 99 percent design temperature. The 97.5 percent design temperature for a given location is a value that the outdoor temperature is *at or above 97.5 percent of the time* during an average year. Viewed another way, the outdoor temperature is below the 97.5 percent design temperature only 2.5 percent of the time in an average year.

The **bin temperature data** for a given location often shows that the outdoor temperature is significantly warmer than the design temperature much of the year. Figure 15-83 plots bin temperature data for Syracuse, NY. The bins are based on increments of 5 °F and range from the coldest possible temperature to a temperature where space heating is no longer required (typically 70 °F).

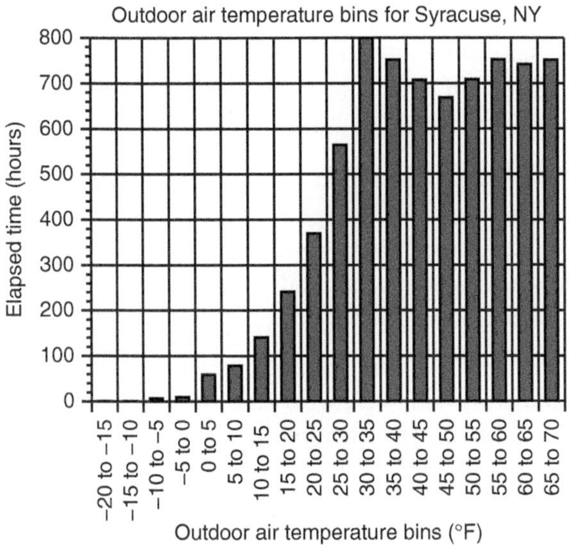

Figure 15-83 | Bin temperature data for Syracuse, NY.

Consider a building in Syracuse, NY, with a design load of 100,000 Btu/hr., based on 70 °F interior temperature and a 97.5 percent design temperature of 2 °F. If the bin temperature data is combined with this load information, another graph can be created that shows the hours during which the heating load equals or exceeds the heat transfer rate shown on the vertical axis. This is shown in Figure 15-84.

The yellow-shaded area under the curve in Figure 15-84 is proportional to the total heating energy required over the heating season.

Consider a boiler that has been sized to provide only 50 percent of the (100,000 Btu/hr.) design heating load building in Syracuse. The green area shown in Figure 15-85 represents the portion of the total seasonal heating energy that would be supplied by this boiler.

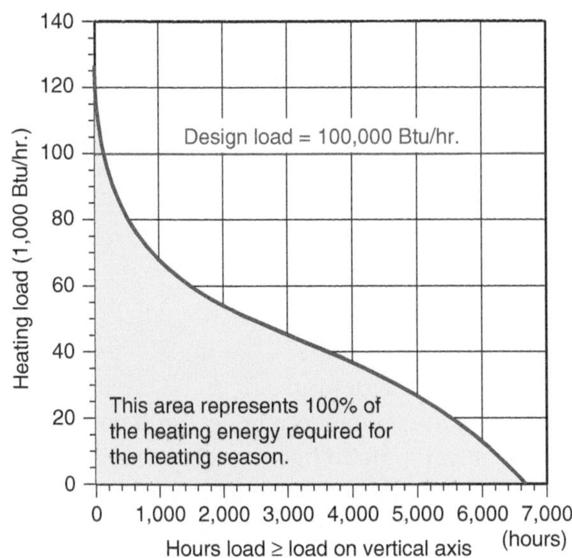

Figure 15-84 | Plot showing the number of hours within a heating season where the heating load equals or exceeds the value on the vertical axis.

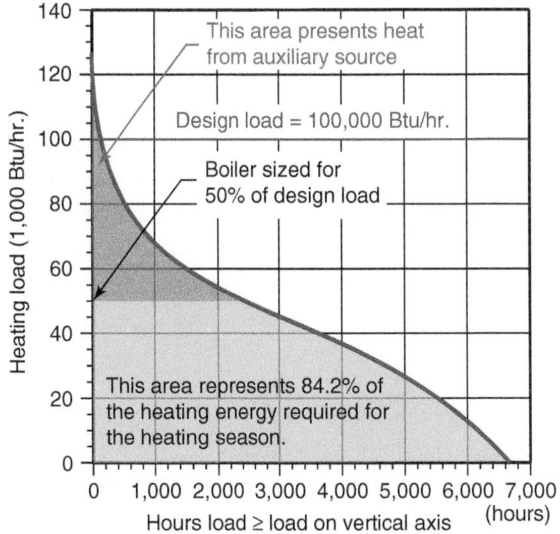

Figure 15-85 | The green shaded area indicates that a boiler sized for 50% of the building's design load supplies approximately 84% of the total seasonal heating energy required by the building.

The green shaded area is 84.2 percent of the total area under the red curve. The implication is that this boiler, which many would say is only half as big as necessary, could still supply about 84.2 percent of the total heating energy required over an entire (average) heating season.

If the boiler were sized for 60 percent of the design heating load, it would supply 90.4 percent of the total seasonal heating energy requirement. If sized for 75 percent of the design load, it would supply 96.3 percent of the total heating energy requirement.

This partial load analysis shows that a boiler that is sized significantly smaller than the design heating load will still supply a high percentage of the total heat energy needed. This is true of pellet-fired boilers as well as boilers burning conventional fuels. This analysis also implies that a boiler sized to the design heating load (or that load plus a safety factor) will spend very little time operating at steady state conditions. It will likely be cycling even when the building is experiencing the 97.5 percent outdoor design temperature.

Automatically controlled pellet-fuel boilers require anywhere from 10 to 30 minutes to reach steady-state operating conditions following an off period. The length of the warm-up period depends on the boiler's water volume relative to combustion rate, and how cool the boiler water is when a call for boiler operation occurs. Under partial load conditions, with limited modulation, and without thermal storage, a boiler sized for design load will often turn off due to reaching a high-limit water temperature *before it achieves combustion conditions that yield the best efficiency and the lowest particulate emissions. This is not a desirable operating characteristic.*

The following guidelines are suggested for sizing a pellet-fired boiler and minimizing the potential for short cycling and its associated lower efficiency and increased emissions. *These guidelines assume that another auxiliary heat source is present to handle the percentage of the load that cannot be supplied by the pellet-fired boiler.*

1. *Don't size the pellet boiler for design load.* Instead, size the pellet-fired boiler for 60 to at most 75 percent of design load. The higher end is for a pellet boiler with wide modulation capabilities such as a 5:1 (or higher) turndown ratio. This allows the pellet boiler to supply "base load" heating, while leaving only 4 to 10 percent of the total seasonal heating energy to be supplied by an auxiliary heat source.
2. *Install thermal storage between the pellet-fired boiler and load.* That storage can range from 1 to 2 gallons of water per 1,000 Btu/hr. of the boiler's full rated capacity. Thus, a pellet boiler rated at 60,000 Btu/hr. would be combined with 60 to 120 gallons of water storage. The upper end of this range (2 gallons/1,000 Btu/hr.) is appropriate when the boiler has limited modulation or no modulation of heat output and the distribution system is highly zoned and has relatively low thermal mass. The lower end of the range (1 gallon per 1,000 Btu/hr.) is appropriate for systems having boilers with a wider modulation range, a high mass distribution system such as a heated concrete slab operated with constant circulation, and limited zoning. This storage sizing criteria produces results that are similar to those generated by using Equation 15.5.

A pellet-fired boiler that is sized based on these guidelines necessitates an auxiliary heat source to meet peak design heating load. In some systems, this heat source will be a second boiler. In other systems, it might be a simple space heater.

If a boiler is used as the auxiliary heat source, it should be sized for 25 to 50 percent of the design heating load. Besides providing the heat output needed to meet peak heating load, a boiler sized to this criteria can supply a significant portion of the load if the pellet-fired boiler is inoperable due to maintenance or other reasons. It is analogous to supplementing a heat pump or solar collector array with an auxiliary heat source. Although the second boiler and its associated fuel and exhaust hardware increases installation cost, use of a smaller pellet-fired boiler offsets a significant portion of that cost. The smaller pellet-fired boiler will also operate at higher average efficiency and thus reduce the fuel required for a given seasonal heating requirement. This **dual-fuel approach** is highly recommended by the author.

Figure 15-86 shows one method for combining a pellet-fired boiler with an auxiliary mod/con boiler and a thermal storage tank.

The pellet-fired boiler is protected against low entering water temperature by a 3-way motorized mixing valve. Many modern pellet-fired boilers contain the control logic to operate this mixing valve. If a boiler without this control logic is used, an external temperature setpoint controller can be used to operate the mixing valve. During a cold start, the mixing valve recirculates water leaving the boiler's outlet port back to the boiler's inlet port. This allows the boiler temperature to rise quickly, since no heat is transferred

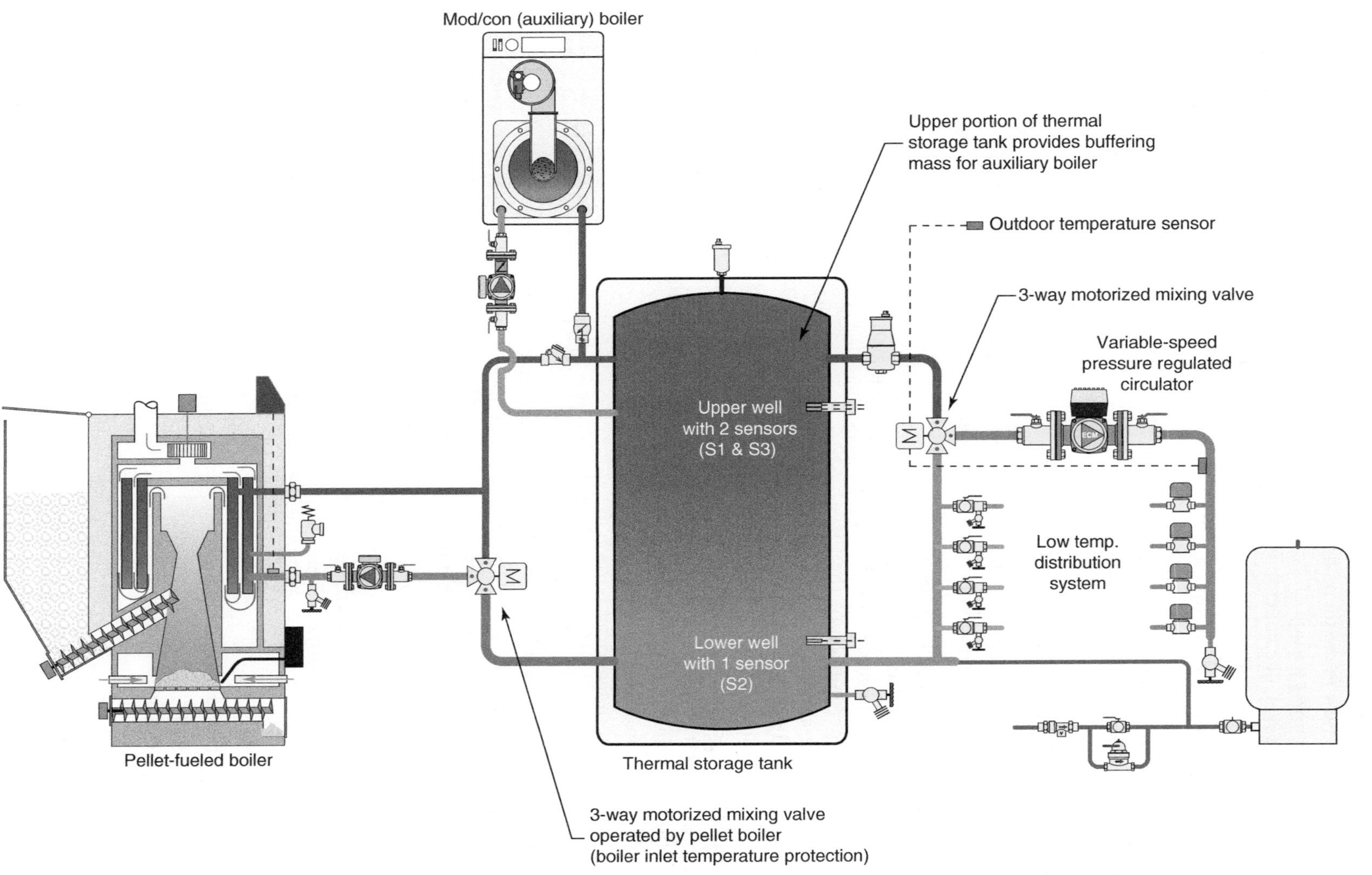

Figure 15-86 | System using a pellet-fired boiler as primary heat source, with mod/con auxiliary boiler.

to the thermal storage tank. As the boiler inlet temperature rises above a threshold where sustained flue gas condensation could occur, the mixing valve smoothly modulates to allow hot water flow to the thermal storage tank. Cool water from the lower portion of the tank is blended with hotter water leaving the boiler to maintain an acceptable boiler inlet temperature.

The swing check valve on the horizontal piping leading into the tank from the pellet-fired boiler prevents reverse thermosyphoning but allows forward thermosyphoning, if necessary, to dissipate a small amount of residual heat from the pellet-fired boiler during a power outage. The spring-loaded check valve in the piping from the auxiliary boiler, and inside the circulator supplying the auxiliary boiler, also prevent thermal migration when the auxiliary boiler is not operating.

The auxiliary boiler is piped so that it interacts with the upper 10 to 20 percent of the water in the thermal storage tank. This detail, in combination with thermal stratification, allows the upper portion of the thermal storage tank to buffer the auxiliary boiler against short cycling. This is especially useful in systems with extensively zoned distribution systems. Temperature stratification within the tank, in combination with careful detailing such as horizontal piping connections carrying flow *into* the tank, minimize mixing that could cause heat transfer into the lower portions of the tank. The thermal mass of the water in the lower portion of the tank is not needed to buffer the auxiliary boiler.

When piped as shown in Figure 15-86, the thermal storage tank also serves as a hydraulic separator between the boiler circulators and the variable-speed distribution circulator. This prevents undesirable interaction between the circulators.

This hybrid approach allows the auxiliary boiler to serve as a supplemental heat source when the system needs it, but still allows the pellet-fired boiler to provide the majority of the seasonal heating energy required by the load.

This approach to sizing is also applicable to wood chip-fired boilers, which require similar, or perhaps slightly longer, warm-up periods once started.

15.10 Example Systems

This section shows several examples of hydronic combisystems designed around wood-fired and pellet-fired boilers. These systems use a variety of subsystems for tasks such as protecting the boiler from low inlet temperature and overheating. They also show an assortment of low-temperature distribution systems and heat emitters. The methods shown for boiler protection and hydronic heating distribution are representative. Other valid methods discussed in this chapter could be used based on the preferences of the designer or the availability of specific equipment. Readers are also encouraged to consider how the combisystems shown in previous chapters for use with solar thermal collectors or heat pumps might be adapted for use with wood-fired or pellet-fired boiler systems.

Readers are also advised that the systems shown are generic "templates," and *not intended as final designs for a specific application*. Specific building and mechanical codes that apply in different locations may require additional safety devices such as low water cut offs, manual reset high limit controllers, or alarm systems on any biomass boilers used in that location. Furthermore, specific boiler manufacturers may require certain safety devices to be installed with their boilers. Designers must determine any code requirements, or specific details required by boiler manufacturer that apply to the systems being designed, and be certain that all required safety devices are installed accordingly.

WOOD-FIRED COMBISYSTEM #1

Several types of unpressurized thermal storage tanks are now available for use with wood-fired boilers. As discussed in Chapter 9, the use of an unpressurized thermal storage tank in combination with a steel or cast-iron boiler, requires a heat exchanger to be installed between the boiler and storage tank. This heat exchanger could be an internal helical coil or an external stainless steel brazed-plate unit. The system shown in Figure 15-87a uses the latter for transferring heat from the boiler to the thermal storage tank and for extracting heat from storage for space heating or domestic water heating.

In this system, the wood-gasification boiler is the sole heat source for space heating. Heat is transferred from the boiler to the brazed-plate heat exchanger (HX1) by circulator (P1). The boiler circuit includes a 3-way thermostatic anti-condensation valve that elevates the boiler's inlet temperature above 130 °F to prevent sustained flue gas condensation. A purging valve is installed upstream of the cold port of this valve to aid in bulk air removal when the system is

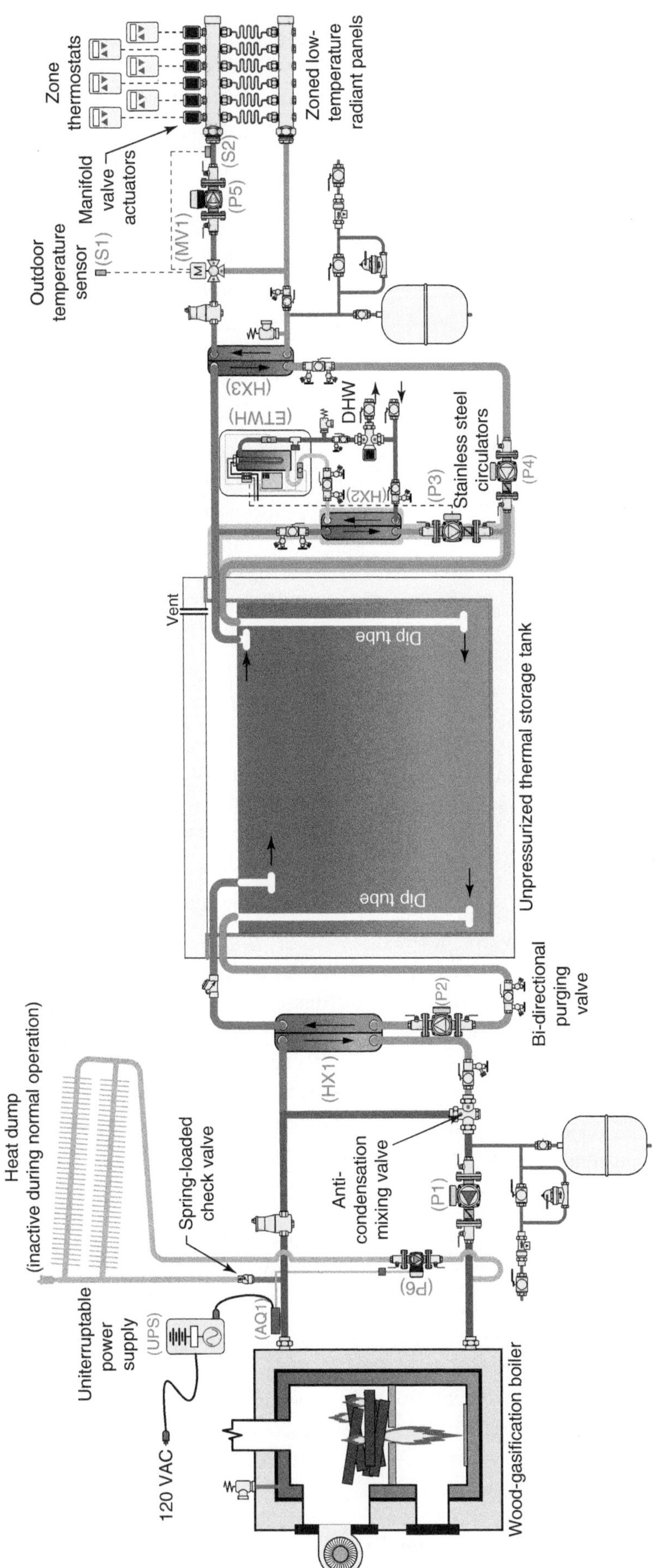

Figure 15-87a | Piping schematic for wood-fired combisystem #1.

commissioned. This system also contains a heat dump subsystem in which flow is created by a small circulator, (P6), supplied by an uninterruptible power supply (UPS). Power from the UPS is routed through a strap on aquastat (AQ1), which closes its contacts if the boiler outlet temperature reaches 180 °F. An AC relay detects the loss of normal AC power and responds by closing a set of contacts, passes line voltage to circulator (P6). Circulator (P6) will remain on until any of the following conditions occur:

a. Normal AC power is restored

b. The battery in the UPS is discharged

c. The boiler outlet temperature drops below 160 °F

The uninterruptible power supply should be sized to operate circulator (P6) for a minimum of 30 minutes.

A stainless steel circulator (P2) operates simultaneously with (P1) to convey heat from heat exchanger (HX1) to the storage tank. Circulator (P2) is located low in the system to maximize static pressure at its inlet. A bidirectional purging valve is provided in the circuit between (HX1) and the thermal storage tank to ensure air displacement from the "gooseneck" piping above the water level in the tank. A swing check is installed to stop reverse thermosyphoning from the tank through the heat exchanger piping.

The piping within the tank uses tees, piped at their side ports, to create horizontal flows that help preserve temperature stratification. The hottest water in the upper portion of the tank is drawn out to heat exchanger (HX2) for domestic water heating or to heat exchanger (HX3) for space heating. Cooler water returning from the distribution system is routed to the bottom of the thermal storage tank.

Domestic water is heated on demand using an external stainless steel brazed-plate heat exchanger (HX2). The details for this approach were discussed in Chapter 9. Whenever there is a demand for domestic hot water of 0.6 gpm or higher, the flow switch inside the electric tankless water heater closes. This closure is used, in combination with a relay, to turn on circulator (P3), which routes water from the upper portion of the thermal storage tank through the primary side of heat exchanger (HX2). Closure of the flow switch also energizes an electrical contactor within the tankless heater closes to supply 240 VAC to triacs, which regulate the power supplied to the heating elements. The electrical power delivered to the elements is limited to that required to provided the desired outgoing water temperature. All heated water leaving the tankless heater flows into an ASSE 1017–rated mixing valve to ensure a safe delivery temperature to the fixtures.

Space heating is provided by a highly zoned low-temperature distribution system. Although the heat emitters shown represent low-temperature radiant panels, the heat emitters could be other devices, such as panel radiators, convectors, towel warmers, low-temperature fin-tube baseboard, fan coils, or a combination of these devices. All heat emitters should be selected and sized to provide design load output at a supply water temperature no higher than 120 °F.

The water temperature supplied to the distribution system is controlled by a 3-way motorized mixing valve. This valve operates based on outdoor reset control. It supplies water to the heat emitters at a temperature that is just warm enough to meet the current heating load. It also protects the low-temperature distribution system from potentially high temperatures in the thermal storage tank.

The electrical wiring diagram for wood-fired combisystem #1 is shown in Figure 15-87b.

The following is a description of operation of the system shown in Figures 15-87a and 15-87b. Readers are encouraged to cross-reference the component designations, shown in parentheses, within both the piping and electrical schematics.

Description of Operation:

1. *Boiler operation:* When the wood-gasification boiler is fired, the operator closes a switch on the boiler that enables 120 VAC to be supplied to the boiler. This switch closure also supplies 120 VAC to circulators (P1) and (P2). Thus, flow between the boiler and heat exchanger (HX1) as well as between (HX1) and the thermal storage tank is enabled. As the temperature at the boiler outlet climbs, the 3-way thermostat valve (MVB) modulates to allow heat flow to heat exchanger (HX1) and onward to thermal storage, while also maintaining the boiler's inlet temperature above 130 °F (or whatever other temperature may be required) to prevent sustained flue gas condensation within the boiler.

2. *Boiler overheat protection mode:* If a power failure occurs, overheat protection is provided by an array of fin-tube elements sized to dissipate the residual heat from the boiler. At the onset of a power failure, 120 VAC is supplied from the uninterruptible power supply (UPS) to a normally open contact in aquastat (A1). This contact closes if the temperature at the boiler outlet rises to 180 °F or above. This passes 120 VAC

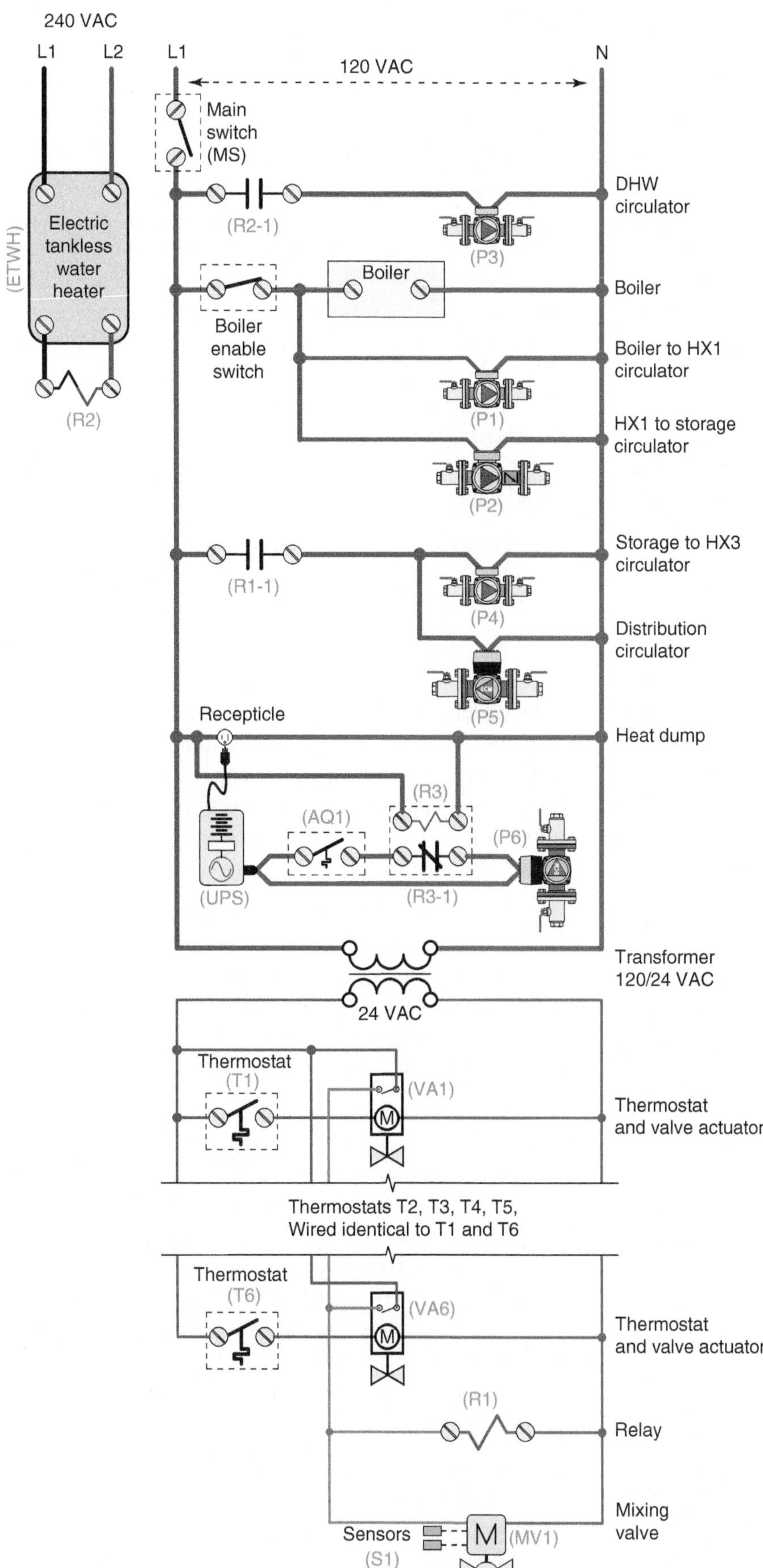

Figure 15-87b | Electrical wiring for wood-fired combisystem #1

from the UPS to a normally closed contact (R3-1). The coil of relay (R3) is energized whenever utility supplied power is available. Thus, contact (R3-1) remains open unless a power outage occurs. The closure of (R3-1) supplies 120 VAC power from the (UPS) to operate circulator (P6) and thus create flow through the heat dump. If the temperature at the boiler outlet drops to 160 °F or lower, the contact in aquastat (A1) opens, turning off circulator (P6). This conserves battery energy within the (UPS) when the boiler temperature is not high enough to justify operating the heat dump.

3. *Space heating mode:* Upon a call from any of the thermostats (T1...T6), 24 VAC is passed to the associated manifold valve actuators (VA1...VA6). When any one or more of these valve actuators reach their fully open position, an end switch within the valve actuator closes to pass 24 VAC to the coil of relay (R1) and motorized mixing valve (MV1). A normally open contact (R1-1) closes to supply 120 VAC to circulators (P4) and (P5). Circulator (P4) creates flow between the thermal storage tank and heat exchanger (HX3). Circulator (P5) is a variable-speed pressure-regulated circulator that adjusts its speed to maintain a constant differential pressure across the distribution manifold as the manifold valve actuators open and close. The motorized mixing valve (MV1) measures the outdoor temperature at sensor (S1) and uses this temperature, along with its settings, to calculate the necessary target supply water temperature to the distribution system. It compares the target supply temperature to the supply temperature measured by sensor (S2) and adjusts the hot water and return water flow rates into the valve to maintain the temperature

at sensor (S2) as close to the target temperature as possible.

4. *Domestic water heating mode:* Whenever there is a demand for domestic hot water of 0.6 gpm or higher, a flow switch within the electric tankless water heater (ETWH) closes. This closure supplies 240 VAC to the coil of relay (R2). The normally open contacts (R2-1) close to turn on circulator (P3), which circulates heated water from the upper portion of the thermal storage tank, through the primary side of the domestic water heat exchanger (HX2). The domestic water leaving (HX2) is either preheated, or fully heated, depending on the temperature in the upper portion of the thermal storage tank. This water passes into the thermostatically controlled tankless water heater (ETWH), which measures its inlet temperature. The electronics within this heater control the electrical power supplied to the heating elements based on the necessary temperature rise (if any) to achieve the set domestic hot water temperature. If the water entering the tankless heater is already at or above the setpoint temperature, the elements are not turned on. All heated water leaving the tankless heater flows into an ASSE 1017–rated mixing valve to ensure a safe delivery temperature to the fixtures. Whenever the demand for domestic hot water drops below 0.4 gpm, circulator (P3) and the tankless electric water heater are turned off.

WOOD-FIRED COMBISYSTEM #2

Some pellet-fired boilers are retrofit into *existing* hydronic heating systems that have a conventionally fueled boiler and existing heat emitters. In many cases, the existing boiler will be fueled by #2 fuel oil or propane. It is also likely that the distribution system was designed around a relatively high supply water temperature. For example, many systems that have an existing oil-fired boiler also have fin-tube baseboard convectors that were sized based on supply water temperatures of 170 °F to 200 °F at design load.

To optimize operation of the pellet-fired boiler, the retrofit should include thermal storage in the range of 1 to 2 gallons of water per 1,000 Btu/hr. of rated pellet-fired boiler capacity. It is also *critically important* that this thermal storage tank can "swing" through a reasonably wide temperature change between operating cycles of the pellet-fired boiler. This lengthens the on-cycle and off-cycle of the pellet-fired boiler, which improves its thermal efficiency and reduces its emissions.

Achieving this temperature swing requires that the distribution system be compatible with lower water temperatures. *In many cases the desired supply water temperature will be significantly lower than the temperature at which the original heat emitters were sized.*

Short of reducing the building's heating load, the only way to lower the supply water temperature required for the distribution system at design load conditions is to add more heat emitter surface area. The additional heat emitters could be fin-tube baseboard, panel radiators, fan-coils, radiant panels, or a combination of these emitters. The suggested approach is to modify the distribution system so that the combination of existing and new heat emitters allows the design load of the building to be met using a supply water temperature no higher than 120 °F.

The system shown in Figure 15-88a demonstrates one way to modify an existing distribution system to achieve lower supply water temperatures.

The distribution system shown in Figure 15-88a is a combination of existing fin-tube baseboard as well as new panel radiators. The new heat emitters have been selected so that the *total* heat output of the modified distribution system, (e.g., the heat output of all existing and new heat emitters), can supply the building's design heat load when operating with a supply water temperature of 120 °F.

The homerun distribution system allows segments of existing baseboards to be reconfigured into independently controlled zones. A thermostatic radiator valve is used to control water flow through each of these new baseboard configurations and thus regulate heat output to each area they serve.

Because the reconfigured distribution system only requires 120 °F supply water temperature at design load, the water temperature in the upper portion of the thermal storage tank could be as low as 115 °F (e.g., 120 °F minus half of the assumed control differential of 10 °F), when the oil-fired boiler is called upon to supplement the pellet-fueled boiler. Under such conditions, the 3-way thermostatic mixing valve maintains the inlet water temperature to the oil-fired boiler at 130 °F, which prevents it from operating with sustained flue gas condensation.

The thermal storage tank is piped with short and generously sized headers at the upper- and lower-left-side piping connections. The space heating circuit is connected across these headers and relatively close to the tank. The supply and return piping from

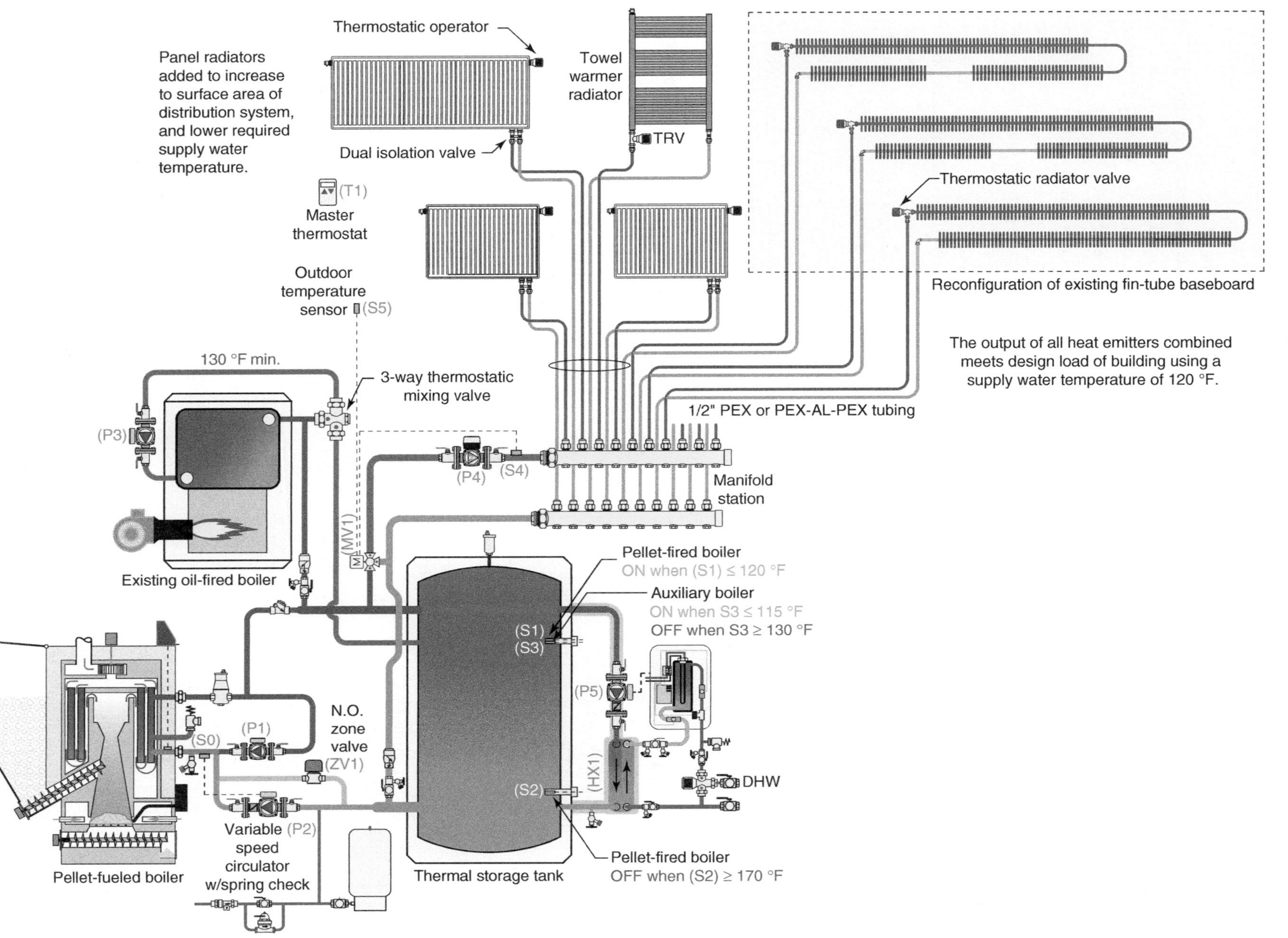

Figure 15-88a | Piping system for wood-fired combisystem #2.

the pellet-fired boiler is also connected across these headers. The supply pipe from the oil-fired boiler is connected to the upper header but not to the lower header. Instead, the piping carrying water from the upper portion of the thermal storage tank to the oil-fired boiler exits from another side wall connection located a few inches below the upper left side header connection. This limits the amount of water that interacts with the oil-fired boiler to the upper 10 to 20 percent of the tank volume. This water volume is sufficient to buffer the oil-fired boiler against the highly zoned distribution, preventing it from short cycling. This piping arrangement also prevents the oil-fired boiler from heating the entire thermal storage tank, which is neither necessary nor desireable. The short/generously sized headers that combine flows from the pellet-fired boiler, the oil-fired boiler, and the thermal storage tank, acting in combination with the low head loss of the thermal storage tank, create good hydraulic separation of circulators (P2), (P3), and (P4).

This piping configuration also allows hot water from either boiler to flow directly to the load without first passing through the thermal storage tank. This allows rapid heat delivery to the load during recovery from thermostat setbacks.

When one or both boilers are operating, and the space heating distribution system is also operating, the flow rate entering the thermal storage tank is the flow rate from the boiler(s), minus the flow rate to the distribution system. This reduces the flow velocity entering the thermal storage tank, which helps to preserve temperature stratification within the tank.

The pellet-fired boiler is protected against sustained flue gas condensation by the variable speed "setpoint" circulator (P2). Figure 15-57 shows an example of such a circulator. The circulator's speed is controlled based on the water temperature entering the pellet-fired boiler. If that temperature is at or below 130 °F, circulator (P2) is off, and no flow passes from the pellet-fired boiler to the thermal storage tank. As the temperature at the boiler inlet climbs above 130 °F, the speed of circulator (P2) increases. If the boiler inlet temperature reaches 135 °F or higher circulator (P2) will operate at full speed.

The piping between the pellet-fired boiler and the thermal storage tank also allows thermosyphon flow from the boiler to the thermal storage tank if a power outage occurs when the boiler is operating. This eliminates the need for any other heat dump.

The swing check valve at the left of the upper tank header prevents reverse thermosyphoning. It also prevents flow reversal from the tank when circulator (P1) is operating, but circulator (P2) is off, or at a very low speed. This piping detail is very effective when used in combination with pellet-fired boilers having low flow resistance, which is typical of most pellet-fired boilers with fire-tube heat exchangers.

The spring-loaded check valve within circulator (P2) provides about 0.5 psi forward opening pressure. This is generally sufficient to prevent undesirable flow of warm water returning from the space heating load through the pellet-fired boiler with that boiler is not firing. If otherwise allowed to occur, this flow would dissipated heat through the jacket of the unfired boiler. It would also induce convective heat loss through the vent connector piping and chimney.

The normally closed zone valve (ZV1) is closed whenever utility power is available to the system. It opens during a power outage to provide a "detour" around the spring-loaded check valve in circulator (P2). This allows for uninhibited thermosyphon flow between the boiler and thermal storage tank during a power outage. This action dissipates any residual heat from the boiler to thermal storage and thus prevents boiler overheating.

Domestic water is heated "on-demand" using an external stainless steel brazed-plate heat exchanger (HX1). Whenever there is a demand for domestic hot water of 0.6 gpm or higher, the flow switch inside the tankless electric water heater closes. This closure is used, in combination with a relay, to turn on the circulator that routes water from the upper portion of the thermal storage tank through the primary side of this heat exchanger. Closure of the flow switch also energizes an electrical contactor within the tankless heater to supply 240 VAC to triacs, which regulate the electrical power supplied to the heating elements. The electrical power delivered is limited to that required to provided the desired outgoing water temperature. All heated water leaving the tankless heater flows into an ASSE 1017–rated mixing valve to ensure a safe delivery temperature to the fixtures. If there is no demand for space heating, and the buffer tank has cooled to room temperature, minimal preheating is provided and the tankless water heater provides most of the energy needed for domestic water heating.

In this system, the boilers are operated by a combination of a 2-stage temperature setpoint controller, and a

1-stage temperature setpoint controller. When the master thermostat creates a space heating demand, the 2-stage controller turns on the pellet-fired boiler if the temperature at the upper tank sensor (S1) is 120 °F or less. If heat output from the pellet-fired boiler is sufficient to meet the load, the temperature at sensor (S1) will remain stable. If the boiler's heat output exceeds the load, the temperature at sensor (S1) will increase. If heat output from the pellet-fired boiler is less than required by the load, the temperature at sensor (S1) will decrease.

The auxiliary boiler will be turned on by the 1-stage temperature controller if the temperature at the upper tank sensor (S3) drops to 115 °F or less. Both boilers will continue operating until the temperature at sensor (S3) climbs to 130 °F, at which point the auxiliary boiler is turned off. This allows the auxiliary boiler to be buffered by the upper portion of the thermal storage tank, but does not allow the auxiliary boiler to heat the entire thermal storage tank.

The pellet-fired boiler keeps running, even if the demand for space heating stops, until the temperature at the *lower* tank sensor (S2) reaches 170 °F. This "stacks" the tank with hot water and lengthens the operating cycle of the pellet-fired boiler. Longer operating cycles improve combustion efficiency and reduce emissions.

Figure 15-88b shows the control temperature ranges along with the on/off setpoints of the boilers.

Figure 15-88c shows the electrical wiring for this system.

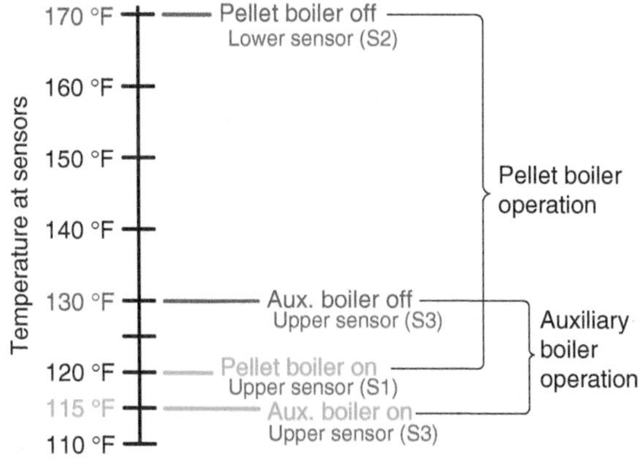

Figure 15-88b Configuration of outdoor reset for determining if the main thermal storage tank can supply space heating, or if auxiliary boiler is required.

The following is a description of operation for combisystem #2. Readers are encouraged to cross-reference the component designations, shown in parentheses, within both the piping and electrical schematics.

Description of Operation:

1. *Boiler operation:* The master thermostat (T1) creates a demand for space heating. This applies 24VAC power to the 2-stage setpoint controller (SP2) and the 1-stage temperature setpoint controller (SP1). The 2-stage controller measures the temperature of sensor (S1) in the upper portion of the storage tank. If the temperature is less than 120 °F, the first stage contact in (SP2) closes. This passes 24 VAC to the coil of relay (R3). One normally open contact in this relay (R3-2) closes to initiate operation of the pellet-fired boiler. Another normally open contact (R3-1) closes to pass 24 VAC to the stage 2 contact in the 2-stage setpoint controller (SP2). A third contact (R3-3) closes in parallel with contact (R1-1). 24 VAC is now simultaneously passing through both stage 1 and stage 2 contacts in (SP2). The stage 1 contact in (SP2) will open when the temperature at the upper sensor (S1) reaches 130 °F. However, 24 VAC can still pass through contact (R3-1) to maintain relay coil (R3) in the on state. This keeps the pellet-fired boiler operating, and allows the storage tank to be fully "stacked" with hot water. The pellet-fired boiler turns off when the temperature at the lower tank sensor (S2) reaches 170 °F, even if the space heating demand ceases.

When started, the pellet-fired boiler applies 120 VAC to circulators (P1) and (P2). Circulator (P1) operates at fixed speed. Circulator (P2) is a variable speed circulator that monitors the temperature at sensor (S0). It begins moving water as the temperature of the water entering the boiler reaches 130 °F. The speed of circulator (P2) increases as the water temperature entering the boiler increases. (P2) will operate at full speed if the boiler inlet temperature reaches 135 °F. As the temperature entering the pellet-fueled boiler decreases the speed of circulator (P2) also decreases.

If the temperature at sensor (S3) in the upper portion of the storage tank drops to or below 115 °F, the normally open contact in the 1-stage setpoint controller (SP1) closes. This enables the auxiliary oil-boiler to fire. The auxiliary boiler supplies 120 VAC to circulator (P3), and transfers heat to the storage tank to supplement the heat output of the pellet-fired boiler. The auxiliary boiler continues to operate until the upper tank sensor (S3) reaches 130 °F, or the heating demand ceases.

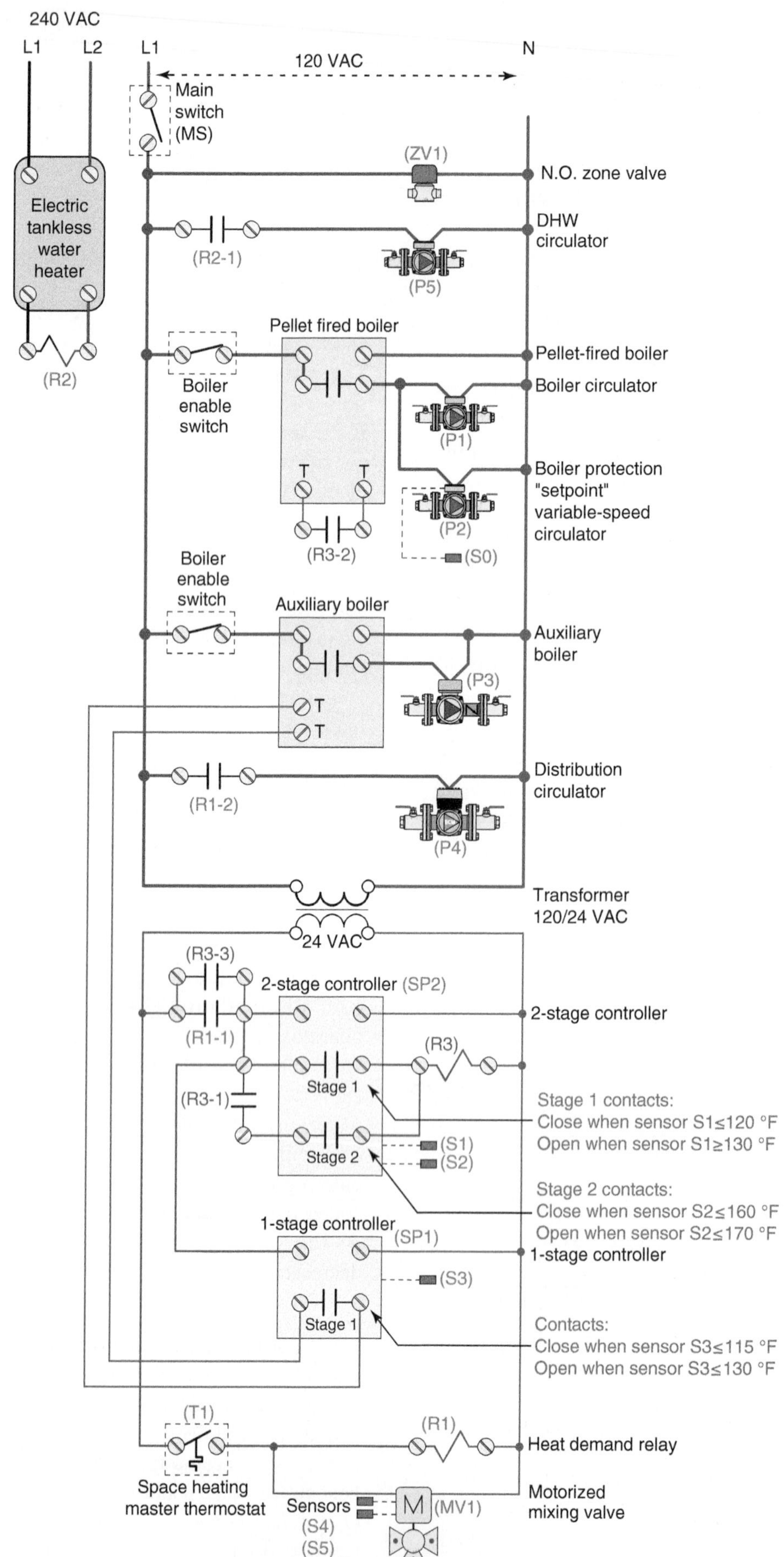

Figure 15-88c | Electrical wiring for combisystem #2.

If a power failure occurs, residual heat from the boiler transfers to storage by thermosyphoning.

2. *Space heating mode:* Space heating is provided by a combination of existing fin-tube baseboard and new panel radiators, each equipped with a wireless thermostatic radiator valve. Space heating is enabled by a master thermostat (T1). When the contacts in the master thermostat close, 24 VAC is passed to the coil of relay (R1). A normally open relay contact (R1-1) closes to supply 120 VAC to circulator (P4). This circulator operates at variable speeds to maintain a constant differential pressure across the distribution system manifold as the thermostatic radiator valves open, close, or modulate.

The motorized mixing valve (MV1) is also turned on by the master thermostat (T1). It measures the outdoor temperature at sensor (S5) and uses this temperature, along with its settings, to calculate the necessary target supply water temperature to the distribution system. It compares the target supply temperature to the actual supply temperature measured by sensor (S4) and adjusts the hot water and return water flow rates into the valve to maintain the temperature at sensor (S4) as close to the target temperature as possible.

3. *Overheating protection mode:* The normally closed zone valve (ZV1) is closed whenever the main switch is closed and utility power is supplied to the system. It opens during a power failure to provide an uninhibited thermosyphon flow path around the spring-loaded check valve in circulator (P2). This allows residual heat from the boiler to be conveyed to the thermal storage tank, and thus prevents boiler overheating.

4. *Domestic water heating mode:* Whenever there is a demand for domestic hot water of 0.6 gpm or higher, a flow switch within the electric tankless water heater (ETWH) closes. This closure supplies 240 VAC to the coil of relay (R2). The normally open contacts (R2-1) close to turn on circulator (P5), which circulates heated water from the upper portion of the thermal storage tank through the primary side of the domestic water heat exchanger (HX1). If the domestic water leaving (HX1) is fully heated, it flows through the tankless heater, but the heating elements remain off. It flows on to an ASSE 1017–rated mixing valve to ensure a safe delivery temperature to the fixtures. Whenever the demand for domestic hot water drops below 0.5 gpm, circulator (P5) and the tankless electric water heater are turned off.

During warm weather, when there is no demand for space heating, the thermal storage tank temperature will drop to room temperature. Because this temperature may still be 10 to 20 °F above entering cold water temperature, the heat exchanger (HX1) can still provide a small preheating effect. The tankless electric water heater will provide the balance of the temperature rise, and thus assume the majority of the domestic water heating load.

WOOD-FIRED COMBISYSTEM #3

The combisystem shown in Figure 15-89a makes use of a natural draft wood-fired boiler as well as a low-mass mod/con auxiliary boiler.

The operating characteristics of a natural draft wood-fired boiler are different from those of a wood-gasification boiler. Unlike the latter, which typically burns a large "charge" of firewood at a rapid rate, a natural draft wood-fired boiler can operate for many hours with less variation in its average heat output, albeit at lower combustion efficiency. Thus, the size of the thermal storage tank can usually be reduced from that required by a wood-gasification boiler.

Another important difference is that most natural draft boilers allow the flow of primary combustion air to continue during a power outage, whereas most wood-gasification boilers will greatly reduce this air flow. If the fire is well established, and a power outage occurs soon after the combustion chamber has been loaded with wood, the boiler will continue producing heat, in some cases for several hours. Thus, it is imperative to equip systems with natural draft wood-fired boilers with a heat dump capable of safely dissipating this residual heat.

The heat dump subsystem shown in Figure 15-89a uses a rack of copper fin tube as the heat dissipater. Flow through the heat dump subassembly is created by a small AC circulator powered by an uninterruptible power supply (UPS). Power from the UPS must pass through an AC relay. The coil of this relay is powered by 120 VAC from the utility. The contacts required to operate circulator (P2) remain open whenever utility power is available. When a power outage occurs, these contacts close to pass 120 VAC from the UPS to circulator (P2). However, it is unnecessary to operate circulator (P2) if the boiler is not at an elevated temperature. This condition is sensed by a strap-on aquastat (AQ1) mounted to the piping at the boiler's outlet. If the temperature at this location reaches 180 °F, and utility electrical power is out, the circuit to circulator (P2) is completed and the heat dump is operational. If the temperature at the boiler outlet drops to 160 °F or less, the

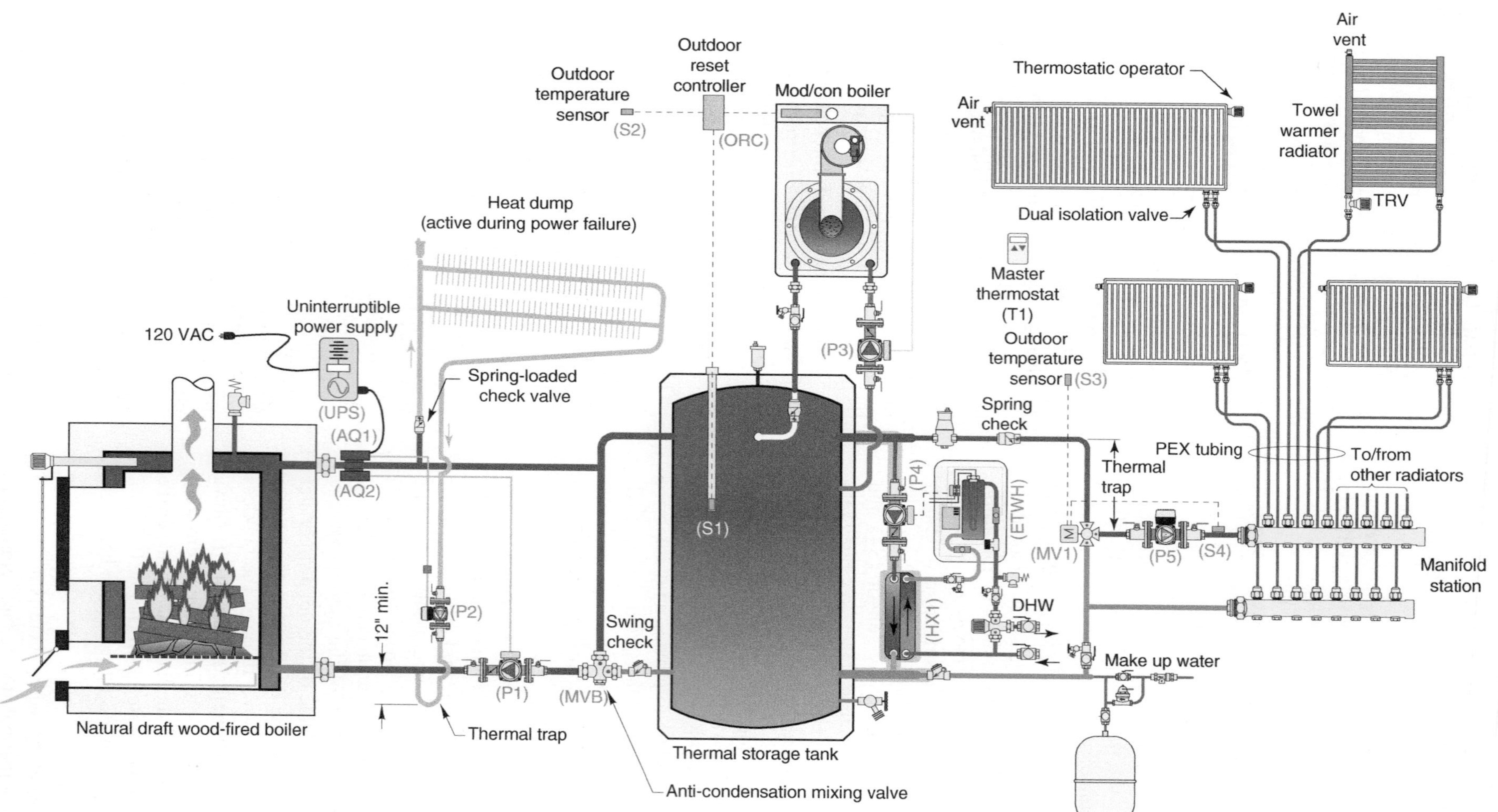

Figure 15-89a | Piping schematic for combisystem #3.

aquastat turns off the heat dump circulator (P2). This helps conserve battery energy in the UPS. The UPS should be sized to operate circulator (P2) for a minimum of 2 hours. A spring-loaded check valve on the supply piping to the heat dump prevents thermosyphoning or reverse flow through the heat dump subassembly during normal system operation. The return side of the heat dump piping contains a thermal trap that further discourages heat migration during normal operation.

The piping between the natural draft wood-fired boiler and thermal storage tank includes a 3-way thermostatic mixing valve. Its purpose is to keep the boiler's inlet temperature at or above 130 °F whenever possible to prevent sustained flue gas condensation within the boiler. As the boiler warms above its minimum inlet water temperature, heated water begins flowing to the storage tank. This hot water can deliver heat for space heating or for heating domestic water. The latter load must be supplied through a heat exchanger.

Upon a call for space heating, circulator (P5) is turned on, along with the 3-way motorized mixing valve (MV1). (P5) is a variable-speed pressure-regulated circulator that responds to the opening, closing, and modulation of the wireless thermostatic radiator valves on each panel radiator and towel warmer. Mixing valve (MV1) operates based on outdoor reset control to regulate the water temperature supplied to the heat emitters. The spring check valve in the piping upstream of the mixing valve, as well as the downward turn in the piping (e.g., the thermal trap), stop heat migration into the space heating subsystem when it is not needed.

The outdoor reset controller (ORC) is also turned on whenever there is a call for space heating from the master thermostat (T1). This controller measures the outdoor temperature and calculates the required supply water temperature to the distribution system. If the water temperature measured by sensor (S1) in the upper portion of the thermal storage tank is at or close to the target temperature, the auxiliary boiler remains off. If the temperature at sensor (S1) is a few degrees below the target temperature, the auxiliary boiler is turned on to heat water in the upper portion of the storage tank to a temperature that can satisfy the current heating load. The auxiliary boiler operates based on a preset high-limit temperature rather than its own internal reset control. It is simply turned on or off by the outdoor reset controller (ORC). The upper portion of the thermal storage tank provides thermal mass to buffer the auxiliary mod/con boiler against short cycling.

Domestic water is heated "on demand" using an external stainless steel brazed-plate heat exchanger (HX1). Whenever there is a demand for domestic hot water of 0.6 gpm or higher, the flow switch inside the tankless electric water heater closes. This closure is used, in combination with a relay, to turn on the circulator that routes water from the upper portion of the thermal storage tank through the primary side of heat exchanger (HX1). Closure of the flow switch also energizes an electrical contactor within the tankless heater to supply 240 VAC to triacs, which regulate the electrical power supplied to the heating elements. The power delivered to the elements is limited to that required to provide the desired outgoing domestic hot water temperature. All heated water leaving the tankless heater flows into an ASSE 1017–rated mixing valve to ensure a safe delivery temperature to the fixtures.

Figure 15-89b shows the electrical wiring for wood-fired combisystem #3.

The following is a description of operation of the system shown in Figures 15-89a and 15-89b. Readers are encouraged to cross-reference the component designations, shown in parentheses, within both the piping and electrical schematics.

Description of Operation:

1. *Wood-fired boiler operation:* When the natural draft wood-fueled boiler is fired, the temperature at its outlet eventually reaches 140 °F. At that point, the normally open contacts in aquastat (AQ2) close to supply 120 VAC to circulator (P1). The 3-way thermostatic valve (MVB) modulates to allow heated water to flow to storage while also maintaining the boiler's inlet temperature above 130 °F (or whatever other temperature may be required) to prevent sustained flue gas condensation within the boiler. When the temperature at aquastat (AQ2) drops below 120 °F, its contacts open, turning off circulator (P1).

2. *Wood-fired boiler overheat protection mode:* During a power failure, 120 VAC is supplied from the uninterruptible power supply (UPS) to a normally open contact in aquastat (AQ1). This contact closes if the temperature at the boiler outlet rises to or above 180 °F. This passes 120 VAC from the UPS to a normally closed contact (R3-1). The coil of relay (R3) is energized whenever utility power is available. Thus, contact (R3-1) remains open unless a power failure occurs. The closure of (R3-1) supplies 120 VAC from the (UPS) to circulator (P2) and creates flow through the heat dump. If the temperature at the boiler outlet drops to 160 °F or

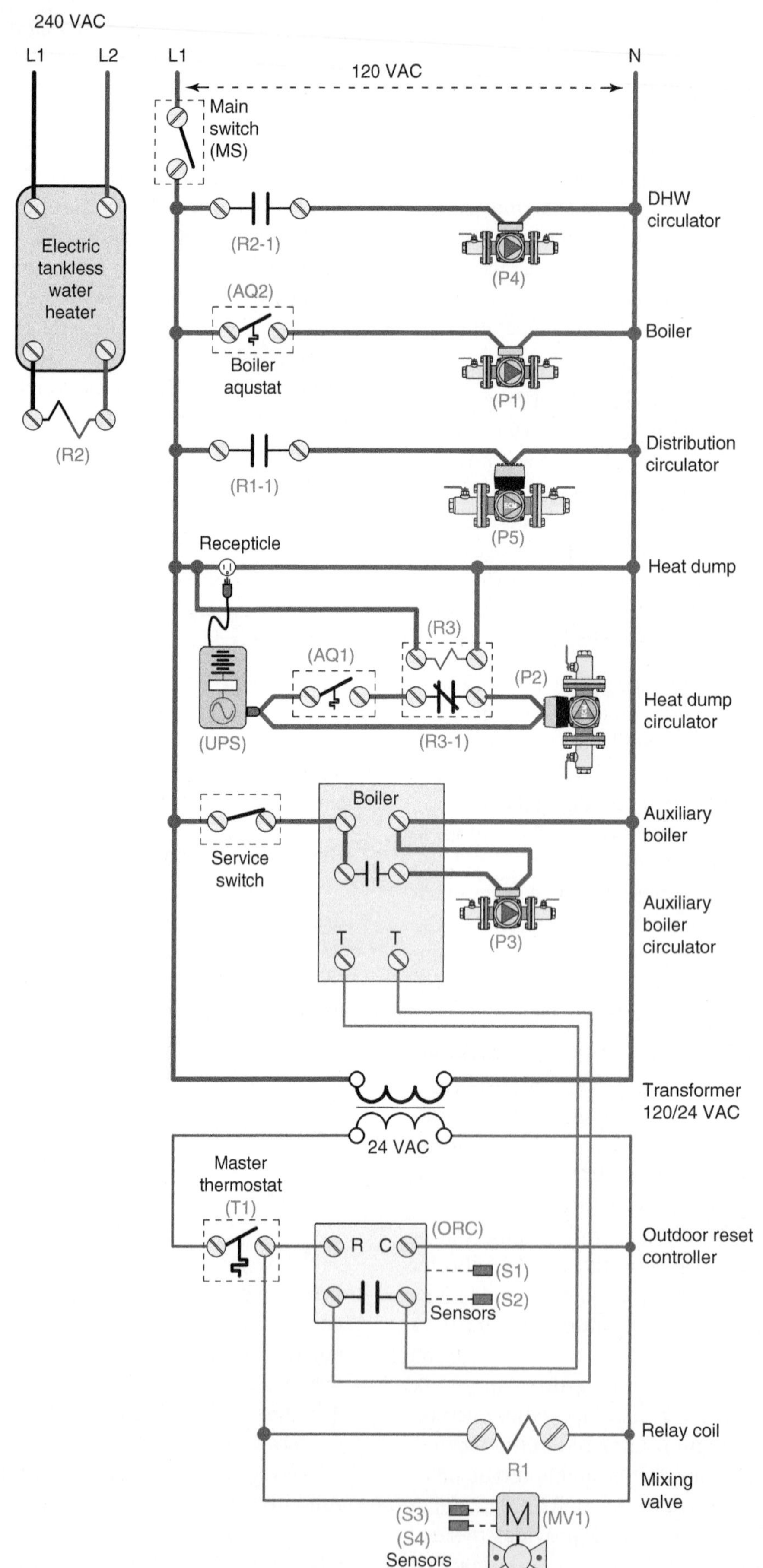

Figure 15-89b | Electrical wiring for combisystem #3

lower, the contacts in aquastat (AQ1) opens, turning off circulator (P2). This conserves battery energy within the (UPS) when the boiler temperature is not high enough to justify operating the heat dump.

3. *Space heating mode:* Space heating is provided by panel radiators, each equipped with a wireless thermostatic radiator valve. Space heating is enabled by a master thermostat (T1). When the contacts in the master thermostat close, 24 VAC is passed to the coil of relay (R1) and to the outdoor reset controller (ORC). A normally open relay contact (R1-1) closes to supply 120 VAC to circulator (P5). This circulator operates at variable speeds to maintain a constant differential pressure across the distribution system manifold as the thermostatic radiator valves open, close, or modulate. When powered on, the (ORC) measures the outdoor temperature using sensor (S2). It uses this temperature, along with its settings, to determine if the water temperature at sensor (S1) in the upper portion of the storage tank is high enough to supply space heating. If it is not, a set of contacts within the (ORC) close to enable the auxiliary boiler to operate. A set of contacts within the boiler close to supply 120 VAC to boiler circulator (P3). The boiler operates under the control of the (ORC) and with a fixed high-limit setting on its internal controls (for safety only). If the wood-fueled boiler starts adding heat to the storage tank, the temperature at sensor (S1) will eventually rise to a value where the (ORC) will turn off the auxiliary boiler.

The motorized mixing valve (MV1) is also turned on by the master thermostat. It measures the outdoor temperature at sensor (S3) and uses this temperature, along with its settings, to calculate the necessary target supply water temperature to the distribution system. It compares the target supply temperature to the actual supply temperature measured by sensor (S4) and adjusts the flow rates into the valve to maintain the temperature at sensor (S4) as close to the target temperature as possible.

4. *Domestic water heating mode:* Whenever there is a demand for domestic hot water of 0.6 gpm or higher, a flow switch within the electric tankless water heater (ETWH) closes. This closure supplies 240 VAC to the coil of relay (R2). The normally open contacts (R2-1) close to turn on circulator (P4), which circulates heated water from the upper portion of the thermal storage tank through the primary side of the domestic water heat exchanger (HX1). The domestic water leaving (HX1) is either preheated or fully heated depending on the temperature in the upper portion of the thermal storage tank. This water passes into the thermostatically controlled tankless water heater (ETWH), which measures its inlet temperature. The electronics within this heater control the electrical power supplied to the heating elements based on the necessary temperature rise (if any) to achieve the set domestic hot water supply temperature. If the water entering the tankless heater is already at or above the setpoint temperature, the elements are not turned on. All heated water leaving the tankless heater flows into an ASSE 1017–rated mixing valve to ensure a safe delivery temperature to the fixtures. Whenever the demand for domestic hot water drops below 0.4 gpm, circulator (P4) and the tankless electric water heater are turned off.

WOOD-FIRED COMBISYSTEM #4

Many people who use a wood-gasification boiler for heating in cold weather don't want to operate it in summer, especially if the only load is domestic water heating. In such situations, solar energy can provide an ideal "summer substitute" for firewood. Figure 15-90a shows one way to combine a wood-gasification boiler with a drainback solar thermal array.

There is no automatically operated auxiliary boiler in this system. All *space heating* must come from either the wood-gasification boiler or the solar collector array.

The wood-gasification boiler supplies heat to a 500-gallon thermal storage tank. A loading unit provides the flow between the boiler and tank. It also protects the boiler from operating with low inlet water temperatures that cause sustained flue gas condensation. At the onset of a power failure, the flapper valve inside the loading unit opens allowing thermosyphon flow to carry residual heat from the boiler into the thermal storage tank.

A drainback solar thermal subsystem is also connected in parallel with the boiler. The common piping they share is very short and sized for a maximum flow velocity of 2 feet per second. This enables the common piping and tank to provide hydraulic separation between the circulator in the loading unit and the solar array circulator. Thus, the boiler and solar subsystem could operate simultaneously without interference between their circulators.

The swing-check valve upstream of the upper-left inlet to the thermal storage tank prevents reverse thermosyphoning when the loading unit is off. It also prevents flow reversal through the inactive boiler.

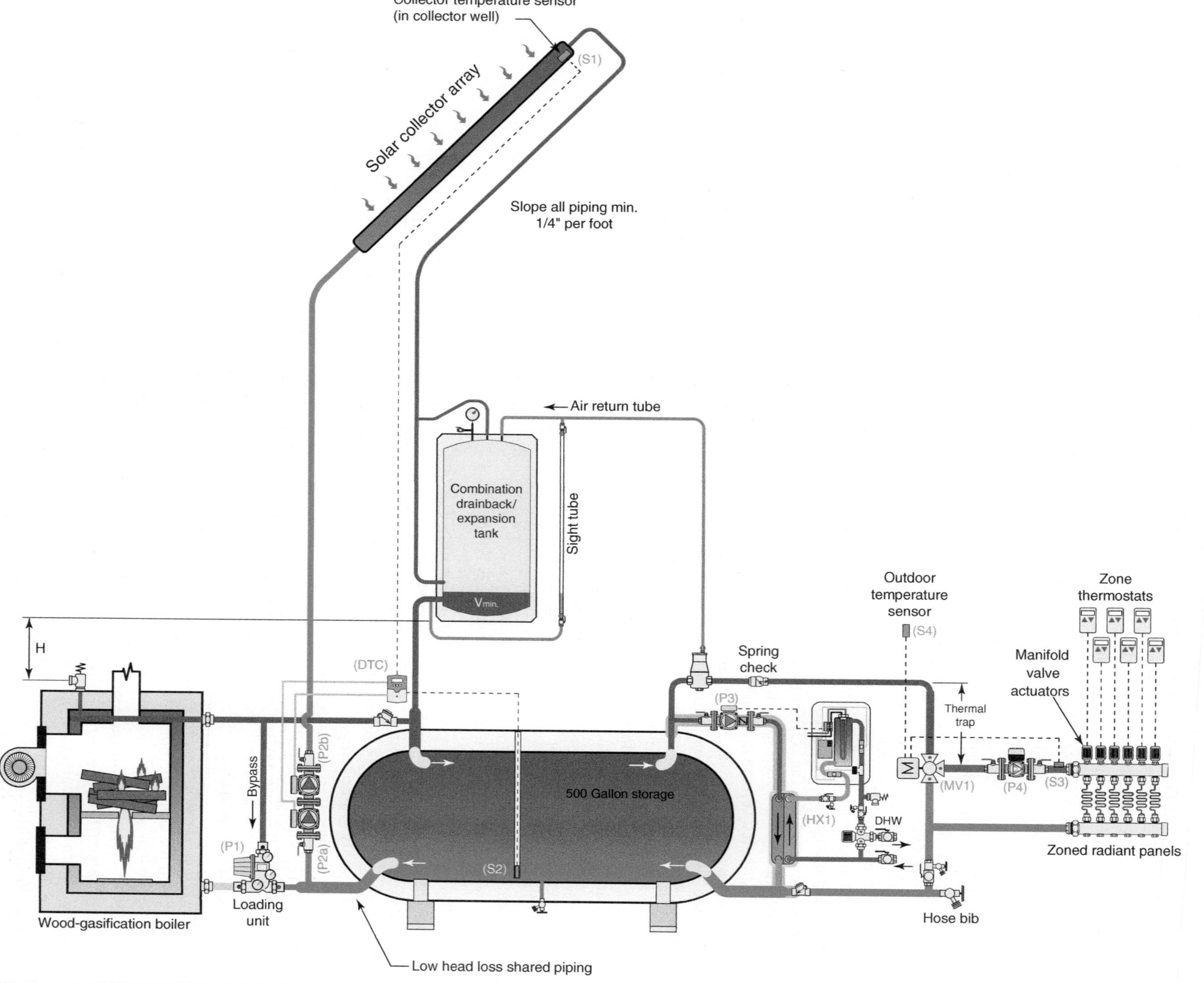

Figure 15-90a | Piping schematic for combisystem #4.

The combination drainback/expansion tank is placed as high within the structure as possible to minimize the initial lift head on the solar circulator. *In all cases, the bottom of this tank should be above the top of the thermal storage tank.* If the bottom of the drainback/expansion tank is at least as high as the highest point in the distribution system, then all locations within the distribution system will be under positive static pressure. The initial air pressure in the drainback/expansion tank could be set as low as zero. However, slightly pressurizing the air in the drainback tank will increase the net positive suction head on the circulators, which reduces noise and helps prevent cavitation.

The volume of the combination drainback/expansion tank is primarily determined by the expansion requirement of a system with a large thermal storage tank that experiences a wide swing in temperature. The volume of this tank can be determined from Equation 15.7.

(Equation 15.7)

$$V_T = \frac{\left(V_T\left(\frac{D_C}{D_H} - 1\right)\right)}{1 - \frac{(P_i + 14.7)(T_h + 458)}{(P_{RV} - 0.42H + 14.7)(T_L + 458)}} + V_{min}$$

where:

V_T = *minimum* volume of the combined drainback/expansion tank (gallons)

D_c = density of cold water used to fill the system (lb./ft.^{3})

D_H = density of water at system's maximum temperature (lb./ft.^{3})

P_i = initial air pressure in drainback/expansion tank (psi) (must be ≥ 0 psi)

P_{RV} = rated opening pressure of relief valve (psi)

H = distance from pressure-relief valve to lowest water level in drainback/expansion tank (feet)

T_h = highest water temperature in system (°F)

T_L = temperature of cold water used to fill system (°F)

V_{min} = minimum volume required at bottom of combined drainback/expansion tank so that no air is entrained into lower piping connections when the collector array and associated piping is completely filled

Example 15.7: A system will be assembled based on the schematic in Figure 15-90a. It uses a 500-gallon pressurized storage tank. The combination drainback/expansion tank will be located with its bottom 1 foot above the 30-psi-rated pressure-relief valve of the boiler. This tank has a minimum volume requirement of 15 gallons to ensure water is drawn from its lower connection without entraining air, when the collector array and associated piping are completely filled. The estimated system volume, including the storage tank, is 560 gallons. The system will be filled with water at a temperature of 50 °F. The maximum temperature of the system water is estimated at 200 °F. Determine the necessary volume of the combined drainback/expansion tank, assuming the initial air pressure in the tank is 0 psi.

Solution: The density of water at both temperature extremes is needed and can be referenced from Figure 3-11.

$$D_{50°F} = 62.4 \text{ lb./ft.}^3$$

$$D_{200°F} = 60.1 \text{ lb./ft.}^3$$

These values, and the remaining data, can now be entered into Equation 15.7:

$$V_T = \frac{560\left(\frac{62.4}{60.1} - 1\right)}{1 - \frac{(0 + 14.7)(200 + 458)}{(30 - 0.42(1) + 14.7)(50 + 458)}} + 15$$

$$= 27.2 \text{ gallon}$$

Discussion: Equation 15.7 is derived from **Charles's law**. The calculated minimum tank volume of 27.2 gallons is conservative, since a portion of the system's volume will be in the low-temperature distribution system and thus not subject to 200 °F temperature. The 15-gallon minimum tank volume is more than half the required volume. This minimum tank volume may be reduced by using a tank with the lowest possible pipe tapping. The suggested minimum volume would be the sum of volumes of the collector array and collector piping multiplied by a safety factor of 1.10. The selected tank should be well insulated, since the hottest water returning from the collector array must pass through it on its way to the main storage tank. Designers are encouraged to enter Equation 15.6 into a spreadsheet and experiment with different values of the variables to determine their effect on the size of the drainback/expansion tank.

Any air captured by the air separator downstream of the thermal storage tank is returned to the drainback/

expansion tank. This is **air control** rather than **air elimination**. There is no automatic make-up water subsystem. The water level in the system is monitored by the sight tube near the drainback/expansion tank. The water level can be adjusted, if necessary, by adding or removing water through a hose connected to the lower hose bib valve. The system should be filled so that the sight tube indicates the minimum water level is present in the combination drainback/expansion tank.

Domestic water is heated "on demand" using an external stainless steel brazed-plate heat exchanger (HX1). Whenever there is a demand for domestic hot water of 0.6 gpm or higher, the flow switch inside the tankless electric water heater closes. This closure is used, in combination with a relay, to turn on the circulator that routes water from the upper portion of the thermal storage tank through the primary side of this heat exchanger. Closure of the flow switch also energizes an electrical contactor within the tankless heater. This closure supplies 240 VAC to triacs, which regulates the electrical power supplied to the heating elements. The power delivered to the elements is limited to that required to provide the desired outgoing water temperature. All heated water leaving the tankless heater flows into an ASSE 1017–rated mixing valve to ensure a safe delivery temperature to the fixtures. This configuration ensures that domestic hot water will always be available, even if the thermal storage tank has cooled down to room temperature after an extended period of no heat input from either the solar collectors, or the wood-gasification boiler.

Space heating is provided by a low-temperature highly zoned radiant panel distribution system. All heat emitters are sized to deliver design load output at a supply temperature not higher than 120 °F. A motorized 3-way mixing valve adjusts the supply water temperature to the heat emitters based on outdoor reset control. The variable-speed pressure-regulated circulator adjusts its speed based on how the manifold valve actuators are opening and closing. The spring-check valve in the piping upstream of the mixing valve, as well as the downward direction of the piping (e.g., the thermal trap), prevents heat migration into the space heating subsystem.

Figure 15-90b shows the electrical wiring for wood-fired combisystem #4.

The following is a description of operation of the system shown in Figures 15-90a and 15-90b. Readers are encouraged to cross-reference the component designations, shown in parentheses, within both the piping and electrical schematics.

Description of Operation:

1. *Wood-gasification boiler operation:* When the wood-gasification boiler is fired, the operator closes a switch that supplies 120 VAC to the boiler's internal electrical components. This switch closure also supplied 120 VAC to circulators (P1) in the loading unit. As the temperature at the boiler outlet climbs, the 3-way thermostat valve within the loading unit modulates to allow heated water to flow to storage while also maintaining the boiler's inlet temperature above 130 °F (or whatever other temperature may be required) to prevent sustained flue gas condensation within the boiler. If a power failure occurs, residual heat from the boiler transfers into the thermal storage by thermosyphoning.

2. *Solar subsystem operating mode:* A differential temperature controller (DTC) measures the collector absorber plate temperature at a sensor (S1) and compares it to the temperature at sensor (S2) in the lower portion of the storage tank. When the temperature at (S1) is at least 12 °F above the temperature at (S2), the collector circulators (P2a and P2b) are both turned on. Both circulators continue to operate for 3 minutes to establish a siphon in the collector return piping. After that, circulator (P2b) is turned off by the (DTC) to reduce power demand. Flow continues in the collector circuit until the temperature at sensor (S1) drops to within 5 °F of sensor (S2). At that point, circulator (P2a) is turned off and water drains from the collector array back to the drainback/expansion tank.

3. *Space heating mode:* Upon a call from any of the thermostats (T1...T6), 24 VAC is passed to the associated manifold valve actuators (VA1...VA6). When any one or more of these valve actuators reach their fully open position, an end switch within the valve actuator closes to pass 24 VAC to the coil of relay (R1) and motorized mixing valve (MV1). Normally open relay contact (R1-1) closes to supply 120 VAC to circulator (P4), which operates at variable speeds to maintain constant differential pressure across the distribution manifold. Mixing valve (MV1) is also turned on when any thermostat calls for heat. It measures the outdoor temperature at sensor (S4) and uses this temperature, along with its settings, to calculate the necessary target supply water temperature to the distribution system. It compares the target supply temperature to the actual supply temperature measured by sensor (S3) and adjusts the hot water and return water flow proportions into the valve to maintain the temperature at sensor (S3) as close to the target temperature as possible.

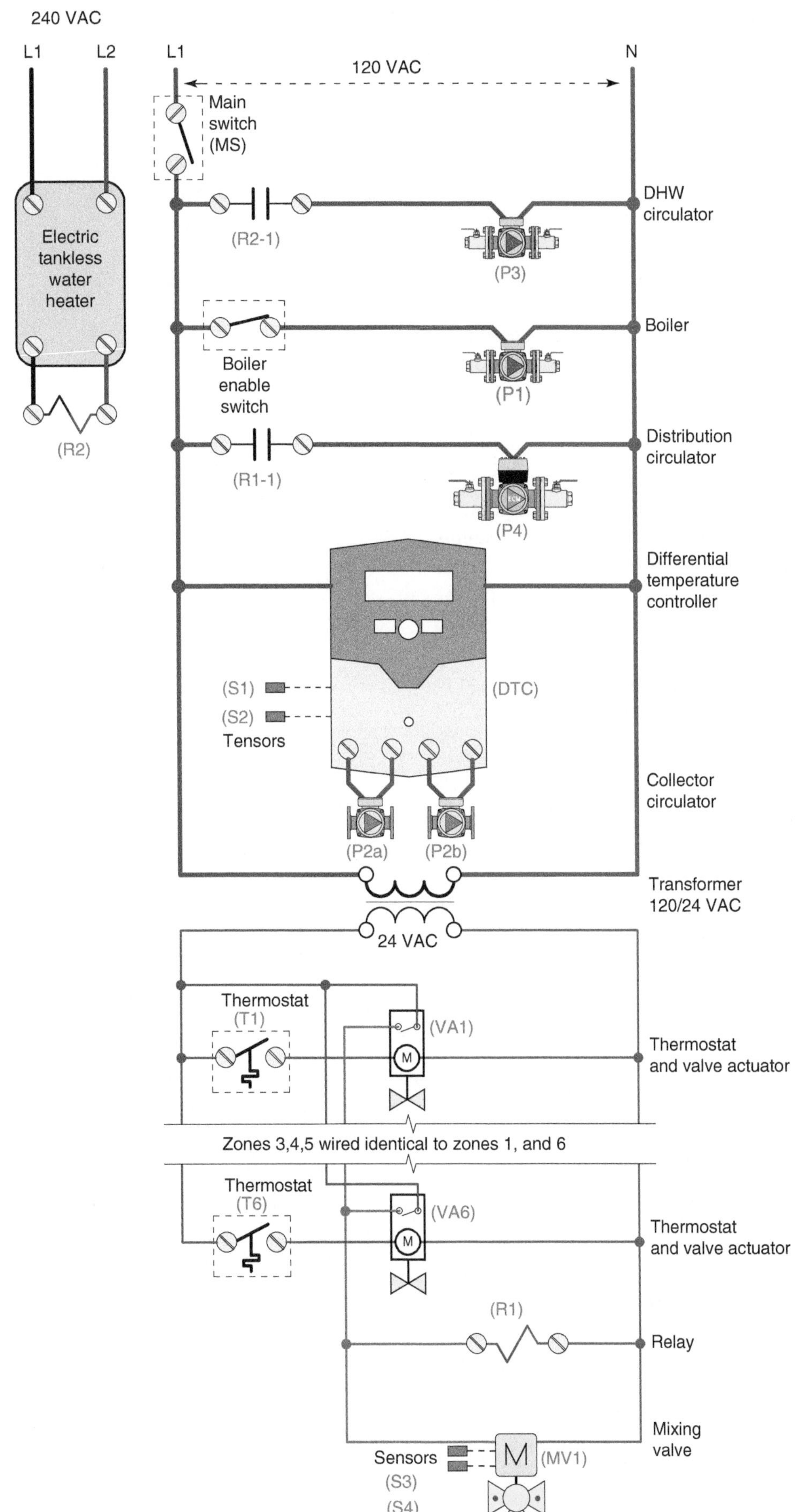

Figure 15-90b | Electrical wiring for combisystem #4.

4. *Domestic water heating mode:* Whenever there is a demand for domestic hot water of 0.6 gpm or higher, a flow switch within the electric tankless water heater (ETWH) closes. This closure supplies 240 VAC to the coil of relay (R2). The normally open contacts (R2-1) close to turn on circulator (P3), which circulates heated water from the upper portion of the thermal storage tank through the primary side of the domestic water heat exchanger (HX1). The domestic water leaving (HX1) is either preheated or fully heated depending on the temperature in the upper portion of the thermal storage tank. This water passes into the thermostatically controlled tankless water heater (ETWH), which measures its inlet temperature. The electronics within this heater control the electrical power supplied to the heating elements based on the necessary temperature rise (if any) to achieve the set domestic hot water supply temperature. If the water entering the tankless heater is already at or above the setpoint temperature, the elements are not turned on. All heated water leaving the tankless heater flows into an ASSE 1017–rated mixing valve to ensure a safe delivery temperature to the fixtures. Whenever the demand for domestic hot water drops below 0.4 gpm, circulator (P3) and the tankless electric water heater are turned off.

WOOD-FIRED COMBISYSTEM #5

As previously discussed, some modern pellet-fired boilers can regulate their heat output by adjusting the rate at which pellets are supplied to their combustion chamber. In systems with minimal zoning, one might assume that the boiler's ability to modulate its heat output, in response to the load, allows it to function without a thermal storage tank. While this may be true in certain circumstances, the author, as well as many others with operating experience with pellet-fired boilers, recommends the use of a thermal storage tank, especially in systems with low-mass heat emitters and extensive zoning. Most modulating pellet-fired boilers cannot adjust their output low enough to match the load of a very small zone, especially when operating under partial load conditions. Furthermore, once started, it is desirable to keep a pellet-fired boiler operating for the longest practical time without overheating the building or otherwise wasting heat. This principle was discussed in Section 15.8. The use of thermal storage in the range of 1 to 2 gallons of water per 1,000 Btu/hr. of rated boiler capacity will help with both of these issues.

The system shown in Figure 15-91a includes a pellet-fired boiler and propane-fired modulating/condensing auxiliary boiler. The pellet-fired boiler will operate as the first stage of heat input. It is sized for 60 percent of the design heating load. As such, it should provide approximately 85 percent of the seasonal space heating energy requirement. The auxiliary boiler is sized for 50 percent of design load. It will supplement the pellet-fired boiler's output when needed to meet peak or near peak loads.

In this system, the boilers are operated by the combination of a 2-stage temperature setpoint controller and a 1-stage temperature setpoint controller. When the master thermostat (T1) creates a space heating demand, the 2-stage controller turns on the pellet-fired boiler if the temperature at the upper tank sensor (S1) is 120 °F or less. If heat output from the pellet-fired boiler matches the heating load, the temperature at sensor (S1) will remain stable. If the boiler's heat output exceeds the load, the temperature at sensor (S1) will increase. If the heat output from the pellet-fired boiler is less than required by the load, the temperature at sensor (S1) will decrease.

With the pellet-fired boiler operating, the auxiliary boiler will be turned on by the 1-stage temperature controller if the temperature at the upper tank sensor (S3) drops to 115 °F or less. Both boilers will continue operating until the temperature at sensor (S3) climbs to 130 °F, at which point the auxiliary boiler is turned off. This allows the auxiliary boiler to be buffered by the upper portion of the thermal storage tank but does not allow the auxiliary boiler to heat the entire tank. The location at which the auxiliary boiler inlet piping connects to the thermal storage tank limits the portion of the tank's volume that interacts with that boiler.

In this system, the pellet-fired boiler keeps running until the temperature at the *lower* tank sensor (S2) reaches 170 °F. The controls action remains in effect even if the space heating demand from the master thermostat stops. This "stacks" the tank with hot water, and lengthens the operating cycle of the pellet-fired boiler. Longer operating cycles improve combustion efficiency and reduce emissions.

Figure 15-91b shows the control temperature ranges along with the on/off setpoints of the boilers.

The pellet-fired boiler is assumed to contain internal control logic that operates a 3-way motorized mixing valve (MV2) to maintain the inlet water temperature high enough to prevent sustained flue gas condensation. If the boiler did not have this control capability, it could be provided by an external

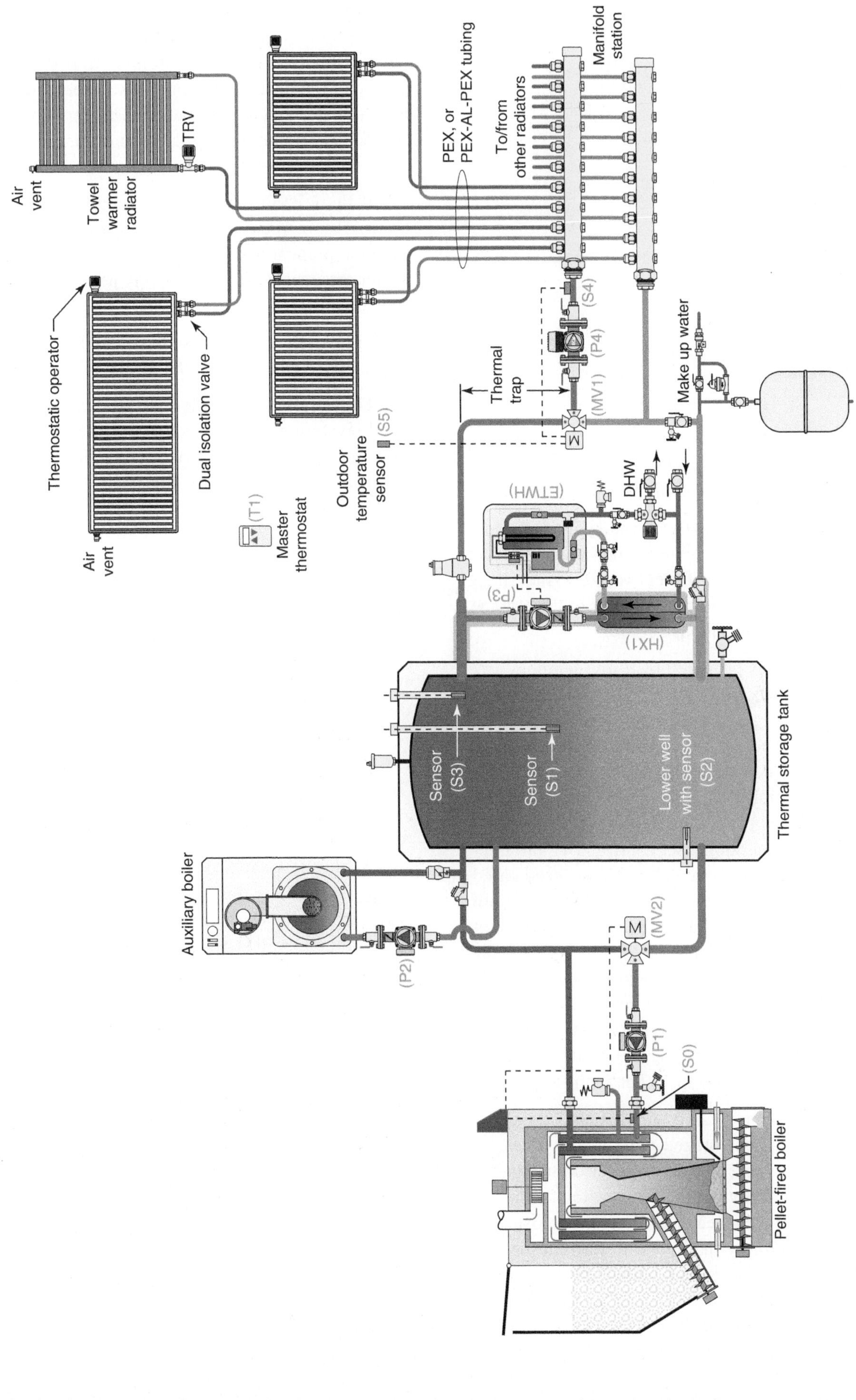

Figure 15-91a | Piping schematic for combisystem #5.

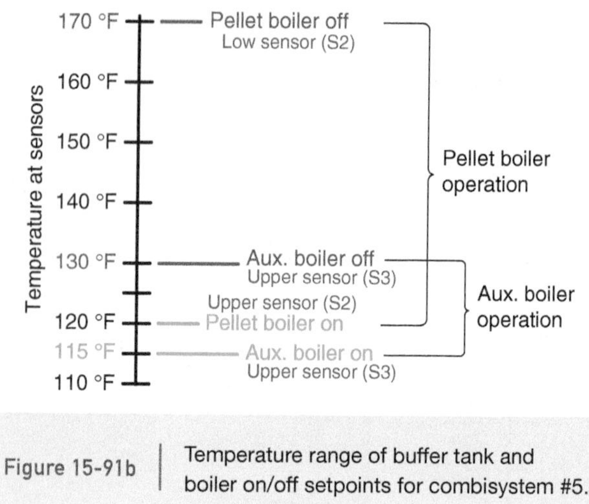

Figure 15-91b | Temperature range of buffer tank and boiler on/off setpoints for combisystem #5.

controller or another form of boiler protection, such as a thermostatic mixing valve could be used.

If a power failure occurs, thermosyphoning flow will develop between pellet-fired boiler and main storage tank. This flow is sufficient to move the small amount of residual heat from the boiler to the storage tank. No external heat dump is needed.

Space heating is provided by a low-temperature, highly zoned distribution system. The panel radiators used are each configured with a wireless thermostatic valve to provide room-by-room temperature control. They are sized to deliver design load output at a supply temperature of 120°F. A motorized 3-way mixing valve adjusts the supply water temperature to the panel radiator based on outdoor reset control. The variable-speed pressure-regulated circulator adjusts speed based on how the wireless thermostatic radiator valves are opening, closing, or modulating. A master thermostat is used to provide a call for space heating. This energizes the distribution circulator and mixing valve. It also enables boiler operation. The downward direction of the piping (e.g., the thermal trap) upstream of the mixing valve is to stop heat migration into the space heating subsystem when no space heating is required.

Domestic water is heated "on demand" using an external stainless steel brazed plate heat exchanger (HX1). Whenever there is a demand for domestic hot water of 0.6 gpm or higher, the flow switch inside the tankless electric water heater closes. This closure is used, in combination with a relay, to turn on the circulator that routes water from the upper portion of the thermal storage tank through the primary side of this heat exchanger. Closure of the flow switch also energizes an electrical contactor within the tankless heater to supply 240 VAC to triacs, which regulate the electrical power supplied to the heating elements. The electrical power delivered is limited to that required to provided the desired outgoing water temperature. All heated water leaving the tankless heater flows into an ASSE 1017–rated mixing valve to ensure a safe delivery temperature to the fixtures.

If there is no demand for space heating, and the thermal storage tank has cooled to room temperature, minimal preheating is provided and the tankless water heater provides most of the energy needed for domestic water heating.

Figure 15-91c shows the electrical wiring for pellet-fueled combisystem #5.

The following is a description of operation for combisystem #5. Readers are encouraged to cross-reference the component designations, shown in parentheses, within both the piping and electrical schematics.

Description of Operation:

1. *Pellet-fire boiler operation:* The master thermostat (T1) creates a demand for space heating. This applies 24 VAC power to the 2-stage setpoint controller (SP2) and the 1-stage temperature setpoint controller (SP1). The 2-stage controller measures the temperature of sensor (S1) in the storage tank. If the temperature is less than 120 °F, the first stage contact in (SP2) closes. This passes 24 VAC to the coil of relay (R3). One normally open contact in this relay (R3-2) closes to initiate operation of the pellet-fired boiler. Another normally open contact (R3-1) closes to pass 24 VAC to the stage 2 contact in the 2-stage setpoint controller (SP2). A third normally open contact (R3-3) closes in parallel with contact (R1-1). 24 VAC is now simultaneously passing through both stage 1 and stage 2 contacts in (SP2). The stage 1 contact in (SP2) will open when the temperature at the upper sensor (S1) reaches 130 °F. However, 24 VAC can still pass through contact (R3-1) to maintain relay coil (R3) in the on state. This keeps the pellet-fired boiler operating and allows the storage tank to be fully "stacked" with hot water. The pellet-fired boiler turns off when the temperature at the lower tank sensor (S2) reaches 170 °F. If the heating demand from master thermostat (T1) ceases before the temperature at sensor (S2) has reached 170 °F, the coil of relay (R1) is turned off, as is the mixing valve (MV1) and the distribution circulator (P4). However, contact (R3-3) maintains 24VAC power to setpoint controller (SP2) allowing the temperature stacking process in the thermal storage tank to continue until sensor (S2) reaches 170 °F.

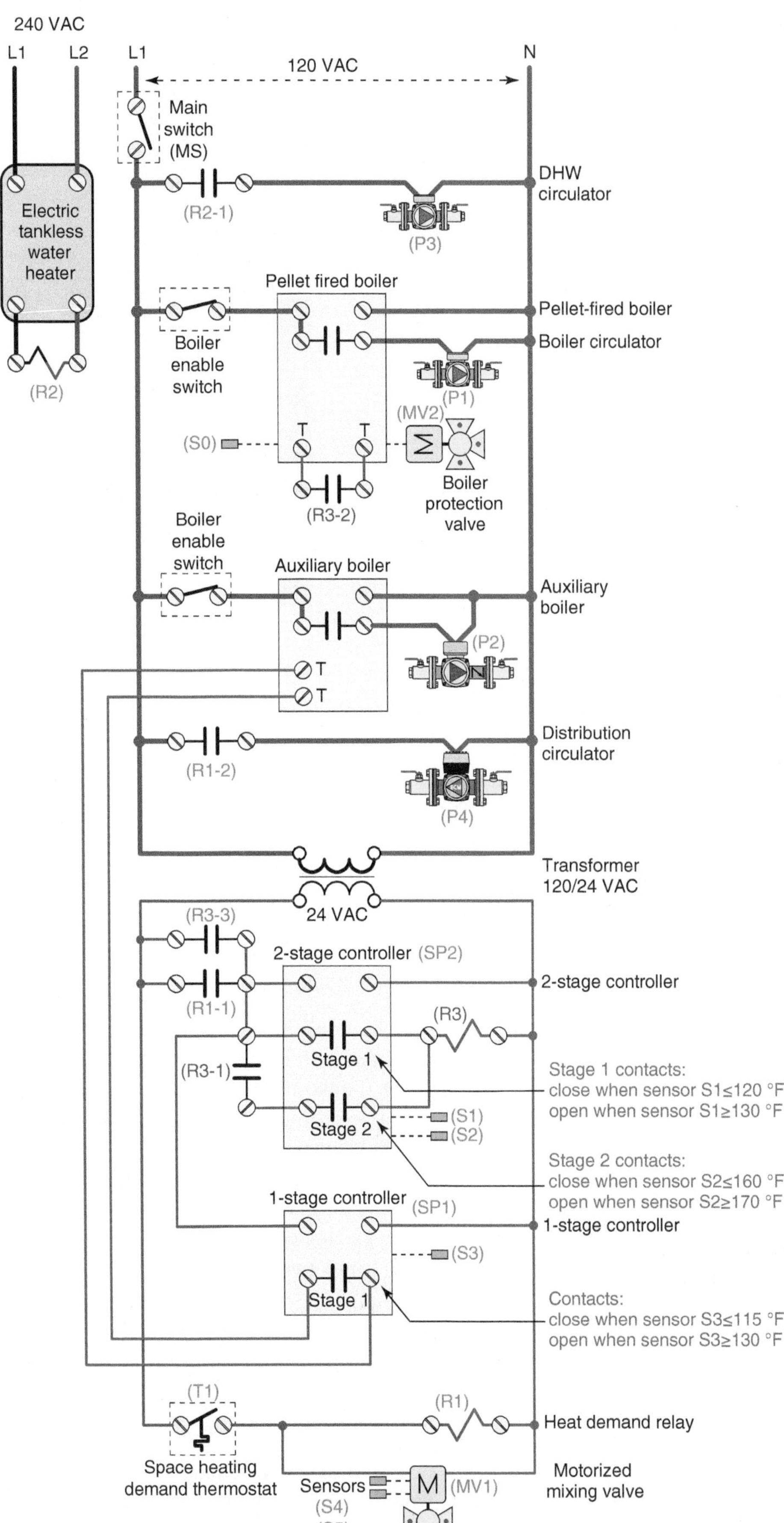

Figure 15-91c | Electrical wiring for combisystem #5.

When started, the pellet-fired boiler applies 120 VAC to circulator (P1). It also measures its entering water temperature at sensor (S0) and operates a 3-way motorized mixing valve (MV2) as necessary to quickly raise the boiler temperature to prevent sustained flue gas condensation. As the temperature at sensor (S0) climbs above the nominal 130 ºF protection limit, the motorized mixing valve (MV2) allows flow from the boiler to the tank, and vice versa. If the temperature at sensor (S0) decreases toward the protection limit, mixing valve (MV2) responds by limiting flow between the pellet-fired boiler and thermal storage tank.

2. *Auxiliary boiler operation:* If there is a demand for space heating, and the temperature at sensor (S3) in the upper portion of the storage tank drops to or below 115 ºF, the normally open contact in the 1-stage setpoint controller (SP1) closes. This enables the auxiliary boiler to fire in setpoint mode (i.e., not based on its own internal outdoor reset control mode). The auxiliary boiler supplies 120 VAC to circulator (P2) and transfers heat to the storage tank to supplement the heat output of the pellet-fired boiler. The auxiliary boiler continues to operate until the upper tank sensor (S3) reaches 130 ºF, or the heating demand ceases.

If a power failure occurs, residual heat from the pellet-fired boiler is conveyed into the thermal storage tank by thermosyphoning.

3. *Space heating mode:* Space heating is provided by panel radiators, each equipped with a wireless thermostatic

radiator valve. Space heating is enabled by a master thermostat (T1). When the contacts in the master thermostat close, 24 VAC is passed to the coil of relay (R1). A normally open relay contact (R1-1) closes to supply 120 VAC to circulator (P4). This circulator operates at variable speeds to maintain a constant differential pressure across the distribution system manifold as the thermostatic radiator valves open, close, or modulate.

The motorized mixing valve (MV1) is also turned on by the master thermostat (T1). It measures the outdoor temperature at sensor (S5) and uses this temperature, along with its settings, to calculate the necessary target supply water temperature to the distribution system. It compares the target supply temperature to the actual supply temperature measured by sensor (S4) and adjusts the hot water and return water flow rates into the valve to maintain the temperature at sensor (S4) as close to the target temperature as possible.

4. *Domestic water heating mode:* Whenever there is a demand for domestic hot water of 0.6 gpm or higher, a flow switch within the electric tankless water heater closes. This closure supplies 240 VAC to the coil of relay (R2). The normally open contacts (R2-1) close to turn on circulator (P3), which circulates heated water from the upper portion of the thermal storage tank through the primary side of the domestic water heat exchanger (HX1). If the domestic water leaving (HX1) is fully heated, it flows through the tankless heater, but the heating elements remain off. The heated domestic water flows on to an ASSE 1017–rated mixing valve, which ensures a safe delivery temperature to the fixtures. Whenever the demand for domestic hot water drops below 0.4 gpm, circulator (P3) and the tankless electric water heater are turned off.

During warm weather, when there is no demand for space heating, the thermal storage tank temperature will drop to room temperature. Because this temperature may still be 10 °F to 20 °F above entering cold water temperature, the heat exchanger (HX1) can still provide a slight preheating effect. The electric tankless water heater provides the balance of the temperature rise and thus assume the majority of the domestic water heating load.

WOOD-FIRED COMBISYSTEM #6

Even small, low-energy-use houses can make use of wood-fueled hydronic heating. However, most wood-gasification boilers have heat outputs several times higher than the design heating loads of such houses. This would require a large storage tank. Smaller homes often cannot accommodate such a large tank, or provide the necessary space for the wood-gasification boiler. Furthermore, the economics of installing such a system would not provide a reasonable return on investment. However, a hydronic wood stove, such as shown in Figure 15-92, and discussed at the beginning of this chapter, provides a well-matched solution to wood heating in a low-energy house.

Figure 15-92 | A small hydronic wood stove.

European hydronic wood stoves typically release 30 to 40 percent of their heat to the space in which they are located. The remaining heat output is captured by a hydronic heat exchanger and carried away from the stove. This heat is available for storage or distribution to other parts of the building. Some hydronic wood stoves can be reconfigured with insulating panels and double or triple glazing on their firebox doors to further increase the hydronic portion of the total heat output.

Figure 15-93a show a piping schematic that integrates a hydronic wood stove along with a modest size thermal storage tank and propane-fired auxiliary boiler. Heat from either heat source can be used to supply space heating or domestic water heating.

This system might be installed in homes where the hydronic wood stove would be located in living space on the first floor, while the thermal storage tank, due to

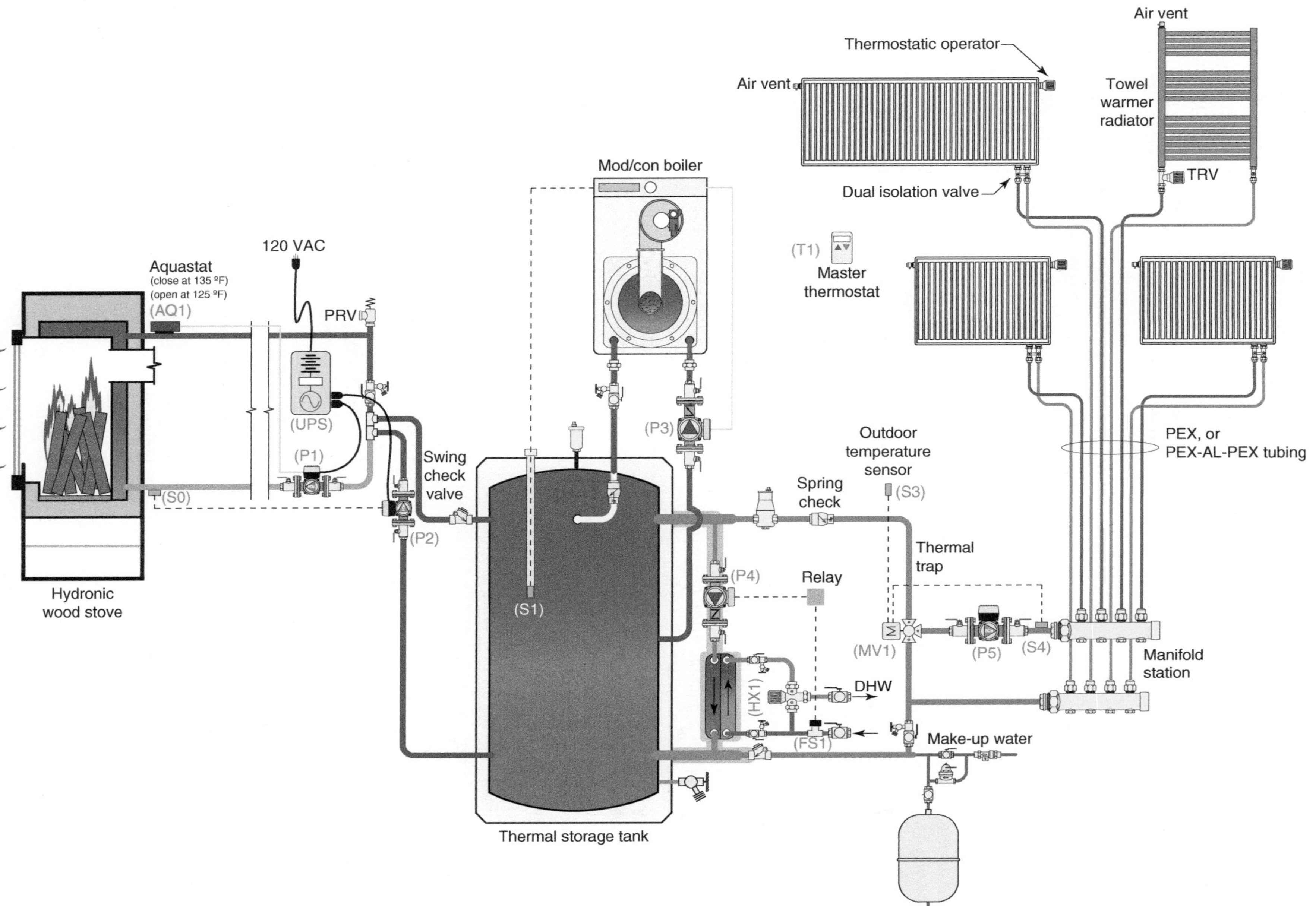

Figure 15-93a | Piping schematic for combisystem #6.

its weight and size, would be located in the basement. Such arrangements do not allow for thermosyphon flow from the hydronic wood stove to thermal storage tank during a power outage. This has been addressed by using two low-power ECM-based circulators, in combination with an uninterruptible power supply (UPS), to transfer heat from the hydronic wood stove to thermal storage. If a power outage occurs while the hydronic wood stove is operating, these circulators remain powered by the (UPS). The (UPS) is sized to allow circulators (P1) and (P2) to operate long enough to transfer the *residual heat* from a well established fire in the wood stove, into the thermal storage tank. *The (UPS) is not intended to allow the wood stove to be refueled during the power outage so that it can continue operating.* Doing so would be of little value, since the remaining AC-powered circulators in the system would not be able to operate. However, if the house was equipped with a standby generator, the entire system could remain in operation during the power outage.

Circulator (P1) is a low power ECM-based circulator that is turned on when the temperature at the outlet of the hydronic wood stove reaches 150 °F. (P1) operates at a fixed speed that provides a nominal 10 °F temperature drop across the heat source loop when there is a well established fire in the wood stove, and circulator (P2) is operating at full speed. Circulator (P2) is a variable-speed setpoint circulator that monitors the temperature of sensor (S0) near the inlet of the wood stove. Its purpose is to protect the hydronic wood stove from inlet water temperatures low enough to cause sustained flue gas condensation. Circulator (P2) begins to operate at a low speed when the temperature at sensor (S0) reaches 130 °F. If the temperature at sensor (S0) continues rise, the speed of (P2) increases. When the temperature at sensor (S0) reaches 135 °F, circulator (P2) is operating at full speed.

The auxiliary boiler continually monitors temperature sensor (S1) in the upper portion of the thermal storage tank. The boiler's internal controller remains enabled in a temperature setpoint mode. As such, the boiler operates when necessary to maintain a setpoint temperature at sensor (S1). The water in the upper portion of the storage tank provides thermal mass to buffer the boiler against short cycling. If the heat exchanger (HX1) were sized for a maximum approach temperature difference of 10 °F, a temperature setpoint of 125 °F at sensor (S1) would enable the domestic water heating subsystem to provide hot domestic water, at the design DHW flow rate and 115 °F whenever needed.

Space heating is provided by a low-temperature, highly zoned distribution system. The panel radiators shown are each configured with a wireless thermostatic valve to provide room-by-room temperature control. They are sized to deliver design load output at a supply temperature not higher than 120 °F. A motorized 3-way mixing valve adjusts the supply water temperature to the heat emitters based on outdoor reset control. The variable-speed pressure-regulated circulator adjusts speed based on how the wireless thermostatic radiator valves are opening, closing, or modulating. A master thermostat is used to provide a call for space heating, which energizes the circulator and mixing valve. The spring-loaded check valve, and downward direction of the piping (e.g., the thermal trap) upstream of the mixing valve stops heat migration into the space heating subsystem when it is not needed.

Domestic water is heated "on demand" using an external stainless steel brazed-plate heat exchanger (HX1). Whenever there is a demand for domestic hot water of 0.6 gpm or higher, the flow switch (FS1) closes. This closure energizes relay (R2), which turns on circulator (P4) to provide flow from the upper portion of the thermal storage tank, through the primary side of heat exchanger (HX1). Domestic cold water flows through the other side of (HX1) and is fully heated. All heated domestic water flows into an ASSE 1017–rated mixing valve to ensure a safe delivery temperature to the fixtures.

Figure 15-93b shows the electrical wiring for combisystem #6.

The following is a description of operation of the system shown in Figures 15-93a and 15-93b. Readers are encouraged to cross-reference the component designations, shown in parentheses, within both the piping and electrical schematics.

Description of Operation:

1. *Heat from hydronic wood stove:* The hydronic wood stove can be fired at any time. When aquastat (AQ1) detects that the temperature at the outlet port has reached 150 °F, its contacts close to energize circulators (P1) and (P2). Circulator (P1) operates at fixed speed. Circulator (P2) is a variable-speed setpoint circulator, which monitors the boiler's inlet water temperature at sensor (S0). When this inlet temperature is below a setpoint that eliminates sustained flue gas condensation (typically about 130 °F), circulator (P2) is off or operating at a very low speed. As the temperature at sensor (S0) climbs above the setpoint, the speed of circulator (P2) increases. This increases the rate of heat transfer to the thermal storage tank. If the temperature at sensor (S0) climbs to, or above, setpoint + 5 °F,

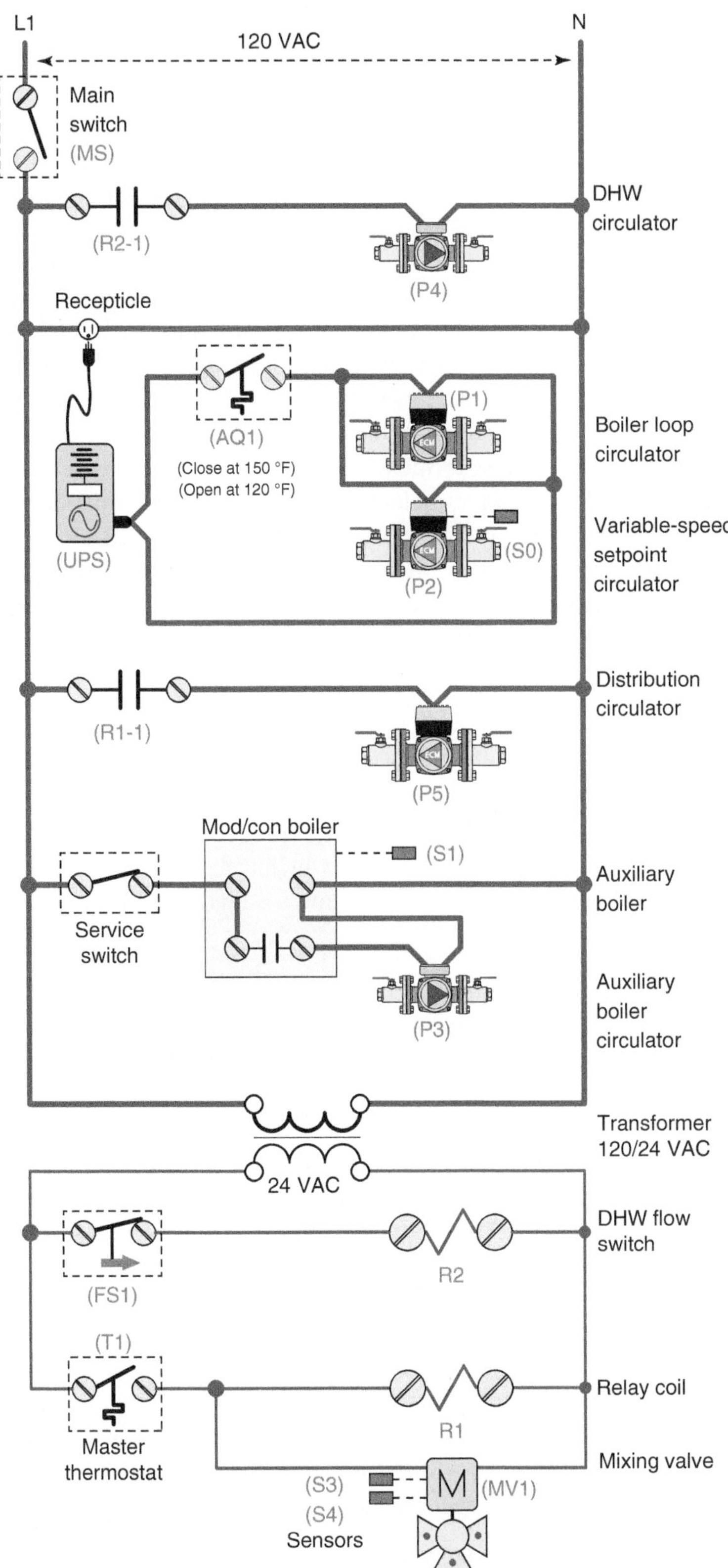

Figure 15-93b | Electrical wiring for wood-fired combisystem #6.

circulator (P2) operates at full speed. If the temperature at sensor (S0) decreases toward 130 °F, circulator (P2) slows down to reduce the rate of heat transfer from the wood stove to storage.

If a power failure occurs, the uninterruptible power supply (UPS) maintains the operation of circulators (P1) and (P2), as described above. The UPS is sized to allow these circulators to operate until the fire in the wood stove has dissipated to a level where heat output is low enough not to be a safety concern. The UPS should be sized so that the minimum run time for circulators (P1) and (P2) during a power outage is 2 hours.

2. *Auxiliary boiler operation:* Auxiliary heat is provided by a propane-fired modulating/condensing boiler. The controls within the boiler are set so that it maintains the temperature at sensor (S1) between 125 °F and 115 °F at all times. This temperature is sufficient for the space heating system as well as for providing domestic hot water. The water in the upper portion of the storage tank buffers the auxiliary boiler against short cycling.

3. *Space heating mode:* Space heating is provided by panel radiators, each equipped with a wireless thermostatic radiator valve. Space heating is enabled by a master thermostat (T1). When the contacts in the master thermostat close, 24 VAC is passed to the coil of relay (R1). A normally open relay contact (R1-1) closes to supply 120 VAC to circulator (P5). This circulator operates at variable speeds to maintain a constant differential pressure across the distribution system manifold as the thermostatic radiator valves open, close, or modulate. The motorized mixing valve (MV1) is also turned on by the master thermostat (T1). It measures the outdoor temperature at sensor (S3), and uses this temperature, along with its settings, to calculate the necessary target supply water temperature of the distribution system. It compares the target supply temperature to the actual supply temperature measured by sensor (S4) and adjusts the hot water and return water flow rates into the valve to maintain the temperature at sensor (S4) as close to the target temperature as possible.

4. *Domestic water heating mode:* Whenever there is a demand for domestic hot water of 0.6 gpm or higher, flow switch (FS1) closes. This provides 24 VAC to the coil of relay (R2). A normally open contact (R2-1) closes to supply 120 VAC to circulator (P4), which moves hot water from the upper portion of the storage tank through the primary side of heat exchanger (HX1).

The domestic water leaving (HX1) is fully heated. This water flows into an ASSE 1017–rated mixing valve to ensure a safe delivery temperature to the fixtures. Whenever the demand for domestic hot water drops below 0.4 gpm, circulator (P4) is turned off. The heat exchanger (HX1) has been sized for a 5 °F approach temperature during maximum domestic water demand.

WOOD-FIRED COMBISYSTEM #7

The system shown in figure 15-94a uses a wood-gasification boiler in combination with a pressurized thermal storage tank, and an auxiliary boiler for full supplemental space heating. It also provides a piping configuration where the distribution system and auxiliary boiler are uniquely coupled to the wood-gasification boiler and thermal storage by a variable speed injection circulator.

Boiler protection is provided by a variable speed setpoint circulator (P2), which increases speed when the boiler inlet water temperature rises above 130 °F. This moves heated water from the boiler to the upper left header at the storage tank. If there is a space heating load, some of this flow will be pulled upward through circulator (P3). The remainder will pass into the thermal storage tank. If there is no space heating load, all flow from the wood-gasification boiler will pass into the thermal storage tank.

When there is a demand for space heating from any of the zone thermostats the variable speed pressure regulated distribution circulator (P4) is turned on. The injection mixing controller (TEK356) and differential temperature controller (TEK156) are also turned on. The differential temperature controller (TEK156) measures the temperature at sensor (S3) on the upper header of the thermal storage tank and compares it to the temperature returning from the distribution system at sensor (S4). If the temperature at (S3) is at least 5 °F above the temperature at sensor (S4), the thermal storage tank can contribute to the space heating load. The (TEK156) then enables operation of the (TEK356) controller, which measures the outdoor temperature and calculates the target supply water temperature required at sensor (S1). If the water temperature at sensor (S1) is below the target supply water temperature, the (TEK356) controller increases the speed of the circulator (P3) to inject heated water from the upper header of the storage tank into the distribution system. If the temperature at (S1) is above the target supply water temperature, the (TEK356) reduces the speed of injection circulator (P3) to reduce heat input to the distribution system. If the (TEK356) controller is unable to establish the target supply water temperature at sensor (S1), it closes a set of dry relay contacts that enable operation of the auxiliary boiler and circulator (P5). Heat produced by the auxiliary boiler is sent to the 25-gallon buffer tank. Modulation of the auxiliary boiler is controlled based on the supply water temperature sensor (S7) and the boiler's outdoor temperature sensor (S8). The auxiliary boiler can operate simultaneously with the injection circulator (P3), provided that the temperature at sensor (S3) remains at least 3 °F above the temperature at sensor (S4). If the temperature at sensor (S3) drops to less than 3 °F above the temperature at sensor (S4), the (TEK156) controller interrupts line voltage to the (TEK356) controller, which turns off injection circulator (P3). This control action allows the storage tank to contribute heat to the distribution system to the lowest practical temperature. Below this temperature, all heat supplied to the distribution system comes from the auxiliary boiler. This also prevents heat generated by the auxiliary boiler from being inadvertently conveyed into the thermal energy storage tank.

If a power failure occurs, the normally open zone valve (ZV1) opens. This allows thermosyphon flow to develop between the thermal storage tank and the wood-gasification boiler, which can dissipate any residual heat from this boiler into the thermal storage tank.

Domestic water is heated on demand using an external stainless steel brazed-plate heat exchanger (HX1). Whenever there is a demand for domestic hot water of 0.6 gpm or higher, the flow switch inside the electric tankless water heater closes. This closure is used, in combination with a relay, to turn on circulator (P6), which routes water from the upper portion of the thermal storage tank through the primary side of heat exchanger (HX1). Closure of the flow switch also energizes an electrical contactor within the tankless heater closes to supply 240 VAC to triacs, which regulate the power supplied to the heating elements. The electrical power delivered to the elements is limited to that required to provided the desired outgoing water temperature. All heated water leaving the tankless heater flows into an ASSE 1017–rated mixing valve to ensure a safe delivery temperature to the fixtures.

The electrical wiring diagram for wood-fired combisystem #7 is shown in Figure 15-94b. This drawing is a slight variation from a typical ladder diagram because it shows two specific embedded controllers (TEK156) and (TEK356), both of which require line voltage and low voltage wiring. Thus, there are conductors between the upper and lower portions of the ladder diagram.

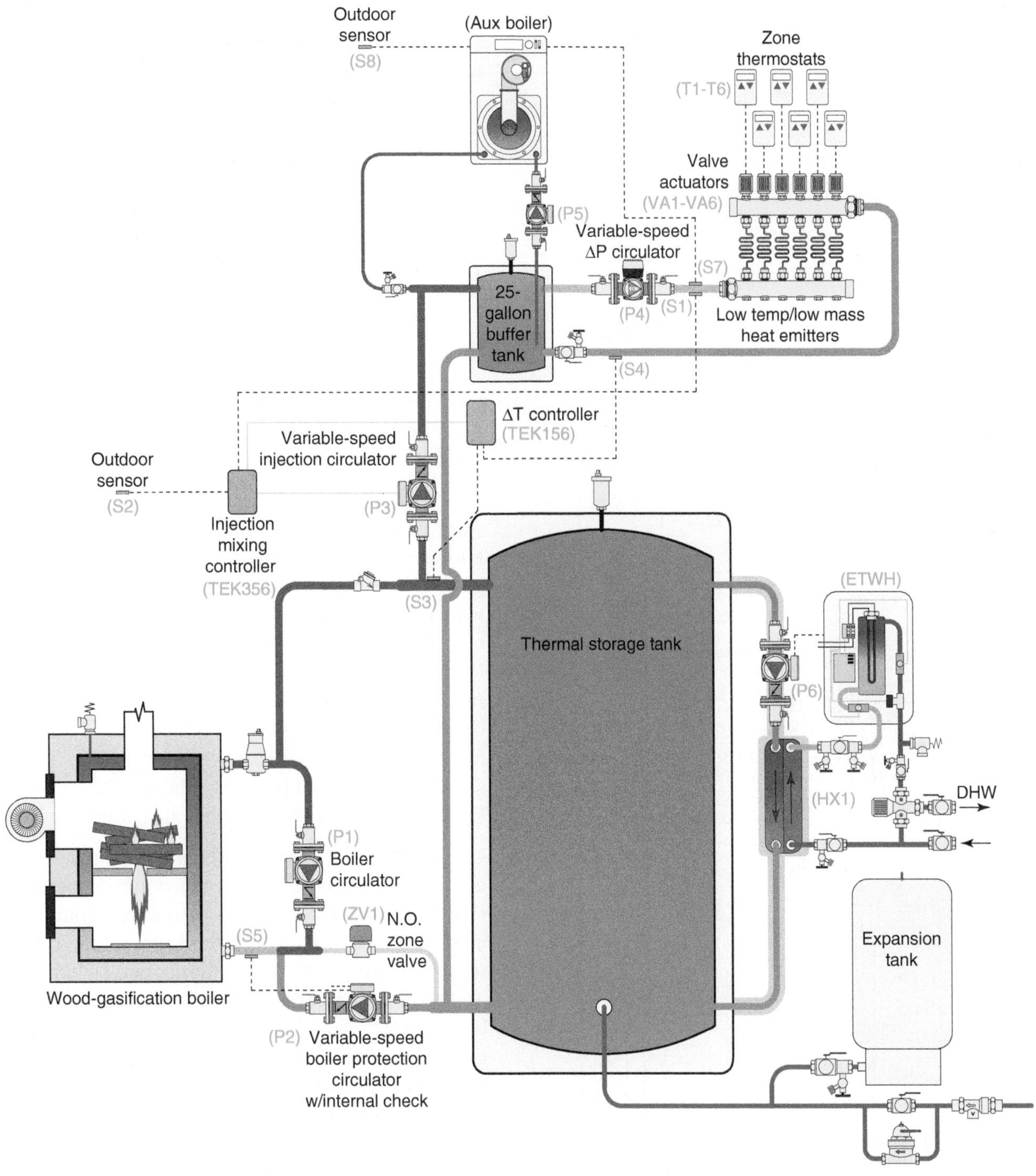

Figure 15-94a | Piping schematic for combisystem #7.

The colors and widths of the lines representing these conductors help make them more recognizeable.

The following is a description of operation of the system shown in Figures 15-94a, and 15-94b. Readers are encouraged to cross-reference the component designations, shown in parentheses, within both the piping and electrical schematics.

Description of Operation:

1. *Wood-gasification boiler operation:* A fire is kindled in the boiler, and after building for a few minutes, the primary combustion chamber is loaded full of wood and the boiler's switch is turned on. This turns on the boiler's combustion blower and circulator (P1) and initiates gasification-mode operation. It also

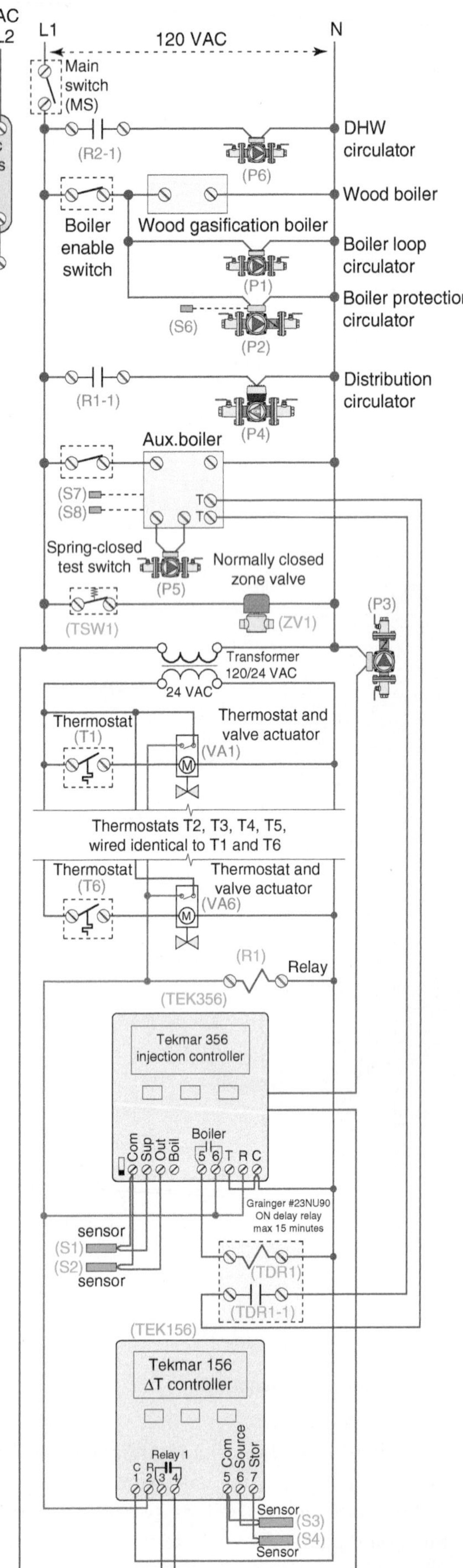

Figure 15-94b | Electrical wiring for combisystem #7.

supplies line voltage to circulator (P2). The electronics within circulator (P2) monitor the temperature of the boiler inlet at sensor (S6). When the temperature at sensor (S6) rises to 130 °F, circulator (P2) increases speed. If the temperature at sensor (S6) drops below 130 °F, the speed of circulator (P2) is reduced to a minimum. The speed of circulator (P2) will continually adjust during the burn cycle to keep the boiler inlet temperature at 130 °F or higher and thus prevent sustained flue gas condensation within the boiler.

Flow created by circulator (P2) passes through the boiler and on to the upper header of the thermal storage tank. If the space heating load is not active, this flow passes into the thermal storage tank. If the space heating load is active, some or all the flow can be drawn through injection circulator (P3).

2. *Overheat protection:* The normally open zone valve (ZV1) opens whenever utility power is lost. This allows an uninhibited flow path between the wood-gasification boiler and thermal storage tank through which thermosyphon flow can develop to move residual heat from the boiler to the storage tank. A spring-loaded, normally closed test switch (TSW1) is wired in series with the zone valve (ZV1). This switch can be pressed to open its contacts, and break line voltage to (ZV1) to verify that it opens correctly. This test should be performed twice each heating season. Switch (TSW1) closes its contacts when its button is released.

3. *Heat transfer to distribution system:* Upon a demand for space heating from any of the thermostats (T1-T6), 24VAC is passed to the associated valve actuator (VA1-VA6). When the valve actuator reaches its fully open position, an internal end switch closes. The end switch passes 24VAC to relay (R1). A normally open contact (R1-1) closes to turn on the distribution circulator (P4), which is set to operate in a constant differential pressure control mode. 24VAC is also sent to the differential temperature controller (TEK156) and injection mixing controller (TEK356). The (TEK156) controller measures the temperature difference between the water returning from the distribution system at sensor (S4) and the storage tank header at sensor (S3). When the temperature at (S3) is at least 5 °F higher than at (S4), a contact in (TEK156) closes to pass line voltage to the injection mixing controller (TEK356). The injection mixing controller (TEK356) measures the outdoor temperature at sensor (S2) and uses this value, along with it settings, to calculate the target supply water temperature to the distribution system. It compares this target temperature to the temperature measured at sensor (S1). If the temperature at (S1) is lower than the target

supply temperature, the (TEK356) controller increases the speed of injection circulator (P3) to inject heated water into the distribution system in an attempt to bring the temperature at (S1) up to the target temperature. If the temperature at (S1) is higher than the target supply temperature, the (TEK356) controller reduces the speed of injection circulator (P3) to bring the temperature at (S1) down to the target temperature. This process continues unless or until the temperature at sensor (S3) become less than 3 °F higher than at sensor (S4). This is typically the result of the fire dwindling in the wood-gasification boiler or the depletion of useful heat in the storage tanks. When this condition is detected, the contact in (TEK156) opens to break line voltage to the (TEK356) controller and thus turns off injection circulator (P3). This prevents heat generated by the auxiliary boiler from being inadvertently conveyed into the thermal energy storage tank.

4. *Auxiliary boiler operation:* The auxiliary boiler is enabled to operate by a relay contact in the (TEK356) controller. Closure of this contact passes 24 VAC to the coil of time delay relay (TDR1), which then begins its timing cycle, which can be set for up to 15 minutes. This adjustable time delay allows sufficient time to warm the 25-gallon buffer tank and establish quasi-steady operation of the distribution system before committing to operation of the auxiliary boiler. When the time delay of relay (TDR1) has elapsed, it closes its normally open contact (TDR1-1), which completes a 24VAC circuit that allow the auxiliary boiler to fire. When turned on, the auxiliary boiler measures the supply water temperature at sensor (S7), and the outdoor temperature at sensor (S8), and operates based on its internal reset controller to keep this temperature close to the calculated target supply temperature.

5. *Domestic water heating mode:* Whenever there is a demand for domestic hot water of 0.6 gpm or higher, a flow switch within the electric tankless water heater (ETWH) closes. This closure supplies 240 VAC to the coil of relay (R2). The normally open contacts (R2-1) close to turn on circulator (P6), which circulates heated water from the upper portion of the thermal storage tank through the primary side of the domestic water heat exchanger (HX1). Closure of the flow switch also energizes an electrical contactor within the electric tankless water heater to supply 240 VAC to triacs, which regulate the power supplied to the heating elements. The electrical power delivered to the elements is limited to that required to provided the desired outgoing water temperature. If the domestic water leaving (HX1) is fully heated, it flows through the tankless heater, but the heating elements remain off. The heated domestic water flows on to an ASSE 1017–rated mixing valve, which ensures a safe delivery temperature to the fixtures. Whenever the demand for domestic hot water drops below 0.4 gpm, circulator (P6) and the tankless electric water heater are turned off.

SUMMARY

This chapter has discussed several heat sources that can burn cordwood, wood chips, or densified wood fuels, such as pellets and briquettes. These heat sources vary in their methods for managing the combustion process as well as the manner in which they interact with the balance of the system.

Thermal storage is critical in managing the heat output of the wood-fueled heat sources, especially wood-gasification boilers that produce a high rate of heat output while burning a fully loaded combustion chamber.

Many of the details used in the example systems of this chapter are similar (or identical) to those used in the example systems from previous chapters and supplied by solar thermal collectors or heat pumps. Examples include thermal storage, variable-speed distribution circulators, hydraulic separation, temperature stratification, distribution systems with valve-based zoning, outdoor reset control, boiler protection, "on-demand" domestic water heating, mixing using 3-way motorized valves, and discriminating use of auxiliary energy. Readers are encouraged to compare these details and fully understand how they can be used across a broad range of systems.

The importance of low-temperature hydronic distribution systems is again evident in this chapter. The lower the water temperature at which the distribution system can operate, the longer the thermal storage tank can supply that load without additional heat input. Over time, this reduces the number of firing cycles required for wood-fired heat sources. Longer firing cycles lead to improved efficiency and reduced emissions. The author recommends that all distribution systems be configured to meet the building's design heating load without exceeding a supply water temperature of 120 °F.

ADDITIONAL RESOURCES

1. (2007). *Planning and Installing Bioenergy Systems*. The German Solar Energy Society. ISBN 978-1-88407-132-6.
2. Jenkins, Dilwyn. (2010). *Wood Pellet Heating Systems*. Earthscan. ISBN 978-1-88407-885-5.

KEY TERMS

2-pipe thermal storage tank
absolute pressure
active heat dump
air control
air elimination
anti-condensation valve
aquastat
ash barrel
ASME-rated [storage tank]
auger
bag silo
barometric damper
bin temperature data
biomass
breach draft
briquette
burn pot
carbon neutral
Charles's law
clinker
cordwood
creosote
damper
day hopper
design load
dewpoint temperature
differential temperature controller
dual-fuel approach
endothermic
energy box
Environmental Protection Agency (EPA)
excess air
exothermic
face cord
failsafe air intake damper
firebox stacking density
fly ash
fuel/air ratio
full cord
gauge pressure
hammermill
heat dump
heat output ratio
hydraulically separate
hydronic wood stove
incomplete combustion
kindling
lambda ratio
lambda sensor
lignins
loading unit
lower heating value
moisture meter
motorized auger
natural draft wood-fired boiler
normally open zone valve
open-loop [hydronic system]
outdoor wood-fired hydronic heater
outgas
overheat protection subsystem
particulate
passive heat dump
Pellet Fuels Institute (PFI)
pellet mill
pellet-fired boiler
pelletization
photosynthesis
$PM_{2.5}$ particulate emission
pneumatic
premium wood pellet
pressure-rated wood-fired boiler
primary combustion air
pyrolytic gas
quality mark [for pellets]
refractory
reverse thermosyphoning
secondary combustion chamber
shunt circulator
single-stage combustion
single-stage wood-fired boiler
stack temperature switch
stand-by mode
steam flash
stoichiometric air requirement
strap-on aquastat
superheated water
sustained flue gas condensation
temperature stacking
thermocline
thermosyphon flow
thermostatically controlled damper
two-stage combustion
uninterruptible power supply (UPS)
vapor pressure
variable frequency drive (VFD)
variable-speed setpoint circulator
water jacket
wood chip
wood-gasification boiler

QUESTIONS AND EXERCISES

1. Describe the factors that led Americans away from using wood as their primary heating fuel during the 1900s.
2. Explain what is meant by a *carbon-neutral* fuel.
3. What were some of the problems associated with early generation hydronic devices that extracted heat from a wood fire?
4. Describe three desirable and three undesirable traits of heating with an outdoor wood-fired hydronic heater.
5. What is a key difference between a natural draft wood-fired boiler and a wood-gasification boiler?
6. What type of hydronic wood burning device is appropriate for a small low-energy house?
7. Compare the fuel energy available *hardwood* at 10, 20, and 50 percent moisture content on the basis of Btu/lb. lower heating value. How does this differ from the fuel energy contained in *softwood* at the same moisture levels and on the same basis of Btu/lb.?
8. Explain what the lower heating value of wood is.

9. Why is it not necessary to add a bonding agent to wood fibers to make wood pellets?

10. Describe the differences between the *endothermic* and *exothermic* phases of wood combustion.

11. Why is creosote accumulation within a vent pipe or chimney dangerous? What can be done to limit creosote formation?

12. What are at least three advantages of installing a heat exchanger between the piping circuit from an outdoor wood-fired hydronic heater and the interior portions of a hydronic heating system?

13. The top of a hydronic distribution system is located 18 feet above the water level in an outdoor wood-fired hydronic heater. The water at the top of the circuit can reach a temperature of 170 °F. There is no heat exchanger between the outdoor heater and the distribution system. Determine if the water will boil at the top of the system when the circulator is turned off.

14. Why should a barometric damper not be installed in the vent connector to a natural draft wood-fired boiler?

15. Why is it import to vent an interior pellet storage container? Be specific.

16. Why is it important to protect wood-fired and pellet-fired boilers from low inlet water temperatures? What happens if this protection is not provided?

17. Why is it usually unnecessary to install a heat dump when a loading unit is used to protect a wood-gasification boiler that connects to a thermal storage tank?

18. Explain the difference between a passive versus active heat dump.

19. Why is it important that flow passes in a specific direction through a helical coil heat exchanger that is suspended within an unpressurized thermal storage tank? How is this flow direction determined?

20. Assume that the firebox of a wood-gasification boiler, when fully loaded, can hold 50 pounds of seasoned (e.g., 20 percent moisture content) firewood. The boiler's average combustion efficiency is 75 percent. Determine the thermal storage tank volume needed, assuming the tank temperature will rise 80 °F as it absorbs heat from burning the full charge of wood.

21. Why is the thermal storage tank for a pellet-fired boiler typically much smaller than the thermal storage tank needed for a wood-gasification boiler?

22. A system uses a wood-gasification boiler combined with a 1,000-gallon pressurized thermal storage tank. It supplies a low-temperature radiant panel distribution system. The water within the thermal storage tank and boiler could reach a maximum temperature of 190 °F. The boiler and piping to the tank have an estimated volume of 25 gallons. The low-temperature portion of the distribution system is estimated to contain 145 gallons of water that will reach a maximum temperature of 105 °F. The upper most portion of the piping system is 15 feet above the connection to the expansion tank. Determine the size of the required expansion tank, assuming the entire system was initially filled with water at 50 °F.

23. Describe a situation in which an auxiliary boiler should heat the upper portion of the same thermal storage tank that is supplied by a wood-gasification boiler.

24. What is a critical design detail that's necessary when using a conventional gas-fired boiler, rather than a modulating/condensing boiler, as the auxiliary heat source to a wood-fired hydronic heating system?

25. An uninterruptible power supply (UPS) has a battery rated to supply 7 amp-hours at 12 volts. Assuming this DC power is converted to AC power with an efficiency of 95 percent, how long could this UPS operate a circulator that requires 30 watts of input power?

26. A system will be assembled based on the schematic in Figure 15-90a. It uses a 1,000-gallon pressurized storage tank. The combination drainback/expansion tank will be located with its bottom 3 foot above the 30 psi rated pressure-relief valve of the boiler. This tank has a minimum volume requirement of 5 gallons to ensure water is drawn from its lower connection without entraining air when the collector array and associated piping are completely filled. The estimated system volume, including the storage tank, is 1,100 gallons. The system will be filled with water at a temperature of 50 °F. The maximum temperature of the system water is estimated at 190 °F. Determine the necessary volume of the combined drainback/expansion tank, assuming the initial air pressure in the tank is 0 psi.

27. Show how the controls of pellet-fired combisystem #5 could be reconfigured to allow the pellet-fired boiler to maintain the thermal storage tank at 120 °F throughout the warm weather season when no space heating is required.

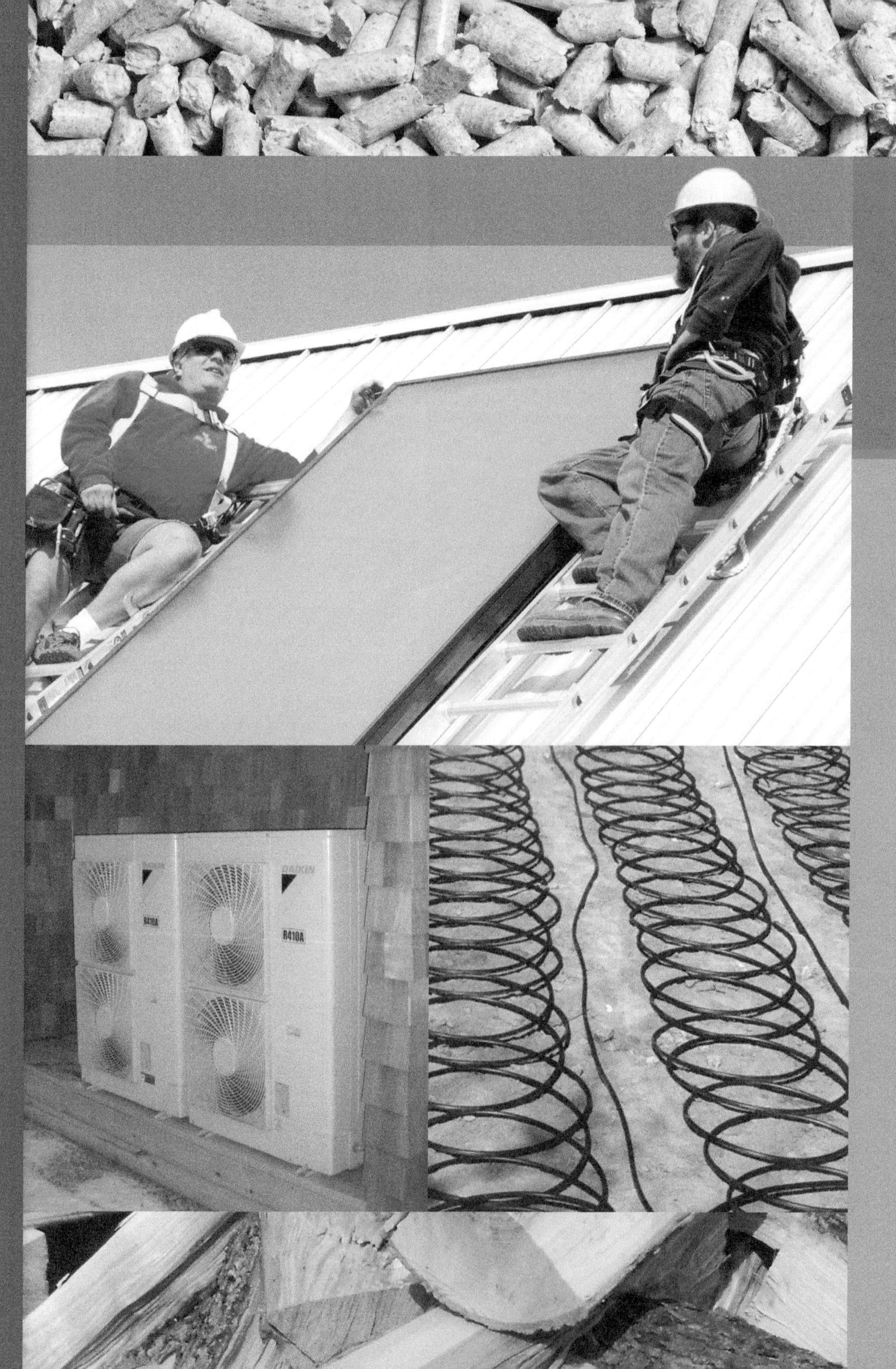
R410A
DAIKIN
R410A

CHAPTER 16

Economic Evaluation Methods and Tools

OBJECTIVES

After studying this chapter, you should be able to:

- Calculate the monthly payments associated with amortizing a loan.
- Explain the difference between annual and monthly compounded interest.
- Calculate the sum of a recurring expense including inflation, over time.
- Calculate the future value of capital invested in an interest bearing account.
- Explain the concept of simple payback.
- Describe some common misapplications of simple payback.
- Explain the concept of *after tax lost opportunity cost.*
- Develop financial comparisons based on total monthly owning and operating cost.
- Calculate the return on investment (ROI) created by competing investments.

16.1 Introduction

Previous chapters have shown that many system using renewable energy heat sources share technical details. This chapter will show that these systems also share financial characteristics. It is common for a solar thermal combisystem, a geothermal heat pump system, or a pellet-fueled boiler system to have an installation cost that is significantly higher than a "standard" approach for providing space heating and domestic hot water. It is also common for the operating cost of the system using the renewable energy heat source to be substantially lower than that of the standard system. Thus, the question often asked is: *Can the reduced operating cost of the system using the renewable energy heat source justify its higher installation cost?* This chapter provides several ways to examine this question, both qualitatively and quantitatively.

There are hundreds of possibilities for what constitutes a "standard" approach for providing space heating and domestic hot water. These approaches vary considerably from one locale to another. In many suburban neighborhoods across the United States and Canada, a standard system is likely to be a natural gas fueled furnace with a forced-air delivery system. The entire house is often treated as a single zone. In other areas, especially those where cooling is considered essential, the standard system might be an air-to-air heat pump combined with a forced-air delivery system. In yet other areas, where electrical energy costs are relatively low, the standard system might be electric resistance baseboard heat.

Figure 16-1 Most North American consumer will only consider the use of renewable energy if it can be provided at a cost comparable to that of other available fuels.

Those who promote the use of renewable energy heating systems should be able to present a clear financial picture of how those systems compare to standard systems. Although some potential clients weigh the environmental benefits of renewable energy heat sources heavily, and are less concerned about the financial consequences, they are in the minority. *The vast majority of potential clients are primarily motivated by the potential of a renewable energy heat source to lower their cost for providing heating and domestic hot water. It is very unlikely that combisystems using renewable energy heat sources will ever gain a significant market share, if a reasonable financial justification cannot be developed and conveyed to potential clients in terms that they can understand.*

This chapter provides the methods and analytical tools necessary to create financial comparisons of the systems described in previous chapters. It discusses financial analysis in relatively simple terms and avoids higher-level approaches, such as net present worth, discount rate, and depreciation. Although higher-level financial analysis methods are valid, provided that the assumptions necessary to execute them are valid, the author has found that few homeowners understand such methods, nor can they relate them to what are typically monthly based household finances. To avoid such confusion, this chapter relies on financial concepts that should be familiar to most homeowners. The goal is to provide tools that allow design professionals to effectively convey the financial implications of their proposed systems to potential clients.

This chapter begins with a summary of the basic financial calculations used in assessing the economic merit of capital investments. It goes on to discuss specific methods for comparing system economics, such as simple payback, total owning and operating cost, and return on investment (ROI).

16.2 Basic Financial Calculations

This section presents several financial calculations that are often used when comparing the cost of a renewable energy system to a standard method of providing space heating and domestic hot water. These calculations convey the time value of **capital** (i.e., available money). They also determine monthly payments when the installation will be financed using a loan.

LOAN AMORTIZATION

The financial analysis of a renewable energy heating system, or even a conventional heating system, often needs to include the effects of a loan. For example, the installation of a renewable energy combisystem might be financed as part of a home mortgage or with a fixed-rate loan from a bank or credit union. The interest rates and other conditions associated with loans are constantly changing based on market conditions. In this book, we limit discussion to fixed-interest rate loans. The loaned amount, which is called the **principal**, will have an associated **annual interest rate** and **term**. The latter is the time over which the loan must be fully repaid.

If the principal, annual interest rate, and term are known, Equation 16.1 can be used to determine the monthly payment.

(Equation 16.1)

$$M = P\left[\frac{\left(\frac{i}{12}\right)\left(\frac{i}{12}+1\right)^{12n}}{\left(\frac{i}{12}+1\right)^{12n}-1}\right]$$

where:

M = fixed monthly payment ($)

P = principal loaned ($)

i = annual interest rate (decimal percent)

n = term of loan (years)

Example 16.1: Determine the fixed monthly payment required to repay a principal of \$50,000 financed using a 30-year loan at an annual interest rate of 5 percent.

Solution: All the necessary inputs are stated. Substituting these numbers into Equation 16.1 and calculating yields:

$$M = P\left[\frac{\left(\frac{i}{12}\right)\left(\frac{i}{12}+1\right)^{12n}}{\left(\frac{i}{12}+1\right)^{12n}-1}\right]$$

$$= \$50{,}000\left[\frac{\left(\frac{.05}{12}\right)\left(\frac{.05}{12}+1\right)^{12(30)}}{\left(\frac{.05}{12}+1\right)^{20(30)}-1}\right]$$

$$= \$50{,}000[0.0053682] = \$268.41/\text{month}$$

Discussion: Be very careful that the annual interest rate is entered as a decimal percent (i.e., 5 percent = 0.05). Also be very careful to multiply the term of the loan, in years, by 12 to determine the exponent in this equation. This allows the equation to be based on monthly rather than annual payments. Also notice that because the quantity $[(i/12)+1]^{12n}$ is the same in the numerator and denominator of Equation 16.1, it only needs to be processed through a calculator once. Its value can then be stored and used as needed in the overall calculation.

Monthly loan payments are sometimes called **amortization schedules**. Because they are used frequently, they can be calculated using several online resources, such as the free online calculator at http://www.myamortizationchart.com/.

FUTURE VALUE

Capital invested in an interest-earning account increases in value over time. Being able to calculate the future value of present-day investment options allows one to assess the **lost opportunity cost** that would occur if the capital were not invested in the interest earning account but instead used to purchase a renewable energy heating system. This will be demonstrated later in this chapter.

The initial capital investment is called the **present value**. The amount it grows to at some future date is called its **future value**. Equation 16.2 relates these values based on the **annually compounded interest** rate:

(Equation 16.2)

$$FV = PV(1+i)^n$$

where:

FV = future value (\$)

PV = present value (\$)

i = annually compounded interest rate (decimal percent)

n = number of years the funds are invested (years)

Example 16.2: Determine the future value of \$5,000 invested in an account that pays 2.5 percent annually compounded interest for 10 years.

Solution: Substituting the data into Equation 16.2 and solving yields:

$$FV = PV(1+i)^n = 5{,}000(1+.025)^{10} = \$6{,}400.40$$

Discussion: This calculation assumes that the accumulated interest is added to the principal once each year. This is called **annual compounding**.

For some investments, such as common savings accounts, the accumulated interest is added to the principal every *month*. This is called **monthly compounded interest**. It increases the future value of an investment relative to annually compounded interest.

Equation 16.3 allows the future value to be calculated when the interest is added to the principal (i.e., compounded) every month.

(Equation 16.3)

$$FV = PV\left(1+\frac{i}{12}\right)^{12n}$$

Where:

FV = future value (\$)

PV = present value (\$)

i = annual interest rate (decimal percent)

n = number of years the funds are invested (years)

Example 16.3: Determine the future value of \$5,000 invested in a savings account that pays 2.5 percent annually, with *monthly* compounding, for 10 years.

Solution:

$$FV = PV\left(1 + \frac{i}{12}\right)^{12n}$$

$$= 5{,}000\left(1 + \frac{0.025}{12}\right)^{12(10)} = \$6{,}418.45$$

Discussion: Comparing this result to that in Example 16.2 shows that monthly compounding increases the future value of the $5,000 by $18.05 ($6,418.45 − $6,400.40). Although not a substantial increase, it does demonstrate that monthly compounding produces more gain when compared to annual compounding.

TIME VALUE OF AN INFLATING RECURRING EXPENSE

The cost of operating a system that provides heating, cooling, or domestic hot water is a **recurring expense.** Although the exact amount of energy used to meet these loads will vary from one year to the next based on weather or changes in demand for heating and domestic hot water, it is common to assume that calculated loads represent long-term averages. Thus, if the calculated heating energy use of a home is 50 MMBtu per year, it will remain at this value, on average, over the life of the system supplying the energy.

The cost of providing energy for space heating and domestic water heating is likely to increase over time. This is especially true of energy supplied by utilities or other commercial providers.

A thorough analysis of energy costs must consider the total cost associated with a recurring expense that inflates as some estimated (or assumed) rate. Equation 16.4 can be used for this purpose.

Equation 16.4

$$C_{total} = C_i\left[\frac{(i + 1)^n - 1}{i}\right]$$

where:

C_{total} = total cost, including inflation, over a period of n years ($)

C_1 = first-year cost ($)

i = annual rate of inflation (decimal percent)

n = number of years over which the cost is to be accumulated (years)

Example 16.4: The heating cost of a home in its first year is $500. Determine the total estimated heating cost over a period of 20 years, assuming the cost of fuel inflates at 4 percent per year.

Solution: The first-year cost is $500. The rate of inflation is 0.04 (e.g., 4 percent), and the number of years is 20. Putting these values into Equation 16.4 yields:

$$C_{total} = C_i\left[\frac{(i + 1)^n - 1}{i}\right] = \$500\left[\frac{(0.04 + 1)^{20} - 1}{0.04}\right]$$

$$= \$500(29.778) = \$14{,}889$$

Discussion: No one can state, with certainty, what the rate of inflation will be 1 year into the future, much less 20 years. It is also not possible to state, with certainty, that the rate of inflation will remain constant, which is an inherent assumption in this equation. Furthermore, this example assumes that weather conditions over the 20 years average out, and that the efficiency of the heat source will not change. Still, even with these assumptions, this equation provides a "what if" tool that can be used to compare alternatives.

Equation 16.4 can also be used to project the total value of accumulating *savings* that may be the result of using renewable energy to offset a conventional fuel that is projected to increase.

Example 16.5: A wood-fired boiler system is expected to generate a savings in conventional fuel of $600 per year. Assuming conventional fuel prices inflate at 5 percent per year, estimate the total savings if the wood-fired boiler is used over the next 15 years.

Solution: Use Equation 16.4, but with the first-year savings inserted instead of the first-year cost.

$$S_{total} = S_i\left[\frac{(i + 1)^n - 1}{i}\right] = \$600\left[\frac{(0.05 + 1)^{15} - 1}{0.05}\right]$$

$$= \$600(21.5786) = \$12{,}947$$

Discussion: Because of the assumptions inherent to Equation 16.4, it is best used in making economic comparisons between competing options instead of as a method to predict, with high certainty, exactly what the total accumulated operating costs will be many years into the future.

Figure 16-2 compares the accumulated savings based on the assumed 5 percent rate of inflation in fuel costs as described in Example 16.5 to a situation in

Figure 16-2 | Accumulated savings over time based on a first year savings of $600, with either 0% or 5% inflation of energy cost.

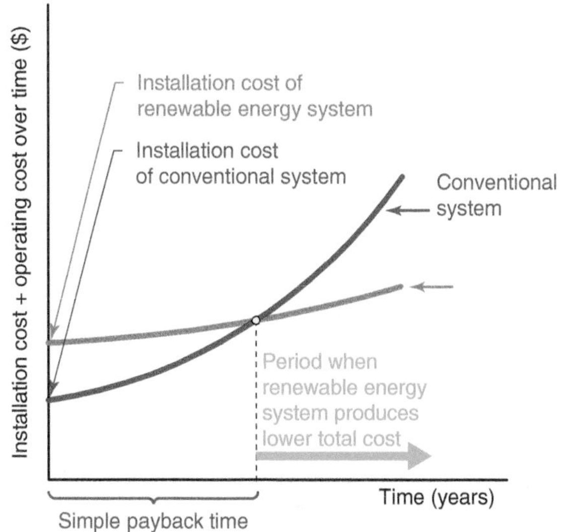

Figure 16-3 | Representative total cost of installation and operation, over time, for a heating systems using either a renewable heat source or a conventional heat source. Simple payback (in years) occurs where the curves cross.

which there is no fuel cost inflation over the 15-year period. In the latter case, the total accumulated savings would simply be the first-year savings ($600) multiplied by 15 years (i.e., $9,000).

Notice how inflation causes the accumulated savings to increase at a higher rate as time goes by. Mathematically, this is similar to the effect of compound interest.

16.3 Simple Payback

For better or for worse, simple payback continues to be the approach many individuals intuitively refer to when assessing the impact of higher installation cost combined with lower operating cost. Although the mathematics of simple payback are "simple," interpretation of the result is often very subjective.

The simple payback of a system that reduces future operating cost is the number of years required for that investment to repay its higher initial cost using the savings it generates, *relative to that of a standard system.* This is illustrated in Figure 16-3.

This graph plots the total of the installation cost plus operating cost over time. The renewable energy system has a higher installation cost but lower operating cost. The reduced operating cost causes the (green) accumulated cost curve to rise at a lower rate than the accumulated cost curve for the conventional system. At the point where the two curves cross, the total accumulated cost of each system is the same. The elapsed time to this point is the simple payback time. Beyond this time, the renewable energy system produces a lower total cost compared to the conventional system. The upward curves in Figure 16-3 are due to inflating conventional energy cost over time. If there were no inflation, both curves would be sloping straight lines.

Simple payback is sometimes *mistakenly* calculated by dividing the installation cost of the renewable energy system by the annual savings that system produces. *This is incorrect because it ignores the fact that some alternative expenditure must be made if the higher cost investment is not made.* For example, if a solar thermal water heating system is *not* installed, there must be some other means of heating water. That alternate means has an associated cost. That cost should include installation, maintenance, and perhaps replacement (if the hardware is likely to require replacement within the expected life of the solar water heating system). Therefore, simple payback should be calculated by dividing the *net* additional cost of the more expensive option by the projected annual savings in operating cost offered by that option. This can be expressed as Equation 16.5.

(Equation 16.5)

$$P = \frac{C_{\text{net}}}{S}$$

where:

P = simple payback time (years)

C_{net} = net installed cost of more expensive system over conventional system ($)

S = annual cost savings realized by using more expensive system ($)

Example 16.6: A solar thermal water heating system has an installed cost (after applicable tax credits and other available incentives) of $8,500. Analysis of the system projects a first-year reduction in fuel cost of $450. The system has a life expectancy of 20 years. Determine the simple payback of this system, *ignoring any cost associated with an alternate system.*

Solution: In this example, the simple payback is based solely on installed cost divided by annual savings.

$$P = \frac{C}{S} = \frac{\$8500}{450\frac{\$}{\text{yr}}} = 18.9 \text{ years}$$

Discussion: If the projected savings remained constant, year after year, and if the system had no maintenance cost, it would take almost 19 years for the solar water heating system to return its initial installation cost. Considering that the anticipated design life of the renewable energy system is 20 years, it's unlikely most homeowners would find this investment worthwhile, especially if the decision was mostly based on financial considerations.

One limitation of simple payback is that it doesn't factor in likely increases in the cost of conventional fuel that would be displaced by the renewable energy heat source. Such **inflation** would reduce the simple payback period.

Another shortcoming of simple payback, one that has been purposely ignored in Example 16.6, is that the cost used in the numerator of the equation should *not* be the full cost of the newly installed renewable energy system. Instead, it should be the *difference* between the cost of that system and the cost of the alternative "conventional" system that would provide the same space heating and hot water requirement. The cost of the latter is essentially unavoidable, assuming that domestic hot water must be provided.

If the building contains an existing heating/domestic hot water (DHW) system, it is prudent to consider the possibility that some or all of that system might have to be replaced within the expected design life of a newly installed renewable energy system. If an evaluation finds that such a replacement is likely, the cost of the replacement should be estimated and subtracted from the installed cost of the renewable energy system.

Example 16.7: The installed cost of a conventional water heating system that would be used instead of the solar thermal water heating system described in Example 16.6 is estimated at $2,000. How does this affect the simple payback of the situation?

Solution: Subtract the cost of the conventional water heating system from that of the solar thermal system and divide by the annual savings.

$$P = \frac{C_{net}}{S} = \frac{(\$8500 - \$2000)}{450\frac{\$}{\text{yr}}} = 14.4 \text{ years}$$

Discussion: Although the simple payback decreases, it is entirely subjective if a 14.4-year simple payback, versus the previous 18.9-year simple payback, would influence the buying decision. Figure 16-4 illustrates the simple payback described in this example. Both lines are straight sloping lines, indicating that no inflation of conventional fuel cost was included.

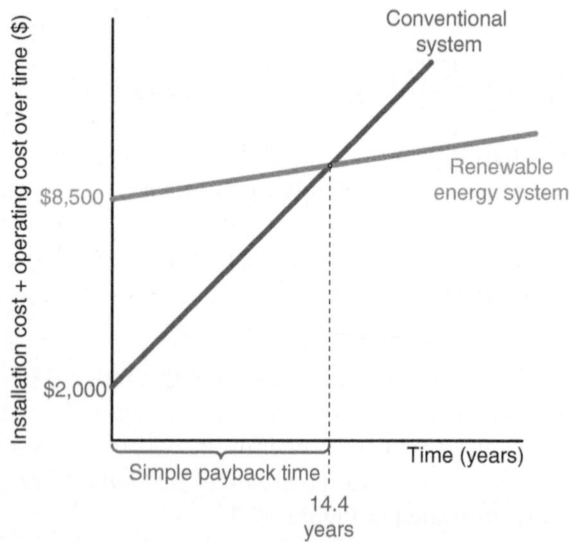

Figure 16-4 Simple payback of 14.4 years for the solar water heating system described in Example 16.7. Straight lines indicate no assumed inflation on fuel cost.

Decisions based on simple payback alone are usually very subjective. Ask yourself this question: Would you invest in a solar thermal water heating system for your home if the simple payback were 9 years rather than 5 years? Perhaps your answer is yes. You may inherently support the use of renewable energy. Another reason might be that you expect the system to last significantly longer than the 9 years required for it to return its additional installation cost over a conventional system. Thus, you perceive a long-term positive return on your investment. Still another reason to answer yes might be that you want to minimize buying energy from the local utility company.

Or, perhaps your answer is no. Maybe you don't see the system as having any merit beyond an improvement to your immediate financial situation. You might also answer *no* because you expect to sell the home within 5 years and thus won't be there to benefit from long-term savings. Another reason could be that you are concerned about the appearance of solar collectors on your house and how that might affect its resale value.

Both the yes and no answers are valid, as are the reasons underlying them. Both answers and the stated justifications are common responses from a wide variety of homeowners.

Using simple payback as the sole means for determining if a renewable energy system is justified precludes other considerations that often significantly influence buying decisions. For example, if a homeowner decides to buy granite rather than laminate countertops for a kitchen, it's highly *unlikely* he or she used simple payback analysis in making that decision. Perhaps the individual wanted the more expensive granite countertops for aesthetic reasons, preferred natural rather than synthetic materials, or was simply trying to impress his or her friends. There could be many reasons other than cost. This reasoning should also be true for purchases involving space heating and domestic water heating systems, but often isn't. The reason is that mechanical systems in buildings are usually viewed as *mundane necessities* rather than *desirable amenities*. Even so, professionals who market renewable energy systems by focusing solely on economics and ignoring influencing factors such as the desire for a low-carbon lifestyle, reduced dependency on imported fuels, or fascination with use of natural resources fail to cultivate what may be very influential factors in the decision-making process of potential customers.

16.4 Monthly Owning and Operating Cost

Another method for comparing the financial merits of a renewable energy system versus those of a conventional system is to consider the **total owning and operating costs** of both systems. This can be done on an annual basis or a monthly basis. The monthly basis is preferred because loans used to finance installation costs are usually repaid on a monthly basis. Utility costs are also paid on a monthly basis. Thus, the financial data for both owning and operating cost are readily available.

The objective of this analysis is to calculate the total cost of owning as well as operating each option for providing space heating and domestic hot water. The option with the lowest total owning and operating cost is, from a financial basis, considered the better investment.

The method used to determine monthly owning cost depends on how the installation will be paid for. If money is withdrawn from a savings account or other interest-bearing account, one should consider the **lost opportunity cost** associated with taking the money out of that account to pay for the installation. One should also assume a reasonable design life for the system and use it as the basis for calculating what would be the accumulated time value of the funds being withdrawn. The net future value of the withdrawn funds can then be divided into the months over which the system is expected to last to determine an average monthly owning cost.

Example 16.8: \$10,000 is withdrawn from an investment account to pay for the installation of a renewable energy heating system. The investment account pays an annual (tax-free) interest rate of 2 percent and is compounded annually. The renewable energy system is expected to last 20 years. Determine the average monthly owning cost associated with self-financing the installation.

Solution: The total value of \$10,000 escalating at 2 percent annually compounded interest over 20 years is found using Equation 16.2:

$$FV = PV(1 + i)^n = 10{,}000(1 + 0.02)^{20} = \$14{,}860$$

The average monthly amount needed to repay this cost over 20 years is \$14,860 divided by 240 months, or \$61.91/month.

Discussion: This simplified analysis does not include the fact that many investment accounts, such as standard savings accounts, are subject to income tax on the interest earnings.

The **"after-tax" lost opportunity cost** would be the total interest income earned multiplied by the percentage of that income that would be retained by the owner. This can be estimated using Equation 16.5.

(Equation 16.5)

$$ATLOC = (FV - PV)(1 - TR)$$

where:

$ATLOC$ = after-tax lost opportunity cost ($)

FV = future value of investment ($)

PV = present value of the investment ($)

TR = tax rate (decimal percent)

Example 16.9: Determine the after-tax lost opportunity cost if an initial investment of $10,000 remains in an interest-bearing account with 2 percent annually compounded interest for 20 years, assuming an effective income tax rate of 30 percent over the term of the investment. What is the monthly amount necessary to repay the initial investment *plus* the after-tax lost opportunity cost?

Solution: The future value of the investment can be found using Equation 16.2:

$$FV = PV(1 + i)^n = 10{,}000(1 + 0.02)^{20} = \$14{,}860$$

The after-tax lost opportunity cost can be found using Equation 16.5:

$$\begin{aligned} ATLOC &= (FV - PV)(1 - TR) \\ &= (14{,}860 - 10{,}000)(1 - 0.3) = \$3402 \end{aligned}$$

Discussion: The ATLOC is an estimate of the "retained" benefit of leaving the initial investment in the account for 20 years. If the $10,000 were withdrawn from the account to pay for the installation of a heating system, the total amount that must be repaid to one's self would be the initial investment plus the ATLOC. In this example, it would be $10,000 + $3,402 = $13,402. The average monthly amount required to repay this total cost over the estimated 20 years system life would be: $13,402/240 months = $55.84 per month.

The calculations in the example are based on several assumptions. First, that the system will last exactly 20 years. Second, that the income tax rate will remain at 30 percent over this period. Third, that the owner chooses to replace the money withdrawn to pay for the system, in equal installments, over 240 months. Still, even with these assumptions, this approach illustrates the estimated monthly *owning* cost of the system and can be used in combination with the estimated month *operating* cost.

If the system will be fully financed using a loan, the monthly owning cost can be calculated using Equation 16.1.

Example 16.10: The $10,000 installed cost of a renewable energy heating system will be fully financed using a loan. Determine the monthly owning cost associated with amortizing this loan over a 20-year term, at an annual interest rate of 4 percent.

Solution: Putting the data into Equation 16.1 and solving yields:

$$\begin{aligned} M &= P\left[\frac{\left(\frac{i}{12}\right)\left(\frac{i}{12}+1\right)^{12n}}{\left(\frac{i}{12}+1\right)^{12n}-1}\right] \\ &= \$10{,}000\left[\frac{\left(\frac{.04}{12}\right)\left(\frac{.04}{12}+1\right)^{12(20)}}{\left(\frac{.04}{12}+1\right)^{12(20)}-1}\right] \\ &= \$10{,}000[0.0060598] = \$60.60/\text{month} \end{aligned}$$

Discussion: Interestingly, the cost of financing the $10,000 installation cost using the 4 percent fixed rate loan is relatively close to the average monthly cost of paying for the system with funds withdrawn from a typical savings account. The difference is only ($60.60 − $55.84) = $4.76 per month. The current availability of low-interest loans to finance renewable energy installations can thus be an effective tool in overcoming the lack of personal capital for such an investment.

Several calculations and cost estimates are needed before the total owning and operating cost of system options can be compared. Some calculations establish the thermal load, some the unit cost of energy, and others the financing cost. It's also necessary to have reasonable estimates (or firm quotes) for the installed cost of each system option. To be accurate, the heating professional must devote time and expertise to develop cost estimates for each system option. The total installed cost of those designs can then be estimated.

Cost estimating is part art and part science. It can vary significantly from one heating professional to another, even when they are estimating the same system. Different professionals use different methods. Some rely on the known installation costs from previous systems. Some use material price estimates from their suppliers, along with mark-ups and labor costs. Still others use annually updated mechanical cost estimating manuals and software, such as those available from the RS Means Company. To arrive at an installed cost, the estimator should include material cost, installation labor estimates, overhead and profit, and warranty reserve allowances.

Example 16.11: A family of five is estimated to use 75 gallons of domestic hot water each day. The average incoming water temperature is 50 °F, and the setpoint for the hot water delivery system is 125 °F. The standby heat loss from the system is estimated at 3 percent of the total energy used to heat the water. An f-Chart analysis indicates that a solar thermal system with a net installed cost of $6,100 can supply 78 percent of the annual domestic water heating energy needs. The alternative is an electric water heater with an installed cost of $900. Assume that both systems will be financed using a 15-year loan with an annual interest rate of 5 percent. Also assume that the local cost of electricity is $0.13/kilowatthour. Determine the following:

a. The average monthly energy required to provide domestic hot water
b. The average monthly cost for providing this hot water using both a conventional electric water heater and the solar thermal system
c. The monthly cost of owning both systems
d. Total monthly owning and operating cost of both systems
e. The simple payback of the solar thermal system

Solution:

a. The average monthly energy required for hot water is determined using a slightly modified form of Equation 2.11, with an additional 3 percent added to account for standby heat losses.

(Equation 2.11 modified)

$$E_{\text{daily}} = \left[\frac{365}{12}\right](G)(8.33)(T_{\text{hot}} - T_{\text{cold}})1.03$$

where:

E_{daily} = daily energy required for DHW production (Btu/day)

G = volume of hot water required per day (gallons)

T_{hot} = hot water temperature supplied to the fixtures (°F)

T_{cold} = cold water temperature supplied to the water heater (°F)

[365/12] = average number of days in a month

1.03 = factor that adds 3 percent to heating load to account for standby heat loss

Entering the data into the modified version of Equation 2.11 yields:

$$E_{\text{monthly}} = \left[\frac{365}{12}\right](G)(8.33)(T_{\text{hot}} - T_{\text{cold}})1.03$$

$$= \left[\frac{365}{12}\right](75)(8.33)(125 - 50)1.03$$

$$= 1.47\ \text{MMBtu/month}$$

Note that the result is expressed in units of MMBtu, where 1 MMBtu = 1,000,000 Btu.

b. The unit cost of electric resistance heating can be found using Figure 1-27:

$$\left(13\frac{\text{cents}}{\text{kwhr}}\right)2.93 = 38.09\frac{\$}{\text{MMBtu}}$$

The average monthly cost for heating the hot water using the electric water heater is:

$$1.47\frac{\text{MMBtu}}{\text{month}}\left(38.09\frac{\$}{\text{MMBtu}}\right) = 55.99\frac{\$}{\text{month}}$$

The average monthly cost for heating the hot water using the solar thermal system is just [1 minus the annual solar fraction], or [1 − 0.78] = 0.22 5 22 percent of the cost of heating the water with electricity:

$$[1 - 0.78]\left(55.99\frac{\$}{\text{month}}\right) = 12.32\frac{\$}{\text{month}}$$

c. The month payment for financing the \$6,100 net installed cost of the solar thermal system, using a 15-year loan at 5 percent annual interest, is:

$$M = P\left[\frac{\left(\frac{i}{12}\right)\left(\frac{i}{12}+1\right)^{12n}}{\left(\frac{i}{12}+1\right)^{12n}-1}\right]$$

$$= 6100\left[\frac{\left(\frac{.05}{12}\right)\left(\frac{.05}{12}+1\right)^{12(15)}}{\left(\frac{.05}{12}+1\right)^{12(15)}-1}\right]$$

$$= 6100(0.0079038) = \$42.21$$

The monthly payment for financing the \$900 electric water heater installation, using a 15-year loan at 5 percent annual interest, is:

$$M = P\left[\frac{\left(\frac{i}{12}\right)\left(\frac{i}{12}+1\right)^{12n}}{\left(\frac{i}{12}+1\right)^{12n}-1}\right]$$

$$= 900\left[\frac{\left(\frac{.05}{12}\right)\left(\frac{.05}{12}+1\right)^{12(15)}}{\left(\frac{.05}{12}+1\right)^{12(15)}-1}\right]$$

$$= 900(0.0079038) = 7.11$$

d. The total monthly owning and operating cost of each system option can now be calculated:

	Standard electric water heater	**Solar thermal water heater**
Monthly owning cost	\$7.11	\$48.21
Monthly operating cost	\$55.99	\$12.32
Total monthly owning and operating cost	\$63.1	\$60.53

e. The simple payback of this solar thermal system would be the installed cost difference divided by the projected annual energy cost savings. One way to calculate the latter is to use the solar heating fraction as a multiplier of the conventional heating cost, as shown below.

$$P = \frac{C_{net}}{S}$$

$$= \frac{(\$6100 - \$900)}{\left[\left(1.47\frac{\text{MMBtu}}{\text{month}}\right)(12\text{ month})\left(38.09\frac{\$}{\text{MMBtu}}\right)(0.78)\right]}$$

$$= 9.2\text{ years}$$

Another way to calculate the simple payback is to divide the difference in installation cost by the calculated difference in monthly energy cost, as shown below.

$$P = \frac{C_{net}}{S} = \frac{(\$6100 - \$900)}{\left[\left(\frac{\$55.99}{\text{month}} - \frac{\$12.32}{\text{month}}\right)\left(12\frac{\text{month}}{\text{year}}\right)\right]}$$

$$= 9.2\text{ years}$$

Discussion: Based on the total monthly owning and operating cost, the solar thermal system *immediately generates a net positive cash flow* of (\$63.10 – \$60.53) = \$2.57 per month. Thus, the owner realizes an *immediate* financial benefit by installing the solar thermal system, even with its installation cost fully financed by the loan. This effect is illustrated in Figure 16-5.

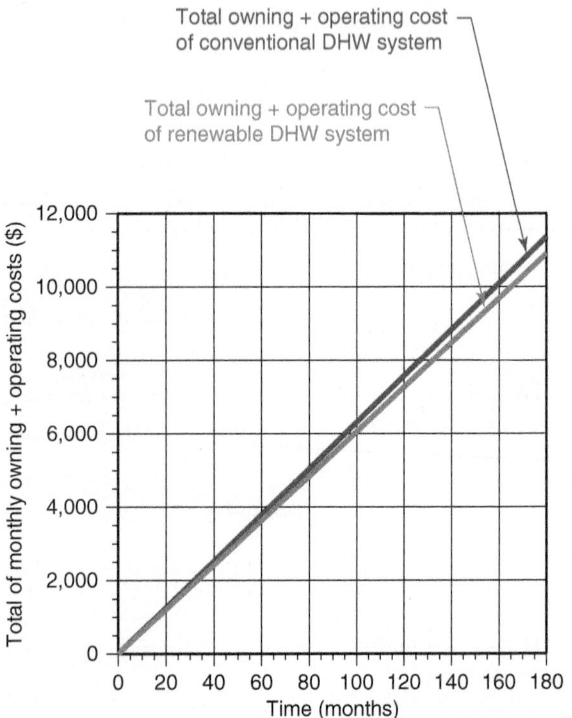

Figure 16-5 | Comparison of total owning and operating cost for systems described in Example 16.11

This analysis ignores any perceived altruistic benefits of the solar thermal system. It also ignores any potential effect that the system has on the resale value of the house. This example also shows that the simple payback of 9.2 years fails to convey the immediate positive cash flow that the solar thermal system generates.

16.5 Return on Investment

Another method of comparing the relative merit of competing investments is called return on investment (ROI) analysis. In a broad sense, ROI can be defined by Equation 16.6.

(Equation 16.6)

$$ROI = \frac{(B - C)}{C}$$

Where:

ROI = return on investment (decimal percent)

B = benefit returned by investment ($)

C = net cost of the investment ($)

ROI is best used as a tool for making *comparisons* between competing alternatives. For purely financial investments, it can be used to compare the merit of investing capital between alternatives such as a money market account, certificate of deposit, or buying shares of a stock. Assuming equal risk factors, the investment with the highest ROI is generally preferred.

In the case of a renewable energy system, the benefit (i.e., the value of B) for an ROI calculation is the total savings achieved through the reduced use of conventional energy. This would typically be calculated over the life of the system. The net cost of the investment would be the life-cycle owning cost of the renewable energy system minus that of a conventional means of providing the same energy.

Example 16.12: A geothermal heat pump system has an installed cost (after applicable tax credits and/or rebates) of $20,000. The cost of a conventional heating/cooling system that could be used instead of the geothermal heat pump is $12,000. An analysis estimates that the geothermal heat pump system would reduce the combined heating and cooling cost to the owners by $850 per year, compared to the conventional system. Assume that both systems would be financed on a home mortgage at a fixed rate of 4 percent for 20 years. Also assume that both systems have the same life expectancy and that electrical energy costs inflate at 3 percent per year. What is the life cycle ROI associated with use of the geothermal heat pump system?

Solution: Before the ROI can be calculated, the total cost of owning both system options needs to be determined as does the total value of savings associated with the use of the geothermal heat pump. These owning costs can be calculated based on the mortgage amortization using Equation 16.1.

The monthly owning cost of the geothermal heat pump system, based on the stated mortgage terms, would be:

$$M = P\left[\frac{\left(\frac{i}{12}\right)\left(\frac{i}{12}+1\right)^{12n}}{\left(\frac{i}{12}+1\right)^{12n}-1}\right]$$

$$= 20{,}000\left[\frac{\left(\frac{0.04}{12}\right)\left(\frac{0.04}{12}+1\right)^{240}}{\left(\frac{0.04}{12}+1\right)^{240}-1}\right]$$

$$= \$20{,}000(0.0060577) = \$121.15\ /\ month$$

The monthly owning cost of the conventional system, based on the stated mortgage terms, would be:

$$M = P\left[\frac{\left(\frac{i}{12}\right)\left(\frac{i}{12}+1\right)^{12n}}{\left(\frac{i}{12}+1\right)^{12n}-1}\right]$$

$$= 12{,}000\left[\frac{\left(\frac{0.04}{12}\right)\left(\frac{0.03}{12}4+1\right)^{240}}{\left(\frac{0.04}{12}+1\right)^{240}-1}\right]$$

$$= \$12{,}000(0.0060577) = \$72.69\ /\ month$$

The net additional owning cost of the geothermal system over that of the conventional system would be 240 (i.e., 20 years × 12 months/year) times the difference in monthly mortgage payments:

$$C_{net} = 240 \text{ months} \times \left[\frac{\$121.15}{\text{month}} - \frac{\$72.69}{\text{month}}\right] = \$11{,}630$$

The benefit (B) would be the total savings in conventional energy over the 20-year design life of the systems, which can be determined using Equation 16.4. The first-year savings is $850. This savings continues to increase based on inflation. The total savings over 20 years, assuming the rate of inflation remains at 3 percent per year, will be:

$$S_{total} = S_i\left[\frac{(i+1)^n - 1}{i}\right] = \$850\left[\frac{(0.03+1)^{20} - 1}{0.03}\right]$$
$$= \$850(26.87) = \$22{,}840$$

Based on life cycle owning cost and savings over the 20-year design life, the ROI will be:

$$ROI = \frac{(B - C_{net})}{C_{net}} = \frac{(\$22{,}840 - \$11{,}630)}{\$11{,}630}$$
$$= 0.964 = 96.4\%$$

Discussion: The 96.4 percent ROI over a 20-year period must be carefully interpreted. It should *not* be compared to any *annual* rate of return, such as investing the capital in a certificate of deposit. Instead, it should be viewed as the financial gain multiplier for the original investment, realized over the 20-year period.

Equation 16.3 can be manipulated to convert a life-cycle ROI to an equivalent fixed annual interest rate that would yield the same financial gain.

Equation 16.3 (repeated):

$$FV = PV\left(1 + \frac{i}{12}\right)^{12n}$$

Where:

FV = future value ($)

PV = present value ($)

i = *annual* interest rate (decimal percent)

n = number of years the funds are invested (years)

Assuming monthly compounding interest and that the future value (FV), after 20 years, is 1.964*(PV), (i.e., a 96.4 percent gain on the original investment), Equation 16.3 can be set up and solved for the equivalent annual interest rate:

$$FV = 1.964(PV) = PV\left(1 + \frac{i}{12}\right)^{20(12)}$$

$$i_{annual} = 12[(1.964)^{1/240} - 1] = 0.03384 = 3.384\%$$

An annual return of 3.384 percent is, in the author's opinion, a reasonable investment in comparison to other readily available low-risk options at the time of this writing. This is especially true since this return will be "tax free." Unlike the situation with a normal interest-bearing account, the owner does not pay income taxes on *savings*. If the capital were instead invested into a savings account paying 3.384 percent annual interest, the owner *would* have to pay income tax on the interest earnings, which would decrease the net return.

FINANCIAL INCENTIVES

Over the last several decades, the U.S. federal government, as well as several state governments and local municipalities, have offered various types of financial incentives to encourage investment in renewable energy systems. The most widely used incentive for residential renewable energy systems has been a federal income **tax credit**.

A tax credit is an amount that is directly subtracted from the total federal income tax owed for a given year. For example, a 25 percent tax credit is equivalent to the federal government paying for 25 percent of the installed cost of the system. Tax credits should not be confused with a **tax deductions**. The latter is an amount subtracted from a person's gross income before his or her federal income tax is calculated.

The U.S. federal government began offering residential renewable energy tax credits on solar thermal energy systems in 1979. These credits subsidized 40 percent of the installed cost of residential solar thermal systems up to a maximum credit of $4,000. Several states also began offering tax credits and other financial incentives during this time. Figure 16-6 shows the rapid growth in annual solar collector sales that occurred when the federal income tax credit became effective in 1979. The market hit a peak during 1984.

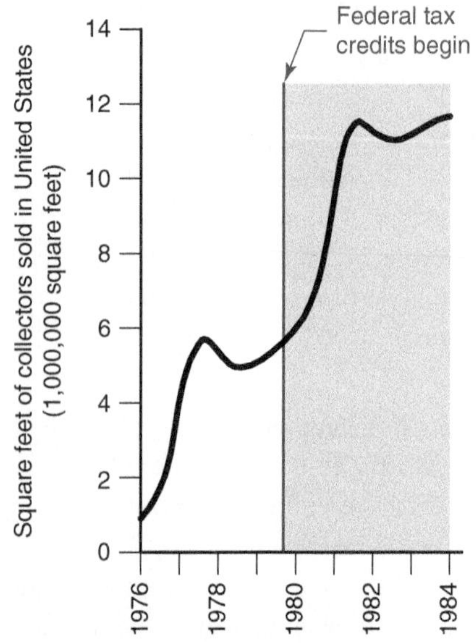

Figure 16-6 | Effect of original 40% federal income tax credit on sales of solar collectors.

The intended purpose of renewable energy tax credits or other rebates is to *temporarily stimulate* market growth. The underlying rationale is that a market temporarily supported by financial incentives should be able to grow to a point where it becomes *self-supporting*. By partially underwriting the installed cost of renewable energy systems, state and federal governments hope to create a market in which the *net* financial investment (e.g., the total owning and operating cost after subtracting tax credits) is competitive, or perhaps even favorable, in comparison to total owning and operating cost associated with conventional energy sources.

There has been considerable debate on the effectiveness of this approach. The historical record of government subsidies on renewable energy systems has shown mixed results. While there is little argument that the availability of subsidies increased sales, there is also evidence that the removal of such subsidies, in combination with lower-priced conventional energy, caused significant drop in sales, as seen in Figure 16-7. The latter is evidence that the intended goal of creating a stable and self-supporting market was not achieved.

As this text is being written, a new period of federal tax credits for investment in specific renewable energy systems is in effect. These tax credits cover 30 percent of the installed cost of *qualified* renewable energy systems. Currently, there is no upper limit on the cost of qualified renewable energy equipment to which the 30 percent federal tax credit applies. These tax credits are scheduled to expire on December 31, 2016.

Two of the systems discussed in this text may qualify for this federal tax credit: solar thermal water heating systems and geothermal heat pump systems. Solar water heating systems must be certified by SRCC to Standard OG-300, or a comparable entity endorsed by the state where the system is installed. To qualify for the federal tax credit, at least half the energy used to heat the dwelling's water must be provided by solar energy. Geothermal heat pumps must meet federal Energy Star criteria to qualify for these tax credits.

As of late 2015, lobbying efforts are underway to extend these federal tax credits to systems using high-performance biomass heat sources such as the wood-gasification boilers and pellet-fueled boilers discussed in Chapter 15.

Some states as well as local municipalities also currently offer financial incentives such as tax credits or rebates on qualifying renewable energy systems. These incentives usually have upper limits on the amount of the tax credit or rebate.

The U.S. Department of Energy maintains a **Database of State Incentives for Renewables & Efficiency**. The information in this database is accessed through the website www.dsireusa.org. This database is frequently updated as incentives are added or removed by state governments or the federal government. Those interested in claiming state or federal tax credits or other financial incentives should make frequent checks of this website to verify currently available incentives as well as specifics on qualifying installation cost and procedures for claiming credits or rebates.

It is the author's opinion that heating systems using renewable energy heat sources will only be accepted by average North American consumers if they are cost-effective, *without financial subsidies*, against competing energy options. Furthermore, they must provide reliable operation and deliver uncompromised thermal comfort.

The debate over government subsidies of renewable energy systems will undoubtedly continue. As it does, designers are encouraged to focus their efforts and apply the concepts discussed in this text to craft systems that can be competitive in an unsubsidized market. Doing so provides greater assurance of long-term market stability and continued use of renewable energy than does continued reliance on financial incentives.

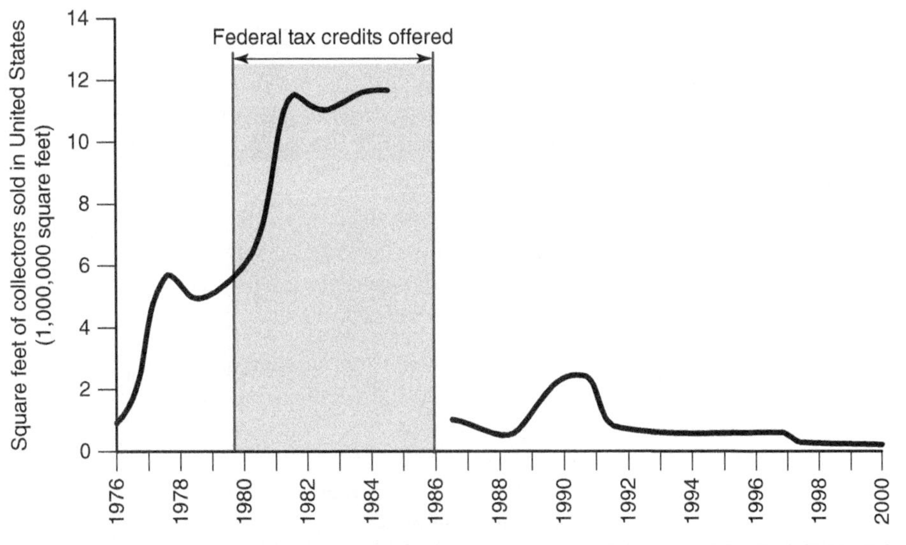

Figure 16-7 | Continuation of graph showing sales of solar collectors from 1976 through 2000.

SUMMARY

This chapter has presented several methods for comparing the financial merit of systems using renewable energy heat sources with those using conventional heat sources. These methods provide perspective on the common situation in which systems using renewable energy heat sources cost more to install than do conventional systems but operate at significantly lower cost.

The concept of simple payback, while valid for comparison purposes, yields numbers that are difficult, or impossible, to compare to other investment options. In contrast, the monthly total owning and operating cost method yields numbers that can be directly compared to alternatives, without subjective interpretation. Similarly, return on investment (ROI) results can be directly compared to alternative investments.

Readers should use discretion with equations that project future costs or future value of investments based on present-day assumptions. While valid for "what if" scenarios, there is no guarantee that assumed conditions will hold for more than a few months, much less two decades or more into the future. It is easy to create financial scenarios, using unrealistically high rates of inflation or energy cost savings that point to a renewable energy system as a "no-brainer" choice. Doing so is at best unprofessional and at worse, deceptive. The market for heating systems using renewable energy heat sources will only be harmed by such unrealistic practices.

Finally, the examples given in this chapter have assumed the net cost of the renewable energy system as a starting point. The net cost is the installed cost minus any applicable tax credits or other rebates. There is no way to know if federal income tax credits or other incentives meant to encourage investment in renewable energy–based systems will continue in the future. The long-term viability of renewable energy systems requires that they can be financially justified without such subsidies.

KEY TERMS

after-tax lost opportunity cost
amortization schedule
annual interest rate
annual compounding
annually compounded interest
capital
Database of State Incentives for Renewables & Efficiency
future value
inflation
lost opportunity cost
monthly compounded interest
present value
principal
recurring expense
return on investment (ROI)
simple payback
tax credit
tax deduction
term
total owning and operating cost

QUESTIONS AND EXERCISES

1. Calculate the monthly payments for a loan of \$25,000 at an annual interest rate of 3.5 percent and a term of:
   a. 10 years
   b. 15 year
   c. 30 years
2. Calculate the future value of an investment of \$8,000 into an account paying 2 percent annual interest, with monthly compounding. Compare this to the same situation, but assuming annually compounded interest.
3. A high-efficiency circulator has been estimated to reduce operating cost by \$160 per year when the currently cost of electricity is \$0.15/kwhr. Estimate the total value of savings, in current dollars, assuming the cost of electricity inflates at 4 percent per year over the next 20 years.
4. The current cost for heating an older home is \$1,500 per year. What is the total estimated cost to provide heat to this house for 15 years, in current dollars, assuming the cost of fuel increases at 5 percent per year, and that no changes occur to the home or its heating system?
5. A person has decided to install a solar thermal combisystem in a new home. The solar heating system has an initial net cost (after tax credits and rebates of

$9,500. Compare the monthly cost of financing this installation using:

a. Money withdrawn from a savings account paying 2 percent annual interest with quarterly compounding. Include the after-tax lost opportunity cost, assuming an effective tax rate of 25 percent. Assume the funds are to be restored using equal monthly payments over a 20-year period.

b. Fully financing the installation with a loan with a 20-year term, and 4 percent interest rate.

6. A family of four is estimated to use 60 gallons of domestic hot water each day. The average incoming water temperature is 45 °F, and the setpoint for the hot water delivery system is 120 °F. The standby heat loss from the system is estimated at 3 percent of the total energy used to heat the water. An f-Chart analysis indicates that a solar thermal system with a net installed cost of $5,800 can supply 72 percent of the annual domestic water heating energy needs. The alternative is an electric water heater with an installed cost of $900. Assume that both systems will be financed using a 15-year loan with an annual interest rate of 5 percent. Also assume that the local cost of electricity is $0.13/kilowatthour. Determine the following:

a. The average monthly energy required to provide domestic hot water

b. The average monthly cost for providing this hot water using both a conventional electric water heater and the solar thermal system

c. The monthly cost of owning both systems

d. The total month owning and operating cost of both systems

e. The simple payback of the solar thermal system

7. A person building a new home is considering use of a wood-gasification boiler system for space heating. That system has an estimated installed cost of $17,500. A conventional heating system for the same home has an estimated installed cost of $11,000. The home's estimated heating energy use is 85 MMBtu per year. The wood-gasification system can provide this heating energy for $10/MMBtu, whereas the conventional system can provide this energy for $22/MMBtu. Determine the simple payback of the wood gasification boiler system.

8. Determine the ROI of the wood-gasification boiler system described in Exercise 7 based on the owning and operating cost of both system options during the first year.

9. Determine the life-cycle ROI of the wood-gasification boiler system described in Exercise 7.

10. Determine the equivalent annual interest rate, assuming annual compounding, for the situation described in Exercise 7.

11. Describe some of the common misapplications of simple payback analysis.

12. Contrast simple payback to a situation in which the monthly total owning and operating cost of a renewable energy heating system is lower than the monthly total owning and operating cost of a conventional system.

13. Make a graph that shows the accumulating value of an investment of $10,000 over a period of 10 years, using the following assumptions:

a. The annual interest rate is 5 percent and is compounded *annually.*

b. The annual interest rate is 5 percent and is compounded *monthly*.

14. Make a graph showing the total owning and operating cost, on a yearly basis, for the situation given in Example 16.11, over the 15-year period.

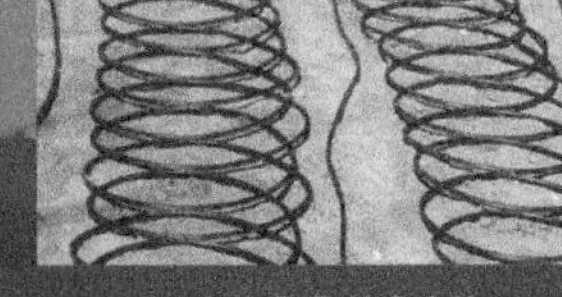

APPENDIX

Schematic Symbols

SCHEMATIC SYMBOLS FOR PIPING COMPONENTS

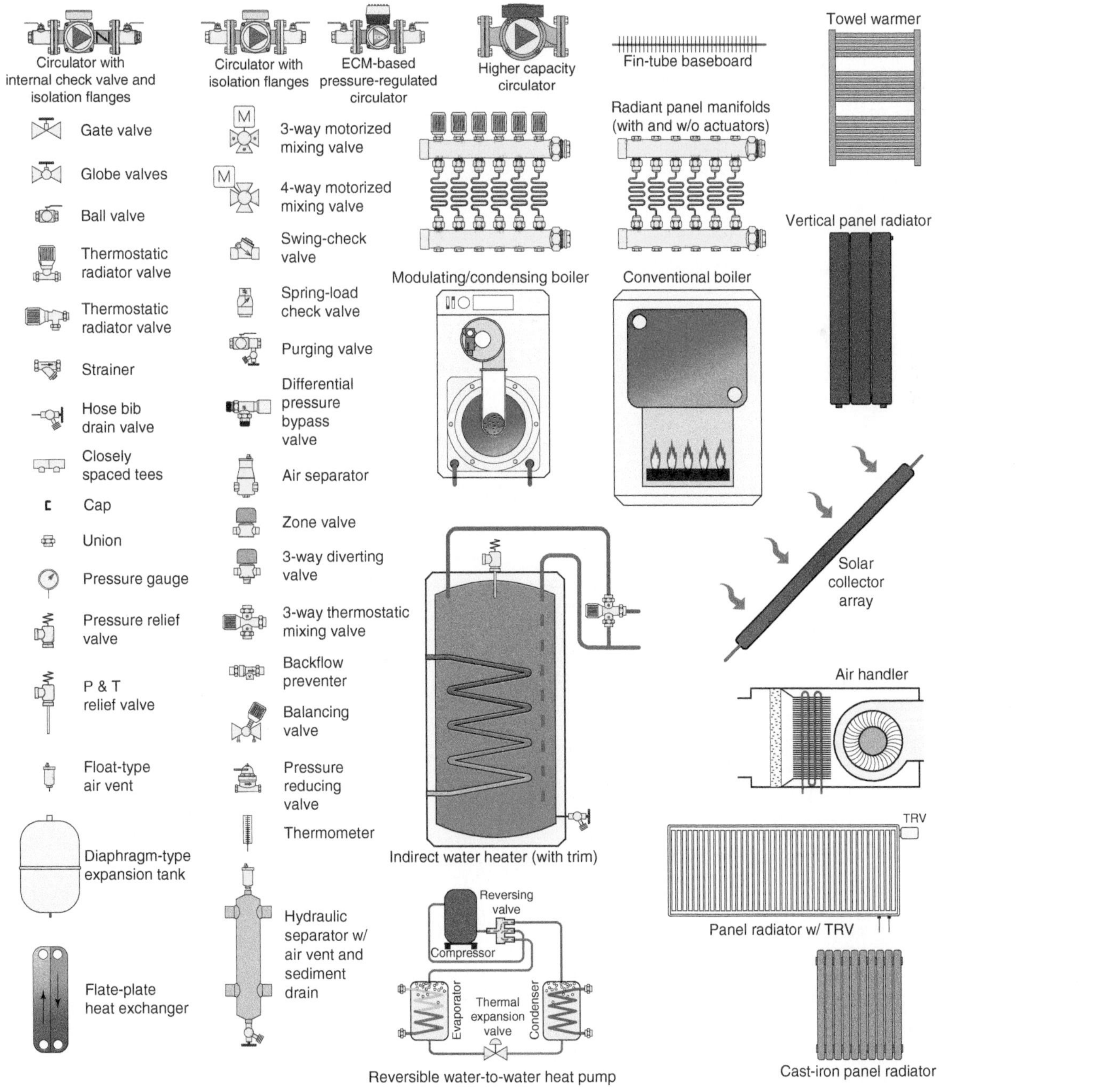

Circulator with internal check valve and isolation flanges

Circulator with isolation flanges

ECM-based pressure-regulated circulator

Higher capacity circulator

Fin-tube baseboard

Towel warmer

Gate valve

Globe valves

Ball valve

Thermostatic radiator valve

Thermostatic radiator valve

Strainer

Hose bib drain valve

Closely spaced tees

Cap

Union

Pressure gauge

Pressure relief valve

P & T relief valve

Float-type air vent

Diaphragm-type expansion tank

Flate-plate heat exchanger

3-way motorized mixing valve

4-way motorized mixing valve

Swing-check valve

Spring-load check valve

Purging valve

Differential pressure bypass valve

Air separator

Zone valve

3-way diverting valve

3-way thermostatic mixing valve

Backflow preventer

Balancing valve

Pressure reducing valve

Thermometer

Hydraulic separator w/ air vent and sediment drain

Radiant panel manifolds (with and w/o actuators)

Modulating/condensing boiler

Conventional boiler

Vertical panel radiator

Solar collector array

Indirect water heater (with trim)

Air handler

TRV

Panel radiator w/ TRV

Reversing valve

Compressor

Evaporator

Thermal expansion valve

Condenser

Reversible water-to-water heat pump

Cast-iron panel radiator

SCHEMATIC SYMBOLS FOR ELECTRICAL COMPONENTS

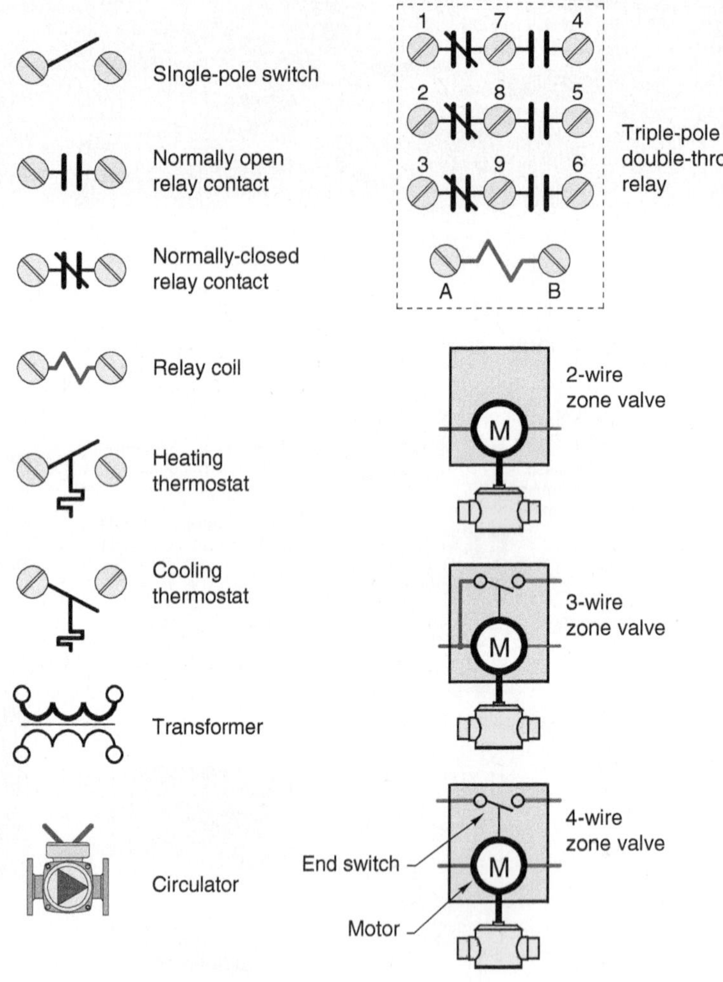

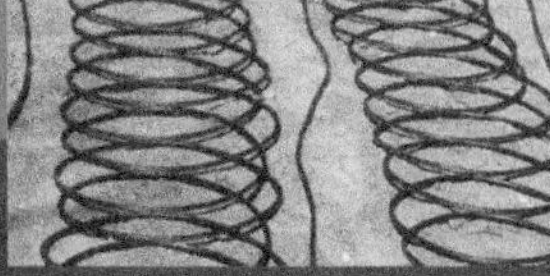

APPENDIX

B R-Values of Common Building Materials

MATERIAL	R-VALUE*	
INSULATIONS		
Fiberglass batts (standard density)	3.17	per inch
Fiberglass batts (high density)	3.5	per inch
Blown fiberglass	2.45	per inch
Blown cellulose fiber	3.1–3.7	per inch
Foam-in-place urethane	5.6–6.3	per inch
Expanded polystyrene panels (beadboard)	3.85	per inch
Extruded polystyrene panels	5.4	per inch
Polyisocyanurate panels (aged)	7.2	per inch
Phenolic foam panels (aged)	8.3	per inch
Vermiculite	2.1	per inch
MASONRY AND CONCRETE		
Concrete	0.10	per inch
8-in concrete block	1.11	for stated thickness
w/vermiculite in cores	2.1	for stated thickness
10-in concrete block	1.20	for stated thickness
w/vermiculite in cores	2.9	for stated thickness
12-in concrete block	1.28	for stated thickness
w/vermiculite in cores	3.7	for stated thickness
Common brick	0.2–0.4	per inch
WOOD AND WOOD PANELS		
Softwoods	0.9–1.1	per inch
Hardwoods	0.8–0.94	per inch
Plywood	1.24	per inch
Waferboard or oriented strand board	1.59	per inch
FLOORING		
Carpet (1/4-in nylon level loop)	1.36	for stated thickness
Carpet (1/2-in polyester plush)	1.92	for stated thickness
Polyurethane foam padding (8 lb density)	4.4	per inch
Vinyl tile or sheet flooring (nominal 1/8 in)	0.21	for stated thickness
Ceramic tile	0.6	per inch

* The R-value for a specific thickness of a material may be obtained by multiplying the R-value per inch by the thickness in inches (or fractions of inches). The units on R-value are the standard U.S. units of °F X hr X ft^2/Btu.

MATERIAL (CONTINUED) — R-VALUE* (CONTINUED)

MISCELLANEOUS		
Drywall	0.9	per inch
Vinyl clapboard siding	0.61	for all thicknesses
Fiberboard sheathing	2.18	per inch
Building felt (15 lb/100 ft^2)	0.06	for stated thickness
Polyolefin housewrap	~0	for all thicknesses
Poly vapor barriers (6-mil)	~0	for stated thickness

The data in this table were taken from a number of sources, including the *ASHRAE Handbook of Fundamentals* and literature from several material suppliers. It represents typical R-values for the various materials. In some cases, a range of R-value is stated due to variability of the material. For more extensive data, consult the *ASHRAE Handbook of Fundamentals* or contact the manufacturer of a specific product.

R-VALUES OF AIR FILMS

INSIDE AIR FILMS	R-VALUE*
Horizontal surface w/upward heat flow (ceiling)	0.61
Horizontal surface w/downward heat flow (floor)	0.92
Vertical surface w/horizontal heat flow (wall)	0.68
45-degree sloped surface w/upward heat flow	0.62
OUTSIDE AIR FILMS	
7.5 mph wind on any surface (summer condition)	0.25
15 mph wind on any surface (winter condition)	0.17

* The R-value for a specific thickness of a material may be obtained by multiplying the R-value per inch by the thickness in inches (or fractions of inches). The units on R-value are the standard U.S. units of °F × hr × ft^2/Btu.

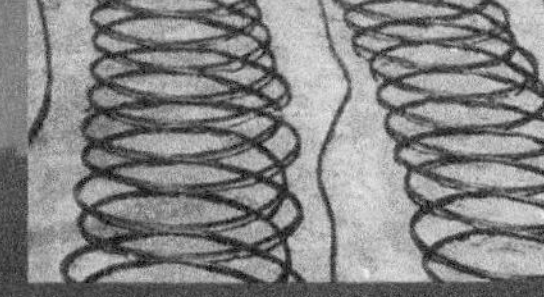

APPENDIX

Useful Conversion Factors and Data C

LENGTH:

1 foot (ft.) = 12 inches (in.) = 0.3048 meter (m)

AREA:

1 square foot ($ft.^2$) = 144 square inches ($in.^2$) = 0.092903 square meters (m^2)

VOLUME:

1 cubic foot ($ft.^3$) = 7.49 gallons (gal.) = 1,728 cubic inches ($in.^3$)
1 gallon (gal.) = 3.7853 liter (L)

FLOW RATE:

1 gallon/minute (gpm) = 0.002225 $ft.^3$/sec. = 0.063008 L/sec. = 0.227126 m^3/hr.

VELOCITY:

1 m/sec. = 3.2808 ft./sec. = 2.2369 mph

PRESSURE:

1 bar = 14.7 (psi)
1 pound per square inch (psi) = 6894.76 N/m^2 = 6894.76 pascal (Pa)
Absolute pressure (psia) = [gauge pressure (psig) + 14.7]

ENERGY:

1 kilowatt hour (kWhr) = 3,413 Btu
1 therm = 100,000 Btu
1 MMBtu = 1,000,000 Btu
1 Joule (J) = 0.7376 ft.·lb. = 0.2388459 calorie (cal.)
1 Kilojoule (KJ) = 0.9478 Btu
1 Btu (British Thermal Unit) = 777.98 ft.·lb.

POWER and HEAT FLOW:

1 kilowatt (kW) = 3,413 Btu/hr.
1 ton = 12,000 Btu/hr.
1 horsepower (hp) = 0.746 kW = 2,546 Btu/hr.

TEMPERATURE (Scale):

°C = (°F − 32)/1.8
°F = (°C × 1.8) + 32

TEMPERATURE (difference):

1 °C = 1.8 °F
1 °F = 0.55555 °C

THERMAL CONDUCTIVITY:

1 Btu/(hr · ft · °F) = 1.7307 W/(m · °C)

HEAT TRANSFER COEFFICIENT:

1 Btu/(hr · ft^2 · °F) = 5.67859 W/(m^2 · °C)

SPECIFIC HEAT:

1 Btu/(lb · °F) = 4185.85 J/(kg · °C)

HEAT FLUX or SOLAR RADIATION INTENSITY:

1 Btu/(hr · ft^2) = 3.15378 W/m^2

INLET FLUID PARAMETER:

1 (°F · hr. · $ft.^2$/Btu) = 0.17615 (°C · m^2/)W

CHEMICAL ENERGY CONTENT OF COMMON FUELS:

1 cubic foot of natural gas = 1,020 Btu (chemical energy content)
1 gallon #2 fuel oil = 140,000 Btu (chemical energy content)
1 gallon propane = 91,200 Btu (chemical energy content)

Glossary

\$/MMBtu The abbreviation for dollars per million Btu.

2-pipe air handler An air handler with a single coil having one supply pipe and one return pipe connection.

2-position spring-return actuator An electrically operated device that produces linear or rotary motion to move the shaft of a valve or damper. The actuator moves in one direction when electrical power is applied and moves in the opposite direction, powered by an internal spring, as soon as the electrical power is removed.

2-stage outdoor reset controller An outdoor reset controller that has two output contacts. One set of contacts closes to operate a heat source. The other set of contacts close to operate a different heat source *if* the output from the first stage heat source is insufficient to meet the load.

2-stage temperature setpoint controller A temperature setpoint controller that has two output contacts. One set of contacts closes to operate a heat source or cooling source. The other set of contacts close to operate a different heat source or cooling source if the output from the first stage heat source, or cooling source, is insufficient to meet the load.

3-way diverting valve An electrically operated valve that is used to route an incoming flow stream to one of two possible outlet streams.

3-way mixing valve Any 3-port valve in which two inlet fluid streams, usually at different temperatures, are blended into a single outlet stream.

3-way motorized mixing valve A 3-port mixing valve having its stem operated by an electrical actuator.

3-way rotary type mixing valve A 3-port mixing valve with a stem that requires approximately 90° of rotation to change a fully open inlet port to a fully closed inlet port.

3-way thermostatic mixing valve A 3-port mixing valve in which the inlet ports are controlled by a temperature sensitive "heat motor." This type of valve requires no electrical power.

4-pipe distribution system A hydronic distribution system commonly used in buildings that can require simultaneous heating and cooling in different areas of the building. The system has a hot water supply and hot water return pipe as well as a chilled water supply and chilled water return pipe. In most cases, these pipes supply two separate coils in each air handler.

4-pipe square A common horizontal earth loop configuration in which the buried piping forms the corners of a square when viewed in cross section. The two upper pipes are directly above the two lower pipes.

4-way valve A type of mixing valve, usually equipped with a motorized actuator, that is specifically designed to interface a conventional (higher-temperature) boiler to a low-temperature hydronic distribution system. This valve creates two internal mixing points: one for supply temperature control, and the other for controlling boiler inlet temperature.

4-wire zone valve A common type of zone valve that has two wires to operate its internal motor and two additional wires connected to an isolated end switch. The end switch closes when the valve reaches its fully open position.

97.5 percent design dry bulb temperature The temperature at which outdoor air is equal to or above 97.5 percent of the year. Often used as the basis for a design heating load calculation.

A

Absolute pressure The pressure of a liquid or gas relative to a pure vacuum. It is often expressed in units of psia (pounds per square inch absolute).

Absolute temperature The temperature expressed relative to absolute zero (–458 °F).

Absorber area The upper surface area of an absorber plate in a flat-plate solar thermal collector.

Absorber plate The plate consisting of sheet metal and bonded metal tubing that absorbs incoming solar radiation, coverts it to heat, and transfers that heat to a fluid passing through the bonded tubing.

Absorber strip Narrow and long strip of copper with bonded tubing that absorbs incoming solar radiation within an evacuated tube solar thermal collector.

Absorptivity The percentage of incoming electromagnetic radiation that is absorbed by the surface it strikes.

Acceptance volume The volume of fluid that can enter a closed expansion tank without the pressure of the air in the tank exceeding a set limit.

Active solar Refers to systems that use circulators or a blower to move heat from solar collectors to storage and from storage to load.

Actuator An electrically or pneumatically driven device that adjusts the position of a valve stem or damper based on the signal it receives from a controller.

AFUE Annual Fuel Utilization Efficiency. An index established by the U.S. Department of Energy to indicate the annual average efficiency at which a boiler converts the energy in fuel to heat.

Aggregate The solid materials in concrete, such as crushed stone (course aggregate) and sand (fine aggregate).

AHRI Abbreviation for the Air-Conditioning, Heating, and Refrigeration Institute

Air and dirt separator A device designed to remove both air bubbles and dirt particles from a fluid stream passing through it.

Air change method A traditional method of expressing the rate of air leakage into a building. One air change per hour means the entire volume of air in the building is replaced with outside air each hour.

Air control Capturing air from areas within a hydronic system where it is undesirable and routing it to other areas in the system where it is desirable (e.g., within an expansion tank).

Air elimination Capturing air within a hydronic system and ejecting that air from the system.

Air film resistance The thermal resistance of the air film along a wall, ceiling, or floor surface.

Air handler A generic name for a device consisting of a blower and a fin-tube coil. It is used to either heat or cool the air passing through it based on the temperature of the fluid circulating through the tubes of the coil.

Air infiltration The process of outside air entering a building as inside air leaks out of the building.

Air separator A device that separates air bubbles from the flowing system fluid and ejects them from the system.

Air-side pressurization The pressure on the air side of a diaphragm-type expansion tank before fluid enters the tank. This pressure is adjusted by adding or releasing air through the Schrader valve on the tank.

Air-source heat pump A heat pump that extracts low-temperature heat from outside air using its evaporator.

Air-to-water heat pump A heat pump that extracts low temperature heat from outside air using its evaporator, increases the temperature of that heat, and transfers it to a fluid flowing through its condenser.

Air vent A manual or automatic device that releases air bubbles from a piping system.

Ampacity The maximum current that can be carried by a conductor in a given application as determined by the National Electrical Code.

Amplitude The distance from the horizontal axis to the farther point on a mathematic curve above or below that axis.

Analog signal An electrical voltage or current that varies over a predetermined range and is typically used to control a modulating device.

Angle of incidence The angle between beam radiation from the sun and a line that is perpendicular to the glazing surface of a solar thermal collector.

Annual solar fraction The percentage of an annual heating load that is met by solar energy.

ANSI/AHRI/ASHRAE/ISO Standard 13256-2 A rating standard often used to express the heating capacity or cooling capacity of a water source heat pump.

Aperture area The area of the glazing on a flat-plate solar collector.

Approach temperature The difference in temperature between the incoming hot fluid and leaving cooler fluid flowing through a heat exchanger.

Aquastat A device that measures the temperature of a liquid at some point in a system and opens or closes electrical contacts based on that temperature and its setpoint temperature.

ASHRAE Standard 93-2010 A testing standard used in the United States to establish the instantaneous thermal efficiency of a solar collector.

ASME pressure vessel code A standard established by the American Society of Mechanical Engineers for testing and certifying acceptable operating conditions for pressure vessels such as tanks.

ASSE 1017–rated thermostatic mixing valve A thermostatic mixing valve that meets the criteria established by the American Society of Sanitary Engineers' Standard 1017. This is a common specified standard for mixing valves that provide anti-scald protection in domestic hot water systems.

ASTM Standard E44 A standard that specifies the acceptable accuracy and verification testing required for heat metering devices.

Atmospheric A term often used to describe gas burners that are designed to operate directly exposed to (unpressurized) room air.

Autumnal equinox The precise moment in September when the sun is directly overhead on the equator at solar noon.

Auxiliary heat source A heat source that supplements the heat output of a renewable energy heat source when needed. One example is a boiler.

Available floor area The floor area in a room that can be used to release heat as part of a floor heating system.

Average floor surface temperature The mean temperature of the upper surface of a heated floor.

Average flow velocity A velocity calculated by dividing the volumetric flow rate of a fluid by the cross sectional area the fluid is flowing through.

B

Backdrafting The reverse flow of combustion gases within a fuel-burning heat source, or its venting system.

Backflow preventer A piping component that contains the functional equivalent of two check valves and a vent port. It prevents any fluid from within a hydronic system from flowing backward and contaminating a fresh water source.

Background ventilation A relatively low but adequate flow rate of outside air for normal ventilation within a building that is not experiencing any condition for which ventilation air flow must be increased, such as an abnormally high interior humidity level due to cooking.

Balance point temperature The outdoor temperature below which a building requires heat input from its space heating system.

Balance-of-system Referring to the portion of a heating system other than the heat source.

Balanced ventilation system A ventilation system in which the flow rate of incoming fresh air is equal to the flow rate of outgoing "stale" air.

Ball valve A valve containing a rotating ball with a hole through it. It can be used for component isolation or limited flow regulation.

Ballast Generally referring to materials used specifically to hold down other devices, such as solar collectors, that are exposed to wind forces of other forces trying to move them.

Bar A unit of pressure. 1 bar = 14.504 pounds per square inch (PSI).

Barium getter A coating applied to the interior end portion of an evacuated tube collector that helps absorb any residual air molecules within the tube and changes color if the vacuum in the tube falls to unacceptable levels.

Baseboard tee A special fitting that resembles a 90°elbow with an additional threaded tapping. This fitting is often used to mount an air vent at the outlet end of a fin-tube baseboard element.

Basic service charge The portion of a monthly utility bill that is a fixed charge associated with having that utility service connected to a building.

Beam (solar) radiation Solar radiation coming directly from the direction of the sun.

Bellows connector A flexible pleated tube designed to connect two devices and allow for slight misalignment as well as expansion and contraction movement.

Bernoulli's equation An equation relating the pressure head, elevation head, and velocity head of a fluid at any point within a flow stream.

Biocide A chemical additive that limits the growth of microbes within a fluid.

Black body A hypothetical body that is a perfect absorber and emitter of thermal radiation.

Blower Different from a fan, a blower creates air motion by using an electric motor to spin a "squirrel cage" assembly of curved metal blades.

Boiler The term commonly used for a pressure-rated hydronic heat source that converts gas, fuel oil, or electricity into heat.

Bond breaker A material used to prevent an adhesive bond between two materials.

Bonded (electrically) Connected to earth ground by a metallic connector.

Boyle's law An equation stating that the product of absolute pressure multiplied by the volume of an ideal gas remains a constant when no temperature change occurs.

Brazed-plate heat exchanger A plate-type heat exchanger typically made of stainless steel plates that have been brazed along their outer edges to form two pressure-rated fluid compartments.

Breather plug Small opening in a solar thermal collector that allows the air pressure inside the collector to equalized with the air pressure outside the collector.

Brine A generic term for a solution of water and antifreeze such as propylene glycol, ethylene glycol, or dissolved salts.

British thermal unit The amount of energy required to raise 1 pound of water by 1°F. Abbreviated as Btu.

Brushless DC motor A motor that has a permanent magnet rotor and electronically commutated stator poles. Sometimes referred to as an ECM motor.

Btu meter A device that senses both flow rate and temperature differential at some location in a hydronic system and uses this information to calculate the total number of Btus that have passed through that location.

Btu metering The process of measuring the total heat passing a given location in a hydronic system by measuring the flow rate and temperature drop of the fluid being used.

Buffer tank An insulated water-filled tank that adds thermal mass to a heating or cooling system. In heating applications is allows the rate of heat generation by the heat source to be different from the rate of heat dissipation from the distribution system.

Building Heat Load Estimator A module with the Hydronics Design Studio software used to estimate the design load of a user-specified building.

Building heating load The rate at which heat must be added to a building to maintain a set indoor temperature.

Building heating load profile A graph of how a building's heating load varies as a function of time.

Building UA The overall heat transfer coefficient of a building. It indicates the rate of heat loss from the building per degree temperature difference between the inside and outside air temperatures. In the United States, the building UA value is typically expressed in units of Btu/hr/°F.

Bulk air All air mixed with the fluid within a hydronic system other than small bubbles. It is imperative to purge the bulk air out of a hydronic piping system when it is commissioned

Bulkhead fitting A low-pressure-rated fitting designed to pass through the wall of an unpressurized tank, allowing a pipe to connect to the tank.

Bullhead tee Any tee configured so that flow enters the side port and passes out through the two end ports.

Buoyancy The upward force that causes a less dense material to rise above a more dense fluid.

Burst point temperature The temperature at which piping filled with an antifreeze solution of a given concentration will burst due to expanding ice crystals in the antifreeze.

Butt fusion machine A fixture designed to hold two ends of polymer piping in precise alignment as their ends are faced, heated, and finally pressed together to create a butt fusion joint.

Butt fusion welding The process of joining two thermoplastic pipe ends by facing, heating, and pressing their ends together.

C

CAD software Software for computer-aided design or computer-aided drafting.

Capacitance rate The product of mass flow rate multiplied by the specific heat for a fluid. This value can be expressed in units of Btu/hr/°F.

Capillary tube A very small diameter tube between a temperature sensing bulb and a controller.

Carnot COP The theoretical maximum COP (coefficient of performance) achievable by any heat pump for a given source and sink temperature.

Cavitation The formation of vapor pockets when the pressure on a liquid drops below its vapor pressure. Cavitation is very undesirable in circulators.

Center off (switch) A switch that does not provide a conducting path when its handle is in the center position.

CFM The abbreviation for cubic feet per minute.

Chain trencher An excavator designed to create narrow trenches several feet deep using a rotating chain with attached scoops.

Chamfer tool A rotating tool that cuts a chamfer (e.g., bevel) on the end of a pipe.

Characterized valve A valve with a specifically shaped flow control element that creates an equal percentage flow characteristic.

Circulator An electrically driven device that adds mechanical energy (e.g., head) to a fluid using a rotating impeller.

Close-coupled Bolting two or more inline circulators together end to end, in the same flow direction, to increase head.

Closed earth loop Piping buried in the ground that forms a complete circuit and can operate with some fluid pressure.

Closed-loop control system A control system configuration in which the controlled variable is sensed and the control action modified (if necessary) based on any deviation between the sensed value and target value of the controlled variable. Also know as a feedback control loop.

Closed-loop system A piping system that is sealed at all points from the atmosphere.

Closely spaced tees Two tees that are installed immediately adjacent to each other to create hydraulic separation between two piping circuits.

Coalesce The process by which molecules of dissolved gases in a liquid come together to form bubbles.

Coalescing media A mesh-type element often used in air separators to enhance the coalescence of dissolved gas molecules.

Coefficient of linear expansion A number indicating the change in length of a material, per unit of original length, and per degree of temperature change.

Coefficient of performance (COP) The ratio of the rate of heat output of a heat pump, to the rate of electrical energy input, in the same units. High COPs are desirable.

Coil A term used in the HVAC industry to describe a fluid-to-air heat exchanger in which copper tubing passes through several rows of aluminum fins. The fluid passes through this tubing while air passes across the aluminum fins.

Cold night fluid shrinkage Describes the fact that an antifreeze in an external piping circuit will thermally contract due reduced night time temperature.

Cold ring clamp A tool used to hold the end of thermoplastic tubing during socket fusion welding.

Collector area An ambiguous term that can be further specified as gross collector area or net collector area. Both are defined in this glossary.

Combisystem A system designed to provide space heating and domestic water heating.

Combustion efficiency The efficiency of a heat source in converting the chemical energy content of its fuel into heat, based on measurements of exhaust gas temperature and carbon dioxide content.

Common piping The piping path in a multiple branch system that is shared with branch circuits and through which the full system flow passes.

Compact style panel radiator A style of panel radiator in which the supply and return piping connect to the bottom of the panel. Compact radiators are also typically supplied with a built-in flow control valve.

Component isolation The use of valves on all piping connections to a given device that allow that device to be isolated for service if necessary.

Compressible fluid A fluid, usually a gas, that can be compressed into a smaller volume when pressure is applied to it.

Compressor The device in which a refrigerant gas is compressed in volume and, at the same time, increased in temperature.

Concrete thin-slab A layer of concrete in the range of 1 to 2 inches thick placed over hydronic tubing to create a radiant floor panel.

Condensate The liquid that is formed as a gas, such as water vapor, condenses.

Condensate drip pan A pan placed under a coil within an air handler to catch and drain away liquid water condensing on the coil.

Condenser The device in which a refrigerant gas releases heat and changes from a vapor into a liquid.

Condenser bulb The end of an evacuated solar collector tube in which a vaporized working fluid condenses to a liquid as it transfers heat to an antifreeze fluid passing through the header of the collector.

Condensing boiler A boiler designed to withstand the effects of sustained flue gas condensation and capable of achieving high thermal efficiency when operated at low incoming water temperatures.

Conduction A process in which heat is transferred by molecular vibrations through a solid or liquid material.

Constant differential pressure control A control algorithm that adjusts the speed of a circulator so that the pressure differential between its inlet and outlet ports remains at a set (fixed) value regardless of flow rate.

Contactor Another name for a relay that is rated to operate at relatively high current and line voltages.

Contacts The metal components within a switch or relay that come together or move apart to determine if electrical current will flow through a given pole of the switch or relay.

Control differential The intended difference in the sensed control variable from where the controller output is turned on to when it is turned off. For example, if a thermostat contact is expected to open at 72 °F and close at 68 °F, the control differential is 4 °F.

Control joint A groove penetrating partway through a concrete slab to force a crack to occur along a predetermined path.

Controlled device The hardware device that is being regulated by the controller.

Controlled variable The physical parameter that is being measured and controlled by the controller. Typical controlled variables in hydronic systems include temperature and pressure.

Controller The device that monitors the controlled variable and provides control outputs that direct operation of a controlled device.

Convection A means by which heat is transferred as the result of fluid movement.

Convector A generic name for heat emitters that transfer the majority of their heat output by convection.

Conventional boiler A boiler that is intended to operate without sustained internal condensation of the fluid gases it creates during combustion.

Cooling capacity The rate at which a device can extract heat from a fluid passing through it. In the United States, this is typically measured in Btu/hr.

Correction factor A number that multiplies a stated value based on differences in operating conditions.

Counterflow A piping arrangement that causes two fluids to flow in opposite directions as they pass through a heat exchanger. This is desirable for maximum heat exchange.

Cover sheet A thin sheet of material (usually plywood or cement board) installed over the top of an above-floor tube and plate radiant panel to serve as a smooth and stable substrate for finish flooring.

Cross-linked polyethylene tubing (PEX) High-density polyethylene tubing that has had its molecular structure altered to change it from a thermoplastic to a thermoset plastic and increase its pressure/temperature ratings.

Cv value A number, sometimes called a "flow coefficient," that indicates the number of gallons per minute of 60 °F water that must pass through a fully open valve to create a 1 psi pressure drop across that valve.

D

Darcy-Weisbach equation A classic equation from fluid mechanics that relates the head loss of a fluid to the flow rate, fluid characteristics, and piping through which the fluid is passing.

Dead loading The load placed on a structure such as a floor because of the weight of the materials used to construct it.

Deadheaded A condition in which a circulator is operating but, due to a blockage or a closed valve, cannot create any flow rate through itself.

Declination angle The angle between the polar north/south axis of the earth and the plane of the earth's orbit around the sun.

Degree day The difference between the outside average temperature and 65 °F over 24 hours. Daily degree days can be added together to obtain monthly and annual total degree days.

Deionized water Water that has been processed to remove a high percentage of chemical ions it would otherwise contain.

Delay-on-break (relay) An operating mode for a time delay relay in which the contacts maintain their operating state until a set time has elapsed from when the input signal to the relay was turned off.

Delay-on-make (relay) An operating mode for a time delay relay in which the contacts maintain their normal (deenergized) state until a period of time has elapsed from when the input signal to the relay was turned on.

Denatured (alcohol) Ethyl alcohol that has been treated to render it undrinkable.

Density In the context of this book, density refers to the weight of a substance divided by its volume. Some readers will recognize this as being the same as the specific weight of a substance. Common English units for density, as defined herein, are lb/ft^3 (pounds per cubic foot).

Description of operation A written description of how devices within a heating or cooling system operate to achieve a desirable result in each of the system's operating modes.

Design heating load The heating load of a building or room when the outside air temperature is at its design dry bulb value.

Design supply water temperature The temperature of water (or other fluid) supplied to a hydronic distribution system under design load conditions.

Desuperheater A refrigerant-to-water heat exchanger that receives superheated refrigerant gas from the discharge of a compressor and cools the refrigerant to near its vapor saturation temperature. The heat extracted from the refrigerant is typically used to heat domestic water.

Dewpoint temperature The temperature at which water vapor in a mixture of other gases begins to condense into liquid droplets. Other compounds in a mixture of gases also have an associated dewpoint temperature.

Diameter ratio (DR) The ratio of the outside diameter of a tube divided by the tube's wall thickness.

Diaphragm-type expansion tank An expansion tank with a built-in air chamber separated from its water chamber by an elastomeric diaphragm.

Differential The difference between two temperatures.

Differential temperature controller A controller that operates its outputs based on the difference between two measured temperatures.

Diffuse radiation Solar radiation that comes from the entire sky dome, but excluding solar radiation coming directly from the position of the sun.

Digital input An input signal to a controller that, at any time, can only have one of two possible states, such as on or off.

Dimensionless number A calculated value, based on several physical quantities, that are combined so that their physical units all cancel out.

DIN 4726 standard An-often cited standard of allowable amounts of oxygen diffusion through polymer tubing.

Dip tube A tube running along the vertical dimension of a tank that allows cooler water to be added to or removed from the lower portion of that tank.

Direct radiation Solar radiation received directly from the solar disk position in the sky. Also referred to as beam radiation.

Dissolved air Molecules of oxygen, nitrogen, or other gases that constitute air and are dissolved within a liquid.

Distilled water Water that has been purified through evaporation and subsequent condensation.

Distribution system The portion of a hydronic system that routes heated or cooled water throughout a building.

Diurnal heat storage Heat storage that is intended to operate on a 24-hour cycle.

Diverter valve A valve (often electrically operated) that directs a single incoming fluid stream in one of two possible directions based on the position of the valve's flow element.

Domestic hot water usage profile A graph showing the quantity of domestic hot water used over a period of time, such as 24 hours.

Domestic water Water that is suitable for human consumption. Sometimes also called potable water.

Domestic water flow switch A switch that closes its contacts when the flow of domestic water passing through it reaches a certain minimum value.

Double rollback bead A desirable deformation of a thermoplastic material around the perimeter of two tube ends being butt fused together.

Double-wall heat exchanger A heat exchanger with an internal channel through which either fluid used in the heat exchanger can flow to the outside of the heat exchanger if a leak occurs.

Double-pumped drainback system A drainback-protected solar thermal system that uses two circulators connected in series to create and maintain flow in the collector circuit.

Drainback freeze protection A solar thermal system in which all water in the collectors and exterior piping drains back to a storage tank within non-freezing space at the end of each collection cycle.

Draw-through fan coil An air handler in which air flows across the coil before it passes through the blower.

Driving ΔT The difference in temperature between the water in a heat emitter and the air surrounding the heat emitter.

Dry contact A switch or relay contact within a controller that does not connect to a power source within that controller.

Dry stagnation A condition in which a solar collector undergoing stagnation does not contain a liquid.

Dual-fuel approach A heating system containing two heat sources that operate on different fuels.

Ductless cooling system A cooling system that delivers indoor air flow from wall or ceiling-mounted cassettes rather than through ducting.

Due south A direction that passes directly through the earth's geographic South Pole. Also sometimes called true south. 180° opposite from true north.

DX The abbreviation for Direct eXpansion refrigeration systems, which use the vapor compression refrigeration cycle.

Dynamic pressure distribution The pressures at all locations within a piping system when the circulator is operating and fluid is flowing.

Dynamic viscosity A physical property that in part determines the drag characteristic of a fluid passing along a surface. Higher viscosity fluids create increased drag.

E

Earth-coupled heat pump A heat pump that extracts low-temperature heat from soil when operating in heating mode and rejects higher temperature heat to soil when operating in cooling mode.

Economically sustainable When the financial merits of a particular approach to heating or cooling will continue to be viewed favorably by mass-market consumers in the absence of financial subsidies.

EEPROM Electrically erasable programmable read-only memory. This type of memory is often used to store "firmware" operating instructions within microprocessor-based controllers.

Effective pump curve The head versus flow rate curve of a circulator contained with an appliance, which has been corrected by subtracting the head loss created by the other components in that appliance.

Effective total R-value The average thermal resistance of a wall, ceiling, floor, etc., after compensating for the presence of framing members.

Effectiveness (of a heat exchanger) The ratio of the actual heat transfer across a heat exchanger to the maximum possible rate of heat transfer. Effectiveness is a unitless number between 0 and 1. Higher values are more desirable.

Electrical contact The metallic parts of a switch or relay that come together or move apart to allow or prevent current flow through the switch or relay.

Electrical noise Any undesirable induced voltage or current within a conductor caused by the electrical or magnetic fields of nearby components.

Electrofusion A process by which thermoplastic pipe is joined using fittings that have embedded electrical filaments that can be electrically heated.

Electromagnetic radiation A means of energy transfer by electrical and magnetic fields. Visible light is one example. Other examples include infrared light, ultraviolet light, X-rays, and microwaves.

Electromagnetic spectrum The entire range of wavelengths of electromagnetic radiation from low-frequency/long-wavelength radio waves to high-frequency/short-wavelength gamma rays.

Electronically commutated motor (ECM) Also known as a brushless DC motor. A motor with a permanent magnet rotor and electronically commutated stator coils. This type of motor must be run by a controller.

Elevation head The mechanical energy contained in a fluid by virtue of its elevation above some established reference elevation.

Embedded controller Any self-contained, dedicated purpose controller that is part of an overall control system.

Emissivity A number between 0 and 1 that indicates the ability of a surface to release thermal radiation. Surfaces with high emissivities release thermal radiation at higher rates compared to surfaces with low emissivities.

EN442 A European standard for estimating the output of panel radiators based on a range of operating conditions. This standard is also used as a basis for certifying heat outputs.

End switch A switch within a zone valve or manifold valve actuator that closes when the zone valve or manifold valve actuator reaches its fully open position

Energy efficiency ratio (EER) An index that indicated the efficiency of a cooling device. Mathematically, it is the device's cooling capacity in Btu/hr divided by its input power in watts.

Energy Star A voluntary program operated by the U.S. Environmental Protection Agency (EPA) to quantify the energy use of appliances as well as new homes and commercial buildings (**www.energystar.gov**).

Entering load water temperature (ELWT) The temperature of water entering the condenser of a water-to-water heat pump.

Entering source water temperature The temperature of the liquid entering the evaporator of a water source heat pump.

Entrained air Air bubbles carried along with the fluid flowing through a piping system.

Entropy A thermodynamic property that indicates the usefulness of energy. Energy in low entropy form (such as electricity) is highly useful because it can be easily converted to several of the forms of energy (e.g., mechanical, electrical, chemical). Energy in high entropy form, such as heat in the soil, is not as useful because of the difficulty in converting it into other forms or using it directly.

EPA Public Law 111–380 Commonly known as the "no-lead law," this amendment to EPA's Safe Drinking Water Act requires all wetted surfaces with components used to deliver potable water to contain less than 0.25 percent lead. This law went into effect on January 4, 2014.

EPDM The abbreviation for ethylene propylene diene monomer. A high-quality rubber compound resistant to many chemicals.

Equal percentage characteristic A valve characteristic in which the flow rate through the valve increases slowly as the stem starts to lift and increases faster the higher the stem lifts. This characteristic gives an equal percentage of change over the previous flow for equal increments of stem movement. This characteristic is desirable when heat output from a heat emitter is being controlled by varying the flow rate through it.

Equilibrium A "balanced" operating condition in which energy input equals energy output.

Equivalent length The concept of replacing a particular piping component by a straight length of pipe, of the same diameter, that would yield the same hydraulic resistance as the component being replaced.

Erosion corrosion A process where metal is removed from the inside surfaces of piping or piping components due to excessively high flow velocity.

Error The difference between a target value and the measured value of a controlled variable.

Escutcheon plate A decorative trim piece that slides over a pipe and covers the hole in the surface the pipe penetrates.

Ethanol A type of alcohol that is sometimes used as an antifreeze in earth loops supplying ground source heat pumps.

Ethylene glycol A type of antifreeze sometimes used in HVAC applications. Ethylene glycol has high oral toxicity.

Evacuated tube A type of solar thermal collector with a glass tube that surrounds the absorber plate. Nearly all air has been evacuated from the glass tube to suppress convective heat loss.

Evacuated tube collector An assembly of several evacuated tubes, along with a manifold and structural support.

Evaporation The process by which a liquid changes into a gas.

Evaporator The component in a refrigeration system inside which a refrigerant absorbs low temperature heat, which causes the refrigerant to evaporate.

Expansion tank A tank specifically designed and sized to accommodate the increased volume of system fluid when that fluid is heated.

External heat exchanger A heat exchanger that is located outside a storage tank.

External VA rating The volt-amps that can safely be supplied to an external electrical load from a controller with a built-in transformer.

Extruded polystyrene foam A type of closed-cell foam insulation board often used for insulation under a concrete slab on grade.

F

F-CHART Software used to implement the f-Chart method of assessing the thermal performance of solar thermal systems.

f-Chart method A simplified method for assessing the performance of solar thermal systems using empirical correlations of simulation result generated by sophisticated computer software called TRNSYS.

Family of circulators A term indicating that a manufacture offers a range of circulators that have different pump curves.

Fan A motor driven propeller that creates air flow, usually against a very low static pressure.

Fan coil A heat emitter that contains an internal blower or fan to force air through a fin-tube coil where it absorbs heat.

Feed water valve A valve that automatically allows cold water to enter a closed hydronic system upon a drop in system pressure.

Feedback A signal sent from a sensor back to a controller that is regulating the controlled variable measured by the sensor.

Feet of head The common English units of expressing the head energy of a fluid. The units of feet represent the total mechanical energy content of each pound of fluid and are derived from simplifying the units of ft•lb/lb.

Fin tube A tube with added fins to enhance convective heat transfer.

Firmware Digital information stored in a special type of solid-state memory that remains valid when power to the overall device is turned off. Sometimes called "non-volatile" memory.

First-hour rating The volume of domestic hot water that a given water heater can deliver within the first hour of a continuous draw. It is based on specified cold water temperature, hot water supply temperature, and rate of heat input.

First law of thermodynamics Energy cannot be created or destroyed, only changed in form.

Flange The common means of connecting a circulator to piping. The circulator's volute has integral flanges on its inlet and discharge ports. These bolt to matching flanges threaded onto the piping. An O-ring provides a pressure tight seal between the flange surfaces.

Flat-plate heat exchanger A modern design for heat exchangers in which fluids flow through alternating spaces between specially shaped stainless steel plates.

Flat-plate solar collector A device that converts solar radiation to heat using a flat, fluid-cooled plate within an insulated and glazed housing.

Float-type air vent A type of automatic air vent that uses an internal float to open and close the air venting valve.

Floating control A common method of controlling an electric actuator that rotates the stem of a mixing valve.

Flow center An assembly of components, including at least one circulator, that prefabricates a portion of the system and thus speeds up installation. Flow centers are commonly used in geothermal heat pump systems between an earth loop and its associated heat pump.

Flow coefficient (Cv) The flow rate, in gallons per minute, of 60 °F water needed to create a pressure drop of 1 psi across a piping component. Cv values for valves are often listed in manufacturers' specifications.

Flow rate The volumetric rate of flow of a fluid. For liquids, it is often expressed in units of gallons per minute (gpm). For gases, it is often expressed in units of cubic feet per minute (cfm).

Flow regulation The action of changing the flow rate through a device, often for controlling heat output or adjusting a mixed fluid temperature.

Flow switch A switch that closes its contacts when flow through it rises above a certain minimum value.

Flow-through evacuated tube A type of evacuated tube solar collector in which the fluid circulated through the collector array also passes through each evacuated tube in that array.

Flow transducer An electronic device that measures flow rate and creates an electrical output signal that can be used by a controller.

Flow velocity The speed of an imaginary fluid particle at some point in a piping system. Common English units for flow velocity are feet per second.

Fluid capacitance rate The product of flow rate multiplied by the specific heat for a fluid stream.

Flush mounting Refers to solar collectors that are mounted parallel to and just above a sloping roof surface.

Foot•pound (ft•lb) A unit of energy that is usually used to express mechanical energy but can be used to indicate any amount of energy. For example, 1 Btu equals approximately 778 foot•pounds of energy.

Foot valve A type of check valve that prevents water from flowing backward through the intake pipe within a water source such as a well.

Forced convection Heat transfer between a solid surface and a moving fluid, where the fluid's motion is created by a fan, blower, or circulator.

Fouling A term describing the undesirable accumulation of particulates, minerals, or other debris on the wetted surfaces of a heat exchanger.

FPT Female pipe thread.

Freezestat A temperature-operated switch often used to turn off a refrigeration process if necessary to prevent a water-filled component from freezing.

Full port ball valve A ball valve in which the hole through the ball is approximately the same diameter as the pipe size of the valve.

Full reset control The ability of an outdoor reset control to reduce the water temperature supplied to the distribution system all the way from some maximum value at design load conditions down to room temperature at no load conditions.

Fully instrumented heat metering A setup for metering heat flow that simultaneously measures flow rate and temperature difference.

Fully modulating (heat source) A heat source that can vary its heat output from 0 to 100 percent of its rated heating output. Most currently available boilers are not fully modulating.

Fusion welding A process used to join thermoplastic tubing and fittings using heat to soften the thermoplastic to it melting point.

G

GPM (gpm) Flow rate expressed in units of gallons per minute.

Galvanic corrosion Corrosion between dissimilar metals caused by the flow of electrons from a less noble metal (on the galvanic chart) to a more noble metal.

Gaseous cavitation Cavitation caused by air bubbles entrained in the fluid rather than vapor pockets.

Gate valve A type of valve designed for component isolation purposes.

Gauge pressure The pressure of a liquid or gas measured relative to atmospheric pressure at sea level. Common English units for gauge pressure are psig (pounds per square inch gauge), or simply psi.

General purpose relay A type of relay that can be used for many different control applications.

Geothermal heat pump A heat pump that extracts low-temperature heat from the earth when operating in heating mode and dissipates higher-temperature heat to the earth when operating in cooling mode.

Geothermal manifold station A piping assembly that allows two or more earth loop circuits to be piping in parallel within a relatively small area.

Glass-lined tank A steel tank that has a heat fused vitreous interior lining that protects the steel from corrosive effects of water in the tank.

Glazing The transparent covering on a solar collector—usually a tempered low-iron content glass.

Globe valve A type of valve designed for flow regulation purposes.

Gooseneck piping Piping that passes through the side wall of an unpressurized thermal storage tank above the wall line and then turns downward both inside and outside the tank.

Graywater heat exchanger A device designed to extract residual heat from water draining from fixtures such as showers and sinks and transfer that heat to cold domestic water on route to a water-heating device.

Gross collector area The projected surface area of a solar collector based on the overall width and height of the collector enclosure or frame.

Ground reflectivity The percentage of incident solar radiation that is reflected from the surface of the earth, or a covering of snow/ice on that surface. Ground reflectivity is a function of the incident angle of the beam solar radiation.

Ground source heat pump A heat pump that extracts low-temperature heat from the earth when operating in heating mode and dissipates higher temperature heat to the earth when operating in cooling mode. This type of heat pump is sometimes called a geothermal heat pump, or an earth-couple heat pump.

H

Hard points Threaded metal studs with associated flashing that have been mounted to a roof structure to support brackets that in turn support solar collectors.

Hardwired logic Specific wiring connections between components that force a control system to operate in a predetermined manner.

Harp-style absorber plate An absorber plate design that uses multiple parallel tubes running between a lower header and upper header.

Head The total mechanical energy content of a fluid at some point in a piping system.

Header well A closed-end tube brazed to the upper header of a solar collector that is sized to allow a temperature sensor to fit snugly inside.

Heat Energy in thermal form. Any material above absolute zero temperature (–458 °F) contains some heat.

Heat capacity A material property indicating the amount of heat required to raise 1 cubic foot of the material by 1 °F. The common English units for heat capacity are Btu/ft^3/°F.

Heat dump An assembly designed to dissipate excess heat from a heat source such as a solar collector or wood-fired boiler.

Heat emitter A generic term for a device that releases heat from a circulating stream of heated water into the space to be heated. Examples include convectors, radiator, and fan coils.

Heat exchanger A device that transfers heat from one fluid to another without allowing the two fluids to make physical contact.

Heat exchanger penalty factor (FR'/FR) A factor that de-rates the performance of a solar collector array based on the presence of a specific heat exchanger between that array and a thermal storage tank.

Heat flux The rate of heat flow across a unit of area. The common English units for heat flux are Btu/hr/ft^2.

Heat meter A device that measures and records the total thermal energy passing across a given location in a hydronic system.

Heat motor A device that moves the stem of a valve when supplied with a suitable voltage. The necessary movement is produced by the expansion and contraction of a wax-filled capsule that is heated by an internal resistor.

Heat pipe A device that moves heat upward using a process in which an internal working fluid is vaporized and subsequently condensed. Many evacuated tube solar collectors operate as heat pipes.

Heat pump A device that uses the refrigeration cycle to move heat from a material at some temperature to a material at a higher temperature.

Heat recovery ventilator (HRV) A device that extracts heat from an exhaust air stream and transfer that heat to an incoming stream of fresh/cooler air.

Heat sink Any material to which heat that is not needed or wanted can be transferred when necessary.

Heat source reset The use of outdoor reset to turn a hydronic heat source on and off based on outdoor temperature.

Heat trap Piping that is shaped to discourage thermosiphoning flow.

Heating capacity The rate of heat output of a heat source.

Heating effect factor An allowance of 15 percent extra heat output, beyond the laboratory tested heat output, allowed by the IBR rating standard for fin-tube baseboard convectors.

Heating run fraction (f_H) The percentage of elapsed time during which a heat pump was operating.

High limit controller An electromechanical or electronic controller that monitors the water temperature in a heat source and turns off the heat source if the measured temperature reaches a set value.

Homerun distribution system A distribution system in which each heat emitter has its own supply and return tube routed from a common manifold station.

Horizontal earth loop Earth loop piping installed in trenches excavated in soil.

Hydraulic equilibrium The condition where the head added by a circulator exactly equals the head dissipated by viscous friction of the fluid flowing through the circuit.

Hydraulic separation A method of coupling two or more hydronic circuits so that the effect created by an operating circulator in one circuit does not influence the effect created by other operating circulators in other circuits.

Hydraulic separator A device used to create hydraulic separation between two or more hydronic circuits.

Hydronic circuit simulator A module in the Hydronics Design Studio software for simulating the flow and heat transfer of a user-defined hydronic piping system.

Hydronic radiant panel Any roof surface such as a floor, wall, or ceiling that contains embedded tubing used to heat the surface.

I

IAPMO/ANSI S1001.1-2013 A code being developed by the International Association of Plumbing and Mechanical Officials that sets minimum installation standards for solar thermal and hydronic systems.

Impeller A rotating disc with curved vanes contained within a centrifugal pump that adds head to a fluid as it is accelerated from its center toward its outer edge.

Implosion A term describing the rapid and violent collapse of vapor pockets as the pressure around them rises above the vapor pressure of the fluid.

Incident angle The angle between beam radiation from the sun and a line perpendicular to the glazing of a solar collector.

Incident angle modifier A number that multiplies the Y-intercept of a solar collector's instantaneous efficiency line to account for sunlight striking the collector at various incident angles.

Incompressible The ability of liquids to transmit large pressures without themselves being compressed into smaller volumes.

Infiltration The unintentional leakage of outside air into heated space.

Infrared radiation Another name for thermal radiation. It is electromagnetic radiation having wavelengths longer than can be seen by the eye.

Infrared thermograph An image taken by a camera that maps the surface temperature of objects seen by the camera into colors that can be displayed along with a scale that relates colors to surface temperature.

Inhibited propylene glycol Propylene glycol to which various chemicals called "inhibitors" have been added to stabilize the solution against thermal and chemical breakdown.

Initial lift head The distance from the static water level in the storage tank, or separate drainback tank, to the top of the collector array in a drainback-protected solar thermal system.

Inlet diffuser Specially designed tube inside a thermal storage tank that allows incoming water to disperse horizontally rather than vertically and thus help maintain temperature stratification within the tank.

Inlet fluid parameter The temperature difference between the fluid entering a solar collector and the outdoor air, divided by the intensity of solar radiator striking the collector. The thermal efficiency of a collector decreases in an approximately linear relationship with increasing inlet fluid parameter.

Instantaneous thermal efficiency The ratio of the thermal energy transferred to a fluid passing through a collector divided by the incident solar radiation on the gross area of that collector, measured and calculated at the same moment.

Intermittent flue gas condensation A condition where some flue gases condense against cold heat transfer surfaces and vent connector piping when combustion begins inside a heat source, but rapidly goes away as the heat transfer surfaces warm above the dewpoint of those flue gases.

intermittent heat source A heat source such as a solar collector or manually operated wood-fired boiler, with an unpredictable short-term heat output based on weather conditions or how the heat source is operated.

Internal heat exchanger A heat exchanger contained within a storage tank and relying on natural convection for heat exchange with the water in the storage tank.

Intumescent coating A fire-resistant coating sprayed over polyurethane insulation.

Inverter drive compressor A compressor used in heat pumps that can operate over a wide range of speed and thus regulate the heat pump's heating and cooling capacity.

ISO 9806 A standard that specifies test procedures for assessing the durability, thermal performance, and safety of solar thermal collectors.

Isogonic chart A map showing magnetic declination (e.g., the difference between north as indicated by a magnetic compass and a true north line).

Isolation flange A special flange that contains a built-in shutoff valve.

Isotherms Lines connecting points having the same temperature. Isotherms are frequently used to display the results of finite element analysis of heat transfer through solid materials.

Iteration A design process that entails making successively refined estimates until a mathematically stable operating condition is determined.

K

KPa (kilopascals of pressure) A pressure equal to 1,000 newtons of force distributed over 1 square meter of area.

L

Ladder diagram Electrical schematic drawing that is used to design and document a control system.

Laminar flow A classification of fluid flow where streamlines remain parallel as the fluid moves along the pipe. Laminar flow is more typical at very low flow velocities.

Latent cooling capacity The ability of a chilled water coil or refrigeration coil to cause water vapor in air passing through the coil to condense.

Latent heat Heat that is added or removed from a substance without any temperature change. This occurs as a material changes phase from solid to liquid, or from liquid to vapor.

Latitude An indication of location on the surface of the earth above the equator (north latitude) or below the equator (south latitude). Zero latitude is the Equator. 90° north latitude is the geographic North Pole.

Leaders The portion of a radiant panel tubing circuit that is not within the radiant panel, but instead connects the tubing within the radiant panel to a manifold station.

Leaving load water temperature The water temperature leaving the condenser of a heat pump operating in heating mode.

Legionella A bacteria that can exist in warm water and under suitable conditions can cause Legionnaires' disease.

Life-cycle cost The total cost of owning and operating a device or system over a period called its "design life" that considers material and installation cost, maintenance cost, and cost of capital.

Lift head The distance from the water level in a storage tank to the top of the collector circuit in a drainback-protected solar thermal system

Line voltage section The upper portion of a ladder diagram.

Liquid A physical state of matter in which a material is pourable and conforms to the shape of its container. Most liquids are also incompressible.

Load heat exchanger A term used in the f-Chart method that describes the overall hydronic heat emitter system that transfers heat from water to a building

Local solar time Time based on the position of the sun. On any day, the sun is always directly above a true north/south line at solar noon. Local clock time can be converted to local solar time using methods described in Chapter 7.

Lockshield valve A valve that mounts on the outlet of a heat emitter and serves as a combination isolation valve, flow balancing valve, and drain valve.

Longitude An indication of location on the surface of the earth. Lines of longitude run from the geographic North Pole to the geographic South Pole and are specified by sector angle relative to the prime meridian (0° longitude), which runs through Greenwich, England.

Low iron glazing A type of glass with low iron-oxide content and thus higher transmissivity to solar radiation compared to standard glass.

Low loss header Another name for a hydraulic separator. It can also mean a header with flow characteristics that result in very little dissipation of head energy.

Low-voltage section The section of a ladder diagram below the secondary winding of a transformer. The primary winding of the transformer is powered by the line voltage section of the ladder diagram.

M

Magnesium hydrolite Granules that are used to neutralize the acidic condensate produced by a condensing boiler burning natural gas or propane

Magnetic declination The deviation between a magnetic north/south line and a true north/south line at a given location on the earth's surface.

Make-up water assembly An assembly of components that automatically adds water to a hydronic system to make up for minor losses. It usually consists of a shut-off valve, feed water valve, and backflow preventer.

Manifold A piping component that serves as a common beginning or ending point for two or more parallel piping circuits.

Manifold station A combination of a supply manifold and return manifold that services two or more circuits.

Manipulated variable The physical quantity that is being regulated by a control process.

Manual air vent A device that, when manually opened, allows air to be released (vented) from a piping system.

Master thermostat An electrical thermostat that controls operation of a circulator in a system using multiple non-electric thermostatically operated valves on the heat emitters.

Maximum siphon height The greatest vertical distance between the water surface in a storage tank and the top of the collector array for which a siphon will remain intact as water flow returns from the collector array to the storage tank. This height depends on the temperature and pressure of the water.

Mean radiant temperature (MRT) The mathematical average of the temperatures of all wall, ceiling, and floor surfaces in a room. In some definitions of MRT, the temperature of each surface has a weighing factor based on the location in the room at which the MRT is being determined and the area of each room surface.

Mechanically compressed collar A collar inserted between a hole in a concrete wall and a pipe passing through that wall. Bolts on the collar are adjusted to create a watertight seal between the hole and pipe

Mercaptan An odorant used in propane to give it a very noticeable "rotten egg" smell.

Methanol A type of alcohol sometimes used as an antifreeze in earth loops serving geothermal heat pumps.

Microbubbles Very small air bubbles formed as gas molecules join together (i.e., coalesce). Individual microbubbles are often too small to be seen and instead appear as a "cloud" within otherwise clear water.

Microbubble air separator A type of air separator that captures microbubbles from the system's fluid. It lowers the dissolved air content of the water, enabling it to absorb any residual air in the system.

Microfan A very small fan sometimes used in panel radiators to enhance convective heat output from the radiator at low-supply water temperatures.

Micron Also called a micrometer. A distance of one-millionth of a meter.

Mineralogy The study of the materials in soil.

Minimum downward pitch The minimum slope of a pipe that allows efficient gravity drainage. Typically a minimum of ¼ inch vertical drop per horizontal foot.

Mixing assembly A collection of hardware that constitutes a complete mixing system and forms a "bridge" between a higher temperature heat source loop and a lower temperature distribution system.

Mixing device Any device in which two or more streams of fluid are intentionally mixed together to create a fluid temperature between the higher and lower entering fluid temperatures.

Mixing reset A method of controlling a mixing assembly in which the mixed water temperature leaving the mixing assembly changes with outdoor temperature.

Mixing valve A valve that blends two fluid streams entering at different temperatures to achieve a desired outlet temperature. The most common types in hydronic heating are 3-way and 4-way mixing valves.

MMBtu One million Btus (British thermal units).

Mod/con boiler An abbreviation commonly used for a modulating and condensing boiler.

Mode selection switch A switch that determines when a system operating in heating mode, cooling mode, or remains off.

Modulation A process by which the rate of energy flow is continuously adjusted as necessary to meet the existing load.

Monthly solar fraction (f) The percentage of a thermal load met by a specific solar thermal system on a monthly basis.

Motorized actuator An assembly that combines an electrically operated motor and a gear train to create slow rotation of a shaft on a valve or damper.

MPT Male pipe thread.

N

National Electrical Code (NEC) A model code created and periodically updated by the National Fire Protection Association (NFPA) that sets minimum installation standards for electrical wiring. This code is widely accepted in most areas of the United States.

National pipe thread (NPT) The standard tapered threads used on piping, fittings, and valves in the United States.

National Renewable Energy Laboratory (NREL) A research laboratory in Golden, Colorado, operated by the U.S. Department of Energy.

Natural convection Heat transfer created by a moving fluid when the fluid's motion is created only by natural processes, such as the tendency of a warm fluid to rise.

Negative efficiency An operating condition where a solar collector is dissipating more heat to its surroundings compared to the solar heat it is absorbing.

Negative temperature coefficient (NTC) thermistor A common type of thermistor temperature sensor used by heating and cooling controllers.

Net collector area The glazed area of a solar collector.

Net positive suction head available (NPSHA) The total head available to push water into a circulator at a given point in a piping system. This indicator is totally determined by the piping system and fluid it contains and does not depend on the circulator. It is expressed in feet of head. To avoid cavitation, the NPSHA must be equal to or greater than the NPSHR of the selected circulator.

Net positive suction head required (NPSHR) The minimum total head required at the inlet of the circulator to prevent cavitation. This value is specified by pump manufacturers.

Nocturnal cooling Using a solar collector or other heat exchanger as a heat-dissipating device during the night.

Nominal cooling capacity The cooling capacity of a heat pump at specific "reference" operating conditions.

Nominal heating capacity The heating capacity of a heat pump at specific "reference" operating conditions.

Nominal inside diameter The approximate inside diameter of a tube. For a copper water tube, the nominal inside diameter is the same as the stated pipe size.

Non-reversible heat pump A heat pump that does not contain a reversing valve.

Normal transmissivity The transmissivity of the glazing on a solar collector when beam solar radiation is perpendicular to the plane of the glazing.

Normally closed contacts Electrical contacts in a relay that remain closed when the relay coil is deenergized.

Normally open contacts Electrical contacts in a relay that remain open when the relay coil is deenergized.

Normally open zone valve A zone valve that remains in its open position when no power is applied to its actuator.

O

Off-cycle losses Heat losses from a heat source that occur when the heat source is not producing heat.

Off-gas To unintentionally release a vaporous substance.

Off-peak periods Set times when electrical energy is sold at a reduced rate by the utility.

Offset A residual error between the target value and actual value of a controlled variable associated with proportional only control.

OG-100 Operating Guideline for Certifying Solar Collectors A standard to which solar thermal collectors are tested and certified by the Solar Rating and Certification Corporation.

Open-loop system A piping system that conveys fresh water or is open to the atmosphere at any point.

Operating differential The actual variation in the controlled variable (often temperature) established by a process using on/off control.

Operating mode A specific condition or state of the controlled devices in a system that allows it to accomplish a specific objective. Hydronic heating systems can have several operating modes, including off, on, priority domestic water heating, heat purging, etc.

Operating point The point on a graph where the pump curve intersects the circuit head loss curve. This point indicates the flow rate at which the circuit achieves hydraulic equilibrium and thus maintains a steady flow rate.

Outdoor reset control A control method that increases the water temperature in a hydronic system as the outdoor temperature drops.

Overshoot The condition in which a controlled variable (often temperature) exceeds the desired temperature.

Oxygen diffusion A process by which oxygen molecules diffuse (travel) through certain materials due to a higher concentration of molecules on one side of that material.

Oxygen diffusion barrier A layer of a special compound or metal on the outer surface or within the cross-section of tubing that significantly reduces the rate at which oxygen can diffuse through the tube wall.

P

Panel radiator A metal heat emitter that mounts on a wall and releases a significant percentage of its heat as thermal radiation.

Parallel piping When two or more branch piping paths originate from a common starting point and end at a common ending point.

Parallel reverse return piping Parallel piping in which the first branch connected on the supply main is the last branch connected to the return main.

Parametric study Varying a single numerical input to a performance model and studying the response of the system to that variation.

Passive solar design Any building or system designed to encourage the absorption and thermal storage of solar energy without use of fans, blowers, or circulators.

Peak hourly demand the maximum amount of domestic hot water required during any single hour of a day.

Peak solar collection window Generally considered to be the time between 9:00 AM and 3:00 PM local solar time, during which a high percentage of the total daily solar energy is available for collection.

Permanent split capacitor (PSC) motor The type of electric motor used in many wet-rotor circulators and blowers. It is relatively inexpensive to build, but not highly efficient.

Pete's Plug A special fitting that accepts a needle-type temperature or pressure probe pushed through an elastomeric seal and reseals itself when the probe is withdrawn.

PEX tubing Cross-linked polyethylene tubing.

PEX-AL-PEX tubing A composite tubing having inner and outer layers of PEX and a center layer of aluminum.

pH buffers Chemical added to glycol-based antifreeze to increase its pH.

Physical quantities Any quantity having a numerical value and units. One example would be a temperature of 100 °F, another would be a flow rate of 5 gpm.

Pilot solenoid valve A small electrically operated valve that creates pressure differentials to move the flow element in a high flow capacity valve.

Pipe size A term that, within the context of hydronics, and for pipes under 12 inches in diameter, refers to the approximate inside diameter of a pipe.

Pitch (of the slinky) The distance each loop of the slinky moves ahead of the previous loop within a trench.

Point of no pressure change (PONPC) The location where an expansion tank is attached to a hydronic circuit. The pressure at this point remains unchanged regardless of whether the circulator is on or off.

Polar axis The rotational axis of the earth passing through the geographic North and South Poles.

Poles (of a switch) The number of independent electrical circuits that can be *simultaneously* passed through a switch or relay.

Potassium acetate An antifreeze material used in some earth loops.

Poured gypsum underlayment A water-based slurry containing gypsum cement, sand, and other admixtures that cures into a hard subflooring. Often poured over hydronic tubing that has been stapled to the subfloor.

PPM Parts per million.

Press-Fit A pipe joining method in which the socket of a special fitting is mechanically compressed against a tube wall (i.e., pressed) using a special tool. An elastomeric O-ring within the fitting forms a pressure-tight seal against the tube wall.

Pressure head The mechanical energy contained in a fluid by virtue of its pressure.

Pressure-regulated circulator A circulator that senses the differential pressure it is producing, and strives to maintain a specific value by adjusting its speed.

Pressure switch A switch that opens and closes it contacts based on the pressure it sensed in a system.

Pressure vessel Any closed container, such as a tank, or heat exchanger, that is intended to operate under positive pressure. Some interpretations of what constitutes a pressure vessel and thus subjects the component to ASME Pressure Vessel Code are subject to interpretation by code enforcement officials.

Pressure-reducing valve A special valve that reduces the pressure of water as it flows from a water supply system into a hydronic system. This valve is also called a feed-water valve.

Pressure-relief valve A spring-load valve that opens to release fluid from the system if the pressure at its location exceeds its rated opening pressure.

Priority load A heating load, such as domestic water heating, that takes priority over other loads when both loads are simultaneously calling for heat. In most cases, all other heating loads served by the same heat source are temporarily turned off whenever the priority load is operating.

Processing algorithm A specific set of programming instructions executed by a controller.

Proportional band The operating range of a controller in which the output signal or control action is proportional to the error between the target setting and the actual (measured) value of the controlled variable.

Proportional differential pressure control An operating mode for a pressure-regulated circulator in which differential pressure decreases linearly with flow rate.

Proportional reset A method of producing a reduced supply temperature in a hydronic distribution circuit by supplying a manually set mixing device with a "hot" fluid temperature that itself is reset by an intelligent mixing device.

Proportional-integral control (PI) A control response where the value of the output signal is based on the error between the target value and the actual (measured) value of the controlled variable as well as how long that error has existed.

Proportional-integral-derivative control (PID) A control response where the value of the output signal is based on the error between the target value and the actual (measured) value of the controlled variable, how long that error has existed, and the time rate of change of error.

Propylene glycol A nontoxic chemical used, in combination with corrosion inhibitors, as a water soluble antifreeze in many HVAC systems.

Pulse width modulation (PWM) A control algorithm in which the length of the on-cycle varies based on the error between the target value and the actual (measured) value of the controlled variable.

Pump (centrifugal) An electrically driven device that uses a rotating impeller to add mechanical energy (i.e., head) to a fluid. In this text, the word circulator refers specifically to a centrifugal pump.

Pump curve A graph indicating the head energy added to the fluid by the pump versus the flow rate of the fluid through the pump. Such curves are provided by pump manufacturers.

Pump efficiency The ratio of the rate of head energy transferred to a flowing fluid divided by the rate of mechanical energy input to the shaft of the pump.

Purge cart An installation tool that uses a high capacity pump to force fluid into an empty hydronic system, and in so doing, fills the system with fluid and purge it of air.

Purging The process of filling a system with water and removing air from the system.

Purging valve A combination of two ball valves within a single body that allows for purging operations that add fluid and remove air from a hydronic system.

Pyranometer An instrument for measuring total solar radiation intensity.

R

R-value (of a material) A value indicating the thermal resistance of the material. The greater the R-value, the slower heat will pass through the material, all other conditions being equal.

Radiant panel A heat emitter that releases 50 percent or more of its heat output as thermal radiation and operates at surface temperatures under 300 °F.

Radiative cooling Heat dissipation by thermal radiation from a surface to a clear nighttime sky

rare earth (neodymium) permanent magnets Powerful magnets used in most brushless DC and ECM motors.

Reciprocating compressor A compressor using a reciprocating piston and rotating crankshaft.

Recirculating hot water system An insulated piping circuit for domestic hot water that contains a small circulator to maintain hot water close to all fixture branches.

Reed switch A small, low-current rated switch with contacts mounted on ferrous metal strips (e.g., reeds) that are contained within a sealed glass tube. The contacts are pulled together by a magnetic field in proximity to the switch.

Refrigerant Any chemical compound that is circulated through a refrigeration cycle.

Relative load heat exchanger size A term used in F-CHART software to quantify the size of the heat exchanger that transfers heat from solar heated fluid to a building.

Relay An electrically operated switch having a specific number of poles and throws.

Relay socket The base a relay plugs into and to which wires leading to the relay are connected.

Reset line A straight line having a specific slope on a graph of supply water temperature versus outdoor air temperature.

Reset ratio The ratio of the change in supply water temperature to the change in outdoor air temperature that a reset control will attempt to maintain. The reset ratio is also the mathematical slope of a reset line.

Resistive loads An electrical load that is primarily resistive, rather than inductive, in nature. Examples include any electrical heating element, or the filament of an incandescent lamp.

Return on investment (ROI) An index used to compare the financial appeal of an investment to other possible investments. Mathematically, ROI is the financial benefit realized by the investment, minus the cost of the investment, divided by the cost of the investment.

Reverse thermosiphoning Undesirable natural convection heat transfer caused by a fluid moving from an area of higher temperature to an area of lower temperature without any assistance by a circulator.

Reversible heat pump A heat pump that can operate in either heating or cooling mode as required.

Reversing valve A special valve used to switch the function of the two refrigerant heat exchangers within a heat pump from serving as the evaporator to serving as the condenser.

Reynold's number A unitless number that can be used to predict whether flow is laminar or turbulent.

Room heating load The rate at which heat must be added to an individual room to maintain interior comfort.

Rotameter A type of flow meter that uses a metal plug sliding vertically along a metal rod within a tapered transparent body.

Rung (of a ladder diagram) Any wiring path between the vertical sides of the line voltage portion of a ladder diagram or between the vertical side of the low-voltage portion of a ladder diagram.

S

S-5 clip A special fastener to clamp solar collector mounting rails to standing seam metal roofing.

Schrader fitting A fitting to allow air to be added or removed from a device or system. The valve on an automotive tire is an example of a Schrader valve.

Sealed combustion A design for a combustion-type heat source in which all air required for combustion is drawn through sealed ducting from the outside of the building. All exhaust gases are also routed through sealed ducting to the outside of the building.

Seasonal average COP The average coefficient of performance of a heat pump measured over an entire heating season. Mathematically, it would be the total useful heat produced by the heat pump (in Btus) divided by the product of the total kilowatt•hours of electrical energy used multiplied by 3,413.

Seasonal efficiency The average efficiency of a heat source measured over an entire season. Mathematically, it would be the total useful heat produced by the heat source divided by the total energy content of the fuel consumed in the process.

Second law of thermodynamics Energy always moves from a state of higher quality (low entropy) to a state of lower quality (higher entropy).

Selective surface A special coating applied to the upper surface of absorber plates or absorber strips in solar collectors that has high absorptivity for incident solar radiation and low emissivity of thermal radiation.

Self-contained air-to-water heat pump A water-to-air heat pump in which the entire refrigeration system is contained in a single enclosure.

Sensible cooling capacity The ability of a coil operating on either chilled water or cold refrigerant to lower the dry bulb temperature of the air passing through it.

Sensible heat Heat that when added or removed from a material is evidenced by a change in the temperature of the material.

Sensible heat quantity equation An equation that relates a quantity of sensible heat to a quantity of water and a specific temperature change in that water.

Sensible heat rate equation An equation that determines the *rate* of sensible heat transfer to or from a flowing fluid, based on the flow rate of that fluid, the temperature change of the fluid, and the fluid's physical characteristics.

Sensor well A metal tube, open at one end and closed at the other, that is installed in a tank or boiler. It serves as a sleeve through which a temperature sensor can be inserted. It holds the sensor in the desired location within a fluid filled device.

Series (piping circuit) An assembly of pipe and piping components that are connected to form a single *closed* loop.

Series (piping path) An assembly of pipe and piping components connected end-to-end to form a single path between two points.

Series circulators Two or more circulators installed in the same piping circuit and having the same flow direction.

Serpentine absorber An absorber plate having a single tubing path that makes several 180° bends as it progresses across the plate area.

Setback mode When a thermostat is set to a reduced temperature, usually during the night or times when the building is not occupied.

Setpoint controller A control that attempts to maintain a preset temperature at some location in a system by controlling the operation of other devices.

Setpoint demand A demand for heat source operation from a load that needs to be supplied with water at a specific set temperature.

Setpoint temperature A desired temperature that is to be maintained if possible and is set on some type of controller, such as a thermostat or setpoint controller.

Shielded cable A cable in which the individual conductors are wrapped with a thin metal sleeve that is connected to earth ground.

Short cycling An undesirable condition in which a heat source turns on and off frequently, but only for short periods.

Shorted (electrically) The unintentional connection of an electrified component to earth ground.

Side-sloped (solar collector) A solar collector that has been sloped sideways within its mounting plane so that it will properly drain in a drainback-protected solar thermal system.

Sight tube A vertically oriented transparent tube with both ends connected to a device such as a tank and used to indicate the height of the liquid within that device.

Sink (heat) A material into which heat is released.

Siphon The ability of a fluid to flow upward through a closed piping path, without the use of a circulator, provided the fluid level at the outlet of that piping path is sufficiently lower than at the inlet.

Sky dome The entire hemisphere above a plane that is tangent to the surface of the earth at a given location.

Slugging The undesirable effect that occurs when liquid refrigerant enters an operating compressor.

Smart circulators Any circulator that contains microprocessor-based electronics that allow it to operate based on an internal instruction set in combination with user settings.

Soft soldering A soldering procedure often used with copper tubing that uses a solder allow that can be melted by a low-temperature torch typically operating on propane, map gas, or acetylene gas.

Soldering A method of metallurgically joining copper or copper alloy pipe and components such as fittings and valves using a molten solder alloy fused to the base metal.

Solar altitude angle (ALT) The angle between the center of the sun and a horizontal plane at a given time and location.

Solar azimuth angle (AZI) The angle between the sun's location, projected onto a horizontal plane, and a true north line. At solar noon, the solar azimuth angle is 180 °F.

Solar circulation station A preassembled collection of hardware designed to interface a solar collector array to a storage tank.

Solar constant The intensity of solar radiation just outside the earth's atmosphere, which is relatively constant but varies slightly due to the elliptical (rather than circulator) orbit of the earth around the sun. Its average value is 432 Btu/hr/ft^2.

Solar heating fraction The percentage of a heating load supplied by solar energy over a specified time, such as a month or year.

Solar noon The exact moment when the solar azimuth is 180 °F, and a line dropped down from the sun's position is aligned with a true north/south line.

Solar Pathfinder™ A device used to assess the effect of shadows on the potential location of solar collectors.

Solar photovoltaic module A panel containing multiple solar cells wired in series, which produces a voltage and direct current when exposed to sunlight.

Solar radiation The entire spectrum of electromagnetic radiation emitted by the sun.

Solar spectrum The entire range of wavelengths of electromagnetic radiation emitted by the sun.

Solar Rating & Certification Corporation (SRCC) A nonprofit corporation established to oversee testing and certification of solar thermal collectors; see www.solar-rating.org.

Solar simulators Artificial light sources that create a spectrum of electromagnetic radiation very similar to that of the sun and are used for indoor testing of solar collectors.

Solar thermal collectors Solar collectors designed to warm a fluid (liquid or air) passing through them.

Solar time Time referenced to the position of the sun rather than local time zone.

Solid-state relay A solid-state device used to turn on a high-amperage AC or DC circuit using a very low-amperage AC or DC signal. Solid-state relays can also be used to amplify low power electrical signals.

Source (of a heat pump) The material from which the low-temperature heat is extracted.

Specific heat A material property indicating the amount of heat required to raise 1 pound of the material by 1 °F, often expressed in Btu/lb/°F.

Split system A refrigeration system, such as used in a heat pump, where the evaporator and condenser are housed in two different enclosures. One enclosure is typically outside and the other inside.

Spool (of a valve) An internal mechanism in a mixing valve that slides or rotates to regulate the amount of hot and cool fluid streams entering the valve.

SRCC rating reports Standardized reports that give the physical characteristic of a solar collector as well as certified thermal and hydraulic performance information.

Stagnation (collector) Any condition in which solar radiation strikes a solar thermal collector but no fluid passes through it to remove heat.

Standard meridians Lines of longitude beginning with 0° for the prime meridian in Greenwich, England, and progressing west in intervals of 15°. For example, the standard meridian for the Eastern time zone in the United States is 75 °F west longitude. Each increment of 15° represents one hour and one time zone.

Standard port ball valve A ball valve design where the orifice through the ball is slightly smaller than the diameter of the pipe to which the valve is connected.

Standardized test day A weather profile including solar radiation and outdoor temperature, which is used in combination with collector performance data for a software simulation as part of the SRCC OG-100 testing standard. The purpose is to estimate the total useful heat output of a specific solar collector based on uniform operating conditions.

Static pressure The pressure at some point in a piping system measured while the fluid is at rest. Static pressure increases from a minimum at the top of a system to a maximum at the bottom. Static pressure is created by the weight of the fluid itself, plus any extra pressurization applied at the top of the fluid.

Stationary air pocket Air that accumulates at the high points within a piping system.

Stator The stationary components in a motor that create a magnetic field around the rotor.

Stator poles Individually controlled electromagnets arranged in a circle around the rotor that interact with the magnetic field of the rotor to generate torque on the rotor.

Steady-state heat transfer A stable, unchanging rate of heat transfer.

Steep pump curve A pump curve that produces relatively high head at low flow rates and operates over a small range of flow rates compared to other circulator designs of comparable motor power.

Strata Hypothetical layers of fluid, each at a specific temperature, that approximate temperature stratification within a storage tank.

Stratification The natural tendency for a warm fluid to rise above a cooler fluid due to density differences. This term is commonly used to describe warm air rising toward the ceiling of a room while cooler air descends toward the floor.

Strut bolt A special bolt with a T-shaped head that is designed to be placed into and retained by a standard channel strut.

Subfloor The structural covering installed over floor joists or floor trusses to create a floor deck.

Subgrade The soil under a concrete floor slab.

Summer solstice Commonly known as the longest day of the year, with the longest time between sunrise and sunset.

Sun path diagram A diagram that allows the altitude angle and azimuth angle of the sun to be quickly determined based on location and time of year, and solar time of day.

Sun path disc A disc used in the Solar Pathfinder device and based on a specific latitude.

Superheat The temperature of a vapor above the temperature at which vaporization begins (e.g., 10 °F superheat means that the temperature of the vapor is 10 °F above the temperature of the substance's liquid/vapor saturation temperature).

Supplemental expansion tank A second expansion tank added to the system to compensate for an inadequately sized expansion tank supplied as part of the heat source or other assembly within the circuit.

Sustained flue gas condensation Continuous formation of liquid condensate from the vaporized compounds within the exhaust stream from a combustion process.

Swing check A check valve with an internal disc that swings on a pivot.

Sydney tube A type of evacuated tube solar collector in which the absorption coating is deposited directly on the inner surface of the inner concentric glass tube.

System head loss curve A curve that indicates the head loss of a hydronic circuit, or an entire distribution system as a function of the flow rate through that circuit or system.

System water Water contained within a hydronic system that is not acceptable for human consumption or other purposes requiring potable or domestic water.

T

Tankless water heater A water-heating device that does not contain a storage tank.

Target supply water temperature The theoretically ideal temperature of water being supplied to a hydronic heating distribution system based on the current heating load.

Target value The desired value of a controlled variable.

Tax credit An amount of money that is directly subtracted from the taxes owed.

Temperature An indication of the average kinetic energy, or molecular vibrations within a material, at a given moment.

Temperature drop The difference between the supply temperature and return temperature in an operating distribution circuit, or part thereof.

Temperature and pressure (T&P) relief valve A relief valve that protects a domestic water heater against both excessive pressure and excessive temperature.

Temperature bin hours The number of hours during which the outdoor temperature is within a specific range, such as between 20 °F and 25 °F. Each range of temperature is called a "bin."

Temperature setpoint controller A controller that monitors a temperature sensor and creates an electrical output signal to operate equipment in an attempt to keep the measured temperature as close as possible to the set (target) temperature.

Thermal break The placement of insulation between two materials to prevent rapid heat transfer between them.

Thermal conductivity A physical property of a material that indicates how fast heat can move through the material by conduction. All other conditions being equal, the higher the thermal conductivity, the faster heat can move through the material.

Thermal efficiency The ratio of heat output rate divided by the energy input rate to the heat source to create that output.

Thermal energy Energy in the form of heat. The term thermal energy is used synonymously with the word heat.

Thermal envelope The enclosure formed by all building surfaces that separate heated from unheated space. Exterior walls, windows, and exposed ceilings are all parts of the thermal envelope of a building.

Thermal equilibrium The condition of a system when the rate of energy loss from the system is exactly the same as the rate of energy input to the system.

Thermal expansion/contraction The movement of a material due to changes in temperature. When heated, nearly all materials expand. When cooled, nearly all materials contract.

Thermal expansion valve (TXV) A valve in a refrigeration system that creates a large drop in pressure and temperature of refrigerant passing through valve.

Thermal grease A special type of grease containing tiny flakes of metal (typically aluminum) that increase conduction heat transfer from the inside of a sensor well to the outer surface of a sensor bulb.

Thermal load The rate of heat transfer required for some specific heating or cooling purpose.

Thermal mass The ability of a material to store sensible heat due to its material properties (e.g., density and specific heat) and temperature.

Thermal penalty Describes a decrease in performance between a heat source, such as a solar collector, and an intended heat sink (such as a storage tank) due to the presence of a heat exchanger between them.

Thermal radiation Heat that is transferred in the infrared portion of the electromagnetic radiation. Thermal radiation cannot be seen by the human eye, but in many other ways behaves like visible light.

Thermal resistance The tendency of a material to resist heat flow. Insulation materials, for example, tend to have high thermal resistance. Other materials such as copper and glass have low thermal resistance.

Thermographic image An image taken by an infrared light–sensing camera in which visible colors are used indicate the surface temperatures of objects in the image.

Thermoplastic A polymer that can be melted and reformed in shape.

Thermoset plastic A polymer that has been cross-linked or otherwise chemically altered so that it will not change back to a formable liquid state, as will a thermoplastic.

Thermosiphoning The tendency of heated water to rise vertically within a hydronic piping system, even when the circulator is off.

Thermostat A temperature-operated switch used to control the operation of a heating or cooling system, typically based on room air temperature.

Thermostatic radiator valve (TRV) The assembly of a thermostatic actuator and a radiator valve.

Three-wire control Another name for floating control in which three wires are required between the controller and actuator.

Thermostatically controlled tankless electric water heater An electric tankless water heater with internal controls that measure incoming water temperature and regulate the power supplied to the heating elements so that the leaving water temperature is as close as possible to the desired setpoint value.

Throws The number of settings of a switch or relay in which an electrical current can pass through the switch or relay contacts.

Time delay relay A relay in which the contacts do not open or close at the same time coil voltage is applied or interrupted.

TiNOX® A type of selective surface coating for the upper surface of the absorber plate in a solar thermal collector.

Ton (of capacity) A term representing a rate of energy flow of 12,000 Btu/hr. It is frequently used to describe the capacity of heat pumps and air conditioners.

Total equivalent length The sum of the equivalent lengths of all piping, fittings, valves, and other components in a piping circuit.

Total head The total mechanical energy content of a fluid based on its pressure, elevation, and speed at some specified location in a piping system.

Total R-value (of an assembly) The overall total thermal resistance of an assembly of materials that are assembled together.

Transmissivity The ratio of the solar radiation intensity transmitted through the glazing of a solar collector to the solar radiation intensity at the outer surface of the glazing. This ratio is usually assumed to be when the beam solar radiation is perpendicular to the plane of the glazing.

Triac A solid-state switching device used to turn an alternating current device on or off, or to vary the AC waveform sent to the device.

TRNSYS A sophisticated computer simulation software that can model the dynamic behavior of a wide range of thermal systems, including many solar thermal systems.

Tube passes (in a coil) The number of planes containing parallel tubes that pass perpendicularly through the fins of the coil.

Turndown ratio The ratio of the maximum heat output from a modulating heat source divided by the minimum stable heat output from that heat source.

Turbulators Twisted metal strips inserted in the firetubes of a boiler that create turbulent flow and enhance convective heat transfer.

U

Ultraviolet Wavelengths in the solar spectrum that are too short to be detected by the human eye.

Undershoot When the value of a controlled variable drops below the intended lower limit of operation.

Underside insulation The insulation placed under a heated floor system.

Unit price (of fuel) The price of fuel expressed on a common basis, such as $/MMBtu (dollar per million Btu).

Unit U-value The area weighted average U-value of a window or door assembly.

Unsaturated state (of air solubility) When the amount of dissolved gases in the system water is low enough to enable the water to absorb residual air trapped in other parts of the system.

Upward heat flux The rate of heat output from a heated floor per unit of floor surface area.

Usable temperature drop The difference between the highest average temperature of water in a storage tank, and the lowest average water temperature that is still useable by the heating distribution system.

Usage profile A graph that shows the quantity of domestic hot water required at any hour of the day.

V

VA Volt-amp. An indication of the ability of a transformer to supply reactive AC power to operate electrically powered devices.

Vaporous cavitation The undesirable formation of vapor pockets within the liquid near or within the impeller of an operating circulator.

Vapor compression refrigeration cycle A refrigeration cycle in which a substance (the refrigerant) is repeatedly compressed and expanded to transfer heat from an area of low temperature to an area of higher temperature.

Vapor pressure The minimum (absolute) pressure that must be maintained on a liquid to prevent it from changing to a vapor. The vapor pressure of most liquids increases as their temperature increases.

Variable refrigerant flow (VRF) Heat pump systems that vary the speed of their compressor and the throttling effect of their expansion device to regulate the flow of refrigerant and thus control its heating or cooling capacity.

Velocity head The mechanical energy possessed by a fluid due to its velocity.

Velocity profile A drawing that shows how the velocity of the fluid varies from the centerline of a pipe out to the pipe wall.

Vernal equinox The precise moment in March when the sun is directly overhead on the equator at solar noon.

Viscosity A means of expressing the natural tendency of a fluid to resist flow.

W

Waterlogged An undesirable condition in which the expansion tank in a hydronic system is completely filled with water.

Water plate The portion of a pressed-steel panel radiator that contains fluid.

Water-to-air heat pump A heat pump that, when operating in heating mode, moves heat from water flowing through its evaporator to air flowing through its condenser coil. When operating in cooling mode heat flow reverses.

Weepage Any undesirable but very minor fluid leakage, such as from the packing around a valve stem.

Wet-rotor circulator A circulator in which the motor's armature and its impeller are an integral unit that is surrounded and cooled by the fluid passing through the circulator.

Winter solstice Commonly known as the shortest day of the year, with the shortest time between sunrise and sunset.

Wire-to-water efficiency The efficiency at which a circulator converts electrical energy input to head (mechanical energy) output.

Wood gasification A process in which wood is heated to the point where it releases pyrolytic gases that are then mixed with secondary air to produce a very hot and efficient combustion process.

Working pressure The allowable upper pressure limit at which a pipe or piping component can safely be used.

Z

Zone An area of a building for which space heating is controlled by a single thermostat.

Zone valve Valves used to allow or prevent the flow of fluid from a heat source through individual zone distribution circuits.

Index

D

E

F

G

H

I

K

L

M

N

O

P

Q

R

S

T

U

V

W

Y

Z